HANDBOOK OF
Special Functions
Derivatives, Integrals, Series and Other Formulas

HANDBOOK OF
Special Functions
Derivatives, Integrals, Series and Other Formulas

Yury A. Brychkov

Computing Center of the Russian
Academy of Sciences

Moscow, Russia

CRC Press
Taylor & Francis Group
Boca Raton London New York

CRC Press is an imprint of the
Taylor & Francis Group, an **informa** business

A CHAPMAN & HALL BOOK

Chapman & Hall/CRC
Taylor & Francis Group
6000 Broken Sound Parkway NW, Suite 300
Boca Raton, FL 33487-2742

© 2008 by Taylor & Francis Group, LLC
Chapman & Hall/CRC is an imprint of Taylor & Francis Group, an Informa business

No claim to original U.S. Government works
Printed in the United States of America on acid-free paper
10 9 8 7 6 5 4 3 2 1

International Standard Book Number-13: 978-1-58488-956-4 (Hardcover)

This book contains information obtained from authentic and highly regarded sources. Reasonable efforts have been made to publish reliable data and information, but the author and publisher cannot assume responsibility for the validity of all materials or the consequences of their use. The authors and publishers have attempted to trace the copyright holders of all material reproduced in this publication and apologize to copyright holders if permission to publish in this form has not been obtained. If any copyright material has not been acknowledged please write and let us know so we may rectify in any future reprint.

Except as permitted under U.S. Copyright Law, no part of this book may be reprinted, reproduced, transmitted, or utilized in any form by any electronic, mechanical, or other means, now known or hereafter invented, including photocopying, microfilming, and recording, or in any information storage or retrieval system, without written permission from the publishers.

For permission to photocopy or use material electronically from this work, please access www.copyright.com (http://www.copyright.com/) or contact the Copyright Clearance Center, Inc. (CCC), 222 Rosewood Drive, Danvers, MA 01923, 978-750-8400. CCC is a not-for-profit organization that provides licenses and registration for a variety of users. For organizations that have been granted a photocopy license by the CCC, a separate system of payment has been arranged.

Trademark Notice: Product or corporate names may be trademarks or registered trademarks, and are used only for identification and explanation without intent to infringe.

Visit the Taylor & Francis Web site at
http://www.taylorandfrancis.com

and the CRC Press Web site at
http://www.crcpress.com

Contents

Preface xix

Chapter 1. The Derivatives 1

1.1.	Elementary Functions	1
1.1.1.	General formulas	1
1.1.2.	Algebraic functions	1
1.1.3.	The exponential function	4
1.1.4.	Hyperbolic functions	6
1.1.5.	Trigonometric functions	7
1.1.6.	The logarithmic function	12
1.1.7.	Inverse trigonometric functions	13
1.2.	The Hurwitz Zeta Function $\zeta(\nu, z)$	15
1.2.1.	Derivatives with respect to the argument	15
1.2.2.	Derivatives with respect to the parameter	15
1.3.	The Exponential Integral $\operatorname{Ei}(z)$	16
1.3.1.	Derivatives with respect to the argument	16
1.4.	The Sine $\operatorname{si}(z)$ and Cosine $\operatorname{ci}(z)$ Integrals	17
1.4.1.	Derivatives with respect to the argument	17
1.5.	The Error Functions $\operatorname{erf}(z)$ and $\operatorname{erfc}(z)$	18
1.5.1.	Derivatives with respect to the argument	18
1.6.	The Fresnel Integrals $S(z)$ and $C(z)$	20
1.6.1.	Derivatives with respect to the argument	20
1.7.	The Generalized Fresnel Integrals $S(z, \nu)$ and $C(z, \nu)$	20
1.7.1.	Derivatives with respect to the argument	20
1.8.	The Incomplete Gamma Functions $\gamma(\nu, z)$ and $\Gamma(\nu, z)$	21
1.8.1.	Derivatives with respect to the argument	21
1.8.2.	Derivatives with respect to the parameter	22
1.9.	The Parabolic Cylinder Function $D_\nu(z)$	22
1.9.1.	Derivatives with respect to the argument	22
1.9.2.	Derivatives with respect to the order	24

1.10.	The Bessel Function $J_\nu(z)$	25
1.10.1.	Derivatives with respect to the argument	25
1.10.2.	Derivatives with respect to the order	28
1.11.	The Bessel Function $Y_\nu(z)$	29
1.11.1.	Derivatives with respect to the argument	29
1.11.2.	Derivatives with respect to the order	31
1.12.	The Hankel Functions $H_\nu^{(1)}(z)$ and $H_\nu^{(2)}(z)$	31
1.12.1.	Derivatives with respect to the argument	31
1.12.2.	Derivatives with respect to the order	32
1.13.	The Modified Bessel Function $I_\nu(z)$	32
1.13.1.	Derivatives with respect to the argument	32
1.13.2.	Derivatives with respect to the order	35
1.14.	The Macdonald Function $K_\nu(z)$	36
1.14.1.	Derivatives with respect to the argument	36
1.14.2.	Derivatives with respect to the order	40
1.15.	The Struve Functions $\mathbf{H}_\nu(z)$ and $\mathbf{L}_\nu(z)$	41
1.15.1.	Derivatives with respect to the argument	41
1.15.2.	Derivatives with respect to the order	42
1.16.	The Anger $\mathbf{J}_\nu(z)$ and Weber $\mathbf{E}_\nu(z)$ Functions	45
1.16.1.	Derivatives with respect to the argument	45
1.16.2.	Derivatives with respect to the order	47
1.17.	The Kelvin Functions $\mathrm{ber}_\nu(z)$, $\mathrm{bei}_\nu(z)$, $\mathrm{ker}_\nu(z)$ and $\mathrm{kei}_\nu(z)$	48
1.17.1.	Derivatives with respect to the argument	48
1.17.2.	Derivatives with respect to the order	51
1.18.	The Legendre Polynomials $P_n(z)$	56
1.18.1.	Derivatives with respect to the argument	56
1.19.	The Chebyshev Polynomials $T_n(z)$ and $U_n(z)$	59
1.19.1.	Derivatives with respect to the argument	59
1.20.	The Hermite Polynomials $H_n(z)$	62
1.20.1.	Derivatives with respect to the argument	62
1.21.	The Laguerre Polynomials $L_n^\lambda(z)$	63
1.21.1.	Derivatives with respect to the argument	63
1.21.2.	Derivatives with respect to the parameter	64
1.22.	The Gegenbauer Polynomials $C_n^\lambda(z)$	64
1.22.1.	Derivatives with respect to the argument	64
1.22.2.	Derivatives with respect to the parameter	66
1.23.	The Jacobi Polynomials $P_n^{(\rho,\sigma)}(z)$	66
1.23.1.	Derivatives with respect to the argument	66
1.23.2.	Derivatives with respect to parameters	69

1.24.	**The Complete Elliptic Integrals $\mathrm{K}(z)$, $\mathrm{E}(z)$ and $\mathrm{D}(z)$**	69
1.24.1.	Derivatives with respect to the argument	69
1.25.	**The Legendre Function $P_\nu^\mu(z)$.**	70
1.25.1.	Derivatives with respect to the argument	70
1.25.2.	Derivatives with respect to parameters	71
1.26.	**The Kummer Confluent Hypergeometric Function $_1F_1(a;b;z)$.**	73
1.26.1.	Derivatives with respect to the argument	73
1.26.2.	Derivatives with respect to parameters	75
1.27.	**The Tricomi Confluent Hypergeometric Function $\Psi(a;b;z)$**	76
1.27.1.	Derivatives with respect to the argument	76
1.27.2.	Derivatives with respect to parameters	77
1.28.	**The Whittaker Functions $M_{\mu,\nu}(z)$ and $W_{\mu,\nu}(z)$**	80
1.28.1.	Derivatives with respect to the argument	80
1.29.	**The Gauss Hypergeometric Function $_2F_1(a,b;c;z)$.**	80
1.29.1.	Derivatives with respect to the argument	80
1.29.2.	Derivatives with respect to parameters	85
1.30.	**The Generalized Hypergeometric Function $_pF_q((a_p);(b_q);z)$**	86
1.30.1.	Derivatives with respect to the argument	86
1.30.2.	Derivatives with respect to parameters	87

Chapter 2. Limits ... 95

2.1.	**Special Functions**	95
2.1.1.	The Bessel functions $J_\nu(z)$, $Y_\nu(z)$, $I_\nu(z)$ and $K_\nu(z)$.	95
2.1.2.	The Struve functions $\mathbf{H}_\nu(z)$ and $\mathbf{L}_\nu(z)$.	95
2.1.3.	The Kelvin functions $\mathrm{ber}_\nu(z)$, $\mathrm{bei}_\nu(z)$, $\mathrm{ker}_\nu(z)$ and $\mathrm{kei}_\nu(z)$	95
2.1.4.	The Legendre polynomials $P_n(z)$.	96
2.1.5.	The Chebyshev polynomials $T_n(z)$ and $U_n(z)$	96
2.1.6.	The Hermite polynomials $H_n(z)$	97
2.1.7.	The Laguerre polynomials $L_n^\lambda(z)$.	97
2.1.8.	The Gegenbauer polynomials $C_n^\lambda(z)$.	98
2.1.9.	The Jacobi polynomials $P_n^{(\rho,\sigma)}(z)$	98
2.1.10.	Hypergeometric functions.	99

Chapter 3. Indefinite Integrals. ... 101

3.1.	**Elementary Functions.**	101
3.1.1.	The logarithmic function	101
3.2.	**Special Functions**	102
3.2.1.	The Bessel functions $J_\nu(x)$, $Y_\nu(x)$, $I_\nu(x)$ and $K_\nu(x)$	102

3.2.2.	The Struve functions $\mathbf{H}_\nu(z)$ and $\mathbf{L}_\nu(z)$	105
3.2.3.	The Airy functions Ai (z) and Bi (z)	105
3.2.4.	Various functions	109

Chapter 4. Definite Integrals 111

4.1.	Elementary Functions.	111
4.1.1.	Algebraic functions	111
4.1.2.	The exponential function	116
4.1.3.	Hyperbolic functions	117
4.1.4.	Trigonometric functions.	121
4.1.5.	The logarithmic function	132
4.1.6.	Inverse trigonometric functions	155
4.2.	The Dilogarithm $\mathrm{Li}_2(z)$.	178
4.2.1.	Integrals containing $\mathrm{Li}_2(z)$ and algebraic functions	178
4.2.2.	Integrals containing $\mathrm{Li}_2(z)$ and trigonometric functions	179
4.2.3.	Integrals containing $\mathrm{Li}_2(z)$ and the logarithmic function	180
4.2.4.	Integrals containing $\mathrm{Li}_2(z)$ and inverse trigonometric functions	180
4.3.	The Sine Si (z) and Cosine ci (z) Integrals	181
4.3.1.	Integrals containing Si (z) and algebraic functions.	181
4.3.2.	Integrals containing Si (z) and trigonometric functions.	181
4.3.3.	Integrals containing Si (z) and the logarithmic function	183
4.3.4.	Integrals containing Si (z) and inverse trigonometric functions	183
4.3.5.	Integrals containing products of Si (z) and ci (z).	184
4.4.	The Error Functions erf (z), erfi (z) and erfc (z).	184
4.4.1.	Integrals containing erf (z) and algebraic functions	184
4.4.2.	Integrals containing erf (z), erfc (z) and the exponential function . . .	185
4.4.3.	Integrals containing erf (z) and trigonometric functions	186
4.4.4.	Integrals containing erf (z) and the logarithmic function	187
4.4.5.	Integrals containing erf (z), erfi (z) and inverse trigonometric functions	187
4.4.6.	Integrals containing products of erf (z), erfc (z) and erfi (z)	188
4.5.	The Fresnel Integrals $S(z)$ and $C(z)$.	189
4.5.1.	Integrals containing $S(z)$ and algebraic functions	189
4.5.2.	Integrals containing $S(z)$ and trigonometric functions	190
4.5.3.	Integrals containing $S(z)$ and the logarithmic function	191
4.5.4.	Integrals containing $C(z)$ and algebraic functions.	191
4.5.5.	Integrals containing $C(z)$ and trigonometric functions.	191
4.5.6.	Integrals containing $C(z)$ and the logarithmic function	192
4.6.	The Incomplete Gamma Function $\gamma(\nu, z)$	192
4.6.1.	Integrals containing $\gamma(\nu, z)$ and algebraic functions.	192
4.6.2.	Integrals containing $\gamma(\nu, z)$ and the exponential function	193
4.6.3.	Integrals containing $\gamma(\nu, z)$ and trigonometric functions	193

4.6.4.	Integrals containing $\gamma(\nu, z)$ and the logarithmic function	195
4.6.5.	Integrals containing $\gamma(\nu, z)$, erf(z) and erfi(z).	195
4.6.6.	Integrals containing products of $\gamma(\nu, z)$	195
4.7.	**The Bessel Function $J_\nu(z)$**	**196**
4.7.1.	Integrals containing $J_\nu(z)$ and algebraic functions	196
4.7.2.	Integrals containing $J_\nu(z)$ and the exponential function.	197
4.7.3.	Integrals containing $J_\nu(z)$ and trigonometric functions	197
4.7.4.	Integrals containing $J_\nu(z)$ and the logarithmic function.	199
4.7.5.	Integrals containing $J_\nu(z)$ and inverse trigonometric functions	199
4.7.6.	Integrals containing $J_\nu(z)$, Si(z) and ci(z)	200
4.7.7.	Integrals containing products of $J_\nu(z)$	201
4.8.	**The Bessel Function $Y_\nu(z)$**	**204**
4.8.1.	Integrals containing $Y_\nu(z)$ and algebraic functions	204
4.8.2.	Integrals containing $Y_\nu(z)$ and $J_\nu(z)$	204
4.9.	**The Modified Bessel Function $I_\nu(z)$**	**205**
4.9.1.	Integrals containing $I_\nu(z)$ and algebraic functions.	205
4.9.2.	Integrals containing $I_\nu(z)$ and the exponential function	206
4.9.3.	Integrals containing $I_\nu(z)$ and trigonometric functions.	208
4.9.4.	Integrals containing $I_\nu(z)$ and the logarithmic function	210
4.9.5.	Integrals containing $I_\nu(z)$ and inverse trigonometric functions	211
4.9.6.	Integrals containing $I_\nu(z)$ and special functions	212
4.9.7.	Integrals containing products of $I_\nu(z)$	213
4.10.	**The Macdonald Function $K_\nu(z)$**	**216**
4.10.1.	Integrals containing $K_\nu(z)$, $J_\nu(z)$, $Y_\nu(z)$ and $I_\nu(z)$	216
4.10.2.	Integrals containing products of $K_\nu(z)$.	216
4.11.	**The Struve Functions $\mathrm{H}_\nu(z)$ and $\mathrm{L}_\nu(z)$**	**217**
4.11.1.	Integrals containing $\mathrm{H}_\nu(z)$, $\mathrm{L}_\nu(z)$ and algebraic functions	217
4.11.2.	Integrals containing $\mathrm{H}_\nu(z)$ and hyperbolic functions	219
4.11.3.	Integrals containing $\mathrm{H}_\nu(z)$, $\mathrm{L}_\nu(z)$ and trigonometric functions	219
4.11.4.	Integrals containing $\mathrm{H}_\nu(z)$, $\mathrm{L}_\nu(z)$ and the logarithmic function.	220
4.11.5.	Integrals containing $\mathrm{H}_\nu(z)$, $\mathrm{L}_\nu(z)$ and inverse trigonometric functions	221
4.12.	**The Kelvin Functions $\mathrm{ber}_\nu(z)$, $\mathrm{bei}_\nu(z)$, $\mathrm{ker}_\nu(z)$ and $\mathrm{kei}_\nu(z)$**	**221**
4.12.1.	Integrals containing $\mathrm{ber}_\nu(z)$, $\mathrm{bei}_\nu(z)$, $\mathrm{ker}_\nu(z)$, $\mathrm{kei}_\nu(z)$ and algebraic functions.	221
4.13.	**The Airy Functions Ai(z) and Bi(z)**	**222**
4.13.1.	Integrals containing products of Ai(z) and Bi(z).	222
4.14.	**The Legendre Polynomials $P_n(z)$**	**223**
4.14.1.	Integrals containing $P_n(z)$ and algebraic functions	223
4.14.2.	Integrals containing $P_n(z)$ and trigonometric functions	226
4.14.3.	Integrals containing $P_n(z)$ and the logarithmic function.	231

4.14.4.	Integrals containing $P_n(z)$, $J_\nu(z)$, $I_\nu(z)$ and $K_\nu(z)$	232
4.14.5.	Integrals containing products of $P_n(z)$	232
4.15.	**The Chebyshev Polynomials $T_n(z)$**	234
4.15.1.	Integrals containing $T_n(z)$ and algebraic functions	234
4.15.2.	Integrals containing $T_n(z)$ and trigonometric functions	236
4.15.3.	Integrals containing $T_n(z)$ and special functions.	241
4.16.	**The Chebyshev Polynomials $U_n(z)$**	241
4.16.1.	Integrals containing $U_n(z)$ and algebraic functions	241
4.16.2.	Integrals containing $U_n(z)$ and trigonometric functions	242
4.16.3.	Integrals containing $U_n(z)$ and $K_\nu(z)$	246
4.16.4.	Integrals containing products of $U_n(z)$	246
4.17.	**The Hermite Polynomials $H_n(z)$**	247
4.17.1.	Integrals containing $H_n(z)$ and algebraic functions	247
4.17.2.	Integrals containing $H_n(z)$ and the exponential function	248
4.17.3.	Integrals containing $H_n(z)$ and trigonometric functions	249
4.17.4.	Integrals containing $H_n(z)$, erf (z) and erfc (z)	251
4.17.5.	Integrals containing $H_n(z)$ and $K_\nu(z)$	251
4.17.6.	Integrals containing products of $H_n(z)$	252
4.18.	**The Laguerre Polynomials $L_n^\lambda(z)$**	254
4.18.1.	Integrals containing $L_n^\lambda(z)$ and algebraic functions	254
4.18.2.	Integrals containing $L_n^\lambda(z)$ and trigonometric functions	255
4.18.3.	Integrals containing $L_n^\lambda(z)$ and erfc (z)	256
4.18.4.	Integrals containing products of $L_n^\lambda(z)$	256
4.19.	**The Gegenbauer Polynomials $C_n^\lambda(z)$**	257
4.19.1.	Integrals containing $C_n^\lambda(z)$ and algebraic functions	257
4.19.2.	Integrals containing $C_n^\lambda(z)$ and trigonometric functions	257
4.19.3.	Integrals containing products of $C_n^\lambda(z)$	261
4.20.	**The Jacobi Polynomials $P_n^{(\rho,\sigma)}(z)$**	262
4.20.1.	Integrals containing $P_n^{(\rho,\sigma)}(z)$ and algebraic functions	262
4.20.2.	Integrals containing $P_n^{(\rho,\sigma)}(z)$ and trigonometric functions	263
4.20.3.	Integrals containing $P_n^{(\rho,\sigma)}(z)$ and $J_\nu(z)$	263
4.20.4.	Integrals containing products of $P_n^{(\rho,\sigma)}(z)$	264
4.21.	**The Complete Elliptic Integral $\mathbf{K}(z)$**	265
4.21.1.	Integrals containing $\mathbf{K}(z)$ and algebraic functions.	265
4.21.2.	Integrals containing $\mathbf{K}(z)$, the exponential, hyperbolic and trigonometric functions	267
4.21.3.	Integrals containing $\mathbf{K}(z)$ and the logarithmic function	268
4.21.4.	Integrals containing $\mathbf{K}(z)$ and inverse trigonometric functions	271
4.21.5.	Integrals containing $\mathbf{K}(z)$ and $\text{Li}_2(z)$	274

4.21.6.	Integrals containing $K(z)$, shi(z) and Si(z)	274
4.21.7.	Integrals containing $K(z)$ and erf(z)	275
4.21.8.	Integrals containing $K(z)$, $S(z)$ and $C(z)$	275
4.21.9.	Integrals containing $K(z)$ and $\gamma(\nu,z)$	275
4.21.10.	Integrals containing $K(z)$, $J_\nu(z)$ and $I_\nu(z)$	276
4.21.11.	Integrals containing $K(z)$, $H_\nu(z)$ and $L_\nu(z)$	277
4.21.12.	Integrals containing $K(z)$ and $L_n^\lambda(z)$	278
4.21.13.	Integrals containing products of $K(z)$	278
4.22.	**The Complete Elliptic Integral $E(z)$**	**279**
4.22.1.	Integrals containing $E(z)$ and algebraic functions	279
4.22.2.	Integrals containing $E(z)$, the exponential, hyperbolic and trigonometric functions	283
4.22.3.	Integrals containing $E(z)$ and the logarithmic function	286
4.22.4.	Integrals containing $E(z)$ and inverse trigonometric functions	289
4.22.5.	Integrals containing $E(z)$ and $\text{Li}_2(z)$	292
4.22.6.	Integrals containing $E(z)$, shi(z) and Si(z)	293
4.22.7.	Integrals containing $E(z)$ and erf(z)	293
4.22.8.	Integrals containing $E(z)$, $S(z)$ and $C(z)$	293
4.22.9.	Integrals containing $E(z)$ and $\gamma(\nu,z)$	294
4.22.10.	Integrals containing $E(z)$, $J_\nu(z)$ and $I_\nu(z)$	294
4.22.11.	Integrals containing $E(z)$, $H_\nu(z)$ and $L_\nu(z)$	296
4.22.12.	Integrals containing $E(z)$ and $L_n^\lambda(z)$	296
4.22.13.	Integrals containing products of $E(z)$ and $K(z)$	296
4.22.14.	Integrals containing products of $E(z)$	299
4.23.	**The Complete Elliptic Integral $D(z)$**	**300**
4.23.1.	Integrals containing $D(z)$ and elementary functions	300
4.23.2.	Integrals containing products of $D(z)$, $K(z)$ and $E(z)$	302
4.24.	**The Generalized Hypergeometric Function $_pF_q((a_p);(b_q);z)$**	**303**
4.24.1.	Integrals containing $_pF_q((a_p);(b_q);z)$ and algebraic functions	303
4.24.2.	Integrals containing $_pF_q((a_p);(b_q);z)$ and trigonometric functions	304
4.24.3.	Integrals containing $_pF_q((a_p);(b_q);z)$ and the logarithmic function	306
4.24.4.	Integrals containing $_pF_q((a_p);(b_q);z)$, $K(z)$ and $E(z)$	306
4.24.5.	Integrals containing products of $_pF_q((a_p);(b_q);z)$	306
Chapter 5.	**Finite Sums**	**309**
5.1.	**The Psi Function $\psi(z)$**	**309**
5.1.1.	Sums containing $\psi(k+a)$	309
5.1.2.	Sums containing products of $\psi(k+a)$	311
5.1.3.	Sums containing $\psi'(k+a,z)$	311

5.2.	The Incomplete Gamma Functions $\gamma(\nu, z)$ and $\Gamma(\nu, z)$	312
5.2.1.	Sums containing $\gamma(nk + \nu, z)$	312
5.2.2.	Sums containing products of $\gamma(\nu \pm k, z)$	313
5.2.3.	Sums containing $\Gamma(\nu \pm k, z)$	313
5.3.	The Bessel Function $J_\nu(z)$	314
5.3.1.	Sums containing $J_{\nu \pm nk}(z)$	314
5.3.2.	Sums containing products of $J_{\nu \pm nk}(z)$	314
5.4.	The Modified Bessel Function $I_\nu(z)$	315
5.4.1.	Sums containing $I_{\nu \pm nk}(z)$	315
5.4.2.	Sums containing products of $J_{\nu \pm nk}(z)$ and $I_{\nu \pm nk}(z)$	315
5.4.3.	Sums containing products of $I_{\nu \pm nk}(z)$	317
5.5.	The Macdonald Function $K_\nu(z)$	318
5.5.1.	Sums containing $K_{\nu \pm nk}(z)$	318
5.5.2.	Sums containing $K_{\nu \pm nk}(z)$ and special functions	318
5.5.3.	Sums containing products of $K_{\nu \pm nk}(z)$	319
5.6.	The Struve Functions $\mathbf{H}_\nu(z)$ and $\mathbf{L}_\nu(z)$	319
5.6.1.	Sums containing $\mathbf{H}_{k+\nu}(z)$ and $\mathbf{L}_{k+\nu}(z)$	319
5.7.	The Legendre Polynomials $P_n(z)$	320
5.7.1.	Sums containing $P_{m \pm nk}(z)$	320
5.7.2.	Sums containing $P_n(z)$ and special functions	325
5.7.3.	Sums containing products of $P_{m \pm nk}(z)$	325
5.7.4.	Sums containing $P_m(\varphi(k, z))$	326
5.7.5.	Sums containing $P_k(\varphi(k, z))$	328
5.7.6.	Sums containing products of $P_m(\varphi(k, z))$	328
5.8.	The Chebyshev Polynomials $T_n(z)$ and $U_n(z)$	329
5.8.1.	Sums containing $T_{m+nk}(z)$	329
5.8.2.	Sums containing products of $T_{m+nk}(z)$	331
5.8.3.	Sums containing $T_n(\varphi(k, z))$	331
5.8.4.	Sums containing $U_{m+nk}(z)$	332
5.8.5.	Sums containing products of $U_n(z)$	335
5.8.6.	Sums containing $U_n(\varphi(k, z))$	335
5.9.	The Hermite Polynomials $H_n(z)$	337
5.9.1.	Sums containing $H_{m \pm nk}(z)$	337
5.9.2.	Sums containing $H_{m \pm nk}(z)$ and special functions	340
5.9.3.	Sums containing products of $H_{m \pm nk}(z)$	341
5.9.4.	Sums containing $H_n(\varphi(k, z))$	342
5.9.5.	Sums containing $H_{m \pm nk}(\varphi(k, z))$	343
5.9.6.	Sums containing products of $H_{m \pm nk}(\varphi(k, z))$	344
5.10.	The Laguerre Polynomials $L_n^\lambda(z)$	344
5.10.1.	Sums containing $L_m^{\lambda \pm nk}(z)$	344

5.10.2. Sums containing $L_{m\pm nk}^{\lambda}(z)$. 346

5.10.3. Sums containing $L_{m\pm pk}^{\lambda\pm nk}(z)$. 348

5.10.4. Sums containing $L_{m\pm pk}^{\lambda\pm nk}(z)$ and special functions 355

5.10.5. Sums containing products of $L_{m\pm pk}^{\lambda\pm nk}(z)$. 357

5.10.6. Sums containing $L_{m\pm pk}^{\lambda\pm nk}(\varphi(k,z))$. 359

5.10.7. Sums containing $L_{m\pm pk}^{\lambda\pm nk}(\varphi(k,z))$ and special functions 361

5.10.8. Sums containing products of $L_{m\pm pk}^{\lambda\pm nk}(\varphi(k,z))$. 362

5.11. **The Gegenbauer Polynomials $C_n^{\lambda}(z)$** 363

5.11.1. Sums containing $C_m^{\lambda\pm nk}(z)$. 363

5.11.2. Sums containing $C_{m\pm pk}^{\lambda}(z)$. 365

5.11.3. Sums containing $C_{m\pm pk}^{\lambda\pm nk}(z)$. 365

5.11.4. Sums containing $C_{m\pm pk}^{\lambda\pm nk}(z)$ and special functions 373

5.11.5. Sums containing products of $C_{m\pm pk}^{\lambda\pm nk}(z)$. 378

5.11.6. Sums containing $C_{mk+n}^{\lambda k+\mu}(\varphi(k,z))$. 381

5.11.7. Sums containing $C_{mk+n}^{\lambda k+\mu}(\varphi(k,z))$ and special functions 388

5.11.8. Sums containing products of $C_{mk+n}^{\lambda k+\mu}(\varphi(k,z))$ 390

5.12. **The Jacobi Polynomials $P_n^{(\rho,\sigma)}(z)$** 391

5.12.1. Sums containing $P_m^{(\rho\pm pk,\sigma\pm qk)}(z)$. 391

5.12.2. Sums containing $P_{m\pm nk}^{(\rho\pm pk,\sigma\pm qk)}(z)$. 392

5.12.3. Sums containing $P_{n\pm mk}^{(\rho\pm pk,\sigma\pm qk)}(z)$ and special functions 402

5.12.4. Sums containing products of $P_{m\pm nk}^{(\rho\pm pk,\sigma\pm qk)}(z)$ 405

5.12.5. Sums containing $P_{m\pm nk}^{(\rho\pm pk,\sigma\pm qk)}(\varphi(k,z))$. 405

5.12.6. Sums containing $P_{n\pm mk}^{(\rho\pm pk,\sigma\pm qk)}(\varphi(k,z))$ and special functions 408

5.12.7. Sums containing products of $P_{n\pm mk}^{(\rho\pm pk,\sigma\pm qk)}(\varphi(k,z))$. 410

5.13. **The Legendre Function $P_\nu^\mu(z)$**. 411

5.13.1. Sums containing $P_{\nu\pm k}^{\mu\pm k}(z)$. 411

5.14. **The Kummer Confluent Hypergeometric Function $_1F_1(a;b;z)$**. 411

5.14.1. Sums containing $_1F_1(a;b;z)$. 411

5.14.2. Sums containing $_1F_1(a;b;z)$ and special functions 412

5.14.3. Sums containing products of $_1F_1(a;b;z)$. 413

5.15. **The Tricomi Confluent Hypergeometric Function $\Psi(a;b;z)$** 413

5.15.1. Sums containing $\Psi(a;b;z)$. 413

5.15.2. Sums containing $\Psi(a;b;z)$ and special functions. 414

5.16. **The Gauss Hypergeometric Function $_2F_1(a,b;c;z)$** 414

5.16.1. Sums containing $_2F_1(a,b;c;z)$ 414

5.16.2.	Sums containing $_2F_1(a,b;c;z)$ and special functions	415
5.16.3.	Sums containing products of $_2F_1(a,b;c;z)$	417
5.17.	**The Generalized Hypergeometric Function $_pF_q((a_p);(b_q);z)$**	418
5.17.1.	Sums containing $_pF_q((a_p)\pm mk;(b_q)\pm nk;z)$	418
5.17.2.	Sums containing $_pF_q((a_p)\pm mk;(b_q)\pm nk;z)$ and special functions	421
5.17.3.	Sums containing $_pF_q((a_p)\pm mk;(b_q)\pm nk;\varphi(k,z))$	422
5.17.4.	Sums containing $_pF_q((a_p)\pm mk;(b_q)\pm nk;\varphi(k,z))$ and special functions	424
5.17.5.	Sums containing products of $_pF_q((a_p)\pm mk;(b_q)\pm nk;\varphi(k,z))$	425
5.17.6.	Various sums containing $_pF_q((a_p)+mk;(b_q)+nk;z)$	425
5.18.	**Multiple Sums**	426
5.18.1.	Sums containing Bessel functions	426
5.18.2.	Sums containing orthogonal polynomials	427

Chapter 6. Infinite Series 429

6.1.	**Elementary Functions**	429
6.1.1.	Series containing algebraic functions	429
6.1.2.	Series containing the exponential function	429
6.1.3.	Series containing hyperbolic functions	430
6.1.4.	Series containing trigonometric functions	431
6.2.	**The Psi Function $\psi(z)$**	431
6.2.1.	Series containing $\psi(ka+b)$	431
6.2.2.	Series containing $\psi(ka+b)$ and trigonometric functions	447
6.2.3.	Series containing products of $\psi(ka+b)$	447
6.2.4.	Series containing $\psi'(ka+b)$	449
6.3.	**The Hurwitz Zeta Function $\zeta(s,z)$**	451
6.3.1.	Series containing $\zeta(k,z)$	451
6.4.	**The Sine Si(z) and Cosine ci(z) Integrals**	451
6.4.1.	Series containing Si$(\varphi(k)x)$	451
6.4.2.	Series containing ci$(\varphi(k)x)$	452
6.4.3.	Series containing Si(kx) and trigonometric functions	453
6.4.4.	Series containing products of Si(kx)	453
6.5.	**The Fresnel Integrals $S(x)$ and $C(x)$**	453
6.5.1.	Series containing $S(\varphi(k)x)$, $C(\varphi(k)x)$ and algebraic functions	453
6.5.2.	Series containing $S(\varphi(k)x)$, $C(\varphi(k)x)$ and trigonometric functions	455
6.5.3.	Series containing $S(kx)$, $C(kx)$ and Si(kx)	458
6.5.4.	Series containing products of $S(kx)$ and $C(kx)$	459
6.6.	**The Incomplete Gamma Function $\gamma(\nu,z)$**	459
6.6.1.	Series containing $\gamma(\nu\pm k,z)$	459
6.6.2.	Series containing products of $\gamma(\nu+k,z)$	460

6.7.	**The Parabolic Cylinder Function $D_\nu(z)$**	460
6.7.1.	Series containing $D_{\nu \pm nk}(z)$ and elementary functions	460
6.8.	**The Bessel Functions $J_\nu(z)$ and $Y_\nu(z)$**	461
6.8.1.	Series containing $J_{nk+\nu}(z)$	461
6.8.2.	Series containing two Bessel functions $J_{nk+\nu}(z)$	462
6.8.3.	Series containing three Bessel functions $J_{nk+\nu}(z)$	465
6.8.4.	Series containing four Bessel functions $J_{nk+\nu}(z)$	466
6.8.5.	Series containing $J_{k+\nu}(z)$ and $\psi(z)$	467
6.8.6.	Series containing $J_\nu(\varphi(k,x))$	467
6.8.7.	Series containing $J_\nu(kx)$ and trigonometric functions	470
6.8.8.	Series containing products of $J_\nu(\varphi(k,x))$	470
6.8.9.	Series containing products of $J_\nu(kx)$ and trigonometric functions	471
6.8.10.	Series containing $J_\nu(kx)$ and $\mathrm{Si}\,(kx)$	472
6.8.11.	Series containing $J_\nu(kx)$, $S(kx)$ and $C(kx)$	472
6.8.12.	Series containing $J_{k\mu+\nu}(\varphi(k,z))$	473
6.8.13.	Various series containing $J_\nu(z)$	474
6.8.14.	Series containing $Y_{k+\nu}(z)$	475
6.9.	**The Modified Bessel Function $I_\nu(z)$**	475
6.9.1.	Series containing $I_{nk+\nu}(z)$	475
6.9.2.	Series containing $I_{k+\nu}(z)$ and $\psi(z)$	476
6.9.3.	Series containing products of $I_{nk+\nu}(z)$	477
6.9.4.	Series containing $I_{nk+\mu}((nk+\nu)z)$	480
6.9.5.	Series containing products of $I_{nk+\nu}((nk+\nu)z)$	480
6.10.	**The Struve Functions $\mathbf{H}_\nu(z)$ and $\mathbf{L}_\nu(z)$**	481
6.10.1.	Series containing $\mathbf{H}_{k+\nu}(z)$ and $\mathbf{L}_{k+\nu}(z)$	481
6.10.2.	Series containing $\mathbf{H}_\nu(\varphi(k)x)$	482
6.10.3.	Series containing $\mathbf{H}_\nu(kx)$ and trigonometric functions	483
6.10.4.	Series containing $\mathbf{H}_\nu(kx)$ and $\mathrm{Si}\,(kx)$	484
6.10.5.	Series containing $\mathbf{H}_\nu(kx)$, $S(kx)$ and $C(kx)$	484
6.10.6.	Series containing $\mathbf{H}_\nu(\varphi(k)x)$ and $J_\mu(kx)$	485
6.10.7.	Series containing product of $\mathbf{H}_\nu(kx)$	485
6.11.	**The Legendre Polynomials $P_n(z)$**	486
6.11.1.	Series containing $P_{nk+m}(z)$	486
6.11.2.	Series containing $P_{nk+m}(z)$ and Bessel functions	486
6.11.3.	Series containing products of $P_{nk+m}(z)$	488
6.11.4.	Series containing $P_{nk+m}(\varphi(k,z))$	488
6.12.	**The Chebyshev Polynomials $T_k(z)$ and $U_k(z)$**	489
6.12.1.	Series containing $T_{nk+m}(\varphi(k,z))$	489
6.12.2.	Series containing $T_{nk+m}(z)$ and Bessel functions	489

6.12.3.	Series containing $U_{nk+m}(\varphi(k,z))$	490
6.12.4.	Series containing $U_{nk+m}(z)$ and Bessel functions	490
6.13.	**Hermite Polynomials $H_n(z)$**	**492**
6.13.1.	Series containing $H_{nk+m}(z)$ and Bessel functions	492
6.13.2.	Series containing products of $H_{nk+m}(z)$	493
6.13.3.	Series containing $H_{nk+m}(\varphi(k,z))$	493
6.13.4.	Series containing $H_{nk+m}(\varphi(k,z))$ and special functions	494
6.13.5.	Series containing products of $H_{nk+m}(\varphi(k,z))$	494
6.14.	**The Laguerre Polynomials $L_n^\lambda(z)$**	**495**
6.14.1.	Series containing $L_{nk+m}^{\lambda \pm lk}(z)$	495
6.14.2.	Series containing $L_{nk+m}^{\lambda \pm lk}(z)$ and special functions	496
6.14.3.	Series containing products of $L_{nk+m}^{\lambda \pm lk}(z)$	497
6.14.4.	Series containing products of $L_n^\lambda(kx)$	497
6.14.5.	Series containing $L_{nk+m}^{\lambda \pm lk}(\varphi(k,z))$	497
6.14.6.	Series containing $L_{nk+m}^{\lambda \pm lk}(\varphi(k,z))$ and special functions	498
6.14.7.	Series containing products of $L_{mk+n}^{\lambda \pm lk}(\varphi(k,z))$	498
6.15.	**The Gegenbauer Polynomials $C_n^\lambda(z)$**	**499**
6.15.1.	Series containing $C_{nk+m}^{\lambda \pm lk}(z)$	499
6.15.2.	Series containing $C_{nk+m}^{\lambda \pm lk}(z)$ and special functions	500
6.15.3.	Series containing products of $C_{nk+m}^{\lambda \pm lk}(z)$	502
6.15.4.	Series containing $C_{nk+m}^{\lambda \pm lk}(\varphi(k,z))$	503
6.15.5.	Series containing $C_{nk+m}^{\lambda \pm lk}(\varphi(k,z))$ and special functions	504
6.16.	**The Jacobi Polynomials $P_n^{(\rho,\sigma)}(z)$**	**505**
6.16.1.	Series containing $P_{m\pm nk}^{(\rho \pm pk, \sigma \pm qk)}(z)$	505
6.16.2.	Series containing $P_{m\pm nk}^{(\rho \pm pk, \sigma \pm qk)}(z)$ and special functions	506
6.16.3.	Series containing products of $P_{m\pm nk}^{(\rho \pm pk, \sigma \pm qk)}(z)$	507
6.16.4.	Series containing $P_{m\pm nk}^{(\rho \pm pk, \sigma \pm qk)}(\varphi(k,z))$	507
6.16.5.	Series containing $P_{m\pm nk}^{(\rho \pm pk, \sigma \pm qk)}(\varphi(k,z))$ and special functions	507
6.17.	**The Generalized Hypergeometric Function $_pF_q((a_p);(b_q);z)$**	**508**
6.17.1.	Series containing $_pF_q((a_p(k));(b_q(k));z)$	508
6.17.2.	Series containing $_pF_q((a_p(k));(b_q(k));z)$ and trigonometric functions	521
6.17.3.	Series containing $_pF_q((a_p(k));(b_q(k));z)$ and special functions	524
6.17.4.	Series containing products of $_pF_q((a_p(k));(b_q(k));z)$	527
6.17.5.	Series containing $_pF_{p+1}((a_p);(b_{p+1});\varphi(k,x))$	527
6.17.6.	Series containing $_pF_{p+1}((a_p(k));(b_{p+1}(k));\varphi(k)z)$	531
6.17.7.	Series containing $_pF_q((a_p(k));(b_q(k));\varphi(k)z)$ and special functions	533
6.17.8.	Series containing products of $_pF_q((a_p(k));(b_q(k));\varphi(k)z)$	536

Chapter 7. The Connection Formulas 537

7.1. Elementary Functions. 537
7.1.1. Trigonometric functions 537

7.2. Special Functions 537
7.2.1. The psi function $\psi(z)$ 537
7.2.2. The incomplete gamma functions $\Gamma(\nu,z)$ and $\gamma(\nu,z)$. 538
7.2.3. The parabolic cylinder function $D_\nu(z)$ 538
7.2.4. The Bessel functions $J_\nu(z)$, $H_\nu^{(1)}(z)$, $H_\nu^{(2)}(z)$, $I_\nu(z)$ and $K_\nu(z)$ 539
7.2.5. The Struve functions $\mathbf{H}_\nu(z)$ and $\mathbf{L}_\nu(z)$. 542
7.2.6. The Anger $\mathbf{J}_\nu(z)$ and Weber $\mathbf{E}_\nu(z)$ functions. 543
7.2.7. The Airy functions $\operatorname{Ai}(z)$ and $\operatorname{Bi}(z)$ 545
7.2.8. The Kelvin functions $\operatorname{ber}_\nu(z)$, $\operatorname{bei}_\nu(z)$, $\operatorname{ker}_\nu(z)$ and $\operatorname{kei}_\nu(z)$ 546
7.2.9. The Legendre polynomials $P_n(z)$. 549
7.2.10. The Chebyshev polynomials $T_n(z)$ and $U_n(z)$ 549
7.2.11. The Hermite polynomials $H_n(z)$ 551
7.2.12. The Laguerre polynomials $L_n^\lambda(z)$. 552
7.2.13. The Gegenbauer polynomials $C_n^\lambda(z)$. 553
7.2.14. The Jacobi polynomials $P_n^{(\rho,\sigma)}(z)$ 555
7.2.15. The polynomials of the imaginary argument 558
7.2.16. The complete elliptic integral $\mathbf{K}(z)$ 559
7.2.17. The complete elliptic integral $\mathbf{E}(z)$. 560
7.2.18. The Legendre function $P_\nu^\mu(z)$ 561

Chapter 8. Representations of Hypergeometric Functions
and of the Meijer G Function. 563

8.1. The Hypergeometric Functions. 563
8.1.1. The Gauss hypergeometric function $_2F_1(a,b;c;z)$ 563
8.1.2. The hypergeometric function $_3F_2(a_1,a_2,a_3;b_1,b_2;z)$ 589
8.1.3. The hypergeometric function $_4F_3(a_1,a_2,a_3,a_4;b_1,b_2,b_3;z)$. 601
8.1.4. The hypergeometric function $_5F_4((a_1,\ldots,a_5);(b_1,\ldots,b_4);z)$ 612
8.1.5. The hypergeometric function $_6F_5(a_1,\ldots,a_6;b_1,\ldots,b_5;z)$ 618
8.1.6. The hypergeometric function $_7F_6(a_1,\ldots,a_7;b_1,\ldots,b_6;z)$ 619
8.1.7. The hypergeometric function $_8F_7(a_1,\ldots,a_8;b_1,\ldots,b_7;z)$ 621
8.1.8. The hypergeometric function $_{10}F_9(a_1,\ldots,a_{10};b_1,\ldots,b_9;z)$ 621
8.1.9. The Kummer confluent hypergeometric function $_1F_1(a;b;z)$ 622
8.1.10. The Tricomi confluent hypergeometric function $\Psi(a;b;z)$ 623
8.1.11. The hypergeometric function $_1F_2(a_1;b_1,b_2;z)$. 625
8.1.12. The hypergeometric function $_2F_2(a_1,a_2;b_1,b_2;z)$ 629
8.1.13. The hypergeometric function $_2F_3(a_1,a_2;b_1,b_2,b_3;z)$ 629
8.1.14. The hypergeometric function $_3F_0(a_1,a_2,a_3;z)$ 631
8.1.15. The hypergeometric function $_5F_0(a_1,a_2,\ldots,a_5;z)$ 631
8.1.16. The hypergeometric function $_4F_1(a_1,\ldots,a_4;b_1;z)$. 631

8.1.17.	The hypergeometric function $_6F_1(a_1,\ldots,a_6;b_1;z)$.	632
8.1.18.	The hypergeometric function $_8F_3(a_1,\ldots,a_8;b_1,b_2,b_3;z)$	632
8.1.19.	The hypergeometric function $_0F_3(b_1,b_2,b_3;z)$.	633
8.1.20.	The hypergeometric function $_0F_7(b_1,\ldots,b_7;z)$.	634
8.1.21.	The hypergeometric function $_2F_5(a_1,a_2;b_1,\ldots,b_5;z)$	635
8.1.22.	The hypergeometric function $_4F_7(a_1,\ldots,a_4;b_1,\ldots,b_7;z)$	636
8.1.23.	The generalized hypergeometric function $_pF_q((a_p);(b_q);z)$.	637
8.2.	**The Meijer Function $G_{p,q}^{m,n}\left(z\,\middle|\,\begin{matrix}(a_p)\\(b_q)\end{matrix}\right)$**	**638**
8.2.1.	General formulas	638
8.2.2.	Various Meijer G functions	639
8.3.	**Representation in Terms of Hypergeometric Functions**	**652**
8.3.1.	Elementary functions.	652
8.3.2.	Special functions.	655

References 669
Index of Notations for Functions and Constants. 673
Index of Notations for Symbols. 679

Preface

The diversity of problems whose solutions require knowledge of properties of elementary and special functions of mathematical physics has given rise to a large number of handbooks in this field of Calculus. Among these are, above all, the books of the Bateman Manuscript Project, namely, *Higher Transcendental Functions*, Vol. 1–3, and *Tables of Integral Transforms*, Vol. 1–2, by A. Erdelyi (Ed.); *Table of Integrals, Series, and Products* by I.S. Gradshteyn and I.M. Ryzhik; *Handbook of Mathematical Functions* by M. Abramowitz and I. Stegun (Eds.); and the 5-volume handbook *Integrals and Series* by A.P. Prudnikov, Yu.A. Brychkov, and O.I. Marichev. Due to numerous applications in science and engineering, the theory of special functions is under permanent development, especially in connection with the requirements of modern computer algebra methods.

The present handbook contains mainly new results. Some known formulas are added for the sake of completeness. Special attention is paid to formulas of derivatives of n-th order with respect to the argument and of the first derivatives with respect to the parameters for most elementary and special functions. A considerable part of the book is devoted to formulas of connection and conversion for elementary and special functions, especially hypergeometric and Meijer G functions.

Chapter 1 contains differentiation formulas for various functions. In Chapter 2 limit formulas are given for the special functions that depend on parameters. Chapters 3 to 6 contain formulas of integration and summation for elementary and special functions, new classes of integrals, finite sums and infinite series being considered. In Chapter 7 connection formulas are given for various elementary and special functions. Chapter 8 is devoted to representations of hypergeometric functions and Meijer G functions in terms of other functions for various values of parameters and arguments.

The notations that are standard are listed at the end of the book. In all chapters, unless other restrictions are indicated, $k, l, m, n, p, q = 0, 1, 2, \ldots$

The author hopes that this handbook will be useful to scientists, engineers, postgraduate students and generally to anybody who uses mathematical methods.

Chapter 1

The Derivatives

1.1. Elementary Functions

1.1.1. General formulas

1. $D^n[f(z^2)] = n! \sum_{k=0}^{[n/2]} \frac{(2z)^{n-2k}}{k!(n-2k)!} f^{(n-k)}(z^2)$.

2. $D^n[f(\sqrt{z})] = \sum_{k=0}^{n-1} (-1)^k \frac{(n+k-1)!}{k!(n-k-1)!} (2\sqrt{z})^{-n-k} f^{(n-k)}(\sqrt{z})$ $[n \geq 1]$.

3. $D^n\left[f\left(\frac{1}{z}\right)\right] = (-1)^n (n-1)! \sum_{k=0}^{n-1} \binom{n}{k} \frac{z^{k-2n}}{(n-k-1)!} f^{(n-k)}\left(\frac{1}{z}\right)$ $[n \geq 1]$.

4. $D^n\left[z^{n-1} f\left(\frac{1}{z}\right)\right] = (-1)^n z^{-n-1} F\left(\frac{1}{z}\right)$, if $D^n[f(z)] = F(z)$.

5. $D^n[f_1(z) f_2(z) \ldots f_m(z)] = \sum_{k_1=0}^{n} \binom{n}{k_1} f_1^{(k_1)}(z) \sum_{k_2=0}^{n-k_1} \binom{n-k_1}{k_2} f_2^{(k_2)}(z) \ldots$
$\times \sum_{k_{m-1}=0}^{n-k_1-\ldots-k_{m-2}} \binom{n-k_1-\ldots-k_{m-2}}{k_{m-1}} f_{m-1}^{(k_{m-1})}(z) f_m^{(n-k_1-\ldots-k_{m-1})}(z)$.

6. $D^n\left[z^{m+n} f\left(\frac{1}{z}\right) D^m\left[z^{m-1} g\left(\frac{1}{z}\right)\right]\right] = (-1)^{m+n} z^{-n-1} F\left(\frac{1}{z}\right)$,
                                                     if $D^n[f(z) D^m[g(z)]] = F(z)$.

1.1.2. Algebraic functions

1. $D^n[z^\lambda] = (-1)^n (-\lambda)_n z^{\lambda-n}$.

2. $D^n[z^\alpha (a-z)^\beta] = n! a^n z^{\alpha-n} (a-z)^{\beta-n} P_n^{(\alpha-n, \beta-n)}\left(1 - \frac{2z}{a}\right)$.

3. $D^n[z^\alpha (a-z)^\beta] = n! z^\alpha (a-z)^{\beta-n} P_n^{(-\alpha-\beta-1, \beta-n)}\left(1 - \frac{2a}{z}\right)$.

4. $D^n[z^\alpha (a-z)^\beta] = n! z^{\alpha-n} (a-z)^\beta P_n^{(\alpha-n, -\alpha-\beta-1)}\left(\frac{a+z}{a-z}\right)$.

1

5. $D^n[z^\lambda(a+z)^\lambda] = \left(\dfrac{a}{4}\right)^n n! \dfrac{(-2\lambda)_n}{\left(\frac{1}{2}-\lambda\right)_n} z^{\lambda-n}(a+z)^{\lambda-n} C_n^{\lambda-n+1/2}\left(1+\dfrac{2z}{a}\right).$

6. $\quad = (-1)^n n!(z^2+az)^{\lambda-n/2} C_n^{-\lambda}\left(\dfrac{2z+a}{2\sqrt{z^2+az}}\right).$

7. $D^n[z^{-1}(a+z)^{-1}]$
$= 2(-1)^{n+1} n!\, a^{-1}(a+z)^{-n-1}\left[1-\left(\dfrac{a+z}{z}\right)^{(n+1)/2} T_{n+1}\left(\dfrac{a+2z}{2\sqrt{az+z^2}}\right)\right].$

8. $D^n[z^{-\lambda-1/2}(a-z)^\lambda]$
$= 2^{-2n} \dfrac{(2n)!}{\left(\frac{1}{2}-\lambda\right)_n} z^{-\lambda-1/2}(a-z)^{\lambda-n} C_{2n}^{\lambda-n+1/2}\left(\sqrt{\dfrac{a}{z}}\right).$

9. $D^n[z^\lambda(a-z)^{n-2\lambda-1}] = (-1)^n n!\, a^{n/2} z^{\lambda-n/2}(a-z)^{-2\lambda-1} C_n^{-\lambda}\left(\dfrac{z+a}{2\sqrt{az}}\right).$

10. $D^n[z^{n-2\lambda-1}(a+z)^\lambda]$
$= \left(-\dfrac{1}{4}\right)^n n! \dfrac{(-2\lambda)_n}{\left(\frac{1}{2}-\lambda\right)_n} z^{n-2\lambda-1}(a+z)^{\lambda-n} C_n^{\lambda-n+1/2}\left(1+\dfrac{2a}{z}\right).$

11. $D^n[z^{-\lambda-3/2}(a-z)^\lambda]$
$= -\dfrac{n!}{2}\dfrac{\left(\frac{3}{2}\right)_n}{\left(-\lambda-\frac{1}{2}\right)_{n+1}} a^{-1/2} z^{-\lambda-1}(a-z)^{\lambda-n} C_{2n+1}^{\lambda-n+1/2}\left(\sqrt{\dfrac{a}{z}}\right).$

12. $D^n[z^{n-1/2}(a-z)^\lambda] = n!\dfrac{\left(\frac{1}{2}\right)_n}{\left(\frac{1}{2}-\lambda\right)_n} a^n z^{-1/2}(a-z)^{\lambda-n} C_{2n}^{\lambda-n+1/2}\left(\sqrt{\dfrac{z}{a}}\right).$

13. $D^n[z^{n+1/2}(a+z)^\lambda]$
$= -2^{-2n-1}\dfrac{(2n+1)!}{(\lambda+1)_{n+1}}(z+a)^{\lambda+1/2} C_{2n+1}^{-\lambda-n-1}\left(\sqrt{\dfrac{z}{z+a}}\right).$

14. $D^n[z^{n-1/2}(a+z)^\lambda] = 2^{-2n}\dfrac{(2n)!}{(\lambda+1)_n} z^{-1/2}(a+z)^\lambda C_{2n}^{-\lambda-n}\left(\sqrt{\dfrac{z}{a+z}}\right).$

15. $D^n[z^n(a+z)^{-1/2}] = n!\, a^{n/2}(a+z)^{-(n+1)/2} P_n\left(\dfrac{2a+z}{2\sqrt{a^2+az}}\right).$

16. $D^n[z^n(a+z)^{n-1/2}] = n!\, a^n(a+z)^{-1/2} P_{2n}\left(\sqrt{1+\dfrac{z}{a}}\right).$

17. $D^n[z^n(a+z)^{n+1/2}] = n!\, a^{n+1/2} P_{2n+1}\left(\sqrt{1+\dfrac{z}{a}}\right).$

18. $D^n[z^n(a+z)^{-n-1/2}] = n!(a+z)^{-n-1/2} P_{2n}\left(\sqrt{\dfrac{a}{a+z}}\right).$

19. $D^n[z^n(a+z)^{-n-3/2}] = n!\, a^{-1/2}(a+z)^{-n-1} P_{2n+1}\left(\sqrt{\dfrac{a}{a+z}}\right).$

20. $D^n[z^n(a+z)^n] = n!\, a^n P_n\left(1+\dfrac{2z}{a}\right).$

21. $D^n[z^{-n-1}(a+z)^n] = (-1)^n n!\, z^{-n-1} P_n\left(1+\dfrac{2a}{z}\right).$

22. $D^n[z^{n-1/2}(a+z)^{-n-1/2}] = (-1)^n n!\, z^{-1/2}(a+z)^{-n-1/2} P_{2n}\left(\sqrt{\dfrac{z}{a+z}}\right).$

23. $D^n[z^{n-1/2}(a+z)^n] = n!\,(-a)^n z^{-1/2} P_{2n}\left(\sqrt{\dfrac{z}{a}}\right).$

24. $D^n[z^{n+1/2}(a+z)^{n+1/2}] = \dfrac{\left(\frac{3}{2}\right)_n}{n+1}\, a^n z^{1/2}(a+z)^{1/2} U_n\left(1+\dfrac{2z}{a}\right).$

25. $\quad = \dfrac{\left(\frac{3}{2}\right)_n}{2(n+1)}\, a^{n+1/2} z^{1/2} U_{2n+1}\left(\sqrt{1+\dfrac{z}{a}}\right).$

26. $D^n[z^{n-1/2}(a+z)^{n+1/2}] = \left(\dfrac{1}{2}\right)_n a^{n+1/2} z^{-1/2} T_{2n+1}\left(\sqrt{1+\dfrac{z}{a}}\right).$

27. $D^n[z^{n+1/2}(a+z)^{n-1/2}] = \left(\dfrac{1}{2}\right)_n a^n z^{1/2}(a+z)^{-1/2} U_{2n}\left(\sqrt{1+\dfrac{z}{a}}\right).$

28. $D^n[z^{n+1/2}(a-z)^{n-1/2}] = (-1)^n \left(\dfrac{1}{2}\right)_n a^{n+1/2}(a-z)^{-1/2} T_{2n+1}\left(\sqrt{\dfrac{z}{a}}\right).$

29. $D^n[z^{n-1/2}(a+z)^{n-1/2}] = \left(\dfrac{1}{2}\right)_n a^n z^{-1/2}(a+z)^{-1/2} T_{2n}\left(\sqrt{1+\dfrac{z}{a}}\right).$

30. $D^n[z^{n-1/2}(a+z)^{-n}] = \left(\dfrac{1}{2}\right)_n z^{-1/2}(a+z)^{-n} T_{2n}\left(\sqrt{\dfrac{a}{a+z}}\right).$

31. $D^n[z^{n+1/2}(a+z)^{-n-1}] = \left(\dfrac{1}{2}\right)_n z^{1/2}(a+z)^{-n-1} U_{2n}\left(\sqrt{\dfrac{a}{a+z}}\right).$

32. $\quad = (-1)^n \left(\dfrac{1}{2}\right)_n (a+z)^{-n-1/2} T_{2n+1}\left(\sqrt{\dfrac{z}{a+z}}\right).$

33. $D^n[z^{n-1/2}(a+z)^{-n-1}]$
$\quad = \left(\dfrac{1}{2}\right)_n (az)^{-1/2}(a+z)^{-n-1/2} T_{2n+1}\left(\sqrt{\dfrac{a}{a+z}}\right).$

34. $\quad = (-1)^n \left(\dfrac{1}{2}\right)_n z^{-1/2}(a+z)^{-n-1} U_{2n}\left(\sqrt{\dfrac{z}{a+z}}\right).$

35. $D^n[z^{n+1/2}(a+z)^{-n-2}]$
$\quad = (-1)^n 2^{-2n} \dfrac{(2n+1)!}{(n+1)!} z^{1/2}(a+z)^{-n-2} U_n\left(\dfrac{z-a}{z+a}\right).$

36. $\quad = \dfrac{\left(\frac{3}{2}\right)_n}{2(n+1)} \left(\dfrac{z}{a}\right)^{1/2} (a+z)^{-n-3/2} U_{2n+1}\left(\sqrt{\dfrac{a}{a+z}}\right).$

37. $D^n[(a^2-z^2)^\lambda] = (-2a)^n n! \dfrac{(-\lambda)_n}{(n-2\lambda)_n} (a^2-z^2)^{\lambda-n} C_n^{\lambda-n+1/2}\left(\dfrac{z}{a}\right).$

38. $D^n[(a^2-z^2)^{-1}]$
$$= n!\, a^{-1}(a-z)^{-n-1}\left[1 - \left(\dfrac{z-a}{z+a}\right)^{(n+1)/2} T_{n+1}\left(\dfrac{z}{\sqrt{z^2-a^2}}\right)\right].$$

39. $D^{2n}[(a^2-z^2)^{-1}] = (2n)!\, a^{-1}(a^2-z^2)^{-n-1/2} T_{2n+1}\left(\dfrac{a}{\sqrt{a^2-z^2}}\right).$

40. $D^{2n+1}[(a^2-z^2)^{-1}] = (2n+1)!\, a^{-1} z (a^2-z^2)^{-n-3/2} U_{2n+1}\left(\dfrac{a}{\sqrt{a^2-z^2}}\right).$

41. $D^n[z^{-1/2}(a+z)^{-1/2}]$
$$= (-1)^n n!\, z^{-(n+1)/2}(a+z)^{-(n+1)/2} P_n\left(\dfrac{2z+a}{2\sqrt{az+z^2}}\right).$$

42. $D^n[(a^2+z^2)^{-1/2}] = (-1)^n n!\, (a^2+z^2)^{-(n+1)/2} P_n\left(\dfrac{z}{\sqrt{a^2+z^2}}\right).$

43. $D^n[(a^2-z^2)^n] = (-2a)^n n!\, P_n\left(\dfrac{z}{a}\right).$

44. $D^n[(a^2-z^2)^{n-1/2}] = (-2a)^n \left(\dfrac{1}{2}\right)_n (a^2-z^2)^{-1/2} T_n\left(\dfrac{z}{a}\right).$

45. $D^n[(az-z^2)^{n-1/2}] = (-a)^n \left(\dfrac{1}{2}\right)_n (az-z^2)^{-1/2} T_{2n}\left(\sqrt{\dfrac{z}{a}}\right).$

46. $D^n[(a^2-z^2)^{n+1/2}] = \dfrac{(-2a)^n}{n+1}\left(\dfrac{3}{2}\right)_n (a^2-z^2)^{1/2} U_n\left(\dfrac{z}{a}\right).$

47. $D^n[z^{-n-1}(a^2-z^2)^n] = (-2a)^n n!\, z^{-n-1} P_n\left(\dfrac{a}{z}\right).$

48. $D^n[z^{n-2\lambda-1}(z^2-a^2)^\lambda]$
$$= (2a)^n n! \dfrac{(-\lambda)_n}{(n-2\lambda)_n} z^{n-2\lambda-1}(z^2-a^2)^{\lambda-n} C_n^{\lambda-n+1/2}\left(\dfrac{a}{z}\right).$$

49. $D^n\left[\dfrac{1}{\sqrt{z^2+a^2}}\left(z+\sqrt{z^2+a^2}\right)^{1/2}\right]$
$$= (-1)^n \dfrac{\sqrt{2}}{a}\left(\dfrac{1}{2}\right)_n (z^2+a^2)^{-(2n+1)/4} \sin\left(\dfrac{2n+1}{2}\operatorname{arccot}\dfrac{z}{a}\right).$$

1.1.3. The exponential function

1. $D^n\left[z^\lambda e^{-az}\right] = n!\, z^{\lambda-n} e^{-az} L_n^{\lambda-n}(az).$

2. $D^n\left[z^\lambda e^{-a/z}\right] = (-1)^n n!\, z^{\lambda-n} e^{-a/z} L_n^{-\lambda-1}\left(\dfrac{a}{z}\right).$

3. $D^n\left[e^{az^2}\right] = (-i)^n a^{n/2} e^{az^2} H_n(i\sqrt{a}\,z).$

4. $D^n\left[z^{n-1}e^{a/z^2}\right] = i^n a^{n/2} z^{-n-1} e^{a/z^2} H_n\left(\dfrac{i\sqrt{a}}{z}\right).$

5. $D^n\left[e^{-az^2}\right] = (-1)^n a^{n/2} e^{-az^2} H_n(\sqrt{a}\,z).$

6. $D^n\left[z^{n-1}e^{-a/z^2}\right] = a^{n/2} z^{-n-1} e^{-a/z^2} H_n\left(\dfrac{\sqrt{a}}{z}\right).$

7. $D^n\left[e^{a\sqrt{z}}\right] = \sqrt{\pi}\left(\dfrac{a}{2}\right)^{n+1/2} z^{(1-2n)/4}[I_{n-1/2}(a\sqrt{z}) + I_{1/2-n}(a\sqrt{z})].$

8. $D^n\left[\dfrac{1}{\sqrt{z}}e^{a\sqrt{z}}\right] = \sqrt{\pi}\left(\dfrac{a}{2}\right)^{n+1/2} z^{-(1+2n)/4}[I_{n+1/2}(a\sqrt{z}) + I_{-n-1/2}(a\sqrt{z})].$

9. $D^n\left[z^{n-1}e^{a/\sqrt{z}}\right]$
$= (-1)^n \sqrt{\pi}\left(\dfrac{a}{2}\right)^{n+1/2} z^{-(5+2n)/4}\left[I_{n-1/2}\left(\dfrac{a}{\sqrt{z}}\right) + I_{1/2-n}\left(\dfrac{a}{\sqrt{z}}\right)\right].$

10. $D^n\left[z^{n-1/2}e^{a/\sqrt{z}}\right]$
$= (-1)^n \sqrt{\pi}\left(\dfrac{a}{2}\right)^{n+1/2} z^{-(3+2n)/4}\left[I_{n+1/2}\left(\dfrac{a}{\sqrt{z}}\right) + I_{-n-1/2}\left(\dfrac{a}{\sqrt{z}}\right)\right].$

11. $D^n\left[e^{-a\sqrt{z}}\right] = (-1)^n \dfrac{a^{n+1/2}}{2^{n-1/2}\sqrt{\pi}} z^{(1-2n)/4} K_{n-1/2}(a\sqrt{z}).$

12. $D^n\left[\dfrac{1}{\sqrt{z}}e^{-a\sqrt{z}}\right] = (-1)^n \dfrac{a^{n+1/2}}{2^{n-1/2}\sqrt{\pi}} z^{-(1+2n)/4} K_{n+1/2}(a\sqrt{z}).$

13. $D^n\left[z^{n-1}e^{-a/\sqrt{z}}\right] = \dfrac{a^{n+1/2}}{2^{n-1/2}\sqrt{\pi}} z^{-(2n+5)/4} K_{n-1/2}\left(\dfrac{a}{\sqrt{z}}\right).$

14. $D^n\left[z^{-n-1/2}e^{-a/\sqrt{z}}\right] = \dfrac{a^{n+1/2}}{2^{n-1/2}\sqrt{\pi}} z^{-(3+2n)/4} K_{n+1/2}\left(\dfrac{a}{\sqrt{z}}\right).$

15. $D^n\left[e^{(-1)^{j+1}ia\sqrt{z}}\right] = (-1)^{j+1} i\sqrt{\pi}\left(\dfrac{a}{2}\right)^{n+1/2} z^{(1-2n)/4} H^{(j)}_{1/2-n}(a\sqrt{z})$ $[j = 1, 2].$

16. $\quad = (-1)^n \sqrt{\pi}\left(\dfrac{a}{2}\right)^{n+1/2} z^{(1-2n)/4} H^{(j)}_{n-1/2}(a\sqrt{z})$ $[j = 1, 2].$

17. $D^n\left[\dfrac{1}{\sqrt{z}}e^{(-1)^{j+1}ia\sqrt{z}}\right]$
$= (-1)^{j+n+1} i\sqrt{\pi}\left(\dfrac{a}{2}\right)^{n+1/2} z^{-(2n+1)/4} H^{(j)}_{n+1/2}(a\sqrt{z})$ $[j = 1, 2].$

18. $\quad = \sqrt{\pi}\left(\dfrac{a}{2}\right)^{n+1/2} z^{-(2n+1)/4} H^{(j)}_{-n-1/2}(a\sqrt{z})$ $[j = 1, 2].$

1.1.4. Hyperbolic functions

1. $D^n[z^\lambda \sinh(az)] = \dfrac{n!}{2} z^{\lambda-n} \left[e^{az} L_n^{\lambda-n}(-az) - e^{-az} L_n^{\lambda-n}(az) \right].$

2. $D^n[z^\lambda \cosh(az)] = \dfrac{n!}{2} z^{\lambda-n} \left[e^{az} L_n^{\lambda-n}(-az) + e^{-az} L_n^{\lambda-n}(az) \right].$

3. $D^n[\operatorname{sech}(az)]$

$$= 2(-a)^n \sum_{k=0}^{n} \dfrac{e^{-kaz} \cos\left[(k+1) \operatorname{arccot} e^{-az}\right]}{(e^{-2az}+1)^{(k+1)/2}} \sum_{m=0}^{k} (-1)^m \binom{k}{m} m^n.$$

4. $D^n[\sinh(az^2)] = \dfrac{(-i)^n}{2} a^{n/2} e^{az^2} H_n\left(i\sqrt{a}\, z\right) - \dfrac{(-1)^n}{2} a^{n/2} e^{-az^2} H_n\left(\sqrt{a}\, z\right).$

5. $D^n[\cosh(az^2)] = \dfrac{(-i)^n}{2} a^{n/2} e^{az^2} H_n\left(i\sqrt{a}\, z\right) + \dfrac{(-1)^n}{2} a^{n/2} e^{-az^2} H_n\left(\sqrt{a}\, z\right).$

6. $D^n[\sinh(a\sqrt{z})] = \sqrt{\pi} \left(\dfrac{a}{2}\right)^{n+1/2} z^{(1-2n)/4} I_{1/2-n}(a\sqrt{z}).$

7. $D^n[\cosh(a\sqrt{z})] = \sqrt{\pi} \left(\dfrac{a}{2}\right)^{n+1/2} z^{(1-2n)/4} I_{n-1/2}(a\sqrt{z}).$

8. $D^n\left[\dfrac{\sinh(a\sqrt{z})}{\sqrt{z}}\right] = \sqrt{\pi} \left(\dfrac{a}{2}\right)^{n+1/2} z^{-(1+2n)/4} I_{n+1/2}(a\sqrt{z}).$

9. $D^n\left[\dfrac{\cosh(a\sqrt{z})}{\sqrt{z}}\right] = \sqrt{\pi} \left(\dfrac{a}{2}\right)^{n+1/2} z^{-(1+2n)/4} I_{-n-1/2}(a\sqrt{z}).$

10. $D^n[\operatorname{sech}\sqrt{z}] = (-1)^n n! \pi \left(\dfrac{4}{\pi^2 + 4z}\right)^{n+1}$

$\times\; {}_{2n+4}F_{2n+3}\!\left(\begin{array}{c} 1, \tfrac{3}{2}, \tfrac{1}{2} - \tfrac{i\sqrt{z}}{\pi}, \ldots, \tfrac{1}{2} - \tfrac{i\sqrt{z}}{\pi}, \tfrac{1}{2} + \tfrac{i\sqrt{z}}{\pi}, \ldots, \tfrac{1}{2} + \tfrac{i\sqrt{z}}{\pi} \\ \tfrac{1}{2}, \tfrac{3}{2} - \tfrac{i\sqrt{z}}{\pi}, \ldots, \tfrac{3}{2} - \tfrac{i\sqrt{z}}{\pi}, \tfrac{3}{2} + \tfrac{i\sqrt{z}}{\pi}, \ldots, \tfrac{3}{2} + \tfrac{i\sqrt{z}}{\pi}; -1 \end{array}\right).$

11. $D^n\left[\dfrac{\operatorname{csch}\sqrt{z}}{\sqrt{z}}\right] = (-1)^{n+1} \dfrac{n!}{z^{n+1}} + \dfrac{2(-1)^n n!}{z^{n+1}}$

$\times\; {}_{2n+3}F_{2n+2}\!\left(\begin{array}{c} 1, -\tfrac{i\sqrt{z}}{\pi}, \ldots, -\tfrac{i\sqrt{z}}{\pi}, \tfrac{i\sqrt{z}}{\pi}, \ldots, \tfrac{i\sqrt{z}}{\pi} \\ 1 - \tfrac{i\sqrt{z}}{\pi}, \ldots, 1 - \tfrac{i\sqrt{z}}{\pi}, 1 + \tfrac{i\sqrt{z}}{\pi}, \ldots, 1 + \tfrac{i\sqrt{z}}{\pi}; -1 \end{array}\right).$

12. $D^n\left[\dfrac{\tanh\sqrt{z}}{\sqrt{z}}\right] = 2(-1)^n n! \left(\dfrac{4}{\pi^2 + 4z}\right)^{n+1}$

$\times\; {}_{2n+3}F_{2n+2}\!\left(\begin{array}{c} 1, \tfrac{1}{2} - \tfrac{i\sqrt{z}}{\pi}, \ldots, \tfrac{1}{2} - \tfrac{i\sqrt{z}}{\pi}, \tfrac{1}{2} + \tfrac{i\sqrt{z}}{\pi}, \ldots, \tfrac{1}{2} + \tfrac{i\sqrt{z}}{\pi} \\ \tfrac{3}{2} - \tfrac{i\sqrt{z}}{\pi}, \ldots, \tfrac{3}{2} - \tfrac{i\sqrt{z}}{\pi}, \tfrac{3}{2} + \tfrac{i\sqrt{z}}{\pi}, \ldots, \tfrac{3}{2} + \tfrac{i\sqrt{z}}{\pi}; 1 \end{array}\right).$

1.1. Elementary Functions

13. $D^n\left[\dfrac{\coth\sqrt{z}}{\sqrt{z}}\right] = (-1)^{n+1}\dfrac{n!}{z^{n+1}}$

$+ \dfrac{2(-1)^n n!}{z^{n+1}}\,{}_{2n+3}F_{2n+2}\left(\begin{array}{c}1,\,-\dfrac{i\sqrt{z}}{\pi},\,\ldots,\,-\dfrac{i\sqrt{z}}{\pi},\,\dfrac{i\sqrt{z}}{\pi},\,\ldots,\,\dfrac{i\sqrt{z}}{\pi}\\[4pt] 1-\dfrac{i\sqrt{z}}{\pi},\,\ldots,\,1-\dfrac{i\sqrt{z}}{\pi},\,1+\dfrac{i\sqrt{z}}{\pi},\,\ldots,\,1+\dfrac{i\sqrt{z}}{\pi}\end{array};\,1\right).$

14. $D^n\left[z^{n-1}\sinh\left(\dfrac{a}{\sqrt{z}}\right)\right] = (-1)^n\sqrt{\pi}\left(\dfrac{a}{2}\right)^{n+1/2} z^{-(5+2n)/4} I_{1/2-n}\left(\dfrac{a}{\sqrt{z}}\right).$

15. $D^n\left[z^{n-1}\cosh\left(\dfrac{a}{\sqrt{z}}\right)\right] = (-1)^n\sqrt{\pi}\left(\dfrac{a}{2}\right)^{n+1/2} z^{-(5+2n)/4} I_{n-1/2}\left(\dfrac{a}{\sqrt{z}}\right).$

16. $D^n\left[z^{n-1/2}\sinh\left(\dfrac{a}{\sqrt{z}}\right)\right] = (-1)^n\sqrt{\pi}\left(\dfrac{a}{2}\right)^{n+1/2} z^{-(3+2n)/4} I_{n+1/2}\left(\dfrac{a}{\sqrt{z}}\right).$

17. $D^n\left[z^{n-1/2}\cosh\left(\dfrac{a}{\sqrt{z}}\right)\right] = (-1)^n\sqrt{\pi}\left(\dfrac{a}{2}\right)^{n+1/2} z^{-(3+2n)/4} I_{-n-1/2}\left(\dfrac{a}{\sqrt{z}}\right).$

18. $D^n[\sinh(a\sqrt[4]{z})]$

$= \dfrac{\sqrt{\pi}}{2^{2n+1/2}} a^{n+1/2} z^{(1-6n)/8} \sum_{k=0}^{n-1} \dfrac{\Gamma(n+k)}{k!\,\Gamma(n-k)} (-a)^{-k} z^{-k/4} I_{k-n+1/2}(a\sqrt[4]{z})$

$[n \geq 1].$

19. $D^n[\cosh(a\sqrt[4]{z})]$

$= \dfrac{\sqrt{\pi}}{2^{2n+1/2}} a^{n+1/2} z^{(1-6n)/8} \sum_{k=0}^{n-1} \dfrac{\Gamma(n+k)}{k!\,\Gamma(n-k)} (-a)^{-k} z^{-k/4} I_{n-k-1/2}(a\sqrt[4]{z})$

$[n \geq 1].$

20. $D^n\left[\dfrac{\sinh(a\sqrt[4]{z})}{\sqrt[4]{z}}\right]$

$= \dfrac{\sqrt{\pi}}{2^{2n+1/2}} a^{n+1/2} z^{-(6n+1)/8} \sum_{k=0}^{n-1} \dfrac{\Gamma(n+k)}{k!\,\Gamma(n-k)} (-a)^{-k} z^{-k/4} I_{n-k+1/2}(a\sqrt[4]{z})$

$[n \geq 1].$

21. $D^n\left[\dfrac{\cosh(a\sqrt[4]{z})}{\sqrt[4]{z}}\right]$

$= \dfrac{\sqrt{\pi}}{2^{2n+1/2}} a^{n+1/2} z^{-(6n+1)/8} \sum_{k=0}^{n-1} \dfrac{\Gamma(n+k)}{k!\,\Gamma(n-k)} (-a)^{-k} z^{-k/4} I_{k-n-1/2}(a\sqrt[4]{z})$

$[n \geq 1].$

1.1.5. Trigonometric functions

1. $D^n[\sin(az)] = a^n \sin\left(az + \dfrac{n\pi}{2}\right).$

2. $D^n[\cos(az)] = a^n \cos\left(az + \dfrac{n\pi}{2}\right)$.

3. $D^n[\sec(az)] = (-1)^{[(n+1)/2]} a^n \sum_{k=1}^{n} \dfrac{(-1)^k}{2^k} \binom{n+1}{k+1} \sec^{k+1}(az)$

$\times \sum_{m=0}^{k} \binom{k}{m}(k-2m)^n \cos\left[(k-2m)az - \dfrac{1-(-1)^n}{4}\pi\right]$ $\quad$ [[13], (47)].

4. $= (-1)^{[(n+1)/2]} a^n \sum_{k=1}^{n} \dfrac{(-1)^k}{2^k} \binom{n+1}{k+1} \sum_{m=0}^{k} \binom{k}{m}(k-2m)^n \sec^{2m+1}(az)$

$\times \sum_{p=0}^{[(k-\gamma)/2]-m} (-1)^p \binom{k-2m}{2p+\gamma} \tan^{2p+\gamma}(az) \quad \left[\gamma = \dfrac{1-(-1)^n}{2};\; [13],\,(52)\right]$.

5. $= (-1)^{[(n+1)/2]} a^n \sec(az) \sum_{k=0}^{n} \dfrac{1}{2^k} \sum_{m=0}^{[(k-\gamma)/2]} (-1)^m \binom{k}{2m+\gamma}$

$\times \tan^{2m+\gamma}(az) \sum_{p=0}^{k} (-1)^p \binom{k}{p}(2p+1)^n \quad \left[\gamma = \dfrac{1-(-1)^n}{2};\; [13],\,(67)\right]$.

6. $= 2(ia)^n \sum_{k=0}^{n} \dfrac{e^{ikaz} \cos\left[(k+1)\arccot e^{iaz}\right]}{(e^{2iaz}+1)^{(k+1)/2}} \sum_{m=0}^{k} (-1)^m \binom{k}{m} m^n$

$\quad$ [[13], (75)].

7. $= (-2)^{n+1} \dfrac{a^n n!}{(2az+\pi)^{n+1}}$

$+ \dfrac{a^n}{(2\pi)^{n+1}} \left[\psi^{(n)}\left(\dfrac{3\pi+2az}{4\pi}\right) - \psi^{(n)}\left(\dfrac{\pi+2az}{4\pi}\right)\right]$

$- (-1)^n \dfrac{a^n}{(2\pi)^{n+1}} \left[\psi^{(n)}\left(\dfrac{\pi-2az}{4\pi}\right) - \psi^{(n)}\left(-\dfrac{\pi+2az}{4\pi}\right)\right]$.

8. $D^n[\csc(az)]$

$= (-1)^{[n/2]} a^n \sum_{k=1}^{n} \dfrac{(-1)^k}{2^k} \binom{n+1}{k+1} \sum_{m=0}^{k} \binom{k}{m}(k-2m)^n \csc^{2m+1}(az)$

$\times \sum_{p=0}^{[(k-\gamma)/2]-m} (-1)^p \binom{k-2m}{2p+\gamma} \cot^{2p+\gamma}(az) \quad \left[\gamma = \dfrac{1-(-1)^n}{2};\; [13],\,(120)\right]$.

9. $= (-1)^{[n/2]} a^n \csc(az) \sum_{k=1}^{n} \dfrac{1}{2^k} \sum_{m=0}^{k} (-1)^m \binom{k}{m}(2m+1)^n$

$\times \sum_{p=0}^{[(k-\gamma)/2]} (-1)^p \binom{k}{2p+\gamma} \cot^{2p+\gamma}(az) \quad \left[\gamma = \dfrac{1-(-1)^n}{2};\; [13],\,(130)\right]$.

1.1. Elementary Functions

10. $= (-1)^{n+1} \dfrac{n!}{az^{n+1}} + \dfrac{a^n}{(2\pi)^{n+1}} \left[\psi^{(n)}\left(\dfrac{\pi + az}{2\pi}\right) - \psi^{(n)}\left(\dfrac{az}{2\pi}\right) \right]$

$\quad - (-1)^n \dfrac{a^n}{(2\pi)^{n+1}} \left[\psi^{(n)}\left(\dfrac{\pi - az}{2\pi}\right) - \psi^{(n)}\left(-\dfrac{az}{2\pi}\right) \right].$

11. $D^n[\tan(az)] = (-1)^{[n/2]+1}(2a)^n \sum_{k=1}^{n} \dfrac{\sec^{k+1}(az)}{2^k}$

$\quad \times \sin\left[(k-1)az + \dfrac{1-(-1)^n}{4}\pi\right] \sum_{m=1}^{k}(-1)^m \binom{k}{m} m^n \quad [[13],(13)].$

12. $= (-1)z^{[n/2]+1}(2a)^n \sec^2(az) \sum_{k=1}^{n}\dfrac{1}{2^k}\sum_{m=1}^{k}(-1)^m\binom{k}{m}m^n$

$\times \sum_{p=0}^{[(k+\gamma-2)/2]} (-1)^p \binom{k-1}{2p-\gamma+1} \tan^{2p-\gamma+1}(az) \quad \left[\gamma = \dfrac{1-(-1)^n}{2};\ [13],(24)\right].$

13. $= \dfrac{a^n}{\pi^{n+1}} \left[\psi^{(n)}\left(\dfrac{1}{2} + \dfrac{az}{\pi}\right) - (-1)^n \psi^{(n)}\left(\dfrac{1}{2} - \dfrac{az}{\pi}\right)\right].$

14. $D^n[\cot(az)] = (-1)^{[(n-1)/2]}(2a)^n \csc^2(az) \sum_{k=0}^{n}\dfrac{1}{2^k}\sum_{m=1}^{k}(-1)^m\binom{k}{m}m^n$

$\times \sum_{p=0}^{[(k+\gamma-2)/2]} (-1)^p \binom{k-1}{2p-\gamma+1} \cot^{2p-\gamma+1}(az) \quad \left[\gamma = \dfrac{1-(-1)^n}{2};\ [13],(97)\right].$

15. $= \dfrac{a^n}{\pi^{n+1}} \left[(-1)^n \psi^{(n)}\left(-\dfrac{az}{\pi}\right) - \psi^{(n)}\left(1 + \dfrac{az}{\pi}\right)\right].$

16. $D^n[z^\lambda \sin(az)] = \dfrac{n!}{2i} z^{\lambda-n} \left[e^{iaz} L_n^{\lambda-n}(-iaz) - e^{-iaz} L_n^{\lambda-n}(iaz)\right].$

17. $D^n[z^\lambda \cos(az)] = \dfrac{n!}{2} z^{\lambda-n} \left[e^{iaz} L_n^{\lambda-n}(-iaz) + e^{-iaz} L_n^{\lambda-n}(iaz)\right].$

18. $D^n[\sin(az^2)]$
$= \dfrac{(-1)^n}{2} a^{n/2} e^{(n-2)\pi i/4} \left[i^n e^{iaz^2} H_n\left(e^{3\pi i/4}\sqrt{a}\,z\right) - e^{-iaz^2} H_n\left(e^{\pi i/4}\sqrt{a}\,z\right)\right].$

19. $D^n[\cos(az^2)]$
$= \dfrac{(-1)^n}{2} a^{n/2} e^{n\pi i/4} \left[i^n e^{iaz^2} H_n\left(e^{3\pi i/4}\sqrt{a}\,z\right) + e^{-iaz^2} H_n\left(e^{\pi i/4}\sqrt{a}\,z\right)\right].$

20. $D^n\left[z^{n-1}\sin\dfrac{a}{z}\right] = (-a)^n z^{-n-1}\sin\left(\dfrac{a}{z} + \dfrac{n\pi}{2}\right).$

21. $D^n\left[z^{n-1}\cos\dfrac{a}{z}\right] = (-a)^n z^{-n-1}\cos\left(\dfrac{a}{z} + \dfrac{n\pi}{2}\right).$

22. $D^n[\sin(a\sqrt{z})] = \sqrt{\pi}\left(\dfrac{a}{2}\right)^{n+1/2} z^{(1-2n)/4} J_{1/2-n}(a\sqrt{z})$.

23. $D^n[\cos(a\sqrt{z})] = (-1)^n\sqrt{\pi}\left(\dfrac{a}{2}\right)^{n+1/2} z^{(1-2n)/4} J_{n-1/2}(a\sqrt{z})$.

24. $D^n[\sin^m(a\sqrt{z})]$

$$= (-1)^{[m/2]} \dfrac{\sqrt{\pi}\, a^{n+1/2}}{2^{m+n-1/2} z^{(2n-1)/4}} \sum_{k=0}^{[m/2]-(1+(-1)^m)/2} (-1)^k \binom{m}{k}$$

$$\times \left\{ \dfrac{1-(-1)^m}{2}(m-2k)^{n+1/2} J_{1/2-n}((m-2k)a\sqrt{z}) \right.$$
$$\left. + (-1)^n \dfrac{1+(-1)^m}{2}(m-2k)^{n+1/2} J_{n-1/2}((m-2k)a\sqrt{z}) \right\} \quad [m \geq 1].$$

25. $D^n[\cos^m(a\sqrt{z})]$

$$= (-1)^n \dfrac{\sqrt{\pi}\, a^{n+1/2}}{2^{m+n+1/2} z^{(2n-1)/4}} \sum_{k=0}^{m} \binom{m}{k}(m-2k)^{n+1/2} J_{n-1/2}((m-2k)a\sqrt{z}).$$

26. $D^n\left[\dfrac{\sin(a\sqrt{z})}{\sqrt{z}}\right] = (-1)^n\sqrt{\pi}\left(\dfrac{a}{2}\right)^{n+1/2} z^{-(1+2n)/4} J_{n+1/2}(a\sqrt{z})$.

27. $D^n\left[\dfrac{\cos(a\sqrt{z})}{\sqrt{z}}\right] = \sqrt{\pi}\left(\dfrac{a}{2}\right)^{n+1/2} z^{-(1+2n)/4} J_{-n-1/2}(a\sqrt{z})$.

28. $D^n[\sec\sqrt{z}] = n!\,\pi\left(\dfrac{4}{\pi^2-4z}\right)^{n+1}$

$$\times\ _{2n+4}F_{2n+3}\left(\begin{array}{c} 1, \dfrac{3}{2}, \dfrac{1}{2}-\dfrac{\sqrt{z}}{\pi},\ldots,\dfrac{1}{2}-\dfrac{\sqrt{z}}{\pi},\dfrac{1}{2}+\dfrac{\sqrt{z}}{\pi},\ldots,\dfrac{1}{2}+\dfrac{\sqrt{z}}{\pi} \\ \dfrac{1}{2},\dfrac{3}{2}-\dfrac{\sqrt{z}}{\pi},\ldots,\dfrac{3}{2}-\dfrac{\sqrt{z}}{\pi},\dfrac{3}{2}+\dfrac{\sqrt{z}}{\pi},\ldots,\dfrac{3}{2}+\dfrac{\sqrt{z}}{\pi};-1\end{array}\right).$$

29. $D^n\left[\dfrac{\csc\sqrt{z}}{\sqrt{z}}\right] = (-1)^{n+1}\dfrac{n!}{z^{n+1}} + \dfrac{2(-1)^n n!}{z^{n+1}}$

$$\times\ _{2n+3}F_{2n+2}\left(\begin{array}{c} 1, -\dfrac{\sqrt{z}}{\pi},\ldots,-\dfrac{\sqrt{z}}{\pi},\dfrac{\sqrt{z}}{\pi},\ldots,\dfrac{\sqrt{z}}{\pi};-1 \\ 1-\dfrac{\sqrt{z}}{\pi},\ldots,1-\dfrac{\sqrt{z}}{\pi},1+\dfrac{\sqrt{z}}{\pi},\ldots,1+\dfrac{\sqrt{z}}{\pi}\end{array}\right).$$

30. $D^n\left[\dfrac{\tan\sqrt{z}}{\sqrt{z}}\right] = 2(n!)\left(\dfrac{4}{\pi^2-4z}\right)^{n+1}$

$$\times\ _{2n+3}F_{2n+2}\left(\begin{array}{c} 1, \dfrac{1}{2}-\dfrac{\sqrt{z}}{\pi},\ldots,\dfrac{1}{2}-\dfrac{\sqrt{z}}{\pi},\dfrac{1}{2}+\dfrac{\sqrt{z}}{\pi},\ldots,\dfrac{1}{2}+\dfrac{\sqrt{z}}{\pi} \\ \dfrac{3}{2}-\dfrac{\sqrt{z}}{\pi},\ldots,\dfrac{3}{2}-\dfrac{\sqrt{z}}{\pi},\dfrac{3}{2}+\dfrac{\sqrt{z}}{\pi},\ldots,\dfrac{3}{2}+\dfrac{\sqrt{z}}{\pi};1\end{array}\right).$$

31. $D^n\left[\dfrac{\cot\sqrt{z}}{\sqrt{z}}\right] = 2(-1)^n n!\, z^{-n-1}$

$$\times\ _{2n+3}F_{2n+2}\left(\begin{array}{c} 1, -\dfrac{\sqrt{z}}{\pi},\ldots,-\dfrac{\sqrt{z}}{\pi},\dfrac{\sqrt{z}}{\pi},\ldots,\dfrac{\sqrt{z}}{\pi};1 \\ 1-\dfrac{\sqrt{z}}{\pi},\ldots,1-\dfrac{\sqrt{z}}{\pi},1+\dfrac{\sqrt{z}}{\pi},\ldots,1+\dfrac{\sqrt{z}}{\pi}\end{array}\right) - (-1)^n \dfrac{n!}{z^{n+1}}.$$

1.1. Elementary Functions

32. $D^n\left[z^{n-1}\sin\dfrac{a}{\sqrt{z}}\right] = (-1)^n\sqrt{\pi}\left(\dfrac{a}{2}\right)^{n+1/2} z^{-(2n+5)/4} J_{1/2-n}\left(\dfrac{a}{\sqrt{z}}\right).$

33. $D^n\left[z^{n-1/2}\sin\dfrac{a}{\sqrt{z}}\right] = \sqrt{\pi}\left(\dfrac{a}{2}\right)^{n+1/2} z^{-(2n+3)/4} J_{n+1/2}\left(\dfrac{a}{\sqrt{z}}\right).$

34. $D^n\left[z^{n-1}\cos\dfrac{a}{\sqrt{z}}\right] = \sqrt{\pi}\left(\dfrac{a}{2}\right)^{n+1/2} z^{-(2n+5)/4} J_{n-1/2}\left(\dfrac{a}{\sqrt{z}}\right).$

35. $D^n\left[z^{n-1/2}\cos\dfrac{a}{\sqrt{z}}\right] = (-1)^n\sqrt{\pi}\left(\dfrac{a}{2}\right)^{n+1/2} z^{-(2n+3)/4} J_{-n-1/2}\left(\dfrac{a}{\sqrt{z}}\right).$

36. $D^n[\sin(a\sqrt[4]{z})] = 2^{-2n-1/2}\sqrt{\pi}\, a^{n+1/2} z^{(1-6n)/8}$
$$\times \sum_{k=0}^{n-1}(-a)^{-k}\dfrac{\Gamma(n+k)}{k!\,\Gamma(n-k)} z^{-k/4} J_{k-n+1/2}(a\sqrt[4]{z}) \quad [n\geq 1].$$

37. $D^n[\cos(a\sqrt[4]{z})] = (-1)^n 2^{-2n-1/2}\sqrt{\pi}\, a^{n+1/2} z^{(1-6n)/8}$
$$\times \sum_{k=0}^{n-1} a^{-k}\dfrac{\Gamma(n+k)}{k!\,\Gamma(n-k)} z^{-k/4} J_{n-k-1/2}(a\sqrt[4]{z}) \quad [n\geq 1].$$

38. $D^n\left[\dfrac{\sin(a\sqrt[4]{z})}{\sqrt[4]{z}}\right] = (-1)^n 2^{-2n-1/2}\sqrt{\pi}\, a^{n+1/2} z^{-(1+6n)/8}$
$$\times \sum_{k=0}^{n-1} a^{-k}\dfrac{\Gamma(n+k)}{k!\,\Gamma(n-k)} z^{-k/4} J_{n-k+1/2}(a\sqrt[4]{z}) \quad [n\geq 1].$$

39. $D^n\left[\dfrac{\cos(a\sqrt[4]{z})}{\sqrt[4]{z}}\right] = 2^{-2n-1/2}\sqrt{\pi}\, a^{n+1/2} z^{-(1+6n)/8}$
$$\times \sum_{k=0}^{n-1}(-1)^k a^{-k}\dfrac{\Gamma(n+k)}{k!\,\Gamma(n-k)} z^{-k/4} J_{k-n-1/2}(a\sqrt[4]{z}) \quad [n\geq 1].$$

40. $D^{4n}\left[\left\{\begin{array}{c}\sinh z\sin z\\ \cosh z\cos z\end{array}\right\}\right] = (-4)^n\left\{\begin{array}{c}\sinh z\sin z\\ \cosh z\cos z\end{array}\right\}.$

41. $D^{4n+1}\left[\left\{\begin{array}{c}\sinh z\sin z\\ \cosh z\cos z\end{array}\right\}\right] = (-4)^n(\pm\cosh z\sin z + \sinh z\cos z).$

42. $D^{4n+2}\left[\left\{\begin{array}{c}\sinh z\sin z\\ \cosh z\cos z\end{array}\right\}\right] = \pm(-1)^n 2^{2n+1}\left\{\begin{array}{c}\cosh z\cos z\\ \sinh z\sin z\end{array}\right\}.$

43. $D^{4n+3}\left[\left\{\begin{array}{c}\sinh z\sin z\\ \cosh z\cos z\end{array}\right\}\right] = (-1)^{n+1} 2^{2n+1}(\cosh z\sin z \mp \sinh z\cos z).$

44. $D^{4n}\left[\left\{\begin{array}{c}\sinh z\cos z\\ \cosh z\sin z\end{array}\right\}\right] = (-4)^n\left\{\begin{array}{c}\sinh z\cos z\\ \cosh z\sin z\end{array}\right\}.$

45. $D^{4n+1}\left[\left\{\begin{array}{c}\sinh z\cos z\\ \cosh z\sin z\end{array}\right\}\right] = (-4)^n(\cosh z\cos z \mp \sinh z\sin z).$

46. $D^{4n+2}\left[\left\{\begin{array}{l}\sinh z \cos z \\ \cosh z \sin z\end{array}\right\}\right] = \mp(-1)^n 2^{2n+1}\left\{\begin{array}{l}\cosh z \sin z \\ \sinh z \cos z\end{array}\right\}.$

47. $D^{4n+3}\left[\left\{\begin{array}{l}\sinh z \cos z \\ \cosh z \sin z\end{array}\right\}\right] = (-1)^{n+1} 2^{2n+1}(\sinh z \sin z \pm \cosh z \cos z).$

48. $D^n\left[\sinh(a\sqrt{z})\left\{\begin{array}{l}\sin(a\sqrt{z}) \\ \cos(a\sqrt{z})\end{array}\right\}\right] = (\mp 1)^n \sqrt{\pi}\left(\dfrac{a}{\sqrt{2}}\right)^{n+1/2} z^{(1-2n)/4}$
$\times \left[\sin\dfrac{(6n+3)\pi}{8} \operatorname{ber}_{\pm n \mp 1/2}(a\sqrt{2z}) + \cos\dfrac{(6n+3)\pi}{8} \operatorname{bei}_{\pm n \mp 1/2}(a\sqrt{2z})\right].$

49. $D^n\left[\cosh(a\sqrt{z})\left\{\begin{array}{l}\sin(a\sqrt{z}) \\ \cos(a\sqrt{z})\end{array}\right\}\right] = (\pm 1)^{n+1} \sqrt{\pi}\left(\dfrac{a}{\sqrt{2}}\right)^{n+1/2} z^{(1-2n)/4}$
$\times \left[\sin\dfrac{(6n+3)\pi}{8} \operatorname{bei}_{\mp n \pm 1/2}(a\sqrt{2z}) - \cos\dfrac{(6n+3)\pi}{8} \operatorname{ber}_{\mp n \pm 1/2}(a\sqrt{2z})\right].$

50. $D^n\left[\dfrac{1}{\sqrt{z}}\sinh(a\sqrt{z})\left\{\begin{array}{l}\sin(a\sqrt{z}) \\ \cos(a\sqrt{z})\end{array}\right\}\right] = (\pm 1)^n \sqrt{\pi}\left(\dfrac{a}{\sqrt{2}}\right)^{n+1/2} z^{-(2n+1)/4}$
$\times \left[\sin\dfrac{(6n+3)\pi}{8} \operatorname{ber}_{\mp n \mp 1/2}(a\sqrt{2z}) + \cos\dfrac{(6n+3)\pi}{8} \operatorname{bei}_{\mp n \mp 1/2}(a\sqrt{2z})\right].$

51. $D^n\left[\dfrac{1}{\sqrt{z}}\cosh(a\sqrt{z})\left\{\begin{array}{l}\sin(a\sqrt{z}) \\ \cos(a\sqrt{z})\end{array}\right\}\right]$
$= (\mp 1)^{n+1} \sqrt{\pi}\left(\dfrac{a}{\sqrt{2}}\right)^{n+1/2} z^{-(2n+1)/4}$
$\times \left[\cos\dfrac{(6n+3)\pi}{8} \operatorname{ber}_{\pm n \pm 1/2}(a\sqrt{2z}) - \sin\dfrac{(6n+3)\pi}{8} \operatorname{bei}_{\pm n \pm 1/2}(a\sqrt{2z})\right].$

1.1.6. The logarithmic function

1. $D^n\left[\ln\left(\sqrt{z} + \sqrt{z+a}\right)\right]$
$= (-1)^{n-1}\dfrac{(n-1)!}{2} z^{-n/2}(z+a)^{-n/2} P_{n-1}\left(\dfrac{2z+a}{2\sqrt{z}\sqrt{z+a}}\right)$ $[n \geq 1].$

2. $D^n\left[z^{n-1}\ln\left(\sqrt{a} + \sqrt{z+a}\right)\right]$
$= \dfrac{(n-1)!}{2z} - \dfrac{(n-1)!}{2z}\left(\dfrac{a}{z+a}\right)^{n/2} P_{n-1}\left(\dfrac{z+2a}{2\sqrt{az+a^2}}\right)$ $[n \geq 1].$

3. $D^n\left[\ln\left(z + \sqrt{z^2+a^2}\right)\right]$
$= (-1)^{n-1}(n-1)!(z^2+a^2)^{-n/2} P_{n-1}\left(\dfrac{z}{\sqrt{z^2+a^2}}\right)$ $[n \geq 1].$

4. $D^n\left[z^{n-1}\ln\left(a + \sqrt{z^2+a^2}\right)\right]$
$= \dfrac{(n-1)!}{z} - \dfrac{(n-1)!}{z} a^n (z^2+a^2)^{-n/2} P_{n-1}\left(\dfrac{a}{\sqrt{z^2+a^2}}\right)$ $[n \geq 1].$

5. $D^{2n}\left[\ln\dfrac{a+z}{a-z}\right] = 2(2n-1)!z(a^2-z^2)^{-n-1/2} U_{2n-1}\left(\dfrac{a}{\sqrt{a^2-z^2}}\right)$ $[n \geq 1].$

6. $D^{2n+1}\left[\ln\dfrac{a+z}{a-z}\right] = 2(2n)!\,(a^2-z^2)^{-n-1/2}T_{2n+1}\left(\dfrac{a}{\sqrt{a^2-z^2}}\right).$

7. $D^n\left[\ln\dfrac{a+\sqrt{z}}{a-\sqrt{z}}\right]$
$= (n-1)!\,a^{2n-1}z^{1/2-n}(a^2-z)^{-n}P_{n-1}^{(1/2-n,\,-n)}\left(1-\dfrac{2z}{a^2}\right)$ $[n\geq 1]$.

8. $D^n\left[z^{n-1}\ln\dfrac{a+\sqrt{z}}{a-\sqrt{z}}\right]$
$= (-1)^n(n-1)!\,az^{n-3/2}(z-a^2)^{-n}P_{n-1}^{(1/2-n,\,-n)}\left(1-\dfrac{2a^2}{z}\right)$ $[n\geq 1]$.

9. $D^n\left[z^{1/2}(a^2-z)^n\,D^n\left[z^{n-1}\ln\dfrac{a+\sqrt{z}}{a-\sqrt{z}}\right]\right] = 0$ $[n\geq 1]$.

10. $D^n\left[z^{n-1/2}(a^2-z)^n\,D^n\left[\ln\dfrac{a+\sqrt{z}}{a-\sqrt{z}}\right]\right] = 0$ $[n\geq 1]$.

11. $D^n\left[z^{-1/2}(a^2-z)^n\,D^n\left[z^{n-1/2}\ln\dfrac{a+\sqrt{z}}{a-\sqrt{z}}\right]\right]$
$= (2n)!\left(-\dfrac{a^2}{4}\right)^n z^{-n-1}\ln\dfrac{a+\sqrt{z}}{a-\sqrt{z}}.$

12. $D^n\left[z^{n+1/2}(a^2-z)^n\,D^n\left[z^{-1/2}\ln\dfrac{a+\sqrt{z}}{a-\sqrt{z}}\right]\right] = \dfrac{(2n)!}{2^{2n}}\ln\dfrac{a+\sqrt{z}}{a-\sqrt{z}}.$

13. $D^n\left[z^{n+1/2}\,D^n\left[z^{-1/2}(a^2-z)^{n-1/2}\ln\dfrac{a+\sqrt{z}}{a-\sqrt{z}}\right]\right]$
$= (-1)^n\left(\dfrac{1}{2}\right)_n^2 a^{2n}(a^2-z)^{-n-1/2}\ln\dfrac{a+\sqrt{z}}{a-\sqrt{z}}.$

14. $D^n\left[z^{n-1/2}\,D^n\left[(a^2-z)^{n-1/2}\ln\dfrac{a+\sqrt{z}}{a-\sqrt{z}}\right]\right]$
$= (-1)^n\left(\dfrac{1}{2}\right)_n^2 a^{2n}z^{-1/2}(a^2-z)^{-n-1/2}\ln\dfrac{a+\sqrt{z}}{a-\sqrt{z}}.$

15. $D^n[z^{n-1}(1-az)^n\,D^n[\ln(1-az)]] = 0$ $[n\geq 1]$.

16. $D^n[(1-az)^n\,D^n[z^{n-1}\ln(1-az)]] = 0$ $[n\geq 1]$.

17. $D^n[z^{n+1}(1-az)^{-1}\,D^n[z^{-1}(1-az)^n\ln(1-az)]]$
$= (-1)^n(n!)^2 a^n(1-az)^{-n-1}\ln(1-az).$

18. $D^n[z^{n+1}(1-az)^n\,D^n[z^{-1}\ln(1-az)]] = (n!)^2 a^n\ln(1-az).$

1.1.7. Inverse trigonometric functions

1. $D^n[\arcsin(az)] = (-i)^{n-1}(n-1)!\,a^n(1-a^2z^2)^{-n/2}P_{n-1}\left(\dfrac{iaz}{\sqrt{1-a^2z^2}}\right)$
$[n\geq 1]$.

2. $D^n[\arccos(az)]$
$$= (-1)^n i^{n-1}(n-1)! a^n (1-a^2z^2)^{-n/2} P_{n-1}\left(\frac{iaz}{\sqrt{1-a^2z^2}}\right) \quad [n \geq 1].$$

3. $D^{2n}[\arctan(az)]$
$$= (-1)^n (2n-1)! a^{2n+1} z(1+a^2z^2)^{-n-1/2} U_{2n-1}\left(\frac{1}{\sqrt{1+a^2z^2}}\right) \quad [n \geq 1].$$

4. $D^{2n+1}[\arctan(az)]$
$$= (-1)^n (2n)! a^{2n+1}(1+a^2z^2)^{-n-1/2} T_{2n+1}\left(\frac{1}{\sqrt{1+a^2z^2}}\right).$$

5. $D^{2n}[\text{arccot}(az)]$
$$= (-1)^{n+1} (2n-1)! a^{2n+1} z(1+a^2z^2)^{-n-1/2} U_{2n-1}\left(\frac{1}{\sqrt{1+a^2z^2}}\right) \quad [n \geq 1].$$

6. $D^{2n+1}[\text{arccot}(az)]$
$$= (-1)^{n+1} (2n)! a^{2n+1}(1+a^2z^2)^{-n-1/2} T_{2n+1}\left(\frac{1}{\sqrt{1+a^2z^2}}\right).$$

7. $D^n[\arcsin(a\sqrt{z})]$
$$= \frac{(-i)^{n-1}}{2}(n-1)! a^n(z-a^2z^2)^{-n/2} P_{n-1}\left(\frac{1-2a^2z}{2a\sqrt{a^2z^2-z}}\right) \quad [n \geq 1].$$

8. $D^n[\arctan(a\sqrt{z})]$
$$= \frac{(n-1)!}{2} az^{1/2-n}(a^2z+1)^{-n} P_{n-1}^{(1/2-n,-n)}(2a^2z+1) \quad [n \geq 1].$$

9. $D^n\left[z^{n-1}\arcsin\frac{a}{\sqrt{z}}\right]$
$$= -\frac{i^{n-1}}{2}(n-1)! a^n z^{-1}(z-a^2)^{-n/2} P_{n-1}\left(\frac{z-2a^2}{2a\sqrt{a^2-z}}\right) \quad [n \geq 1].$$

10. $D^n\left[z^{n-1}\arctan\frac{a}{\sqrt{z}}\right]$
$$= \frac{(-1)^n}{2}(n-1)! az^{n-3/2}(z+a^2)^{-n} P_{n-1}^{(1/2-n,-n)}\left(\frac{2a^2}{z}+1\right) \quad [n \geq 1].$$

11. $D^n\left[z^{n-1/2}(1-a^2z)^{n+1/2} D^n\left[(1-a^2z)^{-1/2}\arcsin(a\sqrt{z})\right]\right]$
$$= \left(\frac{1}{2}\right)_n^2 a^{2n} z^{-1/2} \arcsin(a\sqrt{z}).$$

12. $D^n\left[z^{n+1/2}(1-a^2z)^{n-1/2} D^n\left[z^{-1/2}\arcsin(a\sqrt{z})\right]\right]$
$$= \left(\frac{1}{2}\right)_n^2 a^{2n}(1-a^2z)^{-1/2}\arcsin(a\sqrt{z}).$$

13. $D^n\left[z^{-1/2}(1-a^2z)^{n+1/2} D^n\left[z^{n-1/2}(1-a^2z)^{-1/2}\arcsin(a\sqrt{z})\right]\right]$
$$= (-4)^{-n}(2n)! z^{-n-1}\arcsin(a\sqrt{z}).$$

14. $D^n \left[z^{n+1/2}(1-a^2z)^{n+1/2} D^n \left[z^{-1/2}(1-a^2z)^{-1/2} \arcsin(a\sqrt{z}) \right] \right]$
$$= a^{2n}(n!)^2 \arcsin(a\sqrt{z}).$$

15. $D^n \left[z^{n+1/2}(1-a^2z)^{-1/2} D^n \left[z^{-1/2}(1-a^2z)^{n-1/2} \arcsin(a\sqrt{z}) \right] \right]$
$$= (2n)! \left(-\frac{a^2}{4} \right)^n (1-a^2z)^{-n-1} \arcsin(a\sqrt{z}).$$

16. $D^n \left[z^{1/2}(1-a^2z)^{n-1/2} D^n [z^{n-1} \arcsin(a\sqrt{z})] \right] = 0$ $[n \geq 1]$.

17. $D^n \left[z^{n-1/2}(1-a^2z)^{1/2} D^n [(1-a^2z)^{n-1} \arcsin(a\sqrt{z})] \right] = 0$ $[n \geq 1]$.

18. $D^n \left[z^{n-1/2}(1-a^2z)^{n-1/2} D^n [\arcsin(a\sqrt{z})] \right] = 0$ $[n \geq 1]$.

19. $D^n \left[z^{1/2}(1+a^2z)^n D^n [z^{n-1} \arctan(a\sqrt{z})] \right] = 0$ $[n \geq 1]$.

20. $D^n \left[z^{n-1/2}(1+a^2z)^n D^n [\arctan(a\sqrt{z})] \right] = 0$ $[n \geq 1]$.

21. $D^n \left[z^{-1/2}(1+a^2z)^n D^n \left[z^{n-1/2} \arctan(a\sqrt{z}) \right] \right]$
$$= (2n)! \left(-\frac{1}{4} \right)^n z^{-n-1} \arctan(a\sqrt{z}).$$

22. $D^n \left[z^{n+1/2}(1+a^2z)^n D^n \left[z^{-1/2} \arctan(a\sqrt{z}) \right] \right]$
$$= (2n)! \left(-\frac{a^2}{4} \right)^n \arctan(a\sqrt{z}).$$

23. $D^n \left[z^{n-1/2} D^n \left[(1+a^2z)^{n-1/2} \arctan(a\sqrt{z}) \right] \right]$
$$= \left(\frac{1}{2} \right)_n^2 a^{2n} z^{-1/2} (1+a^2z)^{-n-1/2} \arctan(a\sqrt{z}).$$

24. $D^n \left[z^{n+1/2} D^n \left[z^{-1/2}(1+a^2z)^{n-1/2} \arctan(a\sqrt{z}) \right] \right]$
$$= \left(\frac{1}{2} \right)_n^2 a^{2n} (1+a^2z)^{-n-1/2} \arctan(a\sqrt{z}).$$

1.2. The Hurwitz Zeta Function $\zeta(\nu, z)$

1.2.1. Derivatives with respect to the argument

1. $D^n[\zeta(\nu, az)] = (-a)^n (\nu)_n \zeta(\nu+n, az)$.

2. $D^n \left[z^{n-1} \zeta\left(\nu, \frac{a}{z}\right) \right] = a^n (\nu)_n z^{-n-1} \zeta\left(\nu+n, \frac{a}{z}\right)$.

1.2.2. Derivatives with respect to the parameter

1. $\dfrac{\partial}{\partial s} \zeta(s, z) \big|_{s=0} = \ln \dfrac{\Gamma(z)}{\sqrt{2\pi}}$.

2. $\left.\dfrac{\partial}{\partial s}\zeta\left(s,\dfrac{m}{n}\right)\right|_{s=-2k+1} = \dfrac{1}{2k}\left[\psi(2k)-\ln(2n\pi)\right]B_{2k}\left(\dfrac{m}{n}\right)$

$\qquad - \dfrac{1}{2kn^{2k}}\left[\psi(2k)-\ln(2\pi)\right]B_{2k}$

$\qquad - \dfrac{(-1)^k \pi}{(2n\pi)^{2k}}\sum_{i=1}^{n-1}\sin\dfrac{2im\pi}{n}\psi^{(2k-1)}\left(\dfrac{i}{n}\right) - \dfrac{(-1)^k 2(2k-1)!}{(2n\pi)^{2k}}$

$\qquad \times \sum_{i=1}^{n-1}\cos\dfrac{2im\pi}{n}\zeta'\left(2k,\dfrac{i}{n}\right) + \dfrac{1}{n^{2k}}\zeta'(-2k+1)\qquad [m<n;\ [59],\ (5)]$.

3. $\left.\dfrac{\partial}{\partial s}\zeta\left(s,\dfrac{1}{2}\right)\right|_{s=-2n+1} = -\dfrac{\ln 2}{2^{2n}n}B_{2n} + (2^{1-2n}-1)\zeta'(-2n+1)\qquad [[59],\ (17)]$.

4. $\left.\dfrac{\partial}{\partial s}\zeta\left(s,\dfrac{3\pm 1}{6}\right)\right|_{s=-2n+1} = \pm\dfrac{\sqrt{3}(1-3^{-2n})\pi}{8(1-3^{1-2n})n}B_{2n}$

$\qquad - \dfrac{\ln 3}{2^2\, 3^{2n-1}\,n}B_{2n} \pm \dfrac{(-1)^n}{2^{2n}3^{2n-1/2}\pi^{2n-1}}\psi^{(2n-1)}\left(\dfrac{1}{3}\right)$

$\qquad + \dfrac{3^{1-2n}-1}{2}\zeta'(-2n+1)\qquad [[59],\ (18)]$.

5. $\left.\dfrac{\partial}{\partial s}\zeta\left(s,\dfrac{2\pm 1}{4}\right)\right|_{s=-2n+1} = \pm\dfrac{(1-2^{-2n})\pi}{4n}B_{2n}$

$\qquad - \dfrac{(1-2^{2-2n})\pi}{2^{2n+1}n}B_{2n}\ln 2 \pm \dfrac{(-1)^k}{2^{6n-1}\pi^{2n-1}}\psi^{(2n-1)}\left(\dfrac{1}{4}\right)$

$\qquad - \dfrac{1-2^{1-2n}}{2^{2n-1}}\zeta'(-2n+1)\qquad [[59],\ (19)]$.

6. $\left.\dfrac{\partial}{\partial s}\zeta\left(s,\dfrac{3\pm 2}{6}\right)\right|_{s=-2n+1} = \pm\dfrac{\sqrt{3}(1-3^{-2n})(1+2^{1-2n})\pi}{8n}B_{2n}$

$\qquad + \dfrac{(1-3^{1-2n})\pi}{2^{2n+1}n}B_{2n}\ln 2$

$\qquad + \dfrac{(1-2^{1-2n})\pi}{2^2 3^{2n-1}n}B_{2n}\ln 3 \pm \dfrac{(-1)^k(2^{2k-1}+1)}{2^{4n-1}3^{2n-1/2}\pi^{2n-1}}\psi^{(2n-1)}\left(\dfrac{1}{3}\right)$

$\qquad + \dfrac{(1-2^{1-2n})(1-3^{1-2n})}{2}\zeta'(-2n+1)\qquad [[59],\ (20)]$.

1.3. The Exponential Integral $\operatorname{Ei}(z)$

1.3.1. Derivatives with respect to the argument

1. $D^n[\operatorname{Ei}(-az)] = (-1)^{n-1}(n-1)!\,z^{-n}e^{-az}\sum_{k=0}^{n-1}\dfrac{(az)^k}{k!}\qquad [n\geq 1]$.

2. $\qquad = (n-1)!\,z^{-n}e^{-az}L_{n-1}^{-n}(az)\qquad [n\geq 1]$.

3. $D^n \left[z^{n-1} \operatorname{Ei} \left(-\frac{a}{z} \right) \right] = -(n-1)! \, z^{-1} e^{-a/z} \sum_{k=0}^{n-1} \frac{(a/z)^k}{k!}$ $[n \geq 1]$.

4. $\phantom{D^n \left[z^{n-1} \operatorname{Ei} \left(-\frac{a}{z} \right) \right]} = (-1)^n (n-1)! \, z^{-1} e^{-a/z} L_{n-1}^{-n} \left(\frac{a}{z} \right)$ $[n \geq 1]$.

5. $D^n [e^{az} \operatorname{Ei}(-az)] = a^n e^{az} \operatorname{Ei}(-az) + a^n \sum_{k=0}^{n-1} \frac{(-1)^k k!}{(az)^{k+1}}$ $[n \geq 1]$.

6. $D^n \left[z^{n-1} e^{a/z} \operatorname{Ei} \left(-\frac{a}{z} \right) \right]$
$= (-a)^n z^{-n-1} e^{a/z} \operatorname{Ei} \left(-\frac{a}{z} \right) + (-1)^n a^{n-1} z^{-n} \sum_{k=0}^{n-1} (-1)^k k! \left(\frac{z}{a} \right)^k$ $[n \geq 1]$.

7. $D^n [z^{-1} e^{-az} D^n [z^n e^{az} \operatorname{Ei}(-az)]] = (-1)^n (n!)^2 z^{-n-1} \operatorname{Ei}(-az)$.

8. $D^n \left[z^{2n+1} e^{-a/z} D^n \left[z^{-1} e^{a/z} \operatorname{Ei} \left(-\frac{a}{z} \right) \right] \right] = (-1)^n (n!)^2 \operatorname{Ei} \left(-\frac{a}{z} \right)$.

9. $D^n [z^n e^{-az} D^n [e^{az} \operatorname{Ei}(-az)]] = n! \, a^n \operatorname{Ei}(-az)$.

10. $D^n \left[z^n e^{-a/z} D^n \left[z^{n-1} e^{a/z} \operatorname{Ei} \left(-\frac{a}{z} \right) \right] \right] = n! \, a^n z^{-n-1} \operatorname{Ei} \left(-\frac{a}{z} \right)$.

1.4. The Sine si(z) and Cosine ci(z) Integrals

1.4.1. Derivatives with respect to the argument

1. $D^n [\operatorname{si}(az)] = \frac{(n-1)!}{2i} z^{-n} \left[e^{iaz} L_{n-1}^{-n}(-iaz) - e^{-iaz} L_{n-1}^{-n}(iaz) \right]$ $[n \geq 1]$.

2. $D^n \left[z^{n-1} \operatorname{si} \left(\frac{a}{z} \right) \right]$
$= (-1)^n \frac{(n-1)!}{2iz} \left[e^{ia/z} L_{n-1}^{-n} \left(-\frac{ia}{z} \right) - e^{-ia/z} L_{n-1}^{-n} \left(\frac{ia}{z} \right) \right]$ $[n \geq 1]$.

3. $D^n [\operatorname{ci}(az)] = \frac{(n-1)!}{2} z^{-n} \left[e^{iaz} L_{n-1}^{-n}(-iaz) + e^{-iaz} L_{n-1}^{-n}(iaz) \right]$ $[n \geq 1]$.

4. $D^n \left[z^{n-1} \operatorname{ci} \left(\frac{a}{z} \right) \right]$
$= (-1)^n \frac{(n-1)!}{2z} \left[e^{ia/z} L_{n-1}^{-n} \left(-\frac{ia}{z} \right) + e^{-ia/z} L_{n-1}^{-n} \left(\frac{ia}{z} \right) \right]$ $[n \geq 1]$.

5. $D^n [\sin z \operatorname{si}(z) - \cos z \operatorname{ci}(z)]$
$= (-1)^n \left[\sin \left(z - \frac{n\pi}{2} \right) \operatorname{si}(z) - \cos \left(z - \frac{n\pi}{2} \right) \operatorname{ci}(z) \right.$
$\left. + \frac{1}{z} \sin \frac{n\pi}{2} + \frac{1}{z^{n+2}} \sum_{k=1}^{[n/2]} (n - 2k + 1)! (-z^2)^k \right]$.

6. $D^n[\cos z \operatorname{si}(z) + \sin z \operatorname{ci}(z)] = (-1)^n \left[\cos\left(z - \frac{n\pi}{2}\right)\operatorname{si}(z)\right.$

$$\left. + \sin\left(z - \frac{n\pi}{2}\right)\operatorname{ci}(z) - \frac{1}{z^{n+1}}\sum_{k=1}^{[n/2]}(n-2k)!(-z^2)^k\right].$$

1.5. The Error Functions erf (z) and erfc (z)

1.5.1. Derivatives with respect to the argument

1. $D^n[\operatorname{erf}(az)] = (-1)^{n-1}\dfrac{2a^n}{\sqrt{\pi}}e^{-a^2z^2}H_{n-1}(az)$ $\hspace{2cm}[n \geq 1]$.

2. $D^n[\operatorname{erf}(a\sqrt{z})] = \dfrac{(n-1)!a}{\sqrt{\pi}}z^{1/2-n}e^{-a^2z}L_{n-1}^{1/2-n}(a^2z)$ $\hspace{1cm}[n \geq 1]$.

3. $D^n[z^{n-1}\operatorname{erf}(a\sqrt{z})] = \dfrac{(-1)^{n-1}}{2^{2n-1}\sqrt{\pi z}}e^{-a^2z}H_{2n-1}(a\sqrt{z})$ $\hspace{1cm}[n \geq 1]$.

4. $D^n\left[\dfrac{1}{\sqrt{z}}\operatorname{erf}(a\sqrt{z})\right] = \dfrac{(-1)^n}{\sqrt{\pi}}z^{-n-1/2}\gamma\left(n+\dfrac{1}{2}, a^2z\right)$ $\hspace{1cm}[n \geq 1]$.

5. $D^n\left[z^{n-1}\operatorname{erf}\left(\dfrac{a}{z}\right)\right] = -\dfrac{2a^n}{\sqrt{\pi}}z^{-n-1}e^{-a^2/z^2}H_{n-1}\left(\dfrac{a}{z}\right)$ $\hspace{1cm}[n \geq 1]$.

6. $D^n\left[\operatorname{erf}\left(\dfrac{a}{\sqrt{z}}\right)\right] = -\dfrac{z^{-n}}{2^{2n-1}\sqrt{\pi}}e^{-a^2/z}H_{2n-1}\left(\dfrac{a}{\sqrt{z}}\right)$ $\hspace{1cm}[n \geq 1]$.

7. $D^n\left[z^{n-1/2}\operatorname{erf}\left(\dfrac{a}{\sqrt{z}}\right)\right] = \dfrac{1}{\sqrt{\pi z}}\gamma\left(n+\dfrac{1}{2}, \dfrac{a^2}{z}\right)$ $\hspace{1cm}[n \geq 1]$.

8. $D^n\left[z^{n-1}\operatorname{erf}\left(\dfrac{a}{\sqrt{z}}\right)\right] = (-1)^n\dfrac{(n-1)!a}{\sqrt{\pi}}z^{-1/2}e^{-a^2/z}L_{n-1}^{1/2-n}\left(\dfrac{a^2}{z}\right)$ $\hspace{0.3cm}[n \geq 1]$.

9. $D^n\left[e^{a^2z^2}\operatorname{erf}(az)\right] = (-ia)^n e^{a^2z^2}H_n(iaz)$

$$-\dfrac{2^{(n+1)/2}}{\sqrt{\pi}}n!(-a)^n e^{a^2z^2/2}D_{-n-1}(\sqrt{2}az).$$

10. $D^n\left[z^{n-1}e^{a^2/z^2}\operatorname{erf}\left(\dfrac{a}{z}\right)\right] = (ia)^n z^{-n-1}e^{a^2/z^2}H_n\left(\dfrac{ia}{z}\right)$

$$-\dfrac{2^{(n+1)/2}}{\sqrt{\pi}}n!a^n z^{-n-1}e^{a^2/(2z^2)}D_{-n-1}\left(\dfrac{\sqrt{2}a}{z}\right).$$

11. $D^n\left[e^{a^2z}\operatorname{erf}(a\sqrt{z})\right] = \dfrac{(-a^2)^n}{\sqrt{\pi}}\left(\dfrac{1}{2}\right)_n e^{a^2z}\gamma\left(\dfrac{1}{2}-n, a^2z\right).$

12. $D^n\left[z^{n-1}e^{a^2/z}\operatorname{erf}\left(\frac{a}{\sqrt{z}}\right)\right] = \frac{a^{2n}}{\sqrt{\pi}}\left(\frac{1}{2}\right)_n z^{-n-1}e^{a^2/z}\gamma\left(\frac{1}{2}-n,\frac{a^2}{z}\right).$

13. $D^n\left[z^{n-1/2}e^{-a^2z}D^n\left[e^{a^2z}\operatorname{erf}(a\sqrt{z})\right]\right] = \left(\frac{1}{2}\right)_n a^{2n}z^{-1/2}\operatorname{erf}(a\sqrt{z}).$

14. $D^n\left[z^{n+1/2}e^{-a^2/z}D^n\left[z^{n-1}e^{a^2/z}\operatorname{erf}\left(\frac{a}{\sqrt{z}}\right)\right]\right]$
$$= \left(\frac{1}{2}\right)_n a^{2n}z^{-n-1/2}\operatorname{erf}\left(\frac{a}{\sqrt{z}}\right).$$

15. $D^n\left[z^{n+1/2}e^{a^2z}D^n\left[z^{-1/2}\operatorname{erf}(a\sqrt{z})\right]\right] = \left(\frac{1}{2}\right)_n(-a^2)^n e^{a^2z}\operatorname{erf}(a\sqrt{z}).$

16. $D^n\left[z^{n-1/2}e^{a^2/z}D^n\left[z^{n-1/2}\operatorname{erf}\left(\frac{a}{\sqrt{z}}\right)\right]\right]$
$$= \left(\frac{1}{2}\right)_n(-a^2)^n e^{-a^2/z}\operatorname{erf}\left(\frac{a}{\sqrt{z}}\right).$$

17. $D^n\left[e^{a^2z^2}\operatorname{erfc}(az)\right] = \frac{2^{(n+1)/2}}{\sqrt{\pi}}n!(-a)^n e^{a^2z^2/2}D_{-n-1}(\sqrt{2}az).$

18. $D^n\left[z^{n-1}e^{a^2/z^2}\operatorname{erfc}\left(\frac{a}{z}\right)\right] = \frac{2^{(n+1)/2}}{\sqrt{\pi}}n!a^n z^{-n-1}e^{a^2/(2z^2)}D_{-n-1}\left(\frac{\sqrt{2}a}{z}\right).$

19. $D^n\left[z^{n-1/2}e^{a^2z}\operatorname{erfc}(a\sqrt{z})\right] = \frac{2^{1/2-n}}{\sqrt{\pi}}(2n)!z^{-1/2}e^{a^2z/2}D_{-2n-1}(a\sqrt{2z}).$

20. $D^n\left[z^{-1/2}e^{a^2/z}\operatorname{erfc}\left(\frac{a}{\sqrt{z}}\right)\right]$
$$= (-1)^n\frac{2^{1/2-n}}{\sqrt{\pi}}(2n)!z^{-n-1/2}e^{a^2/(2z)}D_{-2n-1}\left(a\sqrt{\frac{2}{z}}\right).$$

21. $D^n\left[z^{-1/2}e^{-a^2z}D^n\left[z^{n-1/2}e^{a^2z}\operatorname{erfc}(a\sqrt{z})\right]\right] = (-4)^{-n}(2n)!z^{-n-1}\operatorname{erfc}(a\sqrt{z}).$

22. $D^n\left[z^{2n+1/2}e^{-a^2/z}D^n\left[z^{-1/2}e^{a^2/z}\operatorname{erfc}\left(\frac{a}{\sqrt{z}}\right)\right]\right]$
$$= (-4)^{-n}(2n)!\operatorname{erfc}\left(\frac{a}{\sqrt{z}}\right).$$

23. $D^n\left[z^{1/2}e^{a^2z}D^n[z^{n-1}\operatorname{erfc}(a\sqrt{z})]\right] = 0$ $\qquad [n\geq 1].$

24. $D^n\left[z^{n+1/2}e^{-a^2z}D^n\left[z^{-1/2}e^{a^2z}\operatorname{erfc}(a\sqrt{z})\right]\right] = n!a^{2n}\operatorname{erfc}(a\sqrt{z}).$

25. $D^n \left[z^{n-1/2} e^{-a^2/z} D^n \left[z^{n-1/2} e^{a^2/z} \operatorname{erfc} \left(\dfrac{a}{\sqrt{z}} \right) \right] \right]$

$$= n! \, a^{2n} z^{-n-1} \operatorname{erfc} \left(\dfrac{a}{\sqrt{z}} \right).$$

1.6. The Fresnel Integrals $S(z)$ and $C(z)$

1.6.1. Derivatives with respect to the argument

1. $D^n [S(az)]$

$= \sqrt{\dfrac{a}{2\pi}} \, \dfrac{(n-1)!}{2i} z^{1/2-n} \left[e^{iaz} L_{n-1}^{1/2-n}(-iaz) - e^{-iaz} L_{n-1}^{1/2-n}(iaz) \right] \quad [n \geq 1].$

2. $D^n \left[z^{n-1} S\left(\dfrac{a}{z} \right) \right]$

$= (-1)^n \sqrt{\dfrac{a}{2\pi}} \, \dfrac{(n-1)!}{2i} z^{-3/2} \left[e^{ia/z} L_{n-1}^{1/2-n}\left(-\dfrac{ia}{z} \right) - e^{-ia/z} L_{n-1}^{1/2-n}\left(\dfrac{ia}{z} \right) \right]$

$\quad [n \geq 1].$

3. $D^n [C(az)]$

$= \sqrt{\dfrac{a}{2\pi}} \, \dfrac{(n-1)!}{2} z^{1/2-n} \left[e^{iaz} L_{n-1}^{1/2-n}(-iaz) + e^{-iaz} L_{n-1}^{1/2-n}(iaz) \right] \quad [n \geq 1].$

4. $D^n \left[z^{n-1} C\left(\dfrac{a}{z} \right) \right]$

$= (-1)^n \sqrt{\dfrac{a}{2\pi}} \, \dfrac{(n-1)!}{2} z^{-3/2} \left[e^{ia/z} L_{n-1}^{1/2-n}\left(-\dfrac{ia}{z} \right) + e^{-ia/z} L_{n-1}^{1/2-n}\left(\dfrac{ia}{z} \right) \right]$

$\quad [n \geq 1].$

1.7. The Generalized Fresnel Integrals $S(z, \nu)$ and $C(z, \nu)$

1.7.1. Derivatives with respect to the argument

1. $D^n [S(az, \nu)]$

$= -\dfrac{(n-1)!}{2i} a^\nu z^{\nu-n} \left[e^{iaz} L_{n-1}^{\nu-n}(-iaz) - e^{-iaz} L_{n-1}^{\nu-n}(iaz) \right] \quad [n \geq 1].$

2. $D^n \left[z^{n-1} S\left(\dfrac{a}{z}, \nu \right) \right]$

$= (-1)^{n-1} \dfrac{(n-1)!}{2i} a^\nu z^{-\nu-1} \left[e^{ia/z} L_{n-1}^{\nu-n}\left(-\dfrac{ia}{z} \right) - e^{-ia/z} L_{n-1}^{\nu-n}\left(\dfrac{ia}{z} \right) \right]$

$\quad [n \geq 1].$

3. $D^n [C(az, \nu)]$

$= -\dfrac{(n-1)!}{2} a^\nu z^{\nu-n} \left[e^{iaz} L_{n-1}^{\nu-n}(-iaz) + e^{-iaz} L_{n-1}^{\nu-n}(iaz) \right] \quad [n \geq 1].$

4. $D^n\left[z^{n-1}C\left(\frac{a}{z}, \nu\right)\right]$
$$= (-1)^{n-1}\frac{(n-1)!}{2}a^\nu z^{-\nu-1}\left[e^{ia/z}L_{n-1}^{\nu-n}\left(-\frac{ia}{z}\right) - e^{-ia/z}L_{n-1}^{\nu-n}\left(\frac{ia}{z}\right)\right]$$
$$[n \geq 1].$$

1.8. The Incomplete Gamma Functions $\gamma(\nu, z)$ and $\Gamma(\nu, z)$

1.8.1. Derivatives with respect to the argument

1. $D^n[\gamma(\nu, az)] = (n-1)!\, a^\nu z^{\nu-n} e^{-az} L_{n-1}^{\nu-n}(az)$ \hfill $[n \geq 1]$.

2. $D^n\left[z^{n-1}\gamma\left(\nu, \frac{a}{z}\right)\right] = (-1)^n(n-1)!\, a^\nu z^{-\nu-1} e^{-a/z} L_{n-1}^{\nu-n}\left(\frac{a}{z}\right)$ \hfill $[n \geq 1]$.

3. $D^n[z^{-\nu}\gamma(\nu, az)] = (-1)^n z^{-\nu-n}\gamma(\nu+n, az)$.

4. $D^n\left[z^{n+\nu-1}\gamma\left(\nu, \frac{a}{z}\right)\right] = z^{\nu-1}\gamma\left(\nu+n, \frac{a}{z}\right)$.

5. $D^n[e^{az}\gamma(\nu, az)] = (1-\nu)_n(-a)^n e^{az}\gamma(\nu-n, az)$.

6. $D^n\left[z^{n-1}e^{a/z}\gamma\left(\nu, \frac{a}{z}\right)\right] = (1-\nu)_n a^n z^{-n-1} e^{a/z}\gamma\left(\nu-n, \frac{a}{z}\right)$.

7. $D^n[z^{n-\nu}e^{az}\gamma(\nu, az)] = \frac{n!}{\nu}a^\nu \,{}_1F_1\!\left(\begin{matrix}n+1;\ az\\ \nu+1\end{matrix}\right)$.

8. $D^n\left[z^{\nu-1}e^{a/z}\gamma\left(\nu, \frac{a}{z}\right)\right] = (-1)^n \frac{n!}{\nu}a^\nu z^{-n-1} \,{}_1F_1\!\left(\begin{matrix}n+1;\ \frac{a}{z}\\ \nu+1\end{matrix}\right)$.

9. $D^n[z^{\nu+n}e^{-az}D^n[z^{-\nu}e^{az}\gamma(\nu, az)]] = n!\, a^n \gamma(\nu, az)$.

10. $D^n\left[z^{n-\nu}e^{-a/z}D^n\left[z^{n+\nu-1}e^{a/z}\gamma\left(\nu, \frac{a}{z}\right)\right]\right] = n!\, a^n z^{-n-1}\gamma\left(\nu, \frac{a}{z}\right)$.

11. $D^n[z^{\nu+n}e^{az}D^n[z^{-\nu}\gamma(\nu, az)]] = (\nu)_n(-a)^n e^{az}\gamma(\nu, az)$.

12. $D^n\left[z^{n-\nu}e^{a/z}D^n\left[z^{n+\nu-1}\gamma\left(\nu, \frac{a}{z}\right)\right]\right] = (\nu)_n(-a)^n z^{-n-1}e^{a/z}\gamma\left(\nu, \frac{a}{z}\right)$.

13. $D^n[z^{\nu-1}e^{-az}D^n[z^{n-\nu}e^{az}\gamma(\nu, az)]] = (-1)^n n!(1-\nu)_n z^{-n-1}\gamma(\nu, az)$.

14. $D^n\left[z^{2n-\nu+1}e^{-a/z}D^n\left[z^{\nu-1}e^{a/z}\gamma\left(\nu, \frac{a}{z}\right)\right]\right] = (-1)^n n!(1-\nu)_n \gamma\left(\nu, \frac{a}{z}\right)$.

15. $D^n[z^{n-\nu}e^{-az}D^n[e^{az}\gamma(\nu, az)]] = a^n(1-\nu)_n z^{-\nu}\gamma(\nu, az)$.

16. $D^n\left[z^{n+\nu}e^{-a/z}D^n\left[z^{n-1}e^{a/z}\gamma\left(\nu, \frac{a}{z}\right)\right]\right] = a^n(1-\nu)_n z^{\nu-n-1}\gamma\left(\nu, \frac{a}{z}\right)$.

17. $D^n[\Gamma(\nu, az)] = -(n-1)!\, a^\nu z^{\nu-n} e^{-az} L_{n-1}^{\nu-n}(az)$ \hfill $[n \geq 1]$.

18. $D^n\left[z^{n-1}\Gamma\left(\nu, \frac{a}{z}\right)\right] = (-1)^{n-1}(n-1)! a^\nu z^{-\nu-1} e^{-a/z} L_{n-1}^{\nu-n}(az)$ $[n \geq 1]$.

19. $D^n[z^{-\nu}\Gamma(\nu, az)] = (-1)^n z^{-\nu-n}\Gamma(\nu+n, az)$.

20. $D^n\left[z^{n+\nu-1}\Gamma\left(\nu, \frac{a}{z}\right)\right] = z^{\nu-1}\Gamma\left(\nu+n, \frac{a}{z}\right)$.

21. $D^n[e^{az}\Gamma(\nu, az)] = (1-\nu)_n(-a)^n e^{az}\Gamma(\nu-n, az)$.

22. $D^n\left[z^{n-1}e^{a/z}\Gamma\left(\nu, \frac{a}{z}\right)\right] = (1-\nu)_n a^n z^{-n-1} e^{a/z}\Gamma\left(\nu-n, \frac{a}{z}\right)$.

23. $D^n[z^{n-\nu}e^{az}\Gamma(\nu, az)] = n!(1-\nu)_n a^\nu \Psi\left(\begin{matrix}n+1; az\\ \nu+1\end{matrix}\right)$.

24. $D^n\left[z^{\nu-1}e^{a/z}\Gamma\left(\nu, \frac{a}{z}\right)\right] = (-1)^n n!(1-\nu)_n a^\nu z^{-n-1}\Psi\left(\begin{matrix}n+1; \frac{a}{z}\\ \nu+1\end{matrix}\right)$.

25. $D^n[\Gamma(1-n, az)] = (-1)^n \sqrt{\frac{a}{\pi}} z^{1/2-n} e^{-az/2} K_{n-1/2}\left(\frac{az}{2}\right)$.

26. $D^n\left[z^{n-1}\Gamma\left(1-n, \frac{a}{z}\right)\right] = \sqrt{\frac{a}{\pi}} z^{-3/2} e^{-a/(2z)} K_{n-1/2}\left(\frac{a}{2z}\right)$.

1.8.2. Derivatives with respect to the parameter

1. $\dfrac{\partial \gamma(\nu, z)}{\partial \nu} = \gamma(\nu, z)\ln z - \dfrac{z^\nu}{\nu^2} {}_2F_2\left(\begin{matrix}\nu, \nu; -z\\ \nu+1, \nu+1\end{matrix}\right)$.

1.9. The Parabolic Cylinder Function $D_\nu(z)$

1.9.1. Derivatives with respect to the argument

1. $D^n[D_\nu(az)] = \left(-\dfrac{a}{2}\right)^n \sum_{k=0}^n \binom{n}{k} 2^k (-\nu)_k H_{n-k}\left(\dfrac{az}{2}\right) D_{\nu-k}(az)$.

2. $ = \left(-\dfrac{ia}{2}\right)^n \sum_{k=0}^n \binom{n}{k}(-2i)^k H_{n-k}\left(\dfrac{iaz}{2}\right) D_{\nu+k}(az)$.

3. $D^n\left[e^{a^2z^2/4}D_\nu(az)\right] = (-a)^n (-\nu)_n e^{a^2z^2/4} D_{\nu-n}(az)$.

4. $D^n\left[e^{-a^2z^2/4}D_\nu(az)\right] = (-a)^n e^{-a^2z^2/4} D_{\nu+n}(az)$.

5. $D^n\left[z^{n-1}e^{a^2/(4z^2)}D_\nu\left(\dfrac{a}{z}\right)\right] = a^n (-\nu)_n z^{-n-1} e^{a^2/(4z^2)} D_{\nu-n}\left(\dfrac{a}{z}\right)$.

6. $D^n\left[z^{n-1}e^{-a^2/(4z^2)}D_\nu\left(\dfrac{a}{z}\right)\right] = a^n z^{-n-1} e^{-a^2/(4z^2)} D_{\nu+n}\left(\dfrac{a}{z}\right)$.

7. $D^n\left[z^{n-\nu/2-1}e^{a^2z/4}D_\nu(a\sqrt{z})\right] = 2^{-n}(-\nu)_{2n} z^{-\nu/2-1} e^{a^2z/4} D_{\nu-2n}(a\sqrt{z})$.

8. $D^n \left[z^{n+(\nu-1)/2} e^{-a^2 z/4} D_\nu(a\sqrt{z}) \right] = (-2)^{-n} z^{(\nu-1)/2} e^{-a^2 z/4} D_{\nu+2n}(a\sqrt{z}).$

9. $D^n \left[z^{\nu/2} e^{a^2/(4z)} D_\nu\left(\frac{a}{\sqrt{z}}\right) \right]$
$$= (-2)^{-n} (-\nu)_{2n} z^{-n+\nu/2} e^{a^2/(4z)} D_{\nu-2n}\left(\frac{a}{\sqrt{z}}\right).$$

10. $D^n \left[z^{-(\nu+1)/2} e^{-a^2/(4z)} D_\nu\left(\frac{a}{\sqrt{z}}\right) \right]$
$$= 2^{-n} z^{-n-(\nu+1)/2} e^{-a^2/(4z)} D_{\nu+2n}\left(\frac{a}{\sqrt{z}}\right).$$

11. $D^n \left[z^{-1/2} e^{-a^2 z/4} D_{-2n}(a\sqrt{z}) \right] = (-2)^{-n} z^{-n-1/2} e^{-a^2 z/2}.$

12. $D^n \left[z^{n-1/2} e^{-a^2/(4z)} D_{-2n}\left(\frac{a}{\sqrt{z}}\right) \right] = 2^{-n} z^{-1/2} e^{-a^2/(2z)}.$

13. $D^n \left[z^{-1} e^{-a^2 z/4} D_{-2n-1}(a\sqrt{z}) \right] = (-1)^n 2^{-n-1/2} \sqrt{\pi}\, z^{-n-1} \operatorname{erfc}\left(a\sqrt{\frac{z}{2}}\right).$

14. $D^n \left[z^n e^{-a^2/(4z)} D_{-2n-1}\left(\frac{a}{\sqrt{z}}\right) \right] = 2^{-n-1/2} \sqrt{\pi} \operatorname{erfc}\left(\frac{a}{\sqrt{2z}}\right).$

15. $D^n \left[e^{a^2 z/4} D_\nu(a\sqrt{z}) \right]$
$$= (-1)^n e^{a^2 z/4} \sum_{k=0}^{n-1} \frac{\Gamma(n+k)}{\Gamma(n-k) k! (2\sqrt{z})^{n+k}} a^{n-k} (-\nu)_{n-k} D_{\nu-n+k}(a\sqrt{z})$$
$$[n \geq 1].$$

16. $D^n \left[e^{-a^2 z/4} D_\nu(a\sqrt{z}) \right]$
$$= (-1)^n e^{-a^2 z/4} \sum_{k=0}^{n-1} \frac{\Gamma(n+k)}{\Gamma(n-k) k! (2\sqrt{z})^{n+k}} a^{n-k} D_{\nu+n-k}(a\sqrt{z}) \quad [n \geq 1].$$

17. $D^n \left[z^{n+1/2} e^{-a^2 z/2} D^n \left[z^{-1/2} e^{a^2 z/4} D_\nu(a\sqrt{z}) \right] \right]$
$$= \left(\frac{1-\nu}{2}\right)_n \left(\frac{a^2}{2}\right)^n e^{-a^2 z/4} D_\nu(a\sqrt{z}).$$

18. $D^n \left[z^{n-1/2} e^{-a^2/(2z)} D^n \left[z^{n-1/2} e^{a^2/(4z)} D_\nu\left(\frac{a}{\sqrt{z}}\right) \right] \right]$
$$= \left(\frac{1-\nu}{2}\right)_n \left(\frac{a^2}{2}\right)^n z^{-n-1} e^{-a^2/(4z)} D_\nu\left(\frac{a}{\sqrt{z}}\right).$$

19. $D^n \left[z^{n-1/2} e^{-a^2 z/2} D^n \left[e^{a^2 z/4} D_\nu(a\sqrt{z}) \right] \right]$
$$= \left(-\frac{\nu}{2}\right)_n \left(\frac{a^2}{2}\right)^n z^{-1/2} e^{-a^2 z/4} D_\nu(a\sqrt{z}).$$

20. $D^n\left[z^{n+1/2}e^{-a^2/(2z)}D^n\left[z^{n-1}e^{a^2/(4z)}D_\nu\left(\frac{a}{\sqrt{z}}\right)\right]\right]$
$$= \left(-\frac{\nu}{2}\right)_n \left(\frac{a^2}{2}\right)^n z^{-n-1/2} e^{-a^2/(4z)} D_\nu\left(\frac{a}{\sqrt{z}}\right).$$

21. $D^n\left[z^{n-1/2}e^{a^2z/2}D^n\left[e^{-a^2z/4}D_\nu(a\sqrt{z})\right]\right]$
$$= \left(\frac{\nu+1}{2}\right)_n \left(-\frac{a^2}{2}\right)^n z^{-1/2} e^{a^2z/4} D_\nu(a\sqrt{z}).$$

22. $D^n\left[z^{n+1/2}e^{a^2/(2z)}D^n\left[z^{n-1}e^{-a^2/(4z)}D_\nu\left(\frac{a}{\sqrt{z}}\right)\right]\right]$
$$= \left(\frac{\nu+1}{2}\right)_n \left(-\frac{a^2}{2}\right)^n z^{-n-1/2} e^{a^2/(4z)} D_\nu\left(\frac{a}{\sqrt{z}}\right).$$

23. $D^n\left[z^{n+1/2}e^{a^2z/2}D^n\left[z^{-1/2}e^{-a^2z/4}D_\nu(a\sqrt{z})\right]\right]$
$$= \left(\frac{\nu}{2}+1\right)_n \left(-\frac{a^2}{2}\right)^n e^{a^2z/4} D_\nu(a\sqrt{z}).$$

24. $D^n\left[z^{n-1/2}e^{a^2/(2z)}D^n\left[z^{n-1/2}e^{-a^2/(4z)}D_\nu\left(\frac{a}{\sqrt{z}}\right)\right]\right]$
$$= \left(\frac{\nu}{2}+1\right)_n \left(-\frac{a^2}{2}\right)^n z^{-n-1} e^{a^2/(4z)} D_\nu\left(\frac{a}{\sqrt{z}}\right).$$

25. $D^n\left[z^{\nu+1/2}e^{-a^2z/2}D^n\left[z^{n-\nu/2-1}e^{a^2z/4}D_\nu(a\sqrt{z})\right]\right]$
$$= (-4)^{-n}(-\nu)_{2n} z^{(\nu-1)/2-n} e^{-a^2z/4} D_\nu(a\sqrt{z}).$$

26. $D^n\left[z^{2n-\nu-1/2}e^{-a^2/(2z)}D^n\left[z^{\nu/2}e^{a^2/(4z)}D_\nu\left(\frac{a}{\sqrt{z}}\right)\right]\right]$
$$= (-4)^{-n}(-\nu)_{2n} z^{-(\nu+1)/2} e^{-a^2/(4z)} D_\nu\left(\frac{a}{\sqrt{z}}\right).$$

27. $D^n\left[z^{-\nu-1/2}e^{a^2z/2}D^n\left[z^{n+(\nu-1)/2}e^{-a^2z/4}D_\nu(a\sqrt{z})\right]\right]$
$$= (-4)^{-n}(\nu+1)_{2n} z^{-n-\nu/2-1} e^{a^2z/4} D_\nu(a\sqrt{z}).$$

28. $D^n\left[z^{2n+\nu+1/2}e^{a^2/(2z)}D^n\left[z^{-(\nu+1)/2}e^{-a^2/(4z)}D_\nu\left(\frac{a}{\sqrt{z}}\right)\right]\right]$
$$= (-4)^{-n}(\nu+1)_{2n} z^{\nu/2} e^{a^2/(4z)} D_\nu\left(\frac{a}{\sqrt{z}}\right).$$

1.9.2. Derivatives with respect to the order

1. $\dfrac{\partial D_\nu(z)}{\partial \nu}\bigg|_{\nu=2n} = 2^{-n-1} e^{-z^2/4} H_{2n}\left(\dfrac{z}{\sqrt{2}}\right)$
$$\times \left[-C - \ln 2 + \psi\left(\frac{1}{2}-n\right) + \pi\,\mathrm{erfi}\left(\frac{z}{\sqrt{2}}\right) - z^2\,{}_2F_2\!\left(\begin{matrix}1,1;\frac{z^2}{2}\\ \frac{3}{2},2\end{matrix}\right)\right]$$

$$-(-1)^n 2^{n-1} n!\, e^{-z^2/4} \sum_{k=1}^{n} \frac{1}{k} L_{n-k}^{k-1/2}\left(\frac{z^2}{2}\right)$$

$$\times \left[-\sqrt{\frac{\pi}{2}}\, z e^{z^2/2} L_{k-1}^{1/2-k}\left(-\frac{z^2}{2}\right) + \frac{\left(\frac{z^2}{2}\right)^k}{\left(\frac{1}{2}\right)_k}\, {}_1F_1\!\left(\begin{matrix} k;\ \frac{z^2}{2} \\ k+\frac{1}{2} \end{matrix}\right) \right]$$

$$-(-1)^n 2^{n-1} \left(\tfrac{1}{2}\right)_n e^{-z^2/4} \sum_{k=0}^{n} \binom{n}{k} \frac{\left(-\frac{z^2}{2}\right)^k}{\left(\frac{1}{2}\right)_k} \psi\!\left(\tfrac{1}{2}-k\right).$$

2. $\left.\dfrac{\partial D_\nu(z)}{\partial \nu}\right|_{\nu=2n+1} = 2^{-n-3/2} e^{-z^2/4} H_{2n+1}\!\left(\dfrac{z}{\sqrt{2}}\right)$

$$\times \left[C + \ln 2 - \psi\!\left(-\tfrac{1}{2}-n\right) - \pi\,\mathrm{erfi}\!\left(\tfrac{z}{\sqrt{2}}\right) + z^2\,{}_2F_2\!\left(\begin{matrix} 1,1;\ \frac{z^2}{2} \\ \frac{3}{2},2 \end{matrix}\right) \right]$$

$$+ (-1)^n 2^{-n-2}(2n+1)!\, z^{-1} e^{-z^2/4} \sum_{k=1}^{n} \frac{(-2z^2)^k}{(2k-1)!(n-k+1)!} \psi\!\left(\tfrac{1}{2}-k\right)$$

$$+ (-1)^n 2^{n-3/2} n!\, e^{-z^2/4} \sum_{k=1}^{n} \frac{1}{k} \left[L_{n-k}^{k-1/2}\!\left(\tfrac{z^2}{2}\right) + 2(n+1) L_{n-k+1}^{k-3/2}\!\left(\tfrac{z^2}{2}\right) \right]$$

$$\times \left[\frac{z^{2k-1}}{2^{k-1/2} \left(\frac{1}{2}\right)_k}\, {}_1F_1\!\left(\begin{matrix} k;\ \frac{z^2}{2} \\ k+\frac{1}{2} \end{matrix}\right) - \sqrt{\pi}\, e^{z^2/2} L_{k-1}^{1/2-k}\!\left(-\tfrac{z^2}{2}\right) \right]$$

$$- (-1)^n 2^{n-1/2} \sqrt{\pi}\, n!\, e^{z^2/4} L_n^{-n-1/2}\!\left(-\tfrac{z^2}{2}\right)$$

$$+ (-1)^n \frac{n!}{\left(\frac{3}{2}\right)_n} z^{2n+1} e^{-z^2/4}\, {}_1F_1\!\left(\begin{matrix} n+1;\ \frac{z^2}{2} \\ n+\frac{3}{2} \end{matrix}\right) - \frac{z^{2n+1}}{2} e^{-z^2/4} \psi\!\left(-n-\tfrac{1}{2}\right).$$

3. $\left.\dfrac{\partial}{\partial\nu}\left[D_\nu(z) D_\nu(e^{i\pi/2} z)\right]\right|_{\nu=-1/2} = \dfrac{e^{3i\pi/4}}{3} z^3\, {}_2F_3\!\left(\begin{matrix} 1,1;\ \frac{z^4}{16} \\ \frac{5}{4},\frac{3}{2},\frac{7}{4} \end{matrix}\right)$

$$+ \frac{\pi z}{8} \left[2^{3/2} e^{i\pi/4} (C + 2\ln 2)\, I_{-1/4}\!\left(\tfrac{z^2}{4}\right) I_{1/4}\!\left(\tfrac{z^2}{4}\right) \right.$$
$$\left. - e^{i\pi/2}(\pi + 2C + 4\ln 2)\, I_{1/4}^2\!\left(\tfrac{z^2}{4}\right) + (\pi - 2C - 4\ln 2)\, I_{-1/4}^2\!\left(\tfrac{z^2}{4}\right) \right]$$

$$[z > 0].$$

1.10. The Bessel Function $J_\nu(z)$

1.10.1. Derivatives with respect to the argument

1. $D^n[J_\nu(az)] = \left(\pm\dfrac{a}{2}\right)^n \sum_{k=0}^{n} (-1)^k \binom{n}{k} J_{\nu\pm 2k \mp n}(az).$

2. $\quad = n!(-z)^{-n} \sum_{k=0}^{n} \frac{(-az)^k}{(n-k)!} (\nu)_{n-k} \sum_{j=0}^{[k/2]} \frac{(2az)^{-j}}{j!(k-2j)!} J_{\nu+j-k}(az).$

3. $D^n \left[z^{n-1} J_\nu \left(\frac{a}{z} \right) \right] = \left(\mp \frac{a}{2} \right)^n z^{-n-1} \sum_{k=0}^{n} (-1)^k \binom{n}{k} J_{\nu \pm 2k \mp n} \left(\frac{a}{z} \right).$

4. $D^n[J_\nu(a\sqrt{z})] = \left(-\frac{1}{z} \right)^n \sum_{k=0}^{n} (\pm 1)^k \binom{n}{k} \left(\mp \frac{\nu}{2} \right)_{n-k} \left(\frac{a\sqrt{z}}{2} \right)^k J_{\nu \pm k}(a\sqrt{z}).$

5. $D^n[z^{\pm \nu/2} J_\nu(a\sqrt{z})] = \left(\pm \frac{a}{2} \right)^n z^{(\pm \nu - n)/2} J_{\nu \mp n}(a\sqrt{z}).$

6. $D^n[z^{(2n+1)/4} J_{n+1/2}(a\sqrt{z})] = \frac{1}{\sqrt{\pi}} \left(\frac{a}{2} \right)^{n-1/2} \sin(a\sqrt{z}).$

7. $D^n[z^{(2n+1)/4} J_{-n-1/2}(a\sqrt{z})] = \frac{(-1)^n}{\sqrt{\pi}} \left(\frac{a}{2} \right)^{n-1/2} \cos(a\sqrt{z}).$

8. $D^n[z^{-(2n+3)/4} J_{n+1/2}(a\sqrt{z})]$
$\qquad = \sqrt{\pi} \left(\frac{a}{2z} \right)^{n+1/2} J_{n+1/2} \left(\frac{a\sqrt{z}}{2} \right) J_{-n-1/2} \left(\frac{a\sqrt{z}}{2} \right).$

9. $D^n[z^{\pm \nu/4} J_\nu(a\sqrt[4]{z})]$
$\qquad = \left(\pm \frac{a}{4} \right)^n z^{(\pm \nu - 3n)/4} \sum_{k=0}^{n-1} (\mp a)^{-k} \frac{\Gamma(n+k)}{k!\Gamma(n-k)} z^{-k/4} J_{\nu \mp n \pm k}(a\sqrt[4]{z})$
$\qquad\qquad\qquad\qquad\qquad\qquad\qquad\qquad\qquad\qquad [n \geq 1].$

10. $D^n \left[z^{n-1} J_\nu \left(\frac{a}{\sqrt{z}} \right) \right] = \frac{1}{z} \sum_{k=0}^{n} (\pm 1)^k \binom{n}{k} \left(\mp \frac{\nu}{2} \right)_{n-k} \left(\frac{a}{2\sqrt{z}} \right)^k J_{\nu \pm k} \left(\frac{a}{\sqrt{z}} \right).$

11. $D^n \left[z^{n \pm \nu/2 - 1} J_\nu \left(\frac{a}{\sqrt{z}} \right) \right] = \left(\pm \frac{a}{2} \right)^n z^{-(n \mp \nu)/2 - 1} J_{\nu \pm n} \left(\frac{a}{\sqrt{z}} \right).$

12. $D^n \left[z^{(2n-5)/4} J_{n+1/2} \left(\frac{a}{\sqrt{z}} \right) \right] = \frac{(-1)^n}{\sqrt{\pi}} \left(\frac{a}{2} \right)^{n-1/2} z^{-n-1} \sin \left(\frac{a}{\sqrt{z}} \right).$

13. $D^n \left[z^{(2n-5)/4} J_{-n-1/2} \left(\frac{a}{\sqrt{z}} \right) \right] = \frac{1}{\sqrt{\pi}} \left(\frac{a}{2} \right)^{n-1/2} z^{-n-1} \cos \left(\frac{a}{\sqrt{z}} \right).$

14. $D^n \left[z^{(6n-1)/4} J_{n+1/2} \left(\frac{a}{\sqrt{z}} \right) \right]$
$\qquad = (-1)^n \sqrt{\pi} \left(\frac{a}{2} \right)^{n+1/2} z^{-1/2} J_{n+1/2} \left(\frac{a}{2\sqrt{z}} \right) J_{-n-1/2} \left(\frac{a}{2\sqrt{z}} \right).$

15. $D^n[z^{n-1/2} e^{\pm iaz} J_{n-1/2}(az)] = \frac{(2a)^{n-1/2}}{\sqrt{\pi}} z^{n-1} e^{\pm 2iaz} \qquad [n \geq 1].$

1.10. The Bessel Function $J_\nu(z)$ [1.10.1]

16. $D^n\left[z^{-1/2} e^{\pm ia/z} J_{n-1/2}\left(\dfrac{a}{z}\right)\right] = \dfrac{(-1)^n}{\sqrt{\pi}} (2a)^{n-1/2} z^{-2n} e^{\pm 2ia/z}$ $\quad [n \geq 1]$.

17. $D^n[z^{n-1/2} \sin(az) J_{n-1/2}(az)] = \dfrac{(2a)^{n-1/2}}{\sqrt{\pi}} z^{n-1} \sin(2az)$.

18. $D^n[z^{n-1/2} \cos(az) J_{n-1/2}(az)] = \dfrac{(2a)^{n-1/2}}{\sqrt{\pi}} z^{n-1} \cos(2az)$ $\quad [n \geq 1]$.

19. $D^n[z^{n-1/2} \sin(az) J_{1/2-n}(az)] = (-1)^{n+1} \dfrac{(2a)^{n-1/2}}{\sqrt{\pi}} z^{n-1} \cos(2az)$
$\quad [n \geq 1]$.

20. $D^n[z^{n-1/2} \cos(az) J_{1/2-n}(az)] = (-1)^n \dfrac{(2a)^{n-1/2}}{\sqrt{\pi}} z^{n-1} \sin(2az)$.

21. $D^n\left[z^{-1/2} \sin\dfrac{a}{z} J_{n-1/2}\left(\dfrac{a}{z}\right)\right] = \dfrac{(-1)^n}{\sqrt{\pi}} (2a)^{n-1/2} z^{-2n} \sin\dfrac{2a}{z}$.

22. $D^n\left[z^{-1/2} \cos\dfrac{a}{z} J_{n-1/2}\left(\dfrac{a}{z}\right)\right] = \dfrac{(-1)^n}{\sqrt{\pi}} (2a)^{n-1/2} z^{-2n} \cos\dfrac{2a}{z}$ $\quad [n \geq 1]$.

23. $D^n[J_\nu^2(a\sqrt{z})] = \left(\dfrac{a}{2\sqrt{z}}\right)^n \sum_{k=0}^{n} (-1)^k \binom{n}{k} J_{\nu+k}(a\sqrt{z}) J_{\nu-n+k}(a\sqrt{z})$.

24. $D^n[z^{n-1/2} J_{n-1/2}^2(a\sqrt{z})] = \dfrac{a^{n-1/2}}{\sqrt{\pi}} z^{(2n-3)/4} J_{n-1/2}(2a\sqrt{z})$ $\quad [n \geq 1]$.

25. $D^n[z^{n-1/2} J_{1/2-n}^2(a\sqrt{z})] = -\dfrac{a^{n-1/2}}{\sqrt{\pi}} z^{(2n-3)/4} J_{n-1/2}(2a\sqrt{z})$ $\quad [n \geq 1]$.

26. $D^n[z^{n-1/2} J_{n-1/2}(a\sqrt{z}) J_{1/2-n}(a\sqrt{z})]$
$\qquad = \dfrac{a^{n-1/2}}{\sqrt{\pi}} z^{(2n-3)/4} J_{1/2-n}(2a\sqrt{z})$ $\quad [n \geq 1]$.

27. $D^n[z^n J_{n-1/2}(a\sqrt{z}) J_{n+1/2}(a\sqrt{z})] = \dfrac{a^{n-1/2}}{\sqrt{\pi}} z^{(2n-1)/4} J_{n+1/2}(2a\sqrt{z})$.

28. $D^n\left[z^{-1/2} J_{n-1/2}^2\left(\dfrac{a}{\sqrt{z}}\right)\right] = \dfrac{(-1)^n}{\sqrt{\pi}} a^{n-1/2} z^{-(6n+1)/4} J_{n-1/2}\left(\dfrac{2a}{\sqrt{z}}\right)$
$\quad [n \geq 1]$.

29. $D^n\left[z^{-1/2} J_{1/2-n}^2\left(\dfrac{a}{\sqrt{z}}\right)\right] = \dfrac{(-1)^{n+1}}{\sqrt{\pi}} a^{n-1/2} z^{-(6n+1)/4} J_{n-1/2}\left(\dfrac{2a}{\sqrt{z}}\right)$
$\quad [n \geq 1]$.

30. $D^n\left[z^{-1/2} J_{n-1/2}\left(\dfrac{a}{\sqrt{z}}\right) J_{1/2-n}\left(\dfrac{a}{\sqrt{z}}\right)\right]$
$\qquad = \dfrac{(-1)^n}{\sqrt{\pi}} a^{n-1/2} z^{-(6n+1)/4} J_{1/2-n}\left(\dfrac{2a}{\sqrt{z}}\right)$.

31. $D^n \left[z^{-1} J_{n-1/2}\left(\frac{a}{\sqrt{z}}\right) J_{n+1/2}\left(\frac{a}{\sqrt{z}}\right) \right]$
$$= \frac{(-1)^n}{\sqrt{\pi}} a^{n-1/2} z^{-(6n+3)/4} J_{n+1/2}\left(\frac{2a}{\sqrt{z}}\right).$$

32. $D^n \left[z^{n-1} J_\nu^2\left(\frac{a}{\sqrt{z}}\right) \right]$
$$= \left(-\frac{a}{2}\right)^n z^{-n/2-1} \sum_{k=0}^{n} (-1)^k \binom{n}{k} J_{\nu+k}\left(\frac{a}{\sqrt{z}}\right) J_{\nu-n+k}\left(\frac{a}{\sqrt{z}}\right).$$

1.10.2. Derivatives with respect to the order

1. $\left.\dfrac{\partial J_\nu(z)}{\partial \nu}\right|_{\nu=\pm n} = (\pm 1)^n \dfrac{\pi}{2} Y_n(z) \pm (\pm 1)^n \dfrac{n!}{2} \sum_{k=0}^{n-1} \dfrac{\left(\frac{z}{2}\right)^{k-n}}{k!(n-k)} J_k(z).$

2. $\left.\dfrac{\partial J_\nu(z)}{\partial \nu}\right|_{\nu=1/2} = \sqrt{\dfrac{2}{\pi z}} \left[\sin z \operatorname{ci}(2z) - \cos z \operatorname{Si}(2z)\right]$ [[10], 7.9(18)].

3. $\left.\dfrac{\partial J_\nu(z)}{\partial \nu}\right|_{\nu=-1/2} = \sqrt{\dfrac{2}{\pi z}} \left[\sin z \operatorname{Si}(2z) + \cos z \operatorname{ci}(2z)\right]$ [[10], 7.9(19)].

4. $\left.\dfrac{\partial J_\nu(z)}{\partial \nu}\right|_{\nu=n+1/2} = \operatorname{ci}(2z) J_{n+1/2}(z) - (-1)^n \operatorname{Si}(2z) J_{-n-1/2}(z)$
$$+ \frac{n!}{2}\left(\frac{2}{z}\right)^n \sum_{k=0}^{n-1} \frac{\left(\frac{z}{2}\right)^k}{k!(n-k)} J_{k+1/2}(z) - \frac{n!\sqrt{\pi z}}{2} \sum_{k=1}^{n} \frac{\left(\frac{2}{z}\right)^k}{(n-k)!k}$$
$$\times \sum_{p=0}^{k-1} \frac{z^p}{p!} [J_{n-k+1/2}(z) J_{p-1/2}(2z) - (-1)^{n-k-p} J_{k-n-1/2}(z) J_{1/2-p}(2z)].$$

5. $\left.\dfrac{\partial J_\nu(z)}{\partial \nu}\right|_{\nu=1/2-n} = \operatorname{ci}(2z) J_{1/2-n}(z) - (-1)^n \operatorname{Si}(2z) J_{n-1/2}(z)$
$$- \frac{n!}{2} \sum_{k=0}^{n-1} \frac{\left(-\frac{z}{2}\right)^{k-n}}{k!(n-k)} J_{1/2-k}(z) - \frac{n!\sqrt{\pi}}{2} \sum_{k=1}^{n} \frac{2^k}{(n-k)!k}$$
$$\times \sum_{p=0}^{k-1} \frac{z^{p-k+1/2}}{p!} [(-1)^k J_{k-n+1/2}(z) J_{p-1/2}(2z)$$
$$- (-1)^{n+p} J_{n-k-1/2}(z) J_{1/2-p}(2z)].$$

6. $\left.\dfrac{\partial J_\nu(z)}{\partial \nu}\right|_{\nu=\pm n\pm 1/2} = (\mp 1)^{n-1} 2^{n+1/2} \sqrt{\pi}\, z^{-n-1/2}$
$$\times \sum_{k=0}^{n-1} \frac{(n-k)! \left(\frac{z}{2}\right)^{2k}}{k!(n-2k)!\Gamma\left(k+\frac{1}{2}\right)\Gamma\left(k-n+\frac{1}{2}\right)} \left[\left(\psi\left(k+\frac{1}{2}\right) - \psi\left(k-n+\frac{1}{2}\right)\right)\right.$$

$$\times \left\{ \begin{matrix} \sin z \\ \cos z \end{matrix} \right\} \mp \left\{ \begin{matrix} \sin z \\ \cos z \end{matrix} \right\} \operatorname{ci}(2z) + \left\{ \begin{matrix} \cos z \\ \sin z \end{matrix} \right\} \operatorname{Si}(2z) \bigg]$$

$$+ (\mp 1)^n 2^{n-1/2} \sqrt{\pi}\, z^{-n+1/2} \sum_{k=0}^{n-1} \frac{(n-k-1)! \left(\frac{z}{2}\right)^{2k}}{k!(n-2k-1)!\Gamma\left(k+\frac{3}{2}\right)\Gamma\left(k-n+\frac{1}{2}\right)}$$

$$\times \left[\left(\psi\left(k+\frac{3}{2}\right) - \psi\left(k-n+\frac{1}{2}\right) \right) \left\{ \begin{matrix} \cos z \\ \sin z \end{matrix} \right\} \right.$$

$$\left. - \left\{ \begin{matrix} \sin z \\ \cos z \end{matrix} \right\} \operatorname{Si}(2z) \mp \left\{ \begin{matrix} \cos z \\ \sin z \end{matrix} \right\} \operatorname{ci}(2z) \right] \quad [n \geq 1].$$

1.11. The Bessel Function $Y_\nu(z)$

1.11.1. Derivatives with respect to the argument

1. $D^n[Y_\nu(az)] = \left(\pm \frac{a}{2}\right)^n \sum_{k=0}^{n} (-1)^k \binom{n}{k} Y_{\nu \pm 2k \mp n}(az)$.

2. $\quad = n!(-z)^{-n} \sum_{k=0}^{n} \frac{(-az)^k}{(n-k)!} (\nu)_{n-k} \sum_{j=0}^{[k/2]} \frac{(2az)^{-j}}{j!(k-2j)!} Y_{\nu+j-k}(az)$.

3. $D^n\left[z^{n-1} Y_\nu\left(\frac{a}{z}\right)\right] = \left(\mp \frac{a}{2}\right)^n z^{-n-1} \sum_{k=0}^{n} (-1)^k \binom{n}{k} Y_{\nu \pm 2k \mp n}\left(\frac{a}{z}\right)$.

4. $D^n[z^{\pm \nu/2} Y_\nu(a\sqrt{z})] = \left(\pm \frac{a}{2}\right)^n z^{(\pm \nu - n)/2} Y_{\nu \mp n}(a\sqrt{z})$.

5. $D^n[z^{(2n+1)/4} Y_{n+1/2}(a\sqrt{z})] = -\frac{1}{\sqrt{\pi}} \left(\frac{a}{2}\right)^{n-1/2} \cos(a\sqrt{z})$.

6. $D^n[z^{(2n+1)/4} Y_{-n-1/2}(a\sqrt{z})] = \frac{(-1)^n}{\sqrt{\pi}} \left(\frac{a}{2}\right)^{n-1/2} \sin(a\sqrt{z})$.

7. $D^n[z^{-(2n+3)/4} Y_{-n-1/2}(a\sqrt{z})]$
$$= (-1)^{n+1} \sqrt{\pi} \left(\frac{a}{2z}\right)^{n+1/2} Y_{n+1/2}\left(\frac{a\sqrt{z}}{2}\right) Y_{-n-1/2}\left(\frac{a\sqrt{z}}{2}\right).$$

8. $D^n[z^{\pm \nu/4} Y_\nu(a\sqrt[4]{z})]$
$$= \left(\pm \frac{a}{4}\right)^n z^{(\pm \nu - 3n)/4} \sum_{k=0}^{n-1} (\mp a)^{-k} \frac{\Gamma(n+k)}{k!\Gamma(n-k)} z^{-k/4} Y_{\nu \mp n \pm k}(a\sqrt[4]{z}) \quad [n \geq 1].$$

9. $D^n\left[z^{n \pm \nu/2 - 1} Y_\nu\left(\frac{a}{\sqrt{z}}\right)\right] = \left(\pm \frac{a}{2}\right)^n z^{(\pm \nu - n)/2 - 1} Y_{\nu \pm n}\left(\frac{a}{\sqrt{z}}\right)$.

10. $D^n\left[z^{(2n-5)/4} Y_{n+1/2}\left(\frac{a}{\sqrt{z}}\right)\right] = \frac{(-1)^{n+1}}{\sqrt{\pi}} \left(\frac{a}{2}\right)^{n-1/2} z^{-n-1} \cos\left(\frac{a}{\sqrt{z}}\right)$.

11. $D^n\left[z^{(2n-5)/4}Y_{-n-1/2}\left(\dfrac{a}{\sqrt{z}}\right)\right] = \dfrac{1}{\sqrt{\pi}}\left(\dfrac{a}{2}\right)^{n-1/2} z^{-n-1} \sin\left(\dfrac{a}{\sqrt{z}}\right).$

12. $D^n\left[z^{(6n-1)/4}Y_{-n-1/2}\left(\dfrac{a}{\sqrt{z}}\right)\right]$
$= -\sqrt{\pi}\left(\dfrac{a}{2}\right)^{n+1/2} z^{-1/2} Y_{n+1/2}\left(\dfrac{a}{2\sqrt{z}}\right) Y_{-n-1/2}\left(\dfrac{a}{2\sqrt{z}}\right).$

13. $D^n[z^{n-1/2}e^{\pm iaz}Y_{1/2-n}(az)] = (-1)^{n+1}\dfrac{(2a)^{n-1/2}}{\sqrt{\pi}} z^{n-1} e^{\pm 2iaz}$ $\quad [n \geq 1].$

14. $D^n\left[z^{-1/2}e^{\pm ia/z}Y_{1/2-n}\left(\dfrac{a}{z}\right)\right] = -\dfrac{(2a)^{n-1/2}}{\sqrt{\pi}} z^{-2n} e^{\pm 2ia/z}$ $\quad [n \geq 1].$

15. $D^n[z^{n-1/2}Y^2_{1/2-n}(a\sqrt{z})] = (-1)^{n+1}\dfrac{a^{n-1/2}}{\sqrt{\pi}} z^{(2n-3)/4} Y_{1/2-n}(2a\sqrt{z})$
$\quad [n \geq 1].$

16. $D^n[z^{n-1/2}Y_{n-1/2}(a\sqrt{z})Y_{1/2-n}(a\sqrt{z})]$
$= (-1)^{n+1}\dfrac{a^{n-1/2}}{\sqrt{\pi}} z^{(2n-3)/4} Y_{n-1/2}(2a\sqrt{z})$ $\quad [n \geq 1].$

17. $D^n[z^n Y_{n-1/2}(a\sqrt{z})Y_{n+1/2}(a\sqrt{z})]$
$= (-1)^{n+1}\dfrac{a^{n-1/2}}{\sqrt{\pi}} z^{(2n-1)/4} Y_{-n-1/2}(2a\sqrt{z}).$

18. $D^n\left[z^{-1/2}Y^2_{1/2-n}\left(\dfrac{a}{\sqrt{z}}\right)\right] = -\dfrac{a^{n-1/2}}{\sqrt{\pi}} z^{-(6n+1)/4} Y_{1/2-n}\left(\dfrac{2a}{\sqrt{z}}\right)$ $\quad [n \geq 1].$

19. $D^n\left[z^{-1/2}Y_{n-1/2}\left(\dfrac{a}{\sqrt{z}}\right)Y_{1/2-n}\left(\dfrac{a}{\sqrt{z}}\right)\right]$
$= -\dfrac{a^{n-1/2}}{\sqrt{\pi}} z^{-(6n+1)/4} Y_{n-1/2}\left(\dfrac{2a}{\sqrt{z}}\right).$

20. $D^n\left[z^{-1}Y_{n-1/2}\left(\dfrac{a}{\sqrt{z}}\right)Y_{n+1/2}\left(\dfrac{a}{\sqrt{z}}\right)\right]$
$= -\dfrac{1}{\sqrt{\pi}} a^{n-1/2} z^{-(6n+3)/4} Y_{-n-1/2}\left(\dfrac{2a}{\sqrt{z}}\right).$

21. $D^n[z^{n-1/2}J_{n-1/2}(a\sqrt{z})Y_{1/2-n}(a\sqrt{z})]$
$= (-1)^{n+1}\dfrac{a^{n-1/2}}{\sqrt{\pi}} z^{(2n-3)/4} J_{n-1/2}(2a\sqrt{z})$ $\quad [n \geq 1].$

22. $D^n[z^{n-1/2}J_{1/2-n}(a\sqrt{z})Y_{n-1/2}(a\sqrt{z})]$
$= (-1)^{n+1}\dfrac{a^{n-1/2}}{\sqrt{\pi}} z^{(2n-3)/4} J_{n-1/2}(2a\sqrt{z})$ $\quad [n \geq 1].$

23. $D^n[z^{n-1/2}J_{n-1/2}(a\sqrt{z})Y_{n-1/2}(a\sqrt{z})]$
$$= \frac{(-1)^n}{\sqrt{\pi}}a^{n-1/2}z^{(2n-3)/4}J_{1/2-n}(2a\sqrt{z}).$$

24. $D^n[z^n J_{n-1/2}(a\sqrt{z})Y_{-n-1/2}(a\sqrt{z})]$
$$= \frac{(-1)^n}{\sqrt{\pi}}a^{n-1/2}z^{(2n-1)/4}J_{n+1/2}(2a\sqrt{z}).$$

25. $D^n[z^n J_{n+1/2}(a\sqrt{z})Y_{1/2-n}(a\sqrt{z})]$
$$= \frac{(-1)^{n+1}}{\sqrt{\pi}}a^{n-1/2}z^{(2n-1)/4}J_{n+1/2}(2a\sqrt{z}).$$

26. $D^n\left[z^{-1/2}J_{n-1/2}\left(\frac{a}{\sqrt{z}}\right)Y_{n-1/2}\left(\frac{a}{\sqrt{z}}\right)\right]$
$$= \frac{a^{n-1/2}}{\sqrt{\pi}}z^{-(6n+1)/4}J_{1/2-n}\left(\frac{2a}{\sqrt{z}}\right).$$

27. $D^n\left[z^{-1/2}J_{1/2-n}\left(\frac{a}{\sqrt{z}}\right)Y_{n-1/2}\left(\frac{a}{\sqrt{z}}\right)\right]$
$$= -\frac{a^{n-1/2}}{\sqrt{\pi}}z^{-(6n+1)/4}J_{n-1/2}\left(\frac{2a}{\sqrt{z}}\right).$$

28. $D^n\left[z^{-1}J_{n-1/2}\left(\frac{a}{\sqrt{z}}\right)Y_{-n-1/2}\left(\frac{a}{\sqrt{z}}\right)\right] = \frac{a^{n-1/2}}{\sqrt{\pi}}z^{-(6n+3)/4}J_{n+1/2}\left(\frac{2a}{\sqrt{z}}\right).$

29. $D^n\left[z^{-1}J_{n+1/2}\left(\frac{a}{\sqrt{z}}\right)Y_{1/2-n}\left(\frac{a}{\sqrt{z}}\right)\right]$
$$= -\frac{a^{n-1/2}}{\sqrt{\pi}}a^{n-1/2}z^{-(6n+3)/4}J_{n+1/2}\left(\frac{2a}{\sqrt{z}}\right).$$

1.11.2. Derivatives with respect to the order

1. $\left.\dfrac{\partial Y_\nu(z)}{\partial \nu}\right|_{\nu=\pm n} = -(\pm 1)^n \dfrac{\pi}{2} J_n(z) \pm (\pm 1)^n \dfrac{n!}{2} \sum_{k=0}^{n-1} \dfrac{\left(\frac{z}{2}\right)^{k-n}}{k!(n-k)} Y_k(z).$

2. $\left.\dfrac{\partial Y_\nu(z)}{\partial \nu}\right|_{\nu=\pm n\pm 1/2} = -\pi J_{\pm n \pm 1/2}(z) + (-1)^n \left.\dfrac{\partial J_\nu(z)}{\partial \nu}\right|_{\nu=\mp n \mp 1/2}.$

1.12. The Hankel Functions $H_\nu^{(1)}(z)$ and $H_\nu^{(2)}(z)$

1.12.1. Derivatives with respect to the argument

1. $D^n[z^{\pm\nu/2}H_\nu^{(j)}(a\sqrt{z})] = \left(\pm\dfrac{a}{2}\right)^n z^{(\pm\nu-n)/2} H_{\nu\mp n}^{(j)}(a\sqrt{z})$ $\qquad [j=1,2].$

2. $D^n\left[z^{n\pm\nu/2-1}H_\nu^{(j)}\left(\dfrac{a}{\sqrt{z}}\right)\right] = \left(\pm\dfrac{a}{2}\right)^n z^{(\pm\nu-n)/2-1} H_{\nu\pm n}^{(j)}\left(\dfrac{a}{\sqrt{z}}\right)$ $\qquad [j=1,2].$

3. $D^n[z^{(2n+1)/4} H^{(j)}_{n+1/2}(a\sqrt{z})] = \dfrac{(-1)^j i}{\sqrt{\pi}} \left(\dfrac{a}{2}\right)^{n-1/2} e^{(-1)^{j+1} ia\sqrt{z}}$ $[j=1,2]$.

4. $D^n\left[z^{n-1/2} H^{(1)}_{1/2-n}(a\sqrt{z}) H^{(2)}_{1/2-n}(a\sqrt{z})\right] = 0$ $[n \geq 1]$.

1.12.2. Derivatives with respect to the order

1. $\left.\dfrac{\partial H^{(j)}_\nu(z)}{\partial \nu}\right|_{\nu=n} = (-1)^j \dfrac{\pi i}{2} H^{(j)}_n(z) + \dfrac{n!}{2} \sum_{k=0}^{n-1} \dfrac{\left(\frac{z}{2}\right)^{k-n}}{k!(n-k)} H^{(j)}_k(z)$ $[j=1,2]$.

2. $\left.\dfrac{\partial H^{(j)}_\nu(z)}{\partial \nu}\right|_{\nu=-n} = (-1)^{j+n} \dfrac{\pi i}{2} H^{(j)}_n(z) - (-1)^n \dfrac{n!}{2} \sum_{k=0}^{n-1} \dfrac{\left(\frac{z}{2}\right)^{k-n}}{k!(n-k)} H^{(j)}_k(z)$

    $[j=1,2]$.

3. $\left.\dfrac{\partial H^{(j)}_\nu(z)}{\partial \nu}\right|_{\nu=1/2} = \sqrt{\dfrac{2}{\pi z}}$

$\times \{e^{(-1)^j iz}[(-1)^{j+1} i \operatorname{ci}(2z) - \operatorname{Si}(2z)] + (-1)^j i\pi \sin z\}$ $[j=1,2]$.

4. $\left.\dfrac{\partial H^{(j)}_\nu(z)}{\partial \nu}\right|_{\nu=-1/2} = \sqrt{\dfrac{2}{\pi z}}$

$\times \{e^{(-1)^j iz}[\operatorname{ci}(2z) + (-1)^{j+1} i \operatorname{Si}(2z)] + (-1)^j i\pi \cos z\}$ $[j=1,2]$.

1.13. The Modified Bessel Function $I_\nu(z)$

1.13.1. Derivatives with respect to the argument

1. $D^n[I_\nu(az)] = \left(\dfrac{a}{2}\right)^n \sum_{k=0}^{n} \binom{n}{k} I_{\nu \pm 2k \mp n}(az)$.

2. $ = n!(-z)^{-n} \sum_{k=0}^{n} \dfrac{(-az)^k}{(n-k)!} (\nu)_{n-k} \sum_{p=0}^{[k/2]} \dfrac{(2az)^{-j}}{p!(k-2p)!} I_{\nu-k+p}(az)$.

3. $D^n\left[z^{n-1} I_\nu\left(\dfrac{a}{z}\right)\right] = \left(-\dfrac{a}{2}\right)^n z^{-n-1} \sum_{k=0}^{n} \binom{n}{k} I_{\nu \pm 2k \mp n}\left(\dfrac{a}{z}\right)$.

4. $D^n[I_\nu(a\sqrt{z})] = \left(-\dfrac{1}{z}\right)^n \sum_{k=0}^{n} \binom{n}{k} \left(\mp \dfrac{\nu}{2}\right)_{n-k} \left(-\dfrac{a\sqrt{z}}{2}\right)^k I_{\nu \pm k}(a\sqrt{z})$.

5. $D^n[z^{\pm \nu/2} I_\nu(a\sqrt{z})] = \left(\dfrac{a}{2}\right)^n z^{(\pm \nu - n)/2} I_{\nu \mp n}(a\sqrt{z})$.

6. $D^n[z^{(2n+1)/4} I_{n+1/2}(a\sqrt{z})] = \dfrac{1}{\sqrt{\pi}} \left(\dfrac{a}{2}\right)^{n-1/2} \sinh(a\sqrt{z})$.

1.13. The Modified Bessel Function $I_\nu(z)$

7. $D^n[z^{(2n+1)/4}I_{-n-1/2}(a\sqrt{z})] = \dfrac{1}{\sqrt{\pi}}\left(\dfrac{a}{2}\right)^{n-1/2}\cosh(a\sqrt{z}).$

8. $D^n[z^{-(2n+3)/4}I_{n+1/2}(a\sqrt{z})]$
$= \sqrt{\pi}\left(\dfrac{a}{2z}\right)^{n+1/2}I_{n+1/2}\left(\dfrac{a\sqrt{z}}{2}\right)I_{-n-1/2}\left(\dfrac{a\sqrt{z}}{2}\right).$

9. $D^n\left[z^{\pm\nu/4}I_\nu(a\sqrt[4]{z})\right]$
$= \left(\dfrac{a}{4}\right)^n z^{(\pm\nu-3n)/4}\displaystyle\sum_{k=0}^{n-1}(-a)^{-k}\dfrac{\Gamma(n+k)}{k!\,\Gamma(n-k)}z^{-k/4}I_{\nu\mp n\pm k}(a\sqrt[4]{z})\quad [n\geq 1].$

10. $D^n\left[z^{n-1}I_\nu\left(\dfrac{a}{\sqrt{z}}\right)\right] = \dfrac{1}{z}\displaystyle\sum_{k=0}^{n}\binom{n}{k}\left(\mp\dfrac{\nu}{2}\right)_{n-k}\left(-\dfrac{a}{2\sqrt{z}}\right)^k I_{\nu\pm k}\left(\dfrac{a}{\sqrt{z}}\right).$

11. $D^n\left[z^{n\pm\nu/2-1}I_\nu\left(\dfrac{a}{\sqrt{z}}\right)\right] = \left(-\dfrac{a}{2}\right)^n z^{-(n\mp\nu)/2-1}I_{\nu\pm n}\left(\dfrac{a}{\sqrt{z}}\right).$

12. $D^n\left[z^{(2n-5)/4}I_{n+1/2}\left(\dfrac{a}{\sqrt{z}}\right)\right] = \dfrac{(-1)^n}{\sqrt{\pi}}\left(\dfrac{a}{2}\right)^{n-1/2}z^{-n-1}\sinh\left(\dfrac{a}{\sqrt{z}}\right).$

13. $D^n\left[z^{(2n-5)/4}I_{-n-1/2}\left(\dfrac{a}{\sqrt{z}}\right)\right] = \dfrac{(-1)^n}{\sqrt{\pi}}\left(\dfrac{a}{2}\right)^{n-1/2}z^{-n-1}\cosh\left(\dfrac{a}{\sqrt{z}}\right).$

14. $D^n\left[z^{(6n-1)/4}I_{n+1/2}\left(\dfrac{a}{\sqrt{z}}\right)\right]$
$= (-1)^n\sqrt{\pi}\left(\dfrac{a}{2}\right)^{n+1/2}z^{-1/2}I_{n+1/2}\left(\dfrac{a}{2\sqrt{z}}\right)I_{-n-1/2}\left(\dfrac{a}{2\sqrt{z}}\right).$

15. $D^n[z^{n-1/2}e^{\pm az}I_{n-1/2}(az)] = \dfrac{(2a)^{n-1/2}}{\sqrt{\pi}}z^{n-1}e^{\pm 2az}\quad [n\geq 1].$

16. $D^n[z^{n-1/2}e^{-az}I_{n+1/2}(az)] = \dfrac{(2a)^{-n-1/2}}{\sqrt{\pi}}z^{-n-1}\gamma(2n+1,2az).$

17. $D^n[z^{n-1/2}e^{az}I_\nu(az)] = (2a)^\nu z^{\nu-1/2}\dfrac{\Gamma\left(\nu+n+\frac{1}{2}\right)}{\sqrt{\pi}\,\Gamma(2\nu+1)}\,{}_1F_1\left(\begin{array}{c}\nu+n+\frac{1}{2}\\2\nu+1;\,2az\end{array}\right).$

18. $D^n\left[z^{-1/2}e^{\pm a/z}I_{n-1/2}\left(\dfrac{a}{z}\right)\right] = \dfrac{(-1)^n}{\sqrt{\pi}}(2a)^{n-1/2}z^{-2n}e^{\pm 2a/z}\quad [n\geq 1].$

19. $D^n\left[z^{-1/2}e^{-a/z}I_{n+1/2}\left(\dfrac{a}{z}\right)\right] = (-1)^n\dfrac{(2a)^{-n-1/2}}{\sqrt{\pi}}\gamma\left(2n+1,\dfrac{2a}{z}\right).$

20. $D^n\left[\dfrac{1}{\sqrt{z}}e^{a/z}I_\nu\left(\dfrac{a}{z}\right)\right]$
$= (-1)^n(2a)^\nu z^{-\nu-n-1/2}\dfrac{\Gamma\left(\nu+n+\frac{1}{2}\right)}{\sqrt{\pi}\,\Gamma(2\nu+1)}\,{}_1F_1\left(\begin{array}{c}\nu+n+\frac{1}{2}\\2\nu+1;\,\dfrac{2a}{z}\end{array}\right).$

21. $D^n[z^{n-1/2}\sinh(az)I_{n-1/2}(az)] = \dfrac{(2a)^{n-1/2}}{\sqrt{\pi}} z^{n-1}\sinh(2az)$.

22. $D^n[z^{n-1/2}\cosh(az)I_{n-1/2}(az)] = \dfrac{(2a)^{n-1/2}}{\sqrt{\pi}} z^{n-1}\cosh(2az)$ $\quad [n \geq 1]$.

23. $D^n\left[z^{-1/2}\sinh\dfrac{a}{z} I_{n-1/2}\left(\dfrac{a}{z}\right)\right] = \dfrac{(-1)^n}{\sqrt{\pi}}(2a)^{n-1/2}z^{-2n}\sinh\dfrac{2a}{z}$.

24. $D^n\left[z^{-1/2}\cosh\dfrac{a}{z} I_{n-1/2}\left(\dfrac{a}{z}\right)\right] = \dfrac{(-1)^n}{\sqrt{\pi}}(2a)^{n-1/2}z^{-2n}\cosh\dfrac{2a}{z}$ $\quad [n \geq 1]$.

25. $D^n[I_\nu^2(a\sqrt{z})] = \left(\dfrac{a}{2\sqrt{z}}\right)^n \sum\limits_{k=0}^{n}\binom{n}{k} I_{\nu+k}(a\sqrt{z}) I_{\nu-n+k}(a\sqrt{z})$.

26. $D^n[z^{n-1/2}I_{n-1/2}^2(a\sqrt{z})] = \dfrac{a^{n-1/2}}{\sqrt{\pi}} z^{(2n-3)/4} I_{n-1/2}(2a\sqrt{z})$ $\quad [n \geq 1]$.

27. $D^n[z^{n-1/2}I_{1/2-n}^2(a\sqrt{z})] = \dfrac{a^{n-1/2}}{\sqrt{\pi}} z^{(2n-3)/4} I_{n-1/2}(2a\sqrt{z})$ $\quad [n \geq 1]$.

28. $D^n[z^{n-1/2}I_{n-1/2}(a\sqrt{z}) I_{1/2-n}(a\sqrt{z})]$
$\qquad = \dfrac{a^{n-1/2}}{\sqrt{\pi}} z^{(2n-3)/4} I_{1/2-n}(2a\sqrt{z})$ $\quad [n \geq 1]$.

29. $D^n[z^n I_{n-1/2}(a\sqrt{z}) I_{n+1/2}(a\sqrt{z})] = \dfrac{a^{n-1/2}}{\sqrt{\pi}} z^{(2n-1)/4} I_{n+1/2}(2a\sqrt{z})$.

30. $D^n\left[z^{n-1}I_\nu^2\left(\dfrac{a}{\sqrt{z}}\right)\right]$
$\qquad = \left(-\dfrac{a}{2}\right)^n z^{-n/2-1} \sum\limits_{k=0}^{n}\binom{n}{k} I_{\nu+k}\left(\dfrac{a}{\sqrt{z}}\right) I_{\nu-n+k}\left(\dfrac{a}{\sqrt{z}}\right)$.

31. $D^n\left[z^{-1/2}I_{n-1/2}^2\left(\dfrac{a}{\sqrt{z}}\right)\right] = \dfrac{(-1)^n}{\sqrt{\pi}} a^{n-1/2} z^{-(6n+1)/4} I_{n-1/2}\left(\dfrac{2a}{\sqrt{z}}\right)$
$\qquad\qquad\qquad\qquad\qquad\qquad\qquad\qquad\qquad\qquad [n \geq 1]$.

32. $D^n\left[z^{-1/2}I_{1/2-n}^2\left(\dfrac{a}{\sqrt{z}}\right)\right] = \dfrac{(-1)^n}{\sqrt{\pi}} a^{n-1/2} z^{-(6n+1)/4} I_{n-1/2}\left(\dfrac{2a}{\sqrt{z}}\right)$
$\qquad\qquad\qquad\qquad\qquad\qquad\qquad\qquad\qquad\qquad [n \geq 1]$.

33. $D^n\left[z^{-1/2}I_{n-1/2}\left(\dfrac{a}{\sqrt{z}}\right) I_{1/2-n}\left(\dfrac{a}{\sqrt{z}}\right)\right]$
$\qquad = \dfrac{(-1)^n}{\sqrt{\pi}} a^{n-1/2} z^{-(6n+1)/4} I_{1/2-n}\left(\dfrac{2a}{\sqrt{z}}\right)$ $\quad [n \geq 1]$.

34. $D^n\left[z^{-1}I_{n-1/2}\left(\dfrac{a}{\sqrt{z}}\right) I_{n+1/2}\left(\dfrac{a}{\sqrt{z}}\right)\right]$
$\qquad = \dfrac{(-1)^n}{\sqrt{\pi}} a^{n-1/2} z^{-(6n+3)/4} I_{n+1/2}\left(\dfrac{2a}{\sqrt{z}}\right)$.

35. $D^n[z^{2\nu+n}e^{-2az}D^n[z^{-\nu}e^{az}I_\nu(az)]] = \left(\nu+\dfrac{1}{2}\right)_n (2a)^n z^\nu e^{-az} I_\nu(az).$

36. $D^n\left[z^{n-2\nu}e^{-2a/z}D^n\left[z^{\nu+n-1}e^{a/z}I_\nu\left(\dfrac{a}{z}\right)\right]\right]$
$$= \left(\nu+\dfrac{1}{2}\right)_n (2a)^n z^{-\nu-n-1} e^{-a/z} I_\nu\left(\dfrac{a}{z}\right).$$

37. $D^n[z^{n-2\nu}e^{2az}D^n[z^\nu e^{-az}I_\nu(az)]] = \left(\dfrac{1}{2}-\nu\right)_n (-2a)^n z^{-\nu} e^{az} I_\nu(az).$

38. $D^n\left[z^{n+2\nu}e^{2a/z}D^n\left[z^{n-\nu-1}e^{-a/z}I_\nu\left(\dfrac{a}{z}\right)\right]\right]$
$$= \left(\dfrac{1}{2}-\nu\right)_n (-2a)^n z^{\nu-n-1} e^{a/z} I_\nu\left(\dfrac{a}{z}\right).$$

39. $D^n[z^{n-2\nu}e^{-2az}D^n[z^\nu e^{az}I_\nu(az)]] = \left(\dfrac{1}{2}-\nu\right)_n (2a)^n z^{-\nu} e^{-az} I_\nu(az).$

40. $D^n\left[z^{n+2\nu}e^{-2a/z}D^n\left[z^{n-\nu-1}e^{a/z}I_\nu\left(\dfrac{a}{z}\right)\right]\right]$
$$= \left(\dfrac{1}{2}-\nu\right)_n (2a)^n z^{\nu-n-1} e^{-a/z} I_\nu\left(\dfrac{a}{z}\right).$$

41. $D^n[z^{n+2\nu}e^{2az}D^n[z^{-\nu}e^{-az}I_\nu(az)]] = \left(\nu+\dfrac{1}{2}\right)_n (-2a)^n z^\nu e^{az} I_\nu(az).$

42. $D^n\left[z^{n-2\nu}e^{2a/z}D^n\left[z^{n+\nu-1}e^{-a/z}I_\nu\left(\dfrac{a}{z}\right)\right]\right]$
$$= \left(\nu+\dfrac{1}{2}\right)_n (-2a)^n z^{-\nu-n-1} e^{a/z} I_\nu\left(\dfrac{a}{z}\right).$$

1.13.2. Derivatives with respect to the order

1. $\left.\dfrac{\partial I_\nu(z)}{\partial \nu}\right|_{\nu=\pm n} = (-1)^{n+1} K_n(z) \pm \dfrac{n!}{2} \sum_{p=0}^{n-1} \dfrac{\left(-\dfrac{z}{2}\right)^{p-n}}{p!(n-p)} I_p(z).$

2. $\left.\dfrac{\partial I_\nu(z)}{\partial \nu}\right|_{\nu=\pm 1/2} = \sqrt{\dfrac{1}{2\pi z}} \left[e^z \operatorname{Ei}(-2z) \mp e^{-z} \operatorname{Ei}(2z)\right]$ $\qquad$ [[10], 7.8(14)].

3. $= \sqrt{\dfrac{2}{\pi z}} \left[\left\{\begin{matrix}\sinh z \\ \cosh z\end{matrix}\right\} \operatorname{chi}(2z) - \left\{\begin{matrix}\cosh z \\ \sinh z\end{matrix}\right\} \operatorname{shi}(2z)\right].$

4. $\left.\dfrac{\partial I_\nu(z)}{\partial \nu}\right|_{\nu=n+1/2} = \dfrac{1}{2} \operatorname{Ei}(-2z)[I_{-n-1/2}(z) + I_{n+1/2}(z)]$
$\quad - \dfrac{(-1)^n}{\pi} \operatorname{Ei}(2z) K_{n+1/2}(z) + (-1)^n \dfrac{n!}{2} \sum_{k=0}^{n-1} \dfrac{(-1)^k}{k!(n-k)} \left(\dfrac{z}{2}\right)^{k-n} I_{k+1/2}(z)$
$\quad - \dfrac{n!}{2}\sqrt{\dfrac{z}{\pi}} \sum_{k=1}^{n} \dfrac{\left(-\dfrac{2}{z}\right)^k}{(n-k)!k} \sum_{p=0}^{k-1} \dfrac{(-z)^p}{p!}$
$\quad \times \{(-1)^k[I_{n-k+1/2}(z) + I_{k-n-1/2}(z)]K_{p-1/2}(2z)$
$\qquad - (-1)^{n-p} K_{n-k+1/2}(z)[I_{p-1/2}(2z) + I_{1/2-p}(2z)]\}.$

5. $\quad = \frac{1}{2}[\text{chi}(2z) - \text{shi}(2z)][I_{-n-1/2}(z) + I_{n+1/2}(z)] - \frac{(-1)^n}{\pi} K_{n+1/2}(z)$

$\times [\text{chi}(2z) + \text{shi}(2z)] + (-1)^n \frac{n!}{2} \sum_{k=0}^{n-1} \frac{(-1)^k}{k!(n-k)} \left(\frac{z}{2}\right)^{k-n} I_{k+1/2}(z)$

$\frac{n!}{2}\sqrt{\frac{z}{\pi}} \sum_{k=1}^{n} \frac{\left(-\frac{2}{z}\right)^k}{(n-k)!\,k} \sum_{p=0}^{k-1} \frac{(-z)^p}{p!}$

$\times \{(-1)^k [I_{n-k+1/2}(z) + I_{k-n-1/2}(z)] K_{p-1/2}(2z)$
$\qquad - (-1)^{n-p} K_{n-k+1/2}(z)[I_{p-1/2}(2z) + I_{1/2-p}(2z)]\}.$

6. $\left.\dfrac{\partial I_\nu(z)}{\partial \nu}\right|_{\nu=1/2-n} = \dfrac{1}{2}[\text{chi}(2z) - \text{shi}(2z)][I_{-n-1/2}(z) + I_{n+1/2}(z)]$

$\qquad - \dfrac{(-1)^n}{\pi}[\text{chi}(2z) + \text{shi}(2z)]K_{n-1/2}(z)$

$- \dfrac{n!}{2}\sum_{k=0}^{n-1}\dfrac{\left(-\frac{z}{2}\right)^{k-n}}{k!(n-k)} I_{1/2-k}(z) + \dfrac{n!}{2}\sqrt{\dfrac{z}{\pi}}\sum_{k=1}^{n}\dfrac{\left(\frac{2}{z}\right)^k}{(n-k)!\,k}\sum_{p=0}^{k-1}\dfrac{z^p}{p!}$

$\times \{(-1)^{k-1}[I_{n-k-1/2}(z) + I_{k-n+1/2}(z)]K_{p-1/2}(2z)$
$\qquad + (-1)^{n-p} K_{n-k-1/2}(z)[I_{p-1/2}(2z) + I_{1/2-p}(2z)]\}.$

7. $\left.\dfrac{\partial I_\nu(z)}{\partial \nu}\right|_{\nu=\pm n\pm 1/2}$

$= 2^{n-1/2}\sqrt{\pi}\,z^{-n-1/2}\sum_{k=0}^{n-1} \dfrac{(-1)^k (n-k)! \left(\frac{z}{2}\right)^{2k}}{k!(n-2k)!\,\Gamma\!\left(k+\frac{1}{2}\right)\Gamma\!\left(k-n+\frac{1}{2}\right)}$

$\times \left[\mp 2\!\left(\psi\!\left(k+\tfrac{1}{2}\right) - \psi\!\left(k-n+\tfrac{1}{2}\right)\right)\begin{Bmatrix}\sinh z\\ \cosh z\end{Bmatrix} + e^z\,\text{Ei}(-2z) \mp e^{-z}\,\text{Ei}(2z)\right]$

$+ 2^{n-3/2}\sqrt{\pi}\,z^{-n+1/2}\sum_{k=0}^{n-1}\dfrac{(-1)^k (n-k-1)!\left(\frac{z}{2}\right)^{2k}}{k!(n-2k-1)!\,\Gamma\!\left(k+\frac{3}{2}\right)\Gamma\!\left(k-n+\frac{1}{2}\right)}$

$\times \left[\pm 2\!\left(\psi\!\left(k+\tfrac{3}{2}\right) - \psi\!\left(k-n+\tfrac{1}{2}\right)\right)\begin{Bmatrix}\cosh z\\ \sinh z\end{Bmatrix} - e^z\,\text{Ei}(-2z) \mp e^{-z}\,\text{Ei}(2z)\right].$

1.14. The Macdonald Function $K_\nu(z)$

1.14.1. Derivatives with respect to the argument

1. $D^n[K_\nu(az)] = \left(-\dfrac{a}{2}\right)^n \sum_{k=0}^{n} \binom{n}{k} K_{\nu\pm 2k\mp n}(az).$

2. $\quad = n!(-z)^{-n}\sum_{k=0}^{n}\dfrac{(az)^k}{(n-k)!}(\nu)_{n-k}\sum_{p=0}^{[k/2]}\dfrac{(-2az)^{-j}}{p!(k-2p)!}K_{\nu-k+p}(az).$

1.14. The Macdonald Function $K_\nu(z)$

3. $D^n\left[z^{n-1}K_\nu\left(\dfrac{a}{z}\right)\right] = \left(\dfrac{a}{2}\right)^n z^{-n-1} \displaystyle\sum_{k=0}^{n}\binom{n}{k}K_{\nu\pm 2k\mp n}\left(\dfrac{a}{z}\right).$

4. $D^n[z^{\pm\nu/2}K_\nu(a\sqrt{z})] = \left(-\dfrac{a}{2}\right)^n z^{(\pm\nu-n)/2}K_{\nu\mp n}(a\sqrt{z}).$

5. $D^n[z^{(2n+1)/4}K_{n+1/2}(a\sqrt{z})] = (-1)^n \dfrac{\sqrt{\pi}}{2^{n+1/2}} a^{n-1/2} e^{-a\sqrt{z}}.$

6. $D^n[z^{-(2n+3)/4}K_{n+1/2}(a\sqrt{z})] = \dfrac{(-1)^n}{\sqrt{\pi}}\left(\dfrac{a}{2}\right)^{n+1/2} z^{-n-1/2} K^2_{n+1/2}\left(\dfrac{a\sqrt{z}}{2}\right).$

7. $D^n[z^{\pm\nu/4}K_\nu(a\sqrt[4]{z})]$
$= \left(-\dfrac{a}{4}\right)^n z^{(\pm\nu-3n)/4} \displaystyle\sum_{k=0}^{n-1} a^{-k}\dfrac{\Gamma(n+k)}{k!\,\Gamma(n-k)} z^{-k/4} K_{\nu\mp n\pm k}(a\sqrt[4]{z})\quad [n\geq 1].$

8. $D^n\left[z^{n\pm\nu/2-1}K_\nu\left(\dfrac{a}{\sqrt{z}}\right)\right] = \left(\dfrac{a}{2}\right)^n z^{-(n\mp\nu)/2-1} K_{\nu\pm n}\left(\dfrac{a}{\sqrt{z}}\right).$

9. $D^n\left[z^{(6n-1)/4}K_{n+1/2}\left(\dfrac{a}{\sqrt{z}}\right)\right] = \dfrac{1}{\sqrt{\pi}}\left(\dfrac{a}{2}\right)^{n+1/2} z^{-1/2} K^2_{n+1/2}\left(\dfrac{a}{2\sqrt{z}}\right).$

10. $D^n\left[z^{n-1/2}e^{az}K_{n+1/2}(az)\right] = (-1)^n(2n)!\sqrt{\pi}(2a)^{-n-1/2}z^{-n-1}.$

11. $D^n\left[z^{m+1/2}e^{az}K_{m+1/2}(az)\right]$
$= (-1)^{m+n} m!\sqrt{\pi}(2a)^{n-m-1/2} L^{n-2m-1}_{m-n}(2az)\quad [m\geq n].$

12. $D^n\left[z^{n-1/2}e^{az}K_{m+1/2}(az)\right]$
$= (-1)^m(m+n)!\sqrt{\pi}(2a)^{-m-1/2} z^{-m-1} L^{-2m-1}_{m-n}(2az)\quad [m\geq n].$

13. $D^n\left[z^{-m-1/2}e^{az}K_{m+1/2}(az)\right]$
$= (-1)^{m+n}(m+n)!\sqrt{\pi}(2a)^{-m-1/2} z^{-2m-n-1} L^{-2m-1}_{m}(2az).$

14. $D^n\left[z^{m+1/2}e^{-az}K_{m+1/2}(az)\right]$
$= (-1)^{m+n} m!\sqrt{\pi}(2a)^{n-m-1/2} e^{-2az} L^{n-2m-1}_{m}(2az).$

15. $D^n\left[z^{n-1/2}e^{-az}K_{m+1/2}(az)\right]$
$= (-1)^m(m+n)!\sqrt{\pi}(2a)^{-m-1/2} z^{-m-1} e^{-2az} L^{-2m-1}_{m+n}(2az).$

16. $D^n\left[z^{-m-1/2}e^{-az}K_{m+1/2}(az)\right]$
$= (-1)^m(m+n)!\sqrt{\pi}(2a)^{-m-1/2} z^{-2m-n-1} e^{-2az} L^{-2m-n-1}_{m+n}(2az).$

17. $D^n\left[z^{-1/2}e^{a/z}K_{m+1/2}\left(\dfrac{a}{z}\right)\right]$
$= (-1)^{m+n}(m+n)!\sqrt{\pi}(2a)^{-m-1/2} z^{m-n} L^{-2m-1}_{m-n}\left(\dfrac{2a}{z}\right)\quad [m\geq n].$

18. $D^n\left[z^{m+n-1/2}e^{a/z}K_{m+1/2}\left(\dfrac{a}{z}\right)\right]$
$$= (-1)^m(m+n)!\sqrt{\pi}\,(2a)^{-m-1/2}z^{2m}L_m^{-2m-n-1}\left(\dfrac{2a}{z}\right).$$

19. $D^n\left[z^{n-m-3/2}e^{a/z}K_{m+1/2}\left(\dfrac{a}{z}\right)\right]$
$$= (-1)^m m!\sqrt{\pi}\,(2a)^{n-m-1/2}z^{-n-1}L_{m-n}^{n-2m-1}\left(\dfrac{2a}{z}\right) \quad [m \geq n].$$

20. $D^n\left[z^{-1/2}e^{-a/z}K_{m+1/2}\left(\dfrac{a}{z}\right)\right]$
$$= (-1)^{m+n}(m+n)!\sqrt{\pi}\,(2a)^{-m-1/2}z^{m-n}e^{-2a/z}L_{m+n}^{-2m-1}\left(\dfrac{2a}{z}\right).$$

21. $D^n\left[z^{m+n-1/2}e^{-a/z}K_{m+1/2}\left(\dfrac{a}{z}\right)\right]$
$$= (-1)^{m+n}(m+n)!\sqrt{\pi}\,(2a)^{-m-1/2}z^{2m}e^{-2a/z}L_{m+n}^{-2m-n-1}\left(\dfrac{2a}{z}\right).$$

22. $D^n\left[z^{n-m-3/2}e^{-a/z}K_{m+1/2}\left(\dfrac{a}{z}\right)\right]$
$$= (-1)^m m!\sqrt{\pi}\,(2a)^{n-m-1/2}z^{-n-1}e^{-2a/z}L_m^{n-2m-1}\left(\dfrac{2a}{z}\right).$$

23. $D^n[K_\nu^2(a\sqrt{z})] = \left(-\dfrac{a}{2\sqrt{z}}\right)^n \sum_{k=0}^{n}\binom{n}{k}K_{\nu+k}(a\sqrt{z})\,K_{\nu-n+k}(a\sqrt{z}).$

24. $D^n[z^{n-1/2}K_{n-1/2}^2(a\sqrt{z})] = (-1)^n\sqrt{\pi}\,a^{n-1/2}z^{(2n-3)/4}K_{n-1/2}(2a\sqrt{z}).$

25. $D^n[z^n K_{n-1/2}(a\sqrt{z})\,K_{n+1/2}(a\sqrt{z})]$
$$= (-1)^n\sqrt{\pi}\,a^{n-1/2}z^{(2n-1)/4}K_{n+1/2}(2a\sqrt{z}).$$

26. $D^n\left[z^{n-1}K_\nu^2\left(\dfrac{a}{\sqrt{z}}\right)\right]$
$$= \left(\dfrac{a}{2}\right)^n z^{-n/2-1}\sum_{k=0}^{n}\binom{n}{k}K_{\nu+k}\left(\dfrac{a}{\sqrt{z}}\right)K_{\nu-n+k}\left(\dfrac{a}{\sqrt{z}}\right).$$

27. $D^n\left[z^{-1/2}K_{n-1/2}^2\left(\dfrac{a}{\sqrt{z}}\right)\right] = \sqrt{\pi}\,a^{n-1/2}z^{-(6n+1)/4}K_{n-1/2}\left(\dfrac{2a}{\sqrt{z}}\right).$

28. $D^n\left[z^{-1}K_{n-1/2}\left(\dfrac{a}{\sqrt{z}}\right)K_{n+1/2}\left(\dfrac{a}{\sqrt{z}}\right)\right]$
$$= \sqrt{\pi}\,a^{n-1/2}z^{-(6n+3)/4}K_{n+1/2}\left(\dfrac{2a}{\sqrt{z}}\right).$$

29. $D^n[z^{n-1/2}I_{n-1/2}(a\sqrt{z})\,K_{n-1/2}(a\sqrt{z})]$
$$= \dfrac{a^{n-1/2}}{\sqrt{\pi}}z^{(2n-3)/4}K_{n-1/2}(2a\sqrt{z}) \quad [n \geq 1].$$

1.14. The Macdonald Function $K_\nu(z)$

30. $\mathrm{D}^n\left[z^{-1/2}I_{n-1/2}\left(\dfrac{a}{\sqrt{z}}\right)K_{n-1/2}\left(\dfrac{a}{\sqrt{z}}\right)\right]$

$$= \dfrac{(-1)^n}{\sqrt{\pi}}a^{n-1/2}z^{-(6n+1)/4}K_{n-1/2}\left(\dfrac{2a}{\sqrt{z}}\right) \quad [n \geq 1].$$

31. $\mathrm{D}^n[e^{-2az}\,\mathrm{D}^n[z^{n-1/2}e^{az}K_\nu(az)]$

$$= (-1)^n\left(\dfrac{1}{2}-\nu\right)_n\left(\dfrac{1}{2}+\nu\right)_n z^{-n-1/2}e^{-az}K_\nu(az).$$

32. $\mathrm{D}^n\left[z^{2n}e^{-2a/z}\,\mathrm{D}^n\left[z^{-1/2}e^{a/z}K_\nu\left(\dfrac{a}{z}\right)\right]\right]$

$$= (-1)^n\left(\dfrac{1}{2}-\nu\right)_n\left(\dfrac{1}{2}+\nu\right)_n z^{-1/2}e^{-a/z}K_\nu\left(\dfrac{a}{z}\right).$$

33. $\mathrm{D}^n[e^{2az}\,\mathrm{D}^n[z^{n-1/2}e^{-az}K_\nu(az)]$

$$= (-1)^n\left(\dfrac{1}{2}-\nu\right)_n\left(\dfrac{1}{2}+\nu\right)_n z^{-n-1/2}e^{az}K_\nu(az).$$

34. $\mathrm{D}^n\left[z^{2n}e^{2a/z}\,\mathrm{D}^n\left[z^{-1/2}e^{-a/z}K_\nu\left(\dfrac{a}{z}\right)\right]\right]$

$$= (-1)^n\left(\dfrac{1}{2}-\nu\right)_n\left(\dfrac{1}{2}+\nu\right)_n z^{-1/2}e^{a/z}K_\nu\left(\dfrac{a}{z}\right).$$

35. $\mathrm{D}^n[z^{n-2m-1}e^{-2az}\,\mathrm{D}^n[z^{m+1/2}e^{az}K_{m+1/2}(az)]$

$$= (-m)_n(2a)^n z^{-m-1/2}e^{-az}K_{m+1/2}(az).$$

36. $\mathrm{D}^n\left[z^{2m+n+1}e^{-2a/z}\,\mathrm{D}^n\left[z^{n-m-3/2}e^{a/z}K_{m+1/2}\left(\dfrac{a}{z}\right)\right]\right]$

$$= (-m)_n(2a)^n z^{m-n-1/2}e^{-a/z}K_{m+1/2}\left(\dfrac{a}{z}\right).$$

37. $\mathrm{D}^n[z^{n+2m+1}e^{2az}\,\mathrm{D}^n[z^{-m-1/2}e^{-az}K_{m+1/2}(az)]$

$$= \dfrac{(m+n)!}{m!}(-2a)^n z^{m+1/2}e^{az}K_{m+1/2}(az).$$

38. $\mathrm{D}^n\left[z^{n-2m-1}e^{2a/z}\,\mathrm{D}^n\left[z^{n+m-1/2}e^{-a/z}K_{m+1/2}\left(\dfrac{a}{z}\right)\right]\right]$

$$= \dfrac{(m+n)!}{m!}(-2a)^n z^{-m-n-3/2}e^{a/z}K_{m+1/2}\left(\dfrac{a}{z}\right).$$

39. $\mathrm{D}^n[z^{n+2m+1}e^{-2az}\,\mathrm{D}^n[z^{-m-1/2}e^{az}K_{m+1/2}(az)]$

$$= \dfrac{(m+n)!}{m!}(2a)^n z^{m+1/2}e^{-az}K_{m+1/2}(az).$$

40. $\mathrm{D}^n\left[z^{n-2m-1}e^{-2a/z}\,\mathrm{D}^n\left[z^{n+m-1/2}e^{a/z}K_{m+1/2}\left(\dfrac{a}{z}\right)\right]\right]$

$$= \dfrac{(m+n)!}{m!}(2a)^n z^{-m-n-3/2}e^{-a/z}K_{m+1/2}\left(\dfrac{a}{z}\right).$$

41. $\mathrm{D}^n[e^{-2az}\,\mathrm{D}^n[z^{n-1/2}e^{az}K_{m+1/2}(az)]$

$$= \dfrac{(m+n)!}{(m-n)!}z^{-n-1/2}e^{-az}K_{m+1/2}(az).$$

42. $D^n \left[z^{2n} e^{-2a/z} D^n \left[z^{-1/2} e^{a/z} K_{m+1/2}\left(\frac{a}{z}\right) \right] \right]$
$$= \frac{(m+n)!}{(m-n)!} z^{-1/2} e^{-a/z} K_{m+1/2}\left(\frac{a}{z}\right).$$

1.14.2. Derivatives with respect to the order

1. $\left. \dfrac{\partial K_\nu(z)}{\partial \nu} \right|_{\nu=0} = 0.$

2. $\left. \dfrac{\partial K_\nu(z)}{\partial \nu} \right|_{\nu=\pm n} = \pm \dfrac{n!}{2} \sum_{p=0}^{n-1} \dfrac{\left(\frac{z}{2}\right)^{p-n}}{p!(n-p)} K_p(z).$

3. $\left. \dfrac{\partial K_\nu(z)}{\partial \nu} \right|_{\nu=\pm 1/2} = \mp \sqrt{\dfrac{\pi}{2z}}\, e^z \operatorname{Ei}(-2z)$ [[10], 7.8(15)].

4. $\left. \dfrac{\partial K_\nu(z)}{\partial \nu} \right|_{\nu=1/2} = \sqrt{\dfrac{\pi}{2z}}\, e^z [\operatorname{shi}(2z) - \operatorname{chi}(2z)].$

5. $\left. \dfrac{\partial K_\nu(z)}{\partial \nu} \right|_{\nu=n-1/2} = (-1)^n \dfrac{\pi}{2} \operatorname{Ei}(-2z)[I_{n-1/2}(z) + I_{1/2-n}(z)]$

$+ \dfrac{n!}{2} \sum_{k=0}^{n-1} \dfrac{\left(\frac{z}{2}\right)^{k-n}}{k!(n-k)} K_{k-1/2}(z) - (-1)^n \dfrac{n!}{2} \sqrt{\pi} \sum_{k=1}^{n} \dfrac{\left(-\frac{2}{z}\right)^k}{(n-k)!k} [I_{n-k-1/2}(z)$

$+ I_{k-n+1/2}(z)] \sum_{p=0}^{k-1} \dfrac{z^{p+1/2}}{p!} K_{p-1/2}(2z).$

6. $= (-1)^n \dfrac{\pi}{2} [\operatorname{chi}(2z) - \operatorname{shi}(2z)][I_{n-1/2}(z) + I_{1/2-n}(z)]$

$+ \dfrac{n!}{2} \sum_{k=0}^{n-1} \dfrac{\left(\frac{z}{2}\right)^{k-n}}{k!(n-k)} K_{k-1/2}(z) - (-1)^n \dfrac{n!}{2} \sqrt{\pi} \sum_{k=1}^{n} \dfrac{\left(-\frac{2}{z}\right)^k}{(n-k)!k} [I_{n-k-1/2}(z)$

$+ I_{k-n+1/2}(z)] \sum_{p=0}^{k-1} \dfrac{z^{p+1/2}}{p!} K_{p-1/2}(2z).$

7. $\left. \dfrac{\partial K_\nu(z)}{\partial \nu} \right|_{\nu=n+1/2}$

$= (-1)^{n+1} 2^{n-1/2} \pi^{3/2} z^{-n-1/2} \sum_{k=0}^{n-1} \dfrac{(-1)^k (n-k)! \left(\frac{z}{2}\right)^{2k}}{k!(n-2k)! \Gamma\left(k+\frac{1}{2}\right) \Gamma\left(k-n+\frac{1}{2}\right)}$

$\times \left[e^{-z} \left(\psi\left(k+\frac{1}{2}\right) - \psi\left(k-n+\frac{1}{2}\right) \right) + e^z \operatorname{Ei}(-2z) \right]$

$+ (-1)^{n+1} 2^{n-3/2} \pi^{3/2} z^{-n+1/2} \sum_{k=0}^{n-1} \dfrac{(-1)^k (n-k-1)! \left(\frac{z}{2}\right)^{2k}}{k!(n-2k-1)! \Gamma\left(k+\frac{3}{2}\right) \Gamma\left(k-n+\frac{1}{2}\right)}$

$\times \left[e^{-z} \left(\psi\left(k+\frac{3}{2}\right) - \psi\left(k-n+\frac{1}{2}\right) \right) - e^z \operatorname{Ei}(-2z) \right].$

1.15. The Struve Functions $\mathrm{H}_\nu(z)$ and $\mathrm{L}_\nu(z)$

1.15.1. Derivatives with respect to the argument

1. $\mathrm{D}^n[\mathrm{H}_\nu(az)]$
$$= n!\left(-\frac{2}{z}\right)^n \sum_{k=0}^{[n/2]} \frac{\left(-\frac{1}{4}\right)^k}{k!(n-2k)!} \sum_{p=0}^{n-k} \binom{n-k}{p} \left(\frac{\nu}{2}\right)_{n-k-p} \left(-\frac{az}{2}\right)^p \mathrm{H}_{\nu-p}(az).$$

2. $$= n!\left(-\frac{2}{z}\right)^n \sum_{k=0}^{[n/2]} \frac{\left(-\frac{1}{4}\right)^k}{k!(n-2k)!} \sum_{p=0}^{n-k} \binom{n-k}{p} \left(-\frac{\nu}{2}\right)_{n-k-p}$$
$$\times \left[\left(\frac{az}{2}\right)^p \mathrm{H}_{\nu+p}(az) - \frac{1}{\pi}\left(\frac{az}{2}\right)^{\nu+2p-1} \sum_{r=0}^{p-1} \frac{\Gamma\left(r+\frac{1}{2}\right)}{\Gamma\left(\nu+p-r+\frac{1}{2}\right)} \left(\frac{2}{az}\right)^{2r} \right].$$

3. $\mathrm{D}^n[z^{\nu/2}\,\mathrm{H}_\nu(a\sqrt{z})] = \left(\frac{a}{2}\right)^n z^{(\nu-n)/2}\,\mathrm{H}_{\nu-n}(a\sqrt{z}).$

4. $\mathrm{D}^n[z^{-\nu/2}\,\mathrm{H}_\nu(a\sqrt{z})] = \left(-\frac{a}{2}\right)^n z^{-(n+\nu)/2}\,\mathrm{H}_{\nu+n}(a\sqrt{z})$
$$- \frac{(-1)^n}{\pi}\left(\frac{a}{2}\right)^{\nu+2n-1} z^{-1/2} \sum_{k=0}^{n-1} \frac{\Gamma\left(k+\frac{1}{2}\right)}{\Gamma\left(\nu+n-k+\frac{1}{2}\right)} \left(\frac{4}{a^2 z}\right)^k.$$

5. $\mathrm{D}^n[z^{(2n+1)/4}\,\mathrm{H}_{n+1/2}(a\sqrt{z})] = \frac{2}{\sqrt{\pi}}\left(\frac{a}{2}\right)^{n-1/2} \sin^2 \frac{a\sqrt{z}}{2}.$

6. $\mathrm{D}^n\left[z^{n-\nu/2-1}\,\mathrm{H}_\nu\left(\frac{a}{\sqrt{z}}\right)\right] = \left(-\frac{a}{2}\right)^n z^{-(n+\nu)/2-1}\,\mathrm{H}_{\nu-n}\left(\frac{a}{\sqrt{z}}\right).$

7. $\mathrm{D}^n\left[z^{(2n-5)/4}\,\mathrm{H}_{n+1/2}\left(\frac{a}{\sqrt{z}}\right)\right] = (-1)^n \frac{2}{\sqrt{\pi}}\left(\frac{a}{2}\right)^{n-1/2} z^{-n-1} \sin^2 \frac{a}{2\sqrt{z}}.$

8. $\mathrm{D}^n\left[z^{n+\nu/2-1}\,\mathrm{H}_\nu\left(\frac{a}{\sqrt{z}}\right)\right] = \left(\frac{a}{2}\right)^n z^{(\nu-n)/2-1}\,\mathrm{H}_{\nu+n}\left(\frac{a}{\sqrt{z}}\right)$
$$- \frac{1}{\pi}\left(\frac{a}{2}\right)^{\nu+2n-1} z^{-n-1/2} \sum_{k=0}^{n-1} \frac{\Gamma\left(k+\frac{1}{2}\right)}{\Gamma\left(\nu+n-k+\frac{1}{2}\right)} \left(\frac{4z}{a^2}\right)^k.$$

9. $\mathrm{D}^n[\mathrm{L}_\nu(az)]$
$$= n!\left(-\frac{2}{z}\right)^n \sum_{k=0}^{[n/2]} \frac{\left(-\frac{1}{4}\right)^k}{k!(n-2k)!} \sum_{p=0}^{n-k} \binom{n-k}{p} \left(\frac{\nu}{2}\right)_{n-k-p} \left(-\frac{az}{2}\right)^p \mathrm{L}_{\nu-p}(az).$$

10. $\quad = n!\left(-\dfrac{2}{z}\right)^n \sum\limits_{k=0}^{[n/2]} \dfrac{\left(-\dfrac{1}{4}\right)^k}{k!(n-2k)!} \sum\limits_{p=0}^{n-k}(-1)^p\binom{n-k}{p}\left(-\dfrac{\nu}{2}\right)_{n-k-p}$

$\times \left[\left(\dfrac{az}{2}\right)^p \mathbf{L}_{\nu+p}(az) + \dfrac{1}{\pi}\left(\dfrac{az}{2}\right)^{\nu+2p-1}\sum\limits_{r=0}^{p-1}\dfrac{\Gamma\left(r+\dfrac{1}{2}\right)}{\Gamma\left(\nu+p-r+\dfrac{1}{2}\right)}\left(-\dfrac{4}{a^2 z^2}\right)^r\right].$

11. $D^n[z^{\nu/2}\mathbf{L}_\nu(a\sqrt{z})] = \left(\dfrac{a}{2}\right)^n z^{(\nu-n)/2}\mathbf{L}_{\nu-n}(a\sqrt{z}).$

12. $D^n[z^{-\nu/2}\mathbf{L}_\nu(a\sqrt{z})] = \left(\dfrac{a}{2}\right)^n z^{-(n+\nu)/2}\mathbf{L}_{\nu+n}(a\sqrt{z})$

$+ \dfrac{1}{\pi}\left(\dfrac{a}{2}\right)^{\nu+2n-1} z^{-1/2}\sum\limits_{k=0}^{n-1}\dfrac{\Gamma\left(k+\dfrac{1}{2}\right)}{\Gamma\left(\nu+n-k+\dfrac{1}{2}\right)}\left(-\dfrac{4}{a^2 z}\right)^k.$

13. $D^n[z^{(2n+1)/4}\mathbf{L}_{n+1/2}(a\sqrt{z})] = \dfrac{2}{\sqrt{\pi}}\left(\dfrac{a}{2}\right)^{n-1/2}\sinh^2\dfrac{a\sqrt{z}}{2}.$

14. $D^n\left[z^{n-\nu/2-1}\mathbf{L}_\nu\left(\dfrac{a}{\sqrt{z}}\right)\right] = \left(-\dfrac{a}{2}\right)^n z^{-(n+\nu)/2-1}\mathbf{L}_{\nu-n}\left(\dfrac{a}{\sqrt{z}}\right).$

15. $D^n\left[z^{(2n-5)/4}\mathbf{L}_{n+1/2}\left(\dfrac{a}{\sqrt{z}}\right)\right] = (-1)^n \dfrac{2}{\sqrt{\pi}}\left(\dfrac{a}{2}\right)^{n-1/2} z^{-n-1}\sinh^2\dfrac{a}{2\sqrt{z}}.$

16. $D^n\left[z^{n+\nu/2-1}\mathbf{L}_\nu\left(\dfrac{a}{\sqrt{z}}\right)\right] = \left(-\dfrac{a}{2}\right)^n z^{(\nu-n)/2-1}\mathbf{L}_{\nu+n}\left(\dfrac{a}{\sqrt{z}}\right)$

$+ \dfrac{(-1)^n}{\pi}\left(\dfrac{a}{2}\right)^{\nu+2n-1} z^{-n-1/2}\sum\limits_{k=0}^{n-1}\dfrac{\Gamma\left(k+\dfrac{1}{2}\right)}{\Gamma\left(\nu+n-k+\dfrac{1}{2}\right)}\left(-\dfrac{4z}{a^2}\right)^k.$

1.15.2. Derivatives with respect to the order

1. $\left.\dfrac{\partial \mathbf{H}_\nu(z)}{\partial \nu}\right|_{\nu=0} = \dfrac{1}{2\pi} G^{32}_{24}\left(\dfrac{z^2}{4} \,\middle|\, \begin{matrix}\frac{1}{2},\frac{1}{2}\\ \frac{1}{2},\frac{1}{2},0,0\end{matrix}\right) - \dfrac{\pi}{2} J_0(z)$ [Re $z \geq 0$].

2. $\left.\dfrac{\partial \mathbf{H}_\nu(z)}{\partial \nu}\right|_{\nu=n} = -\dfrac{\pi}{2}J_n(z) + \dfrac{2^{n-1}z^{-n}}{\pi} G^{32}_{24}\left(\dfrac{z^2}{4} \,\middle|\, \begin{matrix}\frac{1}{2},\frac{1}{2}\\ \frac{1}{2},\frac{1}{2},n,0\end{matrix}\right)$

$+ \dfrac{1}{\pi}\sum\limits_{k=0}^{n-1}\dfrac{\left(\frac{1}{2}\right)_k}{\left(\frac{1}{2}\right)_{n-k}}\left(\dfrac{z}{2}\right)^{n-2k-1}\left[\log\dfrac{z}{2} - \psi\left(n-k+\dfrac{1}{2}\right)\right]$

$+ \dfrac{n!}{2}\sum\limits_{k=0}^{n-1}(-1)^k\dfrac{\left(\frac{z}{2}\right)^{k-n}}{k!(n-k)}\mathbf{H}_{-k}(z)$ [Re $z \geq 0$].

3. $\left.\dfrac{\partial\,\mathbf{H}_v(z)}{\partial \nu}\right|_{\nu=-n} = (-1)^{n+1}\dfrac{\pi}{2}\,J_n(z)$

$+ (-1)^n\dfrac{2^{n-1}z^{-n}}{\pi}\,G_{24}^{32}\!\left(\dfrac{z^2}{4}\,\bigg|\,\begin{array}{c}\frac{1}{2},\frac{1}{2}\\ \frac{1}{2},\frac{1}{2},n,0\end{array}\right) - \dfrac{n!}{2}\sum_{k=0}^{n-1}\dfrac{\left(-\frac{z}{2}\right)^{k-n}}{k!(n-k)}\,\mathbf{H}_{-k}(z)$

$[\mathrm{Re}\,z \geq 0].$

4. $\left.\dfrac{\partial\,\mathbf{H}_\nu(z)}{\partial \nu}\right|_{\nu=1/2} = \sqrt{\dfrac{2}{\pi z}}$

$\times \left\{ \mathbf{C} + \ln\dfrac{z}{2} + \sin z\,[\mathrm{Si}\,(2z) - 2\,\mathrm{Si}\,(z)] + \cos z\,[\mathrm{ci}\,(2z) - 2\,\mathrm{ci}\,(z)] \right\}$

$[[10], 7.9(22)].$

5. $\left.\dfrac{\partial\,\mathbf{H}_\nu(z)}{\partial \nu}\right|_{\nu=-1/2} = \sqrt{\dfrac{2}{\pi z}}$

$\times \left\{ \cos z\,[\mathrm{Si}\,(2z) - 2\,\mathrm{Si}\,(z)] - \sin z\,[\mathrm{ci}\,(2z) - 2\,\mathrm{ci}\,(z)] \right\}\quad [[10], 7.9(23)].$

6. $\left.\dfrac{\partial\,\mathbf{H}_v(z)}{\partial \nu}\right|_{\nu=n+1/2} = [\mathrm{Si}\,(2z) - 2\,\mathrm{Si}\,(z)]\,J_{n+1/2}(z)$

$+ (-1)^n[\mathrm{ci}\,(2z) - 2\,\mathrm{ci}\,(z)]\,J_{-n-1/2}(z) + \ln\dfrac{z}{2}\,[\mathbf{H}_{n+1/2}(z) - Y_{n+1/2}(z)]$

$+ \dfrac{1}{2\sqrt{\pi}}\left(\dfrac{2}{z}\right)^{n+1/2}\left(\dfrac{1}{2}\right)_n\!\left[3\mathbf{C} + 2\ln 2 + \psi\!\left(\dfrac{1}{2} - n\right)\right]$

$- \dfrac{n!}{2}\left(\dfrac{2}{z}\right)^n\sum_{k=0}^{n-1}\dfrac{\left(-\frac{z}{2}\right)^k}{k!(n-k)}\,J_{-k-1/2}(z) - \dfrac{n!}{2\sqrt{\pi}}\left(\dfrac{2}{z}\right)^{n+1/2}\sum_{k=0}^{n-1}\dfrac{\left(\frac{1}{2}\right)_k}{k!(n-k)}$

$- \dfrac{\left(\frac{z}{2}\right)^{n-1/2}}{\sqrt{\pi}}\sum_{k=0}^{n-1}\dfrac{\left(\frac{1}{2}\right)_k}{(n-k)!}\left(\dfrac{2}{z}\right)^{2k}\psi(n-k+1)$

$- n!\sqrt{\pi}\left(\dfrac{2}{z}\right)^{1/2-n}\sum_{k=0}^{n-1}\dfrac{\left(\frac{z}{2}\right)^k}{k!(n-k)}\sum_{p=0}^{n-k-1}\dfrac{\left(\frac{z}{2}\right)^p}{p!}$

$\times \left\{(-1)^{p+1}J_{k+1/2}(z)\left[J_{1/2-p}(z) - 2^{p-1/2}J_{1/2-p}(2z)\right]\right.$

$\left. - (-1)^k J_{-k-1/2}(z)\left[J_{p-1/2}(z) - 2^{p-1/2}J_{p-1/2}(2z)\right]\right\}.$

7. $\left.\dfrac{\partial\,\mathbf{L}_\nu(z)}{\partial \nu}\right|_{\nu=0} = K_0(z) - \dfrac{1}{\pi^2 z}\,G_{24}^{42}\!\left(\dfrac{z^2}{4}\,\bigg|\,\begin{array}{c}1,1\\ \frac{1}{2},\frac{1}{2},1,1\end{array}\right)\qquad [\mathrm{Re}\,z \geq 0].$

8. $\phantom{\left.\dfrac{\partial\,\mathbf{L}_\nu(z)}{\partial \nu}\right|_{\nu=0}} = -K_0(z) - \dfrac{2}{z}\,G_{46}^{42}\!\left(\dfrac{z^2}{4}\,\bigg|\,\begin{array}{c}1,1,\frac{1}{4},\frac{3}{4}\\ \frac{1}{2},\frac{1}{2},1,1,\frac{1}{4},\frac{3}{4}\end{array}\right)\qquad [\mathrm{Re}\,z \geq 0].$

9. $\left.\dfrac{\partial\,\mathbf{L}_\nu(z)}{\partial \nu}\right|_{\nu=0} = K_0(z) - \dfrac{1}{\pi^2 z}\,G_{24}^{42}\!\left(\dfrac{z^2}{4}\,\bigg|\,\begin{array}{c}1,1\\ \frac{1}{2},\frac{1}{2},1,1\end{array}\right).$

10. $\quad = -K_0(z) - \dfrac{2}{z} G^{42}_{46}\left(\dfrac{z^2}{4} \,\Bigg|\, \begin{matrix} 1,1,\frac{1}{4},\frac{3}{4} \\ \frac{1}{2},\frac{1}{2},1,1,\frac{1}{4},\frac{3}{4} \end{matrix}\right).$

11. $\dfrac{\partial \mathbf{L}_\nu(z)}{\partial \nu}\bigg|_{\nu=n} = (-1)^n K_n(z) + \dfrac{(-2)^{n-1} z^{-n}}{\pi^2} G^{42}_{24}\left(\dfrac{z^2}{4} \,\Bigg|\, \begin{matrix} \frac{1}{2},\frac{1}{2} \\ 0,\frac{1}{2},\frac{1}{2},n \end{matrix}\right)$

$\quad - \dfrac{1}{\pi} \sum_{k=0}^{n-1} (-1)^k \dfrac{\left(\frac{1}{2}\right)_k}{\left(\frac{1}{2}\right)_{n-k}} \left(\dfrac{z}{2}\right)^{n-2k-1} \left[\log \dfrac{z}{2} - \psi\left(n-k+\dfrac{1}{2}\right)\right]$

$\quad + \dfrac{n!}{2} \sum_{k=0}^{n-1} \dfrac{\left(-\frac{z}{2}\right)^{k-n}}{k!(n-k)} \mathbf{L}_{-k}(z) \quad [\operatorname{Re} z \geq 0].$

12. $\dfrac{\partial \mathbf{L}_\nu(z)}{\partial \nu}\bigg|_{\nu=-n} = (-1)^n K_n(z) + \dfrac{(-2)^{n-1} z^{-n}}{\pi^2} G^{42}_{24}\left(\dfrac{z^2}{4} \,\Bigg|\, \begin{matrix} \frac{1}{2},\frac{1}{2} \\ 0,\frac{1}{2},\frac{1}{2},n \end{matrix}\right)$

$\quad - \dfrac{n!}{2} \sum_{k=0}^{n-1} \dfrac{\left(-\frac{z}{2}\right)^{k-n}}{k!(n-k)} \mathbf{L}_{-k}(z) \quad [\operatorname{Re} z \geq 0].$

13. $\dfrac{\partial \mathbf{L}_\nu(z)}{\partial \nu}\bigg|_{\nu=\pm 1/2} = \sqrt{\dfrac{2^{\pm 1}}{\pi z}} \left\{ -\dfrac{1\pm 1}{2}\left(\mathbf{C} + \ln\dfrac{z}{2}\right) \mp e^{-z}[\operatorname{Ei}(2z) - 2\operatorname{Ei}(z)] \right.$

$\quad \left. - e^z [\operatorname{Ei}(-2z) - 2\operatorname{Ei}(-z)] \right\} \quad [[10], 7.8(16)].$

14. $\dfrac{\partial \mathbf{L}_\nu(z)}{\partial \nu}\bigg|_{\nu=1/2} = \sqrt{\dfrac{2}{\pi z}} \left\{ \sinh z [\operatorname{shi}(2z) - 2\operatorname{shi}(z)] \right.$

$\quad \left. - \cosh z [\operatorname{chi}(2z) - 2\operatorname{chi}(z)] - \ln\dfrac{z}{2} - \mathbf{C} \right\}.$

15. $\dfrac{\partial \mathbf{L}_\nu(z)}{\partial \nu}\bigg|_{\nu=-1/2} = \sqrt{\dfrac{2}{\pi z}} \left\{ \cosh z[\operatorname{shi}(2z) - 2\operatorname{shi}(z)] \right.$

$\quad \left. - \sinh z[\operatorname{chi}(2z) - 2\operatorname{chi}(z)] \right\}.$

16. $\dfrac{\partial \mathbf{L}_\nu(z)}{\partial \nu}\bigg|_{\nu=n+1/2} = [\operatorname{shi}(2z) - 2\operatorname{shi}(z)] I_{n+1/2}(z)$

$\quad - [\operatorname{chi}(2z) - 2\operatorname{chi}(z)] I_{-n-1/2}(z) + \ln\dfrac{z}{2} [\mathbf{L}_{n+1/2}(z) - I_{-n-1/2}(z)]$

$\quad + \dfrac{(-1)^{n+1}}{2\pi}\left(\dfrac{2}{z}\right)^{n+1/2} \Gamma\left(n+\dfrac{1}{2}\right)\left[3\mathbf{C} + 2\ln 2 + \psi\left(\dfrac{1}{2} - n\right)\right]$

$\quad + \dfrac{n!}{2}\left(-\dfrac{2}{z}\right)^n \sum_{k=0}^{n-1} \dfrac{\left(-\frac{z}{2}\right)^k}{k!(n-k)} I_{-k-1/2}(z)$

$\quad + (-1)^n \dfrac{n!}{2\sqrt{\pi}}\left(\dfrac{2}{z}\right)^{n+1/2} \sum_{k=0}^{n-1} \dfrac{\left(\frac{1}{2}\right)_k}{k!(n-k)}$

$$+ \frac{\left(\frac{z}{2}\right)^{n-1/2}}{\sqrt{\pi}} \sum_{k=0}^{n-1} (-1)^k \frac{\left(\frac{1}{2}\right)_k}{(n-k)!} \left(\frac{2}{z}\right)^{2k} \psi(n-k+1)$$

$$+ (-1)^n n! \sqrt{\pi} \left(\frac{z}{2}\right)^{1/2-n} \sum_{k=0}^{n-1} \frac{\left(-\frac{z}{2}\right)^k}{k!(n-k)} \sum_{p=0}^{n-k-1} \frac{\left(-\frac{z}{2}\right)^p}{p!}$$

$$\times \left\{ I_{k+1/2}(z) \left[I_{1/2-p}(z) - 2^{p-1/2} I_{1/2-p}(2z) \right] \right.$$
$$\left. - I_{-k-1/2}(z) \left[I_{p-1/2}(z) - 2^{p-1/2} I_{p-1/2}(2z) \right] \right\}.$$

17. $\left. \frac{\partial \mathbf{L}_\nu(z)}{\partial \nu} \right|_{\nu=-n-1/2} = \left[2 \operatorname{chi}(z) - \operatorname{chi}(2z) \right] I_{n+1/2}(z)$
$$+ \left[\operatorname{shi}(2z) - 2 \operatorname{shi}(z) \right] I_{-n-1/2}(z)$$

$$- \frac{n!}{2} \sum_{k=0}^{n-1} \frac{\left(-\frac{z}{2}\right)^{k-n}}{k!(n-k)} I_{k+1/2}(z) + \frac{n!}{2} \sqrt{\pi z} \sum_{k=1}^{n} \frac{\left(-\frac{z}{2}\right)^k}{(n-k)!k} \sum_{p=0}^{k-1} \frac{(-z)^p}{p!}$$

$$\times \left\{ I_{n-k+1/2}(z) \left[2^{1/2-p} I_{p-1/2}(z) - I_{p-1/2}(2z) \right] \right.$$
$$\left. - I_{k-n-1/2}(z) \left[2^{1/2-p} I_{1/2-p}(z) - I_{1/2-p}(2z) \right] \right\}.$$

1.16. The Anger $\mathbf{J}_\nu(z)$ and Weber $\mathbf{E}_\nu(z)$ Functions

1.16.1. Derivatives with respect to the argument

1. $\mathrm{D}^n[\mathbf{J}_\nu(z)] = n! \left(-\frac{z}{2}\right)^{-n} \sum_{k=0}^{[n/2]} \frac{\left(-\frac{1}{4}\right)^k}{k!(n-2k)!} \sum_{p=0}^{n-k} \binom{n-k}{p} \left(\frac{\nu}{2}\right)_{n-k-p} \left(-\frac{z}{2}\right)^p$

$$\times \left\{ \mathbf{J}_{\nu-p}(z) - (-1)^p \frac{\sin(\nu\pi)}{\pi z} \sum_{r=0}^{p-1} \left(\frac{p-r-\nu+1}{2}\right)_r \left(\frac{2}{z}\right)^r \right\}.$$

2. $= n! \left(-\frac{z}{2}\right)^{-n} \sum_{k=0}^{[n/2]} \frac{\left(-\frac{1}{4}\right)^k}{k!(n-2k)!} \sum_{p=0}^{n-k} \binom{n-k}{p} \left(-\frac{\nu}{2}\right)_{n-k-p} \left(\frac{z}{2}\right)^p$

$$\times \left\{ \mathbf{J}_{\nu+p}(z) - (-1)^p \frac{\sin(\nu\pi)}{\pi z} \sum_{r=0}^{p-1} \left(\frac{p-r+\nu+1}{2}\right)_r \left(-\frac{2}{z}\right)^r \right\}.$$

3. $= \left(\pm\frac{1}{2}\right)^n \sum_{k=0}^{n} (-1)^k \binom{n}{k} \mathbf{J}_{\nu \pm 2k \mp n}(z).$

4. $\mathrm{D}^n[z^{\nu/2} \mathbf{J}_\nu(a\sqrt{z})] = \left(\frac{a}{2}\right)^n z^{(\nu-n)/2} \mathbf{J}_{\nu-n}(a\sqrt{z})$

$$- \left(-\frac{a}{2}\right)^n \frac{\sin(\nu\pi)}{\pi a} z^{(\nu-n-1)/2} \sum_{k=0}^{n-1} \left(\frac{n-k-\nu+1}{2}\right)_k \left(\frac{2}{a\sqrt{z}}\right)^k.$$

5. $D^n[z^{-\nu/2}J_\nu(a\sqrt{z})] = \left(-\dfrac{a}{2}\right)^n z^{-(\nu+n)/2} J_{\nu+n}(a\sqrt{z})$

$\qquad - \left(\dfrac{a}{2}\right)^n \dfrac{\sin(\nu\pi)}{\pi a} z^{-(\nu+n+1)/2} \sum_{k=0}^{n-1} \left(\dfrac{n-k+\nu+1}{2}\right)_k \left(-\dfrac{2}{a\sqrt{z}}\right)^k.$

6. $D^n\left[z^{n-\nu/2-1} J_\nu\left(\dfrac{a}{\sqrt{z}}\right)\right] = \left(-\dfrac{a}{2}\right)^n z^{-(\nu+n)/2-1} J_{\nu-n}\left(\dfrac{a}{\sqrt{z}}\right)$

$\qquad - \left(\dfrac{a}{2}\right)^n \dfrac{\sin(\nu\pi)}{\pi a} z^{-(\nu+n+1)/2} \sum_{k=0}^{n-1} \left(\dfrac{n-k-\nu+1}{2}\right)_k \left(\dfrac{2\sqrt{z}}{a}\right)^k.$

7. $D^n\left[z^{n+\nu/2-1} J_\nu\left(\dfrac{a}{\sqrt{z}}\right)\right] = \left(\dfrac{a}{2}\right)^n z^{(\nu-n)/2-1} J_{\nu+n}\left(\dfrac{a}{\sqrt{z}}\right)$

$\qquad - \left(-\dfrac{a}{2}\right)^n \dfrac{\sin(\nu\pi)}{\pi a} z^{(\nu-n-1)/2} \sum_{k=0}^{n-1} \left(\dfrac{n-k+\nu+1}{2}\right)_k \left(-\dfrac{2\sqrt{z}}{a}\right)^k.$

8. $D^n[\mathbf{E}_\nu(z)] = n! \left(-\dfrac{z}{2}\right)^{-n} \sum_{k=0}^{[n/2]} \dfrac{\left(-\dfrac{1}{4}\right)^k}{k!(n-2k)!} \sum_{p=0}^{n-k} \binom{n-k}{p} \left(\dfrac{\nu}{2}\right)_{n-k-p} \left(-\dfrac{z}{2}\right)^p$

$\qquad \times \left\{\mathbf{E}_{\nu-p}(z) + \dfrac{1}{\pi z} \sum_{r=0}^{p-1} [(-1)^r + (-1)^p \cos(\nu\pi)] \left(\dfrac{p-r-\nu+1}{2}\right)_r \left(\dfrac{2}{z}\right)^r\right\}.$

9. $\quad = n! \left(-\dfrac{z}{2}\right)^{-n} \sum_{k=0}^{[n/2]} \dfrac{\left(-\dfrac{1}{4}\right)^k}{k!(n-2k)!} \sum_{p=0}^{n-k} \binom{n-k}{p} \left(-\dfrac{\nu}{2}\right)_{n-k-p} \left(\dfrac{z}{2}\right)^p$

$\qquad \times \left\{\mathbf{E}_{\nu+p}(z) + \dfrac{1}{\pi z} \sum_{r=0}^{p-1} [1 + (-1)^{p+r} \cos(\nu\pi)] \left(\dfrac{p-r+\nu+1}{2}\right)_r \left(\dfrac{2}{z}\right)^r\right\}.$

10. $\quad = \left(\pm\dfrac{1}{2}\right)^n \sum_{k=0}^{n} (-1)^k \binom{n}{k} \mathbf{E}_{\nu\pm 2k\mp n}(z).$

11. $D^n[z^{\nu/2} \mathbf{E}_\nu(a\sqrt{z})] = \left(\dfrac{a}{2}\right)^n z^{(\nu-n)/2} \mathbf{E}_{\nu-n}(a\sqrt{z}) + \dfrac{1}{\pi a} \left(\dfrac{a}{2}\right)^n z^{(\nu-n-1)/2}$

$\qquad \times \sum_{k=0}^{n-1} [(-1)^k + (-1)^n \cos(\nu\pi)] \left(\dfrac{n-k-\nu+1}{2}\right)_k \left(\dfrac{2}{a\sqrt{z}}\right)^k.$

12. $D^n[z^{-\nu/2} \mathbf{E}_\nu(a\sqrt{z})]$

$\quad = \left(-\dfrac{a}{2}\right)^n z^{-(\nu+n)/2} \mathbf{E}_{\nu+n}(a\sqrt{z}) + \dfrac{1}{\pi a} \left(\dfrac{a}{2}\right)^n z^{-(\nu+n+1)/2}$

$\qquad \times \sum_{k=0}^{n-1} [(-1)^n + (-1)^k \cos(\nu\pi)] \left(\dfrac{n-k+\nu+1}{2}\right)_k \left(\dfrac{2}{a\sqrt{z}}\right)^k.$

13. $D^n \left[z^{n-\nu/2-1} \mathbf{E}_\nu \left(\dfrac{a}{\sqrt{z}} \right) \right] = \left(-\dfrac{a}{2} \right)^n z^{-(\nu+n)/2-1} \mathbf{E}_{\nu-n} \left(\dfrac{a}{\sqrt{z}} \right)$

$+ \left(-\dfrac{a}{2} \right)^n \dfrac{z^{-(\nu+n+1)/2}}{\pi a}$

$\times \displaystyle\sum_{k=0}^{n-1} [(-1)^k + (-1)^n \cos(\nu\pi)] \left(\dfrac{n-k-\nu+1}{2} \right)_k \left(\dfrac{2\sqrt{z}}{a} \right)^k.$

14. $D^n \left[z^{n+\nu/2-1} \mathbf{E}_\nu \left(\dfrac{a}{\sqrt{z}} \right) \right]$

$= \left(\dfrac{a}{2} \right)^n z^{(\nu-n)/2-1} \mathbf{E}_{\nu+n} \left(\dfrac{a}{\sqrt{z}} \right) + \left(-\dfrac{a}{2} \right)^n \dfrac{z^{(\nu-n-1)/2}}{\pi a}$

$\times \displaystyle\sum_{k=0}^{n-1} [(-1)^n + (-1)^k \cos(\nu\pi)] \left(\dfrac{n-k+\nu+1}{2} \right)_k \left(\dfrac{2\sqrt{z}}{a} \right)^k.$

1.16.2. Derivatives with respect to the order

1. $\left. \dfrac{\partial \mathbf{J}_\nu(z)}{\partial \nu} \right|_{\nu=n} = \dfrac{n!}{2} \displaystyle\sum_{k=0}^{n-1} \dfrac{\left(\dfrac{2}{z} \right)^{n-k}}{k!(n-k)} J_k(z) + \dfrac{\pi}{2} \mathbf{H}_n(z)$

$- \dfrac{1}{2} \displaystyle\sum_{k=0}^{n-1} \dfrac{\left(\dfrac{1}{2} \right)_k}{\left(\dfrac{1}{2} \right)_{n-k}} \left(\dfrac{z}{2} \right)^{n-2k-1} + \dfrac{(-1)^n}{z} \displaystyle\sum_{k=0}^{n-1} (-1)^k \left(\dfrac{n-k+1}{2} \right)_k \left(\dfrac{2}{z} \right)^k.$

2. $\left. \dfrac{\partial \mathbf{J}_\nu(z)}{\partial \nu} \right|_{\nu=-n} = (-1)^{n-1} \dfrac{n!}{2} \displaystyle\sum_{k=0}^{n-1} \dfrac{\left(\dfrac{2}{z} \right)^{n-k}}{k!(n-k)} J_k(z) + \dfrac{\pi}{2} \mathbf{H}_{-n}(z)$

$+ \dfrac{(-1)^n}{z} \displaystyle\sum_{k=0}^{n-1} \left(\dfrac{n-k+1}{2} \right)_k \left(\dfrac{2}{z} \right)^k.$

3. $\left. \dfrac{\partial \mathbf{E}_\nu(z)}{\partial \nu} \right|_{\nu=n} = \dfrac{\pi}{2} J_n(z)$

$+ \dfrac{n!}{2} \displaystyle\sum_{k=0}^{n-1} \dfrac{\left(\dfrac{2}{z} \right)^{n-k}}{k!(n-k)} \left[-\mathbf{H}_k(z) + \dfrac{1}{\pi} \displaystyle\sum_{r=0}^{k-1} \dfrac{\left(\dfrac{1}{2} \right)_r}{\left(\dfrac{1}{2} \right)_{k-r}} \left(\dfrac{z}{2} \right)^{k-2r-1} \right]$

$+ \dfrac{1}{2\pi} \displaystyle\sum_{k=0}^{n-1} [(-1)^k + (-1)^n] \left(\dfrac{n-k+1}{2} \right)_k \left(-\dfrac{2}{z} \right)^{k+1} \displaystyle\sum_{r=0}^{k-1} \dfrac{1}{2r+n-k+1}.$

4. $\left. \dfrac{\partial \mathbf{E}_\nu(z)}{\partial \nu} \right|_{\nu=-n} = \dfrac{n!}{2} \displaystyle\sum_{k=0}^{n-1} \dfrac{\left(-\dfrac{2}{z} \right)^{n-k}}{k!(n-k)} \mathbf{H}_{-k}(z) + (-1)^n \dfrac{\pi}{2} J_n(z)$

$+ \dfrac{1}{2\pi} \displaystyle\sum_{k=0}^{n-1} [(-1)^k + (-1)^n] \left(\dfrac{n-k+1}{2} \right)_k \left(\dfrac{2}{z} \right)^{k+1} \displaystyle\sum_{r=0}^{k-1} \dfrac{1}{2r+n-k+1}.$

1.17. The Kelvin Functions $\text{ber}_\nu(z)$, $\text{bei}_\nu(z)$, $\text{ker}_\nu(z)$ and $\text{kei}_\nu(z)$

1.17.1. Derivatives with respect to the argument

1. $D^n[z^{\pm\nu/2}\,\text{ber}_\nu(a\sqrt{z})]$
$$= \left(\pm\frac{a}{2}\right)^n z^{(\pm\nu-n)/2}\left[\cos\frac{3n\pi}{4}\,\text{ber}_{\nu\mp n}(a\sqrt{z}) - \sin\frac{3n\pi}{4}\,\text{bei}_{\nu\mp n}(a\sqrt{z})\right].$$

2. $D^n[z^{\pm\nu/2}\,\text{bei}_\nu(a\sqrt{z})]$
$$= \left(\pm\frac{a}{2}\right)^n z^{(\pm\nu-n)/2}\left[\sin\frac{3n\pi}{4}\,\text{ber}_{\nu\mp n}(a\sqrt{z}) + \cos\frac{3n\pi}{4}\,\text{bei}_{\nu\mp n}(a\sqrt{z})\right].$$

3. $D^n[z^{(2n+1)/4}\,\text{ber}_{n+1/2}(a\sqrt{z})] = -\frac{1}{\sqrt{\pi}}\left(\frac{a}{2}\right)^{n-1/2}$
$$\times\left[\sin\frac{3(2n-1)\pi}{8}\,\sinh\!\left(a\sqrt{\frac{z}{2}}\right)\cos\!\left(a\sqrt{\frac{z}{2}}\right)\right.$$
$$\left.+\cos\frac{3(2n-1)\pi}{8}\,\cosh\!\left(a\sqrt{\frac{z}{2}}\right)\sin\!\left(a\sqrt{\frac{z}{2}}\right)\right].$$

4. $D^n[z^{(2n+1)/4}\,\text{bei}_{n+1/2}(a\sqrt{z})] = \frac{1}{\sqrt{\pi}}\left(\frac{a}{2}\right)^{n-1/2}$
$$\times\left[\cos\frac{3(2n-1)\pi}{8}\,\sinh\!\left(a\sqrt{\frac{z}{2}}\right)\cos\!\left(a\sqrt{\frac{z}{2}}\right)\right.$$
$$\left.-\sin\frac{3(2n-1)\pi}{8}\,\cosh\!\left(a\sqrt{\frac{z}{2}}\right)\sin\!\left(a\sqrt{\frac{z}{2}}\right)\right].$$

5. $D^n[z^{(2n+1)/4}\,\text{ber}_{-n-1/2}(a\sqrt{z})] = \frac{(-1)^{n+1}}{\sqrt{\pi}}\left(\frac{a}{2}\right)^{n-1/2}$
$$\times\left[\sin\frac{3(2n-1)\pi}{8}\,\sinh\!\left(a\sqrt{\frac{z}{2}}\right)\sin\!\left(a\sqrt{\frac{z}{2}}\right)\right.$$
$$\left.-\cos\frac{3(2n-1)\pi}{8}\,\cosh\!\left(a\sqrt{\frac{z}{2}}\right)\cos\!\left(a\sqrt{\frac{z}{2}}\right)\right].$$

6. $D^n[z^{(2n+1)/4}\,\text{bei}_{-n-1/2}(a\sqrt{z})] = \frac{(-1)^n}{\sqrt{\pi}}\left(\frac{a}{2}\right)^{n-1/2}$
$$\times\left[\cos\frac{3(2n-1)\pi}{8}\,\sinh\!\left(a\sqrt{\frac{z}{2}}\right)\sin\!\left(a\sqrt{\frac{z}{2}}\right)\right.$$
$$\left.+\sin\frac{3(2n-1)\pi}{8}\,\cosh\!\left(a\sqrt{\frac{z}{2}}\right)\cos\!\left(a\sqrt{\frac{z}{2}}\right)\right].$$

7. $D^n\!\left[z^{-(2n+3)/4}\,\text{ber}_{n+1/2}(a\sqrt{z})\right] = \sqrt{\pi}\left(\frac{a}{2z}\right)^{n+1/2}$
$$\times\left\{\cos\frac{3(2n+1)\pi}{8}\left[\text{ber}_{n+1/2}\!\left(\frac{a\sqrt{z}}{2}\right)\text{ber}_{-n-1/2}\!\left(\frac{a\sqrt{z}}{2}\right)\right.\right.$$
$$\left.\left. - \text{bei}_{n+1/2}\!\left(\frac{a\sqrt{z}}{2}\right)\text{bei}_{-n-1/2}\!\left(\frac{a\sqrt{z}}{2}\right)\right]\right.$$

$$-\sin\frac{3(2n+1)\pi}{8}\Big[\text{ber}_{n+1/2}\Big(\frac{a\sqrt{z}}{2}\Big)\text{bei}_{-n-1/2}\Big(\frac{a\sqrt{z}}{2}\Big)$$
$$+\text{bei}_{n+1/2}\Big(\frac{a\sqrt{z}}{2}\Big)\text{ber}_{-n-1/2}\Big(\frac{a\sqrt{z}}{2}\Big)\Big]\Big\}.$$

8. $D^n\big[z^{-(2n+3)/4}\,\text{bei}_{n+1/2}(a\sqrt{z})\big] = \sqrt{\pi}\Big(\frac{a}{2z}\Big)^{n+1/2}$
$$\times\Big\{\sin\frac{3(2n+1)\pi}{8}\Big[\text{ber}_{n+1/2}\Big(\frac{a\sqrt{z}}{2}\Big)\text{ber}_{-n-1/2}\Big(\frac{a\sqrt{z}}{2}\Big)$$
$$-\text{bei}_{n+1/2}\Big(\frac{a\sqrt{z}}{2}\Big)\text{bei}_{-n-1/2}\Big(\frac{a\sqrt{z}}{2}\Big)\Big]$$
$$+\cos\frac{3(2n+1)\pi}{8}\Big[\text{ber}_{n+1/2}\Big(\frac{a\sqrt{z}}{2}\Big)\text{bei}_{-n-1/2}\Big(\frac{a\sqrt{z}}{2}\Big)$$
$$+\text{bei}_{n+1/2}\Big(\frac{a\sqrt{z}}{2}\Big)\text{ber}_{-n-1/2}\Big(\frac{a\sqrt{z}}{2}\Big)\Big]\Big\}.$$

9. $D^n\Big[z^{n-1/2}e^{az/\sqrt{2}}\Big(\sin\frac{az}{\sqrt{2}}\text{ber}_{n-1/2}(az)+\cos\frac{az}{\sqrt{2}}\text{bei}_{n-1/2}(az)\Big)\Big]$
$$= \frac{(2a)^{n-1/2}}{\sqrt{\pi}}z^{n-1}e^{\sqrt{2}az}\sin\Big[\frac{3(2n-1)\pi}{8}+\sqrt{2}az\Big] \quad [n\geq 1].$$

10. $D^n\Big[z^{n-1/2}e^{az/\sqrt{2}}\Big(\cos\frac{az}{\sqrt{2}}\text{ber}_{n-1/2}(az)-\sin\frac{az}{\sqrt{2}}\text{bei}_{n-1/2}(az)\Big)\Big]$
$$= \frac{(2a)^{n-1/2}}{\sqrt{\pi}}z^{n-1}e^{\sqrt{2}az}\cos\Big[\frac{3(2n-1)\pi}{8}+\sqrt{2}az\Big] \quad [n\geq 1].$$

11. $D^n\Big[z^{n-1/2}\Big(\sinh\frac{az}{\sqrt{2}}\cos\frac{az}{\sqrt{2}}\text{ber}_{n-1/2}(az)$
$$-\cosh\frac{az}{\sqrt{2}}\sin\frac{az}{\sqrt{2}}\text{bei}_{n-1/2}(az)\Big)\Big]$$
$$= \frac{(2a)^{n-1/2}}{\sqrt{\pi}}z^{n-1}\Big[\cos\frac{3(2n-1)\pi}{8}\sinh(\sqrt{2}az)\cos(\sqrt{2}az)$$
$$-\sin\frac{3(2n-1)\pi}{8}\cosh(\sqrt{2}az)\sin(\sqrt{2}az)\Big].$$

12. $D^n\Big[z^{n-1/2}\Big(\cosh\frac{az}{\sqrt{2}}\sin\frac{az}{\sqrt{2}}\text{ber}_{n-1/2}(az)$
$$+\sinh\frac{az}{\sqrt{2}}\cos\frac{az}{\sqrt{2}}\text{bei}_{n-1/2}(az)\Big)\Big]$$
$$= \frac{(2a)^{n-1/2}}{\sqrt{\pi}}z^{n-1}\Big[\sin\frac{3(2n-1)\pi}{8}\sinh(\sqrt{2}az)\cos(\sqrt{2}az)$$
$$+\cos\frac{3(2n-1)\pi}{8}\cosh(\sqrt{2}az)\sin(\sqrt{2}az)\Big].$$

13. $D^n\big[z^{n-1/2}\big(\text{ber}^2_{\pm n\mp 1/2}(a\sqrt{z})-\text{bei}^2_{\pm n\mp 1/2}(a\sqrt{z})\big)\big] = \pm\dfrac{a^{n-1/2}}{\sqrt{\pi}}z^{(2n-3)/4}$
$$\times\Big[\cos\frac{3(2n-1)\pi}{8}\text{ber}_{n-1/2}(2a\sqrt{z})-\sin\frac{3(2n-1)\pi}{8}\text{bei}_{n-1/2}(2a\sqrt{z})\Big]$$
$$[n\geq 1].$$

14. $D^n\left[z^{n-1/2}(\text{ber}_{\pm n \mp 1/2}(a\sqrt{z})\,\text{bei}_{\pm n \mp 1/2}(a\sqrt{z}))\right] = \pm \dfrac{a^{n-1/2}}{2\sqrt{\pi}} z^{(2n-3)/4}$

$\times \left[\sin\dfrac{3(2n-1)\pi}{8}\,\text{ber}_{n-1/2}(2a\sqrt{z}) + \cos\dfrac{3(2n-1)\pi}{8}\,\text{bei}_{n-1/2}(2a\sqrt{z})\right]$

$[n \geq 1]$.

15. $D^n[z^{\pm\nu/2}\,\text{ker}_\nu(a\sqrt{z})]$
$= \left(\pm\dfrac{a}{2}\right)^n z^{(\pm\nu-n)/2}\left[\cos\dfrac{3n\pi}{4}\,\text{ker}_{\nu\mp n}(a\sqrt{z}) - \sin\dfrac{3n\pi}{4}\,\text{kei}_{\nu\mp n}(a\sqrt{z})\right]$.

16. $D^n[z^{\pm\nu/2}\,\text{kei}_\nu(a\sqrt{z})]$
$= \left(\pm\dfrac{a}{2}\right)^n z^{(\pm\nu-n)/2}\left[\sin\dfrac{3n\pi}{4}\,\text{ker}_{\nu\mp n}(a\sqrt{z}) + \cos\dfrac{3n\pi}{4}\,\text{kei}_{\nu\mp n}(a\sqrt{z})\right]$.

17. $D^n[z^{(2n+1)/4}\,\text{ker}_{n+1/2}(a\sqrt{z})]$
$= (-1)^n \dfrac{\sqrt{\pi}}{2^{n+1/2}} a^{n-1/2} e^{-a\sqrt{z/2}} \cos\left[a\sqrt{\dfrac{z}{2}} + \dfrac{(2n+3)\pi}{8}\right]$.

18. $D^n[z^{(2n+1)/4}\,\text{kei}_{n+1/2}(a\sqrt{z})]$
$= (-1)^{n+1} \dfrac{\sqrt{\pi}}{2^{n+1/2}} a^{n-1/2} e^{-a\sqrt{z/2}} \sin\left[a\sqrt{\dfrac{z}{2}} + \dfrac{(2n+3)\pi}{8}\right]$.

19. $D^n\left[z^{-(2n+3)/4}\,\text{ker}_{n+1/2}(a\sqrt{z})\right] = \dfrac{(-1)^n}{\sqrt{\pi}} \left(\dfrac{a}{2z}\right)^{n+1/2}$

$\times \left\{\cos\dfrac{3(2n+1)\pi}{8}\left[\text{ker}^2_{n+1/2}\left(\dfrac{a\sqrt{z}}{2}\right) - \text{kei}^2_{n+1/2}\left(\dfrac{a\sqrt{z}}{2}\right)\right]\right.$

$\left. - 2\sin\dfrac{3(2n+1)\pi}{8}\,\text{ker}_{n+1/2}\left(\dfrac{a\sqrt{z}}{2}\right)\text{kei}_{n+1/2}\left(\dfrac{a\sqrt{z}}{2}\right)\right\}$.

20. $D^n\left[z^{-(2n+3)/4}\,\text{kei}_{n+1/2}(a\sqrt{z})\right] = \dfrac{(-1)^n}{\sqrt{\pi}} \left(\dfrac{a}{2z}\right)^{n+1/2}$

$\times \left\{\sin\dfrac{3(2n+1)\pi}{8}\left[\text{ker}^2_{n+1/2}\left(\dfrac{a\sqrt{z}}{2}\right) - \text{kei}^2_{n+1/2}\left(\dfrac{a\sqrt{z}}{2}\right)\right]\right.$

$\left. + 2\cos\dfrac{3(2n+1)\pi}{8}\,\text{ker}_{n+1/2}\left(\dfrac{a\sqrt{z}}{2}\right)\text{kei}_{n+1/2}\left(\dfrac{a\sqrt{z}}{2}\right)\right\}$.

21. $D^n\left[z^{n-1/2}\,\text{ker}_{n-1/2}(a\sqrt{z})\,\text{kei}_{n-1/2}(a\sqrt{z})\right] = (-1)^n \dfrac{\sqrt{\pi}}{2} a^{n-1/2} z^{(2n-3)/4}$

$\times \left[\cos\dfrac{(2n-1)\pi}{8}\,\text{kei}_{n-1/2}(2a\sqrt{z}) - \sin\dfrac{(2n-1)\pi}{8}\,\text{ker}_{n-1/2}(2a\sqrt{z})\right]$.

22. $D^n\left[z^{n-1/2}\left(\text{ker}^2_{n-1/2}(a\sqrt{z}) - \text{kei}^2_{n-1/2}(a\sqrt{z})\right)\right]$
$= (-1)^n \sqrt{\pi}\, a^{n-1/2} z^{(2n-3)/4}$

$\times \left[\cos\dfrac{(2n-1)\pi}{8}\,\text{ker}_{n-1/2}(2a\sqrt{z}) + \sin\dfrac{(2n-1)\pi}{8}\,\text{kei}_{n-1/2}(2a\sqrt{z})\right]$.

23. $D^n\left[z^{n-1/2}(\text{ber}_{n-1/2}(a\sqrt{z})\,\text{ker}_{n-1/2}(a\sqrt{z})\right.$
$\left.-\,\text{bei}_{n-1/2}(a\sqrt{z})\,\text{kei}_{n-1/2}(a\sqrt{z}))\right]$
$$=\frac{a^{n-1/2}}{\sqrt{\pi}}z^{(2n-3)/4}\left[\cos\frac{3(2n-1)\pi}{8}\,\text{ker}_{n-1/2}(2a\sqrt{z})\right.$$
$$\left.-\sin\frac{3(2n-1)\pi}{8}\,\text{kei}_{n-1/2}(2a\sqrt{z})\right].$$

24. $D^n\left[z^{n-1/2}(\text{ber}_{n-1/2}(a\sqrt{z})\,\text{kei}_{n-1/2}(a\sqrt{z})\right.$
$\left.+\,\text{bei}_{n-1/2}(a\sqrt{z})\,\text{ker}_{n-1/2}(a\sqrt{z}))\right]$
$$=\frac{a^{n-1/2}}{\sqrt{\pi}}z^{(2n-3)/4}\left[\sin\frac{3(2n-1)\pi}{8}\,\text{ker}_{n-1/2}(2a\sqrt{z})\right.$$
$$\left.+\cos\frac{3(2n-1)\pi}{8}\,\text{kei}_{n-1/2}(2a\sqrt{z})\right].$$

1.17.2. Derivatives with respect to the order

1. $\left.\dfrac{\partial\,\text{ber}_\nu(z)}{\partial\nu}\right|_{\nu=n} = -\dfrac{\pi}{2}\,\text{bei}_n(z) - \text{ker}_n(z)$
$$+\frac{n!}{2}\sum_{k=0}^{n-1}\frac{\left(\frac{z}{2}\right)^{k-n}}{k!(n-k)}\left[\cos\frac{5(k-n)\pi}{4}\,\text{ber}_k(z) + \sin\frac{5(k-n)\pi}{4}\,\text{bei}_k(z)\right].$$

2. $\left.\dfrac{\partial\,\text{bei}_\nu(z)}{\partial\nu}\right|_{\nu=n} = \dfrac{\pi}{2}\,\text{ber}_n(z) - \text{kei}_n(z)$
$$+\frac{n!}{2}\sum_{k=0}^{n-1}\frac{\left(\frac{z}{2}\right)^{k-n}}{k!(n-k)}\left[\cos\frac{5(k-n)\pi}{4}\,\text{bei}_k(z) - \sin\frac{5(k-n)\pi}{4}\,\text{ber}_k(z)\right].$$

3. $\left.\dfrac{\partial\,\text{ber}_\nu(z)}{\partial\nu}\right|_{\nu=0} = -\dfrac{\pi}{2}\,\text{bei}_0(z) - \text{ker}_0(z).$

4. $\left.\dfrac{\partial\,\text{bei}_\nu(z)}{\partial\nu}\right|_{\nu=0} = \dfrac{\pi}{2}\,\text{ber}_0(z) - \text{kei}_0(z).$

5. $\left.\dfrac{\partial\,\text{ker}_\nu(z)}{\partial\nu}\right|_{\nu=n} = \dfrac{\pi}{2}\,\text{kei}_n(z)$
$$+\frac{n!}{2}\sum_{k=0}^{n-1}\frac{\left(\frac{z}{2}\right)^{k-n}}{k!(n-k)}\left[\cos\frac{3(k-n)\pi}{4}\,\text{ker}_k(z) - \sin\frac{3(k-n)\pi}{4}\,\text{kei}_k(z)\right].$$

6. $\left.\dfrac{\partial\,\text{kei}_\nu(z)}{\partial\nu}\right|_{\nu=n} = -\dfrac{\pi}{2}\,\text{ker}_n(z)$
$$+\frac{n!}{2}\sum_{k=0}^{n-1}\frac{\left(\frac{z}{2}\right)^{k-n}}{k!(n-k)}\left[\sin\frac{3(k-n)\pi}{4}\,\text{ker}_k(z) + \cos\frac{3(k-n)\pi}{4}\,\text{kei}_k(z)\right].$$

7. $\dfrac{\partial\,\mathrm{ber}_\nu(z)}{\partial \nu}\bigg|_{\nu=n+1/2}$

$$= \frac{1}{6}\,\mathrm{ber}_{n+1/2}(z)\left[6C + 6\ln(2z) - z^4\,{}_2F_5\!\left(\begin{array}{c}1,1;\ -\dfrac{z^4}{16}\\ \frac{5}{4},\frac{3}{2},\frac{7}{4},2,2\end{array}\right)\right]$$

$$- \frac{1}{4}\,\mathrm{bei}_{n+1/2}(z)\left[3\pi + 4z^2\,{}_1F_4\!\left(\begin{array}{c}\frac{1}{2};\ -\dfrac{z^4}{16}\\ \frac{3}{4},\frac{5}{4},\frac{3}{2},\frac{3}{2}\end{array}\right)\right]$$

$$+ (-1)^n\sqrt{2}\,z\,[\mathrm{ber}_{-n-1/2}(z) + \mathrm{bei}_{-n-1/2}(z)]\,{}_1F_4\!\left(\begin{array}{c}\frac{1}{4};\ -\dfrac{z^4}{16}\\ 1,\frac{1}{2},\frac{3}{4},\frac{5}{4},\frac{5}{4}\end{array}\right)$$

$$+ (-1)^n\,\frac{2\sqrt{2}}{9}\,z^3\,[\mathrm{ber}_{-n-1/2}(z) - \mathrm{bei}_{-n-1/2}(z)]\,{}_1F_4\!\left(\begin{array}{c}\frac{3}{4};\ -\dfrac{z^4}{16}\\ \frac{5}{4},\frac{3}{2},\frac{7}{4},\frac{7}{4}\end{array}\right)$$

$$+ \frac{n!}{2}\sum_{k=0}^{n-1}\frac{\left(\frac{z}{2}\right)^{k-n}}{k!(n-k)}\left[\cos\frac{3(n-k)\pi}{4}\,\mathrm{ber}_{k+1/2}(z) + \sin\frac{3(n-k)\pi}{4}\,\mathrm{bei}_{k+1/2}(z)\right]$$

$$- \frac{n!\sqrt{\pi z}}{2}\sum_{k=1}^{n}\frac{\left(\frac{2}{z}\right)^k}{(n-k)!\,k}\sum_{p=0}^{k-1}\frac{z^p}{p!}\left\{\mathrm{ber}_{n-k+1/2}(z)\right.$$

$$\times\left[\cos\frac{3(2k-2p-1)\pi}{8}\,\mathrm{ber}_{p-1/2}(2z) + \sin\frac{3(2k-2p-1)\pi}{8}\,\mathrm{bei}_{p-1/2}(2z)\right]$$

$$+ \mathrm{bei}_{n-k+1/2}(z)\left[\sin\frac{3(2k-2p-1)\pi}{8}\,\mathrm{ber}_{p-1/2}(2z)\right.$$

$$\left. - \cos\frac{3(2k-2p-1)\pi}{8}\,\mathrm{bei}_{p-1/2}(2z)\right]$$

$$- (-1)^{k+n+p}\,\mathrm{ber}_{k-n-1/2}(z)\left[\cos\frac{3(2k-2p-1)\pi}{8}\,\mathrm{ber}_{1/2-p}(2z)\right.$$

$$\left. + \sin\frac{3(2k-2p-1)\pi}{8}\,\mathrm{bei}_{1/2-p}(2z)\right]$$

$$- (-1)^{k+n+p}\,\mathrm{bei}_{k-n-1/2}(z)\left[\sin\frac{3(2k-2p-1)\pi}{8}\,\mathrm{ber}_{1/2-p}(2z)\right.$$

$$\left.\left. - \cos\frac{3(2k-2p-1)\pi}{8}\,\mathrm{bei}_{1/2-p}(2z)\right]\right\}\qquad [|\arg z|\le \pi/4].$$

8. $\dfrac{\partial\,\mathrm{bei}_\nu(z)}{\partial \nu}\bigg|_{\nu=n+1/2} = \dfrac{\partial\,\mathrm{bei}_\nu(z)}{\partial \nu}\bigg|_{\nu=n+1/2}$

$$= \frac{1}{4}\,\mathrm{ber}_{n+1/2}(z)\left[3\pi + 4z^2\,{}_1F_4\!\left(\begin{array}{c}\frac{1}{2};\ -\dfrac{z^4}{16}\\ \frac{3}{4},\frac{5}{4},\frac{3}{2},\frac{3}{2}\end{array}\right)\right]$$

$$+ \frac{1}{6}\,\mathrm{bei}_{n+1/2}(z)\left[6C + 6\ln(2z) - z^4\,{}_2F_5\!\left(\begin{array}{c}1,1;\ -\dfrac{z^4}{16}\\ \frac{5}{4},\frac{3}{2},\frac{7}{4},2,2\end{array}\right)\right]$$

$$+ (-1)^n \sqrt{2}\, z[\mathrm{ber}_{-n-1/2}(z) - \mathrm{bei}_{-n-1/2}(z)]\, {}_1F_4\left(\begin{array}{c}\frac{1}{4};\ -\frac{z^4}{16}\\ \frac{1}{2}, \frac{3}{4}, \frac{5}{4}, \frac{5}{4}\end{array}\right)$$

$$+ (-1)^n \frac{2\sqrt{2}}{9} z^3 [\mathrm{ber}_{-n-1/2}(z) + \mathrm{bei}_{-n-1/2}(z)]\, {}_1F_4\left(\begin{array}{c}\frac{3}{4};\ -\frac{z^4}{16}\\ \frac{5}{4}, \frac{3}{2}, \frac{7}{4}, \frac{7}{4}\end{array}\right)$$

$$- \frac{n!}{2}\sum_{k=0}^{n-1} \frac{\left(\frac{z}{2}\right)^{k-n}}{k!(n-k)}\left[\sin\frac{3(n-k)\pi}{4}\mathrm{ber}_{k+1/2}(z) - \cos\frac{3(n-k)\pi}{4}\mathrm{bei}_{k+1/2}(z)\right]$$

$$+ \frac{n!\sqrt{\pi z}}{2}\sum_{k=1}^{n}\frac{\left(\frac{2}{z}\right)^k}{(n-k)!\,k}\sum_{p=0}^{k-1}\frac{z^p}{p!}\Big\{\mathrm{ber}_{n-k+1/2}(z)$$

$$\times \left[\sin\frac{3(2k-2p-1)\pi}{8}\mathrm{ber}_{p-1/2}(2z) - \cos\frac{3(2k-2p-1)\pi}{8}\mathrm{bei}_{p-1/2}(2z)\right]$$

$$- \mathrm{bei}_{n-k+1/2}(z)\left[\cos\frac{3(2k-2p-1)\pi}{8}\mathrm{ber}_{p-1/2}(2z)\right.$$

$$\left. + \sin\frac{3(2k-2p-1)\pi}{8}\mathrm{bei}_{p-1/2}(2z)\right]$$

$$- (-1)^{k+n+p}\mathrm{ber}_{k-n-1/2}(z)\left[\sin\frac{3(2k-2p-1)\pi}{8}\mathrm{ber}_{1/2-p}(2z)\right.$$

$$\left. - \cos\frac{3(2k-2p-1)\pi}{8}\mathrm{bei}_{1/2-p}(2z)\right]$$

$$+ (-1)^{k+n+p}\mathrm{bei}_{k-n-1/2}(z)\left[\cos\frac{3(2k-2p-1)\pi}{8}\mathrm{ber}_{1/2-p}(2z)\right.$$

$$\left. + \sin\frac{3(2k-2p-1)\pi}{8}\mathrm{bei}_{1/2-p}(2z)\right]\Big\} \qquad [|\arg z| \le \pi/4].$$

9. $\left.\dfrac{\partial\,\mathrm{ber}_\nu(z)}{\partial \nu}\right|_{\nu=1/2-n}$

$$= \frac{1}{6}\mathrm{ber}_{1/2-n}(z)\left[6\mathbf{C} + 6\ln(2z) - z^4\,{}_2F_5\left(\begin{array}{c}1,1;\ -\frac{z^4}{16}\\ \frac{5}{4},\frac{3}{2},\frac{7}{4},2,2\end{array}\right)\right]$$

$$- \frac{1}{4}\mathrm{bei}_{1/2-n}(z)\left[3\pi + 4z^2\,{}_1F_4\left(\begin{array}{c}\frac{1}{2};\ -\frac{z^4}{16}\\ \frac{3}{4},\frac{5}{4},\frac{3}{2},\frac{3}{2}\end{array}\right)\right]$$

$$+ (-1)^n \sqrt{2}\, z[\mathrm{ber}_{n-1/2}(z) + \mathrm{bei}_{n-1/2}(z)]\, {}_1F_4\left(\begin{array}{c}\frac{1}{4};\ -\frac{z^4}{16}\\ \frac{1}{2},\frac{3}{4},\frac{5}{4},\frac{5}{4}\end{array}\right)$$

$$+ (-1)^n \frac{2\sqrt{2}}{9} z^3 [\mathrm{ber}_{n-1/2}(z) - \mathrm{bei}_{n-1/2}(z)]\, {}_1F_4\left(\begin{array}{c}\frac{3}{4};\ -\frac{z^4}{16}\\ \frac{5}{4},\frac{3}{2},\frac{7}{4},\frac{7}{4}\end{array}\right)$$

$$- \frac{n!}{2}\sum_{k=0}^{n-1}\frac{\left(-\frac{z}{2}\right)^{k-n}}{k!(n-k)}\left[\cos\frac{3(n-k)\pi}{4}\mathrm{ber}_{1/2-k}(z) + \sin\frac{3(n-k)\pi}{4}\mathrm{bei}_{1/2-k}(z)\right]$$

$$+ \frac{n!\sqrt{\pi z}}{2} \sum_{k=1}^{n} \frac{\left(\frac{2}{z}\right)^k}{(n-k)!\,k} \sum_{p=0}^{k-1} \frac{z^p}{p!} \left\{ (-1)^{n+p} \operatorname{ber}_{n-k-1/2}(z) \right.$$

$$\times \left[\cos\frac{3(2k-2p-1)\pi}{8} \operatorname{ber}_{1/2-p}(2z) + \sin\frac{3(2k-2p-1)\pi}{8} \operatorname{bei}_{1/2-p}(2z) \right]$$

$$+ (-1)^{n+p} \operatorname{bei}_{n-k-1/2}(z) \left[\sin\frac{3(2k-2p-1)\pi}{8} \operatorname{ber}_{1/2-p}(2z) \right.$$

$$\left. - \cos\frac{3(2k-2p-1)\pi}{8} \operatorname{bei}_{1/2-p}(2z) \right]$$

$$- (-1)^k \operatorname{ber}_{k-n+1/2}(z) \left[\cos\frac{3(2k-2p-1)\pi}{8} \operatorname{ber}_{p-1/2}(2z) \right.$$

$$\left. + \sin\frac{3(2k-2p-1)\pi}{8} \operatorname{bei}_{p-1/2}(2z) \right]$$

$$- (-1)^k \operatorname{bei}_{k-n+1/2}(z) \left[\sin\frac{3(2k-2p-1)\pi}{8} \operatorname{ber}_{p-1/2}(2z) \right.$$

$$\left. \left. - \cos\frac{3(2k-2p-1)\pi}{8} \operatorname{bei}_{p-1/2}(2z) \right] \right\} \quad [|\arg z| \le \pi/4].$$

10. $\left. \dfrac{\partial \operatorname{bei}_\nu(z)}{\partial \nu} \right|_{\nu=1/2-n} = \dfrac{1}{4} \operatorname{ber}_{1/2-n}(z) \left[3\pi + 4z^2 \, {}_1F_4\!\left(\begin{array}{c} \frac{1}{2}; \; -\frac{z^4}{16} \\ \frac{3}{4},\frac{5}{4},\frac{3}{2},\frac{3}{2} \end{array} \right) \right]$

$$+ \frac{1}{6} \operatorname{bei}_{1/2-n}(z) \left[6\,\mathbf{C} + 6\ln(2z) - x^4 \, {}_2F_5\!\left(\begin{array}{c} 1,1;\; -\frac{z^4}{16} \\ \frac{5}{4},\frac{3}{2},\frac{7}{4},2,2 \end{array} \right) \right]$$

$$- (-1)^n \sqrt{2}\, x [\operatorname{ber}_{n-1/2}(z) - \operatorname{bei}_{n-1/2}(z)] \, {}_1F_4\!\left(\begin{array}{c} \frac{1}{4};\; -\frac{z^4}{16} \\ \frac{1}{2},\frac{3}{4},\frac{5}{4},\frac{5}{4} \end{array} \right)$$

$$+ (-1)^n \frac{2\sqrt{2}}{9} z^3 [\operatorname{ber}_{n-1/2}(z) + \operatorname{bei}_{n-1/2}(z)] \, {}_1F_4\!\left(\begin{array}{c} \frac{3}{4};\; -\frac{z^4}{16} \\ \frac{5}{4},\frac{3}{2},\frac{7}{4},\frac{7}{4} \end{array} \right)$$

$$+ \frac{n!}{2} \sum_{k=0}^{n-1} \frac{\left(-\frac{z}{2}\right)^{k-n}}{k!\,(n-k)} \left[\sin\frac{3(n-k)\pi}{4} \operatorname{ber}_{1/2-k}(z) - \cos\frac{3(n-k)\pi}{4} \operatorname{bei}_{1/2-k}(z) \right]$$

$$- \frac{n!\sqrt{\pi z}}{2} \sum_{k=1}^{n} \frac{\left(\frac{2}{z}\right)^k}{(n-k)!\,k} \sum_{p=0}^{k-1} \frac{z^p}{p!} \left\{ (-1)^{n+p} \operatorname{ber}_{n-k-1/2}(z) \right.$$

$$\times \left[\sin\frac{3(2k-2p-1)\pi}{8} \operatorname{ber}_{1/2-p}(2z) - \cos\frac{3(2k-2p-1)\pi}{8} \operatorname{bei}_{1/2-p}(2z) \right]$$

$$- (-1)^{n+p} \operatorname{bei}_{n-k-1/2}(z) \left[\cos\frac{3(2k-2p-1)\pi}{8} \operatorname{ber}_{1/2-p}(2z) \right.$$

$$\left. + \sin\frac{3(2k-2p-1)\pi}{8} \operatorname{bei}_{1/2-p}(2z) \right]$$

$$- (-1)^k \operatorname{ber}_{k-n+1/2}(z) \left[\sin\frac{3(2k-2p-1)\pi}{8} \operatorname{ber}_{p-1/2}(2z) \right.$$

$$\left. - \cos\frac{3(2k-2p-1)\pi}{8} \operatorname{bei}_{p-1/2}(z) \right]$$

1.17.2]   1.17. The Kelvin Functions $\text{ber}_\nu(z)$, $\text{bei}_\nu(z)$, $\text{ker}_\nu(z)$ and $\text{kei}_\nu(z)$

$$+ (-1)^k \text{bei}_{k-n+1/2}(z)\left[\cos\frac{3(2k-2p-1)\pi}{8}\text{ber}_{p-1/2}(2z)\right.$$
$$\left.+ \sin\frac{3(2k-2p-1)\pi}{8}\text{bei}_{p-1/2}(2z)\right]\} \quad [|\arg z| \leq \pi/4].$$

11. $\left.\dfrac{\partial \text{ker}_\nu(z)}{\partial \nu}\right|_{\nu=n-1/2} = \dfrac{\pi}{2}\text{kei}_{n-1/2}(z) - \dfrac{\pi^2}{8}\text{ber}_{n-1/2}(z)$

$$- \frac{\pi}{2}\ln(2z)\text{bei}_{n-1/2}(z) + (-1)^n\frac{\pi}{2}[C + \ln(2z)]\text{ber}_{1/2-n}(z)$$
$$- (-1)^n\frac{\pi^2}{8}\text{bei}_{1/2-n}(z)$$
$$- \frac{\pi z^2}{2}[\text{ber}_{n-1/2}(z) + (-1)^n\text{bei}_{1/2-n}(z)]\,_1F_4\left(\begin{array}{c}\frac{1}{2};\,-\frac{z^4}{16}\\\frac{3}{4},\frac{5}{4},\frac{3}{2},\frac{3}{2}\end{array}\right)$$
$$+ \frac{\pi z}{\sqrt{2}}[\text{ber}_{n-1/2}(z) + \text{bei}_{n-1/2}(z) + (-1)^n\text{ber}_{1/2-n}(z)$$
$$- (-1)^n\text{bei}_{1/2-n}(z)]\,_1F_4\left(\begin{array}{c}\frac{1}{4};\,-\frac{z^4}{16}\\\frac{1}{2},\frac{3}{4},\frac{5}{4},\frac{5}{4}\end{array}\right)$$
$$- \frac{\pi}{18}\left[9C\,\text{bei}_{n-1/2}(z) - 2\sqrt{2}\,z^3[\text{ber}_{n-1/2}(z) - \text{bei}_{n-1/2}(z)\right.$$
$$\left.+ (-1)^n\text{ber}_{1/2-n}(z) + (-1)^n\text{bei}_{1/2-n}(z)]\,_1F_4\left(\begin{array}{c}\frac{3}{4};\,-\frac{z^4}{16}\\\frac{5}{4},\frac{3}{2},\frac{7}{4},\frac{7}{4}\end{array}\right)\right]$$
$$+ \frac{\pi z^4}{12}[\text{bei}_{n-1/2}(z) - (-1)^n\text{ber}_{1/2-n}(z)]\,_2F_5\left(\begin{array}{c}1,1;\,-\frac{z^4}{16}\\\frac{5}{4},\frac{3}{2},\frac{7}{4},2,2\end{array}\right)$$
$$+ \frac{n!}{2}\sum_{k=0}^{n-1}\frac{\left(\frac{z}{2}\right)^{k-n}}{k!(n-k)}\left[\cos\frac{3(n-k)\pi}{4}\text{ker}_{k-1/2}(z) + \sin\frac{3(n-k)\pi}{4}\text{kei}_{k-1/2}(z)\right]$$
$$- \frac{n!\sqrt{\pi z}}{2}\sum_{k=1}^{n}\frac{\left(-\frac{2}{z}\right)^k}{(n-k)!k}\sum_{p=0}^{k-1}\frac{z^p}{p!}\left\{[\text{ber}_{n-k-1/2}(z) + (-1)^{k+n}\text{bei}_{k-n+1/2}(z)]\right.$$
$$\times \left[\cos\frac{3(2k+6p+3)\pi}{8}\text{ker}_{p-1/2}(2z) - \sin\frac{3(2k+6p+3)\pi}{8}\text{kei}_{p-1/2}(2z)\right]$$
$$+ [(-1)^{k+n}\text{ber}_{k-n+1/2}(z) - \text{bei}_{n-k-1/2}(z)]$$
$$\times \left.\left[\sin\frac{3(2k+6p+3)\pi}{8}\text{ker}_{p-1/2}(2z) + \cos\frac{3(2k+6p+3)\pi}{8}\text{kei}_{p-1/2}(2z)\right]\right\}$$
$$[|\arg z| \leq \pi/4].$$

12. $\left.\dfrac{\partial \text{kei}_\nu(z)}{\partial \nu}\right|_{\nu=n-1/2} = -\dfrac{\pi}{2}\text{ker}_{n-1/2}(z)$

$$+ \frac{\pi}{2}[C + \ln(2z)][\text{ber}_{n-1/2}(z) + (-1)^n\text{ber}_{1/2-n}(z)] + (-1)^n\frac{\pi^2}{8}\text{ber}_{1/2-n}(z)$$

$$-\frac{\pi^2}{8}\mathrm{bei}_{n-1/2}(z) + \frac{\pi z^2}{2}[(-1)^n \mathrm{ber}_{1/2-n}(z) - \mathrm{bei}_{n-1/2}(z)]\,_1F_4\left(\begin{array}{c}\frac{1}{2};\,-\frac{z^4}{16}\\ \frac{3}{4},\frac{5}{4},\frac{3}{2},\frac{3}{2}\end{array}\right)$$

$$-\frac{\pi z}{\sqrt{2}}[\mathrm{ber}_{n-1/2}(z) - \mathrm{bei}_{n-1/2}(z) + (-1)^n \mathrm{ber}_{1/2-n}(z)$$

$$+ (-1)^n \mathrm{bei}_{1/2-n}(z)]\,_1F_4\left(\begin{array}{c}\frac{1}{4};\,-\frac{z^4}{16}\\ \frac{1}{2},\frac{3}{4},\frac{5}{4},\frac{5}{4}\end{array}\right)$$

$$+ \frac{\sqrt{2}\,\pi}{9}z^3[\mathrm{ber}_{n-1/2}(z) + \mathrm{bei}_{n-1/2}(z) - (-1)^n \mathrm{ber}_{1/2-n}(z)$$

$$+ (-1)^n \mathrm{bei}_{1/2-n}(z)]\,_1F_4\left(\begin{array}{c}\frac{3}{4};\,-\frac{z^4}{16}\\ \frac{5}{4},\frac{3}{2},\frac{7}{4},\frac{7}{4}\end{array}\right)$$

$$-\frac{\pi z^4}{12}[(-1)^n \mathrm{ber}_{1/2-n}(z) + \mathrm{bei}_{n-1/2}(z)]\,_2F_5\left(\begin{array}{c}1,1;\,-\frac{z^4}{16}\\ \frac{5}{4},\frac{3}{2},\frac{7}{4},2,2\end{array}\right)$$

$$-\frac{n!}{2}\sum_{k=0}^{n-1}\frac{\left(\frac{z}{2}\right)^{k-n}}{k!(n-k)}\left[\sin\frac{3(n-k)\pi}{4}\mathrm{ker}_{k-1/2}(z) - \cos\frac{3(n-k)\pi}{4}\mathrm{kei}_{k-1/2}(z)\right]$$

$$-\frac{n!\sqrt{\pi z}}{2}\sum_{k=1}^{n}\frac{\left(\frac{z}{2}\right)^k}{(n-k)!\,k}\sum_{p=0}^{k-1}\frac{z^p}{p!}$$

$$\times\left\{[(-1)^k \mathrm{ber}_{n-k-1/2}(z) + (-1)^n \mathrm{bei}_{k-n+1/2}(z)]\right.$$
$$\times\left[\sin\frac{3(2k+6p+3)\pi}{8}\mathrm{ker}_{p-1/2}(2z) + \cos\frac{3(2k+6p+3)\pi}{8}\mathrm{kei}_{p-1/2}(2z)\right]$$
$$+ [(-1)^k \mathrm{bei}_{n-k-1/2}(z) - (-1)^n \mathrm{ber}_{k-n+1/2}(z)]$$
$$\times\left[\cos\frac{3(2k+6p+3)\pi}{8}\mathrm{ker}_{p-1/2}(2z) - \sin\frac{3(2k+6p+3)\pi}{8}\mathrm{kei}_{p-1/2}(2z)\right]\right\}$$

$$[|\arg z| \le \pi/4].$$

13. $\left.\dfrac{\partial \mathrm{ker}_\nu(z)}{\partial \nu}\right|_{\nu=1/2-n} = (-1)^{n+1}\pi\,\mathrm{ker}_{n-1/2}(z) + (-1)^{n+1}\left.\dfrac{\partial \mathrm{kei}_\nu(z)}{\partial \nu}\right|_{\nu=n-1/2}.$

14. $\left.\dfrac{\partial \mathrm{kei}_\nu(z)}{\partial \nu}\right|_{\nu=1/2-n} = (-1)^{n+1}\pi\,\mathrm{kei}_{n-1/2}(z) + (-1)^{n}\left.\dfrac{\partial \mathrm{ker}_\nu(z)}{\partial \nu}\right|_{\nu=n-1/2}.$

1.18. The Legendre Polynomials $P_n(z)$

1.18.1. Derivatives with respect to the argument

1. $\mathrm{D}^n[P_m(az)] = (2n-1)!!\,a^n\,C_{m-n}^{n+1/2}(az)$ $\hfill [m \ge n].$

2. $= \dfrac{(m+n)!}{\left(\frac{1}{2}\right)_n (m-n)!}\left(\dfrac{a}{2}\right)^n (1-a^2z^2)^{-n}\,C_{m+n}^{1/2-n}(az)$ $\hfill [m \ge n].$

1.18. THE LEGENDRE POLYNOMIALS $P_n(z)$

3. $D^{2n}[z^{-1/2}(1-a^2z)^{n-1/2}P_{2n}(a\sqrt{z})]$
$$= (-1)^n \left(\frac{1}{2}\right)_n^2 a^{2n}(z-a^2z^2)^{-n-1/2}P_{2n}\left(\frac{1}{a\sqrt{z}}\right).$$

4. $D^{2n}\left[z^n(z-a^2)^{n-1/2}P_{2n}\left(\frac{a}{\sqrt{z}}\right)\right]$
$$= (-1)^n \left(\frac{1}{2}\right)_n^2 a^{2n}(z-a^2)^{-n-1/2}P_{2n}\left(\frac{\sqrt{z}}{a}\right).$$

5. $D^{2n+1}[(1-a^2z)^{n-1/2}P_{2n+1}(a\sqrt{z})]$
$$= (-1)^n \frac{2n+1}{2} \left(\frac{1}{2}\right)_n^2 a^{2n+1}z^{-n-1/2}(1-a^2z)^{-n-3/2}P_{2n}\left(\frac{1}{a\sqrt{z}}\right).$$

6. $D^n\left[z^{-(n+1)/2}(z-a)^{n-1/2}P_n\left(\frac{z+a}{2\sqrt{az}}\right)\right]$
$$= \left(\frac{1}{2}\right)_n a^{n/2}z^{-n-1/2}(z-a)^{-1/2}P_{2n}\left(\sqrt{\frac{a}{z}}\right).$$

7. $D^n\left[z^{n/2}(a-z)^{n-1/2}P_n\left(\frac{z+a}{2\sqrt{az}}\right)\right]$
$$= (-1)^n \left(\frac{1}{2}\right)_n a^{n/2}(a-z)^{-1/2}P_{2n}\left(\sqrt{\frac{z}{a}}\right).$$

8. $D^n\left[z^{m/2}(a-z)^{n-m-1}P_m\left(\frac{z+a}{2\sqrt{az}}\right)\right]$
$$= \frac{m!}{(m-n)!} a^{n/2}z^{(m-n)/2}(a-z)^{-m-1}P_{m-n}\left(\frac{z+a}{2\sqrt{az}}\right) \quad [m \geq n].$$

9. $D^n\left[z^{-(m+1)/2}(a-z)^{m+n}P_m\left(\frac{z+a}{2\sqrt{az}}\right)\right]$
$$= (-1)^n \frac{(m+n)!}{m!} a^{n/2}z^{-(m+n+1)/2}(a-z)^m P_{m+n}\left(\frac{z+a}{2\sqrt{az}}\right).$$

10. $D^n\left[z^{n/2}(z-a)^n P_n\left(\frac{z+a}{2\sqrt{az}}\right)\right] = n! a^{n/2}\left[P_n\left(\sqrt{\frac{z}{a}}\right)\right]^2.$

11. $D^n\left[(z^2-az)^{m/2}P_m\left(\frac{2z-a}{2\sqrt{z^2-az}}\right)\right]$
$$= \frac{(-4)^{-n}}{(1/2-m)_n} \frac{(2m)!}{(2m-2n)!} (z^2-az)^{(m-n)/2}P_{m-n}\left(\frac{2z-a}{2\sqrt{z^2-az}}\right) \quad [m \geq n].$$

12. $D^n\left[z^{n-m-1}(a-z)^{m/2}P_m\left(\frac{2a-z}{2\sqrt{a^2-az}}\right)\right]$
$$= \frac{2^{-2n}a^{n/2}}{\left(\frac{1}{2}-m\right)_n} \frac{(2m)!}{(2m-2n)!} z^{-m-1}(a-z)^{(m-n)/2}P_{m-n}\left(\frac{2a-z}{2\sqrt{a^2-az}}\right) \quad [m \geq n].$$

13. $D^n\left[(z^2-a^2)^{-(m+1)/2}P_m\left(\frac{z}{\sqrt{z^2-a^2}}\right)\right]$
$$= (-1)^n \frac{(m+n)!}{m!} (z^2-a^2)^{-(m+n+1)/2}P_{m+n}\left(\frac{z}{\sqrt{z^2-a^2}}\right).$$

14. $D^{2n}\left[z^n(a-z)^{-1/2}P_{2n}\left(\sqrt{1-\dfrac{a}{z}}\right)\right]$
$$= \left(\dfrac{1}{2}\right)_n^2 (a-z)^{-n-1/2}P_{2n}\left(\sqrt{\dfrac{z}{z-a}}\right).$$

15. $D^{2n+1}\left[z^{n+1/2}P_{2n+1}\left(\sqrt{1-\dfrac{a}{z}}\right)\right]$
$$= (-1)^n \dfrac{2n+1}{2}\left(\dfrac{1}{2}\right)_n^2 (z-a)^{-n-1/2}P_{2n}\left(\sqrt{\dfrac{z}{z-a}}\right).$$

16. $D^{2n}\left[z^{n-1/2}(az-1)^{-1/2}P_{2n}\left(\sqrt{1-az}\right)\right]$
$$= \left(\dfrac{1}{2}\right)_n^2 z^{-n-1/2}(az-1)^{-n-1/2}P_{2n}\left(\dfrac{1}{\sqrt{1-az}}\right).$$

17. $D^{2n}\left[z^{n-1/2}(1-az)^n P_{2n}\left(\dfrac{1}{\sqrt{1-az}}\right)\right]$
$$= (-1)^n \left(\dfrac{1}{2}\right)_n^2 z^{-n-1/2}P_{2n}\left(\sqrt{1-az}\right).$$

18. $D^{2n}\left[(z^2-a^2)^m P_{2m}\left(\dfrac{z}{\sqrt{z^2-a^2}}\right)\right]$
$$= \dfrac{(2m)!}{(2m-2n)!}(z^2-a^2)^{m-n}P_{2m-2n}\left(\dfrac{z}{\sqrt{z^2-a^2}}\right) \quad [m \geq n].$$

19. $D^{2n+1}\left[(z^2-a^2)^m P_{2m}\left(\dfrac{z}{\sqrt{z^2-a^2}}\right)\right]$
$$= \dfrac{(2m)!}{(2m-2n-1)!}(z^2-a^2)^{m-n-1/2}P_{2m-2n-1}\left(\dfrac{z}{\sqrt{z^2-a^2}}\right) \quad [m \geq n+1].$$

20. $D^n\left[z^{-n-1}(a^2-z^2)^n P_{2n}\left(\dfrac{a}{\sqrt{a^2-z^2}}\right)\right]$
$$= (-4)^n \left(\dfrac{1}{2}\right)_n a^n z^{-2n-1}(a^2-z^2)^{n/2}P_n\left(\dfrac{a}{\sqrt{a^2-z^2}}\right).$$

21. $D^{2n}\left[z^{2n-2m-1}(a^2-z^2)^m P_{2m}\left(\dfrac{a}{\sqrt{a^2-z^2}}\right)\right]$
$$= \dfrac{(2m)!}{(2m-2n)!}a^{2n}z^{-2m-1}(a^2-z^2)^{m-n}P_{2m-2n}\left(\dfrac{a}{\sqrt{a^2-z^2}}\right) \quad [m \geq n].$$

22. $D^{2n+1}\left[z^{2n-2m}(a^2-z^2)^m P_{2m}\left(\dfrac{a}{\sqrt{a^2-z^2}}\right)\right]$
$$= -\dfrac{(2m)!}{(2m-2n-1)!}a^{2n+1}z^{-2m-1}(a^2-z^2)^{m-n-1/2}P_{2m-2n-1}\left(\dfrac{a}{\sqrt{a^2-z^2}}\right)$$
$$[m \geq n+1].$$

23. $D^n[z^{n-1/2}(1-a^2z)^{n+1/2}\, D^n[(1-a^2z)^{n-m-1}P_{2m}(a\sqrt{z})]$
$$= \dfrac{(2m)!}{(2m-2n)!}\left(\dfrac{a}{2}\right)^{2n} z^{-1/2}(1-a^2z)^{-m-1/2}P_{2m-2n}(a\sqrt{z}) \quad [m \geq n].$$

24. $D^n\left[(z-a^2)^{n+1/2} D^n\left[z^m(z-a^2)^{n-m-1} P_{2m}\left(\frac{a}{\sqrt{z}}\right)\right]\right]$

$= \frac{(2m)!}{(2m-2n)!}\left(\frac{a}{2}\right)^{2n} z^{m-n}(z-a^2)^{-m-1/2} P_{2m-2n}\left(\frac{a}{\sqrt{z}}\right)$ $[m \geq n]$.

25. $D^n\left[z^n D^n\left[(a-z)^m P_m\left(\frac{a+z}{a-z}\right)\right]\right]$

$= \left[\frac{m!}{(m-n)!}\right]^2 (a-z)^{m-n} P_{m-n}\left(\frac{a+z}{a-z}\right)$ $[m \geq n]$.

26. $D^n\left[z^n D^n\left[z^{m+n}(a-z)^{-m-1} P_m\left(\frac{a+z}{a-z}\right)\right]\right]$

$= \left[\frac{(m+n)!}{m!}\right]^2 a^n z^m (a-z)^{-m-n-1} P_{m+n}\left(\frac{a+z}{a-z}\right)$.

27. $D^n\left[z^n D^n\left[(a-z)^{-m-1} P_m\left(\frac{a+z}{a-z}\right)\right]\right]$

$= \left[\frac{(m+n)!}{m!}\right]^2 (a-z)^{-m-n-1} P_{m+n}\left(\frac{a+z}{a-z}\right)$.

1.19. The Chebyshev Polynomials $T_n(z)$ and $U_n(z)$

1.19.1. Derivatives with respect to the argument

1. $D^n[T_m(az)] = 2^{n-1} m(n-1)! a^n C_{m-n}^n(az)$ $[m \geq n \geq 1]$.

2. $D^n\left[z^{-1/2}(a-z)^n T_n\left(\frac{a+z}{a-z}\right)\right] = (-1)^n \left(\frac{1}{2}\right)_n z^{-n-1/2}(a-z)^n$.

3. $D^{2n}\left[(a^2-z^2)^m T_m\left(\frac{a^2+z^2}{a^2-z^2}\right)\right]$

$= \frac{(2m)!}{(2m-2n)!}(a^2-z^2)^{m-n} T_{m-n}\left(\frac{a^2+z^2}{a^2-z^2}\right)$ $[m \geq n]$.

4. $D^{2n}\left[z^{2n-2m-1}(z^2-a^2)^m T_m\left(\frac{z^2+a^2}{z^2-a^2}\right)\right] = \frac{(2m)!}{(2m-2n)!} a^{2n}$

$\times z^{-2m-1}(z^2-a^2)^{m-n} T_{m-n}\left(\frac{z^2+a^2}{z^2-a^2}\right)$ $[m \geq n]$.

5. $D^{2n}\left[(a^2-z^2)^m T_{2m}\left(\frac{a}{\sqrt{a^2-z^2}}\right)\right]$

$= \frac{(2m)!}{(2m-2n)!}(a^2-z^2)^{m-n} T_{2m-2n}\left(\frac{a}{\sqrt{a^2-z^2}}\right)$ $[m \geq n]$.

6. $D^{2n+1}\left[(a^2-z^2)^m T_{2m}\left(\frac{a}{\sqrt{a^2-z^2}}\right)\right]$

$= \frac{(2m)!}{(2m-2n-1)!} z(a^2-z^2)^{m-n-1} U_{2m-2n-2}\left(\frac{a}{\sqrt{a^2-z^2}}\right)$ $[m \geq n+1]$.

7. $D^{2n}\left[(a^2-z^2)^{m+1/2}T_{2m+1}\left(\dfrac{a}{\sqrt{a^2-z^2}}\right)\right]$

$\qquad = \dfrac{(2m+1)!}{(2m-2n+1)!}(a^2-z^2)^{m-n+1/2}T_{2m-2n+1}\left(\dfrac{a}{\sqrt{a^2-z^2}}\right)$ $\qquad [m \geq n]$.

8. $D^n\left[(z^2-a^2)^{-m/2}T_m\left(\dfrac{z}{\sqrt{z^2-a^2}}\right)\right]$

$\qquad = -\dfrac{(m+n-1)!}{(m-1)!}(z^2-a^2)^{-(m+n)/2}T_{m+n}\left(\dfrac{z}{\sqrt{z^2-a^2}}\right)$ $\qquad [m+n \geq 1]$.

9. $D^{2n}\left[z^{2n-2m-1}(z^2-a^2)^m T_{2m}\left(\dfrac{z}{\sqrt{z^2-a^2}}\right)\right]$

$\qquad = \dfrac{(2m)!}{(2m-2n)!}a^{2n}z^{-2m-1}(z^2-a^2)^{m-n}T_{2m-2n}\left(\dfrac{z}{\sqrt{z^2-a^2}}\right)$ $\qquad [m \geq n]$.

10. $D^{2n+1}\left[z^{2n-2m}(z^2-a^2)^m T_{2m}\left(\dfrac{z}{\sqrt{z^2-a^2}}\right)\right]$

$\qquad = -\dfrac{(2m)!}{(2m-2n-1)!}a^{2n+2}z^{-2m-1}(z^2-a^2)^{m-n-1}U_{2m-2n-2}\left(\dfrac{z}{\sqrt{z^2-a^2}}\right)$

$\qquad\qquad [m \geq n+1]$.

11. $D^{2n}\left[z^{2n-2m-2}(z^2-a^2)^{m+1/2}T_{2m+1}\left(\dfrac{z}{\sqrt{z^2-a^2}}\right)\right]$

$\qquad = \dfrac{(2m+1)!}{(2m-2n+1)!}a^{2n}z^{-2m-2}(z^2-a^2)^{m-n+1/2}T_{2m-2n+1}\left(\dfrac{z}{\sqrt{z^2-a^2}}\right)$

$\qquad\qquad [m \geq n]$.

12. $D^n\left[z^{n-1/2}D^n\left[(a-z)^m T_m\left(\dfrac{a+z}{a-z}\right)\right]\right]$

$\qquad = 2^{-2n}\dfrac{(2m)!}{(2m-2n)!}z^{-1/2}(a-z)^{m-n}T_{m-n}\left(\dfrac{a+z}{a-z}\right)$ $\qquad [m \geq n]$.

13. $D^n\left[z^{n+1/2}D^n\left[z^{n-m-1}(a-z)^m T_m\left(\dfrac{a+z}{a-z}\right)\right]\right]$

$\qquad = \dfrac{(2m)!}{(2m-2n)!}\left(\dfrac{a}{4}\right)^n z^{-m-1/2}(a-z)^{m-n}T_{m-n}\left(\dfrac{a+z}{a-z}\right)$ $\qquad [m \geq n]$.

14. $D^n\left[z^{n+1/2}D^n\left[z^{-1/2}(a-z)^{-m}T_m\left(\dfrac{a+z}{a-z}\right)\right]\right]$

$\qquad = 2^{-2n}(2m)_{2n}(a-z)^{-m-n}T_{m+n}\left(\dfrac{a+z}{a-z}\right)$.

15. $D^n\left[z^{n-1/2}D^n\left[z^{m+n-1/2}(a-z)^{-m}T_m\left(\dfrac{a+z}{a-z}\right)\right]\right]$

$\qquad = 2^{-2n}(2m)_{2n}a^n z^{m-1}(a-z)^{-m-n}T_{m+n}\left(\dfrac{a+z}{a-z}\right)$.

16. $D^n[U_m(az)] = (2a)^n n! C^{n+1}_{m-n}(az)$ $\qquad [m \geq n]$.

1.19. The Chebyshev Polynomials $T_n(z)$ and $U_n(z)$

17. $D^{2n}\left[z(a^2-z^2)^m U_m\left(\dfrac{a^2+z^2}{a^2-z^2}\right)\right]$

    $= \dfrac{(2m+2)!}{(2m-2n+2)!} z(a^2-z^2)^{m-n} U_{m-n}\left(\dfrac{a^2+z^2}{a^2-z^2}\right)$ $\quad [m \geq n]$.

18. $D^{2n}\left[z^{2n-2m-2}(a^2-z^2)^m U_m\left(\dfrac{a^2+z^2}{a^2-z^2}\right)\right]$

    $= \dfrac{(2m+2)!}{(2m-2n+2)!} a^{2n} z^{-2m-2}(a^2-z^2)^{m-n} U_{m-n}\left(\dfrac{a^2+z^2}{a^2-z^2}\right)$ $\quad [m \geq n]$.

19. $D^{2n}\left[z(a^2-z^2)^m U_{2m}\left(\dfrac{a}{\sqrt{a^2-z^2}}\right)\right]$

    $= \dfrac{(2m+1)!}{(2m-2n+1)!} z(a^2-z^2)^{m-n} U_{2m-2n}\left(\dfrac{a}{\sqrt{a^2-z^2}}\right)$ $\quad [m \geq n]$.

20. $D^{2n}\left[z(a^2-z^2)^{m+1/2} U_{2m+1}\left(\dfrac{a}{\sqrt{a^2-z^2}}\right)\right]$

    $= \dfrac{(2m+2)!}{(2m-2n+2)!} z(a^2-z^2)^{m-n+1/2} U_{2m-2n+1}\left(\dfrac{a}{\sqrt{a^2-z^2}}\right)$ $\quad [m \geq n]$.

21. $D^n\left[(z^2-a^2)^{-m/2-1} U_m\left(\dfrac{z}{\sqrt{z^2-a^2}}\right)\right]$

    $= (-1)^n \dfrac{(m+n)!}{m!} (z^2-a^2)^{-(m+n)/2-1} U_{m+n}\left(\dfrac{z}{\sqrt{z^2-a^2}}\right)$.

22. $D^{2n}\left[z^{2n-2m-2}(z^2-a^2)^m U_{2m}\left(\dfrac{z}{\sqrt{z^2-a^2}}\right)\right]$

    $= \dfrac{(2m+1)!}{(2m-2n+1)!} a^{2n} z^{-2m-2}(z^2-a^2)^{m-n} U_{2m-2n}\left(\dfrac{z}{\sqrt{z^2-a^2}}\right)$ $\quad [m \geq n]$.

23. $D^{2n}\left[z^{2n-2m-3}(z^2-a^2)^{m+1/2} U_{2m+1}\left(\dfrac{z}{\sqrt{z^2-a^2}}\right)\right]$

    $= \dfrac{(2m+2)!}{(2m-2n+2)!} a^{2n} z^{-2m-3}(z^2-a^2)^{m-n+1/2} U_{2m-2n+1}\left(\dfrac{z}{\sqrt{z^2-a^2}}\right)$
    $\quad [m \geq n]$.

24. $D^n\left[z^{n-1/2} D^n\left[z^{n-m-1}(a-z)^m U_m\left(\dfrac{a+z}{a-z}\right)\right]\right]$

    $= \dfrac{(2m+2)!}{(2m-2n+2)!}\left(\dfrac{a}{4}\right)^n z^{-m-3/2}(a-z)^{m-n} U_{m-n}\left(\dfrac{a+z}{a-z}\right)$ $\quad [m \geq n]$.

25. $D^n\left[z^{n+1/2} D^n\left[(a-z)^m U_m\left(\dfrac{a+z}{a-z}\right)\right]\right]$

    $= 2^{-2n} \dfrac{(2m+2)!}{(2m-2n+2)!} z^{1/2}(a-z)^{m-n} U_{m-n}\left(\dfrac{a+z}{a-z}\right)$ $\quad [m \geq n]$.

26. $D^n\left[z^{n+1/2} D^n\left[z^{m+n+1/2}(a-z)^{-m-2} U_m\left(\dfrac{a+z}{a-z}\right)\right]\right]$

    $= \dfrac{(2m+2n+1)!}{(2m+1)!}\left(\dfrac{a}{4}\right)^n z^{m+1}(a-z)^{-m-n-2} U_{m+n}\left(\dfrac{a+z}{a-z}\right)$.

27. $D^n\left[z^{n-1/2}\,D^n\left[z^{1/2}(a-z)^{-m-2}U_m\left(\dfrac{a+z}{a-z}\right)\right]\right]$
$\qquad = 2^{-2n}\dfrac{(2m+2n+1)!}{(2m+1)!}(a-z)^{-m-n-2}U_{m+n}\left(\dfrac{a+z}{a-z}\right).$

1.20. The Hermite Polynomials $H_n(z)$

1.20.1. Derivatives with respect to the argument

1. $D^n[H_m(az)] = \dfrac{m!}{(m-n)!}(2a)^n H_{m-n}(az) \qquad [m \geq n].$

2. $D^n[z^{-1/2}H_{2m}(a\sqrt{z})]$
$\qquad = (-1)^{m+n}2^{2m}m!\left(\dfrac{1}{2}-m\right)_n z^{-n-1/2}L_m^{-n-1/2}(a^2 z).$

3. $D^n[H_{2m+1}(a\sqrt{z})]$
$\qquad = (-1)^{m+n}2^{2m+1}m!\left(-\dfrac{1}{2}-m\right)_n az^{-n+1/2}L_m^{-n+1/2}(a^2 z).$

4. $D^n[z^{n-m-1}H_{2m}(a\sqrt{z})] = \dfrac{(2m)!}{(2m-2n)!}z^{-m-1}H_{2m-2n}(a\sqrt{z}) \qquad [m \geq n].$

5. $D^n[z^{n-m-3/2}H_{2m+1}(a\sqrt{z})]$
$\qquad = \dfrac{(2m+1)!}{(2m-2n+1)!}z^{-m-3/2}H_{2m-2n+1}(a\sqrt{z}) \qquad [m \geq n].$

6. $D^n\left[z^{n-1}H_m\left(\dfrac{a}{z}\right)\right] = \dfrac{m!}{(m-n)!}(-2a)^n z^{-n-1}H_{m-n}\left(\dfrac{a}{z}\right) \qquad [m \geq n].$

7. $D^n\left[z^m H_{2m}\left(\dfrac{a}{\sqrt{z}}\right)\right] = (-1)^n \dfrac{(2m)!}{(2m-2n)!}z^{m-n}H_{2m-2n}\left(\dfrac{a}{\sqrt{z}}\right) \qquad [m \geq n].$

8. $D^n\left[z^{m+1/2}H_{2m+1}\left(\dfrac{a}{\sqrt{z}}\right)\right]$
$\qquad = (-1)^n \dfrac{(2m+1)!}{(2m-2n+1)!}z^{m-n+1/2}H_{2m-2n+1}\left(\dfrac{a}{\sqrt{z}}\right) \qquad [m \geq n].$

9. $D^n\left[z^{n-1/2}H_{2m}\left(\dfrac{a}{\sqrt{z}}\right)\right] = (-1)^m 2^{2m}m!\left(\dfrac{1}{2}-m\right)_n z^{-1/2}L_m^{-n-1/2}\left(\dfrac{a^2}{z}\right).$

10. $D^n\left[z^{n-1}H_{2m+1}\left(\dfrac{a}{\sqrt{z}}\right)\right]$
$\qquad = (-1)^m 2^{2m+1}m!\left(-\dfrac{1}{2}-m\right)_n az^{-3/2}L_m^{-n+1/2}\left(\dfrac{a^2}{z}\right).$

11. $D^n\left[e^{-a^2 z^2}H_m(az)\right] = (-a)^n e^{-a^2 z^2}H_{m+n}(az).$

12. $D^n\left[z^{n-1}e^{-a^2/z^2}H_m\left(\dfrac{a}{z}\right)\right] = a^n z^{-n-1}e^{-a^2/z^2}H_{m+n}\left(\dfrac{a}{z}\right).$

13. $D^n\left[z^{-1/2}e^{-a^2z}H_{2m}(a\sqrt{z})\right]$
$$= (-1)^m 2^{2m}(m+n)! z^{-n-1/2} e^{-a^2 z} L_{m+n}^{-n-1/2}(a^2 z).$$

14. $D^n\left[e^{-a^2z}H_{2m+1}(a\sqrt{z})\right]$
$$= (-1)^m 2^{2m+1}(m+n)! a z^{-n+1/2} e^{-a^2 z} L_{m+n}^{-n+1/2}(a^2 z).$$

15. $D^n\left[z^{m+n-1/2}e^{-a^2z}H_{2m}(a\sqrt{z})\right] = \dfrac{(-1)^n}{2^{2n}} z^{m-1/2} e^{-a^2 z} H_{2m+2n}(a\sqrt{z}).$

16. $D^n\left[z^{m+n}e^{-a^2z}H_{2m+1}(a\sqrt{z})\right] = \dfrac{(-1)^n}{2^{2n}} z^m e^{-a^2 z} H_{2m+2n+1}(a\sqrt{z}).$

17. $D^n\left[z^{n-1/2}e^{-a^2/z}H_{2m}\left(\dfrac{a}{\sqrt{z}}\right)\right]$
$$= (-1)^{m+n} 2^{2m}(m+n)! z^{-1/2} e^{-a^2/z} L_{m+n}^{-n-1/2}\left(\dfrac{a^2}{z}\right).$$

18. $D^n\left[z^{n-1}e^{-a^2/z}H_{2m+1}\left(\dfrac{a}{\sqrt{z}}\right)\right]$
$$= (-1)^{m+n} 2^{2m+1}(m+n)! a z^{-3/2} e^{-a^2/z} L_{m+n}^{-n+1/2}\left(\dfrac{a^2}{z}\right).$$

19. $D^n\left[z^{-m-1/2}e^{-a^2/z}H_{2m}\left(\dfrac{a}{\sqrt{z}}\right)\right]$
$$= \dfrac{1}{2^{2n}} z^{-m-n-1/2} e^{-a^2/z} H_{2m+2n}\left(\dfrac{a}{\sqrt{z}}\right).$$

20. $D^n\left[z^{-m-1}e^{-a^2/z}H_{2m+1}\left(\dfrac{a}{\sqrt{z}}\right)\right]$
$$= \dfrac{1}{2^{2n}} z^{-m-n-1} e^{-a^2/z} H_{2m+2n+1}\left(\dfrac{a}{\sqrt{z}}\right).$$

1.21. The Laguerre Polynomials $L_n^\lambda(z)$

1.21.1. Derivatives with respect to the argument

1. $D\left[L_n^\lambda(z)\right] = \dfrac{z-\lambda-n-1}{z} L_n^\lambda(z) + \dfrac{n+1}{z} L_{n+1}^\lambda(z).$

2. $ = L_n^\lambda(z) - L_n^{\lambda+1}(z).$ [54].

3. $D^n[L_m^\lambda(az)] = (-a)^n L_{m-n}^{\lambda+n}(az)$ $[m \geq n]$.

4. $D^n[z^\lambda L_m^\lambda(az)] = (-1)^n(-\lambda-m)_n z^{\lambda-n} L_m^{\lambda-n}(az).$

5. $D^n[z^{n-m-1} L_m^\lambda(az)] = (-\lambda-m)_n z^{-m-1} L_{m-n}^\lambda(az)$ $[m \geq n]$.

6. $D^n\left[z^{n-1} L_m\left(\dfrac{a}{z}\right)\right] = a^n z^{-n-1} L_{m-n}^n\left(\dfrac{a}{z}\right)$ $[m \geq n]$.

7. $D^n\left[z^{n-\lambda-1} L_m^\lambda\left(\frac{a}{z}\right)\right] = (-\lambda - m)_n z^{-\lambda-1} L_m^{\lambda-n}\left(\frac{a}{z}\right).$

8. $D^n\left[z^m L_m^\lambda\left(\frac{a}{z}\right)\right] = (-1)^n (-\lambda - m)_n z^{m-n} L_{m-n}^\lambda\left(\frac{a}{z}\right)$ $\qquad [m \geq n].$

9. $D^n[e^{-az} L_m^\lambda(az)] = (-a)^n e^{-az} L_m^{\lambda+n}(az).$

10. $D^n[z^\lambda e^{-az} L_m^\lambda(az)] = \frac{(m+n)!}{m!} z^{\lambda-n} e^{-az} L_{m+n}^{\lambda-n}(az).$

11. $D^n[z^{\lambda+m+n} e^{-az} L_m^\lambda(az)] = \frac{(m+n)!}{m!} z^{\lambda+m} e^{-az} L_{m+n}^\lambda(az).$

12. $D^n[z^{n-1} e^{-a/z} L_m^\lambda(\frac{a}{z})] = a^n e^{-a/z} L_m^{\lambda+n}\left(\frac{a}{z}\right).$

13. $D^n[z^{n-\lambda-1} e^{-a/z} L_m^\lambda\left(\frac{a}{z}\right)] = (-1)^n \frac{(m+n)!}{m!} z^{-\lambda-1} e^{-a/z} L_{m+n}^{\lambda-n}\left(\frac{a}{z}\right).$

14. $D^n[z^{-\lambda-m-1} e^{-a/z} L_m^\lambda\left(\frac{a}{z}\right)]$
$$= (-1)^n \frac{(m+n)!}{m!} z^{-\lambda-m-n-1} e^{-a/z} L_{m+n}^\lambda\left(\frac{a}{z}\right).$$

1.21.2. Derivatives with respect to the parameter

1. $\dfrac{\partial L_n^\lambda(z)}{\partial \lambda} = \sum\limits_{k=0}^{n-1} \dfrac{1}{n-k} L_k^\lambda(z)$ $\qquad [54].$

1.22. The Gegenbauer Polynomials $C_n^\lambda(z)$

1.22.1. Derivatives with respect to the argument

1. $D^n[C_m^\lambda(az)] = (2a)^n (\lambda)_n C_{m-n}^{\lambda+n}(az)$ $\qquad [m \geq n].$

2. $D^n[z^{\lambda+m+n-1} C_{2m}^\lambda(a\sqrt{z})] = (\lambda)_n z^{\lambda+m-1} C_{2m}^{\lambda+n}(a\sqrt{z}).$

3. $D^n[z^{\lambda+m+n-1/2} C_{2m+1}^\lambda(a\sqrt{z})] = (\lambda)_n z^{\lambda+m-1/2} C_{2m+1}^{\lambda+n}(a\sqrt{z}).$

4. $D^n[z^{n-m-1} C_{2m}^\lambda(a\sqrt{z})] = (\lambda)_n z^{-m-1} C_{2m-2n}^{\lambda+n}(a\sqrt{z})$ $\qquad [m \geq n].$

5. $D^n[z^{n-m-3/2} C_{2m+1}^\lambda(a\sqrt{z})] = (\lambda)_n z^{-m-3/2} C_{2m-2n+1}^{\lambda+n}(a\sqrt{z})$ $\qquad [m \geq n].$

6. $D^n[z^{m+n-1/2}(1-a^2 z)^{\lambda-1/2} C_{2m}^\lambda(a\sqrt{z})] = 2^{-n} \dfrac{(m+n)!}{m!} \dfrac{(2m+2n-1)!!}{(2m-1)!!}$
$$\times \frac{1}{(1-\lambda)_n} z^{m-1/2}(1-a^2 z)^{\lambda-n-1/2} C_{2m+2n}^{\lambda-n}(a\sqrt{z}).$$

7. $D^n[z^{m+n}(1-a^2 z)^{\lambda-1/2} C_{2m+1}^\lambda(a\sqrt{z})] = 2^{-n} \dfrac{(m+n)!}{m!} \dfrac{(2m+2n+1)!!}{(2m+1)!!}$
$$\times \frac{1}{(1-\lambda)_n} z^m (1-a^2 z)^{\lambda-n-1/2} C_{2m+2n+1}^{\lambda-n}(a\sqrt{z}).$$

1.22. The Gegenbauer Polynomials $C_n^\lambda(z)$

8. $D^n\left[z^{n-1}C_m^\lambda\left(\dfrac{a}{z}\right)\right] = (-2a)^n(\lambda)_n z^{-n-1}C_{m-n}^{\lambda+n}\left(\dfrac{a}{z}\right)$ $\qquad [m \geq n]$.

9. $D^n\left[z^{n-2\lambda}(a-2z)^{\lambda-1/2}C_m^\lambda\left(1-\dfrac{a}{z}\right)\right]$
$= 2^{-n}\dfrac{(m+n)!}{m!}\dfrac{(1-2\lambda-m)_n}{(1-\lambda)_n} z^{n-2\lambda}(a-2z)^{\lambda-n-1/2}C_{m+n}^{\lambda-n}\left(1-\dfrac{a}{z}\right).$

10. $D^n\left[(a-z)^{n-1}C_m^\lambda\left(\dfrac{a+z}{a-z}\right)\right] = 2^{2n}(\lambda)_n a^n(a-z)^{-n-1}C_{m-n}^{\lambda+n}\left(\dfrac{a+z}{a-z}\right)]$
$\qquad [m \geq n]$.

11. $D^n\left[z^{\lambda-1/2}(a-z)^{n-2\lambda}C_m^\lambda\left(\dfrac{a+z}{a-z}\right)\right]$
$= (-1)^n \dfrac{(m+n)!}{m!}\dfrac{(2\lambda)_m\left(\frac{1}{2}-\lambda\right)_n}{(2\lambda-2n)_{m+n}} z^{\lambda-n-1/2}(a-z)^{n-2\lambda}C_{m+n}^{\lambda-n}\left(\dfrac{a+z}{a-z}\right).$

12. $D^n\left[z^m C_{2m}^\lambda\left(\dfrac{a}{\sqrt{z}}\right)\right] = (-1)^m(\lambda)_n z^{m-n}C_{2m-2n}^{\lambda+n}\left(\dfrac{a}{\sqrt{z}}\right)$ $\qquad [m \geq n]$.

13. $D^n\left[z^{m+1/2}C_{2m+1}^\lambda\left(\dfrac{a}{\sqrt{z}}\right)\right] = (-1)^n(\lambda)_n z^{m-n+1/2}C_{2m-2n+1}^{\lambda+n}\left(\dfrac{a}{\sqrt{z}}\right)$
$\qquad [m \geq n]$.

14. $D^n\left[z^{-\lambda-m}(z-a^2)^{\lambda-1/2}C_{2m}^\lambda\left(\dfrac{a}{\sqrt{z}}\right)\right]$
$= (-2)^{-n}\dfrac{(m+n)!}{m!}\dfrac{(2m+2n-1)!!}{(2m-1)!!}\dfrac{1}{(1-\lambda)_n} z^{-m-\lambda}$
$\times (z-a^2)^{\lambda-n-1/2}C_{2m+2n}^{\lambda-n}\left(\dfrac{a}{\sqrt{z}}\right).$

15. $D^n\left[z^{-\lambda-m-1/2}(z-a^2)^{\lambda-1/2}C_{2m+1}^\lambda\left(\dfrac{a}{\sqrt{z}}\right)\right]$
$= (-2)^{-n}\dfrac{(m+n)!}{m!}\dfrac{(2m+2n+1)!!}{(2m+1)!!}\dfrac{1}{(1-\lambda)_n} z^{-\lambda-m-1/2}$
$\times (a^2-z)^{\lambda-n-1/2}C_{2m+2n+1}^{\lambda-n}\left(\dfrac{a}{\sqrt{z}}\right).$

16. $D^n\left[z^{m/2}(a-z)^{n-m-1}C_m^\lambda\left(\dfrac{z+a}{2\sqrt{az}}\right)\right]$
$= \dfrac{(\lambda)_m(1-2\lambda-m)_n}{(1-\lambda-m)_n(\lambda)_{m-n}} a^{n/2}z^{(m-n)/2}(a-z)^{-m-1}C_{m-n}^\lambda\left(\dfrac{z+a}{2\sqrt{az}}\right)$ $\qquad [m \geq n]$.

17. $D^n\left[z^{-\lambda-m/2}(a-z)^{2\lambda+m+n-1}C_m^\lambda\left(\dfrac{z+a}{2\sqrt{az}}\right)\right]$
$= (-1)^n \dfrac{(m+n)!}{m!} a^{n/2}z^{-\lambda-(m+n)/2}(a-z)^{2\lambda+m-1}C_{m+n}^\lambda\left(\dfrac{z+a}{2\sqrt{az}}\right).$

18. $D^n\left[(z^2-a^2)^{-\lambda-m/2}C_m^\lambda\left(\dfrac{z}{\sqrt{z^2-a^2}}\right)\right]$
$= (-1)^n \dfrac{(m+n)!}{m!} (z^2-a^2)^{-\lambda-(m+n)/2}C_{m+n}^\lambda\left(\dfrac{z}{\sqrt{z^2-a^2}}\right).$

1.22.2. Derivatives with respect to the parameter

1. $\dfrac{\partial C_n^\lambda(z)}{\partial \lambda} = [\psi(n+\lambda) - \psi(\lambda)] C_n^\lambda(z) + \sum_{k=1}^{[n/2]} \dfrac{\lambda + n - 2k}{k(\lambda + n - k)} C_{n-2k}^\lambda(z)$

[[77], (50)].

2. $= \left[\psi\!\left(\lambda + \dfrac{1}{2}\right) - \psi\!\left(\lambda + n + \dfrac{1}{2}\right) - 2\psi(2\lambda) + 2\psi(2\lambda + 2n)\right] C_n^\lambda(z)$

$+ 2 \sum_{k=0}^{n-1} \dfrac{[1 + (-1)^{n-k}](k + \lambda)}{(n-k)(2\lambda + k + n)} C_k^\lambda(z)$ [54].

3. $= 2[\psi(n + 2\lambda) - \psi(2\lambda)] C_n^\lambda(z)$

$- \sum_{k=1}^{[n/2]} \dfrac{(\lambda)_{2k}}{(2\lambda + n)_{2k}} \left(\dfrac{1}{k} + \dfrac{2}{2k + 2\lambda - 1}\right) 2^{2k}(z^2 - 1)^k C_{n-2k}^{\lambda + 2k}(z)$ [[77], (52)].

4. $\dfrac{\partial}{\partial \lambda}\left[\dfrac{C_n^\lambda(z)}{(\lambda)_n}\right] = \dfrac{1}{(\lambda)_n} \sum_{k=1}^{[n/2]} \dfrac{\lambda + n - 2k}{k(\lambda + n - k)} C_{n-2k}^\lambda(z)$

[[77], (49)].

1.23. The Jacobi Polynomials $P_n^{(\rho,\sigma)}(z)$

1.23.1. Derivatives with respect to the argument

1. $D\left[P_n^{(\rho,\sigma)}(z)\right] = \dfrac{\rho + \sigma + n + 1}{1 + z}\left[P_n^{(\rho+1,\sigma)}(z) - P_n^{(\rho,\sigma)}(z)\right]$.

2. $= -\dfrac{2(\rho + n)}{1 - z^2} P_n^{(\rho-1,\sigma)}(z) + \dfrac{2\rho + n - nz}{1 - z^2} P_n^{(\rho,\sigma)}(z)$.

3. $= \dfrac{\rho + \sigma + n + 1}{1 - z}\left[P_n^{(\rho,\sigma)}(z) - P_n^{(\rho,\sigma+1)}(z)\right]$.

4. $= \dfrac{2(\sigma + n)}{1 - z^2} P_n^{(\rho,\sigma-1)}(z) - \dfrac{2\sigma + n + nz}{1 - z^2} P_n^{(\rho,\sigma)}(z)$ [54].

5. $D^n\left[P_m^{(\rho,\sigma)}(az)\right] = (\rho + \sigma + m + 1)_n \left(\dfrac{a}{2}\right)^n P_{m-n}^{(\rho+n,\sigma+n)}(az)$ $[m \geq n]$.

6. $D^n\left[(1 + az)^\sigma P_m^{(\rho,\sigma)}(az)\right]$
$= (-a)^n(-\sigma - m)_n (1 + az)^{\sigma - n} P_m^{(\rho+n,\sigma-n)}(az)$.

7. $D^n\left[(1 + az)^{n-m-1} P_m^{(\rho,\sigma)}(az)\right]$
$= (-a)^n(-\sigma - m)_n (1 + az)^{-m-1} P_{m-n}^{(\rho+n,\sigma)}(az)$ $[m \geq n]$.

8. $D^n\left[(1 + az)^{\rho + \sigma + m + n} P_m^{(\rho,\sigma)}(az)\right]$
$= a^n(\rho + \sigma + m + 1)_n (1 + az)^{\rho + \sigma + m} P_m^{(\rho+n,\sigma)}(az)$.

1.23.1]   1.23. The Jacobi Polynomials $P_n^{(\rho,\sigma)}(z)$

9. $D^n\left[(1-az)^\rho(1+az)^\sigma P_m^{(\rho,\sigma)}(az)\right]$
$$= (-2a)^n \frac{(m+n)!}{m!}(1-az)^{\rho-n}(1+az)^{\sigma-n} P_{m+n}^{(\rho-n,\sigma-n)}(az).$$

10. $D^n\left[(1-az)^{m+n+\rho}(1+az)^\sigma P_m^{(\rho,\sigma)}(az)\right]$
$$= (-2a)^n \frac{(m+n)!}{m!}(1-az)^{\rho+m}(1+az)^{\sigma-n} P_{m+n}^{(\rho,\sigma-n)}(az).$$

11. $D^n\left[(az-1)^{2\rho+n} P_n^{(\rho,-n-1/2)}(az)\right]$
$$= n!\frac{(\rho+1)_n}{(2\rho+1)_n} a^n (az-1)^{2\rho}\left[C_n^{\rho+1/2}\left(\sqrt{\frac{az+1}{2}}\right)\right]^2.$$

12. $D^{2n}\left[z^{2\rho+1} P_m^{(\rho,1/2-m-n)}(1+az^2)\right]$
$$= \frac{(\rho+1)_m(-2\rho-1)_{2n}}{(\rho-n+1)_m} z^{2\rho-2n+1} P_m^{(\rho-n,1/2-m+n)}(1+az^2).$$

13. $D^n\left[z^{n-1} P_m^{(\rho,\sigma)}\left(\frac{a}{z}\right)\right]$
$$= (\rho+\sigma+n+1)_n\left(-\frac{a}{2}\right)^n z^{-n-1} P_{m-n}^{(\rho+n,\sigma+n)}\left(\frac{a}{z}\right) \quad [m\geq n].$$

14. $D^n[z^{n-\sigma-1}(z+a)^\sigma P_m^{(\rho,\sigma)}\left(\frac{a}{z}\right)]$
$$= a^n(-\sigma-m)_n z^{-\sigma-1}(z+a)^{\sigma-n} P_m^{(\rho+n,\sigma-n)}\left(\frac{a}{z}\right).$$

15. $D^n\left[z^m(z+a)^{n-m-1} P_m^{(\rho,\sigma)}\left(\frac{a}{z}\right)\right]$
$$= a^n(-\sigma-m)_n z^{m-n}(z+a)^{-m-1} P_{m-n}^{(\rho+n,\sigma)}\left(\frac{a}{z}\right) \quad [m\geq n].$$

16. $D^n\left[z^{-\rho-\sigma-m-1}(z+a)^{\rho+\sigma+m+n} P_m^{(\rho,\sigma)}\left(\frac{a}{z}\right)\right]$
$$= (-a)^n(\rho+\sigma+m+1)_n z^{-\rho-\sigma-m-n-1}(z+a)^{\rho+\sigma+m} P_m^{(\rho+n,\sigma)}\left(\frac{a}{z}\right).$$

17. $D^n\left[z^{-\rho-\sigma+n-1}(z-a)^\rho(z+a)^\sigma P_m^{(\rho,\sigma)}\left(\frac{a}{z}\right)\right]$
$$= (2a)^n \frac{(m+n)!}{m!} z^{-\rho-\sigma+n-1}(z-a)^{\rho-n}(z+a)^{\sigma-n} P_{m+n}^{(\rho-n,\sigma-n)}\left(\frac{a}{z}\right).$$

18. $D^n\left[z^{-\rho-\sigma-m-1}(z-a)^{m+n+\rho}(z+a)^\sigma P_m^{(\rho,\sigma)}\left(\frac{a}{z}\right)\right]$
$$= (2a)^n \frac{(m+n)!}{m!} z^{-\rho-\sigma-m-1}(z-a)^{\rho+m}(z+a)^{\sigma-n} P_{m+n}^{(\rho,\sigma-n)}\left(\frac{a}{z}\right).$$

19. $D^n\left[z^{n-1} P_m^{(\rho,\sigma)}\left(1-\frac{a}{z}\right)\right]$
$$= \left(\frac{a}{2}\right)^n (\rho+\sigma+m+1)_n z^{-n-1} P_{m-n}^{(\rho+n,\sigma+n)}\left(1-\frac{a}{z}\right) \quad [m\geq n].$$

20. $D^n\left[z^m P_m^{(\rho,\sigma)}\left(1-\dfrac{a}{z}\right)\right] = (-1)^n(-\rho-m)_n z^{m-n} P_{m-n}^{(\rho,\sigma+n)}\left(1-\dfrac{a}{z}\right)$
$$[m \geq n].$$

21. $D^n\left[z^{-\rho-\sigma-m-1} P_m^{(\rho,\sigma)}\left(1-\dfrac{a}{z}\right)\right]$
$$= (-1)^n(\rho+\sigma+m+1)_n z^{-\rho-\sigma-m-n-1} P_m^{(\rho,\sigma+n)}\left(1-\dfrac{a}{z}\right).$$

22. $D^n\left[z^{n-\rho-1} P_m^{(\rho,\sigma)}\left(1-\dfrac{a}{z}\right)\right]$
$$= (-1)^n(\rho+m-n+1)_n z^{-\rho-1} P_m^{(\rho-n,\sigma+n)}\left(1-\dfrac{a}{z}\right).$$

23. $D^n\left[z^m(a-2z)^\sigma P_m^{(\rho,\sigma)}\left(1-\dfrac{a}{z}\right)\right]$
$$= 2^n(-\sigma-m)_n z^m (a-2z)^{\sigma-n} P_m^{(\rho,\sigma-n)}\left(1-\dfrac{a}{z}\right).$$

24. $D^n\left[z^m(a-2z)^{n-m-1} P_m^{(\rho,\sigma)}\left(1-\dfrac{a}{z}\right)\right]$
$$= a^n(-\sigma-m)_n z^{m-n}(a-2z)^{-m-1} P_{m-n}^{(\rho+n,\sigma)}\left(1-\dfrac{a}{z}\right) \quad [m \geq n].$$

25. $D^n\left[z^m(a-2z)^{n-m-\rho-1} P_m^{(\rho,\sigma)}\left(1-\dfrac{a}{z}\right)\right]$
$$= (-2)^n(-\rho-m)_n z^m (a-2z)^{-\rho-m-1} P_m^{(\rho-n,\sigma)}\left(1-\dfrac{a}{z}\right).$$

26. $D^n\left[z^{-\rho-\sigma-m-1}(a-2z)^\sigma P_m^{(\rho,\sigma)}\left(1-\dfrac{a}{z}\right)\right]$
$$= 2^n \dfrac{(m+n)!}{m!} z^{-\rho-\sigma-m-1}(a-2z)^{\sigma-n} P_{m+n}^{(\rho,\sigma-n)}\left(1-\dfrac{a}{z}\right).$$

27. $D^n\left[z^{n-\rho-\sigma-1}(a-2z)^\sigma P_m^{(\rho,\sigma)}\left(1-\dfrac{a}{z}\right)\right]$
$$= 2^n \dfrac{(m+n)!}{m!} z^{n-\rho-\sigma-1}(a-2z)^{\sigma-n} P_{m+n}^{(\rho-n,\sigma-n)}\left(1-\dfrac{a}{z}\right).$$

28. $D^n\left[z^{-\rho-\sigma-m-1}(a-2z)^{\sigma+m+n} P_m^{(\rho,\sigma)}\left(1-\dfrac{a}{z}\right)\right]$
$$= 2^n \dfrac{(m+n)!}{m!} z^{-\rho-\sigma-m-1}(a-2z)^{\sigma+m} P_{m+n}^{(\rho-n,\sigma)}\left(1-\dfrac{a}{z}\right).$$

29. $D^n\left[z^{2\rho+n}(a-z)^n P_n^{(\rho,-\rho-n-1/2)}\left(\dfrac{a+z}{a-z}\right)\right]$
$$= n!\,\dfrac{(\rho+1)_n}{(2\rho+1)_n} a^n z^{2\rho}\left[C_n^{\rho+1/2}\left(\sqrt{1-\dfrac{z}{a}}\right)\right]^2.$$

30. $D^n\left[z^{-2\rho-n-1}(z-a)^n P_n^{(\rho,-\rho-n-1/2)}\left(\dfrac{z+a}{z-a}\right)\right]$
$$= (-1)^n n!\,\dfrac{(\rho+1)_n}{(2\rho+1)_n} z^{-2\rho-n-1}\left[C_n^{\rho+1/2}\left(\sqrt{1-\dfrac{a}{z}}\right)\right]^2.$$

1.23.2. Derivatives with respect to parameters

1. $\dfrac{\partial P_n^{(\rho,\sigma)}(z)}{\partial \rho} = [\psi(\rho+\sigma+2n+1) - \psi(\rho+\sigma+n+1)] P_n^{(\rho,\sigma)}(z)$

$+ \sum_{k=0}^{n-1} \dfrac{\rho+\sigma+2k+1}{(n-k)(\rho+\sigma+k+n+1)} \dfrac{(\sigma+k+1)_{n-k}}{(\rho+\sigma+k+1)_{n-k}} P_k^{(\rho,\sigma)}(z)$ [54].

2. $\dfrac{\partial P_n^{(\rho,\sigma)}(z)}{\partial \sigma} = [\psi(\rho+\sigma+2n+1) - \psi(\rho+\sigma+n+1)] P_n^{(\rho,\sigma)}(z)$

$+ \sum_{k=0}^{n-1} (-1)^{n-k} \dfrac{\rho+\sigma+2k+1}{(n-k)(\rho+\sigma+k+n+1)} \dfrac{(\rho+k+1)_{n-k}}{(\rho+\sigma+k+1)_{n-k}} P_k^{(\rho,\sigma)}(z)$ [54].

3. $\dfrac{\partial}{\partial \rho}\left[\dfrac{P_n^{(\rho,\sigma)}(z)}{(\rho+1)_n}\right]$

$= \dfrac{1}{(\rho+1)_n} \sum_{k=1}^{n} (-1)^{k+1} \dfrac{(-\sigma-n)_k}{(\rho+n+1)_k} \left(\dfrac{1}{k} + \dfrac{1}{\rho+k}\right)\left(\dfrac{z-1}{2}\right)^k P_{n-k}^{(\rho+2k,\sigma)}(z)$

[[77], (51)].

4. $\dfrac{\partial}{\partial \rho}\left[\dfrac{P_n^{(\rho,\sigma)}(z)}{(\rho+\sigma+n+1)_n}\right]$

$= \dfrac{(\sigma+1)_n}{(\rho+\sigma+1)_{2n}} \sum_{k=0}^{n-1} \dfrac{\rho+\sigma+2k+1}{(n-k)(\rho+\sigma+k+n+1)} \dfrac{(\rho+\sigma+1)_k}{(\sigma+1)_k} P_k^{(\rho,\sigma)}(z)$

[[77], (48)].

1.24. The Complete Elliptic Integrals $K(z)$, $E(z)$ and $D(z)$

1.24.1. Derivatives with respect to the argument

1. $D^n[z^n(1-a^2z)^n D^n[K(a\sqrt{z})]] = \left(\dfrac{1}{2}\right)_n^2 a^{2n} K(a\sqrt{z})$.

2. $D^n\left[(1-a^2z)^n D^n\left[z^{n-1/2} K(a\sqrt{z})\right]\right] = (-1)^n \left(\dfrac{1}{2}\right)_n^2 z^{-n-1/2} K(a\sqrt{z})$.

3. $D^n\left[z^n D^n\left[(1-a^2z)^{n-1/2} K(a\sqrt{z})\right]\right]$
$= (-1)^n \left(\dfrac{1}{2}\right)_n^2 a^{2n}(1-a^2z)^{-n-1/2} K(a\sqrt{z})$.

4. $D^n[z^n(1-a^2z)^{n-1} D^n[E(a\sqrt{z})]] = \left(-\dfrac{1}{2}\right)_n \left(\dfrac{1}{2}\right)_n \dfrac{a^{2n}}{1-a^2z} E(a\sqrt{z})$.

5. $D^n\left[z^2(1-a^2z)^{n-1} D^n\left[z^{n-3/2} E(a\sqrt{z})\right]\right]$
$= (-1)^n \left(-\dfrac{1}{2}\right)_n^2 \dfrac{z^{1/2-n}}{1-a^2z} E(a\sqrt{z})$.

6. $D^n \left[z^n (1 - a^2 z)^{-1} D^n [(1 - a^2 z)^{n-1/2} E(a\sqrt{z})] \right]$
$$= (-1)^n \left(\frac{1}{2} \right)_n \left(\frac{3}{2} \right)_n a^{2n} (1 - a^2 z)^{-n-3/2} E(a\sqrt{z}).$$

7. $D^n \left[z^n (1 - a^2 z) D^n [(1 - a^2 z)^{n-3/2} E(a\sqrt{z})] \right]$
$$= (-1)^n \left(-\frac{1}{2} \right)_n \left(\frac{1}{2} \right)_n a^{2n} (1 - a^2 z)^{-n-1/2} E(a\sqrt{z}).$$

8. $D^n [z^n (1 - a^2 z)^{n+1} D^n [(1 - a^2 z)^{-1} E(a\sqrt{z})]] = \left(\frac{1}{2} \right)_n \left(\frac{3}{2} \right)_n a^{2n} E(a\sqrt{z}).$

9. $D^n \left[(1 - a^2 z)^{n+1} D^n [z^{n-1/2} (1 - a^2 z)^{-1} E(a\sqrt{z})] \right]$
$$= (-1)^n \left(\frac{1}{2} \right)_n^2 z^{-n-1/2} E(a\sqrt{z}).$$

10. $D^n [z^{n+1} (1 - a^2 z)^n D^n [D(a\sqrt{z})]] = \left(\frac{1}{2} \right)_n \left(\frac{3}{2} \right)_n a^{2n} z \, D(a\sqrt{z}).$

11. $D^n [z^{n-1} (1 - a^2 z)^n D^n [z \, D(a\sqrt{z})]] = \left(-\frac{1}{2} \right)_n \left(\frac{1}{2} \right)_n a^{2n} D(a\sqrt{z}).$

12. $D^n \left[z (1 - a^2 z)^n D^n [z^{n-1/2} D(a\sqrt{z})] \right]$
$$= (-1)^n \left(-\frac{1}{2} \right)_n \left(\frac{1}{2} \right)_n a^{2n} z^{1/2-n} D(a\sqrt{z}).$$

13. $D^n \left[z^{-1} (1 - a^2 z)^n D^n [z^{n+1/2} D(a\sqrt{z})] \right]$
$$= (-1)^n \left(\frac{1}{2} \right)_n \left(\frac{3}{2} \right)_n z^{-n-1/2} D(a\sqrt{z}).$$

14. $D^n [z^{n+1} D^n [(1 - a^2 z)^{n-1/2} D(a\sqrt{z})]]$
$$= (-1)^n \left(\frac{1}{2} \right)_n^2 a^{2n} z (1 - a^2 z)^{-n-1/2} D(a\sqrt{z}).$$

1.25. The Legendre Function $P_\nu^\mu(z)$

1.25.1. Derivatives with respect to the argument

1. $D^n \left[P_\nu^\mu \left(\frac{z}{a} \right) \right] = n! (-2a)^n (a^2 - z^2)^{-n} \sum_{k=0}^n \frac{(2a)^{-k}}{k!} \frac{\left(\frac{\mu}{2} \right)_{n-k}}{(n - k + \mu)_{n-k}}$
$$\times (\nu - \mu + 1)_k (-\mu - \nu)_k (a^2 - z^2)^{k/2} C_{n-k}^{k-n+(1-\mu)/2} \left(\frac{z}{a} \right) P_\nu^{\mu-k} \left(\frac{z}{a} \right).$$

2. $D^n \left[P_\nu^\mu \left(\frac{z}{a} \right) \right] = n! (-2a)^n (a^2 - z^2)^{-n} \sum_{k=0}^n \frac{(2a)^{-k}}{k!} \frac{\left(-\frac{\mu}{2} \right)_{n-k}}{(n - k - \mu)_{n-k}}$
$$\times (a^2 - z^2)^{k/2} C_{n-k}^{k-n+(\mu+1)/2} \left(\frac{z}{a} \right) P_\nu^{\mu+k} \left(\frac{z}{a} \right).$$

3. $\mathrm{D}^n\left[(a^2-z^2)^{\mu/2}P_\nu^\mu\left(\dfrac{z}{a}\right)\right]$
$\qquad = (-1)^n(\nu-\mu+1)_n(-\mu-\nu)_n(a^2-z^2)^{(\mu-n)/2}P_\nu^{\mu-n}\left(\dfrac{z}{a}\right).$

4. $\mathrm{D}^n\left[(a^2-z^2)^{-\mu/2}P_\nu^\mu\left(\dfrac{z}{a}\right)\right] = (-1)^n(a^2-z^2)^{-(\mu+n)/2}P_\nu^{\mu+n}\left(\dfrac{z}{a}\right).$

5. $\mathrm{D}^n\left[z^{n-(\mu+\nu)/2-1}(a-z)^{\mu/2}P_\nu^\mu\left(\sqrt{\dfrac{z}{a}}\right)\right]$
$\qquad = 2^{-n}a^{n/2}(-\mu-\nu)_{2n}z^{-(\mu+\nu)/2-1}(a-z)^{(\mu-n)/2}P_{\nu-n}^{\mu-n}\left(\sqrt{\dfrac{z}{a}}\right).$

6. $\mathrm{D}^n\left[z^{n+(\mu+\nu-1)/2}(a-z)^{-\mu/2}P_\nu^\mu\left(\sqrt{\dfrac{z}{a}}\right)\right]$
$\qquad = (-2)^{-n}a^{n/2}z^{(\mu+\nu-1)/2}(a-z)^{-(\mu+n)/2}P_{\nu+n}^{\mu+n}\left(\sqrt{\dfrac{z}{a}}\right).$

7. $\mathrm{D}^n\left[z^{n+(\nu-\mu-1)/2}(a-z)^{\mu/2}P_\nu^\mu\left(\sqrt{\dfrac{z}{a}}\right)\right]$
$\qquad = 2^{-n}a^{n/2}(\nu-\mu+1)_{2n}z^{(\nu-\mu-1)/2}(a-z)^{(\mu-n)/2}P_{\nu+n}^{\mu-n}\left(\sqrt{\dfrac{z}{a}}\right).$

8. $\mathrm{D}^n\left[z^{n+(\mu-\nu)/2-1}(a-z)^{-\mu/2}P_\nu^\mu\left(\sqrt{\dfrac{z}{a}}\right)\right]$
$\qquad = (-2)^{-n}a^{n/2}z^{(\mu-\nu)/2-1}(a-z)^{-(\mu+n)/2}P_{\nu-n}^{\mu+n}\left(\sqrt{\dfrac{z}{a}}\right).$

9. $\mathrm{D}^n\left[z^{\nu/2}(z-a)^{\mu/2}P_\nu^\mu\left(\sqrt{\dfrac{a}{z}}\right)\right]$
$\qquad = (-2)^{-n}(-\mu-\nu)_{2n}z^{(\nu-n)/2}(z-a)^{(\mu-n)/2}P_{\nu-n}^{\mu-n}\left(\sqrt{\dfrac{a}{z}}\right).$

10. $\mathrm{D}^n\left[z^{\nu/2}(z-a)^{-\mu/2}P_\nu^\mu\left(\sqrt{\dfrac{a}{z}}\right)\right]$
$\qquad = 2^{-n}z^{(\nu-n)/2}(z-a)^{-(\mu+n)/2}P_{\nu-n}^{\mu+n}\left(\sqrt{\dfrac{a}{z}}\right).$

11. $\mathrm{D}^n\left[z^{-(\nu+1)/2}(z-a)^{\mu/2}P_\nu^\mu\left(\sqrt{\dfrac{a}{z}}\right)\right]$
$\qquad = (-2)^{-n}(\nu-\mu+1)_{2n}z^{-(\nu+n+1)/2}(z-a)^{(\mu-n)/2}P_{\nu+n}^{\mu-n}\left(\sqrt{\dfrac{a}{z}}\right).$

12. $\mathrm{D}^n\left[z^{-(\nu+1)/2}(z-a)^{-\mu/2}P_\nu^\mu\left(\sqrt{\dfrac{a}{z}}\right)\right]$
$\qquad = 2^{-n}z^{-(\nu+n+1)/2}(z-a)^{-(\mu+n)/2}P_{\nu+n}^{\mu+n}\left(\sqrt{\dfrac{a}{z}}\right).$

1.25.2. Derivatives with respect to parameters

1. $\mathrm{D}_\nu[P_\nu(z)]|_{\nu=n} = -\ln\dfrac{z+1}{2}P_n(z)$
$\qquad\qquad -n!\displaystyle\sum_{k=1}^n 2^{k+1}\dfrac{\left(\frac{1}{2}\right)_k}{(k+n)!\,k}(1-z)^k C_{n-k}^{k+1/2}(z)$ [[73], (5.9)].

2. $D_\nu[P_\nu(z)]|_{\nu=-n-1} = - D_\nu[P_\nu(z)]|_{\nu=n}$.

3. $D_\nu[P_\nu^\mu(z)]|_{\nu=n-1/2}$
$$= -\ln z\, P_{n-1/2}^\mu(z) + (-1)^n 2^{n-1}\left(\frac{2\mu-2n+1}{4}\right)_n (1-z^2)^{n/2}$$
$$\times \sum_{k=0}^n \binom{n}{k} \frac{z^{2k}}{\left(\frac{2\mu-2n+1}{4}\right)_k}\left[2\ln z + \psi\left(\frac{2\mu+2n+1}{4}\right) - \psi\left(\frac{2\mu-2n+4k+1}{4}\right)\right]$$
$$\times \left[\delta_{k,0} P_{-1/2}^{\mu-n}(z) + (-2z)^{-k}(1-z^2)^{-k/2}\right.$$
$$\left.\times \sum_{p=0}^{k-1} \frac{(k+p-1)!}{p!(k-p-1)!}(2z)^{-p}(1-z^2)^{p/2}\, P_{-1/2}^{\mu+k-n-p}(z)\right].$$

4. $\quad = \left[\psi\left(\frac{1}{2}-\mu-n\right) - \psi\left(\frac{1}{2}-\mu+n\right) - \ln z\right] P_{n-1/2}^\mu(z)$
$$+ \frac{2^{n-1}}{\left(\frac{1}{2}-\mu\right)_n \left(\frac{1}{2}+\mu\right)_n}(1-z^2)^{n/2}$$
$$\times \sum_{k=0}^n \binom{n}{k}(-z^2)^k \left[2\left(\frac{2\mu-2n+3}{4}\right)_{n-k}\ln z - (n-k)!\sum_{p=0}^{n-k-1}\frac{\left(\frac{2\mu-2n+3}{4}\right)_p}{p!(n-k-p)}\right]$$
$$\times \left[\delta_{k,0} P_{-1/2}^{\mu+n}(z) + (-2z)^{-k}(1-z^2)^{-k/2}\sum_{p=0}^{k-1}\frac{(k+p-1)!}{p!(k-p-1)!}\left(\frac{1}{2}-\mu-n\right)_{k-p}^2\right.$$
$$\left.\times (2z)^{-p}(1-z^2)^{p/2}\, P_{-1/2}^{\mu-k+n+p}(z)\right].$$

5. $D_\nu[P_\nu^\mu(z)]|_{\nu=-n-1/2} = - D_\nu[P_\nu^\mu(z)]|_{\nu=n-1/2}$.

6. $D_\mu[P_\nu^\mu(z)]|_{\mu=0}$
$$= \left\{\frac{\pi\csc(\nu\pi)}{4(\nu+2)}\left[(2\nu+5)\cos(\nu\pi)+1\right]P_\nu(z) - [2\nu+5+\cos(\nu\pi)]P_\nu(-z)\right\}$$
$$+ \frac{\pi\cot\frac{\nu\pi}{2}}{4(\nu+1)(\nu+2)z}\left\{[\nu+2-(2\nu+3)z^2][P_{\nu+1}(z)+P_{\nu+1}(-z)]\right.$$
$$\left. + \sqrt{1-z^2}[P_{\nu+2}^1(z)+P_{\nu+2}^1(-z)]\right\}.$$

7. $D_\mu\left[P_{1/2}^\mu(z)\right]\Big|_{\mu=1/2} = \frac{(1-z^2)^{-1/4}}{\sqrt{2\pi}}\left[-2(C+\ln 2)z - \pi\sqrt{1-z^2}\right.$
$$\left. + \left(z+i\sqrt{1-z^2}\right)\ln\left(1-\frac{iz}{\sqrt{1-z^2}}\right) + \left(z+i\sqrt{1-z^2}\right)\ln\left(1-\frac{iz}{\sqrt{1-z^2}}\right)\right]$$
$$[|\arg(1\pm z)|<\pi].$$

1.26. The Kummer Confluent Hypergeometric Function $_1F_1(a; b; z)$

1.26.1. Derivatives with respect to the argument

1. $\mathrm{D}^n \left[{}_1F_1\left(\begin{matrix} a;\ cz \\ b \end{matrix} \right) \right] = c^n \dfrac{(a)_n}{(b)_n} {}_1F_1\left(\begin{matrix} a+n;\ cz \\ b+n \end{matrix} \right)$ $\hfill$ [[6], 6.4.10].

2. $\mathrm{D}^n \left[z^{a+n-1} {}_1F_1\left(\begin{matrix} a;\ cz \\ b \end{matrix} \right) \right] = (a)_n z^{a-1} {}_1F_1\left(\begin{matrix} a+n \\ b;\ cz \end{matrix} \right)$ $\hfill$ [[6], 6.4.11].

3. $\mathrm{D}^n \left[z^{b-1} {}_1F_1\left(\begin{matrix} a;\ cz \\ b \end{matrix} \right) \right] = (-1)^n (1-b)_n z^{b-n-1} {}_1F_1\left(\begin{matrix} a;\ cz \\ b-n \end{matrix} \right)$ $\hfill$ [[6], 6.4.12].

4. $\mathrm{D}^n \left[e^{-cz} {}_1F_1\left(\begin{matrix} a;\ cz \\ b \end{matrix} \right) \right] = (-c)^n \dfrac{(b-a)_n}{(b)_n} e^{-cz} {}_1F_1\left(\begin{matrix} a;\ cz \\ b+n \end{matrix} \right)$ $\hfill$ [[6], 6.4.13].

5. $\mathrm{D}^n \left[z^{b-1} e^{-cz} {}_1F_1\left(\begin{matrix} a;\ cz \\ b \end{matrix} \right) \right] = (-1)^n (1-b)_n z^{b-n-1} e^{-cz} {}_1F_1\left(\begin{matrix} a-n;\ cz \\ b-n \end{matrix} \right).$

6. $\mathrm{D}^n \left[z^{b-a+n-1} e^{-cz} {}_1F_1\left(\begin{matrix} a;\ cz \\ b \end{matrix} \right) \right] = (b-a)_n z^{b-a-1} e^{-cz} {}_1F_1\left(\begin{matrix} a-n \\ b;\ cz \end{matrix} \right)$ $\hfill$ [[6], 6.4.14].

7. $\mathrm{D}^n \left[z^{n-1} {}_1F_1\left(\begin{matrix} a;\ \frac{c}{z} \\ b \end{matrix} \right) \right] = (-c)^n \dfrac{(a)_n}{(b)_n} z^{-n-1} {}_1F_1\left(\begin{matrix} a+n;\ \frac{c}{z} \\ b+n \end{matrix} \right).$

8. $\mathrm{D}^n \left[z^{-a} {}_1F_1\left(\begin{matrix} a;\ \frac{c}{z} \\ b \end{matrix} \right) \right] = (-1)^n (a)_n z^{-a-n} {}_1F_1\left(\begin{matrix} a+n \\ b;\ \frac{c}{z} \end{matrix} \right).$

9. $\mathrm{D}^n \left[z^{n-b} {}_1F_1\left(\begin{matrix} a;\ \frac{c}{z} \\ b \end{matrix} \right) \right] = (1-b)_n z^{-b} {}_1F_1\left(\begin{matrix} a;\ \frac{c}{z} \\ b-n \end{matrix} \right).$

10. $\mathrm{D}^n \left[z^{n-1} e^{-c/z} {}_1F_1\left(\begin{matrix} a;\ \frac{c}{z} \\ b \end{matrix} \right) \right] = c^n \dfrac{(b-a)_n}{(b)_n} z^{-n-1} e^{-c/z} {}_1F_1\left(\begin{matrix} a;\ \frac{c}{z} \\ b+n \end{matrix} \right).$

11. $\mathrm{D}^n \left[z^{a-b} e^{-c/z} {}_1F_1\left(\begin{matrix} a;\ \frac{c}{z} \\ b \end{matrix} \right) \right] = (-1)^n (b-a)_n z^{a-b-n} e^{-c/z} {}_1F_1\left(\begin{matrix} a-n \\ b;\ \frac{c}{z} \end{matrix} \right).$

12. $\mathrm{D}^{2n+\sigma} \left[{}_1F_1\left(\begin{matrix} \frac{1}{2} - n \\ b;\ z^2 \end{matrix} \right) \right] = (-4)^n \dfrac{\left(\frac{1}{2}\right)_n \left(\sigma + \frac{1}{2}\right)_n}{(b)_{n+\sigma}} z^\sigma {}_1F_1\left(\begin{matrix} n+\sigma+\frac{1}{2} \\ b+n+\sigma;\ z^2 \end{matrix} \right)$
$\hfill [\sigma = 0 \text{ or } 1;\ [78]].$

13. $\mathrm{D}^{2n+\sigma} \left[z \, {}_1F_1\left(\begin{matrix} \frac{3}{2} - \sigma - n \\ b;\ z^2 \end{matrix} \right) \right] = (-4)^n \dfrac{\left(\frac{3}{2}\right)_n \left(\sigma - \frac{1}{2}\right)_n}{(b)_n} z^{1-\sigma} {}_1F_1\left(\begin{matrix} n+\frac{3}{2} \\ b+n;\ z^2 \end{matrix} \right)$
$\hfill [\sigma = 0 \text{ or } 1;\ [78]].$

14. $D^{2n+\sigma}\left[z^{2b-1}\,_1F_1\left(\begin{matrix}b-\sigma-n+\frac{1}{2}\\b;\,z^2\end{matrix}\right)\right]$

$= (-1)^\sigma(1-2b)_{2n+\sigma}z^{2b-\sigma-2n-1}\,_1F_1\left(\begin{matrix}b+\frac{1}{2}\\b-n;\,z^2\end{matrix}\right)$ $[\sigma = 0 \text{ or } 1; [78]]$.

15. $D^{2n+\sigma}\left[z^{2b-2}\,_1F_1\left(\begin{matrix}b-n-\frac{1}{2}\\b;\,z^2\end{matrix}\right)\right]$

$= (-1)^\sigma(2-2b)_{2n+\sigma}z^{2b-\sigma-2n-2}\,_1F_1\left(\begin{matrix}b-\frac{1}{2};\,z^2\\b-\sigma-n\end{matrix}\right)$ $[\sigma = 0 \text{ or } 1; [78]]$.

16. $D^{2n+\sigma}\left[\,_1F_1\left(\begin{matrix}a;\,z^2\\\frac{1}{2}\end{matrix}\right)\right] = 2^{2n+2\sigma}(a)_{n+\sigma}z^\sigma\,_1F_1\left(\begin{matrix}a+n+\sigma\\\sigma+\frac{1}{2};\,z^2\end{matrix}\right)$

 $[\sigma = 0 \text{ or } 1; [78]]$.

17. $D^{2n+\sigma}\left[z\,_1F_1\left(\begin{matrix}a;\,z^2\\\frac{3}{2}\end{matrix}\right)\right] = 2^{2n}(a)_n z^{1-\sigma}\,_1F_1\left(\begin{matrix}a+n;\,z^2\\\frac{3}{2}-\sigma\end{matrix}\right)$

 $[\sigma = 0 \text{ or } 1; [78]]$.

18. $D^{2n+\sigma}\left[e^{-z^2}\,_1F_1\left(\begin{matrix}a+n-\frac{1}{2}\\a;\,z^2\end{matrix}\right)\right]$

$= (-1)^\sigma 2^{2n}\dfrac{\left(\frac{1}{2}\right)_n\left(\sigma+\frac{1}{2}\right)_n}{(a)_{n+\sigma}}z^\sigma e^{-z^2}\,_1F_1\left(\begin{matrix}a-\frac{1}{2};\,z^2\\a+n+\sigma\end{matrix}\right)$ $[\sigma = 0 \text{ or } 1; [78]]$.

19. $D^{2n+\sigma}\left[ze^{-z^2}\,_1F_1\left(\begin{matrix}a+n+\sigma-\frac{3}{2}\\a;\,z^2\end{matrix}\right)\right]$

$= 2^{2n}\dfrac{\left(\frac{3}{2}\right)_n\left(\sigma-\frac{1}{2}\right)_n}{(a)_n}z^{1-\sigma}e^{-z^2}\,_1F_1\left(\begin{matrix}a-\frac{3}{2}\\a+n;\,z^2\end{matrix}\right)$ $[\sigma = 0 \text{ or } 1; [78]]$.

20. $D^{2n+\sigma}\left[z^{2a-1}e^{-z^2}\,_1F_1\left(\begin{matrix}n+\sigma-\frac{1}{2}\\a;\,z^2\end{matrix}\right)\right]$

$= (-1)^\sigma(1-2a)_{2n+\sigma}z^{2a-2n-\sigma-1}e^{-z^2}\,_1F_1\left(\begin{matrix}-n-\frac{1}{2}\\a-n;\,z^2\end{matrix}\right)$ $[\sigma = 0 \text{ or } 1; [78]]$.

21. $D^{2n+\sigma}\left[z^{2a-2}e^{-z^2}\,_1F_1\left(\begin{matrix}n+\sigma-\frac{1}{2}\\a;\,z^2\end{matrix}\right)\right] = (-1)^\sigma(2-2a)_{2n+\sigma}$

$\times z^{2a-2n-\sigma-2}e^{-z^2}\,_1F_1\left(\begin{matrix}\frac{1}{2}-n-\sigma\\a-n-\sigma;\,z^2\end{matrix}\right)$ $[\sigma = 0 \text{ or } 1; [78]]$.

22. $D^{2n+\sigma}\left[e^{-z^2} {}_1F_1\begin{pmatrix} a; z^2 \\ \frac{1}{2} \end{pmatrix}\right]$

$= (-4)^{n+\sigma}\left(\frac{1}{2} - a\right)_{n+\sigma} z^\sigma e^{-z^2} {}_1F_1\begin{pmatrix} a-n; z^2 \\ \sigma + \frac{1}{2} \end{pmatrix}$ $[\sigma = 0 \text{ or } 1; [78]]$.

23. $D^{2n+\sigma}\left[ze^{-z^2} {}_1F_1\begin{pmatrix} a; z^2 \\ \frac{3}{2} \end{pmatrix}\right]$

$= (-4)^n \left(\frac{3}{2} - a\right)_n z^{1-\sigma} e^{-z^2} {}_1F_1\begin{pmatrix} a - n - \sigma \\ \frac{3}{2} - \sigma; z^2 \end{pmatrix}$ $[\sigma = 0 \text{ or } 1; [78]]$.

1.26.2. Derivatives with respect to parameters

1. $D_a\left[{}_1F_1\begin{pmatrix} a; z \\ n+1 \end{pmatrix}\right]\bigg|_{a=m+n+1}$

$= \frac{m!\,n!}{(m+n)!} [\psi(m+1) - \psi(m+n+1)] e^z L_m^{-n}(-z)$

$+ n!\, z^{-n} \bigg\{ [C + 2\ln z + \text{shi}(z) - \text{chi}(z)] L_{m+n}^{-n}(-z)$

$\quad - \sum_{k=1}^{m+n} \frac{1}{k} L_{m+n-k}^{k-n}(-z) [2(-1)^k e^z + L_{k-1}^{-k}(z)] \bigg\}$

$\quad - \frac{m!\,(n!)^2}{(m+n)!} z^{-n} e^z \sum_{k=0}^{m} \binom{m}{k} \frac{z^k}{k!}$

$\times \bigg\{ [C + \ln z + \psi(m+1)] L_n^{k-n}(-z) - \sum_{p=1}^{n} \frac{(-1)^p}{p} L_{n-p}^{k-n+p}(-z) \bigg\}$.

2. $D_a\left[{}_1F_1\begin{pmatrix} a; z \\ \frac{1}{2} - n \end{pmatrix}\right]\bigg|_{a=m+1/2}$

$= \frac{e^z}{\left(\frac{1}{2}\right)_m \left(\frac{1}{2}\right)_n} \bigg\{ 2(-1)^n m!\, z^{n+1} \sum_{k=0}^{n} \binom{n}{k}\left(-\frac{1}{2}\right)_k (-z)^{-k}$

$\times \sum_{p=0}^{m} \frac{(-1)^p}{p!} \left(k - \frac{1}{2}\right)_p L_{m-p}^{n+p}(-z) {}_2F_2\begin{pmatrix} 1, 1; -z \\ \frac{3}{2} - k - p, 2 \end{pmatrix}$

$- (-1)^n (m+n)! \left[\psi\left(m+\frac{1}{2}\right) + \ln 2 + C\right] L_{m+n}^{-n-1/2}(-z)$

$- (-1)^{m+n} \sum_{k=0}^{m-1} \frac{1}{k!(m-k)} \sum_{p=0}^{k} \binom{k}{p}\left(\frac{1}{2} - m\right)_{k-p} L_n^{p-n-1/2}(-z) \bigg\}$.

3. $D_a\left[{}_1F_1\left(\begin{matrix}a;\ z\\ \frac{3}{2}-n\end{matrix}\right)\right]\Big|_{a=3/2}$

$$= \frac{(-z)^{n+1}}{\left(-\frac{3}{2}\right)_{n+1}} e^z \sum_{k=0}^{n} \binom{n}{k}\left(-\frac{3}{2}\right)_k (-z)^{-k} {}_2F_2\left(\begin{matrix}1,\ 1;\ -z\\ 2,\ \frac{5}{2}-k\end{matrix}\right).$$

4. $D_a\left[{}_1F_1\left(\begin{matrix}n+1;\ z\\ a\end{matrix}\right)\right]$

$$= z^{1-a} e^z L_n^{1-a}(-z)\gamma(a-1,\ z) + \sum_{k=1}^{n} \frac{1}{k} L_{n-k}^{k-a+1}(-z) L_{k-1}^{a-k-1}(z)$$

$$- (a-1) e^z \sum_{k=0}^{n} \frac{(-z)^k}{k!(a+k-1)^2} L_{n-k}^k(-z) {}_2F_2\left(\begin{matrix}a+k-1,\ a+k-1\\ a+k,\ a+k;\ -z\end{matrix}\right)$$

$$[a \neq 0,\ \pm 1,\ \pm 2,\ \dots].$$

5. $D_a\left[{}_1F_1\left(\begin{matrix}a;\ z\\ 2a+b\end{matrix}\right)\right]\Big|_{a=(n-b+1)/2} = \left[2\psi(n+1) - \psi\left(\frac{n-b+1}{2}\right) - \ln z\right]$

$$\times {}_1F_1\left(\begin{matrix}\frac{n-b+1}{2}\\ n+1;\ z\end{matrix}\right) + (-1)^{n+1} n! z^{-n} \Gamma\left(\frac{1-b-n}{2}\right) \Psi\left(\begin{matrix}\frac{1-b-n}{2}\\ 1-n;\ z\end{matrix}\right)$$

$$+ \frac{(-1)^n (n!)^2 z^{-n}}{\Gamma\left(\frac{n-b+1}{2}\right)} \sum_{k=0}^{n-1} \frac{\Gamma\left(k+\frac{1-n-b}{2}\right)}{(k!)^2 (n-k)} (-z)^k {}_1F_1\left(\begin{matrix}k+\frac{1-b-n}{2}\\ k+1;\ z\end{matrix}\right).$$

6. $D_a\left[{}_1F_1\left(\begin{matrix}a;\ -z\\ 2a+b\end{matrix}\right)\right]\Big|_{a=(n-b+1)/2} = \frac{1}{2}\left[2\psi(n+1) - \psi\left(\frac{n-b+1}{2}\right) - \ln z\right]$

$$\times {}_1F_1\left(\begin{matrix}\frac{n-b+1}{2};\ -z\\ n+1\end{matrix}\right) + \frac{n! z^{-n}}{\Gamma\left(\frac{n-b+1}{2}\right)}$$

$$\times \left[\pi G_{23}^{21}\left(z\ \Big|\ \begin{matrix}\frac{b+n+1}{2},\ -\frac{1}{2}\\ 0,\ n,\ -\frac{1}{2}\end{matrix}\right) + \Gamma(2a+b) \sum_{k=0}^{n-1} \frac{\Gamma\left(k+\frac{1-n-b}{2}\right)}{(k!)^2 (n-k)} z^k\right.$$

$$\left. \times {}_1F_1\left(\begin{matrix}k+\frac{1-b-n}{2}\\ k+1;\ -z\end{matrix}\right)\right].$$

1.27. The Tricomi Confluent Hypergeometric Function $\Psi(a;\ b;\ z)$

1.27.1. Derivatives with respect to the argument

1. $D^n\left[\Psi\left(\begin{matrix}a;\ cz\\ b\end{matrix}\right)\right] = (-c)^n (a)_n \Psi\left(\begin{matrix}a+n;\ cz\\ b+n\end{matrix}\right)$ [[6], 6.6.11].

2. $D^n\left[z^{a+n-1}\Psi\left(\begin{matrix}a;\ cz\\ b\end{matrix}\right)\right] = (a)_n (a-b+1)_n z^{a-1} \Psi\left(\begin{matrix}a+n;\ cz\\ b\end{matrix}\right)$ [[6], 6.6.13].

1.27.2]   1.27. The Tricomi Confluent Hypergeometric Function $\Psi(a;b;z)$

3. $D^n \left[z^{b-1} \Psi\left(\begin{matrix} a;\ cz \\ b \end{matrix} \right) \right] = (-1)^n (a-b+1)_n z^{b-n-1} \Psi\left(\begin{matrix} a;\ cz \\ b-n \end{matrix} \right)$ [[6], 6.6.12].

4. $D^n \left[e^{-cz} \Psi\left(\begin{matrix} a;\ cz \\ b \end{matrix} \right) \right] = (-c)^n e^{-cz} \Psi\left(\begin{matrix} a;\ cz \\ b+n \end{matrix} \right)$ [[6], 6.6.14].

5. $D^n \left[z^{b-a+n-1} e^{-cz} \Psi\left(\begin{matrix} a;\ cz \\ b \end{matrix} \right) \right] = (-1)^n z^{b-a-1} e^{-cz} \Psi\left(\begin{matrix} a-n;\ cz \\ b \end{matrix} \right)$ [[6], 6.6.15].

6. $D^n \left[z^{n-1} \Psi\left(\begin{matrix} a;\ \frac{c}{z} \\ b \end{matrix} \right) \right] = c^n (a)_n z^{-n-1} \Psi\left(\begin{matrix} a+n;\ \frac{c}{z} \\ b+n \end{matrix} \right)$.

7. $D^n \left[z^{-a} \Psi\left(\begin{matrix} a;\ \frac{c}{z} \\ b \end{matrix} \right) \right] = (-1)^n (a)_n (a-b+1)_n z^{-a-n} \Psi\left(\begin{matrix} a+n;\ \frac{c}{z} \\ b \end{matrix} \right)$.

8. $D^n \left[z^{n-b} \Psi\left(\begin{matrix} a;\ \frac{c}{z} \\ b \end{matrix} \right) \right] = (a-b+1)_n z^{-b} \Psi\left(\begin{matrix} a;\ \frac{c}{z} \\ b-n \end{matrix} \right)$.

9. $D^n \left[z^{n-1} e^{-c/z} \Psi\left(\begin{matrix} a;\ \frac{c}{z} \\ b \end{matrix} \right) \right] = c^n e^{-c/z} z^{-n-1} \Psi\left(\begin{matrix} a;\ \frac{c}{z} \\ b+n \end{matrix} \right)$.

10. $D^n \left[z^{a-b} e^{-c/z} \Psi\left(\begin{matrix} a;\ \frac{c}{z} \\ b \end{matrix} \right) \right] = z^{a-b-n} e^{-c/z} \Psi\left(\begin{matrix} a-n;\ \frac{c}{z} \\ b \end{matrix} \right)$.

1.27.2. Derivatives with respect to parameters

1. $D_a \left[\Psi\left(\begin{matrix} a;\ z \\ b+n \end{matrix} \right) \right]\bigg|_{a=b}$
$= -[\psi(b) + \ln z] \Psi\left(\begin{matrix} b;\ z \\ b+n \end{matrix} \right) + \sum_{k=0}^{n-1} \binom{n-1}{k} (b)_k\, z^{-b-k} \psi(b+k)$
$\quad - \frac{1}{\Gamma(b)} \sum_{k=0}^{n-1} \binom{n-1}{k} \left\{ (-1)^k \frac{\pi \csc(b\pi)}{b+k}\, {}_1F_1\left(\begin{matrix} b+k;\ z \\ b+k+1 \end{matrix} \right) \right.$
$\quad + \left. \frac{\Gamma(b+k)}{z^{b+k}} \left[\frac{z}{b+k-1}\, {}_2F_2\left(\begin{matrix} 1,1;\ z \\ 2, 2-b-k \end{matrix} \right) + \psi(b+k) - \ln z \right] \right\}$ $[n \geq 1]$.

2. $D_a \left[\Psi\left(\begin{matrix} a;\ z \\ b+1 \end{matrix} \right) \right]\bigg|_{a=b} = -z^{-b} \psi(b) - \Gamma(1-b)(-z)^{-b} \gamma(b,-z)$
$\quad + \dfrac{z^{1-b}}{1-b}\, {}_2F_2\left(\begin{matrix} 1,1;\ z \\ 2-b, 2 \end{matrix} \right)$.

3. $D_a \left[\Psi\left(\begin{matrix} a;\ z \\ n \end{matrix} \right) \right]\bigg|_{a=m} = -[\psi(m) + \ln z] \Psi\left(\begin{matrix} m;\ z \\ n \end{matrix} \right)$
$\quad + \dfrac{z^{-m}}{(m-1)!} \sum_{k=0}^{n-m-1} \binom{n-m-1}{k} (k+m-1)!\, z^{-k} \psi(k+m)$
$\quad - \dfrac{(-z)^{-m}}{(m-1)!} \sum_{k=0}^{n-m-1} \binom{n-m-1}{k} (k+m-1)!\, (-z)^{-k}$

77

$$\times \left\{ \sum_{p=1}^{k+m-1} \frac{1}{p} L_{p-1}^{-p}(z) L_{k+m-p-1}^{-k-m+p}(-z) - e^z [\operatorname{shi}(z) - \operatorname{chi}(z)] L_{k+m-1}^{-k-m}(-z) \right\}$$

$$[n > m \geq 1].$$

4. $D_a \left[\Psi \begin{pmatrix} a;\ z \\ n+\frac{1}{2} \end{pmatrix} \right]\bigg|_{a=-m} = \frac{(-1)^m}{z^n} \left(-\frac{1}{2}\right)_n \sum_{k=0}^{n} \binom{n}{k} \frac{1}{\left(\frac{3}{2}-n\right)_k}$

$$\times \left\{ (k+m)!\, L_{k+m}^{-k-1/2}(z) \right.$$

$$\times \left[2z\, {}_2F_2\!\left(\begin{matrix}1,1;\ z\\ \frac{3}{2},2\end{matrix}\right) - \pi\operatorname{erfi}(\sqrt{z}) - \psi\!\left(\frac{1}{2}-m\right) - \psi\!\left(n-k-\frac{1}{2}\right) + 2 \right]$$

$$- (k+m)! \sum_{p=1}^{k+m} \frac{1}{p} L_{k+m-p}^{p-k-1/2}(z)$$

$$\times \left[\sqrt{\pi z}\, e^z L_{p-1}^{1/2-p}(-z) - \frac{z^p}{\left(\frac{1}{2}\right)_p}\, {}_1F_1\!\left(\begin{matrix}p;\ z\\ p+\frac{1}{2}\end{matrix}\right) \right]$$

$$+ (-1)^k \left(\frac{1}{2}\right)_k \left(\frac{1}{2}\right)_m \sum_{p=0}^{k+m} \binom{k+m}{p} \frac{(-z)^p}{\left(\frac{1}{2}-k\right)_p} \psi\!\left(k-p+\frac{1}{2}\right) \bigg\}.$$

5. $D_b \left[\Psi\!\begin{pmatrix}a;\ z\\ b\end{pmatrix}\right]\bigg|_{b=a+n+1} = (-1)^n n!\, z^{-a-n}$

$$\times \sum_{k=0}^{n} \left\{ (-1)^k [\psi(k+1) + \psi(a) + \mathbf{C} - \ln z] \right.$$

$$- \frac{z^{k+1}}{(1-a)_{k+1}(k+1)}\, {}_2F_2\!\left(\begin{matrix}k+1,k+1;\ z\\ k-a+2,k+2\end{matrix}\right)$$

$$- \Gamma(-a)z^a \left[(e^z-1)\delta_{n,0} + L_k^{a-k-1}(-z) \right.$$

$$\left. + \frac{(-1)^k}{k!} \sum_{p=0}^{k} \binom{k}{p}(-a)_{k-p}(-z)^{-a}\gamma(a+p+1,-z) \right]\bigg\}.$$

6. $D_b \left[\Psi\!\begin{pmatrix}m+1;\ z\\ b\end{pmatrix}\right]\bigg|_{b=n+3/2} = \frac{2^{2m}}{(2m)!} \left[\mathbf{C} + 2\ln 2 + \psi\!\left(m+\frac{1}{2}\right)\right] \sum_{k=0}^{n} \binom{n}{k}(k+m)!(-z)^{-k}$

$$\times \left[\sqrt{\pi}\, z^{-1/2} e^z L_{k+m}^{-k-1/2}(-z)\operatorname{erfc}(\sqrt{z}) - \sum_{p=1}^{k+m} \frac{1}{p} L_{k+m-p}^{p-k-1/2}(-z) L_{p-1}^{1/2-p}(z) \right]$$

1.27.2]   1.27. The Tricomi Confluent Hypergeometric Function $\Psi(a;b;z)$

$$+ \frac{2^{2m}\sqrt{\pi}}{(2m)!} z^{-1/2} e^z \sum_{k=0}^{n} \binom{n}{k}(k+m)!(-z)^{-k}$$

$$\times \left\{ \frac{4z^{1/2}}{\sqrt{\pi}} \sum_{p=0}^{k+m} \frac{(-z)^p}{p!(2p+1)^2} L_{k+m-p}^{p-k}(-z) \, _2F_2\!\left(\begin{array}{c} p+\frac{1}{2}, p+\frac{1}{2}; -z \\ p+\frac{3}{2}, p+\frac{3}{2} \end{array} \right) \right.$$

$$\left. - [C + \ln(4z)] L_{k+m}^{-k-1/2}(-z) + \sum_{p=1}^{k+m} \frac{(-1)^p}{p} L_{k+m-p}^{p-k-1/2}(-z) \right\}.$$

7. $D_b\!\left[\Psi\!\left(\begin{array}{c} m+1 \\ b; z \end{array} \right) \right]\Bigg|_{b=3/2-n} = \frac{2^{2m} \left(\frac{1}{2}\right)_m \left(-\frac{1}{2}\right)_n}{(2m)!\left(m+\frac{1}{2}\right)_n} \sum_{k=0}^{n} \binom{n}{k} \frac{(k+m)!}{\left(\frac{3}{2}-n\right)_k}$

$$\times \left[2 + \psi\!\left(m+n+\frac{1}{2}\right) - \psi\!\left(n-k-\frac{1}{2}\right) \right] \sum_{p=0}^{k+m} \binom{m}{k+m-p}$$

$$\times \left[\sqrt{\pi} z^{-1/2} e^z L_p^{-p-1/2}(-z) \operatorname{erfc}(\sqrt{z}) - \sum_{r=1}^{p} \frac{1}{r} L_{p-r}^{r-p-1/2}(-z) L_{r-1}^{1/2-r}(z) \right]$$

$$+ \frac{2^{2m} m! \left(-\frac{1}{2}\right)_n}{(2m)!\left(m+\frac{1}{2}\right)_n} e^z \sum_{k=0}^{n} \binom{n}{k} \frac{1}{\left(\frac{3}{2}-n\right)_k} \sum_{p=k}^{k+m} \binom{k+m}{p} \frac{1}{(p-k)!}$$

$$\times \left\{ 4z^p \sum_{r=0}^{p} \binom{p}{r} \frac{(-1)^r}{(2r+1)^2} \, _2F_2\!\left(\begin{array}{c} r+\frac{1}{2}, r+\frac{1}{2}; -z \\ r+\frac{3}{2}, r+\frac{3}{2} \end{array} \right) \right.$$

$$- p!\sqrt{\pi} z^{-1/2}[C+\ln(4z)]L_p^{-p-1/2}(-z)$$

$$\left. + (-1)^p p!\sqrt{\pi} z^{-1/2} \sum_{r=0}^{p-1} \frac{(-1)^r}{p-r} L_r^{-r-1/2}(-z) \right\}.$$

8. $D_a\!\left[\Psi\!\left(\begin{array}{c} a; z \\ 2a+n \end{array} \right) \right]\Bigg|_{a=m} = -[\ln z + \psi(m)] \Psi\!\left(\begin{array}{c} m; z \\ 2m+n \end{array} \right)$

$$+ \frac{z^{-m}}{(m-1)!} \sum_{k=0}^{m+n-1} \binom{m+n-1}{k}(k+m-1)! z^{-k} \psi(k+m)$$

$$- \frac{(-z)^{-m}}{(m-1)!} \sum_{k=0}^{m+n-1} \binom{m+n-1}{k}(k+m-1)!(-z)^{-k}$$

$$\times \left\{ e^z [\operatorname{shi}(z) - \operatorname{chi}(z)] L_{k+m-1}^{-k-m}(-z) - \sum_{p=1}^{k+m-1} \frac{1}{p} L_{p-1}^{-p}(z) L_{k+m-p-1}^{p-k-m}(-z) \right\}$$

$$[m \geq 1].$$

9. $D_a\left[\Psi\binom{a;\ z}{2a+b}\right]\bigg|_{a=n+(1-b)/2} = -\ln z\,\Psi\binom{n+\frac{1-b}{2};\ z}{2n+1}$

$$+2^{2n-1}n!\,z^{-n}e^z\sum_{k=0}^{n-1}\frac{2^{-2k}}{k!(n-k)}G_{23}^{30}\left(z\left|\begin{array}{c}\frac{1-b}{2},\ k-n+\frac{1}{2}\\ \frac{1}{2},\ 2k-n,\ -n\end{array}\right.\right).$$

1.28. The Whittaker Functions $M_{\mu,\nu}(z)$ and $W_{\mu,\nu}(z)$

1.28.1. Derivatives with respect to the argument

1. $D^n[z^{n-\mu-1}e^{\pm az/2}M_{\mu,\nu}(az)]$
$$= (\pm 1)^n \left(\frac{1}{2}-\mu+\nu\right)_n z^{-\mu-1}e^{\pm az/2}M_{\mu\mp n,\nu}(az).$$

2. $D^n[z^{\nu-1/2}e^{\pm az/2}M_{\mu,\nu}(az)]$
$$= (-1)^n(-2\nu)_n a^{n/2}z^{\nu-(n+1)/2}e^{\pm az/2}M_{\mu\mp n/2,\nu-n/2}(az).$$

3. $D^n[z^{-\nu-1/2}e^{\pm az/2}M_{\mu,\nu}(az)]$
$$= (\pm 1)^n \frac{\left(\nu-\mu+\frac{1}{2}\right)_n}{(2\nu+1)_n} a^{n/2}z^{-\nu-(n+1)/2}e^{\pm az/2}M_{\mu\mp n/2,\nu+n/2}(az).$$

4. $D^n[z^{\pm\nu-1/2}e^{\pm az/2}W_{\mu,\nu}(az)]$
$$= (-1)^n \left(\frac{1}{2}-\mu-\nu\right)_n a^{n/2}z^{\pm\nu-(n+1)/2}e^{\pm az/2}W_{\mu-n/2,\nu\mp n}(az).$$

5. $D^n[z^{\pm\nu-1/2}e^{az/2}W_{\mu,\nu}(az)]$
$$= (-1)^n \left(\frac{1}{2}-\mu\mp\nu\right)_n a^{n/2}z^{\pm\nu-(n+1)/2}e^{az/2}W_{\mu-n/2,\nu\mp n/2}(az).$$

6. $D^n[z^{n-\mu-1}e^{az/2}W_{\mu,\nu}(az)]$
$$= \left(\frac{1}{2}-\mu-\nu\right)_n \left(\frac{1}{2}-\mu+\nu\right)_n z^{-\mu-1}e^{az/2}W_{\mu-n,\nu}(az).$$

7. $D^n[z^{\pm\nu-1/2}e^{-az/2}W_{\mu,\nu}(az)]$
$$= (-1)^n a^{n/2}z^{\pm\nu-(n+1)/2}e^{-az/2}W_{\mu+n/2,\nu\mp n/2}(az).$$

1.29. The Gauss Hypergeometric Function $_2F_1(a, b;\ c;\ z)$

1.29.1. Derivatives with respect to the argument

1. $D^n\left[{}_2F_1\binom{a,\ b}{c;\ z}\right] = \frac{(a)_n(b)_n}{(c)_n}\,{}_2F_1\binom{a+n,\ b+n}{c+n;\ z}$ [[6], 2.8.20].

2. $D^n\left[z^{a+n-1}\,{}_2F_1\binom{a,\ b}{c;\ z}\right] = (a)_n z^{a-1}\,{}_2F_1\binom{a+n,\ b}{c;\ z}$ [[6], 2.8.21].

1.29. The Gauss Hypergeometric Function $_2F_1(a,b;c;z)$

3. $D^n \left[z^{c-1} {}_2F_1\left(\begin{matrix} a,\,b \\ c;\,z \end{matrix}\right) \right] = (-1)^n (1-c)_n z^{c-n-1} {}_2F_1\left(\begin{matrix} a,\,b \\ c-n;\,z \end{matrix}\right)$ [[6], 2.8.22].

4. $D^n \left[(1-z)^{a+n-1} {}_2F_1\left(\begin{matrix} a,\,b \\ c;\,z \end{matrix}\right) \right]$
$= (-1)^n \dfrac{(a)_n (c-b)_n}{(c)_n} (1-z)^{a-1} {}_2F_1\left(\begin{matrix} a+n,\,b \\ c+n;\,z \end{matrix}\right)$ [[6], 2.8.25].

5. $D^n \left[(1-z)^{a+b-c} {}_2F_1\left(\begin{matrix} a,\,b \\ c;\,z \end{matrix}\right) \right]$
$= \dfrac{(c-a)_n (c-b)_n}{(c)_n} (1-z)^{a+b-c-n} {}_2F_1\left(\begin{matrix} a,\,b \\ c+n;\,z \end{matrix}\right)$ [[6], 2.8.24].

6. $D^n \left[z^{c-1}(1-z)^{b-c+n} {}_2F_1\left(\begin{matrix} a,\,b \\ c;\,z \end{matrix}\right) \right]$
$= (-1)^n (1-c)_n z^{c-n-1}(1-z)^{b-c} {}_2F_1\left(\begin{matrix} a-n,\,b \\ c-n;\,z \end{matrix}\right)$ [[6], 2.8.26].

7. $D^n \left[z^{c-1}(1-z)^{a+b-c} {}_2F_1\left(\begin{matrix} a,\,b \\ c;\,z \end{matrix}\right) \right]$
$= (-1)^n (1-c)_n z^{c-n-1}(1-z)^{a+b-c-n} {}_2F_1\left(\begin{matrix} a-n,\,b-n \\ c-n;\,z \end{matrix}\right)$ [[6], 2.8.27].

8. $D^n \left[z^{c-a+n-1}(1-z)^{a+b-c} {}_2F_1\left(\begin{matrix} a,\,b \\ c;\,z \end{matrix}\right) \right]$
$= (c-a)_n z^{c-a-1}(1-z)^{a+b-c-n} {}_2F_1\left(\begin{matrix} a-n,\,b \\ c;\,z \end{matrix}\right)$ [[6], 2.8.23].

9. $D^n \left[z^{n-1} {}_2F_1\left(\begin{matrix} a,\,b \\ c;\,\frac{1}{z} \end{matrix}\right) \right] = (-1)^n \dfrac{(a)_n (b)_n}{(c)_n} z^{-n-1} {}_2F_1\left(\begin{matrix} a+n,\,b+n \\ c+n;\,\frac{1}{z} \end{matrix}\right)$.

10. $D^n \left[z^{-a} {}_2F_1\left(\begin{matrix} a,\,b \\ c;\,\frac{1}{z} \end{matrix}\right) \right] = (-1)^n (a)_n z^{-a-n} {}_2F_1\left(\begin{matrix} a+n,\,b \\ c;\,\frac{1}{z} \end{matrix}\right)$.

11. $D^n \left[z^{n-c} {}_2F_1\left(\begin{matrix} a,\,b \\ c;\,\frac{1}{z} \end{matrix}\right) \right] = (1-c)_n z^{-c} {}_2F_1\left(\begin{matrix} a,\,b \\ c-n;\,\frac{1}{z} \end{matrix}\right)$.

12. $D^n \left[z^{-a}(z-1)^{a+n-1} {}_2F_1\left(\begin{matrix} a,\,b \\ c;\,\frac{1}{z} \end{matrix}\right) \right]$
$= \dfrac{(a)_n (c-b)_n}{(c)_n} z^{-a-n}(z-1)^{a-1} {}_2F_1\left(\begin{matrix} a+n,\,b \\ c+n;\,\frac{1}{z} \end{matrix}\right)$.

13. $D^n \left[z^{c-a-b+n-1}(z-1)^{a+b-c} {}_2F_1 \begin{pmatrix} a, b \\ c; \frac{1}{z} \end{pmatrix} \right]$

$$= (-1)^n \frac{(c-a)_n (c-b)_n}{(c)_n} z^{c-a-b-1}(z-1)^{a+b-c-n} {}_2F_1 \begin{pmatrix} a, b \\ c+n; \frac{1}{z} \end{pmatrix}.$$

14. $D^n \left[z^{-b}(z-1)^{b-c+n} {}_2F_1 \begin{pmatrix} a, b \\ c; \frac{1}{z} \end{pmatrix} \right]$

$$= (1-c)_n z^{-b}(z-1)^{b-c} {}_2F_1 \begin{pmatrix} a-n, b \\ c-n; \frac{1}{z} \end{pmatrix}.$$

15. $D^n \left[z^{n-a-b}(z-1)^{a+b-c} {}_2F_1 \begin{pmatrix} a, b \\ c; \frac{1}{z} \end{pmatrix} \right]$

$$= (1-c)_n z^{n-a-b}(z-1)^{a+b-c-n} {}_2F_1 \begin{pmatrix} a-n, b-n \\ c-n; \frac{1}{z} \end{pmatrix}.$$

16. $D^n \left[z^{-b}(z-1)^{a+b-c} {}_2F_1 \begin{pmatrix} a, b \\ c; \frac{1}{z} \end{pmatrix} \right]$

$$= (-1)^n (c-a)_n z^{-b}(z-1)^{a+b-c-n} {}_2F_1 \begin{pmatrix} a-n, b \\ c; \frac{1}{z} \end{pmatrix}.$$

17. $D^{2n+\sigma} \left[{}_2F_1 \begin{pmatrix} -n+\frac{1}{2}, b \\ c; z^2 \end{pmatrix} \right]$

$$= (-4)^n \left(\frac{1}{2}\right)_n \left(\sigma+\frac{1}{2}\right)_n \frac{(b)_{n+\sigma}}{(c)_{n+\sigma}} z^\sigma {}_2F_1 \begin{pmatrix} n+\sigma+\frac{1}{2}, b+n+\sigma \\ c+n+\sigma; z^2 \end{pmatrix}$$

$$[\sigma = 0 \text{ or } 1; [79], (36, 40)].$$

18. $D^{2n+\sigma} \left[z \, {}_2F_1 \begin{pmatrix} -n-\sigma+\frac{3}{2}, b \\ c; z^2 \end{pmatrix} \right]$

$$= (-4)^n \left(\frac{3}{2}\right)_n \left(\sigma-\frac{1}{2}\right)_n \frac{(b)_{n+\sigma}}{(c)_{n+\sigma}} z^{1-\sigma} {}_2F_1 \begin{pmatrix} n+\frac{3}{2}, b+n \\ c+n; z^2 \end{pmatrix}$$

$$[\sigma = 0 \text{ or } 1; [79], (34, 41)].$$

19. $D^{2n+\sigma} \left[z^{2c-1} {}_2F_1 \begin{pmatrix} c-n-\sigma+\frac{1}{2}, b \\ c; z^2 \end{pmatrix} \right] = (-1)^\sigma (1-2c)_{2n+\sigma}$

$$\times z^{2c-2n-\sigma-1} {}_2F_1 \begin{pmatrix} c+\frac{1}{2}, b \\ c-n; z^2 \end{pmatrix} \quad [\sigma = 0 \text{ or } 1; [79], (33, 37)].$$

1.29. The Gauss Hypergeometric Function $_2F_1(a,b;c;z)$

20. $\mathrm{D}^{2n+\sigma}\left[z^{2c-2}\,_2F_1\!\left(\begin{array}{c}c-n-\frac{1}{2},\,b\\c;\,z^2\end{array}\right)\right]$

$$= (-1)^\sigma (2-2c)_{2n+\sigma}\, z^{2c-2n-\sigma-2}\,_2F_1\!\left(\begin{array}{c}c-\frac{1}{2},\,b\\c-n-\sigma;\,z^2\end{array}\right)$$

$[\sigma = 0 \text{ or } 1;\ [79],\ (35,\ 39)].$

21. $\mathrm{D}^{2n+\sigma}\left[(1-z^2)^{b+n-1/2}\,_2F_1\!\left(\begin{array}{c}c+n-\frac{1}{2},\,b\\c;\,z^2\end{array}\right)\right] = (-4)^n\left(\frac{1}{2}\right)_n\left(\sigma+\frac{1}{2}\right)_n$

$$\times\,\frac{(c-b)_{n+\sigma}}{(c)_{n+\sigma}}\,z^\sigma(1-z^2)^{b-n-\sigma-1/2}\,_2F_1\!\left(\begin{array}{c}c-\frac{1}{2},\,b\\c+n+\sigma;\,z^2\end{array}\right)$$

$[\sigma = 0 \text{ or } 1;\ [79],\ (48,\ 52)].$

22. $\mathrm{D}^{2n+\sigma}\left[z(1-z^2)^{b+n+\sigma-3/2}\,_2F_1\!\left(\begin{array}{c}c+n+\sigma-\frac{3}{2},\,b\\c;\,z^2\end{array}\right)\right]$

$$= (-4)^n\left(\frac{3}{2}\right)_n\left(\sigma-\frac{1}{2}\right)_n\frac{(c-b)_n}{(c)_n}\,z^{1-\sigma}(1-z^2)^{b-n-3/2}\,_2F_1\!\left(\begin{array}{c}c-\frac{3}{2},\,b\\c+n;\,z^2\end{array}\right)$$

$[\sigma = 0 \text{ or } 1;\ [79],\ (46,\ 50)].$

23. $\mathrm{D}^{2n+\sigma}\left[z^{2c-1}(1-z^2)^{b-c+n+\sigma-1/2}\,_2F_1\!\left(\begin{array}{c}n+\sigma-\frac{1}{2},\,b\\c;\,z^2\end{array}\right)\right]$

$$= (-1)^\sigma(1-2c)_{2n+\sigma}\,z^{2c-2n-\sigma-1}(1-z^2)^{b-c-n-1/2}\,_2F_1\!\left(\begin{array}{c}-n-\frac{1}{2},\,b-n\\c-n;\,z^2\end{array}\right)$$

$[\sigma = 0 \text{ or } 1;\ [79],\ (45,\ 49)].$

24. $\mathrm{D}^{2n+\sigma}\left[z^{2c-2}(1-z^2)^{b-c+n+1/2}\,_2F_1\!\left(\begin{array}{c}n+\frac{1}{2},\,b\\c;\,z^2\end{array}\right)\right] = (-1)^\sigma(2-2c)_{2n+\sigma}$

$$\times\,z^{2c-2n-\sigma-2}(1-z^2)^{b-c-n-\sigma+1/2}\,_2F_1\!\left(\begin{array}{c}-n-\sigma+\frac{1}{2},\,b-n-\sigma\\c-n-\sigma;\,z^2\end{array}\right)$$

$[\sigma = 0 \text{ or } 1;\ [79],\ (47,\ 51)].$

25. $\mathrm{D}^{2n+\sigma}\left[\,_2F_1\!\left(\begin{array}{c}a,\,b\\\frac{1}{2};\,z^2\end{array}\right)\right]$

$$= 2^{2n+2\sigma}(a)_{n+\sigma}(b)_{n+\sigma}\,z^\sigma\,_2F_1\!\left(\begin{array}{c}a+n+\sigma,\,b+n+\sigma\\\frac{1}{2}+\sigma;\,z^2\end{array}\right)$$

$[\sigma = 0 \text{ or } 1;\ [79],\ (42,\ 44)].$

26. $D^{2n+\sigma}\left[z\,{}_2F_1\left(\begin{matrix}a,b\\\frac{3}{2};z^2\end{matrix}\right)\right] = 2^{2n}(a)_n(b)_n\,z^{1-\sigma}\,{}_2F_1\left(\begin{matrix}a+n,b+n\\\frac{3}{2}-\sigma;z^2\end{matrix}\right)$

$[\sigma = 0 \text{ or } 1; [79], (41, 44)].$

27. $D^{2n+\sigma}\left[(1-z^2)^{a+b-1/2}\,{}_2F_1\left(\begin{matrix}a,b\\\frac{1}{2};z^2\end{matrix}\right)\right] = 2^{2n+2\sigma}\left(\frac{1}{2}-a\right)_{n+\sigma}$

$\times \left(\frac{1}{2}-b\right)_{n+\sigma} z^\sigma(1-z^2)^{a+b-2n-\sigma-1/2}\,{}_2F_1\left(\begin{matrix}a-n,b-n\\\frac{1}{2}+\sigma;z^2\end{matrix}\right)$

$[\sigma = 0 \text{ or } 1; [79], (54, 55)].$

28. $D^{2n+\sigma}\left[z(1-z^2)^{a+b-1/2}\,{}_2F_1\left(\begin{matrix}a,b\\\frac{3}{2};z^2\end{matrix}\right)\right] = 2^{2n}\left(\frac{3}{2}-a\right)_n\left(\frac{3}{2}-b\right)_n$

$\times z^{1-\sigma}(1-z^2)^{a+b-2n-\sigma-3/2}\,{}_2F_1\left(\begin{matrix}a-n-\sigma,b-n-\sigma\\\frac{3}{2}-\sigma;z^2\end{matrix}\right)$

$[\sigma = 0 \text{ or } 1; [79], (53, 56)].$

29. $D^{2n+\sigma}\left[(1-z^2)^{a+n+\sigma-1}\,{}_2F_1\left(\begin{matrix}a,a+\sigma-\frac{1}{2}\\\frac{1}{2};z^2\end{matrix}\right)\right]$

$= (-4)^n(a)_{n+\sigma}(1-a-\sigma)_{n+\sigma}\,z^\sigma(1-z^2)^{a-n-1}\,{}_2F_1\left(\begin{matrix}a+\sigma,a+\sigma-\frac{1}{2}\\\sigma+\frac{1}{2};z^2\end{matrix}\right)$

$[\sigma = 0 \text{ or } 1; [79], (58, 60)].$

30. $D^{2n+\sigma}\left[z(1-z^2)^{a+n-1}\,{}_2F_1\left(\begin{matrix}a,a-\sigma+\frac{1}{2}\\\frac{3}{2};z^2\end{matrix}\right)\right] = (-4)^n(a)_n(1-a+\sigma)_n$

$\times z^{1-\sigma}(1-z^2)^{a-n-\sigma-1}\,{}_2F_1\left(\begin{matrix}a-\sigma,a-\sigma+\frac{1}{2}\\\frac{3}{2}-\sigma;z^2\end{matrix}\right)\quad [\sigma = 0 \text{ or } 1; [79], (58, 62)].$

31. $D^{2n+\sigma}\left[(1-z^2)^{n+\sigma}\,{}_2F_1\left(\begin{matrix}1,a\\\frac{1}{2};z^2\end{matrix}\right)\right]$

$= (-4)^{n+\sigma}(n+\sigma)!\left(\frac{1}{2}-a\right)_{n+\sigma} z^\sigma\,{}_2F_1\left(\begin{matrix}n+\sigma+1,a\\\sigma+\frac{1}{2};z^2\end{matrix}\right)$

$[\sigma = 0 \text{ or } 1; [79], (57, 59)].$

1.29.2. Derivatives with respect to parameters

1. $D_a\left[{}_2F_1\binom{n+1, a;\ z}{a+b}\right]$

$$= bz(1-z)^{b-n-1} \sum_{k=0}^{n} \frac{(b+1)_k}{k!(a+b+k)^2} z^k (1-z)^k P_{n-k}^{(k+1, b+k-n-1)}(1-2z)$$

$$\times\ {}_3F_2\binom{a+b+k,\ a+b+k,\ b+k+1;\ z}{a+b+k+1,\ a+b+k+1}.$$

2. $D_b\left[{}_2F_1\binom{n+1, a}{b;\ z}\right] = -az(1-z)^{-a-1} \sum_{k=0}^{n} \frac{(a+1)_k}{k!(b+k)^2} (-z)^k (1-z)^{-2k}$

$$\times P_{n-k}^{(k+1, a+k-n-1)}\left(\frac{1+z}{1-z}\right) {}_3F_2\binom{a+k+1,\ b+k,\ b+k;\ \frac{z}{z-1}}{b+k+1,\ b+k+1}.$$

3. $D_b\left[{}_2F_1\binom{\frac{1}{2}, 1}{b;\ z}\right]\bigg|_{b=1/2}$

$$= \frac{1}{2(z-1)}\left[(1+2\sqrt{z})\ln(1+\sqrt{z}) + (1-2\sqrt{z})\ln(1-\sqrt{z}) - \ln(1-z)\right].$$

4. $D_a\left[{}_2F_1\binom{a, b}{b+1;\ z}\right]$

$$= bz^{-b}(1-z)^{1-a}\left[(1-z)^{a-1}\,\mathrm{B}(1-a, b)[\psi(b-a+1)-\psi(1-a)]\right.$$

$$+ \frac{1}{1-a}\ln(1-z)\,{}_2F_1\binom{1-a, 1-b}{2-a;\ 1-z} - \frac{1}{(1-a)^2}\,{}_3F_2\binom{1-a, 1-a, 1-b}{2-a, 2-a;\ 1-z}\bigg].$$

5. $D_a\left[{}_2F_1\binom{a, a+b;\ -z}{2a+c}\right]\bigg|_{a=(n-c+1)/2}$

$$= \left[-\ln z + 2\psi(n+1) - \psi\left(\frac{n-c+1}{2}\right) - \psi\left(b + \frac{n-c+1}{2}\right)\right]$$

$$\times\ {}_2F_1\binom{\frac{n-c+1}{2}, b+\frac{n-c+1}{2}}{n+1;\ -z}$$

$$+ \frac{1}{\Gamma\left(\frac{n-c+1}{2}\right)\Gamma\left(b+\frac{n-c+1}{2}\right)}\left\{(-1)^n \frac{2^{n-b} n!\,(1+z)^{(c-n-1)/2-b}}{\Gamma(n-b+1)}\right.$$

$$\times\left[-\sin\left(\frac{c+n}{2}\pi\right)\Gamma(b)\Gamma\left(\frac{1+n-c}{2}\right)\Gamma\left(\frac{1-n-c}{2}\right)(1+z)^b\right.$$

$$\times\ {}_2F_1\binom{\frac{1-n-c}{2},\ \frac{1+n-c}{2}}{1-b;\ \frac{1}{1+z}}$$

$$+ \sin\left(\frac{2b-c-n}{2}\pi\right)\Gamma(-b)\Gamma\left(b+\frac{1-n-c}{2}\right)\Gamma\left(b+\frac{n-c+1}{2}\right)$$

$$\times {}_2F_1\left(\begin{array}{c}\frac{n+c+1}{2},\, b+\frac{n-c+1}{2}\\ b+1;\, \frac{1}{1+z}\end{array}\right)\Bigg] + (n!)^2 z^{-n}$$

$$\times \sum_{k=0}^{n-1} \frac{z^k \Gamma\left(k+\frac{1-n-c}{2}\right)\Gamma\left(k+b+\frac{1-n-c}{2}\right)}{(k!)^2 (n-k)}$$

$$\times {}_2F_1\left(\begin{array}{c}k+\frac{1-n-c}{2},\, k+b+\frac{1-n-c}{2}\\ k+1;\, -z\end{array}\right)\Bigg\}.$$

1.30. The Generalized Hypergeometric Function ${}_pF_q((a_p);\,(b_q);\,z)$

1.30.1. Derivatives with respect to the argument

1. $\displaystyle D^n\left[{}_pF_q\left(\begin{array}{c}(a_p);\,z\\(b_q)\end{array}\right)\right] = \frac{\prod(a_p)_n}{\prod(b_q)_n}\,{}_pF_q\left(\begin{array}{c}(a_p)+n;\,z\\(b_q)+n\end{array}\right).$

2. $\displaystyle D^n\left[z^r\,{}_pF_q\left(\begin{array}{c}(a_p);\,z^m\\(b_q)\end{array}\right)\right]$

$$= (-1)^n (-r)_n\, z^{r-n}\,{}_{p+m}F_{q+m}\left(\begin{array}{c}(a_p),\,\Delta(m,\,r+1);\,z^m\\(b_q),\,\Delta(m,\,r-n+1)\end{array}\right)$$
$$[r \neq n-1,\,n-2,\,n-3,\,\ldots].$$

3. $\displaystyle D^n\left[z^r\,{}_pF_q\left(\begin{array}{c}(a_p);\,z\\(b_q)\end{array}\right)\right]$

$$= \frac{n!}{(n-r)!}\frac{\prod(a_p)_{n-r}}{\prod(b_q)_{n-r}}\,{}_{p+1}F_{q+1}\left(\begin{array}{c}(a_p)+n-r,\,n+1;\,z\\(b_q)+n-r,\,n-r+1\end{array}\right)$$

$$+ \sum_{k=0}^{-r-1}\frac{(k-n+r+1)_n}{k!}\frac{\prod(a_p)_k}{\prod(b_q)_k}\,z^{k-n+r} \quad [r = n-1,\,n-2,\,n-3,\,\ldots].$$

4. $\displaystyle D^{2n}\left[{}_pF_q\left(\begin{array}{c}(a_p);\,z^2\\(b_q)\end{array}\right)\right] = 2^{2n}\left(\frac{1}{2}\right)_n\frac{\prod(a_p)_n}{\prod(b_q)_n}\,{}_{p+1}F_{q+1}\left(\begin{array}{c}(a_p)+n,\,n+\frac{1}{2}\\(b_q)+n,\,\frac{1}{2};\,z^2\end{array}\right).$

5. $\displaystyle D^{2n+1}\left[{}_pF_q\left(\begin{array}{c}(a_p);\,z^2\\(b_q)\end{array}\right)\right]$

$$= 2^{2n+1} z \left(\frac{3}{2}\right)_n \frac{\prod(a_p)_{n+1}}{\prod(b_q)_{n+1}}\,{}_{p+1}F_{q+1}\left(\begin{array}{c}(a_p)+n+1,\,n+\frac{3}{2}\\(b_q)+n+1,\,\frac{3}{2};\,z^2\end{array}\right).$$

1.30.2. Derivatives with respect to parameters

1. $\mathrm{D}_a^n \left[{}_{p+1}F_q \left(\begin{matrix} a, (a_p) \\ (b_q); z \end{matrix} \right) \right] \bigg|_{a=0} = (-1)^n n! \sum_{k=n}^{\infty} S_n^k \frac{(-z)^k}{k!} \frac{\prod (a_p)_k}{\prod (b_q)_k}.$

2. $\mathrm{D}_a \left[{}_{p+2}F_{q+1} \left(\begin{matrix} -n, a, (a_p) \\ b, (b_q); z \end{matrix} \right) \right]$

   $= [\psi(b-a) - \psi(b-a+n)] \, {}_{p+2}F_q \left(\begin{matrix} -n, a, (a_p) \\ b, (b_q); z \end{matrix} \right)$

   $- \frac{n!}{(a-b+1)_n} \sum_{k=0}^{n-1} \frac{(-1)^k (2k-n+b-a)(b-a-n)_k}{k!(n-k)(k-a+b)}$

   $\times \sum_{j=0}^{n-k} \binom{n-k}{j} (-z)^j \frac{\prod (a_p)}{\prod (b_q)} \, {}_{p+2}F_{q+1} \left(\begin{matrix} -k, a+n-k, (a_p)+j \\ b, (b_q)+j; z \end{matrix} \right).$

3. $\mathrm{D}_b \left[{}_{p+1}F_{q+1} \left(\begin{matrix} -n, (a_p); z \\ b, (b_q) \end{matrix} \right) \right] = [\psi(b) - \psi(b+n)] \, {}_{p+1}F_{q+1} \left(\begin{matrix} -n, (a_p) \\ b, (b_q); z \end{matrix} \right)$

   $+ \frac{n!}{(b)_n} \sum_{k=0}^{n-1} \frac{(b)_k}{k!(n-k)} \, {}_{p+1}F_{q+1} \left(\begin{matrix} -k, (a_p) \\ b, (b_q); z \end{matrix} \right).$

4. $\mathrm{D}_a \left[{}_{p+2}F_q \left(\begin{matrix} (a_p), a, b-a \\ (b_q); z \end{matrix} \right) \right] \bigg|_{a=n+b/2}$

   $= \left[\psi\left(\frac{b}{2} - n\right) - \psi\left(\frac{b}{2} + n\right) \right] {}_{p+2}F_q \left(\begin{matrix} (a_p), \frac{b}{2}-n, \frac{b}{2}+n \\ (b_q); z \end{matrix} \right)$

   $+ 2^{2n-1} n! \frac{\Gamma\left(\frac{b+1}{2}\right)}{\Gamma\left(\frac{b}{2}+n\right)} \sum_{k=0}^{n-1} \frac{2^{-2k} \Gamma\left(\frac{b}{2}+2k-n\right)}{k!(n-k) \Gamma\left(\frac{b+1}{2}+k-n\right)}$

   $\times {}_{p+3}F_{q+1} \left(\begin{matrix} (a_p), \frac{b+1}{2}, \frac{b}{2}-n, \frac{b}{2}+2k-n \\ (b_q), \frac{b+1}{2}+k-n; z \end{matrix} \right).$

5. $\mathrm{D}_a \left[{}_{p+2}F_{q+1} \left(\begin{matrix} -n, a, (a_p) \\ a+b, (b_q); z \end{matrix} \right) \right]$

   $= [\psi(a+n) - \psi(a) + \psi(a+b) - \psi(a+b+n)]$

   $\times {}_{p+2}F_{q+1} \left(\begin{matrix} -n, a, (a_p) \\ a+b, (b_q); z \end{matrix} \right)$

   $+ \frac{n!(b)_n}{(1-a)_n (a+b)_n} \sum_{k=0}^{n-1} (2k-n+a) \frac{(a-n)_k (a+b)_k}{k!(n-k)(a+k)(1-b-n)_k}$

   $\times {}_{p+2}F_{q+1} \left(\begin{matrix} -k, a+k-n, (a_p) \\ a+b, (b_q); z \end{matrix} \right).$

6. $D_a \left[{}_3F_2 \left(\begin{matrix} \frac{1}{2}, 1, 1 \\ a, 2; z \end{matrix} \right) \right] \Big|_{a=1/2}$

$$= \frac{1}{2z} \left[4 \ln(1-z) + 4\sqrt{z} \ln \frac{1+\sqrt{z}}{1+\sqrt{z}} - \ln^2 \frac{1+\sqrt{z}}{1+\sqrt{z}} \right].$$

7. $D_a \left[{}_3F_2 \left(\begin{matrix} a, a, a; z \\ a + \frac{1}{2}, 2a \end{matrix} \right) \right] \Big|_{a=1/2} = -\frac{4}{\pi^2} \mathbf{K} \left(\sqrt{\frac{1-\sqrt{1-z}}{2}} \right)$

$$\times \left[\pi \mathbf{K} \left(\sqrt{\frac{1+\sqrt{1-z}}{2}} \right) + 2 \ln \frac{\sqrt{z}}{8} \mathbf{K} \left(\sqrt{\frac{1-\sqrt{1-z}}{2}} \right) \right].$$

8. $D_a \left[{}_3F_2 \left(\begin{matrix} a, a, a; -z \\ a + 1/2, 2a \end{matrix} \right) \right] \Big|_{a=1/2} = -\frac{8}{\pi^2} \mathbf{K} \left(\frac{\sqrt{z}}{\sqrt{z+1}+1} \right)$

$$\times \left[\frac{\sqrt{2}\pi}{(\sqrt{z+1}+1)^{1/2}(\sqrt{z+1}+\sqrt{z})^{1/2}} \mathbf{K} \left(\frac{1}{\sqrt{z+1}+\sqrt{z}} \right) \right.$$

$$\left. + \frac{2\ln \frac{\sqrt{z}}{8}}{\sqrt{z+1}+1} \mathbf{K} \left(\frac{\sqrt{z}}{\sqrt{z+1}+1} \right) \right].$$

9. $D_d \left[{}_4F_3 \left(\begin{matrix} a + \frac{1}{2}, a+1, a-b+c, a+b+c \\ 2a+1, 2d+1, 2a-2d+1; z \end{matrix} \right) \right] \Big|_{d=a} =$

$$- \left[\psi(a+b+c) + \psi(b-a-c+1) + 2\mathbf{C} + \ln \frac{z}{4} \right]$$

$$\times {}_4F_3 \left(\begin{matrix} a + \frac{1}{2}, a+1, a-b+c, a+b+c \\ 2a+1, 2a+1, 1; z \end{matrix} \right) + 2^{2a} \sqrt{\pi} \frac{\Gamma(2a+1)\Gamma(b-a-c+1)}{\Gamma(a+b+c)}$$

$$\times G^{23}_{55} \left(z \left| \begin{matrix} \frac{1}{2} - a, -a, 1-a-b-c, 1-a+b-c, -\frac{1}{2} \\ 0, 0, -\frac{1}{2}, -2a, -2a \end{matrix} \right. \right).$$

10. $D_a \left[{}_4F_3 \left(\begin{matrix} a, a + \frac{1}{2}, a+b-c, a+b+c \\ 2a, d, 2a-d+1; z \end{matrix} \right) \right] \Big|_{a=d/2}$

$$= - \left[\ln \frac{z}{4} + \psi \left(c+b+\frac{d}{2} \right) + \psi \left(c-b-\frac{d}{2}+1 \right) + 2\mathbf{C} \right]$$

$$\times {}_4F_3 \left(\begin{matrix} \frac{d}{2}, \frac{d+1}{2}, b-c+\frac{d}{2}, b+c+\frac{d}{2} \\ 1, d, d; z \end{matrix} \right) + 2^{d-1}\sqrt{\pi} \frac{\Gamma(d)\Gamma\left(c-b-\frac{d}{2}+1\right)}{\Gamma\left(c+b+\frac{d}{2}\right)}$$

$$\times G^{23}_{55} \left(z \left| \begin{matrix} \frac{1-d}{2}, \frac{2-2b-2c-d}{2}, 1-\frac{d}{2}, -\frac{1}{2}, c-b-\frac{d}{2}+1 \\ 0, 0, -\frac{1}{2}, 1-d, 1-d \end{matrix} \right. \right).$$

11. $D_a \left[{}_1F_2 \left(\begin{matrix} a; z \\ 1, 1 \end{matrix} \right) \right] \Big|_{a=1} = K_0(2\sqrt{z}) + \left(\frac{1}{2} \ln z + \mathbf{C} \right) I_0(2\sqrt{z}).$

1.30.2]   1.30. The Generalized Hypergeometric Function $_pF_q((a_p);(b_q);z)$

12. $D_a\left[_1F_2\left(\begin{array}{c}a;\ -z\\1,\ 1\end{array}\right)\right]\bigg|_{a=1} = -\dfrac{\pi}{2}Y_0(2\sqrt{z}) + \left(C + \dfrac{1}{2}\ln z\right)J_0(2\sqrt{z}).$

13. $D_a\left[_1F_2\left(\begin{array}{c}a;\ z\\ \frac{3}{2},\ \frac{3}{2}\end{array}\right)\right]\bigg|_{a=3/2}$
$$= \dfrac{1}{4\sqrt{z}}\{\sinh(2\sqrt{z})[2C + \ln(16z) - 2\,\mathrm{chi}\,(4\sqrt{z}) - 4]$$
$$+ 2\cosh(2\sqrt{z})\,\mathrm{shi}\,(4\sqrt{z})\}.$$

14. $D_a\left[_1F_2\left(\begin{array}{c}n+1;\ z\\a,\ 2-a\end{array}\right)\right]\bigg|_{a=1/2} = -2 + \dfrac{\left(\frac{1}{2}\right)_n}{2(n!)\,z}$

$$\times \sum_{k=0}^{n}\binom{n}{k}\dfrac{z^k}{\left(\frac{1}{2}\right)_k}\left[\pi z^{(1-k)/2}\,\mathbf{L}_k(2\sqrt{z}) + \sum_{p=0}^{k-1}\dfrac{\Gamma\left(p+\frac{1}{2}\right)}{\Gamma\left(k-p+\frac{1}{2}\right)}(-z)^{-p}\right]$$

$$- \sum_{k=0}^{n}\binom{n}{k}\dfrac{(-1)^k}{k!}\sum_{p=0}^{k}(-1)^p\binom{k}{p}\left(-\dfrac{1}{2}\right)_{k-p}z^{p/2}$$

$$\times\left[\pi\,\mathbf{L}_{p+1}(2\sqrt{z}) + z^{p/2}\sum_{r=0}^{p-1}\dfrac{\Gamma\left(r+\frac{1}{2}\right)}{\Gamma\left(p-r+\frac{3}{2}\right)}(-z)^{-r}\right].$$

15. $D_a\left[_1F_2\left(\begin{array}{c}n+1;\ -z\\a,\ 2-a\end{array}\right)\right]\bigg|_{a=1/2} = -2 + \dfrac{\left(\frac{1}{2}\right)_n}{2(n!)\,z}$

$$\times \sum_{k=0}^{n}\binom{n}{k}\dfrac{(-z)^k}{\left(\frac{1}{2}\right)_k}\left[\pi z^{(1-k)/2}\,\mathbf{H}_k(2\sqrt{z}) - \sum_{r=0}^{k-1}\dfrac{\Gamma\left(r+\frac{1}{2}\right)}{\Gamma\left(k-r+\frac{3}{2}\right)}\left(\dfrac{1}{z}\right)^r\right]$$

$$+ \sum_{k=0}^{n}\dfrac{(-1)^k}{k!}\binom{n}{k}\sum_{p=0}^{k}\binom{k}{p}z^{p/2}\left(-\dfrac{1}{2}\right)_{k-p}$$

$$\times\left[\pi\,\mathbf{H}_{p+1}(2\sqrt{z}) - z^{p/2}\sum_{r=0}^{p-1}\dfrac{\Gamma\left(r+\frac{1}{2}\right)}{\Gamma\left(p-r+\frac{3}{2}\right)}\left(\dfrac{1}{z}\right)^r\right].$$

16. $D_a\left[_1F_2\left(\begin{array}{c}1;\ z\\a,\ a+\frac{1}{2}\end{array}\right)\right]\bigg|_{a=3/2} = -\dfrac{1}{2z}\{3 + 2\sinh(2\sqrt{z})\,\mathrm{shi}\,(2\sqrt{z})$
$$+ \cosh(2\sqrt{z})[-3 + 2C + 2\ln(2\sqrt{z}) - 2\,\mathrm{chi}\,(2\sqrt{z})]\}.$$

17. $D_a\left[_1F_2\left(\begin{array}{c}a;\ -z\\a+b,\ 2a+c\end{array}\right)\right]\bigg|_{a=(n-c+1)/2}$
$$= \left[2\psi(n+1) - \ln z - \psi\left(\dfrac{n-c+1}{2}\right) + \psi\left(b + \dfrac{n-c+1}{2}\right)\right]$$
$$\times\,_1F_2\left(\begin{array}{c}\frac{n-c+1}{2};\ -z\\b+\frac{n-c+1}{2},\ n+1\end{array}\right)$$

$$+ n! \frac{\Gamma\left(b + \frac{n-c+1}{2}\right)}{\Gamma\left(\frac{n-c+1}{2}\right)} \left[\pi G_{24}^{21}\left(z \left|\begin{array}{c} \frac{c-n+1}{2}, -n - \frac{1}{2} \\ 0, -n, -n - \frac{1}{2}, \frac{c-n+1}{2} - b \end{array}\right.\right)\right.$$

$$+ n! \sum_{k=0}^{n-1} \frac{z^{k-n}}{(k!)^2 (n-k)} \frac{\Gamma\left(k + \frac{1-n-c}{2}\right)}{\Gamma\left(k + b + \frac{1-n-c}{2}\right)} {}_1F_2\left(\begin{array}{c} k + \frac{1-n-c}{2}; -z \\ k+1, k+b+\frac{1-n-c}{2} \end{array}\right)\right].$$

18. $D_a \left[{}_1F_2\left(\begin{array}{c} a;\ z \\ a+b,\ 2a+c \end{array}\right)\right]\bigg|_{a=(n-c+1)/2}$

$$= \frac{1}{2}\left[2\psi(n+1) - \ln z - \psi\left(\frac{n-c+1}{2}\right) + \psi\left(b + \frac{n-c+1}{2}\right)\right]$$

$$\times {}_1F_2\left(\begin{array}{c} \frac{n-c+1}{2};\ z \\ b + \frac{n-c+1}{2},\ n+1 \end{array}\right) + (-1)^n n!\, z^{-n/2} \frac{\Gamma\left(b + \frac{n-c+1}{2}\right)}{2\Gamma\left(\frac{n-c+1}{2}\right)}$$

$$\times \left[G_{13}^{21}\left(z \left|\begin{array}{c} \frac{c+1}{2} \\ -\frac{n}{2},\ \frac{n}{2},\ \frac{c-2b+1}{2} \end{array}\right.\right)\right.$$

$$- n! \sum_{k=0}^{n-1} (-1)^k \frac{z^{k-n/2}}{(k!)^2 (n-k)} \frac{\Gamma\left(k + \frac{1-n-c}{2}\right)}{\Gamma\left(k + b + \frac{1-n-c}{2}\right)}$$

$$\times {}_1F_2\left(\begin{array}{c} k + \frac{1-n-c}{2};\ z \\ k+1,\ k+b+\frac{1-n-c}{2} \end{array}\right)\right].$$

19. $D_a[{}_0F_2(a, 2a+b;\ z)]|_{a=(1-b)/2}$

$$= \left[\psi\left(\frac{b+1}{2}\right) - \ln z - 2\mathbf{C}\right] {}_0F_2\left(\begin{array}{c} z \\ 1,\ \frac{1-b}{2} \end{array}\right) + \frac{\pi}{\Gamma\left(\frac{b+1}{2}\right)\Gamma\left(\frac{b+3}{2}\right)}$$

$$\times \left\{ z^{(b+1)/2} \tan \frac{b\pi}{2} \Gamma\left(-\frac{b+1}{2}\right) {}_0F_2\left(\begin{array}{c} z \\ \frac{b+3}{2},\ \frac{b+3}{2} \end{array}\right)\right.$$

$$\left. + \Gamma\left(\frac{b+3}{2}\right) G_{14}^{30}\left(z \left|\begin{array}{c} -\frac{1}{2} \\ 0,\ 0,\ \frac{b+1}{2},\ -\frac{1}{2} \end{array}\right.\right)\right\}.$$

20. $D_a\left[{}_2F_2\left(\begin{array}{c} a,\ a;\ z \\ 1,\ a+1 \end{array}\right)\right]\bigg|_{a=1}$

$$= \frac{1}{z}\left[e^z - 1 + (2 + e^z)(\mathbf{C} + \ln z) - 2\operatorname{Ei}(z) - e^z \operatorname{Ei}(-z)\right] \quad [z > 0].$$

21. $D_a\left[{}_2F_2\left(\begin{array}{c} a,\ a;\ -z \\ 1,\ a+1 \end{array}\right)\right]\bigg|_{a=1}$

$$= \frac{1}{z}\left[1 - e^{-z} - (2 + e^{-z})(\mathbf{C} + \ln z) + 2\operatorname{Ei}(-z) + e^{-z}\operatorname{Ei}(z)\right] \quad [z > 0].$$

22. $D_a \left[{}_2F_2 \begin{pmatrix} a, a+\frac{1}{2}; z \\ b, b+\frac{1}{2} \end{pmatrix} \right] \bigg|_{a=b}$

$$= 2ze^z \left[\frac{1}{2b} {}_2F_2 \begin{pmatrix} 1, 1; -z \\ 2, b+1 \end{pmatrix} + \frac{1}{2b+1} {}_2F_2 \begin{pmatrix} 1, 1; -z \\ 2, b+\frac{3}{2} \end{pmatrix} \right].$$

23. $D_a \left[{}_2F_2 \begin{pmatrix} a, a+b; z \\ a+\frac{b+1}{2}, 2a+c \end{pmatrix} \right] \bigg|_{a=(n-c+1)/2}$

$$= \left[-\ln z - \psi\left(\frac{n-c+1}{2}\right) - \psi\left(\frac{n-c+1}{2}+b\right) \right.$$

$$+ \psi\left(\frac{n+b-c}{2}+1\right) + 2\psi(n+1) \right] {}_2F_2 \begin{pmatrix} \frac{n-c+1}{2}, \frac{n-c+1}{2}+b; z \\ \frac{n+b-c}{2}+1, n+1 \end{pmatrix}$$

$$- (-1)^n \frac{n!\,\Gamma\left(\frac{n+b-c}{2}+1\right) z^{-n/2}}{\Gamma\left(\frac{n-c+1}{2}\right) \Gamma\left(\frac{n-c+1}{2}+b\right)} \left[G_{23}^{22}\left(z \,\middle|\, \begin{matrix} \frac{c+1}{2}, \frac{c+1}{2}-b \\ -\frac{n}{2}, \frac{n}{2}, \frac{c-b}{2} \end{matrix} \right) \right.$$

$$- n!\,z^{-n/2} \sum_{k=0}^{n-1} \frac{(-z)^k \Gamma\left(k+\frac{1-c-n}{2}\right) \Gamma\left(k+b+\frac{1-c-n}{2}\right)}{(k!)^2 (n-k) \Gamma\left(k+1+\frac{b-c-n}{2}\right)}$$

$$\times {}_2F_2 \begin{pmatrix} k+\frac{1-c-n}{2}, k+b+\frac{1-c-n}{2}; z \\ k+1, k+1+\frac{b-n-c}{2} \end{pmatrix} \bigg].$$

24. $D_a \left[{}_2F_2 \begin{pmatrix} a, a+b; -z \\ a+\frac{b+1}{2}, 2a+c \end{pmatrix} \right] \bigg|_{a=(n-c+1)/2}$

$$= \left[-\ln z - \psi\left(\frac{n-c+1}{2}\right) - \psi\left(\frac{n-c+1}{2}+b\right) \right.$$

$$+ \psi\left(\frac{n+b-c}{2}+1\right) + 2\psi(n+1) \right] \times {}_2F_2 \begin{pmatrix} \frac{n-c+1}{2}, \frac{n-c+1}{2}+b; -z \\ \frac{n+b-c}{2}+1, n+1 \end{pmatrix}$$

$$+ \frac{n!\,\Gamma\left(\frac{n+b-c}{2}+1\right) z^{-n/2}}{\Gamma\left(\frac{n-c+1}{2}\right) \Gamma\left(\frac{n-c+1}{2}+b\right)} \left[\pi G_{34}^{22}\left(z \,\middle|\, \begin{matrix} \frac{c+1}{2}, \frac{c+1}{2}-b, -\frac{n+1}{2} \\ -\frac{n}{2}, \frac{n}{2}, -\frac{n+1}{2}, \frac{c-b}{2} \end{matrix} \right) \right.$$

$$+ n!\,z^{-n/2} \sum_{k=0}^{n-1} \frac{z^k \Gamma\left(k+\frac{1-c-n}{2}\right) \Gamma\left(k+b+\frac{1-c-n}{2}\right)}{(k!)^2 (n-k) \Gamma\left(k+1+\frac{b-c-n}{2}\right)}$$

$$\times {}_2F_2 \begin{pmatrix} k+\frac{1-c-n}{2}, k+b+\frac{1-c-n}{2}; -z \\ k+1, k+1+\frac{b-n-c}{2} \end{pmatrix} \bigg].$$

25. $D_a\left[{}_2F_2\left({a, a; z \atop a+1, 2a+b}\right)\right]\Big|_{a=(n-b+1)/2}$

$= \left[\dfrac{2}{n-b+1} - \ln z - \psi\left(\dfrac{n-b+1}{2}\right) + 2\psi(n+1)\right]$

$\times {}_2F_2\left({\dfrac{n-b+1}{2}, \dfrac{n-b+1}{2}; z \atop \dfrac{n-b+3}{2}, n+1}\right)$

$- (-1)^n \dfrac{n!(n-b+1)}{\Gamma\left(\dfrac{n-b+1}{2}\right)} z^{-n/2} \left[\dfrac{1}{2} G^{22}_{23}\left(z \left|{\dfrac{b+1}{2}, \dfrac{b+1}{2} \atop -\dfrac{n}{2}, \dfrac{n}{2}, \dfrac{b-1}{2}}\right.\right)\right.$

$- n! z^{-n/2} \sum_{k=0}^{n-1} \dfrac{(-z)^k \Gamma\left(k + \dfrac{1-n-b}{2}\right)}{(k!)^2 (n-k)(2k-n-b+1)}$

$\left. \times {}_2F_2\left({k + \dfrac{1-b-n}{2}, k + \dfrac{1-n-b}{2}; z \atop k+1, k + \dfrac{3-n-b}{2}}\right)\right]$.

26. $D_a\left[{}_2F_2\left({a, a; -z \atop a+1, 2a+b}\right)\right]\Big|_{a=(n-b+1)/2}$

$= \left[\dfrac{2}{n-b+1} - \ln z - \psi\left(\dfrac{n-b+1}{2}\right) + 2\psi(n+1)\right]$

$\times {}_2F_2\left({\dfrac{n-b+1}{2}, \dfrac{n-b+1}{2}; -z \atop \dfrac{n-b+3}{2}, n+1}\right)$

$+ \dfrac{n!(n-b+1)z^{-n/2}}{\Gamma\left(\dfrac{n-b+1}{2}\right)} \left[\dfrac{\pi}{2} G^{22}_{34}\left(z \left|{\dfrac{b+1}{2}, \dfrac{b+1}{2}, -\dfrac{n+1}{2} \atop -\dfrac{n}{2}, \dfrac{n}{2}, \dfrac{b-1}{2}, -\dfrac{n+1}{2}}\right.\right)\right.$

$+ n! z^{-n/2} \sum_{k=0}^{n-1} \dfrac{z^k \Gamma\left(k + \dfrac{1-b-n}{2}\right)}{(k!)^2 (n-k)(2k-n-b+1)}$

$\left. \times {}_2F_2\left({k + \dfrac{1-b-n}{2}, k + \dfrac{1-b-n}{2}; -z \atop k+1, k + \dfrac{3-b-n}{2}}\right)\right]$.

27. $D_a\left[{}_2F_2\left({a, a+\tfrac{1}{2}; z \atop a+1, 2a+b}\right)\right]\Big|_{a=(n-b+1)/2}$

$= \left[\dfrac{2}{n-b+1} - \ln z - \psi\left(\dfrac{n-b}{2} + 1\right) + 2\psi(n+1)\right]$

$\times {}_2F_2\left({\dfrac{n-b+1}{2}, \dfrac{n-b}{2} + 1; z \atop \dfrac{n-b+3}{2}, n+1}\right)$

$- \dfrac{(-1)^n n!(n-b+1)z^{-n/2}}{2\Gamma\left(\dfrac{n-b}{2} + 1\right)} \left[G^{22}_{23}\left(z \left|{\dfrac{b}{2}, \dfrac{b+1}{2} \atop -\dfrac{n}{2}, \dfrac{n}{2}, \dfrac{b-1}{2}}\right.\right)\right.$

$$- 2(n!)\, z^{-n/2} \sum_{k=0}^{n-1} \frac{(-z)^k \Gamma\!\left(k - \frac{n+b}{2} + 1\right)}{(k!)^2 (n-k)(2k - n - b + 1)}$$

$$\times\, {}_2F_2\!\left(\begin{matrix} k + \frac{1-b-n}{2},\, k - \frac{n+b}{2} + 1;\, z \\ k+1,\, k + \frac{3-n-b}{2} \end{matrix}\right)\Bigg].$$

28. $D_a \left[{}_2F_3\!\left(\begin{matrix} a,\, a + \frac{1}{2};\, z \\ a+1,\, a+1,\, 2a+b \end{matrix}\right) \right]\bigg|_{a=(n-b+1)/2}$

$$= \left[\frac{4}{n-b+1} - \ln z + \psi\!\left(\frac{n-b+1}{2}\right) - \psi\!\left(\frac{n-b}{2} + 1\right) + 2\psi(n+1) \right]$$

$$\times\, {}_2F_3\!\left(\begin{matrix} \frac{n-b+1}{2},\, \frac{n-b}{2} + 1;\, z \\ \frac{n-b+3}{2},\, \frac{n-b+3}{2},\, n+1 \end{matrix}\right)$$

$$- (-1)^n n!\,(n-b+1) \frac{\Gamma\!\left(\frac{n-b+3}{2}\right)}{\Gamma\!\left(\frac{n-b}{2}+1\right)}\, z^{-n/2} \left[\frac{1}{2} G^{22}_{24}\!\left(z \,\Bigg|\, \begin{matrix} \frac{b}{2},\, \frac{b+1}{2} \\ -\frac{n}{2},\, \frac{n}{2},\, \frac{b-1}{2},\, \frac{b-1}{2} \end{matrix}\right) \right.$$

$$- n! \sum_{k=0}^{n-1} \frac{(-1)^k z^{k-n/2}}{(k!)^2 (n-k)(2k-n-b+1)} \frac{\Gamma\!\left(k - \frac{n+b}{2} + 1\right)}{\Gamma\!\left(k + \frac{3-n-b}{2}\right)}$$

$$\times\, {}_2F_3\!\left(\begin{matrix} k + \frac{1-b-n}{2},\, k - \frac{n+b}{2} + 1;\, z \\ k+1,\, k + \frac{3-n-b}{2},\, k + \frac{3-n-b}{2} \end{matrix}\right)\Bigg].$$

29. $D_a \left[{}_2F_3\!\left(\begin{matrix} a,\, a + \frac{1}{2};\, z \\ a+1,\, a+1,\, 2a+b \end{matrix}\right) \right]\bigg|_{a=(n-b+1)/2} = \frac{4}{n-b+1} - \ln z$

$$+ \psi\!\left(\frac{n-b+1}{2}\right) - \psi\!\left(\frac{n-b}{2} + 1\right)$$

$$+ 2\psi(n+1)\, {}_2F_3\!\left(\begin{matrix} \frac{n-b+1}{2},\, \frac{n-b}{2} + 1;\, z \\ \frac{n-b+3}{2},\, \frac{n-b+3}{2},\, n+1 \end{matrix}\right)$$

$$- (-1)^n n!\,(n-b+1) \frac{\Gamma\!\left(\frac{n-b+3}{2}\right)}{2\Gamma\!\left(\frac{n-b}{2}+1\right)}\, z^{-n/2} \left[\frac{1}{2} G^{22}_{24}\!\left(z \,\Bigg|\, \begin{matrix} \frac{b}{2},\, \frac{b+1}{2} \\ -\frac{n}{2},\, \frac{n}{2},\, \frac{b-1}{2},\, \frac{b-1}{2} \end{matrix}\right) \right.$$

$$- v - n! \sum_{k=0}^{n-1} \frac{(-1)^k z^{k-n/2}}{(k!)^2 (n-k)(2k-n-b+1)} \frac{\Gamma\!\left(k - \frac{n+b}{2} + 1\right)}{\Gamma\!\left(k + \frac{3-n-b}{2}\right)}$$

$$\times\, {}_2F_3\!\left(\begin{matrix} k + \frac{1-b-n}{2},\, k - \frac{n+b}{2} + 1;\, z \\ k+1,\, k + \frac{3-n-b}{2},\, k + \frac{3-n-b}{2} \end{matrix}\right)\Bigg].$$

30. $D_a \left[{}_2F_3 \left(\begin{matrix} a, a + \frac{1}{2}; z \\ a+b, a+b+\frac{1}{2}, 2a+c \end{matrix} \right) \right] \Big|_{a=(n-c+1)/2}$

$= [-\ln z + 2\psi(n + 2b - c + 1) - 2\psi(n - c + 1) + 2\psi(n+1)]$

$\times {}_2F_3 \left(\begin{matrix} \frac{n-c+1}{2}, \frac{n-c}{2} + 1; z \\ \frac{n-c+1}{2} + b, \frac{n-c}{2} + b + 1, n+1 \end{matrix} \right) + (-1)^n n! \, z^{-n/2} \frac{\Gamma(2b - c + n + 1)}{\Gamma(n - c + 1)}$

$\times \left[-2^{-2b} G_{24}^{22} \left(z \Big| \begin{matrix} \frac{c}{2}, \frac{c+1}{2} \\ -\frac{n}{2}, \frac{n}{2}, \frac{c-2b}{2}, \frac{c-2b+1}{2} \end{matrix} \right) \right.$

$+ n! \sum_{k=0}^{n-1} \frac{(-1)^k z^{k-n/2}}{(k!)^2 (n-k)} \frac{\Gamma(2k - n - c + 1)}{\Gamma(2k - n + 2b - c + 1)}$

$\left. \times {}_2F_3 \left(\begin{matrix} k + \frac{1-c-n}{2}, k - \frac{n+c}{2} + 1; z \\ k+1, k + \frac{1-n-c}{2} + b, k - \frac{n+c}{2} + b + 1 \end{matrix} \right) \right].$

31. $D_a \left[{}_2F_3 \left(\begin{matrix} a, a + \frac{1}{2}; -z \\ a+b, a+b+\frac{1}{2}, 2a+c \end{matrix} \right) \right] \Big|_{a=(n-c+1)/2}$

$= [-\ln z - 2\psi(n - c + 1) + 2\psi(n + 2b - c + 1) + 2\psi(n+1)]$

$\times {}_2F_3 \left(\begin{matrix} \frac{n-c+1}{2}, \frac{n-c}{2} + 1; -z \\ \frac{n-c+1}{2} + b, \frac{n-c}{2} + b + 1, n+1 \end{matrix} \right) + n! \, z^{-n/2} \frac{\Gamma(n + 2b - c + 1)}{\Gamma(n - c + 1)}$

$\times \left[2^{-2b} \pi G_{35}^{22} \left(z \Big| \begin{matrix} \frac{c}{2}, \frac{c+1}{2}, -\frac{n+1}{2} \\ -\frac{n}{2}, \frac{n}{2}, \frac{c-2b}{2}, \frac{c-2b+1}{2}, -\frac{n+1}{2} \end{matrix} \right) \right.$

$+ n! \, z^{-n/2} \sum_{k=0}^{n-1} \frac{z^k}{(k!)^2 (n-k)} \frac{\Gamma(2k - n - c + 1)}{\Gamma(2k - n + 2b - c + 1)}$

$\left. \times {}_2F_3 \left(\begin{matrix} k + \frac{1-c-n}{2}, k - \frac{n+c}{2} + 1; -z \\ k+1, k + \frac{1-n-c}{2} + b, k - \frac{n+c}{2} + b + 1 \end{matrix} \right) \right].$

Chapter 2

Limits

2.1. Special Functions

2.1.1. The Bessel functions $J_\nu(z)$, $Y_\nu(z)$, $I_\nu(z)$ and $K_\nu(z)$

1. $\lim\limits_{\nu\to\infty} \nu^{\nu+1/2} \left(\dfrac{2}{ez}\right)^\nu \left\{\begin{matrix} J_\nu(z) \\ I_\nu(z) \end{matrix}\right\} = \dfrac{1}{\sqrt{2\pi}}$.

2. $\lim\limits_{\nu\to\infty} \nu^{\nu+1/2} \left(\dfrac{2}{ez}\right)^\nu \left\{\begin{matrix} J_\nu(z\sqrt{\nu}) \\ I_\nu(z\sqrt{\nu}) \end{matrix}\right\} = \dfrac{1}{\sqrt{2\pi}} e^{\mp z^2/4}$.

3. $\lim\limits_{n\to\infty} (-1)^n n \left[\sin\left(n^2 z + \dfrac{n\pi}{2}\right) J_{n+1/2}(n^2 z) \right.$

    $\left. + \cos\left(n^2 z - \dfrac{n\pi}{2}\right) J_{-n-1/2}(n^2 z)\right] = \sqrt{\dfrac{2}{\pi z}} \cos\dfrac{1}{2z}$.

4. $\lim\limits_{n\to\infty} \left(\dfrac{ez}{2n}\right)^n [\sin z\, J_{n+1/2}(z) - \cos z\, Y_{n+1/2}(z)] = \dfrac{2}{\sqrt{\pi z}} \cos z$.

5. $\lim\limits_{n\to\infty} \left(\dfrac{ez}{2n}\right)^n [\cos z\, J_{n+1/2}(z) - \sin z\, Y_{n+1/2}(z)] = \dfrac{2}{\sqrt{\pi z}} \sin z$.

6. $\lim\limits_{n\to\infty} n[J_{n+1/2}^2(nz) + Y_{n+1/2}^2(nz)] = \dfrac{2\operatorname{sgn}(z)}{\pi\sqrt{z^2-1}}$ $\qquad [z^2 > 1]$.

7. $\lim\limits_{\nu\to\infty} \left(\dfrac{z}{2\nu}\right)^{\nu-1/2} e^\nu K_\nu(z) = \pm\sqrt{\dfrac{\pi}{z}}$ $\qquad \left[\left\{\begin{matrix}|\arg z| < \pi \\ z < 0\end{matrix}\right\}\right]$.

2.1.2. The Struve functions $H_\nu(z)$ and $L_\nu(z)$

1. $\lim\limits_{\nu\to\infty} \nu^{(\nu+1)/2} \left(\dfrac{2}{ez}\right)^\nu H_\nu(z\sqrt{\nu}) = \dfrac{1}{\sqrt{2\pi}} e^{-z^2/4} \operatorname{erfi}\left(\dfrac{z}{2}\right)$.

2. $\lim\limits_{\nu\to\infty} \nu^{(\nu+1)/2} \left(\dfrac{2}{ez}\right)^\nu L_\nu(z\sqrt{\nu}) = \dfrac{1}{\sqrt{2\pi}} e^{z^2/4} \operatorname{erf}\left(\dfrac{z}{2}\right)$.

2.1.3. The Kelvin functions $\operatorname{ber}_\nu(z)$, $\operatorname{bei}_\nu(z)$, $\operatorname{ker}_\nu(z)$ and $\operatorname{kei}_\nu(z)$

1. $\lim\limits_{\nu\to\infty} \left(\dfrac{2}{ez}\right)^\nu \nu^{\nu+1/2} \left[\cos\dfrac{3\nu\pi}{4} \operatorname{ber}_\nu(z) + \sin\dfrac{3\nu\pi}{4} \operatorname{bei}_\nu(z)\right] = \dfrac{1}{\sqrt{2\pi}}$.

2. $\lim\limits_{\nu\to\infty} \left(\dfrac{z}{2\nu}\right)^{\nu-1/2} e^{\nu} \left[\sin\dfrac{(6\nu-1)\pi}{8}\,\mathrm{ker}_{\nu}(z) + \cos\dfrac{(6\nu-1)\pi}{8}\,\mathrm{kei}_{\nu}(z)\right]$
$$= -\sqrt{\dfrac{\pi}{z}}\,\sin\dfrac{\pi}{8}.$$

2.1.4. The Legendre polynomials $P_n(z)$

1. $\lim\limits_{n\to\infty} n^{1/2} z^{n/2} P_n\left(\dfrac{z+1}{2\sqrt{z}}\right) = \dfrac{1}{\sqrt{\pi(1-z)}}$ $\qquad [|z|<1]$.

2. $\lim\limits_{n\to\infty} n^{1/2-n}(2z)^{-n} P_n(1+nz) = \dfrac{1}{\sqrt{\pi}}\,e^{1/z}$.

3. $\lim\limits_{n\to\infty} n^{(1-n)/2}(2z)^{-n} P_n(\sqrt{n}\,z) = \dfrac{1}{\sqrt{\pi}}\,e^{-z^{-2}/4}$.

4. $\lim\limits_{n\to\infty} (-1)^n n^{1/2} P_{2n}\left(\dfrac{z}{n}\right) = \dfrac{\cos(2z)}{\sqrt{\pi}}$.

5. $\lim\limits_{n\to\infty} (-1)^n n^{1/2} P_{2n+1}\left(\dfrac{z}{n}\right) = \dfrac{\sin(2z)}{\sqrt{\pi}}$.

6. $\lim\limits_{n\to\infty} P_n\left(1+\dfrac{z}{n^2}\right) = I_0(\sqrt{2z})$.

7. $\lim\limits_{n\to\infty} P_n\left(\sqrt{1+\dfrac{z^2}{n^2}}\right) = I_0(z)$.

8. $\lim\limits_{n\to\infty} P_n\left(\dfrac{n}{\sqrt{n^2+z^2}}\right) = J_0(z)$.

9. $\lim\limits_{n\to\infty} P_n\left(\dfrac{n+z}{\sqrt{n(n+2z)}}\right) = I_0(z)$.

2.1.5. The Chebyshev polynomials $T_n(z)$ and $U_n(z)$

1. $\lim\limits_{n\to\infty} \left(z-\sqrt{z^2-1}\right)^n T_n(z) = \dfrac{1}{2}$ $\qquad [\mathrm{Re}\,z>1]$.

2. $\lim\limits_{n\to\infty} (2zn)^{-n} T_n(1+nz) = \dfrac{1}{2}\,e^{1/z}$.

3. $\lim\limits_{n\to\infty} (-1)^n T_{2n}\left(\dfrac{z}{n}\right) = \cos(2z)$.

4. $\lim\limits_{n\to\infty} (-1)^n T_{2n+1}\left(\dfrac{z}{n}\right) = \sin(2z)$.

5. $\lim\limits_{n\to\infty} T_n\left(1+\dfrac{z}{n^2}\right) = \cosh\sqrt{2z}$.

6. $\lim\limits_{n\to\infty} T_n\left(\dfrac{n^2+z^2}{n^2-z^2}\right) = \cosh(2z)$.

7. $\lim_{n\to\infty} (2z)^{-n} n^{-n/2} T_n(\sqrt{n}\, z) = \dfrac{1}{2} e^{-z^{-2}/4}$.

8. $\lim_{n\to\infty} T_n\left(\sqrt{1 + \dfrac{z^2}{n^2}}\right) = \cosh z$.

9. $\lim_{n\to\infty} T_n\left(\dfrac{n}{\sqrt{n^2 + z^2}}\right) = \cos z$.

10. $\lim_{n\to\infty} \left(z - \sqrt{z^2 - 1}\right)^n U_n(z) = \left(2 - 2z^2 + 2z\sqrt{z^2 - 1}\right)^{-1}$ [Re $z > 1$].

11. $\lim_{n\to\infty} (2zn)^{-n} U_n(1 + nz) = e^{1/z}$.

12. $\lim_{n\to\infty} (-1)^n U_{2n}\left(\dfrac{z}{n}\right) = \cos(2z)$.

13. $\lim_{n\to\infty} (-1)^n U_{2n+1}\left(\dfrac{z}{n}\right) = \sin(2z)$.

14. $\lim_{n\to\infty} \dfrac{1}{n} U_n\left(1 + \dfrac{z}{n^2}\right) = \dfrac{\sinh\sqrt{2z}}{\sqrt{2z}}$.

15. $\lim_{n\to\infty} \dfrac{1}{n} U_n\left(\dfrac{n^2 + z^2}{n^2 - z^2}\right) = \dfrac{\sinh(2z)}{2z}$.

16. $\lim_{n\to\infty} (2z)^{-n} n^{-n/2} U_n(\sqrt{n}\, z) = e^{-z^{-2}/4}$.

17. $\lim_{n\to\infty} \dfrac{1}{n} U_n\left(\sqrt{1 + \dfrac{z^2}{n^2}}\right) = \dfrac{\sinh z}{z}$.

18. $\lim_{n\to\infty} \dfrac{1}{n} U_n\left(\dfrac{n}{\sqrt{n^2 + z^2}}\right) = \dfrac{\sin z}{z}$.

2.1.6. The Hermite polynomials $H_n(z)$

1. $\lim_{n\to\infty} \dfrac{1}{(2nz)^n} H_n(nz) = e^{-z^{-2}/4}$.

2. $\lim_{n\to\infty} \left(-\dfrac{e}{4n}\right)^n H_{2n}\left(\dfrac{z}{\sqrt{n}}\right) = 2^{1/2} \cos(2z)$.

3. $\lim_{n\to\infty} \dfrac{(-e)^n}{(4n)^{n+1/2}} H_{2n+1}\left(\dfrac{z}{\sqrt{n}}\right) = 2^{1/2} \sin(2z)$.

2.1.7. The Laguerre polynomials $L_n^\lambda(z)$

1. $\lim_{n\to\infty} n^{-\lambda} L_n^\lambda\left(\dfrac{z}{n}\right) = z^{-\lambda/2} J_\lambda(2\sqrt{z})$ [82].

2. $\lim_{\lambda\to\infty} \lambda^{-n/2} L_n^\lambda(\lambda - \sqrt{\lambda}\, z) = \dfrac{2^{-n/2}}{n!} H_n\left(\dfrac{z}{\sqrt{2}}\right)$ [75].

3. $\lim\limits_{s\to\infty} s^{-n/2} L_n^{\lambda+s}\left(s\dfrac{\sqrt{s}-z}{\sqrt{s}-t}\right) = \dfrac{2^{-n/2}}{n!} H_n\left(\dfrac{z-t}{\sqrt{2}}\right)$ [[80], (3)].

2.1.8. The Gegenbauer polynomials $C_n^\lambda(z)$

1. $\lim\limits_{\lambda\to 0} \lambda^{-1} C_n^\lambda(z) = \dfrac{2}{n} T_n(z).$

2. $\lim\limits_{\lambda\to\infty} \lambda^{-n} C_n^\lambda(z) = \dfrac{(-2z)^n}{n!}.$

3. $\lim\limits_{n\to\infty} n^{1-\lambda} z^{n/2} C_n^\lambda\left(\dfrac{z+1}{2\sqrt{z}}\right) = \dfrac{(1-z)^{-\lambda}}{\Gamma(\lambda)}$ $[|z|<1]$.

4. $\lim\limits_{\lambda\to\infty} \lambda^{-n/2} C_n^\lambda\left(\dfrac{z}{\sqrt{\lambda}}\right) = \dfrac{1}{n!} H_n(z).$

5. $\lim\limits_{n\to\infty} (-2z)^{-n} n^{\lambda-n} C_n^{\lambda-n}(nz) = (2z)^{-\lambda} I_{-\lambda}\left(\dfrac{1}{z}\right).$

6. $\lim\limits_{n\to\infty} (-2z)^{-n} n^{\lambda-n} C_n^\lambda(1+nz) = (2z)^{-\lambda} e^{1/z} I_{-\lambda}\left(\dfrac{1}{z}\right).$

7. $\lim\limits_{n\to\infty} (-1)^n n^{1-\lambda} C_{2n}^\lambda\left(\dfrac{z}{n}\right) = \dfrac{\cos(2z)}{\Gamma(\lambda)}.$

8. $\lim\limits_{n\to\infty} (-1)^n n^{1-\lambda} C_{2n+1}^\lambda\left(\dfrac{z}{n}\right) = \dfrac{\sin(2z)}{\Gamma(\lambda)}.$

9. $\lim\limits_{n\to\infty} (-1)^n n^{1/2} \lambda^{-n\lambda} (\lambda-1)^{n\lambda-n} C_{2n}^{n\lambda-n+\mu}\left(\dfrac{z}{n}\right)$
   $= \dfrac{1}{\sqrt{2\pi}} \left(\dfrac{\lambda}{\lambda-1}\right)^{\mu-1/2} \cos(2\sqrt{\lambda}\,z)$ $[\lambda\notin[0,1]]$.

10. $\lim\limits_{n\to\infty} (-1)^n n^{1/2} \lambda^{-n\lambda} (\lambda-1)^{n\lambda-n} C_{2n+1}^{n\lambda-n+\mu}\left(\dfrac{z}{n}\right)$
    $= \dfrac{1}{\sqrt{2\pi}} \dfrac{\lambda^\mu}{(\lambda-1)^{\mu-1/2}} \sin(2\sqrt{\lambda}\,z)$ $[\lambda\notin[0,1]]$.

11. $\lim\limits_{n\to\infty} \left(\dfrac{n}{2}\right)^\lambda C_n^{\lambda-n/2}\left(\dfrac{z}{\sqrt{n}}\right) = \dfrac{1}{\Gamma(1-\lambda)}\, {}_1F_1\!\left(\begin{matrix}\lambda;\ -\dfrac{z^2}{2}\\ \dfrac{1}{2}\end{matrix}\right).$

12. $\lim\limits_{n\to\infty} n^{1-2\lambda} C_n^\lambda\left(\dfrac{n}{\sqrt{n^2+z^2}}\right) = \dfrac{\sqrt{\pi}}{\Gamma(\lambda)}(2z)^{1/2-\lambda} J_{\lambda-1/2}(z).$

2.1.9. The Jacobi polynomials $P_n^{(\rho,\sigma)}(z)$

1. $\lim\limits_{\rho\to\infty} \rho^{-n} P_n^{(\rho,\sigma-\rho-n)}(\rho z) = \left(-\dfrac{z}{2}\right)^n L_n^{-\sigma-n-1}\left(\dfrac{2}{z}\right).$

2. $\lim\limits_{\sigma\to\infty} P_n^{(\rho,\sigma+a)}\left(1+\dfrac{z}{\sigma}\right) = L_n^\rho\left(-\dfrac{z}{2}\right).$

2.1. Special Functions

3. $\lim\limits_{\sigma\to\infty} P_n^{(\rho,-\sigma+a)}\left(\dfrac{\sigma+z}{\sigma-z}\right) = L_n^\rho(z).$

4. $\lim\limits_{\rho\to\infty} \rho^{-n/2} P_n^{(\rho,\rho)}\left(\dfrac{z}{\sqrt{\rho}}\right) = \dfrac{2^{-n}}{n!} H_n(z).$

5. $\lim\limits_{n\to\infty} n^{-\rho} P_n^{(\rho,\sigma-n)}\left(1+\dfrac{z}{n}\right) = \dfrac{1}{\Gamma(\rho+1)}\, {}_1F_1\!\left(\begin{array}{c}\rho+\sigma+1\\ \rho+1;\ \dfrac{z}{2}\end{array}\right).$

6. $\lim\limits_{n\to\infty} n^{-\rho} P_n^{(\rho,-n-\rho/2-1/2)}\left(1+\dfrac{z}{n}\right) = \dfrac{\sqrt{\pi}}{\Gamma\!\left(\dfrac{\rho+1}{2}\right)}\left(\dfrac{2}{z}\right)^{\rho/2} e^{z/4} I_{\rho/2}\!\left(\dfrac{z}{4}\right).$

7. $\lim\limits_{n\to\infty} n^{-1/2} P_n^{(1/2,-n-1)}\left(1+\dfrac{z^2}{n}\right) = \dfrac{\sqrt{2}}{z}\, \operatorname{erfi}\!\left(\dfrac{z}{\sqrt{2}}\right).$

8. $\lim\limits_{n\to\infty} n^{-1/2} P_n^{(1/2,-n-1/2)}\left(1+\dfrac{z^2}{n}\right) = \dfrac{\sqrt{2}}{z} e^{z^2/2} \operatorname{erf}\!\left(\dfrac{z}{\sqrt{2}}\right).$

9. $\lim\limits_{n\to\infty} n^{-\rho} P_n^{(\rho,-n-\rho)}\left(1+\dfrac{z}{n}\right) = \dfrac{\left(\dfrac{2}{z}\right)^{\rho}}{\Gamma(\rho)} e^{z/2} \gamma\!\left(\rho,\dfrac{z}{2}\right).$

10. $\lim\limits_{n\to\infty} n^{-\rho} P_n^{(\rho,-n-m-\rho-1)}\left(1+\dfrac{z}{n}\right) = \dfrac{m!}{\Gamma(\rho+m+1)} L_m^\rho\!\left(\dfrac{z}{2}\right).$

11. $\lim\limits_{n\to\infty} n^{-\rho} P_n^{(\rho,n\sigma-\rho-1)}\left(1+\dfrac{z}{n^2}\right) = \left(\dfrac{2}{\sigma z+z}\right)^{\rho/2} I_\rho\!\left(\sqrt{2(\sigma+1)z}\right).$

2.1.10. Hypergeometric functions

1. $\lim\limits_{s\to\infty} s^{-a}\, {}_1F_1\!\left(\begin{array}{c}a;\ s^2-sz\\ b+s^2\end{array}\right) = e^{z^2/4} D_{-a}(z)$ [33].

2. $\lim\limits_{s\to\infty} s^{-a/2}\, {}_2F_1\!\left(\begin{array}{c}a, b+s\\ c+\dfrac{s}{2};\ \dfrac{1}{2}-\dfrac{z}{2\sqrt{s}}\end{array}\right) = e^{z^2/4} D_{-a}(z)$ [[75], (6)].

Chapter 3

Indefinite Integrals

3.1. Elementary Functions

3.1.1. The logarithmic function

1. $\displaystyle\int \frac{\ln x \ln^n (ax+b)}{ax+b}\, dx = -\frac{1}{(n+1)a} \ln\left(-\frac{a}{b}\right) \ln^{n+1}(ax+b)$
$\displaystyle\qquad - \frac{n!}{a} \sum_{k=0}^{n} \frac{(-1)^k}{(n-k)!} \ln^{n-k}(ax+b)\, \text{Li}_{k+2}\left(\frac{ax}{b}+1\right).$

2. $\displaystyle\int \frac{\ln^2 x \ln(ax+b)}{(cx+d)^{n+1}}\, dx = \frac{a}{nc}\int \frac{\ln^2 x}{(ax+b)(cx+d)^n}\, dx$
$\displaystyle\qquad + \frac{2}{nc}\int \frac{\ln x \ln(ax+b)}{x(cx+d)^n}\, dx - \frac{1}{nc}\frac{\ln^2 x \ln(ax+b)}{(cx+d)^n} \quad [n \geq 1].$

3. $\displaystyle\int \frac{\ln x \ln^n(ax+b)}{(ax+b)^{m+1}}\, dx = \frac{n}{m}\int \frac{\ln x \ln^{n-1}(ax+b)}{(ax+b)^{m+1}}\, dx$
$\displaystyle\qquad + \frac{1}{ma}\int \frac{\ln x \ln^n(ax+b)}{x(ax+b)^m}\, dx - \frac{1}{ma}\frac{\ln x \ln^n(ax+b)}{(ax+b)^m} \quad [n \geq 1].$

4. $\displaystyle\int \frac{\ln x \ln(ax+b)\ln(cx+d)}{x^{n+1}}\, dx = \frac{c}{n}\int \frac{\ln x \ln(ax+b)}{x^n(cx+d)}\, dx$
$\displaystyle\qquad + \frac{a}{n}\int \frac{\ln x \ln(cx+d)}{x^n(ax+b)}\, dx$
$\displaystyle\qquad + \frac{1}{n}\int \frac{\ln(ax+b)\ln(cx+d)}{x^{n+1}}\, dx - \frac{\ln x}{nx^n}\ln(ax+b)\ln(cx+d) \quad [n \geq 1].$

5. $\displaystyle\int \frac{\ln(ax+b)\ln(cx+d)}{x^{n+1}}\, dx = \frac{c}{n}\int \frac{\ln(ax+b)}{x^n(cx+d)}\, dx$
$\displaystyle\qquad + \frac{a}{n}\int \frac{\ln(cx+d)}{x^n(ax+b)}\, dx - \frac{1}{nx^n}\ln(ax+b)\ln(cx+d) \quad [n \geq 1].$

6. $\displaystyle\int \frac{\ln^2 x \ln^2(ax+b)}{(ax+b)^{n+1}}\, dx = \frac{2}{n}\int \frac{\ln^2 x \ln(ax+b)}{(ax+b)^{n+1}}\, dx$
$\displaystyle\qquad + \frac{2}{na}\int \frac{\ln x \ln^2(ax+b)}{x(ax+b)^n}\, dx - \frac{1}{na(ax+b)^n}\ln^2 x \ln^2(ax+b) \quad [n \geq 1].$

7. $\int \ln x \ln^n(ax+b)\,dx = \dfrac{ax+b}{a} \ln x \ln^n(ax+b)$
$\qquad\qquad - n \int \ln x \ln^{n-1}(ax+b)\,dx - \dfrac{1}{a} \int \dfrac{ax+b}{x} \ln^n(ax+b)\,dx.$

3.2. Special Functions

3.2.1. The Bessel functions $J_\nu(x)$, $Y_\nu(x)$, $I_\nu(x)$ and $K_\nu(x)$

Notation: $Z_\nu(x)$, $\overline{Z}(x) = J_\nu(x)$ or $Y_\nu(x)$.

1. $\int x^{\nu+2n+1} J_\nu(x)\,dx = (-2)^n n!\, x^{\nu+n+1} \sum\limits_{k=0}^{n} \dfrac{(-1)^k}{k!} \left(\dfrac{x}{2}\right)^k J_{\nu+n-k+1}(x).$

2. $\int x^n J_\nu(x)\,dx = \sum\limits_{k=0}^{n} 2^k \left(\dfrac{\nu-n+1}{2}\right)_k x^{n-k} J_{\nu+k+1}(x)$
$\qquad\qquad + 2^n \left(\dfrac{\nu-n+1}{2}\right)_n \int J_{\nu+n}(x)\,dx.$

3. $\int x^{\nu+1} \ln x\, J_0(x) J_\nu(x)\,dx = \dfrac{x^{\nu+1}}{4(\nu+1)^2} \{2 J_0(x)[(\nu+1)J_{\nu+1}(x)$
$\qquad + x(\nu \ln x + \ln x - 1) J_\nu(x)] + x[\pi(\nu+1)[J_\nu(x)Y_0(x) + J_{\nu+1}(x)Y_1(x)]$
$\qquad + 2(\nu \ln x + \ln x - 1) J_1(x) J_{\nu+1}(x)]\}$
$\qquad - \dfrac{\sqrt{\pi}}{4} G^{22}_{35}\left(x^2 \left|\begin{array}{c} \frac{\nu}{2}+1, \frac{\nu+3}{2}, \nu+\frac{1}{2} \\ \nu+1, \nu+1, 0, 1, \nu+\frac{1}{2} \end{array}\right.\right).$

4. $\int \sin x \ln x\, J_0(x)\,dx = (x \sin x \ln x - \cos x - 2x \sin x) J_0(x)$
$\qquad\qquad + x \cos x (2 - \ln x) J_1(x).$

5. $\int \cos x \ln x\, J_0(x)\,dx = (\sin x - 2x \cos x + x \cos x \ln x) J_0(x)$
$\qquad\qquad + x \sin x (\ln x - 2) J_1(x).$

6. $\int \dfrac{1}{x J_\nu^2(x)} f\left(\dfrac{J_{-\nu}(x)}{J_\nu(x)}\right) dx = -\dfrac{\pi}{2} \csc(\nu\pi) F\left(\dfrac{J_{-\nu}(x)}{J_\nu(x)}\right)$ $\qquad [f(x) = F'(x)].$

7. $\int \dfrac{1}{x J_\nu^2(x)} f\left(\dfrac{Y_\nu(x)}{J_\nu(x)}\right) dx = \dfrac{\pi}{2} F\left(\dfrac{Y_\nu(x)}{J_\nu(x)}\right)$ $\qquad [f(x) = F'(x)].$

8. $\int \dfrac{[J_\nu(x)]^{n-1}}{x\,[J_{-\nu}(x)]^{n+1}}\,dx = \dfrac{\pi}{2n \sin(\nu\pi)} \left[\dfrac{J_\nu(x)}{J_{-\nu}(x)}\right]^n.$

9. $\int \dfrac{[J_\nu(x)]^{n-1}}{x\,[Y_\nu(x)]^{n+1}}\,dx = -\dfrac{\pi}{2n} \left[\dfrac{J_\nu(x)}{Y_\nu(x)}\right]^n.$

10. $\int x^{n\nu} Z_\nu^{n-1}(x) Z_{\nu-1}(x)\,dx = \dfrac{1}{n} x^{n\nu} Z_\nu^n(x).$

11. $\displaystyle\int x^{-n\nu} Z_\nu^{n-1}(x) Z_{\nu+1}(x)\, dx = -\frac{1}{n} x^{-n\nu} Z_\nu^n(x).$

12. $\displaystyle\int \frac{Z_{\nu-1}(x)}{Z_\nu(x)}\, dx = \ln Z_\nu(x) + \nu \ln x.$

13. $\displaystyle\int \frac{Z_{\nu+1}(x)}{Z_\nu(x)}\, dx = -\ln Z_\nu(x) + \nu \ln x.$

14. $\displaystyle\int x^{2n\nu-1} [Z_\nu(x)]^2 \frac{[Z_{\nu+1}(x)]^{n-1}}{[Z_{\nu-1}(x)]^{n+1}}\, dx = \frac{x^{2n\nu}}{2n\nu} \left[\frac{Z_{\nu+1}(x)}{Z_{\nu-1}(x)}\right]^n.$

15. $\displaystyle\int \frac{1}{x} \left[\frac{Z_0(x)}{Z_1(x)}\right]^2 dx = \frac{Z_0(x)}{x Z_1(x)} - \ln x.$

16. $\displaystyle\int \frac{1}{x} \frac{[Z_\nu(x)]^2}{Z_{\nu-1}(x) Z_{\nu+1}(x)}\, dx = \frac{1}{2\nu} \ln \frac{Z_{\nu+1}(x)}{Z_{\nu-1}(x)} + \ln x.$

17. $\displaystyle\int \frac{[Z_\nu(x)]^{n-1}}{x \, [\overline{Z}_\nu(x)]^{n+1}}\, dx = \frac{1}{nx \left[Z'_\nu(x)\overline{Z}_\nu(x) - Z_\nu(x)\overline{Z}'_\nu(x)\right]} \left(\frac{Z_\nu(x)}{\overline{Z}_\nu(x)}\right)^n.$

18. $\displaystyle\int \frac{1}{x J_\nu^2(x)} f\!\left(\frac{H_\nu^{(1)}(x)}{J_\nu(x)}\right) dx = -\frac{\pi i}{2} \csc(\nu\pi) F\!\left(\frac{H_\nu^{(1)}(x)}{J_\nu(x)}\right) \qquad [f(x) = F'(x)].$

19. $\displaystyle\int \frac{1}{x J_\nu^2(x)} f\!\left(\frac{H_\nu^{(2)}(x)}{J_\nu(x)}\right) dx = \frac{\pi i}{2} \csc(\nu\pi) F\!\left(\frac{H_\nu^{(2)}(x)}{J_\nu(x)}\right) \qquad [f(x) = F'(x)].$

20. $\displaystyle\int \frac{1}{x [H_\nu^{(1)}(x)]^2} f\!\left(\frac{H_\nu^{(2)}(x)}{H_\nu^{(1)}(x)}\right) dx = \frac{\pi i}{4} F\!\left(\frac{H_\nu^{(2)}(x)}{H_\nu^{(1)}(x)}\right) \qquad [f(x) = F'(x)].$

21. $\displaystyle\int x^\mu I_\nu(x)\, dx = \sum_{k=0}^{n-1} 2^k \left(\frac{\nu-\mu+1}{2}\right)_k z^{\mu-k} I_{\nu+k+1}(x)$
$\qquad\qquad\qquad\qquad + 2^n \left(\frac{\nu-\mu+1}{2}\right)_n \displaystyle\int x^{\mu-n} I_{\nu+n}(x)\, dx.$

22. $\displaystyle\int x^\mu I_\nu(x)\, dx = \frac{(-2)^{-n}}{\left(\frac{1+\mu-\nu}{2}\right)_n}$
$\qquad \times \left[\displaystyle\int x^{\mu+n} I_{\nu-n}(x)\, dx - \sum_{k=1}^{n} 2^{n-k} \left(\frac{\nu-\mu+1}{2}-n\right)_{n-k} z^{\mu+k} I_{\nu-k+1}(x)\right].$

23. $\displaystyle\int x \ln x\, I_0^2(x)\, dx = \frac{x^2}{2} \ln x [I_0^2(x) - I_1^2(x)]$
$\qquad\qquad\qquad\qquad - \frac{x^2}{2}\left[I_0^2(x) - I_1^2(x) - \frac{1}{x} I_0(x) I_1(x)\right].$

24. $\displaystyle\int \frac{1}{x I_\nu^2(x)} f\!\left[\frac{I_{-\nu}(x)}{I_\nu(x)}\right] dx = -\frac{\pi}{2} \csc(\nu\pi) F\!\left[\frac{I_{-\nu}(x)}{I_\nu(x)}\right] \qquad [f(x) = F'(x)].$

25. $\int x^{n\nu} I_\nu^{n-1}(x) I_{\nu-1}(x) \, dx = \dfrac{1}{n} x^{n\nu} I_\nu^n(x).$

26. $\int x^{-n\nu} I_\nu^{n-1}(x) I_{\nu+1}(x) \, dx = \dfrac{1}{n} x^{-n\nu} I_\nu^n(x).$

27. $\int \dfrac{I_{\nu+1}(x)}{I_\nu(x)} \, dx = \ln I_\nu(x) - \nu \ln x.$

28. $\int \dfrac{I_{\nu-1}(x)}{I_\nu(x)} \, dx = \ln I_\nu(x) + \nu \ln x.$

29. $\int e^{\pm x} \ln x \, K_0(x) \, dx = x e^{\pm x} (\ln x - 2)[K_0(x) \pm K_1(x)] \pm e^{\pm x} K_0(x).$

30. $\int \dfrac{1}{x I_\nu^2(x)} f\left[\dfrac{K_\nu(x)}{I_\nu(x)}\right] dx = -F\left[\dfrac{K_\nu(x)}{I_\nu(x)}\right] \qquad [f(x) = F'(x)].$

31. $\int \dfrac{[I_\nu(x)]^{n-1}}{x \, [K_\nu(x)]^{n+1}} dx = \dfrac{1}{nx \, [I_\nu'(x) K_\nu(x) - I_\nu(x) K_\nu'(x)]} \left(\dfrac{I_\nu(x)}{K_\nu(x)}\right)^n.$

32. $\int \dfrac{K_{\nu-1}(x)}{K_\nu(x)} dx = -\ln K_\nu(x) - \nu \ln x.$

33. $\int \dfrac{K_{\nu+1}(x)}{K_\nu(x)} dx = -\ln K_\nu(x) + \nu \ln x.$

34. $\int x^{n\nu} K_\nu^{n-1}(x) K_{\nu-1}(x) \, dx = -\dfrac{1}{n} x^{n\nu} K_\nu^n(x).$

35. $\int x^{-n\nu} K_\nu^{n-1}(x) K_{\nu+1}(x) \, dx = -\dfrac{1}{n} x^{-n\nu} K_\nu^n(x).$

36. $\int K_n(x) K_{n+1}(x) \, dx = \dfrac{(-1)^{n+1}}{2} K_0^2(x) + (-1)^{n+1} \sum_{j=1}^{n} (-1)^j K_j^2(x).$

37. $\int x^{-p} K_\mu(x) K_\nu(x) \, dx = I(p, \mu, \nu),$

$I(p, \mu, \nu) = -\dfrac{1}{\mu + \nu + p - 1}[I(p-1, \mu-1, \nu) + I(p-1, \mu, \nu-1)$
$\qquad\qquad\qquad + x^{1-p} K_\mu(x) K_\nu(x)] \quad [\mu + \nu + p - 1 \neq 0].$

38. $I(p, \mu, \nu) = \dfrac{1}{\mu + \nu - p + 1}[I(p-1, \mu+1, \nu) + I(p-1, \mu, \nu+1)$
$\qquad\qquad\qquad + x^{1-p} K_\mu(x) K_\nu(x)] \quad [\mu + \nu - p + 1 \neq 0].$

39. $I(1, \mu, \nu) = -\dfrac{x}{\mu^2 - \nu^2}[K_\mu(x) K_\nu'(x) + K_\mu'(x) K_\nu(x)].$

40. $\quad = -\dfrac{1}{\mu+\nu} K_\mu(x) K_\nu(x) - \dfrac{x}{\mu^2-\nu^2}[K_{\mu-1}(x) K_\nu(x) - K_\mu(x) K_{\nu-1}(x)],$

$$I(1, m, m) = -\dfrac{1}{2m}\Big[(-1)^m K_0^2(x) + 2\sum_{j=1}^{m-1}(-1)^{j+m} K_j^2(x) + K_m^2(x)\Big] \quad [m \geq 1].$$

3.2.2. The Struve functions $H_\nu(z)$ and $L_\nu(z)$

1. $\displaystyle \int x^\mu\, L_\nu(x)\,dx = \sum_{k=0}^{n-1} 2^k \binom{\dfrac{1-\mu-\nu}{2}}{k} z^{\mu-k}\, L_{\nu-k-1}(x)$

$$-\dfrac{1}{\sqrt{\pi}} \sum_{k=0}^{n-1} \dfrac{2^{2k-\nu+1}\binom{\dfrac{1-\mu-\nu}{2}}{k} z^{\mu+\nu-2k}}{(\mu+\nu-2k)\Gamma\!\left(\nu - k + \dfrac{1}{2}\right)}$$

$$+ 2^n \binom{\dfrac{1-\mu-\nu}{2}}{n}\int x^{\mu-n}\, L_{\nu-n}(x)\,dx.$$

2. $\displaystyle \int x^{2n-\nu+1}\, L_\nu(x)\,dx = n!\, z^{2n-\nu+1}\sum_{k=0}^{n-1}\dfrac{\left(-\dfrac{2}{z}\right)^k}{(n-k)!}\, L_{\nu-k-1}(x)$

$$-\dfrac{n!\, z^{2n+1}}{2^{\nu-1}\sqrt{\pi}}\sum_{k=0}^{n-1}(-1)^k \dfrac{\left(\dfrac{2}{z}\right)^{2k}}{(n-k)!\,(2n-2k+1)\Gamma\!\left(\nu-k+\dfrac{1}{2}\right)}$$

$$+ (-2)^n n!\left[z^{n-\nu+1}\, L_{\nu-n-1}(x) - \dfrac{2^{n-\nu+1} z}{\sqrt{\pi}\,\Gamma\!\left(\nu-n+\dfrac{1}{2}\right)}\right].$$

3. $\displaystyle \int x\ln x\, H_0(x)\,dx = x\ln x\, H_1(x) + H_0(x) - \dfrac{2x}{\pi}.$

4. $\displaystyle \int x\ln x\, L_0(x)\,dx = x\ln x\, L_1(x) - L_0(x) + \dfrac{2x}{\pi}.$

3.2.3. The Airy functions $\mathrm{Ai}(z)$ and $\mathrm{Bi}(z)$

Notation: $y = a\,\mathrm{Ai}(x) + b\,\mathrm{Bi}(x)$, a, b are constants.

1. $\displaystyle \int x^n y\,dx = x^{n-1}y' - (n-1)x^{n-2}y + (n-1)(n-2)\int x^{n-3}y\,dx$

[[53], (4)].

2. $\displaystyle \int xy\,dx = y'$ \hfill [[53], (6)].

3. $\displaystyle \int x^2 y\,dx = xy' - y$ \hfill [[53], (7)].

4. $\displaystyle \int x^3 y\,dx = x^2 y' - 2xy + 2\int y\,dx$ \hfill [[53], (8)].

5. $\int x^4 y \, dx = x^3 y' - 3x^2 y + 6y'$ [[53], (9)].

6. $\int x^n y^2 \, dx = \dfrac{1}{2n+1} \Big[x^{n+1} y^2 - x^n y'^2 + nx^{n-1} yy' - \dfrac{n}{2}(n-1)x^{n-2} y^2$
$\qquad\qquad\qquad + \dfrac{n}{2}(n-1)(n-2) \int x^{n-3} y^2 \, dx \Big]$ [[53], (11)].

7. $\int y^2 \, dx = xy^2 - y'^2$ [[53], (12)].

8. $\int xy^2 \, dx = \dfrac{1}{3}\left(x^2 y^2 - xy'^2 + yy'\right)$ [[53], (13)).

9. $\int x^2 y^2 \, dx = \dfrac{1}{5}\left(x^3 y^2 - x^2 y'^2 + 2xyy' - y^2\right)$ [[53], (14)].

10. $\int x^n y^3 \, dx = x^{n-1} y^2 y'$
$\qquad - \dfrac{n-1}{3} x^{n-2} y^3 - \dfrac{2}{3} x^{n-2} y'^3 + \dfrac{1}{3}(n-1)(n-2) \int x^{n-3} y^3 \, dx$
$\qquad\qquad + \dfrac{2}{3}(n-2) \int x^{n-3} y'^3 \, dx$ [[53], (24)].

11. $\int x^n y^3 \, dx = x^{n-1} y^2 y'$
$\qquad - \dfrac{1}{9}(7n-11) x^{n-2} y^3 - \dfrac{2}{3} x^{n-2} y'^3 + \dfrac{2}{3}(n-2) x^{n-3} yy'^2$
$\qquad - \dfrac{1}{3}(n-2)(n-3) x^{n-4} y^2 y' + \dfrac{1}{9}(n-2)(n-3)(n-4) x^{n-5} y^3$
$+ \dfrac{10}{9}(n-2)^2 \int x^{n-3} y^3 \, dx - \dfrac{1}{9}(n-2)(n-3)(n-4)(n-5) \int x^{n-6} y^3 \, dx$
$\qquad\qquad\qquad\qquad\qquad\qquad$ [[53], (27)].

12. $\int x^2 y^3 \, dx = xy^2 y' - \dfrac{1}{3} y^3 - \dfrac{2}{3} y'^3$ [[53], (25)].

13. $\int x^3 y^3 \, dx = x^2 y^2 y' - \dfrac{10}{9} xy^3 - \dfrac{2}{3} xy'^3 + \dfrac{2}{3} yy'^2 + \dfrac{10}{9} \int y^3 \, dx$ [[53], (28)].

14. $\int x^4 y^3 \, dx = x^3 y^2 y'$
$\qquad - \dfrac{17}{9} x^2 y^3 - \dfrac{2}{3} x^2 y'^3 + \dfrac{4}{3} xyy'^2 - \dfrac{2}{3} y^2 y' + \dfrac{40}{9} \int xy^3 \, dx$ [[53], (29)].

15. $\int x^5 y^3 \, dx = x^4 y^2 y'$
$\qquad - \dfrac{8}{3} x^3 y^3 - \dfrac{2}{3} x^3 y'^3 + 2x^2 yy'^2 + 8xy^2 y' - \dfrac{8}{3} y^3 - \dfrac{20}{3} y'^3$ [[53], (30)].

16. $\int x^n y^4\, dx = \dfrac{1}{5n+3}\bigg[3x^{n+1}y^4 - 6x^n y^2 y'^2$
$\qquad + 2nx^{n-1}y^3 y' + 3x^{n-1}y'^4 - \dfrac{n}{2}(n-1)x^{n-2}y^4$
$\qquad + \dfrac{n}{2}(n-1)(n-2)\int x^{n-3}y^4\, dx - 3(n-1)\int x^{n-2}y'^4\, dx\bigg]$ [[53], (38)].

17. $\int x^n y^4\, dx = \dfrac{1}{8n}\bigg[3x^{n+1}y^4 - 6x^n y^2 y'^2 + (5n-3)x^{n-1}y^3 y'$
$\qquad + 3x^{n-1}y'^4 - 3(n-1)x^{n-2}yy'^3 - \dfrac{1}{4}(n-1)(5n-3)x^{n-2}y^4$
$\qquad + \dfrac{3}{4}(n-1)(n-2)x^{n-4}y'^4 - \dfrac{3}{4}(n-1)(n-2)(n-4)\int x^{n-5}y'^4\, dx$
$\qquad + \dfrac{1}{4}(n-1)(n-2)(5n-3)\int x^{n-3}y^4\, dx\bigg]$ [[53], (39)].

18. $\int xy^4\, dx = \dfrac{1}{8}(3x^2 y^4 - 6xy^2 y'^2 + 2xy^3 y' + 3y'^4)$ [[53], (40)].

19. $\int x^2 y^4\, dx = \dfrac{1}{16}\left(3x^3 y^4 - 6x^2 y^2 y'^2 + 7xy^3 y' + 3xy'^4 - 3yy'^3 - \dfrac{7}{4}y^4\right)$
[[53], (41)].

20. $\int x^n y'\, dx = x^n y - n\int x^{n-1} y\, dx$ [[53], (10)].

21. $\int yy'\, dx = \dfrac{1}{2}y^2$ [[53], (16)].

22. $\int xyy'\, dx = \dfrac{1}{2}y'^2$ [[53], (17)].

23. $\int x^2 yy'\, dx = \dfrac{1}{3}\left(\dfrac{1}{2}x^2 y^2 + xy'^2 - yy'\right)$ [[53], (18)].

24. $\int x^n y^2 y'\, dx = \dfrac{1}{3}x^n y^3 - \dfrac{n}{3}\int x^{n-1}y^3\, dx$ [[53], (34)].

25. $\int x^n y^3 y'\, dx = \dfrac{1}{4}x^n y^4 - \dfrac{n}{4}\int x^{n-1}y^4\, dx$ [[53], (10)].

26. $\int x^n y'^2\, dx = \dfrac{1}{2n+3}\bigg[-x^{n+2}y^2 + x^{n+1}y'^2 + (n+2)x^n yy'$
$\qquad - \dfrac{n}{2}(n+2)x^{n-1}y^2 + \dfrac{1}{2}(n-1)n(n+2)\int x^{n-2}y^2\, dx\bigg]$ [[53], (20)].

27. $\int y'^2\, dx = \dfrac{1}{3}(-x^2 y^2 + xy'^2 + 2yy')$ [[53], (21)].

28. $\int xy'^2\, dx = \dfrac{1}{5}\left(-x^3 y^2 + x^2 y'^2 + 3xyy' - \dfrac{3}{2}y^2\right)$ [[53], (22)].

29. $\int x^2 y'^2 \, dx = \dfrac{1}{7} \left(-x^4 y^2 + x^3 y'^2 + 4x^2 yy' - 4y'^2 \right)$ \hfill [[53], (23)].

30. $\int x^n yy'^2 \, dx = \dfrac{1}{2} x^n y^2 y' - \dfrac{n}{6} x^{n-1} y^3 - \dfrac{1}{2} \int x^{n+1} y^3 \, dx$
$\qquad\qquad + \dfrac{n}{6}(n-1) \int x^{n-2} y^3 \, dx$ \hfill [[53], (34)].

31. $\int x^3 y^2 y' \, dx = \dfrac{1}{3} x^3 y^3 - xy^2 y' + \dfrac{1}{3} y^3 + \dfrac{2}{3} y'^3$ \hfill [[53], (36)].

32. $\int xyy'^2 \, dx = \dfrac{1}{3} y'^3$ \hfill [[53], (37)].

33. $\int x^n y^2 y'^2 \, dx = \dfrac{1}{3} x^n y^3 y' - \dfrac{n}{12} x^{n-1} y^4 - \dfrac{1}{3} \int x^{n+1} y^4 \, dx$
$\qquad\qquad - \dfrac{n}{12}(n-1) \int x^{n-2} y^4 \, dx$ \hfill [[53], (47)].

34. $\int x^n y'^3 \, dx = -\dfrac{2}{3} x^{n+1} y^3 + x^n yy'^2 - \dfrac{n}{2} x^{n-1} y^2 y' + \dfrac{n}{6}(n-1) x^{n-2} y^3$
$\qquad + \dfrac{1}{6}(7n+4) \int x^n y^3 \, dx - \dfrac{n}{6}(n-1)(n-2) \int x^{n-3} y^3 \, dx$ \hfill [[53], (26)].

35. $\int y'^3 \, dx = -\dfrac{2}{3} xy^3 + yy'^2 + \dfrac{2}{3} \int y^3 \, dx$ \hfill [[53], (31)].

36. $\int xy'^3 \, dx = -\dfrac{2}{3} x^2 y^3 + xyy'^2 - \dfrac{1}{2} y^2 y' + \dfrac{11}{6} \int xy^3 \, dx$ \hfill [[53], (32)].

37. $\int x^2 y'^3 \, dx = -\dfrac{2}{3} x^3 y^3 + x^2 yy'^2 + 2xy^2 y' - \dfrac{2}{3} y^3 - 2y'^3$ \hfill [[53], (33)].

38. $\int x^n yy'^3 \, dx = \dfrac{1}{2} x^n y^2 y'^2 - \dfrac{n}{6} x^{n-1} y^3 y' - \dfrac{1}{4} x^{n+1} y^4 + \dfrac{n}{24}(n-1) x^{n-2} y^4$
$\qquad + \dfrac{5n+3}{12} \int x^n y^4 \, dx - \dfrac{n}{24}(n-1)(n-2) \int x^{n-3} y^4 \, dx$ \hfill [[53], (48)].

39. $\int x^n y'^4 \, dx = \dfrac{1}{3(n+1)} \Big[3x^{n+3} y^4 - 6x^{n+2} y^2 y'^2 + 2(n+2) x^{n+1} y^3 y'$
$\qquad + 3x^{n+1} y'^4 - \dfrac{1}{2}(n+1)(n+2) x^n y^4 - (5n+13) \int x^{n+2} y^4 \, dx$
$\qquad\qquad + \dfrac{n}{2}(n+1)(n+2) \int x^{n-1} y^4 \, dx$ \hfill [[53], (42)].

40. $\int y'^4 \, dx = \dfrac{3}{16} x^3 y^4 - \dfrac{3}{8} x^2 y^2 y'^2 - \dfrac{9}{16} xy^3 y' + \dfrac{3}{16} xy'^4$
$\qquad\qquad + \dfrac{13}{16} yy'^3 + \dfrac{9}{64} y^4$ \hfill [[53], (45)].

41. $\int \dfrac{1}{\mathrm{Ai}^2(x)} f\!\left(\dfrac{\mathrm{Bi}(x)}{\mathrm{Ai}(x)} \right) dx = \pi F\!\left(\dfrac{\mathrm{Bi}(x)}{\mathrm{Ai}(x)} \right)$ \hfill $[f(x) = F'(x)]$.

42. $\int \dfrac{1}{\text{Bi}^2(x)} f\left(\dfrac{\text{Ai}(x)}{\text{Bi}(x)}\right) dx = -\pi F\left(\dfrac{\text{Ai}(x)}{\text{Bi}(x)}\right)$ \qquad $[f(x) = F'(x)]$.

3.2.4. Various functions

1. $\int x^n \zeta(s,\, x)\, dx = -\dfrac{n!\, x^n}{s-1} \sum_{k=0}^{n} (-1)^k \dfrac{(2-s)_k}{(n-k)!}\, x^{-k} \zeta(s-k-1,\, x)$.

2. $\int e^x \Gamma^2(\nu,\, x)\, dx = \dfrac{2}{\nu} \left[x^\nu \Gamma(\nu,\, x) - \Gamma(2\nu,\, x)\right] + e^x \Gamma^2(\nu,\, x)$.

3. $\int \dfrac{x^{\nu-1} e^{-x}}{\gamma^2(\nu,\, x)} f\left[\dfrac{\Gamma(\nu,\, x)}{\gamma(\nu,\, x)}\right] dx = -\dfrac{1}{\Gamma(\nu)} F\left[\dfrac{\Gamma(\nu,\, x)}{\gamma(\nu,\, x)}\right]$ \qquad $[f(x) = F'(x)]$.

4. $\int \dfrac{1}{D_\nu^2(x)} f\left[\dfrac{D_\nu(-x)}{D_\nu(x)}\right] dx = \dfrac{\Gamma(-\nu)}{\sqrt{2\pi}} F\left[\dfrac{D_\nu(-x)}{D_\nu(x)}\right]$ \qquad $[f(x) = F'(x)]$.

5. $\int \dfrac{1}{D_\nu^2(x)} f\left[\dfrac{D_{-\nu-1}(ix)}{D_\nu(x)}\right] dx = i e^{i\nu\pi/2} F\left[\dfrac{D_{-\nu-1}(ix)}{D_\nu(x)}\right]$ \qquad $[f(x) = F'(x)]$.

6. $\int P_\mu(x) P_\nu(x)\, dx = -\dfrac{1}{(\mu-\nu)(\mu+\nu+1)}$
$\times \left[\mu P_{\mu-1}(x) P_\nu(x) - \nu P_\mu(x) P_{\nu-1}(x) - (\mu-\nu) z P_\mu(x) P_\nu(x)\right]$.

7. $\int [P_\nu(x)]^2\, dx = -\dfrac{1}{2\nu+1}$
$\times \left[P_{\nu-1}(x) P_\nu(x) - z[P_\nu(x)]^2 + \nu P_\nu(x) \dfrac{\partial P_{\nu-1}(x)}{\partial \nu} - \nu P_{\nu-1}(x) \dfrac{\partial P_\nu(x)}{\partial \nu}\right]$.

8. $\int x \ln x\, \mathbf{K}(x)\, dx = (x^2 \ln x - \ln x - x^2 + 1)\, \mathbf{K}(x) + (\ln x - 2)\, \mathbf{E}(x)$.

9. $\int x \ln x\, \mathbf{E}(x)\, dx = \dfrac{1}{3}\left(x^2 \ln x - \ln x - \dfrac{4}{3} x^2 + \dfrac{4}{3}\right) \mathbf{K}(x)$
$\hspace{2cm} + \dfrac{1}{3}\left(x^2 \ln x + \ln x - \dfrac{1}{3} x^2 - \dfrac{7}{3}\right) \mathbf{E}(x)$.

Chapter 4

Definite Integrals

4.1. Elementary Functions

4.1.1. Algebraic functions

Condition: $a > 0$.

1. $\displaystyle\int_0^a x^{s-1}(a-x)^{t-1}[1-bx(a-x)]^\nu \, dx = a^{s+t-1} \, \mathrm{B}(s,t)$

$$\times {}_3F_2\left(\begin{array}{c} s,\, t,\, -\nu;\, \dfrac{a^2 b}{4} \\ \dfrac{s+t}{2},\, \dfrac{s+t+1}{2} \end{array}\right) \qquad [\mathrm{Re}\, s,\, \mathrm{Re}\, t > 0;\, |\arg(4 - a^2 b)| < \pi].$$

2. $\displaystyle\int_0^a x^{-\nu-3/2}(a-x)^{-\nu-1/2}[1 - bx(a-x)]^\nu \, dx$

$$= \sqrt{\pi}\, \frac{\Gamma\!\left(-\nu - \dfrac{1}{2}\right)}{\Gamma(-\nu)} \left(\frac{4}{a^2} - b\right)^{\nu+1/2} \qquad [\mathrm{Re}\,\nu < -1/2;\, |\arg(4 - a^2 b)| < \pi].$$

3. $\displaystyle\int_0^a x^{-1/2}(a-x)^{-1/2}[1 - bx(a-x)]^n \, dx$

$$= \pi \left(1 - \frac{a^2 b}{4}\right)^{n/2} P_n\!\left(\frac{8 - a^2 b}{4\sqrt{4 - a^2 b}}\right) \qquad [a > 0].$$

4. $\displaystyle\int_0^a x^{-1/2}(a-x)^{1/2}[1 - bx(a-x)]^{1/2} \, dx = a\, \mathrm{E}\!\left(\frac{a\sqrt{b}}{2}\right)$

$$[|\arg(4 - a^2 b)| < \pi].$$

5. $\displaystyle\int_0^a x^{-1/2}(a-x)^{-1/2}[1 - bx(a-x)]^{1/2} \, dx = 2\, \mathrm{E}\!\left(\frac{a\sqrt{b}}{2}\right)$

$$[|\arg(4 - a^2 b)| < \pi].$$

6. $\displaystyle\int_0^a x^{1/2}(a-x)^{1/2}[1-bx(a-x)]^{-1/2}dx = \frac{2}{b}\left[\mathbf{K}\left(\frac{a\sqrt{b}}{2}\right) - \mathbf{E}\left(\frac{a\sqrt{b}}{2}\right)\right]$

$[|\arg(4-a^2b)| < \pi]$.

7. $\displaystyle\int_0^a x^{-1/2}(a-x)^{-1/2}[1-bx(a-x)]^{-1/2}dx = 2\,\mathbf{K}\left(\frac{a\sqrt{b}}{2}\right)$

$[|\arg(4-a^2b)| < \pi]$.

8. $\displaystyle\int_0^a x^{-1/4}(a-x)^{-3/4}[1-bx(a-x)]^{-1/2}dx$

$\displaystyle = \frac{4}{\sqrt{2+a\sqrt{b}}}\,\mathbf{K}\left(\frac{2^{1/2}a^{1/2}b^{1/4}}{\sqrt{2+a\sqrt{b}}}\right)$ $[|\arg(4-a^2b)| < \pi]$.

9. $\displaystyle\int_0^a x^{-1/4}(a-x)^{-3/4}[1-bx(a-x)]^{-1}dx$

$\displaystyle = \frac{\pi\sqrt{2}}{\sqrt{4-a^2b}}\left[\left(1+\frac{a\sqrt{b}}{2}\right)^{1/2} + \left(1-\frac{a\sqrt{b}}{2}\right)^{1/2}\right]$ $[|\arg(4-a^2b)| < \pi]$.

10. $\displaystyle\int_0^a x^{-1/2}(a-x)^{-1/2}[1-bx(a-x)]^{-1}dx = \frac{2\pi}{\sqrt{4-a^2b}}$ $[|\arg(4-a^2b)| < \pi]$.

11. $\displaystyle\int_0^a x^{1/2}(a-x)^{1/2}[1-bx(a-x)]^{-1}dx = \frac{\pi}{b}\left[\left(1-\frac{a^2b}{4}\right)^{-1/2} - 1\right]$

$[|\arg(4-a^2b)| < \pi]$.

12. $\displaystyle\int_0^a x^{1/2}(a-x)^{1/2}[1-bx(a-x)]^{-3/2}dx$

$\displaystyle = \frac{8}{b(4-a^2b)}\,\mathbf{E}\left(\frac{a\sqrt{b}}{2}\right) - \frac{2}{b}\,\mathbf{K}\left(\frac{a\sqrt{b}}{2}\right)$ $[|\arg(4-a^2b)| < \pi]$.

13. $\displaystyle\int_0^a \frac{x^{-1/2}(a-x)^{-1/2}}{1+\sqrt{1-bx(a-x)}}\,dx = 2\,\mathbf{K}\left(\frac{a\sqrt{b}}{2}\right) - 2\,\mathbf{D}\left(\frac{a\sqrt{b}}{2}\right)$

$[|\arg(4-a^2b)| < \pi]$.

14. $\displaystyle\int_0^a \frac{dx}{1+\sqrt{1-bx(a-x)}} = \frac{1}{\sqrt{b}}\ln\frac{2+a\sqrt{b}}{2-a\sqrt{b}} + \frac{2}{ab}\ln\left(1-\frac{a^2b}{4}\right)$

$[|\arg(4-a^2b)| < \pi]$.

15. $\displaystyle\int_0^a \frac{x^{1/2}(a-x)^{1/2}}{1+\sqrt{1-bx(a-x)}}\,dx = \frac{\pi}{b} - \frac{2}{b}\,\mathrm{E}\!\left(\frac{a\sqrt{b}}{2}\right)$ $[|\arg(4-a^2b)|<\pi]$.

16. $\displaystyle\int_0^a \frac{x^{-1/2}(a-x)^{1/2}}{1+\sqrt{1-bx(a-x)}}\,dx = a\,\mathrm{K}\!\left(\frac{a\sqrt{b}}{2}\right) - a\,\mathrm{D}\!\left(\frac{a\sqrt{b}}{2}\right)$

$[|\arg(4-a^2b)|<\pi]$.

17. $\displaystyle\int_0^a x^s(a-x)^{s+1/2}\bigl[1+b\sqrt{x(a-x)}\bigr]^\nu dx = 2^{-2s-1}\sqrt{\pi}\,a^{2s+3/2}\,\frac{\Gamma(2s+2)}{\Gamma\!\left(2s+\tfrac{5}{2}\right)}$

$\times\,_2F_1\!\left(\begin{matrix}-\nu,\,2s+2\\ 2s+\tfrac{5}{2};\,-\tfrac{ab}{2}\end{matrix}\right)$ $[\mathrm{Re}\,s>-1;\;|\arg(2+ab)|<\pi]$.

18. $\displaystyle\int_0^a x^{-3/4}(a-x)^{-1/4}\bigl[1+b\sqrt{x(a-x)}\bigr]^n dx$

$= \sqrt{2}\,\pi\!\left(1+\frac{ab}{2}\right)^{n/2} P_n\!\left(\frac{4+ab}{2^{3/2}\sqrt{2+ab}}\right)$ $[|\arg(2+ab)|<\pi]$.

19. $\displaystyle\int_0^a x^{1/2}\bigl[1-b\sqrt{x(a-x)}\bigr]^{1/2} dx$

$= \dfrac{\sqrt{a}}{4b}\!\left[\dfrac{1}{\sqrt{2ab}}\!\left(1+\dfrac{3ab}{2}\right)\!\left(1-\dfrac{ab}{2}\right)\ln\dfrac{\sqrt{2}+\sqrt{ab}}{\sqrt{2}-\sqrt{ab}} + \dfrac{3a}{2} - 1\right]$

$[|\arg(2+ab)|<\pi]$.

20. $\displaystyle\int_0^a x^{-1/2}\bigl[1-b\sqrt{x(a-x)}\bigr]^{1/2} dx$

$= \sqrt{a} + \dfrac{1}{\sqrt{2b}}\!\left(1-\dfrac{ab}{2}\right)\ln\dfrac{\sqrt{2}+\sqrt{ab}}{\sqrt{2}-\sqrt{ab}}$ $[|\arg(2-ab)|<\pi]$.

21. $\displaystyle\int_0^a x^{-1/4}(a-x)^{1/4}\bigl[1-b\sqrt{x(a-x)}\bigr]^{1/2} dx$

$= \dfrac{\sqrt{2}}{3b}\!\left[(2-ab)\,\mathrm{K}\!\left(\sqrt{\dfrac{ab}{2}}\right) + 2(ab-1)\,\mathrm{E}\!\left(\sqrt{\dfrac{ab}{2}}\right)\right]$ $[|\arg(2-ab)|<\pi]$.

22. $\displaystyle\int_0^a x^{-3/4}(a-x)^{-1/4}\bigl[1-b\sqrt{x(a-x)}\bigr]^{1/2} dx = 2^{3/2}\,\mathrm{E}\!\left(\sqrt{\dfrac{ab}{2}}\right)$

$[|\arg(2-ab)|<\pi]$.

23. $\int_0^a x^{1/2}\bigl[1-b\sqrt{x(a-x)}\bigr]^{-1/2}dx$

$$=\frac{1}{\sqrt{2}\,b^{3/2}}\left(1+\frac{ab}{2}\right)\ln\frac{\sqrt{2}+\sqrt{ab}}{\sqrt{2}-\sqrt{ab}}-\frac{\sqrt{a}}{b}\quad [|\arg(2+ab)|<\pi].$$

24. $\int_0^a x^{-1/2}\bigl[1+b\sqrt{x(a-x)}\bigr]^{-1/2}dx = \sqrt{\dfrac{8}{b}}\,\arctan\sqrt{\dfrac{ab}{2}}$

$$[|\arg(2-ab)|<\pi].$$

25. $\int_0^a x^{-1/2}\bigl[1-b\sqrt{x(a-x)}\bigr]^{-1/2}dx = \sqrt{\dfrac{2}{b}}\,\ln\dfrac{\sqrt{2}+\sqrt{ab}}{\sqrt{2}-\sqrt{ab}}$

$$[|\arg(2+ab)|<\pi].$$

26. $\int_0^a x^{-1/4}(a-x)^{1/4}\bigl[1-b\sqrt{x(a-x)}\bigr]^{-1/2}dx = \sqrt{2}\,a\,\mathrm{D}\!\left(\sqrt{\dfrac{ab}{2}}\right)$

$$[|\arg(2+ab)|<\pi].$$

27. $\int_0^a x^{-3/4}(a-x)^{-1/4}\bigl[1-b\sqrt{x(a-x)}\bigr]^{-1/2}dx = 2^{3/2}\,\mathrm{K}\!\left(\sqrt{\dfrac{ab}{2}}\right)$

$$[|\arg(2-ab)|<\pi].$$

28. $\int_0^a x^{1/2}\bigl[1-b\sqrt{x(a-x)}\bigr]^{-1}dx$

$$=\frac{2\sqrt{a}}{b}\left[\frac{2}{\sqrt{ab(2-ab)}}\arcsin\sqrt{\dfrac{ab}{2}}-1\right]\quad[|\arg(2-ab)|<\pi].$$

29. $\int_0^a x^{-1/2}\bigl[1-b\sqrt{x(a-x)}\bigr]^{-1}dx = \dfrac{4}{\sqrt{b(2-ab)}}\arcsin\sqrt{\dfrac{ab}{2}}$

$$[|\arg(2-ab)|<\pi].$$

30. $\int_0^a x^{-1/4}(a-x)^{1/4}\bigl[1+b\sqrt{x(a-x)}\bigr]^{-1}dx$

$$=\frac{\sqrt{2}\,\pi}{b}\left[1-\left(1+\frac{ab}{2}\right)^{-1/2}\right]\quad[|\arg(2+ab)|<\pi].$$

31. $\int_0^a x^{-3/4}(a-x)^{-1/4}\bigl[1+b\sqrt{x(a-x)}\bigr]^{-1}dx = \dfrac{2\pi}{\sqrt{ab+2}}$

$$[|\arg(2+ab)|<\pi].$$

32. $\int_0^a x^{-1/2}\left[1+b\sqrt{x(a-x)}\right]^{-3/2} dx = \dfrac{4\sqrt{a}}{ab+2}$ $\qquad [|\arg(2+ab)|<\pi].$

33. $\int_0^a x^{-1/4}(a-x)^{1/4}\left[1-b\sqrt{x(a-x)}\right]^{-3/2} dx$

$= -\dfrac{2\sqrt{2}}{b(ab-2)}\left[(ab-2)\mathbf{K}\left(\sqrt{\dfrac{ab}{2}}\right)+2\mathbf{E}\left(\sqrt{\dfrac{ab}{2}}\right)\right]$ $\qquad [|\arg(2+ab)|<\pi].$

34. $\int_0^a x^{-3/4}(a-x)^{-1/4}\left[1-b\sqrt{x(a-x)}\right]^{-3/2} dx = \dfrac{2^{5/2}}{2-ab}\mathbf{E}\left(\sqrt{\dfrac{ab}{2}}\right)$

$\qquad [|\arg(2-ab)|<\pi].$

35. $\int_0^a x^{1/2}\left[1-b\sqrt{x(a-x)}\right]^{-2} dx$

$= \dfrac{2\sqrt{a}}{b(2-ab)}+\dfrac{4(ab-1)}{b^{3/2}(2-ab)^{3/2}}\arcsin\sqrt{\dfrac{ab}{2}}$ $\qquad [|\arg(2-ab)|<\pi].$

36. $\int_0^a x^{-1/2}\left[1-b\sqrt{x(a-x)}\right]^{-2} dx$

$= \dfrac{2\sqrt{a}}{2-ab}+\dfrac{4}{\sqrt{b}\,(2-ab)^{3/2}}\arcsin\sqrt{\dfrac{ab}{2}}$ $\qquad [|\arg(2-ab)|<\pi].$

37. $\int_0^a x^{-1/4}(a-x)^{1/4}\left[1-b\sqrt{x(a-x)}\right]^{-2} dx = \dfrac{\pi a}{(2-ab)^{3/2}}$

$\qquad [|\arg(2-ab)|<\pi].$

38. $\int_0^a x^{-3/4}(a-x)^{-1/4}\left[1-b\sqrt{x(a-x)}\right]^{-2} dx = \dfrac{\pi(4-ab)}{(2-ab)^{3/2}}$

$\qquad [|\arg(2-ab)|<\pi].$

39. $\int_0^a x^{-1/2}\left[1-b\sqrt{x(a-x)}\right]^{-5/2} dx = \dfrac{4\sqrt{a}\,(6-ab)}{3(2-ab)^2}$ $\qquad [|\arg(2-ab)|<\pi].$

40. $\int_{-1}^1 \dfrac{(1-x^2)^{s-1/2}}{(1+2ax+a^2)^s(1+2bx+b^2)^s} dx = \dfrac{\sqrt{\pi}\,\Gamma\!\left(s+\dfrac{1}{2}\right)}{\Gamma(s+1)}\,{}_2F_1\!\left({s,\,2s \atop s+1;\,ab}\right)$

$\qquad [\operatorname{Re} s > -1/2].$

4.1.2. The exponential function

Condition: $a > 0$.

1. $\int_0^\infty x^n e^{-bx^2-cx}\,dx = 2^{-n}i^n b^{-(n+1)/2}\left[\frac{\sqrt{\pi}}{2}e^{c^2/(4b)}\operatorname{erfc}\left(\frac{c}{2\sqrt{b}}\right)H_n\left(\frac{ic}{2\sqrt{b}}\right)\right.$

$\left.+\sum_{k=1}^n \binom{n}{k}(-i)^k H_{n-k}\left(\frac{ic}{2\sqrt{b}}\right)H_{k-1}\left(\frac{c}{2\sqrt{b}}\right)\right]$ [Re $b > 0$].

2. $\int_0^a x^{s-1}(a-x)^{t-1}e^{bx(a-x)}\,dx = a^{s+t-1}\mathrm{B}(s,t)\,{}_2F_2\left(\begin{array}{c} s,\,t;\ \frac{a^2 b}{4} \\ \frac{s+t}{2},\ \frac{s+t+1}{2} \end{array}\right)$

[Re s, Re $t > 0$].

3. $\int_0^a e^{bx(a-x)}\,dx = \sqrt{\frac{\pi}{b}}\,e^{a^2 b/4}\operatorname{erf}\left(\frac{a\sqrt{b}}{2}\right).$

4. $\int_0^a x e^{bx(a-x)}\,dx = \frac{a}{2}\sqrt{\frac{\pi}{b}}\,e^{a^2 b/4}\operatorname{erf}\left(\frac{a\sqrt{b}}{2}\right).$

5. $\int_0^a x^2 e^{bx(a-x)}\,dx = \frac{a}{2b}\left[\frac{a^2 b+2}{2a}\sqrt{\frac{\pi}{b}}\,e^{a^2 b/4}\operatorname{erf}\left(\frac{a\sqrt{b}}{2}\right)-1\right].$

6. $\int_0^a x^{-1/2}(a-x)^{-1/2}e^{bx(a-x)}\,dx = \pi e^{a^2 b/8} I_0\left(\frac{a^2 b}{8}\right).$

7. $\int_0^a x^{1/2}(a-x)^{-1/2}e^{bx(a-x)}\,dx = \frac{\pi a}{2}e^{a^2 b/8} I_0\left(\frac{a^2 b}{8}\right).$

8. $\int_0^a x^{1/2}(a-x)^{1/2}e^{bx(a-x)}\,dx = \frac{\pi a^2}{8}e^{a^2 b/8}\left[I_0\left(\frac{a^2 b}{8}\right)+I_1\left(\frac{a^2 b}{8}\right)\right].$

9. $\int_0^a x^{3/2}(a-x)^{-1/2}e^{bx(a-x)}\,dx = \frac{\pi a^2}{8}e^{a^2 b/8}\left[3I_0\left(\frac{a^2 b}{8}\right)-I_1\left(\frac{a^2 b}{8}\right)\right].$

10. $\int_0^\infty x^{-1/2}\exp\left(ax - \frac{x^3}{12}\right)dx = \pi^{3/2}[\operatorname{Ai}^2(a)+\operatorname{Bi}^2(a)]$ [[65], (2.21)].

11. $\displaystyle\int_0^a x^s(a-x)^{s+1/2} e^{b\sqrt{x(a-x)}}\,dx$

$$= 2^{-2s-1}\sqrt{\pi}\, a^{2s+3/2}\, \frac{\Gamma(2s+2)}{\Gamma\left(2s+\frac{5}{2}\right)}\, {}_1F_1\left(\begin{array}{c}2s+2;\\ 2s+\frac{5}{2}\end{array}\frac{ab}{2}\right) \qquad [\mathrm{Re}\,s > -1].$$

12. $\displaystyle\int_0^a x^{1/2} e^{b\sqrt{x(a-x)}}\,dx = \frac{\sqrt{a}}{b} + (ab-1)\sqrt{\frac{\pi}{2b^3}}\, e^{ab/2}\,\mathrm{erf}\left(\sqrt{\frac{ab}{2}}\right).$

13. $\displaystyle\int_0^a x^{-1/2} e^{b\sqrt{x(a-x)}}\,dx = \sqrt{\frac{2\pi}{b}}\, e^{ab/2}\,\mathrm{erf}\left(\sqrt{\frac{ab}{2}}\right).$

14. $\displaystyle\int_0^a x^{-1/4}(a-x)^{1/4} e^{b\sqrt{x(a-x)}}\,dx = \frac{\pi a}{2^{3/2}}\, e^{ab/4}\left[I_0\left(\frac{ab}{4}\right) + I_1\left(\frac{ab}{4}\right)\right].$

15. $\displaystyle\int_0^a x^{-1/4}(a-x)^{-3/4} e^{b\sqrt{x(a-x)}}\,dx = \sqrt{2}\,\pi\, e^{ab/4}\, I_0\left(\frac{ab}{4}\right).$

16. $\displaystyle\int_{-\infty}^{\infty} e^{i(x^3/3+xz)}\,dx = 2\pi\,\mathrm{Ai}(z) \qquad [\mathrm{Im}\,z = 0;\ [66]].$

17. $\displaystyle\int_0^{\infty} \frac{x}{(x^2+z^2)(e^{2\pi x}+1)}\,dx = \frac{1}{2}\psi\left(z+\frac{1}{2}\right) - \frac{1}{2}\ln z \qquad [\mathrm{Re}\,z > 0].$

18. $\displaystyle\int_0^{\infty} \frac{x}{(x^2+z^2)^{n+1}(e^{2\pi x}+1)}\,dx = \frac{1}{4nz^{2n}}$

$$+ \frac{(-1)^n}{2(n!)}\sum_{k=0}^{n-1}(-1)^k \frac{(n+k-1)!}{k!(n-k-1)!}(2z)^{-n-k}\psi^{(n-k)}\left(z+\frac{1}{2}\right) \qquad [\mathrm{Re}\,z > 0].$$

4.1.3. Hyperbolic functions

Condition: $a > 0$.

1. $\displaystyle\int_0^a x^{s-1}(a-x)^{t-1}\sinh\left(b\sqrt{x(a-x)}\right)dx = a^{s+t} b\, \mathrm{B}\left(s+\frac{1}{2},\, t+\frac{1}{2}\right)$

$$\times {}_2F_3\left(\begin{array}{c}s+\frac{1}{2},\, t+\frac{1}{2};\\ \frac{3}{2},\, \frac{s+t+1}{2},\, \frac{s+t}{2}+1\end{array}\frac{a^2 b^2}{16}\right) \qquad [\mathrm{Re}\,s,\,\mathrm{Re}\,t > -1/2].$$

2. $\int_0^a \sinh\big(b\sqrt{x(a-x)}\big)\,dx = \dfrac{\pi a}{2} I_1\!\left(\dfrac{ab}{2}\right).$

3. $\int_0^a x^{-1/2} \sinh\big(b\sqrt{x(a-x)}\big)\,dx$

$$= \sqrt{\dfrac{\pi}{2b}}\left[e^{ab/2}\,\mathrm{erf}\!\left(\sqrt{\dfrac{ab}{2}}\right) - e^{-ab/2}\,\mathrm{erfi}\!\left(\sqrt{\dfrac{ab}{2}}\right)\right].$$

4. $\int_0^a x^{-1} \sinh\big(b\sqrt{x(a-x)}\big)\,dx$

$$= \dfrac{\pi ab}{4}\left\{-\pi I_1\!\left(\dfrac{ab}{2}\right)\mathbf{L}_0\!\left(\dfrac{ab}{2}\right) + I_0\!\left(\dfrac{ab}{2}\right)\left[2 + \pi \mathbf{L}_1\!\left(\dfrac{ab}{2}\right)\right]\right\}.$$

5. $\int_0^a x^{1/2}(a-x)^{1/2} \sinh\big(b\sqrt{x(a-x)}\big)\,dx = \dfrac{\pi a}{4b}\left[ab\,\mathbf{L}_0\!\left(\dfrac{ab}{2}\right) - 2\mathbf{L}_1\!\left(\dfrac{ab}{2}\right)\right].$

6. $\int_0^a x^{1/2}(a-x)^{-1/2} \sinh\big(b\sqrt{x(a-x)}\big)\,dx = \dfrac{\pi a}{2}\mathbf{L}_0\!\left(\dfrac{ab}{2}\right).$

7. $\int_0^a x^{-1/2}(a-x)^{-1/2} \sinh\big(b\sqrt{x(a-x)}\big)\,dx = \pi\,\mathbf{L}_0\!\left(\dfrac{ab}{2}\right).$

8. $\int_0^a x^{-1/2}(a-x)^{-1} \sinh\big(b\sqrt{x(a-x)}\big)\,dx = \dfrac{\pi}{a^{1/2}}\,\mathrm{erf}\!\left(\sqrt{\dfrac{ab}{2}}\right)\mathrm{erfi}\!\left(\sqrt{\dfrac{ab}{2}}\right).$

9. $\int_0^a x^{-3/4}(a-x)^{-3/4} \sinh\big(b\sqrt{x(a-x)}\big)\,dx = \sqrt{\dfrac{\pi^3 b}{2}}\,I_{1/4}^2\!\left(\dfrac{ab}{4}\right).$

10. $\int_0^a x^{-1}(a-x)^{-1} \sinh\big(b\sqrt{x(a-x)}\big)\,dx$

$$= \dfrac{\pi b}{2}\left\{-\pi I_1\!\left(\dfrac{ab}{2}\right)\mathbf{L}_0\!\left(\dfrac{ab}{2}\right) + I_0\!\left(\dfrac{ab}{2}\right)\left[2 + \pi \mathbf{L}_1\!\left(\dfrac{ab}{2}\right)\right]\right\}.$$

11. $\int_0^a x^{-5/4}(a-x)^{-5/4} \sinh\big(b\sqrt{x(a-x)}\big)\,dx$

$$= \sqrt{\dfrac{\pi^3 b^3}{2}}\left[I_{-1/4}^2\!\left(\dfrac{ab}{4}\right) - I_{3/4}^2\!\left(\dfrac{ab}{4}\right)\right].$$

4.1.3] 4.1. ELEMENTARY FUNCTIONS

12. $\int_0^a x^s (a-x)^{s+1/2} \sinh\left(b\sqrt[4]{x(a-x)}\right) dx$

$$= 2^{-2s-3/2}\sqrt{\pi}\, a^{2s+2} b\, \frac{\Gamma\left(2s+\frac{5}{2}\right)}{\Gamma(2s+3)} \,_1F_2\left(\begin{array}{c} 2s+\frac{5}{2}; \frac{ab^2}{8} \\ \frac{3}{2}, 2s+3 \end{array}\right) \quad [\operatorname{Re} s > -5/4].$$

13. $\int_0^a x^{-1/4}(a-x)^{1/4} \sinh\left(b\sqrt[4]{x(a-x)}\right) dx = \frac{\pi a}{\sqrt{2}} \mathbf{L}_0\left(b\sqrt{\frac{a}{2}}\right)$

$$- \frac{\pi a^{1/2}}{b}\mathbf{L}_1\left(b\sqrt{\frac{a}{2}}\right).$$

14. $\int_0^a x^{-1/2} \sinh\left(b\sqrt[4]{x(a-x)}\right) dx = \pi a^{1/2} I_1\left(b\sqrt{\frac{a}{2}}\right).$

15. $\int_0^a x^{1/2} \sinh\left(b\sqrt[4]{x(a-x)}\right) dx = \frac{\pi a}{2b}\left[3\sqrt{2}\,I_2\left(b\sqrt{\frac{a}{2}}\right) + \sqrt{a}\,b\,I_3\left(b\sqrt{\frac{a}{2}}\right)\right].$

16. $\int_0^a x^{-1/4}(a-x)^{-3/4} \sinh\left(b\sqrt[4]{x(a-x)}\right) dx = \sqrt{2}\,\pi\, \mathbf{L}_0\left(b\sqrt{\frac{a}{2}}\right).$

17. $\int_0^a x^{-1/2}(a-x)^{-1} \sinh\left(b\sqrt[4]{x(a-x)}\right) dx = \frac{\pi b}{\sqrt{2}}$

$$\times \left\{2I_0\left(b\sqrt{\frac{a}{2}}\right) + \pi\left[I_0\left(b\sqrt{\frac{a}{2}}\right)\mathbf{L}_1\left(b\sqrt{\frac{a}{2}}\right) - I_1\left(b\sqrt{\frac{a}{2}}\right)\mathbf{L}_0\left(b\sqrt{\frac{a}{2}}\right)\right]\right\}.$$

18. $\int_0^a x^{s-1}(a-x)^{t-1} \cosh\left(b\sqrt{x(a-x)}\right) dx$

$$= a^{s+t-1} \mathrm{B}(s,t) \,_2F_3\left(\begin{array}{c} s, t; \frac{a^2 b^2}{16} \\ \frac{1}{2}, \frac{s+t}{2}, \frac{s+t+1}{2} \end{array}\right) \quad [\operatorname{Re} s, \operatorname{Re} t > 0].$$

19. $\int_0^a \cosh\left(b\sqrt{x(a-x)}\right) dx = \frac{\pi a}{2} \mathbf{L}_{-1}\left(\frac{ab}{2}\right).$

20. $\int_0^a x \cosh\left(b\sqrt{x(a-x)}\right) dx = \frac{\pi a^2}{4} \mathbf{L}_{-1}\left(\frac{ab}{2}\right).$

21. $\displaystyle\int_0^a x^2 \cosh\left(b\sqrt{x(a-x)}\right) dx$

$$= \frac{a}{8b^2}\left[2\pi ab\,\mathbf{L}_0\left(\frac{ab}{2}\right) + \pi(a^2b^2-8)\,\mathbf{L}_1\left(\frac{ab}{2}\right) + 2a^2b^2\right].$$

22. $\displaystyle\int_0^a x^{1/2} \cosh\left(b\sqrt{x(a-x)}\right) dx$

$$= \sqrt{\frac{\pi}{8b^3}}\left[(ab-1)e^{ab/2}\,\mathrm{erf}\left(\sqrt{\frac{ab}{2}}\right) + (ab+1)e^{-ab/2}\,\mathrm{erfi}\left(\sqrt{\frac{ab}{2}}\right)\right].$$

23. $\displaystyle\int_0^a x^{-1/2} \cosh\left(b\sqrt{x(a-x)}\right) dx$

$$= \sqrt{\frac{\pi}{2b}}\left[e^{ab/2}\,\mathrm{erf}\left(\sqrt{\frac{ab}{2}}\right) + e^{-ab/2}\,\mathrm{erfi}\left(\sqrt{\frac{ab}{2}}\right)\right].$$

24. $\displaystyle\int_0^a x^{1/2}(a-x)^{1/2}\cosh\left(b\sqrt{x(a-x)}\right) dx = \frac{\pi a}{4b}\left[2I_1\left(\frac{ab}{2}\right) + ab\,I_2\left(\frac{ab}{2}\right)\right].$

25. $\displaystyle\int_0^a x^{1/2}(a-x)^{-1/2}\cosh\left(b\sqrt{x(a-x)}\right) dx = \frac{\pi a}{2}I_0\left(\frac{ab}{2}\right).$

26. $\displaystyle\int_0^a x^{-1/2}(a-x)^{-1/2}\cosh\left(b\sqrt{x(a-x)}\right) dx = \pi I_0\left(\frac{ab}{2}\right).$

27. $\displaystyle\int_0^a x^{-1/4}(a-x)^{-1/4}\cosh\left(b\sqrt{x(a-x)}\right) dx$

$$= \sqrt{\frac{\pi^3}{8b}}\,I_{1/4}\left(\frac{ab}{4}\right)\left[2I_{1/4}\left(\frac{ab}{4}\right) + ab\,I_{5/4}\left(\frac{ab}{4}\right)\right].$$

28. $\displaystyle\int_0^a x^{-1/4}(a-x)^{-3/4}\cosh\left(b\sqrt{x(a-x)}\right) dx = \sqrt{2}\,\pi\cosh\frac{ab}{4}\,I_0\left(\frac{ab}{4}\right).$

29. $\displaystyle\int_0^a x^{-3/4}(a-x)^{-3/4}\cosh\left(b\sqrt{x(a-x)}\right) dx = \sqrt{\frac{\pi^3 b}{2}}\,I_{-1/4}^2\left(\frac{ab}{4}\right).$

30. $\displaystyle\int_0^a x^s(a-x)^{s+1/2}\cosh\left(b\sqrt[4]{x(a-x)}\right)dx$

$$= 2^{-2s-1}\sqrt{\pi}\, a^{2s+3/2}\frac{\Gamma(2s+2)}{\Gamma\left(2s+\frac{5}{2}\right)}\, _1F_2\left(\begin{array}{c}2s+2;\ \frac{ab^2}{8}\\ \frac{1}{2},\, 2s+\frac{5}{2}\end{array}\right)\qquad [\mathrm{Re}\, s>-1].$$

31. $\displaystyle\int_0^a x^{1/2}\cosh\left(b\sqrt[4]{x(a-x)}\right)dx$

$$=\frac{\sqrt{a}\,\pi}{2b^2}\left[ab^2\,\mathbf{L}_{-1}\left(b\sqrt{\frac{a}{2}}\right)-\sqrt{2a}\,b\,\mathbf{L}_0\left(b\sqrt{\frac{a}{2}}\right)+4\,\mathbf{L}_1\left(b\sqrt{\frac{a}{2}}\right)\right].$$

32. $\displaystyle\int_0^a x^{-1/2}\cosh\left(b\sqrt[4]{x(a-x)}\right)dx=\pi\sqrt{a}\,\mathbf{L}_{-1}\left(b\sqrt{\frac{a}{2}}\right).$

33. $\displaystyle\int_0^a x^{1/4}(a-x)^{-1/4}\cosh\left(b\sqrt[4]{x(a-x)}\right)dx$

$$=\frac{\pi\sqrt{a}}{b}I_1\left(b\sqrt{\frac{a}{2}}\right)+\frac{\pi a}{\sqrt{2}}I_2\left(b\sqrt{\frac{a}{2}}\right).$$

34. $\displaystyle\int_0^a x^{-1/4}(a-x)^{-3/4}\cosh\left(b\sqrt[4]{x(a-x)}\right)dx=\sqrt{2}\,\pi I_0\left(b\sqrt{\frac{a}{2}}\right).$

35. $\displaystyle\int_0^\infty\frac{x^{2n}}{e^x-1}\left[2^{2n+1}\sinh(2bx)-\sinh(bx)\right]dx$

$$=\frac{1}{2}\left[\psi^{(2n)}\left(\frac{1}{2}+b\right)-\psi^{(2n)}\left(\frac{1}{2}-b\right)\right]\qquad[\mathrm{Re}\,b<1/2].$$

36. $\displaystyle\int_0^\infty\frac{x^{2n+1}}{e^x-1}\left[2^{2n+1}\cosh(2bx)-\cosh(bx)\right]dx$

$$=\frac{1}{2}\left[\psi^{(2n+1)}\left(\frac{1}{2}+b\right)+\psi^{(2n+1)}\left(\frac{1}{2}-b\right)\right]\qquad[\mathrm{Re}\,b<1/2].$$

4.1.4. Trigonometric functions

1. $\displaystyle\int_0^\infty\frac{\sin(ax)}{x\sqrt{x^2+1}}dx=\frac{\pi a}{2}[K_0(a)\,\mathbf{L}_{-1}(a)+K_1(a)\,\mathbf{L}_0(a)]\qquad[a>0].$

2. $\displaystyle\int_0^{\pi/2}\frac{\sin^{2n}x}{\sin^{2n}x+\cos^{2n}x}dx=\frac{\pi}{4}.$

3. $\displaystyle\int_0^a \sin^{\nu-2}(a-x)\sin(\nu x)\,dx = \frac{1}{\nu-1}\sin^\nu a$ $\qquad [a>0;\ \operatorname{Re}\nu>1].$

4. $\displaystyle\int_0^a \sin^{\nu-2}(a-x)\cos(\nu x)\,dx = \frac{1}{\nu-1}\sin^{\nu-1}a\cos a$ $\qquad [a>0;\ \operatorname{Re}\nu>1].$

5. $\displaystyle\int_0^a (\cos x-\cos a)^{\nu-1}\cos(\nu x)\,dx = 2^{\nu-1}\,\mathrm{B}(\nu,\nu)\sin^{2\nu-1}a$ $\qquad [a,\ \operatorname{Re}\nu>0].$

6. $\displaystyle\int_0^{\pi/2}\frac{\sin^6(nx)}{\sin^6 x}\,dx = \frac{n\pi}{40}(11n^4+5n^2+4).$

7. $\displaystyle\int_0^{\pi/2}\sin^2[(n+1)x]\,\frac{\sin^4(nx)}{\sin^4 x}\,dx = \frac{n^2\pi}{8}(n+1).$

8. $\displaystyle\int_0^{\pi/2}\frac{\sin^3[(2n+1)x]}{\sin^3 x}\,dx = \frac{\pi}{2}+\frac{3n\pi}{2}(n+1).$

9. $\displaystyle\int_0^{\pi/2}\frac{\sin[(2n+1)x]}{\sin^3 x}\sin^2(nx)\sin^2[(n+1)x]\,dx = \frac{n\pi}{8}(n+1).$

10. $\displaystyle\int_0^\infty x^{-1/2}\cos\left(\frac{x^3}{12}+ax-\frac{b}{x}+\frac{\pi}{4}\right)dx$
$\qquad\qquad = 2\pi^{3/2}\,\mathrm{Ai}\left(a+\sqrt{b}\right)\mathrm{Ai}\left(a-\sqrt{b}\right)$ $\qquad [[65],\,(7)].$

11. $\displaystyle\int_0^a x^{s-1}(a-x)^{t-1}\sin\left(b\sqrt{x(a-x)}\right)dx = a^{s+t}b\,\mathrm{B}\!\left(s+\tfrac{1}{2},\,t+\tfrac{1}{2}\right)$
$\qquad\qquad \times {}_2F_3\!\left(\begin{array}{c} s+\tfrac{1}{2},\,t+\tfrac{1}{2};\ -\tfrac{a^2b^2}{16} \\ \tfrac{3}{2},\,\tfrac{s+t+1}{2},\,\tfrac{s+t}{2}+1 \end{array}\right)$ $\qquad [a>0;\ \operatorname{Re}s,\operatorname{Re}t>-1/2].$

12. $\displaystyle\int_0^a \sin\left(b\sqrt{x(a-x)}\right)dx = \frac{\pi a}{2}J_1\!\left(\frac{ab}{2}\right)$ $\qquad [a>0].$

13. $\displaystyle\int_0^a x^{1/2}\sin\left(b\sqrt{x(a-x)}\right)dx = \sqrt{\frac{\pi}{b^3}}\left[\left(ab\sin\frac{ab}{2}+\cos\frac{ab}{2}\right)C\!\left(\frac{ab}{2}\right)\right.$
$\qquad\qquad \left. -\left(ab\cos\frac{ab}{2}-\sin\frac{ab}{2}\right)S\!\left(\frac{ab}{2}\right)-\sqrt{\frac{ab}{\pi}}\,\right]$ $\qquad [a>0].$

14. $\displaystyle\int_0^a x^{-1/2} \sin\left(b\sqrt{x(a-x)}\right) dx$

$$= 2\sqrt{\frac{\pi}{b}} \left[\sin\frac{ab}{2} C\left(\frac{ab}{2}\right) - \cos\frac{ab}{2} S\left(\frac{ab}{2}\right)\right] \quad [a > 0].$$

15. $\displaystyle\int_0^a x^{-1} \sin\left(b\sqrt{x(a-x)}\right) dx$

$$= \frac{\pi ab}{4}\left\{\pi J_1\left(\frac{ab}{2}\right)\mathbf{H}_0\left(\frac{ab}{2}\right) + J_0\left(\frac{ab}{2}\right)\left[2 - \pi \mathbf{H}_1\left(\frac{ab}{2}\right)\right]\right\} \quad [a > 0].$$

16. $\displaystyle\int_0^a x^{1/2}(a-x)^{1/2} \sin\left(b\sqrt{x(a-x)}\right) dx$

$$= \frac{\pi a}{4b}\left[ab\,\mathbf{H}_0\left(\frac{ab}{2}\right) - 2\mathbf{H}_1\left(\frac{ab}{2}\right)\right] \quad [a > 0].$$

17. $\displaystyle\int_0^a x^{1/2}(a-x)^{-1/2} \sin\left(b\sqrt{x(a-x)}\right) dx = \frac{\pi a}{2} \mathbf{H}_0\left(\frac{ab}{2}\right) \quad [a > 0].$

18. $\displaystyle\int_0^a x^{-1/2}(a-x)^{-1/2} \sin\left(b\sqrt{x(a-x)}\right) dx = \pi \mathbf{H}_0\left(\frac{ab}{2}\right) \quad [a > 0].$

19. $\displaystyle\int_0^a x^{-3/4}(a-x)^{-3/4} \sin\left(b\sqrt{x(a-x)}\right) dx = \sqrt{\frac{\pi^3 b}{2}} J_{1/4}^2\left(\frac{ab}{4}\right) \quad [a > 0].$

20. $\displaystyle\int_0^a x^{-1}(a-x)^{-1} \sin\left(b\sqrt{x(a-x)}\right) dx$

$$= \frac{\pi b}{2}\left\{\pi J_1\left(\frac{ab}{2}\right)\mathbf{H}_0\left(\frac{ab}{2}\right) + J_0\left(\frac{ab}{2}\right)\left[2 - \pi \mathbf{H}_1\left(\frac{ab}{2}\right)\right]\right\} \quad [a > 0].$$

21. $\displaystyle\int_0^a x^{-5/4}(a-x)^{-5/4} \sin\left(b\sqrt{x(a-x)}\right) dx$

$$= \sqrt{\frac{\pi^3 b^3}{2}} \left[J_{-1/4}^2\left(\frac{ab}{4}\right) + J_{3/4}^2\left(\frac{ab}{4}\right)\right] \quad [a > 0].$$

22. $\displaystyle\int_0^a x^s(a-x)^{s+1/2} \sin\left(b\sqrt[4]{x(a-x)}\right) dx = 2^{-2s-3/2}\sqrt{\pi}\, a^{2s+2} b$

$$\times \frac{\Gamma\left(2s+\frac{5}{2}\right)}{\Gamma(2s+3)} {}_1F_2\left(\begin{matrix} 2s+\frac{5}{2}; & -\frac{ab^2}{8} \\ \frac{3}{2}, 2s+3 & \end{matrix}\right) \quad [a > 0;\ \text{Re}\,s > -5/4].$$

23. $\int_0^a x^{1/2} \sin\left(b \sqrt[4]{x(a-x)}\right) dx = \dfrac{\pi a}{2b} \left[3\sqrt{2}\, J_2\!\left(b\sqrt{\dfrac{a}{2}}\right) - \sqrt{a}\, b\, J_3\!\left(b\sqrt{\dfrac{a}{2}}\right)\right]$ $\qquad [a > 0]$.

24. $\int_0^a x^{-1/2} \sin\left(b \sqrt[4]{x(a-x)}\right) dx = \pi a^{1/2} J_1\!\left(b\sqrt{\dfrac{a}{2}}\right)$ $\qquad [a > 0]$.

25. $\int_0^a x^{-1/4}(a-x)^{1/4} \sin\left(b \sqrt[4]{x(a-x)}\right) dx$

$\qquad = \dfrac{\pi a}{\sqrt{2}}\, \mathbf{H}_0\!\left(b\sqrt{\dfrac{a}{2}}\right) - \dfrac{\pi a^{1/2}}{b}\, \mathbf{H}_1\!\left(b\sqrt{\dfrac{a}{2}}\right)$ $\qquad [a > 0]$.

26. $\int_0^a x^{-1/4}(a-x)^{-3/4} \sin\left(b \sqrt[4]{x(a-x)}\right) dx = \sqrt{2}\,\pi\, \mathbf{H}_0\!\left(b\sqrt{\dfrac{a}{2}}\right)$ $\qquad [a > 0]$.

27. $\int_0^a x^{-1/2}(a-x)^{-1} \sin\left(b \sqrt[4]{x(a-x)}\right) dx = \dfrac{\pi b}{\sqrt{2}}\left\{ 2 J_0\!\left(b\sqrt{\dfrac{a}{2}}\right)\right.$

$\qquad \left. + \pi \left[J_1\!\left(b\sqrt{\dfrac{a}{2}}\right) \mathbf{H}_0\!\left(b\sqrt{\dfrac{a}{2}}\right) - J_0\!\left(b\sqrt{\dfrac{a}{2}}\right) \mathbf{H}_1\!\left(b\sqrt{\dfrac{a}{2}}\right) \right]\right\}$ $\qquad [a > 0]$.

28. $\int_0^a x^{s-1}(a-x)^{t-1} \cos\left(b\sqrt{x(a-x)}\right) dx$

$\qquad = a^{s+t-1}\, \mathrm{B}(s,t)\, {}_2F_3\!\left(\begin{matrix} s,\,t;\, -\dfrac{a^2 b^2}{16} \\ \dfrac{1}{2},\, \dfrac{s+t}{2},\, \dfrac{s+t+1}{2} \end{matrix}\right)$ $\qquad [a,\, \operatorname{Re} s,\, \operatorname{Re} t > 0]$.

29. $\int_0^a \cos\left(b\sqrt{x(a-x)}\right) dx = \dfrac{\pi a}{2}\, \mathbf{H}_{-1}\!\left(\dfrac{ab}{2}\right)$ $\qquad [a > 0]$.

30. $\int_0^a x \cos\left(b\sqrt{x(a-x)}\right) dx = \dfrac{\pi a^2}{4}\, \mathbf{H}_{-1}\!\left(\dfrac{ab}{2}\right)$ $\qquad [a > 0]$.

31. $\int_0^a x^2 \cos\left(b\sqrt{x(a-x)}\right) dx$

$\qquad = \dfrac{a}{8b^2} \left[2\pi ab\, \mathbf{H}_0\!\left(\dfrac{ab}{2}\right) - \pi(a^2 b^2 + 8)\, \mathbf{H}_1\!\left(\dfrac{ab}{2}\right) + 2a^2 b^2 \right]$ $\qquad [a > 0]$.

32. $\int_0^a x^{-1/2} \cos\left(b\sqrt{x(a-x)}\right) dx$

$$= 2\sqrt{\frac{\pi}{b}} \left[\cos \frac{ab}{2} C\left(\frac{ab}{2}\right) + \sin \frac{ab}{2} S\left(\frac{ab}{2}\right)\right] \quad [a>0].$$

33. $\int_0^a x^{1/2}(a-x)^{1/2} \cos\left(b\sqrt{x(a-x)}\right) dx = \frac{\pi a}{4b}\left[2J_1\left(\frac{ab}{2}\right) - ab J_2\left(\frac{ab}{2}\right)\right]$

$[a>0].$

34. $\int_0^a x^{1/2}(a-x)^{-1/2} \cos\left(b\sqrt{x(a-x)}\right) dx = \frac{\pi a}{2} J_0\left(\frac{ab}{2}\right) \quad [a>0].$

35. $\int_0^a x^{-1/2}(a-x)^{-1/2} \cos\left(b\sqrt{x(a-x)}\right) dx = \pi J_0\left(\frac{ab}{2}\right) \quad [a>0].$

36. $\int_0^a x^{-1/4}(a-x)^{-1/4} \cos\left(b\sqrt{x(a-x)}\right) dx$

$$= \sqrt{\frac{\pi^3}{8b}} J_{1/4}\left(\frac{ab}{4}\right) \left[2J_{1/4}\left(\frac{ab}{4}\right) - ab J_{5/4}\left(\frac{ab}{4}\right)\right] \quad [a>0].$$

37. $\int_0^a x^{-1/4}(a-x)^{-3/4} \cos\left(b\sqrt{x(a-x)}\right) dx = \sqrt{2}\pi \cos\frac{ab}{4} J_0\left(\frac{ab}{4}\right)$

$[a>0].$

38. $\int_0^a x^{-3/4}(a-x)^{-3/4} \cos\left(b\sqrt{x(a-x)}\right) dx = \sqrt{\frac{\pi^3 b}{2}} J_{-1/4}^2\left(\frac{ab}{4}\right) \quad [a>0].$

39. $\int_0^a x^s(a-x)^{s+1/2} \cos\left(b\sqrt[4]{x(a-x)}\right) dx = 2^{-2s-1}\sqrt{\pi}\, a^{2s+3/2}$

$$\times \frac{\Gamma(2s+2)}{\Gamma\left(2s+\frac{5}{2}\right)} {}_1F_2\left(\begin{array}{c}2s+2;\ -\frac{ab^2}{8}\\ \frac{1}{2},\ 2s+\frac{5}{2}\end{array}\right) \quad [a>0;\ \operatorname{Re} s>-1].$$

40. $\int_0^a x^{1/2} \cos\left(b\sqrt[4]{x(a-x)}\right) dx = \frac{a^{1/2}\pi}{2b^2}$

$$\times \left[ab^2 \mathbf{H}_{-1}\left(b\sqrt{\frac{a}{2}}\right) - \sqrt{2a}\, b\, \mathbf{H}_0\left(b\sqrt{\frac{a}{2}}\right) + 4\mathbf{H}_1\left(b\sqrt{\frac{a}{2}}\right)\right] \quad [a>0].$$

41. $\int_0^a x^{-1/2} \cos\left(b\sqrt[4]{x(a-x)}\right) dx = \pi\sqrt{a}\, \mathbf{H}_{-1}\left(b\sqrt{\frac{a}{2}}\right) \quad [a>0].$

42. $\int\limits_0^a x^{1/4}(a-x)^{-1/4}\cos\left(b\sqrt[4]{x(a-x)}\right)dx$

$$=\frac{\pi\sqrt{a}}{b}J_1\left(b\sqrt{\frac{a}{2}}\right)-\frac{\pi a}{\sqrt{2}}J_2\left(b\sqrt{\frac{a}{2}}\right)\qquad [a>0].$$

43. $\int\limits_0^a x^{-1/4}(a-x)^{-3/4}\cos\left(b\sqrt[4]{x(a-x)}\right)dx=\sqrt{2}\pi J_0\left(b\sqrt{\frac{a}{2}}\right)\qquad [a>0].$

44. $\int\limits_0^{\pi/2}\cos^\mu x\cos(ax)(1+b\cos^2 x)^\nu dx$

$$=\frac{2^{-\mu-1}\pi\Gamma(\mu+1)}{\Gamma\left(\frac{\mu-a}{2}+1\right)\Gamma\left(\frac{\mu+a}{2}+1\right)}\,{}_3F_2\left(\begin{array}{c}\frac{\mu+1}{2},\frac{\mu}{2}+1,-\nu;\,-b\\ \frac{\mu-a}{2}+1,\frac{\mu+a}{2}+1\end{array}\right)$$

$$[\operatorname{Re}\mu>-1;\,|\arg(1+b)|<\pi].$$

45. $\int\limits_0^{\pi/2}\cos(2nx)(1+a\sin^2 x)^\nu dx$

$$=\frac{2^{-2n-1}}{n!}\pi a^n(-\nu)_n\,{}_2F_1\left(\begin{array}{c}n+\frac{1}{2},n-\nu;\,-a\\ 2n+1\end{array}\right)\qquad [|\arg(1+a)|<\pi].$$

46. $\int\limits_0^\pi \cos(nx)(1+a\cos^2 x)^\nu dx$

$$=\frac{2^{-n}\pi a^{n/2}\Gamma\left(\frac{n}{2}-\nu\right)}{\Gamma\left(\frac{n}{2}+1\right)\Gamma(-\nu)}\cos\frac{n\pi}{2}\,{}_2F_1\left(\begin{array}{c}\frac{n+1}{2},\frac{n}{2}-\nu;\,-a\\ n+1\end{array}\right)\qquad [|\arg(1+a)|<\pi].$$

47. $\int\limits_0^\pi \sin^\mu x\left\{\begin{array}{c}\sin(ax)\\ \cos(ax)\end{array}\right\}(1+b\sin^2 x)^\nu dx$

$$=\frac{2^{-\mu}\pi\Gamma(\mu+1)}{\Gamma\left(\frac{\mu-a}{2}+1\right)\Gamma\left(\frac{\mu+a}{2}+1\right)}\left\{\begin{array}{c}\sin(a\pi/2)\\ \cos(a\pi/2)\end{array}\right\}{}_3F_2\left(\begin{array}{c}\frac{\mu+1}{2},\frac{\mu}{2}+1,-\nu;\,-b\\ \frac{\mu-a}{2}+1,\frac{\mu+a}{2}+1\end{array}\right)$$

$$[\operatorname{Re}\mu>-1;\,|\arg(1+b)|<\pi].$$

48. $\int\limits_0^{m\pi}\left\{\begin{array}{c}\sin(ax)\\ \cos(ax)\end{array}\right\}(1+b\sin^2 x)^\nu dx$

$$=\frac{2\sin(m\pi a/2)}{a}\left\{\begin{array}{c}\sin(m\pi a/2)\\ \cos(m\pi a/2)\end{array}\right\}{}_3F_2\left(\begin{array}{c}-\nu,\frac{1}{2},1;\,-b\\ 1-\frac{a}{2},1+\frac{a}{2}\end{array}\right).$$

49. $\displaystyle\int_0^{m\pi} \left\{\begin{array}{c}\sin(ax)\\\cos(ax)\end{array}\right\} \cos(b\sin x)\,dx$

$$= \frac{2\sin(m\pi a/2)}{a} \left\{\begin{array}{c}\sin(m\pi a/2)\\\cos(m\pi a/2)\end{array}\right\} {}_1F_2\left(\begin{array}{c}1;\ -\frac{b^2}{4}\\1-\frac{a}{2},\ 1+\frac{a}{2}\end{array}\right).$$

50. $\displaystyle\int_0^{m\pi} \frac{1}{\sin x} \left\{\begin{array}{c}\sin(ax)\\\cos(ax)\end{array}\right\} \sin(b\sin x)\,dx$

$$= \frac{2b\sin(m\pi a/2)}{a} \left\{\begin{array}{c}\sin(m\pi a/2)\\\cos(m\pi a/2)\end{array}\right\} {}_2F_3\left(\begin{array}{c}\frac{1}{2},\ 1;\ -\frac{b^2}{4}\\\frac{3}{2},\ 1-\frac{a}{2},\ 1+\frac{a}{2}\end{array}\right).$$

51. $\displaystyle\int_0^{\pi} x\sin(nx - z\sin x)\,dx$

$$= \frac{\pi}{2}\sum_{k=0}^{n-1}\left(\frac{1}{2}\right)_k\left(\frac{1}{2}\right)_{n-k}\left(\frac{z}{2}\right)^{n-2k-1} - (-1)^n\frac{\pi}{z}\sum_{k=0}^{n-1}\left(\frac{n-k+1}{2}\right)_k\left(-\frac{2}{z}\right)^k$$

$$- \frac{n!\,\pi}{2}\sum_{k=0}^{n-1}\frac{\left(\frac{2}{z}\right)^{n-k}}{k!(n-k)}J_k(z) - \frac{\pi^2}{2}\mathbf{H}_n(z).$$

52. $\displaystyle\int_0^{\pi} x\cos(nx - z\sin x)\,dx = \frac{1}{2z}\sum_{k=0}^{n-1}\left[1+(-1)^{k+n}\right]$

$$\times \left(\frac{n-k+1}{2}\right)_k\left(\frac{2}{z}\right)^k\left[\psi\left(\frac{n-k+1}{2}\right) - \psi\left(\frac{n+k+1}{2}\right)\right]$$

$$+ \frac{\pi^2}{2}J_n(z) - \frac{n!}{2}\sum_{k=0}^{n-1}\frac{\left(\frac{2}{z}\right)^{n-k}}{k!(n-k)}\left[\pi\mathbf{H}_k(z) - \sum_{p=0}^{k-1}\left(\frac{1}{2}\right)_p\left(\frac{1}{2}\right)_{k-p}\left(\frac{z}{2}\right)^{k-2p-1}\right].$$

53. $\displaystyle\int_0^{\infty} e^{-ax}\sin^m(bx)\sin^n(cx)\,dx = \frac{m!}{b}\left(\frac{i}{2}\right)^{m+n+1}$

$$\times \sum_{k=0}^{n}(-1)^k\binom{n}{k}\left(-\frac{m}{2}+\frac{kc}{b}-\frac{nc}{2b}+\frac{ia}{2b}\right)^{-1}_{m+1} \quad [\mathrm{Re}\,a>0].$$

54. $\displaystyle\int_0^{\infty} e^{-ax}\cos^m(bx)\cos^n(cx)\,dx = 2^{-m-n}m!\sum_{k=0}^{n}\binom{n}{k}\frac{1}{a+imb-2ikc+inc}$

$$\times {}_2F_1\left(\begin{array}{c}-m,\ -\frac{m}{2}+\frac{kc}{b}-\frac{nc}{2b}+\frac{ia}{2b};\ -1\\1-\frac{m}{2}+\frac{kc}{b}-\frac{nc}{2b}+\frac{ia}{2b}\end{array}\right) \quad [\mathrm{Re}\,a>0].$$

55. $$\int_0^\infty e^{-ax}\sin^m(bx)\cos^n(cx)\,dx = 2^{-m-n-1}i^{m+1}\frac{m!}{b}$$
$$\times \sum_{k=0}^n \binom{n}{k}\left(-\frac{m}{2} + \frac{kc}{b} - \frac{nc}{2b} + \frac{ia}{2b}\right)^{-1}_{m+1} \quad [\operatorname{Re} a > 0].$$

56. $$\int_0^{m\pi} e^{-ax}(1+b\sin^2 x)^\nu\,dx = \frac{1-e^{-m\pi a}}{a}\,_3F_2\left(\begin{array}{c}-\nu,\frac{1}{2},1;\,-b\\1-\frac{ia}{2},1+\frac{ia}{2}\end{array}\right)$$
$$[|\arg(1+b)| < \pi \text{ for } \nu \neq 0,1,2,\ldots].$$

57. $$\int_0^{m\pi} e^{-ax}\sin(b\sin x)\,dx = b\frac{1-(-1)^m e^{-m\pi a}}{a^2+1}\,_2F_3\left(\begin{array}{c}1;\,-\frac{b^2}{4}\\ \frac{3-ia}{2},\frac{3+ia}{2}\end{array}\right).$$

58. $$\int_0^{m\pi} e^{-ax}\cos(b\sin x)\,dx = \frac{1-e^{-m\pi a}}{a}\,_1F_2\left(\begin{array}{c}1;\,-\frac{b^2}{4}\\1-\frac{ia}{2},1+\frac{ia}{2}\end{array}\right).$$

59. $$\int_0^{m\pi} \frac{e^{-ax}}{\sin x}\sin(b\sin x)\,dx = \frac{b}{a}(1-e^{-m\pi a})\,_1F_2\left(\begin{array}{c}\frac{1}{2},1;\,-\frac{b^2}{4}\\ \frac{3}{2},1-\frac{ia}{2},1+\frac{ia}{2}\end{array}\right).$$

60. $$\int_0^{m\pi} e^{-ax+b\sin^2 x}\,dx = \frac{1-e^{-m\pi a}}{a}\,_2F_2\left(\begin{array}{c}\frac{1}{2},1;\,b\\1-\frac{ia}{2},1+\frac{ia}{2}\end{array}\right).$$

61. $$\int_0^{m\pi} \left\{\begin{array}{c}\sin(ax)\\ \cos(ax)\end{array}\right\} e^{b\sin^2 x}\,dx$$
$$= \frac{2}{a}\sin\frac{m\pi a}{2}\left\{\begin{array}{c}\sin(m\pi a/2)\\ \cos(m\pi a/2)\end{array}\right\}\,_2F_2\left(\begin{array}{c}\frac{1}{2},1;\,b\\1-\frac{a}{2},1+\frac{a}{2}\end{array}\right).$$

62. $$\int_0^\infty e^{-ax}(1+b\sin^2 x)^\nu\,x = \frac{1}{a}\,_3F_2\left(\begin{array}{c}-\nu,\frac{1}{2},1;\,-b\\1-\frac{ia}{2},1+\frac{ia}{2}\end{array}\right)$$
$$[\operatorname{Re} a > 0;\,|\arg(1+b)| < \pi \text{ for } \nu \neq 0,1,2,\ldots].$$

63. $$\int_0^\infty e^{-ax}\sin(b\sin x)\,dx = \frac{b}{a^2+1}\,_1F_2\left(\begin{array}{c}1;\,-\frac{b^2}{4}\\ \frac{3-ia}{2},\frac{3+ia}{2}\end{array}\right) \quad [\operatorname{Re} a > 0].$$

64. $$\int_0^\infty e^{-ax}\cos(b\sin x)\,dx = \frac{1}{a}\,_1F_2\left(\begin{array}{c}1;\,-\frac{b^2}{4}\\1-\frac{ia}{2},1+\frac{ia}{2}\end{array}\right) \quad [\operatorname{Re} a > 0].$$

65. $$\int_0^\infty \frac{e^{-ax}}{\sin x}\sin(b\sin x)\,dx = \frac{b}{a}\,_2F_3\left(\begin{array}{c}\frac{1}{2},1;\,-\frac{b^2}{4}\\ \frac{3}{2},1-\frac{ia}{2},1+\frac{ia}{2}\end{array}\right) \quad [\operatorname{Re} a > 0].$$

66. $\int\limits_0^\infty e^{-ax+b\sin^2 x}\,dx = \frac{1}{a}\,_2F_2\left(\begin{array}{c}\frac{1}{2},\,1;\,b\\1-\frac{ia}{2},\,1+\frac{ia}{2}\end{array}\right)$ \hfill $[\operatorname{Re} a > 0]$.

67. $\int\limits_0^\pi \sin^\mu x \left\{\begin{array}{c}\sin(ax)\\\cos(ax)\end{array}\right\} e^{b\sin^2 x}\,dx = \frac{2^{-\mu}\pi\Gamma(\mu+1)}{\Gamma\left(\frac{\mu-a}{2}+1\right)\Gamma\left(\frac{\mu+a}{2}+1\right)}$

$$\times \left\{\begin{array}{c}\sin(a\pi/2)\\\cos(a\pi/2)\end{array}\right\}\,_2F_2\left(\begin{array}{c}\frac{\mu+1}{2},\,\frac{\mu}{2}+1;\,b\\\frac{\mu-a}{2}+1,\,\frac{\mu+a}{2}+1\end{array}\right).$$

68. $\int\limits_0^\infty \cosh^\nu x\, e^{a\operatorname{sech}^2 x}\cos(bx)\,dx = \frac{2^{-\nu-2}}{\Gamma(-\nu)}\Gamma\!\left(\frac{-ib-\nu}{2}\right)\Gamma\!\left(\frac{ib-\nu}{2}\right)$

$$\times\,_2F_2\left(\begin{array}{c}\frac{-ib-\nu}{2},\,\frac{ib-\nu}{2}\\-\frac{\nu}{2},\,\frac{1-\nu}{2};\,a\end{array}\right) \quad [\operatorname{Re}\nu < 0].$$

69. $\int\limits_0^{m\pi} \left\{\begin{array}{c}\sin(ax)\\\cos(ax)\end{array}\right\}\cosh(b\sin x)\,dx$

$$= \frac{2\sin(m\pi a/2)}{a}\left\{\begin{array}{c}\sin(m\pi a/2)\\\cos(m\pi a/2)\end{array}\right\}\,_1F_2\left(\begin{array}{c}1;\,\frac{b^2}{4}\\1-\frac{a}{2},\,1+\frac{a}{2}\end{array}\right).$$

70. $\int\limits_0^{m\pi} \frac{1}{\sin x}\left\{\begin{array}{c}\sin(ax)\\\cos(ax)\end{array}\right\}\sinh(b\sin x)\,dx$

$$= \frac{2b\sin(m\pi a/2)}{a}\left\{\begin{array}{c}\sin(m\pi a/2)\\\cos(m\pi a/2)\end{array}\right\}\,_2F_3\left(\begin{array}{c}\frac{1}{2},\,1;\,\frac{b^2}{4}\\\frac{3}{2},\,1-\frac{a}{2},\,1+\frac{a}{2}\end{array}\right).$$

71. $\int\limits_0^{m\pi} \frac{1}{\sin x}\left\{\begin{array}{c}\sin(ax)\\\cos(ax)\end{array}\right\}\sinh\left(b\sqrt{\sin x}\right)\sin\left(b\sqrt{\sin x}\right)\,dx$

$$= \frac{2b^2\sin(m\pi a/2)}{a}\left\{\begin{array}{c}\sin(m\pi a/2)\\\cos(m\pi a/2)\end{array}\right\}\,_2F_5\left(\begin{array}{c}\frac{1}{2},\,1;\,-\frac{b^4}{64}\\\frac{3}{4},\,\frac{5}{4},\,\frac{3}{2},\,1-\frac{a}{2},\,1+\frac{a}{2}\end{array}\right).$$

72. $\int\limits_0^{m\pi} \left\{\begin{array}{c}\sin(ax)\\\cos(ax)\end{array}\right\}\cosh\left(b\sqrt{\sin x}\right)\cos\left(b\sqrt{\sin x}\right)\,dx$

$$= \frac{2\sin(m\pi a/2)}{a}\left\{\begin{array}{c}\sin(m\pi a/2)\\\cos(m\pi a/2)\end{array}\right\}\,_1F_4\left(\begin{array}{c}1;\,-\frac{b^4}{64}\\\frac{1}{4},\,\frac{3}{4},\,1-\frac{a}{2},\,1+\frac{a}{2}\end{array}\right).$$

73. $\displaystyle\int_0^{m\pi} e^{-ax}\cosh(b\sin x)\,dx = \frac{1-e^{-m\pi a}}{a}\,{}_1F_2\left(\begin{array}{c}1;\ \frac{b^2}{4}\\ 1-\frac{ia}{2},1+\frac{ia}{2}\end{array}\right).$

74. $\displaystyle\int_0^{m\pi} \frac{e^{-ax}}{\sin x}\sinh(b\sin x)\,dx = \frac{b}{a}(1-e^{-m\pi a})\,{}_2F_3\left(\begin{array}{c}\frac{1}{2},1;\ \frac{b^2}{4}\\ \frac{3}{2},1-\frac{ia}{2},1+\frac{ia}{2}\end{array}\right).$

75. $\displaystyle\int_0^{m\pi} \frac{e^{-ax}}{\sin x}\sinh\!\left(b\sqrt{\sin x}\right)\sin\!\left(b\sqrt{\sin x}\right)dx$

$$= \frac{b^2}{a}(1-e^{-m\pi a})\,{}_2F_5\left(\begin{array}{c}\frac{1}{2},1;\ -\frac{b^4}{64}\\ \frac{3}{4},\frac{5}{4},\frac{3}{2},1-\frac{ia}{2},1+\frac{ia}{2}\end{array}\right).$$

76. $\displaystyle\int_0^{m\pi} e^{-ax}\cosh\!\left(b\sqrt{\sin x}\right)\cos\!\left(b\sqrt{\sin x}\right)dx$

$$= \frac{1-e^{-m\pi a}}{a}\,{}_1F_4\left(\begin{array}{c}1;\ -\frac{b^4}{64}\\ \frac{1}{4},\frac{3}{4},1-\frac{ia}{2},1+\frac{ia}{2}\end{array}\right).$$

77. $\displaystyle\int_0^{\infty} e^{-ax}\cos(b\sin x)\,dx = \frac{1}{a}\,{}_1F_2\left(\begin{array}{c}1;\ \frac{b^2}{4}\\ 1-\frac{ia}{2},1+\frac{ia}{2}\end{array}\right)\qquad [\mathrm{Re}\,a>0].$

78. $\displaystyle\int_0^{\infty} \frac{e^{-ax}}{\sin x}\sinh\!\left(b\sqrt{\sin x}\right)\sin\!\left(b\sqrt{\sin x}\right)dx$

$$= \frac{b^2}{a}\,{}_2F_5\left(\begin{array}{c}\frac{1}{2},1;\ -\frac{b^4}{64}\\ \frac{3}{4},\frac{5}{4},\frac{3}{2},1-\frac{ia}{2},1+\frac{ia}{2}\end{array}\right)\qquad [\mathrm{Re}\,a>0].$$

79. $\displaystyle\int_0^{\infty} e^{-ax}\cosh\!\left(b\sqrt{\sin x}\right)\cos\!\left(b\sqrt{\sin x}\right)dx$

$$= \frac{1}{a}\,{}_1F_4\left(\begin{array}{c}1;\ -\frac{b^4}{64}\\ \frac{1}{4},\frac{3}{4},1-\frac{ia}{2},1+\frac{ia}{2}\end{array}\right)\qquad [\mathrm{Re}\,a>0].$$

80. $\displaystyle\int_0^{\infty}\cosh^\nu x\cos(bx)\cos(c\,\mathrm{sech}\,x)\,dx = \frac{2^{-\nu-2}}{\Gamma(-\nu)}\Gamma\!\left(\frac{-ib-\nu}{2}\right)\Gamma\!\left(\frac{ib-\nu}{2}\right)$

$$\times {}_2F_3\left(\begin{array}{c}\frac{-ib-\nu}{2},\frac{ib-\nu}{2};\ -\frac{c^2}{4}\\ \frac{1}{2},-\frac{\nu}{2},\frac{1-\nu}{2}\end{array}\right)\qquad [\mathrm{Re}\,\nu<0].$$

81. $\displaystyle\int_0^\infty \cosh^\nu x \cos(bx)\sin(c\,\text{sech}\,x)\,dx$

$$= \frac{2^{-\nu-1}c}{\Gamma(1-\nu)}\Gamma\!\left(\frac{1-ib-\nu}{2}\right)\Gamma\!\left(\frac{1+ib-\nu}{2}\right)$$

$$\times {}_2F_3\!\left(\begin{array}{c}\frac{1-ib-\nu}{2},\frac{1+ib-\nu}{2};\\ \frac{3}{2},\frac{1-\nu}{2},1-\frac{\nu}{2}\end{array}\;-\frac{c^2}{4}\right)\qquad [\text{Re}\,\nu<1].$$

82. $\displaystyle\int_0^{\pi/2}\cos(b\cos x)\cos(c\sin(2nx))\,dx = \frac{\pi}{2}J_0(b)J_0(c).$

83. $\displaystyle\int_0^a \sinh(b\sqrt{x})\sin(b\sqrt{a-x})\,dx = \frac{\pi}{8}(ab)^2.$

84. $\displaystyle\int_0^a (a-x)^{-1/2}\sinh(b\sqrt{x})\cos(b\sqrt{a-x})\,dx = \frac{\pi ab}{2}.$

85. $\displaystyle\int_0^a x^{-1/2}(a-x)^{-1/2}\cosh(b\sqrt{x})\cos(b\sqrt{a-x})\,dx = \pi.$

86. $\displaystyle\int_0^{\pi/2}\sin(2nx)\sinh(a\sin x)\sin(a\cos x)\,dx = (-1)^{n+1}\frac{\pi a^{2n}}{4(2n)!}\qquad [n\geq 1].$

87. $\displaystyle\int_0^{\pi/2}\sin(2nx)\sinh(a\cos x)\sin(a\sin x)\,dx = \frac{\pi a^{2n}}{4(2n)!}\qquad [n\geq 1].$

88. $\displaystyle\int_0^{\pi/2}\sinh^2(a\sin x)\sin^2(a\cos x)\,dx = \frac{\pi}{8}[I_0(2a)+J_0(2a)]-\frac{\pi}{4}.$

89. $\displaystyle\int_0^{\pi/2}\cos^2(a\sin x)\cos^2(a\cos x)\,dx = \frac{\pi}{8}\left[2J_0(2a)+J_0(2\sqrt{2}\,a)+1\right].$

90. $\displaystyle\int_0^{\pi/2}\cosh^2(a\cos x)\sin^2(a\sin x)\,dx = \frac{\pi}{8}[I_0(2a)-J_0(2a)].$

91. $\displaystyle\int_0^{\pi/2}\sinh^2(a\cos x)\cos^2(a\sin x)\,dx = \frac{\pi}{8}[I_0(2a)-J_0(2a)].$

92. $\displaystyle\int_0^{\pi/2} \sinh^2(a\sin x)\cos^2(a\cos x)\,dx = \frac{\pi}{8}\left[I_0(2a) - J_0(2a)\right].$

93. $\displaystyle\int_0^{\pi} \cos^2(a\sin x)\sin^2(a\cos x)\,dx = \frac{\pi}{4}\left[1 - J_0(2\sqrt{2}\,a)\right].$

94. $\displaystyle\int_0^{\pi} e^{2a\cos(nx)}\left\{\begin{array}{l}\sin(a\sin(nx))\\ \cos(a\sin(nx))\end{array}\right\}^2 dx = \mp\frac{\pi}{2} + \frac{\pi}{2}I_0(2a).$

4.1.5. The logarithmic function

1. $\displaystyle\int_0^1 \frac{\ln x}{\sqrt{4-x^2}}\,dx = \frac{1}{32\sqrt{3}}\left[-\varsigma\left(2,\frac{1}{6}\right) + \varsigma\left(2,\frac{1}{3}\right) - \varsigma\left(2,\frac{2}{3}\right) + \varsigma\left(2,\frac{5}{6}\right)\right].$

2. $\displaystyle\int_0^1 \frac{\ln^2 x}{\sqrt{4-x^2}}\,dx = \frac{7\pi^3}{216}.$

3. $\displaystyle\int_0^1 \frac{\ln^3 x}{\sqrt{4-x^2}}\,dx$
$= -\dfrac{\pi}{4}\varsigma(3) - \dfrac{\sqrt{3}}{1024}\left[\varsigma\left(4,\dfrac{1}{6}\right) - \varsigma\left(4,\dfrac{1}{3}\right) + \varsigma\left(4,\dfrac{2}{3}\right) - \varsigma\left(4,\dfrac{5}{6}\right)\right].$

4. $\displaystyle\int_0^1 \frac{x^{1/2}}{x^3+1}\ln(1-x)\,dx = -\frac{2}{9}\mathbf{G} - \frac{\pi}{6}\ln 2.$

5. $\displaystyle\int_0^1 \frac{x}{x^4+1}\ln(1-x)\,dx = \frac{1}{32}\left[\pi\ln 2 - 8\mathbf{G} - 4\pi\ln\left(1+\sqrt{2}\right)\right].$

6. $\displaystyle\int_0^1 \frac{(1-x)^{s-1}}{(1-xz)^{s+1}}\ln(1-x)\,dx = \frac{(z-1)^{-1}}{s^2}\,{}_2F_1\!\left(\begin{array}{c}1,\,s;\,z\\ s+1\end{array}\right)$

[Re $s > 0$; $|1-z| < \pi$].

7. $\displaystyle\int_0^1 \frac{1}{1+x}\ln\left(1+x^{2+\sqrt{3}}\right)dx = \frac{\pi^2}{12}(1-\sqrt{3}) + \ln 2\ln(1+\sqrt{3})$ \hfill [60].

8. $\displaystyle\int_0^1 \frac{1}{1+x}\ln\left(1+x^{3+\sqrt{8}}\right)dx = \frac{\pi^2}{24}(3-\sqrt{32})$
$+ \dfrac{1}{2}\ln 2\left[\ln 2 + \dfrac{3}{2}\ln(3+\sqrt{8})\right]$ \hfill [60].

9. $\int_0^1 \frac{1}{1+x} \ln\left(1 + x^{4+\sqrt{15}}\right) dx$

$$= \frac{\pi^2}{12}(2-\sqrt{15}) + \ln\frac{1+\sqrt{5}}{2} \ln(2+\sqrt{3}) + \ln 2 \ln(\sqrt{3}+\sqrt{5}) \quad [60].$$

10. $\int_0^1 \frac{1}{1+x} \ln\left(1 + x^{5+\sqrt{24}}\right) dx = \frac{\pi^2}{24}(5-\sqrt{96}) + \frac{1}{2}\ln(1+\sqrt{2})\ln(2+\sqrt{3})$

$$+ \frac{1}{2}\ln 2\left[\ln 2 + \frac{3}{2}\ln(5+\sqrt{24})\right] \quad [60].$$

11. $\int_0^1 \frac{1}{1+x} \ln\left(1 + x^{6+\sqrt{35}}\right) dx = \frac{\pi^2}{12}(3-\sqrt{35}) + \ln\frac{1+\sqrt{5}}{2}\ln(8+3\sqrt{7})$

$$+ \ln 2 \ln(\sqrt{5}+\sqrt{7}) \quad [60].$$

12. $\int_0^1 \frac{1}{1+x} \ln\left(1 + x^{8+\sqrt{63}}\right) dx = \frac{\pi^2}{12}(4-\sqrt{63}) + \ln\frac{5+\sqrt{21}}{2}\ln(2+\sqrt{3})$

$$+ \ln 2 \ln(3+\sqrt{7}) \quad [60].$$

13. $\int_0^1 \frac{1}{1+x} \ln\left(1 + x^{11+\sqrt{120}}\right) dx = \frac{\pi^2}{24}\left(11-\sqrt{480}\right)$

$$+ \frac{1}{2}\ln(1+\sqrt{2})\ln(4+\sqrt{15}) + \frac{1}{2}\ln(2+\sqrt{3})\ln(3+\sqrt{10})$$
$$+ \frac{1}{2}\ln\frac{1+\sqrt{5}}{2}\ln(5+\sqrt{24}) + \frac{1}{2}\ln 2\left[\ln 2 + \frac{3}{2}\ln(11+\sqrt{120})\right] \quad [60].$$

14. $\int_0^1 \frac{1}{1+x} \ln\left(1 + x^{13+\sqrt{168}}\right) dx = \frac{\pi^2}{24}\left(13-\sqrt{672}\right)$

$$+ \frac{1}{2}\ln(1+\sqrt{2})\ln(5+\sqrt{21}) + \frac{1}{4}\ln(2+\sqrt{3})\ln(15+\sqrt{224})$$
$$+ \frac{1}{4}\ln(5+\sqrt{24})\ln(8+\sqrt{63}) + \frac{1}{2}\ln 2\left[\ln 2 + \frac{3}{2}\ln(13+\sqrt{168})\right] \quad [60].$$

15. $\int_0^1 \frac{1}{1+x} \ln\left(1 + x^{14+\sqrt{195}}\right) dx = \frac{\pi^2}{12}\left(7-\sqrt{195}\right)$

$$+ \ln\frac{1+\sqrt{5}}{2}\ln(25+4\sqrt{39}) + \ln\frac{3+\sqrt{13}}{2}\ln(4+\sqrt{15})$$
$$+ \ln 2 \ln(\sqrt{15}+\sqrt{13}) \quad [60].$$

16. $\int_0^a x^{s-1}(a-x)^{t-1} \ln(1-bx(a-x))\, dx$

$$= -a^{s+t+1} b\, B(s+1, t+1)\, {}_4F_3\left(\begin{array}{c} 1,1,s+1,t+1;\\ 2, \frac{s+t+1}{2}, \frac{s+t}{2}+1 \end{array} \frac{a^2 b}{4}\right)$$

$$[a > 0;\ \operatorname{Re} s,\ \operatorname{Re} t > -1;\ |\arg(4 - a^2 b)| < \pi].$$

17. $\int_0^a \ln(1 - bx(a-x))\, dx = \dfrac{4}{\sqrt{b}}\left(1 - \dfrac{a^2 b}{4}\right)^{1/2} \arcsin\left(\dfrac{a\sqrt{b}}{2}\right) - 2a$

$$[a > 0;\ |\arg(4 - a^2 b)| < \pi].$$

18. $\int_0^a x^{-1} \ln(1 - bx(a-x))\, dx = -2\arcsin^2 \dfrac{a\sqrt{b}}{2} \qquad [a > 0;\ |\arg(4 - a^2 b)| < \pi].$

19. $\int_0^a \dfrac{1}{x(a-x)} \ln(1 - bx(a-x))\, dx = -\dfrac{4}{a} \arcsin^2 \dfrac{a\sqrt{b}}{2}$

$$[a > 0;\ |\arg(4 - a^2 b)| < \pi].$$

20. $\int_0^a x^{-1/2}(a-x)^{-1/2} \ln(1 - bx(a-x))\, dx = 2\pi \ln \dfrac{2 + \sqrt{4 - a^2 b}}{4}$

$$[a > 0;\ |\arg(4 - a^2 b)| < \pi].$$

21. $\int_0^a x^{-3/2}(a-x)^{-3/2} \ln(1 - bx(a-x))\, dx = \dfrac{4\pi}{a^2}\left(\sqrt{4 - a^2 b} - 2\right)$

$$[a > 0;\ |\arg(4 - a^2 b)| < \pi].$$

22. $\int_0^a x^{-1/2}(a-x)^{-3/2} \ln(1 - bx(a-x))\, dx = \dfrac{2\pi}{a}\left(\sqrt{4 - a^2 b} - 2\right)$

$$[a > 0;\ |\arg(4 - a^2 b)| < \pi].$$

23. $\int_0^\infty \dfrac{x}{x^2 + a^2} \ln \dfrac{x^2 + 2bx + c}{x^2 - 2bx + c}\, dx = 2\pi \arctan \dfrac{b}{|a| + \sqrt{c - b^2}}$

$$[b^2 \leq c;\ [40], (42)].$$

24. $\int_0^\infty \dfrac{1}{x} \ln \dfrac{x^2 + 2bx + c}{x^2 - 2bx + c}\, dx = 2\pi \arctan \dfrac{b}{\sqrt{c}} \qquad [b^2 \leq c;\ [40], (43)].$

25. $\displaystyle\int_0^\infty \frac{1}{x} \ln \frac{x^2 + 2bx + c}{x^2 + 2dx + c}\, dx = \arccos^2 \frac{d}{\sqrt{c}} - \arccos^2 \frac{b}{\sqrt{c}}$

$[b^2 \leq c;\ b^2 \leq d;\ [40],\ (47)].$

26. $\displaystyle\int_0^a x^{s+1/2}(a-x)^s \ln\left(1 - b\sqrt{x(a-x)}\right) dx$

$= -2^{-2s-2}\pi^{1/2} a^{2s+5/2} b\, \frac{\Gamma(2s+3)}{\Gamma\left(2s+\frac{7}{2}\right)}\, {}_3F_2\left(\begin{array}{c} 1,1,2s+3 \\ 2, 2s+\frac{7}{2};\ \frac{ab}{2} \end{array}\right)$

$[a > 0;\ \mathrm{Re}\, s > -1;\ |\arg(2-ab)| < \pi].$

27. $\displaystyle\int_0^a x^{1/2} \ln\left(1 - b\sqrt{x(a-x)}\right) dx = \frac{4}{3b^{3/2}}(ab+1)\sqrt{2-ab}\, \arcsin\left(\sqrt{\frac{ab}{2}}\right)$

$\displaystyle - \frac{2\sqrt{a}}{9b}(5ab+6) \quad [a>0;\ |\arg(2-ab)| < \pi].$

28. $\displaystyle\int_0^a x^{-1/2} \ln\left(1 - b\sqrt{x(a-x)}\right) dx$

$= \displaystyle\frac{4}{\sqrt{b}}\left[\sqrt{2-ab}\, \arcsin\left(\sqrt{\frac{ab}{2}}\right) - \sqrt{ab}\right] \quad [a>0;\ |\arg(2-ab)| < \pi].$

29. $\displaystyle\int_0^a x^{-1/2}(a-x)^{-1} \ln\left(1 - b\sqrt{x(a-x)}\right) dx = -\frac{4}{\sqrt{a}} \arcsin^2 \sqrt{\frac{ab}{2}}$

$[a>0;\ |\arg(2-ab)| < \pi].$

30. $\displaystyle\int_0^a x^{1/4}(a-x)^{-1/4} \ln\left(1 - b\sqrt{x(a-x)}\right) dx$

$= \displaystyle\frac{\pi}{\sqrt{2b}}\left[ab \ln\left(1 + \sqrt{1 - \frac{ab}{2}}\right) + 2\sqrt{1 - \frac{ab}{2}} + \frac{ab}{2}(1 - 2\ln 2) - 2\right]$

$[a>0;\ |\arg(2-ab)| < \pi].$

31. $\displaystyle\int_0^a x^{-1/4}(a-x)^{-3/4} \ln\left(1 - b\sqrt{x(a-x)}\right) dx$

$= \displaystyle 2^{3/2}\pi \ln\left(\frac{1}{2} + \frac{1}{2}\sqrt{1 - \frac{ab}{2}}\right) \quad [a>0;\ |\arg(2-ab)| < \pi].$

32. $\displaystyle\int_0^a x^{-3/4}(a-x)^{-5/4} \ln\left(1 - b\sqrt{x(a-x)}\right) dx = \frac{4\pi}{a}\left(\sqrt{2-ab} - \sqrt{2}\right)$

$[a>0;\ |\arg(2-ab)| < \pi].$

33. $\int_0^a x^{s-1}(a-x)^{t-1} \ln\left(b\sqrt{x(a-x)} + \sqrt{1+b^2 x(a-x)}\right) dx$

$$= a^{s+t} b\, \mathrm{B}\left(s+\frac{1}{2}, t+\frac{1}{2}\right) {}_4F_3\left(\begin{array}{c} \frac{1}{2}, \frac{1}{2}, s+\frac{1}{2}, t+\frac{1}{2}; -\frac{a^2 b^2}{4} \\ \frac{3}{2}, \frac{s+t+1}{2}, \frac{s+t}{2}+1 \end{array}\right)$$

$[a, \operatorname{Re} s, \operatorname{Re} t > 0;\ |\arg(4+a^2 b^2)| < \pi]$.

34. $\int_0^a x \ln\left(b\sqrt{x(a-x)} + \sqrt{1+b^2 x(a-x)}\right) dx$

$$= \frac{a}{b}\sqrt{1+\frac{a^2 b^2}{4}}\left[\mathrm{K}\left(\frac{ab}{\sqrt{4+a^2 b^2}}\right) - \mathrm{E}\left(\frac{ab}{\sqrt{4+a^2 b^2}}\right)\right]$$

$[a > 0;\ |\arg(4+a^2 b^2)| < \pi]$.

35. $\int_0^a x^{-1/2}(a-x)^{-1/2} \ln\left(b\sqrt{x(a-x)} + \sqrt{1+b^2 x(a-x)}\right) dx$

$$= i\left[\mathrm{Li}_2\left(-\frac{iab}{2}\right) - \mathrm{Li}_2\left(\frac{iab}{2}\right)\right] \quad [a > 0;\ |\arg(4+a^2 b^2)| < \pi].$$

36. $\int_0^a x^{1/2}(a-x)^{-1/2} \ln\left(b\sqrt{x(a-x)} + \sqrt{1+b^2 x(a-x)}\right) dx$

$$= \frac{ia}{2}\left[\mathrm{Li}_2\left(-\frac{iab}{2}\right) - \mathrm{Li}_2\left(\frac{iab}{2}\right)\right] \quad [a > 0;\ |\arg(4+a^2 b^2)| < \pi].$$

37. $\int_0^a x^{s+1/2}(a-x)^s \ln\left(b\sqrt[4]{x(a-x)} + \sqrt{1+b^2\sqrt{x(a-x)}}\right) dx$

$$= 2^{-2s-3/2} \pi^{1/2} a^{2s+2} b \frac{\Gamma\left(2s+\frac{5}{2}\right)}{\Gamma(2s+3)} {}_3F_2\left(\begin{array}{c} \frac{1}{2}, \frac{1}{2}, 2s+\frac{5}{2} \\ \frac{3}{2}, 2s+3; -\frac{ab^2}{2} \end{array}\right)$$

$[a > 0;\ \operatorname{Re} s > -5/4;\ |\arg(2+ab^2)| < \pi]$.

38. $\int_0^a x^{-1/2} \ln\left(b\sqrt[4]{x(a-x)} + \sqrt{1+b^2\sqrt{x(a-x)}}\right) dx$

$$= \frac{2^{3/2}}{b}\sqrt{1+\frac{ab^2}{2}}\left[\mathrm{K}\left(\frac{b\sqrt{a}}{\sqrt{2+ab^2}}\right) - \mathrm{E}\left(\frac{b\sqrt{a}}{\sqrt{2+ab^2}}\right)\right]$$

$[a > 0;\ |\arg(2+ab^2)| < \pi]$.

39. $\int_0^a x^{-1/4}(a-x)^{-3/4} \ln\left(b\sqrt[4]{x(a-x)} + \sqrt{1+b^2\sqrt{x(a-x)}}\right) dx$

$$= \sqrt{2}\,i\left[\mathrm{Li}_2\left(-ib\sqrt{\frac{a}{2}}\right) - \mathrm{Li}_2\left(ib\sqrt{\frac{a}{2}}\right)\right] \quad [a > 0;\ |\arg(2+ab^2)| < \pi].$$

40. $\displaystyle\int_0^a \frac{x^{1/2}}{\sqrt{1+b^2\sqrt{x(a-x)}}} \ln\left(b\sqrt[4]{x(a-x)} + \sqrt{1+b^2\sqrt{x(a-x)}}\right) dx$

$$= \frac{\pi}{4\sqrt{2}\,b^3}\left[ab^2 + (ab^2 - 2)\ln\left(1 + \frac{ab^2}{2}\right)\right] \quad [a>0;\ |\arg(2+ab^2)|<\pi].$$

41. $\displaystyle\int_0^a \frac{x^{-1/2}}{\sqrt{1+b^2\sqrt{x(a-x)}}} \ln\left(b\sqrt[4]{x(a-x)} + \sqrt{1+b^2\sqrt{x(a-x)}}\right) dx$

$$= \frac{\pi}{\sqrt{2}\,b}\ln\left(1 + \frac{ab^2}{2}\right) \quad [a>0;\ |\arg(2+ab^2)|<\pi].$$

42. $\displaystyle\int_0^a \frac{x(a-x)^{1/2}}{\sqrt{1+b^2\sqrt{x(a-x)}}} \ln\left(b\sqrt[4]{x(a-x)} + \sqrt{1+b^2\sqrt{x(a-x)}}\right) dx$

$$= \frac{\pi}{64\sqrt{2}\,b^5}\left[ab^2(7ab^2 - 12) + 2(3a^2b^4 - 4ab^2 + 12)\ln\left(1 + \frac{ab^2}{2}\right)\right]$$

$$[a>0;\ |\arg(2+ab^2)|<\pi].$$

43. $\displaystyle\int_0^a \frac{x^{-1/2}}{\sqrt{1+b^2\sqrt{x(a-x)}}} \ln\left(b\sqrt[4]{x(a-x)} + \sqrt{1+b^2\sqrt{x(a-x)}}\right) dx$

$$= \frac{\pi}{\sqrt{2}\,b}\ln\left(1 + \frac{ab^2}{2}\right) \quad [a>0;\ |\arg(2+ab^2)|<\pi].$$

44. $\displaystyle\int_0^a \frac{x^{-1/2}(a-x)^{-1}}{\sqrt{1+b^2\sqrt{x(a-x)}}} \ln\left(b\sqrt[4]{x(a-x)} + \sqrt{1+b^2\sqrt{x(a-x)}}\right) dx$

$$= \frac{2\pi}{\sqrt{a}} \arctan\left(b\sqrt{\frac{a}{2}}\right) \quad [a>0;\ |\arg(2+ab^2)|<\pi].$$

45. $\displaystyle\int_0^1 \frac{x(1-x^2)^{s-1}}{(1-ax^2)^{s+1/2}} \ln\frac{1+x}{1-x}\, dx$

$$= \frac{\sqrt{\pi}}{2}\,\frac{\Gamma(s)}{s\Gamma\left(s+\frac{1}{2}\right)}(1-a)^{-1/2}\,{}_2F_1\!\left(\begin{array}{c}1,\,s\\s+1;\,a\end{array}\right) \quad [\mathrm{Re}\,s > 0;\ |1-a|<\pi].$$

46. $\displaystyle\int_0^a x^{s-1}(1+bx)^\nu \ln\frac{\sqrt{a}+\sqrt{a-x}}{\sqrt{x}}\, dx$

$$= \frac{1}{2}\sqrt{\pi}\,a^s\,\frac{\Gamma(s)}{s\Gamma\left(s+\frac{1}{2}\right)}\,{}_3F_2\!\left(\begin{array}{c}-\nu,\,s,\,s;\,-ab\\s+\frac{1}{2},\,s+1\end{array}\right)$$

$$[a,\,\mathrm{Re}\,s > 0;\ \mathrm{Re}\,\nu > -1\text{ for }b<0;\ |\arg(1+ab)|<\pi].$$

47. $\int_0^a \dfrac{1}{1+bx} \ln \dfrac{\sqrt{a}+\sqrt{a-x}}{\sqrt{x}}\, dx = \dfrac{1}{b} \ln^2\left(\sqrt{ab}+\sqrt{ab+1}\right)$

$[a>0;\ |\arg(1+ab)|<\pi]$.

48. $\int_0^a \dfrac{1}{1-bx} \ln \dfrac{\sqrt{a}+\sqrt{a-x}}{\sqrt{x}}\, dx = \dfrac{1}{b} \arcsin^2 \sqrt{ab}$ $\quad [a>0;\ |\arg(1-ab)|<\pi]$.

49. $\int_0^a \dfrac{1}{(1+bx)^2} \ln \dfrac{\sqrt{a}+\sqrt{a-x}}{\sqrt{x}}\, dx = \sqrt{\dfrac{a}{b(ab+1)}} \ln\left(\sqrt{ab}+\sqrt{ab+1}\right)$

$[a>0;\ |\arg(1+ab)|<\pi]$.

50. $\int_0^a \dfrac{1}{(1-bx)^2} \ln \dfrac{\sqrt{a}+\sqrt{a-x}}{\sqrt{x}}\, dx = \sqrt{\dfrac{a}{b(1-ab)}} \arcsin \sqrt{ab}$

$[a>0;\ |\arg(1-ab)|<\pi]$.

51. $\int_0^a \dfrac{1}{\sqrt{1+bx}} \ln \dfrac{\sqrt{a}+\sqrt{a-x}}{\sqrt{x}}\, dx = 2\sqrt{\dfrac{a}{b}} \arctan\left(\sqrt{ab}\right) - \dfrac{1}{b}\ln(1+ab)$

$[a>0;\ |\arg(1+ab)|<\pi]$.

52. $\int_0^a \dfrac{1}{\sqrt{1-bx}} \ln \dfrac{\sqrt{a}+\sqrt{a-x}}{\sqrt{x}}\, dx = \sqrt{\dfrac{a}{b}} \ln \dfrac{1+\sqrt{ab}}{1-\sqrt{ab}} + \dfrac{1}{b}\ln(1-ab)$

$[a>0;\ |\arg(1-ab)|<\pi]$.

53. $\int_0^a \dfrac{x}{\sqrt{1+bx}} \ln \dfrac{\sqrt{a}+\sqrt{a-x}}{\sqrt{x}}\, dx$

$= \dfrac{1}{3b^2}\left[\sqrt{ab}(ab-3)\arctan\sqrt{ab} + 2\ln(1+ab) + ab\right]$

$[a>0;\ |\arg(1+ab)|<\pi]$.

54. $\int_0^a \dfrac{x}{\sqrt{1-bx}} \ln \dfrac{\sqrt{a}+\sqrt{a-x}}{\sqrt{x}}\, dx$

$= \dfrac{1}{6b^2}\left[\sqrt{ab}(ab+3)\ln \dfrac{1+\sqrt{ab}}{1-\sqrt{ab}} + 4\ln(1-ab) - 2ab\right]$

$[a>0;\ |\arg(1-ab)|<\pi]$.

55. $\int_0^a \dfrac{\sqrt{x}}{1+bx} \ln \dfrac{\sqrt{a}+\sqrt{a-x}}{\sqrt{x}}\, dx$

$= \dfrac{\pi\sqrt{a}}{b} - \dfrac{2\pi}{b^{3/2}} \ln \dfrac{\sqrt{(ab+1)^{1/2}-1}+\sqrt{(ab+1)^{1/2}+1}}{\sqrt{2}}$

$[a>0;\ |\arg(1+ab)|<\pi]$.

4.1. Elementary Functions

56. $\displaystyle\int_0^a \frac{\sqrt{x}}{1-bx} \ln\frac{\sqrt{a}+\sqrt{a-x}}{\sqrt{x}}\,dx = -\frac{\pi\sqrt{a}}{b} + \frac{2\pi}{b^{3/2}}\arcsin\sqrt{\frac{1-(1-ab)^{1/2}}{2}}$

$[a>0;\ |\arg(1-ab)|<\pi]$.

57. $\displaystyle\int_0^a \frac{x^{-1/2}}{1+bx} \ln\frac{\sqrt{a}+\sqrt{a-x}}{\sqrt{x}}\,dx = \frac{\pi}{\sqrt{b}}\ln\left(\sqrt{ab}+\sqrt{ab+1}\right)$

$[a>0;\ |\arg(1+ab)|<\pi]$.

58. $\displaystyle\int_0^a \frac{x^{-1/2}}{1-bx} \ln\frac{\sqrt{a}+\sqrt{a-x}}{\sqrt{x}}\,dx = \frac{\pi}{\sqrt{b}}\arcsin\sqrt{ab}$ $\quad [a>0;\ |\arg(1-ab)|<\pi]$.

59. $\displaystyle\int_0^a \frac{x^{-1/2}}{(1+bx)^2} \ln\frac{\sqrt{a}+\sqrt{a-x}}{\sqrt{x}}\,dx$

$= \dfrac{\pi}{2}\left[\sqrt{\dfrac{a}{ab+1}} + \dfrac{1}{\sqrt{b}}\ln\left(\sqrt{ab}+\sqrt{ab+1}\right)\right]$

$[a>0;\ |\arg(1+ab)|<\pi]$.

60. $\displaystyle\int_0^a \frac{x^{-1/2}}{(1-bx)^2} \ln\frac{\sqrt{a}+\sqrt{a-x}}{\sqrt{x}}\,dx = \frac{\pi}{2}\sqrt{\frac{a}{1-ab}} + \frac{\pi}{2\sqrt{b}}\arcsin\sqrt{ab}$

$[a>0;\ |\arg(1-ab)|<\pi]$.

61. $\displaystyle\int_0^a x^{s-1}(a-x)^{t-1}\ln\frac{1+b\sqrt{x(a-x)}}{1-b\sqrt{x(a-x)}}\,dx = 2a^{s+t}b\,\mathrm{B}\left(s+\frac{1}{2},\,t+\frac{1}{2}\right)$

$\times\,{}_4F_3\!\left(\begin{array}{c}\frac{1}{2},\,1,\,s+\frac{1}{2},\,t+\frac{1}{2};\,\frac{a^2b^2}{4}\\ \frac{3}{2},\,\frac{s+t+1}{2},\,\frac{s+t}{2}+1\end{array}\right)$ $\quad [a,\,\mathrm{Re}\,s,\,\mathrm{Re}\,t > -1;\ |\arg(4-a^2b^2)|<\pi]$.

62. $\displaystyle\int_0^a \ln\frac{1+b\sqrt{x(a-x)}}{1-b\sqrt{x(a-x)}}\,dx = \frac{\pi a^2 b}{2+\sqrt{4-a^2b^2}}$ $\quad [a>0;\ |\arg(4-a^2b^2)|<\pi]$.

63. $\displaystyle\int_0^a x\ln\frac{1+b\sqrt{x(a-x)}}{1-b\sqrt{x(a-x)}}\,dx = \frac{\pi a^3 b}{2}\left(2+\sqrt{4-a^2b^2}\right)^{-1}$

$[a>0;\ |\arg(4-a^2b^2)|<\pi]$.

64. $\displaystyle\int_0^a x^2 \ln\frac{1+b\sqrt{x(a-x)}}{1-b\sqrt{x(a-x)}}\,dx = \frac{\pi}{12b^3}\left[9a^2b^2 - 4(a^2b^2-1)\sqrt{4-a^2b^2} - 8\right]$

$[a>0;\ |\arg(4-a^2b^2)|<\pi]$.

65. $\displaystyle\int_0^a x^{-1} \ln \frac{1+b\sqrt{x(a-x)}}{1-b\sqrt{x(a-x)}}\, dx = 2\pi \arcsin \frac{ab}{2}$ $\quad [a>0;\ |\arg(4-a^2b^2)|<\pi]$.

66. $\displaystyle\int_0^a \frac{1}{x(a-x)} \ln \frac{1+b\sqrt{x(a-x)}}{1-b\sqrt{x(a-x)}}\, dx = \frac{4\pi}{a} \arcsin \frac{ab}{2}$

$\quad [a>0;\ |\arg(4-a^2b^2)|<\pi]$.

67. $\displaystyle\int_a^1 \frac{x}{\sqrt{x^2-a^2}} \ln \frac{\sqrt{1+x}+\sqrt{1-x}}{\sqrt{1+x}-\sqrt{1-x}}\, dx = \frac{\pi}{2}(1-a)$ $\quad [0\le a\le 1]$.

68. $\displaystyle\int_0^a x^{s+1/2}(a-x)^s \ln \frac{1+b\sqrt[4]{x(a-x)}}{1-b\sqrt[4]{x(a-x)})}\, dx$

$= 2^{-2s-1/2}\pi^{1/2} a^{2s+2} b\, \frac{\Gamma\!\left(2s+\frac{5}{2}\right)}{\Gamma(2s+3)}\, {}_3F_2\!\left(\begin{array}{c}\frac{1}{2},\,1,\,2s+\frac{5}{2}\\ \frac{3}{2},\,2s+3;\ \frac{ab^2}{2}\end{array}\right)$

$\quad [a>0;\ \operatorname{Re} s > -1;\ |\arg(2-ab^2)|<\pi]$.

69. $\displaystyle\int_0^a x^{-1/2} \ln \frac{1+b\sqrt[4]{x(a-x)}}{1-b\sqrt[4]{x(a-x)}}\, dx = \frac{2\pi}{b}\left(\sqrt{2}-\sqrt{2-ab^2}\right)$

$\quad [a>0;\ |\arg(2-ab^2)|<\pi]$.

70. $\displaystyle\int_0^a x^{1/2} \ln \frac{1+b\sqrt[4]{x(a-x)}}{1-b\sqrt[4]{x(a-x)}}\, dx$

$= \frac{\pi}{3\sqrt{2}\,b^3}\left[3ab^2 - 2(ab^2+1)\sqrt{4-2ab^2}+4\right]$ $\quad [a>0;\ |\arg(2-ab^2)|<\pi]$.

71. $\displaystyle\int_0^a x^{-1/2}(a-x)^{-1} \ln \frac{1+b\sqrt[4]{x(a-x)}}{1-b\sqrt[4]{x(a-x)}}\, dx = \frac{4\pi}{\sqrt{a}} \arcsin\!\left(b\sqrt{\frac{a}{2}}\right)$

$\quad [a>0;\ |\arg(2-ab^2)|<\pi]$.

72. $\displaystyle\int_0^\infty \frac{e^{-ax}}{[x(x^2+z^2)(\sqrt{x^2+z^2}+x)]^{1/2}} \ln\!\left(\sqrt{x^2+z^2}+x\right) dx$

$= \dfrac{\pi}{2^{5/2}\,z}\left\{\pi \sin \dfrac{az}{2} J_0\!\left(\dfrac{az}{2}\right) - \cos \dfrac{az}{2} Y_0\!\left(\dfrac{az}{2}\right)\right.$

$\qquad + 4\cos \dfrac{az}{2} \ln z\, J_0\!\left(\dfrac{az}{2}\right) + 4\sin \dfrac{az}{2} \ln z\, Y_0\!\left(\dfrac{az}{2}\right)$

$\qquad + 2J_0\!\left(\dfrac{az}{2}\right)\left[\cos \dfrac{az}{2}\operatorname{ci}(az) + \sin \dfrac{az}{2}\operatorname{Si}(az)\right]$

$\qquad \left. - 2Y_0\!\left(\dfrac{az}{2}\right)\left[\sin \dfrac{az}{2}\operatorname{ci}(pz) - \cos \dfrac{az}{2}\operatorname{Si}(az)\right]\right\}$ $\quad [\operatorname{Re} a,\ \operatorname{Re} z > 0]$.

73. $\displaystyle\int_0^a x\sinh(bx)\ln\frac{a+\sqrt{a^2-x^2}}{x}\,dx = \frac{\pi^2 a}{4b}\left[I_1(ab)\,\mathbf{L}_0(ab) - I_0(ab)\,\mathbf{L}_1(ab)\right]$

$$[a>0].$$

74. $\displaystyle\int_0^\infty \left(\frac{1}{4}\operatorname{csch}^2\frac{x}{2} - \frac{1}{x^2}\right)\ln x\,dx = \frac{\mathbf{C}}{2} - \frac{1}{2}\ln(2\pi).$

75. $\displaystyle\int_0^a \cosh(bx)\ln\frac{a+\sqrt{a^2-x^2}}{x}\,dx$

$$= \frac{\pi a}{4}\left[\pi I_0(ab)\,\mathbf{L}_1(ab) - \pi I_1(ab)\,\mathbf{L}_0(ab) + 2I_0(ab)\right] \quad [a>0].$$

76. $\displaystyle\int_0^a x^2\cosh(bx)\ln\frac{a+\sqrt{a^2-x^2}}{x}\,dx$

$$= \frac{\pi a}{2b^2}\left[\pi I_0(ab)\,\mathbf{L}_1(ab) - \pi I_1(ab)\mathbf{L}_0(ab) + ab\,I_1(ab)\right] \quad [a>0].$$

77. $\displaystyle\int_0^a \cos bx\,\ln(a^2-x^2)\,dx$

$$= \frac{1}{b}\left[\sin(ab)\left(\ln\frac{2a}{b} - \mathbf{C}\right) - \cos(ab)\operatorname{Si}(2ab) + \sin(ab)\operatorname{ci}(2ab)\right] \quad [a>0].$$

78. $\displaystyle\int_0^a \frac{\cos(bx)}{\sqrt{a^2-x^2}}\ln(a^2-x^2)\,dx = \frac{\pi^2}{4}Y_0(ab) - \frac{\pi}{2}\left(\ln\frac{2b}{a} + \mathbf{C}\right)J_0(ab)$

$$[a>0].$$

79. $\displaystyle\int_0^\infty \frac{\cos(ax)}{x^2+z^2}\ln(x^2+z^2)\,dx$

$$= \frac{\pi}{2z}e^{az}\left[e^{-2az}\left(\ln\frac{2z}{a} - \mathbf{C}\right) - \operatorname{shi}(2az) + \operatorname{chi}(2az)\right] \quad [a,\operatorname{Re} z>0].$$

80. $\displaystyle\int_0^\infty \frac{\sin(ax)}{\sqrt{x^2+z^2}}\ln(x^2+z^2)\,dx = \frac{\pi}{2}\left(\ln\frac{z}{2a} - \mathbf{C}\right)[I_0(az) - \mathbf{L}_0(az)]$

$$+ \frac{1}{4\pi}G^{32}_{24}\left(\frac{a^2 z^2}{4}\,\middle|\,\begin{matrix}\frac{1}{2},\frac{1}{2}\\ 0,0,\frac{1}{2},\frac{1}{2}\end{matrix}\right) \quad [a,\operatorname{Re} z>0].$$

81. $\displaystyle\int_0^\infty \frac{\cos(ax)}{\sqrt{x^2+z^2}}\ln(x^2+z^2)\,dx = \left(\ln\frac{z}{2a} - \mathbf{C}\right)K_0(az) \quad [a,\operatorname{Re} z>0].$

82. $\displaystyle\int_a^\infty \frac{\sin bx}{\sqrt{x^2-a^2}} \ln(x^2-a^2)\,dx = -\frac{\pi^2}{4} Y_0(ab) - \frac{\pi}{2}\left(\ln\frac{2b}{a} + \mathbf{C}\right) J_0(ab)$

$[a, b > 0]$.

83. $\displaystyle\int_a^\infty \frac{\cos bx}{\sqrt{x^2-a^2}} \ln(x^2-a^2)\,dx = -\frac{\pi^2}{4} J_0(ab) + \frac{\pi}{2}\left(\ln\frac{2b}{a} + \mathbf{C}\right) Y_0(ab)$

$[a, b > 0]$.

84. $\displaystyle\int_0^a \sqrt{x(a-x)} \begin{Bmatrix}\sin bx\\ \cos bx\end{Bmatrix} \ln(ax-x^2)\,dx = \frac{\pi}{4b^2} \begin{Bmatrix}\sin(ab/2)\\ \cos(ab/2)\end{Bmatrix}$

$\times \left\{ 4 J_0\!\left(\dfrac{ab}{2}\right) + ab\left[\pi Y_1\!\left(\dfrac{ab}{2}\right) - 2\left(\ln\dfrac{4b}{a} + \mathbf{C} - 2\right) J_1\!\left(\dfrac{ab}{2}\right)\right]\right\}$ $[a, b > 0]$.

85. $\displaystyle\int_0^a \frac{1}{\sqrt{x(a-x)}} \begin{Bmatrix}\sin bx\\ \cos bx\end{Bmatrix} \ln(ax-x^2)\,dx$

$= \dfrac{\pi}{2}\begin{Bmatrix}\sin(ab/2)\\ \cos(ab/2)\end{Bmatrix} \left[\pi Y_0\!\left(\dfrac{ab}{2}\right) - 2\left(\ln\dfrac{4b}{a} + \mathbf{C}\right) J_0\!\left(\dfrac{ab}{2}\right)\right]$ $[a, b > 0]$.

86. $\displaystyle\int_0^\infty \frac{\left(\sqrt{x^2+z^2}+x\right)^{1/2}}{\sqrt{x(x^2+z^2)}} \sin(ax) \ln\!\left(\sqrt{x^2+z^2}+x\right) dx$

$= \dfrac{\pi}{2^{3/2}} e^{-az/2} K_0\!\left(\dfrac{az}{2}\right) + \dfrac{\pi}{2^{1/2}} e^{-az/2} \ln z\, I_0\!\left(\dfrac{az}{2}\right)$

$\qquad - \dfrac{\pi}{2^{3/2}} e^{az/2} \operatorname{Ei}(-az) I_0\!\left(\dfrac{az}{2}\right)$ $[\operatorname{Re} z > 0]$.

87. $\displaystyle\int_0^\infty \frac{\left(\sqrt{x^2+z^2}+x\right)^{-1/2}}{\sqrt{x(x^2+z^2)}} \cos(ax) \ln\!\left(\sqrt{x^2+z^2}+x\right) dx$

$= \dfrac{\pi}{2^{3/2} a^{1/2} z} e^{-az/2} K_0\!\left(\dfrac{az}{2}\right) + \dfrac{\pi}{2^{1/2} z} e^{-az/2} \ln z\, I_0\!\left(\dfrac{az}{2}\right)$

$\qquad + \dfrac{\pi}{2^{3/2} z} e^{az/2} \operatorname{Ei}(-az) I_0\!\left(\dfrac{az}{2}\right)$ $[\operatorname{Re} z > 0]$.

88. $\displaystyle\int_0^{\pi/2} \frac{\cos(2nx)}{\sin x} \ln\frac{1+a\sin x}{1-a\sin x}\,dx$

$= (-1)^n \dfrac{2^{-2n}\pi a^{2n+1}}{2n+1}\, {}_3F_2\!\left(\begin{array}{c} n+\dfrac{1}{2}, n+\dfrac{1}{2}, n+1 \\ n+\dfrac{3}{2}, 2n+1;\ a^2 \end{array}\right)$ $[|\arg(1-a^2)| < \pi]$.

4.1.5] 4.1. ELEMENTARY FUNCTIONS

89. $\displaystyle\int_0^{\pi/2} \cos^\nu x \cos(ax) \ln\frac{1+b\cos x}{1-b\cos x}\, dx = \frac{2^{-\nu-1}\pi b\, \Gamma(\nu+2)}{\Gamma\!\left(\frac{\nu-a+3}{2}\right)\Gamma\!\left(\frac{\nu+a+3}{2}\right)}$

$\times\; {}_4F_3\!\left(\begin{array}{c}\frac12,\,1,\,\frac{\nu}{2}+1,\,\frac{\nu+3}{2};\,b^2\\ \frac32,\,\frac{\nu-a+3}{2},\,\frac{\nu+a+3}{2}\end{array}\right)\qquad [\operatorname{Re}\nu>-2;\ |\arg(1-b^2)|<\pi].$

90. $\displaystyle\int_0^{\pi}\sin^\nu x\left\{\begin{array}{c}\sin(ax)\\ \cos(ax)\end{array}\right\}\ln\frac{1+b\sin x}{1-b\sin x}\, dx = \frac{2^{-\nu}\pi b\,\Gamma(\nu+2)}{\Gamma\!\left(\frac{\nu-a+3}{2}\right)\Gamma\!\left(\frac{\nu+a+3}{2}\right)}$

$\times\left\{\begin{array}{c}\sin(\pi a/2)\\ \cos(\pi a/2)\end{array}\right\}{}_4F_3\!\left(\begin{array}{c}\frac12,\,1,\,\frac{\nu}{2}+1,\,\frac{\nu+3}{2};\,b^2\\ \frac32,\,\frac{\nu-a+3}{2},\,\frac{\nu+a+3}{2}\end{array}\right)\qquad [\operatorname{Re}\nu>-2;\ |\arg(1-b^2)|<\pi].$

91. $\displaystyle\int_0^{m\pi}\frac{1}{\sin^2 x}\left\{\begin{array}{c}\sin(ax)\\ \cos(ax)\end{array}\right\}\ln(1+b\sin^2 x)\, dx = \frac{2b\sin(m\pi a/2)}{a}$

$\times\left\{\begin{array}{c}\sin(m\pi a/2)\\ \cos(m\pi a/2)\end{array}\right\}{}_4F_3\!\left(\begin{array}{c}\frac12,\,1,\,1,\,1;\,-b\\ 2,\,1-\frac{a}{2},\,1+\frac{a}{2}\end{array}\right)\qquad [|\arg(1+b)|<\pi].$

92. $\displaystyle\int_0^{\pi/2}\cos^\nu x\cos(ax)\ln(1+b\cos^2 x)\, dx = \frac{2^{-\nu-3}\pi b\,\Gamma(\nu+3)}{\Gamma\!\left(\frac{\nu-a}{2}+2\right)\Gamma\!\left(\frac{\nu+a}{2}+2\right)}$

$\times\; {}_4F_3\!\left(\begin{array}{c}1,\,1,\,\frac{\nu+3}{2},\,\frac{\nu}{2}+2;\,-b\\ 2,\,\frac{\nu-a}{2}+2,\,\frac{\nu+a}{2}+2\end{array}\right)\qquad [\operatorname{Re}\nu>-3;\ |\arg(1+b)|<\pi].$

93. $\displaystyle\int_0^{\pi/2}\frac{\cos(ax)}{\sqrt{1+b^2\cos^2 x}}\ln\!\left(b\cos x+\sqrt{1+b^2\cos^2 x}\right)dx$

$=\dfrac{\cos(a\pi/2)}{1-a^2}\,{}_3F_2\!\left(\begin{array}{c}1,\,1,\,1;\,-b^2\\ \frac{3-a}{2},\,\frac{3+a}{2}\end{array}\right)\qquad [|\arg(1+b^2)|<\pi].$

94. $\displaystyle\int_0^{\pi}\sin^\nu x\left\{\begin{array}{c}\sin(ax)\\ \cos(ax)\end{array}\right\}\ln(1+b\sin^2 x)\, dx = \frac{2^{-\nu-2}\pi b\,\Gamma(\nu+3)}{\Gamma\!\left(\frac{\nu-a}{2}+2\right)\Gamma\!\left(\frac{\nu+a}{2}+2\right)}$

$\times\left\{\begin{array}{c}\sin(a\pi/2)\\ \cos(a\pi/2)\end{array}\right\}{}_4F_3\!\left(\begin{array}{c}1,\,1,\,\frac{\nu+3}{2},\,\frac{\nu}{2}+2;\,-b\\ 2,\,\frac{\nu-a}{2}+2,\,\frac{\nu+a}{2}+2\end{array}\right)\qquad [\operatorname{Re}\nu>-1;\ |\arg(1+b)|<\pi].$

95. $\displaystyle\int_0^\pi \frac{\sin^\nu x}{\sqrt{1+b^2\sin^2 x}} \begin{Bmatrix} \sin(ax) \\ \cos(ax) \end{Bmatrix} \ln\left(b\sin x + \sqrt{1+b^2\sin^2 x}\right) dx$

$= \dfrac{2^{-\nu-1}\pi b\, \Gamma(\nu+2)}{\Gamma\left(\frac{\nu-a+3}{2}\right)\Gamma\left(\frac{\nu+a+3}{2}\right)} \begin{Bmatrix} \sin(a\pi/2) \\ \cos(a\pi/2) \end{Bmatrix}\; {}_4F_3\left(\begin{matrix} 1, 1, \frac{\nu}{2}+1, \frac{\nu+3}{2}; -b^2 \\ \frac{3}{2}, \frac{\nu-a+3}{2}, \frac{\nu+a+3}{2} \end{matrix}\right)$

$[\operatorname{Re}\nu > -1;\ |\arg(1+b^2)| < \pi]$.

96. $\displaystyle\int_0^\pi \sin^\nu x \begin{Bmatrix} \sin(ax) \\ \cos(ax) \end{Bmatrix} \ln\left(b\sin x + \sqrt{1+b^2\sin^2 x}\right) dx$

$= \dfrac{2^{-\nu-1}\pi b\, \Gamma(\nu+2)}{\Gamma\left(\frac{\nu-a+3}{2}\right)\Gamma\left(\frac{\nu+a+3}{2}\right)} \begin{Bmatrix} \sin(a\pi/2) \\ \cos(a\pi/2) \end{Bmatrix}\; {}_4F_3\left(\begin{matrix} \frac{1}{2}, \frac{1}{2}, \frac{\nu}{2}+1, \frac{\nu+3}{2}; -b^2 \\ \frac{3}{2}, \frac{\nu-a+3}{2}, \frac{\nu+a+3}{2} \end{matrix}\right)$

$[\operatorname{Re}\nu > -2;\ |\arg(1+b^2)| < \pi]$.

97. $\displaystyle\int_0^{m\pi} \frac{1}{\sin x} \begin{Bmatrix} \sin(ax) \\ \cos(ax) \end{Bmatrix} \ln \frac{1+b\sin x}{1-b\sin x}\, dx = \frac{4b\sin(m\pi a/2)}{a}$

$\times \begin{Bmatrix} \sin(m\pi a/2) \\ \cos(m\pi a/2) \end{Bmatrix}\; {}_4F_3\left(\begin{matrix} \frac{1}{2}, \frac{1}{2}, 1, 1;\ b^2 \\ \frac{3}{2}, 1-\frac{a}{2}, 1+\frac{a}{2} \end{matrix}\right)$ $[|\arg(1-b^2)| < \pi]$.

98. $\displaystyle\int_0^a x^{s-1}\sin(bx)\ln\frac{a+\sqrt{a^2-x^2}}{x}\, dx = \frac{\pi^{1/2} a^{s+1} b\, \Gamma\left(\frac{s+1}{2}\right)}{2s(s+1)\Gamma\left(\frac{s}{2}\right)}$

$\times {}_2F_3\left(\begin{matrix} \frac{s+1}{2}, \frac{s+1}{2};\ -\frac{a^2 b^2}{4} \\ \frac{3}{2}, \frac{s}{2}+1, \frac{s+3}{2} \end{matrix}\right)$ $[a > 0;\ \operatorname{Re} s > -1]$.

99. $\displaystyle\int_0^a x^{s-1}\cos(bx)\ln\frac{a+\sqrt{a^2-x^2}}{x}\, dx$

$= \dfrac{\pi^{1/2} a^s \Gamma\left(\frac{s}{2}\right)}{2s\,\Gamma\left(\frac{s+1}{2}\right)}\; {}_2F_3\left(\begin{matrix} \frac{s}{2}, \frac{s}{2};\ -\frac{a^2 b^2}{4} \\ \frac{1}{2}, \frac{s+1}{2}, \frac{s}{2}+1 \end{matrix}\right)$ $[a, \operatorname{Re} s > 0]$.

100. $\displaystyle\int_0^a x\sin(bx)\ln\frac{a+\sqrt{a^2-x^2}}{a-\sqrt{a^2-x^2}}\, dx$

$= \dfrac{\pi^2 a}{4b}\left[J_1(ab)\mathbf{H}_0(ab) - J_0(ab)\mathbf{H}_1(ab)\right]$ $[a > 0]$.

101. $\displaystyle\int_0^a \cos(bx)\ln\frac{a+\sqrt{a^2-x^2}}{x}\, dx$

$= \dfrac{\pi a}{4}\left[\pi J_1(ab)\mathbf{H}_0(ab) - \pi J_0(ab)\mathbf{H}_1(ab) + 2J_0(ab)\right]$ $[a > 0]$.

102. $\displaystyle\int_0^a x^2 \cos(bx) \ln\frac{a+\sqrt{a^2-x^2}}{x}\,dx$

$\displaystyle = \frac{\pi a}{2b^2}\left[\pi J_0(ab)\,\mathbf{H}_1(ab) - \pi J_1(ab)\,\mathbf{H}_0(ab) + ab\,J_1(ab)\right] \quad [a>0].$

103. $\displaystyle\int_0^{m\pi} \frac{e^{-ax}}{\sin^2 x} \ln(1+b\sin^2 x)\,dx = \frac{b}{a}(1-e^{-m\pi a})$

$\displaystyle \times\ {}_4F_3\!\left(\begin{array}{c}\frac{1}{2},1,1,1;\ -b\\ 2,\ 1-\frac{ia}{2},\ 1+\frac{ia}{2}\end{array}\right)\quad [\mathrm{Re}\,a>0;\ |\arg(1+b)|<\pi].$

104. $\displaystyle\int_0^{m\pi} \frac{e^{-ax}}{\sin x} \ln\frac{1+b\sin x}{1-b\sin x}\,dx = \frac{2b}{a}(1-e^{-m\pi a})\ {}_4F_3\!\left(\begin{array}{c}\frac{1}{2},\frac{1}{2},1,1;\ b^2\\ \frac{3}{2},\ 1-\frac{ia}{2},\ 1+\frac{ia}{2}\end{array}\right)$

$[|\arg(1+b^2)|<\pi].$

105. $\displaystyle\int_0^\infty \frac{e^{-ax}}{\sin x}\ln(1+b\sin x)\,dx = \frac{b}{a}\ {}_4F_3\!\left(\begin{array}{c}\frac{1}{2},\frac{1}{2},1,1;\ b^2\\ \frac{3}{2},\ \frac{1-ia}{2},\ \frac{1+ia}{2}\end{array}\right)$

$\displaystyle - \frac{b^2}{2(a^2+1)}\ {}_4F_3\!\left(\begin{array}{c}1,1,1,\frac{3}{2};\ b^2\\ 2,\ \frac{3-ia}{2},\ \frac{3+ia}{2}\end{array}\right)\quad [\mathrm{Re}\,a>0;\ |\arg(1-b^2)|<\pi].$

106. $\displaystyle\int_0^\infty e^{-ax}\ln\frac{1+b\sin x}{1-b\sin x}\,dx = \frac{2b}{a^2+1}\ {}_3F_2\!\left(\begin{array}{c}\frac{1}{2},1,1;\ b^2\\ \frac{3-ia}{2},\ \frac{3+ia}{2}\end{array}\right)$

$[\mathrm{Re}\,a>0;\ |\arg(1-b^2)|<\pi].$

107. $\displaystyle\int_0^\infty \frac{e^{-ax}}{\sin x}\ln\frac{1+b\sin x}{1-b\sin x}\,dx = \frac{2b}{a}\ {}_4F_3\!\left(\begin{array}{c}\frac{1}{2},\frac{1}{2},1,1;\ b^2\\ \frac{3}{2},\ 1-\frac{ia}{2},\ 1+\frac{ia}{2}\end{array}\right)$

$[\mathrm{Re}\,a>0;\ |\arg(1-b^2)|<\pi].$

108. $\displaystyle\int_0^\infty \frac{e^{-ax}}{\sin^2 x}\ln(1+b\sin^2 x)\,dx = \frac{b}{a}\ {}_4F_3\!\left(\begin{array}{c}\frac{1}{2},1,1,1;\ -b\\ 2,\ 1-\frac{ia}{2},\ 1+\frac{ia}{2}\end{array}\right)$

$[\mathrm{Re}\,a>0;\ |\arg(1+b)|<\pi].$

109. $\displaystyle\int_0^\infty \frac{dx}{(x+z)^{n+1}(\ln^2 x+\pi^2)} = \frac{(-1)^n}{n!}\,\mathrm{D}_z^n\!\left[\frac{1}{\ln z}\right] - \frac{1}{(z-1)^{n+1}}$

$[|\arg z|<\pi;\ z\ne 1].$

110. $\displaystyle\int_0^1 \left[\left(1 - \frac{x}{2}\ln\frac{1+x}{1-x}\right)^2 + \frac{\pi^2 x^2}{4}\right]^{-1} dx = \frac{4}{5}$ [40].

111. $\displaystyle\int_0^1 x^2\left[\left(1 - \frac{x}{2}\ln\frac{1+x}{1-x}\right)^2 + \frac{\pi^2 x^2}{4}\right]^{-1} dx = \frac{36}{175}$ [40].

112. $\displaystyle\int_0^1 x^4\left[\left(1 - \frac{x}{2}\ln\frac{1+x}{1-x}\right)^2 + \frac{\pi^2 x^2}{4}\right]^{-1} dx = \frac{92}{875}$ [40].

113. $\displaystyle\int_0^1 \frac{1}{a^2 x^2 + 1}\left[\left(1 - \frac{x}{2}\ln\frac{1+x}{1-x}\right)^2 + \frac{\pi^2 x^2}{4}\right]^{-1} dx = \frac{\arctan a}{a - \arctan a} - \frac{3}{a^2}$

[Re $a > 0$; [40]].

114. $\displaystyle\int_0^1 x^6\left[\left(1 - \frac{x}{2}\ln\frac{1+x}{1-x}\right)^2 + \frac{\pi^2 x^2}{4}\right]^{-1} dx = \frac{22548}{336875}$ [40].

115. $\displaystyle\int_{-1}^1 \frac{1}{\sqrt{1-x^2}} \ln\left(1 + 2ax + a^2\right)\ln\left(1 + 2bx + b^2\right) dx = 2\pi\,\text{Li}_2(ab)$.

116. $\displaystyle\int_0^a x^{s-1}(a-x)^{t-1} \ln^2\left(b\sqrt{x(a-x)} + \sqrt{1 + b^2 x(a-x)}\right) dx$

$= a^{s+t+1} b^2\, \text{B}(s+1, t+1)\, {}_5F_4\!\left(\begin{array}{c} 1, 1, 1, s+1, t+1;\ -\dfrac{a^2 b^2}{4} \\ \dfrac{3}{2},\, 2,\, \dfrac{s+t}{2}+1,\, \dfrac{s+t+3}{2} \end{array}\right)$

$[a > 0;\ \text{Re}\,s,\,\text{Re}\,t > -1;\ |\arg(4 + a^2 b^2)| < \pi]$.

117. $\displaystyle\int_0^a x^{1/2}(a-x)^{1/2} \ln^2\left(b\sqrt{x(a-x)} + \sqrt{1 + b^2 x(a-x)}\right) dx$

$= \dfrac{\pi}{16 b^2}\left[(4 + a^2 b^2)\ln\left(1 + \dfrac{a^2 b^2}{4}\right) - a^2 b^2\,\text{Li}_2\!\left(-\dfrac{a^2 b^2}{4}\right) - a^2 b^2\right]$

$[a > 0;\ |\arg(4 + a^2 b^2)| < \pi]$.

118. $\displaystyle\int_0^a x^{1/2}(a-x)^{-1/2} \ln^2\left(b\sqrt{x(a-x)} + \sqrt{1 + b^2 x(a-x)}\right) dx$

$= -\dfrac{\pi a}{4}\,\text{Li}_2\!\left(-\dfrac{a^2 b^2}{4}\right)$ $[a > 0;\ |\arg(4 + a^2 b^2)| < \pi]$.

4.1.5]   4.1. Elementary Functions

119. $\int_0^a x^{1/2}(a-x)^{-3/2}\ln^2\left(b\sqrt{x(a-x)}+\sqrt{1+b^2 x(a-x)}\right)dx$

$$= \frac{\pi}{2}\left[4ab\arctan\frac{ab}{2}-4\ln\left(1+\frac{a^2b^2}{4}\right)-\operatorname{Li}_2\left(-\frac{a^2b^2}{4}\right)\right]$$

$$[a>0;\ |\arg(4+a^2b^2)|<\pi].$$

120. $\int_0^a x^{-1/2}(a-x)^{-1/2}\ln^2\left(b\sqrt{x(a-x)}+\sqrt{1+b^2 x(a-x)}\right)dx$

$$= -\frac{\pi}{2}\operatorname{Li}_2\left(-\frac{a^2b^2}{4}\right) \qquad [a>0;\ |\arg(4+a^2b^2)|<\pi].$$

121. $\int_0^a x^{-1/2}(a-x)^{-3/2}\ln^2\left(b\sqrt{x(a-x)}+\sqrt{1+b^2 x(a-x)}\right)dx$

$$= \frac{2\pi}{a}\left[ab\arctan\frac{ab}{2}-\ln\left(1+\frac{a^2b^2}{4}\right)\right] \qquad [a>0;\ |\arg(4+a^2b^2)|<\pi].$$

122. $\int_0^a x^{-3/2}(a-x)^{-3/2}\ln^2\left(b\sqrt{x(a-x)}+\sqrt{1+b^2 x(a-x)}\right)dx$

$$= \frac{4\pi}{a^2}\left[ab\arctan\frac{ab}{2}-\ln\left(1+\frac{a^2b^2}{4}\right)\right] \qquad [a>0;\ |\arg(4+a^2b^2)|<\pi].$$

123. $\int_0^a x^s(a-x)^{s+1/2}\ln^2\left(b\sqrt[4]{x(a-x)}+\sqrt{1+b^2\sqrt{x(a-x)}}\right)dx$

$$= \frac{2^{-2s-2}\sqrt{\pi}\,a^{2s+5/2}b^2\Gamma(2s+3)}{\Gamma\left(2s+\frac{7}{2}\right)}\,_4F_3\left(\begin{array}{c}1,1,1,2s+3;\\ \frac{3}{2},2,2s+\frac{7}{2}\end{array};-\frac{ab^2}{2}\right)$$

$$[a>0;\ \operatorname{Re} s>-3/2;\ |\arg(2+ab^2)|<\pi].$$

124. $\int_0^a x^{-1/4}(a-x)^{-3/4}\ln^2\left(b\sqrt[4]{x(a-x)}+\sqrt{1+b^2\sqrt{x(a-x)}}\right)dx$

$$= -\frac{\pi}{\sqrt{2}}\operatorname{Li}_2\left(-\frac{ab^2}{2}\right) \qquad [a>0;\ |\arg(2+ab^2)|<\pi].$$

125. $\int_0^a x^{1/4}(a-x)^{-1/4}\ln^2\left(b\sqrt[4]{x(a-x)}+\sqrt{1+b^2\sqrt{x(a-x)}}\right)dx$

$$= -\frac{\pi}{4\sqrt{2}\,b^2}\left[ab^2-(2+ab^2)\ln\left(1+\frac{ab^2}{2}\right)+ab^2\operatorname{Li}_2\left(-\frac{ab^2}{2}\right)\right]$$

$$[a>0;\ |\arg(2+ab^2)|<\pi].$$

126. $\displaystyle\int_0^1 x^{-1} \ln x \ln \frac{a+x}{a-x}\, dx = \text{Li}_3\left(-\frac{1}{a}\right) - \text{Li}_3\left(\frac{1}{a}\right)$ $\quad [a > 1]$.

127. $\displaystyle\int_0^a x^{-1/2} \ln(1-bx) \ln \frac{\sqrt{a}+\sqrt{a-x}}{\sqrt{x}}\, dx = 2\pi\sqrt{a}\left(\ln\frac{1+\sqrt{1-ab}}{2} - 1\right)$

$\quad + \dfrac{4\pi}{\sqrt{b}} \arcsin\sqrt{\dfrac{1-(1-ab)^{1/2}}{2}}$ $\quad [a > 0;\ |\arg(1-ab)| < \pi]$.

128. $\displaystyle\int_0^a \ln(1+bx) \ln \frac{\sqrt{a}+\sqrt{a-x}}{\sqrt{x}}\, dx$

$= -3a + 2\sqrt{\dfrac{a(1+ab)}{b}}\, \ln\left(\sqrt{ab}+\sqrt{1+ab}\right)$

$\quad + \dfrac{1}{b}\ln^2\left(\sqrt{ab}+\sqrt{1+ab}\right)$ $\quad [a > 0;\ |\arg(1+ab)| < \pi]$.

129. $\displaystyle\int_0^1 \frac{1}{x} \ln(1-x) \ln \frac{1+\sqrt{1-x}}{\sqrt{x}}\, dx = \frac{1}{4}\left[7\zeta(3) - 2\pi^2 \ln 2\right]$.

130. $\displaystyle\int_0^a x^{-3/2} \ln(1+bx) \ln \frac{\sqrt{a}+\sqrt{a-x}}{\sqrt{x}}\, dx$

$= \dfrac{2\pi}{\sqrt{a}} \left[1 - \sqrt{ab+1} + \sqrt{ab}\, \ln\left(\sqrt{ab}+\sqrt{ab+1}\right)\right]$

$\quad [a > 0;\ |\arg(1+ab)| < \pi]$.

131. $\displaystyle\int_0^a x^{-3/2} \ln(1-bx) \ln \frac{\sqrt{a}+\sqrt{a-x}}{\sqrt{x}}\, dx$

$= \dfrac{2\pi}{\sqrt{a}}\left(1 - \sqrt{1-ab} - \sqrt{ab}\, \arcsin\sqrt{ab}\right)$ $\quad [a > 0;\ |\arg(1-ab)| < \pi]$.

132. $\displaystyle\int_0^a x^{s-1} \ln\frac{a+\sqrt{a^2-x^2}}{x} \ln\frac{1+bx}{1-bx}\, dx = \dfrac{2\pi^{1/2} a^{s+1} b\, \Gamma\left(\frac{s+1}{2}\right)}{s^2(s+1)\Gamma\left(\frac{s}{2}\right)}$

$\times \left[(s+1)\,_3F_2\!\left(\begin{array}{c} \frac{1}{2},\, 1,\, \frac{s+1}{2};\ a^2b^2 \\ \frac{3}{2},\, \frac{s}{2}+1 \end{array}\right) - \,_3F_2\!\left(\begin{array}{c} 1,\, \frac{s+1}{2},\, \frac{s+1}{2};\ a^2b^2 \\ \frac{s}{2}+1,\, \frac{s+3}{2} \end{array}\right)\right]$

$\quad [a > 0;\ \text{Re}\, s > -1;\ |\arg(1-a^2b^2)| < \pi]$.

133. $\displaystyle\int_0^a x \ln\frac{a+\sqrt{a^2-x^2}}{x}\ln\frac{1+bx}{1-bx}\,dx$

$\displaystyle = \frac{\pi}{2b^2}\left[\arcsin(ab)-4\arcsin\sqrt{\frac{1-(1-a^2b^2)^{1/2}}{2}}+ab\left(2-\sqrt{1-a^2b^2}\right)\right]$

$[a>0;\ |\arg(1-a^2b^2)|<\pi]$.

134. $\displaystyle\int_0^1 \frac{1}{x}\ln\frac{1+\sqrt{1-x^2}}{x}\ln\frac{1+x}{1-x}\,dx = \frac{\pi^2}{2}\ln 2$.

135. $\displaystyle\int_0^a x^{s-1}\ln\left(bx+\sqrt{1+b^2x^2}\right)\ln\frac{a+\sqrt{a^2-x^2}}{x}\,dx = \frac{\pi^{1/2}a^{s+1}b\,\Gamma\!\left(\frac{s+1}{2}\right)}{s^2(s+1)\Gamma\!\left(\frac{s}{2}\right)}$

$\displaystyle \times\left[(s+1)\,_3F_2\!\left(\begin{array}{c}\frac{1}{2},\frac{1}{2},\frac{s+1}{2}\\ \frac{3}{2},\frac{s}{2}+1\end{array};-a^2b^2\right) - {}_3F_2\!\left(\begin{array}{c}\frac{1}{2},\frac{s+1}{2},\frac{s+1}{2}\\ \frac{s}{2}+1,\frac{s+3}{2}\end{array};-a^2b^2\right)\right]$

$[a>0;\ \operatorname{Re} s>-1/2;\ |\arg(1+a^2b^2)|<\pi]$.

136. $\displaystyle\int_0^a \ln\left(bx+\sqrt{b^2x^2+1}\right)\ln\frac{a+\sqrt{a^2-x^2}}{x}\,dx$

$\displaystyle = \frac{ia}{2b}\left[\operatorname{Li}_2(-iab)-\operatorname{Li}_2(iab)\right] + \frac{1}{2b}\ln(1+a^2b^2) - a\arctan(ab)$

$[a>0;\ |\arg(1+a^2b^2)|<\pi]$.

137. $\displaystyle\int_0^1 \ln\left(x+\sqrt{1+x^2}\right)\ln\frac{1+\sqrt{1-x^2}}{x}\,dx = \mathbf{G} - \frac{\pi}{4} + \frac{1}{2}\ln 2$.

138. $\displaystyle\int_0^a x^2\ln\left(bx+\sqrt{b^2x^2+1}\right)\ln\frac{a+\sqrt{a^2-x^2}}{x}\,dx$

$\displaystyle = \frac{ia^3}{12}\left[\operatorname{Li}_2(-iab)-\operatorname{Li}_2(iab)\right]$

$\displaystyle + \frac{1}{36b^3}\left[ab(a^2b^2+9)\arctan(ab) - 4\ln(1+a^2b^2) - 5a^2b^2\right]$

$[a>0;\ |\arg(1+a^2b^2)|<\pi]$.

139. $\displaystyle\int_0^1 x^2\ln\left(x+\sqrt{x^2+1}\right)\ln\frac{1+\sqrt{1-x^2}}{x}\,dx$

$\displaystyle = \frac{1}{72}(12\mathbf{G}+5\pi-8\ln 2-10)$.

140. $\displaystyle\int_0^a \frac{x^{s-1}}{\sqrt{1+b^2x^2}} \ln\left(bx+\sqrt{1+b^2x^2}\right) \ln\frac{a+\sqrt{a^2-x^2}}{x}\,dx$

$$= \frac{\pi^{1/2}a^{s+1}b\,\Gamma\left(\frac{s+1}{2}\right)}{2(s+1)\Gamma\left(\frac{s}{2}+1\right)}\,{}_4F_3\left(\begin{array}{c}1,1,\frac{s+1}{2},\frac{s+1}{2};\\ \frac{3}{2},\frac{s}{2}+1,\frac{s+3}{2}\end{array};-a^2b^2\right)$$

$[a>0;\ \operatorname{Re} s>-1;\ |\arg(1+a^2b^2)|<\pi].$

141. $\displaystyle\int_0^a \frac{x}{\sqrt{b^2x^2+1}} \ln\left(bx+\sqrt{b^2x^2+1}\right) \ln\frac{a+\sqrt{a^2-x^2}}{x}\,dx$

$= \dfrac{\pi}{2b^2}\arctan(ab) + \dfrac{\pi a}{4b}[\ln(a^2b^2+1)-2] \quad [a>0;\ |\arg(1+a^2b^2)|<\pi].$

142. $\displaystyle\int_0^a \frac{x^{-1}}{\sqrt{b^2x^2+1}} \ln\left(bx+\sqrt{b^2x^2+1}\right) \ln\frac{a+\sqrt{a^2-x^2}}{x}\,dx$

$= \dfrac{\pi i}{8}[\operatorname{Li}_2(-a^2b^2) - 4\operatorname{Li}_2(iab)] \quad [a>0;\ |\arg(1+a^2b^2)|<\pi].$

143. $\displaystyle\int_0^1 \frac{1}{x\sqrt{x^2+1}} \ln\left(x+\sqrt{x^2+1}\right) \ln\frac{1+\sqrt{1-x^2}}{x}\,dx = \frac{\pi\mathbf{G}}{2}.$

144. $\displaystyle\int_0^1 x^{s-1}\ln\left(ax+\sqrt{1+a^2x^2}\right)\ln\frac{1+\sqrt{1-x^2}}{1-\sqrt{1-x^2}}\,dx = \frac{2\pi^{1/2}a\,\Gamma\left(\frac{s+1}{2}\right)}{s^2(s+1)\Gamma\left(\frac{s}{2}\right)}$

$\times\left[(s+1)\,{}_3F_2\left(\begin{array}{c}\frac{1}{2},\frac{1}{2},\frac{s+1}{2};\\ \frac{3}{2},\frac{s}{2}+1;\end{array}-a^2\right) - {}_3F_2\left(\begin{array}{c}\frac{1}{2},\frac{s+1}{2},\frac{s+1}{2};\\ \frac{s}{2}+1,\frac{s+3}{2};\end{array}-a^2\right)\right]$

$[|\arg(1+a^2)|<\pi;\ \operatorname{Re} s>0].$

145. $\displaystyle\int_0^1 \ln\left(ax+\sqrt{a^2x^2+1}\right)\ln\frac{1+\sqrt{1-x^2}}{1-\sqrt{1-x^2}}\,dx = i[\operatorname{Li}_2(-ia)-\operatorname{Li}_2(ia)]$

$\qquad - 2\arctan a + \dfrac{1}{a}\ln(a^2+1) \quad [|\arg(1+a^2)|<\pi].$

146. $\displaystyle\int_0^1 x^{s-1}\ln\frac{a+x}{a-x}\ln\frac{1+\sqrt{1-x^2}}{1-\sqrt{1-x^2}}\,dx = \frac{4\pi^{1/2}\Gamma\left(\frac{s+1}{2}\right)}{as^2(s+1)\Gamma\left(\frac{s}{2}\right)}$

$\times\left[(s+1)\,{}_3F_2\left(\begin{array}{c}\frac{1}{2},1,\frac{s+1}{2};\\ \frac{3}{2},\frac{s}{2}+1;\end{array}a^{-2}\right) - {}_3F_2\left(\begin{array}{c}1,\frac{s+1}{2},\frac{s+1}{2};\\ \frac{s}{2}+1,\frac{s+3}{2};\end{array}a^{-2}\right)\right]$

$[|\arg(1+a^{-2})|<\pi;\ \operatorname{Re} s>0].$

147. $\displaystyle\int_0^1 \ln\frac{1+x}{1-x} \ln\frac{1+\sqrt{1-x^2}}{1-\sqrt{1-x^2}}\, dx = 8\mathrm{G} - \frac{\pi^2}{2}.$

148. $\displaystyle\int_0^1 x \ln\frac{1+x}{1-x} \ln\frac{1+\sqrt{1-x^2}}{1-\sqrt{1-x^2}}\, dx = 2\pi - \frac{\pi^2}{2}.$

149. $\displaystyle\int_0^1 \frac{1}{x} \ln\frac{1+x}{1-x} \ln\frac{1+\sqrt{1-x^2}}{1-\sqrt{1-x^2}}\, dx = \pi^2 \ln 2.$

150. $\displaystyle\int_0^1 \frac{1}{\sqrt{x}} \ln\left(\sqrt{x}+\sqrt{x+1}\right) \ln\frac{1+\sqrt{1-x}}{1-\sqrt{1-x}}\, dx = 4\mathrm{G} - \pi + 2\ln 2.$

151. $\displaystyle\int_0^a x^{s-1} \ln\frac{a+\sqrt{a^2-x^2}}{x} \ln^2\left(bx+\sqrt{1+b^2x^2}\right) dx$

$$= \frac{\pi^{1/2} a^{s+2} b^2 \Gamma\!\left(\frac{s}{2}+1\right)}{2(s+2)\Gamma\!\left(\frac{s+3}{2}\right)}\, {}_5F_4\!\left(\begin{array}{c}1,1,1,\frac{s}{2}+1,\frac{s}{2}+1 \\ \frac{3}{2},2,\frac{s+3}{2},\frac{s}{2}+2;\, -a^2b^2\end{array}\right)$$

$[a>0;\ \mathrm{Re}\, s > -2;\ |\arg(1+a^2b^2)|<\pi].$

152. $\displaystyle\int_0^a \ln^2\left(bx+\sqrt{b^2x^2+1}\right) \ln\frac{a+\sqrt{a^2-x^2}}{x}\, dx = \frac{\pi a}{4}[4 - 2\ln(a^2b^2+1)$

$- \mathrm{Li}_2(-a^2b^2)] - \frac{\pi}{b}\arctan(ab) \quad [a>0;\ |\arg(1+a^2b^2)|<\pi].$

153. $\displaystyle\int_0^1 \ln^2\left(x+\sqrt{x^2+1}\right) \ln\frac{1+\sqrt{1-x^2}}{x}\, dx$

$$= \frac{\pi}{48}(48 - 12\pi + \pi^2 - 24\ln 2) \quad [a>0].$$

154. $\displaystyle\int_0^1 x^2 \ln^2\left(bx+\sqrt{b^2x^2+1}\right) \ln\frac{a+\sqrt{a^2-x^2}}{x}\, dx$

$$= \frac{\pi}{216b^3}[48\arctan(ab) + 3ab(a^2b^2+9)\ln(1+a^2b^2) - 9a^3b^3 \mathrm{Li}_2(-a^2b^2)$$

$- ab(11a^2b^2+48)] \quad [a>0;\ |\arg(1+a^2b^2)|<\pi].$

155. $\displaystyle\int_0^1 x^2 \ln^2\left(x+\sqrt{x^2+1}\right) \ln\frac{1+\sqrt{1-x^2}}{x}\, dx$

$$= \frac{\pi}{864}(48\pi + 3\pi^2 + 120\ln 2 - 236).$$

156. $\int_0^a x^{-2} \ln^2\left(bx + \sqrt{b^2x^2+1}\right) \ln \dfrac{a + \sqrt{a^2 - x^2}}{x} \, dx$

$= \dfrac{i\pi b}{2}[\operatorname{Li}_2(-iab) - \operatorname{Li}_2(iab)] - \pi b \arctan(ab) + \dfrac{\pi}{2a} \ln(a^2b^2 + 1)$

$\qquad\qquad\qquad\qquad\qquad [a > 0;\ |\arg(1 + a^2b^2)| < \pi].$

157. $\int_0^a x^{-2} \ln^2\left(x + \sqrt{x^2+1}\right) \ln \dfrac{1 + \sqrt{1 - x^2}}{x} \, dx = \dfrac{\pi}{4}(4G - \pi + 2\ln 2).$

158. $\int_0^1 x^a(1-x)^{-1/2}(x+3)^{(a-1)/2} \ln^n(x^3 + 3x^2) \, dx$

$= \dfrac{\sqrt{\pi}}{3} \mathbf{D}_a^n \left[2^{a+n} \dfrac{\Gamma\left(\dfrac{a+1}{2}\right)}{\Gamma\left(\dfrac{a}{2}+1\right)} \right] \qquad [\operatorname{Re} a > 0].$

159. $\int_0^1 \dfrac{x^a(1-x)^{-a/2-3/4}}{\sqrt{2 + (\sqrt{2}-1)x}} \ln^n \dfrac{x^2}{1-x} \, dx$

$= -\mathbf{D}_a^n \left[2^{(a+1)/2+n} \cos \dfrac{(1-2a)\pi}{4} \dfrac{\Gamma(a+1)\Gamma\left(\dfrac{2a+5}{4}\right)\Gamma\left(-a-\dfrac{1}{2}\right)}{\Gamma\left(\dfrac{a+3}{4}\right)\Gamma\left(\dfrac{a}{4}+1\right)} \right]$

$\qquad\qquad\qquad\qquad\qquad [-1 < \operatorname{Re} a < 1/2].$

160. $\int_0^1 x^a(2-x)^{-a-1} \ln^n \dfrac{x}{2-x} \, dx = \dfrac{1}{2} \mathbf{D}_a^n \left[\psi\left(\dfrac{a}{2}+1\right) - \psi\left(\dfrac{a+1}{2}\right) \right]$

$\qquad\qquad\qquad\qquad\qquad [-1 < \operatorname{Re} a < 0].$

161. $\int_0^1 x^a(1-x)^{-3a/2}(3-x)^{a/2} \ln^n \dfrac{x^2(3-x)}{(1-x)^3} \, dx$

$= \mathbf{D}_a^n \left[\dfrac{2^{n-a} 3^{3a/2} \Gamma(a+1) \Gamma\left(1 - \dfrac{3a}{2}\right)}{\Gamma\left(1 - \dfrac{a}{2}\right)} \right] \qquad [\operatorname{Re} a > -1].$

162. $\int_0^1 x^a(1-x)^{-3a-3/2}(4-x)^{-1/2} \ln^n \dfrac{x}{(1-x)^3} \, dx$

$= -\mathbf{D}_a^n \left[\dfrac{\cos(a\pi)\Gamma(2a+2)\Gamma\left(-3a - \dfrac{3}{2}\right)}{\Gamma\left(\dfrac{1}{2} - a\right)} \right] \qquad [-1 < \operatorname{Re} a < -1/6].$

4.1.5] 4.1. ELEMENTARY FUNCTIONS

163. $\displaystyle\int_0^1 x^{-1/2}(1-x)^{2a+1}(4-x)^a \ln^n \frac{(1-x)^2}{4-x}\, dx$

$$= \frac{2\sqrt{\pi}}{3} D_a^n \left[\frac{2^{2a}\Gamma(a+1)}{\Gamma\left(a+\frac{3}{2}\right)}\right] \qquad [\mathrm{Re}\, a > -1].$$

164. $\displaystyle\int_0^1 x^a(1-x)^{-2a-5/3}(9-x)^{2a+1} \ln^n \frac{x(9-x)^2}{(1-x)^2}\, dx$

$$= \frac{1}{\Gamma\left(\frac{2}{3}\right)} D_a^n\left[3^{3a+1}\Gamma(a+1)\Gamma\left(-a-\frac{1}{3}\right)\right] \qquad [-1 < \mathrm{Re}\, a < -1/3].$$

165. $\displaystyle\int_0^1 x^a(1-x)^{-a/2-2/3}(9-x)^{a/2} \ln^n \frac{x^2(9-x)}{1-x}\, dx$

$$= -\frac{2^{n+1}}{\Gamma\left(\frac{2}{3}\right)} D_a^n\left[3^{3a/2}\cos\frac{(2\pi-3a)\pi}{6}\Gamma(a+1)\Gamma\left(-a-\frac{1}{3}\right)\right]$$

$$[-1 < \mathrm{Re}\, a < -1/3].$$

166. $\displaystyle\int_0^1 x^a(1-x)^{-a/3-1/2}(x+8)^{a/3} \ln^n \frac{x^3(x+8)}{1-x}\, dx$

$$= 3^n\sqrt{\pi}\, D_a^n\left[\frac{2^{2a}\sec\frac{a\pi}{3}\Gamma(a+1)}{\Gamma\left(a+\frac{3}{2}\right)}\right] \qquad [-1 < \mathrm{Re}\, a < 3/2].$$

167. $\displaystyle\int_0^1 x^a(1-x)^{-2a-5/3}(x+8)^{-a-4/3} \ln^n \frac{x}{(1-x)^2(x+8)}\, dx$

$$= \frac{1}{\Gamma\left(\frac{5}{3}\right)} D_a^n\left[3^{-3a-4}\Gamma(a+1)\Gamma\left(-a-\frac{1}{3}\right)\right] \qquad [-1 < \mathrm{Re}\, a < -1/3].$$

168. $\displaystyle\int_0^1 x^a(1-x)^{-2a-4/3}(x+8)^{-a-2/3} \ln^n \frac{x}{(1-x)^2(x+8)}\, dx$

$$= \frac{2}{9\Gamma\left(\frac{2}{3}\right)} D_a^n\left[3^{-3a}\cos(a\pi)\Gamma(2a+1)\Gamma\left(-2a-\frac{1}{3}\right)\right] \qquad [-1 < \mathrm{Re}\, a < -1/6].$$

169. $\displaystyle\int_0^1 \frac{x^{-1/2}(1-x)^{a-3/4}}{(2+(\sqrt{2}-1)x)^{2a}} \ln^n \frac{1-x}{(2+(\sqrt{2}-1)x)^2}\, dx$

$$= \pi D_a^n\left[2^{-3a}\frac{\Gamma\left(a+\frac{1}{4}\right)}{\Gamma\left(\frac{2a+3}{4}\right)\Gamma\left(\frac{a+1}{2}\right)}\right] \qquad [-1 < \mathrm{Re}\, a < 1/2].$$

170. $\displaystyle\int_0^1 \frac{x^a(1-x)^{-3a-3/2}}{\sqrt{4-x}} \ln^n \frac{x}{(1-x)^3}\, dx$

$$= \frac{\sqrt{\pi}}{3} \mathrm{D}_a^n \left[2^{2a+1} \frac{\Gamma(a+1)\Gamma\left(-\frac{1}{2}-3a\right)}{\Gamma^2\left(\frac{1}{2}-a\right)} \right] \qquad [-1 < \mathrm{Re}\, a < -1/6].$$

171. $\displaystyle\int_0^1 x^a (1-x)^{-3a-3/2}(x+3)^{2a} \ln^n \frac{x^3+3x^2}{(1-x)^3}\, dx$

$$= \sqrt{\pi}\, \mathrm{D}_a^n \left[3^{3a} \frac{\Gamma(a+1)\Gamma\left(-3a-\frac{1}{2}\right)}{\Gamma^2\left(\frac{1}{2}-a\right)} \right] \qquad [-1 < \mathrm{Re}\, a < -1/6].$$

172. $\displaystyle\int_0^1 x^a (1-x)^{-3(2a+1)/2}(x+3)^{-a-1/2} \ln^n \frac{x}{(1-x)^3(x+3)}\, dx$

$$= \frac{\Gamma\left(\frac{4}{3}\right)}{\sqrt{\pi}}\, \mathrm{D}_a^n \left[2^{-2-12a} 3^{3a+3/2} \cos(2a\pi) \frac{\Gamma(4a+1)\Gamma\left(\frac{1}{2}-2a\right)\Gamma\left(-\frac{1}{2}-3a\right)}{\Gamma\left(\frac{1}{3}-2a\right)\Gamma\left(\frac{1}{2}+a\right)} \right]$$
$$[-1 < \mathrm{Re}\, a < -1/6].$$

173. $\displaystyle\int_0^1 x^a (1-x)^{3a}(9-8x)^{-a-1/2} \ln^n \frac{x(1-x)^3}{9-8x}\, dx$

$$= \frac{\sqrt{\pi}}{4}\, \mathrm{D}_a^n \left[\frac{2^{-6a}\Gamma(3a+1)}{\Gamma\left(3a+\frac{3}{2}\right)} \right] \qquad [\mathrm{Re}\, a > -1].$$

174. $\displaystyle\int_0^1 x^a (1-x)^{(a-2)/3}(9-8x)^{-a-1/2} \ln^n \frac{x^3(1-x)}{(9-8x)^3}\, dx$

$$= \frac{3^n \sqrt{\pi}}{2}\, \mathrm{D}_a^n \left[\frac{2^{-2a}\Gamma\left(\frac{a+1}{3}\right)}{\Gamma\left(\frac{2a+5}{6}\right)} \right] \qquad [\mathrm{Re}\, a > -1].$$

175. $\displaystyle\int_0^1 x^a (1-x)^{a/3}(9-8x)^{-a-3/2} \ln^n \frac{x^3(1-x)}{(9-8x)^3}\, dx$

$$= \frac{\sqrt{\pi}}{12}\, \mathrm{D}_a^n \left[\frac{2^{-2a}\Gamma(a+1)}{\Gamma\left(a+\frac{3}{2}\right)} \right] \qquad [\mathrm{Re}\, a > -1].$$

176. $\int_0^1 x^a (1-x)^{-(a+b)/2-1} (1+\sqrt{1-x})^{2b} \ln^n\left(\frac{x}{\sqrt{1-x}}\right) dx$

$$= D_a^n \left[2^{2a-2b+1} B\left(\frac{a+1}{2}, b-a\right)\right] \quad [-1 < \operatorname{Re} a < -\operatorname{Re} b].$$

177. $\int_0^1 x^a (1-x)^b (2-x)^{-a-2b-2} \ln^m \frac{x}{2-x} \ln^n \frac{1-x}{(2-x)^2} dx$

$$= D_a^m D_b^n \left[\frac{2^{-2b-2} \Gamma\left(\frac{a+1}{2}\right) \Gamma(b+1)}{\Gamma\left(\frac{a+3}{2}+b\right)} \right] \quad [\operatorname{Re} a, \operatorname{Re} b > -1].$$

178. $\int_0^1 x^{a-1} (1-x)^{b-2a} (1+x)^{-b} \ln^m \frac{1-x}{1+x} \ln^n \frac{x}{(1-x)^2} dx$

$$= D_b^m D_a^n \left[\frac{\Gamma(a) \Gamma\left(\frac{b-2a+1}{2}\right)}{2^{2a} \Gamma\left(\frac{b+1}{2}\right)} \right] \quad [0 < \operatorname{Re} a < \operatorname{Re}(b+1)/2].$$

179. $\int_0^1 \frac{x^{a-1}}{(x^b+1)^{n+1}} \ln^{m+n} x \ln \ln \frac{1}{x} dx$

$$= \frac{(-1)^n}{n!} D_a^m D_b^n \left[\frac{C + \ln(2b)}{2b} \left\{ \psi\left(\frac{a}{2b}\right) - \psi\left(\frac{a+b}{2b}\right)\right\} \right.$$
$$\left. + \frac{1}{2b} \left\{ \zeta'\left(1, \frac{a}{2b}\right) - \zeta'\left(1, \frac{a+b}{2b}\right)\right\}\right] \quad [\operatorname{Re} a > 0;\ [A2],\ (21)].$$

4.1.6. Inverse trigonometric functions

1. $\int_0^a (a^2 - x^2)^{-1/2} \arcsin(bx) dx = \frac{1}{2} [\operatorname{Li}_2(ab) + \operatorname{Li}_2(-ab)]$

$$[|\arg(1-a^2 b^2)| < \pi].$$

2. $\int_0^a (a^2 - x^2)^{1/2} \arcsin(bx) dx$

$$= \frac{a}{4b} \left\{ \frac{1-a^2 b^2}{2ab} \ln \frac{1+ab}{1-ab} + ab[\operatorname{Li}_2(ab) - \operatorname{Li}_2(-ab)] - 1 \right\}$$

$$[|\arg(1-a^2 b^2)| < \pi].$$

3. $\int_0^a x(a^2 - x^2)^{1/2} \arcsin(bx) dx$

$$= \frac{a^2}{9b} [2(1 - 2a^2 b^2) D(ab) - (1 - 3a^2 b^2) K(ab)] \quad [|\arg(1-a^2 b^2)| < \pi].$$

4. $\displaystyle\int_0^a x(a^2-x^2)^{-1/2}\arcsin(bx)\,dx = a^2 b[\mathbf{K}(ab) - \mathbf{D}(ab)]$

$$[|\arg(1-a^2b^2)| < \pi].$$

5. $\displaystyle\int_0^1 \frac{(1-x^2)^{1/2}}{x(1-b^2x^2)^{1/2}}\arcsin(bx)\,dx = \frac{\pi}{4}\left[\ln\frac{1+b}{1-b} + \frac{1}{b}\ln(1-b^2)\right]$

$$[|\arg(1-b^2)| < \pi].$$

6. $\displaystyle\int_0^1 \frac{(1-x^2)^{-1/2}}{x(1-b^2x^2)^{1/2}}\arcsin(bx)\,dx = \frac{\pi}{4}\ln\frac{1+b}{1-b}$ $\qquad [|\arg(1-b^2)| < \pi].$

7. $\displaystyle\int_0^1 \frac{x(1-x^2)^{s-1}}{(1-x^2z)^{s+1}}\arcsin x\,dx = \frac{\sqrt{\pi}}{4}\frac{\Gamma\left(s+\frac{1}{2}\right)}{s^2\Gamma(s)}(1-z)^{-1}\,{}_2F_1\!\left(\begin{array}{c}\frac{1}{2},\,s\\s+1;\,z\end{array}\right)$

$$[\mathrm{Re}\,s > 0;\ |1-z| < \pi].$$

8. $\displaystyle\int_0^a x^{s-1}(a-x)^{t-1}\arcsin\!\left(b\sqrt{x(a-x)}\right)dx$

$= a^{s+t}b\,\mathrm{B}\!\left(s+\tfrac{1}{2},\,t+\tfrac{1}{2}\right){}_4F_3\!\left(\begin{array}{c}\frac{1}{2},\frac{1}{2},\,s+\frac{1}{2},\,t+\frac{1}{2};\,\frac{a^2b^2}{4}\\\frac{3}{2},\,\frac{s+t+1}{2},\,\frac{s+t}{2}+1\end{array}\right)$

$$[a,\,\mathrm{Re}\,s,\,\mathrm{Re}\,t > -1/2;\ |\arg(4-a^2b^2)| < \pi].$$

9. $\displaystyle\int_0^a \arcsin\!\left(b\sqrt{x(a-x)}\right)dx = \frac{a^2 b}{2}\left[\mathbf{K}\!\left(\frac{ab}{2}\right) - \mathbf{D}\!\left(\frac{ab}{2}\right)\right]$

$$[a > 0;\ |\arg(4-a^2b^2)| < \pi].$$

10. $\displaystyle\int_0^a x\arcsin\!\left(b\sqrt{x(a-x)}\right)dx = \frac{a^3 b}{4}\left[\mathbf{K}\!\left(\frac{ab}{2}\right) - \mathbf{D}\!\left(\frac{ab}{2}\right)\right]$

$$[a > 0;\ |\arg(4-a^2b^2)| < \pi].$$

11. $\displaystyle\int_0^a x^{1/2}(a-x)^{-1/2}\arcsin\!\left(b\sqrt{x(a-x)}\right)dx$

$= \dfrac{a}{2}\left[\mathrm{Li}_2\!\left(\dfrac{ab}{2}\right) - \mathrm{Li}_2\!\left(-\dfrac{ab}{2}\right)\right]\qquad [a > 0;\ |\arg(4-a^2b^2)| < \pi].$

12. $\displaystyle\int_0^a x^{-1/2}(a-x)^{-1/2}\arcsin\!\left(b\sqrt{x(a-x)}\right)dx = \mathrm{Li}_2\!\left(\frac{ab}{2}\right) - \mathrm{Li}_2\!\left(-\frac{ab}{2}\right)$

$$[a > 0;\ |\arg(4-a^2b^2)| < \pi].$$

13. $$\int_0^a \frac{1}{\sqrt{1-b^2x(a-x)}} \arcsin\left(b\sqrt{x(a-x)}\right) dx = -\frac{\pi}{2b} \ln\left(1 - \frac{a^2b^2}{4}\right)$$
$$[a>0;\ |\arg(4-a^2b^2)|<\pi].$$

14. $$\int_0^a \frac{x^{-1}}{\sqrt{1-b^2x(a-x)}} \arcsin\left(b\sqrt{x(a-x)}\right) dx = \frac{\pi}{2} \ln\frac{2+ab}{2-ab}$$
$$[a>0;\ |\arg(4-a^2b^2)|<\pi].$$

15. $$\int_0^a \frac{x^{-1}(a-x)^{-1}}{\sqrt{1-b^2x(a-x)}} \arcsin\left(b\sqrt{x(a-x)}\right) dx = \frac{\pi}{a} \ln\frac{2+ab}{2-ab}$$
$$[a>0;\ |\arg(4-a^2b^2)|<\pi].$$

16. $$\int_{-a}^a \frac{(x+a)^{-1}}{(x^2+a^2)^{1/2}} \arcsin\frac{b(x+a)}{\sqrt{x^2+a^2}}\, dx = \frac{\pi b}{2a}\, _3F_2\!\left(\begin{array}{c}\frac{1}{2},\frac{1}{2},\frac{1}{2}\\ 1,\frac{3}{2};\ 2b^2\end{array}\right)$$
$$[a>0;\ |\arg(1-2b^2)|<\pi].$$

17. $$\int_{-a}^a \frac{x+a}{(x^2+a^2)^{3/2}} \arcsin\frac{b(x+a)}{\sqrt{x^2+a^2}}\, dx = \frac{2b}{a}\left[\mathbf{K}\!\left(\sqrt{2}b\right) - \mathbf{D}\!\left(\sqrt{2}b\right)\right]$$
$$[a>0;\ |\arg(1-2b^2)|<\pi].$$

18. $$\int_0^a x^{s+1/2}(a-x)^s \arcsin\left(b\sqrt[4]{x(a-x)}\right) dx$$
$$= 2^{-2s-3/2}\pi^{1/2}a^{2s+2}b\,\frac{\Gamma\!\left(2s+\frac{5}{2}\right)}{\Gamma(2s+3)}\, _3F_2\!\left(\begin{array}{c}\frac{1}{2},\frac{1}{2},2s+\frac{5}{2}\\ \frac{3}{2},2s+3;\ \frac{ab^2}{2}\end{array}\right)$$
$$[a>0;\ \operatorname{Re} s>-1;\ |\arg(2-ab^2)|<\pi].$$

19. $$\int_0^a x^{1/2} \arcsin\left(b\sqrt[4]{x(a-x)}\right) dx$$
$$= \frac{2^{1/2}}{9b^3}\left[(3a^2b^4-4ab^2-4)\,\mathbf{K}\!\left(b\sqrt{\frac{a}{2}}\right) + (5ab^2+4)\,\mathbf{E}\!\left(b\sqrt{\frac{a}{2}}\right)\right]$$
$$[a>0;\ |\arg(2-ab^2)|<\pi].$$

20. $$\int_0^a x^{-1/2} \arcsin\left(b\sqrt[4]{x(a-x)}\right) dx$$
$$= \frac{2^{1/2}}{b}\left[(ab^2-2)\,\mathbf{K}\!\left(b\sqrt{\frac{a}{2}}\right) + 2\,\mathbf{E}\!\left(b\sqrt{\frac{a}{2}}\right)\right] \quad [a>0;\ |\arg(2-ab^2)|<\pi].$$

21. $\displaystyle\int_0^a x^{-1/4}(a-x)^{1/4} \arcsin\left(b\sqrt[4]{x(a-x)}\right) dx$

$$= \frac{a^{1/2}}{2b}\left\{1 - \frac{2-ab^2}{2^{3/2}a^{1/2}b} \ln\frac{\sqrt{2}+\sqrt{a}\,b}{\sqrt{2}-\sqrt{a}\,b}\right.$$
$$\left.+ b\sqrt{\frac{a}{2}}\left[\text{Li}_2\left(b\sqrt{\frac{a}{2}}\right) - \text{Li}_2\left(-b\sqrt{\frac{a}{2}}\right)\right]\right\} \qquad [a>0;\ |\arg(2-ab^2)|<\pi].$$

22. $\displaystyle\int_0^a x^{-1/4}(a-x)^{-3/4} \arcsin\left(b\sqrt[4]{x(a-x)}\right) dx$

$$= \sqrt{2}\left[\text{Li}_2\left(b\sqrt{\frac{a}{2}}\right) - \text{Li}_2\left(-b\sqrt{\frac{a}{2}}\right)\right] \qquad [a>0;\ |\arg(2-ab^2)|<\pi].$$

23. $\displaystyle\int_0^a \frac{x(a-x)^{1/2}}{\sqrt{1-b^2\sqrt{x(a-x)}}} \arcsin\left(b\sqrt[4]{x(a-x)}\right) dx$

$$= -\frac{\pi}{64\sqrt{2}\,b^5}\left[ab^2(7ab^2+12) + 2(3a^2b^4+4ab^2+12)\ln\left(1-\frac{ab^2}{2}\right)\right]$$
$$[a>0;\ |\arg(2-ab^2)|<\pi].$$

24. $\displaystyle\int_0^a \frac{x^{1/2}}{\sqrt{1-b^2\sqrt{x(a-x)}}} \arcsin\left(b\sqrt[4]{x(a-x)}\right) dx$

$$= -\frac{\pi}{2^{5/2}b^3}\left[ab^2 + (ab^2+2)\ln\left(1-\frac{ab^2}{2}\right)\right] \qquad [a>0;\ |\arg(2-ab^2)|<\pi].$$

25. $\displaystyle\int_0^a \frac{x^{-1/2}}{\sqrt{1-b^2\sqrt{x(a-x)}}} \arcsin\left(b\sqrt[4]{x(a-x)}\right) dx = -\frac{\pi}{\sqrt{2}\,b}\ln\left(1-\frac{ab^2}{2}\right)$

$$[a>0;\ |\arg(2-ab^2)|<\pi].$$

26. $\displaystyle\int_0^a \frac{x^{-1/2}(a-x)^{-1}}{\sqrt{1-b^2\sqrt{x(a-x)}}} \arcsin\left(b\sqrt[4]{x(a-x)}\right) dx = \frac{\pi}{\sqrt{a}}\ln\frac{\sqrt{2}+\sqrt{a}\,b}{\sqrt{2}-\sqrt{a}\,b}$

$$[a>0;\ |\arg(2-ab^2)|<\pi].$$

27. $\displaystyle\int_0^{\pi/2} \cos^\nu x \cos(ax) \arcsin(b\cos x)\, dx = \frac{2^{-\nu-2}\pi b\,\Gamma(\nu+2)}{\Gamma\left(\frac{\nu-a+3}{2}\right)\Gamma\left(\frac{\nu+a+3}{2}\right)}$

$$\times\ {}_4F_3\left(\begin{matrix}\frac{1}{2},\frac{1}{2},1+\frac{\nu}{2},\frac{\nu+3}{2};\ b^2\\ \frac{3}{2},\frac{\nu-a+3}{2},\frac{\nu+a+3}{2}\end{matrix}\right) \qquad [\text{Re}\,\nu>-2;\ |\arg(1-b^2)|<\pi].$$

28. $\displaystyle\int_0^{\pi/2} \frac{\cos(2nx)}{\sin x} \arcsin(a\sin x)\,dx = (-1)^n \frac{2^{-2n-1}\sqrt{\pi}\,a^{2n+1}\Gamma\left(n+\frac{1}{2}\right)}{n!(2n+1)}$

$$\times {}_3F_2\left(\begin{array}{c} n+\frac{1}{2},\, n+\frac{1}{2},\, n+\frac{1}{2} \\ n+\frac{3}{2},\, 2n+1;\, a^2 \end{array}\right) \qquad [|\arg(1-a^2)| < \pi].$$

29. $\displaystyle\int_0^{\pi/2} \frac{\cos(2nx)}{\sin x\sqrt{1-a^2\sin^2 x}} \arcsin(a\sin x)\,dx = (-1)^n \frac{n!\,\pi^{3/2}\,a^{2n+1}}{2^{2n+2}\Gamma\left(n+\frac{3}{2}\right)}$

$$\times {}_3F_2\left(\begin{array}{c} n+\frac{1}{2},\, n+1,\, n+1 \\ n+\frac{3}{2},\, 2n+1;\, a^2 \end{array}\right) \qquad [|\arg(1-a^2)| < \pi].$$

30. $\displaystyle\int_0^{m\pi} \frac{1}{\sin x} \left\{\begin{array}{c}\sin(ax)\\ \cos(ax)\end{array}\right\} \arcsin(b\sin x)\,dx = \frac{2b}{a}\sin\frac{m\pi a}{2}$

$$\times \left\{\begin{array}{c}\sin(m\pi a/2)\\ \cos(m\pi a/2)\end{array}\right\} {}_4F_3\left(\begin{array}{c} \frac{1}{2},\frac{1}{2},\frac{1}{2},1;\, b^2 \\ \frac{3}{2},\, 1-\frac{a}{2},\, 1+\frac{a}{2} \end{array}\right) \qquad [|\arg(1-b^2)| < \pi].$$

31. $\displaystyle\int_0^{m\pi} \frac{1}{\sin x\sqrt{1-b^2\sin^2 x}} \left\{\begin{array}{c}\sin(ax)\\ \cos(ax)\end{array}\right\} \arcsin(b\sin x)\,dx = \frac{2b}{a}\sin\frac{m\pi a}{2}$

$$\times \left\{\begin{array}{c}\sin(m\pi a/2)\\ \cos(m\pi a/2)\end{array}\right\} {}_4F_3\left(\begin{array}{c} \frac{1}{2},1,1,1;\, b^2 \\ \frac{3}{2},\, 1-\frac{a}{2},\, 1+\frac{a}{2} \end{array}\right) \qquad [|\arg(1-b^2)| < \pi].$$

32. $\displaystyle\int_0^{\pi} \sin^\nu x \left\{\begin{array}{c}\sin(ax)\\ \cos(ax)\end{array}\right\} \arcsin(b\sin x)\,dx = \frac{2^{-\nu-1}\pi b\,\Gamma(\nu+2)}{\Gamma\left(\frac{\nu-a+3}{2}\right)\Gamma\left(\frac{\nu+a+3}{2}\right)}$

$$\times \left\{\begin{array}{c}\sin(a\pi/2)\\ \cos(a\pi/2)\end{array}\right\} {}_4F_3\left(\begin{array}{c} \frac{1}{2},\frac{1}{2},\, 1+\frac{\nu}{2},\, \frac{\nu+3}{2};\, b^2 \\ \frac{3}{2},\, \frac{\nu-a+3}{2},\, \frac{\nu+a+3}{2} \end{array}\right) \qquad [\mathrm{Re}\,\nu > -2;\; |\arg(1-b^2)| < \pi].$$

33. $\displaystyle\int_0^{\pi} \frac{\cos(nx)}{\cos x} \arcsin(a\cos x)\,dx$

$$= \frac{2^{-n}\sqrt{\pi}\,a^{n+1}\Gamma\left(\frac{n+1}{2}\right)}{(n+1)\Gamma\left(\frac{n}{2}+1\right)} \cos^2\frac{n\pi}{2}\, {}_3F_2\left(\begin{array}{c} \frac{n+1}{2},\, \frac{n+1}{2},\, \frac{n+1}{2} \\ \frac{n+3}{2},\, n+1;\, a^2 \end{array}\right)$$

$$[|\arg(1-a^2)| < \pi].$$

34. $\displaystyle\int_0^\pi \frac{\sin^\nu x}{\sqrt{1-b^2\sin^2 x}}\begin{Bmatrix}\sin(ax)\\ \cos(ax)\end{Bmatrix}\arcsin(b\sin x)\,dx$

$= \dfrac{2^{-\nu-1}\pi b\,\Gamma(\nu+2)}{\Gamma\!\left(\dfrac{\nu-a+3}{2}\right)\Gamma\!\left(\dfrac{\nu+a+3}{2}\right)} \begin{Bmatrix}\sin(a\pi/2)\\ \cos(a\pi/2)\end{Bmatrix} {}_4F_3\!\left(\begin{array}{c}1,1,1+\dfrac{\nu}{2},\dfrac{\nu+3}{2};\,b^2\\ \dfrac{3}{2},\dfrac{\nu-a+3}{2},\dfrac{\nu+a+3}{2}\end{array}\right)$

$[\operatorname{Re}\nu > -2;\ |\arg(1-b^2)| < \pi].$

35. $\displaystyle\int_0^{m\pi} \frac{e^{-ax}}{\sin x}\arcsin(b\sin x)\,dx = \frac{b}{a}\left(1-e^{-m\pi a}\right){}_4F_3\!\left(\begin{array}{c}\dfrac{1}{2},\dfrac{1}{2},\dfrac{1}{2},1;\,b^2\\ \dfrac{3}{2},1-\dfrac{ia}{2},1+\dfrac{ia}{2}\end{array}\right)$

$[|\arg(1-b^2)| < \pi].$

36. $\displaystyle\int_0^{m\pi} \frac{e^{-ax}}{\sin x\sqrt{1-b^2\sin^2 x}}\arcsin(b\sin x)\,dx$

$= \dfrac{b}{a}\left(1-e^{-m\pi a}\right){}_4F_3\!\left(\begin{array}{c}\dfrac{1}{2},1,1,1;\,b^2\\ \dfrac{3}{2},1-\dfrac{ia}{2},1+\dfrac{ia}{2}\end{array}\right)$ $[|\arg(1-b^2)| < \pi].$

37. $\displaystyle\int_0^\infty e^{-ax}\arcsin(b\sin x)\,dx = \frac{b}{a^2+1}\,{}_3F_2\!\left(\begin{array}{c}\dfrac{1}{2},\dfrac{1}{2},1;\,b^2\\ \dfrac{3-ia}{2},\dfrac{3+ia}{2}\end{array}\right)$

$\bigl[\operatorname{Re} a > 0;\ |\arg(1-b^2)| < \pi\bigr].$

38. $\displaystyle\int_0^\infty \frac{e^{-ax}}{\sin x}\arcsin(b\sin x)\,dx = \frac{b}{a}\,{}_4F_3\!\left(\begin{array}{c}\dfrac{1}{2},\dfrac{1}{2},\dfrac{1}{2},1;\,b^2\\ \dfrac{3}{2},1-\dfrac{ia}{2},1+\dfrac{ia}{2}\end{array}\right)$

$\bigl[\operatorname{Re} a > 0;\ |\arg(1-b^2)| < \pi\bigr].$

39. $\displaystyle\int_0^\infty \frac{e^{-ax}}{\sin x\sqrt{1-b^2\sin^2 x}}\arcsin(b\sin x)\,dx = \frac{b}{a}\,{}_4F_3\!\left(\begin{array}{c}\dfrac{1}{2},1,1,1;\,b^2\\ \dfrac{3}{2},1-\dfrac{ia}{2},1+\dfrac{ia}{2}\end{array}\right)$

$\bigl[\operatorname{Re} a > 0;\ |\arg(1-b^2)| < \pi\bigr].$

40. $\displaystyle\int_0^1 \ln x\,\arcsin(ax)\,dx = \frac{1}{a}\left(2-2\sqrt{1-a^2} - a\arcsin a + \ln\frac{1+\sqrt{1-a^2}}{2}\right)$

$\bigl[|\arg(1-a^2)| < \pi\bigr].$

41. $\displaystyle\int_0^1 \ln x\,\arcsin x\,dx = 2 - \frac{\pi}{2} - \ln 2.$

42. $\displaystyle\int_0^1 x\ln x\,\arcsin x\,dx = \frac{\pi}{8}(\ln 2 - 1).$

43. $\int_0^1 x^2 \ln x \arcsin x \, dx = \dfrac{1}{54}(14 - 3\pi - 12\ln 2).$

44. $\int_0^1 \dfrac{1}{x} \ln x \arcsin x \, dx = -\dfrac{\pi}{48}(\pi^2 + 12\ln^2 2).$

45. $\int_0^1 x(x^2 - x^4)^\nu \ln^n(x^2 - x^4) \arcsin x \, dx = \dfrac{\pi^{3/2}}{16} D_\nu^n \left[\dfrac{2^{-2\nu}\Gamma(\nu+1)}{\Gamma\left(\nu + \dfrac{3}{2}\right)} \right]$

$\qquad\qquad\qquad\qquad\qquad\qquad\qquad\qquad\qquad\qquad\qquad\qquad [\operatorname{Re}\nu > -1].$

46. $\int_0^a x^{s-1} \ln \dfrac{a + \sqrt{a^2 - x^2}}{x} \arcsin(bx)\, dx = \dfrac{\pi^{1/2} a^{s+1} b \Gamma\left(\dfrac{s+1}{2}\right)}{s(s+1)\Gamma\left(\dfrac{s}{2}\right)}$

$\qquad \times \left[(s+1)\,{}_3F_2\!\left(\begin{matrix} \tfrac{1}{2}, \tfrac{1}{2}, \tfrac{s+1}{2}; a^2 b^2 \\ \tfrac{3}{2}, \tfrac{s}{2}+1 \end{matrix} \right) - {}_3F_2\!\left(\begin{matrix} \tfrac{1}{2}, \tfrac{s+1}{2}, \tfrac{s+1}{2} \\ \tfrac{s}{2}+1, \tfrac{s+3}{2}; a^2 b^2 \end{matrix} \right) \right]$

$\qquad\qquad\qquad\qquad\qquad [a > 0;\ \operatorname{Re} s > -1;\ |\arg(1 - a^2 b^2)| < \pi].$

47. $\int_0^a \ln \dfrac{a + \sqrt{a^2 - x^2}}{x} \arcsin(bx)\, dx = \dfrac{a}{2}[\operatorname{Li}_2(ab) - \operatorname{Li}_2(-ab)] - \dfrac{a}{2} \ln \dfrac{1+ab}{1-ab}$

$\qquad\qquad\qquad - \dfrac{1}{2b} \ln(1 - a^2 b^2) \quad [a > 0;\ |\arg(1 - a^2 b^2)| < \pi].$

48. $\int_0^a x^2 \ln \dfrac{a + \sqrt{a^2 - x^2}}{x} \arcsin(bx)\, dx = \dfrac{a^3}{12}[\operatorname{Li}_2(ab) - \operatorname{Li}_2(-ab)]$

$\qquad + \dfrac{1}{72b^3}\left[ab(a^2 b^2 - 9) \ln \dfrac{1+ab}{1-ab} - 8\ln(1 - a^2 b^2) + 10 a^2 b^2 \right]$

$\qquad\qquad\qquad\qquad\qquad [a > 0;\ |\arg(1 - a^2 b^2)| < \pi].$

49. $\int_0^1 x^2 \ln \dfrac{1 + \sqrt{1 - x^2}}{x} \arcsin x \, dx = \dfrac{1}{144}(20 + 3\pi^2 - 32\ln 2).$

50. $\int_0^a \dfrac{x^{-1}}{\sqrt{1 - b^2 x^2}} \ln \dfrac{a + \sqrt{a^2 - x^2}}{x} \arcsin(bx)\, dx = \dfrac{\pi}{4}[\operatorname{Li}_2(ab) - \operatorname{Li}_2(-ab)]$

$\qquad\qquad\qquad\qquad\qquad [a > 0;\ |\arg(1 - a^2 b^2)| < \pi].$

51. $\int_0^a \dfrac{x}{\sqrt{1 - b^2 x^2}} \ln \dfrac{a + \sqrt{a^2 - x^2}}{x} \arcsin(bx)\, dx = \dfrac{\pi a}{2b} - \dfrac{\pi}{4b^2} \ln \dfrac{1+ab}{1-ab}$

$\qquad\qquad - \dfrac{\pi a}{4b} \ln(1 - a^2 b^2) \quad [a > 0;\ |\arg(1 - a^2 b^2)| < \pi].$

52. $\int_0^1 \dfrac{x}{\sqrt{1-x^2}} \ln \dfrac{1+\sqrt{1-x^2}}{x} \arcsin x\, dx = \dfrac{\pi}{2}(1-\ln 2).$

53. $\int_0^1 \dfrac{x^{-1}}{\sqrt{1-x^2}} \ln \dfrac{1+\sqrt{1-x^2}}{x} \arcsin x\, dx = \dfrac{\pi^3}{16}.$

54. $\int_0^1 \ln \dfrac{1+\sqrt{1-x^2}}{1-\sqrt{1-x^2}} \arcsin x\, dx = \dfrac{\pi^2}{4} - 2\ln 2.$

55. $\int_0^1 x^2 \ln \dfrac{1+\sqrt{1-x^2}}{1-\sqrt{1-x^2}} \arcsin x\, dx = \dfrac{1}{72}\left(20 + 3\pi^2 - 32\ln 2\right).$

56. $\int_0^1 \dfrac{x^{-1}}{\sqrt{1-a^2 x^2}} \ln \dfrac{1+\sqrt{1-x^2}}{1-\sqrt{1-x^2}} \arcsin(ax)\, dx = \dfrac{\pi}{2}\left[\operatorname{Li}_2(a) - \operatorname{Li}_2(-a)\right]$

$[|\arg(1+a)| < \pi].$

57. $\int_0^1 x(a^2 - x^2)^{-1/2} \arccos x\, dx = \dfrac{\pi a}{2} - a\,\mathbf{E}\left(\dfrac{1}{a}\right)$ $\quad [|\arg(a^2-1)| < \pi].$

58. $\int_0^1 (1-a^2 x^2)^{-1/2} \arccos x\, dx = \dfrac{1}{2a}\left[\operatorname{Li}_2(a) - \operatorname{Li}_2(-a)\right]\quad [|\arg(1-a^2)|<\pi].$

59. $\int_0^1 (1-a^2 x^2)^{-3/2} \arccos x\, dx = \dfrac{1}{2a} \ln \dfrac{1+a}{1-a}\quad [|\arg(1-a^2)|<\pi].$

60. $\int_0^1 x^{s-1} \sinh(ax) \arccos x\, dx = \dfrac{\pi^{1/2} a\, \Gamma\!\left(\frac{s}{2}+1\right)}{(s+1)^2 \Gamma\!\left(\frac{s+1}{2}\right)}\, {}_2F_3\!\left(\begin{array}{c} \frac{s+1}{2},\, \frac{s}{2}+1;\, \frac{a^2}{4} \\ \frac{3}{2},\, \frac{s+3}{2},\, \frac{s+3}{2} \end{array}\right)$

$[\operatorname{Re} s > -1].$

61. $\int_0^1 \sinh(ax) \arccos x\, dx = \dfrac{\pi}{2a}[I_0(a) - 1].$

62. $\int_0^1 x \sinh(ax) \arccos x\, dx = \dfrac{\pi}{2a^2}\left[a\,\mathbf{L}_{-1}(a) - \mathbf{L}_0(a)\right].$

63. $\int_0^1 x^{s-1} \cosh(ax) \arccos x \, dx = \dfrac{\pi^{1/2} \Gamma\left(\frac{s+1}{2}\right)}{s^2 \Gamma\left(\frac{s}{2}\right)} {}_2F_3\left(\begin{array}{c} \frac{s}{2}, \frac{s+1}{2}; \frac{a^2}{4} \\ \frac{1}{2}, \frac{s}{2}+1, \frac{s}{2}+1 \end{array}\right)$

$\quad$ [Re $s > 0$].

64. $\int_0^1 \cosh(ax) \arccos x \, dx = \dfrac{\pi}{2a} \mathbf{L}_0(a).$

65. $\int_0^1 x \cosh(ax) \arccos x \, dx = \dfrac{\pi}{2a^2} [1 - I_0(a) + a\, I_1(a)].$

66. $\int_0^1 x^{s-1} \sin(ax) \arccos x \, dx = \dfrac{\pi^{1/2} a \Gamma\left(\frac{s}{2}+1\right)}{(s+1)^2 \Gamma\left(\frac{s+1}{2}\right)} {}_2F_3\left(\begin{array}{c} \frac{s+1}{2}, \frac{s}{2}+1; -\frac{a^2}{4} \\ \frac{3}{2}, \frac{s+3}{2}, \frac{s+3}{2} \end{array}\right)$

$\quad$ [Re $s > -1$].

67. $\int_0^1 \sin(ax) \arccos x \, dx = \dfrac{\pi}{2a} [1 - J_0(a)].$

68. $\int_0^1 x \sin(ax) \arccos x \, dx = \dfrac{\pi}{2a^2} [\mathbf{H}_0(a) - a\, \mathbf{H}_{-1}(a)].$

69. $\int_0^1 x^{s-1} \cos(ax) \arccos x \, dx = \dfrac{\pi^{1/2} \Gamma\left(\frac{s+1}{2}\right)}{s^2 \Gamma\left(\frac{s}{2}\right)} {}_2F_3\left(\begin{array}{c} \frac{s}{2}, \frac{s+1}{2}; -\frac{a^2}{4} \\ \frac{1}{2}, \frac{s}{2}+1, \frac{s}{2}+1 \end{array}\right)$

$\quad$ [Re $s > 0$].

70. $\int_0^1 \cos(ax) \arccos x \, dx = \dfrac{\pi}{2a} \mathbf{H}_0(a).$

71. $\int_0^1 x \cos(ax) \arccos x \, dx = \dfrac{\pi}{2a^2} [J_0(a) + a\, J_1(a) - 1].$

72. $\int_0^1 x^{s-1} \ln(1+ax^2) \arccos x \, dx = \dfrac{\pi^{1/2} a \Gamma\left(\frac{s+3}{2}\right)}{s(s+2)^2 \Gamma\left(\frac{s}{2}+1\right)}$

$\quad \times \left[(s+2)\, {}_3F_2\left(\begin{array}{c} 1, 1, \frac{s+3}{2} \\ 2, \frac{s}{2}+2; -a \end{array}\right) - 2\, {}_3F_2\left(\begin{array}{c} 1, \frac{s}{2}+1, \frac{s+3}{2} \\ \frac{s}{2}+2, \frac{s}{2}+2; -a \end{array}\right)\right]$

$\quad$ [Re $s > -2$; $|\arg(1+a)| < \pi$].

73. $\displaystyle\int_0^1 x\ln(1+ax^2)\arccos x\,dx$

$$= \frac{\pi}{4a}\left[1 - \sqrt{1+a} + (a+2)\ln\frac{1+\sqrt{1+a}}{2}\right] \qquad [|\arg(1+a)| < \pi].$$

74. $\displaystyle\int_0^1 x\ln(1-x^2)\arccos x\,dx = \frac{\pi}{4}(\ln 2 - 1).$

75. $\displaystyle\int_0^1 \frac{1}{x}\ln(1+ax^2)\arccos x\,dx$

$$= \frac{\pi}{4}\left[\ln^2\frac{1+\sqrt{1+a}}{2} - 2\operatorname{Li}_2\left(\frac{1-\sqrt{1+a}}{2}\right)\right] \qquad [|\arg(1+a)| < \pi].$$

76. $\displaystyle\int_0^1 \frac{1}{x}\ln(1-x^2)\arccos x\,dx = \frac{\pi}{24}(12\ln^2 2 - \pi^2).$

77. $\displaystyle\int_0^1 \frac{1}{x^2}\ln(1-x^2)\arccos x\,dx = \frac{\pi^2}{4} - 4G.$

78. $\displaystyle\int_0^1 x^{s-1}\ln\frac{a+x}{a-x}\arccos x\,dx = \frac{\pi^{1/2}\Gamma\left(\frac{s}{2}\right)}{2a(s+1)\Gamma\left(\frac{s+3}{2}\right)}$

$$\times\left[(s+1)\,_3F_2\left(\begin{matrix}\frac{1}{2},1,\frac{s}{2}+1\\ \frac{3}{2},\frac{s+3}{2}\end{matrix};a^{-2}\right) - {}_3F_2\left(\begin{matrix}1,\frac{s+1}{2},\frac{s}{2}+1\\ \frac{s+3}{2},\frac{s+3}{2}\end{matrix};a^{-2}\right)\right]$$

$$[\operatorname{Re} s > -1;\ |\arg(a^2-1)| < \pi].$$

79. $\displaystyle\int_0^1 \ln\frac{a+x}{a-x}\arccos x\,dx = \pi\left(a - \sqrt{a^2-1} + a\ln\frac{a-\sqrt{a^2-1}}{2a}\right)$

$$[|\arg(a^2-1)| < \pi].$$

80. $\displaystyle\int_0^1 x^2\arccos x\ln\frac{a+x}{a-x}\,dx$

$$= \frac{\pi}{36}\left[4a^3 + 9a - 4(a^2+2)\sqrt{a^2-1} + 12a^3\ln\frac{a+\sqrt{a^2-1}}{2a}\right]$$

$$[|\arg(a^2-1)| < \pi].$$

81. $\int_0^1 x^{s-1} \ln\left(ax + \sqrt{1+a^2x^2}\right) \arccos x\, dx = \dfrac{\pi^{1/2} a\, \Gamma\left(\frac{s}{2}\right)}{4(s+1)\Gamma\left(\frac{s+3}{2}\right)}$

$$\times \left[(s+1)\,{}_3F_2\left(\begin{array}{c}\frac{1}{2},\frac{1}{2},\frac{s}{2}+1\\ \frac{3}{2},\frac{s+3}{2}\end{array};-a^2\right) - {}_3F_2\left(\begin{array}{c}\frac{1}{2},\frac{s+1}{2},\frac{s}{2}+1\\ \frac{s+3}{2},\frac{s+3}{2}\end{array};-a^2\right)\right]$$

$[\operatorname{Re} s > -1;\ |\arg(1+a^2)| < \pi]$.

82. $\int_0^1 \ln\left(ax + \sqrt{1+a^2x^2}\right) \arccos x\, dx$

$= \dfrac{\sqrt{1+a^2}}{a}\left[\mathbf{K}\left(\dfrac{a}{\sqrt{1+a^2}}\right) - 2\mathbf{E}\left(\dfrac{a}{\sqrt{1+a^2}}\right)\right] + \dfrac{\pi}{2a}$ $[|\arg(1+a^2)| < \pi]$.

83. $\int_0^1 \ln\left(x + \sqrt{1+x^2}\right) \arccos x\, dx = \dfrac{\pi}{2} - \dfrac{2}{\sqrt{\pi}}\Gamma^2\left(\dfrac{3}{4}\right)$.

84. $\int_0^1 x \ln\left(ax + \sqrt{1+a^2x^2}\right) \arccos x\, dx$

$= \dfrac{1}{8a}\left\{-2a^2 + ia(a^2+1)\left[\operatorname{Li}_2(-ia) - \operatorname{Li}_2(ia)\right]\right\}$ $[|\arg(1+a^2)| < \pi]$.

85. $\int_0^1 x \ln\left(x + \sqrt{1+x^2}\right) \arccos x\, dx = \dfrac{\mathbf{G}}{2} - \dfrac{1}{4}$.

86. $\int_0^1 x^2 \ln\left(ax + \sqrt{1+a^2x^2}\right) \arccos x\, dx = \dfrac{6a^4 + 5a^2 - 1}{27a^3\sqrt{1+a^2}}\mathbf{K}\left(\dfrac{a}{\sqrt{1+a^2}}\right)$

$+ \dfrac{7}{27a^3}(1-a^2)\sqrt{1+a^2}\,\mathbf{E}\left(\dfrac{a}{\sqrt{1+a^2}}\right) - \dfrac{\pi}{9a^3}$ $[|\arg(1+a^2)| < \pi]$.

87. $\int_0^1 x^2 \ln\left(x + \sqrt{1+x^2}\right) \arccos x\, dx = \dfrac{5}{54\sqrt{2\pi}}\Gamma^2\left(\dfrac{1}{4}\right) - \dfrac{\pi}{9}$.

88. $\int_0^1 \dfrac{1}{x} \ln\left(ax + \sqrt{1+a^2x^2}\right) \arccos x\, dx = \dfrac{a}{8}\Phi\left(-a^2, 3, \dfrac{1}{2}\right)$

$[|\arg(1+a^2)| < \pi]$.

89. $\int_0^1 \dfrac{1}{x} \ln\left(x + \sqrt{1+x^2}\right) \arccos x\, dx = \dfrac{\pi^3}{32}$.

90. $\int_0^1 \dfrac{x^{s-1}}{\sqrt{1+a^2x^2}} \ln\left(ax+\sqrt{1+a^2x^2}\right) \arccos x \, dx = \dfrac{\pi^{1/2} a \Gamma\left(\frac{s}{2}+1\right)}{2(s+1)\Gamma\left(\frac{s+3}{2}\right)}$

$\times {}_4F_3\left(\begin{matrix} 1,1,\frac{s+1}{2},\frac{s}{2}+1 \\ \frac{3}{2}, \frac{s+3}{2}, \frac{s+3}{2} \end{matrix}; -a^2\right)$ $\quad [\operatorname{Re} s > 0;\ |\arg(1+a^2)| < \pi]$.

91. $\int_0^1 \dfrac{1}{\sqrt{1+a^2x^2}} \ln\left(ax+\sqrt{1+a^2x^2}\right) \arccos x \, dx = -\dfrac{\pi}{8a} \operatorname{Li}_2(-a^2)$

$[|\arg(1+a^2)| < \pi]$.

92. $\int_0^1 \dfrac{1}{\sqrt{1+x^2}} \ln\left(x+\sqrt{1+x^2}\right) \arccos x \, dx = \dfrac{\pi^3}{96}$.

93. $\int_0^1 \dfrac{x^2}{\sqrt{1+a^2x^2}} \ln\left(ax+\sqrt{1+a^2x^2}\right) \arccos x \, dx$

$= \dfrac{\pi}{16a^3}\left[(a^2+1)\ln(a^2+1) + \operatorname{Li}_2(-a^2)\right] \quad [|\arg(1+a^2)| < \pi]$.

94. $\int_0^1 \dfrac{x^2}{\sqrt{1+x^2}} \ln\left(x+\sqrt{1+x^2}\right) \arccos x \, dx = \dfrac{\pi}{192}(24\ln 2 - \pi^2)$.

95. $\int_0^1 x^{s-1} \ln^2\left(ax+\sqrt{1+a^2x^2}\right) \arccos x \, dx = \dfrac{\pi^{1/2} a^2 \Gamma\left(\frac{s+3}{2}\right)}{2s(s+2)\Gamma\left(\frac{s}{2}+2\right)}$

$\times \left[(s+2)\, {}_4F_3\left(\begin{matrix} 1,1,1,\frac{s+3}{2}; -a^2 \\ \frac{3}{2}, 2, \frac{s}{2}+2 \end{matrix}\right) - 2\, {}_4F_3\left(\begin{matrix} 1,1,\frac{s}{2}+1,\frac{s+3}{2}; -a^2 \\ \frac{3}{2}, \frac{s}{2}+2, \frac{s}{2}+2 \end{matrix}\right)\right]$

$[\operatorname{Re} s > 0;\ |\arg(1+a^2)| < \pi]$.

96. $\int_0^1 x \ln^2\left(ax+\sqrt{1+a^2x^2}\right) \arccos x \, dx$

$= -\dfrac{\pi}{16a^2}\left[a^2 + (a^2+1)\operatorname{Li}_2(-a^2)\right] \quad [|\arg(1+a^2)| < \pi]$.

97. $\int_0^1 x \ln^2\left(x+\sqrt{1+x^2}\right) \arccos x \, dx = \dfrac{\pi^3}{96} - \dfrac{\pi}{16}$.

98. $\displaystyle\int_0^1 x^3 \ln^2\left(ax + \sqrt{1 + a^2 x^2}\right) \arccos x\, dx$

$\displaystyle = \frac{\pi}{512 a^4}\left[4a^2 - 15a^4 + 8(a^2 + 1)^2 \ln(1 + a^2) - 12(a^4 - 1)\operatorname{Li}_2(-a^2)\right]$

$[|\arg(1 + a^2)| < \pi]$.

99. $\displaystyle\int_0^1 x^3 \ln^2\left(x + \sqrt{1 + x^2}\right) \arccos x\, dx = \frac{\pi}{512}(32\ln 2 - 11)$.

100. $\displaystyle\int_0^1 \frac{1}{x} \ln^2\left(ax + \sqrt{1 + a^2 x^2}\right) \arccos x\, dx = -\frac{\pi}{8}\operatorname{Li}_3(-a^2)$

$[|\arg(1 + a^2)| < \pi]$.

101. $\displaystyle\int_0^1 \frac{1}{x} \ln^2\left(x + \sqrt{x^2 + 1}\right) \arccos x\, dx = \frac{3\pi}{32}\zeta(3)$.

102. $\displaystyle\int_0^1 x^{s-1} \arcsin(ax) \arccos x\, dx$

$\displaystyle = \frac{\pi^{1/2} a\, \Gamma\left(\frac{s}{2}\right)}{4(s+1)\Gamma\left(\frac{s+3}{2}\right)}\left[(s+1)\,_3F_2\left(\begin{matrix}\frac{1}{2}, \frac{1}{2}, \frac{s}{2}+1 \\ \frac{3}{2}, \frac{s+3}{2}\end{matrix}; a^2\right) - \,_3F_2\left(\begin{matrix}\frac{1}{2}, \frac{s+1}{2}, \frac{s}{2}+1 \\ \frac{s+3}{2}, \frac{s+3}{2}\end{matrix}; a^2\right)\right]$

$[\operatorname{Re} s > 0;\ |\arg(1 - a^2)| < \pi]$.

103. $\displaystyle\int_0^1 \arcsin(ax)\arccos x\, dx = \frac{1}{2a}\left[2(a^2 - 1)\mathbf{K}(a) + 4\mathbf{E}(a) - \pi\right]$

$[|\arg(1 - a^2)| < \pi]$.

104. $\displaystyle\int_0^1 \arcsin(x)\arccos x\, dx = 2 - \frac{\pi}{2}$.

105. $\displaystyle\int_0^1 x \arcsin(ax)\arccos x\, dx = \frac{a}{16}\left[\Phi\left(a^2, 2, \frac{1}{2}\right) - \Phi\left(a^2, 2, \frac{3}{2}\right)\right]$

$[|\arg(1 - a^2)| < \pi]$.

106. $\displaystyle\int_0^1 x \arcsin x \arccos x\, dx = \frac{1}{4}$.

107. $\displaystyle\int_0^1 x^2 \arcsin(ax) \arccos x \, dx$

$$= \frac{1}{27a^3} \left[(6a^4 - 5a^2 - 1) \, \mathbf{K}(a) + 7(a^2 + 1) \, \mathbf{E}(a) - 3\pi\right]$$

$$[|\arg(1-a^2)| < \pi].$$

108. $\displaystyle\int_0^1 x^2 \arcsin x \arccos x \, dx = \frac{14}{27} - \frac{\pi}{9}.$

109. $\displaystyle\int_0^1 \frac{1}{x} \arcsin(ax) \arccos x \, dx = \frac{a}{8} \Phi\left(a^2, 3, \frac{1}{2}\right)$ $\qquad [|\arg(1-a^2)| < \pi].$

110. $\displaystyle\int_0^1 \frac{1}{x} \arcsin x \arccos x \, dx = \frac{7}{8}\zeta(3).$

111. $\displaystyle\int_0^1 \frac{x^{s-1}}{\sqrt{1-a^2x^2}} \arcsin(ax) \arccos x \, dx = \frac{\pi^{1/2} a \, \Gamma\left(\frac{s}{2}+1\right)}{2(s+1)\Gamma\left(\frac{s+3}{2}\right)}$

$$\times {}_4F_3\left(\begin{matrix} 1, 1, \frac{s+1}{2}, \frac{s}{2}+1 \\ \frac{3}{2}, \frac{s+3}{2}, \frac{s+3}{2} \end{matrix} ; a^2\right) \qquad [\text{Re } s > 0; \, |\arg(1-a^2)| < \pi].$$

112. $\displaystyle\int_0^1 \frac{1}{\sqrt{1-a^2x^2}} \arcsin(ax) \arccos x \, dx = \frac{\pi}{8a} \text{Li}_2(a^2)$ $\qquad [|\arg(1-a^2)| < \pi].$

113. $\displaystyle\int_0^1 \frac{1}{\sqrt{1-x^2}} \arcsin x \arccos x \, dx = \frac{\pi^3}{48}.$

114. $\displaystyle\int_0^1 \frac{x^2}{\sqrt{1-a^2x^2}} \arcsin(ax) \arccos x \, dx$

$$= \frac{\pi}{16a^3} \left[(1-a^2)\ln(1-a^2) + \text{Li}_2(a^2)\right] \quad [|\arg(1-a^2)| < \pi].$$

115. $\displaystyle\int_0^1 \frac{x^2}{\sqrt{1-x^2}} \arcsin x \arccos x \, dx = \frac{\pi^3}{96}.$

116. $\displaystyle\int_{-1}^1 \frac{1}{\sqrt{1-x^2}} \ln(1 + 2ax + a^2) \arccos^2 x \, dx = -4\pi \, \text{Li}_3(a)$

$$[|\arg(1-a^2)| < \pi].$$

117. $\int_0^a x^{-1}(a^2-x^2)^{1/2} \arctan(bx)\, dx$

$$= \frac{\pi}{2a}\left[1 - \sqrt{a^2b^2+1} + ab\ln\left(ab+\sqrt{a^2b^2+1}\right)\right]$$

$$[a>0;\ |\arg(1+a^2b^2)|<\pi].$$

118. $\int_0^a x^{-1}(a^2-x^2)^{-1/2} \arctan(bx)\, dx = \frac{\pi}{2a}\ln\left(ab+\sqrt{a^2b^2+1}\right)$

$$[a>0;\ |\arg(1+a^2b^2)|<\pi].$$

119. $\int_0^\infty \frac{1}{x(x^2+a^2)}\arctan x\, dx = \frac{\pi}{2a^2}\ln(1+a)\quad [\operatorname{Re} a>0;\ |\arg(1+a)|<\pi].$

120. $\int_0^1 \frac{1}{1+x^2}\arctan\left(x^{3+\sqrt{8}}\right) dx = \frac{1}{16}\ln 2\ln\left(3+\sqrt{8}\right)$ [60].

121. $\int_0^1 \frac{1}{1+x^2}\arctan\left(x^{5+\sqrt{24}}\right) dx = \frac{1}{8}\ln\left(1+\sqrt{2}\right)\ln\left(2+\sqrt{3}\right)$

$$-\frac{1}{16}\ln 2\ln\left(5+\sqrt{24}\right) \quad [60].$$

122. $\int_0^1 \frac{1}{1+x^2}\arctan\left(x^{11+\sqrt{120}}\right) dx = -\frac{1}{8}\ln\left(1+\sqrt{2}\right)\ln\left(4+\sqrt{15}\right)$

$$-\frac{1}{8}\ln\left(2+\sqrt{3}\right)\ln\left(3+\sqrt{10}\right) + \frac{3}{8}\ln\frac{1+\sqrt{5}}{2}\ln\left(5+\sqrt{24}\right)$$

$$+\frac{1}{16}\ln 2\ln\left(11+\sqrt{120}\right) \quad [60].$$

123. $\int_0^1 \frac{1}{1+x^2}\arctan\left(x^{13+\sqrt{168}}\right) dx$

$$= -\frac{3}{8}\ln\left(1+\sqrt{2}\right)\ln\frac{5+\sqrt{21}}{2} - \frac{1}{16}\ln 2\ln\left(13+\sqrt{168}\right)$$

$$+\frac{1}{16}\ln\left(2+\sqrt{3}\right)\ln\left(15+\sqrt{224}\right) + \frac{1}{16}\ln\left(5+\sqrt{24}\right)\ln\left(8+\sqrt{63}\right) \quad [60].$$

124. $\int_0^\infty \frac{1}{x(x^2+b^2)}\arctan\frac{ax}{x^2+b^2}\, dx = \frac{\pi}{2b^2}\ln\frac{a+\sqrt{a^2+4b^2}}{2b}\quad [\operatorname{Re} b>0].$

125. $\displaystyle\int\limits_0^\infty \frac{1}{x(a^2x^2+1)}\arctan[(a^2+1)x+a^2x^3]\,dx$

$$= \frac{\pi}{2}\ln\frac{(1+a)(a+\sqrt{4+a^2})}{2a} \qquad [\operatorname{Re} a > 0].$$

126. $\displaystyle\int\limits_0^a x^{s-1}(a-x)^{t-1}\arctan\left(b\sqrt{x(a-x)}\right)dx$

$$= a^{s+t}b\,\mathrm{B}\left(s+\frac{1}{2}, t+\frac{1}{2}\right){}_3F_2\left(\begin{array}{c}\frac{1}{2},\,1,\,s+\frac{1}{2},\,t+\frac{1}{2};\,-\frac{a^2b^2}{4}\\ \frac{3}{2},\,\frac{s+t+1}{2},\,\frac{s+t}{2}+1\end{array}\right)$$

$$[a > 0;\ \operatorname{Re} s,\,\operatorname{Re} t > -1/2;\ |\arg(4+a^2b^2)| < \pi].$$

127. $\displaystyle\int\limits_0^a \arctan\left(b\sqrt{x(a-x)}\right)dx = \frac{\pi}{2b}\left(\sqrt{4+a^2b^2}-2\right)$

$$[a > 0;\ |\arg(4+a^2b^2)| < \pi].$$

128. $\displaystyle\int\limits_0^a x\arctan\left(b\sqrt{x(a-x)}\right)dx = \frac{\pi a}{4b}\left(\sqrt{4+a^2b^2}-2\right)$

$$[a > 0;\ |\arg(4+a^2b^2)| < \pi].$$

129. $\displaystyle\int\limits_0^a x^{-1}\arctan\left(b\sqrt{x(a-x)}\right)dx = \pi\ln\left(\frac{ab}{2}+\sqrt{1+\frac{a^2b^2}{4}}\right)$

$$[a > 0;\ |\arg(4+a^2b^2)| < \pi].$$

130. $\displaystyle\int\limits_0^a x^{-1}(a-x)^{-1}\arctan\left(b\sqrt{x(a-x)}\right)dx$

$$= \frac{2\pi}{a}\ln\left(\frac{ab}{2}+\sqrt{1+\frac{a^2b^2}{4}}\right) \qquad [a > 0;\ |\arg(4+a^2b^2)| < \pi].$$

131. $\displaystyle\int\limits_0^a x^{s+1/2}(a-x)^s \arctan\left(b\sqrt[4]{x(a-x)}\right)dx = 2^{-2s-3/2}\pi^{1/2}a^{2s+2}b$

$$\times \frac{\Gamma\left(2s+\frac{5}{2}\right)}{\Gamma(2s+3)}\,{}_3F_2\left(\begin{array}{c}\frac{1}{2},\,1,\,2s+\frac{5}{2}\\ \frac{3}{2},\,2s+3;\,-\frac{ab^2}{2}\end{array}\right) \qquad [a > 0;\ \operatorname{Re} s > -1;\ |\arg(2+ab^2)| < \pi].$$

132. $\int\limits_0^a x^{1/2} \arctan\left(b\sqrt[4]{x(a-x)}\right) dx$

$$= \frac{\pi}{6\sqrt{2}\,b^3}\left[2(ab^2-1)\sqrt{2ab^2+4} - 3ab^2 + 4\right]$$

$$[a > 0;\ |\arg(2+ab^2)| < \pi].$$

133. $\int\limits_0^a x^{-1/2} \arctan\left(b\sqrt[4]{x(a-x)}\right) dx = \frac{\pi}{b}\left(\sqrt{ab^2+2} - \sqrt{2}\right)$

$$[a > 0;\ |\arg(2+ab^2)| < \pi].$$

134. $\int\limits_0^a x^{-1/2}(a-x)^{-1} \arctan\left(b\sqrt[4]{x(a-x)}\right) dx$

$$= \frac{2\pi}{\sqrt{a}} \ln\left(b\sqrt{\frac{a}{2}} + \sqrt{1+\frac{ab^2}{2}}\right) \quad [a > 0;\ |\arg(2+ab^2)| < \pi].$$

135. $\int\limits_0^\infty \frac{1}{e^{2\pi x}+1} \arctan x\, dx = \frac{3}{4}\ln 2 - \frac{1}{2}.$

136. $\int\limits_0^{\pi/2} \cos^\nu x \cos(ax) \arctan(b\cos x)\, dx = \frac{2^{-\nu-2}\pi b\, \Gamma(\nu+2)}{\Gamma\left(\frac{\nu-a+3}{2}\right)\Gamma\left(\frac{\nu+a+3}{2}\right)}$

$$\times {}_4F_3\left(\begin{array}{c}\frac{1}{2},1,1+\frac{\nu}{2},\frac{\nu+3}{2};\ -b^2\\ \frac{3}{2},\frac{\nu-a+3}{2},\frac{\nu+a+3}{2}\end{array}\right) \quad [\mathrm{Re}\,\nu > -1;\ |\arg(1+b^2)| < \pi].$$

137. $\int\limits_0^{\pi/2} \frac{\cos(2nx)}{\sin x} \arctan(a\sin x)\, dx$

$$= \frac{2^{-2n-1}\pi a^{2n+1}}{2n+1} {}_3F_2\left(\begin{array}{c} n+\frac{1}{2}, n+\frac{1}{2}, n+1 \\ n+\frac{3}{2}, 2n+1;\ -a^2 \end{array}\right) \quad [|\arg(1+a^2)| < \pi].$$

138. $\int\limits_0^{m\pi} \frac{1}{\sin x}\left\{\begin{array}{c}\sin(ax)\\ \cos(ax)\end{array}\right\} \arctan(b\sin x)\, dx$

$$= \frac{2b}{a}\sin\frac{m\pi a}{2}\left\{\begin{array}{c}\sin(m\pi a/2)\\ \cos(m\pi a/2)\end{array}\right\} {}_4F_3\left(\begin{array}{c}\frac{1}{2},\frac{1}{2},1,1;\ -b^2\\ \frac{3}{2},1-\frac{a}{2},1+\frac{a}{2}\end{array}\right) \quad [|\arg(1+b^2)| < \pi].$$

139. $\int_0^\pi \sin^\nu x \begin{Bmatrix} \sin(ax) \\ \cos(ax) \end{Bmatrix} \arctan(b\sin x)\,dx = \dfrac{2^{-\nu-1}\pi b\,\Gamma(\nu+2)}{\Gamma\left(\dfrac{\nu-a+3}{2}\right)\Gamma\left(\dfrac{\nu+a+3}{2}\right)}$

$\times \begin{Bmatrix} \sin(a\pi/2) \\ \cos(a\pi/2) \end{Bmatrix} {}_4F_3\left(\begin{array}{c} \tfrac{1}{2}, 1, \tfrac{\nu}{2}+1, \tfrac{\nu+3}{2};\ -b^2 \\ \tfrac{3}{2}, \tfrac{\nu-a+3}{2}, \tfrac{\nu+a+3}{2} \end{array}\right)$

$[\operatorname{Re}\nu > -1;\ |\arg(1+b^2)| < \pi].$

140. $\int_0^{m\pi} \dfrac{e^{-ax}}{\sin x} \arctan(b\sin x)\,dx = \dfrac{b}{a}\left(1 - e^{-m\pi a}\right) {}_4F_3\left(\begin{array}{c} \tfrac{1}{2}, \tfrac{1}{2}, 1, 1;\ -b^2 \\ \tfrac{3}{2}, 1-\tfrac{ia}{2}, 1+\tfrac{ia}{2} \end{array}\right).$

141. $\int_0^\infty e^{-ax} \arctan(b\sin x)\,dx = \dfrac{b}{a^2+1}\,{}_3F_2\left(\begin{array}{c} \tfrac{1}{2}, 1, 1;\ -b^2 \\ \tfrac{3-ia}{2}, \tfrac{3+ia}{2} \end{array}\right)$

$\left[\operatorname{Re} a > 0;\ |\arg(1+b^2)| < \pi\right].$

142. $\int_0^\infty \dfrac{e^{-ax}}{\sin x} \arctan(b\sin x)\,dx = \dfrac{b}{a}\,{}_4F_3\left(\begin{array}{c} \tfrac{1}{2}, \tfrac{1}{2}, 1, 1;\ -b^2 \\ \tfrac{3}{2}, 1-\tfrac{ia}{2}, 1+\tfrac{ia}{2} \end{array}\right)$

$\left[\operatorname{Re} a > 0;\ |\arg(1+b^2)| < \pi\right].$

143. $\int_0^\infty \dfrac{\sin(z\arctan x)}{(1+x^2)^{z/2}(e^{ax}-1)}\,dx = \dfrac{\left(\tfrac{a}{2\pi}\right)^{z-1}}{2}\zeta\left(z, \dfrac{a}{2\pi}\right) - \dfrac{1}{2(z-1)} - \dfrac{\pi}{2a}$

$[\operatorname{Re} a > 0].$

144. $\int_0^a x^{s-1} \ln\dfrac{a+\sqrt{a^2-x^2}}{x} \arctan(bx)\,dx = \dfrac{\pi^{1/2} a^{s+1} b\,\Gamma\left(\tfrac{s+1}{2}\right)}{s(s+1)\Gamma\left(\tfrac{s}{2}\right)}$

$\times {}_4F_3\left(\begin{array}{c} \tfrac{1}{2}, 1, \tfrac{s+1}{2}, \tfrac{s+1}{2} \\ \tfrac{3}{2}, \tfrac{s}{2}+1, \tfrac{s+3}{2};\ -a^2 b^2 \end{array}\right)$ $[a > 0;\ \operatorname{Re} s > -1;\ |\arg(1+a^2 b^2)| < \pi].$

145. $\int_0^1 x^{s-1} \ln\dfrac{1+\sqrt{1-x^2}}{1-\sqrt{1-x^2}} \arctan(ax)\,dx$

$= \dfrac{2\pi^{1/2} a\,\Gamma\left(\tfrac{s+1}{2}\right)}{s^2(s+1)\Gamma\left(\tfrac{s}{2}\right)}\left[(s+1)\,{}_3F_2\left(\begin{array}{c} \tfrac{1}{2}, 1, \tfrac{s+1}{2} \\ \tfrac{3}{2}, \tfrac{s}{2}+1;\ -a^2 \end{array}\right) - {}_3F_2\left(\begin{array}{c} 1, \tfrac{s+1}{2}, \tfrac{s+1}{2} \\ \tfrac{s}{2}, \tfrac{s+3}{2};\ -a^2 \end{array}\right)\right]$

$[\operatorname{Re} a > 0;\ \operatorname{Re} s > -1].$

146. $\int_0^1 x \ln\dfrac{1+\sqrt{1-x^2}}{1-\sqrt{1-x^2}} \arctan(ax)\,dx = \dfrac{\pi}{2a^3}\left[a^2\left(\sqrt{a^2+1}-2\right)\right.$

$\left. + 2a \ln\left(\sqrt{\tfrac{1}{2}\sqrt{a^2+1}-\tfrac{1}{2}} + \sqrt{\tfrac{1}{2}\sqrt{a^2+1}+\tfrac{1}{2}}\right)\right]$ $[|\arg(1+a^2)| < \pi].$

147. $\displaystyle\int_0^\infty \frac{\ln(x^2+1)}{e^{ax}-1}\arctan x\,dx = \frac{a-\pi}{2a}\ln^2\frac{a}{2\pi} - \frac{\pi}{a}\ln\frac{a}{2\pi}\ln\left[\frac{1}{2\pi}\Gamma^2\left(\frac{a}{2\pi}\right)\right]$

$$-1 - \frac{\pi}{a}\frac{\partial^2}{\partial z^2}\zeta\left(z,\frac{a}{2\pi}\right)\bigg|_{z=0} \quad [\operatorname{Re} a > 0].$$

148. $\displaystyle\int_0^a x^{s-1}(a^2-x^2)^{t-1}\arcsin^2(bx)\,dx$

$$= \frac{1}{2}a^{s+2t}b^2\,\mathrm{B}\left(\frac{s}{2}+1,\,t\right)\,{}_4F_3\!\left(\begin{array}{c}1,1,1,\frac{s}{2}+1;\,a^2b^2\\ \frac{3}{2},2,\frac{s}{2}+t+1\end{array}\right)$$

$$[a,\,\operatorname{Re} t > 0;\ \operatorname{Re} s > -2;\ |\arg(1-a^2b^2)| < \pi].$$

149. $\displaystyle\int_0^a (a^2-x^2)^{1/2}\arcsin^2(bx)\,dx$

$$= \frac{\pi a^2}{8}\left[\operatorname{Li}_2(a^2b^2) - \left(\frac{1}{a^2b^2}-1\right)\ln(1-a^2b^2) - 1\right]$$

$$[a > 0;\ |\arg(1-a^2b^2)| < \pi].$$

150. $\displaystyle\int_0^a (a^2-x^2)^{-1/2}\arcsin^2(bx)\,dx = \frac{\pi}{4}\operatorname{Li}_2(a^2b^2)$

$$[a > 0;\ |\arg(1-a^2b^2)| < \pi].$$

151. $\displaystyle\int_0^a x^{s-1}(a-x)^{t-1}\arcsin^2(b\sqrt{x(a-x)})\,dx$

$$= a^{s+t+1}b^2\,\mathrm{B}(s+1,t+1)\,{}_5F_4\!\left(\begin{array}{c}1,1,1,s+1,t+1;\,\frac{a^2b^2}{4}\\ \frac{3}{2},2,\frac{s+t}{2}+1,\frac{s+t+3}{2}\end{array}\right)$$

$$[a > 0;\ \operatorname{Re} s,\operatorname{Re} t > -1;\ |\arg(4-a^2b^2)| < \pi].$$

152. $\displaystyle\int_0^a x^{1/2}(a-x)^{1/2}\arcsin^2(b\sqrt{x(a-x)})\,dx$

$$= \frac{\pi}{16b^2}\left[a^2b^2 + (4-a^2b^2)\ln\left(1-\frac{a^2b^2}{4}\right) + a^2b^2\operatorname{Li}_2\left(\frac{a^2b^2}{4}\right)\right]$$

$$[a > 0;\ |\arg(4-a^2b^2)| < \pi].$$

153. $\displaystyle\int_0^a x^{1/2}(a-x)^{-1/2}\arcsin^2(b\sqrt{x(a-x)})\,dx = \frac{\pi a}{4}\operatorname{Li}_2\left(\frac{a^2b^2}{4}\right)$

$$[a > 0;\ |\arg(4-a^2b^2)| < \pi].$$

Ch. 4. Definite Integrals [4.1.6

154. $\displaystyle\int_0^a x^{1/2}(a-x)^{-3/2}\arcsin^2(b\sqrt{x(a-x)})\,dx$

$$= \frac{\pi}{2}\left[2ab\ln\frac{2+ab}{2-ab} + 4\ln\left(1-\frac{a^2b^2}{4}\right) - \operatorname{Li}_2\left(\frac{a^2b^2}{4}\right)\right]$$

$$[a>0;\ |\arg(4-a^2b^2)|<\pi].$$

155. $\displaystyle\int_0^a x^{-1/2}(a-x)^{-1/2}\arcsin^2(b\sqrt{x(a-x)})\,dx = \frac{\pi}{2}\operatorname{Li}_2\left(\frac{a^2b^2}{4}\right)$

$$[a>0;\ |\arg(4-a^2b^2)|<\pi].$$

156. $\displaystyle\int_0^a x^{-1/2}(a-x)^{-3/2}\arcsin^2(b\sqrt{x(a-x)})\,dx$

$$= \frac{\pi}{a}\left[ab\ln\frac{2+ab}{2-ab} + 2\ln\left(1-\frac{a^2b^2}{4}\right)\right] \quad [a>0;\ |\arg(4-a^2b^2)|<\pi].$$

157. $\displaystyle\int_0^a x^{-3/2}(a-x)^{-3/2}\arcsin^2(b\sqrt{x(a-x)})\,dx$

$$= \frac{2\pi}{a^2}\left[ab\ln\frac{2+ab}{2-ab} + 2\ln\left(1-\frac{a^2b^2}{4}\right)\right] \quad [a>0;\ |\arg(4-a^2b^2)|<\pi].$$

158. $\displaystyle\int_0^a x^s(a-x)^{s+1/2}\arcsin^2(b\sqrt[4]{x(a-x)})\,dx$

$$= \frac{2^{-2s-2}\sqrt{\pi}\,a^{2s+5/2}b^2\Gamma(2s+3)}{\Gamma\left(2s+\frac{7}{2}\right)}\,{}_4F_3\left(\begin{array}{c}1,1,1,2s+3;\\ \frac{3}{2},2,2s+\frac{7}{2}\end{array}\frac{ab^2}{2}\right)$$

$$[a>0;\ \operatorname{Re} s>-3/2;\ |\arg(2-ab^2)|<\pi].$$

159. $\displaystyle\int_0^a x^{1/4}(a-x)^{-1/4}\arcsin^2(b\sqrt[4]{x(a-x)})\,dx$

$$= \frac{\pi}{4\sqrt{2}\,b^2}\left[ab^2 + (2-ab^2)\ln\left(1-\frac{ab^2}{2}\right) + ab^2\operatorname{Li}_2\left(\frac{ab^2}{2}\right)\right]$$

$$[a>0;\ |\arg(2-ab^2)|<\pi].$$

160. $\displaystyle\int_0^a x^{-1/4}(a-x)^{-3/4}\arcsin^2(b\sqrt[4]{x(a-x)})\,dx = \frac{\pi}{\sqrt{2}}\operatorname{Li}_2\left(\frac{ab^2}{2}\right)$

$$[a>0;\ |\arg(2-ab^2)|<\pi].$$

161. $\int_0^a x^{-5/4}(a-x)^{-3/4} \arcsin^2(b\sqrt[4]{x(a-x)})\,dx = \dfrac{2\pi b}{\sqrt{a}} \ln \dfrac{\sqrt{2}+\sqrt{ab}}{\sqrt{2}-\sqrt{ab}}$

$\qquad + \dfrac{2^{3/2}\pi}{a} \ln\left(1 - \dfrac{ab^2}{2}\right)$ $\quad [a>0;\ |\arg(2-ab^2)| < \pi]$.

162. $\displaystyle\int_{-a}^{a} \dfrac{1}{(x+a)^2} \arcsin^2 \dfrac{b(x+a)}{\sqrt{x^2+a^2}}\,dx$

$= \dfrac{\pi b^2}{2a}\left[\dfrac{1}{\sqrt{2}b}\ln\dfrac{1+\sqrt{2}b}{1-\sqrt{2}b} + \dfrac{1}{2b^2}\ln(1-2b^2)\right]$ $\quad [a>0;\ |\arg(1-b^2)| < \pi]$.

163. $\displaystyle\int_0^a x^{s-1} \ln\dfrac{a+\sqrt{a^2-x^2}}{x} \arcsin^2(bx)\,dx$

$= \dfrac{\sqrt{\pi}}{2(s+2)} a^{s+2} \dfrac{\Gamma\!\left(\dfrac{s}{2}+1\right)b^2}{\Gamma\!\left(\dfrac{s+3}{2}\right)} {}_5F_4\!\left(\begin{array}{c} 1,1,1,\dfrac{s}{2}+1,\dfrac{s}{2}+1;\ a^2b^2 \\ \dfrac{3}{2},2,\dfrac{s+3}{2},\dfrac{s}{2}+2 \end{array}\right)$

$\qquad\qquad\qquad [a>0;\ \operatorname{Re} s > -2;\ |\arg(1-a^2b^2)| < \pi]$.

164. $\displaystyle\int_0^{m\pi} \dfrac{e^{-ax}}{\sin^2 x} \arcsin^2(b\sin x)\,dx = \dfrac{b^2}{a}(1-e^{-m\pi a})$

$\qquad \times {}_5F_4\!\left(\begin{array}{c} \dfrac{1}{2},1,1,1,1;\ b^2 \\ \dfrac{3}{2},2,1-\dfrac{ia}{2},1+\dfrac{ia}{2} \end{array}\right)$ $\quad [\operatorname{Re} a > 0;\ |\arg(1-b^2)| < \pi]$.

165. $\displaystyle\int_0^{\infty} \dfrac{e^{-ax}}{\sin^2 x} \arcsin^2(b\sin x)\,dx = \dfrac{b^2}{a}\, {}_5F_4\!\left(\begin{array}{c} \dfrac{1}{2},1,1,1;\ b^2 \\ \dfrac{3}{2},2,1-\dfrac{ia}{2},1+\dfrac{ia}{2} \end{array}\right)$

$\qquad\qquad\qquad [\operatorname{Re} a > 0;\ |\arg(1-b^2)| < \pi]$.

166. $\displaystyle\int_0^a \ln\dfrac{a+\sqrt{a^2-x^2}}{x} \arcsin^2(bx)\,dx = \dfrac{\pi a}{4}\left[\operatorname{Li}_2(a^2b^2) + 2\ln(1-a^2b^2) - 4\right]$

$\qquad + \dfrac{\pi}{2b}\ln\dfrac{1+ab}{1-ab}$ $\quad [a>0;\ |\arg(1-a^2b^2)| < \pi]$.

167. $\displaystyle\int_0^1 \ln\dfrac{1+\sqrt{1-x^2}}{x} \arcsin^2 x\,dx = \dfrac{\pi^3}{24} + (\ln 2 - 1)\pi$.

168. $\displaystyle\int_0^a x^2 \ln\dfrac{a+\sqrt{a^2-x^2}}{x} \arcsin^2(bx)\,dx = \dfrac{\pi a^3}{24}\operatorname{Li}_2(a^2b^2)$

$\qquad - \dfrac{\pi}{216b^3}\left[3ab(a^2b^2-9)\ln(1-a^2b^2) - 24\ln\dfrac{1+ab}{1-ab} + 48ab - 11a^3b^3\right]$

$\qquad\qquad\qquad [a>0;\ |\arg(1-a^2b^2)| < \pi]$.

169. $\displaystyle\int_0^1 x^2 \ln \frac{1+\sqrt{1-x^2}}{x} \arcsin^2 x \, dx = \frac{\pi}{432}(3\pi^2 + 96\ln 2 - 74).$

170. $\displaystyle\int_0^a \frac{1}{x^2} \ln \frac{a+\sqrt{a^2-x^2}}{x} \arcsin^2(bx) \, dx$

$\quad - \dfrac{b\pi}{2}\left[\mathrm{Li}_2(ab) - \mathrm{Li}_2(-ab) - \ln\dfrac{1+ab}{1-ab}\right] - \dfrac{\pi}{2a}\ln(1-a^2b^2)$

$\qquad\qquad\qquad\qquad\qquad\qquad [a > 0;\; |\arg(1-a^2b^2)| < \pi].$

171. $\displaystyle\int_0^{\pi/2} \cos^\nu x \cos(ax) \arcsin^2(b\cos x) \, dx = \frac{2^{-\nu-3}\pi b\,\Gamma(\nu+3)}{\Gamma\!\left(\frac{\nu-a}{2}+2\right)\Gamma\!\left(\frac{\nu+a}{2}+2\right)}$

$\quad \times {}_5F_4\!\left(\begin{array}{c}1,1,1,\frac{\nu+3}{2},\frac{\nu}{2}+2;\;b\\ \frac{3}{2},2,\frac{\nu-a}{2}+2,\frac{\nu+a}{2}+2\end{array}\right) \quad [\mathrm{Re}\,\nu > 0;\; |\arg(1-b)| < \pi].$

172. $\displaystyle\int_0^\pi \sin^\nu x \left\{\begin{array}{c}\sin(ax)\\ \cos(ax)\end{array}\right\} \arcsin^2(b\sin x) \, dx = \frac{2^{-\nu-2}\pi b^2\,\Gamma(\nu+3)}{\Gamma\!\left(\frac{\nu-a}{2}+2\right)\Gamma\!\left(\frac{\nu+a}{2}+2\right)}$

$\quad \times \left\{\begin{array}{c}\sin(a\pi/2)\\ \cos(a\pi/2)\end{array}\right\} {}_5F_4\!\left(\begin{array}{c}1,1,1,\frac{\nu+3}{2},\frac{\nu}{2}+2;\;b\\ \frac{3}{2},2,\frac{\nu-a}{2}+2,\frac{\nu+a}{2}+2\end{array}\right)$

$\qquad\qquad\qquad\qquad\qquad [\mathrm{Re}\,\nu > -1;\; |\arg(1-b)| < \pi].$

173. $\displaystyle\int_0^{m\pi} \frac{1}{\sin^2 x}\left\{\begin{array}{c}\sin(ax)\\ \cos(ax)\end{array}\right\} \arcsin^2(b\sin x)\, dx$

$= \dfrac{2b^2}{a}\sin\dfrac{m\pi a}{2}\left\{\begin{array}{c}\sin(m\pi a/2)\\ \cos(m\pi a/2)\end{array}\right\} {}_5F_4\!\left(\begin{array}{c}\frac{1}{2},1,1,1,1;\;b^2\\ \frac{3}{2},2,1-\frac{a}{2},1+\frac{a}{2}\end{array}\right)$

$\qquad\qquad\qquad\qquad\qquad\qquad [|\arg(1-b^2)| < \pi].$

174. $\displaystyle\int_0^1 \frac{1}{x^2} \ln\frac{1+\sqrt{1-x^2}}{x} \arcsin^2(x) \, dx = \frac{\pi^3}{8} - \pi\ln 2.$

175. $\displaystyle\int_0^1 x^{s-1} \arcsin^2(ax) \arccos x \, dx = \frac{\pi^{1/2} a^2 \Gamma\!\left(\frac{s+3}{2}\right)}{2s(s+2)\Gamma\!\left(\frac{s}{2}+2\right)}$

$\quad \times \left[(s+2)\,{}_4F_3\!\left(\begin{array}{c}1,1,1,\frac{s+3}{2}\\ \frac{3}{2},2,\frac{s}{2}+2;\;a^2\end{array}\right) - 2\,{}_4F_3\!\left(\begin{array}{c}1,1,\frac{s}{2}+1,\frac{s+3}{2}\\ \frac{3}{2},\frac{s}{2}+2,\frac{s}{2}+2;\;a^2\end{array}\right)\right]$

$\qquad\qquad\qquad\qquad\qquad\qquad [\mathrm{Re}\,s > -1;\; |\arg(1-a^2)| < \pi].$

176. $\displaystyle\int_0^1 \frac{1}{x} \arccos x \arcsin^2(ax)\, dx = \frac{\pi}{8} \operatorname{Li}_3(a^2)$ $\qquad [|\arg(1-a^2)| < \pi]$.

177. $\displaystyle\int_0^1 \frac{1}{x} \arccos x \arcsin^2 x\, dx = \frac{\pi}{8} \zeta(3)$.

178. $\displaystyle\int_0^1 x \arccos x \arcsin^2(ax)\, dx = \frac{\pi}{16a^2}[a^2 + (a^2-1)\operatorname{Li}_2(a^2)]$

$\qquad [|\arg(1-a^2)| < \pi]$.

179. $\displaystyle\int_0^1 x^3 \arccos x \arcsin^2(ax)\, dx$

$= \dfrac{\pi}{512 a^4}[4a^2 + 15a^4 - 8(a^2-1)^2 \ln(1-a^2) + 12(a^4-1)\operatorname{Li}_2(a^2)]\operatorname{Li}$

$\qquad [|\arg(1-a^2)| < \pi]$.

180. $\displaystyle\int_0^1 x^3 \arccos x \arcsin^2 x\, dx = \frac{19\pi}{512}$.

181. $\displaystyle\int_0^1 x^{s-1} \arccos x \arctan(ax)\, dx = \frac{\pi^{1/2} a \Gamma\!\left(\frac{s}{2}\right)}{4(s+1)\Gamma\!\left(\frac{s+3}{2}\right)}$

$\times \left[(s+1)\,{}_3F_2\!\left(\begin{array}{c}\frac{1}{2},1,\frac{s}{2}+1\\ \frac{3}{2},\frac{s+3}{2}\end{array};-a^2\right) - {}_3F_2\!\left(\begin{array}{c}1,\frac{s+1}{2},\frac{s}{2}+1\\ \frac{s+3}{2},\frac{s+3}{2}\end{array};-a^2\right)\right]$

$\qquad [\operatorname{Re} s > -1;\ |\arg(1+a^2)| < \pi]$.

182. $\displaystyle\int_0^1 \arccos x \arctan(ax)\, dx = \frac{\pi}{2a}\left(\sqrt{1+a^2} - \ln\frac{1+\sqrt{1+a^2}}{2} - 1\right)$

$\qquad [|\arg(1+a^2)| < \pi]$.

183. $\displaystyle\int_0^1 x^2 \arccos x \arctan(ax)\, dx$

$= \dfrac{\pi}{72a^3}(4-9a^2) + \dfrac{\pi}{18a^3\sqrt{1+a^2}}(2a^4+a^2-1) + \dfrac{\pi}{6a^3}\ln\dfrac{1+\sqrt{1+a^2}}{2}\ln$

$\qquad [|\arg(1+a^2)| < \pi]$.

184. $\int\limits_0^\infty \left(\dfrac{\arctan x}{x}\right)^n dx = I_n,\ \ I_2 = \pi \ln 2,\ \ I_3 = \dfrac{3\pi}{2} \ln 2 - \dfrac{\pi^3}{16},$

$I_4 = 2\pi \left(1 - \dfrac{\pi^2}{12}\right) \ln 2 - \dfrac{\pi^3}{12} + \dfrac{3\pi}{4} \zeta(3),\ \ I_5 = \dfrac{5\pi}{2} \left(1 - \dfrac{\pi^2}{3}\right) \ln 2 - \dfrac{5\pi^3}{48}$

$$+ \dfrac{\pi^5}{128} + \dfrac{15\pi}{4} \zeta(3).$$

4.2. The Dilogarithm $\mathrm{Li}_2(z)$

4.2.1. Integrals containing $\mathrm{Li}_2(z)$ and algebraic functions

1. $\int\limits_0^a x^{s-1}(a-x)^{t-1} \mathrm{Li}_2(bx(a-x))\, dx$

$$= a^{s+t+1} b\, \mathrm{B}(s+1, t+1)\, {}_5F_4\!\left(\begin{matrix} 1, 1, 1, s+1, t+1;\ \dfrac{a^2 b}{4} \\ 2, 2, \dfrac{s+t}{2}+1, \dfrac{s+t+3}{2} \end{matrix}\right)$$

$$[a > 0;\ \mathrm{Re}\, s,\, \mathrm{Re}\, t > -1;\ |\arg(4 - a^2 b)| < \pi].$$

2. $\int\limits_0^a x^{1/2}(a-x)^{-1/2} \mathrm{Li}_2(bx(a-x))\, dx = \pi a\, \mathrm{Li}_2\!\left(\dfrac{2 - \sqrt{4 - a^2 b}}{4}\right)$

$$- \dfrac{\pi a}{2} \ln^2 \dfrac{2 + \sqrt{4 - a^2 b}}{4} \quad [a > 0;\ |\arg(4 - a^2 b)| < \pi].$$

3. $\int\limits_0^a x^{-1/2}(a-x)^{-1/2} \mathrm{Li}_2(bx(a-x))\, dx = 2\pi\, \mathrm{Li}_2\!\left(\dfrac{2 - \sqrt{4 - a^2 b}}{4}\right)$

$$- \pi \ln^2 \dfrac{2 + \sqrt{4 - a^2 b}}{4} \quad [a > 0;\ |\arg(4 - a^2 b)| < \pi].$$

4. $\int\limits_0^a x^{-1/2}(a-x)^{-3/2} \mathrm{Li}_2(bx(a-x))\, dx$

$$= \dfrac{8\pi}{a} \left(1 - \sqrt{1 - \dfrac{a^2 b}{4}} + \ln \dfrac{2 + \sqrt{4 - a^2 b}}{4}\right) \quad [a > 0;\ |\arg(4 - a^2 b)| < \pi].$$

5. $\int\limits_0^a x^s (a-x)^{s+1/2} \mathrm{Li}_2\!\left(b\sqrt{x(a-x)}\right) dx = \dfrac{2^{-2s-2} \sqrt{\pi}\, a^{2s+5/2} b\, \Gamma(2s+3)}{\Gamma\!\left(2s + \dfrac{7}{2}\right)}$

$$\times {}_4F_3\!\left(\begin{matrix} 1, 1, 1, 2s+3 \\ 2, 2, 2s + \dfrac{7}{2};\ \dfrac{ab}{2} \end{matrix}\right) \quad [a > 0;\ \mathrm{Re}\, s > 3/2;\ |\arg(2 - ab)| < \pi].$$

6. $\displaystyle\int_0^a x^{-1/2}\, \mathrm{Li}_2\bigl(b\sqrt{x(a-x)}\bigr)\, dx$

$$= 4a^{1/2}\left(\arcsin^2\sqrt{\frac{ab}{2}} + 2\sqrt{\frac{2}{ab}-1}\,\arcsin\sqrt{\frac{ab}{2}} - 2\right)$$

$$[a>0;\ |\arg(2-ab)|<\pi].$$

7. $\displaystyle\int_0^a x^{-3/4}(a-x)^{-5/4}\,\mathrm{Li}_2\bigl(b\sqrt{x(a-x)}\bigr)\,dx$

$$= \frac{2^{7/2}\pi}{a}\left[1-\sqrt{1-\frac{ab}{2}}+\ln\!\left(1+\sqrt{1-\frac{ab}{2}}\right)-\ln 2\right]$$

$$[a>0;\ |\arg(2-ab)|<\pi].$$

8. $\displaystyle\int_0^a x^{-1/4}(a-x)^{-3/4}\,\mathrm{Li}_2\bigl(b\sqrt{x(a-x)}\bigr)\,dx$

$$= 2^{1/2}\pi\left[2\,\mathrm{Li}_2\!\left(\frac{1}{2}-\frac{1}{2}\sqrt{1-\frac{ab}{2}}\right) - \ln^2\!\left(\frac{1}{2}+\frac{1}{2}\sqrt{1-\frac{ab}{2}}\right)\right]$$

$$[a>0;\ |\arg(2-ab)|<\pi].$$

4.2.2. Integrals containing $\mathrm{Li}_2(z)$ and trigonometric functions

1. $\displaystyle\int_0^{m\pi}\frac{1}{\sin^2 x}\left\{\begin{matrix}\sin(ax)\\ \cos(ax)\end{matrix}\right\}\mathrm{Li}_2(b\sin^2 x)\, dx$

$$= \frac{2b}{a}\sin\frac{m\pi a}{2}\left\{\begin{matrix}\sin(m\pi a/2)\\ \cos(m\pi a/2)\end{matrix}\right\}\,{}_5F_4\!\left(\begin{matrix}\frac{1}{2},1,1,1,1;\ b\\ 2,2,1-\frac{a}{2},1+\frac{a}{2}\end{matrix}\right)$$

$$[a>0;\ |\arg(1-b)|<\pi].$$

2. $\displaystyle\int_0^{\pi}\sin^\nu x\left\{\begin{matrix}\sin(ax)\\ \cos(ax)\end{matrix}\right\}\mathrm{Li}_2(b\sin^2 x)\, dx = \frac{2^{-\nu-2}\pi b\,\Gamma(\nu+3)}{\Gamma\!\left(\frac{\nu-a}{2}+2\right)\Gamma\!\left(\frac{\nu+a}{2}+2\right)}$

$$\times\left\{\begin{matrix}\sin(a\pi/2)\\ \cos(a\pi/2)\end{matrix}\right\}\,{}_5F_4\!\left(\begin{matrix}1,1,1,\frac{\nu+3}{2},\frac{\nu}{2}+2;\ b\\ 2,2,\frac{\nu-a}{2}+2,\frac{\nu+a}{2}+2\end{matrix}\right)$$

$$[\mathrm{Re}\,\nu>-1;\ |\arg(1-b)|<\pi].$$

3. $\displaystyle\int_0^{\pi/2}\cos^\nu x\cos(ax)\,\mathrm{Li}_2(b\cos^2 x)\, dx$

$$= \frac{2^{-\nu-3}\pi b\,\Gamma(\nu+3)}{\Gamma\!\left(\frac{\nu-a}{2}+2\right)\Gamma\!\left(\frac{\nu+a}{2}+2\right)}\,{}_5F_4\!\left(\begin{matrix}1,1,1,\frac{\nu+3}{2},\frac{\nu}{2}+2;\ b\\ 2,2,\frac{\nu-a}{2}+2,\frac{\nu+a}{2}+2\end{matrix}\right)$$

$$[\mathrm{Re}\,\nu>-1;\ |\arg(1-b)|<\pi].$$

4. $\displaystyle\int_0^{m\pi} \frac{e^{-ax}}{\sin^2 x}\,\mathrm{Li}_2(b\sin^2 x)\,dx = \frac{b}{a}(1-e^{-m\pi a})\,_5F_4\!\left(\begin{matrix}\frac{1}{2},1,1,1,1;\,b\\ 2,2,1-\frac{ia}{2},1+\frac{ia}{2}\end{matrix}\right).$

5. $\displaystyle\int_0^\infty \frac{e^{-ax}}{\sin x}\,\mathrm{Li}_2(b\sin x)\,dx$

$\displaystyle = \frac{b}{a}\,_5F_4\!\left(\begin{matrix}\frac{1}{2},\frac{1}{2},\frac{1}{2},1,1;\,b^2\\ \frac{3}{2},\frac{3}{2},\frac{3-ia}{2},\frac{3+ia}{2}\end{matrix}\right) + \frac{b^2}{4(a^2+1)}\,_5F_4\!\left(\begin{matrix}1,1,1,1,\frac{3}{2};\,b^2\\ 2,2,\frac{3-ia}{2},\frac{3+ia}{2}\end{matrix}\right)$

$\qquad\qquad\qquad\qquad\qquad\qquad [\mathrm{Re}\,a>0;\ |\arg(1-b^2)|<\pi].$

6. $\displaystyle\int_0^\infty \frac{e^{-ax}}{\sin^2 x}\,\mathrm{Li}_2(b\sin^2 x)\,dx = \frac{b}{a}\,_5F_4\!\left(\begin{matrix}\frac{1}{2},1,1,1,1;\,b\\ 2,2,1-\frac{ia}{2},1+\frac{ia}{2}\end{matrix}\right)$

$\qquad\qquad\qquad\qquad\qquad\qquad [\mathrm{Re}\,a>0;\ |\arg(1-b^2)|<\pi].$

4.2.3. Integrals containing $\mathrm{Li}_2(z)$ and the logarithmic function

1. $\displaystyle\int_0^a x^{s-1} \ln\frac{\sqrt{a}+\sqrt{a-x}}{\sqrt{x}}\,\mathrm{Li}_2(bx)\,dx = \frac{\pi^{1/2} a^{s+1} b\,\Gamma(s)}{2s(s+1)\Gamma\!\left(s+\frac{3}{2}\right)}$

$\displaystyle \times \left[\,_3F_2\!\left(\begin{matrix}1,s+1,s+1;\,ab\\ s+\frac{3}{2},s+2\end{matrix}\right) - (s+1)\,_3F_2\!\left(\begin{matrix}1,1,s+1;\,ab\\ 2,s+\frac{3}{2}\end{matrix}\right)\right.$

$\displaystyle \left. + s(s+1)\,_4F_3\!\left(\begin{matrix}1,1,1,s+1;\,ab\\ 2,2,s+\frac{3}{2}\end{matrix}\right)\right] \quad [a>0;\ \mathrm{Re}\,s>-2;\ |\arg(1-ab)|<\pi].$

2. $\displaystyle\int_0^a \ln\frac{\sqrt{a}+\sqrt{a-x}}{\sqrt{x}}\,\mathrm{Li}_2(bx)\,dx = \left(2a+\frac{1}{b}\right)\arcsin^2(\sqrt{ab})$

$\displaystyle + 6a\sqrt{\frac{1}{ab}-1}\,\arcsin(\sqrt{ab}) - 7a \quad [a>0;\ |\arg(1-ab)|<\pi].$

3. $\displaystyle\int_0^a x^{-3/2} \ln\frac{\sqrt{a}+\sqrt{a-x}}{\sqrt{x}}\,\mathrm{Li}_2(bx)\,dx$

$\displaystyle = \frac{4\pi}{\sqrt{a}}\left(\sqrt{ab}\,\arcsin\sqrt{ab} - \ln\frac{1+\sqrt{1-ab}}{2} + 2\sqrt{1-ab} - 2\right)$

$\qquad\qquad\qquad\qquad\qquad\qquad [a>0;\ |\arg(1-ab)|<\pi].$

4.2.4. Integrals containing $\mathrm{Li}_2(z)$ and inverse trigonometric functions

1. $\displaystyle\int_0^1 \frac{1}{x}\arccos\sqrt{x}\,\mathrm{Li}_2(x)\,dx = \frac{\pi}{12}[8\ln^3 2 - 2\pi^2 \ln 2 + 12\zeta(3)].$

2. $\displaystyle\int_0^1 x^{s-1} \arccos\sqrt{x}\, \text{Li}_2(ax)\, dx = \frac{\pi^{1/2} a \Gamma\left(s+\frac{3}{2}\right)}{2s^2(s+1)\Gamma(s+2)}$

$$\times \left[{}_3F_2\!\left(\begin{matrix}1,\, s+1,\, s+\frac{3}{2}\\ s+2,\, s+2;\, a\end{matrix}\right) - (s+1)\, {}_3F_2\!\left(\begin{matrix}1,\, 1,\, s+\frac{3}{2}\\ 2,\, s+2;\, a\end{matrix}\right)\right.$$

$$\left. + s(s+1)\, {}_4F_3\!\left(\begin{matrix}1,\, 1,\, 1,\, s+\frac{3}{2}\\ 2,\, 2,\, s+2;\, a\end{matrix}\right)\right] \qquad [|\arg(1-a)| < \pi].$$

4.3. The Sine Si(z) and Cosine ci(z) Integrals

4.3.1. Integrals containing Si(z) and algebraic functions

1. $\displaystyle\int_0^a x^{-1/2} \text{Si}\left(b\sqrt[4]{x(a-x)}\right) dx = \frac{\pi\sqrt{a}}{2^{3/2}}\left\{2\sqrt{a}\, J_0\!\left(b\sqrt{\frac{a}{2}}\right) - 2\sqrt{2}\, J_1\!\left(b\sqrt{\frac{a}{2}}\right)\right.$

$$\left. + \pi\sqrt{a}\, b\left[J_1\!\left(b\sqrt{\frac{a}{2}}\right)\mathbf{H}_0\!\left(b\sqrt{\frac{a}{2}}\right) - J_0\!\left(b\sqrt{\frac{a}{2}}\right)\mathbf{H}_1\!\left(b\sqrt{\frac{a}{2}}\right)\right]\right\} \quad [a>0].$$

4.3.2. Integrals containing Si(z) and trigonometric functions

1. $\displaystyle\int_0^a \frac{1}{\sqrt{x(a-x)}}\, [\sin x\, \text{Si}(2x) + \cos x\, \text{ci}(2x)]\, dx = \frac{\pi^2}{4} \cos\frac{a}{2}\, Y_0\!\left(\frac{a}{2}\right)$

$$+ \frac{\pi}{2}\left[\sin\frac{a}{2}\,\text{Si}(a) + \cos\frac{a}{2}\,\text{ci}(a)\right] J_0\!\left(\frac{a}{2}\right) \quad [a>0].$$

2. $\displaystyle\int_0^a \frac{1}{\sqrt{x(a-x)}}\, [\cos x\, \text{Si}(2x) - \sin x\, \text{ci}(2x)]\, dx = -\frac{\pi^2}{4} \sin\frac{a}{2}\, Y_0\!\left(\frac{a}{2}\right)$

$$+ \frac{\pi}{2}\left[\cos\frac{a}{2}\,\text{Si}(a) - \sin\frac{a}{2}\,\text{ci}(a)\right] J_0\!\left(\frac{a}{2}\right) \quad [a>0].$$

3. $\displaystyle\int_0^{\pi/2} \cos^\nu x \cos(ax)\, \text{Si}(b\cos x)\, dx$

$$= \frac{2^{-\nu-2}\pi b\,\Gamma(\nu+2)}{\Gamma\!\left(\frac{\nu-a+3}{2}\right)\Gamma\!\left(\frac{\nu+a+3}{2}\right)}\, {}_3F_4\!\left(\begin{matrix}\frac{1}{2},\, \frac{\nu}{2}+1,\, \frac{\nu+3}{2};\, -\frac{b^2}{4}\\ \frac{3}{2},\, \frac{3}{2},\, \frac{\nu-a+3}{2},\, \frac{\nu+a+3}{2}\end{matrix}\right) \quad [\text{Re}\,\nu > -2].$$

4. $\displaystyle\int_0^{\pi/2} \frac{\cos(2nx)}{\sin x}\, \text{Si}(a\sin x)\, dx$

$$= \frac{2^{-2n-1}\pi a^{2n+1}}{(2n+1)!(2n+1)}\, {}_2F_3\!\left(\begin{matrix}n+\frac{1}{2},\, n+\frac{1}{2};\, -\frac{a^2}{4}\\ n+\frac{3}{2},\, n+\frac{3}{2},\, 2n+1\end{matrix}\right).$$

5. $\displaystyle\int_0^\pi \sin^\nu x \begin{Bmatrix} \sin(ax) \\ \cos(ax) \end{Bmatrix} \operatorname{Si}(b\sin x)\,dx = \dfrac{2^{-\nu-1}\pi b\,\Gamma(\nu+2)}{\Gamma\left(\dfrac{\nu-a+3}{2}\right)\Gamma\left(\dfrac{\nu+a+3}{2}\right)}$

$\times \begin{Bmatrix} \sin(a\pi/2) \\ \cos(a\pi/2) \end{Bmatrix} {}_3F_4\left(\begin{array}{c} \dfrac{1}{2},\dfrac{\nu}{2}+1,\dfrac{\nu+3}{2};\ -\dfrac{b^2}{4} \\ \dfrac{3}{2},\dfrac{3}{2},\dfrac{\nu-a+3}{2},\dfrac{\nu+a+3}{2} \end{array}\right)$ [Re $\nu > -2$].

6. $\displaystyle\int_0^\pi \dfrac{\cos(nx)}{\cos x}\operatorname{Si}(a\cos x)\,dx$

$= \dfrac{2^{-n}\pi a^{n+1}}{(n+1)!\,(n+1)}\cos\dfrac{n\pi}{2}\,{}_2F_3\left(\begin{array}{c} \dfrac{n+1}{2},\dfrac{n+1}{2};\ -\dfrac{a^2}{4} \\ \dfrac{n+3}{2},\dfrac{n+3}{2},n+1 \end{array}\right).$

7. $\displaystyle\int_0^{m\pi} \dfrac{1}{\sin x}\begin{Bmatrix} \sin(ax) \\ \cos(ax) \end{Bmatrix}\operatorname{Si}(b\sin x)\,dx$

$= \dfrac{2b}{a}\sin\dfrac{m\pi a}{2}\begin{Bmatrix} \sin(m\pi a/2) \\ \cos(m\pi a/2) \end{Bmatrix}{}_3F_4\left(\begin{array}{c} \dfrac{1}{2},\dfrac{1}{2},1;\ -\dfrac{b^2}{4} \\ \dfrac{3}{2},\dfrac{3}{2},1-\dfrac{a}{2},1+\dfrac{a}{2} \end{array}\right).$

8. $\displaystyle\int_0^\infty \dfrac{1}{\sqrt{x}}e^{-ax}\left[\sin x\,\operatorname{Si}(2x)+\cos x\,\operatorname{ci}(2x)\right]dx =$

$-\sqrt{\dfrac{\pi}{2}}\dfrac{(a+\sqrt{1+a^2})^{1/2}}{(1+a^2)^{1/2}}\ln\left(a+\sqrt{1+a^2}\right)$ [Re $a > 0$].

9. $\displaystyle\int_0^\infty \dfrac{e^{-ax}}{\sqrt{x}}\left[\sin x\,\operatorname{ci}(2x)-\cos x\,\operatorname{Si}(2x)\right]dx =$

$-\sqrt{\dfrac{\pi}{2}}\dfrac{(1+a^2)^{-1/2}}{(a+\sqrt{1+a^2})^{1/2}}\ln\left(a+\sqrt{1+a^2}\right)$ [Re $a > 0$].

10. $\displaystyle\int_0^{m\pi} \dfrac{e^{-ax}}{\sin x}\operatorname{Si}(b\sin x)\,dx = \dfrac{b}{a}(1-e^{-m\pi a})\,{}_3F_4\left(\begin{array}{c} \dfrac{1}{2},\dfrac{1}{2},1;\ -\dfrac{b^2}{4} \\ \dfrac{3}{2},\dfrac{3}{2},1-\dfrac{ia}{2},1+\dfrac{ia}{2} \end{array}\right).$

11. $\displaystyle\int_0^\infty \dfrac{e^{-ax}}{\sin x}\operatorname{Si}(b\sin x)\,dx = \dfrac{b}{a}\,{}_3F_4\left(\begin{array}{c} \dfrac{1}{2},\dfrac{1}{2},1;\ -\dfrac{b^2}{4} \\ \dfrac{3}{2},\dfrac{3}{2},1-\dfrac{ia}{2},1+\dfrac{ia}{2} \end{array}\right)$ [Re $a > 0$].

12. $\displaystyle\int_0^\infty \cosh^\nu x \cos(bx) \operatorname{Si}(c \operatorname{sech} x)\, dx = \frac{2^{-\nu-1} c}{\Gamma(1-\nu)}$

$$\times \Gamma\left(\frac{1-\nu-ib}{2}\right)\Gamma\left(\frac{1-\nu+ib}{2}\right) {}_3F_4\left(\begin{array}{c}\frac{1}{2},\frac{1-\nu-ib}{2},\frac{1-\nu+ib}{2};\\ \frac{3}{2},\frac{3}{2},\frac{1-\nu}{2},1-\frac{\nu}{2}\end{array} -\frac{c^2}{4}\right)$$

$[\operatorname{Re}\nu < 1]$.

4.3.3. Integrals containing $\operatorname{Si}(z)$ and the logarithmic function

1. $\displaystyle\int_0^a x^{s-1} \ln\frac{a+\sqrt{a^2-x^2}}{x} \operatorname{Si}(bx)\, dx = \frac{\sqrt{\pi}}{2} \frac{a^{s+1} b\, \Gamma\left(\frac{s+1}{2}\right)}{s(s+1)\Gamma\left(\frac{s}{2}+1\right)}$

$$\times\left[(s+1){}_2F_3\left(\begin{array}{c}\frac{1}{2},\frac{s+1}{2};\\ \frac{3}{2},\frac{3}{2},\frac{s}{2}+1\end{array} -\frac{a^2b^2}{4}\right) - 2{}_2F_3\left(\begin{array}{c}\frac{s+1}{2},\frac{s+1}{2};\\ \frac{3}{2},\frac{s}{2}+1,\frac{s+3}{2}\end{array} -\frac{a^2b^2}{4}\right)\right]$$

$[a>0;\ \operatorname{Re} s > -1]$.

2. $\displaystyle\int_0^a x \ln\frac{a+\sqrt{a^2-x^2}}{x} \operatorname{Si}(bx)\, dx = \frac{\pi a}{8b}\{2ab[ab\, J_0(ab) - J_1(ab)]$

$$+ \pi(a^2b^2 - 1)[J_1(ab)\mathbf{H}_0(ab) - J_0(ab)\mathbf{H}_1(ab)]\} \quad [a>0].$$

3. $\displaystyle\int_0^a x^3 \ln\frac{a+\sqrt{a^2-x^2}}{x} \operatorname{Si}(bx)\, dx$

$$= \frac{\pi a}{24b^3}\{2ab[ab(a^2b^2+1)J_0(ab) - (a^2b^2+5)J_1(ab)]\}$$

$$+ \frac{\pi^2 a}{24b^3}(a^4b^4 + 9)[J_1(ab)\mathbf{H}_0(ab) - J_0(ab)\mathbf{H}_1(ab)] \quad [a>0].$$

4.3.4. Integrals containing $\operatorname{Si}(z)$ and inverse trigonometric functions

1. $\displaystyle\int_0^1 x^{s-1} \arccos x\, \operatorname{Si}(ax)\, dx$

$$= \frac{\pi^{1/2} a\, \Gamma\left(\frac{s}{2}+1\right)}{2(s+1)\Gamma\left(\frac{s+3}{2}\right)} {}_3F_4\left(\begin{array}{c}\frac{1}{2},\frac{s+1}{2},\frac{s}{2}+1;\\ \frac{3}{2},\frac{3}{2},\frac{s+3}{2},\frac{s+3}{2}\end{array} -\frac{a^2}{4}\right) \quad [\operatorname{Re} s > -1].$$

2. $\displaystyle\int_0^1 \arccos x\, \operatorname{Si}(ax)\, dx = \frac{\pi}{4a}\{-2 + 2(a^2+1)J_0(a) - 2a\, J_1(a)$

$$+ \pi a^2[J_1(a)\mathbf{H}_0(a) - J_0(a)\mathbf{H}_1(a)]\}.$$

4.3.5. Integrals containing products of Si(z) and ci(z)

1. $\displaystyle\int_0^\infty x^{s-1}[\sin(x)\operatorname{ci}(2x) - \cos(x)\operatorname{Si}(2x)]^2\, dx$

$$= \frac{2^{-s-4}}{s}\Gamma(s)\left\{\pi^2 s[3-\cos(\pi s)]\sec\frac{\pi s}{2} + 4\pi[1+\cos(\pi s)]\csc\frac{\pi s}{2}\right.$$
$$\left. + 4s\cos\frac{\pi s}{2}\left[\psi'\!\left(\frac{1+s}{2}\right) - \psi'\!\left(\frac{s}{2}\right)\right]\right\} \qquad [-2 < \operatorname{Re} s < 0].$$

2. $\displaystyle\int_0^\infty x^{s-1}[\sin(x)\operatorname{ci}(2x) - \cos(x)\operatorname{Si}(2x)][\cos(x)\operatorname{ci}(2x) + \sin(x)\operatorname{Si}(2x)]\,dx$

$$= 2^{-s-3}\Gamma(s)\left\{\frac{\pi^2}{2}[\cos(\pi s) + 3]\csc\frac{\pi s}{2}\right.$$
$$\left. + \sin\frac{\pi s}{2}\left[3\psi'\!\left(\frac{1+s}{2}\right) - 4\psi'(s) - \psi'\!\left(\frac{s}{2}\right)\right]\right\} \qquad [-1 < \operatorname{Re} s < 1].$$

3. $\displaystyle\int_0^a \operatorname{shi}(b\sqrt{x})\operatorname{Si}(b\sqrt{a-x})\,dx = \frac{\pi(ab)^2}{8}\,{}_2F_5\!\left(\begin{array}{c}\frac{1}{2},\frac{1}{2};\ \frac{a^2 b^4}{256}\\ \frac{3}{4},\frac{5}{4},\frac{3}{2},\frac{3}{2},2\end{array}\right) \qquad [a>0].$

4.4. The Error Functions erf(z), erfi(z) and erfc(z)

4.4.1. Integrals containing erf(z) and algebraic functions

1. $\displaystyle\int_0^a x^{s-1}(a-x)^{t-1}\operatorname{erf}\!\left(b\sqrt{x(a-x)}\right)dx = \frac{2}{\sqrt{\pi}}a^{s+t}b\,\mathrm{B}\!\left(s+\frac{1}{2},t+\frac{1}{2}\right)$

$$\times {}_3F_3\!\left(\begin{array}{c}\frac{1}{2},s+\frac{1}{2},t+\frac{1}{2};\ -\frac{a^2 b^2}{4}\\ \frac{3}{2},\frac{s+t+1}{2},\frac{s+t}{2}+1\end{array}\right) \qquad [a>0;\ \operatorname{Re} s, \operatorname{Re} t > -1/2].$$

2. $\displaystyle\int_0^a x^{s+1/2}(a-x)^s \operatorname{erf}\!\left(b\sqrt[4]{x(a-x)}\right)dx = 2^{-2s-1/2}a^{2s+2}b\,\frac{\Gamma\!\left(2s+\frac{5}{2}\right)}{\Gamma(2s+3)}$

$$\times {}_2F_2\!\left(\begin{array}{c}\frac{1}{2},2s+\frac{5}{2};\ -\frac{ab^2}{2}\\ \frac{3}{2},2s+3\end{array}\right) \qquad [a>0;\ \operatorname{Re} s > -5/4].$$

3. $\displaystyle\int_0^a x^{1/2}\operatorname{erf}\!\left(b\sqrt[4]{x(a-x)}\right)dx$

$$= \sqrt{\frac{\pi}{2}}\,\frac{a}{3b}\,e^{-ab^2/4}\left[ab^2 I_0\!\left(\frac{ab^2}{4}\right) + (ab^2+1)I_1\!\left(\frac{ab^2}{4}\right)\right] \qquad [a>0].$$

4.4.2] **4.4. The Error Functions erf (z), erfi (z) and erfc (z)**

4. $\int_0^a x^{-1/2} \operatorname{erf}\left(b\sqrt[4]{x(a-x)}\right) dx = \sqrt{\dfrac{\pi}{2}}\, abe^{-ab^2/4}\left[I_0\left(\dfrac{ab^2}{4}\right) + I_1\left(\dfrac{ab^2}{4}\right)\right]$

$[a > 0]$.

4.4.2. Integrals containing erf (z), erfc (z) and the exponential function

1. $\int_0^a x^{s-1}(a-x)^{t-1} e^{b^2 x(a-x)} \operatorname{erf}\left(b\sqrt{x(a-x)}\right) dx$

$= \dfrac{2}{\sqrt{\pi}} a^{s+t} b\, \mathrm{B}\left(s+\dfrac{1}{2},\, t+\dfrac{1}{2}\right)\, {}_3F_3\left(\begin{array}{c} 1,\ s+\dfrac{1}{2},\ t+\dfrac{1}{2};\ \dfrac{a^2 b^2}{4} \\ \dfrac{3}{2},\ \dfrac{s+t+1}{2},\ \dfrac{s+t}{2}+1 \end{array}\right)$

$[a > 0;\ \operatorname{Re} s,\ \operatorname{Re} t > -1/2]$.

2. $\int_0^a e^{b^2 x(a-x)} \operatorname{erf}\left(b\sqrt{x(a-x)}\right) dx = \dfrac{\sqrt{\pi}}{b}\left(e^{a^2 b^2/4} - 1\right)$ $\qquad [a > 0]$.

3. $\int_0^a x^{-1} e^{b^2 x(a-x)} \operatorname{erf}\left(b\sqrt{x(a-x)}\right) dx = \pi\, \operatorname{erfi}\left(\dfrac{ab}{2}\right)$ $\qquad [a > 0]$.

4. $\int_0^a x^{-1}(a-x)^{-1} e^{b^2 x(a-x)} \operatorname{erf}\left(b\sqrt{x(a-x)}\right) dx = \dfrac{2\pi}{a}\, \operatorname{erfi}\left(\dfrac{ab}{2}\right)$ $\qquad [a > 0]$.

5. $\int_0^a x^{s+1/2}(a-x)^s e^{b^2 \sqrt{x(a-x)}} \operatorname{erf}\left(b\sqrt[4]{x(a-x)}\right) dx = 2^{-2s-1/2} a^{2s+2} b$

$\times \dfrac{\Gamma\left(2s+\dfrac{5}{2}\right)}{\Gamma(2s+3)}\, {}_2F_2\left(\begin{array}{c} 1,\ 2s+\dfrac{5}{2};\ \dfrac{ab^2}{2} \\ \dfrac{3}{2},\ 2s+3 \end{array}\right)$ $\qquad [a > 0;\ \operatorname{Re} s > -5/4]$.

6. $\int_0^a x^{1/2} e^{b^2 \sqrt{x(a-x)}} \operatorname{erf}\left(b\sqrt[4]{x(a-x)}\right) dx$

$= \dfrac{1}{2b^3}\sqrt{\dfrac{\pi}{2}}\left[2 - ab^2 + 2(ab^2 - 1)e^{ab^2/2}\right]$ $\qquad [a > 0]$.

7. $\int_0^a x^{-1/2} e^{b^2 \sqrt{x(a-x)}} \operatorname{erf}\left(b\sqrt[4]{x(a-x)}\right) dx = \dfrac{\sqrt{2\pi}}{b}\left(e^{ab^2/2} - 1\right)$ $\qquad [a > 0]$.

8. $\int_0^a x^{-1/2}(a-x)^{-1}e^{b^2\sqrt{x(a-x)}}\operatorname{erf}\left(b\sqrt[4]{x(a-x)}\right)dx = \dfrac{2\pi}{\sqrt{a}}\operatorname{erfi}\left(b\sqrt{\dfrac{a}{2}}\right)$

$[a>0]$.

4.4.3. Integrals containing erf (z) and trigonometric functions

1. $\displaystyle\int_0^\pi \sin(ax)\operatorname{erf}(b\sin x)\,dx = \dfrac{2b\sin(\pi a)}{\sqrt{\pi}(1-a^2)}\,{}_2F_2\!\left(\begin{array}{c}\frac{1}{2},\,1;\,-b^2\\[2pt]\frac{3-a}{2},\,\frac{3+a}{2}\end{array}\right).$

2. $\displaystyle\int_0^\pi \dfrac{\cos(nx)}{\cos x}\operatorname{erf}(a\cos x)\,dx$

$= \dfrac{2^{1-n}\sqrt{\pi}\,a^{n+1}}{(n+1)\Gamma\!\left(\frac{n}{2}+1\right)}\cos\dfrac{n\pi}{2}\,{}_2F_2\!\left(\begin{array}{c}\frac{n+1}{2},\,\frac{n+1}{2};\,-a^2\\[2pt]\frac{n+3}{2},\,n+1\end{array}\right).$

3. $\displaystyle\int_0^{\pi/2}\dfrac{\cos(2nx)}{\sin x}\operatorname{erf}(a\sin x)\,dx$

$= \dfrac{2^{-2n}\sqrt{\pi}\,a^{2n+1}\Gamma\!\left(n+\frac{1}{2}\right)}{n!\,(2n+1)}\,{}_2F_2\!\left(\begin{array}{c}n+\frac{1}{2},\,n+\frac{1}{2};\,-a^2\\[2pt]n+\frac{3}{2},\,2n+1\end{array}\right).$

4. $\displaystyle\int_0^{\pi/2}\cos^\nu x\cos(ax)\operatorname{erf}(b\cos x)\,dx = \dfrac{2^{-\nu-1}\sqrt{\pi}\,b\,\Gamma(\nu+2)}{\Gamma\!\left(\frac{\nu-a+3}{2}\right)\Gamma\!\left(\frac{\nu+a+3}{2}\right)}$

$\times\,{}_3F_3\!\left(\begin{array}{c}\frac{1}{2},\,1+\frac{\nu}{2},\,\frac{\nu+3}{2};\,b^2\\[2pt]\frac{3}{2},\,\frac{\nu-a+3}{2},\,\frac{\nu+a+3}{2}\end{array}\right)$ $[\operatorname{Re}\nu>-2]$.

5. $\displaystyle\int_0^\pi \sin^\nu x\left\{\begin{array}{c}\sin(ax)\\\cos(ax)\end{array}\right\}\operatorname{erf}(b\sin x)\,dx = \dfrac{2^{-\nu}\sqrt{\pi}\,b\,\Gamma(\nu+2)}{\Gamma\!\left(\frac{\nu-a+3}{2}\right)\Gamma\!\left(\frac{\nu+a+3}{2}\right)}$

$\times\left\{\begin{array}{c}\sin(a\pi/2)\\\cos(a\pi/2)\end{array}\right\}{}_3F_3\!\left(\begin{array}{c}\frac{1}{2},\,1+\frac{\nu}{2},\,\frac{\nu+3}{2};\,-b^2\\[2pt]\frac{3}{2},\,\frac{\nu-a+3}{2},\,\frac{\nu+a+3}{2}\end{array}\right)$ $[\operatorname{Re}\nu>-2]$.

6. $\displaystyle\int_0^{\pi/2}\cos^\nu x\cos(ax)e^{b^2\cos^2 x}\operatorname{erf}(b\cos x)\,dx = \dfrac{2^{-\nu-1}\sqrt{\pi}\,b\,\Gamma(\nu+2)}{\Gamma\!\left(\frac{\nu-a+3}{2}\right)\Gamma\!\left(\frac{\nu+a+3}{2}\right)}$

$\times\,{}_3F_3\!\left(\begin{array}{c}1,\,1+\frac{\nu}{2},\,\frac{\nu+3}{2};\,b^2\\[2pt]\frac{3}{2},\,\frac{\nu-a+3}{2},\,\frac{\nu+a+3}{2}\end{array}\right)$ $[\operatorname{Re}\nu>-2]$.

4.4.5] **4.4. THE ERROR FUNCTIONS erf (z), erfi (z) AND erfc (z)**

7. $\displaystyle\int_0^\pi \sin^\nu x \left\{\begin{matrix}\sin(ax)\\\cos(ax)\end{matrix}\right\} e^{b^2\sin^2 x} \operatorname{erf}(b\sin x)\, dx = \dfrac{2^{-\nu}\sqrt{\pi}\,b\,\Gamma(\nu+2)}{\Gamma\left(\dfrac{\nu-a+3}{2}\right)\Gamma\left(\dfrac{\nu+a+3}{2}\right)}$

$\times \left\{\begin{matrix}\sin(a\pi/2)\\\cos(a\pi/2)\end{matrix}\right\} {}_3F_3\left(\begin{matrix}1,1+\dfrac{\nu}{2},\dfrac{\nu+3}{2};\,b^2\\\dfrac{3}{2},\dfrac{\nu-a+3}{2},\dfrac{\nu+a+3}{2}\end{matrix}\right)$ [Re $\nu > -1$].

8. $\displaystyle\int_0^\infty e^{-ax}\operatorname{erf}(b\sin x)\, dx = \dfrac{2b}{\sqrt{\pi}\,(a^2+1)}\,{}_2F_2\left(\begin{matrix}\dfrac{1}{2},1;\,-b^2\\\dfrac{3-ia}{2},\dfrac{3+ia}{2}\end{matrix}\right)$ [Re $a > 0$].

9. $\displaystyle\int_0^\infty e^{-ax+b^2\sin^2 x}\operatorname{erf}(b\sin x)\, dx = \dfrac{2b}{\sqrt{\pi}\,(a^2+1)}\,{}_2F_2\left(\begin{matrix}1,1;\,b^2\\\dfrac{3-ia}{2},\dfrac{3+ia}{2}\end{matrix}\right)$

[Re $a > 0$].

4.4.4. Integrals containing erf (z) and the logarithmic function

1. $\displaystyle\int_0^a x^{s-1}\ln\dfrac{a+\sqrt{a^2-x^2}}{x}\operatorname{erf}(bx)\, dx = \dfrac{2a^{s+1}b\,\Gamma\left(\dfrac{s+1}{2}\right)}{s(s+1)\Gamma\left(\dfrac{s}{2}\right)}$

$\times {}_3F_3\left(\begin{matrix}\dfrac{1}{2},\dfrac{s+1}{2},\dfrac{s+1}{2};\,-a^2b^2\\\dfrac{3}{2},\dfrac{s}{2}+1,\dfrac{s+3}{2}\end{matrix}\right)$ [$a > 0$; Re $s > -1$].

4.4.5. Integrals containing erf (z), erfi (z) and inverse trigonometric functions

1. $\displaystyle\int_0^1 \arccos x\,\operatorname{erf}(ax)\, dx$

$= \dfrac{\sqrt{\pi}}{2a}\left\{e^{-a^2/2}\left[(a^2+1)I_0\left(\dfrac{a^2}{2}\right)+a^2 I_1\left(\dfrac{a^2}{2}\right)\right]-1\right\}.$

2. $\displaystyle\int_0^1 x^2 \arccos x\,\operatorname{erf}(ax)\, dx = \dfrac{\sqrt{\pi}}{36a^3}$

$\times \left[(4a^4+3a^2+6)e^{-a^2/2}I_0\left(\dfrac{a^2}{2}\right)+a^2(4a^2-1)e^{-a^2/2}I_1\left(\dfrac{a^2}{2}\right)-6\right].$

3. $\displaystyle\int_0^1 e^{a^2 x^2}\arccos x\,\operatorname{erf}(ax)\, dx = \dfrac{\sqrt{\pi}}{4a}[\operatorname{Ei}(a^2)-2\ln a - C].$

4. $\int_0^1 x^2 e^{a^2 x^2} \arccos x \, \text{erf}(ax) \, dx$

$$= \frac{\sqrt{\pi}}{8a^3} \left[2e^{a^2} - a^2 - \text{Ei}(a^2) + 2\ln a + C - 2 \right].$$

5. $\int_0^1 \arccos x \, \text{erfi}(ax) \, dx$

$$= \frac{\sqrt{\pi}}{2a} \left\{ 1 + e^{a^2/2} \left[(a^2 - 1) I_0\left(\frac{a^2}{2}\right) - a^2 I_1\left(\frac{a^2}{2}\right) \right] \right\}.$$

6. $\int_0^1 x^2 \arccos x \, \text{erfi}(ax) \, dx$

$$= \frac{\sqrt{\pi}}{36a^3} \left[(4a^4 - 3a^2 + 6) e^{a^2/2} I_0\left(\frac{a^2}{2}\right) - a^2(4a^2 + 1) e^{a^2/2} I_1\left(\frac{a^2}{2}\right) - 6 \right].$$

7. $\int_0^1 e^{-a^2 x^2} \arccos x \, \text{erfi}(ax) \, dx = \frac{\sqrt{\pi}}{4a} [C + 2\ln a - \text{Ei}(-a^2)].$

8. $\int_0^1 x^2 e^{-a^2 x^2} \arccos x \, \text{erfi}(ax) \, dx$

$$= \frac{\sqrt{\pi}}{8a^3} \left[2e^{-a^2} + a^2 - \text{Ei}(-a^2) + 2\ln a + C - 2 \right].$$

4.4.6. Integrals containing products of $\text{erf}(z)$, $\text{erfc}(z)$ and $\text{erfi}(z)$

1. $\int_0^\infty x \, \text{erfi}(ax) \, \text{erfc}(bx) \, dx = \frac{a^2 + b^2}{4\pi a^2 b^2} \ln \frac{b + a}{b - a} - \frac{1}{2\pi ab}$ $\qquad [a < b]$.

2. $\int_0^\infty \frac{1}{x} \, \text{erfi}(ax) \, \text{erfc}(bx) \, dx = \frac{1}{\pi} \left[\text{Li}_2\left(\frac{a}{b}\right) - \text{Li}_2\left(-\frac{a}{b}\right) \right]$ $\qquad [|a| < |b|]$.

3. $\int_0^\infty \frac{1}{x} \, \text{erfi}(x) \, \text{erfc}(x) \, dx = \frac{\pi}{4}.$

4. $\int_0^a \text{erf}(b\sqrt{x}) \, \text{erf}(b\sqrt{a - x}) \, dx = \frac{1}{b^2} \left(e^{-ab^2} + ab^2 - 1 \right)$ $\qquad [a > 0]$.

5. $\int_0^a \text{erfi}(b\sqrt{x}) \, \text{erfi}(b\sqrt{a - x}) \, dx = \frac{1}{b^2} \left(e^{ab^2} - ab^2 - 1 \right)$ $\qquad [a > 0]$.

6. $\displaystyle\int_0^a \operatorname{erfi}(b\sqrt{x})\operatorname{erf}(b\sqrt{a-x})\,dx = a^2 b^2$

$\times \left[I_0(ab^2) - \dfrac{1}{ab^2} I_1(ab^2) + \dfrac{\pi}{2} I_0(ab^2)\mathbf{L}_1(ab^2) - \dfrac{\pi}{2} I_1(ab^2)\mathbf{L}_0(ab^2) \right]$

$[a > 0]$.

7. $\displaystyle\int_0^a e^{-b^2 x} \operatorname{erfi}(b\sqrt{x})\operatorname{erf}(b\sqrt{a-x})\,dx = 2a e^{-ab^2/2} I_1\!\left(\dfrac{ab^2}{2}\right)$ $\quad [a > 0]$.

8. $\displaystyle\int_0^a x^{s-1}(a-x)^{t-1}\operatorname{erf}\!\left(b\sqrt[4]{x(a-x)}\right)\operatorname{erfi}\!\left(b\sqrt[4]{x(a-x)}\right)dx$

$= \dfrac{4}{\pi} a^{s+t} b^2 \,\mathrm{B}\!\left(s+\dfrac{1}{2}, t+\dfrac{1}{2}\right)\, {}_4F_5\!\left(\begin{array}{c}\tfrac{1}{2},\,1,\,s+\tfrac{1}{2},\,t+\tfrac{1}{2};\\ \tfrac{3}{4},\,\tfrac{3}{2},\,\tfrac{5}{4},\,\tfrac{s+t+1}{2},\,\tfrac{s+t}{2}+1\end{array}\;\dfrac{a^2 b^2}{16}\right)$

$[a > 0;\ \operatorname{Re} s,\ \operatorname{Re} t > -1/2]$.

9. $\displaystyle\int_0^a e^{b^2 x}\operatorname{erf}(b\sqrt{x})\operatorname{erfi}(b\sqrt{a-x})\,dx = 2a e^{ab^2/2} I_1\!\left(\dfrac{ab^2}{2}\right)$ $\quad [a > 0]$.

10. $\displaystyle\int_0^a e^{2b^2 x}\operatorname{erf}(b\sqrt{x})\operatorname{erfi}(b\sqrt{a-x})\,dx = \dfrac{e^{ab^2}}{b^2}\left[\cosh(ab^2) - 1\right]$ $\quad [a > 0]$.

4.5. The Fresnel Integrals $S(z)$ and $C(z)$

4.5.1. Integrals containing $S(z)$ and algebraic functions

1. $\displaystyle\int_0^a x^{s-1}(a-x)^{t-1} S\!\left(b\sqrt{x(a-x)}\right)dx$

$= \dfrac{a^{s+t+1/2}}{3}\sqrt{\dfrac{2b^3}{\pi}}\,\mathrm{B}\!\left(s+\dfrac{3}{4}, t+\dfrac{3}{4}\right)\, {}_3F_4\!\left(\begin{array}{c}\tfrac{3}{4},\,s+\tfrac{3}{4},\,t+\tfrac{3}{4};\\ \tfrac{3}{2},\,\tfrac{7}{4},\,\tfrac{2s+2t+3}{4},\,\tfrac{2s+2t+5}{4}\end{array}\;-\dfrac{a^2 b^2}{16}\right)$

$[a > 0;\ \operatorname{Re} s,\ \operatorname{Re} t > -3/4]$.

2. $\displaystyle\int_0^a x^{s+1/2}(a-x)^s S\!\left(b\sqrt[4]{x(a-x)}\right)dx$

$= \dfrac{2^{-2s-5/4}}{3} a^{2s+9/4} b^{3/2}\,\dfrac{\Gamma\!\left(2s+\tfrac{11}{4}\right)}{\Gamma\!\left(2s+\tfrac{13}{4}\right)}\, {}_2F_3\!\left(\begin{array}{c}\tfrac{3}{4},\,2s+\tfrac{11}{4};\\ \tfrac{3}{2},\,\tfrac{7}{4},\,2s+\tfrac{13}{4}\end{array}\;-\dfrac{ab^2}{8}\right)$

$[a > 0;\ \operatorname{Re} s > -11/8]$.

4.5.2. Integrals containing $S(z)$ and trigonometric functions

1. $\displaystyle\int_0^{\pi/2} \frac{\cos(2nx)}{\sin^{3/2} x} S(a\sin x)\, dx$

$$= \frac{2^{-2n-1/2}\sqrt{\pi}\, a^{2n+3/2}}{(2n+1)!(4n+3)}\, {}_2F_3\left(\begin{array}{c} n+\frac{1}{2},\, n+\frac{3}{4};\\ n+\frac{3}{2},\, n+\frac{7}{4},\, 2n+1 \end{array};\, -\frac{a^2}{4}\right).$$

2. $\displaystyle\int_0^{\pi} \sin^\mu x \begin{Bmatrix} \sin(ax) \\ \cos(ax) \end{Bmatrix} S(b\sin x)\, dx = \frac{2^{-\mu-1}\Gamma\!\left(\mu+\frac{5}{2}\right)\sqrt{b^3\pi}}{3\Gamma\!\left(\frac{2\mu-2a+5}{4}\right)\Gamma\!\left(\frac{2\mu+2a+5}{4}\right)}$

$$\times \begin{Bmatrix} \sin(a\pi/2) \\ \cos(a\pi/2) \end{Bmatrix} {}_3F_4\left(\begin{array}{c} \frac{3}{4},\, \frac{2\mu+5}{4},\, \frac{2\mu+7}{4};\, -\frac{b^2}{4}\\ \frac{3}{2},\, \frac{7}{4},\, \frac{2\mu-2a+7}{4},\, \frac{2\mu+2a+7}{4} \end{array}\right) \qquad [\mathrm{Re}\,\mu > -5/2].$$

3. $\displaystyle\int_0^{m\pi} \frac{1}{\sin^{3/2} x} \begin{Bmatrix} \sin(ax) \\ \cos(ax) \end{Bmatrix} S(b\sin x)\, dx$

$$= \frac{2^{3/2}}{3a}\sin\frac{m\pi a}{2}\begin{Bmatrix} \sin(m\pi a/2) \\ \cos(m\pi a/2) \end{Bmatrix}\left(\frac{b^3}{\pi}\right)^{1/2} {}_3F_4\left(\begin{array}{c} \frac{1}{2},\, \frac{3}{4},\, 1;\, -\frac{b^2}{4}\\ \frac{3}{2},\, \frac{7}{4},\, 1-\frac{a}{2},\, 1+\frac{a}{2} \end{array}\right).$$

4. $\displaystyle\int_0^{m\pi} \frac{e^{-ax}}{\sin^{3/2} x} S(b\sin x)\, dx$

$$= \frac{1-e^{-m\pi a}}{3a}\left(\frac{2b^3}{\pi}\right)^{1/2} {}_3F_4\left(\begin{array}{c} \frac{1}{2},\, \frac{3}{4},\, 1;\, -\frac{b^2}{4}\\ \frac{3}{2},\, \frac{7}{4},\, 1-\frac{ia}{2},\, 1+\frac{ia}{2} \end{array}\right).$$

5. $\displaystyle\int_0^{\infty} \frac{e^{-ax}}{\sin^{3/2} x} S(b\sin x)\, dx = \frac{1}{3a}\left(\frac{2b^3}{\pi}\right)^{1/2} {}_3F_4\left(\begin{array}{c} \frac{1}{2},\, \frac{3}{4},\, 1;\, -\frac{b^2}{4}\\ \frac{3}{2},\, \frac{7}{4},\, 1-\frac{ia}{2},\, 1+\frac{ia}{2} \end{array}\right)$

$$[\mathrm{Re}\,a > 0].$$

6. $\displaystyle\int_0^{\infty} \cosh^\nu x \cos(bx) S(c\,\mathrm{sech}\,x)\, dx = \frac{2^{-\nu} c^{3/2}}{3\sqrt{\pi}\,\Gamma\!\left(\frac{3}{2}-\nu\right)}$

$$\times \Gamma\!\left(\frac{3-2\nu-2ib}{4}\right)\Gamma\!\left(\frac{3-2\nu+2ib}{4}\right) {}_3F_4\left(\begin{array}{c} \frac{3}{4},\, \frac{3-2\nu-ib}{4},\, \frac{3-2\nu+2ib}{4}\\ \frac{3}{2},\, \frac{7}{4},\, \frac{3-2\nu}{4},\, \frac{5-2\nu}{4} \end{array};\, -\frac{c^2}{4}\right)$$

$$[\mathrm{Re}\,\nu < 3/2].$$

4.5.3. Integrals containing $S(z)$ and the logarithmic function

1. $\displaystyle\int_0^a x^{s-1} \ln\frac{a+\sqrt{a^2-x^2}}{x} S(bx)\,dx = \frac{1}{3}a^{s+3/2}\left(\frac{b}{2}\right)^{3/2}\frac{\Gamma\!\left(\frac{2s+3}{4}\right)}{s(2s+3)\Gamma\!\left(\frac{2s+5}{4}\right)}$

$\displaystyle\times\left[(2s+3)\,_2F_3\!\left(\begin{array}{c}\frac{3}{4},\frac{2s+3}{4};\,-\frac{a^2b^2}{4}\\ \frac{3}{2},\frac{7}{4},\frac{2s+5}{4}\end{array}\right) - \,_2F_3\!\left(\begin{array}{c}\frac{2s+3}{4},\frac{2s+3}{4};\,-\frac{a^2b^2}{4}\\ \frac{3}{2},\frac{2s+5}{4},\frac{2s+7}{4}\end{array}\right)\right]$

$[a>0;\ \operatorname{Re} s > -3/2].$

4.5.4. Integrals containing $C(z)$ and algebraic functions

1. $\displaystyle\int_0^a x^{s-1}(a-x)^{t-1}C\!\left(b\sqrt{x(a-x)}\right)dx$

$\displaystyle = a^{s+t-1/2}\sqrt{\frac{2b}{\pi}}\,B\!\left(s+\frac{1}{4},t+\frac{1}{4}\right)\,_3F_4\!\left(\begin{array}{c}\frac{1}{4},s+\frac{1}{4},t+\frac{1}{4};\,-\frac{a^2b^2}{16}\\ \frac{1}{2},\frac{5}{4},\frac{2s+2t+1}{4},\frac{2s+2t+3}{4}\end{array}\right)$

$[a>0;\ \operatorname{Re} s,\operatorname{Re} t > -1/4].$

2. $\displaystyle\int_0^a x^{s+1/2}(a-x)^s C\!\left(b\sqrt[4]{x(a-x)}\right)dx$

$\displaystyle = 2^{-2s-3/4}a^{2s+7/4}b^{1/2}\,\frac{\Gamma\!\left(2s+\frac{9}{4}\right)}{\Gamma\!\left(2s+\frac{11}{4}\right)}\,_2F_3\!\left(\begin{array}{c}\frac{1}{4},2s+\frac{9}{4};\,-\frac{ab^2}{8}\\ \frac{1}{2},\frac{5}{4},2s+\frac{11}{4}\end{array}\right)$

$[a>0;\ \operatorname{Re} s > -9/8].$

4.5.5. Integrals containing $C(z)$ and trigonometric functions

1. $\displaystyle\int_0^{\pi/2}\frac{\cos(2nx)}{\sqrt{\sin x}}C(a\sin x)\,dx = \frac{\sqrt{\pi}\left(\frac{a}{2}\right)^{2n+1/2}}{(2n)!(4n+1)}\,_1F_2\!\left(\begin{array}{c}n+\frac{1}{4};\,-\frac{a^2}{4}\\ n+\frac{5}{4},2n+1\end{array}\right).$

2. $\displaystyle\int_0^\pi \sin^\mu x\left\{\begin{array}{c}\sin(ax)\\ \cos(ax)\end{array}\right\}C(b\sin x)\,dx = \frac{2^{-\mu}\Gamma\!\left(\mu+\frac{3}{2}\right)\sqrt{b\pi}}{\Gamma\!\left(\frac{2\mu-2a+5}{2}\right)\Gamma\!\left(\frac{2\mu+2a+5}{2}\right)}\sqrt{\frac{b}{\pi}}$

$\displaystyle\times\left\{\begin{array}{c}\sin(a\pi/2)\\ \cos(a\pi/2)\end{array}\right\}\,_3F_4\!\left(\begin{array}{c}\frac{1}{4},\frac{2\mu+3}{4},\frac{2\mu+5}{4};\,-\frac{b^2}{4}\\ \frac{1}{2},\frac{5}{4},\frac{2\mu-2a+5}{4},\frac{2\mu+2a+5}{4}\end{array}\right)\quad [\operatorname{Re}\mu > -3/2].$

3. $\displaystyle\int_0^{m\pi}\frac{1}{\sqrt{\sin x}}\left\{\begin{array}{c}\sin(ax)\\ \cos(ax)\end{array}\right\}C(b\sin x)\,dx$

$\displaystyle = \frac{2^{3/2}}{a}\sin\frac{m\pi a}{2}\left\{\begin{array}{c}\sin(m\pi a/2)\\ \cos(m\pi a/2)\end{array}\right\}\left(\frac{b}{\pi}\right)^{1/2}\,_2F_3\!\left(\begin{array}{c}\frac{1}{4},1;\,-\frac{b^2}{4}\\ \frac{5}{4},1-\frac{a}{2},1+\frac{a}{2}\end{array}\right).$

4. $\int_0^{m\pi} \dfrac{e^{-ax}}{\sqrt{\sin x}} C(b\sin x)\,dx = \dfrac{1}{a}\sqrt{\dfrac{2b}{\pi}}\,(1-e^{-m\pi a})\,{}_2F_3\!\left(\begin{array}{c}\frac{1}{4},1;\ -\frac{b^2}{4}\\ \frac{5}{4},1-\frac{ia}{2},1+\frac{ia}{2}\end{array}\right).$

5. $\int_0^\infty \dfrac{e^{-ax}}{\sqrt{\sin x}} C(b\sin x)\,dx = \dfrac{1}{a}\left(\dfrac{2b}{\pi}\right)^{1/2}{}_2F_3\!\left(\begin{array}{c}\frac{1}{4},1;\ -\frac{b^2}{4}\\ \frac{5}{4},1-\frac{ia}{2},1+\frac{ia}{2}\end{array}\right)$ $\quad[\operatorname{Re}a>0].$

6. $\int_0^\infty \cosh^\nu x\,\cos(bx)\,C(c\operatorname{sech} x)\,dx = \dfrac{2^{-\nu-1}c^{1/2}}{\sqrt{\pi}\,\Gamma\!\left(\frac{1}{2}-\nu\right)}$

$\times\Gamma\!\left(\dfrac{1-2\nu-2ib}{4}\right)\Gamma\!\left(\dfrac{1-2\nu+2ib}{4}\right){}_3F_4\!\left(\begin{array}{c}\frac{1}{4},\ \frac{1-2\nu-2ib}{4},\ \frac{1-2\nu+2ib}{4}\\ \frac{1}{2},\frac{5}{4},\frac{1-2\nu}{4},\frac{3-2\nu}{4};\ -\frac{c^2}{4}\end{array}\right)$

$[\operatorname{Re}\nu<1].$

4.5.6. Integrals containing $C(z)$ and the logarithmic function

1. $\int_0^a x^{s-1}\ln\dfrac{a+\sqrt{a^2-x^2}}{x}\,C(bx)\,dx = \sqrt{\dfrac{b}{8}}\,\dfrac{a^{s+1/2}\Gamma\!\left(\frac{2s+1}{4}\right)}{s(2s+1)\Gamma\!\left(\frac{2s+3}{4}\right)}$

$\times\left[(2s+1)\,{}_2F_3\!\left(\begin{array}{c}\frac{1}{4},\frac{2s+1}{4};\ -\frac{a^2b^2}{4}\\ \frac{1}{2},\frac{5}{4},\frac{2s+3}{4}\end{array}\right)-{}_2F_3\!\left(\begin{array}{c}\frac{2s+1}{4},\frac{2s+1}{4};\ -\frac{a^2b^2}{4}\\ \frac{1}{2},\frac{2s+3}{4},\frac{2s+5}{4}\end{array}\right)\right]$

$[a>0;\ \operatorname{Re}s>1/2].$

4.6. The Incomplete Gamma Function $\gamma(\nu,z)$

4.6.1. Integrals containing $\gamma(\nu,z)$ and algebraic functions

1. $\int_0^a x^{s-1}(a-x)^{t-1}\gamma(\nu,bx(a-x))\,dx = \dfrac{a^{s+t+2\nu-1}b^\nu}{\nu}\,\mathrm{B}(s+\nu,t+\nu)$

$\times\,{}_3F_3\!\left(\begin{array}{c}\nu,s+\nu,t+\nu;\ -\frac{a^2b}{4}\\ \nu+1,\frac{s+t}{2}+\nu,\frac{s+t+1}{2}+\nu\end{array}\right)\quad[a,\operatorname{Re}\nu,\operatorname{Re}(s+\nu),\operatorname{Re}(t+\nu)>0].$

2. $\int_0^a x^{s+1/2}(a-x)^s\gamma\!\left(\nu,b\sqrt{x(a-x)}\right)dx$

$= 2^{-2s-\nu-1}a^{2s+\nu+3/2}b^\nu\,\dfrac{\sqrt{\pi}\,\Gamma(2s+\nu+2)}{\nu\Gamma\!\left(2s+\nu+\frac{5}{2}\right)}\,{}_2F_2\!\left(\begin{array}{c}\nu,2s+\nu+2;\ -\frac{ab}{2}\\ \nu+1,2s+\nu+\frac{5}{2}\end{array}\right)$

$[a,\operatorname{Re}\nu,\operatorname{Re}(s+\nu)>0].$

4.6.2. Integrals containing $\gamma(\nu, z)$ and the exponential function

1. $\displaystyle\int_0^a x^{s-1}(a-x)^{t-1} e^{bx(a-x)} \gamma(\nu, bx(a-x))\, dx$

$$= \frac{a^{s+t+2\nu-1} b^\nu}{\nu} B(s+\nu, t+\nu)\, {}_3F_3\left(\begin{array}{c} 1,\, s+\nu,\, t+\nu;\, \dfrac{a^2 b}{4} \\ \nu+1,\, \dfrac{s+t}{2}+\nu,\, \dfrac{s+t+1}{2}+\nu \end{array}\right)$$

$[a,\, \operatorname{Re}\nu,\, \operatorname{Re}(s+\nu),\, \operatorname{Re}(t+\nu) > 0].$

2. $\displaystyle\int_0^a e^{bx(a-x)} \gamma(\nu, bx(a-x))\, dx = \frac{\sqrt{\pi}\,\Gamma(\nu)}{\Gamma\left(\nu+\frac{1}{2}\right)} b^{-1/2} e^{a^2 b/4} \gamma\left(\nu+\frac{1}{2},\, \frac{a^2 b}{4}\right)$

$[a,\, \operatorname{Re}\nu > 0].$

3. $\displaystyle\int_0^a x^{s+1/2}(a-x)^s e^{b\sqrt{x(a-x)}} \gamma(\nu, b\sqrt{x(a-x)})\, dx$

$$= 2^{-2s-\nu-1} a^{2s+\nu+3/2} b^\nu \frac{\sqrt{\pi}\,\Gamma(2s+\nu+2)}{\nu\Gamma\left(2s+\nu+\frac{5}{2}\right)}\, {}_2F_2\left(\begin{array}{c} 1,\, 2s+\nu+2;\, \dfrac{ab}{2} \\ \nu+1,\, 2s+\nu+\dfrac{5}{2} \end{array}\right)$$

$[a,\, \operatorname{Re}\nu > 0;\, \operatorname{Re}(s+\nu/2) > -1].$

4. $\displaystyle\int_0^a x^{-1/2} e^{b\sqrt{x(a-x)}} \gamma(\nu, b\sqrt{x(a-x)})\, dx$

$$= \sqrt{\frac{2\pi}{b}}\, e^{ab/2} \frac{\Gamma(\nu)}{\Gamma\left(\nu+\frac{1}{2}\right)} \gamma\left(\nu+\frac{1}{2},\, \frac{ab}{2}\right) \quad [a > 0;\, \operatorname{Re}\nu > -1/2].$$

4.6.3. Integrals containing $\gamma(\nu, z)$ and trigonometric functions

1. $\displaystyle\int_0^{m\pi} \left\{\begin{array}{c} \sin(ax) \\ \cos(ax) \end{array}\right\} \sin^{-2\nu} x\, \gamma(\nu, b\sin^2 x)\, dx$

$$= \frac{2b^\nu}{\nu a} \sin\frac{m\pi a}{2} \left\{\begin{array}{c} \sin(m\pi a/2) \\ \cos(m\pi a/2) \end{array}\right\}\, {}_3F_3\left(\begin{array}{c} \nu,\, \dfrac{1}{2},\, 1;\, -b \\ \nu+1,\, 1-\dfrac{a}{2},\, 1+\dfrac{a}{2} \end{array}\right) \quad [\operatorname{Re}\nu > 0].$$

2. $\displaystyle\int_0^{\pi/2} \cos^\mu x \cos(ax) e^{b\cos^2 x} \gamma(\nu, b\cos^2 x)\, dx$

$$= \frac{2^{-\mu-2\nu-1} \pi b^\nu \Gamma(\mu+2\nu+1)}{\nu\Gamma\left(\dfrac{\mu+2\nu-a+2}{2}\right) \Gamma\left(\dfrac{\mu+2\nu+a+2}{2}\right)}$$

$$\times\, {}_3F_3\left(\begin{array}{c} 1,\, \dfrac{1+\mu}{2}+\nu,\, \dfrac{\mu}{2}+\nu+1;\, b \\ \nu+1,\, \dfrac{\mu-a}{2}+\nu+1,\, \dfrac{\mu+a}{2}+\nu+1 \end{array}\right) \quad [\operatorname{Re}\nu,\, \operatorname{Re}(\mu+\nu+1) > 0].$$

3. $\displaystyle\int_0^\pi \sin^\mu x \begin{Bmatrix} \sin(ax) \\ \cos(ax) \end{Bmatrix} e^{b\sin^2 x} \gamma(\nu, b\sin^2 x)\, dx$

$= \dfrac{2^{-\mu-2\nu}\pi b^\nu \Gamma(\mu+2\nu+1)}{\nu\Gamma\left(\frac{\mu+2\nu-a}{2}+1\right)\Gamma\left(\frac{\mu+2\nu+a}{2}+1\right)} \begin{Bmatrix} \sin(a\pi/2) \\ \cos(a\pi/2) \end{Bmatrix}$

$\times {}_3F_3\left(\begin{array}{c} 1, \frac{\mu+1}{2}+\nu, \frac{\mu}{2}+\nu+1;\ b \\ \nu+1, \frac{\mu-a}{2}+\nu+1, \frac{\mu+a}{2}+\nu+1 \end{array}\right)$

$[\operatorname{Re}\nu,\ \operatorname{Re}(\mu+\nu+(3\pm 1)/2) > 0]$.

4. $\displaystyle\int_0^{m\pi} \sin^{-2\nu} x \begin{Bmatrix} \sin(ax) \\ \cos(ax) \end{Bmatrix} e^{b\sin^2 x} \gamma(\nu, b\sin^2 x)\, dx$

$= \dfrac{2b^\nu}{\nu a}\sin\dfrac{m\pi a}{2} \begin{Bmatrix} \sin(m\pi a/2) \\ \cos(m\pi a/2) \end{Bmatrix} {}_3F_3\left(\begin{array}{c} \frac{1}{2}, 1, 1;\ b \\ \nu+1, 1-\frac{a}{2}, 1+\frac{a}{2} \end{array}\right)$ $[\operatorname{Re}\nu > 0]$.

5. $\displaystyle\int_0^{m\pi} e^{-ax+b\sin^2 x} \sin^{-2\nu} x\, \gamma(\nu, b\sin^2 x)\, dx$

$= (1-e^{-m\pi a})\dfrac{b^\nu}{\nu a} {}_3F_3\left(\begin{array}{c} \frac{1}{2}, 1, 1;\ b \\ \nu+1, 1-\frac{ia}{2}, 1+\frac{ia}{2} \end{array}\right)$ $[\operatorname{Re}\nu > 0]$.

6. $\displaystyle\int_0^{m\pi} e^{-ax} \sin^{-2\nu} x\, \gamma(\nu, b\sin^2 x)\, dx$

$= (1-e^{-m\pi a})\dfrac{b^\nu}{\nu a} {}_3F_3\left(\begin{array}{c} \nu, \frac{1}{2}, 1;\ b \\ \nu+1, 1-\frac{ia}{2}, 1+\frac{ia}{2} \end{array}\right)$ $[\operatorname{Re}\nu > 0]$.

7. $\displaystyle\int_0^\infty e^{-ax} \sin^{-2\nu} x\, \gamma(\nu, b\sin^2 x)\, dx = \dfrac{b^\nu}{\nu a} {}_3F_3\left(\begin{array}{c} \nu, \frac{1}{2}, 1;\ b \\ \nu+1, 1-\frac{ia}{2}, 1+\frac{ia}{2} \end{array}\right)$

$[\operatorname{Re}a,\ \operatorname{Re}\nu > 0]$.

8. $\displaystyle\int_0^\infty e^{-ax+b\sin^2 x} \sin^{-2\nu} x\, \gamma(\nu, b\sin^2 x)\, dx$

$= \dfrac{b^\nu}{\nu a} {}_3F_3\left(\begin{array}{c} \frac{1}{2}, 1, 1;\ b \\ \nu+1, 1-\frac{ia}{2}, 1+\frac{ia}{2} \end{array}\right)$ $[\operatorname{Re}a,\ \operatorname{Re}\nu > 0]$.

4.6.4. Integrals containing $\gamma(\nu, z)$ and the logarithmic function

1. $\displaystyle\int_0^a x^{s-1} \ln \frac{\sqrt{a} + \sqrt{a-x}}{\sqrt{x}} \gamma(\nu, bx)\, dx = \frac{\pi^{1/2} a^{s+\nu} b^\nu \Gamma(s+\nu)}{2\nu(s+\nu)\Gamma\left(s+\nu+\frac{1}{2}\right)}$

$\times {}_3F_3\left(\begin{matrix} \nu, s+\nu, s+\nu; -ab \\ \nu+1, s+\nu+\frac{1}{2}, s+\nu+1 \end{matrix}\right)$ $[a, \operatorname{Re}\nu, \operatorname{Re}(s+\nu) > 0]$.

2. $\displaystyle\int_0^a x^{s-1} e^{bx} \ln \frac{\sqrt{a}+\sqrt{a-x}}{\sqrt{x}} \gamma(\nu, bx)\, dx = \frac{\pi^{1/2} a^{s+\nu} b^\nu \Gamma(s+\nu)}{2\nu(s+\nu)\Gamma\left(s+\nu+\frac{1}{2}\right)}$

$\times {}_3F_3\left(\begin{matrix} 1, s+\nu, s+\nu; ab \\ \nu+1, s+\nu+\frac{1}{2}, s+\nu+1 \end{matrix}\right)$ $[a, \operatorname{Re}\nu, \operatorname{Re}(s+\nu) > 0]$.

4.6.5. Integrals containing $\gamma(\nu, z)$, $\operatorname{erf}(z)$ and $\operatorname{erfi}(z)$

1. $\displaystyle\int_0^a \operatorname{erf}\left(\sqrt{b(a-x)}\right) \gamma(\nu, bx)\, dx$

$= \dfrac{b^{-1}\Gamma(\nu)}{2\Gamma\left(\nu+\frac{3}{2}\right)} \left[2(ab)^{\nu+3/2} e^{-ab} + (2ab - 2\nu - 1)\gamma\left(\nu+\frac{3}{2}, ab\right)\right]$

$[a, \operatorname{Re}\nu > 0]$.

2. $\displaystyle\int_0^a e^{bx} \operatorname{erfi}\left(\sqrt{b(a-x)}\right) \gamma(\nu, bx)\, dx$

$= \dfrac{\Gamma(\nu)}{\Gamma\left(\nu+\frac{5}{2}\right)} a^{\nu+3/2} b^{\nu+1/2} {}_1F_1\left(\begin{matrix} \frac{3}{2}; ab \\ \nu+\frac{5}{2} \end{matrix}\right)$ $[a, \operatorname{Re}\nu > 0]$.

3. $\displaystyle\int_0^a e^{2bx} \operatorname{erfi}\left(\sqrt{b(a-x)}\right) \gamma(\nu, bx)\, dx$

$= \dfrac{\Gamma(\nu)}{\Gamma\left(\nu+\frac{5}{2}\right)} a^{\nu+3/2} b^{\nu+1/2} e^{ab} {}_1F_2\left(\begin{matrix} 1; \frac{a^2 b^2}{4} \\ \frac{2\nu+5}{4}, \frac{2\nu+7}{4} \end{matrix}\right)$ $[a, \operatorname{Re}\nu > 0]$.

4.6.6. Integrals containing products of $\gamma(\nu, z)$

1. $\displaystyle\int_0^a e^{2bx} \gamma(\mu, bx) \gamma(\nu, b(a-x))\, dx = \frac{\Gamma(\mu)\Gamma(\nu)}{\Gamma(\mu+\nu+2)} a^{\mu+\nu+1} b^{\mu+\nu} e^{ab+\mu\pi i}$

$\times {}_1F_2\left(\begin{matrix} 1; \frac{a^2 b^2}{4} \\ \frac{\mu+\nu}{2}+1, \frac{\mu+\nu+3}{2} \end{matrix}\right)$ $[a, b, \operatorname{Re}\mu, \operatorname{Re}\nu > 0]$.

4.7. The Bessel Function $J_\nu(z)$

4.7.1. Integrals containing $J_\nu(z)$ and algebraic functions

1. $\displaystyle\int_0^\infty \frac{x^{n+2p+1}}{(x^4+ax^2+b)^{m+p+1/2}} J_n(cx)\, dx$

$\displaystyle = \frac{(-1)^{m+n+p}}{\left(\frac{1}{2}\right)_m \left(m+\frac{1}{2}\right)_p} \sum_{k=0}^n (-1)^k \binom{n}{k} \mathrm{D}_a^p \mathrm{D}_b^m \left[u_+^k u_-^{n-k} I_k(cu_+) I_{n-k}(cu_-)\right]$

$\left[n < 4m+2p;\ u_\pm = 2^{-1}\left(a \pm 2\sqrt{b}\right)^{1/2}\right].$

2. $\displaystyle\int_0^a x^s (a-x)^{s+1/2} J_\nu\left(b\sqrt[4]{x(a-x)}\right) dx = 2^{-2s-3\nu/2-1}\sqrt{\pi}\, a^{2s+(\nu+3)/2} b^\nu$

$\displaystyle \times \frac{\Gamma\left(2s+\frac{\nu}{2}+2\right)}{\Gamma(\nu+1)\Gamma\left(2s+\frac{\nu+5}{2}\right)} {}_1F_2\left(\begin{matrix} 2s+\frac{\nu}{2}+2;\ -\frac{ab^2}{8} \\ \nu+1,\ 2s+\frac{\nu+5}{2} \end{matrix}\right)$

$[a > 0;\ \mathrm{Re}(s+\nu/4) > -1].$

3. $\displaystyle\int_0^a x^{-\nu/4}(a-x)^{-(\nu+2)/4} J_\nu\left(b\sqrt[4]{x(a-x)}\right) dx$

$\displaystyle = 2^{(2\nu+3)/4}\sqrt{\pi}\, a^{(1-2\nu)/4} b^{-1/2} \mathbf{H}_{\nu-1/2}\left(b\sqrt{\frac{a}{2}}\right)\quad [a>0].$

4. $\displaystyle\int_0^a x^{\nu/4}(a-x)^{(\nu-2)/4} J_\nu\left(b\sqrt[4]{x(a-x)}\right) dx$

$\displaystyle = 2^{(3-2\nu)/4}\sqrt{\pi}\, a^{(2\nu+1)/4} b^{-1/2} J_{\nu+1/2}\left(b\sqrt{\frac{a}{2}}\right)\quad [a>0;\ \mathrm{Re}\,\nu > -2].$

5. $\displaystyle\int_0^a x^{-1/4}(a-x)^{-3/4} J_\nu\left(b\sqrt[4]{x(a-x)}\right) dx = 2^{1/2}\pi J_{\nu/2}^2\left(b\sqrt{\frac{a}{8}}\right)$

$[a>0;\ \mathrm{Re}\,\nu > -1].$

6. $\displaystyle\int_0^a x^{1/2} J_0\left(b\sqrt[4]{x(a-x)}\right) dx$

$\displaystyle = \frac{1}{b^3}\left[\sqrt{2}(ab^2-2)\sin\left(b\sqrt{\frac{a}{2}}\right) + 2\sqrt{a}\,b\cos\left(b\sqrt{\frac{a}{2}}\right)\right]$

$[a>0].$

7. $\displaystyle\int_0^a x^{-1/2} J_0\left(b\sqrt[4]{x(a-x)}\right) dx = \frac{2^{3/2}}{b}\sin\left(b\sqrt{\frac{a}{2}}\right) \qquad [a>0].$

8. $\int_0^a x^{-1/4}(a-x)^{1/4} J_0\left(b\sqrt[4]{x(a-x)}\right) dx$

$$= \frac{\pi a}{2^{3/2}} \left[J_0^2\left(b\sqrt{\frac{a}{8}}\right) - J_1^2\left(b\sqrt{\frac{a}{8}}\right) \right] \quad [a > 0].$$

9. $\int_0^a x^{-1/4}(a-x)^{1/4} J_1\left(b\sqrt[4]{x(a-x)}\right) dx$

$$= \frac{2}{b^2} \left[\sqrt{2} \sin\left(b\sqrt{\frac{a}{2}}\right) - \sqrt{a} b \cos\left(b\sqrt{\frac{a}{2}}\right) \right] \quad [a > 0].$$

10. $\int_0^a x^{-3/4}(a-x)^{-1/4} J_1\left(b\sqrt[4]{x(a-x)}\right) dx = \frac{4}{\sqrt{ab}} \left[1 - \cos\left(b\sqrt{\frac{a}{2}}\right) \right]$

$$[a > 0].$$

4.7.2. Integrals containing $J_\nu(z)$ and the exponential function

1. $\int_0^\infty x^{m+(n-1)/2} (x+z)^{(n-1)/2} e^{-ax} J_\nu\left(b\sqrt{x^2 + xz}\right) dx$

$$= (-1)^m 2^n b^{-\nu} D_h^n D_a^m \left[h^{(\nu+n)/2} e^{az/2} I_{(\nu+n)/2}(u_-) K_{(\nu+n)/2}(u_+) \right]\big|_{h=c^2}$$

$$\left[u_\pm = z\left(\sqrt{a^2 + h} \pm a\right)/4; \; \operatorname{Re}(2\nu + 2m + n) > 2; \; \operatorname{Re} a > |\operatorname{Im} b|; \; |\arg z| < \pi \right].$$

4.7.3. Integrals containing $J_\nu(z)$ and trigonometric functions

1. $\int_0^a x^{m+\nu/2} (a-x)^n \sin\left(b\sqrt{a-x}\right) J_\nu(c\sqrt{x}) dx = (-1)^{m+n} 2^{m+1/2} \sqrt{\pi}$

$$\times a^{(2m+2\nu+3)/4} c^\nu \sum_{k=0}^m (-1)^k \binom{m}{k} (-m-\nu)_{m-k} \left(-\frac{\sqrt{a}c^2}{2} \right)^k$$

$$\times D_b^{2n} \left[b(b^2 + c^2)^{-(2m+2k+2\nu+3)/4} J_{m+k+\nu+3/2}\left(\sqrt{a(b^2+c^2)}\right) \right] \quad [a > 0].$$

2. $\int_0^\pi \sin^\mu x \begin{Bmatrix} \sin(ax) \\ \cos(ax) \end{Bmatrix} J_\nu(b \sin x) dx$

$$= \frac{2^{-\mu-2\nu} \pi b^\nu \Gamma(\mu + \nu + 1)}{\Gamma(\nu+1)\Gamma\left(\frac{\mu+\nu-a}{2}+1\right)\Gamma\left(\frac{\mu+\nu+a}{2}+1\right)} \begin{Bmatrix} \sin(a\pi/2) \\ \cos(a\pi/2) \end{Bmatrix}$$

$$\times {}_2F_3\left(\frac{\mu+\nu+1}{2}, \frac{\mu+\nu}{2}+1; -\frac{b^2}{4} \atop \nu+1, \frac{\mu+\nu-a}{2}+1, \frac{\mu+\nu+a}{2}+1 \right) \quad [\operatorname{Re}(\mu+\nu) > -1].$$

3. $$\int_0^{\pi/2} \cos^\mu x \cos(ax) J_\nu(b\cos x)\, dx$$

$$= \frac{2^{-\mu-2\nu-1}\pi b^\nu \Gamma(\mu+\nu+1)}{\Gamma(\nu+1)\Gamma\left(\frac{\mu+\nu-a}{2}+1\right)\Gamma\left(\frac{\mu+\nu+a}{2}+1\right)}$$

$$\times {}_2F_3\left(\begin{array}{c}\frac{\mu+\nu+1}{2},\ \frac{\mu+\nu}{2}+1;\ -\frac{b^2}{4}\\ \nu+1,\ \frac{\mu+\nu-a}{2}+1,\ \frac{\mu+\nu+a}{2}+1\end{array}\right) \qquad [\operatorname{Re}(\mu\mid\nu)>-1].$$

4. $$\int_0^{\pi/2} \cos(2nx) \sin^{-\nu} x\, J_\nu(a\sin x)\, dx$$

$$= \frac{2^{-4n-\nu-1}\pi a^{2n+\nu}}{n!\,\Gamma(n+\nu+1)}\, {}_1F_2\left(\begin{array}{c}n+\frac{1}{2};\ -\frac{a^2}{4}\\ n+\nu+1,\ 2n+1\end{array}\right).$$

5. $$\int_0^{m\pi} e^{-ax} \sin^{-\nu} x\, J_\nu(b\sin x)\, dx$$

$$= \frac{\left(\frac{b}{2}\right)^\nu}{\Gamma(\nu+1)a}\left(1-e^{-m\pi a}\right) {}_2F_3\left(\begin{array}{c}\frac{1}{2},\ 1;\ -\frac{b^2}{4}\\ \nu+1,\ 1-\frac{ia}{2},\ 1+\frac{ia}{2}\end{array}\right).$$

6. $$\int_0^\infty e^{-ax} \sin^{-\nu} x\, J_\nu(b\sin x)\, dx = \frac{\left(\frac{b}{2}\right)^\nu}{\Gamma(\nu+1)a}\, {}_2F_3\left(\begin{array}{c}\frac{1}{2},\ 1;\ -\frac{b^2}{4}\\ \nu+1,\ 1-\frac{ia}{2},\ 1+\frac{ia}{2}\end{array}\right)$$

$$[\operatorname{Re} a>0].$$

7. $$\int_0^\pi \cos 2mx\, (a^2-b^2\sin^2 x)^{n/2} J_n\left(\sqrt{a^2-b^2\sin^2 x}\right) dx$$

$$= (-1)^{m+n} \frac{\pi}{2^n} \sum_{k=0}^n (-1)^k \binom{n}{k} b^{2n-2k} \left(a-\sqrt{a^2-b^2}\right)^{2k-n}$$

$$\times J_{m+k}\left(\frac{a-\sqrt{a^2-b^2}}{2}\right) J_{m-n+k}\left(\frac{a+\sqrt{a^2-b^2}}{2}\right) \qquad [0<b\leq a].$$

8. $$\int_0^{2\pi} \cos mx\, (a^2+b^2+2ab\cos x)^{n/2} J_n\left(\sqrt{a^2+b^2+2ab\cos x}\right) dx$$

$$= (-1)^{m+n} 2\pi \sum_{k=0}^n (-1)^k \binom{n}{k} a^k b^{n-k} J_{m+k}(a) J_{m-n+k}(b).$$

4.7.4. Integrals containing $J_\nu(z)$ and the logarithmic function

1. $\displaystyle\int_0^1 x \ln x\, J_0(ax)\, dx = \frac{1}{a^2}[J_0(a) - 1]$.

2. $\displaystyle\int_0^a \frac{x \ln x}{\sqrt{a^2 - x^2}} J_0(bx)\, dx = \frac{1}{b}\{\sin(ab)\ln a + \sin(ab)[\operatorname{ci}(2ab) - \operatorname{ci}(ab)]$
$\hspace{5cm} + \cos(ab)[\operatorname{Si}(ab) - \operatorname{Si}(2ab)]\}$ $\quad [a > 0]$.

3. $\displaystyle\int_0^a x^{s-1} \ln \frac{a + \sqrt{a^2 - x^2}}{x} J_\nu(bx)\, dx = \frac{\pi^{1/2} a^{s+\nu} \left(\frac{b}{2}\right)^\nu \Gamma\left(\frac{s+\nu}{2}\right)}{2(s+\nu)\Gamma\left(\frac{s+\nu+1}{2}\right)\Gamma(\nu+1)}$

$\hspace{2cm} \times {}_2F_3\left(\begin{array}{c} \frac{s+\nu}{2}, \frac{s+\nu}{2}; -\frac{a^2 b^2}{4} \\ \frac{s+\nu+1}{2}, \frac{s+\nu}{2}+1, \nu+1 \end{array}\right)$ $\quad [a, \operatorname{Re}(s+\nu) > 0]$.

4. $\displaystyle\int_0^a \ln \frac{a + \sqrt{a^2 - x^2}}{x} J_1(bx)\, dx = \frac{1}{b}[\mathbf{C} + \ln(ab) - \operatorname{ci}(ab)]$ $\quad [a > 0]$.

5. $\displaystyle\int_0^a x \ln \frac{a + \sqrt{a^2 - x^2}}{x} J_0(bx)\, dx = \frac{1}{b^2}[1 - \cos(ab)]$ $\quad [a > 0]$.

6. $\displaystyle\int_0^a x^3 \ln \frac{a + \sqrt{a^2 - x^2}}{x} J_0(bx)\, dx$
$\hspace{2cm} = \frac{1}{b^4}[3ab\sin(ab) + (4 - a^2 b^2)\cos(ab) - 4]$ $\quad [a > 0]$.

7. $\displaystyle\int_0^\infty x^{m+n} e^{-px} \ln x\, J_n(ax)\, dx$

$\hspace{1cm} = (-1)^{m+n}(2a)^n D_p^m D_u^n \left[(p^2 + u)^{-1/2} \left(\ln \frac{p + \sqrt{p^2 + u}}{2(p^2 + u)} - \mathbf{C}\right)\right]\bigg|_{u=a^2}$
$\hspace{10cm} [\operatorname{Re} p > |\operatorname{Im} a|]$.

4.7.5. Integrals containing $J_\nu(z)$ and inverse trigonometric functions

1. $\displaystyle\int_0^1 x^{s-1} \arccos x\, J_\nu(ax)\, dx$

$\hspace{1cm} = \frac{\pi^{1/2}\left(\frac{a}{2}\right)^\nu \Gamma\left(\frac{s+\nu+1}{2}\right)}{(s+\nu)^2 \Gamma(\nu+1)\Gamma\left(\frac{s+\nu}{2}\right)} {}_2F_3\left(\begin{array}{c} \frac{s+\nu}{2}, \frac{s+\nu+1}{2}; -\frac{a^2}{4} \\ \frac{s+\nu}{2}+1, \frac{s+\nu}{2}+1, \nu+1 \end{array}\right)$
$\hspace{10cm} [\operatorname{Re}(s+\nu) > 0]$.

2. $\int_0^1 \arccos x \, J_0(ax) \, dx = \dfrac{1}{a} \operatorname{Si}(a).$

3. $\int_0^1 x \arccos x \, J_0(ax) \, dx = \dfrac{\pi}{2a} J_0\left(\dfrac{a}{2}\right) J_1\left(\dfrac{a}{2}\right).$

4. $\int_0^1 x^2 \arccos x \, J_0(ax) \, dx = \dfrac{1}{a^3} [2\sin a - a\cos a - \operatorname{Si}(a)].$

5. $\int_0^1 \arccos x \, J_1(ax) \, dx = \dfrac{\pi}{2a} \left[1 - J_0^2\left(\dfrac{a}{2}\right)\right].$

6. $\int_0^1 x \arccos x \, J_1(ax) \, dx = \dfrac{1}{a^2} [\operatorname{Si}(a) - \sin a].$

7. $\int_0^1 \dfrac{1}{x} \arccos x \, J_1(ax) \, dx = \operatorname{Si}(a) + \dfrac{1}{a}(\cos a - 1).$

4.7.6. Integrals containing $J_\nu(z)$, $\operatorname{Si}(z)$ and $\operatorname{ci}(z)$

1. $\int_0^\infty x^{s-1}[\sin(x)\operatorname{Si}(2x) + \cos(x)\operatorname{ci}(2x)] J_\nu(x) \, dx$

$$= -\dfrac{2^{-s-1}\Gamma(s+\nu)}{\pi^{1/2}\Gamma(\nu-s+1)} \Gamma\left(\dfrac{1}{2} - s\right) \cos\left(\dfrac{s+\nu}{2}\pi\right)$$

$$\times \left[\psi\left(\dfrac{1-s-\nu}{2}\right) + \psi\left(\dfrac{1-s+\nu}{2}\right) - \psi\left(1 - \dfrac{s-\nu}{2}\right) - \psi\left(\dfrac{s+\nu}{2}\right)\right]$$

$$[0 < \operatorname{Re}(s+\nu) < 3/2].$$

2. $\int_0^\infty x^{s-1}[\sin(x)\operatorname{ci}(2x) - \cos(x)\operatorname{Si}(2x)] J_\nu(x) \, dx$

$$= -\dfrac{2^{-s-1}\Gamma(s+\nu)}{\pi^{1/2}\Gamma(\nu-s+1)} \Gamma\left(\dfrac{1}{2} - s\right) \sin\left(\dfrac{s+\nu}{2}\pi\right)$$

$$\times \left[\psi\left(1 - \dfrac{s+\nu}{2}\right) - \psi\left(\dfrac{1-s+\nu}{2}\right) + \psi\left(1 + \dfrac{\nu-s}{2}\right) - \psi\left(\dfrac{1+s+\nu}{2}\right)\right]$$

$$[-1/2 < \operatorname{Re}(s+\nu) < 3/2].$$

4.7.7. Integrals containing products of $J_\nu(z)$

1. $\displaystyle\int_0^a J_\mu(x)\, J_\nu(a-x)\, dx = \frac{a}{\mu+\nu}\left\{ J_{\mu+\nu}(a) - \frac{(a/2)^{\mu+\nu}}{\Gamma(\mu+\nu+1)}\right.$

$$\times \left[\frac{\cos a}{\mu+\nu+1}\, {}_3F_4\left(\begin{array}{c}\frac{2\mu+2\nu+1}{4}, \frac{\mu+\nu+1}{2}, \frac{2\mu+2\nu+3}{4}; -a^2\\ \frac{1}{2}, \frac{\mu+\nu+3}{2}, \mu+\nu+\frac{1}{2}, \mu+\nu+1\end{array}\right)\right.$$

$$\left.\left.+\frac{a\sin a}{\mu+\nu+2}\, {}_3F_4\left(\begin{array}{c}\frac{2\mu+2\nu+3}{4}, \frac{\mu+\nu}{2}+1, \frac{2\mu+2\nu+5}{4}; -a^2\\ \frac{3}{2}, \frac{\mu+\nu}{2}+2, \mu+\nu+1, \mu+\nu+\frac{3}{2}\end{array}\right)\right]\right\}$$

$$[a>0;\ \operatorname{Re}\mu,\operatorname{Re}\nu>-1].$$

2. $\displaystyle\int_0^a \frac{1}{x^2} J_\mu(x)\, J_\nu(a-x)\, dx = \frac{1}{2\mu}\left[\frac{1}{\mu-1} J_{\mu+\nu-1}(a) + \frac{1}{\mu+1} J_{\mu+\nu+1}(a)\right]$

$$[a>0;\ \operatorname{Re}\mu>1;\ \operatorname{Re}\nu>-1].$$

3. $\displaystyle\int_0^a \frac{1}{x(a-x)^2} J_\mu(x)\, J_\nu(a-x)\, dx$

$$= \frac{1}{2\mu\nu a}\left[\frac{\mu+\nu-1}{\nu-1} J_{\mu+\nu-1}(a) + \frac{\mu+\nu+1}{\nu+1} J_{\mu+\nu+1}(a)\right]$$

$$[a,\operatorname{Re}\mu>0;\ \operatorname{Re}\nu>1].$$

4. $\displaystyle\int_0^a \frac{1}{x^2(a-x)^2} J_\mu(x)\, J_\nu(a-x)\, dx$

$$= \frac{1}{2\mu\nu a^2}\left[\frac{(\mu+\nu-1)(\mu+\nu-2)}{(\mu-1)(\nu-1)} J_{\mu+\nu-1}(a)\right.$$

$$\left.+\frac{(\mu+\nu+1)(\mu+\nu+2)}{(\mu+1)(\nu+1)} J_{\mu+\nu+1}(a)\right] \quad [a>0;\ \operatorname{Re}\mu,\operatorname{Re}\nu>1].$$

5. $\displaystyle\int_0^a x^{(2n-1)/4}(a-x)^{\nu/2} J_{n-1/2}(b\sqrt{x})\, J_\nu(c\sqrt{a-x})\, dx$

$$= 2b^{n-1/2} c^\nu \left(\frac{a}{b^2+c^2}\right)^{(2\nu+2n+1)/4} J_{\nu+n+1/2}\left(\sqrt{ab^2+ac^2}\right)$$

$$[a>0;\ \operatorname{Re}\nu>-1].$$

6. $\displaystyle\int_0^a x^{m+\mu/2}(a-x)^{n+\nu/2} J_\mu(b\sqrt{x})\, J_\nu(c\sqrt{a-x})\, dx = (-1)^{m+n} 2^{m+n+1} b^\mu c^\nu$

$$\times \left(\frac{a}{b^2+c^2}\right)^{(\mu+\nu+m+n+1)/2} \sum_{j=0}^{m}\sum_{k=0}^{n} \binom{m}{j}\binom{n}{k} (-\mu-m)_{m-j}(-\nu-n)_{n-k}$$

$$\times \frac{b^{2j}c^{2k}}{2^{j+k}} \left(\frac{a}{b^2+c^2}\right)^{(j+k)/2} J_{\mu+\nu+j+k+m+n+1}\left(\sqrt{ab^2+ac^2}\right)$$

$$[a > 0;\ \operatorname{Re}\mu > -m-1;\ \operatorname{Re}\nu > -n-1].$$

7. $\displaystyle\int_0^a x^{n-1/2}(a-x)^{\nu/2} J_{n-1/2}(b\sqrt{x})\, J_{1/2-n}(b\sqrt{x})\, J_\nu\left(c\sqrt{a-x}\right)\, dx$

$$= (-1)^{n-1}\frac{2^{n+1/2}}{\sqrt{\pi}} c^\nu \left(\frac{a}{4b^2+c^2}\right)^{(2\nu+2n+1)/4} \sum_{k=0}^{n-1}\binom{n-1}{k}\left(\frac{1}{2}\right)_{n-k-1}$$

$$\times \left(\frac{4ab^4}{4b^2+c^2}\right)^{k/2} J_{\nu+n+k+1/2}\left(\sqrt{4ab^2+ac^2}\right) \quad [n \geq 1;\ a > 0;\ \operatorname{Re}\nu > -1].$$

8. $\displaystyle\int_a^b \frac{1}{b-x} J_1(b-x)\, J_0\left(\sqrt{x^2-a^2}\right)\, dx = \frac{b-a}{\sqrt{b^2-a^2}} J_1\left(\sqrt{b^2-a^2}\right)$

$$[b > a > 0].$$

9. $\displaystyle\int_a^b \frac{1}{\sqrt{x^2-a^2}} J_0(b-x)\, J_1\left(\sqrt{x^2-a^2}\right)\, dx$

$$= \frac{1}{a}\left[J_0(b-a) - J_0\left(\sqrt{b^2-a^2}\right)\right] \quad [b > a > 0].$$

10. $\displaystyle\int_0^1 e^{2ax} J_0^2\left(a\sqrt{x-x^2}\right)\, dx = e^a I_0(a)\left[1+\frac{\pi}{2}\mathbf{L}_1(a)\right] - \frac{\pi}{2}e^a I_1(a)\mathbf{L}_0(a).$

11. $\displaystyle\int_0^1 \frac{x}{\sqrt{1-x^2}}\cosh\left[(a+b)\sqrt{1-x^2}\right] J_0(ax)\, J_0(bx)\, dx$

$$= I_0(2\sqrt{ab})\left[1+\frac{\pi}{2}\mathbf{L}_1(2\sqrt{ab})\right] - \frac{\pi}{2}I_1(2\sqrt{ab})\mathbf{L}_0(2\sqrt{ab}).$$

12. $\displaystyle\int_0^{\pi/2}\cos(2nx)\sin^{-\mu-\nu}x\, J_\mu(a\sin x)\, J_\nu(a\sin x)\, dx$

$$= \frac{2^{-4n-\mu-\nu-1}\pi a^{2n+\mu+\nu}\Gamma(2n+\mu+\nu+1)}{n!\,\Gamma(n+\mu+1)\Gamma(n+\nu+1)}$$

$$\times \frac{1}{\Gamma(n+\mu+\nu+1)}\, {}_3F_4\left(\begin{array}{c} n+\frac{\mu+\nu+1}{2},\, n+\frac{\mu+\nu}{2}+1,\, n+\frac{1}{2};\, -a^2 \\ n+\mu+1,\, n+\nu+1,\, n+\mu+\nu+1,\, 2n+1 \end{array}\right).$$

13. $\displaystyle\int_0^\pi \cos(nx) \cos^{-\mu-\nu} x \, J_\mu(a\cos x) J_\nu(a\cos x) \, dx$

$$= \frac{2^{-2n-\mu-\nu} \pi a^{n+\mu+\nu} \Gamma(n+\mu+\nu+1)}{\Gamma\left(\frac{n}{2}+1\right) \Gamma\left(\frac{n}{2}+\mu+1\right) \Gamma\left(\frac{n}{2}+\nu+1\right)}$$

$$\times \frac{\cos(n\pi/2)}{\Gamma\left(\frac{n}{2}+\mu+\nu+1\right)} \, _3F_4\left(\begin{array}{c} \frac{n+\mu+\nu+1}{2}, \frac{n+\mu+\nu}{2}+1, \frac{n+1}{2}; -a^2 \\ \frac{n}{2}+\mu+1, \frac{n}{2}+\nu+1, \frac{n}{2}+\mu+\nu+1, n+1 \end{array}\right).$$

14. $\displaystyle\int_0^\pi \sin(ax) \sin^{1-\mu-\nu} x \, J_\mu(b\sin x) J_\nu(b\sin x) \, dx$

$$= \frac{\left(\frac{b}{2}\right)^{\mu+\nu} \sin(\pi a)}{\Gamma(\mu+1)\Gamma(\nu+1)(1-a^2)} \, _4F_5\left(\begin{array}{c} \frac{\mu+\nu+1}{2}, \frac{\mu+\nu}{2}+1, 1, \frac{3}{2}; -b^2 \\ \mu+1, \nu+1, \mu+\nu+1, \frac{3-a}{2}, \frac{3+a}{2} \end{array}\right).$$

15. $\displaystyle\int_0^{m\pi} e^{-ax} \sin^{-\mu-\nu} x \, J_\mu(b\sin x) J_\nu(b\sin x) \, dx$

$$= \frac{\left(\frac{b}{2}\right)^{\mu+\nu} (1-e^{-m\pi a})}{\Gamma(\mu+1)\Gamma(\nu+1)a} \, _4F_5\left(\begin{array}{c} \frac{\mu+\nu+1}{2}, \frac{\mu+\nu}{2}+1, \frac{1}{2}, 1; -b^2 \\ \mu+1, \nu+1, \mu+\nu+1, 1-\frac{ia}{2}, 1+\frac{ia}{2} \end{array}\right).$$

16. $\displaystyle\int_0^\infty e^{-ax} \sin^{-\mu-\nu} x \, J_\mu(b\sin x) J_\nu(b\sin x) \, dx$

$$= \frac{\left(\frac{b}{2}\right)^{\mu+\nu}}{\Gamma(\mu+1)\Gamma(\nu+1)a} \, _4F_5\left(\begin{array}{c} \frac{\mu+\nu+1}{2}, \frac{\mu+\nu}{2}+1, \frac{1}{2}, 1; -b^2 \\ \mu+1, \nu+1, \mu+\nu+1, 1-\frac{ia}{2}, 1+\frac{ia}{2} \end{array}\right) \quad [\operatorname{Re} a > 0].$$

17. $\displaystyle\int_0^a x \ln \frac{a+\sqrt{a^2-x^2}}{x} J_0^2(bx) \, dx = \frac{a}{2b} [2ab \, J_0(2ab) - J_1(2ab)]$

$$+ \frac{\pi a^2}{2} [J_1(2ab) \mathbf{H}_0(2ab) - J_0(2ab) \mathbf{H}_1(2ab)] \quad [a > 0].$$

18. $\displaystyle\int_0^a x^3 \ln \frac{a+\sqrt{a^2-x^2}}{x} J_0^2(bx) \, dx$

$$= \frac{a}{12b^3} [2ab(a^2b^2-2) J_0(2ab) - (a^2b^2-4) J_1(2ab)]$$

$$+ \frac{\pi a^2}{48b^2} (4a^2b^2-3) [J_1(2ab) \mathbf{H}_0(2ab) - J_0(2ab) \mathbf{H}_1(2ab)] \quad [a > 0].$$

19. $\displaystyle\int_0^a (a^2-x^2)^{(2n-1)/4} J_{n-1/2}\left(b\sqrt{a^2-x^2}\right) J_\nu(cx)\,dx$

$$= (-1)^n 2^{n-1/2}\sqrt{\pi}\,b^{n-1/2}$$
$$\times \left. D_u^n\left[J_{\nu/2}\left(\frac{a}{2}\sqrt{u+c^2}+\frac{a\sqrt{u}}{2}\right) J_{\nu/2}\left(\frac{a}{2}\sqrt{u+c^2}-\frac{a\sqrt{u}}{2}\right)\right]\right|_{u=b^2}$$
$$[a>0;\ \operatorname{Re}\nu>-1].$$

20. $\displaystyle\int_0^1 x\ln x\, J_0^2(ax)\,dx = -\frac{1}{2}\left[J_0^2(a)+J_1^2(a)-\frac{1}{a}J_0(a)J_1(a)\right].$

21. $\displaystyle\int_0^1 x\ln x\, J_1^2(ax)\,dx = \frac{1}{2a^2}[1-(a^2+1)J_0^2(a)+a\,J_0(a)J_1(a)-a^2 J_1^2(a)].$

22. $\displaystyle\int_0^1 \frac{1}{x^2}\arccos x\, J_1^2(ax)\,dx$

$$= -\frac{1}{6a}\left[3a - 2a(4a^2+1)J_0(2a)+(4a^2-1)J_1(2a)\right]$$
$$+\frac{2\pi a^2}{3}[J_1(2a)\mathbf{H}_0(2a)-J_0(2a)\mathbf{H}_1(2a)].$$

23. $\displaystyle\int_{-1}^1 J_0\!\left(a\sqrt{x-1}\right) J_0\!\left(a\sqrt{x+1}\right) J_0\!\left(b\sqrt{x-1}\right) J_0\!\left(b\sqrt{x+1}\right)dx$

$$= 2\,{}_1F_4\!\left(\begin{array}{c}\frac{1}{2};\ \frac{a^2b^2}{4}\\ 1,1,1,\frac{3}{2}\end{array}\right).$$

4.8. The Bessel Function $Y_\nu(z)$

4.8.1. Integrals containing $Y_\nu(z)$ and algebraic functions

1. $\displaystyle\int_0^a \frac{1}{\sqrt{a^2-x^2}}\,Y_0(cx)\,dx = \frac{\pi}{2}J_0\!\left(\frac{ac}{2}\right)Y_0\!\left(\frac{ac}{2}\right)$ $[a>0].$

4.8.2. Integrals containing $Y_\nu(z)$ and $J_\nu(z)$

1. $\displaystyle\int_0^a \frac{1}{x}J_1(x)Y_0(a-x)\,dx = \frac{2}{\pi a}J_0(a)+Y_1(a)$ $[a>0].$

2. $\int_0^a e^{-ax}\left[\dfrac{2}{\pi bx}J_0(bx)+Y_1(bx)\right]dx$

$$=\dfrac{2}{\pi b}\left(1-\dfrac{a}{\sqrt{a^2+b^2}}\right)\ln\dfrac{a+\sqrt{a^2+b^2}}{b}\qquad [\operatorname{Re}a>|\operatorname{Im}b|].$$

4.9. The Modified Bessel Function $I_\nu(z)$

4.9.1. Integrals containing $I_\nu(z)$ and algebraic functions

1. $\int_0^a x^{s-1}(a-x)^{t-1}I_\nu\left(b\sqrt{x(a-x)}\right)dx$

$$=B\left(s+\dfrac{\nu}{2},t+\dfrac{\nu}{2}\right)\dfrac{a^{s+t+\nu-1}}{\Gamma(\nu+1)}\left(\dfrac{b}{2}\right)^\nu {}_2F_3\left(\begin{array}{c}s+\dfrac{\nu}{2},\,t+\dfrac{\nu}{2};\,\dfrac{a^2b^2}{16}\\ \nu+1,\,\dfrac{s+t+\nu}{2},\,\dfrac{s+t+\nu+1}{2}\end{array}\right).$$

2. $\int_0^a x^{\nu/2}(a-x)^{\nu/2}I_\nu\left(b\sqrt{x(a-x)}\right)dx$

$$=\sqrt{\pi}\left(\dfrac{a}{2}\right)^{\nu+1/2}\left(\dfrac{2}{b}\right)^{1/2}I_{\nu+1/2}\left(\dfrac{ab}{2}\right)\qquad [a>0;\ \operatorname{Re}\nu>-1].$$

3. $\int_0^a I_0\left(b\sqrt{x(a-x)}\right)dx=\dfrac{2}{b}\sinh\left(\dfrac{ab}{2}\right)\qquad [a>0].$

4. $\int_0^a x^{-1/2}(a-x)^{-1/2}I_0\left(b\sqrt{x(a-x)}\right)dx=\pi I_0^2\left(\dfrac{ab}{4}\right)\qquad [a>0].$

5. $\int_0^a x^s(a-x)^{s+1/2}I_\nu\left(b\sqrt[4]{x(a-x)}\right)dx=2^{-2s-3\nu/2-1}\sqrt{\pi}\,a^{2s+(\nu+3)/2}b^\nu$

$$\times\dfrac{\Gamma\left(2s+\dfrac{\nu}{2}+2\right)}{\Gamma(\nu+1)\Gamma\left(2s+\dfrac{\nu+5}{2}\right)}\,{}_1F_2\left(\begin{array}{c}2s+\dfrac{\nu}{2}+2;\,\dfrac{ab^2}{8}\\ \nu+1,\,2s+\dfrac{\nu+5}{2}\end{array}\right)\quad [a>0;\ \operatorname{Re}(s+\nu/4)>-1].$$

6. $\int_0^a x^{\nu/4}(a-x)^{(\nu-2)/4}I_\nu\left(b\sqrt[4]{x(a-x)}\right)dx$

$$=2^{(3-2\nu)/4}\sqrt{\pi}\,a^{(2\nu+1)/4}b^{-1/2}I_{\nu+1/2}\left(b\sqrt{\dfrac{a}{2}}\right)\qquad [a>0;\ \operatorname{Re}\nu>-3].$$

7. $\int_0^a x^{-\nu/4}(a-x)^{-(\nu+2)/4} I_\nu\left(b\sqrt[4]{x(a-x)}\right) dx$

$$= 2^{(2\nu+3)/4}\sqrt{\pi}\, a^{(1-2\nu)/4} b^{-1/2}\, \mathbf{L}_{\nu-1/2}\left(b\sqrt{\frac{a}{2}}\right) \quad [a>0;\ \mathrm{Re}\,\nu > -2].$$

8. $\int_0^a x^{-1/4}(a-x)^{-3/4} I_\nu\left(b\sqrt[4]{x(a-x)}\right) dx = 2^{1/2}\pi I_{\nu/2}^2\left(b\sqrt{\frac{a}{8}}\right)$

$$[a>0;\ \mathrm{Re}\,\nu > -3/2].$$

9. $\int_0^a x^{1/2} I_0\left(b\sqrt[4]{x(a-x)}\right) dx$

$$= \frac{1}{b^3}\left[\sqrt{2}(ab^2+2)\sinh\left(b\sqrt{\frac{a}{2}}\right) - 2\sqrt{a}\,b\cosh\left(b\sqrt{\frac{a}{2}}\right)\right] \quad [a>0].$$

10. $\int_0^a x^{-1/2} I_0\left(b\sqrt[4]{x(a-x)}\right) dx = \frac{2^{3/2}}{b}\sinh\left(b\sqrt{\frac{a}{2}}\right) \quad [a>0].$

11. $\int_0^a x^{-1/4}(a-x)^{1/4} I_0\left(b\sqrt[4]{x(a-x)}\right) dx$

$$= \frac{\pi a}{2^{3/2}}\left[I_0^2\left(b\sqrt{\frac{a}{8}}\right) + I_1^2\left(b\sqrt{\frac{a}{8}}\right)\right] \quad [a>0].$$

12. $\int_0^a x^{-1/4}(a-x)^{1/4} I_1\left(b\sqrt[4]{x(a-x)}\right) dx$

$$= \frac{2}{b^2}\left[\sqrt{a}\,b\cosh\left(b\sqrt{\frac{a}{2}}\right) - \sqrt{2}\sinh\left(b\sqrt{\frac{a}{2}}\right)\right] \quad [a>0].$$

13. $\int_0^a x^{-3/4}(a-x)^{-1/4} I_1\left(b\sqrt[4]{x(a-x)}\right) dx = \frac{4}{\sqrt{a}\,b}\left[\cosh\left(b\sqrt{\frac{a}{2}}\right) - 1\right]$

$$[a>0].$$

4.9.2. Integrals containing $I_\nu(z)$ and the exponential function

1. $\int_0^a x^{s-1}(a-x)^{t-1} e^{bx(a-x)} I_\nu(bx(a-x))\, dx$

$$= \mathrm{B}(s+\nu, t+\nu)\frac{a^{s+t+2\nu-1}}{\Gamma(\nu+1)}\left(\frac{b}{2}\right)^\nu {}_3F_3\left(\begin{array}{c}\nu+\frac{1}{2},\ s+\nu,\ t+\nu;\ \frac{a^2 b}{2}\\ 2\nu+1,\ \frac{s+t}{2}+\nu,\ \frac{s+t+1}{2}+\nu\end{array}\right)$$

$$[a,\ \mathrm{Re}\,(s+\nu),\ \mathrm{Re}\,(t+\nu) > 0].$$

2. $\int_0^a e^{bx(a-x)} I_0(bx(a-x))\, dx = \sqrt{\dfrac{\pi}{2b}}\, \mathrm{erfi}\left(a\sqrt{\dfrac{b}{2}}\right)$ $\qquad [a > 0]$.

3. $\int_0^a x e^{bx(a-x)} I_0(bx(a-x))\, dx = \dfrac{a}{2}\sqrt{\dfrac{\pi}{2b}}\, \mathrm{erfi}\left(a\sqrt{\dfrac{b}{2}}\right)$ $\qquad [a > 0]$.

4. $\int_0^a x^2 e^{bx(a-x)} I_0(bx(a-x))\, dx$

$= \dfrac{3a^2b+1}{8}\sqrt{\dfrac{\pi}{2b^3}}\, \mathrm{erfi}\left(a\sqrt{\dfrac{b}{2}}\right) - \dfrac{a}{8b} e^{a^2b/2}$ $\qquad [a > 0]$.

5. $\int_0^a x^{-1} e^{bx(a-x)} I_\nu(bx(a-x))\, dx$

$= \dfrac{a}{\nu}\sqrt{\dfrac{\pi b}{8}}\, e^{a^2b/4}\left[I_{\nu-1/2}\left(\dfrac{a^2b}{4}\right) - I_{\nu+1/2}\left(\dfrac{a^2b}{4}\right)\right]$ $\qquad [a,\ \mathrm{Re}\,\nu > 0]$.

6. $\int_0^a x^{-1} e^{bx(a-x)} I_1(bx(a-x))\, dx = \dfrac{1}{a^2b}(2e^{a^2b/2} - a^2b - 2)$ $\qquad [a > 0]$.

7. $\int_0^a x^{-1}(a-x)^{-1} e^{bx(a-x)} I_\nu(bx(a-x))\, dx$

$= \dfrac{1}{\nu}\sqrt{\dfrac{\pi b}{2}}\, e^{a^2b/4}\left[I_{\nu-1/2}\left(\dfrac{a^2b}{4}\right) - I_{\nu+1/2}\left(\dfrac{a^2b}{4}\right)\right]$ $\qquad [a,\ \mathrm{Re}\,\nu > 0]$.

8. $\int_0^a x^{-1}(a-x)^{-1} e^{bx(a-x)} I_1(bx(a-x))\, dx = \dfrac{2}{a^3b}(2e^{a^2b/2} - a^2b - 2)$

$\qquad [a > 0]$.

9. $\int_0^a x^{1/2} e^{b\sqrt{x(a-x)}} I_0\left(b\sqrt{x(a-x)}\right) dx$

$= \dfrac{a^{1/2}}{4b}\left[e^{ab} - \left(\dfrac{1}{2} - ab\right)\sqrt{\dfrac{\pi}{ab}}\, \mathrm{erfi}\left(\sqrt{ab}\right)\right]$ $\qquad [a > 0]$.

10. $\int_0^a x^{-1/2} e^{b\sqrt{x(a-x)}} I_0\left(b\sqrt{x(a-x)}\right) dx = \sqrt{\dfrac{\pi}{b}}\, \mathrm{erfi}\left(\sqrt{ab}\right)$ $\qquad [a > 0]$.

11. $\displaystyle\int_0^a x^{-1/2} e^{b\sqrt{x(a-x)}} I_1\left(b\sqrt{x(a-x)}\right) dx$

$$= \frac{1}{a^{1/2}b}\left[2e^{ab} - \sqrt{\pi ab}\,\operatorname{erfi}\left(\sqrt{ab}\right) - 2\right] \quad [a>0].$$

12. $\displaystyle\int_0^a x^{-1}(a-x)^{-1/2} e^{b\sqrt{x(a-x)}} I_1\left(b\sqrt{x(a-x)}\right) dx$

$$= \frac{2}{a^{3/2}b}\left(e^{ab} - ab - 1\right) \quad [a>0].$$

4.9.3. Integrals containing $I_\nu(z)$ and trigonometric functions

1. $\displaystyle\int_0^{\pi/2} \cos^\mu x \cos(ax) I_\nu(b\cos x)\,dx$

$$= \frac{2^{-\mu-2\nu-1}\pi b^\nu \Gamma(\mu+\nu+1)}{\Gamma(\nu+1)\Gamma\!\left(\frac{\mu+\nu-a}{2}+1\right)\Gamma\!\left(\frac{\mu+\nu+a}{2}+1\right)}$$

$$\times\; {}_2F_3\!\left(\begin{array}{c}\frac{\mu+\nu+1}{2},\,\frac{\mu+\nu}{2}+1;\,\frac{b^2}{4}\\ \nu+1,\,\frac{\mu+\nu-a}{2}+1,\,\frac{\mu+\nu+a}{2}+1\end{array}\right) \quad [\operatorname{Re}(\mu+\nu)>-1].$$

2. $\displaystyle\int_0^\pi \sin^\mu x \left\{\begin{array}{c}\sin(ax)\\ \cos(ax)\end{array}\right\} I_\nu(b\sin x)\,dx$

$$= \frac{2^{-\mu-2\nu}\pi b^\nu \Gamma(\mu+\nu+1)}{\Gamma(\nu+1)\Gamma\!\left(\frac{\mu+\nu-a}{2}+1\right)\Gamma\!\left(\frac{\mu+\nu+a}{2}+1\right)}$$

$$\times \left\{\begin{array}{c}\sin(a\pi/2)\\ \cos(a\pi/2)\end{array}\right\} {}_2F_3\!\left(\begin{array}{c}\frac{\mu+\nu+1}{2},\,\frac{\mu+\nu}{2}+1;\,\frac{b^2}{4}\\ \nu+1,\,\frac{\mu+\nu-a}{2}+1,\,\frac{\mu+\nu+a}{2}+1\end{array}\right) \quad [\operatorname{Re}(\mu+\nu)>-1].$$

3. $\displaystyle\int_0^{m\pi} \sin^{-\nu} x \left\{\begin{array}{c}\sin(ax)\\ \cos(ax)\end{array}\right\} I_\nu(b\sin x)\,dx$

$$= \frac{2}{a}\sin\frac{m\pi a}{2}\left\{\begin{array}{c}\sin(m\pi a/2)\\ \cos(m\pi a/2)\end{array}\right\} \frac{\left(\frac{b}{2}\right)^\nu}{\Gamma(\nu+1)} \,{}_2F_3\!\left(\begin{array}{c}\frac{1}{2},\,1;\,\frac{b^2}{4}\\ \nu+1,\,1-\frac{a}{2},\,1+\frac{a}{2}\end{array}\right).$$

4. $\displaystyle\int_0^{m\pi} \sin^{-2\nu} x \left\{\begin{array}{c}\sin(ax)\\ \cos(ax)\end{array}\right\} I_\nu(b\sin^2 x)\,dx = \frac{2}{a}\sin\frac{m\pi a}{2}\left\{\begin{array}{c}\sin(m\pi a/2)\\ \cos(m\pi a/2)\end{array}\right\}$

$$\times \frac{\left(\frac{b}{2}\right)^\nu}{\Gamma(\nu+1)} \,{}_4F_5\!\left(\begin{array}{c}\frac{1}{4},\,\frac{1}{2},\,\frac{3}{4},\,1;\,b\\ \nu+1,\,1-\frac{a}{4},\,1-\frac{a}{4},\,1+\frac{a}{4},\,1+\frac{a}{4}\end{array}\right).$$

4.9. The Modified Bessel Function $I_\nu(z)$

5. $\displaystyle\int_0^{m\pi} e^{-ax} \sin^{-\nu} x\, I_\nu(b\sin x)\, dx$

$$= (1 - e^{-m\pi a}) \frac{\left(\frac{b}{2}\right)^\nu}{\Gamma(\nu+1)a}\, {}_2F_3\left(\begin{array}{c} \frac{1}{2},\, 1;\, \frac{b^2}{4} \\ \nu+1,\, 1-\frac{ia}{2},\, 1+\frac{ia}{2} \end{array}\right).$$

6. $\displaystyle\int_0^{m\pi} e^{-ax+b\sin^2 x} \sin^{-2\nu} x\, I_\nu(b\sin^2 x)\, dx$

$$= (1 - e^{-m\pi a}) \frac{\left(\frac{b}{2}\right)^\nu}{\Gamma(\nu+1)a}\, {}_3F_3\left(\begin{array}{c} \frac{1}{2},\, 1,\, \nu+\frac{1}{2};\, 2b \\ 2\nu+1,\, 1-\frac{ia}{2},\, 1+\frac{ia}{2} \end{array}\right).$$

7. $\displaystyle\int_0^\infty e^{-ax} \sin^{-\nu} x\, I_\nu(b\sin x)\, dx = \frac{\left(\frac{b}{2}\right)^\nu}{\Gamma(\nu+1)a}\, {}_2F_3\left(\begin{array}{c} \frac{1}{2},\, 1;\, \frac{b^2}{4} \\ \nu+1,\, 1-\frac{ia}{2},\, 1+\frac{ia}{2} \end{array}\right)$

$\qquad\qquad\qquad\qquad\qquad\qquad\qquad\qquad\qquad\qquad\qquad$ [Re $a > 0$].

8. $\displaystyle\int_0^\infty e^{-ax+b\sin^2 x} \sin^{-2\nu} x\, I_\nu(b\sin^2 x)\, dx$

$$= \frac{\left(\frac{b}{2}\right)^\nu}{\Gamma(\nu+1)a}\, {}_3F_3\left(\begin{array}{c} \nu+\frac{1}{2},\, \frac{1}{2},\, 1;\, 2b \\ 2\nu+1,\, 1-\frac{ia}{2},\, 1+\frac{ia}{2} \end{array}\right) \qquad [\text{Re}\, a > 0].$$

9. $\displaystyle\int_0^\pi \sin^\nu x \left\{\begin{array}{c} \sin(ax) \\ \cos(ax) \end{array}\right\} \cos(b\sin x)\, I_0(b\sin x)\, dx$

$$= \frac{2^{-\nu}\pi\Gamma(\nu+1)}{\Gamma\!\left(\frac{\nu-a}{2}+1\right)\Gamma\!\left(\frac{\nu+a}{2}+1\right)} \left\{\begin{array}{c} \sin(a\pi/2) \\ \cos(a\pi/2) \end{array}\right\} {}_4F_4\left(\begin{array}{c} \frac{1}{4},\, \frac{3}{4},\, \frac{\nu+1}{2},\, 1+\frac{\nu}{2};\, b \\ \frac{1}{2},\, \frac{1}{2},\, 1+\frac{\nu-a}{2},\, 1+\frac{\nu+a}{2} \end{array}\right)$$

$\qquad\qquad\qquad\qquad\qquad\qquad\qquad\qquad\qquad\qquad\qquad$ [Re $\nu > -1$].

10. $\displaystyle\int_0^\pi \sin^\mu x \left\{\begin{array}{c} \sin(ax) \\ \cos(ax) \end{array}\right\} \sinh(b\sin x)\, I_\nu(b\sin x)\, dx$

$$= \frac{2^{-\mu-2\nu-1}\pi b^{\nu+1}\Gamma(\mu+\nu+2)}{\Gamma(\nu+1)\Gamma\!\left(\frac{\mu+\nu-a+3}{2}\right)\Gamma\!\left(\frac{\mu+\nu+a+3}{2}\right)}$$

$$\times \left\{\begin{array}{c} \sin(a\pi/2) \\ \cos(a\pi/2) \end{array}\right\} {}_4F_5\left(\begin{array}{c} \frac{2\nu+3}{4},\, \frac{2\nu+5}{4},\, \frac{\mu+\nu}{2}+1,\, \frac{\mu+\nu+3}{2};\, b^2 \\ \frac{3}{2},\, \nu+1,\, \nu+\frac{3}{2},\, \frac{\mu+\nu-a+3}{2},\, \frac{\mu+\nu+a+3}{2} \end{array}\right)$$

$\qquad\qquad\qquad\qquad\qquad\qquad\qquad\qquad\qquad\qquad\qquad$ [Re $(\mu+\nu) > -3$].

11. $\displaystyle\int_0^{\pi} \sin^\mu x \begin{Bmatrix} \sin(ax) \\ \cos(ax) \end{Bmatrix} \cosh(b\sin x) I_\nu(b\sin x)\, dx$

$$= \frac{2^{-\mu-2\nu}\pi b^\nu \Gamma(\mu+\nu+1)}{\Gamma(\nu+1)\Gamma\left(\frac{\mu+\nu-a}{2}+1\right)\Gamma\left(\frac{\mu+\nu+a}{2}+1\right)} \begin{Bmatrix} \sin(a\pi/2) \\ \cos(a\pi/2) \end{Bmatrix}$$

$$\times\, _4F_5\left(\begin{matrix} \frac{2\nu+3}{4}, \frac{2\nu+5}{4}, \frac{\mu+\nu}{2}+1, \frac{\mu+\nu+3}{2}; b^2 \\ \frac{3}{2}, \nu+1, \nu+\frac{3}{2}, \frac{\mu+\nu-a+3}{2}, \frac{\mu+\nu+a+3}{2} \end{matrix}\right) \qquad [\mathrm{Re}\,(\mu+\nu) > -2].$$

12. $\displaystyle\int_0^{\pi/2} \cos^\mu x \cos(ax) \sinh(b\cos x) I_\nu(b\cos x)\, dx$

$$= \frac{2^{-\mu-2\nu-2}\pi b^{\nu+1}\Gamma(\mu+\nu+2)}{\Gamma(\nu+1)\Gamma\left(\frac{\mu+\nu-a+3}{2}\right)\Gamma\left(\frac{\mu+\nu+a+3}{2}\right)}$$

$$\times\, _4F_5\left(\begin{matrix} \frac{2\nu+3}{4}, \frac{2\nu+5}{4}, \frac{\mu+\nu}{2}+1, \frac{\mu+\nu+3}{2}; b^2 \\ \frac{3}{2}, \nu+1, \nu+\frac{3}{2}, \frac{\mu+\nu-a+3}{2}, \frac{\mu+\nu+a+3}{2} \end{matrix}\right) \qquad [\mathrm{Re}\,(\mu+\nu) > -3].$$

13. $\displaystyle\int_0^{\pi/2} \cos^\mu x \cos(ax) \cosh(b\cos x) I_\nu(b\cos x)\, dx$

$$= \frac{2^{-\mu-2\nu-1}\pi b^\nu \Gamma(\mu+\nu+1)}{\Gamma(\nu+1)\Gamma\left(\frac{\mu+\nu-a}{2}+1\right)\Gamma\left(\frac{\mu+\nu+a}{2}+1\right)}$$

$$\times\, _4F_5\left(\begin{matrix} \frac{2\nu+1}{4}, \frac{2\nu+3}{4}, \frac{\mu+\nu+1}{2}, \frac{\mu+\nu}{2}+1; b^2 \\ \frac{1}{2}, \nu+\frac{1}{2}, \nu+1, \frac{\mu+\nu-a}{2}+1, \frac{\mu+\nu+a}{2}+1 \end{matrix}\right) \qquad [\mathrm{Re}\,(\mu+\nu) > -2].$$

14. $\displaystyle\int_0^{\infty} \cosh^\mu(b\cos x)\cos(bx) I_\nu(c\,\mathrm{sech}\, x)\, dx = \frac{2^{-\mu-2}c^\nu}{\Gamma(\nu+1)\Gamma(\nu-\mu)}$

$$\times \Gamma\left(\frac{\nu-\mu-ib}{2}\right)\Gamma\left(\frac{\nu-\mu+ib}{2}\right)\, _2F_3\left(\begin{matrix} \frac{\nu-\mu-ib}{2}, \frac{\nu-\mu+ib}{2}; \frac{c^2}{4} \\ \nu+1, \frac{\nu-\mu}{2}, \frac{\mu-\nu+1}{2} \end{matrix}\right)$$

$$[\mathrm{Re}\,(\nu-\mu) > 0].$$

4.9.4. Integrals containing $I_\nu(z)$ and the logarithmic function

1. $\displaystyle\int_0^{a} x^{s-1}\ln\frac{a+\sqrt{a^2-x^2}}{x} I_\nu(bx)\, dx = \frac{\pi^{1/2} a^{s+\nu} \left(\frac{b}{2}\right)^\nu \Gamma\left(\frac{s+\nu}{2}\right)}{2(s+\nu)\Gamma\left(\frac{s+\nu+1}{2}\right)\Gamma(\nu+1)}$

$$\times\, _2F_3\left(\begin{matrix} \frac{s+\nu}{2}, \frac{s+\nu}{2}; \frac{a^2b^2}{4} \\ \frac{s+\nu+1}{2}, \frac{s+\nu}{2}+1, \nu+1 \end{matrix}\right) \qquad [a, \mathrm{Re}\,(s+\nu) > 0].$$

2. $\displaystyle\int_0^a \ln\frac{a+\sqrt{a^2-x^2}}{x} I_1(bx)\,dx = \frac{1}{b}[\operatorname{chi}(ab) - \ln(ab) - C]$ $\quad [a>0]$.

3. $\displaystyle\int_0^a x \ln\frac{a+\sqrt{a^2-x^2}}{x} I_0(bx)\,dx = \frac{\cosh(ab)-1}{b^2}$ $\quad [a>0]$.

4. $\displaystyle\int_0^a x^3 \ln\frac{a+\sqrt{a^2-x^2}}{x} I_0(bx)\,dx = \frac{1}{b^4}[(a^2b^2+4)\cosh(ab)$
$\qquad\qquad\qquad\qquad\qquad\qquad\qquad\qquad - 3ab\sinh(ab) - 4]$ $\quad [a>0]$.

5. $\displaystyle\int_0^1 x \ln\frac{1+\sqrt{1-x^2}}{1-\sqrt{1-x^2}} I_0(ax)\,dx = \frac{4}{a^2}\sinh^2\frac{a}{2}$.

6. $\displaystyle\int_0^a x^{s-1} e^{bx} \ln\frac{\sqrt{a}+\sqrt{a-x}}{\sqrt{x}} I_\nu(bx)\,dx = \frac{2^{-\nu-1}\pi^{1/2}a^{s+\nu}b^\nu \Gamma(s+\nu)}{(s+\nu)\Gamma\left(s+\nu+\frac{1}{2}\right)\Gamma(\nu+1)}$

$\qquad\qquad\qquad\times {}_3F_3\left(\begin{array}{c}\nu+\frac{1}{2},\, s+\nu,\, s+\nu;\, 2ab\\ 2\nu+1,\, s+\nu+\frac{1}{2},\, s+\nu+1\end{array}\right)$ $\quad [a, \operatorname{Re}(s+\nu)>0]$.

4.9.5. Integrals containing $I_\nu(z)$ and inverse trigonometric functions

1. $\displaystyle\int_0^1 x^{s-1} \arccos x\, I_\nu(ax)\,dx$

$\qquad = \dfrac{\pi^{1/2}\left(\frac{a}{2}\right)^\nu \Gamma\left(\frac{s+\nu+1}{2}\right)}{(s+\nu)^2 \Gamma(\nu+1)\Gamma\left(\frac{s+\nu}{2}\right)} {}_2F_3\left(\begin{array}{c}\frac{s+\nu}{2},\, \frac{s+\nu+1}{2};\, \frac{a^2}{4}\\ \frac{s+\nu}{2}+1,\, \frac{s+\nu}{2}+1,\, \nu+1\end{array}\right)$

$\qquad\qquad\qquad\qquad\qquad\qquad\qquad\qquad\qquad\qquad [\operatorname{Re}(s+\nu)>0]$.

2. $\displaystyle\int_0^1 \arccos x\, I_0(ax)\,dx = \frac{1}{a}\operatorname{shi}(a)$.

3. $\displaystyle\int_0^1 x \arccos x\, I_0(ax)\,dx = \frac{\pi}{2a} I_0\left(\frac{a}{2}\right) I_1\left(\frac{a}{2}\right)$.

4. $\displaystyle\int_0^1 x^2 \arccos x\, I_0(ax)\,dx = \frac{1}{a^3}[a\cosh a - 2\sinh a + \operatorname{shi}(a)]$.

5. $\displaystyle\int_0^1 \frac{1}{x}\arccos x\, I_1(ax)\,dx = \operatorname{shi}(a) + \frac{1-\cosh a}{a}$.

6. $\int_0^1 \arccos x \, I_1(ax) \, dx = \dfrac{\pi}{2a} \left[I_0^2\left(\dfrac{a}{2}\right) - 1 \right].$

7. $\int_0^1 x \arccos x \, I_1(ax) \, dx = \dfrac{1}{a^2} \left[\sinh a - \operatorname{shi}(a) \right].$

4.9.6. Integrals containing $I_\nu(z)$ and special functions

1. $\displaystyle\int_0^\infty x^{s-1} e^x \operatorname{Ei}(-2x) I_\nu(x) \, dx$

$$= -\dfrac{2^{-s} \pi^{1/2}}{s+\nu} \sec(\nu\pi) \dfrac{\Gamma(s+\nu)}{\Gamma\left(\frac{1}{2}-\nu\right)\Gamma(1+2\nu)} {}_3F_2\!\left(\begin{array}{c} \nu+\frac{1}{2},\, s+\nu,\, s+\nu \\ s+\nu+1,\, 2\nu+1;\, 1 \end{array}\right)$$

$[\operatorname{Re}(s-\nu) > 0].$

2. $\displaystyle\int_0^a e^{bx} \operatorname{erf}\!\left(\sqrt{2b(a-x)}\right) I_0(bx) \, dx$

$$= (\pi a^3 b)^{1/2} \left[I_{-1/4}(ab) I_{1/4}(ab) - I_{-3/4}(ab) I_{3/4}(ab) \right]$$

$[\operatorname{Re} a > 0].$

3. $\displaystyle\int_0^a e^{-bx} \operatorname{erf}\!\left(\sqrt{2b(a-x)}\right) I_0(bx) \, dx = \sqrt{\dfrac{2a}{b\pi}} - \dfrac{e^{-2ab}}{2b} \operatorname{erfi}\!\left(\sqrt{2ab}\right).$

4. $\displaystyle\int_0^a e^{bx} \operatorname{erfi}\!\left(\sqrt{2b(a-x)}\right) I_0(bx) \, dx = \dfrac{e^{2ab}}{2b} \operatorname{erf}\!\left(\sqrt{2ab}\right) - \sqrt{\dfrac{2a}{b\pi}}.$

5. $\displaystyle\int_0^a x^{\mu/2}(a-x)^{\nu/2} J_\mu(b\sqrt{x}) I_\nu(b\sqrt{a-x}) \, dx = \dfrac{a^{\mu+\nu+1} b^{\mu+\nu}}{2^{\mu+\nu} \Gamma(\mu+\nu+2)}$

$[a > 0;\ \operatorname{Re}\mu,\ \operatorname{Re}\nu > -1].$

6. $\displaystyle\int_0^{m\pi} \left\{\begin{array}{c} \sin(ax) \\ \cos(ax) \end{array}\right\} \sin^{-\nu} x \, J_\nu(b\sqrt{\sin x}) I_\nu(b\sqrt{\sin x}) \, dx = 2\sin\dfrac{m\pi a}{2}$

$$\times \left\{\begin{array}{c} \sin(m\pi a/2) \\ \cos(m\pi a/2) \end{array}\right\} \dfrac{\left(\frac{b}{2}\right)^{2\nu}}{\Gamma^2(\nu+1)a} \, {}_2F_5\!\left(\begin{array}{c} \frac{1}{2},\, 1;\, -\frac{b^4}{64} \\ \frac{\nu+1}{2},\, \frac{\nu}{2}+1,\, \nu+1,\, 1-\frac{a}{2},\, 1+\frac{a}{2} \end{array}\right).$$

7. $\displaystyle\int_0^{m\pi} e^{-ax}\sin^{-\nu}x\, J_\nu(b\sqrt{\sin x})\, I_\nu(b\sqrt{\sin x})\, dx$

$$= (1 - e^{-m\pi a})\frac{\left(\frac{b}{2}\right)^{2\nu}}{\Gamma^2(\nu+1)a}\, {}_2F_5\left(\begin{array}{c}\frac{1}{2},\, 1;\, -\frac{b^4}{64}\\ \frac{\nu+1}{2},\, \frac{\nu}{2}+1,\, \nu+1,\, 1-\frac{ia}{2},\, 1+\frac{ia}{2}\end{array}\right).$$

8. $\displaystyle\int_0^{\infty} e^{-ax}\sin^{-\nu}x\, J_\nu(b\sqrt{\sin x})\, I_\nu(b\sqrt{\sin x})\, dx$

$$= \frac{\left(\frac{b}{2}\right)^{2\nu}}{\Gamma^2(\nu+1)a}\, {}_2F_5\left(\begin{array}{c}\frac{1}{2},\, 1;\, -\frac{b^4}{64}\\ \frac{\nu+1}{2},\, \frac{\nu}{2}+1,\, \nu+1,\, 1-\frac{ia}{2},\, 1+\frac{ia}{2}\end{array}\right) \quad [\operatorname{Re} a > 0].$$

9. $\displaystyle\int_0^a x^{s-1}\ln\frac{a^2 + \sqrt{a^4 - x^4}}{x^2}\, J_\nu(bx)\, I_\nu(bx)\, dx$

$$= \frac{2^{-2\nu-3}\pi^{1/2}a^{s+2\nu}b^{2\nu}\Gamma\left(\frac{s+2\nu}{4}\right)}{(s+2\nu)\Gamma\left(\frac{s+2\nu+2}{4}\right)\Gamma^2(\nu+1)}$$

$$\times {}_2F_5\left(\begin{array}{c}\frac{s+2\nu}{4},\, \frac{s+2\nu}{4};\, -\frac{a^4 b^4}{64}\\ \frac{s+2\nu+2}{2},\, \frac{s+2\nu}{2}+1,\, \frac{\nu+1}{2},\, \frac{\nu}{2}+1,\, \nu+1\end{array}\right) \quad [\operatorname{Re}(s+2\nu) > 0].$$

10. $\displaystyle\int_{-1}^1 e^{ax} J_0(a\sqrt{1-x^2})\, J_0(b\sqrt{1-x})\, I_0(b\sqrt{1+x})\, dx = 2\, {}_1F_3\left(\begin{array}{c}\frac{1}{2};\, \frac{ab^2}{2}\\ 1,\, 1,\, \frac{3}{2}\end{array}\right).$

4.9.7. Integrals containing products of $I_\nu(z)$

1. $\displaystyle\int_0^a x^\mu(a-x)^\nu I_\mu(bx)\, I_\nu(b(a-x))\, dx$

$$= a^{\mu+\nu+1/2}\sqrt{\frac{b}{2\pi}}\, B\left(\mu + \frac{1}{2},\, \nu + \frac{1}{2}\right) I_{\mu+\nu+1/2}(ab)$$

$$[a > 0;\ \operatorname{Re}\mu,\, \operatorname{Re}\nu > -1/2].$$

2. $\displaystyle\int_0^a I_0^2(b\sqrt{x(a-x)})\, dx = a\, I_0(ab) + \frac{\pi a}{2}[I_0(ab)\, \mathbf{L}_1(ab) - I_1(ab)\, \mathbf{L}_0(ab)]$

$$[a > 0].$$

3. $\displaystyle\int_0^a x\, I_0^2(b\sqrt{x(a-x)})\, dx = \frac{a^2}{2} I_0(ab) + \frac{\pi a^2}{4}[I_0(ab)\, \mathbf{L}_1(ab) - I_1(ab)\, \mathbf{L}_0(ab)]$

$$[a > 0].$$

4. $\displaystyle\int_0^a I_1^2\bigl(b\sqrt{x(a-x)}\bigr)\,dx = \frac{2}{b}I_1(ab) - a\,I_0(ab)$

$\displaystyle\qquad\qquad\qquad - \frac{\pi a}{2}[I_0(ab)\,\mathbf{L}_1(ab) - I_1(ab)\,\mathbf{L}_0(ab)]\qquad [a>0].$

5. $\displaystyle\int_0^a x^{-1/2} I_0^2\bigl(b\sqrt[4]{x(a-x)}\bigr)\,dx = 2a^{1/2} I_0(b\sqrt{2a})$

$\displaystyle\qquad\qquad + \pi a^{1/2}\bigl[I_0(b\sqrt{2a})\,\mathbf{L}_1(b\sqrt{2a}) - I_1(b\sqrt{2a})\,\mathbf{L}_0(b\sqrt{2a})\bigr]\qquad [a>0].$

6. $\displaystyle\int_0^a x^{-1/2} I_1^2\bigl(b\sqrt[4]{x(a-x)}\bigr)\,dx = -2a^{1/2} I_0(b\sqrt{2a}) + \frac{2^{3/2}}{b}I_1(b\sqrt{2a})$

$\displaystyle\qquad\qquad +\pi a^{1/2}\bigl[I_1(b\sqrt{2a})\,\mathbf{L}_0(b\sqrt{2a}) - I_0(b\sqrt{2a})\,\mathbf{L}_1(b\sqrt{2a})\bigr]\qquad [a>0].$

7. $\displaystyle\int_0^a x^{-1/4}(a-x)^{1/4} I_0\bigl(b\sqrt[4]{x(a-x)}\bigr)\,I_1\bigl(b\sqrt[4]{x(a-x)}\bigr)\,dx$

$\displaystyle\qquad = \frac{\pi a^{1/2}}{2b}\bigl[I_1(b\sqrt{2a})\,\mathbf{L}_0(b\sqrt{2a}) - I_0(b\sqrt{2a})\,\mathbf{L}_1(b\sqrt{2a})\bigr]\qquad [a>0].$

8. $\displaystyle\int_0^a x^{-3/4}(a-x)^{-1/4} I_0\bigl(b\sqrt[4]{x(a-x)}\bigr)\,I_1\bigl(b\sqrt[4]{x(a-x)}\bigr)\,dx$

$\displaystyle\qquad\qquad\qquad = \frac{2}{a^{1/2}b}\bigl[I_0(b\sqrt{2a}) - 1\bigr]\qquad [a>0].$

9. $\displaystyle\int_0^a x^{-1}(a-x)^{-1/2} I_1^2\bigl(b\sqrt[4]{x(a-x)}\bigr)\,dx = \frac{2^{3/2}}{ab}I_1(b\sqrt{2a}) - \frac{2}{a^{1/2}}\qquad [a>0].$

10. $\displaystyle\int_0^a x^\nu (a-x)^\nu e^{2bx} I_\nu(bx)\,I_\nu(b(a-x))\,dx$

$\displaystyle\qquad = \left(\frac{a}{2}\right)^{4\nu+1} b^{2\nu} e^{ab}\,\frac{\Gamma\bigl(\nu+\tfrac{1}{2}\bigr)}{\Gamma(\nu+1)\Gamma\bigl(2\nu+\tfrac{3}{2}\bigr)}\,{}_1F_2\!\left(\begin{array}{c}\nu+\tfrac{1}{2};\ a^2b^2\\ 2\nu+1,\ 2\nu+\tfrac{3}{2}\end{array}\right)$

$\displaystyle\qquad\qquad\qquad\qquad\qquad\qquad\qquad\qquad\qquad [a>0;\ \operatorname{Re}\nu>-1/2].$

11. $\displaystyle\int_0^a e^{2bx} I_0(bx)\,I_0(b(a-x))\,dx$

$\displaystyle\qquad = ae^{ab}\Bigl[I_0(2ab) + \frac{\pi}{2}I_0(2ab)\,\mathbf{L}_1(2ab) - \frac{\pi}{2}I_1(2ab)\mathbf{L}_0(2ab)\Bigr].$

12. $\displaystyle\int_0^1 x\ln x\,I_0^2(ax)\,dx = \frac{1}{2}[I_1^2(a) - I_0^2(a) + \frac{1}{a}I_0(a)\,I_1(a)].$

13. $\int_0^1 x \ln x \, I_1^2(ax) \, dx = \dfrac{1}{2a^2} [1 + (a^2 - 1) I_0^2(a) - a\, I_0(a) I_1(a) - a^2 I_1^2(a)]$.

14. $\int_0^a x \ln \dfrac{a + \sqrt{a^2 - x^2}}{x} I_0^2(bx) \, dx = \dfrac{a}{2b} [2ab\, I_0(2ab) - I_1(2ab)]$

$\qquad\qquad + \dfrac{\pi a^2}{2} [I_0(2ab)\, \mathbf{L}_1(2ab) - I_1(2ab)\, \mathbf{L}_0(2ab)] \quad [a > 0]$.

15. $\int_0^a x^3 \ln \dfrac{a + \sqrt{a^2 - x^2}}{x} I_0^2(bx) \, dx$

$\qquad = \dfrac{a}{12b^3} [2ab(a^2 b^2 + 2) I_0(2ab) - (a^2 b^2 + 4) I_1(2ab)]$

$\qquad + \dfrac{\pi a^2}{48 b^2} (4a^2 b^2 + 3) [I_0(2ab) \mathbf{L}_1(2ab) - I_1(2ab) \mathbf{L}_0(2ab)] \quad [a > 0]$.

16. $\int_0^a x^{(\nu-1)/2} (a-x)^\nu J_\nu(2b\sqrt{x}) I_\nu^2(b\sqrt{a-x}) \, dx$

$= \dfrac{\Gamma^2\left(\nu + \frac{1}{2}\right) a^{3\nu + 1/2} b^{3\nu}}{\pi^{1/2} \Gamma^2(\nu+1) \Gamma\left(3\nu + \frac{3}{2}\right)} \; {}_2F_5\left(\begin{array}{c} \frac{1}{2}, \nu + \frac{1}{2}; \; \dfrac{a^2 b^4}{16} \\ \dfrac{\nu+1}{2}, \dfrac{\nu}{2}+1, \dfrac{6\nu+3}{4}, \dfrac{6\nu+5}{4}, \nu+1 \end{array}\right) \quad [a > 0]$.

17. $\int_0^a x^\nu (a-x)^\nu J_\nu^2(b\sqrt{x}) I_\nu^2(b\sqrt{a-x}) \, dx = \dfrac{2^{-6\nu - 1} a^{4\nu + 1} b^{4\nu} \Gamma\left(\nu + \frac{1}{2}\right)}{\Gamma^3(\nu+1) \Gamma\left(2\nu + \frac{3}{2}\right)}$

$\qquad \times {}_2F_5\left(\begin{array}{c} \frac{1}{2}, \nu + \frac{1}{2}; \; \dfrac{a^2 b^4}{16} \\ \dfrac{\nu+1}{2}, \dfrac{\nu}{2}+1, \nu+1, 2\nu+1, 2\nu+\dfrac{3}{2} \end{array}\right) \quad [a > 0]$.

18. $\int_{-1}^1 J_0(a\sqrt{1+x}) J_0(b\sqrt{1+x}) I_0(a\sqrt{1-x}) I_0(b\sqrt{1-x}) \, dx$

$\qquad\qquad\qquad\qquad = 2\, {}_1F_4\left(\begin{array}{c} \frac{1}{2}; \; \dfrac{a^2 b^2}{4} \\ 1, 1, 1, \dfrac{3}{2} \end{array}\right)$.

19. $\int_0^a x^{-\nu}(a-x)^{-\nu} J_\nu(b\sqrt{x}) J_{-\nu}(b\sqrt{x}) I_\nu(b\sqrt{a-x}) I_{-\nu}(b\sqrt{a-x}) \, dx$

$= \dfrac{\sin^2(\nu\pi) \Gamma(1-\nu)}{\nu^2 \pi^{3/2} \Gamma\left(\frac{3}{2} - \nu\right)} \left(\dfrac{2}{a}\right)^{2\nu - 1} {}_2F_5\left(\begin{array}{c} \frac{1}{2}, \nu + \frac{1}{2}; \; \dfrac{a^2 b^4}{16} \\ 1-\nu, \dfrac{3}{2}-\nu, \dfrac{\nu+1}{2}, \dfrac{\nu}{2}+1, \nu+1 \end{array}\right)$

$\qquad\qquad\qquad\qquad\qquad\qquad\qquad\qquad\qquad\qquad\qquad [a > 0]$.

4.10. The Macdonald Function $K_\nu(z)$

4.10.1. Integrals containing $K_\nu(z)$, $J_\nu(z)$, $Y_\nu(z)$ and $I_\nu(z)$

1. $\displaystyle\int_0^a \frac{1}{x} I_1(x) K_0(a-x)\,dx = \frac{2}{\pi a} I_0(a) - K_1(a)$.

2. $\displaystyle\int_0^\infty e^{-ax}\left[\frac{1}{bx} I_0(bx) - K_1(bx)\right] dx = \frac{1}{b}\left(\frac{a}{\sqrt{a^2-b^2}} - 1\right)\ln\frac{a+\sqrt{a^2-b^2}}{b}$

   $[\operatorname{Re} a > |\operatorname{Re} b|]$.

3. $\displaystyle\int_0^\infty x\, J_1(ax) I_1(ax) Y_0(bx) K_0(bx)\,dx = -\frac{1}{2\pi a^2}\ln\left(1 - \frac{a^4}{b^4}\right)$ $\quad [0 < a < b]$.

4.10.2. Integrals containing products of $K_\nu(z)$

1. $\displaystyle\int_0^\infty \frac{x^{\nu-2}}{(x+a)^\nu} K_0(x) K_\nu(x+a)\,dx$

   $= 2^{-\nu}\sqrt{\pi}\,\frac{\Gamma^2(\nu-1)}{\Gamma(\nu+\frac{1}{2})} a^{-\nu-1}\left[\nu(\nu-1) K_0(a) + (2\nu-1)a K_1(a)\right]$

   $[|\arg a| < \pi,\ \operatorname{Re}\nu > 1]$.

2. $\displaystyle\int_0^\infty x K_0^3(x)\,dx = \frac{1}{6}\psi'\!\left(\frac{1}{3}\right) - \frac{\pi^2}{9}$ $\hfill$ [37].

3. $\displaystyle\int_0^\infty x K_0^4(x)\,dx = \frac{7}{8}\zeta(3)$ $\hfill$ [37].

4. $\displaystyle\int_0^\infty x^3 K_0^4(x)\,dx = -\frac{3}{16} + \frac{7}{32}\zeta(3)$ $\hfill$ [37].

5. $\displaystyle\int_0^\infty x^5 K_0^4(x)\,dx = -\frac{27}{64} + \frac{49}{128}\zeta(3)$ $\hfill$ [37].

6. $\displaystyle\int_0^\infty x^7 K_0^4(x)\,dx = -\frac{37}{16} + \frac{63}{32}\zeta(3)$ $\hfill$ [37].

7. $\displaystyle\int_0^\infty x^3 K_0^2(x) K_1^2(x)\,dx = \frac{1}{16} + \frac{7}{32}\zeta(3)$ $\hfill$ [37].

8. $\int_0^\infty x^2 K_0^3(x) K_1(x) \, dx = \dfrac{7}{16} \zeta(3)$ [37].

9. $\int_0^\infty x^4 K_0^3(x) K_1(x) \, dx = -\dfrac{3}{16} + \dfrac{7}{32} \zeta(3)$ [37].

10. $\int_0^\infty x^4 K_0(x) K_1^3(x) \, dx = \dfrac{1}{4}$ [37].

11. $\int_0^\infty x^7 K_1^4(x) \, dx = \dfrac{201}{64} - \dfrac{315}{128} \zeta(3)$ [37].

4.11. The Struve Functions $\mathbf{H}_\nu(z)$ and $\mathbf{L}_\nu(z)$

4.11.1. Integrals containing $\mathbf{H}_\nu(z)$, $\mathbf{L}_\nu(z)$ and algebraic functions

1. $\displaystyle\int_0^a x^{s-1}(a-x)^{t-1} \left\{ \begin{array}{l} \mathbf{H}_\nu\!\left(b\sqrt{x(a-x)}\right) \\ \mathbf{L}_\nu\!\left(b\sqrt{x(a-x)}\right) \end{array} \right\} dx$

$= 2^{-\nu} a^{s+t+\nu} b^{\nu+1} \dfrac{\Gamma(s+\frac{\nu+1}{2})\Gamma(t+\frac{\nu+1}{2})}{\sqrt{\pi}\,\Gamma\!\left(\nu+\frac{3}{2}\right)\Gamma(s+t+\nu+1)}$

$\times \; {}_3F_4\!\left(\begin{array}{c} 1,\, s+\frac{\nu+1}{2},\, t+\frac{\nu+1}{2};\, \mp\frac{a^2 b^2}{16} \\ \frac{3}{2},\, \nu+\frac{3}{2},\, \frac{s+t+\nu+1}{2},\, \frac{s+t+\nu}{2}+1 \end{array}\right)$

$[a>0;\; \operatorname{Re}(s+\nu),\, \operatorname{Re}(s+\nu) > -1].$

2. $\displaystyle\int_0^a x^{-\nu/2}(a-x)^{-\nu/2} \mathbf{L}_\nu\!\left(b\sqrt{x(a-x)}\right) dx$

$= 2^\nu \sqrt{\pi}\, a^{1/2-\nu} b^{-1/2} I_{\nu-1/2}\!\left(\dfrac{ab}{2}\right) - \dfrac{\sqrt{\pi}}{\Gamma(\nu+1/2)}\left(\dfrac{b}{2}\right)^{\nu-1}$ $[a>0].$

3. $\displaystyle\int_0^a \mathbf{L}_0\!\left(b\sqrt{x(a-x)}\right) dx = \dfrac{2}{b}\left[\cosh\!\left(\dfrac{ab}{2}\right) - 1\right]$ $[a>0].$

4. $\displaystyle\int_0^a x^{s+1/2}(a-x)^s \left\{ \begin{array}{l} \mathbf{H}_\nu\!\left(b\sqrt[4]{x(a-x)}\right) \\ \mathbf{L}_\nu\!\left(b\sqrt[4]{x(a-x)}\right) \end{array} \right\} dx = 2^{-2s-3(\nu+1)/2} a^{2s+\nu/2+2} b^{\nu+1}$

$\times \; \dfrac{\Gamma\!\left(2s+\frac{\nu+5}{2}\right)}{\Gamma\!\left(2s+\frac{\nu}{2}+3\right)\Gamma\!\left(\nu+\frac{3}{2}\right)} \, {}_2F_3\!\left(\begin{array}{c} 1,\, 2s+\frac{\nu+5}{2};\, \mp\frac{ab^2}{8} \\ \frac{3}{2},\, \nu+\frac{3}{2},\, 2s+\frac{\nu}{2}+3 \end{array}\right)$

$[a>0;\; \operatorname{Re}(4s+\nu) > -5].$

5. $\int_0^a x^{-1/2} \mathbf{H}_0\left(b\sqrt[4]{x(a-x)}\right) dx = \frac{2^{3/2}}{b}\left[1 - \cos\left(b\sqrt{\frac{a}{2}}\right)\right]$ $[a > 0]$.

6. $\int_0^a x^{-1/2}(a-x)^{-1} \mathbf{H}_0\left(b\sqrt[4]{x(a-x)}\right) dx = \frac{4}{\sqrt{a}} \operatorname{Si}\left(b\sqrt{\frac{a}{2}}\right)$ $[a > 0]$.

7. $\int_0^a x^{-1/4}(a-x)^{1/4} \mathbf{H}_1\left(b\sqrt[4]{x(a-x)}\right) dx$
$$= \frac{2^{3/2}}{b^2}\left[1 - \cos\left(b\sqrt{\frac{a}{2}}\right) - \frac{ab^2}{4}\sin\left(b\sqrt{\frac{a}{2}}\right)\right] + \frac{a}{2^{1/2}} \quad [a > 0].$$

8. $\int_0^a x^{-3/4}(a-x)^{-1/4} \mathbf{H}_1\left(b\sqrt[4]{x(a-x)}\right) dx = -\frac{4}{\sqrt{ab}}\sin\left(b\sqrt{\frac{a}{2}}\right) + 2^{3/2}$
$[a > 0]$.

9. $\int_0^a x^{-3/4}(a-x)^{-5/4} \mathbf{H}_1\left(b\sqrt[4]{x(a-x)}\right) dx$
$$= \frac{2}{a^{3/2}b}\left[-2\sin\left(b\sqrt{\frac{a}{2}}\right) + \sqrt{2a}\,b\cos\left(b\sqrt{\frac{a}{2}}\right) + ab^2 \operatorname{Si}\left(b\sqrt{\frac{a}{2}}\right)\right] \quad [a > 0].$$

10. $\int_0^a x^{-1/2} \mathbf{L}_0\left(b\sqrt[4]{x(a-x)}\right) dx = \frac{2^{3/2}}{b}\left[\cosh\left(b\sqrt{\frac{a}{2}}\right) - 1\right]$ $[a > 0]$.

11. $\int_0^a x^{-1/2}(a-x)^{-1} \mathbf{L}_0\left(b\sqrt[4]{x(a-x)}\right) dx = \frac{4}{\sqrt{a}} \operatorname{shi}\left(b\sqrt{\frac{a}{2}}\right)$ $[a > 0]$.

12. $\int_0^a x^{-1/4}(a-x)^{1/4} \mathbf{L}_1\left(b\sqrt[4]{x(a-x)}\right) dx$
$$= \frac{2^{3/2}}{b^2}\left[1 - \cosh\left(b\sqrt{\frac{a}{2}}\right) + \frac{ab^2}{4}\sinh\left(b\sqrt{\frac{a}{2}}\right)\right] - \frac{a}{2^{1/2}} \quad [a > 0].$$

13. $\int_0^a x^{-3/4}(a-x)^{-1/4} \mathbf{L}_1\left(b\sqrt[4]{x(a-x)}\right) dx = \frac{4}{\sqrt{ab}}\sinh\left(b\sqrt{\frac{a}{2}}\right) - 2^{3/2}$
$[a > 0]$.

14. $\int_0^a x^{-3/4}(a-x)^{-5/4} \mathbf{L}_1\left(b\sqrt[4]{x(a-x)}\right) dx$
$$= \frac{2}{a^{3/2}b}\left[2\sinh\left(b\sqrt{\frac{a}{2}}\right) - \sqrt{2a}\,b\cosh\left(b\sqrt{\frac{a}{2}}\right) + ab^2 \operatorname{shi}\left(b\sqrt{\frac{a}{2}}\right)\right] \quad [a > 0].$$

4.11.2. Integrals containing $\mathbf{H}_\nu(z)$ and hyperbolic functions

1. $\displaystyle\int_0^a \frac{x^{\nu/2}}{\sqrt{a-x}}\sinh(b\sqrt{a-x})\,\mathbf{H}_\nu(b\sqrt{x})\,dx$

$$= \frac{a^{\nu+3/2}b^{\nu+2}}{2^\nu\sqrt{\pi}\,\Gamma\!\left(\nu+\dfrac{5}{2}\right)}\,{}_2F_5\!\left(\begin{array}{c}\dfrac{1}{2},\,1;\,\dfrac{a^2b^4}{256}\\[2pt]\dfrac{3}{4},\,\dfrac{5}{4},\,\dfrac{3}{2},\,\dfrac{2\nu+5}{4},\,\dfrac{2\nu+7}{4}\end{array}\right)\quad [a>0;\ \operatorname{Re}\nu>-5/2].$$

4.11.3. Integrals containing $\mathbf{H}_\nu(z)$, $\mathbf{L}_\nu(z)$ and trigonometric functions

1. $\displaystyle\int_0^{\pi/2}\sin^{-\nu-1}x\cos(2nx)\,\mathbf{H}_\nu(a\sin x)\,dx$

$$= \frac{2^{-4n-\nu-2}\pi a^{2n+\nu+1}}{\Gamma\!\left(n+\dfrac{3}{2}\right)\Gamma\!\left(n+\nu+\dfrac{3}{2}\right)}\,{}_2F_3\!\left(\begin{array}{c}n+\dfrac{1}{2},\,n+1;\,-\dfrac{a^2}{4}\\[2pt]n+\nu+\dfrac{3}{2},\,n+\dfrac{3}{2},\,2n+1\end{array}\right).$$

2. $\displaystyle\int_0^\pi \cos^{-\nu-1}x\cos(nx)\,\mathbf{H}_\nu(a\cos x)\,dx$

$$= \frac{2^{-2n-\nu-1}\pi a^{n+\nu+1}}{\Gamma\!\left(\dfrac{n+3}{2}\right)\Gamma\!\left(\dfrac{n+3}{2}+\nu\right)}\cos\frac{n\pi}{2}\,{}_2F_3\!\left(\begin{array}{c}\dfrac{n+1}{2},\,\dfrac{n}{2}+1;\,-\dfrac{a^2}{4}\\[2pt]\dfrac{n+3}{2}+\nu,\,\dfrac{n+3}{2},\,n+1\end{array}\right).$$

3. $\displaystyle\int_0^\pi \sin^\mu x\left\{\begin{array}{c}\sin(ax)\\\cos(ax)\end{array}\right\}\mathbf{H}_\nu(b\sin x)\,dx$

$$= \frac{2^{-\mu-2\nu-1}\sqrt{\pi}\,b^{\nu+1}\Gamma(\mu+\nu+2)}{\Gamma\!\left(\dfrac{\mu+\nu-a+3}{2}\right)\Gamma\!\left(\dfrac{\mu+\nu+a+3}{2}\right)\Gamma\!\left(\nu+\dfrac{3}{2}\right)}$$

$$\times\left\{\begin{array}{c}\sin(a\pi/2)\\\cos(a\pi/2)\end{array}\right\}{}_3F_4\!\left(\begin{array}{c}1,\,\dfrac{\mu+\nu}{2}+1,\,\dfrac{\mu+\nu+3}{2};\,-\dfrac{b^2}{4}\\[2pt]\dfrac{3}{2},\,\dfrac{\mu+\nu-a+3}{2},\,\dfrac{\mu+\nu+a+3}{2},\,\nu+\dfrac{3}{2}\end{array}\right)$$

$$[\operatorname{Re}(\mu+\nu)>-(5\pm 1)/2].$$

4. $\displaystyle\int_0^\pi \sin^\mu x\left\{\begin{array}{c}\sin(ax)\\\cos(ax)\end{array}\right\}\mathbf{L}_\nu(b\sin x)\,dx$

$$= \frac{2^{-\mu-2\nu-1}\sqrt{\pi}\,b^{\nu+1}\Gamma(\mu+\nu+2)}{\Gamma\!\left(\dfrac{\mu+\nu-a+3}{2}\right)\Gamma\!\left(\dfrac{\mu+\nu+a+3}{2}\right)\Gamma\!\left(\nu+\dfrac{3}{2}\right)}$$

$$\times \begin{Bmatrix} \sin(a\pi/2) \\ \cos(a\pi/2) \end{Bmatrix} {}_3F_4\left(\begin{array}{c} 1, \frac{\mu+\nu}{2}+1, \frac{\mu+\nu+3}{2}; \frac{b^2}{4} \\ \frac{3}{2}, \frac{\mu+\nu-a+3}{2}, \frac{\mu+\nu+a+3}{2}, \nu+\frac{3}{2} \end{array}\right)$$

$$[\operatorname{Re}(\mu+\nu) > -(5\pm 1)/2].$$

5. $\displaystyle\int_0^{m\pi} \sin^{-\nu-1} x \sin(ax) \begin{Bmatrix} \mathbf{H}_\nu(b\sin x) \\ \mathbf{L}_\nu(b\sin x) \end{Bmatrix} dx$

$$= [1-\cos(m\pi a)] \frac{2^{-\nu}\pi^{-1/2} b^{\nu+1}}{\Gamma\left(\nu+\frac{3}{2}\right)a} {}_3F_4\left(\begin{array}{c} \frac{1}{2}, 1, 1; \mp\frac{b^2}{4} \\ \frac{3}{2}, \nu+\frac{3}{2}, 1-\frac{a}{2}, 1+\frac{a}{2} \end{array}\right).$$

6. $\displaystyle\int_0^{m\pi} \sin^{-\nu-1} x \cos(ax) \begin{Bmatrix} \mathbf{H}_\nu(b\sin x) \\ \mathbf{L}_\nu(b\sin x) \end{Bmatrix} dx$

$$= \sin(m\pi a) \frac{2^{-\nu}\pi^{-1/2} b^{\nu+1}}{\Gamma\left(\nu+\frac{3}{2}\right)a} {}_3F_4\left(\begin{array}{c} \frac{1}{2}, 1, 1; \mp\frac{b^2}{4} \\ \frac{3}{2}, \nu+\frac{3}{2}, 1-\frac{a}{2}, 1+\frac{a}{2} \end{array}\right).$$

7. $\displaystyle\int_0^{m\pi} e^{-ax} \sin^{-\nu-1} x \begin{Bmatrix} \mathbf{H}_\nu(b\sin x) \\ \mathbf{L}_\nu(b\sin x) \end{Bmatrix} dx$

$$= (1-e^{-m\pi a}) \frac{2^{-\nu}\pi^{-1/2} b^{\nu+1}}{\Gamma\left(\nu+\frac{3}{2}\right)a} {}_3F_4\left(\begin{array}{c} \frac{1}{2}, 1, 1; \mp\frac{b^2}{4} \\ \frac{3}{2}, \nu+\frac{3}{2}, 1-\frac{ia}{2}, 1+\frac{ia}{2} \end{array}\right).$$

8. $\displaystyle\int_0^\infty e^{-ax} \sin^{-\nu-1} x \begin{Bmatrix} \mathbf{H}_\nu(b\sin x) \\ \mathbf{L}_\nu(b\sin x) \end{Bmatrix} dx$

$$= \frac{2^{-\nu}\pi^{-1/2} b^{\nu+1}}{\Gamma(\nu+3/2)a} {}_3F_4\left(\begin{array}{c} \frac{1}{2}, 1, 1; \mp\frac{b^2}{4} \\ \frac{3}{2}, \nu+\frac{3}{2}, 1-\frac{ia}{2}, 1+\frac{ia}{2} \end{array}\right) \quad [\operatorname{Re} a > 0].$$

4.11.4. Integrals containing $\mathbf{H}_\nu(z)$, $\mathbf{L}_\nu(z)$ and the logarithmic function

1. $\displaystyle\int_0^a x^{s-1} \ln\frac{a+\sqrt{a^2-x^2}}{x} \begin{Bmatrix} \mathbf{H}_\nu(bx) \\ \mathbf{L}_\nu(bx) \end{Bmatrix} dx = \frac{a^{s+\nu+1}\left(\frac{b}{2}\right)^{\nu+1} \Gamma\left(\frac{s+\nu+1}{2}\right)}{(s+\nu+1)\Gamma\left(\frac{s+\nu}{2}+1\right)\Gamma\left(\nu+\frac{3}{2}\right)}$

$$\times {}_3F_4\left(\begin{array}{c} 1, \frac{s+\nu+1}{2}, \frac{s+\nu+1}{2}; \mp\frac{a^2 b^2}{4} \\ \frac{3}{2}, \nu+\frac{3}{2}, \frac{s+\nu}{2}+1, \frac{s+\nu+3}{2} \end{array}\right) \quad [a>0;\ \operatorname{Re}(s+\nu) > -1].$$

2. $\displaystyle\int_0^a x \ln\frac{a+\sqrt{a^2-x^2}}{x} \mathbf{H}_0(bx)\, dx = \frac{1}{b^2}[ab - \sin(ab)] \quad [a>0].$

3. $\int_0^a x \ln \frac{a + \sqrt{a^2 - x^2}}{x} \mathbf{L}_0(bx)\, dx = \frac{1}{b^2} [\sinh(ab) - ab]$ $\qquad [a > 0]$.

4.11.5. Integrals containing $\mathbf{H}_\nu(z)$, $\mathbf{L}_\nu(z)$ and inverse trigonometric functions

1. $\int_0^1 x^{s-1} \arccos x \left\{ \begin{array}{c} \mathbf{H}_\nu(ax) \\ \mathbf{L}_\nu(ax) \end{array} \right\} dx$

$= \frac{a^{\nu+1} \Gamma\left(\frac{s+\nu}{2} + 1\right)}{2^\nu (s+\nu+1)^2 \Gamma\left(\nu + \frac{3}{2}\right) \Gamma\left(\frac{s+\nu+1}{2}\right)} {}_3F_3\left(\begin{array}{c} 1, \frac{s+\nu+1}{2}, \frac{s+\nu}{2} + 1; \mp \frac{a^2}{4} \\ \frac{3}{2}, \nu + \frac{3}{2}, \frac{s+\nu+3}{2}, \frac{s+\nu+3}{2} \end{array}\right)$

$\qquad [\operatorname{Re}(s+\nu) > -1]$.

2. $\int_0^1 \arccos x\, \mathbf{H}_0(ax)\, dx = \frac{1}{a}[\mathbf{C} - \operatorname{ci}(a) + \ln a]$ $\qquad [|\arg a| < \pi]$.

3. $\int_0^1 \frac{1}{x} \arccos x\, \mathbf{H}_1(ax)\, dx = \mathbf{C} - 1 - \operatorname{ci}(a) + \ln a + \frac{1}{a} \sin a$ $\qquad [|\arg a| < \pi]$.

4. $\int_0^1 \arccos x\, \mathbf{L}_0(ax)\, dx = \frac{1}{a}[\operatorname{chi}(a) - \mathbf{C} - \ln a]$ $\qquad [|\arg a| < \pi]$.

5. $\int_0^1 \frac{1}{x} \arccos x\, \mathbf{L}_1(ax)\, dx = 1 - \mathbf{C} + \operatorname{chi}(a) - \ln a - \frac{1}{a} \sinh a$ $\qquad [|\arg a| < \pi]$.

6. $\int_0^a x^{\mu/2} (a-x)^{\nu/2} \mathbf{H}_\mu(b\sqrt{x}) \mathbf{L}_\nu(b\sqrt{a-x})\, dx$

$= \frac{(ab)^{\mu+\nu+2}}{2^{\mu+\nu} \pi \Gamma(\mu+\nu+3)} {}_2F_5\left(\begin{array}{c} \frac{1}{2}, 1; \frac{a^2 b^4}{256} \\ \frac{3}{4}, \frac{5}{4}, \frac{3}{2}, \mu+\nu+\frac{3}{2}, \mu+\nu+2 \end{array}\right)$

$\qquad [a > 0;\ \operatorname{Re}\mu,\ \operatorname{Re}\nu > -3/2]$.

4.12. The Kelvin Functions $\operatorname{ber}_\nu(z)$, $\operatorname{bei}_\nu(z)$, $\operatorname{ker}_\nu(z)$ and $\operatorname{kei}_\nu(z)$

4.12.1. Integrals containing $\operatorname{ber}_\nu(z)$, $\operatorname{bei}_\nu(z)$, $\operatorname{ker}_\nu(z)$, $\operatorname{kei}_\nu(z)$ and algebraic functions

1. $\int_0^a \frac{1}{\sqrt{a^2 - x^2}} \operatorname{ber}_\nu(bx)\, dx = \frac{\pi}{2} \left[\operatorname{ber}^2_{\nu/2}\left(\frac{ab}{2}\right) - \operatorname{bei}^2_{\nu/2}\left(\frac{ab}{2}\right) \right]$ $\qquad [a > 0]$.

2. $\displaystyle\int_0^a \frac{1}{\sqrt{a^2-x^2}}\operatorname{bei}_\nu(bx)\,dx = \pi\operatorname{ber}_{\nu/2}\left(\frac{ab}{2}\right)\operatorname{bei}_{\nu/2}\left(\frac{ab}{2}\right)$ \hfill $[a>0]$.

3. $\displaystyle\int_a^\infty \frac{1}{\sqrt{x^2-a^2}}\operatorname{ker}_\nu(bx)\,dx = \frac{1}{2}\left[\operatorname{ker}^2_{\nu/2}\left(\frac{ab}{2}\right) - \operatorname{kei}^2_{\nu/2}\left(\frac{ab}{2}\right)\right]$ \hfill $[a>0]$.

4. $\displaystyle\int_a^\infty \frac{1}{\sqrt{x^2-a^2}}\operatorname{kei}_\nu(bx)\,dx = \operatorname{ker}_{\nu/2}\left(\frac{ab}{2}\right)\operatorname{kei}_{\nu/2}\left(\frac{ab}{2}\right)$ \hfill $[a>0]$.

4.13. The Airy Functions Ai(z) and Bi(z)

4.13.1. Integrals containing products of Ai(z) and Bi(z)

1. $\displaystyle\int_0^\infty x^{s-1}\operatorname{Ai}(x)\operatorname{Bi}(x)\,dx = \frac{2}{\pi^{1/2}}\,12^{-(2s+5)/6}\,\frac{\Gamma(s)\,\Gamma\left(\frac{1-2s}{6}\right)}{\Gamma\left(\frac{1+s}{3}\right)\Gamma\left(\frac{2-s}{3}\right)}$

\hfill $[0<\operatorname{Re} s<1/2;\;[65],(2.16)]$.

2. $\displaystyle\int_0^\infty x^{s-1}\operatorname{Ai}(x)\operatorname{Bi}(-x)\,dx = \frac{12^{(s-5)/6}}{\pi^{1/2}}\,\frac{\Gamma\left(\frac{s}{2}\right)\Gamma\left(\frac{s+1}{6}\right)}{\Gamma\left(\frac{s+4}{6}\right)\Gamma\left(\frac{2-s}{6}\right)}$

\hfill $[\operatorname{Re} s>0;\;[65],(2.7)]$.

3. $\displaystyle\int_0^\infty x^{s-1}\operatorname{Ai}^2(x)\,dx = \frac{2^{-2(s+1)/3}3^{-(2s+5)/6}\Gamma(s)}{\pi^{1/2}\Gamma\left(\frac{2s+5}{6}\right)}$ \hfill $[\operatorname{Re} s>0]$.

4. $\displaystyle\int_0^\infty x^{s-1}\operatorname{Ai}\left(xe^{\pi i/6}\right)\operatorname{Ai}\left(xe^{-\pi i/6}\right)\,dx = \frac{2^{(s-8)/3}3^{(s-5)/6}}{\pi^{3/2}}\Gamma\left(\frac{s}{2}\right)\Gamma\left(\frac{s+1}{6}\right)$

\hfill $[\operatorname{Re} s>0;\;[65],(4.5)]$.

5. $\displaystyle\int_0^\infty x^{s-1}\left[\operatorname{Ai}^2(-x) + \operatorname{Bi}^2(-x)\right]\,dx = \frac{4}{\pi^{3/2}}\,12^{-(2s+5)/6}\Gamma(s)\,\Gamma\left(\frac{1-2s}{6}\right)$

\hfill $[0<\operatorname{Re} s<1/2;\;[65],(2.28)]$.

6. $\displaystyle\int_0^\infty \operatorname{Ai}^3(x)\,dx = \frac{\Gamma^2\left(\frac{1}{3}\right)}{12\pi^2} - \frac{\Gamma\left(\frac{2}{3}\right)}{2^{5/3}\pi^2}\,{}_2F_1\!\left(\begin{array}{c}\frac{1}{6},\frac{1}{3}\\ \frac{7}{6};\frac{1}{4}\end{array}\right)$ \hfill $[[66],(2.11)]$.

7. $\displaystyle\int_0^\infty \operatorname{Bi}^3(-x)\,dx = -\frac{\Gamma^2\left(\frac{1}{3}\right)}{2\sqrt{3}\,\pi^2} + \frac{3^{3/2}\,\Gamma\left(\frac{2}{3}\right)}{2^{5/3}\pi^2}\,{}_2F_1\!\left(\begin{array}{c}\frac{1}{6},\frac{1}{3}\\ \frac{7}{6};\frac{1}{4}\end{array}\right)$ \hfill $[[66],(3.15)]$.

8. $\int_0^\infty \mathrm{Ai}^2(x)\,\mathrm{Bi}(x)\,dx = \dfrac{\Gamma^2\left(\frac{1}{3}\right)}{12\sqrt{3}\,\pi^2} + \dfrac{\Gamma\left(\frac{2}{3}\right)}{2^{5/3}\sqrt{3}\,\pi^2}\,{}_2F_1\left(\begin{array}{c}\frac{1}{6},\frac{1}{3}\\ \frac{7}{6};\frac{1}{4}\end{array}\right)$ \hfill [[66], (2.27)].

9. $\int_0^\infty \mathrm{Ai}^2(-x)\,\mathrm{Bi}(-x)\,dx = \dfrac{\Gamma^2\left(\frac{1}{3}\right)}{6\sqrt{3}\,\pi^2} - \dfrac{\Gamma\left(\frac{2}{3}\right)}{2^{5/3}\sqrt{3}\,\pi^2}\,{}_2F_1\left(\begin{array}{c}\frac{1}{6},\frac{1}{3}\\ \frac{7}{6};\frac{1}{4}\end{array}\right)$ \hfill [[66], (2.34)].

10. $\int_0^\infty \mathrm{Ai}(-x)\,\mathrm{Bi}^2(-x)\,dx = \dfrac{\Gamma^2\left(\frac{1}{3}\right)}{6\pi^2} - \dfrac{\Gamma\left(\frac{2}{3}\right)}{2^{5/3}\pi^2}\,{}_2F_1\left(\begin{array}{c}\frac{1}{6},\frac{1}{3}\\ \frac{7}{6};\frac{1}{4}\end{array}\right)$ \hfill [[66], (3.7)].

11. $\int_0^\infty \mathrm{Ai}^3(x)\,\mathrm{Bi}(x)\,dx = \dfrac{1}{24\pi}$ \hfill [[67], (3.7)].

12. $\int_0^\infty \mathrm{Ai}^3(-x)\,\mathrm{Bi}(-x)\,dx = \dfrac{1}{12\pi}$ \hfill [[67], (2.18)].

13. $\int_0^\infty \mathrm{Ai}(-x)\,\mathrm{Bi}^3(-x)\,dx = \dfrac{1}{12\pi}$ \hfill [[67], (4.7)].

14. $\int_0^\infty \mathrm{Ai}^4(x)\,dx = \dfrac{\ln 3}{24\pi^2}$ \hfill [57].

15. $\int_0^\infty x\,\mathrm{Ai}^4(x)\,dx = \dfrac{3^{-5/6}}{32\pi^3}\Gamma^2\left(\frac{1}{3}\right) - \dfrac{3^{5/3}}{128\pi^4}\Gamma^4\left(\frac{2}{3}\right)$ \hfill [57].

16. $\int_0^\infty x^2\,\mathrm{Ai}^4(x)\,dx = \dfrac{7}{1024\sqrt[3]{9}\,\pi^4}\Gamma^4\left(\frac{1}{3}\right) - \dfrac{3^{5/6}}{128\pi^3}\Gamma^2\left(\frac{2}{3}\right)$ \hfill [57].

4.14. The Legendre Polynomials $P_n(z)$

4.14.1. Integrals containing $P_n(z)$ and algebraic functions

1. $\int_a^1 (x-a)^{-1/2} P_n(x)\,dx = \dfrac{2}{2n+1}\sqrt{1-a}\,U_{2n}\left(\sqrt{\dfrac{a+1}{2}}\right)$ \hfill $[\operatorname{Re} a > 0]$.

2. $\int_0^a (a^2 - x^2)^{-1/2} P_{2n}(bx)\,dx = (-1)^n \dfrac{\left(\frac{1}{2}\right)_n \pi}{n!\,2}\,P_{2n}\left(\sqrt{1-a^2 b^2}\right)$ \hfill $[\operatorname{Re} a > 0]$.

3. $\int_{-1}^{1} \dfrac{1}{(1 - 2ax + a^2)^{3/2}} P_n(x)\, dx = \dfrac{2a^n}{1 - a^2}.$

4. $\int_{-1}^{1} \dfrac{1}{(2a - x)^{n+2}} P_n(x)\, dx = \dfrac{1}{n+1} \left(\dfrac{2}{4a^2 - 1} \right)^{n+1}$ $\quad [|\arg(4a^2 - 1)| < \pi].$

5. $\int_{0}^{1} \dfrac{1}{(x^2 + a^2)^{n+3/2}} P_{2n}(x)\, dx = \dfrac{(-1)^n}{2n + 1} a^{-2n-2}(a^2 + 1)^{-n-1/2}$ $\quad [\operatorname{Re} a > 0].$

6. $\int_{0}^{1} \dfrac{x}{(x^2 + a^2)^{n+5/2}} P_{2n+1}(x)\, dx = \dfrac{(-1)^n}{2n + 3} a^{-2n-2}(a^2 + 1)^{-n-3/2}$ $\quad [\operatorname{Re} a > 0].$

7. $\int_{0}^{a} (a^2 - x^2)^{s-1} (1 - b^2 x^2)^{-s-n-1/2} P_{2n}(bx)\, dx$

$= \dfrac{(-1)^n}{2} \operatorname{B}\left(n + \dfrac{1}{2}, s\right) a^{2s-1} (1 - a^2 b^2)^{-n-1/2} P_n^{(s-1/2, 0)}(1 - 2a^2 b^2)$

$\quad [a > 0].$

8. $\int_{0}^{a} x(a^2 - x^2)^{s-1} (1 - b^2 x^2)^{-s-n-3/2} P_{2n+1}(bx)\, dx$

$= \dfrac{(-1)^n}{2} \operatorname{B}\left(n + \dfrac{3}{2}, s\right) a^{2s+1} b (1 - a^2 b^2)^{-n-3/2} P_n^{(s+1/2, 0)}(1 - 2a^2 b^2)$

$\quad [a > 0].$

9. $\int_{0}^{a} (1 - b^2 x^2)^{-n-3/2} P_{2n}(bx)\, dx = \dfrac{b^{-1}}{2n + 1} (1 - a^2 b^2)^{-n-1/2} P_{2n+1}(ab).$

10. $\int_{-1}^{1} \dfrac{1}{1 - x} [1 - P_n(x)]\, dx = 2\mathrm{C} + 2\psi(n + 1).$

11. $\int_{-1}^{1} \dfrac{1}{(1 - x)^{3/2}} [1 - P_n(x)]\, dx = 2^{3/2} n.$

12. $\int_{0}^{a} P_n(1 + bx(a - x))\, dx = \dfrac{a}{2n + 1} U_{2n}\left(\sqrt{1 + \dfrac{a^2 b}{8}} \right).$

4.14.1]   4.14. THE LEGENDRE POLYNOMIALS $P_n(z)$

13. $\displaystyle\int_0^a (a-x)^{s-1}(1+bx)^{-s-n-1} P_{2n}\left(\sqrt{1+bx}\right) dx$

$$= \frac{2(2n)!}{(2s)_{2n+1}} a^s (1+ab)^{-n-1} C_{2n}^{s+1/2}\left(\sqrt{1+ab}\right) \quad [a>0].$$

14. $\displaystyle\int_0^a (a-x)^{-1/2}(1+bx)^{-n-3/2} P_{2n}\left(\sqrt{1+bx}\right) dx$

$$= \frac{2a^{1/2}}{2n+1}(1+ab)^{-n-1} U_{2n}\left(\sqrt{1+ab}\right) \quad [a>0].$$

15. $\displaystyle\int_0^a x^{-1/2}(a-x)^{-1/2} P_{2n}\left(b\sqrt{x(a-x)}\right) dx$

$$= (-1)^n \pi \frac{\left(\frac{1}{2}\right)_n}{n!} P_{2n}\left(\sqrt{1-\frac{a^2 b^2}{4}}\right) \quad [a>0].$$

16. $\displaystyle\int_0^a x^{s-1}(a-x)^{s-1/2} P_n\left(1+b\sqrt{x(a-x)}\right) dx$

$$= \frac{2^{-2s+1}\sqrt{\pi}\, a^{2s-1/2}\Gamma(2s)}{\Gamma\left(2s+\frac{1}{2}\right)} \,_3F_2\left(\begin{array}{c}-n,\, n+1,\, 2s\\ 1,\, 2s+\frac{1}{2}; \end{array} -\frac{ab}{4}\right) \quad [a>0;\ \operatorname{Re}\nu>0].$$

17. $\displaystyle\int_0^a x^{-1/2} P_n\left(1+b\sqrt{x(a-x)}\right) dx = \frac{2a^{1/2}}{2n+1} U_{2n}\left(\sqrt{1+\frac{ab}{4}}\right) \quad [a>0].$

18. $\displaystyle\int_0^a x^{-1/4}(a-x)^{-3/4} P_n\left(1+b\sqrt{x(a-x)}\right) dx$

$$= \sqrt{2}\,\pi \left[P_n\left(\sqrt{1+\frac{ab}{4}}\right)\right]^2 \quad [a>0].$$

19. $\displaystyle\int_0^a P_{2n}\left(\sqrt{1-bx(a-x)}\right) dx$

$$= (-1)^n \frac{n!}{\left(\frac{3}{2}\right)_n} ab^{-1/2}\left[a\sqrt{b}\, P_{2n}\left(\frac{a\sqrt{b}}{2}\right) - 2P_{2n-1}\left(\frac{a\sqrt{b}}{2}\right)\right] \quad [a>0;\ n\geq 1].$$

20. $\displaystyle\int_0^a \frac{1}{\sqrt{1-bx(a-x)}} P_{2n+1}\left(\sqrt{1-bx(a-x)}\right) dx$

$$= \frac{2(-1)^n n!}{\left(\frac{3}{2}\right)_n b^{1/2}} P_{2n+1}\left(\frac{a\sqrt{b}}{2}\right) \quad [a>0].$$

21. $\int_0^a x^{-1/2} P_{2n}\left(\sqrt{1-b\sqrt{x(a-x)}}\right) dx$

$= (-1)^{n+1} \dfrac{(n-1)!}{\left(\frac{1}{2}\right)_n} \left(\dfrac{2}{b}\right)^{1/2} \left[\sqrt{\dfrac{ab}{2}} P_{2n}\left(\sqrt{\dfrac{ab}{2}}\right) - P_{2n+1}\left(\sqrt{\dfrac{ab}{2}}\right)\right]$ $\quad [a > 0]$.

22. $\int_0^a \dfrac{x^{-1/2}}{\sqrt{1-b\sqrt{x(a-x)}}} P_{2n+1}\left(\sqrt{1-b\sqrt{x(a-x)}}\right) dx$

$= (-1)^n \dfrac{n!}{\left(\frac{3}{2}\right)_n} \dfrac{2^{3/2}}{b^{1/2}} P_{2n+1}\left(\sqrt{\dfrac{ab}{2}}\right)$ $\quad [a > 0]$.

23. $\int_0^a x^{-1/2}\left[1-b\sqrt{x(a-x)}\right]^n P_n\left(\dfrac{1+b\sqrt{x(a-x)}}{1-b\sqrt{x(a-x)}}\right) dx$

$= -\dfrac{2^{1-n} n!}{\left(\frac{3}{2}\right)_n} b^{-1/2} (ab-2)^{n+1/2} P_{2n+1}\left(\sqrt{\dfrac{ab}{ab-2}}\right)$ $\quad [a > 0]$.

24. $\int_0^a x^{-1/4}(a-x)^{-3/4} P_{2n}\left(b\sqrt[4]{x(a-x)}\right) dx$

$= (-1)^n \sqrt{2}\,\pi\, \dfrac{\left(\frac{1}{2}\right)_n}{n!} P_{2n}\left(\sqrt{1-\dfrac{ab^2}{2}}\right)$ $\quad [a > 0]$.

25. $\int_0^b x^{-n-3} e^{-a/x^2} P_n\left(\dfrac{x}{b}\right) dx = \dfrac{a^{-n/2-1}}{2^{n+1}} e^{-a/b^2} H_n\left(\dfrac{\sqrt{a}}{b}\right)$ $\quad [\operatorname{Re} a > 0]$.

4.14.2. Integrals containing $P_n(z)$ and trigonometric functions

1. $\int_0^\pi \sin^\mu x \begin{Bmatrix} \sin(ax) \\ \cos(ax) \end{Bmatrix} P_{2n}(b\sin x)\, dx$

$= (-1)^n \dfrac{2^{-\mu}\pi\Gamma(\mu+1)\left(\frac{1}{2}\right)_n}{n!\,\Gamma\left(\frac{\mu-a}{2}+1\right)\Gamma\left(\frac{\mu+a}{2}+1\right)} \begin{Bmatrix} \sin(a\pi/2) \\ \cos(a\pi/2) \end{Bmatrix}$

$\times\, {}_4F_3\left(\begin{matrix} -n,\, n+\frac{1}{2},\, \frac{\mu+1}{2},\, \frac{\mu}{2}+1;\, b^2 \\ \frac{1}{2},\, \frac{\mu-a}{2}+1,\, \frac{\mu+a}{2}+1 \end{matrix}\right)$ $\quad [\operatorname{Re}\mu > -1]$.

4.14. The Legendre Polynomials $P_n(z)$

2. $\displaystyle\int_0^\pi \sin^\mu x \begin{Bmatrix} \sin(ax) \\ \cos(ax) \end{Bmatrix} P_{2n+1}(b\sin x)\,dx = (-1)^n \dfrac{2^{-\mu-1}\pi b\,\Gamma(\mu+2)\left(\frac{3}{2}\right)_n}{n!\,\Gamma\!\left(\frac{\mu-a+3}{2}\right)\Gamma\!\left(\frac{\mu+a+3}{2}\right)}$

$\times \begin{Bmatrix} \sin(a\pi/2) \\ \cos(a\pi/2) \end{Bmatrix} {}_4F_3\!\left(\begin{array}{c} -n,\ n+\frac{3}{2},\ \frac{\mu}{2}+1,\ \frac{\mu+3}{2};\ b^2 \\ \frac{3}{2},\ \frac{\mu-a+3}{2},\ \frac{\mu+a+3}{2} \end{array}\right)$ $\quad$ [$\operatorname{Re}\mu > -2$].

3. $\displaystyle\int_0^\pi \sin^\mu x \begin{Bmatrix} \sin(ax) \\ \cos(ax) \end{Bmatrix} P_n\!\left(\dfrac{b}{\sin x}\right) dx = \dfrac{2^{2n-\mu}\pi b^n \Gamma(\mu-n+1)\left(\frac{1}{2}\right)_n}{n!\,\Gamma\!\left(\frac{\mu-a-n}{2}+1\right)\Gamma\!\left(\frac{\mu+a-n}{2}+1\right)}$

$\times \begin{Bmatrix} \sin(a\pi/2) \\ \cos(a\pi/2) \end{Bmatrix} {}_4F_3\!\left(\begin{array}{c} -\frac{n}{2},\ \frac{1-n}{2},\ \frac{\mu-n+1}{2},\ \frac{\mu-n}{2}+1;\ b^{-2} \\ \frac{1}{2}-n,\ \frac{\mu-a-n}{2}+1,\ \frac{\mu+a-n}{2}+1 \end{array}\right)$ $\quad$ [$\operatorname{Re}\mu > n-1$].

4. $\displaystyle\int_0^\pi \sin x \sin(ax) P_{2n}\!\left(\sqrt{1+b\sin^2 x}\right) dx = \dfrac{\sin(\pi a)}{1-a^2}\,{}_3F_2\!\left(\begin{array}{c} -n,\ n+\frac{1}{2},\ \frac{3}{2} \\ \frac{3-a}{2},\ \frac{3+a}{2};\ -b \end{array}\right).$

5. $\displaystyle\int_0^{m\pi} \begin{Bmatrix} \sin(ax) \\ \cos(ax) \end{Bmatrix} P_{2n}(b\sin x)\,dx$

$= (-1)^n \dfrac{2\left(\frac{1}{2}\right)_n}{n!\,a}\sin\dfrac{m\pi a}{2}\begin{Bmatrix} \sin(m\pi a/2) \\ \cos(m\pi a/2) \end{Bmatrix} {}_3F_2\!\left(\begin{array}{c} -n,\ n+\frac{1}{2},\ 1;\ b^2 \\ 1-\frac{a}{2},\ 1+\frac{a}{2} \end{array}\right).$

6. $\displaystyle\int_0^{m\pi} \begin{Bmatrix} \sin(ax) \\ \cos(ax) \end{Bmatrix} P_n(\cos x)\,dx$

$= (-1)^n \dfrac{2}{a}\sin\dfrac{m\pi a}{2}\begin{Bmatrix} \sin(m\pi a/2) \\ \cos(m\pi a/2) \end{Bmatrix} {}_3F_2\!\left(\begin{array}{c} -n,\ n+1,\ \frac{1}{2};\ 1 \\ 1-a,\ 1+a \end{array}\right).$

7. $\displaystyle\int_0^{m\pi} \dfrac{1}{\sin x}\begin{Bmatrix} \sin(ax) \\ \cos(ax) \end{Bmatrix} P_{2n+1}(b\sin x)\,dx$

$= (-1)^n \dfrac{2b}{n!\,a}\left(\dfrac{3}{2}\right)_n \sin\dfrac{m\pi a}{2}\begin{Bmatrix} \sin(m\pi a/2) \\ \cos(m\pi a/2) \end{Bmatrix} {}_4F_3\!\left(\begin{array}{c} -n,\ n+\frac{3}{2},\ \frac{1}{2},\ 1;\ b^2 \\ \frac{3}{2},\ 1-\frac{a}{2},\ 1+\frac{a}{2} \end{array}\right).$

8. $\displaystyle\int_0^{m\pi} \dfrac{1}{\cos x}\begin{Bmatrix} \sin(ax) \\ \cos(ax) \end{Bmatrix} P_{2n+1}(\cos x)\,dx$

$= \dfrac{2}{a}\sin\dfrac{m\pi a}{2}\begin{Bmatrix} \sin(m\pi a/2) \\ \cos(m\pi a/2) \end{Bmatrix} {}_3F_2\!\left(\begin{array}{c} -n,\ n+\frac{3}{2},\ \frac{1}{2};\ 1 \\ 1-\frac{a}{2},\ 1+\frac{a}{2} \end{array}\right).$

9. $\int_0^{m\pi} \begin{Bmatrix} \sin(ax) \\ \cos(ax) \end{Bmatrix} P_n(1+b\sin^2 x)\, dx$

$$= \frac{2}{a}\sin\frac{m\pi a}{2} \begin{Bmatrix} \sin(m\pi a/2) \\ \cos(m\pi a/2) \end{Bmatrix} {}_3F_2\left(\begin{array}{c} -n, n+1, \frac{1}{2}; -\frac{b}{2} \\ 1-\frac{a}{2}, 1+\frac{a}{2} \end{array}\right).$$

10. $\int_0^{m\pi} \begin{Bmatrix} \sin(ax) \\ \cos(ax) \end{Bmatrix} P_{2n}\left(\sqrt{1+b\sin^2 x}\right) dx$

$$= \frac{2}{a}\sin\frac{m\pi a}{2} \begin{Bmatrix} \sin(m\pi a/2) \\ \cos(m\pi a/2) \end{Bmatrix} {}_3F_2\left(\begin{array}{c} -n, n+\frac{1}{2}, \frac{1}{2}; -b \\ 1-\frac{a}{2}, 1+\frac{a}{2} \end{array}\right).$$

11. $\int_0^{m\pi} \frac{1}{\sqrt{1+b\sin^2 x}} \begin{Bmatrix} \sin(ax) \\ \cos(ax) \end{Bmatrix} P_{2n+1}\left(\sqrt{1+b\sin^2 x}\right) dx$

$$= \frac{2}{a}\sin\frac{m\pi a}{2} \begin{Bmatrix} \sin(m\pi a/2) \\ \cos(m\pi a/2) \end{Bmatrix} {}_3F_2\left(\begin{array}{c} -n, n+\frac{3}{2}, \frac{1}{2}; -b \\ 1-\frac{a}{2}, 1+\frac{a}{2} \end{array}\right).$$

12. $\int_0^{m\pi} \begin{Bmatrix} \sin(ax) \\ \cos(ax) \end{Bmatrix} (1+b\sin^2 x)^{n/2} P_n\left(\frac{1}{\sqrt{1+b\sin^2 x}}\right) dx$

$$= \frac{2}{a}\sin\frac{m\pi a}{2} \begin{Bmatrix} \sin(m\pi a/2) \\ \cos(m\pi a/2) \end{Bmatrix} {}_3F_2\left(\begin{array}{c} -\frac{n}{2}, \frac{1-n}{2}, \frac{1}{2}; -b \\ 1-\frac{a}{2}, 1+\frac{a}{2} \end{array}\right).$$

13. $\int_0^{m\pi} \begin{Bmatrix} \sin(ax) \\ \cos(ax) \end{Bmatrix} \sin^n x\, P_n\left(\frac{b}{\sin x}\right) dx$

$$= 2\sin\frac{m\pi a}{2} \begin{Bmatrix} \sin(m\pi a/2) \\ \cos(m\pi a/2) \end{Bmatrix} \frac{\left(\frac{1}{2}\right)_n}{n!\,a} (2b)^n\, {}_4F_3\left(\begin{array}{c} -\frac{n}{2}, \frac{1-n}{2}, \frac{1}{2}, 1;\ b^{-2} \\ \frac{1}{2}-n, 1-\frac{a}{2}, 1+\frac{a}{2} \end{array}\right).$$

14. $\int_0^{m\pi} \begin{Bmatrix} \sin(ax) \\ \cos(ax) \end{Bmatrix} \cos^n x\, P_n\left(\frac{1}{\cos x}\right) dx$

$$= \frac{2}{a}\sin\frac{m\pi a}{2} \begin{Bmatrix} \sin(m\pi a/2) \\ \cos(m\pi a/2) \end{Bmatrix} {}_3F_2\left(\begin{array}{c} -\frac{n}{2}, \frac{1-n}{2}, \frac{1}{2}; 1 \\ 1-\frac{a}{2}, 1+\frac{a}{2} \end{array}\right).$$

15. $\int_0^{m\pi} \begin{Bmatrix} \sin(ax) \\ \cos(ax) \end{Bmatrix} \sin^{2n} x\, P_n(\cot^2 x)\, dx$

$$= \frac{2^{n+1}\left(\frac{1}{2}\right)_n}{n!\,a} \sin\frac{m\pi a}{2} \begin{Bmatrix} \sin(m\pi a/2) \\ \cos(m\pi a/2) \end{Bmatrix} {}_4F_3\left(\begin{array}{c} -n, -n, \frac{1}{2}, 1; 2 \\ -2n, 1-\frac{a}{2}, 1+\frac{a}{2} \end{array}\right).$$

4.14. The Legendre Polynomials $P_n(z)$

16. $\displaystyle\int_0^{m\pi} e^{-ax} P_{2n}(b\sin x)\, dx$

$$= (-1)^n \frac{\left(\frac{1}{2}\right)_n}{n!\,a} (1 - e^{-m\pi a})\, {}_3F_2\left(\begin{array}{c} -n,\, n+\frac{1}{2},\, 1;\, b^2 \\ 1 - \frac{ia}{2},\, 1 + \frac{ia}{2} \end{array}\right).$$

17. $\displaystyle\int_0^{m\pi} \frac{e^{-ax}}{\sin x} P_{2n+1}(b\sin x)\, dx$

$$= (-1)^n \frac{\left(\frac{3}{2}\right)_n b}{n!\,a} (1 - e^{-m\pi a})\, {}_4F_3\left(\begin{array}{c} -n,\, n+\frac{3}{2},\, \frac{1}{2},\, 1;\, b^2 \\ \frac{3}{2},\, 1 - \frac{ia}{2},\, 1 + \frac{ia}{2} \end{array}\right).$$

18. $\displaystyle\int_0^{m\pi} \frac{e^{-ax}}{\cos x} P_{2n+1}(\cos x)\, dx = \frac{1}{a}(1 - e^{-m\pi a})\, {}_3F_2\left(\begin{array}{c} -n,\, n+\frac{3}{2},\, \frac{1}{2};\, 1 \\ 1 - \frac{ia}{2},\, 1 + \frac{ia}{2} \end{array}\right).$

19. $\displaystyle\int_0^{m\pi} e^{-ax} P_n(1 + b\sin^2 x)\, dx = \frac{1}{a}(1 - e^{-m\pi a})\, {}_3F_2\left(\begin{array}{c} -n,\, n+1,\, \frac{1}{2};\, -\frac{b}{2} \\ 1 - \frac{ia}{2},\, 1 + \frac{ia}{2} \end{array}\right).$

20. $\displaystyle\int_0^{m\pi} e^{-ax} P_{2n}\left(\sqrt{1 + b\sin^2 x}\right) dx$

$$= \frac{1}{a}(1 - e^{-m\pi a})\, {}_3F_2\left(\begin{array}{c} -n,\, n+\frac{1}{2},\, \frac{1}{2};\, -b \\ 1 - \frac{ia}{2},\, 1 + \frac{ia}{2} \end{array}\right).$$

21. $\displaystyle\int_0^{m\pi} \frac{e^{-ax}}{\sqrt{1 + b\sin^2 x}} P_{2n+1}\left(\sqrt{1 + b\sin^2 x}\right) dx$

$$= \frac{1}{a}(1 - e^{-m\pi a})\, {}_3F_2\left(\begin{array}{c} -n,\, n+\frac{3}{2},\, \frac{1}{2};\, -b \\ 1 - \frac{ia}{2},\, 1 + \frac{ia}{2} \end{array}\right).$$

22. $\displaystyle\int_0^{m\pi} e^{-ax}(1 + b\sin^2 x)^{n/2} P_n\left(\frac{1}{\sqrt{1 + b\sin^2 x}}\right) dx$

$$= \frac{1}{a}(1 - e^{-m\pi a})\, {}_3F_2\left(\begin{array}{c} -\frac{n}{2},\, \frac{1-n}{2},\, \frac{1}{2};\, -b \\ 1 - \frac{ia}{2},\, 1 + \frac{ia}{2} \end{array}\right).$$

23. $\displaystyle\int_0^{m\pi} e^{-ax}\sin^n x\, P_n\!\left(\frac{b}{\sin x}\right) dx$

$$= \frac{\left(\frac{1}{2}\right)_n}{n!\,a}(2b)^n\left(1-e^{-m\pi a}\right){}_4F_3\!\left(\begin{array}{c}-\frac{n}{2},\,\frac{1-n}{2},\,\frac{1}{2},\,1;\,b^{-2}\\ \frac{1}{2}-n,\,1-\frac{ia}{2},\,1+\frac{ia}{2}\end{array}\right).$$

24. $\displaystyle\int_0^{m\pi} e^{-ax}\cos^n x\, P_n\!\left(\frac{1}{\cos x}\right) dx = \frac{1}{a}\left(1-e^{-m\pi a}\right){}_3F_2\!\left(\begin{array}{c}-\frac{n}{2},\,\frac{1-n}{2},\,\frac{1}{2};\,1\\ 1-\frac{ia}{2},\,1+\frac{ia}{2}\end{array}\right).$

25. $\displaystyle\int_0^{m\pi} e^{-ax}\sin^{2n} x\, P_n(\cot^2 x)\, dx$

$$= 2^n(1-e^{-m\pi a})\frac{\left(\frac{1}{2}\right)_n}{n!\,a}\,{}_4F_3\!\left(\begin{array}{c}-n,\,-n,\,\frac{1}{2},\,1;\,2\\ -2n,\,1-\frac{ia}{2},\,1+\frac{ia}{2}\end{array}\right).$$

26. $\displaystyle\int_0^{\infty} e^{-ax} P_{2n}(b\sin x)\, dx = (-1)^n\frac{\left(\frac{1}{2}\right)_n}{n!\,a}\,{}_3F_2\!\left(\begin{array}{c}-n,\,n+\frac{1}{2},\,1;\,b^2\\ 1-\frac{ia}{2},\,1+\frac{ia}{2}\end{array}\right)$ [Re $a > 0$].

27. $\displaystyle\int_0^{\infty} e^{-ax} P_{2n+1}(b\sin x)\, dx = (-1)^n\frac{\left(\frac{3}{2}\right)_n}{n!}\frac{b}{a^2+1}\,{}_3F_2\!\left(\begin{array}{c}-n,\,n+\frac{3}{2},\,1;\,b^2\\ \frac{3-ia}{2},\,\frac{3+ia}{2}\end{array}\right)$

[Re $a > 0$].

28. $\displaystyle\int_0^{\infty}\frac{e^{-ax}}{\sin x} P_{2n+1}(b\sin x)\, dx = (-1)^n\frac{\left(\frac{3}{2}\right)_n}{n!\,a}b\,{}_4F_3\!\left(\begin{array}{c}-n,\,n+\frac{3}{2},\,\frac{1}{2},\,1;\,b^2\\ \frac{3}{2},\,1-\frac{ia}{2},\,1+\frac{ia}{2}\end{array}\right)$

[Re $a > 0$].

29. $\displaystyle\int_0^{\infty}\frac{e^{-ax}}{\cos x} P_{2n+1}(\cos x)\, dx = \frac{1}{a}\,{}_3F_2\!\left(\begin{array}{c}-n,\,n+\frac{3}{2},\,\frac{1}{2};\,b\\ 1-\frac{ia}{2},\,1+\frac{ia}{2}\end{array}\right)$ [Re $a > 0$].

30. $\displaystyle\int_0^{\infty} e^{-ax} P_n(1+b\sin^2 x)\, dx = \frac{1}{a}\,{}_3F_2\!\left(\begin{array}{c}-n,\,n+1,\,\frac{1}{2};\,-\frac{b}{2}\\ 1-\frac{ia}{2},\,1+\frac{ia}{2}\end{array}\right)$ [Re $a > 0$].

31. $\displaystyle\int_0^{\infty} e^{-ax} P_{2n}\!\left(\sqrt{1+b\sin^2 x}\right) dx = \frac{1}{a}\,{}_3F_2\!\left(\begin{array}{c}-n,\,n+\frac{1}{2},\,\frac{1}{2};\,-b\\ 1-\frac{ia}{2},\,1+\frac{ia}{2}\end{array}\right)$ [Re $a > 0$].

32. $\displaystyle\int_0^\infty \frac{e^{-ax}}{\sqrt{1+b\sin^2 x}} P_{2n+1}\left(\sqrt{1+b\sin^2 x}\right) dx$

$$= \frac{1}{a}\,{}_3F_2\left(\begin{matrix}-n, n+\frac{3}{2}, \frac{1}{2}; -b\\ 1-\frac{ia}{2}, 1+\frac{ia}{2}\end{matrix}\right) \qquad [\operatorname{Re} a > 0].$$

33. $\displaystyle\int_0^\infty e^{-ax}(1+b\sin^2 x)^{n/2} P_n\left(\frac{1}{\sqrt{1+b\sin^2 x}}\right) dx$

$$= \frac{1}{a}\,{}_3F_2\left(\begin{matrix}-\frac{n}{2}, \frac{1-n}{2}, \frac{1}{2}; -b\\ 1-\frac{ia}{2}, 1+\frac{ia}{2}\end{matrix}\right) \qquad [\operatorname{Re} a > 0].$$

34. $\displaystyle\int_0^\infty e^{-ax} \sin^n x\, P_n\left(\frac{b}{\sin x}\right) dx = \frac{\left(\frac{1}{2}\right)_n}{n!\,a}(2b)^n\,{}_4F_3\left(\begin{matrix}-\frac{n}{2}, \frac{1-n}{2}, \frac{1}{2}, 1;\, b^{-2}\\ \frac{1}{2}-n, 1-\frac{ia}{2}, 1+\frac{ia}{2}\end{matrix}\right)$

$$[\operatorname{Re} a > 0].$$

35. $\displaystyle\int_0^\infty e^{-ax} \cos^n x\, P_n\left(\frac{1}{\cos x}\right) dx = \frac{1}{a}\,{}_3F_2\left(\begin{matrix}-\frac{n}{2}, \frac{1-n}{2}, \frac{1}{2};\, 1\\ 1-\frac{ia}{2}, 1+\frac{ia}{2}\end{matrix}\right) \qquad [\operatorname{Re} a > 0].$

36. $\displaystyle\int_0^\infty e^{-ax} \sin^{2n} x\, P_n(\cot^2 x)\, dx = 2^n \frac{\left(\frac{1}{2}\right)_n}{n!\,a}\,{}_4F_3\left(\begin{matrix}-n, -n, \frac{1}{2}, 1;\, 2\\ -2n, 1-\frac{ia}{2}, 1+\frac{ia}{2}\end{matrix}\right)$

$$[\operatorname{Re} a > 0].$$

4.14.3. Integrals containing $P_n(z)$ and the logarithmic function

Condition: $a > 0$.

1. $\displaystyle\int_{-a}^{a} (a-x)^{n-1} \ln\left(\frac{x+a}{2a}\right) P_n\left(\frac{x}{a}\right) dx = \frac{\left(-\frac{a}{2}\right)^n}{n^2}\left[\frac{n!}{\left(\frac{1}{2}\right)_n} - 2^{2n}\right].$

2. $\displaystyle\int_0^a \ln\frac{\sqrt{a}+\sqrt{a-x}}{\sqrt{x}} P_n(1+bx)\, dx$

$$= \frac{1}{n(n+1)b}\left[\frac{(n+1)!}{\left(\frac{1}{2}\right)_{n+1}} P_{n+1}^{(-1/2,\,-3/2)}(1+ab) - 1\right].$$

3. $\displaystyle\int_0^a \ln\frac{\sqrt{a}+\sqrt{a-x}}{\sqrt{x}} P_{2n}(\sqrt{1-bx})\, dx$

$$= \frac{1}{(n+1)(2n-1)b}\left[1 - \frac{(n+1)!}{\left(\frac{1}{2}\right)_{n+1}} P_{n+1}^{(-1/2,\,-2)}(1-2ab)\right].$$

4. $\displaystyle\int_0^a \ln\frac{\sqrt{a}+\sqrt{a-x}}{\sqrt{x}}\,\frac{P_{2n+1}(\sqrt{1-bx})}{\sqrt{1-bx}}\,dx$

$$= \frac{1}{(n+1)(2n+1)b}\left[1-\frac{(n+1)!}{\left(\frac{1}{2}\right)_{n+1}}P_{n+1}^{(-1/2,-1)}(1-2ab)\right].$$

4.14.4. Integrals containing $P_n(z)$, $J_\nu(z)$, $I_\nu(z)$ and $K_\nu(z)$

1. $\displaystyle\int_{-1}^1 e^{ax}J_0\!\left(a\sqrt{1-x^2}\right)P_n(x)\,dx = \frac{2a^n}{n!(2n+1)}.$

2. $\displaystyle\int_{-1}^1 J_0\!\left(a\sqrt{1-x}\right)I_0\!\left(b\sqrt{1+x}\right)P_n(x)\,dx = \frac{2^{1-n}a^n}{(n!)^2(2n+1)}.$

3. $\displaystyle\int_0^1 x^{-5/2}K_{n+1/2}\!\left(\frac{a}{x}\right)P_n(x)\,dx = \frac{\sqrt{\pi}}{4}\left(\frac{2}{a}\right)^{3/2}e^{-a}$ [Re $a>0$].

4. $\displaystyle\int_0^1 x(1-x^2)^{-7/4}K_{2n+3/2}\!\left(\frac{a}{\sqrt{1-x^2}}\right)P_{2n+1}(x)\,dx$

$$= (-1)^n\frac{2^{1/2}a^{-3/2}}{n!}\Gamma\!\left(n+\frac{3}{2}\right)K_0(a) \quad \text{[Re }a>0\text{]}.$$

5. $\displaystyle\int_a^\infty K_0(2\sqrt{x})\,P_{2n+1}\!\left(i\sqrt{\frac{x}{a}-1}\right)dx$

$$= (-1)^n i\,\frac{\Gamma\!\left(n+\frac{3}{2}\right)}{n!}\,a^{1/4}K_{2n+3/2}(2\sqrt{a}) \quad [a>0].$$

6. $\displaystyle\int_{2a}^\infty K_0(\sqrt{x})\,P_n\!\left(\frac{x}{a}-1\right)dx = \sqrt{8a}\,K_{2n+1}(\sqrt{2a})$ $[a>0]$.

4.14.5. Integrals containing products of $P_n(z)$

1. $\displaystyle\int_1^a P_{2n}(x)P_{2n}\!\left(\frac{x}{a}\right)dx = \frac{1}{4n+1}(a^{2n+1}-a^{-2n}).$

2. $\displaystyle\int_1^a P_{2n+1}(x)P_{2n+1}\!\left(\frac{x}{a}\right)dx = \frac{1}{4n+3}(a^{2n+2}-a^{-2n-1}).$

3. $\int_1^a P_n(1-2x) P_n\left(1 - \frac{2x}{a}\right) dx = \frac{1}{2n+1}(a^{n+1} - a^{-n})$.

4. $\int_a^1 \frac{1}{x} P_n(1-2x) P_n\left(1 - \frac{2x}{a}\right) dx = -\ln a$ $[a > 0]$.

5. $\int_a^1 \frac{1}{x^2} P_n(x) P_n\left(\frac{x}{a}\right) dx = \frac{1}{a} - 1$.

6. $\int_0^{\pi/2} \cos^\nu x \cos(ax) \left[P_n\left(\sqrt{1 + b\cos^2 x}\right)\right]^2 dx$

$$= \frac{2^{-\nu-1} \pi \Gamma(\nu+1)}{\Gamma\left(\frac{\nu-a}{2}+1\right) \Gamma\left(\frac{\nu+a}{2}+1\right)} \; {}_5F_4\left(\begin{array}{c} -n, n+1, \frac{1}{2}, \frac{\nu+1}{2}, \frac{\nu}{2}+1 \\ 1, 1, \frac{\nu-a}{2}+1, \frac{\nu+a}{2}+1; -b \end{array}\right)$$

$[\operatorname{Re} \nu > -1]$.

7. $\int_0^\pi \sin^\nu x \left\{\begin{array}{c}\sin(ax)\\ \cos(ax)\end{array}\right\} \left[P_n\left(\sqrt{1 + b\sin^2 x}\right)\right]^2 dx = \frac{2^{-\nu} \pi \Gamma(\nu+1)}{\Gamma\left(\frac{\nu-a}{2}+1\right) \Gamma\left(\frac{\nu+a}{2}+1\right)}$

$$\times \left\{\begin{array}{c}\sin(a\pi/2)\\ \cos(a\pi/2)\end{array}\right\} \; {}_5F_4\left(\begin{array}{c} -n, n+1, \frac{1}{2}, \frac{\nu+1}{2}, \frac{\nu}{2}+1 \\ 1, 1, \frac{\nu-a}{2}+1, \frac{\nu+a}{2}+1; -b \end{array}\right)$$ $[\operatorname{Re} \nu > -1]$.

8. $\int_0^{m\pi} \left\{\begin{array}{c}\sin(ax)\\ \cos(ax)\end{array}\right\} [P_n(\cos x)]^2 dx$

$$= \frac{2}{a} \sin \frac{m\pi a}{2} \left\{\begin{array}{c}\sin(m\pi a/2)\\ \cos(m\pi a/2)\end{array}\right\} \; {}_4F_3\left(\begin{array}{c} -n, n+1, \frac{1}{2}, \frac{1}{2}; 1 \\ 1, 1-\frac{a}{2}, 1+\frac{a}{2} \end{array}\right).$$

9. $\int_0^{m\pi} \left\{\begin{array}{c}\sin(ax)\\ \cos(ax)\end{array}\right\} \left[P_n\left(\sqrt{1 + b\sin^2 x}\right)\right]^2 dx$

$$= \frac{2}{a} \sin \frac{m\pi a}{2} \left\{\begin{array}{c}\sin(m\pi a/2)\\ \cos(m\pi a/2)\end{array}\right\} \; {}_4F_3\left(\begin{array}{c} -n, n+1, \frac{1}{2}, \frac{1}{2}; -b \\ 1, 1-\frac{a}{2}, 1+\frac{a}{2} \end{array}\right).$$

10. $\int_0^{m\pi} e^{-ax} [P_n(\cos x)]^2 dx = \frac{1}{a}(1 - e^{-m\pi a}) \; {}_4F_3\left(\begin{array}{c} -n, n+1, \frac{1}{2}, \frac{1}{2}; 1 \\ 1, 1-\frac{ia}{2}, 1+\frac{ia}{2} \end{array}\right).$

11. $$\int_0^{m\pi} e^{-ax}\left[P_n\left(\sqrt{1+b\sin^2 x}\right)\right]^2 dx$$
$$= \frac{1}{a}\left(1-e^{-m\pi a}\right){}_4F_3\left(\begin{array}{c}-n, n+1, \frac{1}{2}, \frac{1}{2}; -b\\ 1, 1-\frac{ia}{2}, 1+\frac{ia}{2}\end{array}\right).$$

12. $$\int_0^\infty e^{-ax}[P_n(\cos x)]^2 dx = \frac{1}{a}{}_4F_3\left(\begin{array}{c}-n, n+1, \frac{1}{2}, \frac{1}{2}; 1\\ 1, 1-\frac{ia}{2}, 1+\frac{ia}{2}\end{array}\right) \qquad [\operatorname{Re} a > 0].$$

13. $$\int_0^\infty e^{-ax}\left[P_n\left(\sqrt{1+b\sin^2 x}\right)\right]^2 dx = \frac{1}{a}{}_4F_3\left(\begin{array}{c}-n, n+1, \frac{1}{2}, \frac{1}{2}; -b\\ 1, 1-\frac{ia}{2}, 1+\frac{ia}{2}\end{array}\right)$$
$$[\operatorname{Re} a > 0].$$

14. $$\int_0^1 [P_m(x)]^2[P_n(x)]^2 dx$$
$$= \frac{1}{(2m+1)(2n+1)}{}_9F_8\left(\begin{array}{c}-m, m+1, -n, n+1, \frac{1}{2}, \frac{1}{2}, \frac{1}{2}, \frac{1}{2}, \frac{5}{4}; 1\\ \frac{1}{2}-m, \frac{3}{2}+m, \frac{1}{2}-n, \frac{3}{2}+n, \frac{1}{4}, 1, 1, 1\end{array}\right).$$

4.15. The Chebyshev Polynomials $T_n(z)$

4.15.1. Integrals containing $T_n(z)$ and algebraic functions

1. $$\int_0^1 \frac{(1-x^2)^{-1/2}}{(x^2+a^2)^{n+1}} T_{2n}(x)\, dx = (-1)^n \pi \frac{\left(\frac{1}{2}\right)_n}{2(n!)} a^{-2n-1}(a^2+1)^{-n-1/2}$$
$$[\operatorname{Re} a > 0].$$

2. $$\int_0^1 \frac{x(1-x^2)^{-1/2}}{(x^2+a^2)^{n+2}} T_{2n+1}(x)\, dx = (-1)^n \pi \frac{\left(\frac{3}{2}\right)_n}{4(n+1)!} a^{-2n-1}(a^2+1)^{-n-3/2}$$
$$[\operatorname{Re} a > 0].$$

3. $$\int_0^a (a^2-x^2)^{-1/2}(1-b^2x^2)^{-n-1} T_{2n}(bx)\, dx$$
$$= (-1)^n \frac{\pi}{2}(1-a^2b^2)^{-n-1/2} P_{2n}\left(\sqrt{1-a^2b^2}\right) \qquad [a > 0].$$

4. $$\int_0^a x^{-1/2}(1-bx)^{-n-3/2} T_{2n}\left(\sqrt{1+bx}\right) dx$$
$$= \frac{2a^{1/2}}{2n+1}(1+ab)^{-n-1/2} U_{2n}\left(\sqrt{1+ab}\right) \qquad [a > 0].$$

5. $\displaystyle\int_0^a x^{-1/2}(a-x)^{-1/2}(1+bx)^{-n-1}T_{2n}\left(\sqrt{1+bx}\right)dx$

$$= \pi(1+ab)^{-n-1/2}P_{2n}\left(\sqrt{1+ab}\right) \quad [a>0].$$

6. $\displaystyle\int_0^a x^{-1/2}(a-x)^{-1/2}T_{2n}\left(b\sqrt{x(a-x)}\right)dx$

$$= (-1)^n \frac{\pi}{2}\left[P_n\left(1-\frac{a^2b^2}{2}\right)+P_{n-1}\left(1-\frac{a^2b^2}{2}\right)\right] \quad [n\geq 1;\ a>0].$$

7. $\displaystyle\int_0^a \frac{x^{-1/2}(a-x)^{-1/2}}{\sqrt{1+bx(a-x)}}T_{2n+1}\left(\sqrt{1+bx(a-x)}\right)dx = \pi P_n\left(1+\frac{a^2b}{2}\right) \quad [a>0].$

8. $\displaystyle\int_0^a x^s(a-x)^{s-1/2}T_{2n}\left(b\sqrt[4]{x(a-x)}\right)dx$

$$= (-1)^n \frac{\pi^{1/2}a^{2s+1/2}\Gamma(2s+1)}{2^{2s}\Gamma\left(2s+\frac{3}{2}\right)}\,{}_3F_2\left(\begin{array}{c}-n,\,n,\,2s+1\\ \frac{1}{2},\,2s+\frac{3}{2};\,\frac{ab^2}{2}\end{array}\right) \quad [a>0;\ \mathrm{Re}\,s>-1/2].$$

9. $\displaystyle\int_0^a x^s(a-x)^{s-1/2}T_{2n+1}\left(b\sqrt[4]{x(a-x)}\right)dx$

$$= (-1)^n\frac{(2n+1)\pi^{1/2}a^{2s+1}b\,\Gamma\left(2s+\frac{3}{2}\right)}{2^{2s+1/2}\Gamma(2s+2)}\,{}_3F_2\left(\begin{array}{c}-n,\,n+1,\,2s+\frac{3}{2}\\ \frac{3}{2},\,2s+2;\,\frac{ab^2}{2}\end{array}\right)$$
$$[a>0;\ \mathrm{Re}\,s>-3/4].$$

10. $\displaystyle\int_0^a x^s(a-x)^{s-1/2}T_{2n}\left(\sqrt{1+b\sqrt{x(a-x)}}\right)dx$

$$= \frac{\pi^{1/2}a^{2s+1/2}\Gamma(2s+1)}{2^{2s}\Gamma\left(2s+\frac{3}{2}\right)}\,{}_3F_2\left(\begin{array}{c}-n,\,n,\,2s+1\\ \frac{1}{2},\,2s+\frac{3}{2};\,-\frac{ab}{2}\end{array}\right) \quad [a>0;\ \mathrm{Re}\,s>-1/2].$$

11. $\displaystyle\int_0^a \frac{x^s(a-x)^{s-1/2}}{\sqrt{1+b\sqrt{x(a-x)}}}T_{2n+1}\left(\sqrt{1+b\sqrt{x(a-x)}}\right)dx$

$$= \frac{\pi^{1/2}a^{2s+1/2}\Gamma(2s+1)}{2^{2s}\Gamma\left(2s+\frac{3}{2}\right)}\,{}_3F_2\left(\begin{array}{c}-n,\,n+1,\,2s+1\\ \frac{1}{2},\,2s+\frac{3}{2};\,-\frac{ab}{2}\end{array}\right) \quad [a>0;\ \mathrm{Re}\,s>-1/2].$$

4.15.2. Integrals containing $T_n(z)$ and trigonometric functions

1. $\displaystyle\int_0^\pi \sin^\mu x \begin{Bmatrix} \sin(ax) \\ \cos(ax) \end{Bmatrix} T_{2n}(b \sin x)\, dx$

$$= (-1)^n \frac{2^{-\mu}\pi\Gamma(\mu+1)}{\Gamma\left(\frac{\mu-a}{2}+1\right)\Gamma\left(\frac{\mu+a}{2}+1\right)} \begin{Bmatrix} \sin(a\pi/2) \\ \cos(a\pi/2) \end{Bmatrix}$$

$$\times {}_4F_3\left(\begin{array}{c} -n,\, n,\, \frac{\mu+1}{2},\, \frac{\mu}{2}+1;\, b^2 \\ \frac{1}{2},\, \frac{\mu-a}{2}+1,\, \frac{\mu+a}{2}+1 \end{array}\right) \quad [\operatorname{Re}\mu > -1].$$

2. $\displaystyle\int_0^\pi \sin^\mu x \begin{Bmatrix} \sin(ax) \\ \cos(ax) \end{Bmatrix} T_{2n+1}(b \sin x)\, dx$

$$= (-1)^n \frac{2^{-\mu-1}(2n+1)\pi b\,\Gamma(\mu+2)}{\Gamma\left(\frac{\mu-a+3}{2}\right)\Gamma\left(\frac{\mu+a+3}{2}\right)} \begin{Bmatrix} \sin(a\pi/2) \\ \cos(a\pi/2) \end{Bmatrix}$$

$$\times {}_4F_3\left(\begin{array}{c} -n,\, n+1,\, \frac{\mu}{2}+1,\, \frac{\mu+3}{2};\, b^2 \\ \frac{3}{2},\, \frac{\mu-a+3}{2},\, \frac{\mu+a+3}{2} \end{array}\right) \quad [\operatorname{Re}\mu > -2].$$

3. $\displaystyle\int_0^\pi \sin x \sin(ax) T_{2n}\left(\sqrt{1+b\sin^2 x}\right) dx = \frac{\sin(\pi a)}{1-a^2}\, {}_4F_3\left(\begin{array}{c} -n,\, n,\, 1,\, \frac{3}{2};\, -b \\ \frac{1}{2},\, \frac{3-a}{2},\, \frac{3+a}{2} \end{array}\right).$

4. $\displaystyle\int_0^\pi \frac{\sin x \sin(ax)}{\sqrt{1+b\sin^2 x}} T_{2n+1}\left(\sqrt{1+b\sin^2 x}\right) dx$

$$= \frac{\sin(\pi a)}{1-a^2}\, {}_4F_3\left(\begin{array}{c} -n,\, n+1,\, 1,\, \frac{3}{2};\, -b \\ \frac{1}{2},\, \frac{3-a}{2},\, \frac{3+a}{2} \end{array}\right).$$

5. $\displaystyle\int_0^\pi \sin^\mu x \begin{Bmatrix} \sin(ax) \\ \cos(ax) \end{Bmatrix} T_n\left(\frac{b}{\sin x}\right) dx = \frac{2^{2n-\mu}\pi b^n \Gamma(\mu-n+1)}{\Gamma\left(\frac{\mu-a-n}{2}+1\right)\Gamma\left(\frac{\mu+a-n}{2}+1\right)}$

$$\times \begin{Bmatrix} \sin(a\pi/2) \\ \cos(a\pi/2) \end{Bmatrix} {}_4F_3\left(\begin{array}{c} -\frac{n}{2},\, \frac{1-n}{2},\, \frac{\mu-n+1}{2},\, \frac{\mu-n}{2}+1;\, b^{-2} \\ 1-n,\, \frac{\mu-a-n}{2}+1,\, \frac{\mu+a-n}{2}+1 \end{array}\right) \quad [\operatorname{Re}\mu > -1].$$

6. $\displaystyle\int_0^{m\pi} \begin{Bmatrix} \sin(ax) \\ \cos(ax) \end{Bmatrix} T_{2n}(b\sin x)\, dx$

$$= (-1)^n \frac{2}{a}\sin\frac{m\pi a}{2}\begin{Bmatrix} \sin(m\pi a/2) \\ \cos(m\pi a/2) \end{Bmatrix} {}_3F_2\left(\begin{array}{c} -n,\, n,\, 1;\, b^2 \\ 1-\frac{a}{2},\, 1+\frac{a}{2} \end{array}\right).$$

4.15. The Chebyshev Polynomials $T_n(z)$

7. $\displaystyle\int_0^{m\pi} \frac{1}{\sin x}\begin{Bmatrix}\sin(ax)\\\cos(ax)\end{Bmatrix} T_{2n+1}(b\sin x)\,dx$

$= 2(-1)^n \sin\dfrac{m\pi a}{2}\begin{Bmatrix}\sin(m\pi a/2)\\\cos(m\pi a/2)\end{Bmatrix}\dfrac{2n+1}{a}b\,{}_4F_3\!\left(\begin{array}{c}-n,\,n+1,\,\frac{1}{2},\,1;\,b^2\\\frac{3}{2},\,1-\frac{a}{2},\,1+\frac{a}{2}\end{array}\right).$

8. $\displaystyle\int_0^{m\pi}\begin{Bmatrix}\sin(ax)\\\cos(ax)\end{Bmatrix} T_{2n}(\cos x)\,dx$

$=\dfrac{2}{a}\sin\dfrac{m\pi a}{2}\begin{Bmatrix}\sin(m\pi a/2)\\\cos(m\pi a/2)\end{Bmatrix}{}_3F_2\!\left(\begin{array}{c}-n,\,n,\,1;\,1\\1-\frac{a}{2},\,1+\frac{a}{2}\end{array}\right).$

9. $\displaystyle\int_0^{m\pi}\frac{1}{\cos x}\begin{Bmatrix}\sin(ax)\\\cos(ax)\end{Bmatrix} T_{2n+1}(\cos x)\,dx$

$=\dfrac{2}{a}\sin\dfrac{m\pi a}{2}\begin{Bmatrix}\sin(m\pi a/2)\\\cos(m\pi a/2)\end{Bmatrix}{}_3F_2\!\left(\begin{array}{c}-n,\,n+1,\,1;\,1\\1-\frac{a}{2},\,1+\frac{a}{2}\end{array}\right).$

10. $\displaystyle\int_0^{m\pi}\begin{Bmatrix}\sin(ax)\\\cos(ax)\end{Bmatrix} T_n(1+b\sin^2 x)\,dx$

$=\dfrac{2}{a}\sin\dfrac{m\pi a}{2}\begin{Bmatrix}\sin(m\pi a/2)\\\cos(m\pi a/2)\end{Bmatrix}{}_3F_2\!\left(\begin{array}{c}-n,\,n,\,1;\,-\frac{b}{2}\\1-\frac{a}{2},\,1+\frac{a}{2}\end{array}\right).$

11. $\displaystyle\int_0^{m\pi}\begin{Bmatrix}\sin(ax)\\\cos(ax)\end{Bmatrix} T_{2n}\!\left(\sqrt{1+b\sin^2 x}\right)dx$

$=\dfrac{2}{a}\sin\dfrac{m\pi a}{2}\begin{Bmatrix}\sin(m\pi a/2)\\\cos(m\pi a/2)\end{Bmatrix}{}_3F_2\!\left(\begin{array}{c}-n,\,n,\,1;\,-b\\1-\frac{a}{2},\,1+\frac{a}{2}\end{array}\right).$

12. $\displaystyle\int_0^{m\pi}\frac{1}{\sqrt{1+b\sin^2 x}}\begin{Bmatrix}\sin(ax)\\\cos(ax)\end{Bmatrix} T_{2n+1}\!\left(\sqrt{1+b\sin^2 x}\right)dx$

$=\dfrac{2}{a}\sin\dfrac{m\pi a}{2}\begin{Bmatrix}\sin(m\pi a/2)\\\cos(m\pi a/2)\end{Bmatrix}{}_3F_2\!\left(\begin{array}{c}-n,\,n+1,\,1;\,-b\\1-\frac{a}{2},\,1+\frac{a}{2}\end{array}\right).$

13. $\displaystyle\int_0^{m\pi}\begin{Bmatrix}\sin(ax)\\\cos(ax)\end{Bmatrix}(1+b\sin^2 x)^{n/2}T_n\!\left(\dfrac{1}{\sqrt{1+b\sin^2 x}}\right)dx$

$=\dfrac{2}{a}\sin\dfrac{m\pi a}{2}\begin{Bmatrix}\sin(m\pi a/2)\\\cos(m\pi a/2)\end{Bmatrix}{}_3F_2\!\left(\begin{array}{c}-\frac{n}{2},\,\frac{1-n}{2},\,1;\,-b\\1-\frac{a}{2},\,1+\frac{a}{2}\end{array}\right).$

14. $\displaystyle\int_0^{m\pi} \begin{Bmatrix} \sin(ax) \\ \cos(ax) \end{Bmatrix} \sin^n x \, T_n\left(\frac{b}{\sin x}\right) dx$

$= \sin\frac{m\pi a}{2} \begin{Bmatrix} \sin(m\pi a/2) \\ \cos(m\pi a/2) \end{Bmatrix} \frac{2^n b^n}{a} {}_4F_3\left(\begin{matrix} -\frac{n}{2}, \frac{1-n}{2}, \frac{1}{2}, 1; b^{-2} \\ 1-n, 1-\frac{a}{2}, 1+\frac{a}{2} \end{matrix}\right)$ $[n \geq 1]$.

15. $\displaystyle\int_0^{m\pi} \begin{Bmatrix} \sin(ax) \\ \cos(ax) \end{Bmatrix} \cos^n x \, T_n\left(\frac{1}{\cos x}\right) dx$

$= \frac{2}{a} \sin\frac{m\pi a}{2} \begin{Bmatrix} \sin(m\pi a/2) \\ \cos(m\pi a/2) \end{Bmatrix} {}_3F_2\left(\begin{matrix} -\frac{n}{2}, \frac{1-n}{2}, 1; 1 \\ 1-\frac{a}{2}, 1+\frac{a}{2} \end{matrix}\right).$

16. $\displaystyle\int_0^{m\pi} \begin{Bmatrix} \sin(ax) \\ \cos(ax) \end{Bmatrix} \cos^{2n} x \, T_{2n}(i\tan x) \, dx$

$= (-1)^n \frac{2}{a} \sin\frac{m\pi a}{2} \begin{Bmatrix} \sin(m\pi a/2) \\ \cos(m\pi a/2) \end{Bmatrix} {}_3F_2\left(\begin{matrix} -n, \frac{1}{2}-n, 1; 1 \\ 1-\frac{a}{2}, 1+\frac{a}{2} \end{matrix}\right).$

17. $\displaystyle\int_0^{m\pi} \begin{Bmatrix} \sin(ax) \\ \cos(ax) \end{Bmatrix} \sin^{2n} x \, T_n(\cot^2 x) \, dx$

$= \frac{2^n}{a} \sin\frac{m\pi a}{2} \begin{Bmatrix} \sin(m\pi a/2) \\ \cos(m\pi a/2) \end{Bmatrix} {}_4F_3\left(\begin{matrix} -n, \frac{1}{2}-n, \frac{1}{2}, 1; 2 \\ 1-2n, 1-\frac{a}{2}, 1+\frac{a}{2} \end{matrix}\right)$ $[n \geq 2]$.

18. $\displaystyle\int_0^{m\pi} e^{-ax} T_{2n}(b\sin x) \, dx = \frac{(-1)^n}{a}\left(1-e^{-m\pi a}\right) {}_3F_2\left(\begin{matrix} -n, n, 1; b^2 \\ 1-\frac{ia}{2}, 1+\frac{ia}{2} \end{matrix}\right).$

19. $\displaystyle\int_0^{m\pi} \frac{e^{-ax}}{\sin x} T_{2n+1}(b\sin x) \, dx$

$= (-1)^n (2n+1)\frac{b}{a}\left(1-e^{-m\pi a}\right) {}_4F_3\left(\begin{matrix} -n, n+1, \frac{1}{2}, 1; b^2 \\ \frac{3}{2}, 1-\frac{ia}{2}, 1+\frac{ia}{2} \end{matrix}\right).$

20. $\displaystyle\int_0^{m\pi} e^{-ax} T_{2n}(\cos x) \, dx = \frac{1}{a}\left(1-e^{-m\pi a}\right) {}_3F_2\left(\begin{matrix} -n, n, 1; 1 \\ 1-\frac{ia}{2}, 1+\frac{ia}{2} \end{matrix}\right).$

21. $\displaystyle\int_0^{m\pi} \frac{e^{-ax}}{\cos x} T_{2n+1}(\cos x) \, dx = \frac{1}{a}\left(1-e^{-m\pi a}\right) {}_3F_2\left(\begin{matrix} -n, n+1, 1; 1 \\ 1-\frac{ia}{2}, 1+\frac{ia}{2} \end{matrix}\right).$

22. $\displaystyle\int_0^{m\pi} e^{-ax} T_n(1+b\sin^2 x) \, dx = \frac{1}{a}\left(1-e^{-m\pi a}\right) {}_3F_2\left(\begin{matrix} -n, n, 1; -\frac{b}{2} \\ 1-\frac{ia}{2}, 1+\frac{ia}{2} \end{matrix}\right).$

4.15. The Chebyshev Polynomials $T_n(z)$

23. $\displaystyle\int_0^{m\pi} e^{-ax} T_{2n}\left(\sqrt{1+b\sin^2 x}\right) dx = \frac{1}{a}(1-e^{-m\pi a})\, {}_3F_2\!\left(\begin{array}{c}-n,\, n,\, 1;\, -b\\ 1-\frac{ia}{2},\, 1+\frac{ia}{2}\end{array}\right).$

24. $\displaystyle\int_0^{m\pi} \frac{e^{-ax}}{\sqrt{1+b\sin^2 x}} T_{2n+1}\left(\sqrt{1+b\sin^2 x}\right) dx$

$$= \frac{1}{a}(1-e^{-m\pi a})\, {}_3F_2\!\left(\begin{array}{c}-n,\, n+1,\, 1;\, -b\\ 1-\frac{ia}{2},\, 1+\frac{ia}{2}\end{array}\right).$$

25. $\displaystyle\int_0^{m\pi} e^{-ax}(1+b\sin^2 x)^{n/2} T_n\!\left(\frac{1}{\sqrt{1+b\sin^2 x}}\right) dx$

$$= \frac{1}{a}(1-e^{-m\pi a})\, {}_3F_2\!\left(\begin{array}{c}-\frac{n}{2},\, \frac{1-n}{2},\, 1;\, -b\\ 1-\frac{ia}{2},\, 1+\frac{ia}{2}\end{array}\right).$$

26. $\displaystyle\int_0^{m\pi} e^{-ax} \sin^n x\, T_n\!\left(\frac{b}{\sin x}\right) dx$

$$= (1-e^{-m\pi a})(1+\delta_{0,n})\frac{2^{n-1} b^n}{a}\, {}_4F_3\!\left(\begin{array}{c}-\frac{n}{2},\, \frac{1-n}{2},\, \frac{1}{2},\, 1;\, b^{-2}\\ 1-n,\, 1-\frac{ia}{2},\, 1+\frac{ia}{2}\end{array}\right).$$

27. $\displaystyle\int_0^{m\pi} e^{-ax} \cos^n x\, T_n\!\left(\frac{1}{\cos x}\right) dx = \frac{1}{a}(1-e^{-m\pi a})\, {}_4F_3\!\left(\begin{array}{c}-\frac{n}{2},\, \frac{1-n}{2},\, 1;\, 1\\ 1-\frac{ia}{2},\, 1+\frac{ia}{2}\end{array}\right).$

28. $\displaystyle\int_0^{m\pi} e^{-ax} \sin^{2n} x\, T_n(\cot^2 x)\, dx$

$$= (1-e^{-m\pi a})(1+\delta_{0,n})\frac{2^{n-1}}{a}\, {}_4F_3\!\left(\begin{array}{c}-n,\, \frac{1}{2}-n,\, \frac{1}{2},\, 1;\, 2\\ 1-2n,\, 1-\frac{ia}{2},\, 1+\frac{ia}{2}\end{array}\right).$$

29. $\displaystyle\int_0^{\infty} e^{-ax} T_{2n}(b\sin x)\, dx = \frac{(-1)^n}{a}\, {}_3F_2\!\left(\begin{array}{c}-n,\, n,\, 1;\, b^2\\ 1-\frac{ia}{2},\, 1+\frac{ia}{2}\end{array}\right)$ [Re $a > 0$].

30. $\displaystyle\int_0^{\infty} e^{-ax} T_n(\cos x)\, dx = \frac{1}{a}\, {}_3F_2\!\left(\begin{array}{c}-n,\, n,\, 1;\, 1\\ 1-ia,\, 1+ia\end{array}\right)$ [Re $a > 0$].

31. $\displaystyle\int_0^{\infty} e^{-ax} T_{2n}(\cos x)\, dx = \frac{1}{a}\, {}_3F_2\!\left(\begin{array}{c}-n,\, n,\, 1;\, 1\\ 1-\frac{ia}{2},\, 1+\frac{ia}{2}\end{array}\right)$ [Re $a > 0$].

32. $\int_0^\infty e^{-ax} T_{2n+1}(b\sin x)\,dx = (-1)^n \dfrac{(2n+1)b}{a^2+1}\,{}_3F_2\!\left(\begin{matrix}-n,\,n+1,\,1;\,b^2\\ \frac{3-ia}{2},\,\frac{3+ia}{2}\end{matrix}\right)$ $\qquad [\operatorname{Re} a > 0].$

33. $\int_0^\infty \dfrac{e^{-ax}}{\sin x} T_{2n+1}(b\sin x)\,dx$

$\qquad\qquad = (-1)^n (2n+1)\dfrac{b}{a}\,{}_4F_3\!\left(\begin{matrix}-n,\,n+1,\,\frac{1}{2},\,1;\,b^2\\ \frac{3}{2},\,1-\frac{ia}{2},\,1+\frac{ia}{2}\end{matrix}\right)\quad [\operatorname{Re} a > 0].$

34. $\int_0^\infty \dfrac{e^{-ax}}{\cos x} T_{2n+1}(\cos x)\,dx = \dfrac{1}{a}\,{}_3F_2\!\left(\begin{matrix}-n,\,n+1,\,1;\,1\\ 1-\frac{ia}{2},\,1+\frac{ia}{2}\end{matrix}\right) \qquad [\operatorname{Re} a > 0].$

35. $\int_0^\infty e^{-ax} T_n(1+b\sin^2 x)\,dx = \dfrac{1}{a}\,{}_3F_2\!\left(\begin{matrix}-n,\,n,\,1;\,-\frac{b}{2}\\ 1-\frac{ia}{2},\,1+\frac{ia}{2}\end{matrix}\right) \qquad [\operatorname{Re} a > 0].$

36. $\int_0^\infty e^{-ax}(1+b\sin^2 x)^{n/2}\, T_n\!\left(\dfrac{1}{\sqrt{1+b\sin^2 x}}\right)dx$

$\qquad\qquad = \dfrac{1}{a}\,{}_3F_2\!\left(\begin{matrix}-\frac{n}{2},\,\frac{1-n}{2},\,1;\,-b\\ 1-\frac{ia}{2},\,1+\frac{ia}{2}\end{matrix}\right) \qquad [\operatorname{Re} a > 0].$

37. $\int_0^\infty e^{-ax} T_{2n}\!\left(\sqrt{1+b\sin^2 x}\right)dx = \dfrac{1}{a}\,{}_3F_2\!\left(\begin{matrix}-n,\,n,\,1;\,-b\\ 1-\frac{ia}{2},\,1+\frac{ia}{2}\end{matrix}\right) \qquad [\operatorname{Re} a > 0].$

38. $\int_0^\infty \dfrac{e^{-ax}}{\sqrt{1+b\sin^2 x}}\,T_{2n+1}\!\left(\sqrt{1+b\sin^2 x}\right)dx$

$\qquad\qquad = \dfrac{1}{a}\,{}_3F_2\!\left(\begin{matrix}-n,\,n+1,\,1;\,-b\\ 1-\frac{ia}{2},\,1+\frac{ia}{2}\end{matrix}\right) \qquad [\operatorname{Re} a > 0].$

39. $\int_0^\infty e^{-ax} \sin^n x\, T_n\!\left(\dfrac{b}{\sin x}\right)dx$

$\qquad\qquad = (1+\delta_{0,n})\dfrac{2^{n-1} b^n}{a}\,{}_4F_3\!\left(\begin{matrix}-\frac{n}{2},\,\frac{1-n}{2},\,\frac{1}{2},\,1;\,b^{-2}\\ 1-n,\,1-\frac{ia}{2},\,1+\frac{ia}{2}\end{matrix}\right) \quad [n \ge 2;\ \operatorname{Re} a > 0].$

40. $\int_0^\infty e^{-ax} \cos^n x\, T_n\!\left(\dfrac{1}{\cos x}\right)dx = \dfrac{1}{a}\,{}_4F_3\!\left(\begin{matrix}-\frac{n}{2},\,\frac{1-n}{2},\,1;\,1\\ 1-\frac{ia}{2},\,1+\frac{ia}{2}\end{matrix}\right) \qquad [\operatorname{Re} a > 0].$

41. $\int_0^\infty e^{-ax} \sin^{2n} x \, T_n(\cot^2 x) \, dx = \frac{2^{n-1}}{a} {}_4F_3\left(\begin{array}{c} -n, \frac{1}{2}-n, \frac{1}{2}, 1; 2 \\ 1-2n, 1-\frac{ia}{2}, 1+\frac{ia}{2} \end{array}\right)$

$$[n \geq 2; \operatorname{Re} a > 0].$$

4.15.3. Integrals containing $T_n(z)$ and special functions

1. $\int_0^\infty K_1(ax) T_{2n+1}(bx) \, dx = \frac{(2n+1)\pi^2}{8a} \left[I_{-n-1/2}^2\left(\frac{a}{2b}\right) - I_{n+1/2}^2\left(\frac{a}{2b}\right) \right]$

$$[\operatorname{Re} a > 0].$$

2. $\int_a^1 x^{-1/2}(x-a)^{-1/2} P_{2n+1}(\sqrt{x}) T_{2n}\left(\sqrt{\frac{x}{a}}\right) dx = 2a^n(1-a)^{1/2}$

$$[0 < a < 1].$$

3. $\int_a^1 x^{-1/2}(x-a)^{-1/2} P_{2n+2}(\sqrt{x}) T_{2n+1}\left(\sqrt{\frac{x}{a}}\right) dx = 2a^{n+1/2}(1-a)^{1/2}$

$$[0 < a < 1].$$

4.16. The Chebyshev Polynomials $U_n(z)$

4.16.1. Integrals containing $U_n(z)$ and algebraic functions

1. $\int_0^a (a^2-x^2)^{-1/2} U_{2n}(bx) \, dx = (-1)^n \frac{\pi}{2} P_n(1-2a^2b^2)$ $\qquad [a > 0].$

2. $\int_0^a (a^2-x^2)^{-1/2}(b^2-x^2)^{-n-1} U_{2n}\left(\frac{x}{b}\right) dx$

$$= (-1)^n \frac{\pi}{2} (b^2-a^2)^{-n-1} P_{2n+1}\left(\sqrt{1-\frac{a^2}{b^2}}\right)$$

$$[a > 0; |a/b| < 1].$$

3. $\int_0^1 \frac{(1-x^2)^{1/2}}{(x^2+a^2)^{n+2}} U_{2n}(x) \, dx = (-1)^n \pi \frac{\left(\frac{3}{2}\right)_n}{4(n+1)!} a^{-2n-3}(a^2+1)^{-n-1/2}$

$$[\operatorname{Re} a > 0].$$

4. $\int_0^1 \frac{x(1-x^2)^{1/2}}{(x^2+a^2)^{n+3}} U_{2n+1}(x) \, dx = (-1)^n \pi \frac{\left(\frac{3}{2}\right)_n}{4n!(n+2)} a^{-2n-3}(a^2+1)^{-n-3/2}$

$$[\operatorname{Re} a > 0].$$

5. $\int_0^a x^{-1/2}(a-x)^{-1/2} U_{2n}\left(b\sqrt{x(a-x)}\right) dx = (-1)^n \pi P_n\left(1 - \frac{a^2 b^2}{2}\right)$

$\qquad [a > 0].$

4.16.2. Integrals containing $U_n(z)$ and trigonometric functions

1. $\int_0^\pi \sin^\mu x \begin{Bmatrix} \sin(ax) \\ \cos(ax) \end{Bmatrix} U_{2n}(b \sin x)\, dx = (-1)^n \dfrac{2^{-\mu}\pi\Gamma(\mu+1)}{\Gamma\left(\frac{\mu-a}{2}+1\right)\Gamma\left(\frac{\mu+a}{2}+1\right)}$

$\times \begin{Bmatrix} \sin(a\pi/2) \\ \cos(a\pi/2) \end{Bmatrix}\ {}_4F_3\left(\begin{matrix} -n,\, n+1,\, \frac{\mu+1}{2},\, \frac{\mu}{2}+1;\, b^2 \\ \frac{1}{2},\, \frac{\mu-a}{2}+1,\, \frac{\mu+a}{2}+1 \end{matrix}\right)$ $\qquad [\operatorname{Re}\mu > -1].$

2. $\int_0^\pi \sin^\mu x \begin{Bmatrix} \sin(ax) \\ \cos(ax) \end{Bmatrix} U_{2n+1}(b \sin x)\, dx$

$= \dfrac{2^{1-\mu}(n+1)\pi b\, \Gamma(\mu+2)}{n!\,\Gamma\left(\frac{\mu-a+3}{2}\right)\Gamma\left(\frac{\mu+a+3}{2}\right)} \cos\dfrac{a\pi}{2} \begin{Bmatrix} \sin(a\pi/2) \\ \cos(a\pi/2) \end{Bmatrix}$

$\times {}_4F_3\left(\begin{matrix} -n,\, n+2,\, \frac{\mu}{2}+1,\, \frac{\mu+3}{2};\, b^2 \\ \frac{3}{2},\, \frac{\mu-a+3}{2},\, \frac{\mu+a+3}{2} \end{matrix}\right)$ $\qquad [\operatorname{Re}\mu > -2].$

3. $\int_0^\pi \sin x \sin(ax) U_{2n}\left(\sqrt{1+b\sin^2 x}\right) dx$

$= (2n+1)\dfrac{\sin(\pi a)}{1-a^2}\ {}_3F_2\left(\begin{matrix} -n,\, n+1,\, 1;\, -b \\ \frac{3-a}{2},\, \frac{3+a}{2} \end{matrix}\right).$

4. $\int_0^\pi \dfrac{\sin x \sin(ax)}{\sqrt{1+b\sin^2 x}} U_{2n+1}\left(\sqrt{1+b\sin^2 x}\right) dx$

$= 2(n+1)\dfrac{\sin(\pi a)}{1-a^2}\ {}_3F_2\left(\begin{matrix} -n,\, n+2,\, 1;\, -b \\ \frac{3-a}{2},\, \frac{3+a}{2} \end{matrix}\right).$

5. $\int_0^\pi \sin^\mu x \begin{Bmatrix} \sin(ax) \\ \cos(ax) \end{Bmatrix} U_n\left(\dfrac{b}{\sin x}\right) dx = \dfrac{2^{2n-\mu}\pi\Gamma(\mu-n+1)}{\Gamma\left(\frac{\mu-a-n}{2}+1\right)\Gamma\left(\frac{\mu+a-n}{2}+1\right)}$

$\times \begin{Bmatrix} \sin(a\pi/2) \\ \cos(a\pi/2) \end{Bmatrix}\ {}_4F_3\left(\begin{matrix} -\frac{n}{2},\, \frac{1-n}{2},\, \frac{\mu-n+1}{2},\, \frac{\mu-n}{2}+1;\, b^{-2} \\ -n,\, \frac{\mu-a-n}{2}+1,\, \frac{\mu+a-n}{2}+1 \end{matrix}\right)$ $\qquad [\operatorname{Re}\mu > n-1].$

6. $\int_0^{m\pi} \begin{Bmatrix} \sin(ax) \\ \cos(ax) \end{Bmatrix} U_{2n}(b \sin x)\, dx$

$= (-1)^n \dfrac{2}{a} \sin\dfrac{m\pi a}{2} \begin{Bmatrix} \sin(m\pi a/2) \\ \cos(m\pi a/2) \end{Bmatrix}\ {}_3F_2\left(\begin{matrix} -n,\, n+1,\, 1;\, b^2 \\ 1-\frac{a}{2},\, 1+\frac{a}{2} \end{matrix}\right).$

4.16. The Chebyshev Polynomials $U_n(z)$

7. $\displaystyle\int_0^{m\pi} \left\{\begin{array}{c} \sin(ax) \\ \cos(ax) \end{array}\right\} U_{2n}(\cos x)\, dx$

$$= 2\sin\frac{m\pi a}{2}\left\{\begin{array}{c} \sin(m\pi a/2) \\ \cos(m\pi a/2) \end{array}\right\}\frac{2n+1}{a}\,{}_4F_3\left(\begin{array}{c} -n,\, n+1,\, \frac{1}{2},\, 1;\, 1 \\ \frac{3}{2},\, 1-\frac{a}{2},\, 1+\frac{a}{2} \end{array}\right).$$

8. $\displaystyle\int_0^{m\pi} \frac{1}{\sin x}\left\{\begin{array}{c} \sin(ax) \\ \cos(ax) \end{array}\right\} U_{2n+1}(b\sin x)\, dx$

$$= 4(-1)^n\sin\frac{m\pi a}{2}\left\{\begin{array}{c} \sin(m\pi a/2) \\ \cos(m\pi a/2) \end{array}\right\}\frac{n+1}{a}\,b\,{}_4F_3\left(\begin{array}{c} -n,\, n+2,\, \frac{1}{2},\, 1;\, b^2 \\ \frac{3}{2},\, 1-\frac{a}{2},\, 1+\frac{a}{2} \end{array}\right).$$

9. $\displaystyle\int_0^{m\pi} \frac{1}{\cos x}\left\{\begin{array}{c} \sin(ax) \\ \cos(ax) \end{array}\right\} U_{2n+1}(\cos x)\, dx$

$$= 4\sin\frac{m\pi a}{2}\left\{\begin{array}{c} \sin(m\pi a/2) \\ \cos(m\pi a/2) \end{array}\right\}\frac{n+1}{a}\,{}_4F_3\left(\begin{array}{c} -n,\, n+2,\, \frac{1}{2},\, 1;\, 1 \\ \frac{3}{2},\, 1-\frac{a}{2},\, 1+\frac{a}{2} \end{array}\right).$$

10. $\displaystyle\int_0^{m\pi} \left\{\begin{array}{c} \sin(ax) \\ \cos(ax) \end{array}\right\} U_n(1+b\sin^2 x)\, dx$

$$= 2\sin\frac{m\pi a}{2}\left\{\begin{array}{c} \sin(m\pi a/2) \\ \cos(m\pi a/2) \end{array}\right\}\frac{n+1}{a}\,{}_4F_3\left(\begin{array}{c} -n,\, n+2,\, \frac{1}{2},\, 1;\, -\frac{b}{2} \\ \frac{3}{2},\, 1-\frac{a}{2},\, 1+\frac{a}{2} \end{array}\right).$$

11. $\displaystyle\int_0^{m\pi} \left\{\begin{array}{c} \sin(ax) \\ \cos(ax) \end{array}\right\} \sin^n x\, U_n\left(\frac{b}{\sin x}\right) dx$

$$= 2\sin\frac{m\pi a}{2}\left\{\begin{array}{c} \sin(m\pi a/2) \\ \cos(m\pi a/2) \end{array}\right\}\frac{(2b)^n}{a}\,{}_4F_3\left(\begin{array}{c} -\frac{n}{2},\, \frac{1-n}{2},\, \frac{1}{2},\, 1;\, b^{-2} \\ -n,\, 1-\frac{ia}{2},\, 1+\frac{ia}{2} \end{array}\right).$$

12. $\displaystyle\int_0^{m\pi} \left\{\begin{array}{c} \sin(ax) \\ \cos(ax) \end{array}\right\} U_{2n}\left(\sqrt{1+b\sin^2 x}\right) dx$

$$= 2\sin\frac{m\pi a}{2}\left\{\begin{array}{c} \sin(m\pi a/2) \\ \cos(m\pi a/2) \end{array}\right\}\frac{2n+1}{a}\,{}_4F_3\left(\begin{array}{c} -n,\, n+1,\, \frac{1}{2},\, 1;\, -b \\ \frac{3}{2},\, 1-\frac{a}{2},\, 1+\frac{a}{2} \end{array}\right).$$

13. $\displaystyle\int_0^{m\pi} \frac{1}{\sqrt{1+b\sin^2 x}}\left\{\begin{array}{c} \sin(ax) \\ \cos(ax) \end{array}\right\} U_{2n+1}\left(\sqrt{1+b\sin^2 x}\right) dx$

$$= 4\sin\frac{m\pi a}{2}\left\{\begin{array}{c} \sin(m\pi a/2) \\ \cos(m\pi a/2) \end{array}\right\}\frac{n+1}{a}\,{}_4F_3\left(\begin{array}{c} -n,\, n+2,\, \frac{1}{2},\, 1;\, -b \\ \frac{3}{2},\, 1-\frac{a}{2},\, 1+\frac{a}{2} \end{array}\right).$$

14. $\displaystyle\int_0^{m\pi} \begin{Bmatrix} \sin(ax) \\ \cos(ax) \end{Bmatrix} (1+b\sin^2 x)^{n/2} U_n\left(\frac{1}{\sqrt{1+b\sin^2 x}}\right) dx$

$$= 2\sin\frac{m\pi a}{2} \begin{Bmatrix} \sin(m\pi a/2) \\ \cos(m\pi a/2) \end{Bmatrix} \frac{n+1}{a} \, {}_4F_3\left(\begin{array}{c} -\frac{n}{2}, \frac{1-n}{2}, \frac{1}{2}, 1; -b \\ \frac{3}{2}, 1-\frac{a}{2}, 1+\frac{a}{2} \end{array}\right).$$

15. $\displaystyle\int_0^{m\pi} \begin{Bmatrix} \sin(ax) \\ \cos(ax) \end{Bmatrix} \sin^{2n} x\, U_n(\cot^2 x)\, dx$

$$= \sin\frac{m\pi a}{2} \begin{Bmatrix} \sin(m\pi a/2) \\ \cos(m\pi a/2) \end{Bmatrix} \frac{2^{n+1}}{a} \, {}_4F_3\left(\begin{array}{c} -n, -\frac{1}{2}-n, \frac{1}{2}, 1;\, 2 \\ -2n-1, 1-\frac{a}{2}, 1+\frac{a}{2} \end{array}\right).$$

16. $\displaystyle\int_0^{m\pi} e^{-ax} U_{2n}(b\sin x)\, dx = \frac{(-1)^n}{a}(1-e^{-m\pi a})\, {}_3F_2\left(\begin{array}{c} -n, n+1, 1;\, b^2 \\ 1-\frac{ia}{2}, 1+\frac{ia}{2} \end{array}\right).$

17. $\displaystyle\int_0^{m\pi} \frac{e^{-ax}}{\sin x} U_{2n+1}(b\sin x)\, dx$

$$= 2(-1)^n(n+1)\frac{b}{a}(1-e^{-m\pi a})\, {}_4F_3\left(\begin{array}{c} -n, n+2, \frac{1}{2}, 1;\, b^2 \\ \frac{3}{2}, 1-\frac{ia}{2}, 1+\frac{ia}{2} \end{array}\right).$$

18. $\displaystyle\int_0^{m\pi} e^{-ax} U_{2n}(\cos x)\, dx = \frac{2n+1}{a}(1-e^{-m\pi a})\, {}_4F_3\left(\begin{array}{c} -n, n+1, \frac{1}{2}, 1;\, 1 \\ \frac{3}{2}, 1-\frac{ia}{2}, 1+\frac{ia}{2} \end{array}\right).$

19. $\displaystyle\int_0^{m\pi} \frac{e^{-ax}}{\cos x} U_{2n+1}(\cos x)\, dx = 2(n+1)\frac{1-e^{-m\pi a}}{a}\, {}_4F_3\left(\begin{array}{c} -n, n+2, \frac{1}{2}, 1;\, 1 \\ \frac{3}{2}, 1-\frac{ia}{2}, 1+\frac{ia}{2} \end{array}\right).$

20. $\displaystyle\int_0^{m\pi} e^{-ax} U_n(1+b\sin^2 x)\, dx$

$$= (n+1)\frac{1-e^{-m\pi a}}{a}\, {}_4F_3\left(\begin{array}{c} -n, n+2, \frac{1}{2}, 1;\, -\frac{b}{2} \\ \frac{3}{2}, 1-\frac{ia}{2}, 1+\frac{ia}{2} \end{array}\right).$$

21. $\displaystyle\int_0^{m\pi} e^{-ax} U_{2n}\left(\sqrt{1+b\sin^2 x}\right) dx$

$$= (2n+1)\frac{1-e^{-m\pi a}}{a}\, {}_4F_3\left(\begin{array}{c} -n, n+1, \frac{1}{2}, 1;\, -b \\ \frac{3}{2}, 1-\frac{ia}{2}, 1+\frac{ia}{2} \end{array}\right).$$

22. $\displaystyle\int_0^{m\pi} \frac{e^{-ax}}{\sqrt{1+b\sin^2 x}} U_{2n+1}\left(\sqrt{1+b\sin^2 x}\right) dx$

$$= 2(n+1)\frac{1-e^{-m\pi a}}{a}\,{}_4F_3\left(\begin{array}{c}-n, n+2, \frac{1}{2}, 1; -b\\ \frac{3}{2}, 1-\frac{ia}{2}, 1+\frac{ia}{2}\end{array}\right).$$

23. $\displaystyle\int_0^{m\pi} e^{-ax}(1+b\sin^2 x)^{n/2} U_n\left(\frac{1}{\sqrt{1+b\sin^2 x}}\right) dx$

$$= \frac{n+1}{a}(1-e^{-m\pi a})\,{}_4F_3\left(\begin{array}{c}-\frac{n}{2}, \frac{1-n}{2}, \frac{1}{2}, 1; -b\\ \frac{3}{2}, 1-\frac{ia}{2}, 1+\frac{ia}{2}\end{array}\right).$$

24. $\displaystyle\int_0^{m\pi} e^{-ax} \sin^n x\, U_n\left(\frac{b}{\sin x}\right) dx$

$$= \frac{(2b)^n}{a}(1-e^{-m\pi a})\,{}_4F_3\left(\begin{array}{c}-\frac{n}{2}, \frac{1-n}{2}, \frac{1}{2}, 1; b^{-2}\\ -n, 1-\frac{ia}{2}, 1+\frac{ia}{2}\end{array}\right).$$

25. $\displaystyle\int_0^{m\pi} e^{-ax} \sin^{2n} x\, U_n(b\cot^2 x)\, dx$

$$= \frac{2^n}{a}(1-e^{-m\pi a})\,{}_4F_3\left(\begin{array}{c}-n, -n-\frac{1}{2}, \frac{1}{2}, 1; 2\\ -2n-1, 1-\frac{ia}{2}, 1+\frac{ia}{2}\end{array}\right).$$

26. $\displaystyle\int_0^\infty e^{-ax} U_{2n}(b\sin x)\, dx = \frac{(-1)^n}{a}\,{}_3F_2\left(\begin{array}{c}-n, n+1, 1; b^2\\ 1-\frac{ia}{2}, 1+\frac{ia}{2}\end{array}\right)$ $\quad[\operatorname{Re} a > 0]$.

27. $\displaystyle\int_0^\infty e^{-ax} U_{2n+1}(b\sin x)\, dx = (-1)^n \frac{2(n+1)b}{a^2+1}\,{}_3F_2\left(\begin{array}{c}-n, n+2, 1; b^2\\ \frac{3-ia}{2}, \frac{3+ia}{2}\end{array}\right)$

$\quad[\operatorname{Re} a > 0]$.

28. $\displaystyle\int_0^\infty e^{-ax} U_{2n}(\cos x)\, dx = \frac{2n+1}{a}\,{}_4F_3\left(\begin{array}{c}-n, n+1, \frac{1}{2}, 1; 1\\ \frac{3}{2}, 1-\frac{ia}{2}, 1+\frac{ia}{2}\end{array}\right)$ $\quad[\operatorname{Re} a > 0]$.

29. $\displaystyle\int_0^\infty \frac{e^{-ax}}{\sin x} U_{2n+1}(b\sin x)\, dx = 2(-1)^n \frac{n+1}{a} b\,{}_4F_3\left(\begin{array}{c}-n, n+2, \frac{1}{2}, 1; b^2\\ \frac{3}{2}, 1-\frac{ia}{2}, 1+\frac{ia}{2}\end{array}\right)$

$\quad[\operatorname{Re} a > 0]$.

30. $\displaystyle\int_0^\infty \frac{e^{-ax}}{\cos x} U_{2n+1}(\cos x)\, dx = \frac{2(n+1)}{a}\,{}_4F_3\left(\begin{array}{c}-n, n+2, \frac{1}{2}, 1; 1\\ \frac{3}{2}, 1-\frac{ia}{2}, 1+\frac{ia}{2}\end{array}\right)$ $\quad[\operatorname{Re} a > 0]$.

31. $\displaystyle\int_0^\infty e^{-ax} U_n(1+b\sin^2 x)\, dx = \frac{n+1}{a}\, {}_4F_3\left(\begin{array}{c}-n, n+2, \frac{1}{2}, 1; -\frac{b}{2}\\ \frac{3}{2}, 1-\frac{ia}{2}, 1+\frac{ia}{2}\end{array}\right)$

[Re $a > 0$].

32. $\displaystyle\int_0^\infty e^{-ax} U_{2n}\left(\sqrt{1+b\sin^2 x}\right) dx = \frac{2n+1}{a}\, {}_4F_3\left(\begin{array}{c}-n, n+1, \frac{1}{2}, 1; -b\\ \frac{3}{2}, 1-\frac{ia}{2}, 1+\frac{ia}{2}\end{array}\right)$

[Re $a > 0$].

33. $\displaystyle\int_0^\infty \frac{e^{-ax}}{\sqrt{1+b\sin^2 x}}\, U_{2n+1}\left(\sqrt{1+b\sin^2 x}\right) dx$

$\displaystyle = \frac{2(n+1)}{a}\, {}_4F_3\left(\begin{array}{c}-n, n+2, \frac{1}{2}, 1; -b\\ \frac{3}{2}, 1-\frac{ia}{2}, 1+\frac{ia}{2}\end{array}\right)$ [Re $a > 0$].

34. $\displaystyle\int_0^\infty e^{-ax}(1+b\sin^2 x)^{n/2}\, U_n\left(\frac{1}{\sqrt{1+b\sin^2 x}}\right) dx$

$\displaystyle = \frac{n+1}{a}\, {}_4F_3\left(\begin{array}{c}-\frac{n}{2}, \frac{1-n}{2}, \frac{1}{2}, 1; -b\\ \frac{3}{2}, 1-\frac{ia}{2}, 1+\frac{ia}{2}\end{array}\right)$ [Re $a > 0$].

35. $\displaystyle\int_0^\infty e^{-ax}\sin^n x\, U_n\left(\frac{b}{\sin x}\right) dx = \frac{(2b)^n}{a}\, {}_4F_3\left(\begin{array}{c}-\frac{n}{2}, \frac{1-n}{2}, \frac{1}{2}, 1; b^{-2}\\ -n, 1-\frac{ia}{2}, 1+\frac{ia}{2}\end{array}\right)$

[Re $a > 0$].

36. $\displaystyle\int_0^\infty e^{-ax}\sin^{2n} x\, U_n(b\cot^2 x)\, dx = \frac{2^n}{a}\, {}_4F_3\left(\begin{array}{c}-n, -n-\frac{1}{2}, \frac{1}{2}, 1; 2\\ -2n-1, 1-\frac{ia}{2}, 1+\frac{ia}{2}\end{array}\right)$

[Re $a > 0$].

4.16.3. Integrals containing $U_n(z)$ and $K_\nu(z)$

1. $\displaystyle\int_0^\infty K_0(ax) U_{2n}(bx)\, dx = \frac{\pi^2}{8b}\left[I_{-n-1/2}^2\left(\frac{a}{2b}\right) - I_{n+1/2}^2\left(\frac{a}{2b}\right)\right]$ [Re $a > 0$].

4.16.4. Integrals containing products of $U_n(z)$

1. $\displaystyle\int_0^{m\pi}\left\{\begin{array}{c}\sin(ax)\\ \cos(ax)\end{array}\right\}[U_n(\cos x)]^2\, dx = 2\sin\frac{m\pi a}{2}\left\{\begin{array}{c}\sin(m\pi a/2)\\ \cos(m\pi a/2)\end{array}\right\}\frac{(n+1)^2}{a}$

$\displaystyle \times\left[2\,{}_4F_3\left(\begin{array}{c}-n, n+2, \frac{1}{2}, 1; 1\\ \frac{3}{2}, 1-\frac{a}{2}, 1+\frac{a}{2}\end{array}\right) - {}_4F_3\left(\begin{array}{c}-n, n+2, 1, 1; 1\\ 2, 1-\frac{a}{2}, 1+\frac{a}{2}\end{array}\right)\right].$

2. $\int_0^{m\pi} e^{-ax}[U_n(\cos x)]^2\, dx = (1 - e^{-m\pi a})\dfrac{(n+1)^2}{a}$

$$\times \left[{}_2 4F_3\!\left(\begin{matrix} -n,\, n+2,\, \tfrac{1}{2},\, 1;\, 1 \\ \tfrac{3}{2},\, 1-\tfrac{ia}{2},\, 1+\tfrac{ia}{2} \end{matrix}\right) - {}_4F_3\!\left(\begin{matrix} -n,\, n+2,\, 1,\, 1;\, 1 \\ 2,\, 1-\tfrac{ia}{2},\, 1+\tfrac{ia}{2} \end{matrix}\right) \right].$$

3. $\int_0^{\infty} e^{-ax}[U_n(\cos x)]^2\, dx$

$$= \dfrac{(n+1)^2}{a}\left[{}_2 4F_3\!\left(\begin{matrix} -n,\, n+2,\, \tfrac{1}{2},\, 1;\, 1 \\ \tfrac{3}{2},\, 1-\tfrac{ia}{2},\, 1+\tfrac{ia}{2} \end{matrix}\right) - {}_4F_3\!\left(\begin{matrix} -n,\, n+2,\, 1,\, 1;\, 1 \\ 2,\, 1-\tfrac{ia}{2},\, 1+\tfrac{ia}{2} \end{matrix}\right) \right]$$

[Re $a > 0$].

4.17. The Hermite Polynomials $H_n(z)$

4.17.1. Integrals containing $H_n(z)$ and algebraic functions

1. $\int_0^a x^{s-1}(a-x)^{t-1} H_{2n}\!\left(b\sqrt{x(a-x)}\right) dx$

$$= (-1)^n \dfrac{(2n)!}{n!} a^{s+t-1} \mathrm{B}(s,t)\, {}_3F_3\!\left(\begin{matrix} -n,\, s,\, t;\, \tfrac{a^2 b^2}{4} \\ \tfrac{1}{2},\, \tfrac{s+t}{2},\, \tfrac{s+t+1}{2} \end{matrix}\right) \quad [a,\, \mathrm{Re}\, s,\, \mathrm{Re}\, t > 0].$$

2. $\int_0^a x^{s-1}(a-x)^{t-1} H_{2n+1}\!\left(b\sqrt{x(a-x)}\right) dx$

$$= 2(-1)^n \dfrac{(2n+1)!}{n!} a^{s+t} b\, \mathrm{B}\!\left(s+\tfrac{1}{2},\, t+\tfrac{1}{2}\right){}_3F_3\!\left(\begin{matrix} -n,\, s+\tfrac{1}{2},\, t+\tfrac{1}{2};\, \tfrac{a^2 b^2}{4} \\ \tfrac{3}{2},\, \tfrac{s+t+1}{2},\, \tfrac{s+t}{2}+1 \end{matrix}\right)$$

[$a > 0$; Re s, Re $t > -1/2$].

3. $\int_0^a x^{1/2}(a-x)^{1/2} H_{2n}\!\left(b\sqrt{x(a-x)}\right) dx$

$$= (-1)^n \dfrac{(2n-1)!}{(n+1)!}\dfrac{\pi a^2}{8}\left[2n L_n^1\!\left(\dfrac{a^2 b^2}{4}\right) - n a^2 b^2 L_{n-1}^2\!\left(\dfrac{a^2 b^2}{4}\right) \right]$$

[$n \geq 1$; $a > 0$].

4. $\int_0^a x^{1/2}(a-x)^{-1/2} H_{2n}\!\left(b\sqrt{x(a-x)}\right) dx = (-1)^n \dfrac{(2n)!}{n!}\dfrac{\pi a}{2} L_n\!\left(\dfrac{a^2 b^2}{4}\right)$

[$a > 0$].

5. $\int_0^a x^{-1/2}(a-x)^{-1/2} H_{2n}\left(b\sqrt{x(a-x)}\right) dx = (-1)^n \dfrac{(2n)!}{n!} \pi L_n\left(\dfrac{a^2 b^2}{4}\right)$

$$[a > 0].$$

6. $\int_0^a H_{2n+1}\left(b\sqrt{x(a-x)}\right) dx = (-1)^n \dfrac{(2n+1)!}{(n+1)!} \dfrac{\pi a^2 b}{4} L_n^1\left(\dfrac{a^2 b^2}{4}\right) \quad [a > 0].$

7. $\int_0^a x H_{2n+1}\left(b\sqrt{x(a-x)}\right) dx = (-1)^n \dfrac{(2n+1)!}{(n+1)!} \dfrac{\pi a^3 b}{8} L_n^1\left(\dfrac{a^2 b^2}{4}\right) \quad [a > 0].$

8. $\int_0^a x^s (a-x)^{s+1/2} H_{2n}\left(b\sqrt[4]{x(a-x)}\right) dx$

$= (-1)^n 2^{-2s-1} \sqrt{\pi}\, a^{2s+3/2} \dfrac{(2n)!}{n!} \dfrac{\Gamma(2s+2)}{\Gamma\left(2s+\tfrac{5}{2}\right)}\, {}_2F_2\left(\begin{array}{c} -n,\, 2s+2;\, \tfrac{ab^2}{2} \\ \tfrac{1}{2},\, 2s+\tfrac{5}{2} \end{array}\right)$

$$[a > 0;\ \operatorname{Re} s > -1].$$

9. $\int_0^a x^s (a-x)^{s+1/2} H_{2n+1}\left(b\sqrt[4]{x(a-x)}\right) dx$

$= (-1)^n 2^{-2s-1/2} \sqrt{\pi}\, a^{2s+2} b\, \dfrac{(2n+1)!}{n!} \dfrac{\Gamma\left(2s+\tfrac{5}{2}\right)}{\Gamma(2s+3)}\, {}_2F_2\left(\begin{array}{c} -n,\, 2s+\tfrac{5}{2};\, \tfrac{ab^2}{2} \\ \tfrac{3}{2},\, 2s+3 \end{array}\right)$

$$[a > 0;\ \operatorname{Re} s > -5/4].$$

10. $\int_0^a x^{-1/2} H_{2n+1}\left(b\sqrt[4]{x(a-x)}\right) dx = (-1)^n \dfrac{(2n+1)!}{(n+1)!} \dfrac{\pi ab}{\sqrt{2}} L_n^1\left(\dfrac{ab^2}{2}\right)$

$$[a > 0].$$

11. $\int_0^a x^{-1/4}(a-x)^{-3/4} H_{2n}\left(b\sqrt[4]{x(a-x)}\right) dx$

$= (-1)^n \dfrac{(2n)!}{n!} \sqrt{2}\, \pi L_n\left(\dfrac{ab^2}{2}\right) \quad [a > 0].$

4.17.2. Integrals containing $H_n(z)$ and the exponential function

1. $\int_a^\infty x(x^2-a^2)^{n-3/2} e^{-b^2 x^2} H_{2n}(bx)\, dx = 2^{2n-1} a^{2n} b\, \Gamma\left(n-\dfrac{1}{2}\right) e^{-a^2 b^2}$

$$[a > 0;\ |\arg b| < \pi/4;\ n \geq 1].$$

2. $\displaystyle\int_a^\infty (x^2-a^2)^{n-1/2} e^{-b^2 x^2} H_{2n+1}(bx)\,dx = (2a)^{2n}\Gamma\!\left(n+\frac{1}{2}\right) e^{-a^2 b^2}$

$\qquad\qquad\qquad\qquad\qquad\qquad\qquad\qquad\qquad\qquad [a>0;\ |\arg b|<\pi/4].$

3. $\displaystyle\int_0^\infty x^{n-1} e^{-ax^2} H_n(bx)\,dx = (n-1)!\,a^{-n/2} T_n\!\left(\frac{b}{\sqrt{a}}\right) \qquad [n\ge 1;\ \operatorname{Re} a>0].$

4.17.3. Integrals containing $H_n(z)$ and trigonometric functions

1. $\displaystyle\int_0^\pi \sin(ax) H_{2n+1}(b\sin x)\,dx$

$\displaystyle = (-1)^n \frac{2^{2n+1} b \sin(\pi a)}{1-a^2}\left(\frac{3}{2}\right)_n {}_2F_2\!\left(\begin{array}{c}-n,\,1;\,b^2\\ \frac{3-a}{2},\,\frac{3+a}{2}\end{array}\right).$

2. $\displaystyle\int_0^\pi \sin x \sin(ax) H_{2n}(b\sin x)\,dx$

$\displaystyle = (-4)^n \frac{\sin(\pi a)}{1-a^2}\left(\frac{1}{2}\right)_n {}_3F_3\!\left(\begin{array}{c}-n,\,1,\,\frac{3}{2};\,b^2\\ \frac{1}{2},\,\frac{3-a}{2},\,\frac{3+a}{2}\end{array}\right).$

3. $\displaystyle\int_0^{m\pi} \left\{\begin{array}{c}\sin(ax)\\ \cos(ax)\end{array}\right\} H_{2n}(b\sin x)\,dx$

$\displaystyle = 2\sin\frac{m\pi a}{2}\left\{\begin{array}{c}\sin(m\pi a/2)\\ \cos(m\pi a/2)\end{array}\right\}\frac{(-4)^n}{a}\left(\frac{1}{2}\right)_n {}_2F_2\!\left(\begin{array}{c}-n,\,1;\,b^2\\ 1-\frac{a}{2},\,1+\frac{a}{2}\end{array}\right).$

4. $\displaystyle\int_0^{m\pi} \frac{1}{\sin x}\left\{\begin{array}{c}\sin(ax)\\ \cos(ax)\end{array}\right\} H_{2n+1}(b\sin x)\,dx$

$\displaystyle = (-1)^n 2^{2n+1}\frac{b}{a}\sin\frac{m\pi a}{2}\left\{\begin{array}{c}\sin(m\pi a/2)\\ \cos(m\pi a/2)\end{array}\right\}\left(\frac{3}{2}\right)_n {}_3F_3\!\left(\begin{array}{c}-n,\,\frac{1}{2},\,1;\,b^2\\ \frac{3}{2},\,1-\frac{a}{2},\,1+\frac{a}{2}\end{array}\right).$

5. $\displaystyle\int_0^{m\pi} \left\{\begin{array}{c}\sin(ax)\\ \cos(ax)\end{array}\right\} \sin^{2n} x\, H_{2n}\!\left(\frac{b}{\sin x}\right)dx$

$\displaystyle = 2\sin\frac{m\pi a}{2}\left\{\begin{array}{c}\sin(m\pi a/2)\\ \cos(m\pi a/2)\end{array}\right\}\frac{(2b)^{2n}}{a} {}_4F_2\!\left(\begin{array}{c}-n,\,\frac{1}{2}-n,\,\frac{1}{2},\,1\\ 1-\frac{a}{2},\,1+\frac{a}{2};\,-b^{-2}\end{array}\right).$

6. $\displaystyle\int_0^{m\pi} \left\{\begin{matrix}\sin(ax)\\ \cos(ax)\end{matrix}\right\} \sin^{2n+1} x\, H_{2n+1}\left(\frac{b}{\sin x}\right) dx$

$= 2\sin\dfrac{m\pi a}{2} \left\{\begin{matrix}\sin(m\pi a/2)\\ \cos(m\pi a/2)\end{matrix}\right\} \dfrac{(2b)^{2n+1}}{a}\, {}_4F_2\left(\begin{matrix}-n, -n-\frac{1}{2}, \frac{1}{2}, 1\\ 1-\frac{a}{2}, 1+\frac{a}{2};\ -b^{-2}\end{matrix}\right).$

7. $\displaystyle\int_0^{m\pi} e^{-ax} H_{2n}(b\sin x)\, dx = \dfrac{(-4)^n}{a}\left(\dfrac{1}{2}\right)_n (1-e^{-m\pi a})\, {}_2F_2\left(\begin{matrix}-n, 1; b^2\\ 1-\frac{ia}{2}, 1+\frac{ia}{2}\end{matrix}\right).$

8. $\displaystyle\int_0^{m\pi} \dfrac{e^{-ax}}{\sin x} H_{2n+1}(b\sin x)\, dx$

$= (-1)^n 2^{2n+1}(1-e^{-m\pi a})\left(\dfrac{3}{2}\right)_n \dfrac{b}{a}\, {}_3F_3\left(\begin{matrix}-n, \frac{1}{2}, 1; b^2\\ \frac{3}{2}, 1-\frac{ia}{2}, 1+\frac{ia}{2}\end{matrix}\right).$

9. $\displaystyle\int_0^{m\pi} e^{-ax} \sin^{2n} x\, H_{2n}\left(\dfrac{b}{\sin x}\right) dx$

$= \dfrac{(2b)^{2n}}{a}(1-e^{-m\pi a})\, {}_4F_2\left(\begin{matrix}-n, \frac{1}{2}-n, \frac{1}{2}, 1;\ -b^{-2}\\ 1-\frac{ia}{2}, 1+\frac{ia}{2}\end{matrix}\right).$

10. $\displaystyle\int_0^{m\pi} e^{-ax} \sin^{2n+1} x\, H_{2n+1}\left(\dfrac{b}{\sin x}\right) dx$

$= \dfrac{(2b)^{2n+1}}{a}(1-e^{-m\pi a})\, {}_4F_2\left(\begin{matrix}-n, -n-\frac{1}{2}, \frac{1}{2}, 1;\ -b^{-2}\\ 1-\frac{ia}{2}, 1+\frac{ia}{2}\end{matrix}\right).$

11. $\displaystyle\int_0^\infty e^{-ax} H_{2n}(b\sin x)\, dx = \dfrac{(-4)^n}{a}\left(\dfrac{1}{2}\right)_n\, {}_2F_2\left(\begin{matrix}-n, 1; b^2\\ 1-\frac{ia}{2}, 1+\frac{ia}{2}\end{matrix}\right)\quad [\operatorname{Re} a > 0].$

12. $\displaystyle\int_0^\infty \dfrac{e^{-ax}}{\sin x} H_{2n+1}(b\sin x)\, dx$

$= (-1)^{n+1} 2^{2n+1}\left(\dfrac{3}{2}\right)_n \dfrac{b}{a}\, {}_3F_3\left(\begin{matrix}-n, \frac{1}{2}, 1; b^2\\ \frac{3}{2}, 1-\frac{ia}{2}, 1+\frac{ia}{2}\end{matrix}\right)\quad [\operatorname{Re} a > 0].$

13. $\displaystyle\int_0^\infty e^{-ax} \sin^{2n} x\, H_{2n}\left(\dfrac{b}{\sin x}\right) dx = \dfrac{(2b)^{2n}}{a}\, {}_4F_2\left(\begin{matrix}-n, \frac{1}{2}-n, \frac{1}{2}, 1;\ -b^{-2}\\ 1-\frac{ia}{2}, 1+\frac{ia}{2}\end{matrix}\right)$

$[\operatorname{Re} a > 0].$

14. $\displaystyle\int_0^\infty e^{-ax}\sin^{2n+1}x\,H_{2n+1}\left(\frac{b}{\sin x}\right)dx$

$$= \frac{(2b)^{2n+1}}{a}\,{}_4F_2\left(\begin{array}{c}-n,\,-n-\frac{1}{2},\,\frac{1}{2},\,1\\ 1-\frac{ia}{2},\,1+\frac{ia}{2};\,-b^{-2}\end{array}\right) \qquad [\mathrm{Re}\,a > 0].$$

4.17.4. Integrals containing $H_n(z)$, erf(z) and erfc(z)

1. $\displaystyle\int_0^\infty \mathrm{erfc}\,(ax)H_{2n}(bx)\,dx = \frac{(-1)^n}{\sqrt{\pi}\,a}\frac{(2n)!}{\left(\frac{3}{2}\right)_n}P_n^{(1/2,\,-n-1/2)}\left(1-\frac{2b^2}{a^2}\right)$

$$[|\arg a| < \pi/4].$$

2. $\displaystyle\int_0^\infty \mathrm{erfc}\,(ax)H_{2n+1}(bx)\,dx = \frac{(-1)^n b}{2a^2}\frac{(2n+1)!}{(n+1)!}P_n^{(1,\,-n-1)}\left(1-\frac{2b^2}{a^2}\right)$

$$[|\arg a| < \pi/4].$$

3. $\displaystyle\int_0^\infty x\,\mathrm{erfc}\,(ax)H_{2n+1}(bx)\,dx$

$$= (-1)^n\frac{2b}{3\sqrt{\pi}\,a^3}\frac{(2n+1)!}{\left(\frac{5}{2}\right)_n}P_n^{(3/2,\,-n-1/2)}\left(1-\frac{2b^2}{a^2}\right) \qquad [|\arg a| < \pi/4].$$

4. $\displaystyle\int_0^\infty e^{-2x^2}\mathrm{erf}\,(ax)H_{2n+1}(x)\,dx = (-1)^n\frac{2^{n-1}n!\,a}{\sqrt{a^2+2}}P_n^{(1/2,\,-n-1)}\left(\frac{3a^2+2}{a^2+2}\right).$

4.17.5. Integrals containing $H_n(z)$ and $K_\nu(z)$

1. $\displaystyle\int_0^\infty K_0(ax)H_{2n}(bx)\,dx = (-1)^n(2n)!\pi\frac{2^{2n-1}b^{2n}}{a^{2n+1}}L_n^{-n-1/2}\left(-\frac{a^2}{4b^2}\right)$

$$[\mathrm{Re}\,a > 0].$$

2. $\displaystyle\int_0^\infty x\,K_1(ax)H_{2n}(bx)\,dx = (-1)^n(2n)!\pi\frac{2^{2n-1}b^{2n}}{a^{2n+2}}L_n^{-n-3/2}\left(-\frac{a^2}{4b^2}\right)$

$$[\mathrm{Re}\,a > 0].$$

3. $\displaystyle\int_0^\infty K_1(ax)H_{2n+1}(bx)\,dx = (-1)^n(2n+1)!\pi\frac{2^{2n}b^{2n+1}}{a^{2n+2}}L_n^{-n-1/2}\left(-\frac{a^2}{4b^2}\right)$

$$[\mathrm{Re}\,a > 0].$$

4.17.6. Integrals containing products of $H_n(z)$

1. $\displaystyle\int_0^a H_{2m+1}(b\sqrt{x})\, H_{2n+1}(b\sqrt{a-x})\, dx$

$$= \frac{(-2)^{m+n}\pi(2m+1)!!(2n+1)!!}{(m+n+1)(m+n+2)}\, a^2 b^2 L_{m+n}^2(ab^2) \quad [a>0].$$

2. $\displaystyle\int_0^a H_{2n+1}(b\sqrt{x})\, H_{2n+1}(ib\sqrt{a-x})\, dx$

$$= 2^{4n+1} i\, \Gamma^2\!\left(n+\tfrac{3}{2}\right) a^2 b^2 \, {}_1F_2\!\left(\begin{array}{c} -n;\ \dfrac{a^2 b^4}{4} \\ \dfrac{3}{2},\ 2 \end{array}\right) \quad [a>0].$$

3. $\displaystyle\int_0^a (a-x)^{-1/2}\, H_{2m+1}(b\sqrt{x})\, H_{2n}(b\sqrt{a-x})\, dx$

$$= (-2)^{m+n}\pi\, \frac{(2m+1)!!(2n-1)!!}{m+n+1}\, ab\, L_{m+n}^1(ab^2) \quad [a>0].$$

4. $\displaystyle\int_0^a x^{-1/2}(a-x)^{-1/2}\, H_{2m}(b\sqrt{x})\, H_{2n}(b\sqrt{a-x})\, dx$

$$= (-2)^{m+n}(2m-1)!!(2n-1)!!\, L_{m+n}(ab^2) \quad [a>0].$$

5. $\displaystyle\int_0^a x^{-1/2}(a-x)^{-1/2}\, H_{2n}(b\sqrt{x})\, H_{2n}(ib\sqrt{a-x})\, dx$

$$= 2^{4n}\, \Gamma^2\!\left(n+\tfrac{1}{2}\right)\, {}_1F_2\!\left(\begin{array}{c} -n;\ \dfrac{a^2 b^4}{4} \\ \dfrac{1}{2},\ 1 \end{array}\right) \quad [a>0].$$

6. $\displaystyle\int_0^a (a-x)^{-1/2}\, H_{2n+1}(b\sqrt{x})\, H_{2n}(ib\sqrt{a-x})\, dx$

$$= 2^{4n+1}\, \Gamma\!\left(n+\tfrac{1}{2}\right)\Gamma\!\left(n+\tfrac{3}{2}\right) ab\, {}_1F_2\!\left(\begin{array}{c} -n;\ \dfrac{a^2 b^4}{4} \\ 1,\ \dfrac{3}{2} \end{array}\right) \quad [a>0].$$

7. $\displaystyle\int_0^{m\pi} \left\{\begin{array}{c}\sin(ax)\\ \cos(ax)\end{array}\right\} H_{2n}(b\sqrt{\sin x})\, H_{2n}(ib\sqrt{\sin x})\, dx$

$$= \frac{2^{4n+1}}{a}\left(\tfrac{1}{2}\right)_n^2 \sin\frac{m\pi a}{2}\left\{\begin{array}{c}\sin(m\pi a/2)\\ \cos(m\pi a/2)\end{array}\right\} {}_4F_5\!\left(\begin{array}{c} -n,\, n+\dfrac{1}{2},\, \dfrac{1}{2},\, 1;\ \dfrac{b^4}{4} \\ \dfrac{1}{4},\, \dfrac{1}{2},\, \dfrac{3}{4},\, 1-\dfrac{a}{2},\, 1+\dfrac{a}{2} \end{array}\right).$$

8. $\displaystyle\int_0^{m\pi} \frac{1}{\sin x}\left\{\begin{array}{c}\sin(ax)\\ \cos(ax)\end{array}\right\} H_{2n+1}(b\sqrt{\sin x})\, H_{2n+1}(ib\sqrt{\sin x})\, dx$

$\displaystyle = 2^{4n+3} i \frac{b^2}{a}\left(\frac{3}{2}\right)_n^2 \sin\frac{m\pi a}{2}\left\{\begin{array}{c}\sin(m\pi a/2)\\ \cos(m\pi a/2)\end{array}\right\}{}_4F_5\!\left(\begin{array}{c}-n, n+\frac{3}{2}, \frac{1}{2}, 1;\ \frac{b^4}{4}\\ \frac{3}{2}, \frac{3}{4}, \frac{5}{4}, 1-\frac{a}{2}, 1+\frac{a}{2}\end{array}\right).$

9. $\displaystyle\int_0^{m\pi} e^{-ax} H_{2n}(b\sqrt{\sin x})\, H_{2n}(ib\sqrt{\sin x})\, dx$

$\displaystyle = (1-e^{-m\pi a})\frac{2^{4n}}{a}\left(\frac{1}{2}\right)_n^2 {}_4F_5\!\left(\begin{array}{c}-n, n+\frac{1}{2}, \frac{1}{2}, 1;\ \frac{b^4}{4}\\ \frac{1}{4}, \frac{1}{2}, \frac{3}{4}, 1-\frac{ia}{2}, 1+\frac{ia}{2}\end{array}\right).$

10. $\displaystyle\int_0^{m\pi} \frac{e^{-ax}}{\sin x} H_{2n+1}(b\sqrt{\sin x})\, H_{2n+1}(ib\sqrt{\sin x})\, dx$

$\displaystyle = (1-e^{-m\pi a}) 2^{4n+2} i\frac{b^2}{a}\left(\frac{3}{2}\right)_n^2 {}_4F_5\!\left(\begin{array}{c}-n, n+\frac{3}{2}, \frac{1}{2}, 1;\ \frac{b^4}{4}\\ \frac{3}{2}, \frac{3}{4}, \frac{5}{4}, 1-\frac{ia}{2}, 1+\frac{ia}{2}\end{array}\right).$

11. $\displaystyle\int_0^{\infty} e^{-ax} H_{2n}(b\sqrt{\sin x})\, H_{2n}(ib\sqrt{\sin x})\, dx$

$\displaystyle = \frac{2^{4n}}{a}\left(\frac{1}{2}\right)_n^2 {}_4F_5\!\left(\begin{array}{c}-n, n+\frac{1}{2}, \frac{1}{2}, 1;\ \frac{b^4}{4}\\ \frac{1}{4}, \frac{1}{2}, \frac{3}{4}, 1-\frac{ia}{2}, 1+\frac{ia}{2}\end{array}\right)\quad [\operatorname{Re} a > 0].$

12. $\displaystyle\int_0^{\infty} \frac{e^{-ax}}{\sin x} H_{2n+1}(b\sqrt{\sin x})\, H_{2n+1}(ib\sqrt{\sin x})\, dx$

$\displaystyle = 2^{4n+2} i\frac{b^2}{a}\left(\frac{3}{2}\right)_n^2 {}_4F_5\!\left(\begin{array}{c}-n, n+\frac{3}{2}, \frac{1}{2}, 1;\ \frac{b^4}{4}\\ \frac{3}{2}, \frac{3}{4}, \frac{5}{4}, 1-\frac{ia}{2}, 1+\frac{ia}{2}\end{array}\right)\quad [\operatorname{Re} a > 0].$

13. $\displaystyle\int_0^{\infty} \operatorname{erfc}(x) H_n(x) H_{n+1}(x)\, dx$

$\displaystyle = 2^{n-1} n! + [(-1)^n - 1]\frac{n!(n+1)!}{8}\left(\left[\frac{n+1}{2}\right]!\right)^{-2}.$

14. $\displaystyle\int_0^{\infty} \operatorname{erfc}(\sqrt{2}\, x) H_n(x) H_{n+1}(x)\, dx$

$\displaystyle = 2^{n-2}\frac{\left(\frac{3}{2}\right)_n}{n+1} + [(-1)^n - 1]\frac{n!(n+1)!}{8}\left(\left[\frac{n+1}{2}\right]!\right)^{-2}.$

15. $\displaystyle\int_0^\infty e^{-x^2} H_{2m}(x) H_n^2\left(\frac{x}{\sqrt{2}}\right) dx = \frac{2^{n-1}\sqrt{\pi}(2m)!(-n)_m \left(\frac{1}{2}\right)_n}{m!\left(\frac{1}{2}-n\right)_m}.$

16. $\displaystyle\int_0^\infty e^{-x^2} H_m^2(x) H_n^2(x)\, dx = 2^{m+n-1} m!\, n!\, \sqrt{\pi}\; {}_3F_2\!\left(\begin{matrix} -m,\, -n,\, \frac{1}{2} \\ 1,\, 1;\, 4 \end{matrix}\right).$

17. $\displaystyle\int_0^\infty e^{-2x^2} H_m^2(x) H_n^2(x)\, dx = \sqrt{\frac{\pi}{8}}\, (2m-1)!!\,(2n-1)!!\; {}_3F_2\!\left(\begin{matrix} -m,\, -n,\, \frac{1}{2};\, 1 \\ \frac{1}{2}-m,\, \frac{1}{2}-n \end{matrix}\right).$

18. $\displaystyle\int_0^\infty \frac{1}{x} e^{-2x^2} H_n^2(x) H_{n+1}^2(x)\, dx = 2^{2n} (n!)^2 \sum_{k=0}^{[n/2]} \frac{\left(\frac{1}{2}\right)_k^2}{(k!)^2}.$

4.18. The Laguerre Polynomials $L_n^\lambda(z)$

4.18.1. Integrals containing $L_n^\lambda(z)$ and algebraic functions

1. $\displaystyle\int_0^a x^{s-1}(a-x)^{t-1} L_n^\lambda(bx(a-x))\, dx = \frac{(\lambda+1)_n}{n!} a^{s+t-1} \mathrm{B}(s,t)$

$\times {}_3F_3\!\left(\begin{matrix} -n,\, s,\, t;\, \frac{a^2 b}{4} \\ \lambda+1,\, \frac{s+t}{2},\, \frac{s+t+1}{2} \end{matrix}\right) \qquad [\mathrm{Re}\, s,\, \mathrm{Re}\, t > 0].$

2. $\displaystyle\int_0^a x^\lambda (a-x)^\lambda L_n^\lambda(bx(a-x))\, dx$

$= \pi^{1/2} \left(\frac{a}{2}\right)^{2\lambda+1} \frac{\Gamma(n+\lambda+1)}{\Gamma\left(n+\lambda+\frac{3}{2}\right)} L_n^{\lambda+1/2}\!\left(\frac{a^2 b}{4}\right) \qquad [\mathrm{Re}\,\lambda > -1].$

3. $\displaystyle\int_0^a L_n(bx(a-x))\, dx = (-1)^n \frac{n!}{(2n+1)!\, b^{1/2}} H_{2n+1}\!\left(\frac{a\sqrt{b}}{2}\right).$

4. $\displaystyle\int_0^a x^s (a-x)^{s+1/2} L_n^\lambda\!\left(b\sqrt{x(a-x)}\right) dx = \frac{\sqrt{\pi}\, a^{2s+3/2}}{2^{2s+1}} \frac{(\lambda+1)_n \Gamma(2s+2)}{n!\, \Gamma\left(2s+\frac{5}{2}\right)}$

$\times {}_2F_2\!\left(\begin{matrix} -n,\, 2s+2;\, \frac{ab}{2} \\ \lambda+1,\, 2s+\frac{5}{2} \end{matrix}\right) \qquad [a > 0;\; \mathrm{Re}\, s > -1].$

5. $\displaystyle\int_0^a x^{-1/2} L_n\!\left(b\sqrt{x(a-x)}\right) dx = (-1)^n \frac{n!}{(2n+1)!} \sqrt{\frac{2}{b}}\, H_{2n+1}\!\left(\sqrt{\frac{ab}{2}}\right)$

$[a > 0].$

4.18.2. Integrals containing $L_n^\lambda(z)$ and trigonometric functions

1. $$\int_0^{m\pi} \begin{Bmatrix} \sin(ax) \\ \cos(ax) \end{Bmatrix} L_n^\lambda(b\sin^2 x)\,dx = \frac{2}{a}\sin\frac{m\pi a}{2}$$
$$\times \begin{Bmatrix} \sin(m\pi a/2) \\ \cos(m\pi a/2) \end{Bmatrix} \frac{(\lambda+1)_n}{n!}\,_3F_3\left(\begin{array}{c} -n,\,\frac{1}{2},\,1;\,b \\ \lambda+1,\,1-\frac{a}{2},\,1+\frac{a}{2} \end{array}\right).$$

2. $$\int_0^\pi \sin^\mu x \begin{Bmatrix} \sin(ax) \\ \cos(ax) \end{Bmatrix} L_n^\lambda(b\sin^2 x)\,dx = \frac{2^{-\mu}\pi\Gamma(\mu+1)(\lambda+1)_n}{n!\,\Gamma\left(\frac{\mu-a}{2}+1\right)\Gamma\left(\frac{\mu+a}{2}+1\right)}$$
$$\times \begin{Bmatrix} \sin(a\pi/2) \\ \cos(a\pi/2) \end{Bmatrix} {}_3F_3\left(\begin{array}{c} -n,\,\frac{\mu+1}{2},\,\frac{\mu}{2}+1;\,b \\ \frac{\mu-a}{2}+1,\,\frac{\mu+a}{2}+1,\,\lambda+1 \end{array}\right).$$

3. $$\int_0^{m\pi} \begin{Bmatrix} \sin(ax) \\ \cos(ax) \end{Bmatrix} \sin^{2n} x\, L_n^\lambda\left(\frac{b}{\sin^2 x}\right) dx = \frac{2(-b)^n}{n!\,a}\sin\frac{m\pi a}{2}$$
$$\times \begin{Bmatrix} \sin(m\pi a/2) \\ \cos(m\pi a/2) \end{Bmatrix} {}_4F_2\left(\begin{array}{c} -n,\,-\lambda-n,\,\frac{1}{2},\,1 \\ 1-\frac{a}{2},\,1+\frac{a}{2};\,-\frac{1}{b} \end{array}\right).$$

4. $$\int_0^{m\pi} e^{-ax} L_n^\lambda(b\sin^2 x)\,dx = (1-e^{-m\pi a})\frac{(\lambda+1)_n}{n!\,a}\,_3F_3\left(\begin{array}{c} -n,\,\frac{1}{2},\,1;\,b \\ \lambda+1,\,1-\frac{ia}{2},\,1+\frac{ia}{2} \end{array}\right).$$

5. $$\int_0^{m\pi} e^{-ax} \sin^{2n} x\, L_n^\lambda\left(\frac{b}{\sin^2 x}\right) dx$$
$$= \frac{(-b)^n}{n!\,a}\left(1-e^{-m\pi a}\right) {}_4F_2\left(\begin{array}{c} -n,\,-\lambda-n,\,\frac{1}{2},\,1 \\ 1-\frac{ia}{2},\,1+\frac{ia}{2};\,-\frac{1}{b} \end{array}\right).$$

6. $$\int_0^\infty e^{-ax} L_n^\lambda(b\sin^2 x)\,dx = \frac{(\lambda+1)_n}{n!\,a}\,_3F_3\left(\begin{array}{c} -n,\,\frac{1}{2},\,1;\,b \\ \lambda+1,\,1-\frac{ia}{2},\,1+\frac{ia}{2} \end{array}\right) \quad [\operatorname{Re} a > 0].$$

7. $$\int_0^\infty e^{-ax} \sin^{2n} x\, L_n^\lambda\left(\frac{b}{\sin^2 x}\right) dx = \frac{(-b)^n}{n!\,a}\,_4F_2\left(\begin{array}{c} -n,\,-\lambda-n,\,\frac{1}{2},\,1 \\ 1-\frac{ia}{2},\,1+\frac{ia}{2};\,-\frac{1}{b} \end{array}\right).$$
$$[\operatorname{Re} a > 0].$$

8. $$\int_0^a x^{s-1} \ln\frac{\sqrt{a}+\sqrt{a-x}}{\sqrt{x}} L_n^\lambda(bx)\,dx$$
$$= \frac{\pi^{1/2} a^s (\lambda+1)_n \Gamma(s)}{n!\,2s\,\Gamma\left(s+\frac{1}{2}\right)}\,_3F_3\left(\begin{array}{c} -n,\,s,\,s;\,ab \\ \lambda+1,\,s+\frac{1}{2},\,s+1 \end{array}\right) \quad [a > 0;\,\operatorname{Re} s > -1/2].$$

4.18.3. Integrals containing $L_n^\lambda(z)$ and erfc(z)

1. $\displaystyle\int_0^\infty \mathrm{erfc}\,(a\sqrt{x})\,L_n(bx)\,dx = \frac{1}{2(n+1)a^2}\,P_n^{(1,\,-n-1/2)}\left(1-\frac{2b}{a^2}\right)$ $\qquad$ [Re $a > 0$].

4.18.4. Integrals containing products of $L_n^\lambda(z)$

1. $\displaystyle\int_0^a x^\lambda(a-x)^\mu L_m^\lambda(bx)L_n^\mu(b(a-x))\,dx = \frac{(m+n)!}{m!\,n!}$

$\qquad\times \mathrm{B}(\lambda+m+1,\mu+n+1)a^{\lambda+\mu+1}L_{m+n}^{\lambda+\mu+1}(ab)$ $\qquad$ [Re λ, Re $\mu > -1$].

2. $\displaystyle\int_0^a x^\lambda(a-x)^\mu L_n^\lambda(bx)L_n^\mu(-b(a-x))\,dx$

$= \dfrac{a^{\lambda+\mu+1}}{(n!)^2}\,\dfrac{\Gamma(\lambda+n+1)\Gamma(\mu+n+1)}{\Gamma(\lambda+\mu+2)}\,{}_1F_2\!\left(\begin{array}{c}-n;\ \dfrac{a^2b^2}{4}\\[2pt]\dfrac{\lambda+\mu}{2}+1,\ \dfrac{\lambda+\mu+3}{2}\end{array}\right)$

$\qquad$ [Re λ, Re $\mu > -1$].

3. $\displaystyle\int_0^{m\pi}\left\{\begin{array}{c}\sin(ax)\\ \cos(ax)\end{array}\right\}L_n^\lambda(-b\sin x)L_n^\lambda(b\sin x)\,dx = \frac{2}{a}\sin\frac{m\pi a}{2}\left\{\begin{array}{c}\sin(m\pi a/2)\\ \cos(m\pi a/2)\end{array}\right\}$

$\qquad\times\left[\dfrac{(\nu+1)_n}{n!}\right]^2{}_4F_5\!\left(\begin{array}{c}-n,\ \lambda+n+1,\ \frac{1}{2},\ 1;\ \dfrac{b^2}{4}\\[2pt]\dfrac{\lambda+1}{2},\ \dfrac{\lambda}{2}+1,\ \lambda+1,\ 1-\dfrac{a}{2},\ 1+\dfrac{a}{2}\end{array}\right).$

4. $\displaystyle\int_0^{m\pi} e^{-ax}L_n^\lambda(-b\sin x)\,L_n^\lambda(b\sin x)\,dx$

$= \dfrac{(\lambda+1)_n^2}{(n!)^2\,a}(1-e^{-m\pi a})\,{}_4F_5\!\left(\begin{array}{c}-n,\ \lambda+n+1,\ \frac{1}{2},\ 1;\ \dfrac{b^2}{4}\\[2pt]\dfrac{\lambda+1}{2},\ \dfrac{\lambda}{2}+1,\ \lambda+1,\ 1-\dfrac{ia}{2},\ 1+\dfrac{ia}{2}\end{array}\right).$

5. $\displaystyle\int_0^\infty e^{-ax}L_n^\lambda(-b\sin x)\,L_n^\lambda(b\sin x)\,dx$

$= \dfrac{(\lambda+1)_n^2}{(n!)^2\,a}\,{}_4F_5\!\left(\begin{array}{c}-n,\ \lambda+n+1,\ \frac{1}{2},\ 1;\ \dfrac{b^2}{4}\\[2pt]\dfrac{\lambda+1}{2},\ \dfrac{\lambda}{2}+1,\ \lambda+1,\ 1-\dfrac{ia}{2},\ 1+\dfrac{ia}{2}\end{array}\right)$ $\qquad$ [Re $a > 0$].

6. $\displaystyle\int_0^\infty x^{2\lambda} e^{-2x} \left[L_m^\lambda(x)\right]^2 \left[L_n^\lambda(x)\right]^2 dx = \frac{2^{-2\lambda-1}}{(m!\,n!)^2} \Gamma(2\lambda+1) \left(\frac{1}{2}\right)_m \left(\frac{1}{2}\right)_n$

$\displaystyle\times (\lambda+1)_m (\lambda+1)_n \, {}_4F_3\!\left(\begin{array}{c} -m,\, -n,\, \lambda+\frac{1}{2},\, \frac{1}{2};\, 1 \\ \frac{1}{2}-m,\, \frac{1}{2}-n,\, \lambda+1 \end{array}\right)$ $\quad[\operatorname{Re}\lambda > -1/2].$

4.19. The Gegenbauer Polynomials $C_n^\lambda(z)$

4.19.1. Integrals containing $C_n^\lambda(z)$ and algebraic functions

1. $\displaystyle\int_0^a x^{-1/2}(a-x)^{-1/2} C_{2n}^\lambda\!\left(b\sqrt{x(a-x)}\right) dx$

$\displaystyle = (-1)^n \pi \frac{(\lambda)_n}{n!} P_n^{(0,\lambda-1)}\!\left(1 - \frac{a^2 b^2}{2}\right).$

2. $\displaystyle\int_0^a C_{2n+1}^\lambda\!\left(b\sqrt{x(a-x)}\right) dx = (-1)^n \pi \frac{(\lambda)_{n+1}}{4(n+1)!} a^2 b P_n^{(1,\lambda-1)}\!\left(1 - \frac{a^2 b^2}{2}\right).$

4.19.2. Integrals containing $C_n^\lambda(z)$ and trigonometric functions

1. $\displaystyle\int_0^\pi \sin^\mu x \left\{\begin{array}{c}\sin(ax)\\ \cos(ax)\end{array}\right\} C_{2n}^\lambda(b\sin x) \, dx$

$\displaystyle = (-1)^n \frac{2^{-\mu} \pi \Gamma(\mu+1)(\lambda)_n}{n!\,\Gamma\!\left(\frac{\mu-a}{2}+1\right)\Gamma\!\left(\frac{\mu+a}{2}+1\right)} \left\{\begin{array}{c}\sin(a\pi/2)\\ \cos(a\pi/2)\end{array}\right\}$

$\displaystyle \times {}_4F_3\!\left(\begin{array}{c} -n,\, \lambda+n,\, \frac{\mu+1}{2},\, \frac{\mu}{2}+1;\, b^2 \\ \frac{1}{2},\, \frac{\mu-a}{2}+1,\, \frac{\mu+a}{2}+1 \end{array}\right)$ $\quad[\operatorname{Re}\mu > -1].$

2. $\displaystyle\int_0^\pi \sin^\mu x \left\{\begin{array}{c}\sin(ax)\\ \cos(ax)\end{array}\right\} C_{2n+1}^\lambda(b\sin x) \, dx$

$\displaystyle = (-1)^n \frac{2^{-\mu} \pi b \, \Gamma(\mu+2)(\lambda)_{n+1}}{n!\,\Gamma\!\left(\frac{\mu-a+3}{2}\right)\Gamma\!\left(\frac{\mu+a+3}{2}\right)} \left\{\begin{array}{c}\sin(a\pi/2)\\ \cos(a\pi/2)\end{array}\right\}$

$\displaystyle \times {}_4F_3\!\left(\begin{array}{c} -n,\, \lambda+n+1,\, \frac{\mu}{2}+1,\, \frac{\mu+3}{2};\, b^2 \\ \frac{3}{2},\, \frac{\mu-a+3}{2},\, \frac{\mu+a+3}{2} \end{array}\right)$ $\quad[\operatorname{Re}\mu > -2].$

3. $\displaystyle\int_0^{m\pi} \begin{Bmatrix} \sin(ax) \\ \cos(ax) \end{Bmatrix} C_{2n}^{\lambda}(b\sin x)\, dx$

$\displaystyle = (-1)^n \frac{2(\lambda)_n}{n!\, a} \sin\frac{m\pi a}{2} \begin{Bmatrix} \sin(m\pi a/2) \\ \cos(m\pi a/2) \end{Bmatrix} {}_3F_2\!\left(\begin{matrix} -n,\ \lambda+n,\ 1;\ b^2 \\ 1-\frac{a}{2},\ 1+\frac{a}{2} \end{matrix}\right).$

4. $\displaystyle\int_0^{m\pi} \frac{1}{\sin x}\begin{Bmatrix} \sin(ax) \\ \cos(ax) \end{Bmatrix} C_{2n+1}^{\lambda}(b\sin x)\, dx$

$\displaystyle = (-1)^n \frac{4(\lambda)_{n+1}}{n!\, a} \sin\frac{m\pi a}{2} \begin{Bmatrix} \sin(m\pi a/2) \\ \cos(m\pi a/2) \end{Bmatrix} {}_3F_2\!\left(\begin{matrix} -n,\ \lambda+n+1,\ 1;\ b^2 \\ 1-\frac{a}{2},\ 1+\frac{a}{2} \end{matrix}\right).$

5. $\displaystyle\int_0^{m\pi} \frac{1}{\cos x}\begin{Bmatrix} \sin(ax) \\ \cos(ax) \end{Bmatrix} C_{2n+1}^{\lambda}(\cos x)\, dx$

$\displaystyle = 2\sin\frac{m\pi a}{2} \begin{Bmatrix} \sin(m\pi a/2) \\ \cos(m\pi a/2) \end{Bmatrix} \frac{(2\lambda)_{2n+1}}{(2n+1)!\, a}\, {}_4F_3\!\left(\begin{matrix} -n,\ \lambda+n+1,\ \frac{1}{2},\ 1;\ 1 \\ \lambda+\frac{1}{2},\ 1-\frac{a}{2},\ 1+\frac{a}{2} \end{matrix}\right).$

6. $\displaystyle\int_0^{m\pi} \begin{Bmatrix} \sin(ax) \\ \cos(ax) \end{Bmatrix} C_{2n}^{\lambda}\!\left(\sqrt{1+b\sin^2 x}\right) dx$

$\displaystyle = 2\sin\frac{m\pi a}{2} \begin{Bmatrix} \sin(m\pi a/2) \\ \cos(m\pi a/2) \end{Bmatrix} \frac{(2\lambda)_{2n}}{(2n)!\, a}\, {}_4F_3\!\left(\begin{matrix} -n,\ \lambda+n,\ \frac{1}{2},\ 1;\ -b \\ \lambda+\frac{1}{2},\ 1-\frac{a}{2},\ 1+\frac{a}{2} \end{matrix}\right).$

7. $\displaystyle\int_0^{m\pi} \frac{1}{\sqrt{1+b\sin^2 x}}\begin{Bmatrix} \sin(ax) \\ \cos(ax) \end{Bmatrix} C_{2n+1}^{\lambda}\!\left(\sqrt{1+b\sin^2 x}\right) dx$

$\displaystyle = 2\sin\frac{m\pi a}{2} \begin{Bmatrix} \sin(m\pi a/2) \\ \cos(m\pi a/2) \end{Bmatrix} \frac{(2\lambda)_{2n+1}}{(2n+1)!\, a}\, {}_4F_3\!\left(\begin{matrix} -n,\ \lambda+n+1,\ \frac{1}{2},\ 1;\ -b \\ \lambda+\frac{1}{2},\ 1-\frac{a}{2},\ 1+\frac{a}{2} \end{matrix}\right).$

8. $\displaystyle\int_0^{m\pi} \begin{Bmatrix} \sin(ax) \\ \cos(ax) \end{Bmatrix} (1+b\sin^2 x)^{n/2} C_n^{\lambda}\!\left(\frac{1}{\sqrt{1+b\sin^2 x}}\right) dx$

$\displaystyle = 2\sin\frac{m\pi a}{2} \begin{Bmatrix} \sin(m\pi a/2) \\ \cos(m\pi a/2) \end{Bmatrix} \frac{(2\lambda)_n}{n!\, a}\, {}_4F_3\!\left(\begin{matrix} -\frac{n}{2},\ \frac{1-n}{2},\ \frac{1}{2},\ 1;\ -b \\ \lambda+\frac{1}{2},\ 1-\frac{a}{2},\ 1+\frac{a}{2} \end{matrix}\right).$

9. $\displaystyle\int_0^{m\pi} \begin{Bmatrix} \sin(ax) \\ \cos(ax) \end{Bmatrix} \sin^n x\, C_n^{\lambda}\!\left(\frac{b}{\sin x}\right) dx$

$\displaystyle = 2\sin\frac{m\pi a}{2} \begin{Bmatrix} \sin(m\pi a/2) \\ \cos(m\pi a/2) \end{Bmatrix} \frac{(\lambda)_n}{n!\, a} (2b)^n\, {}_4F_3\!\left(\begin{matrix} -\frac{n}{2},\ \frac{1-n}{2},\ \frac{1}{2},\ 1;\ b^{-2} \\ 1-\lambda-n,\ 1-\frac{a}{2},\ 1+\frac{a}{2} \end{matrix}\right).$

4.19. The Gegenbauer Polynomials $C_n^\lambda(z)$

10. $\displaystyle\int_0^{m\pi} e^{-ax} C_{2n}^\lambda(b\sin x)\, dx$

$$= (-1)^n \frac{(\lambda)_n}{n!\,a} \left(1 - e^{-m\pi a}\right) {}_3F_2\!\left(\begin{matrix} -n,\, \lambda+n,\, 1;\, b^2 \\ 1 - \dfrac{ia}{2},\, 1 + \dfrac{ia}{2} \end{matrix}\right).$$

11. $\displaystyle\int_0^{m\pi} \frac{e^{-ax}}{\sin x} C_{2n+1}^\lambda(b\sin x)\, dx$

$$= 2(-1)^n \frac{b(\lambda)_{n+1}}{n!\,a} \left(1 - e^{-m\pi a}\right) {}_4F_3\!\left(\begin{matrix} -n,\, \lambda+n+1,\, \dfrac{1}{2},\, 1;\, b^2 \\ \dfrac{3}{2},\, 1 - \dfrac{ia}{2},\, 1 + \dfrac{ia}{2} \end{matrix}\right).$$

12. $\displaystyle\int_0^{m\pi} e^{-ax} C_{2n}^\lambda(\cos x)\, dx$

$$= \frac{(2\lambda)_n}{(2n)!\,a} \left(1 - e^{-m\pi a}\right) {}_4F_3\!\left(\begin{matrix} -n,\, \lambda+n,\, \dfrac{1}{2},\, 1;\, 1 \\ \lambda+\dfrac{1}{2},\, 1 - \dfrac{ia}{2},\, 1 + \dfrac{ia}{2} \end{matrix}\right).$$

13. $\displaystyle\int_0^{m\pi} \frac{e^{-ax}}{\cos x} C_{2n+1}^\lambda(\cos x)\, dx$

$$= \left(1 - e^{-m\pi a}\right) \frac{(2\lambda)_{2n+1}}{(2n+1)!\,a}\, {}_4F_3\!\left(\begin{matrix} -n,\, \lambda+n+1,\, \dfrac{1}{2},\, 1;\, 1 \\ \lambda+\dfrac{1}{2},\, 1 - \dfrac{ia}{2},\, 1 + \dfrac{ia}{2} \end{matrix}\right).$$

14. $\displaystyle\int_0^{m\pi} e^{-ax} \sin^n x\, C_n^\lambda\!\left(\frac{b}{\sin x}\right) dx$

$$= \frac{(\lambda)_n}{n!\,a}(2b)^n \left(1 - e^{-m\pi a}\right) {}_4F_3\!\left(\begin{matrix} -\dfrac{n}{2},\, \dfrac{1-n}{2},\, \dfrac{1}{2},\, 1;\, b^{-2} \\ 1-\lambda-n,\, 1 - \dfrac{ia}{2},\, 1 + \dfrac{ia}{2} \end{matrix}\right).$$

15. $\displaystyle\int_0^{m\pi} e^{-ax} C_n^\lambda(1 + b\sin^2 x)\, dx$

$$= \frac{(2\lambda)_n}{n!\,a} \left(1 - e^{-m\pi a}\right) {}_4F_3\!\left(\begin{matrix} -n,\, 2\lambda+n,\, \dfrac{1}{2},\, 1;\, -\dfrac{b}{2} \\ \lambda+\dfrac{1}{2},\, 1 - \dfrac{ia}{2},\, 1 + \dfrac{ia}{2} \end{matrix}\right).$$

16. $\displaystyle\int_0^{m\pi} e^{-ax} C_{2n}^\lambda\!\left(\sqrt{1 + b\sin^2 x}\right) dx$

$$= \frac{(2\lambda)_{2n}}{(2n)!\,a} \left(1 - e^{-m\pi a}\right) {}_4F_3\!\left(\begin{matrix} -n,\, \lambda+n,\, \dfrac{1}{2},\, 1;\, -b \\ \lambda+\dfrac{1}{2},\, 1 - \dfrac{ia}{2},\, 1 + \dfrac{ia}{2} \end{matrix}\right).$$

17. $\displaystyle\int_0^{m\pi} \frac{e^{-ax}}{\sqrt{1+b\sin^2 x}} C_{2n+1}^\lambda\left(\sqrt{1+b\sin^2 x}\right) dx$

$\displaystyle = \frac{(2\lambda)_{2n+1}}{(2n+1)!\,a}\left(1 - e^{-m\pi a}\right) {}_4F_3\left(\begin{array}{c} -n, \lambda+n+1, \frac{1}{2}, 1; -b \\ \lambda+\frac{1}{2}, 1-\frac{ia}{2}, 1+\frac{ia}{2} \end{array}\right).$

18. $\displaystyle\int_0^{m\pi} e^{-ax}(1+b\sin^2 x)^{n/2} C_n^\lambda\left(\frac{1}{\sqrt{1+b\sin^2 x}}\right) dx$

$\displaystyle = \frac{(2\lambda)_n}{n!\,a}\left(1 - e^{-m\pi a}\right) {}_4F_3\left(\begin{array}{c} -\frac{n}{2}, \frac{1-n}{2}, \frac{1}{2}, 1; -b \\ \lambda+\frac{1}{2}, 1-\frac{ia}{2}, 1+\frac{ia}{2} \end{array}\right).$

19. $\displaystyle\int_0^\infty e^{-ax} C_{2n}^\lambda(b\sin x)\, dx = (-1)^n \frac{(\lambda)_n}{n!\,a} {}_3F_2\left(\begin{array}{c} -n, \lambda+n, 1; b^2 \\ 1-\frac{ia}{2}, 1+\frac{ia}{2} \end{array}\right)$ [Re $a > 0$].

20. $\displaystyle\int_0^\infty e^{-ax} C_{2n+1}^\lambda(b\sin x)\, dx$

$\displaystyle = (-1)^n \frac{2b(\lambda)_{n+1}}{n!\,(a^2+1)} {}_3F_2\left(\begin{array}{c} -n, \lambda+n+1, 1; b^2 \\ \frac{3-ia}{2}, \frac{3+ia}{2} \end{array}\right)$ [Re $a > 0$].

21. $\displaystyle\int_0^\infty \frac{e^{-ax}}{\sin x} C_{2n+1}^\lambda(b\sin x)\, dx$

$\displaystyle = 2(-1)^n \frac{(\lambda)_{n+1} b}{n!\,a} {}_4F_3\left(\begin{array}{c} -n, \lambda+n+1, \frac{1}{2}, 1; b^2 \\ \frac{3}{2}, 1-\frac{ia}{2}, 1+\frac{ia}{2} \end{array}\right)$ [Re $a > 0$].

22. $\displaystyle\int_0^\infty e^{-ax} C_{2n}^\lambda(\cos x)\, dx = \frac{(2\lambda)_{2n}}{(2n)!\,a} {}_4F_3\left(\begin{array}{c} -n, \lambda+n, \frac{1}{2}, 1; 1 \\ \lambda+\frac{1}{2}, 1-\frac{ia}{2}, 1+\frac{ia}{2} \end{array}\right)$ [Re $a > 0$].

23. $\displaystyle\int_0^\infty \frac{e^{-ax}}{\cos x} C_{2n+1}^\lambda(\cos x)\, dx = \frac{(2\lambda)_{2n+1}}{(2n+1)!\,a} {}_4F_3\left(\begin{array}{c} -n, \lambda+n+1, \frac{1}{2}, 1; 1 \\ \lambda+\frac{1}{2}, 1-\frac{ia}{2}, 1+\frac{ia}{2} \end{array}\right)$

[Re $a > 0$].

24. $\displaystyle\int_0^\infty e^{-ax} \sin^n x \, C_n^\lambda\left(\frac{b}{\sin x}\right) dx$

$\displaystyle = \frac{(\lambda)_n}{n!\,a}(2b)^n {}_4F_3\left(\begin{array}{c} -\frac{n}{2}, \frac{1-n}{2}, \frac{1}{2}, 1; b^{-2} \\ 1-\lambda-n, 1-\frac{ia}{2}, 1+\frac{ia}{2} \end{array}\right)$ [Re $a > 0$].

4.19. The Gegenbauer Polynomials $C_n^\lambda(z)$

25. $\displaystyle\int_0^\infty e^{-ax} C_n^\lambda(1 + b\sin^2 x)\, dx = \frac{(2\lambda)_n}{n!\, a}\, {}_4F_3\left(\begin{array}{c} -n,\ 2\lambda + n,\ \frac{1}{2},\ 1;\ -\frac{b}{2} \\ \lambda + \frac{1}{2},\ 1 - \frac{ia}{2},\ 1 + \frac{ia}{2} \end{array}\right)$

[Re $a > 0$].

26. $\displaystyle\int_0^\infty e^{-ax} C_{2n}^\lambda\left(\sqrt{1 + b\sin^2 x}\right) dx = \frac{(2\lambda)_{2n}}{(2n)!\, a}\, {}_4F_3\left(\begin{array}{c} -n,\ \lambda + n,\ \frac{1}{2},\ 1;\ -b \\ \lambda + \frac{1}{2},\ 1 - \frac{ia}{2},\ 1 + \frac{ia}{2} \end{array}\right)$

[Re $a > 0$].

27. $\displaystyle\int_0^\infty e^{-ax}(1 + b\sin^2 x)^{n/2} C_n^\lambda\left(\frac{1}{\sqrt{1 + b\sin^2 x}}\right) dx$

$= \dfrac{(2\lambda)_n}{n!\, a}\, {}_4F_3\left(\begin{array}{c} -\frac{n}{2},\ \frac{1-n}{2},\ \frac{1}{2},\ 1;\ -b \\ \lambda + \frac{1}{2},\ 1 - \frac{ia}{2},\ 1 + \frac{ia}{2} \end{array}\right)$ [Re $a > 0$].

4.19.3. Integrals containing products of $C_n^\lambda(z)$

1. $\displaystyle\int_0^{\pi/2} \cos^\nu x \cos(ax) \left[C_n^\lambda\left(\sqrt{1 + b\cos^2 x}\right)\right]^2 dx$

$= \dfrac{2^{-\nu-1}\pi \Gamma(\nu + 1)(2\lambda)_n^2}{(n!)^2\, \Gamma\!\left(\frac{\nu-a}{2} + 1\right)\Gamma\!\left(\frac{\nu+a}{2} + 1\right)}\, {}_5F_4\left(\begin{array}{c} -n,\ \lambda,\ 2\lambda + n,\ \frac{\nu+1}{2},\ 1 + \frac{\nu}{2};\ -b \\ \lambda + \frac{1}{2},\ 2\lambda,\ 1 + \frac{\nu-a}{2},\ 1 + \frac{\nu+a}{2} \end{array}\right)$

[Re $\nu > -1$].

2. $\displaystyle\int_0^\pi \sin^\nu x \left\{\begin{array}{c}\sin(ax) \\ \cos(ax)\end{array}\right\}\left[C_n^\lambda\left(\sqrt{1 + b\sin^2 x}\right)\right]^2 dx$

$= \dfrac{2^{-\nu}\pi\Gamma(\nu+1)(2\lambda)_n^2}{(n!)^2\, \Gamma\!\left(\frac{\nu-a}{2}+1\right)\Gamma\!\left(\frac{\nu+a}{2}+1\right)} \left\{\begin{array}{c}\sin(a\pi/2) \\ \cos(a\pi/2)\end{array}\right\}$

$\times {}_5F_4\left(\begin{array}{c} -n,\ \lambda,\ 2\lambda + n,\ \frac{\nu+1}{2},\ 1 + \frac{\nu}{2};\ -b \\ \lambda + \frac{1}{2},\ 2\lambda,\ 1 + \frac{\nu-a}{2},\ 1 + \frac{\nu+a}{2} \end{array}\right)$ [Re $\nu > -1$].

3. $\displaystyle\int_0^{m\pi} \left\{\begin{array}{c}\sin(ax) \\ \cos(ax)\end{array}\right\} [C_n^\lambda(\cos x)]^2\, dx = \frac{2}{a}\sin\frac{m\pi a}{2}\left\{\begin{array}{c}\sin(m\pi a/2) \\ \cos(m\pi a/2)\end{array}\right\}$

$\times \left[\dfrac{(2\lambda)_n}{n!}\right]^2 {}_5F_4\left(\begin{array}{c} -n,\ n + 2\lambda,\ \lambda,\ \frac{1}{2},\ 1;\ 1 \\ 2\lambda,\ \lambda + \frac{1}{2},\ 1 - \frac{a}{2},\ 1 + \frac{a}{2} \end{array}\right).$

4. $\displaystyle\int_0^{m\pi}\begin{Bmatrix}\sin(ax)\\\cos(ax)\end{Bmatrix}\left[C_n^\lambda\left(\sqrt{1+b\sin^2 x}\right)\right]^2 dx$

$\displaystyle = \frac{2}{a}\sin\frac{m\pi a}{2}\begin{Bmatrix}\sin(m\pi a/2)\\\cos(m\pi a/2)\end{Bmatrix}\left[\frac{(2\lambda)_n}{n!}\right]^2 {}_5F_4\left(\begin{array}{c}-n,\,n+2\lambda,\,\lambda,\,\frac{1}{2},\,1;\,-b\\2\lambda,\,\lambda+\frac{1}{2},\,1-\frac{a}{2},\,1+\frac{a}{2}\end{array}\right).$

5. $\displaystyle\int_0^{m\pi} e^{-ax}\left[C_n^\lambda(\cos x)\right]^2 dx$

$\displaystyle = \frac{1-e^{-m\pi a}}{a}\left[\frac{(2\lambda)_n}{n!}\right]^2 {}_5F_4\left(\begin{array}{c}-n,\,\lambda,\,2\lambda+n,\,\frac{1}{2},\,1;\,1\\\lambda+\frac{1}{2},\,2\lambda,\,1-\frac{ia}{2},\,1+\frac{ia}{2}\end{array}\right).$

6. $\displaystyle\int_0^{m\pi} e^{-ax}\left[C_n^\lambda\left(\sqrt{1+b\sin^2 x}\right)\right]^2 dx$

$\displaystyle = \frac{1-e^{-m\pi a}}{a}\left[\frac{(2\lambda)_n}{n!}\right]^2 {}_5F_4\left(\begin{array}{c}-n,\,n+2\lambda,\,\lambda,\,\frac{1}{2},\,1;\,-b\\\lambda+\frac{1}{2},\,2\lambda,\,1-\frac{ia}{2},\,1+\frac{ia}{2}\end{array}\right).$

7. $\displaystyle\int_0^\infty e^{-ax}\left[C_n^\lambda(\cos x)\right]^2 dx = \frac{1}{a}\left[\frac{(2\lambda)_n}{n!}\right]^2 {}_5F_4\left(\begin{array}{c}-n,\,n+2\lambda,\,\lambda,\,\frac{1}{2},\,1;\,1\\\lambda+\frac{1}{2},\,2\lambda,\,1-\frac{ia}{2},\,1+\frac{ia}{2}\end{array}\right)$

[Re $a > 0$].

8. $\displaystyle\int_0^\infty e^{-ax}\left[C_n^\lambda\left(\sqrt{1+b\sin^2 x}\right)\right]^2 dx$

$\displaystyle = \frac{1}{a}\left[\frac{(2\lambda)_n}{n!}\right]^2 {}_5F_4\left(\begin{array}{c}-n,\,n+2\lambda,\,\lambda,\,\frac{1}{2},\,1;\,-b\\\lambda+\frac{1}{2},\,2\lambda,\,1-\frac{ia}{2},\,1+\frac{ia}{2}\end{array}\right)$ [Re $a > 0$].

4.20. The Jacobi Polynomials $P_n^{(\rho,\sigma)}(z)$

4.20.1. Integrals containing $P_n^{(\rho,\sigma)}(z)$ and algebraic functions

1. $\displaystyle\int_0^a x^{s-1}(a-x)^{t-1} P_n^{(\rho,\sigma)}(1+bx(a-x))\,dx = \frac{(\rho+1)_n}{n!}\,\mathrm{B}(s,t)$

$\displaystyle\times a^{s+t-1}\,{}_4F_3\left(\begin{array}{c}-n,\,\rho+\sigma+n+1,\,s,\,t\\\rho+1,\,\frac{s+t}{2},\,\frac{s+t+1}{2};\,-\frac{a^2 b}{8}\end{array}\right)$ [Re s, Re $t > 0$].

2. $\displaystyle\int_0^a x^s(a-x)^{s+1/2} P_n^{(\rho,\sigma)}\left(1+b\sqrt{x(a-x)}\right) dx = 2^{-2s-1}\frac{(\rho+1)_n}{n!}$

$\displaystyle \times B\left(\frac{1}{2}, 2s+2\right) a^{2s+3/2}\, {}_3F_2\left(\begin{array}{c}-n, \rho+\sigma+n+1, 2s+2\\ \rho+1, 2s+\frac{5}{2};\ -\frac{ab}{4}\end{array}\right)$ [Re $s > -1$].

4.20.2. Integrals containing $P_n^{(\rho,\sigma)}(z)$ and trigonometric functions

1. $\displaystyle\int_0^{2m\pi} \left\{\begin{array}{c}\sin(ax)\\ \cos(ax)\end{array}\right\} P_n^{(\rho,\sigma)}(\cos x)\, dx$

$\displaystyle = 2\sin(m\pi a)\left\{\begin{array}{c}\sin(m\pi a)\\ \cos(m\pi a)\end{array}\right\}\frac{(\rho+1)_n}{n!\, a}\, {}_4F_3\left(\begin{array}{c}-n, \rho+\sigma+n+1, \frac{1}{2}, 1\\ \rho+1, 1-a, 1+a;\ 1\end{array}\right)$.

2. $\displaystyle\int_0^{m\pi} \left\{\begin{array}{c}\sin(ax)\\ \cos(ax)\end{array}\right\} P_n^{(\rho,\sigma)}(\cos^2 x)\, dx$

$\displaystyle = 2\sin\frac{m\pi a}{2}\left\{\begin{array}{c}\sin(m\pi a/2)\\ \cos(m\pi a/2)\end{array}\right\}\frac{(\rho+1)_n}{n!\, a}\, {}_4F_3\left(\begin{array}{c}-n, \rho+\sigma+n+1, \frac{1}{2}, 1\\ \rho+1, 1-\frac{a}{2}, 1+\frac{a}{2};\ \frac{1}{2}\end{array}\right)$.

3. $\displaystyle\int_0^{m\pi} e^{-ax} P_n^{(\rho,\sigma)}(\cos^2 x)\, dx$

$\displaystyle = (1-e^{-m\pi a})\frac{(\rho+1)_n}{n!\, a}\, {}_4F_3\left(\begin{array}{c}-n, \rho+\sigma+n+1, \frac{1}{2}, 1\\ \rho+1, 1-\frac{ia}{2}, 1+\frac{ia}{2};\ \frac{1}{2}\end{array}\right)$.

4. $\displaystyle\int_0^{\infty} e^{-ax} P_n^{(\rho,\sigma)}(\cos^2 x)\, dx = \frac{(\rho+1)_n}{n!\, a}\, {}_4F_3\left(\begin{array}{c}-n, \rho+\sigma+n+1, \frac{1}{2}, 1\\ \rho+1, 1-\frac{ia}{2}, 1+\frac{ia}{2};\ \frac{1}{2}\end{array}\right)$

[Re $a > 0$].

5. $\displaystyle\int_0^{\infty} e^{-ax} P_n^{(\rho,\sigma)}(1+b\sin^2 x)\, dx$

$\displaystyle = \frac{(\rho+1)_n}{n!\, a}\, {}_4F_3\left(\begin{array}{c}-n, \rho+\sigma+n+1, \frac{1}{2}, 1\\ \rho+1, 1-\frac{ia}{2}, 1+\frac{ia}{2};\ -\frac{b}{2}\end{array}\right)$ [Re $a > 0$].

4.20.3. Integrals containing $P_n^{(\rho,\sigma)}(z)$ and $J_\nu(z)$

1. $\displaystyle\int_{-1}^{1} (1-x)^{\rho/2}(1+x)^\sigma J_\rho\left(a\sqrt{1-x}\right) P_n^{(\rho,\sigma)}(x)\, dx$

$\displaystyle = 2^{(\rho+3\sigma+3)/2} a^{-\sigma-1}\frac{\Gamma(\sigma+n+1)}{n!} J_{\rho+\sigma+2n+1}\left(\sqrt{2}a\right)$ [Re ρ, Re $\sigma > -1$].

4.20.4. Integrals containing products of $P_n^{(\rho,\sigma)}(z)$

1. $\displaystyle\int_0^{m\pi} \begin{Bmatrix} \sin(ax) \\ \cos(ax) \end{Bmatrix} P_n^{(\rho,\sigma)}(-\cos x) P_n^{(\rho,\sigma)}(\cos x)\, dx$

$$= (-1)^n 2\sin\frac{m\pi a}{2} \begin{Bmatrix} \sin(m\pi a/2) \\ \cos(m\pi a/2) \end{Bmatrix} \frac{(\rho+1)_n(\sigma+1)_n}{(n!)^2\, a}$$

$$\times\ _6F_5\left(\begin{array}{c} -n,\ \frac{\rho+\sigma+1}{2},\ \frac{\rho+\sigma}{2}+1,\ \rho+\sigma+n+1,\ \frac{1}{2},\ 1 \\ \rho+1,\ \sigma+1,\ \rho+\sigma+1,\ 1-\frac{a}{2},\ 1+\frac{a}{2};\ 1 \end{array}\right).$$

2. $\displaystyle\int_0^{m\pi} e^{-ax} P_n^{(\rho,\sigma)}(-\cos x) P_n^{(\rho,\sigma)}(\cos x)\, dx$

$$= (-1)^n(1-e^{-m\pi a})\,\frac{(\rho+1)_n(\sigma+1)_n}{(n!)^2\, a}$$

$$\times\ _6F_5\left(\begin{array}{c} -n,\ \frac{\rho+\sigma+1}{2},\ \frac{\rho+\sigma}{2}+1,\ \rho+\sigma+n+1,\ \frac{1}{2},\ 1;\ 1 \\ \rho+1,\ \sigma+1,\ \rho+\sigma+1,\ 1-\frac{ia}{2},\ 1+\frac{ia}{2} \end{array}\right).$$

3. $\displaystyle\int_0^{m\pi} e^{-ax} P_n^{(\rho,\sigma)}\left(-\sqrt{1+b\sin^2 x}\right) P_n^{(\rho,\sigma)}\left(\sqrt{1+b\sin^2 x}\right) dx$

$$= (-1)^n(1-e^{-m\pi a})\,\frac{(\rho+1)_n(\sigma+1)_n}{(n!)^2\, a}$$

$$\times\ _6F_5\left(\begin{array}{c} -n,\ \frac{\rho+\sigma+1}{2},\ \frac{\rho+\sigma}{2}+1,\ \rho+\sigma+n+1,\ \frac{1}{2},\ 1 \\ \rho+1,\ \sigma+1,\ \rho+\sigma+1,\ 1-\frac{ia}{2},\ 1+\frac{ia}{2};\ -b \end{array}\right).$$

4. $\displaystyle\int_0^{\infty} e^{-ax} P_n^{(\rho,\sigma)}(-\cos x) P_n^{(\rho,\sigma)}(\cos x)\, dx = (-1)^n \frac{\Gamma(\rho+1)\Gamma(\sigma+1)}{(n!)^2\, a}$

$$\times\ _6F_5\left(\begin{array}{c} -n,\ \frac{\rho+\sigma+1}{2},\ \frac{\rho+\sigma}{2}+1,\ \rho+\sigma+n+1,\ \frac{1}{2},\ 1 \\ \rho+1,\ \sigma+1,\ \rho+\sigma+1,\ 1-\frac{ia}{2},\ 1+\frac{ia}{2};\ 1 \end{array}\right) \quad [\operatorname{Re} a > 0].$$

5. $\displaystyle\int_0^{\infty} e^{-ax} P_n^{(\rho,\sigma)}\left(-\sqrt{1+b\sin^2 x}\right) P_n^{(\rho,\sigma)}\left(\sqrt{1+b\sin^2 x}\right) dx$

$$= (-1)^n \frac{\Gamma(\rho+1)\Gamma(\sigma+1)}{(n!)^2\, a}\ _6F_5\left(\begin{array}{c} -n,\ \frac{\rho+\sigma+1}{2},\ \frac{\rho+\sigma}{2}+1,\ \rho+\sigma+n+1,\ \frac{1}{2},\ 1 \\ \rho+1,\ \sigma+1,\ \rho+\sigma+1,\ 1-\frac{ia}{2},\ 1+\frac{ia}{2};\ -b \end{array}\right)$$

$$[\operatorname{Re} a > 0].$$

4.21. The Complete Elliptic Integral $\mathbf{K}(z)$

4.21.1. Integrals containing $\mathbf{K}(z)$ and algebraic functions

1. $\displaystyle\int_0^1 \frac{x(1-x^2)^{s-1}}{(1-ax^2)^{s+1/2}} \mathbf{K}(x)\,dx = \frac{\pi}{4}\frac{\Gamma^2(s)}{\Gamma^2\left(s+\frac{1}{2}\right)}(1-a)^{-1/2}\,{}_2F_1\left(\begin{array}{c}\frac{1}{2},\,s \\ s+\frac{1}{2};\,a\end{array}\right)$
$\quad\quad [\operatorname{Re} s > 0;\ |\arg(1-a)| < \pi].$

2. $\displaystyle\int_0^1 x(1-x^2)^{-1/2}\mathbf{K}(ax)\,dx = \frac{\pi}{2a}\arcsin a \quad\quad [|\arg(1-a^2)| < \pi].$

3. $\displaystyle\int_0^1 x^{s-1}(1+ax)^\nu \mathbf{K}(\sqrt{1-x})\,dx = \frac{\pi\Gamma^2(s)}{2\Gamma^2\left(s+\frac{1}{2}\right)}\,{}_3F_2\left(\begin{array}{c}-\nu,\,s,\,s;\,a \\ s+\frac{1}{2},\,s+\frac{1}{2}\end{array}\right)$
$\quad\quad [\operatorname{Re} s > 0;\ |\arg(1-a)| < \pi].$

4. $\displaystyle\int_0^a x^{s-1}(a-x)^{t-1}\mathbf{K}\left(b\sqrt{x(a-x)}\right)dx$

$\displaystyle\quad = \frac{\pi}{2}\mathrm{B}(s,t)\,a^{s+t-1}\,{}_4F_3\left(\begin{array}{c}\frac{1}{2},\,\frac{1}{2},\,s,\,t;\,\frac{a^2b^2}{4} \\ 1,\,\frac{s+t}{2},\,\frac{s+t+1}{2}\end{array}\right)$
$\quad\quad [\operatorname{Re} s,\,\operatorname{Re} t > 0;\ |\arg(4-a^2b^2)| < \pi].$

5. $\displaystyle\int_0^a \mathbf{K}\left(b\sqrt{x(a-x)}\right)dx = \frac{\pi}{b}\arcsin\frac{ab}{2} \quad\quad [|\arg(4-a^2b^2)| < \pi].$

6. $\displaystyle\int_0^a x\,\mathbf{K}\left(b\sqrt{x(a-x)}\right)dx = \frac{\pi a}{2b}\arcsin\frac{ab}{2} \quad\quad [|\arg(4-a^2b^2)| < \pi].$

7. $\displaystyle\int_0^a x^2\,\mathbf{K}\left(b\sqrt{x(a-x)}\right)dx = \frac{\pi}{8b^3}\left[ab\sqrt{1-\frac{a^2b^2}{4}}+(3a^2b^2-2)\arcsin\frac{ab}{2}\right]$
$\quad\quad [|\arg(4-a^2b^2)| < \pi].$

8. $\displaystyle\int_0^a x^{-1/2}(a-x)^{1/2}\mathbf{K}\left(b\sqrt{x(a-x)}\right)dx = \frac{\pi^2 a}{4}\psi_1\left(\frac{a^2b^2}{4}\right)$
$\quad\quad [|\arg(4-a^2b^2)| < \pi].$

9. $\displaystyle\int_0^a x^{-1/2}(a-x)^{-1/2}\mathbf{K}\left(b\sqrt{x(a-x)}\right)dx = \frac{\pi^2}{2}\psi_1\left(\frac{a^2b^2}{4}\right)$
$\quad\quad [|\arg(4-a^2b^2)| < \pi].$

10. $\displaystyle\int_0^a x^{s+1/2}(a-x)^s \operatorname{K}\left(b\sqrt[4]{x(a-x)}\right) dx = 2^{-2s-2}\pi^{3/2} a^{2s+3/2} \frac{\Gamma(2s+2)}{\Gamma\left(2s+\frac{5}{2}\right)}$

$\times {}_3F_2\!\left(\begin{array}{c}\frac{1}{2},\frac{1}{2},2s+2\\ 1,2s+\frac{5}{2};\ \frac{ab^2}{2}\end{array}\right)\quad [\operatorname{Re} s > -1;\ |\arg(2-ab^2)| < \pi].$

11. $\displaystyle\int_0^a x^{1/2} \operatorname{K}\left(b\sqrt[4]{x(a-x)}\right) dx = -\frac{\pi a^{1/2}}{4b^2}\sqrt{1-\frac{ab^2}{2}}$

$+ \dfrac{\pi}{2^{3/2} b^3}(ab^2+1)\arcsin\!\left(b\sqrt{\dfrac{a}{2}}\right)\quad [|\arg(2-ab^2)| < \pi].$

12. $\displaystyle\int_0^a x^{-1/2} \operatorname{K}\left(b\sqrt[4]{x(a-x)}\right) dx = \frac{2^{1/2}\pi}{b}\arcsin\!\left(b\sqrt{\frac{a}{2}}\right)$

$[|\arg(2-ab^2)| < \pi].$

13. $\displaystyle\int_0^a x^{1/4}(a-x)^{-1/4}\operatorname{K}\left(b\sqrt[4]{x(a-x)}\right) dx$

$= \dfrac{a\pi^2}{4\sqrt{2}}\left[2\psi_1\!\left(\dfrac{ab^2}{2}\right) - \psi_2\!\left(\dfrac{ab^2}{2}\right)\right]\quad [|\arg(2-ab^2)| < \pi].$

14. $\displaystyle\int_0^a x^{-3/4}(a-x)^{-1/4}\operatorname{K}\left(b\sqrt[4]{x(a-x)}\right) dx = \frac{\pi^2}{\sqrt{2}}\psi_1\!\left(\frac{ab^2}{2}\right)$

$[|\arg(2-ab^2)| < \pi].$

15. $\displaystyle\int_0^2 x^{1/4}(2-x)^{-1/4}\operatorname{K}\left(\sqrt[4]{x(2-x)}\right) dx = \frac{1}{8\sqrt{2\pi}}\left[\Gamma^4\!\left(\frac{1}{4}\right) + 16\Gamma^4\!\left(\frac{3}{4}\right)\right].$

16. $\displaystyle\int_0^2 x^{-1/4}(2-x)^{-3/4}\operatorname{K}\left(\sqrt[4]{x(2-x)}\right) dx = \frac{\pi^3}{\sqrt{2}}\Gamma^{-4}\!\left(\frac{3}{4}\right).$

17. $\displaystyle\int_0^a \frac{x^{-1/2}}{\sqrt{1+b^2\sqrt{x(a-x)}}}\operatorname{K}\!\left(\frac{b\sqrt[4]{x(a-x)}}{\sqrt{1+b^2\sqrt{x(a-x)}}}\right) dx$

$= \dfrac{2^{1/2}\pi}{b}\ln\!\left(b\sqrt{\dfrac{a}{2}} + \sqrt{1+\dfrac{ab^2}{2}}\right)\quad [|\arg(2+ab^2)| < \pi].$

4.21.2. Integrals containing $K(z)$, the exponential, hyperbolic and trigonometric functions

1. $\displaystyle\int_0^1 x^{s-1} e^{ax} \, K\!\left(\sqrt{1-x}\right) dx = \frac{\pi \Gamma^2(s)}{2\Gamma^2\!\left(s+\frac{1}{2}\right)} \, {}_2F_2\!\left(\begin{matrix} s, s; a \\ s+\frac{1}{2}, s+\frac{1}{2} \end{matrix}\right)$ [Re $s > 0$].

2. $\displaystyle\int_0^1 x^{s-1} \left\{\begin{matrix}\sinh(a\sqrt{x}) \\ \sin(a\sqrt{x})\end{matrix}\right\} K\!\left(\sqrt{1-x}\right) dx$

$= \dfrac{\pi a \Gamma^2\!\left(s+\frac{1}{2}\right)}{2\Gamma^2(s+1)} \, {}_2F_3\!\left(\begin{matrix} s+\frac{1}{2}, s+\frac{1}{2}; \pm\frac{a^2}{4} \\ \frac{3}{2}, s+1, s+1 \end{matrix}\right)$ [Re $s > -1/2$].

3. $\displaystyle\int_0^1 x^{s-1} \left\{\begin{matrix}\cosh(a\sqrt{x}) \\ \cos(a\sqrt{x})\end{matrix}\right\} K\!\left(\sqrt{1-x}\right) dx$

$= \dfrac{\pi \Gamma^2(s)}{2\Gamma^2\!\left(s+\frac{1}{2}\right)} \, {}_2F_3\!\left(\begin{matrix} s, s; \pm\frac{a^2}{4} \\ \frac{1}{2}, s+\frac{1}{2}, s+\frac{1}{2} \end{matrix}\right)$ [Re $s > 0$].

4. $\displaystyle\int_0^1 \sinh(a\sqrt{x}) \, K\!\left(\sqrt{1-x}\right) dx = \frac{\pi^2}{2} I_0\!\left(\frac{a}{2}\right) I_1\!\left(\frac{a}{2}\right)$.

5. $\displaystyle\int_0^1 x^{-1/2} \cosh(a\sqrt{x}) \, K\!\left(\sqrt{1-x}\right) dx = \frac{\pi^2}{2} I_0^2\!\left(\frac{a}{2}\right)$.

6. $\displaystyle\int_0^1 \sin(a\sqrt{x}) \, K\!\left(\sqrt{1-x}\right) dx = \frac{\pi^2}{2} J_0\!\left(\frac{a}{2}\right) J_1\!\left(\frac{a}{2}\right)$.

7. $\displaystyle\int_0^1 x^{1/2} \cos(a\sqrt{x}) \, K\!\left(\sqrt{1-x}\right) dx$

$= \dfrac{\pi^2}{4}\left[J_0^2\!\left(\frac{a}{2}\right) - \frac{2}{a} J_0\!\left(\frac{a}{2}\right) J_1\!\left(\frac{a}{2}\right) - J_1^2\!\left(\frac{a}{2}\right)\right]$.

8. $\displaystyle\int_0^1 x^{-1/2} \cos(a\sqrt{x}) \, K\!\left(\sqrt{1-x}\right) dx = \frac{\pi^2}{2} J_0^2\!\left(\frac{a}{2}\right)$.

9. $\displaystyle\int_0^{\pi/2} \cos(2nx) \, K(a \sin x) \, dx = 2^{-2n-2} \pi (-a^2)^n$

$\times \dfrac{\Gamma^2\!\left(n+\frac{1}{2}\right)}{(n!)^2} \, {}_3F_2\!\left(\begin{matrix} \frac{n+1}{2}, \frac{n+1}{2}, \frac{n+1}{2} \\ n+1, 2n+1; a^2 \end{matrix}\right)$ $[|\arg(1-a^2)| < \pi]$.

10. $\displaystyle\int_0^{\pi/2} \cos^\nu x \cos(ax) \, \mathrm{K}(b\cos x) \, dx = \dfrac{2^{-\nu-2}\pi^2 \Gamma(\nu+1)}{\Gamma\left(\dfrac{\nu-a}{2}+1\right)\Gamma\left(\dfrac{\nu+a}{2}+1\right)}$

$\times {}_4F_3\left(\begin{array}{c} \dfrac{1}{2}, \dfrac{1}{2}, \dfrac{\nu+1}{2}, 1+\dfrac{\nu}{2}; \, b^2 \\ 1, 1+\dfrac{\nu-a}{2}, 1+\dfrac{\nu+a}{2} \end{array}\right)$ $\quad [\operatorname{Re}\nu > -1; \ |\arg b| < \pi].$

11. $\displaystyle\int_0^{\pi} \sin^\nu x \left\{\begin{array}{c}\sin(ax)\\ \cos(ax)\end{array}\right\} \mathrm{K}(b\sin x)\, dx$

$= \dfrac{2^{-\nu-1}\pi^2 \Gamma(\nu+1)}{\Gamma\left(\dfrac{\nu-a}{2}+1\right)\Gamma\left(\dfrac{\nu+a}{2}+1\right)} \left\{\begin{array}{c}\sin(a\pi/2)\\ \cos(a\pi/2)\end{array}\right\}$

$\times {}_4F_3\left(\begin{array}{c}\dfrac{1}{2},\dfrac{1}{2},\dfrac{\nu+1}{2},1+\dfrac{\nu}{2};\, b^2 \\ 1, 1+\dfrac{\nu-a}{2}, 1+\dfrac{\nu+a}{2}\end{array}\right)$ $\quad[\operatorname{Re}\nu > -1; \ |\arg b| < \pi].$

12. $\displaystyle\int_0^{m\pi} \left\{\begin{array}{c}\sin(ax)\\ \cos(ax)\end{array}\right\} \mathrm{K}(b\sin x)\, dx = \dfrac{\pi}{a}\sin\dfrac{m\pi a}{2}$

$\times \left\{\begin{array}{c}\sin(m\pi a/2)\\ \cos(m\pi a/2)\end{array}\right\} {}_3F_2\left(\begin{array}{c}\dfrac{1}{2},\dfrac{1}{2},\dfrac{1}{2};\, b^2 \\ 1-\dfrac{a}{2}, 1+\dfrac{a}{2}\end{array}\right)$ $\quad[|\arg(1-b^2)| < \pi].$

13. $\displaystyle\int_0^{m\pi} e^{-ax}\mathrm{K}(b\sin x)\, dx = \dfrac{\pi}{2a}(1-e^{-m\pi a})\, {}_3F_2\left(\begin{array}{c}\dfrac{1}{2},\dfrac{1}{2},\dfrac{1}{2};\, b^2 \\ 1-\dfrac{ia}{2}, 1+\dfrac{ia}{2}\end{array}\right)$

$[|\arg(1-b^2)| < \pi].$

14. $\displaystyle\int_0^\infty e^{-ax}\mathrm{K}(b\sin x)\, dx = \dfrac{\pi}{2a}\, {}_3F_2\left(\begin{array}{c}\dfrac{1}{2},\dfrac{1}{2},\dfrac{1}{2};\, b^2 \\ 1-\dfrac{ia}{2}, 1+\dfrac{ia}{2}\end{array}\right)$

$\bigl[\operatorname{Re} a > 0; \ |\arg(1-b^2)| < \pi\bigr].$

4.21.3. Integrals containing $\mathrm{K}(z)$ and the logarithmic function

1. $\displaystyle\int_0^1 x(1-x^2)^{-3/2} \ln x \, \mathrm{K}(x)\, dx = -\dfrac{\pi}{2}.$

2. $\displaystyle\int_0^1 \dfrac{x^3}{(2-x^2)^{5/2}} \ln\dfrac{4x^4(1-x^2)}{(2-x^2)^4} \mathrm{K}(x)\, dx = -\dfrac{32}{9} + \dfrac{\sqrt{2}\pi}{3} + 2\ln 2.$

3. $\displaystyle\int_0^1 x^{s-1} \ln(1+ax)\, \mathbf{K}\left(\sqrt{1-x}\right) dx = \frac{\pi a \Gamma^2(s+1)}{2\Gamma^2\left(s+\frac{3}{2}\right)}$

$$\times {}_4F_3\left(\begin{array}{c} 1, 1, s+1, s+1 \\ 2, s+\frac{3}{2}, s+\frac{3}{2};\ -a \end{array}\right) \quad [\operatorname{Re} s > -1;\ |\arg(1+a)| < \pi].$$

4. $\displaystyle\int_0^1 x^{-3/2} \ln(1-ax)\, \mathbf{K}\left(\sqrt{1-x}\right) dx = 2\pi[(1-a)\,\mathbf{K}(\sqrt{a}) - \mathbf{E}(\sqrt{a})]$

$$[|\arg(1-a)| < \pi].$$

5. $\displaystyle\int_0^1 x^{-3/2} \ln(1+ax)\, \mathbf{K}\left(\sqrt{1-x}\right) dx$

$$= 2\pi(a+1)^{1/2}\left[\mathbf{K}\left(\sqrt{\frac{a}{a+1}}\right) - \mathbf{E}\left(\sqrt{\frac{a}{a+1}}\right)\right] \quad [|\arg(1+a)| < \pi].$$

6. $\displaystyle\int_0^1 x^{s-1} \ln\left(a\sqrt{x} + \sqrt{1+a^2 x}\right) \mathbf{K}\left(\sqrt{1-x}\right) dx = \frac{\pi a \Gamma^2\left(s+\frac{1}{2}\right)}{2\Gamma^2(s+1)}$

$$\times {}_4F_3\left(\begin{array}{c} \frac{1}{2}, \frac{1}{2}, s+\frac{1}{2}, s+\frac{1}{2} \\ \frac{3}{2}, s+1, s+1;\ -a^2 \end{array}\right) \quad [\operatorname{Re} s > -1/2;\ |\arg(1+a^2)| < \pi].$$

7. $\displaystyle\int_0^1 \ln\left(a\sqrt{x}+\sqrt{1+a^2 x}\right) \mathbf{K}\left(\sqrt{1-x}\right) dx = \frac{\pi a}{8}\left[2\psi_2(-a^2) - \psi_3(-a^2)\right]$

$$[|\arg(1+a^2)| < \pi].$$

8. $\displaystyle\int_0^1 \frac{x^{s-1}}{\sqrt{1+a^2 x}} \ln\left(a\sqrt{x}+\sqrt{1+a^2 x}\right) \mathbf{K}\left(\sqrt{1-x}\right) dx = \frac{\pi a \Gamma^2\left(s+\frac{1}{2}\right)}{2\Gamma^2(s+1)}$

$$\times {}_4F_3\left(\begin{array}{c} 1, 1, s+\frac{1}{2}, s+\frac{1}{2} \\ \frac{3}{2}, s+1, s+1;\ -a^2 \end{array}\right) \quad [\operatorname{Re} s > -1/2;\ |\arg(1+a^2)| < \pi].$$

9. $\displaystyle\int_0^1 \frac{1}{\sqrt{1+a^2 x}} \ln\left(a\sqrt{x}+\sqrt{1+a^2 x}\right) \mathbf{K}\left(\sqrt{1-x}\right) dx = \frac{\pi^2}{2a} \ln\frac{a+\sqrt{1+a^2}}{2}$

$$[|\arg(1+a^2)| < \pi].$$

10. $\displaystyle\int_0^1 \frac{x^{-1}}{\sqrt{1+a^2 x}} \ln\left(a\sqrt{x}+\sqrt{1+a^2 x}\right) \mathbf{K}\left(\sqrt{1-x}\right) dx$

$$= \frac{\pi^2}{2} \ln\left(a+\sqrt{1+a^2}\right) \quad [|\arg(1+a^2)| < \pi].$$

11. $\int_0^1 x^{s-1} \ln \dfrac{1+a\sqrt{x}}{1-a\sqrt{x}} \, \mathbf{K}(\sqrt{1-x}) \, dx = \dfrac{\pi a \Gamma^2\left(s+\dfrac{1}{2}\right)}{\Gamma^2(s+1)}$

$\times {}_4F_3\left(\begin{matrix} \dfrac{1}{2}, 1, s+\dfrac{1}{2}, s+\dfrac{1}{2} \\ \dfrac{3}{2}, s+1, s+1; \, a^2 \end{matrix}\right)$ $\quad [\operatorname{Re} s > -1/2; \, |\arg(1-a^2)| < \pi]$.

12. $\int_0^1 \ln \dfrac{1+a\sqrt{x}}{1-a\sqrt{x}} \, \mathbf{K}(\sqrt{1-x}) \, dx = \dfrac{\pi}{a}\left[\pi - 2\,\mathbf{E}(a)\right] \qquad [|\arg(1-a^2)| < \pi]$.

13. $\int_0^1 \dfrac{1}{x} \ln \dfrac{1+\sqrt{x}}{1-\sqrt{x}} \, \mathbf{K}(\sqrt{1-x}) \, dx = 4\pi \mathbf{G}$.

14. $\int_0^1 x^{s-1} \ln \dfrac{1+\sqrt{1-x^2}}{1-\sqrt{1-x^2}} \, \mathbf{K}(ax) \, dx = \dfrac{\pi^{3/2} \Gamma\left(\dfrac{s}{2}\right)}{2s\,\Gamma\left(\dfrac{s+1}{2}\right)} \, {}_4F_3\left(\begin{matrix} \dfrac{1}{2}, \dfrac{1}{2}, \dfrac{s}{2}, \dfrac{s}{2}; \, a^2 \\ 1, \dfrac{s+1}{2}, \dfrac{s}{2}+1 \end{matrix}\right)$

$[\operatorname{Re} s > 0; \, |\arg(1-a^2)| < \pi]$.

15. $\int_0^1 x \ln \dfrac{1+\sqrt{1-x^2}}{1-\sqrt{1-x^2}} \, \mathbf{K}(ax) \, dx = \dfrac{\pi}{a^2}\left(a \arcsin a + \sqrt{1-a^2} - 1\right)$

$[|\arg(1-a^2)| < \pi]$.

16. $\int_0^a x^{s-1} \ln \dfrac{a+\sqrt{a^2-x^2}}{x} \, \mathbf{K}(bx) \, dx = \dfrac{\pi^{3/2} a^s \Gamma\left(\dfrac{s}{2}\right)}{4s\,\Gamma\left(\dfrac{s+1}{2}\right)} \, {}_4F_3\left(\begin{matrix} \dfrac{1}{2}, \dfrac{1}{2}, \dfrac{s}{2}, \dfrac{s}{2}; \, a^2 b^2 \\ 1, \dfrac{s+1}{2}, \dfrac{s}{2}+1 \end{matrix}\right)$

$[\operatorname{Re} s > 0; \, |\arg(1-a^2 b^2)| < \pi]$.

17. $\int_0^a x \ln \dfrac{a+\sqrt{a^2-x^2}}{x} \, \mathbf{K}(bx) \, dx = \dfrac{\pi}{2b^2}\left[ab \arcsin(ab) + \sqrt{1-a^2 b^2} - 1\right]$

$[|\arg(1-a^2 b^2)| < \pi]$.

18. $\int_0^a x^3 \ln \dfrac{a+\sqrt{a^2-x^2}}{x} \, \mathbf{K}(bx) \, dx = \dfrac{\pi}{72 b^4 \sqrt{1-a^2 b^2}}\,(a^4 b^4 - 17 a^2 b^2 + 16)$

$+ \dfrac{\pi}{72 b^4}\left[3ab(2a^2 b^2 + 3)\arcsin(ab) - 16\right] \quad [|\arg(1-a^2 b^2)| < \pi]$.

19. $\int_0^a \dfrac{x^3}{\sqrt{b^2 x^2 + 1}} \ln \dfrac{a+\sqrt{a^2-x^2}}{x} \, \mathbf{K}\left(\dfrac{bx}{\sqrt{b^2 x^2 + 1}}\right) dx = \dfrac{\pi \sqrt{a^2 b^2 + 1}}{72 b^4}$

$\times (a^2 b^2 + 16) + \dfrac{\pi}{72 b^4}\left[3ab(2a^2 b^2 - 3)\ln\left(ab + \sqrt{a^2 b^2 + 1}\right) - 16\right]$

$[|\arg(1+a^2 b^2)| < \pi]$.

20. $\displaystyle\int_0^a \frac{x}{\sqrt{b^2x^2+1}} \ln\frac{a+\sqrt{a^2-x^2}}{x} \mathbf{K}\left(\frac{bx}{\sqrt{b^2x^2+1}}\right) dx$

$\displaystyle = \frac{\pi}{2b^2}\left[ab\ln\left(ab+\sqrt{a^2b^2+1}\right) - \sqrt{a^2b^2+1}+1\right]$ $\quad [|\arg(1+a^2b^2)|<\pi].$

21. $\displaystyle\int_0^1 x \ln\frac{1+a\sqrt{1-x^2}}{1-a\sqrt{1-x^2}} \mathbf{K}(x)\, dx = \frac{\pi}{a}\left[\frac{\pi}{2} - \mathbf{E}(a)\right]$ $\quad [|\arg(1-a)|<\pi].$

22. $\displaystyle\int_0^1 x^{s-1} \ln^2\left(a\sqrt{x}+\sqrt{1+a^2x}\right) \mathbf{K}(\sqrt{1-x})\, dx = \frac{\pi a^2 \Gamma^2(s+1)}{2\Gamma^2\left(s+\frac{3}{2}\right)}$

$\displaystyle \times {}_5F_4\left(\begin{array}{c}1,1,1,s+1,s+1 \\ \frac{3}{2},2,s+\frac{3}{2},s+\frac{3}{2};\end{array} -a^2\right)$ $\quad [\operatorname{Re} s > -1;\ |\arg(1+a^2)|<\pi].$

23. $\displaystyle\int_0^1 x^{-1/2} \ln^2\left(a\sqrt{x}+\sqrt{1+a^2x}\right) \mathbf{K}(\sqrt{1-x})\, dx$

$\displaystyle = \frac{\pi^2}{4}\left[\ln^2 2 + \ln\left(1+\sqrt{1+a^2}\right)\ln\frac{1+\sqrt{1+a^2}}{4} - 2\operatorname{Li}_2\left(\frac{1-\sqrt{1+a^2}}{2}\right)\right]$

$\quad [|\arg(1+a^2)|<\pi].$

24. $\displaystyle\int_0^1 x^{-3/2} \ln^2\left(a\sqrt{x}+\sqrt{1+a^2x}\right) \mathbf{K}(\sqrt{1-x})\, dx$

$\displaystyle = \pi^2\left[1 - \sqrt{1-a^2} + a\ln\left(a+\sqrt{1+a^2}\right)\right]$ $\quad [|\arg(1+a^2)|<\pi].$

25. $\displaystyle\int_0^1 (x-x^2)^s \ln^n(x-x^2) \mathbf{K}(\sqrt{x})\, dx = \frac{\pi^2}{\Gamma^2(3/4)} D_s^n\left[\frac{2^{-2s-2}\Gamma^2(s+1)}{\Gamma^2\left(s+\frac{5}{4}\right)}\right]$

$\quad [\operatorname{Re} s > -1].$

4.21.4. Integrals containing $\mathbf{K}(z)$ and inverse trigonometric functions

1. $\displaystyle\int_0^1 \arcsin(a\sqrt{x}) \mathbf{K}(\sqrt{1-x})\, dx = \frac{\pi^2 a}{8}[2\psi_2(a^2) - \psi_3(a^2)]$

$\quad [|\arg(1-a^2)|<\pi].$

2. $\displaystyle\int_0^1 \arcsin(\sqrt{x}) \mathbf{K}(\sqrt{1-x})\, dx = \frac{2}{\pi}\Gamma^4\left(\frac{3}{4}\right).$

3. $\int_0^1 x \arcsin(\sqrt{1-x})\, \mathbf{K}(\sqrt{x})\, dx = \frac{1}{\pi}\Gamma^4\left(\frac{3}{4}\right) - \frac{1}{432\pi}\Gamma^4\left(\frac{1}{4}\right).$

4. $\int_0^1 \frac{x^{s-1}}{\sqrt{1-a^2 x}} \arcsin(a\sqrt{x})\, \mathbf{K}(\sqrt{1-x})\, dx = \frac{\pi a \Gamma^2\left(s+\frac{1}{2}\right)}{2\Gamma^2(s+1)}$

$\times\ {}_4F_3\left(\begin{array}{c} 1, 1, s+\frac{1}{2}, s+\frac{1}{2} \\ \frac{3}{2}, s+1, s+1;\ a^2 \end{array}\right)$ $\quad [\operatorname{Re} s > -1/2;\ |\arg(1-a^2)| < \pi].$

5. $\int_0^1 \frac{x^{-1}}{\sqrt{1-a^2 x}} \arcsin(a\sqrt{x})\, \mathbf{K}(\sqrt{1-x})\, dx = \frac{\pi^2}{2} \arcsin a$

$\quad\quad\quad\quad\quad\quad\quad [|\arg(1-a^2)| < \pi].$

6. $\int_0^1 \frac{1}{\sqrt{1-a^2 x}} \arcsin(a\sqrt{x})\, \mathbf{K}(\sqrt{1-x})\, dx = -\frac{\pi^2}{2a} \ln \frac{1+\sqrt{1-a^2}}{2}$

$\quad\quad\quad\quad\quad\quad\quad [|\arg(1-a^2)| < \pi].$

7. $\int_0^1 x^{s-1} \arccos x\, \mathbf{K}(ax)\, dx = \frac{\pi^{3/2}\Gamma\left(\frac{s+1}{2}\right)}{2s^2\Gamma\left(\frac{s}{2}\right)}\, {}_4F_3\left(\begin{array}{c} \frac{1}{2}, \frac{1}{2}, \frac{s}{2}, \frac{s+1}{2};\ a^2 \\ 1, \frac{s}{2}+1, \frac{s}{2}+1 \end{array}\right)$

$\quad\quad\quad\quad\quad\quad\quad [\operatorname{Re} s > 0;\ |\arg(1-a^2)| < \pi].$

8. $\int_0^1 \arccos x\, \mathbf{K}(x)\, dx = \frac{\pi^2}{4} \ln 2.$

9. $\int_0^1 x \arccos x\, \mathbf{K}(ax)\, dx = \frac{\pi^2}{16}[2\psi_2(a^2) - \psi_3(a^2)]$ $\quad [|\arg(1-a^2)| < \pi].$

10. $\int_0^1 x \arccos x\, \mathbf{K}(x)\, dx = 4\pi^3 \Gamma^{-4}\left(\frac{1}{4}\right).$

11. $\int_0^1 x^2 \arccos x\, \mathbf{K}(x)\, dx = \frac{\pi^2}{16} \ln 2.$

12. $\int_0^1 \frac{\arccos x}{x^2}\left[\frac{\pi}{2} - \mathbf{K}(ax)\right] dx = \frac{\pi}{2}\left(\sqrt{1-a^2} - \ln\frac{1+\sqrt{1-a^2}}{2} - 1\right)$

$\quad\quad\quad\quad\quad\quad\quad [|\arg(1-a^2)| < \pi].$

13. $\displaystyle\int_0^1 \frac{x^{s-1}}{\sqrt{a^2x^2+1}} \arccos x\, \mathbf{K}\left(\frac{ax}{\sqrt{a^2x^2+1}}\right) dx = \frac{\pi^{3/2}\Gamma\left(\frac{s+1}{2}\right)}{2s^2\Gamma\left(\frac{s}{2}\right)}$

$$\times {}_4F_3\left(\begin{matrix}\frac{1}{2},\frac{1}{2},\frac{s}{2},\frac{s+1}{2};\ -a^2\\ 1,\frac{s}{2}+1,\frac{s}{2}+1\end{matrix}\right) \qquad \left[\operatorname{Re} s > -1/2;\ |\arg(1+a^2)| < \pi\right].$$

14. $\displaystyle\int_0^1 \frac{x}{\sqrt{a^2x^2+1}} \arccos x\, \mathbf{K}\left(\frac{ax}{\sqrt{a^2x^2+1}}\right) dx = \frac{\pi^2}{16}\left[2\psi_2(-a^2) - \psi_3(-a^2)\right]$

$$\left[|\arg(1+a^2)| < \pi\right].$$

15. $\displaystyle\int_0^1 x^{s-1}\arctan(a\sqrt{x})\,\mathbf{K}(\sqrt{1-x})\,dx$

$$= \frac{\pi a\, \Gamma^2\left(s+\frac{1}{2}\right)}{2\Gamma^2(s+1)}\, {}_4F_3\left(\begin{matrix}\frac{1}{2},1,s+\frac{1}{2},s+\frac{1}{2};\ -a^2\\ \frac{3}{2},s+1,s+1\end{matrix}\right)$$

$$\left[\operatorname{Re} s > -1/2;\ |\arg(1+a^2)| < \pi\right].$$

16. $\displaystyle\int_0^1 \arctan(a\sqrt{x})\,\mathbf{K}(\sqrt{1-x})\,dx = \frac{\pi}{2a}\left[2\sqrt{1+a^2}\,\mathbf{E}\left(\frac{a}{\sqrt{1+a^2}}\right) - \pi\right]$

$$\left[|\arg(1+a^2)| < \pi\right].$$

17. $\displaystyle\int_0^1 \arctan(\sqrt{x})\,\mathbf{K}(\sqrt{1-x})\,dx = \frac{\sqrt{2\pi}}{8}\left[\Gamma^2\left(\frac{1}{4}\right) + 8\pi^2\Gamma^{-2}\left(\frac{1}{4}\right) - (2\pi)^{3/2}\right].$

18. $\displaystyle\int_0^1 \arctan(a\sqrt{x})\,\mathbf{K}(\sqrt{1-x})\,dx = \frac{\pi}{a}\left[(1+a^2)^{1/2}\mathbf{E}\left(\frac{a}{\sqrt{1+a^2}}\right) - \frac{\pi}{2}\right]$

$$\left[|\arg(1+a^2)| < \pi\right].$$

19. $\displaystyle\int_0^1 x^{s-1}\arcsin^2(a\sqrt{x})\,\mathbf{K}(\sqrt{1-x})\,dx = \frac{\pi a^2 \Gamma^2(s+1)}{2\Gamma^2\left(s+\frac{3}{2}\right)}$

$$\times {}_5F_4\left(\begin{matrix}1,1,1,s+1,s+1;\ a^2\\ \frac{3}{2},2,s+\frac{3}{2},s+\frac{3}{2}\end{matrix}\right) \qquad \left[\operatorname{Re} s > -1;\ |\arg(1-a^2)| < \pi\right].$$

20. $\displaystyle\int_0^1 x^{-1/2}\arcsin^2(a\sqrt{x})\,\mathbf{K}(\sqrt{1-x})\,dx$

$$= \frac{\pi^2}{4}\left[2\operatorname{Li}_2\left(\frac{1-\sqrt{1-a^2}}{2}\right) - \ln^2\left(\frac{1+\sqrt{1-a^2}}{2}\right)\right] \qquad \left[|\arg(1-a^2)| < \pi\right].$$

21. $\int_0^1 x^{-3/2} \arcsin^2(a\sqrt{x}) \, \mathrm{K}\left(\sqrt{1-x}\right) dx = \pi^2 \left(\sqrt{1-a^2} + a \arcsin a - 1\right)$

$$\left[|\arg(1-a^2)| < \pi\right].$$

22. $\int_0^1 x^{-1/2} \arcsin^2 \sqrt{x} \, \mathrm{K}\left(\sqrt{1-x}\right) dx = \dfrac{\pi^2}{24}\left(\pi^2 - 12 \ln^2 2\right).$

23. $\int_0^1 x^{-3/2} \arcsin^2(\sqrt{x}) \, \mathrm{K}\left(\sqrt{1-x}\right) dx = \dfrac{\pi^2}{2}(\pi - 2).$

4.21.5. Integrals containing $\mathrm{K}(z)$ and $\mathrm{Li}_2(z)$

1. $\int_0^1 x^{s-1} \mathrm{Li}_2(ax) \, \mathrm{K}\left(\sqrt{1-x}\right) dx = \dfrac{\pi a \Gamma^2(s+1)}{2 \Gamma^2\left(s+\dfrac{3}{2}\right)}$

$$\times {}_5F_4\left(\begin{matrix} 1, 1, 1, s+1, s+1; a \\ 2, 2, s+\dfrac{3}{2}, s+\dfrac{3}{2} \end{matrix}\right) \quad [\mathrm{Re}\, s > -1;\ |\arg(1-a^2)| < \pi].$$

2. $\int_0^1 x^{-3/2} \mathrm{Li}_2(ax) \, \mathrm{K}\left(\sqrt{1-x}\right) dx = 2\pi \left[2(a-1)\, \mathrm{K}(\sqrt{a}) + 4\,\mathrm{E}(\sqrt{a}) - \pi\right]$

$$\left[|\arg(1-a^2)| < \pi\right].$$

3. $\int_0^1 x^{-3/2} \mathrm{Li}_2(-ax) \, \mathrm{K}\left(\sqrt{1-x}\right) dx = 4\pi \sqrt{a+1}$

$$\times \left[2\,\mathrm{E}\left(\sqrt{\dfrac{a}{a+1}}\right) - \mathrm{K}\left(\sqrt{\dfrac{a}{a+1}}\right)\right] - 2\pi^2 \quad [|\arg(1+a^2)| < \pi].$$

4. $\int_0^1 x^{-3/2} \mathrm{Li}_2(x) \, \mathrm{K}\left(\sqrt{1-x}\right) dx = 2\pi(4 - \pi).$

5. $\int_0^1 x^{-3/2} \mathrm{Li}_2(-x) \, \mathrm{K}\left(\sqrt{1-x}\right) dx = 2^{5/2} \sqrt{\pi}\, \Gamma^2\left(\dfrac{3}{4}\right) - 2\pi^2.$

4.21.6. Integrals containing $\mathrm{K}(z)$, $\mathrm{shi}(z)$ and $\mathrm{Si}(z)$

1. $\int_0^1 x^{s-1} \left\{\begin{matrix} \mathrm{shi}(a\sqrt{x}) \\ \mathrm{Si}(a\sqrt{x}) \end{matrix}\right\} \mathrm{K}\left(\sqrt{1-x}\right) dx$

$$= \dfrac{\pi a \Gamma^2\left(s+\dfrac{1}{2}\right)}{2\Gamma^2(s+1)} \,{}_3F_4\left(\begin{matrix} \dfrac{1}{2}, s+\dfrac{1}{2}, s+\dfrac{1}{2}; \pm\dfrac{a^2}{4} \\ \dfrac{3}{2}, \dfrac{3}{2}, s+1, s+1 \end{matrix}\right) \quad [\mathrm{Re}\, s > -1/2].$$

2. $\displaystyle\int_0^1 \text{Si}(a\sqrt{x})\,\mathbf{K}(\sqrt{1-x})\,dx = \frac{\pi^2}{4}\left[a J_0^2\left(\frac{a}{2}\right) - 2 J_0\left(\frac{a}{2}\right) J_1\left(\frac{a}{2}\right) + a J_1^2\left(\frac{a}{2}\right)\right].$

4.21.7. Integrals containing $\mathbf{K}(z)$ and $\text{erf}(z)$

1. $\displaystyle\int_0^1 x^{s-1}\,\text{erf}(a\sqrt{x})\,\mathbf{K}(\sqrt{1-x})\,dx$

$$= \frac{\sqrt{\pi}\,a\,\Gamma^2\left(s+\frac{1}{2}\right)}{\Gamma^2(s+1)}\,{}_3F_3\left(\begin{array}{c}\frac{1}{2},\,s+\frac{1}{2},\,s+\frac{1}{2};\,-a^2\\ \frac{3}{2},\,s+1,\,s+1\end{array}\right) \qquad [\text{Re}\,s > -1/2].$$

2. $\displaystyle\int_0^1 x^{s-1} e^{a^2 x}\,\text{erf}(a\sqrt{x})\,\mathbf{K}(\sqrt{1-x})\,dx$

$$= \frac{\sqrt{\pi}\,a\,\Gamma^2\left(s+\frac{1}{2}\right)}{\Gamma^2(s+1)}\,{}_3F_3\left(\begin{array}{c}1,\,s+\frac{1}{2},\,s+\frac{1}{2};\,a^2\\ \frac{3}{2},\,s+1,\,s+1\end{array}\right) \qquad [\text{Re}\,s > -1/2].$$

4.21.8. Integrals containing $\mathbf{K}(z)$, $S(z)$ and $C(z)$

1. $\displaystyle\int_0^1 x^{s-1} S(a\sqrt{x})\,\mathbf{K}(\sqrt{1-x})\,dx$

$$= \frac{1}{3}\sqrt{\frac{\pi a^3}{2}}\,\frac{\Gamma^2\left(s+\frac{3}{4}\right)}{\Gamma^2\left(s+\frac{5}{4}\right)}\,{}_3F_4\left(\begin{array}{c}\frac{3}{4},\,s+\frac{3}{4},\,s+\frac{3}{4};\,-\frac{a^2}{4}\\ \frac{3}{2},\,\frac{7}{4},\,s+\frac{5}{4},\,s+\frac{5}{4}\end{array}\right) \qquad [\text{Re}\,s > -3/4].$$

2. $\displaystyle\int_0^1 x^{s-1} C(a\sqrt{x})\,\mathbf{K}(\sqrt{1-x})\,dx$

$$= \sqrt{\frac{\pi a}{2}}\,\frac{\Gamma^2\left(s+\frac{1}{4}\right)}{\Gamma^2\left(s+\frac{3}{4}\right)}\,{}_3F_4\left(\begin{array}{c}\frac{1}{4},\,s+\frac{1}{4},\,s+\frac{1}{4};\,-\frac{a^2}{4}\\ \frac{1}{2},\,\frac{5}{4},\,s+\frac{3}{4},\,s+\frac{3}{4}\end{array}\right) \qquad [\text{Re}\,s > -1/4].$$

4.21.9. Integrals containing $\mathbf{K}(z)$ and $\gamma(\nu,\,z)$

1. $\displaystyle\int_0^1 x^{s-1}\gamma(\nu,\,ax)\,\mathbf{K}(\sqrt{1-x})\,dx$

$$= \frac{\pi a^\nu \Gamma^2(s+\nu)}{2\nu\,\Gamma^2\left(s+\nu+\frac{1}{2}\right)}\,{}_3F_3\left(\begin{array}{c}\nu,\,s+\nu,\,s+\nu;\,-a\\ \nu+1,\,s+\nu+\frac{1}{2},\,s+\nu+\frac{1}{2}\end{array}\right) \qquad [\text{Re}\,(s+\nu) > 0].$$

2. $\int_0^1 x^{s-1} e^{ax} \gamma(\nu, ax) \, \mathrm{K}\left(\sqrt{1-x}\right) dx$

$$= \frac{\pi a^\nu \Gamma^2(s+\nu)}{2\nu \Gamma^2\left(s+\nu+\frac{1}{2}\right)} {}_3F_3\left(\begin{matrix} 1, s+\nu, s+\nu; a \\ \nu+1, s+\nu+\frac{1}{2}, s+\nu+\frac{1}{2} \end{matrix}\right) \quad [\mathrm{Re}\,(s+\nu) > 0].$$

4.21.10. Integrals containing $\mathrm{K}(z)$, $J_\nu(z)$ and $I_\nu(z)$

1. $\int_0^1 x^{s-1} \left\{\begin{matrix} J_\nu(a\sqrt{x}) \\ I_\nu(a\sqrt{x}) \end{matrix}\right\} \mathrm{K}\left(\sqrt{1-x}\right) dx = \frac{\pi a^\nu \Gamma^2\left(s+\frac{\nu}{2}\right)}{2^{\nu+1} \Gamma^2\left(s+\frac{\nu+1}{2}\right) \Gamma(\nu+1)}$

$$\times {}_2F_3\left(\begin{matrix} s+\frac{\nu}{2}, s+\frac{\nu}{2}; \mp\frac{a^2}{4} \\ s+\frac{\nu+1}{2}, s+\frac{\nu+1}{2}, \nu+1 \end{matrix}\right) \quad [\mathrm{Re}\,(2s+\nu) > 0].$$

2. $\int_0^1 \left\{\begin{matrix} I_0(a\sqrt{x}) \\ J_0(a\sqrt{x}) \end{matrix}\right\} \mathrm{K}\left(\sqrt{1-x}\right) dx = \frac{\pi}{a} \left\{\begin{matrix} \mathbf{L}_0(a) \\ \mathbf{H}_0(a) \end{matrix}\right\}.$

3. $\int_0^1 \sqrt{x}\, J_1(a\sqrt{x}) \, \mathrm{K}\left(\sqrt{1-x}\right) dx = \frac{\pi}{a^2} [\mathbf{H}_0(a) - a\,\mathbf{H}_{-1}(a)].$

4. $\int_0^1 \sqrt{x}\, I_1(a\sqrt{x}) \, \mathrm{K}\left(\sqrt{1-x}\right) dx = \frac{\pi}{a^2} [a\,\mathbf{L}_{-1}(a) - \mathbf{L}_0(a)].$

5. $\int_0^1 x^{s-1} e^{ax} I_\nu(ax) \, \mathrm{K}\left(\sqrt{1-x}\right) dx = \frac{\pi \Gamma^2(s+\nu)}{2\Gamma^2\left(s+\nu+\frac{1}{2}\right) \Gamma(\nu+1)} \left(\frac{a}{2}\right)^\nu$

$$\times {}_3F_3\left(\begin{matrix} \nu+\frac{1}{2}, s+\nu, s+\nu; 2a \\ 2\nu+1, s+\nu+\frac{1}{2}, s+\nu+\frac{1}{2} \end{matrix}\right) \quad [\mathrm{Re}\,(s+\nu) > 0].$$

6. $\int_0^1 \left\{\begin{matrix} I_0(a\sqrt{x}) \\ J_0(a\sqrt{x}) \end{matrix}\right\}^2 \mathrm{K}\left(\sqrt{1-x}\right) dx = \frac{1}{a} \left\{\begin{matrix} \mathrm{shi}\,(2a) \\ \mathrm{Si}\,(2a) \end{matrix}\right\}.$

7. $\int_0^1 J_1^2(a\sqrt{x}) \, \mathrm{K}\left(\sqrt{1-x}\right) dx = \frac{1}{a^2} [\cos(2a) + a\,\mathrm{Si}\,(2a) - 1].$

8. $\int_0^1 I_1^2(a\sqrt{x}) \, \mathrm{K}\left(\sqrt{1-x}\right) dx = \frac{1}{a^2} [2\sinh^2 a - a\,\mathrm{shi}\,(2a)].$

9. $\int_0^1 x J_0^2(a\sqrt{x}) K(\sqrt{1-x}) dx = \dfrac{1}{16a^3} [5 \sin(2a) - 6a \cos(2a)$

$\hspace{6cm} + 2(2a^2 - 1) \operatorname{Si}(2a)].$

10. $\int_0^1 x I_0^2(a\sqrt{x}) K(\sqrt{1-x}) dx = \dfrac{1}{16a^3} [-5 \sinh(2a) + 6a \cosh(2a)$

$\hspace{6cm} + 2(2a^2 + 1) \operatorname{shi}(2a)].$

11. $\int_0^1 \dfrac{1}{x} J_1^2(a\sqrt{x}) K(\sqrt{1-x}) dx = \mathbf{C} - \dfrac{1}{2} - \dfrac{1}{4a^2} + \dfrac{1}{2a} \sin(2a)$

$\hspace{6cm} + \dfrac{1}{4a^2} \cos(2a) + \ln(2a) - \operatorname{ci}(2a).$

12. $\int_0^1 \dfrac{1}{x} I_1^2(a\sqrt{x}) K(\sqrt{1-x}) dx$

$\hspace{1cm} = -\mathbf{C} + \dfrac{1}{2} - \dfrac{1}{4a^2} - \dfrac{1}{2a} \sinh(2a) + \dfrac{1}{4a^2} \cosh(2a) - \ln(2a) + \operatorname{chi}(2a).$

13. $\int_0^1 x J_1^2(a\sqrt{x}) K(\sqrt{1-x}) dx = \dfrac{1}{16a^3} [10a \cos(2a) - 11 \sin(2a)$

$\hspace{6cm} + (4a^2 + 6) \operatorname{Si}(2a)].$

14. $\int_0^1 x I_1^2(a\sqrt{x}) K(\sqrt{1-x}) dx = \dfrac{1}{16a^3} [10a \cosh(2a) - 11 \sinh(2a)$

$\hspace{6cm} + (6 - 4a^2) \operatorname{shi}(2a)].$

4.21.11. Integrals containing $K(z)$, $\mathbf{H}_\nu(z)$ and $\mathbf{L}_\nu(z)$

1. $\int_0^1 x^{s-1} \left\{ \begin{array}{c} \mathbf{H}_\nu(a\sqrt{x}) \\ \mathbf{L}_\nu(a\sqrt{x}) \end{array} \right\} K(\sqrt{1-x}) dx = \dfrac{\sqrt{\pi} \Gamma^2 \left(s + \frac{\nu+1}{2}\right) \left(\frac{a}{2}\right)^{\nu+1}}{2\Gamma^2 \left(s + \frac{\nu}{2} + 1\right) \Gamma\left(\nu + \frac{3}{2}\right)}$

$\hspace{2cm} \times {}_3F_4 \left(\begin{array}{c} 1, s + \frac{\nu+1}{2}, s + \frac{\nu+1}{2}; \mp \frac{a^2}{4} \\ \frac{3}{2}, \nu + \frac{3}{2}, s + \frac{\nu}{2} + 1, s + \frac{\nu}{2} + 1 \end{array} \right) \quad [\operatorname{Re}(2s + \nu) > -1].$

2. $\int_0^1 \mathbf{H}_0(a\sqrt{x}) K(\sqrt{1-x}) dx = \dfrac{\pi}{a} [1 - J_0(a)].$

3. $\int_0^1 \mathbf{L}_0(a\sqrt{x}) K(\sqrt{1-x}) dx = \dfrac{\pi}{a} [I_0(a) - 1].$

4. $\int_0^1 x\, \mathbf{H}_0(a\sqrt{x})\, \mathbf{K}\left(\sqrt{1-x}\right) dx = \dfrac{\pi}{4a^3}[a^2 - 4 + 4(1-a^2)J_0(a) + 8a\, J_1(a)]$.

5. $\int_0^1 x\, \mathbf{L}_0(a\sqrt{x})\, \mathbf{K}\left(\sqrt{1-x}\right) dx = \dfrac{\pi}{4a^3}[4(1+a^2)I_0(a) - 8a\, I_1(a) - a^2 - 4]$.

4.21.12. Integrals containing $\mathbf{K}(z)$ and $L_n^\lambda(z)$

1. $\int_0^1 x^{s-1} L_n^\lambda(ax)\, \mathbf{K}\left(\sqrt{1-x}\right) dx$

$$= \frac{\pi \Gamma^2(s)(\lambda+1)_n}{n!\, 2\Gamma^2\left(s+\dfrac{1}{2}\right)}\, {}_3F_3\!\left(\begin{matrix} -n,\, s,\, s;\, a \\ \lambda+1,\, s+\dfrac{1}{2},\, s+\dfrac{1}{2} \end{matrix}\right) \quad [\operatorname{Re} s > 0].$$

4.21.13. Integrals containing products of $\mathbf{K}(z)$

1. $\int_0^1 \dfrac{1}{\sqrt{1-x^2}}\, \mathbf{K}^2(x)\, dx = \dfrac{\pi^{-3}}{64}\Bigg[-\dfrac{1}{2}\Gamma^8\!\left(\dfrac{1}{4}\right) + \Gamma^8\!\left(\dfrac{1}{4}\right)\, {}_5F_4\!\left(\begin{matrix} \tfrac{1}{4},\tfrac{1}{4},\tfrac{1}{4},\tfrac{1}{4},1 \\ \tfrac{3}{4},\tfrac{3}{4},\tfrac{3}{4},\tfrac{3}{4};1 \end{matrix}\right)$
$\qquad + 256\,\Gamma^8\!\left(\dfrac{3}{4}\right)\, {}_5F_4\!\left(\begin{matrix} \tfrac{3}{4},\tfrac{3}{4},\tfrac{3}{4},\tfrac{3}{4},1 \\ \tfrac{5}{4},\tfrac{5}{4},\tfrac{5}{4},\tfrac{5}{4};1 \end{matrix}\right)\Bigg]$.

2. $\int_0^1 \dfrac{x}{\sqrt{1-x^2}}\, \mathbf{K}^2(x)\, dx = \dfrac{\pi^4}{16}\, {}_7F_6\!\left(\begin{matrix} \tfrac{1}{2},\tfrac{1}{2},\tfrac{1}{2},\tfrac{1}{2},\tfrac{1}{2},\tfrac{1}{2},\tfrac{5}{4} \\ \tfrac{1}{4},1,1,1,1,1;\,1 \end{matrix}\right)$.

3. $\int_0^1 x^{s-1}\, \mathbf{K}(a\sqrt{x})\, \mathbf{K}\left(\sqrt{1-x}\right) dx = \dfrac{\pi^2}{4}\, \dfrac{\Gamma^2(s)}{\Gamma^2\!\left(s+\dfrac{1}{2}\right)}\, {}_4F_3\!\left(\begin{matrix} \tfrac{1}{2},\tfrac{1}{2},s,s;\,a^2 \\ 1,s+\tfrac{1}{2},s+\tfrac{1}{2} \end{matrix}\right)$
$\qquad\qquad [\operatorname{Re} s > 0;\ |\arg(1-a^2)| < \pi]$.

4. $\int_0^1 \mathbf{K}(a\sqrt{x})\, \mathbf{K}\left(\sqrt{1-x}\right) dx = \dfrac{\pi}{2a}\,[\operatorname{Li}_2(a) - \operatorname{Li}_2(-a)] \qquad [|\arg(1-a^2)| < \pi]$.

5. $\int_0^1 \mathbf{K}(\sqrt{x})\, \mathbf{K}\left(\sqrt{1-x}\right) dx = \dfrac{\pi^3}{8}$.

6. $\int_0^1 x\, \mathbf{K}(a\sqrt{x})\, \mathbf{K}\left(\sqrt{1-x}\right) dx = \dfrac{\pi}{8a^3}\Bigg\{(1-a^2)\ln\dfrac{1-a}{1+a}$
$\qquad\qquad + (1+a^2)[\operatorname{Li}_2(a) - \operatorname{Li}_2(-a)]\Bigg\} \qquad [|\arg(1-a^2)| < \pi]$.

7. $\int_0^1 x\, K(\sqrt{x})\, K(\sqrt{1-x})\, dx = \dfrac{\pi^3}{16}$.

8. $\int_0^1 x^2\, K(\sqrt{x})\, K(\sqrt{1-x})\, dx = \dfrac{11\pi^3}{256}$.

9. $\displaystyle\int_0^1 \dfrac{x^{s-1}}{\sqrt{1+a^2x}}\, K\!\left(\dfrac{a\sqrt{x}}{\sqrt{1+a^2x}}\right) K(\sqrt{1-x})\, dx$

$= \dfrac{\pi^2}{4}\, \dfrac{\Gamma^2(s)}{\Gamma^2\!\left(s+\tfrac{1}{2}\right)}\, {}_4F_3\!\left(\begin{matrix}\tfrac{1}{2},\tfrac{1}{2},s,s;\ -a^2\\ 1,\, s+\tfrac{1}{2},\, s+\tfrac{1}{2}\end{matrix}\right)$ $\quad [\operatorname{Re} s > 0;\ |\arg(1+a^2)| < \pi]$.

10. $\displaystyle\int_0^1 \dfrac{1}{\sqrt{1+x}}\, K\!\left(\sqrt{\dfrac{x}{1+x}}\right) K(\sqrt{1-x})\, dx = \pi\, \mathbf{G}$.

11. $\displaystyle\int_0^1 \dfrac{x}{\sqrt{1+x}}\, K\!\left(\sqrt{\dfrac{x}{1+x}}\right) K(\sqrt{1-x})\, dx = \dfrac{\pi^2}{8}$.

12. $\displaystyle\int_0^1 \dfrac{x^2}{\sqrt{1+x}}\, K\!\left(\sqrt{\dfrac{x}{1+x}}\right) K(\sqrt{1-x})\, dx = \dfrac{\pi}{64}(3 + 14\,\mathbf{G})$.

13. $\displaystyle\int_0^{\pi/2} K(\sin x)\, K(\sin(2x))\, dx = -\dfrac{\pi^{-3}}{128}\, \Gamma^8\!\left(\dfrac{1}{4}\right)$

$+ \dfrac{\pi^{-3}}{64}\,\Gamma^8\!\left(\dfrac{1}{4}\right) {}_7F_6\!\left(\begin{matrix}\tfrac{1}{8},\tfrac{1}{8},\tfrac{1}{4},\tfrac{1}{4},\tfrac{5}{8},\tfrac{5}{8},1\\ \tfrac{3}{8},\tfrac{3}{8},\tfrac{3}{4},\tfrac{3}{4},\tfrac{7}{8},\tfrac{7}{8};\,1\end{matrix}\right) - \dfrac{\pi}{9}\,{}_7F_6\!\left(\begin{matrix}\tfrac{5}{8},\tfrac{5}{8},\tfrac{3}{4},\tfrac{3}{4},1,\tfrac{9}{8},\tfrac{9}{8}\\ \tfrac{7}{8},\tfrac{7}{8},\tfrac{5}{4},\tfrac{5}{4},\tfrac{11}{8},\tfrac{11}{8};\,1\end{matrix}\right)$.

4.22. The Complete Elliptic Integral $E(z)$

4.22.1. Integrals containing $E(z)$ and algebraic functions

1. $\displaystyle\int_0^1 \dfrac{x(1-x^2)^{s-1}}{(1-ax^2)^{s+1/2}}\, E(x)\, dx$

$= \dfrac{\pi}{2}\, \dfrac{s\,\Gamma^2(s)}{(2s+1)\Gamma^2\!\left(s+\tfrac{1}{2}\right)}\,(1-a)^{-1/2}\, {}_2F_1\!\left(\begin{matrix}1/2,\, s\\ s+3/2;\ a\end{matrix}\right)$

$\quad [\operatorname{Re} s > 0;\ |\arg(1-a)| < \pi]$.

2. $\displaystyle\int_0^1 \frac{x}{\sqrt{1-x^2}}\, \mathbf{E}(ax)\, dx$

$\displaystyle = \frac{\pi}{96a^3}\left[3a(2a^2+1)\sqrt{1-a^2} + 3(4a^2-1)\arcsin a\right]$ $\quad [|\arg(1-a^2)| < \pi]$.

3. $\displaystyle\int_0^a x^{s-1}(a-x)^{t-1}\, \mathbf{E}\left(b\sqrt{x(a-x)}\right) dx$

$\displaystyle = \frac{\pi}{2}\,\mathrm{B}(s,t)\, a^{s+t-1}\, {}_4F_3\!\left(\begin{array}{c} -\frac{1}{2}, \frac{1}{2}, s, t;\ \frac{a^2b^2}{4} \\ 1,\ \frac{s+t}{2},\ \frac{s+t+1}{2} \end{array}\right)$

$\quad [a, \operatorname{Re} s, \operatorname{Re} t > 0;\ |\arg(4-a^2b^2)| < \pi]$.

4. $\displaystyle\int_0^a \mathbf{E}\left(b\sqrt{x(a-x)}\right) dx = \frac{\pi a}{4}\left(\frac{2}{ab}\arcsin\frac{ab}{2} + \sqrt{1-\frac{a^2b^2}{4}}\right)$

$\quad [a > 0;\ |\arg(4-a^2b^2)| < \pi]$.

5. $\displaystyle\int_0^a x\, \mathbf{E}\left(b\sqrt{x(a-x)}\right) dx = \frac{\pi a^2}{8}\left(\frac{2}{ab}\arcsin\frac{ab}{2} + \sqrt{1-\frac{a^2b^2}{4}}\right)$

$\quad [a > 0;\ |\arg(4-a^2b^2)| < \pi]$.

6. $\displaystyle\int_0^a x^2\, \mathbf{E}\left(b\sqrt{x(a-x)}\right) dx$

$\displaystyle = \frac{\pi}{64b^3}\left[4(3a^2b^2-1)\arcsin\frac{ab}{2} + ab(5a^2b^2+2)\sqrt{1-\frac{a^2b^2}{4}}\right]$

$\quad [a > 0;\ |\arg(4-a^2b^2)| < \pi]$.

7. $\displaystyle\int_0^a x^{1/2}(a-x)^{1/2}\, \mathbf{E}\left(b\sqrt{x(a-x)}\right) dx = \frac{\pi^2 a^2}{48}\left[(4-a^2b^2)\psi_1\!\left(\frac{a^2b^2}{4}\right)\right.$

$\displaystyle \left. + (2+a^2b^2)\psi_2\!\left(\frac{a^2b^2}{4}\right) - \frac{a^2b^2}{4}\psi_3\!\left(\frac{a^2b^2}{4}\right)\right]$ $\quad [a > 0;\ |\arg(4-a^2b^2)| < \pi]$.

8. $\displaystyle\int_0^a x^{-1/2}(a-x)^{1/2}\, \mathbf{E}\left(b\sqrt{x(a-x)}\right) dx = \frac{\pi^2 a}{16}\left[(4-a^2b^2)\psi_1\!\left(\frac{a^2b^2}{4}\right)\right.$

$\displaystyle \left. + a^2b^2\psi_2\!\left(\frac{a^2b^2}{4}\right) - \frac{a^2b^2}{4}\psi_3\!\left(\frac{a^2b^2}{4}\right)\right]$ $\quad [a > 0;\ |\arg(4-a^2b^2)| < \pi]$.

4.22. The Complete Elliptic Integral $E(z)$

9. $\displaystyle\int_0^a x^{-1/2}(a-x)^{-1/2}\, E\left(b\sqrt{x(a-x)}\right) dx = \frac{\pi^2}{8}\left[(4-a^2b^2)\psi_1\left(\frac{a^2b^2}{4}\right)\right.$

$\left. + a^2b^2\psi_2\left(\frac{a^2b^2}{4}\right) - \frac{a^2b^2}{4}\psi_3\left(\frac{a^2b^2}{4}\right)\right]$ $\quad [a>0;\ |\arg(4-a^2b^2)|<\pi]$.

10. $\displaystyle\int_0^a \frac{x^{s-1}(a-x)^{t-1}}{1-b^2 x(a-x)}\, E\left(b\sqrt{x(a-x)}\right) dx$

$= \dfrac{\pi}{2}\, B(s,t)\, a^{s+t-1}\, {}_4F_3\!\left(\begin{array}{c} \frac{1}{2},\frac{3}{2},s,t;\ \frac{a^2b^2}{4} \\ 1,\ \frac{s+t}{2},\ \frac{s+t+1}{2} \end{array}\right)$

$[a,\ \mathrm{Re}\,s,\ \mathrm{Re}\,t>0;\ |\arg(4-a^2b^2)|<\pi]$.

11. $\displaystyle\int_0^a \frac{1}{1-b^2 x(a-x)}\, E\left(b\sqrt{x(a-x)}\right) dx = \frac{\pi a}{\sqrt{4-a^2b^2}}$

$[a>0;\ |\arg(4-a^2b^2)|<\pi]$.

12. $\displaystyle\int_0^a \frac{x}{1-b^2 x(a-x)}\, E\left(b\sqrt{x(a-x)}\right) dx = \frac{\pi a^2}{2\sqrt{4-a^2b^2}}$

$[a>0;\ |\arg(4-a^2b^2)|<\pi]$.

13. $\displaystyle\int_0^a \frac{x^{1/2}(a-x)^{1/2}}{1-b^2 x(a-x)}\, E\left(b\sqrt{x(a-x)}\right) dx$

$= \dfrac{\pi^2 a^2}{8(4-a^2b^2)}\left[(2+a^2b^2)\psi_2\!\left(\frac{a^2b^2}{4}\right) - a^2b^2\psi_1\!\left(\frac{a^2b^2}{4}\right) - \frac{a^2b^2}{4}\psi_3\!\left(\frac{a^2b^2}{4}\right)\right]$

$[a>0;\ |\arg(4-a^2b^2)|<\pi]$.

14. $\displaystyle\int_0^a \frac{x^{-1/2}(a-x)^{1/2}}{1-b^2 x(a-x)}\, E\left(b\sqrt{x(a-x)}\right) dx$

$= \dfrac{\pi^2 a}{4(4-a^2b^2)}\left[2(2-a^2b^2)\psi_1\!\left(\frac{a^2b^2}{4}\right) + \frac{3a^2b^2}{2}\psi_2\!\left(\frac{a^2b^2}{4}\right) - \frac{a^2b^2}{4}\psi_3\!\left(\frac{a^2b^2}{4}\right)\right]$

$[a>0;\ |\arg(4-a^2b^2)|<\pi]$.

15. $\displaystyle\int_0^a \frac{x^{-1/2}(a-x)^{-1/2}}{1-b^2 x(a-x)}\, E\left(b\sqrt{x(a-x)}\right) dx$

$= \dfrac{\pi^2}{2(4-a^2b^2)}\left[2(2-a^2b^2)\psi_1\!\left(\frac{a^2b^2}{4}\right) + \frac{3a^2b^2}{2}\psi_2\!\left(\frac{a^2b^2}{4}\right) - \frac{a^2b^2}{4}\psi_3\!\left(\frac{a^2b^2}{4}\right)\right]$

$[a>0;\ |\arg(4-a^2b^2)|<\pi]$.

16. $\int_0^a x^{s+1/2}(a-x)^s \, \mathbf{E}\left(b\sqrt[4]{x(a-x)}\right) dx = 2^{-2s-2}\pi^{3/2} a^{2s+3/2} \dfrac{\Gamma(2s+2)}{\Gamma\left(2s+\frac{5}{2}\right)}$

$\times \; {}_3F_2\left(\begin{array}{c} -\frac{1}{2},\, \frac{1}{2},\, 2s+2 \\ 1,\, 2s+\frac{5}{2}; \end{array} \dfrac{ab^2}{2}\right) \qquad [a>0;\; \operatorname{Re} s > -1;\; |\arg(2-a^2b^2)| < \pi].$

17. $\int_0^a x^{1/2} \, \mathbf{E}\left(b\sqrt[4]{x(a-x)}\right) dx$

$= \dfrac{\pi a^{1/2}}{16 b^3}\left[\sqrt{\dfrac{2}{a}}(2ab^2+1)\arcsin\left(b\sqrt{\dfrac{a}{2}}\right) + b(3ab^2-1)\sqrt{1-\dfrac{ab^2}{2}}\right]$

$[a>0;\; |\arg(2-a^2b^2)|<\pi].$

18. $\int_0^a x^{-1/2} \, \mathbf{E}\left(b\sqrt[4]{x(a-x)}\right) dx$

$= \dfrac{\pi\sqrt{a}}{2}\left[\dfrac{1}{b}\sqrt{\dfrac{2}{a}}\arcsin\left(b\sqrt{\dfrac{a}{2}}\right) + \sqrt{1-\dfrac{ab^2}{2}}\right] \qquad [|\arg(2-a^2b^2)|<\pi].$

19. $\int_0^a x^{1/4}(a-x)^{-1/4} \, \mathbf{E}\left(b\sqrt[4]{x(a-x)}\right) dx$

$= \dfrac{\pi^2 a}{12\sqrt{2}}\left[4\left(1-\dfrac{ab^2}{2}\right)\psi_1\left(\dfrac{ab^2}{2}\right) - (1-2ab^2)\psi_2\left(\dfrac{ab^2}{2}\right) - \dfrac{ab^2}{2}\psi_3\left(\dfrac{ab^2}{2}\right)\right]$

$[a>0;\; |\arg(2-a^2b^2)|<\pi].$

20. $\int_0^a x^{-1/4}(a-x)^{-3/4} \, \mathbf{E}\left(b\sqrt[4]{x(a-x)}\right) dx$

$= \dfrac{\pi^2}{\sqrt{2}}\left[\left(1-\dfrac{ab^2}{2}\right)\psi_1\left(\dfrac{ab^2}{2}\right) + \dfrac{ab^2}{2}\psi_2\left(\dfrac{ab^2}{2}\right) - \dfrac{ab^2}{8}\psi_3\left(\dfrac{ab^2}{2}\right)\right]$

$[a>0;\; |\arg(2-a^2b^2)|<\pi].$

21. $\int_0^a \dfrac{x^{s+1/2}(a-x)^s}{1-b^2\sqrt{x(a-x)}} \, \mathbf{E}\left(b\sqrt[4]{x(a-x)}\right) dx = \dfrac{\pi^{3/2} a^{2s+3/2}\Gamma(2s+2)}{2^{2s+2}\Gamma\left(2s+\frac{5}{2}\right)}$

$\times \; {}_3F_2\left(\begin{array}{c} \frac{1}{2},\, \frac{3}{2},\, 2s+2 \\ 1,\, 2s+\frac{5}{2}; \end{array} \dfrac{ab^2}{2}\right) \qquad [a>0;\; \operatorname{Re} s > -1;\; |\arg(2-a^2b^2)| < \pi].$

22. $\int_0^a \dfrac{x^{1/2}}{1-b^2\sqrt{x(a-x)}} \, \mathbf{E}\left(b\sqrt[4]{x(a-x)}\right) dx = \dfrac{\pi\sqrt{a}(ab^2+2)}{4b^2}\left(1-\dfrac{ab^2}{2}\right)^{-1/2}$

$\quad - \dfrac{\pi}{\sqrt{2}\,b^3}\arcsin\left(b\sqrt{\dfrac{a}{2}}\right) \qquad [a>0;\; |\arg(2-a^2b^2)|<\pi].$

23. $\int_0^a \dfrac{x^{-1/2}}{1 - b^2\sqrt{x(a-x)}} \mathbf{E}\left(b\sqrt[4]{x(a-x)}\right) dx = \pi\sqrt{\dfrac{2a}{2 - a^2 b^2}}$

$\qquad\qquad\qquad\qquad\qquad\qquad\qquad\qquad [a > 0;\ |\arg(2 - a^2 b^2)| < \pi].$

24. $\int_0^a \dfrac{x^{1/4}(a-x)^{-1/4}}{1 - b^2\sqrt{x(a-x)}} \mathbf{E}\left(b\sqrt[4]{x(a-x)}\right) dx = \dfrac{a\pi^2}{2^{5/2}}\left(1 - \dfrac{ab^2}{2}\right)^{-1}$

$\qquad \times \left[(1 + ab^2)\,\psi_2\!\left(\dfrac{ab^2}{2}\right) - ab^2 \psi_2\!\left(\dfrac{ab^2}{2}\right) - \dfrac{ab^2}{4}\psi_3\!\left(\dfrac{ab^2}{2}\right)\right]$

$\qquad\qquad\qquad\qquad\qquad\qquad\qquad\qquad [a > 0;\ |\arg(2 - a^2 b^2)| < \pi].$

25. $\int_0^a \dfrac{x^{-1/4}(a-x)^{-3/4}}{1 - b^2\sqrt{x(a-x)}} \mathbf{E}\left(b\sqrt[4]{x(a-x)}\right) dx$

$\qquad = \dfrac{\pi^2}{2^{5/2}}\left(1 - \dfrac{ab^2}{2}\right)^{-1}\left[4(1 - ab^2)\,\psi_1\!\left(\dfrac{ab^2}{2}\right)\right.$

$\qquad \left. + 3ab^2\psi_2\!\left(\dfrac{ab^2}{2}\right) - \dfrac{ab^2}{2}\psi_3\!\left(\dfrac{ab^2}{2}\right)\right] \quad [a > 0;\ |\arg(2 - a^2 b^2)| < \pi].$

26. $\int_0^a \dfrac{x^{-1/2}}{[1 + b^2\sqrt{x(a-x)}]^{1/2}} \mathbf{E}\!\left(\dfrac{b\sqrt[4]{x(a-x)}}{\sqrt{1 + b^2\sqrt{x(a-x)}}}\right) dx = \pi\sqrt{\dfrac{2a}{2 + ab^2}}$

$\qquad\qquad\qquad\qquad\qquad\qquad\qquad\qquad [a > 0;\ |\arg(2 + a^2 b^2)| < \pi].$

27. $\int_0^1 x^{s-1}(1 + ax)^\nu \mathbf{E}\left(\sqrt{1-x}\right) dx = \dfrac{\pi s^2 \Gamma(s)}{(2s + 1)\Gamma^2\!\left(s + \dfrac{1}{2}\right)}$

$\qquad \times {}_3F_2\!\left(\begin{matrix} -\nu,\, s,\, s+1;\ a \\ s + \dfrac{1}{2},\, s + \dfrac{3}{2} \end{matrix}\right) \quad [\operatorname{Re} s > 0;\ |\arg(1 - a)| < \pi].$

4.22.2. Integrals containing $\mathbf{E}(z)$, the exponential, hyperbolic and trigonometric functions

1. $\int_0^1 x^{s-1} e^{ax} \mathbf{E}\left(\sqrt{1-x}\right) dx = \dfrac{\pi \Gamma(s)\Gamma(s+1)}{2\Gamma\!\left(s + \dfrac{1}{2}\right)\Gamma\!\left(s + \dfrac{3}{2}\right)} {}_2F_3\!\left(\begin{matrix} s,\, s+1;\ a \\ s + \dfrac{1}{2},\, s + \dfrac{3}{2} \end{matrix}\right)$

$\qquad\qquad\qquad\qquad\qquad\qquad\qquad\qquad [\operatorname{Re} s > 0].$

2. $\int_0^1 x^{s-1} \left\{\begin{matrix} \sinh(a\sqrt{x}) \\ \sin(a\sqrt{x}) \end{matrix}\right\} \mathbf{E}\left(\sqrt{1-x}\right) dx$

$\qquad = \dfrac{\pi a(2s+1)\Gamma^2\!\left(s+\dfrac{1}{2}\right)}{4(s+1)\Gamma^2(s+1)} {}_2F_3\!\left(\begin{matrix} s + \dfrac{1}{2},\, s + \dfrac{3}{2};\ \mp\dfrac{a^2}{4} \\ \dfrac{3}{2},\, s+1,\, s+2 \end{matrix}\right) \quad [\operatorname{Re} s > -1/2].$

3. $\int_0^1 x^{s-1} \left\{ \begin{array}{c} \cosh(a\sqrt{x}) \\ \cos(a\sqrt{x}) \end{array} \right\} \mathbf{E}\left(\sqrt{1-x}\right) dx$

$$= \frac{\pi s \Gamma^2(s)}{(2s+1)\Gamma^2\left(s+\frac{1}{2}\right)} \; {}_2F_3\left(\begin{array}{c} s, s+1; \pm\frac{a^2}{4} \\ \frac{1}{2}, s+\frac{1}{2}, s+\frac{3}{2} \end{array}\right) \quad [\operatorname{Re} s > 0].$$

4. $\int_0^1 \sinh(a\sqrt{x}) \mathbf{E}\left(\sqrt{1-x}\right) dx = \frac{\pi^2}{2a}\left[I_0\left(\frac{a}{2}\right)I_1\left(\frac{a}{2}\right) - I_1^2\left(\frac{a}{2}\right)\right].$

5. $\int_0^1 \frac{1}{x}\sinh(a\sqrt{x}) \mathbf{E}\left(\sqrt{1-x}\right) dx = \frac{\pi^2 a}{4}\left[I_0^2\left(\frac{a}{2}\right) - I_1^2\left(\frac{a}{2}\right)\right].$

6. $\int_0^1 \frac{1}{\sqrt{x}}\cosh(a\sqrt{x}) \mathbf{E}\left(\sqrt{1-x}\right) dx = \frac{\pi^2}{4}\left[I_0^2\left(\frac{a}{2}\right) + I_1^2\left(\frac{a}{2}\right)\right].$

7. $\int_0^1 \sin(a\sqrt{x}) \mathbf{E}\left(\sqrt{1-x}\right) dx = \frac{\pi^2}{2a}\left[J_0\left(\frac{a}{2}\right)J_1\left(\frac{a}{2}\right) - J_1^2\left(\frac{a}{2}\right)\right].$

8. $\int_0^1 \frac{1}{x}\sin(a\sqrt{x}) \mathbf{E}\left(\sqrt{1-x}\right) dx = \frac{\pi^2 a}{4}\left[J_0^2\left(\frac{a}{2}\right) + J_1^2\left(\frac{a}{2}\right)\right].$

9. $\int_0^1 \frac{1}{\sqrt{x}}\cos(a\sqrt{x}) \mathbf{E}\left(\sqrt{1-x}\right) dx = \frac{\pi^2}{4}\left[J_0^2\left(\frac{a}{2}\right) - J_1^2\left(\frac{a}{2}\right)\right].$

10. $\int_0^{\pi/2} \cos(2nx) \mathbf{E}(a\sin x) dx = -2^{-2n-3}\pi(-a^2)^n \frac{\Gamma\left(n-\frac{1}{2}\right)\Gamma\left(n+\frac{1}{2}\right)}{(n!)^2}$

$$\times {}_3F_2\left(\begin{array}{c} n-\frac{1}{2}, n+\frac{1}{2}, n+\frac{1}{2} \\ n+1, 2n+1; \; a^2 \end{array}\right) \quad [|\arg(1-a^2)| < \pi].$$

11. $\int_0^{\pi/2} \cos^\nu x \cos(ax) \mathbf{E}(b\cos x) dx = \frac{2^{-\nu-2}\pi^2\Gamma(\nu+1)}{\Gamma\left(\frac{\nu-a}{2}+1\right)\Gamma\left(\frac{\nu+a}{2}+1\right)}$

$$\times {}_4F_3\left(\begin{array}{c} -\frac{1}{2}, \frac{1}{2}, \frac{\nu+1}{2}, 1+\frac{\nu}{2}; \; b^2 \\ 1, 1+\frac{\nu-a}{2}, 1+\frac{\nu+a}{2} \end{array}\right) \quad [\operatorname{Re}\nu > -1; \; |\arg(1-b^2)| < \pi].$$

4.22. The Complete Elliptic Integral $\mathbf{E}(z)$

12. $\displaystyle\int_0^{\pi/2} \frac{\cos^\nu x}{1-b^2\cos^2 x} \cos(ax)\, \mathbf{E}(b\cos x)\, dx = \frac{2^{-\nu-2}\pi^2 \Gamma(\nu+1)}{\Gamma\left(\frac{\nu-a}{2}+1\right)\Gamma\left(\frac{\nu+a}{2}+1\right)}$

$\times {}_4F_3\left(\begin{array}{c}\frac{1}{2},\frac{3}{2},\frac{\nu+1}{2},\frac{\nu}{2}+1;\,b^2 \\ 1,\frac{\nu-a}{2}+1,\frac{\nu+a}{2}+1\end{array}\right)$ $[\operatorname{Re}\nu > -1;\ |\arg(1-b^2)| < \pi]$.

13. $\displaystyle\int_0^\pi \sin^\nu x \begin{Bmatrix}\sin(ax)\\ \cos(ax)\end{Bmatrix} \mathbf{E}(b\sin x)\, dx$

$= \dfrac{2^{-\nu-1}\pi^2 \Gamma(\nu+1)}{\Gamma\left(\frac{\nu-a}{2}+1\right)\Gamma\left(\frac{\nu+a}{2}+1\right)} \begin{Bmatrix}\sin(a\pi/2)\\ \cos(a\pi/2)\end{Bmatrix} {}_4F_3\left(\begin{array}{c}-\frac{1}{2},\frac{1}{2},\frac{\nu+1}{2},\frac{\nu}{2}+1;\,b^2 \\ 1,\frac{\nu-a}{2}+1,\frac{\nu+a}{2}+1\end{array}\right)$

$[\operatorname{Re}\nu > -1;\ |\arg(1-b^2)| < \pi]$.

14. $\displaystyle\int_0^\pi \frac{\sin^\nu x}{1-b^2 \sin^2 x} \begin{Bmatrix}\sin(ax)\\ \cos(ax)\end{Bmatrix} \mathbf{E}(b\sin x)\, dx$

$= \dfrac{2^{-\nu-1}\pi^2 \Gamma(\nu+1)}{\Gamma\left(\frac{\nu-a}{2}+1\right)\Gamma\left(\frac{\nu+a}{2}+1\right)} \begin{Bmatrix}\sin(a\pi/2)\\ \cos(a\pi/2)\end{Bmatrix} {}_4F_3\left(\begin{array}{c}\frac{1}{2},\frac{3}{2},\frac{\nu+1}{2},\frac{\nu}{2}+1;\,b^2 \\ 1,\frac{\nu-a}{2}+1,\frac{\nu+a}{2}+1\end{array}\right)$

$[\operatorname{Re}\nu > -1;\ |\arg(1-b^2)| < \pi]$.

15. $\displaystyle\int_0^\pi \cos(nx)\, \mathbf{E}(a\cos x)\, dx = -2^{-n-2}\pi a^n \frac{\Gamma\left(\frac{n-1}{2}\right)\Gamma\left(\frac{n+1}{2}\right)}{\Gamma^2\left(\frac{n}{2}+1\right)} \cos^2\frac{n\pi}{2}$

$\times {}_3F_2\left(\begin{array}{c}\frac{n-1}{2},\frac{n+1}{2},\frac{n+1}{2} \\ \frac{n}{2}+1,\, n+1;\, a^2\end{array}\right)$ $[|\arg(1-a^2)| < \pi]$.

16. $\displaystyle\int_0^{m\pi} \begin{Bmatrix}\sin(ax)\\ \cos(ax)\end{Bmatrix} \mathbf{E}(b\sin x)\, dx = \frac{\pi}{a}\sin\frac{m\pi a}{2} \begin{Bmatrix}\sin(m\pi a/2)\\ \cos(m\pi a/2)\end{Bmatrix}$

$\times {}_3F_2\left(\begin{array}{c}-\frac{1}{2},\frac{1}{2},\frac{1}{2};\,b^2 \\ 1-\frac{a}{2},\,1+\frac{a}{2}\end{array}\right)$ $[|\arg(1-b^2)| < \pi]$.

17. $\displaystyle\int_0^{m\pi} e^{-ax}\, \mathbf{E}(b\sin x)\, dx = \frac{\pi}{2a}(1-e^{-m\pi a})\, {}_3F_2\left(\begin{array}{c}-\frac{1}{2},\frac{1}{2},\frac{1}{2};\,b^2 \\ 1-\frac{ia}{2},\,1+\frac{ia}{2}\end{array}\right)$

$[|\arg(1-b^2)| < \pi]$.

18. $\displaystyle\int_0^{m\pi} \frac{e^{-ax}}{1-b^2\sin^2 x}\, \mathbf{E}(b\sin x)\, dx = \frac{\pi}{2a}(1-e^{-m\pi a})\, {}_3F_2\left(\begin{array}{c}\frac{1}{2},\frac{1}{2},\frac{3}{2};\,b^2 \\ 1-\frac{ia}{2},\,1+\frac{ia}{2}\end{array}\right)$

$[|\arg(1-b^2)| < \pi]$.

19. $\int_0^\infty e^{-ax} \mathbf{E}(b \sin x) \, dx = \dfrac{\pi}{2a} \, {}_3F_2\left(\begin{array}{c} -\frac{1}{2}, \frac{1}{2}, \frac{1}{2}; \ b^2 \\ 1 - \frac{ia}{2}, 1 + \frac{ia}{2} \end{array}\right)$

$\left[\operatorname{Re} a > 0; \ |\arg(1 - b^2)| < \pi\right].$

4.22.3. Integrals containing $\mathbf{E}(z)$ and the logarithmic function

1. $\displaystyle\int_0^1 x(1-x^2)^{-3/2} \ln x \, \mathbf{E}(x) \, dx = \dfrac{\pi}{2} - \dfrac{\pi^2}{4}.$

2. $\displaystyle\int_0^1 x^{s-1} \ln(1+ax) \, \mathbf{E}\left(\sqrt{1-x}\right) dx$

$= \dfrac{\pi a (s+1) \Gamma^2(s+1)}{(2s+3)\Gamma^2\left(s+\frac{3}{2}\right)} \, {}_4F_3\left(\begin{array}{c} 1, 1, s+1, s+2; \ -a \\ 2, s+\frac{3}{2}, s+\frac{5}{2} \end{array}\right)$

$\left[\operatorname{Re} s > -1; \ |\arg(1+a)| < \pi\right].$

3. $\displaystyle\int_0^1 x^{-3/2} \ln(1+ax) \, \mathbf{E}\left(\sqrt{1-x}\right) dx = 2\pi (a+1)^{1/2} \, \mathbf{E}\left(\sqrt{\dfrac{a}{a+1}}\right) - \pi^2$

$\left[|\arg(1+a)| < \pi\right].$

4. $\displaystyle\int_0^1 x^{-3/2} \ln(1-ax) \, \mathbf{E}\left(\sqrt{1-x}\right) dx = 2\pi \, \mathbf{E}(a) - \pi^2 \qquad \left[|\arg(1-a)| < \pi\right].$

5. $\displaystyle\int_0^1 x^{s-1} \ln\left(a\sqrt{x} + \sqrt{1+a^2 x}\right) \mathbf{E}\left(\sqrt{1-x}\right) dx = \dfrac{\pi a (2s+1) \Gamma^2\left(s+\frac{1}{2}\right)}{4(s+1)\Gamma^2(s+1)}$

$\times {}_4F_3\left(\begin{array}{c} \frac{1}{2}, \frac{1}{2}, s+\frac{1}{2}, s+\frac{3}{2} \\ \frac{3}{2}, s+1, s+2; \ -a^2 \end{array}\right) \qquad \left[\operatorname{Re} s > -1/2; \ |\arg(1+a^2)| < \pi\right].$

6. $\displaystyle\int_0^1 \ln\left(a\sqrt{x} + \sqrt{1+a^2 x}\right) \mathbf{E}\left(\sqrt{1-x}\right) dx$

$= \dfrac{\pi^2}{72a} \left[4(a^2+1)\psi_1(-a^2) - 4(1-2a^2)\psi_2(-a^2) - 5a^2 \psi_3(-a^2)\right]$

$\left[|\arg(1+a^2)| < \pi\right].$

7. $\displaystyle\int_0^1 \dfrac{1}{\sqrt{1+a^2 x}} \ln\left(a\sqrt{x} + \sqrt{1+a^2 x}\right) \mathbf{E}\left(\sqrt{1-x}\right) dx$

$= \dfrac{\pi^2}{8a^3} \left(2 + a^2 - 2\sqrt{1+a^2} + 2a^2 \ln \dfrac{1+\sqrt{1+a^2}}{2}\right) \qquad \left[|\arg(1+a^2)| < \pi\right].$

8. $\int_0^1 \frac{x^{-1}}{\sqrt{1+a^2x}} \ln\left(a\sqrt{x} + \sqrt{1+a^2x}\right) \mathbf{E}\left(\sqrt{1-x}\right) dx$

$$= \frac{\pi^2}{2a}\left(\sqrt{1+a^2} - 1\right) \quad [|\arg(1+a^2)| < \pi].$$

9. $\int_0^1 x^{s-1} \ln\frac{1+a\sqrt{x}}{1-a\sqrt{x}} \mathbf{E}\left(\sqrt{1-x}\right) dx = \frac{\pi a(2s+1)\Gamma^2\left(s+\frac{1}{2}\right)}{2(s+1)\Gamma^2(s+1)}$

$$\times {}_4F_3\left(\begin{matrix}\frac{1}{2}, 1, s+\frac{1}{2}, s+\frac{3}{2} \\ \frac{3}{2}, s+1, s+2; a^2\end{matrix}\right) \quad [\operatorname{Re} s > -1/2; |\arg(1-a^2)| < \pi].$$

10. $\int_0^1 \ln\frac{1+a\sqrt{x}}{1-a\sqrt{x}} \mathbf{E}\left(\sqrt{1-x}\right) dx$

$$= \frac{\pi}{6a^3}\left[3\pi a^2 + 4(a^2-1)\mathbf{K}(a) + 4(1-2a^2)\mathbf{E}(a)\right] \quad [|\arg(1-a^2)| < \pi].$$

11. $\int_0^1 \ln\frac{1+\sqrt{x}}{1-\sqrt{x}} \mathbf{E}\left(\sqrt{1-x}\right) dx = \frac{\pi^2}{2} - \frac{2\pi}{3}.$

12. $\int_0^1 \frac{1}{x} \ln\frac{1+a\sqrt{x}}{1-a\sqrt{x}} \mathbf{E}\left(\sqrt{1-x}\right) dx = \frac{2\pi}{a}\left[(a^2-1)\mathbf{K}(a) + \mathbf{E}(a)\right]$

$$[|\arg(1-a^2)| < \pi].$$

13. $\int_0^1 \frac{1}{x} \ln\frac{1+\sqrt{x}}{1-\sqrt{x}} \mathbf{E}\left(\sqrt{1-x}\right) dx = 2\pi.$

14. $\int_0^1 x^{s-1} \ln\frac{1+\sqrt{1-x^2}}{1-\sqrt{1-x^2}} \mathbf{E}(ax) dx = \frac{\pi^{3/2}\Gamma\left(\frac{s}{2}\right)}{2s\Gamma\left(\frac{s+1}{2}\right)} {}_4F_3\left(\begin{matrix}-\frac{1}{2}, \frac{1}{2}, \frac{s}{2}, \frac{s}{2}; a^2 \\ 1, \frac{s+1}{2}, \frac{s}{2}+1\end{matrix}\right)$

$$[\operatorname{Re} s > 0; |\arg(1-a^2)| < \pi].$$

15. $\int_0^1 x \ln\frac{1+\sqrt{1-x^2}}{1-\sqrt{1-x^2}} \mathbf{E}(ax) dx$

$$= \frac{\pi}{6a^2}\left[3a\arcsin a + (a^2+2)\sqrt{1-a^2} - 2\right] \quad [|\arg(1-a^2)| < \pi].$$

16. $\int_0^1 \frac{x}{1-a^2x^2} \ln\frac{1+\sqrt{1-x^2}}{1-\sqrt{1-x^2}} \mathbf{E}(ax) dx = \frac{\pi}{a^2}\left(1 - \sqrt{1-a^2}\right)$

$$[|\arg(1-a^2)| < \pi].$$

17. $\displaystyle\int_0^1 \frac{x}{\sqrt{1+a^2x^2}} \ln\frac{1+\sqrt{1-x^2}}{1-\sqrt{1-x^2}} \mathbf{E}\left(\frac{ax}{\sqrt{1+a^2x^2}}\right) dx = \frac{\pi}{a^2}\left(\sqrt{1+a^2}-1\right)$

$$[|\arg(1+a^2)| < \pi].$$

18. $\displaystyle\int_0^a x^{s-1} \ln\frac{a+\sqrt{a^2-x^2}}{x} \mathbf{E}(bx)\, dx$

$$= \frac{\pi^{3/2} a^s \Gamma\left(\frac{s}{2}\right)}{4s\,\Gamma\left(\frac{s+1}{2}\right)} {}_4F_3\left(\begin{array}{c}-\frac{1}{2},\frac{1}{2},\frac{s}{2},\frac{s}{2};\ a^2b^2\\ 1,\frac{s+1}{2},\frac{s}{2}+1\end{array}\right)$$

$$[a,\,\operatorname{Re} s>0;\ |\arg(1-a^2b^2)|<\pi].$$

19. $\displaystyle\int_0^a x \ln\frac{a+\sqrt{a^2-x^2}}{x} \mathbf{E}(bx)\, dx$

$$= \frac{\pi}{12b^2}\left[3ab\arcsin(ab) + (a^2b^2+2)\sqrt{1-a^2b^2} - 2\right]$$

$$[a>0;\ |\arg(1-a^2b^2)|<\pi].$$

20. $\displaystyle\int_0^a x^3 \ln\frac{a+\sqrt{a^2-x^2}}{x} \mathbf{E}(bx)\, dx = \frac{\pi}{1440 b^4}$

$$\times\left[15ab(4a^2b^2+3)\arcsin(ab) + (54a^4b^4 - 13a^2b^2 + 64)\sqrt{1-a^2b^2} - 64\right]$$

$$[a>0;\ |\arg(1-a^2b^2)|<\pi].$$

21. $\displaystyle\int_0^a \frac{x^{s-1}}{1-b^2x^2} \ln\frac{a+\sqrt{a^2-x^2}}{x} \mathbf{E}(bx)\, dx$

$$= \frac{\pi^{3/2} a^s \Gamma\left(\frac{s}{2}\right)}{4s\,\Gamma\left(\frac{s+1}{2}\right)} {}_4F_3\left(\begin{array}{c}\frac{1}{2},\frac{3}{2},\frac{s}{2},\frac{s}{2};\ a^2b^2\\ 1,\frac{s+1}{2},\frac{s}{2}+1\end{array}\right)$$

$$[a,\,\operatorname{Re} s>0;\ |\arg(1-a^2b^2)|<\pi].$$

22. $\displaystyle\int_0^a \frac{1}{1-b^2x^2} \ln\frac{a+\sqrt{a^2-x^2}}{x} \mathbf{E}(bx)\, dx = \frac{\pi^2 a}{4}\psi_1(a^2b^2)$

$$[a>0;\ |\arg(1-a^2b^2)|<\pi].$$

23. $\displaystyle\int_0^a \frac{x}{1-b^2x^2}\ln\frac{a+\sqrt{a^2-x^2}}{x}\mathbf{E}(bx)\,dx = \frac{\pi}{2b^2}\left(1-\sqrt{1-a^2b^2}\right)$

$$[a>0;\ |\arg(1-a^2b^2)|<\pi].$$

24. $\displaystyle\int_0^1 \frac{1}{1-x^2}\ln\frac{1+\sqrt{1-x^2}}{x}\mathbf{E}(x)\,dx = \frac{1}{16\pi}\Gamma^4\left(\frac{1}{4}\right).$

25. $\displaystyle\int_0^a x\sqrt{b^2x^2+1}\,\ln\frac{a+\sqrt{a^2-x^2}}{x}\,\mathbf{E}\!\left(\frac{bx}{\sqrt{b^2x^2+1}}\right)dx$

$\quad = \dfrac{\pi}{12b^2}\left[3ab\ln\left(ab+\sqrt{a^2b^2+1}\right) + (a^2b^2-2)\sqrt{a^2b^2+1}+2\right]$

$$[a>0;\ |\arg(1+a^2b^2)|<\pi].$$

26. $\displaystyle\int_0^a x^3\sqrt{b^2x^2+1}\,\ln\frac{a+\sqrt{a^2-x^2}}{x}\,\mathbf{E}\!\left(\frac{bx}{\sqrt{b^2x^2+1}}\right)dx$

$\quad = \dfrac{\pi}{1440b^4}\Big[15ab(4a^2b^2-3)\ln\left(ab+\sqrt{a^2b^2+1}\right)$
$\quad + (54a^4b^4+13a^2b^2+64)\sqrt{a^2b^2+1}-64\Big]\quad [a>0;\ |\arg(1+a^2b^2)|<\pi].$

27. $\displaystyle\int_0^1 x^{s-1}\ln^2\!\left(a\sqrt{x}+\sqrt{1+a^2x}\right)\mathbf{E}(\sqrt{1-x})\,dx$

$\quad = \dfrac{\pi a^2(s+1)\Gamma^2(s+1)}{(2s+3)\Gamma^2\!\left(s+\frac{3}{2}\right)}\,{}_5F_4\!\left(\begin{array}{c}1,1,1,s+1,s+2\\ \frac{3}{2},2,s+\frac{3}{2},s+\frac{5}{2};\ -a^2\end{array}\right)$

$$[a>0;\ \mathrm{Re}\,s>-1;\ |\arg(1+a^2)|<\pi].$$

28. $\displaystyle\int_0^1 x^{-3/2}\ln^2\!\left(a\sqrt{x}+\sqrt{1+a^2x}\right)\mathbf{E}(\sqrt{1-x})\,dx$

$\quad = \pi^2\!\left(\sqrt{1+a^2}-\ln\frac{1+\sqrt{1+a^2}}{2}-1\right)\quad [a>0;\ |\arg(1+a^2)|<\pi].$

4.22.4. Integrals containing $\mathbf{E}(z)$ and inverse trigonometric functions

1. $\displaystyle\int_0^1 x^{s-1}\arcsin(a\sqrt{x})\,\mathbf{E}(\sqrt{1-x})\,dx = \dfrac{\pi a\Gamma\!\left(s+\frac{1}{2}\right)\Gamma\!\left(s+\frac{3}{2}\right)}{2(s+1)\Gamma^2(s+1)}$

$\quad \times {}_4F_3\!\left(\begin{array}{c}\frac{1}{2},\frac{1}{2},s+\frac{1}{2},s+\frac{3}{2}\\ \frac{3}{2},s+1,s+2;\ a^2\end{array}\right)\quad [\mathrm{Re}\,s>-1/2;\ |\arg(1-a^2)|<\pi].$

2. $\int_0^1 \arcsin(a\sqrt{x})\, \mathbf{E}(\sqrt{1-x})\, dx$

$$= \frac{\pi^2}{72a}[4(a^2-1)\psi_1(a^2) + 4(2a^2+1)\psi_2(a^2) - 5a^2\psi_3(a^2)]$$
$$[|\arg(1-a^2)| < \pi].$$

3. $\int_0^1 \arcsin(a\sqrt{x})\, \mathbf{E}(\sqrt{1-x})\, dx = \frac{1}{144\pi}\Gamma^4\left(\frac{1}{4}\right) + \frac{1}{\pi}\Gamma^4\left(\frac{3}{4}\right)$
$$[|\arg(1-a^2)| < \pi].$$

4. $\int_0^1 \frac{1}{x}\arcsin(a\sqrt{x})\, \mathbf{E}(\sqrt{1-x})\, dx = \frac{\pi^2 a}{4}\psi_2(a^2).$

5. $\int_0^1 \frac{1}{x}\arcsin(\sqrt{x})\, \mathbf{E}(\sqrt{1-x})\, dx = \frac{1}{16\pi}\Gamma^4\left(\frac{1}{4}\right) - \frac{1}{\pi}\Gamma^4\left(\frac{3}{4}\right).$

6. $\int_0^1 \frac{x^{s-1}}{\sqrt{1-a^2 x}}\arcsin(a\sqrt{x})\, \mathbf{E}(\sqrt{1-x})\, dx = \frac{\pi a \Gamma\left(s+\frac{1}{2}\right)\Gamma\left(s+\frac{3}{2}\right)}{2(s+1)\Gamma^2(s+1)}$

$$\times \,_4F_3\left(\begin{array}{c} 1,1,s+\frac{1}{2},s+\frac{3}{2} \\ \frac{3}{2}, s+1, s+2; a^2 \end{array}\right) \quad [\operatorname{Re} s > -1/2;\ |\arg(1-a^2)| < \pi].$$

7. $\int_0^1 \frac{1}{\sqrt{1-a^2 x}}\arcsin(a\sqrt{x})\, \mathbf{E}(\sqrt{1-x})\, dx$

$$= \frac{\pi^2}{8a^3}\left(2 - a^2 - 2\sqrt{1-a^2} - 2a^2 \ln\frac{1+\sqrt{1-a^2}}{2}\right) \quad [|\arg(1-a^2)| < \pi].$$

8. $\int_0^1 \frac{x^{-1}}{\sqrt{1-a^2 x}}\arcsin(a\sqrt{x})\, \mathbf{E}(\sqrt{1-x})\, dx = \frac{\pi^2}{2a}\left(1 - \sqrt{1-a^2}\right)$
$$[|\arg(1-a^2)| < \pi].$$

9. $\int_0^1 x^{s-1}\arccos x\, \mathbf{E}(ax)\, dx = \frac{\pi^{3/2}\Gamma\left(\frac{s+1}{2}\right)}{2s^2\Gamma\left(\frac{s}{2}\right)}\,_4F_3\left(\begin{array}{c} -\frac{1}{2}, \frac{1}{2}, \frac{s}{2}, \frac{s+1}{2}; a^2 \\ 1, \frac{s}{2}+1, \frac{s}{2}+1 \end{array}\right)$
$$[\operatorname{Re} s > 0;\ |\arg(1-a^2)| < \pi].$$

10. $\int_0^1 \arccos x\, \mathbf{E}(x)\, dx = \frac{\pi^2}{16}(1+2\ln 2).$

11. $\displaystyle\int_0^1 x\arccos x\, \mathbf{E}(x)\, dx = \frac{1}{288\pi}\Gamma^4\!\left(\frac{1}{4}\right) + \frac{1}{2\pi}\Gamma^4\!\left(\frac{3}{4}\right).$

12. $\displaystyle\int_0^1 x^2 \arccos x\, \mathbf{E}(x)\, dx = \frac{\pi^2}{256}(5 + 4\ln 2).$

13. $\displaystyle\int_0^1 x^{s-1}\sqrt{1+a^2x^2}\, \arccos x\, \mathbf{E}\!\left(\frac{ax}{\sqrt{1+a^2x^2}}\right) dx = \frac{\pi^{3/2}\Gamma\!\left(\frac{s+1}{2}\right)}{2s^2\Gamma\!\left(\frac{s}{2}\right)}$

$\times\, {}_4F_3\!\left(\begin{matrix}-\frac{1}{2},\frac{1}{2},\frac{s}{2},\frac{s+1}{2}\\ 1,\frac{s}{2}+1,\frac{s}{2}+1;\ -a^2\end{matrix}\right)\quad [\operatorname{Re} s > 0;\ |\arg(1+a^2)|<\pi].$

14. $\displaystyle\int_0^1 \frac{x^{s-1}}{1-a^2x^2}\arccos x\, \mathbf{E}(ax)\, dx$

$=\dfrac{\pi^{3/2}\Gamma\!\left(\frac{s+1}{2}\right)}{2s^2\Gamma\!\left(\frac{s}{2}\right)}\, {}_4F_3\!\left(\begin{matrix}\frac{1}{2},\frac{3}{2},\frac{s}{2},\frac{s+1}{2}\\ 1,\frac{s}{2}+1,\frac{s}{2}+1;\ a^2\end{matrix}\right)\quad [\operatorname{Re} s > 0;\ |\arg(1-a^2)|<\pi].$

15. $\displaystyle\int_0^1 \frac{1}{1-a^2x^2}\arccos x\, \mathbf{E}(ax)\, dx = \frac{\pi}{2a}\arcsin a\qquad [|\arg(1-a^2)|<\pi].$

16. $\displaystyle\int_0^1 \frac{x}{1-x^2}\arccos x\, \mathbf{E}(x)\, dx = \frac{1}{32\pi}\left[\Gamma^4\!\left(\frac{1}{4}\right) - 16\Gamma^4\!\left(\frac{3}{4}\right)\right].$

17. $\displaystyle\int_0^1 \frac{x^2}{1-x^2}\arccos x\, \mathbf{E}(x)\, dx = \frac{\pi^2}{16}(3 - 2\ln 2).$

18. $\displaystyle\int_0^1 \frac{1}{\sqrt{1+a^2x^2}}\arccos x\, \mathbf{E}\!\left(\frac{ax}{\sqrt{1+a^2x^2}}\right) dx = \frac{\pi}{2a}\ln\!\left(a+\sqrt{1+a^2}\right)$

$[|\arg(1+a^2)|<\pi].$

19. $\displaystyle\int_0^1 x^{s-1}\arctan(a\sqrt{x})\, \mathbf{E}(\sqrt{1-x})\, dx = \frac{\pi a\,\Gamma\!\left(s+\frac{1}{2}\right)\Gamma\!\left(s+\frac{3}{2}\right)}{2(s+1)\Gamma^2(s+1)}$

$\times\, {}_4F_3\!\left(\begin{matrix}\frac{1}{2},1,s+\frac{1}{2},s+\frac{3}{2}\\ \frac{3}{2},s+1,s+2;\ -a^2\end{matrix}\right)\quad [\operatorname{Re} s > 0;\ |\arg(1+a^2)|<\pi].$

20. $\displaystyle\int_0^1 \arctan(a\sqrt{x})\, \mathrm{E}\left(\sqrt{1-x}\right) dx$

$\displaystyle = \frac{\pi(a^2+1)^{1/2}}{12a^3}\left[4(2a^2+1)\,\mathrm{E}\left(\frac{a}{\sqrt{a^2+1}}\right) - 4\,\mathrm{K}\left(\frac{a}{\sqrt{a^2+1}}\right) - \frac{3\pi a^2}{\sqrt{a^2+1}}\right]$

$[|\arg(1+a^2)| < \pi].$

21. $\displaystyle\int_0^1 \arctan\sqrt{x}\, \mathrm{E}\left(\sqrt{1-x}\right) dx = -\frac{\pi^2}{4} + \frac{1}{12}\sqrt{\frac{\pi}{2}}\left[\Gamma^4\!\left(\frac{1}{4}\right) + 12\Gamma^4\!\left(\frac{3}{4}\right)\right].$

22. $\displaystyle\int_0^1 \frac{1}{x}\arctan(a\sqrt{x})\, \mathrm{E}\left(\sqrt{1-x}\right) dx = \frac{\pi}{a}(a^2+1)^{1/2}$

$\displaystyle\times \left[\mathrm{K}\left(\frac{a}{\sqrt{a^2+1}}\right) - \mathrm{E}\left(\frac{a}{\sqrt{a^2+1}}\right)\right]\qquad [|\arg(1+a^2)| < \pi].$

23. $\displaystyle\int_0^1 x^{s-1}\arcsin^2(a\sqrt{x})\,\mathrm{E}\left(\sqrt{1-x}\right) dx$

$\displaystyle = \frac{\pi a^2(s+1)\Gamma^2(s+1)}{(2s+3)\Gamma^2\!\left(s+\frac{3}{2}\right)}\,{}_5F_4\!\left(\begin{array}{c}1,1,1,s+1,s+2\\ \frac{3}{2},2,s+\frac{3}{2},s+\frac{5}{2};\ a^2\end{array}\right)$

$[\operatorname{Re} s > -1;\ |\arg(1-a^2)| < \pi].$

24. $\displaystyle\int_0^1 x^{-3/2}\arcsin^2(a\sqrt{x})\,\mathrm{E}\left(\sqrt{1-x}\right) dx$

$\displaystyle = \pi^2\left(1 - \sqrt{1-a^2} + \ln\frac{1+\sqrt{1-a^2}}{2}\right)\qquad [|\arg(1-a^2)| < \pi].$

25. $\displaystyle\int_0^1 x^{-3/2}\arcsin^2\sqrt{x}\,\mathrm{E}\left(\sqrt{1-x}\right) dx = \pi^2(1-\ln 2).$

4.22.5. Integrals containing $\mathrm{E}(z)$ and $\mathrm{Li}_2(z)$

1. $\displaystyle\int_0^1 x^{s-1}\,\mathrm{Li}_2(ax)\,\mathrm{E}\left(\sqrt{1-x}\right) dx$

$\displaystyle = \frac{\pi a(s+1)\Gamma^2(s+1)}{(2s+3)\Gamma^2\!\left(s+\frac{3}{2}\right)}\,{}_4F_3\!\left(\begin{array}{c}1,1,1,s+1,s+2\\ 2,2,s+\frac{3}{2},s+\frac{5}{2};\ a\end{array}\right)$

$[\operatorname{Re} s > -1;\ |\arg(1-a)| < \pi].$

4.22.6. Integrals containing $E(z)$, shi(z) and Si(z)

1. $\displaystyle\int_0^1 x^{s-1} \left\{ \begin{array}{l} \text{shi}(a\sqrt{x}) \\ \text{Si}(a\sqrt{x}) \end{array} \right\} E(\sqrt{1-x})\,dx$

$\displaystyle = \frac{\pi a(2s+1)\Gamma^2\left(s+\frac{1}{2}\right)}{4(s+1)\Gamma^2(s+1)} \, {}_3F_4\left(\begin{array}{c} \frac{1}{2}, s+\frac{1}{2}, s+\frac{3}{2}; \pm\frac{a^2}{4} \\ \frac{3}{2}, \frac{3}{2}, s+1, s+2 \end{array}\right)$ [Re $s > -1/2$].

2. $\displaystyle\int_0^1 \text{Si}(a\sqrt{x}) E(\sqrt{1-x})\,dx$

$\displaystyle = \frac{\pi^2}{6a}\left[a^2 J_0^2\left(\frac{a}{2}\right) - 2a J_0\left(\frac{a}{2}\right) J_1\left(\frac{a}{2}\right) + (1+a^2) J_1^2\left(\frac{a}{2}\right)\right].$

4.22.7. Integrals containing $E(z)$ and erf(z)

1. $\displaystyle\int_0^1 x^{s-1} \text{erf}(a\sqrt{x}) E(\sqrt{1-x})\,dx$

$\displaystyle = \frac{\sqrt{\pi}\, a(2s+1)\Gamma^2\left(s+\frac{1}{2}\right)}{2(s+1)\Gamma^2(s+1)} \, {}_3F_3\left(\begin{array}{c} \frac{1}{2}, s+\frac{1}{2}, s+\frac{3}{2} \\ \frac{3}{2}, s+1, s+2; -a^2 \end{array}\right)$ [Re $s > -1/2$].

2. $\displaystyle\int_0^1 x^{s-1} e^{a^2 x} \text{erf}(a\sqrt{x}) E(\sqrt{1-x})\,dx$

$\displaystyle = \frac{\sqrt{\pi}\, a(2s+1)\Gamma^2\left(s+\frac{1}{2}\right)}{2(s+1)\Gamma^2(s+1)} \, {}_3F_3\left(\begin{array}{c} 1, s+\frac{1}{2}, s+\frac{3}{2} \\ \frac{3}{2}, s+1, s+2; a^2 \end{array}\right)$ [Re $s > -1/2$].

4.22.8. Integrals containing $E(z)$, $S(z)$ and $C(z)$

1. $\displaystyle\int_0^1 x^{s-1} S(a\sqrt{x}) E(\sqrt{1-x})\,dx$

$\displaystyle = \frac{1}{3}\sqrt{\frac{\pi a^3}{2}} \, \frac{(4s+3)\Gamma^2\left(s+\frac{3}{4}\right)}{(4s+5)\Gamma^2\left(s+\frac{5}{4}\right)} \, {}_3F_4\left(\begin{array}{c} \frac{3}{4}, s+\frac{3}{4}, s+\frac{7}{4}; -\frac{a^2}{4} \\ \frac{3}{2}, \frac{7}{4}, s+\frac{5}{4}, s+\frac{9}{4} \end{array}\right)$ [Re $s > -3/4$].

2. $\displaystyle\int_0^1 x^{s-1} C(a\sqrt{x}) E(\sqrt{1-x})\,dx$

$\displaystyle = \sqrt{\frac{\pi a}{2}} \, \frac{(4s+1)\Gamma^2\left(s+\frac{1}{4}\right)}{(4s+3)\Gamma^2\left(s+\frac{3}{4}\right)} \, {}_3F_4\left(\begin{array}{c} \frac{1}{4}, s+\frac{1}{4}, s+\frac{5}{4}; -\frac{a^2}{4} \\ \frac{1}{2}, \frac{5}{4}, s+\frac{3}{4}, s+\frac{7}{4} \end{array}\right)$ [Re $s > -1/4$].

4.22.9. Integrals containing $\mathrm{E}(z)$ and $\gamma(\nu, z)$

1. $\displaystyle\int_0^1 x^{s-1} \gamma(\nu, ax) \, \mathrm{E}\left(\sqrt{1-x}\right) dx$

$$= \frac{\pi a^\nu (s+\nu) \Gamma^2(s+\nu)}{\nu(2s+\nu+1)\Gamma^2\left(s+\nu+\frac{1}{2}\right)} \,_3F_3\left(\begin{array}{c} \nu,\, s+\nu,\, s+\nu+1;\, -a \\ \nu+1,\, s+\nu+\frac{1}{2},\, s+\nu+\frac{3}{2} \end{array}\right)$$

$[\mathrm{Re}\,(s+\nu) > 0]$.

2. $\displaystyle\int_0^1 x^{s-1} e^{ax} \gamma(\nu, ax) \, \mathrm{E}\left(\sqrt{1-x}\right) dx$

$$= \frac{\pi a^\nu (s+\nu) \Gamma^2(s+\nu)}{\nu(2s+\nu+1)\Gamma^2\left(s+\nu+\frac{1}{2}\right)} \,_3F_3\left(\begin{array}{c} 1,\, s+\nu,\, s+\nu+1;\, a \\ \nu+1,\, s+\nu+\frac{1}{2},\, s+\nu+\frac{3}{2} \end{array}\right)$$

$[\mathrm{Re}\,(s+\nu) > 0]$.

4.22.10. Integrals containing $\mathrm{E}(z)$, $J_\nu(z)$ and $I_\nu(z)$

1. $\displaystyle\int_0^1 x^{s-1} \left\{\begin{array}{c} J_\nu(a\sqrt{x}) \\ I_\nu(a\sqrt{x}) \end{array}\right\} \mathrm{E}\left(\sqrt{1-x}\right) dx$

$$= \frac{\pi a^\nu (2s+\nu) \Gamma^2\left(s+\frac{\nu}{2}\right)}{2^{\nu+1}(2s+\nu+1)\Gamma^2\left(s+\frac{\nu}{2}+1\right)\Gamma(\nu+1)} \,_2F_3\left(\begin{array}{c} s+\frac{\nu}{2},\, s+\frac{\nu}{2}+1;\, \mp\frac{a^2}{4} \\ s+\frac{\nu+1}{2},\, s+\frac{\nu+3}{2},\, \nu+1 \end{array}\right)$$

$[\mathrm{Re}\,(s+\nu/2) > 0]$.

2. $\displaystyle\int_0^1 J_0(a\sqrt{x}) \, \mathrm{E}\left(\sqrt{1-x}\right) dx = \frac{\pi}{a^2} [a\,\mathbf{H}_0(a) - \mathbf{H}_1(a)]$.

3. $\displaystyle\int_0^1 \frac{1}{\sqrt{x}} J_1(a\sqrt{x}) \, \mathrm{E}\left(\sqrt{1-x}\right) dx = \frac{\pi}{a} \mathbf{H}_1(a)$.

4. $\displaystyle\int_0^1 I_0(a\sqrt{x}) \, \mathrm{E}\left(\sqrt{1-x}\right) dx = \frac{\pi}{a^2} [a\,\mathbf{L}_0(a) - \mathbf{L}_1(a)]$.

5. $\displaystyle\int_0^1 \frac{1}{\sqrt{x}} I_1(a\sqrt{x}) \, \mathrm{E}\left(\sqrt{1-x}\right) dx = \frac{\pi}{a} \mathbf{L}_1(a)$.

6. $\int_0^1 x^{s-1} e^{ax} I_\nu(ax) \mathbf{E}\left(\sqrt{1-x}\right) dx$

$$= \frac{\pi(s+\nu)\Gamma^2(s+\nu)}{(2s+2\nu+1)\Gamma^2\left(s+\nu+\frac{1}{2}\right)\Gamma(\nu+1)} \left(\frac{a}{2}\right)^\nu$$

$$\times {}_3F_3\left(\begin{array}{c} \nu+\frac{1}{2},\ s+\nu,\ s+\nu+1;\ 2a \\ 2\nu+1,\ s+\nu+\frac{1}{2},\ s+\nu+\frac{3}{2} \end{array}\right) \quad [\operatorname{Re}(s+\nu)>0].$$

7. $\int_0^1 x^{s-1} \left\{\begin{array}{c} J_\mu(a\sqrt{x})\,J_\nu(a\sqrt{x}) \\ I_\mu(a\sqrt{x})\,I_\nu(a\sqrt{x}) \end{array}\right\} \mathbf{E}\left(\sqrt{1-x}\right) dx$

$$= \frac{(2s+\mu+\nu)\Gamma^2\left(s+\frac{\mu+\nu}{2}\right)}{(2s+\mu+\nu+1)\Gamma^2\left(s+\frac{\mu+\nu+1}{2}\right)} \frac{2^{-\mu-\nu-1}\pi a^{\mu+\nu}}{\Gamma(\mu+1)\Gamma(\nu+1)}$$

$$\times {}_4F_5\left(\begin{array}{c} \frac{\mu+\nu+1}{2},\ \frac{\mu+\nu}{2}+1,\ s+\frac{\mu+\nu}{2},\ s+\frac{\mu+\nu}{2}+1;\ \mp a^2 \\ \mu+1,\ \nu+1,\ \mu+\nu+1,\ s+\frac{\mu+\nu+1}{2},\ s+\frac{\mu+\nu+3}{2} \end{array}\right)$$

$$[\operatorname{Re}(2s+\mu+\nu)>0].$$

8. $\int_0^1 J_0^2(a\sqrt{x}) \mathbf{E}\left(\sqrt{1-x}\right) dx = \frac{1}{8a^3}[\sin(2a) - 2a\cos(2a) + 4a^2 \operatorname{Si}(2a)].$

9. $\int_0^1 J_1^2(a\sqrt{x}) \mathbf{E}\left(\sqrt{1-x}\right) dx = \frac{1}{8a^3}[6a\cos(2a) - 3\sin(2a) + 4a^2 \operatorname{Si}(2a)].$

10. $\int_0^1 \frac{1}{x} J_1^2(a\sqrt{x}) \mathbf{E}\left(\sqrt{1-x}\right) dx = \frac{1}{2a^2}[\cos(2a) + 2a^2 - 1].$

11. $\int_0^1 I_0^2(a\sqrt{x}) \mathbf{E}\left(\sqrt{1-x}\right) dx = \frac{1}{8a^3}[2a\cosh(2a) - \sinh(2a) + 4a^2 \operatorname{shi}(2a)].$

12. $\int_0^1 I_1^2(a\sqrt{x}) \mathbf{E}\left(\sqrt{1-x}\right) dx$

$$= \frac{1}{8a^3}[6a\cosh(2a) - 3\sinh(2a) - 4a^2 \operatorname{shi}(2a)].$$

13. $\int_0^1 \frac{1}{x} I_1^2(a\sqrt{x}) \mathbf{E}\left(\sqrt{1-x}\right) dx = \frac{1}{2a^2}[\cosh(2a) - 2a^2 - 1].$

4.22.11. Integrals containing $\mathbf{E}(z)$, $\mathbf{H}_\nu(z)$ and $\mathbf{L}_\nu(z)$

1. $\displaystyle\int_0^1 x^{s-1} \left\{ \begin{matrix} \mathbf{H}_\nu(a\sqrt{x}) \\ \mathbf{L}_\nu(a\sqrt{x}) \end{matrix} \right\} \mathbf{E}\left(\sqrt{1-x}\right) dx$

$$= \frac{\sqrt{\pi}(2s+\nu+1)\Gamma^2\left(s+\frac{\nu+1}{2}\right)}{(2s+\nu+2)\Gamma^2\left(s+\frac{\nu}{2}+1\right)\Gamma\left(\nu+\frac{3}{2}\right)} \left(\frac{a}{2}\right)^{\nu+1}$$

$$\times {}_3F_4\left(\begin{matrix} 1, s+\frac{\nu+1}{2}, s+\frac{\nu+3}{2}; \mp\frac{a^2}{4} \\ \frac{3}{2}, \nu+\frac{3}{2}, s+\frac{\nu}{2}+1, s+\frac{\nu}{2}+2 \end{matrix}\right) \qquad [\operatorname{Re}(2s+\nu) > -1].$$

2. $\displaystyle\int_0^1 \mathbf{H}_0(a\sqrt{x})\, \mathbf{E}\left(\sqrt{1-x}\right) dx = \frac{\pi}{a}[1 - J_0(a)].$

3. $\displaystyle\int_0^1 \mathbf{L}_0(a\sqrt{x})\, \mathbf{E}\left(\sqrt{1-x}\right) dx = \frac{\pi}{a}[I_0(a) - 1].$

4. $\displaystyle\int_0^1 \frac{1}{x} \mathbf{H}_0(a\sqrt{x})\, \mathbf{E}\left(\sqrt{1-x}\right) dx = \pi[a\, J_0(a) - J_1(a)]$

$$+ \frac{\pi^2 a}{2}[J_1(a)\, \mathbf{H}_0(a) - J_0(a)\, \mathbf{H}_1(a)].$$

5. $\displaystyle\int_0^1 \frac{1}{x} \mathbf{L}_0(a\sqrt{x})\, \mathbf{E}\left(\sqrt{1-x}\right) dx = \pi[a\, I_0(a) - I_1(a)]$

$$- \frac{\pi^2 a}{2}[I_1(a)\, \mathbf{L}_0(a) - I_0(a)\, \mathbf{L}_1(a)].$$

4.22.12. Integrals containing $\mathbf{E}(z)$ and $L_n^\lambda(z)$

1. $\displaystyle\int_0^1 x^{s-1} L_n^\lambda(ax)\, \mathbf{E}\left(\sqrt{1-x}\right) dx$

$$= \frac{\pi s\, \Gamma^2(s)\, (\lambda+1)_n}{n!\,(2s+1)\Gamma^2\left(s+\frac{1}{2}\right)} {}_3F_3\left(\begin{matrix} -n, s, s+1; a \\ \nu+1, s+\frac{1}{2}, s+\frac{3}{2} \end{matrix}\right) \qquad [\operatorname{Re} s > 0].$$

4.22.13. Integrals containing products of $\mathbf{E}(z)$ and $\mathbf{K}(z)$

1. $\displaystyle\int_0^1 \frac{x}{\sqrt{1-x^2}}\, \mathbf{K}(x)\, \mathbf{E}(x)\, dx = \frac{\pi^4}{32}\, {}_7F_6\left(\begin{matrix} -\frac{1}{2}, \frac{1}{2}, \frac{1}{2}, \frac{1}{2}, \frac{1}{2}, \frac{1}{2}, \frac{5}{4} \\ \frac{1}{4}, 1, 1, 1, 1, 2; 1 \end{matrix}\right).$

2. $\int_0^1 x^{s-1} \mathrm{K}(a\sqrt{x}) \mathrm{E}(\sqrt{1-x}) \, dx$

$$= \frac{\pi^2}{2} \frac{s\Gamma^2(s)}{(2s+1)\Gamma^2\left(s+\frac{1}{2}\right)} {}_4F_3\left(\begin{array}{c} \frac{1}{2}, \frac{1}{2}, s, s+1; a^2 \\ 1, s+\frac{1}{2}, s+\frac{3}{2} \end{array}\right)$$

$$[\operatorname{Re} s > 0; \; |\arg(1-a^2)| < \pi].$$

3. $\int_0^1 x^{s-1} \mathrm{K}(\sqrt{1-x}) \mathrm{E}(a\sqrt{x}) \, dx = \frac{\pi^2}{4} \frac{\Gamma^2(s)}{\Gamma^2\left(s+\frac{1}{2}\right)} {}_4F_3\left(\begin{array}{c} -\frac{1}{2}, \frac{1}{2}, s, s; a^2 \\ 1, s+\frac{1}{2}, s+\frac{1}{2} \end{array}\right)$

$$[\operatorname{Re} s > 0; \; |\arg(1-a^2)| < \pi].$$

4. $\int_0^1 \mathrm{K}(\sqrt{1-x}) \mathrm{E}(a\sqrt{x}) \, dx$

$$= \frac{\pi}{4a}\left[a + \frac{1-a^2}{2} \ln \frac{1+a}{1-a} + \operatorname{Li}_2(a) - \operatorname{Li}_2(-a)\right] \quad [|\arg(1-a^2)| < \pi].$$

5. $\int_0^1 \mathrm{K}(a\sqrt{x}) \mathrm{E}(\sqrt{1-x}) \, dx$

$$= \frac{\pi}{4a}\left[\frac{1}{a} + \frac{1-a^2}{2a^2} \ln \frac{1-a}{1+a} + \operatorname{Li}_2(a) - \operatorname{Li}_2(-a)\right] \quad [|\arg(1-a^2)| < \pi].$$

6. $\int_0^1 \mathrm{K}(\sqrt{x}) \mathrm{E}(\sqrt{1-x}) \, dx = \frac{\pi^3}{16} + \frac{\pi}{4}.$

7. $\int_0^1 x \, \mathrm{K}(\sqrt{x}) \mathrm{E}(\sqrt{1-x}) \, dx = \frac{5\pi^3}{128} + \frac{\pi}{8}.$

8. $\int_0^1 x^2 \mathrm{K}(\sqrt{x}) \mathrm{E}(\sqrt{1-x}) \, dx = \frac{15\pi^3}{512} + \frac{\pi}{12}.$

9. $\int_0^1 \frac{x^{s-1}}{1-a^2 x} \mathrm{E}(a\sqrt{x}) \mathrm{K}(\sqrt{1-x}) \, dx = \frac{\pi^2}{4} \frac{\Gamma^2(s)}{\Gamma^2\left(s+\frac{1}{2}\right)}$

$$\times {}_4F_3\left(\begin{array}{c} \frac{1}{2}, \frac{3}{2}, s, s; a^2 \\ 1, s+\frac{1}{2}, s+\frac{1}{2} \end{array}\right) \quad [\operatorname{Re} s > 0; \; |\arg(1-a^2)| < \pi].$$

10. $\int_0^1 \frac{1}{1-a^2 x} \mathrm{E}(a\sqrt{x}) \mathrm{K}(\sqrt{1-x}) \, dx = \frac{\pi}{2a} \ln \frac{1+a}{1-a} \quad [|\arg(1-a^2)| < \pi].$

11. $\displaystyle\int_0^1 \frac{x^{s-1}}{\sqrt{1+a^2x}}\, \mathbf{E}\left(\frac{a\sqrt{x}}{\sqrt{1+a^2x}}\right) \mathbf{K}\left(\sqrt{1-x}\right) dx = \frac{\pi^2}{4}\, \frac{\Gamma^2(s)}{\Gamma^2\left(s+\frac{1}{2}\right)}$

$\times\, {}_4F_3\!\left(\begin{array}{c}\frac{1}{2},\frac{3}{2},s,s;\ -a^2\\ 1,s+\frac{1}{2},s+\frac{1}{2}\end{array}\right)$ $\quad\left[\operatorname{Re} s > 0;\ |\arg(1+a^2)| < \pi\right].$

12. $\displaystyle\int_0^1 \frac{1}{\sqrt{1+a^2x}}\, \mathbf{E}\left(\frac{a\sqrt{x}}{\sqrt{1+a^2x}}\right) \mathbf{K}\left(\sqrt{1-x}\right) dx = \frac{\pi}{a}\arctan a.$

13. $\displaystyle\int_0^1 \frac{x^{s-1}}{\sqrt{1+a^2x}}\, \mathbf{K}\left(\frac{a\sqrt{x}}{\sqrt{1+a^2x}}\right) \mathbf{E}\left(\sqrt{1-x}\right) dx = \frac{\pi^2}{4}\,\frac{\Gamma(s)\Gamma(s+1)}{\Gamma\left(s+\frac{1}{2}\right)\Gamma\left(s+\frac{3}{2}\right)}$

$\times\, {}_4F_3\!\left(\begin{array}{c}\frac{1}{2},\frac{1}{2},s,s+1;\ -a^2\\ 1,s+\frac{1}{2},s+\frac{3}{2}\end{array}\right)$ $\quad\left[\operatorname{Re} s > 0;\ |\arg(1+a^2)| < \pi\right].$

14. $\displaystyle\int_0^1 \frac{1}{\sqrt{1+a^2x}}\, \mathbf{K}\left(\frac{a\sqrt{x}}{\sqrt{1+a^2x}}\right) \mathbf{E}\left(\sqrt{1-x}\right) dx$

$= -\dfrac{\pi}{4a^2} + \dfrac{\pi}{4a^3}(a^2-1)\arctan a - \dfrac{\pi i}{4a}[\operatorname{Li}_2(ia)-\operatorname{Li}_2(-ia)]$

$\left[|\arg(1+a^2)| < \pi\right].$

15. $\displaystyle\int_0^1 \frac{1}{\sqrt{1+x}}\, \mathbf{K}\left(\sqrt{\frac{x}{1+x}}\right) \mathbf{E}\left(\sqrt{1-x}\right) dx = \frac{\pi}{8}(\pi + 4\mathbf{G} - 2).$

16. $\displaystyle\int_0^1 \sqrt{1+x}\, \mathbf{K}\left(\sqrt{1-x}\right) \mathbf{E}\left(\sqrt{\frac{x}{1+x}}\right) dx = \frac{\pi}{8}(\pi + 4\mathbf{G} + 2).$

17. $\displaystyle\int_0^1 \frac{x}{\sqrt{1+x}}\, \mathbf{K}\left(\sqrt{\frac{x}{1+x}}\right) \mathbf{E}\left(\sqrt{1-x}\right) dx = \frac{\pi}{8}(\pi + 2\mathbf{G} + 5).$

18. $\displaystyle\int_0^1 x\sqrt{1+x}\, \mathbf{K}\left(\sqrt{1-x}\right) \mathbf{E}\left(\sqrt{\frac{x}{1+x}}\right) dx = \frac{\pi}{32}(3\pi + 2\mathbf{G} + 5).$

19. $\displaystyle\int_0^1 x^{s-1}(1+a^2x)^{1/2}\, \mathbf{K}\left(\sqrt{1-x}\right) \mathbf{E}\left(\frac{a\sqrt{x}}{\sqrt{1+a^2x}}\right) dx = \frac{\pi^2}{4}\,\frac{\Gamma^2(s)}{\Gamma^2\left(s+\frac{1}{2}\right)}$

$\times\, {}_4F_3\!\left(\begin{array}{c}-\frac{1}{2},\frac{1}{2},s,s;\ -a^2\\ 1,s+\frac{1}{2},s+\frac{1}{2}\end{array}\right)$ $\quad\left[\operatorname{Re} s > 0;\ |\arg(1+a^2)| < \pi\right].$

4.22.14. Integrals containing products of $\mathbf{E}(z)$

1. $\displaystyle\int_0^1 \frac{x}{\sqrt{1-x^2}}\, \mathbf{E}^2(x)\, dx = \frac{\pi^4}{64}\, {}_7F_6\!\left(\begin{array}{c} -\frac{1}{2}, -\frac{1}{2}, \frac{1}{2}, \frac{1}{2}, \frac{1}{2}, \frac{1}{2}, \frac{5}{4} \\ \frac{1}{4}, 1, 1, 1, 2, 2;\, 1 \end{array}\right).$

2. $\displaystyle\int_0^1 x^{s-1}\, \mathbf{E}(a\sqrt{x})\, \mathbf{E}(\sqrt{1-x})\, dx$

$$= \frac{\pi^2}{4}\, \frac{\Gamma(s)\Gamma(s+1)}{\Gamma\!\left(s+\frac{1}{2}\right)\Gamma\!\left(s+\frac{3}{2}\right)}\, {}_4F_3\!\left(\begin{array}{c} -\frac{1}{2}, \frac{1}{2}, s, s+1;\, a^2 \\ 1, s+\frac{1}{2}, s+\frac{3}{2} \end{array}\right)$$

$$\left[\operatorname{Re} s > 0;\ |\arg(1-a^2)| < \pi\right].$$

3. $\displaystyle\int_0^1 \mathbf{E}(a\sqrt{x})\, \mathbf{E}(\sqrt{1-x})\, dx$

$$= \frac{\pi}{8a}\left[\frac{3a^2+1}{2a} - \frac{4a^2 - 3a^4 - 1}{2a^2}\ln\frac{1-a}{1+a} + \operatorname{Li}_2(a) - \operatorname{Li}_2(-a)\right]$$

$$\left[|\arg(1-a^2)| < \pi\right].$$

4. $\displaystyle\int_0^1 \mathbf{E}(\sqrt{x})\, \mathbf{E}(\sqrt{1-x})\, dx = \frac{\pi^3}{32} + \frac{\pi}{4}.$

5. $\displaystyle\int_0^1 x\, \mathbf{E}(\sqrt{x})\, \mathbf{E}(\sqrt{1-x})\, dx = \frac{\pi^3}{64} + \frac{\pi}{8}.$

6. $\displaystyle\int_0^1 x^2\, \mathbf{E}(\sqrt{x})\, \mathbf{E}(\sqrt{1-x})\, dx = \frac{21\pi^3}{2048} + \frac{\pi}{12}.$

7. $\displaystyle\int_0^1 \frac{x^{s-1}}{1-a^2 x}\, \mathbf{E}(a\sqrt{x})\, \mathbf{E}(\sqrt{1-x})\, dx = \frac{\pi^2}{4}\, \frac{\Gamma(s)\Gamma(s+1)}{\Gamma\!\left(s+\frac{1}{2}\right)\Gamma\!\left(s+\frac{3}{2}\right)}$

$$\times\, {}_4F_3\!\left(\begin{array}{c} \frac{1}{2}, \frac{3}{2}, s, s+1;\, a^2 \\ 1, s+\frac{1}{2}, s+\frac{3}{2} \end{array}\right) \qquad \left[\operatorname{Re} s > 0;\ |\arg(1-a^2)| < \pi\right].$$

8. $\displaystyle\int_0^1 \frac{1}{1-a^2 x}\, \mathbf{E}(a\sqrt{x})\, \mathbf{E}(\sqrt{1-x})\, dx = \frac{\pi}{4a^3}\left[(a^2+1)\ln\frac{1+a}{1-a} - 2a\right]$

$$\left[|\arg(1-a^2)| < \pi\right].$$

9. $\int_0^1 (1+a^2x)^{1/2} \operatorname{E}\left(\sqrt{1-x}\right) \operatorname{E}\left(\dfrac{a\sqrt{x}}{\sqrt{1+a^2x}}\right) dx$

$= \dfrac{\pi}{16}\left\{3 - \dfrac{1}{a^2} + \dfrac{3a^4 + 4a^2 + 1}{a^3}\arctan a - \dfrac{2i}{a}\left[\operatorname{Li}_2(ia) - \operatorname{Li}_2(-ia)\right]\right\}$

$\left[|\arg(1-a^2)| < \pi\right].$

10. $\int_0^1 \dfrac{1}{1+a^2x} \operatorname{E}\left(\dfrac{a\sqrt{x}}{\sqrt{1+a^2x}}\right) \operatorname{E}\left(\sqrt{1-x}\right) dx = \dfrac{\pi}{2a^3}\left[a + (a^2-1)\arctan a\right]$

$\left[|\arg(1-a^2)| < \pi\right].$

11. $\int_0^1 (1+x)^{1/2} \operatorname{E}\left(\sqrt{\dfrac{x}{1+x}}\right) \operatorname{E}\left(\sqrt{1-x}\right) dx = \dfrac{\pi}{8}(\pi + 2\operatorname{G} + 1).$

12. $\int_0^1 x(1+x)^{1/2} \operatorname{E}\left(\sqrt{\dfrac{x}{1+x}}\right) \operatorname{E}\left(\sqrt{1-x}\right) dx = \dfrac{\pi}{32}(2\pi + 2\operatorname{G} + 5).$

4.23. The Complete Elliptic Integral $\operatorname{D}(z)$

4.23.1. Integrals containing $\operatorname{D}(z)$ and elementary functions

1. $\int_0^a x^{s-1}(a-x)^{t-1} \operatorname{D}\left(b\sqrt{x(a-x)}\right) dx$

$= \dfrac{\pi}{4} \operatorname{B}(s,t) a^{s+t-1} {}_4F_3\left(\begin{array}{c}\frac{1}{2}, \frac{3}{2}, s, t; \\ 2, \frac{s+t}{2}, \frac{s+t+1}{2}\end{array}\bigg| \dfrac{a^2b^2}{4}\right)$

$\left[\operatorname{Re} s, \operatorname{Re} t > 0;\ |\arg(4 - a^2b^2)| < \pi\right].$

2. $\int_0^a \operatorname{D}\left(b\sqrt{x(a-x)}\right) dx = \dfrac{\pi}{ab^2}\left(2 - \sqrt{4-a^2b^2}\right)$ $\left[|\arg(4-a^2b^2)| < \pi\right].$

3. $\int_0^a x \operatorname{D}\left(b\sqrt{x(a-x)}\right) dx = \dfrac{\pi}{2b^2}\left(2 - \sqrt{4-a^2b^2}\right)$ $\left[|\arg(4-a^2b^2)| < \pi\right].$

4. $\int_0^a x^{1/2}(a-x)^{1/2} \operatorname{D}\left(b\sqrt{x(a-x)}\right) dx$

$= \dfrac{\pi^2 a^2}{32}\left[4\psi_1\left(\dfrac{a^2b^2}{4}\right) - 4\psi_2\left(\dfrac{a^2b^2}{4}\right) + \psi_3\left(\dfrac{a^2b^2}{4}\right)\right]$

$\left[|\arg(4-a^2b^2)| < \pi\right].$

4.23.1]   4.23. The Complete Elliptic Integral $D(z)$

5. $\displaystyle\int_0^a x^{-1/2}(a-x)^{1/2}\, D\!\left(b\sqrt{x(a-x)}\right)dx = \frac{\pi^2 a}{8}\left[2\psi_1\!\left(\frac{a^2 b^2}{4}\right)-\psi_2\!\left(\frac{a^2 b^2}{4}\right)\right]$

$$[|\arg(4-a^2 b^2)|<\pi].$$

6. $\displaystyle\int_0^a x^{-1/2}(a-x)^{-1/2}\, D\!\left(b\sqrt{x(a-x)}\right)dx = \frac{\pi^2}{4}\left[2\psi_1\!\left(\frac{a^2 b^2}{4}\right)-\psi_2\!\left(\frac{a^2 b^2}{4}\right)\right]$

$$[|\arg(4-a^2 b^2)|<\pi].$$

7. $\displaystyle\int_0^a x^{s+1/2}(a-x)^s\, D\!\left(b\sqrt[4]{x(a-x)}\right)dx = 2^{-2s-3}\pi^{3/2} a^{2s+3/2}\,\frac{\Gamma(2s+2)}{\Gamma\!\left(2s+\frac{5}{2}\right)}$

$$\times\ {}_3F_2\!\left(\begin{array}{c}\frac{1}{2},\,\frac{3}{2},\,2s+2;\ \frac{ab^2}{2}\\ 2,\,2s+\frac{5}{2}\end{array}\right)\quad [\mathrm{Re}\,s>-1;\ |\arg(2-ab^2)|<\pi].$$

8. $\displaystyle\int_0^a x^{1/2}\, D\!\left(b\sqrt[4]{x(a-x)}\right)dx = \frac{\pi}{\sqrt{2}\,b^3}\arcsin\!\left(b\sqrt{\frac{a}{2}}\right)-\frac{\pi\sqrt{a}}{2b^2}\sqrt{1-\frac{ab^2}{2}}$

$$[|\arg(2-ab^2)|<\pi].$$

9. $\displaystyle\int_0^a x^{-1/2}\, D\!\left(b\sqrt[4]{x(a-x)}\right)dx = \pi a^{1/2}\left(1+\sqrt{1-\frac{ab^2}{2}}\right)^{-1}$

$$[|\arg(2-ab^2)|<\pi].$$

10. $\displaystyle\int_0^a x^{-1/4}(a-x)^{1/4}\, D\!\left(b\sqrt[4]{x(a-x)}\right)dx = 2^{-7/2}\pi^2 a$

$$\times\left[4\psi_1\!\left(\frac{ab^2}{2}\right)-4\psi_2\!\left(\frac{ab^2}{2}\right)+\psi_3\!\left(\frac{ab^2}{2}\right)\right]\quad [|\arg(2-ab^2)|<\pi].$$

11. $\displaystyle\int_0^a x^{-3/4}(a-x)^{-1/4}\, D\!\left(b\sqrt[4]{x(a-x)}\right)dx$

$$= 2^{-3/2}\pi^2\left[2\psi_1\!\left(\frac{ab^2}{2}\right)-\psi_2\!\left(\frac{ab^2}{2}\right)\right]\quad [|\arg(2-ab^2)|<\pi].$$

12. $\displaystyle\int_0^{m\pi} e^{-ax}\, D(b\sin x)\,dx = \frac{\pi a}{2b^2}(1-e^{-m\pi a})\left[{}_3F_2\!\left(\begin{array}{c}\frac{1}{2},\,\frac{1}{2},\,\frac{3}{2};\ b^2\\ 1-\frac{ia}{2},\,1+\frac{ia}{2}\end{array}\right)-1\right]$

$$[|\arg(1-b^2)|<\pi].$$

13. $\displaystyle\int_0^\infty e^{-ax}\,\mathbf{D}(b\sin x)\,dx = \frac{\pi a}{2b^2}\left[{}_3F_2\!\left(\begin{matrix}-\tfrac{1}{2},\,-\tfrac{1}{2},\,\tfrac{1}{2};\,b^2\\ -\tfrac{ia}{2},\,\tfrac{ia}{2}\end{matrix}\right)-1\right]$

$$[\operatorname{Re} a > 0;\ |\arg(1-b^2)|<\pi].$$

14. $\displaystyle\int_0^a x^{s-1}\ln\frac{a+\sqrt{a^2-x^2}}{x}\,\mathbf{D}(bx)\,dx = \frac{\pi^{3/2}a^s\Gamma\!\left(\tfrac{s}{2}\right)}{8s\,\Gamma\!\left(\tfrac{s+1}{2}\right)}$

$$\times\,{}_4F_3\!\left(\begin{matrix}\tfrac{1}{2},\,\tfrac{3}{2},\,\tfrac{s}{2},\,\tfrac{s}{2};\,a^2b^2\\ 2,\,\tfrac{s+1}{2},\,\tfrac{s}{2}+1\end{matrix}\right)\qquad [a,\,\operatorname{Re} s > 0;\ |\arg(1-a^2b^2)|<\pi].$$

15. $\displaystyle\int_0^a \ln\frac{a+\sqrt{a^2-x^2}}{x}\,\mathbf{D}(bx)\,dx = \frac{\pi^2 a}{8}\psi_2(a^2b^2)$

$$[a>0;\ |\arg(1-a^2b^2)|<\pi].$$

16. $\displaystyle\int_0^a x\ln\frac{a+\sqrt{a^2-x^2}}{x}\,\mathbf{D}(bx)\,dx$

$$= \frac{\pi}{2b^2}\left(1-\sqrt{1-a^2b^2}+\ln\frac{1+\sqrt{1-a^2b^2}}{2}\right)$$

$$[a>0;\ |\arg(1-a^2b^2)|<\pi].$$

17. $\displaystyle\int_0^1 x^{s-1}\arccos x\,\mathbf{D}(ax)\,dx = \frac{\pi^{3/2}\Gamma\!\left(\tfrac{s+1}{2}\right)}{4s^2\Gamma\!\left(\tfrac{s}{2}\right)}\,{}_4F_3\!\left(\begin{matrix}\tfrac{1}{2},\,\tfrac{3}{2},\,\tfrac{s}{2},\,\tfrac{s+1}{2};\,a^2\\ 2,\,\tfrac{s}{2}+1,\,\tfrac{s}{2}+1\end{matrix}\right)$

$$[\operatorname{Re} s > 0;\ |\arg(1-a^2)|<\pi].$$

18. $\displaystyle\int_0^1 \arccos x\,\mathbf{D}(ax)\,dx = \frac{\pi}{2a^2}\left(a\arcsin a+\sqrt{1-a^2}-1\right)$

$$[|\arg(1-a^2)|<\pi].$$

4.23.2. Integrals containing products of $\mathbf{D}(z)$, $\mathbf{K}(z)$ and $\mathbf{E}(z)$

1. $\displaystyle\int_0^1 x^{s-1}\mathbf{K}\!\left(\sqrt{1-x}\right)\mathbf{D}(a\sqrt{x})\,dx = \frac{\pi^2\Gamma^2(s)}{8\Gamma^2\!\left(s+\tfrac{1}{2}\right)}\,{}_4F_3\!\left(\begin{matrix}\tfrac{1}{2},\,\tfrac{3}{2},\,s,\,s;\,a^2\\ 2,\,s+\tfrac{1}{2},\,s+\tfrac{1}{2}\end{matrix}\right)$

$$[\operatorname{Re} s > 0;\ |\arg(1-a^2)|<\pi].$$

2. $\displaystyle\int_0^1 \mathbf{K}\!\left(\sqrt{1-x}\right)\mathbf{D}(a\sqrt{x})\,dx = \frac{\pi}{2a^2}\left[\ln(1-a^2)+a\ln\frac{1+a}{1-a}\right]$

$$[|\arg(1-a^2)|<\pi].$$

3. $\int_0^1 K\left(\sqrt{1-x}\right) D\left(\sqrt{x}\right) dx = \pi \ln 2.$

4. $\int_0^1 x^{s-1} E\left(\sqrt{1-x}\right) D\left(a\sqrt{x}\right) dx$

$$= \frac{\pi^2 \Gamma(s)\Gamma(s+1)}{8\Gamma\left(s+\frac{1}{2}\right)\Gamma\left(s+\frac{3}{2}\right)} {}_4F_3\left(\begin{array}{c} \frac{1}{2}, \frac{3}{2}, s, s+1; a^2 \\ 2, s+\frac{1}{2}, s+\frac{3}{2} \end{array}\right)$$

$\left[\operatorname{Re} s > 0;\ |\arg(1-a^2)| < \pi\right].$

5. $\int_0^1 E\left(\sqrt{1-x}\right) D\left(a\sqrt{x}\right) dx = \frac{\pi}{2a^3}\left(a + \frac{a^2-1}{2} \ln\frac{1+a}{1-a}\right)$

$\left[|\arg(1-a^2)| < \pi\right].$

6. $\int_0^1 E\left(\sqrt{1-x}\right) D\left(\sqrt{x}\right) dx = \frac{\pi}{2}.$

7. $\int_0^1 x\, E\left(\sqrt{1-x}\right) D\left(\sqrt{x}\right) dx = \frac{\pi^3}{32}.$

4.24. The Generalized Hypergeometric Function ${}_pF_q((a_p);\ (b_q);\ z)$

4.24.1. Integrals containing ${}_pF_q((a_p);\ (b_q);\ z)$ and algebraic functions

1. $\int_0^a x^{s-1}(a-x)^{t-1}\, {}_pF_{q+1}\left(\begin{array}{c}(a_p);\ bx(a-x)\\ (b_q)\end{array}\right) dx = B(s,t) a^{s+t-1}$

$\times\ {}_{p+2}F_{q+2}\left(\begin{array}{c}(a_p),\ s,\ t;\ \frac{a^2 b}{4}\\ (b_q),\ \frac{s+t}{2},\ \frac{s+t+1}{2}\end{array}\right)$ $\left[a,\ \operatorname{Re} s,\ \operatorname{Re} t > 0;\ |\arg(4-a^2 b)| < \pi\right].$

2. $\int_0^a x^{s-1}(a-x)^{s-1/2}\, {}_pF_{q+1}\left(\begin{array}{c}(a_p);\ b\sqrt{x(a-x)}\\ (b_q)\end{array}\right) dx = \frac{\sqrt{\pi}\, \Gamma(2s)\, a^{2s-1/2}}{2^{2s-1}\Gamma\left(2s+\frac{1}{2}\right)}$

$\times\ {}_{p+1}F_{q+1}\left(\begin{array}{c}(a_p),\ 2s;\ \frac{ab}{2}\\ (b_q),\ 2s+\frac{1}{2}\end{array}\right)$ $\left[a,\ \operatorname{Re} s > 0;\ |\arg(2-ab)| < \pi\right].$

4.24.2. Integrals containing $_pF_q((a_p); (b_q); z)$ and trigonometric functions

1. $$\int_0^{m\pi} \begin{Bmatrix} \sin(ax) \\ \cos(ax) \end{Bmatrix} {}_pF_q\left(\begin{matrix}(a_p); b\sin^2 x \\ (b_q)\end{matrix}\right) dx$$

$$= \frac{2}{a}\sin\frac{m\pi a}{2}\begin{Bmatrix}\sin(m\pi a/2) \\ \cos(m\pi a/2)\end{Bmatrix} {}_{p+2}F_{q+2}\left(\begin{matrix}(a_p), \frac{1}{2}, 1; b \\ (b_q), 1-\frac{a}{2}, 1+\frac{a}{2}\end{matrix}\right).$$

2. $$\int_0^{\pi/2} \cos(2nx) {}_pF_q\left(\begin{matrix}(a_p); a\sin^2 x \\ (b_q)\end{matrix}\right) dx$$

$$= \frac{\pi(-a)^n}{2^{2n+1}n!}\frac{\prod(a_p)_n}{\prod(b_q)_n} {}_{p+1}F_{q+1}\left(\begin{matrix}(a_p)+n, n+\frac{1}{2}; a \\ (b_q)+n, 2n+1\end{matrix}\right).$$

3. $$\int_0^{\pi/2} \sin x \sin(2n+1)x \, {}_pF_q\left(\begin{matrix}(a_p); b\sin^2 x \\ (b_q)\end{matrix}\right) dx$$

$$= \frac{2^{-2n-2}\pi}{n!}(-b)^n \frac{\prod(a_p)_n}{\prod(b_q)_n} {}_{p+1}F_{q+1}\left(\begin{matrix}(a_p)+n, n+\frac{3}{2}; b \\ (b_q)+n, 2n+2\end{matrix}\right).$$

4. $$\int_0^\pi \sin^\nu x \begin{Bmatrix}\sin(ax) \\ \cos(ax)\end{Bmatrix} {}_pF_q\left(\begin{matrix}(a_p); b\sin^2 x \\ (b_q)\end{matrix}\right) dx$$

$$= \frac{2^{-\nu}\pi\Gamma(\nu+1)}{\Gamma\left(\frac{\nu+a}{2}+1\right)\Gamma\left(\frac{\nu-a}{2}+1\right)}\begin{Bmatrix}\sin(a\pi/2) \\ \cos(a\pi/2)\end{Bmatrix}$$

$$\times {}_{p+1}F_{q+2}\left(\begin{matrix}(a_p), \frac{\nu+1}{2}, \frac{\nu}{2}+1; b \\ (b_q), \frac{\nu-a}{2}+1, \frac{\nu+a}{2}+1\end{matrix}\right) \quad [\mathrm{Re}\,\nu > -1].$$

5. $$\int_0^\pi \cos(nx) {}_pF_q\left(\begin{matrix}(a_p); a\cos^2 x \\ (b_q)\end{matrix}\right) dx$$

$$= \frac{2^{-n}\pi(-a)^{n/2}}{\Gamma\left(\frac{n}{2}+1\right)}\cos\frac{n\pi}{2}\frac{\prod(a_p)_{n/2}}{\prod(b_q)_{n/2}} {}_pF_q\left(\begin{matrix}(a_p)+\frac{n}{2}, \frac{n+1}{2}; a \\ (b_q)+\frac{n}{2}, n+1\end{matrix}\right).$$

6. $$\int_0^{\pi/2} \cos^\nu x \cos(ax) {}_pF_q\left(\begin{matrix}(a_p); b\cos^2 x \\ (b_q)\end{matrix}\right) dx$$

$$= \frac{2^{-\nu-1}\pi\Gamma(\nu+1)}{\Gamma\left(\frac{\nu+a}{2}+1\right)\Gamma\left(\frac{\nu-a}{2}+1\right)} {}_{p+1}F_{q+2}\left(\begin{matrix}(a_p), \frac{\nu+1}{2}, \frac{\nu}{2}+1; b \\ (b_q), \frac{\nu-a}{2}+1, \frac{\nu+a}{2}+1\end{matrix}\right)$$

$$[\mathrm{Re}\,\nu > -1].$$

7. $\int_0^{m\pi} e^{-ax} {}_pF_q\!\left(\!\begin{array}{c}(a_p); \, b\sin x\\(b_q)\end{array}\!\right) dx$

$$= \frac{1-e^{-m\pi a}}{a} \, {}_{2p+1}F_{2q+2}\!\left(\!\begin{array}{c}\frac{(a_p)}{2}, \frac{(a_p)+1}{2}, 1; \frac{b^2}{4^{q-p+1}}\\ \frac{(b_q)}{2}, \frac{(b_q)+1}{2}, 1-\frac{ia}{2}, 1+\frac{ia}{2}\end{array}\!\right)$$

$$+ \frac{b\left(1-e^{-m\pi}\right)}{a^2+1} \frac{\prod_{i=1}^p a_i}{\prod_{j=1}^q b_j} \, {}_{2p+1}F_{2q+2}\!\left(\!\begin{array}{c}\frac{(a_p)+1}{2}, \frac{(a_p)}{2}+1, 1; \frac{b^2}{4^{q-p+1}}\\ \frac{(b_q)+1}{2}, \frac{(b_q)}{2}+1, \frac{3}{2}-\frac{ia}{2}, \frac{3}{2}+\frac{ia}{2}\end{array}\!\right)$$

$$[\text{Re}\, a > 0].$$

8. $\int_0^{m\pi} e^{-ax} {}_pF_q\!\left(\!\begin{array}{c}(a_p); \, b\sin^2 x\\(b_q)\end{array}\!\right) dx$

$$= \frac{1-e^{-m\pi a}}{a} \, {}_{p+2}F_{q+2}\!\left(\!\begin{array}{c}(a_p), \frac{1}{2}, 1; b\\(b_q), 1-\frac{ia}{2}, 1+\frac{ia}{2}\end{array}\!\right).$$

9. $\int_0^{m\pi} e^{-ax} \sin x \, {}_pF_q\!\left(\!\begin{array}{c}(a_p); \, b\sin^2 x\\(b_q)\end{array}\!\right) dx$

$$= \frac{1-(-1)^m e^{-m\pi a}}{a^2+1} \, {}_{p+2}F_{q+2}\!\left(\!\begin{array}{c}(a_p), 1, \frac{3}{2}; b\\(b_q), \frac{3-ia}{2}, \frac{3+ia}{2}\end{array}\!\right).$$

10. $\int_0^\infty e^{-ax} {}_pF_q\!\left(\!\begin{array}{c}(a_p); \, b\sin x\\(b_q)\end{array}\!\right) dx$

$$= \frac{1}{a} \, {}_{2p+1}F_{2q+2}\!\left(\!\begin{array}{c}\frac{(a_p)}{2}, \frac{(a_p)+1}{2}, 1; \frac{b^2}{4^{q-p+1}}\\ \frac{(b_q)}{2}, \frac{(b_q)+1}{2}, 1-\frac{ia}{2}, 1+\frac{ia}{2}\end{array}\!\right)$$

$$+ \frac{b}{a^2+1} \frac{\prod_{i=1}^p a_i}{\prod_{j=1}^q b_j} \, {}_{2p+1}F_{2q+2}\!\left(\!\begin{array}{c}\frac{(a_p)+1}{2}, \frac{(a_p)}{2}+1, 1; \frac{b^2}{4^{q-p+1}}\\ \frac{(b_q)+1}{2}, \frac{(b_q)}{2}+1, \frac{3-ia}{2}, \frac{3+ia}{2}\end{array}\!\right) \quad [\text{Re}\, a > 0].$$

11. $\int_0^\infty e^{-ax} {}_pF_q\!\left(\!\begin{array}{c}(a_p); \, b\sin^2 x\\(b_q)\end{array}\!\right) dx = \frac{1}{a} \, {}_{p+2}F_{q+2}\!\left(\!\begin{array}{c}(a_p), \frac{1}{2}, 1; b\\(b_q); 1-\frac{ia}{2}, 1+\frac{ia}{2}\end{array}\!\right)$

$$[\text{Re}\, a > 0].$$

12. $\int_0^\infty e^{-ax} \sin x \, {}_pF_q\left(\begin{array}{c}(a_p);\ b\sin^2 x\\(b_q)\end{array}\right) dx$

$$= \frac{1}{a^2+1} {}_{p+2}F_{q+2}\left(\begin{array}{c}(a_p), 1, \frac{3}{2};\ b\\(b_q), \frac{3-ia}{2}, \frac{3+ia}{2}\end{array}\right) \qquad [\operatorname{Re} a > 0].$$

4.24.3. Integrals containing ${}_pF_q((a_p); (b_q); z)$ and the logarithmic function

1. $\int_0^a x^{s-1} \ln \frac{\sqrt{a}+\sqrt{a-x}}{\sqrt{x}} \, {}_pF_q\left(\begin{array}{c}(a_p);\ bx\\(b_q)\end{array}\right) dx$

$$= \frac{\sqrt{\pi}}{2s} a^s \frac{\Gamma(s)}{\Gamma\left(s+\frac{1}{2}\right)} {}_{p+2}F_{q+2}\left(\begin{array}{c}(a_p), s, s;\ ab\\(b_q), s+\frac{1}{2}, s+1\end{array}\right) \qquad [a, \operatorname{Re} s > 0].$$

4.24.4. Integrals containing ${}_pF_q((a_p); (b_q); z)$, $K(z)$ and $E(z)$

1. $\int_0^1 x^{s-1} K\left(\sqrt{1-x}\right) {}_pF_q\left(\begin{array}{c}(a_p);\ ax\\(b_q)\end{array}\right) dx$

$$= \frac{\pi}{2} \frac{\Gamma^2(s)}{\Gamma^2\left(s+\frac{1}{2}\right)} {}_{p+2}F_{q+2}\left(\begin{array}{c}(a_p), s, s;\ a\\(b_q), s+\frac{1}{2}, s+\frac{1}{2}\end{array}\right) \qquad [\operatorname{Re} s > 0].$$

2. $\int_0^1 x^{s-1} E\left(\sqrt{1-x}\right) {}_pF_q\left(\begin{array}{c}(a_p);\ ax\\(b_q)\end{array}\right) dx$

$$= \frac{\pi}{2} \frac{\Gamma(s)\Gamma(s+1)}{\Gamma\left(s+\frac{1}{2}\right)\Gamma\left(s+\frac{3}{2}\right)} {}_{p+2}F_{q+2}\left(\begin{array}{c}(a_p), s, s+1;\ a\\(b_q), s+\frac{1}{2}, s+\frac{3}{2}\end{array}\right) \qquad [\operatorname{Re} s > 0].$$

4.24.5. Integrals containing products of ${}_pF_q((a_p); (b_q); z)$

1. $\int_0^a x^{s-1}(a-x)^{t-1} {}_1F_1\left(\begin{array}{c}b;\ wx\\s\end{array}\right) {}_1F_1\left(\begin{array}{c}c;\ w(a-x)\\t\end{array}\right) dx$

$$= B(s,t) a^{s+t-1} {}_1F_1\left(\begin{array}{c}b+c;\ aw\\s+t\end{array}\right)$$

$$[a, \operatorname{Re} s, \operatorname{Re} t > 0].$$

2. $\int_0^a x^{s-1}(a-x)^{t-1} {}_1F_1\left(\begin{array}{c}b;\ wx\\s\end{array}\right) {}_1F_1\left(\begin{array}{c}b;\ -w(a-x)\\t\end{array}\right) dx$

$$= B(s,t) a^{s+t-1} {}_1F_2\left(\begin{array}{c}b;\ \frac{a^2w^2}{4}\\\frac{s+t}{2}, \frac{s+t+1}{2}\end{array}\right) \qquad [a, \operatorname{Re} s, \operatorname{Re} t > 0].$$

3. $\displaystyle\int_0^a x^{s-1}(a-x)^{t-1} {}_1F_2\!\left(\begin{matrix}b;\ wx\\ c,\ s\end{matrix}\right){}_1F_2\!\left(\begin{matrix}b;\ -w(a-x)\\ c,\ t\end{matrix}\right)dx$

$\displaystyle = \mathrm{B}(s,t)\,a^{s+t-1}\,{}_2F_5\!\left(\begin{matrix}b,\ c-b;\ \dfrac{a^2 w^2}{16}\\ c,\ \dfrac{c}{2},\ \dfrac{c+1}{2},\ \dfrac{s+t}{2},\ \dfrac{s+t+1}{2}\end{matrix}\right)\qquad [a,\ \mathrm{Re}\,s,\ \mathrm{Re}\,t > 0].$

4. $\displaystyle\int_0^a x^{\mu+m-1}(a-x)^{\nu+n-1}\,{}_0F_1(\mu;\ -bx)\,{}_0F_1(\nu;\ -c(a-x))\,dx$

$\displaystyle = (-1)^{m+n}\Gamma(\mu)\Gamma(\nu)\left(\frac{a}{b+c}\right)^{(\mu+\nu+m+n-1)/2}$

$\displaystyle \times \sum_{j=0}^{m}\sum_{k=0}^{n}\binom{m}{j}\binom{n}{k}(1-\mu-m)_{m-j}(1-\nu-n)_{n-k}\,b^j c^k \left(\frac{a}{b+c}\right)^{(j+k)/2}$

$\displaystyle \qquad\qquad\qquad\qquad\times J_{\mu+\nu+j+k+m+n-1}\!\left[2\sqrt{a(b+c)}\right]\qquad [a>0].$

5. $\displaystyle\int_0^a x^{n+1/2}(a-x)^{(m+\nu)/2-1}\,{}_0F_1(\nu;\ -b(a-x))\,{}_1F_2\!\left(\begin{matrix}\tfrac{1}{2};\ -cx\\ n+\tfrac{3}{2},\ \tfrac{1}{2}-n\end{matrix}\right)$

$\displaystyle = (-1)^n\left(n+\tfrac{1}{2}\right)\pi\Gamma(\nu)(1-\nu)_m \sum_{k=0}^{n}\binom{n}{k}\left(\tfrac{1}{2}\right)_{n-k} c^k$

$\displaystyle \times D_u^m D_w^{k+n+1}\!\left[b^{(1-\nu)/2} J_{(\nu-1)/2}\!\left(\sqrt{a}\sqrt{u+w}+\sqrt{aw}\right)\right.$

$\displaystyle \qquad\qquad\qquad \left.\times J_{(\nu-1)/2}\!\left(\sqrt{a}\sqrt{u+w}-\sqrt{aw}\right)\right]\bigg|_{\substack{u=b\\ w=c}}\qquad [a>0].$

6. $\displaystyle\int_1^\infty x^{-m-n-2}\,{}_2F_2\!\left(\begin{matrix}-m,\ -m,\ -m-\tfrac{1}{2};\ x\\ \tfrac{1}{2}-m,\ -2m-1\end{matrix}\right){}_2F_2\!\left(\begin{matrix}-n,\ -n,\ -n-\tfrac{1}{2};\ x\\ \tfrac{1}{2}-n,\ -2n-1\end{matrix}\right)dx$

$\displaystyle = (-1)^{m+n}\frac{(m!)^2(n!)^2}{2(2m)!(2n)!}\left[\mathrm{C}+2\ln 2+\psi\!\left(m+\tfrac{1}{2}\right)\right]\qquad [m\le n].$

7. $\displaystyle\int_a^\infty (a-x)^{\nu-1}\,{}_0F_1(\nu;\ ab-bx)\,{}_pF_{p+1}\!\left(\begin{matrix}(a_p);\ -cx\\ (b_{p+1})\end{matrix}\right)=0$

$\displaystyle \qquad\qquad \left[a>0;\ 0<c<4b;\ \mathrm{Re}\!\left(\nu+\sum a_p-\sum b_{p+1}\right)<-1\right].$

Chapter 5

Finite Sums

5.1. The Psi Function $\psi(z)$

5.1.1. Sums containing $\psi(k+a)$

1. $\displaystyle\sum_{k=1}^{n} \psi(k) = n\psi(n+1) - n.$

2. $\displaystyle\sum_{k=0}^{n} \psi(k+a) = (a-1)\left[1 - \psi(a-1)\right] - (a+n)\left[1 - \psi(a+n)\right].$

3. $\displaystyle\sum_{k=0}^{n} k\psi(k+a) = \frac{a-1}{4}\left[2 - 3a + 2a\psi(a)\right]$
$\qquad\qquad\qquad + \frac{a+n}{4}\left[3a - 3 - n + 2(1 - a + n)\psi(a+n+1)\right].$

4. $\displaystyle\sum_{k=0}^{n} k^2\psi(k+a) = \frac{n}{36}\left[1 + 12a(1-a) - n(3 - 6a + 4n)\right]$
$\qquad - \frac{(a-1)a(2a-1)}{6}\psi(a) + \frac{1}{6}\left[(a-1)a(2a-1) + n(n+1)(2n+1)\right]\psi(a+n).$

5. $\displaystyle\sum_{k=0}^{n} k^3\psi(k+a) = \frac{n+1}{48}\left[12a^3 - 6(n+4)a^2\right.$
$\qquad\qquad \left. + 2(n+2)(2n+3)a - n(n+2)(3n+5)\right] + \frac{1}{4}(a-1)^2 a^2 \psi(a)$
$\qquad\qquad\qquad + \frac{1}{4}\left[n^2(n+1)^2 - (a-1)^2 a^2\right]\psi(a+n+1).$

6. $\displaystyle\sum_{k=1}^{n} \frac{k}{2^k}\psi(k+1) = 2(1 - C + \ln 2) - 2^{-n}(n+2)\psi(n+3)$
$\qquad\qquad\qquad - \frac{2^{-n}}{n+2}\,{}_2F_1\!\left(\begin{matrix}1,\,1;\,-1\\n+3\end{matrix}\right) \quad [n \geq 1].$

7. $\displaystyle\sum_{k=1}^{n} 2^{-k}(k+a-2)\psi(k+a) = 2 + a\psi(a) - 2^{-n}\left[2 + (a+n)\psi(a+n)\right].$

8. $\displaystyle\sum_{k=1}^{n}(-1)^k \binom{n}{k}\psi(k) = \left(2 - \frac{1}{n}\right)\mathbf{C} + \left(1 - \frac{1}{n}\right)\psi(n+1)$ $\qquad [n \geq 1]$.

9. $\displaystyle\sum_{k=0}^{n}\frac{(a)_k}{k!}\psi(k+1) = \frac{1}{a} + \frac{(a+1)_n}{n!\,a}[a\psi(n+1) - 1]$.

10. $\displaystyle\sum_{k=0}^{n}\frac{(b)_k}{k!}\psi(k+b) = \frac{(b+1)_n}{n!}[\psi(b) - \psi(b+1) + \psi(b+n+1)]$.

11. $\displaystyle\sum_{k=0}^{n}\frac{(a)_k}{(b)_k}\psi(k+a)$
$= \dfrac{1}{(a-b+1)^2}\left\{b - 1 - \dfrac{a(a+1)_n}{(b)_n} + (a-b+1)\left[1 - b + \dfrac{a(a+1)_n}{(b)_n}\right]\psi(a)\right.$
$\left. + \dfrac{(b-a-1)(a+1)_n}{(b)_n}[a\psi(a+1) - a\psi(a+n+1) - 1]\right\}$.

12. $\displaystyle\sum_{k=0}^{n}\frac{(b)_k}{(a)_k}\psi(k+a) = \dfrac{a-1}{(b-a+1)^2}[1 + (a-b-1)\psi(a-1)]$
$\qquad\qquad - \dfrac{(b)_{n+1}}{(b-a+1)^2(a)_n}[1 + (a-b-1)\psi(a+n)]$.

13. $\displaystyle\sum_{k=0}^{n}\frac{(a)_k}{k!\,(n-k+1)}\psi(k+a)$
$= \dfrac{(a)_{n+1}}{(n+1)!}\{\psi(a+n+1)[\psi(-a-n) - \psi(1-a)]$
$\qquad\qquad + \psi'(1-a) - \psi'(-a-n)\}$.

14. $\displaystyle\sum_{k=0}^{n}2^k\binom{n}{k}(2n-k)!\,(a)_k\,\psi(k+a)$
$= 2^{2n-1}n!\left(\dfrac{a+1}{2}\right)_n\left[2\ln 2 + \psi\left(\dfrac{a}{2}\right) + \psi\left(n + \dfrac{a+1}{2}\right)\right]$.

15. $\displaystyle\sum_{k=1}^{n}\frac{(2n-k-1)!}{(n-k)!}\psi(k)$
$= \dfrac{2^{2n-1}\left(\frac{1}{2}\right)_n}{n}\left[2\ln 2 - \mathbf{C} - \dfrac{1}{n} + \psi\left(n + \dfrac{1}{2}\right) - \psi(2n)\right]$ $\qquad [n \geq 1]$.

16. $\displaystyle\sum_{k=0}^{n}2^k\frac{(2n-k)!}{(n-k)!}\psi(k+1) = 2^{2n-1}n!\,[\psi(n+1) - \mathbf{C}]$.

17. $\displaystyle\sum_{k=0}^{n} \binom{n}{k}\binom{2n}{k}^{-1} \frac{(a-1)_k}{k!} \psi(k+a)$

$$= \frac{2^{-2n}(a)_{2n}}{\left(\frac{1}{2}\right)_n (a)_n} \left[2\psi(a) - \psi(a+n) - \psi\left(\frac{a+1}{2}\right) + \psi\left(\frac{a+n+1}{2}\right)\right].$$

5.1.2. Sums containing products of $\psi(k+a)$

1. $\displaystyle\sum_{k=0}^{n} \psi^2(k+a) = 2n + (2a-1)\psi(a) + (1-a)\psi^2(a)$

$$+ (1 - 2n - 2a)\psi(n+a) + (n+a)\psi^2(n+a).$$

2. $\displaystyle\sum_{k=0}^{n} \frac{(b)_k}{(a)_k} \psi(k+a)\,\psi(k+b)$

$$= \frac{a-1}{(a-b-1)^2}\left\{\frac{2}{a-b-1} + \psi(b) + \psi(a-1)\left[1 + (a-b-1)\psi(b)\right]\right.$$

$$- \frac{(b)_{n+1}}{(a)_n (a-b-1)^2}\left[\frac{2}{a-b-1} + \psi(b+n+1)\right.$$

$$\left.\left. + \psi(a+n)\left(1 + (a-b-1)\psi(b+n+1)\right)\right]\right\}.$$

3. $\displaystyle\sum_{k=0}^{n} \psi^3(k+a) = -6n + 3(1-2a)\psi(a) + \frac{3}{2}(2a-1)\psi^2(a)$

$$+ (1-a)\psi^3(a) + 3(2n+2a-1)\psi(n+a) + \frac{3}{2}(1-2n-2a)\psi^2(n+a)$$

$$+ (n+a)\psi^3(n+a) + \frac{1}{2}\psi'(a) - \frac{1}{2}\psi'(n+a).$$

5.1.3. Sums containing $\psi'(k+a, z)$

1. $\displaystyle\sum_{k=1}^{n} (-1)^k \binom{n}{k} \psi'\left(k+\frac{1}{2}\right) = -\frac{\pi^2}{2} + \frac{(n-1)!}{\left(\frac{1}{2}\right)_n}\left[C + 2\ln 2 + \psi\left(n+\frac{1}{2}\right)\right]$

$$[n \geq 1].$$

2. $\displaystyle\sum_{k=1}^{n} \frac{(-1)^k}{2k+1} \binom{n}{k} \psi'\left(k+\frac{1}{2}\right) = \frac{n!\,\pi^2}{2\left(\frac{3}{2}\right)_n} - \frac{\pi^2}{2}$

$$+ \frac{2n}{\left(\frac{3}{2}\right)_n}\, n!\,{}_4F_3\!\left(\begin{array}{c}1-n,1,1,1\\ \frac{3}{2},2,2;\,1\end{array}\right) \quad [n \geq 1].$$

3. $\sum_{k=0}^{n} \psi(k+a)\psi'(k+a) = \psi(a) - \frac{1}{2}\psi^2(a) - \psi(n+a)$

$$+ \frac{1}{2}\psi^2(n+a) + \frac{1}{2}\left[2a - 1 + 2(1-a)\psi(a)\right]\psi'(a)$$

$$+ \frac{1}{2}\left[1 - 2n - 2a + 2(n+a)\psi(n+a)\right]\psi'(n+a).$$

5.2. The Incomplete Gamma Functions $\gamma(\nu, z)$ and $\Gamma(\nu, z)$

5.2.1. Sums containing $\gamma(nk + \nu, z)$

1. $\sum_{k=0}^{n} \frac{z^{-k}}{\nu + k - 1} \gamma(\nu + k, z) = \frac{z^\nu e^{-z}}{\nu - 1} {}_2F_2\!\left(\begin{matrix} 1, 1;\ z \\ \nu, 2 \end{matrix}\right)$

$$- \frac{z^\nu e^{-z}}{\nu + n} {}_2F_2\!\left(\begin{matrix} 1, 1;\ z \\ \nu + n + 1, 2 \end{matrix}\right).$$

2. $\sum_{k=0}^{n} \frac{(-1)^k}{k!} \gamma(k+1, z) = \frac{1 + (-1)^n}{2} - \frac{(-z)^n}{n!} e^{-z} {}_3F_0\!\left(\begin{matrix} -\frac{n}{2}, \frac{1-n}{2}, 1 \\ 4z^{-2} \end{matrix}\right).$

3. $\sum_{k=0}^{n} \binom{n}{k}(-z)^{-k} \gamma(\nu + k, z) = \frac{n! z^\nu e^{-z}}{(\nu)_{n+1}} {}_1F_1\!\left(\begin{matrix} n+1;\ z \\ \nu + n + 1 \end{matrix}\right).$

4. $\sum_{k=0}^{n} (-1)^k \binom{n}{k} \frac{1}{(\nu)_k (k+1)} \gamma(\nu + k, z)$

$$= \frac{1}{n+1}\left[(\nu - 1)\gamma(\nu - 1, z) - \frac{n!}{(\nu)_n} z^{\nu-1} e^{-z} L_n^{\nu-1}(z)\right].$$

5. $\sum_{k=0}^{n} (-1)^k \binom{n}{k} \frac{(a-1)_k}{(a-n)_k (\nu)_k} \gamma(\nu + k, z) = \frac{z^\nu e^{-z}}{\nu} {}_2F_2\!\left(\begin{matrix} 1-n, a;\ z \\ \nu + 1, a - n \end{matrix}\right).$

6. $\sum_{k=0}^{n} \frac{(-1)^k}{(k+m)!} \binom{n}{k} \gamma(k + m + 1, z) = \delta_{n,0} - (-1)^m e^{-z} L_{m+n}^{-m-1}(z).$

7. $\sum_{k=0}^{n} \binom{n}{k} t^k \gamma(k+1, z) = t^n e^{1/t}\left[\gamma\!\left(n+1, z+\frac{1}{t}\right) - \gamma\!\left(n+1, \frac{1}{t}\right)\right].$

8. $\sum_{k=0}^{n} \binom{n}{k} \frac{(k+n+1)!}{(k+1)!} (-z)^{-k} \gamma(k+1, z) = \frac{n! z}{n+1}$

$$- \frac{(n!)^2}{(2n+2)!} z^{n+2} e^{-z} {}_1F_1\!\left(\begin{matrix} n+2;\ z \\ 2n + 3 \end{matrix}\right).$$

5.2.3]   5.2. The Incomplete Gamma Functions $\gamma(\nu, z)$ and $\Gamma(\nu, z)$

9. $\displaystyle\sum_{k=0}^{n} \binom{n}{k} \frac{(a)_k}{(b)_k} (-z)^{-k} \gamma(k+1, z) = {}_3F_1\left(\begin{array}{c} -n, a, 1 \\ b; z^{-1} \end{array}\right)$

$$- \frac{(b-a)_n}{(b)_n} e^{-z} \, {}_3F_1\left(\begin{array}{c} -n, a, 1; -z^{-1} \\ a-b-n+1 \end{array}\right).$$

10. $\displaystyle\sum_{k=0}^{n} \binom{n}{k} \frac{(a+m)_k}{(k+m)!} (-z)^{-k} \gamma(k+m+1, z) = (-z)^{a+m} \Psi\left(\begin{array}{c} a+m; -z \\ a+m+n+1 \end{array}\right)$

$$- \frac{(1-a)_n}{(m+n)!} z^m e^{-z} \, {}_3F_1\left(\begin{array}{c} -m-n, a, 1 \\ a-n; -z^{-1} \end{array}\right).$$

11. $\displaystyle\sum_{k=0}^{n} (-1)^k \binom{n}{k} \frac{(\nu+n)_k}{(\nu+1)_k (\nu)_{2k}} \gamma(\nu + 2k, z)$

$$= \frac{(n-1)! n!}{(\nu+1)_n (\nu+1)_{n-1}} \frac{z^\nu e^{-z}}{\nu} L_n^\nu(-z) L_{n-1}^\nu(z) \quad [n \geq 1].$$

5.2.2. Sums containing products of $\gamma(\nu \pm k, z)$

1. $\displaystyle\sum_{k=0}^{2n} \binom{2n}{k} \gamma(\nu+k, z) \gamma(\nu+2n-k, -z)$

$$= (2n)! \frac{e^{\nu \pi i} z^{2\nu + 2n}}{(\nu)_{2n+1} (\nu + n)} {}_3F_4\left(\begin{array}{c} n+\frac{1}{2}, n+1, \nu+n; \frac{z^2}{4} \\ \frac{\nu+1}{2} + n, \frac{\nu}{2} + n + 1, \nu + n + 1, \frac{1}{2} \end{array}\right).$$

2. $\displaystyle\sum_{k=0}^{2n+1} \binom{2n+1}{k} \gamma(\nu+k, z) \gamma(\nu+2n+1-k, -z)$

$$= (2n+2)! \frac{e^{(\nu+1)\pi i} z^{2\nu+2n+2}}{(\nu)_{2n+3}(\nu+n+1)} {}_3F_4\left(\begin{array}{c} n+\frac{3}{2}, n+2, \nu+n+1; \frac{z^2}{4} \\ \frac{\nu+3}{2}+n, \frac{\nu}{2}+n+2, \nu+n+2, \frac{3}{2} \end{array}\right).$$

5.2.3. Sums containing $\Gamma(\nu \pm k, z)$

1. $\displaystyle\sum_{k=0}^{n} \binom{n}{k} (-z)^k \Gamma(\nu - k, z) = n! \, e^{-z} \Psi\left(\begin{array}{c} 1-\nu+n \\ 1-\nu; z \end{array}\right).$

2. $\displaystyle\sum_{k=0}^{n} \binom{n}{k} (1-\nu)_k \Gamma(\nu - k, z) = (-1)^{n-1} (n-1)! z^{\nu-n} e^{-z} L_{n-1}^{\nu-n}(z) \quad [n \geq 1].$

3. $\displaystyle\sum_{k=0}^{n} \binom{n}{k} \frac{(1-\nu)_k}{(1-\nu-n)_k} (-z)^k \Gamma(\nu - k, z) = \frac{n!}{(\nu)_n} z^{\nu+n} e^{-z} \Psi\left(\begin{array}{c} n+1; z \\ \nu+n+1 \end{array}\right).$

4. $\displaystyle\sum_{k=0}^{n} (-1)^k \binom{n}{k} \frac{1}{(\nu)_k} \Gamma(\nu+k, z) = -\frac{(n-1)!}{(\nu)_n} z^\nu e^{-z} L_{n-1}^\nu(z) \quad [n \geq 1].$

5.3. The Bessel Function $J_\nu(z)$

5.3.1. Sums containing $J_{\nu \pm nk}(z)$

1. $\displaystyle\sum_{k=0}^{[n/2]} \frac{(-n)_{2k}}{k!} (2z)^{-k} J_{k-n+1/2}(z) = \sqrt{\frac{2}{\pi z}} \sin\left(z + \frac{n\pi}{2}\right).$

2. $\displaystyle\sum_{k=0}^{[n/2]} \frac{(-n)_{2k}}{k!} (-2z)^{-k} J_{n-k-1/2}(z) = (-1)^n \sqrt{\frac{2}{\pi z}} \cos\left(z + \frac{n\pi}{2}\right).$

3. $\displaystyle\sum_{k=0}^{n} \binom{n}{k} (2k+\nu) \frac{(\nu)_k}{(\nu+n+1)_k} J_{2k+\nu}(z) = (\nu)_{n+1} \left(\frac{2}{z}\right)^n J_{\nu+n}(z).$

4. $\displaystyle\sum_{k=0}^{n} (-1)^k \binom{n}{k} (2k+\nu) \frac{(\nu)_k}{(\nu+n+1)_k} J_{4k+2\nu}(z)$

$$= \frac{2^{-2\nu-1} z^{2\nu}}{\Gamma(2\nu)} {}_1F_2\!\left(\begin{matrix} \nu+n+\frac{1}{2}; -\frac{z^2}{4} \\ \nu+\frac{1}{2}, 2\nu+2n+1 \end{matrix}\right).$$

5. $\displaystyle\sum_{k=0}^{n} \binom{n}{k} \frac{\left(-\frac{z}{2}\right)^k}{(\nu-n+1)_k} J_{k+\nu}(z) = \frac{\left(-\frac{z}{2}\right)^n}{(-\nu)_n} J_{\nu-n}(z).$

5.3.2. Sums containing products of $J_{\nu \pm nk}(z)$

1. $\displaystyle\sum_{k=1}^{n} (-1)^k (k+\nu) J_{k+\nu}^2(z) = \frac{1}{2} J_{\nu+1}(z) \left[z J_{\nu+2}(z) - 2(\nu+1) J_{\nu+1}(z)\right]$

$\qquad + \dfrac{(-1)^n}{2} J_{\nu+n+1}(z) \left[2(\nu+n+1) J_{\nu+n+1}(z) - z J_{\nu+n+2}(z)\right].$

2. $\displaystyle\sum_{k=0}^{n} \binom{n}{k} (k+\nu) \frac{(2\nu)_k}{(2\nu+n+1)_k} J_{k+\nu}^2(z) = \frac{\left(\frac{z}{2}\right)^{2\nu}}{\nu \Gamma^2(\nu)} {}_1F_2\!\left(\begin{matrix} \nu+\frac{1}{2}; -z^2 \\ \nu+1, 2\nu+n+1 \end{matrix}\right).$

3. $\displaystyle\sum_{k=0}^{2n+1} (2n-2k+1) \binom{2n+1}{k} J_{k-n-1/2}(w) J_{k-n-1/2}(z)$

$$= -\frac{2^{n+3/2}}{\Gamma\!\left(-n-\frac{1}{2}\right)} \left(\frac{w+z}{wz}\right)^{n+1/2} J_{-n-1/2}(w+z).$$

5.4. The Modified Bessel Function $I_\nu(z)$

5.4.1. Sums containing $I_{\nu\pm nk}(z)$

1. $\displaystyle\sum_{k=0}^{n} \binom{n}{k} \frac{\left(\frac{z}{2}\right)^k}{(\nu - n + 1)_k} I_{k+\nu}(z) = \frac{\left(-\frac{z}{2}\right)^n}{(-\nu)_n} I_{\nu-n}(z).$

2. $\displaystyle\sum_{k=0}^{[n/2]} \frac{(-n)_{2k}}{k!} (2z)^{-k} I_{\pm n \mp k \mp 1/2}(z) = \frac{1}{\sqrt{2\pi z}} [e^z \pm (-1)^n e^{-z}].$

3. $\displaystyle\sum_{k=1}^{n} (2k + \nu) I_{2k+\nu}(z) = \frac{z}{2} [I_{\nu+1}(z) - I_{\nu+2n+1}(z)].$

4. $\displaystyle\sum_{k=0}^{n} (-1)^k \binom{n}{k} (2k + \nu) \frac{(\nu)_k}{(\nu + n + 1)_k} I_{2k+\nu}(z) = (\nu)_{n+1} \left(\frac{2}{z}\right)^n I_{\nu+n}(z).$

5.4.2. Sums containing products of $J_{\nu\pm nk}(z)$ and $I_{\nu\pm nk}(z)$

1. $\displaystyle\sum_{k=0}^{n} \binom{n}{k} J_{\nu-n+k}(z) I_{\nu+k}(z)$

$$= \frac{(-1)^n(-\nu)_n}{\Gamma^2(\nu+1)} \left(\frac{z}{2}\right)^{2\nu-n} {}_0F_3\left(\nu+1, \frac{\nu-n+1}{2}, \frac{\nu-n}{2}+1; -\frac{z^4}{64}\right).$$

2. $\displaystyle\sum_{k=0}^{n} \binom{n}{k} J_{\nu-n+k}(z) I_{\nu-k}(z)$

$$= \frac{(-1)^n(-2\nu)_n}{\Gamma^2(\nu+1)} \left(\frac{z}{2}\right)^{2\nu-n} {}_1F_4\left(\nu+\frac{1}{2}; \frac{\nu+1}{2}, \frac{\nu}{2}+1, \nu-\frac{n+1}{2}, \nu-\frac{n}{2}+1; -\frac{z^4}{64}\right).$$

3. $\displaystyle\sum_{k=0}^{n} \binom{n}{k}\left(\frac{z}{w}\right)^k J_{k-n+1/2}(w) I_{k-1/2}(z)$

$$= \frac{w^{-n-1/2}}{\sqrt{2\pi z}} [(w+iz)^{n+1/2} J_{1/2-n}(w+iz) + (w-iz)^{n+1/2} J_{1/2-n}(w-iz)].$$

4. $\displaystyle\sum_{k=0}^{n} (-1)^k \binom{n}{k}\left(\frac{z}{w}\right)^k J_{n-k-1/2}(w) I_{k-1/2}(z)$

$$= \frac{w^{-n-1/2}}{\sqrt{2\pi z}} [(w+iz)^{n+1/2} J_{n-1/2}(w+iz) + (w-iz)^{n+1/2} J_{n-1/2}(w-iz).$$

5. $\displaystyle\sum_{k=0}^{n} (-1)^k \binom{n}{k} J_{k-1/2}(z) I_{n-k+1/2}(z) = (-1)^n \frac{2^{(2n+3)/4}}{\sqrt{\pi z}}$

$$\times \left[\sin\frac{(6n+3)\pi}{8} \operatorname{ber}_{n+1/2}(\sqrt{2}\,z) + \cos\frac{(6n+3)\pi}{8} \operatorname{bei}_{n+1/2}(\sqrt{2}\,z)\right].$$

6. $\displaystyle\sum_{k=0}^{n}(-1)^k\binom{n}{k}J_{k-1/2}(z)\,I_{k-n+1/2}(z) = \frac{2^{(2n+3)/4}}{\sqrt{\pi z}}$

$\displaystyle\times\left[\sin\frac{(6n+3)\pi}{8}\,\mathrm{ber}_{1/2-n}(\sqrt{2}\,z) + \cos\frac{(6n+3)\pi}{8}\,\mathrm{bei}_{1/2-n}(\sqrt{2}\,z)\right].$

7. $\displaystyle\sum_{k=0}^{n}\binom{n}{k}J_{-k-1/2}(z)\,I_{k-n+1/2}(z) = (-1)^n\,\frac{2^{(2n+3)/4}}{\sqrt{\pi z}}$

$\displaystyle\times\left[\sin\frac{(6n+3)\pi}{8}\,\mathrm{ber}_{n+1/2}(\sqrt{2}\,z) + \cos\frac{(6n+3)\pi}{8}\,\mathrm{bei}_{n+1/2}(\sqrt{2}\,z)\right].$

8. $\displaystyle\sum_{k=0}^{n}(-1)^k\binom{n}{k}J_{k+1/2}(z)\,I_{k-n+1/2}(z) = \frac{2^{(2n+3)/4}}{\sqrt{\pi z}}$

$\displaystyle\times\left[\sin\frac{(6n+3)\pi}{8}\,\mathrm{ber}_{-n-1/2}(\sqrt{2}\,z) + \cos\frac{(6n+3)\pi}{8}\,\mathrm{bei}_{-n-1/2}(\sqrt{2}\,z)\right].$

9. $\displaystyle\sum_{k=0}^{n}(-1)^k\binom{n}{k}J_{k-1/2}(z)\,I_{k-n-1/2}(z) = \frac{2^{(2n+3)/4}}{\sqrt{\pi z}}$

$\displaystyle\times\left[\cos\frac{(6n+3)\pi}{8}\,\mathrm{ber}_{-n-1/2}(\sqrt{2}\,z) - \sin\frac{(6n+3)\pi}{8}\,\mathrm{bei}_{-n-1/2}(\sqrt{2}\,z)\right].$

10. $\displaystyle\sum_{k=0}^{n}\binom{n}{k}J_{1/2-k}(z)\,I_{n-k+1/2}(z) = \frac{2^{(2n+3)/4}}{\sqrt{\pi z}}$

$\displaystyle\times\left[\sin\frac{(6n+3)\pi}{8}\,\mathrm{ber}_{-n-1/2}(\sqrt{2}\,z) + \cos\frac{(6n+3)\pi}{8}\,\mathrm{bei}_{-n-1/2}(\sqrt{2}\,z)\right].$

11. $\displaystyle\sum_{k=0}^{n}\binom{n}{k}J_{-k-1/2}(z)\,I_{n-k-1/2}(z) = \frac{2^{(2n+3)/4}}{\sqrt{\pi z}}$

$\displaystyle\times\left[\cos\frac{(6n+3)\pi}{8}\,\mathrm{ber}_{-n-1/2}(\sqrt{2}\,z) - \sin\frac{(6n+3)\pi}{8}\,\mathrm{bei}_{-n-1/2}(\sqrt{2}\,z)\right].$

12. $\displaystyle\sum_{k=0}^{n}(-1)^k\binom{n}{k}J_{k+1/2}(z)\,I_{n-k-1/2}(z) = (-1)^n\,\frac{2^{(2n+3)/4}}{\sqrt{\pi z}}$

$\displaystyle\times\left[\sin\frac{(6n+3)\pi}{8}\,\mathrm{bei}_{n+1/2}(\sqrt{2}\,z) - \cos\frac{(6n+3)\pi}{8}\,\mathrm{ber}_{n+1/2}(\sqrt{2}\,z)\right].$

13. $\sum_{k=0}^{n} \binom{n}{k} J_{1/2-k}(z) I_{n-k-1/2}(z) = \dfrac{2^{(2n+3)/4}}{\sqrt{\pi z}}$

$\times \left[\sin \dfrac{(6n+3)\pi}{8} \operatorname{bei}_{1/2-n}(\sqrt{2}\,z) - \cos \dfrac{(6n+3)\pi}{8} \operatorname{ber}_{1/2-n}(\sqrt{2}\,z) \right].$

14. $\sum_{k=0}^{n} \binom{n}{k} J_{1/2-k}(z) I_{k-n-1/2}(z) = (-1)^n \dfrac{2^{(2n+3)/4}}{\sqrt{\pi z}}$

$\times \left[\sin \dfrac{(6n+3)\pi}{8} \operatorname{bei}_{n+1/2}(\sqrt{2}\,z) - \cos \dfrac{(6n+3)\pi}{8} \operatorname{ber}_{n+1/2}(\sqrt{2}\,z) \right].$

5.4.3. Sums containing products of $I_{\nu \pm nk}(z)$

1. $\sum_{k=0}^{n} (-1)^k \binom{n}{k} (\nu+k) \dfrac{(2\nu)_k}{(2\nu+n+1)_k} I_{\nu+k}^2(z)$

$= \dfrac{\left(\dfrac{z}{2}\right)^{2\nu}}{\nu \Gamma^2(\nu)} {}_1F_2\left(\begin{matrix} \nu + \dfrac{1}{2};\ z^2 \\ \nu+1,\ 2\nu+n+1 \end{matrix} \right).$

2. $\sum_{k=0}^{n} \binom{n}{k} I_{\mu-k}(z) I_{\nu+k}(z) = (-1)^n \dfrac{(\mu+\nu-n+1)_{2n}}{\Gamma(\mu+1)\Gamma(\nu+n+1)(-\mu-\nu)_n} \left(\dfrac{z}{2}\right)^{\mu+\nu}$

$\times {}_2F_3\left(\begin{matrix} \dfrac{\mu+\nu+n+1}{2},\ \dfrac{\mu+\nu+n}{2}+1;\ z^2 \\ \mu+1,\ \nu+n+1,\ \mu+\nu+1 \end{matrix} \right).$

3. $\sum_{k=0}^{n} \binom{n}{k} I_{\mu-k}(z) I_{\nu-k}(z)$

$= \dfrac{(-\mu)_n(-\nu)_n}{\Gamma(\mu+1)\Gamma(\nu+1)} \left(\dfrac{z}{2}\right)^{\mu+\nu-2n} {}_2F_3\left(\begin{matrix} \dfrac{\mu+\nu-n+1}{2},\ \dfrac{\mu+\nu-n}{2}+1;\ z^2 \\ \mu-n+1,\ \nu-n+1,\ \mu+\nu-n+1 \end{matrix} \right).$

4. $\sum_{k=0}^{2n+1} (-1)^k (2n-2k+1) \binom{2n+1}{k} I_{k-n-1/2}(w) I_{k-n-1/2}(z)$

$= -\dfrac{2^{n+3/2}}{\Gamma\left(-n-\dfrac{1}{2}\right)} \left(\dfrac{w+z}{wz}\right)^{n+1/2} I_{-n-1/2}(w+z).$

5. $\sum_{k=1}^{n} \dfrac{(-1)^k}{k^{1/2}} \binom{n}{k} \left[I_{-m-1/2}^2\left(\dfrac{z}{\sqrt{k}}\right) - I_{m+1/2}^2\left(\dfrac{z}{\sqrt{k}}\right) \right]$

$= (-1)^n 2^{1-2n} z^{-2n-1} \dfrac{[(2n)!]^2}{n!\,\pi} \delta_{m,n} - (-1)^m \dfrac{2}{\pi z} \quad [m \le n].$

6. $\sum_{k=0}^{n} \dfrac{(-1)^k}{I_k(z) I_{k+1}(z)} = (-1)^n z \dfrac{K_{n+1}(z)}{I_{n+1}(z)} + z \dfrac{K_0(z)}{I_0(z)}.$

5.5. The Macdonald Function $K_\nu(z)$

5.5.1. Sums containing $K_{\nu \pm nk}(z)$

1. $\displaystyle\sum_{k=0}^{n} \frac{(n+k)!}{k!(n-k)!} t^k K_{k+1/2}(z)$

$$= \frac{1}{\sqrt{\pi t}} K_{n+1/2}\left(\frac{z}{2} - \frac{1}{2}\sqrt{z^2 - \frac{2z}{t}}\right) K_{n+1/2}\left(\frac{z}{2} + \frac{1}{2}\sqrt{z^2 - \frac{2z}{t}}\right) \quad [68].$$

2. $\displaystyle\sum_{k=0}^{[n/2]} \frac{(-2z)^{-k}}{k!(n-2k)!} K_{n-k-1/2}(z) = \frac{1}{n!}\sqrt{\frac{\pi}{2z}} e^{-z}.$

3. $\displaystyle\sum_{k=0}^{n} \frac{\left(\frac{1}{2} - a - n\right)_k}{(n-k)!(1-a-2n)_k} \left(\frac{2}{z}\right)^k K_{k+1/2}(z)$

$$= \Gamma\left(n + \frac{1}{2}\right)\left[\frac{2^{-a}\sqrt{\pi}\,\Gamma(2a+2n)}{\Gamma(a+2n)} z^{-a-2n} I_{a-1/2}(z)\right.$$
$$\left. - \frac{a+2n}{(2n+1)!(a+n)} 2^{2n-1/2} z^{1/2} e^{-z} {}_2F_2\!\left(\begin{matrix}a+2n+1,\,1;\,2z\\2a+2n+1,\,2n+2\end{matrix}\right)\right].$$

4. $\displaystyle\sum_{k=0}^{n} \binom{n}{k} \frac{\left(-\frac{z}{2}\right)^k}{\left(\frac{1}{2} - m - n\right)_k} K_{m-k+1/2}(z) = (-1)^{m+n} \frac{2^{-n-1}\pi z^n}{\left(m+\frac{1}{2}\right)_n} I_{-m-n-1/2}(z).$

5. $\displaystyle\sum_{k=0}^{n} \binom{n}{k}(\nu+n)_k \left(\frac{2}{z}\right)^k K_{\nu+k}(z) = K_{\nu+2n}(z).$

5.5.2. Sums containing $K_{\nu \pm nk}(z)$ and special functions

1. $\displaystyle\sum_{k=0}^{n} (-1)^k \binom{n}{k} J_{\nu+k-n}(z) K_{\nu+k}(z)$

$$= \frac{\Gamma\!\left(\frac{\nu-n}{2}\right)}{4\Gamma\!\left(\frac{\nu+n}{2}+1\right)} \left(-\frac{z}{2}\right)^n {}_0F_3\!\left(\begin{matrix}-\frac{z^4}{64}\\ \frac{n-\nu}{2}+1,\,\frac{n+\nu}{2}+1,\,\frac{1}{2}\end{matrix}\right)$$

$$- \frac{\Gamma\!\left(\frac{\nu-n-1}{2}\right)}{4\Gamma\!\left(\frac{\nu+n+3}{2}\right)} \left(-\frac{z}{2}\right)^{n+2} {}_0F_3\!\left(\begin{matrix}-\frac{z^4}{64}\\ \frac{n-\nu+3}{2},\,\frac{n+\nu+3}{2},\,\frac{3}{2}\end{matrix}\right)$$

$$+ (-1)^n \frac{\Gamma(n-\nu)}{2\Gamma(\nu+1)} \left(\frac{z}{2}\right)^{2\nu-n} {}_0F_3\!\left(\begin{matrix}-\frac{z^4}{64}\\ \frac{\nu-n+1}{2},\,\frac{\nu-n}{2}+1,\,\nu+1\end{matrix}\right).$$

2. $\displaystyle\sum_{k=0}^{n}(-1)^k\binom{n}{k}\left(\frac{z}{w}\right)^k I_{\pm n\mp k\mp 1/2}(w)\,K_{k-1/2}(z)$

$$= (-1)^n \frac{w^{-n-1/2}}{\sqrt{2\pi z}}[(z-w)^{n+1/2}K_{n-1/2}(z-w)$$
$$\pm (z+w)^{n+1/2}K_{n-1/2}(z+w)].$$

3. $\displaystyle\sum_{k=0}^{n}(-1)^k k\, I_{k+\nu}(z)\,K_{k-\nu}(z) = -\frac{z}{4}[I_\nu(z)\,K_{\nu-1}(z) - I_{\nu+1}(z)\,K_\nu(z)]$

$$+ (-1)^n \frac{z}{4}[I_{n+\nu}(z)\,K_{n-\nu+1}(z) - I_{n+\nu+1}(z)\,K_{n-\nu}(z)].$$

5.5.3. Sums containing products of $K_{\nu\pm nk}(z)$

1. $\displaystyle\sum_{k=0}^{n}(2k+1)K_{k+1/2}(w)\,K_{k+1/2}(z)$

$$= (n+1)\sqrt{\pi}\sum_{k=0}^{n}\frac{(k+n+1)!}{(k+1)!\,(n-k)!}\left(\frac{w+z}{2wz}\right)^{k+1/2}K_{k+1/2}(w+z) \quad [68].$$

2. $\displaystyle\sum_{k=0}^{n}(-1)^k \frac{2k+1}{(n-k)!\,(k+n+1)!} K_{k+1/2}(w)\,K_{k+1/2}(z)$

$$= (-1)^n \frac{\sqrt{\pi}}{n!}\left(\frac{w+z}{2wz}\right)^{n+1/2} K_{n+1/2}(w+z).$$

5.6. The Struve Functions $\mathbf{H}_\nu(z)$ and $\mathbf{L}_\nu(z)$

5.6.1. Sums containing $\mathbf{H}_{k+\nu}(z)$ and $\mathbf{L}_{k+\nu}(z)$

1. $\displaystyle\sum_{k=0}^{n}\frac{(-n)_k(n)_k}{k!}\left(\frac{2}{z}\right)^k \mathbf{H}_{k-1/2}(z)$

$$= (-1)^n \sqrt{\frac{\pi}{2z}}\,J_{n+1/2}\left(\frac{z}{2}\right)\left[2(2n+1)J_{n+1/2}\left(\frac{z}{2}\right) - z\,J_{n+3/2}\left(\frac{z}{2}\right)\right].$$

2. $\displaystyle\sum_{k=0}^{n}\frac{(-n)_k(n)_k}{k!}\left(\frac{2}{z}\right)^k \mathbf{L}_{k-1/2}(z)$

$$= \sqrt{\frac{\pi}{2z}}\,I_{n+1/2}\left(\frac{z}{2}\right)\left[2(2n+1)I_{n+1/2}\left(\frac{z}{2}\right) + z\,I_{n+3/2}\left(\frac{z}{2}\right)\right].$$

3. $\displaystyle\sum_{k=0}^{n}\binom{n}{k}(a)_k\left(-\frac{2}{z}\right)^k \mathbf{H}_{k+\nu}(z)$

$$= \frac{2^{-\nu}z^{\nu+1}\left(\nu-a+\frac{3}{2}\right)_n}{\sqrt{\pi}\,\Gamma\left(\nu+n+\frac{3}{2}\right)}\,{}_2F_3\left(\begin{array}{c}\nu-a+n+\frac{3}{2},\,1;\,-\frac{z^2}{4}\\ \nu+n+\frac{3}{2},\,\nu-a+\frac{3}{2},\,\frac{3}{2}\end{array}\right).$$

4. $\displaystyle\sum_{k=0}^{n}\binom{n}{k}(a)_k\left(-\frac{2}{z}\right)^k \mathbf{L}_{k+\nu}(z)$

$$= \frac{2^{-\nu} z^{\nu+1}\left(\nu - a + \frac{3}{2}\right)_n}{\sqrt{\pi}\,\Gamma\left(\nu + n + \frac{3}{2}\right)}\, {}_2F_3\left(\begin{array}{c}\nu - a + n + \frac{3}{2},\, 1;\, \frac{z^2}{4} \\ \nu + n + \frac{3}{2},\, \nu - a + \frac{3}{2},\, \frac{3}{2}\end{array}\right).$$

5.7. The Legendre Polynomials $P_n(z)$

5.7.1. Sums containing $P_{m\pm nk}(z)$

1. $\displaystyle\sum_{k=0}^{n} P_k(z) = \frac{(2n)!}{(n!)^2}\left(\frac{z-1}{2}\right)^n {}_3F_2\left(\begin{array}{c}-n,\,-n,\,-n-\frac{1}{2};\,\frac{2}{1-z} \\ \frac{1}{2}-n,\,-2n-1\end{array}\right).$

2. $\displaystyle\sum_{k=0}^{n}(2k+1)\,P_k(z) = \frac{n+1}{z-1}\left[P_{n+1}(z) - P_n(z)\right].$

3. $\displaystyle\sum_{k=0}^{n}\binom{n}{k} t^k P_k(z) = (1 + 2tz + t^2)^{n/2}\, P_n\!\left(\frac{1+tz}{\sqrt{1+2tz+t^2}}\right).$

4. $\displaystyle\sum_{k=1}^{n}\frac{k\,\Gamma(2n-k)}{(n-k)!}(2z)^k P_k(z) = \left(\frac{1}{2}\right)_n (2z)^n \qquad [n \geq 1].$

5. $\displaystyle\sum_{k=0}^{n}\frac{(-1)^k}{(n-k)!\,(a)_k}\left(z - \sqrt{z^2-1}\right)^k P_k(z)$

$$= \frac{1}{(a)_n}\, P_n^{(a-3/2,\,1-a-n)}\!\left(3 - 4z^2 + 4z\sqrt{z^2-1}\right).$$

6. $\displaystyle\sum_{k=0}^{n-1}(-1)^k\binom{n}{k}(n-k)^n\left(z + \sqrt{z^2-1}\right)^k P_k(z)$

$$= \frac{2^n \left(\frac{1}{2}\right)_n}{n!}\left(1 - \frac{z}{\sqrt{z^2-1}}\right)^{-n}\sum_{k=0}^{n} 2^{-k} \sigma_n^k \frac{(-n)_k^2}{(1/2-n)_k}\left(\frac{z}{\sqrt{z^2-1}} - 1\right)^k.$$

7. $\displaystyle\sum_{k=0}^{n}\binom{n}{k}\frac{(a)_k}{k!}\left(\frac{2}{1-z}\right)^k P_k(z)$

$$= \frac{(a)_n}{n!}\left(\frac{1+z}{1-z}\right)^n {}_4F_3\left(\begin{array}{c}-n,\,-n,\,\frac{1-a-n}{2},\,1-\frac{a+n}{2} \\ 1-a-n,\,1-a-n,\,1;\,\frac{4z-4}{z+1}\end{array}\right).$$

8. $\displaystyle\sum_{k=0}^{n}(-1)^k\frac{2k+1}{(n-k)!\,(k+n+1)!}\, P_k(z) = \frac{2^{-n}}{(n!)^2}(1-z)^n.$

5.7.1]   5.7. The Legendre Polynomials $P_n(z)$

9. $\displaystyle\sum_{k=0}^{n}\binom{n}{k}\frac{2^{-k}}{(n-k)!\left(\frac{1}{2}-n\right)_k}\left(\frac{\sqrt{z^2-1}-z}{1-z^2+z\sqrt{z^2-1}}\right)^k P_k(z)$

$\displaystyle = \frac{2^{-n}}{\left(\frac{1}{2}\right)_n}\left(1-z^2+z\sqrt{z^2-1}\right)^{-n}\,_3F_2\left(\begin{array}{c}-\frac{n}{2},\,\frac{1-n}{2},\,\frac{1}{2}\\ 1,\,1;\,4\left(1-z^2+z\sqrt{z^2-1}\right)^2\end{array}\right).$

10. $\displaystyle\sum_{k=0}^{n}\frac{(a)_k}{(n-k)!(k+n+1)!(2-a)_k}P_k(z)$

$\displaystyle = \frac{(a)_n}{(n!)^2(2n+1)(2-a)_n}\left(\frac{z+1}{2}\right)^n\,_3F_2\left(\begin{array}{c}-n,\,-n,\,\frac{3}{2}-a;\,\frac{2}{z+1}\\ \frac{1}{2}-n,\,1-a-n\end{array}\right).$

11. $\displaystyle\sum_{k=0}^{n}(2k+1)\frac{(a)_k(b)_k}{(n-k)!(k+n+1)!(2-a)_k(2-b)_k}P_k(z)$

$\displaystyle = \frac{(a)_n(b)_n}{(n!)^2(2-a)_n(2-b)_n}\left(\frac{1+z}{2}\right)^n\,_3F_2\left(\begin{array}{c}-n,\,-n,\,2-a-b;\,\frac{2}{1+z}\\ 1-a-n,\,1-b-n\end{array}\right).$

12. $\displaystyle\sum_{k=0}^{n}(-1)^k\binom{n}{k}\left(\sqrt{z^2-1}-z\right)^{2k}P_{2k}(z)$

$\displaystyle = (1-z^2+z\sqrt{z^2-1})^n P_n^{(n,-n-1/2)}\left(2z^2-2z\sqrt{z^2-1}-1\right).$

13. $\displaystyle\sum_{k=0}^{n}(-1)^k\binom{n}{k}\frac{(a)_k}{\left(\frac{1}{2}\right)_k}z^{-2k}P_{2k}(z)$

$\displaystyle = \frac{(a)_n}{\left(\frac{1}{2}\right)_n}z^{-2n}(1-z^2)^n\,_4F_3\left(\begin{array}{c}-n,\,-n,\,\frac{1-2a-2n}{4},\,\frac{3-2a-2n}{4}\\ \frac{1}{2}-a-n,\,1-a-n,\,\frac{1}{2};\,\frac{4z^2}{z^2-1}\end{array}\right).$

14. $\displaystyle\sum_{k=0}^{n}\binom{n}{k}\frac{(a)_k}{\left(\frac{1}{2}\right)_k}(1-z^2)^{-k}P_{2k}(z)$

$\displaystyle = \frac{(a)_n}{\left(\frac{1}{2}\right)_n}z^{2n}(1-z^2)^{-n}\,_4F_3\left(\begin{array}{c}-n,\,\frac{1}{2}-n,\,\frac{1-a-n}{2},\,1-\frac{a+n}{2}\\ 1-a-n,\,1-a-n,\,1;\,4-4z^{-2}\end{array}\right).$

15. $\displaystyle\sum_{k=0}^{n}\frac{4k+1}{(n-k)!\left(n+\frac{3}{2}\right)_k}P_{2k}(z) = \frac{2n+1}{n!}z^{2n}.$

16. $\displaystyle\sum_{k=0}^{n}(-1)^k\binom{n}{k}(4k+1)\frac{\left(\frac{1}{2}\right)_k}{\left(n+\frac{3}{2}\right)_k}P_{2k}(z) = \frac{2\left(\frac{1}{2}\right)_{n+1}}{n!}(1-z^2)^n.$

17. $\sum_{k=0}^{n}(-1)^k \dfrac{(4k+1)(n)_k}{(n-k)!\left(\frac{3}{2}-n\right)_k\left(\frac{3}{2}+n\right)_k} P_{2k}(z) = \dfrac{1-4n^2}{n!} T_{2n}(z).$

18. $\sum_{k=0}^{n}(-1)^k \binom{n}{k} \dfrac{(a)_k}{\left(\frac{1}{2}\right)_k} z^{-2k} P_{2k}(z)$

$= \dfrac{(a)_n}{n!} z^{-2n}(1-z^2)^n \, {}_4F_3\left(\begin{array}{c} -n,\,-n,\,\frac{1-2a-2n}{4},\,\frac{3-2a-2n}{4} \\ \frac{1}{2}-a-n,\,1-a-n,\,\frac{1}{2};\,\frac{4z^2}{z^2-1} \end{array}\right).$

19. $\sum_{k=0}^{n} \binom{n}{k} \dfrac{(a)_k}{\left(\frac{1}{2}\right)_k} (1-z^2)^{-k} P_{2k}(z)$

$= \dfrac{(a)_n}{\left(\frac{1}{2}\right)_n} z^{2n}(1-z^2)^{-n} \, {}_4F_3\left(\begin{array}{c} -n,\,\frac{1}{2}-n,\,\frac{1-a-n}{2},\,1-\frac{a+n}{2} \\ 1-a-n,\,1-a-n,\,1;\,4-4z^{-2} \end{array}\right).$

20. $\sum_{k=0}^{n} \dfrac{4k+1}{(n-k)!(4k-1)(4k+3)\left(n+\frac{3}{2}\right)_k} P_{2k}(z)$

$= \dfrac{2n+1}{n!(4n-1)(4n+3)} z^{2n} \, {}_3F_2\left(\begin{array}{c} -n,\,\frac{1}{2}-n,\,1;\,z^{-2} \\ \frac{1}{4}-n,\,\frac{5}{4}-n \end{array}\right).$

21. $\sum_{k=0}^{n}(-1)^k \binom{n}{k}(4k+1) \dfrac{(k+n)!\left(\frac{1}{2}\right)_k^2}{k!\left(\frac{1}{2}-n\right)_k\left(\frac{3}{2}+n\right)_k} P_{2k}(z)$

$= n!(2n+1)[P_n(z)]^2.$

22. $\sum_{k=0}^{n}(-1)^k \binom{n}{k} \dfrac{4k+1}{(4k-1)(4k+3)} \dfrac{\left(\frac{1}{2}\right)_k}{\left(n+\frac{3}{2}\right)_k} P_{2k}(z)$

$= \dfrac{\left(\frac{3}{2}\right)_n}{n!(4n-1)(4n+3)} (1-z^2)^n \, {}_3F_2\left(\begin{array}{c} -n,\,-n,\,1;\,\frac{1}{1-z^2} \\ \frac{1}{4}-n,\,\frac{5}{4}-n \end{array}\right).$

23. $\sum_{k=0}^{n}(-1)^k \binom{n}{k}(4k+1) \dfrac{(a)_k (b)_k \left(\frac{1}{2}\right)_k}{\left(n+\frac{3}{2}\right)_k\left(\frac{3}{2}-a\right)_k\left(\frac{3}{2}-b\right)_k} P_{2k}(z)$

$= \dfrac{2(a)_n (b)_n \left(\frac{1}{2}\right)_{n+1}}{n!\left(\frac{3}{2}-a\right)_n\left(\frac{3}{2}-b\right)_n} (1-z^2)^n \, {}_3F_2\left(\begin{array}{c} -n,\,-n,\,\frac{3}{2}-a-b;\,\frac{1}{1-z^2} \\ 1-a-n,\,1-b-n \end{array}\right).$

24. $\sum_{k=0}^{n} \dfrac{4k+3}{(n-k)!\left(n+\frac{5}{2}\right)_k} P_{2k+1}(z) = \dfrac{2n+3}{n!} z^{2n+1}.$

25. $\sum_{k=0}^{n} \binom{n}{k} \frac{(a)_k}{\left(\frac{3}{2}\right)_k} (1-z^2)^{-k} P_{2k+1}(z)$

$$= \frac{(a)_n}{\left(\frac{3}{2}\right)_n} z^{2n+1}(1-z^2)^{-n} \,_4F_3\left(\begin{array}{c} -n, -\frac{1}{2}-n, \frac{1-a-n}{2}, 1-\frac{a+n}{2} \\ 1-a-n, 1-a-n, 1; \, 4-4z^{-2} \end{array}\right).$$

26. $\sum_{k=0}^{n}(-1)^k \binom{n}{k} \frac{(a)_k}{\left(\frac{3}{2}\right)_k} z^{-2k} P_{2k+1}(z)$

$$= \frac{(a)_n}{n!} z^{1-2n}(1-z^2)^n \,_4F_3\left(\begin{array}{c} -n, -n, \frac{3-2a-2n}{4}, \frac{5-2a-2n}{4} \\ 1-a-n, \frac{3}{2}-a-n, \frac{3}{2}; \, \frac{4z^2}{z^2-1} \end{array}\right).$$

27. $\sum_{k=0}^{n} \binom{n}{k} \frac{(a)_k}{\left(\frac{3}{2}\right)_k} (1-z^2)^{-k} P_{2k+1}(z)$

$$= \frac{(a)_n}{\left(\frac{3}{2}\right)_n} z^{2n+1}(1-z^2)^{-n} \,_4F_3\left(\begin{array}{c} -n, -n-\frac{1}{2}, \frac{1-a-n}{2}, 1-\frac{a+n}{2} \\ 1-a-n, 1-a-n, 1; \, 4-4z^{-2} \end{array}\right).$$

28. $\sum_{k=0}^{n}(-1)^k \binom{n}{k}\left(2k+\frac{3}{2}\right) \frac{\left(\frac{3}{2}\right)_k}{\left(n+\frac{5}{2}\right)_k} P_{2k+1}(z) = \frac{\left(\frac{3}{2}\right)_{n+1}}{n!} z(1-z^2)^n.$

29. $\sum_{k=0}^{n} \frac{4k+3}{(n-k)!(4k+1)(4k+5)} \frac{1}{\left(n+\frac{5}{2}\right)_k} P_{2k+1}(z)$

$$= \frac{2n+3}{n!(4n+1)(4n+5)} z^{2n+1} \,_3F_2\left(\begin{array}{c} -n, -\frac{1}{2}-n, 1; \, z^{-2} \\ -\frac{1}{4}-n, \frac{3}{4}-n \end{array}\right).$$

30. $\sum_{k=0}^{n}(-1)^k \binom{n}{k} \frac{4k+3}{(4k+1)(4k+5)} \frac{\left(\frac{3}{2}\right)_k}{\left(n+\frac{5}{2}\right)_k} P_{2k+1}(z)$

$$= \frac{2\left(\frac{3}{2}\right)_{n+1} z}{n!(4n+1)(4n+5)} (1-z^2)^n \,_3F_2\left(\begin{array}{c} -n, -n, 1; \, \frac{1}{1-z^2} \\ -\frac{1}{4}-n, \frac{3}{4}-n \end{array}\right).$$

31. $\sum_{k=0}^{n}(-1)^k \binom{n}{k}(4k+3) \frac{(a)_k (b)_k \left(\frac{3}{2}\right)_k}{\left(n+\frac{5}{2}\right)_k \left(\frac{5}{2}-a\right)_k \left(\frac{5}{2}-b\right)_k} P_{2k+1}(z)$

$$= \frac{2(a)_n (b)_n \left(\frac{3}{2}\right)_{n+1}}{n! \left(\frac{5}{2}-a\right)_n \left(\frac{5}{2}-b\right)_n} z(1-z^2)^n \,_3F_2\left(\begin{array}{c} -n, -n, \frac{5}{2}-a-b; \, \frac{1}{1-z^2} \\ 1-a-n, 1-b-n \end{array}\right).$$

32. $\displaystyle\sum_{k=0}^{[n/2]} (-1)^k \binom{m}{k} (2n - 4k + 1) \frac{\left(-n - \frac{1}{2}\right)_k}{\left(m - n + \frac{1}{2}\right)_k} P_{n-2k}(z)$

$\displaystyle = -2^{2m+1} \frac{(n-2m)!}{n!} \left(\frac{1}{2}\right)_m \left(-n - \frac{1}{2}\right)_{m+1} (1 - z^2)^m C_{n-2m}^{m+1/2}(z)$

$[2m \le n].$

33. $\displaystyle\sum_{k=0}^{[n/2]} (-1)^k (2n - 4k + 1) \frac{\left(-n - \frac{1}{2}\right)_k}{k!} P_{n-2k}(z) = \left(\frac{3}{2}\right)_n \frac{(2z)^n}{n!}.$

34. $\displaystyle\sum_{k=0}^{[n/2]} (2n - 4k + 1) \frac{(a)_k \left(-\frac{1}{2} - n\right)_k}{k! \left(\frac{1}{2} - a - n\right)_k} P_{n-2k}(z) = \frac{\left(\frac{3}{2}\right)_n}{\left(a + \frac{1}{2}\right)_n} C_n^{a+1/2}(z).$

35. $\displaystyle\sum_{k=0}^{[n/3]} \frac{2n - 6k + 1}{k!} \left(-\frac{2n+1}{3}\right)_k P_{n-3k}(z) = 3^n P_n^{\left(\frac{1-n}{3}, \frac{1-4n}{6}\right)}\!\left(\frac{4z-1}{3}\right).$

36. $\displaystyle\sum_{k=0}^{[n/3]} (4n - 12k + 1) \frac{(-n)_{3k} \left(-\frac{4n+1}{6}\right)_k}{k! \left(-n + \frac{1}{2}\right)_{3k}} P_{2n-6k}(z)$

$\displaystyle = 2^{2n} \frac{\left(\frac{3}{2}\right)_{2n}}{(2n)!} (z^2 - 1)^n \,_3F_2\!\left(\begin{array}{c} -n, -n, -\frac{4n+1}{6}; \\ -n - \frac{1}{4}, -n + \frac{1}{4} \end{array} \frac{3}{4 - 4z^2}\right).$

37. $\displaystyle\sum_{k=0}^{[n/3]} (4n - 12k + 3) \frac{(-n)_{3k} \left(-\frac{4n+3}{6}\right)_k}{k! \left(-n - \frac{1}{2}\right)_{3k}} P_{2n-6k+1}(z)$

$\displaystyle = 2^{2n+1} \frac{\left(\frac{3}{2}\right)_{2n+1}}{(2n+1)!} z(z^2 - 1)^n \,_3F_2\!\left(\begin{array}{c} -n, -n, -\frac{4n+3}{6}; \\ -n - \frac{3}{4}, -n - \frac{1}{4} \end{array} \frac{3}{4 - 4z^2}\right).$

38. $\displaystyle\sum_{k=0}^{n} (-1)^k \binom{n}{k} \frac{(-m)_k}{(a)_k} \left(z + \sqrt{z^2 - 1}\right)^k P_{m-k}(z)$

$\displaystyle = \frac{(a+m)_n}{(a)_n} \left(z + \sqrt{z^2 - 1}\right)^m$

$\displaystyle \times \,_3F_2\!\left(\begin{array}{c} -m, 1 - a - m, \frac{1}{2}; \; 2(1 - z^2 + z\sqrt{z^2 - 1}) \\ 1 - a - m - n, 1 \end{array}\right) \quad [m \ge n].$

39. $\displaystyle\sum_{k=0}^{n} (-1)^k \binom{n}{k} \frac{(m+1)_k}{\left(m - n + \frac{3}{2}\right)_k} \left(z + \sqrt{z^2 - 1}\right)^{-k} P_{k+m}(z)$

$\displaystyle = \frac{\left(\frac{1}{2}\right)_n}{\left(-m - \frac{1}{2}\right)_n} \left(z + \sqrt{z^2 - 1}\right)^{-m-2n}$

$\displaystyle \times P_m^{(0, -m-n-1/2)}\!\left(4z^2 + 4z\sqrt{z^2 - 1} - 3\right).$

5.7.2. Sums containing $P_n(z)$ and special functions

1. $\displaystyle\sum_{k=0}^{n}\binom{2n}{2k}\frac{2^{2n-2k}-1}{(2z)^{2k}}B_{2n-2k}P_{2k}(z)=\frac{n}{(2z)^{2n-1}}P_{2n-1}(z)$ $\qquad [n\geq 1].$

2. $\displaystyle\sum_{k=0}^{n}\binom{2n+1}{2k+1}(2z)^{-2k}B_{2n-2k}P_{2k+1}(z)=\frac{2n+1}{2^{2n}z^{2n-1}}z^{1-2n}P_{2n}(z).$

3. $\displaystyle\sum_{k=0}^{n}\left(z-\sqrt{z^2-1}\right)^{k}\psi(k+1)P_k(z)$

$\qquad = \psi(n+2)\,P_n^{(1,\,-n-3/2)}\!\left(4z^2-3-4z\sqrt{z^2-1}\right)$

$\qquad +\left(1-z^2+z\sqrt{z^2-1}\right)^{-1}\left[1-P_{n+1}^{(0,\,-n-5/2)}\!\left(4z^2-3-4z\sqrt{z^2-1}\right)\right].$

5.7.3. Sums containing products of $P_{m\pm nk}(z)$

1. $\displaystyle\sum_{k=0}^{n}(2k+1)\,[P_k(z)]^2$

$\qquad = \dfrac{(n+1)^2}{1-z^2}\left\{[P_n(z)]^2-2zP_n(z)P_{n+1}(z)+[P_{n+1}(z)]^2\right\}.$

2. $\displaystyle\sum_{k=0}^{n}(-1)^{k}\frac{2k+1}{(n-k)!(k+n+1)!}\,[P_k(z)]^2 = \frac{\left(\frac{1}{2}\right)_n}{(n!)^3}(1-z^2)^{n}.$

3. $\displaystyle\sum_{k=0}^{n}(-1)^{k}\frac{(2k+1)\left(n+\frac{1}{2}\right)_k}{(n-k)!(k+n+1)!\left(\frac{3}{2}-n\right)_k}\,[P_k(z)]^2 = \frac{1-2n}{(n!)^2}P_{2n}(z).$

4. $\displaystyle\sum_{k=0}^{n}(-1)^{k}(4k+1)\frac{\left(\frac{1}{2}\right)_k}{k!}P_{2k}\!\left(\sqrt{\frac{1-z}{2}}\right)P_{2k}\!\left(\sqrt{\frac{1+z}{2}}\right)$

$\qquad = 2(-1)^{n+1}(n+1)\frac{\left(\frac{3}{2}\right)_n}{n!}P_{2n+2}\!\left(\sqrt{\frac{1-z}{2}}\right)P_{2n+2}\!\left(\sqrt{\frac{1+z}{2}}\right)$

$\qquad +\dfrac{9\left(\frac{5}{2}\right)_n^2}{8(n!)^2}(1-z^2)\,{}_4F_3\!\left(\begin{array}{c}-n,\,n+\frac{5}{2},\,\frac{5}{4},\,\frac{7}{4}\\ \frac{3}{2},\,2,\,\frac{5}{2};\,1-z^2\end{array}\right).$

5. $\displaystyle\sum_{k=0}^{n}(-1)^k(4k+3)\frac{\left(\frac{3}{2}\right)_k}{k!}P_{2k+1}\left(\sqrt{\frac{1-z}{2}}\right)P_{2k+1}\left(\sqrt{\frac{1+z}{2}}\right)$

$\displaystyle = 2(-1)^{n+1}(n+1)\frac{\left(\frac{5}{2}\right)_n}{n!}P_{2n+3}\left(\sqrt{\frac{1-z}{2}}\right)P_{2n+3}\left(\sqrt{\frac{1+z}{2}}\right)$

$\displaystyle + \frac{25\left(\frac{7}{2}\right)_n^2}{16(n!)^2}(1-z^2)^{3/2}\,{}_4F_3\!\left(\begin{array}{c}-n,\,n+\frac{7}{2},\,\frac{7}{4},\,\frac{9}{4}\\ 2,\,\frac{5}{2},\,\frac{7}{2};\,1-z^2\end{array}\right).$

5.7.4. Sums containing $P_m(\varphi(k,z))$

1. $\displaystyle\sum_{k=0}^{n}(-1)^k\binom{n}{k}P_m(w+kz)=0$ $\qquad [m<n].$

2. $\displaystyle\sum_{k=1}^{n}(-1)^k\binom{n}{k}P_m(1+kz)=(-2z)^m\left(\frac{1}{2}\right)_m\delta_{m,n}-1$ $\qquad [n\geq m].$

3. $\displaystyle\sum_{k=0}^{n}(-1)^k\binom{n}{k}P_{m+n}(w+kz)$

$\displaystyle = (2n)!\left(-\frac{z}{2}\right)^n\sum_{k=0}^{m}\sigma_{k+n}^n\frac{\left(n+\frac{1}{2}\right)_k}{(k+n)!}(2z)^k C_{m-k}^{k+n+1/2}(w).$

4. $\displaystyle\sum_{k=0}^{n}(-1)^k\binom{n}{k}P_{2m}\!\left(\sqrt{k}\,z\right)=(-1)^m z^{2m}\frac{\left(\frac{1}{2}\right)_{2m}}{\left(\frac{1}{2}\right)_m}\delta_{m,n}$ $\qquad [n\geq m].$

5. $\displaystyle\sum_{k=1}^{n}(-1)^k\binom{n}{k}k^{-1/2}P_{2m+1}\!\left(\sqrt{k}\,z\right)=(-1)^m z^{2m+1}\frac{\left(\frac{3}{2}\right)_{2m}}{\left(\frac{3}{2}\right)_m}\delta_{m,n}$

$\displaystyle\qquad\qquad\qquad\qquad\qquad\qquad\qquad -(-1)^m\frac{\left(\frac{3}{2}\right)_m}{m!}z \quad [n\geq m].$

6. $\displaystyle\sum_{k=0}^{n}(-1)^k\binom{n}{k}P_{2m}\!\left(\sqrt{1+kz}\right)=(-z)^m\frac{\left(\frac{1}{2}\right)_{2m}}{\left(\frac{1}{2}\right)_m}\delta_{m,n}$ $\qquad [n\geq m].$

7. $\displaystyle\sum_{k=1}^{n}\frac{(-1)^k}{\sqrt{1+kz}}\binom{n}{k}P_{2m+1}\!\left(\sqrt{1+kz}\right)=(-4z)^m\frac{m!\left(\frac{3}{2}\right)_{2m}}{(2m+1)!}\delta_{m,n}-1$ $\;[n\geq m].$

8. $\displaystyle\sum_{k=1}^{n}(-1)^k\binom{n}{k}k^m P_m\!\left(1+\frac{z}{k}\right)=(-1)^m n!\,\delta_{m,n}-\frac{\left(\frac{1}{2}\right)_m}{m!}(2z)^m$ $\qquad [n\geq m].$

9. $\sum_{k=1}^{n}(-1)^k\binom{n}{k}(k+z)^m P_{2m}\left(\sqrt{\dfrac{z}{k+z}}\right) = \left(\dfrac{1}{2}\right)_m \delta_{m,n} - z^m$ $\quad [n \geq m].$

10. $\sum_{k=1}^{n}(-1)^k\binom{n}{k}(k+z)^{m+1/2} P_{2m+1}\left(\sqrt{\dfrac{z}{k+z}}\right) = \sqrt{z}\left(\dfrac{3}{2}\right)_m \delta_{m,n}$
$\qquad - z^{m+1/2} \quad [n \geq m].$

11. $\sum_{k=1}^{n}(-1)^k\binom{n}{k}(k-z)^m P_m\left(\dfrac{k+z}{k-z}\right) = (-1)^m m!\,\delta_{m,n} - z^m$ $\quad [n \geq m].$

12. $\sum_{k=1}^{n}(-1)^k\binom{n}{k}k^m P_{2m}\left(\dfrac{z}{\sqrt{k}}\right) = \left(\dfrac{1}{2}\right)_m \delta_{m,n} - \dfrac{\left(\dfrac{1}{2}\right)_{2m}}{(2m)!}(2z)^{2m}$ $\quad [n \geq m].$

13. $\sum_{k=1}^{n}(-1)^k\binom{n}{k}k^{m+1/2} P_{2m+1}\left(\dfrac{z}{\sqrt{k}}\right) = z\left(\dfrac{3}{2}\right)_m \delta_{m,n}$
$\qquad - \dfrac{\left(\dfrac{3}{2}\right)_{2m}}{(2m+1)!}2^{2m} z^{2m+1} \quad [n \geq m].$

14. $\sum_{k=1}^{n}(-1)^k\binom{n}{k}k^m P_{2m}\left(\sqrt{1+\dfrac{z}{k}}\right) = (-1)^m m!\,\delta_{m,n} - \dfrac{\left(\dfrac{1}{2}\right)_{2m}}{(2m)!}(4z)^m$
$\qquad [n \geq m].$

15. $\sum_{k=1}^{n}(-1)^k\binom{n}{k}\dfrac{k^{m+1/2}}{\sqrt{k+z}} P_{2m+1}\left(\sqrt{1+\dfrac{z}{k}}\right) = (-1)^m m!\,\delta_{m,n}$
$\qquad - \dfrac{\left(\dfrac{3}{2}\right)_{2m}}{(2m+1)!}(4z)^m \quad [n \geq m].$

16. $\sum_{k=1}^{n}(-1)^k\binom{n}{k}(k+z)^m P_{2m}\left(\sqrt{\dfrac{k}{k+z}}\right) = (-1)^m m!\,\delta_{m,n}$
$\qquad - \dfrac{\left(\dfrac{1}{2}\right)_m}{m!}(-z)^m \quad [n \geq m].$

17. $\sum_{k=1}^{n}(-1)^k\binom{n}{k}k^{-1/2}(k+z)^{m+1/2} P_{2m+1}\left(\sqrt{\dfrac{k}{k+z}}\right)$
$\qquad = (-1)^m m!\,\delta_{m,n} - \dfrac{\left(\dfrac{3}{2}\right)_m}{m!}(-z)^m \quad [n \geq m].$

18. $\sum_{k=1}^{n}(-1)^k\binom{n}{k}k^{m/2}P_m\left(\dfrac{k+z}{2\sqrt{kz}}\right)=(-1)^m\left(\dfrac{1}{2}\right)_m z^{-m/2}\delta_{m,n}$

$\qquad\qquad\qquad\qquad\qquad\qquad\qquad\qquad -\dfrac{\left(\frac{1}{2}\right)_m}{m!}z^{m/2}\qquad [n\geq m].$

19. $\sum_{k=1}^{n}(-1)^k\binom{n}{k}k^{m/2}(k+z)^{m/2}P_m\left(\dfrac{2k+z}{2\sqrt{k(k+z)}}\right)=-z^m\dfrac{\left(\frac{1}{2}\right)_m}{m!}$

$\qquad +(-1)^n m!\,n!\,z^{m-n}\dfrac{\left(\frac{1}{2}\right)_{m-n}}{(m-n)!^2}\sum_{k=0}^{m-n}\sigma_{k+n}^n\dfrac{(n-m)_k^2}{(k+n)!}\dfrac{(-z)^{-k}}{\left(n-m+\frac{1}{2}\right)_k}.$

5.7.5. Sums containing $P_k(\varphi(k,z))$

1. $\sum_{k=1}^{n}(-1)^k\binom{n}{k}a^k(ka+b)^{n-k-1}(k^2+kz)^{k/2}P_k\left(\dfrac{2k+z}{2\sqrt{k^2+kz}}\right)$

$\qquad\qquad\qquad = -b^{n-1}+\dfrac{(b^2-abz)^{n/2}}{na+b}P_n\left(\dfrac{2b-az}{2\sqrt{b^2-abz}}\right).$

2. $\sum_{k=0}^{n}(-1)^k\binom{n}{k}(ka+1)^{n-k/2-1}(ka+z+1)^{k/2}$

$\qquad\qquad\times P_k\left(\dfrac{2ka+z+2}{2\sqrt{(ka+1)(ka+z+1)}}\right)=\dfrac{\left(\frac{1}{2}\right)_n(-z)^n}{n!(na+1)}.$

5.7.6. Sums containing products of $P_m(\varphi(k,z))$

1. $\sum_{k=0}^{n}(-1)^k\binom{n}{k}[P_m(kw+z)]^2=0\qquad [2m<n].$

2. $\sum_{k=1}^{n}(-1)^k\binom{n}{k}k^m P_{2m}\left(\sqrt{1+\dfrac{z}{k}}+\sqrt{\dfrac{z}{k}}\right)P_{2m}\left(\sqrt{1+\dfrac{z}{k}}-\sqrt{\dfrac{z}{k}}\right)$

$\qquad\qquad\qquad = m!\,(-1)^m\delta_{m,n}-\dfrac{\left(\frac{1}{2}\right)_{2m}}{(m!)^2}(-4z)^m\qquad [n\geq m].$

3. $\sum_{k=1}^{n}(-1)^k\binom{n}{k}k^m P_{2m+1}\left(\sqrt{1+\dfrac{z}{k}}+\sqrt{\dfrac{z}{k}}\right)P_{2m+1}\left(\sqrt{1+\dfrac{z}{k}}-\sqrt{\dfrac{z}{k}}\right)$

$\qquad\qquad\qquad = m!\,(-1)^m\delta_{m,n}-\dfrac{\left(\frac{3}{2}\right)_{2m}}{(m!)^2}(-4z)^m\qquad [n\geq m].$

5.8. The Chebyshev Polynomials $T_n(z)$ and $U_n(z)$

5.8.1. Sums containing $T_{m+nk}(z)$

1. $\displaystyle\sum_{k=0}^{n}\binom{n}{k}t^k T_k(z) = (1+2tz+t^2)^{n/2}\, T_n\!\left(\dfrac{1+tz}{\sqrt{1+2tz+t^2}}\right).$

2. $\displaystyle\sum_{k=0}^{n}\binom{n}{k}\dfrac{(a)_k}{\left(\frac{1}{2}\right)_k}\left(\dfrac{2}{1-z}\right)^k T_k(z)$
$= \dfrac{(a)_n}{\left(\frac{1}{2}\right)_n}\left(\dfrac{1+z}{1-z}\right)^n{}_4F_3\!\left(\begin{array}{c}-n,\ \frac{1}{2}-n,\ \dfrac{1-2a-2n}{4},\ \dfrac{3-2a-2n}{4}\\[2pt]\frac{1}{2}-a-n,\ 1-a-n,\ \frac{1}{2};\ \dfrac{4z-4}{z+1}\end{array}\right).$

3. $\displaystyle\sum_{k=0}^{n} T_{2k}(z) = \dfrac{1}{2}U_{2n}(z) + \dfrac{1}{2}.$

4. $\displaystyle\sum_{k=0}^{n}(n^2-k^2)T_{2k}(z) = \dfrac{1}{4(z^2-1)}\left[2n^2(z^2-1)+2nT_{2n}(z)-zU_{2n-1}(z)\right]$
$\hfill [n\geq 1].$

5. $\displaystyle\sum_{k=0}^{n}\binom{2n}{n-k}T_{2k}(z) = 2^{2n-1}z^{2n}+\dfrac{1}{2}\binom{2n}{n}.$

6. $\displaystyle\sum_{k=0}^{n}\dfrac{1}{(n-k)!(n+k)!}T_{2k}(z) = \dfrac{(2z)^{2n}}{2(2n)!}+\dfrac{1}{2(n!)^2}.$

7. $\displaystyle\sum_{k=0}^{n}(-1)^k\binom{n}{k}\dfrac{(a)_k}{\left(\frac{1}{2}\right)_k}z^{-2k}T_{2k}(z)$
$= \dfrac{(a)_n}{\left(\frac{1}{2}\right)_n}z^{-2n}(1-z^2)^n\,{}_4F_3\!\left(\begin{array}{c}-n,\ \frac{1}{2}-n,\ \dfrac{1-2a-2n}{4},\ \dfrac{3-2a-2n}{4}\\[2pt]\frac{1}{2}-a-n,\ 1-a-n,\ \frac{1}{2};\ \dfrac{4z^2}{z^2-1}\end{array}\right).$

8. $\displaystyle\sum_{k=0}^{n}\binom{n}{k}\dfrac{(a)_k}{\left(\frac{1}{2}\right)_k}(1-z^2)^{-k}T_{2k}(z)$
$= \dfrac{(a)_n}{\left(\frac{1}{2}\right)_n}z^{2n}(1-z^2)^{-n}\,{}_4F_3\!\left(\begin{array}{c}-n,\ \frac{1}{2}-n,\ \dfrac{1-2a-2n}{4},\ \dfrac{3-2a-2n}{4}\\[2pt]\frac{1}{2}-a-n,\ 1-a-n,\ \frac{1}{2};\ 4-4z^{-2}\end{array}\right).$

9. $\displaystyle\sum_{k=0}^{n} T_{2k+1}(z) = \dfrac{1}{2}U_{2n-1}(z)\hfill [n\geq 1].$

10. $\displaystyle\sum_{k=0}^{n}\binom{2n+1}{n-k}T_{2k+1}(z) = 2^{2n}z^{2n+1}.$

11. $\displaystyle\sum_{k=0}^{n} \frac{1}{k+n+1}\binom{2n}{n-k} T_{2k+1}(z) = \frac{2^{2n} z^{2n+1}}{2n+1}.$

12. $\displaystyle\sum_{k=0}^{n} (n-k)(n+k+1) T_{2k+1}(z)$
$$= \frac{1}{4(z^2-1)} \left[(2n+1) T_{2n+1}(z) - z U_{2n}(z)\right].$$

13. $\displaystyle\sum_{k=0}^{n} \frac{1}{(n-k)!(k+n+1)!} T_{2k+1}(z) = \frac{2^{2n} z^{2n+1}}{(2n+1)!}.$

14. $\displaystyle\sum_{k=0}^{n} (-1)^k \frac{2k+1}{(n-k)!(k+n+1)!} T_{2k+1}(z) = \frac{2^{2n} z}{(2n)!} (1-z^2)^n.$

15. $\displaystyle\sum_{k=0}^{n} \binom{n}{k} \frac{(a)_k}{\left(\frac{3}{2}\right)_k} (1-z^2)^{-k} T_{2k+1}(z)$
$$= \frac{(a)_n}{\left(\frac{3}{2}\right)_n} z^{2n+1} (1-z^2)^{-n} \, {}_4F_3\left(\begin{array}{c} -n, -n-\frac{1}{2}, \frac{1-2a-2n}{4}, \frac{3-2a-2n}{4} \\ \frac{1}{2}-a-n, 1-a-n, \frac{1}{2}; \; 4-4z^{-2} \end{array}\right).$$

16. $\displaystyle\sum_{k=0}^{n} (-1)^k \frac{(a)_k}{(n-k)!(2k+1)!} \left(\frac{2}{z}\right)^{2k} T_{2k+1}(z)$
$$= \frac{(a)_n}{(2n)!} z^{1-2n} \left(\frac{1-z^2}{4}\right)^n \, {}_4F_3\left(\begin{array}{c} -n, \frac{1}{2}-n, \frac{3-2a-2n}{4}, \frac{5-2a-2n}{4} \\ 1-a-n, \frac{3}{2}-a-n, \frac{3}{2}; \; \frac{4z^2}{z^2-1} \end{array}\right).$$

17. $\displaystyle\sum_{k=0}^{n} (-1)^k \frac{(1+a+n)_k}{(n-k)!(k+n+1)!(1-a-n)_k} T_{2k+1}(z)$
$$= \frac{1}{2(a+n)(a)_n^2} C_{2n+1}^a(z).$$

18. $\displaystyle\sum_{k=0}^{[n/2]} (-1)^k \binom{n}{k} \frac{(a)_k}{(1-a-n)_k} T_{n-2k}(z) = \frac{n!}{2(a)_n} C_n^a(z)$
$$+ \frac{(-1)^{[n/2]}}{2} \delta_{2[n/2],n} \binom{n}{\left[\frac{n}{2}\right]} \frac{(a)_{[n/2]}}{(1-a-n)_{[n/2]}}.$$

19. $\displaystyle\sum_{k=0}^{[n/3]} \frac{\left(-\frac{2n}{3}\right)_k}{k!} T_{n-3k}(z) = \frac{3^n}{2} P_n^{(-n/3,\,-2n/3)}\left(\frac{4z-1}{3}\right)$
$$+ \frac{1}{2} \delta_{3[n/3],n} \frac{\left(-\frac{2n}{3}\right)_{[n/3]}}{\left[\frac{n}{3}\right]!}.$$

20. $\sum_{k=0}^{[n/2]} (-1)^k \dfrac{\left(\frac{1}{2}-n\right)_k}{k!(2n-2k+1)} T_{2n-4k+1}(z) = \dfrac{(-2)^n z}{2n+1} P_n^{(1/2,\,-n-1/2)}(1-4z^2).$

21. $\sum_{k=0}^{n} \binom{m}{k} T_{m+n-2k}(z) = (2z)^m T_n(z).$

22. $\sum_{k=0}^{n} \binom{2n+1}{2k+1}(2z)^{-2k} B_{2n-2k} T_{2k+1}(z) = \dfrac{2n+1}{2^{2n} z^{2n-1}} T_{2n}(z).$

23. $\sum_{k=0}^{n} \binom{2n}{2k} \dfrac{2^{2n-2k}-1}{(2z)^{2k}} B_{2n-2k} T_{2k}(z) = \dfrac{n}{(2z)^{2n-1}} z^{1-2n} T_{2n-1}(z) \qquad [n \geq 1].$

5.8.2. Sums containing products of $T_{m+nk}(z)$

1. $\sum_{k=0}^{n} (-1)^k \dfrac{\left(n-\frac{1}{2}\right)_k}{(n-k)!(k+n)!\left(\frac{3}{2}-n\right)_k} [T_k(z)]^2$

$= \dfrac{1}{2(n!)^2} + \dfrac{1}{4\left(-\frac{1}{2}\right)_n \left(\frac{1}{2}\right)_n} [P_{2n}(z) - z P_{2n-1}(z)] \qquad [n \geq 1].$

2. $\sum_{k=0}^{n} (-1)^k T_{2k+1}\left(\sqrt{\dfrac{1-z}{2}}\right) T_{2k+1}\left(\sqrt{\dfrac{1+z}{2}}\right)$

$= (-1)^{n+1} \dfrac{(n+1)^2}{2n+3} T_{2n+3}\left(\sqrt{\dfrac{1-z}{2}}\right) T_{2n+3}\left(\sqrt{\dfrac{1+z}{2}}\right)$

$+ (-1)^n \dfrac{(n+1)(n+2)}{2(2n+3)} T_{2n+3}\left(\sqrt{1-z^2}\right)$

$+ \dfrac{1}{4\sqrt{1-z^2}} \left[1 - {}_2F_1\left(\begin{array}{c}-n-2,\,n+1\\ \frac{1}{2}\end{array};\,1-z^2\right)\right].$

5.8.3. Sums containing $T_n(\varphi(k,z))$

1. $\sum_{k=0}^{n} (-1)^k \binom{n}{k} T_m(w+kz) = 0 \qquad\qquad [m<n].$

2. $\sum_{k=0}^{n} (-1)^k \binom{n}{k} T_{m+n}(w+kz)$

$= 2^{n-1} n!\,(m+n)(-z)^n \sum_{k=0}^{m} \sigma_{k+n}^{n} \dfrac{(2z)^k}{k+n} C_{m-k}^{k+n}(w).$

3. $\sum_{k=1}^{n} (-1)^k \binom{n}{k} k^m T_m\left(1+\dfrac{z}{k}\right) = (-1)^m m!\,\delta_{m,n} - 2^{m-1} z^m \qquad [n \geq m \geq 1].$

4. $\displaystyle\sum_{k=0}^{n}(-1)^k\binom{n}{k}T_{2m}(\sqrt{k}\,z) = (-1)^m m!\,2^{2m-1}z^{2m}\delta_{m,n}$ $\qquad [n \geq m \geq 1]$.

5. $\displaystyle\sum_{k=1}^{n}(-1)^k\binom{n}{k}k^{-1/2}T_{2m+1}(\sqrt{k}\,z) = (-1)^m 2^{2m}z^{2m+1}m!\,\delta_{m,n} - (-1)^m(2m+1)z$ $\qquad [n \geq m]$.

6. $\displaystyle\sum_{k=1}^{n}(-1)^k\binom{n}{k}k^m T_{2m}\!\left(\frac{z}{\sqrt{k}}\right) = m!\,\delta_{m,n} - 2^{2m-1}z^{2m}$ $\qquad [n \geq m \geq 1]$.

7. $\displaystyle\sum_{k=1}^{n}(-1)^k\binom{n}{k}k^{m+1/2}T_{2m+1}\!\left(\frac{z}{\sqrt{k}}\right) = m!(2m+1)z\,\delta_{m,n} - 2^{2m}z^{2m+1}$ $\qquad [n \geq m]$.

8. $\displaystyle\sum_{k=1}^{n}(-1)^k\binom{n}{k}k^m T_{2m}\!\left(\sqrt{1+\frac{z}{k}}\right) = (-1)^m m!\,\delta_{m,n} - 2^{2m-1}z^m$ $\qquad [n \geq m \geq 1]$.

9. $\displaystyle\sum_{k=1}^{n}(-1)^k\binom{n}{k}\frac{k^{m+1/2}}{\sqrt{k+z}}T_{2m+1}\!\left(\sqrt{1+\frac{z}{k}}\right) = (-1)^m m!\,\delta_{m,n} - (4z)^m$ $\qquad [n \geq m]$.

10. $\displaystyle\sum_{k=1}^{n}(-1)^k\binom{n}{k}(k+z)^m T_{2m}\!\left(\sqrt{\frac{z}{k+z}}\right) = m!\,\delta_{m,n} - z^m$ $\qquad [n \geq m]$.

11. $\displaystyle\sum_{k=1}^{n}(-1)^k\binom{n}{k}(k+z)^{m+1/2}T_{2m+1}\!\left(\sqrt{\frac{z}{k+z}}\right)$
$= (-1)^{m+1}m!(2m+1)\sqrt{z}\,\delta_{m,n} - z^{m+1/2}$ $\qquad [n \geq m]$.

12. $\displaystyle\sum_{k=1}^{n}(-1)^k\binom{n}{k}(k+z)^m T_{2m}\!\left(\sqrt{\frac{k}{k+z}}\right) = (-1)^m m!\,\delta_{m,n} - (-z)^m$ $\qquad [n \geq m]$.

13. $\displaystyle\sum_{k=1}^{n}(-1)^k\binom{n}{k}k^{-1/2}(k+z)^{m+1/2}T_{2m+1}\!\left(\sqrt{\frac{k}{k+z}}\right) = (-1)^m m!\,\delta_{m,n}$
$- (2m+1)(-z)^m$ $\qquad [n \geq m]$.

5.8.4. Sums containing $U_{m+nk}(z)$

1. $\displaystyle\sum_{k=0}^{n}\frac{t^k}{k+1}\binom{n}{k}U_k(z) = \frac{(1+2tz+t^2)^{n/2}}{n+1}U_n\!\left(\frac{1+tz}{\sqrt{1+2tz+t^2}}\right).$

2. $\sum_{k=0}^{n}(-1)^k(k+1)\binom{2n+2}{n-k}U_k(z) = 2^n(n+1)(1-z)^n.$

3. $\sum_{k=0}^{n}(-1)^k \frac{k+1}{(n-k)!(k+n+2)!}U_k(z) = \frac{2^{n-1}}{(2n+1)!}(1-z)^n.$

4. $\sum_{k=0}^{n} \frac{2^{3k}(a)_k}{(n-k)!(2k+2)!}(1-z)^{-k}U_k(z)$

$= \frac{2^{2n-1}(a)_n}{(2n+1)!}\left(\frac{1+z}{1-z}\right)^n {}_4F_3\left(\begin{matrix} -n, -\frac{1}{2}-n, \frac{3-2a-2n}{4}, \frac{5-2a-2n}{4} \\ 1-a-n, \frac{3}{2}-a-n, \frac{3}{2}; \frac{4z-4}{z+1} \end{matrix}\right).$

5. $\sum_{k=0}^{n} \frac{(a)_k}{(n-k)!(2k+2)!}\left(\frac{8}{1-z}\right)^k U_k(z)$

$= \frac{2^{2n-1}(a)_n}{(2n+1)!}\left(\frac{1+z}{1-z}\right)^n {}_4F_3\left(\begin{matrix} -n, -n-\frac{1}{2}, \frac{3-2a-2n}{4}, \frac{5-2a-2n}{4} \\ 1-a-n, \frac{3}{2}-a-n, \frac{3}{2}; \frac{4z-4}{z+1} \end{matrix}\right).$

6. $\sum_{k=0}^{n} \frac{k+1}{(n-k)!(k+n+2)!(2k+1)(2k+3)}U_k(z)$

$= -\frac{2^{-1/2}}{\left(\frac{3}{2}\right)_n^2 (2n+3)\sqrt{1-z}} C_{2n+1}^{-n-1/2}\left(\sqrt{\frac{1-z}{2}}\right).$

7. $\sum_{k=0}^{n}(k+1)\frac{(a)_k(b)_k}{(n-k)!(k+n+2)!(3-a)_k(3-b)_k}U_k(z)$

$= \frac{2^{n-1}(a)_n(b)_n}{(2n+1)!(3-a)_n(3-b)_n}(1+z)^n {}_3F_2\left(\begin{matrix} -n, -n-\frac{1}{2}, 3-a-b \\ 1-a-n, 1-b-n; \frac{2}{1+z} \end{matrix}\right).$

8. $\sum_{k=0}^{n}(-1)^k U_{2k}(z) = U_n(1-2z^2).$

9. $\sum_{k=0}^{n}(-1)^k(2k+1)U_{2k}(z)$

$= -\frac{1}{2z^2} + (-1)^n(n+1)U_{2n+2}(z) - \frac{2(n+2)!}{\left(\frac{3}{2}\right)_n} z^{-2} P_{n+2}^{(-3/2,-1/2)}(1-2z^2).$

10. $\sum_{k=0}^{n} \frac{2k+1}{(n-k)!(k+n+1)!}U_{2k}(z) = \frac{(2z)^{2n}}{(2n)!}.$

11. $\displaystyle\sum_{k=0}^{n} \frac{(a)_k}{(n-k)!(2k+1)!} \left(\frac{4}{1-z^2}\right)^k U_{2k}(z)$

$\displaystyle = \frac{(a)_n}{(2n)!} (2z)^{2n}(1-z^2)^{-n} \,{}_4F_3\left(\begin{array}{c} -n,\, \frac{1}{2}-n,\, \frac{3-2a-2n}{4},\, \frac{5-2a-2n}{4} \\ 1-a-n,\, \frac{3}{2}-a-n,\, \frac{3}{2};\, 4-4z^{-2} \end{array}\right).$

12. $\displaystyle\sum_{k=0}^{n} (-1)^k \binom{n}{k} \frac{(a)_k}{\left(\frac{3}{2}\right)_k} z^{-2k} U_{2k}(z)$

$\displaystyle = \frac{(a)_n}{\left(\frac{3}{2}\right)_n} z^{-2n}(1-z^2)^n \,{}_4F_3\left(\begin{array}{c} -n,\, -n-\frac{1}{2},\, \frac{1-2a-2n}{4},\, \frac{3-2a-2n}{4} \\ \frac{1}{2}-a-n,\, 1-a-n,\, \frac{1}{2};\, \frac{4z^2}{z^2-1} \end{array}\right).$

13. $\displaystyle\sum_{k=0}^{n} \frac{k+1}{(n-k)!(k+n+2)!} U_{2k+1}(z) = \frac{2^{2n} z^{2n+1}}{(2n+1)!}.$

14. $\displaystyle\sum_{k=0}^{n} (-1)^k \frac{k+1}{(n-k)!(k+n+2)!} U_{2k+1}(z) = \frac{2^{2n} z}{(2n+1)!} (1-z^2)^n.$

15. $\displaystyle\sum_{k=0}^{n} \frac{(a)_k}{(n-k)!(2k+2)!} \left(\frac{4}{1-z^2}\right)^k U_{2k+1}(z)$

$\displaystyle = \frac{(a)_n}{(2n+1)!} z^{2n+1} \left(\frac{4}{1-z^2}\right)^n \,{}_4F_3\left(\begin{array}{c} -n,\, -n-\frac{1}{2},\, \frac{3-2a-2n}{2},\, \frac{5-2a-2n}{2} \\ 1-a-n,\, \frac{3}{2}-a-n,\, \frac{3}{2};\, 4-4z^{-2} \end{array}\right).$

16. $\displaystyle\sum_{k=0}^{n} (-1)^k \frac{(a)_k}{(n-k)!(2k+2)!} \left(\frac{2}{z}\right)^{2k} U_{2k+1}(z)$

$\displaystyle = \frac{2^{2n}(a)_n}{(2n+1)!} z^{1-2n}(1-z^2)^n \,{}_4F_3\left(\begin{array}{c} -n,\, -\frac{1}{2}-n,\, \frac{3-2a-2n}{4},\, \frac{5-2a-2n}{4} \\ 1-a-n,\, \frac{3}{2}-a-n,\, \frac{3}{2};\, \frac{4z^2}{z^2-1} \end{array}\right).$

17. $\displaystyle\sum_{k=0}^{n} (2k+1) \frac{(a)_k (b)_k}{(n-k)!(k+n+1)!(2-a)_k(2-b)_k} U_{2k}(z)$

$\displaystyle = \frac{(a)_n (b)_n}{(2n)!(2-a)_n(2-b)_n} (2z)^{2n} \,{}_3F_2\left(\begin{array}{c} -n,\, \frac{1}{2}-n,\, 2-a-b;\, z^{-2} \\ 1-a-n,\, 1-b-n \end{array}\right).$

18. $\displaystyle\sum_{k=0}^{[n/2]} \binom{n}{k} \frac{n-2k+1}{n-k+1} U_{n-2k}(z) = (2z)^n.$

19. $\displaystyle\sum_{k=0}^{[n/2]} (-1)^k (n-2k+1) \frac{(a-1)_k}{k!(n-k+1)!(1-a-n)_k} U_{n-2k}(z) = \frac{1}{(a)_n} C_n^a(z).$

20. $\displaystyle\sum_{k=0}^{[n/2]} \binom{n+1}{k} \frac{1}{(m-k)!(m-n)_k} U_{n-2k}(z)$

$$= -2^{2m} \frac{(n-2m)!}{(n+1)!} (-n-1)_{m+1} (1-z^2)^m C_{n-2m}^{m+1}(z) \quad [2m \leq n].$$

21. $\displaystyle\sum_{k=0}^{[n/2]} (-1)^k (2n-4k+1) \frac{\left(-n-\frac{1}{2}\right)_k}{k!} U_{2n-4k}(z)$

$$= (-2)^n (2n+1) P_n^{(-1/2,\, 1/2-n)}(1-4z^2).$$

22. $\displaystyle\sum_{k=0}^{[n/3]} \frac{n-3k+1}{k!} \left(-\frac{2n+2}{3}\right)_k U_{n-3k}(z) = 3^n P_n^{((2-n)/3,\,(1-n)/3)}\left(\frac{4z-1}{3}\right).$

23. $\displaystyle\sum_{k=0}^{n} U_{m+n-2k}(z) = U_m(z) U_n(z).$

24. $\displaystyle\sum_{k=0}^{n} \binom{2n+2}{2k+2} (2z)^{-2k} B_{2n-2k} U_{2k+1}(z) = (n+1)(2z)^{1-2n} U_{2n}(z).$

25. $\displaystyle\sum_{k=0}^{n} \binom{2n+1}{2k+1} \frac{2^{2n-2k}-1}{(2z)^{2k}} B_{2n-2k} U_{2k}(z) = \frac{2n+1}{2^{2n} z^{2n-1}} U_{2n-1}(z) \quad [n \geq 1].$

5.8.5. Sums containing products of $U_n(z)$

1. $\displaystyle\sum_{k=0}^{n} (k+1) [U_k(z)]^2$

$$= \frac{1}{4(1-z^2)} \left\{ (n+2)^2 [U_n(z)]^2 - 2(n+1)(n+2) z U_n(z) U_{n+1}(z) \right.$$
$$\left. + (n+1)^2 [U_{n+1}(z)]^2 \right\}.$$

2. $\displaystyle\sum_{k=0}^{n} (-1)^k \frac{\left(\frac{3}{2}+n\right)_k}{(n-k)!(k+n+2)!\left(\frac{3}{2}-n\right)_k} [U_k(z)]^2$

$$= \frac{2^{2n} n! (1-z^2)^{-1}}{(2n+2)! \left(-\frac{1}{2}\right)_n} [P_{2n}(z) - z P_{2n+1}(z)].$$

5.8.6. Sums containing $U_n(\varphi(k,z))$

1. $\displaystyle\sum_{k=0}^{n} (-1)^k \binom{n}{k} U_m(w+kz) = 0 \qquad [m < n].$

2. $\displaystyle\sum_{k=0}^{n} (-1)^k \binom{n}{k} U_{m+n}(w+kz) = n! (-2z)^n \sum_{k=0}^{m} \sigma_{k+n}^n (2z)^k C_{m-k}^{k+n+1}(w).$

335

3. $\sum_{k=0}^{n}(-1)^{k}\binom{n}{k}U_{m}(1+kz) = n!(-2z)^{m}\delta_{m,n}$ $\qquad [n \geq m].$

4. $\sum_{k=1}^{n}(-1)^{k}\binom{n}{k}k^{m}U_{m}\left(1+\frac{z}{k}\right) = (-1)^{m}(m+1)!\delta_{m,n} - (2z)^{m}$ $\qquad [n \geq m].$

5. $\sum_{k=1}^{n}(-1)^{k}\binom{n}{k}(k-z)^{m}U_{m}\left(\frac{k+z}{k-z}\right) = (-1)^{m}(m+1)!n!\delta_{m,n}$
$\qquad - (m+1)z^{m}$ $\qquad [n \geq m].$

6. $\sum_{k=0}^{n}(-1)^{k}\binom{n}{k}U_{2m}(\sqrt{k}\,z) = (-1)^{m}m!(2z)^{2m}\delta_{m,n}$ $\qquad [n \geq m].$

7. $\sum_{k=1}^{n}(-1)^{k}\binom{n}{k}k^{-1/2}U_{2m+1}(\sqrt{k}\,z) = (-1)^{m}m!(2z)^{2m+1}\delta_{m,n}$
$\qquad - 2(-1)^{m}(m+1)z$ $\qquad [n \geq m].$

8. $\sum_{k=1}^{n}(-1)^{k}\binom{n}{k}k^{m}U_{2m}\left(\frac{z}{\sqrt{k}}\right) = m!\delta_{m,n} - (2z)^{2m}$ $\qquad [n \geq m].$

9. $\sum_{k=1}^{n}(-1)^{k}\binom{n}{k}k^{m+1/2}U_{2m+1}\left(\frac{z}{\sqrt{k}}\right) = 2(m+1)!z\,\delta_{m,n} - (2z)^{2m+1}$
$\qquad [n \geq m].$

10. $\sum_{k=1}^{n}(-1)^{k}\binom{n}{k}k^{m}U_{2m}\left(\sqrt{1+\frac{z}{k}}\right) = (-1)^{m}m!(2m+1)\delta_{m,n} - (4z)^{m}$
$\qquad [n \geq m].$

11. $\sum_{k=1}^{n}(-1)^{k}\binom{n}{k}\frac{k^{m+1/2}}{\sqrt{k+z}}U_{2m+1}\left(\sqrt{1+\frac{z}{k}}\right) = 2(-1)^{m}(m+1)!\delta_{m,n}$
$\qquad - 2^{2m+1}z^{m}$ $\qquad [n \geq m].$

12. $\sum_{k=1}^{n}(-1)^{k}\binom{n}{k}(k+z)^{m}U_{2m}\left(\sqrt{\frac{z}{k+z}}\right) = (-1)^{m+1}m!\delta_{m,n}$
$\qquad - (2m+1)z^{m}$ $\qquad [n \geq m].$

13. $\sum_{k=1}^{n}(-1)^{k}\binom{n}{k}(k+z)^{m+1/2}U_{2m+1}\left(\sqrt{\frac{z}{k+z}}\right)$
$\qquad = 2(-1)^{m}(m+1)!\sqrt{z}\,\delta_{m,n} - 2(-1)^{m}(m+1)z^{m+1/2}$ $\qquad [n \geq m].$

14. $\sum_{k=1}^{n}(-1)^k\binom{n}{k}(k+z)^m U_{2m}\left(\sqrt{\dfrac{k}{k+z}}\right) = (-1)^m m!(2m+1)\delta_{m,n}$
$\qquad - (-z)^m \quad [n \geq m].$

15. $\sum_{k=1}^{n}(-1)^k\binom{n}{k}k^{-1/2}(k+z)^{m+1/2}U_{2m+1}\left(\sqrt{\dfrac{k}{k+z}}\right)$
$\qquad = 2(-1)^m(m+1)!\delta_{m,n} - 2(m+1)(-z)^m \quad [n \geq m].$

16. $\sum_{k=0}^{2n}\dfrac{(-1)^k}{k+1}\binom{2n}{k}(ka+1)^{2n-k-1}[(ka+1)^2+z]^{k/2}U_k\left(\dfrac{ka+1}{\sqrt{(ka+1)^2+z}}\right)$
$\qquad = \dfrac{(-z)^n}{(2n+1)(2na+1)}.$

5.9. The Hermite Polynomials $H_n(z)$

5.9.1. Sums containing $H_{m\pm nk}(z)$

1. $\sum_{k=0}^{n}\dfrac{t^k}{(n-k)!(2k)!}H_{2k}(z) = \dfrac{(t-1)^n}{(2n)!}(t-1)^n H_{2n}\left(z\sqrt{\dfrac{t}{t-1}}\right).$

2. $\sum_{k=0}^{n-1}\dfrac{(n-k)^n}{(n-k)!(2k)!}H_{2k}(z) = (2z)^{2n}\sum_{k=0}^{n}\dfrac{\sigma_n^k}{(2n-2k)!}(2z)^{-2k}.$

3. $\sum_{k=0}^{n}\binom{n}{k}\dfrac{2^{-2k}}{(a)_k}H_{2k}(z) = (-1)^n\dfrac{n!}{(a)_n}L_n^{1/2-a-n}(z^2).$

4. $\sum_{k=0}^{n}\dfrac{\left(\frac{1}{2}-n\right)_k}{(n-k)!(2k)!}z^{-2k}H_{2k}(z) = \dfrac{\left(\frac{1}{2}\right)_n}{n!}z^{-2n}\,{}_2F_3\left(\begin{array}{c}-\frac{n}{2},\frac{1-n}{2};-z^4\\\frac{1}{4},\frac{1}{2},\frac{3}{4}\end{array}\right).$

5. $\sum_{k=0}^{n}\dfrac{(a)_k}{(n-k)!(2k)!}(-z^2)^{-k}H_{2k}(z)$
$\qquad = \dfrac{(a)_n}{n!}z^{-2n}\,{}_3F_3\left(\begin{array}{c}-n,\frac{1-2a-2n}{4},\frac{3-2a-2n}{4}\\1-a-n,\frac{1}{2}-a-n,\frac{1}{2};4z^2\end{array}\right).$

6. $\sum_{k=0}^{n}\binom{n}{k}\binom{2n}{k}^{-1}\dfrac{1}{(n-k)!(2k)!}H_{2k}(z) = \dfrac{n!}{(2n)!}\,{}_2F_2\left(\begin{array}{c}-n,-n\\\frac{1}{2},1;z^2\end{array}\right).$

7. $\sum_{k=0}^{n}\dfrac{(a)_k}{(n-k)!(2k)!(b)_k}H_{2k}(z) = \dfrac{(b-a)_n}{n!(b)_n}\,{}_2F_2\left(\begin{array}{c}-n,a;z^2\\a-b-n+1,\frac{1}{2}\end{array}\right).$

8. $\displaystyle\sum_{k=0}^{n}(-1)^k \frac{\sigma_m^{n-k+1}}{(2k)!} H_{2k}(z) = (-1)^n \frac{(2z)^{2n}}{(2n)!}\, {}_{m+1}F_{m-1}\!\left(\begin{array}{c}-n,\, \frac{1}{2}-n,\, 2,\, \ldots,\, 2\\ 1,\, \ldots,\, 1;\, -z^{-2}\end{array}\right).$

9. $\displaystyle\sum_{k=0}^{n} \frac{t^k}{(n-k)!(2k+1)!} H_{2k+1}(z)$

$$= \frac{t^{-1/2}}{(2n+1)!}(t-1)^{n+1/2} H_{2n+1}\!\left(z\sqrt{\frac{t}{t-1}}\right).$$

10. $\displaystyle\sum_{k=0}^{n} \frac{(n-k)^{n+1}}{(n-k)!(2k+1)!} H_{2k+1}(z) = (2z)^{2n+1}\sum_{k=0}^{n}\frac{\sigma_{n+1}^k}{(2n-2k+1)!}(2z)^{-2k}.$

11. $\displaystyle\sum_{k=0}^{n}(-1)^k \frac{(a)_k}{(n-k)!(2k+1)!} z^{-2k} H_{2k+1}(z)$

$$= 2\frac{(a)_n}{n!} z^{1-2n}\, {}_3F_3\!\left(\begin{array}{c}-n,\, \frac{3-2a-2n}{4},\, \frac{5-2a-2n}{4}\\ 1-a-n,\, \frac{3}{2}-a-n,\, \frac{3}{2};\, 4z^2\end{array}\right).$$

12. $\displaystyle\sum_{k=0}^{n} \frac{(a)_k}{(n-k)!(2k+1)!} (-z^2)^{-k} H_{2k+1}(z)$

$$= 2\frac{(a)_n}{n!} z^{1-2n}\, {}_3F_3\!\left(\begin{array}{c}-n,\, \frac{3-2a-2n}{4},\, \frac{5-2a-2n}{4}\\ 1-a-n,\, \frac{3}{2}-a-n,\, \frac{3}{2};\, 4z^2\end{array}\right).$$

13. $\displaystyle\sum_{k=0}^{n}\binom{n}{k}\binom{2n}{k}^{-1}\frac{1}{(n-k)!(2k+1)!} H_{2k+1}(z) = 2z\frac{n!}{(2n)!}\, {}_2F_2\!\left(\begin{array}{c}-n,\, -n\\ 1,\, \frac{3}{2};\, z^2\end{array}\right).$

14. $\displaystyle\sum_{k=0}^{n}\binom{n}{k}\frac{2^{-2k-1}}{(a)_k} H_{2k+1}(z) = (-1)^n\frac{n!\, z}{(a)_n} L_n^{3/2-a-n}(z^2).$

15. $\displaystyle\sum_{k=0}^{n}\frac{\left(-n-\frac{1}{2}\right)_k}{(n-k)!(2k+1)!} z^{-2k} H_{2k+1}(z)$

$$= 2\frac{\left(\frac{3}{2}\right)_n}{n!} z^{1-2n}\, {}_2F_3\!\left(\begin{array}{c}-\frac{n}{2},\, \frac{1-n}{2};\, -z^4\\ \frac{3}{4},\, \frac{5}{4},\, \frac{3}{2}\end{array}\right).$$

16. $\displaystyle\sum_{k=0}^{n}\frac{(a)_k}{(n-k)!(2k+1)!(b)_k} H_{2k+1}(z) = 2z\frac{(b-a)_n}{n!(b)_n}\, {}_2F_2\!\left(\begin{array}{c}-n,\, a;\, z^2\\ a-b-n+1,\, \frac{3}{2}\end{array}\right).$

5.9. The Hermite Polynomials $H_n(z)$

17. $\sum_{k=0}^{n}(-1)^k \dfrac{\sigma_m^{n-k+1}}{(2k+1)!} H_{2k+1}(z)$

$$= (-1)^n \dfrac{(2z)^{2n+1}}{(2n+1)!} {}_{m+1}F_{m-1}\!\left(\begin{array}{c} -n, -\tfrac{1}{2}-n, 2, \ldots, 2 \\ 1, \ldots, 1;\ -z^{-2} \end{array}\right).$$

18. $\sum_{k=0}^{n}\binom{n}{k}\dfrac{2^{-2k}}{(k+m)!} H_{2k+2m}(z) = (-1)^{m+n}\, 2^{2m}\, L_{m+n}^{-n-1/2}(z^2).$

19. $\sum_{k=0}^{n}\binom{n}{k}\dfrac{(a+m)_k}{(2k+2m)!} H_{2k+2m}(z) = (-1)^m \dfrac{(1-a)_n}{(m+n)!}\, {}_2F_2\!\left(\begin{array}{c} -m-n,\ a \\ a-n,\ \tfrac{1}{2};\ z^2 \end{array}\right).$

20. $\sum_{k=0}^{n}\binom{n}{k}\dfrac{(a+m)_k}{(2k+2m+1)!} H_{2k+2m+1}(z)$

$$= 2(-1)^m \dfrac{(1-a)_n}{(m+n)!} z\, {}_2F_2\!\left(\begin{array}{c} -m-n,\ a \\ a-n,\ \tfrac{3}{2};\ z^2 \end{array}\right).$$

21. $\sum_{k=0}^{n}\binom{n}{k}\dfrac{2^{-2k}}{(k+m)!} H_{2k+2m+1}(z) = (-1)^{m+n}\, 2^{2m+1}\, z\, L_{m+n}^{1/2-n}(z^2).$

22. $\sum_{k=0}^{n}(-1)^k \binom{n}{k}\dfrac{(2k)!}{(4k)!} H_{4k}(z) = n!\dfrac{(2z^2)^n}{\left(\tfrac{1}{2}\right)_{2n}} L_n^{-1/2}\!\left(\dfrac{z^2}{2}\right).$

23. $\sum_{k=0}^{n}(-1)^k \binom{n}{k}\dfrac{(2k)!}{(4k+1)!} H_{4k+1}(z) = \dfrac{2^{n+1} n!}{\left(\tfrac{3}{2}\right)_{2n}} z^{2n+1} L_n^{1/2}\!\left(\dfrac{z^2}{2}\right).$

24. $\sum_{k=0}^{[n/2]}\dfrac{2^{-2k}}{k!(2n-4k)!(a)_k} H_{2n-4k}(z) = \dfrac{(2z)^{2n}}{(2n)!}\, {}_3F_1\!\left(\begin{array}{c} -n,\ \tfrac{1}{2}-n,\ a-\tfrac{1}{2} \\ 2a-1;\ -2z^{-2} \end{array}\right).$

25. $\sum_{k=0}^{[n/2]}\dfrac{2^{-2k}}{k!(2n-4k+1)!(a)_k} H_{2n-4k+1}(z)$

$$= \dfrac{(2z)^{2n+1}}{(2n+1)!}\, {}_3F_1\!\left(\begin{array}{c} -n,\ -\tfrac{1}{2}-n,\ a-\tfrac{1}{2} \\ 2a-1;\ -2z^{-2} \end{array}\right).$$

26. $\sum_{k=0}^{n}\binom{n}{k}\dfrac{2^{2k}}{k+1}\left(\tfrac{1}{2}-n\right)_k H_{2n-2k}(z) P_k^{(\rho-k,\,1)}(3)$

$$= (2z)^{2n}\, {}_3F_1\!\left(\begin{array}{c} -n,\ \tfrac{1}{2}-n,\ \rho+2 \\ 2;\ -2z^{-2} \end{array}\right).$$

27. $\sum_{k=0}^{n} \binom{n}{k} \frac{2^{2k}}{k+1} \left(-n-\frac{1}{2}\right)_k H_{2n-2k+1}(z) P_k^{(\rho-k,1)}(3)$

$$= (2z)^{2n+1} {}_3F_1\left(\begin{array}{c} -n, -n-\frac{1}{2}, \rho+2 \\ 2; -2z^{-2} \end{array}\right).$$

28. $\sum_{k=0}^{n} \binom{m}{k} \frac{2^k}{(n-k)!} H_{m+n-2k}(z) = \frac{1}{n!} H_m(z) H_n(z) \qquad [m \geq n;\ [34],\ (44)].$

5.9.2. Sums containing $H_{m\pm nk}(z)$ and special functions

1. $\sum_{k=0}^{n} \binom{2n+1}{2k+1} (4z)^{-2k} B_{2n-2k} H_{2k+1}(z) = \frac{2n+1}{2^{4n-1} z^{2n-1}} H_{2n}(z).$

2. $\sum_{k=0}^{n} \binom{2n}{2k} \frac{2^{2n-2k}-1}{(4z)^{2k}} B_{2n-2k} H_{2k}(z) = \frac{n}{(4z)^{2n-1}} H_{2n-1}(z) \qquad [n \geq 1].$

3. $\sum_{k=0}^{n} \frac{1}{(2k)!(n-k+1)!} \psi(a-k) H_{2k}(z) = \frac{1}{(2n+2)!} \psi(a-n-1)$

$\times [(2z)^{2n+2} - H_{2n+2}(z)] + \frac{(2z)^{2n}}{(2n)!(a-n-1)} {}_4F_2\left(\begin{array}{c} -n, -n+\frac{1}{2}, 1, 1 \\ a-n, 2; -z^{-2} \end{array}\right).$

4. $\sum_{k=0}^{n} \frac{1}{(2k+1)!(n-k+1)!} \psi(a-k) H_{2k+1}(z) = \frac{1}{(2n+3)!} \psi(a-n-1)$

$\times [(2z)^{2n+3} - H_{2n+3}(z)] + \frac{(2z)^{2n+1}}{(2n+1)!(a-n-1)} {}_4F_2\left(\begin{array}{c} -n, -n-\frac{1}{2}, 1, 1 \\ a-n, 2; -z^{-2} \end{array}\right).$

5. $\sum_{k=0}^{n} (-1)^k \frac{(1-a)_k}{(2k)!} \psi(a-k) H_{2k}(z) = (-1)^{n+1} \frac{(1-a)_k}{(2n)!(n-a+1)} (2z)^{2n}$

$\times \left[(a-n-1)\psi(a-n-1) {}_3F_1\left(\begin{array}{c} -n, -n+\frac{1}{2}, a-n-1;\ -z^{-2} \\ a-n \end{array}\right) \right.$

$\left. + {}_4F_2\left(\begin{array}{c} -n, -n+\frac{1}{2}, a-n-1, a-n-1 \\ a-n, a-n;\ -z^{-2} \end{array}\right)\right].$

6. $\sum_{k=0}^{n} (-1)^k \frac{(1-a)_k}{(2k+1)!} \psi(a-k) H_{2k+1}(z)$

$= (-1)^{n+1} \frac{(1-a)_n}{(2n+1)!(n-a+1)} (2z)^{2n+1}$

$$\times \left[(a-n-1)\psi(a-n-1) \, {}_3F_1\left(\begin{matrix} -n, -n-\frac{1}{2}, a-n-1 \\ a-n; \, -z^{-2} \end{matrix} \right) \right.$$

$$\left. + {}_4F_2\left(\begin{matrix} -n, -n-\frac{1}{2}, a-n-1, a-n-1 \\ a-n, a-n; \, -z^{-2} \end{matrix} \right) \right].$$

7. $\sum_{k=0}^{n}(-1)^k \binom{n}{k} \dfrac{2^{-k/2}}{(\nu+1)_k} D_{\nu+k}(\sqrt{2}z) H_k(z) = \dfrac{(-1)^n}{(\nu+1)_n} D_{\nu+2n}(\sqrt{2}z).$

8. $\sum_{k=0}^{n} \binom{n}{k} \left(\dfrac{i}{\sqrt{2}}\right)^k D_{\nu-k}(\sqrt{2}z) H_k(iz) = (n-\nu)_n D_{\nu-2n}(\sqrt{2}z).$

9. $\sum_{k=0}^{n} \dfrac{(w-\sqrt{w^2-1})^k}{(n-k)!(2k)!} P_{n-k}(w) H_{2k}(z)$

$$= 2^n \dfrac{\left(\frac{1}{2}\right)_n}{(n!)^2} (w^2-1)^{n/2} \, {}_2F_2\left(\begin{matrix} -n, -n; \, \frac{z^2}{2} - \frac{wz^2}{2\sqrt{w^2-1}} \\ \frac{1}{2}-n, \frac{1}{2} \end{matrix} \right).$$

10. $\sum_{k=0}^{n} \dfrac{(w-\sqrt{w^2-1})^k}{(n-k)!(2k+1)!} P_{n-k}(w) H_{2k+1}(z)$

$$= 2^{n+1} \dfrac{\left(\frac{1}{2}\right)_n}{(n!)^2} (w^2-1)^{n/2} z \, {}_2F_2\left(\begin{matrix} -n, -n; \, \frac{z^2}{2} - \frac{wz^2}{2\sqrt{w^2-1}} \\ \frac{1}{2}-n, \frac{3}{2} \end{matrix} \right).$$

11. $\sum_{k=0}^{[n/2]} \dfrac{1}{(n-2k)!(\lambda+1)_k} L_k^\lambda(-1) H_{n-2k}(z) = \dfrac{(-1)^n}{(\lambda+1)_n} C_n^{-\lambda-n}(z).$

5.9.3. Sums containing products of $H_{m\pm nk}(z)$

1. $\sum_{k=0}^{n} \binom{n}{k} \dfrac{\left(-\frac{1}{4}\right)^k}{(2k)!} [H_{2k}(z)]^2 = (-4)^n L_{2n}^{-1/2-n}(z^2).$

2. $\sum_{k=0}^{n} \binom{n}{k} \dfrac{\left(-\frac{1}{4}\right)^k}{(2k+1)!} [H_{2k+1}(z)]^2 = \dfrac{(-1)^n 2^{2n+2}}{2n+1} z^2 L_{2n}^{1/2-n}(z^2).$

3. $\sum_{k=1}^{n}(-i)^k \binom{n}{k} H_{n-k}(iz) H_{k-1}(z)$

$$= 2^{(n-1)/2} n! (-i)^n e^{z^2/2} D_{-n-1}(\sqrt{2}z) - \dfrac{\sqrt{\pi}}{2} e^{z^2} \operatorname{erfc}(z) H_n(iz).$$

4. $\sum_{k=0}^{n} \binom{n}{k} H_{2k}(w) H_{2n-2k}(z) = (-4)^n n! L_n(w^2+z^2).$

5. $\sum_{k=0}^{n} \binom{n}{k} H_{2n-2k+1}(w) H_{2k+1}(z) = (-1)^n 2^{2n+2} n! \, wz L_n^2(w^2 + z^2).$

6. $\sum_{k=0}^{n} \binom{n}{k} H_{2n-2k}(w) H_{2k+1}(z) = (-1)^n 2^{2n+1} n! \, z L_n^1(w^2 + z^2).$

7. $\sum_{k=0}^{n} 2^{2k} \dfrac{\left(\frac{1}{2} - n\right)_k}{(2k)!(n-k)!} H_{2k}(w) H_{2n-2k}(z) = \dfrac{2^{2n}}{n!} (w^2 + z^2)^n T_n\left(\dfrac{z^2 - w^2}{z^2 + w^2}\right).$

8. $\sum_{k=0}^{n} 2^{2k} \dfrac{\left(\frac{1}{2} - n\right)_k}{(2k+1)!(n-k)!} H_{2k+1}(w) H_{2n-2k}(z)$
$$= \dfrac{(-1)^n 2^{2n+1}}{n!(2n+1)} (w^2 + z^2)^{n+1/2} T_{2n+1}\left(\dfrac{w}{\sqrt{w^2 + z^2}}\right).$$

9. $\sum_{k=0}^{n} 2^{2k} \dfrac{\left(-\frac{1}{2} - n\right)_k}{(2k+1)!(n-k)!} H_{2k+1}(w) H_{2n-2k+1}(z)$
$$= \dfrac{(-1)^n 2^{2n+1} z}{(n+1)!} (w^2 + z^2)^{n+1/2} U_{2n+1}\left(\dfrac{w}{\sqrt{w^2 + z^2}}\right).$$

5.9.4. Sums containing $H_n(\varphi(k, z))$

1. $\sum_{k=0}^{n} (-1)^k \binom{n}{k} H_m(w + kz) = 0 \qquad [m < n].$

2. $\sum_{k=0}^{n} (-1)^k \binom{n}{k} H_{2m}(\sqrt{k}\, z)$
$$= (-1)^m (2m)! \, n! \, (2z)^{2n} \sum_{k=0}^{m-n} \sigma_{k+n}^n \dfrac{(-4z^2)^k}{(m-n-k)!(2k+2n)!}.$$

3. $\sum_{k=1}^{n} \dfrac{(-1)^k}{\sqrt{k}} \binom{n}{k} H_{2m+1}(\sqrt{k}\, z) = (-1)^{m+1} 2^{2m+1} \left(\dfrac{3}{2}\right)_m z$
$$+ (-1)^m (2m+1)! \, n! \, (2z)^{2n+1} \sum_{k=0}^{m-n} \sigma_{k+n}^n \dfrac{(-4z^2)^k}{(m-n-k)!(2k+2n+1)!}.$$

4. $\sum_{k=0}^{n} (-1)^k \binom{n}{k} H_{2m+2n}(\sqrt{w + kz})$
$$= (-4)^{m+n} n! \, (m+n)! \, z^n \sum_{k=0}^{m} \sigma_{k+n}^n \dfrac{(-z)^k}{(k+n)!} L_{m-k}^{k+n-1/2}(w).$$

5. $\displaystyle\sum_{k=0}^{n} \frac{(-1)^k}{\sqrt{kw+z}} \binom{n}{k} H_{2m+2n+1}\left(\sqrt{w+kz}\right)$

$$= (-1)^{m+n} 2^{2m+2n+1} n! (m+n)! z^n \sum_{k=0}^{m} \sigma_{k+n}^{n} \frac{(-z)^k}{(k+n)!} L_{m-k}^{k+n+1/2}(w).$$

6. $\displaystyle\sum_{k=1}^{n} (-1)^k \binom{n}{k} k^m H_{2m}\left(\frac{z}{\sqrt{k}}\right) = -(2z)^{2m}$

$$+ (2m)! \, n! \, (2z)^{2m-2n} \sum_{k=0}^{m-n} (-1)^k \frac{\sigma_{k+n}^n}{(k+n)!(2m-2n-2k)!} (2z)^{-2k}.$$

7. $\displaystyle\sum_{k=1}^{n} (-1)^k \binom{n}{k} k^{m+1/2} H_{2m+1}\left(\frac{z}{\sqrt{k}}\right) = -(2z)^{2m+1}$

$$+ (2m+1)! \, n! \, (2z)^{2m-2n+1} \sum_{k=0}^{m-n} (-1)^k \frac{\sigma_{k+n}^n}{(k+n)!(2m-2n-2k+1)!} (2z)^{-2k}.$$

8. $\displaystyle\sum_{k=0}^{n} (-1)^k \binom{n}{k} (ka+b)^{m+n} H_{2m+2n}\left(\frac{z}{\sqrt{ka+b}}\right)$

$$= \frac{2^{-2m} n! (2m+2n)!}{\left(\frac{1}{2}\right)_m} a^n b^m \sum_{k=0}^{m} \sigma_{k+n}^{n} \frac{\left(\frac{1}{2}-m\right)_k}{(n+k)!(m-k)!} \left(\frac{4a}{b}\right)^k H_{2m-2k}\left(\frac{z}{\sqrt{b}}\right).$$

9. $\displaystyle\sum_{k=0}^{n} (-1)^k \binom{n}{k} (ka+b)^{m+n+1/2} H_{2m+2n+1}\left(\frac{z}{\sqrt{ka+b}}\right)$

$$= \frac{2^{-2m} n! (2m+2n+1)!}{\left(\frac{3}{2}\right)_m} a^n b^{m+1/2}$$

$$\times \sum_{k=0}^{m} \sigma_{k+n}^{n} \frac{\left(-\frac{1}{2}-m\right)_k}{(n+k)!(m-k)!} \left(\frac{4a}{b}\right)^k H_{2m-2k+1}\left(\frac{z}{\sqrt{b}}\right).$$

5.9.5. Sums containing $H_{m \pm nk}(\varphi(k,z))$

1. $\displaystyle\sum_{k=1}^{n} \frac{(ka)^k (ka-b)^{n-k-1}}{(n-k)!(2k)!} H_{2k}\left(\frac{z}{\sqrt{k}}\right)$

$$= \frac{(-1)^n b^{n-1}}{n!} + \frac{b^n}{(2n)!(na-b)} H_{2n}\left(z\sqrt{\frac{a}{b}}\right).$$

2. $\displaystyle\sum_{k=1}^{n} \frac{a^k k^{k+1/2} (ka-b)^{n-k-1}}{(n-k)!(2k+1)!} H_{2k+1}\left(\frac{z}{\sqrt{k}}\right)$

$$= \frac{2(-1)^n b^{n-1} z}{n!} + \frac{a^{-1/2} b^{n+1/2}}{(2n+1)!(na-b)} H_{2n+1}\left(z\sqrt{\frac{a}{b}}\right).$$

3. $\sum_{k=1}^{n} \dfrac{k^{n-1}}{k!(2n-2k)!} H_{2n-2k}\left(\dfrac{z}{\sqrt{k}}\right) = \dfrac{(2z)^{2n-2}}{(2n-2)!}$ $\qquad [n \geq 1]$.

4. $\sum_{k=1}^{n} \dfrac{k^{n-k}(k+1)^{k-1}}{k!(2n-2k)!} H_{2n-2k}\left(\dfrac{z}{\sqrt{k}}\right) = -\dfrac{(2z)^{2n}}{(2n)!} + \dfrac{(-1)^n}{(2n)!} H_{2n}(iz)$.

5. $\sum_{k=1}^{n} \dfrac{k^{n-k+1/2}(k+1)^{k-1}}{k!(2n-2k+1)!} H_{2n-2k+1}\left(\dfrac{z}{\sqrt{k}}\right)$
$$= -\dfrac{(2z)^{2n+1}}{(2n+1)!} - \dfrac{(-1)^n i}{(2n+1)!} H_{2n+1}(iz).$$

6. $\sum_{k=1}^{n} \dfrac{k^{n-1/2}}{k!(2n-2k+1)!} H_{2n-2k+1}\left(\dfrac{z}{\sqrt{k}}\right) = \dfrac{(2z)^{2n-1}}{(2n-1)!}$ $\qquad [n \geq 1]$.

7. $\sum_{k=0}^{n} \dfrac{(ka+1)^{n-1}}{(n-k)!(2k)!} H_{2k}\left(\dfrac{z}{\sqrt{ka+1}}\right) = \dfrac{(2z)^{2n}}{(2n)!(na+1)}$.

8. $\sum_{k=0}^{n} \dfrac{(ka+1)^{n-1/2}}{(n-k)!(2k+1)!} H_{2k+1}\left(\dfrac{z}{\sqrt{ka+1}}\right) = \dfrac{(2z)^{2n+1}}{(2n+1)!(na+1)}$.

5.9.6. Sums containing products of $H_{m \pm nk}(\varphi(k,z))$

1. $\sum_{k=0}^{n} (-1)^k \binom{n}{k} [H_m(w+kz)]^2 = 0$ $\qquad [2m < n]$.

2. $\sum_{k=0}^{n} \dfrac{(-1)^k (k+1)^{n-1/2}}{(2k+1)!(2n-2k)!} H_{2n-2k}\left(\dfrac{w}{\sqrt{k+1}}\right) H_{2k+1}\left(\dfrac{z}{\sqrt{k+1}}\right)$
$$= \dfrac{2^{2n+2}}{(2n+2)!\, z} \left[w^{2n+2} - (w^2+z^2)^{n+1} T_{n+1}\left(\dfrac{w^2-z^2}{w^2+z^2}\right)\right].$$

3. $\sum_{k=0}^{n} \dfrac{(-1)^k (k+1)^n}{(2k+1)!(2n-2k+1)!} H_{2n-2k+1}\left(\dfrac{w}{\sqrt{k+1}}\right) H_{2k+1}\left(\dfrac{z}{\sqrt{k+1}}\right)$
$$= \dfrac{2^{2n+3} w}{(2n+3)!\, z} \left[w^{2n+2} + (-1)^n (w^2+z^2)^{n+1} U_{2n+2}\left(\dfrac{z}{\sqrt{w^2+z^2}}\right)\right].$$

5.10. The Laguerre Polynomials $L_n^\lambda(z)$

5.10.1. Sums containing $L_m^{\lambda \pm nk}(z)$

1. $\sum_{k=0}^{n} \dfrac{(-n)_k (n)_k}{k!(k+m)!} L_m^k(z) = \dfrac{(-z)^n}{(m+n)!} L_{m-n}^{2n}(z)$ $\qquad [m \geq n]$.

5.10. The Laguerre Polynomials $L_n^\lambda(z)$

2. $\displaystyle\sum_{k=0}^{n}(-1)^k\binom{n}{k}L_m^{\lambda+k}(z)=(-1)^n L_{m-n}^{\lambda+n}(z)$ $\qquad [m\geq n]$.

3. $\displaystyle\sum_{k=0}^{n}\binom{n}{k}\frac{k^r}{(\lambda+m+1)_k}(-z)^k L_m^{\lambda+k}(z)$

$$=\frac{1}{m!(\lambda+m+1)_n}\sum_{k=1}^{r}\sigma_r^k(-n)_k(m+n-k)!\,z^k L_{m+n-k}^{\lambda+k}(z).$$

4. $\displaystyle\sum_{k=0}^{n}(-1)^k\binom{n}{k}\frac{(a)_k}{(\lambda+m+1)_k}L_m^{\lambda+k}(z)$

$$=\frac{(\lambda+1)_m(\lambda-a+1)_n}{m!(\lambda+1)_n}\,{}_2F_2\!\left(\begin{array}{c}-m,\lambda-a+n+1;\,z\\ \lambda+n+1,\lambda-a+1\end{array}\right).$$

5. $\displaystyle\sum_{k=0}^{n}\binom{n}{k}\frac{(-z)^k}{(\lambda-n+1)_k}L_m^{\lambda+k}(z)=(-1)^n\frac{(m+n)!}{m!(-\lambda)_n}L_{m+n}^{\lambda-n}(z).$

6. $\displaystyle\sum_{k=0}^{n}\binom{n}{k}\frac{(-z)^k}{(\lambda+m+1)_k}L_m^{\lambda+k}(z)=\frac{(m+n)!}{m!(\lambda+m+1)_n}L_{m+n}^{\lambda}(z).$

7. $\displaystyle\sum_{k=0}^{n}\binom{n}{k}\frac{(a+m+n)_k}{(a)_k(\lambda+m+1)_k}(-z)^k L_m^{\lambda+k}(z)=\frac{(\lambda+1)_m}{m!}\,{}_2F_2\!\left(\begin{array}{c}-m-n,a+n\\ \lambda+1,a;\,z\end{array}\right).$

8. $\displaystyle\sum_{k=0}^{n}\binom{n}{k}\frac{(\lambda+m+n+1)_k}{(\lambda+1)_k(\lambda+m+1)_{2k}}(-z^2)^k L_m^{\lambda+2k}(z)$

$$=\frac{n!(m+n)!(\lambda+1)_m}{m!(\lambda+1)_n(\lambda+1)_{m+n}}L_n^{\lambda}(-z)\,L_{m+n}^{\lambda}(z).$$

9. $\displaystyle\sum_{k=0}^{n}L_m^{\lambda-k}(z)=L_{m+1}^{\lambda}(z)-L_{m+1}^{\lambda-n-1}(z).$

10. $\displaystyle\sum_{k=0}^{n}(-1)^k\binom{n}{k}L_m^{\lambda-k}(z)=L_{m-n}^{\lambda}(z)$ $\qquad [m\geq n]$.

11. $\displaystyle\sum_{k=0}^{n}(-1)^k\binom{n}{k}\frac{(-\lambda-m)_k}{(1-\lambda-n)_k}L_m^{\lambda-k}(z)=\frac{z^n}{(\lambda)_n}L_{m-n}^{\lambda+n}(z)$ $\qquad [m\geq n]$.

12. $\displaystyle\sum_{k=0}^{n}(-1)^k\binom{n}{k}\frac{(-\lambda-m)_k}{(1-a-n)_k}L_m^{\lambda-k}(z)$

$$=\frac{(\lambda+1)_m(a-\lambda)_n}{m!(a)_n}\,{}_2F_2\!\left(\begin{array}{c}-m,\lambda-a+1;\,z\\ \lambda+1,\lambda-a-n+1\end{array}\right).$$

13. $\displaystyle\sum_{k=0}^{n}\binom{n}{k}\frac{(-\lambda-m)_k(a+m+n)_k}{(a)_k}z^{-k}L_m^{\lambda-k}(z)$

$$=\frac{(-z)^m}{m!}\,_3F_1\!\left(\begin{matrix}-m-n,\,-\lambda-m,\,a+n\\a;\,-z^{-1}\end{matrix}\right).$$

14. $\displaystyle\sum_{k=0}^{n}(-1)^k\binom{n}{k}\frac{(-\lambda-m)_k}{(a)_k}L_m^{\lambda-k}(z)$

$$=\frac{(\lambda+a+m)_n(-z)^m}{m!(a)_n}\,_3F_1\!\left(\begin{matrix}-m,\,-\lambda-m,\,1-\lambda-a-m\\1-\lambda-a-m-n;\,-z^{-1}\end{matrix}\right).$$

15. $\displaystyle\sum_{k=0}^{n}\binom{n}{k}(n-\lambda)_k z^{-k}L_m^{\lambda-k}(z)=\frac{(m+n)!}{m!}(-z)^{-n}L_{m+n}^{\lambda-2n}(z).$

16. $\displaystyle\sum_{k=0}^{n}\binom{n}{k}\frac{(-z)^k}{(\lambda+m+1)_k(k+1)}L_m^{\lambda+k}(z)$

$$=\frac{z^{-1}}{n+1}\left[(\lambda+m)L_m^{\lambda-1}(z)-\frac{(m+n+1)!z}{m!(\lambda+m+1)_n}L_{m+n+1}^{\lambda-1}(z)\right].$$

5.10.2. Sums containing $L_{m\pm nk}^{\lambda}(z)$

1. $\displaystyle\sum_{k=0}^{n}\frac{t^k}{(n-k)!(\lambda+1)_k}L_k^{\lambda}(z)=\frac{(t+1)^n}{(\lambda+1)_n}L_n^{\lambda}\!\left(\frac{tz}{t+1}\right).$

2. $\displaystyle\sum_{k=0}^{n-1}(-1)^k\frac{(n-k)^n}{(n-k)!(\lambda+1)_k}L_k^{\lambda}(z)=\frac{z^n}{n!(\lambda+1)_n}\sum_{k=0}^{n}\sigma_n^k(-n)_k(-\lambda-n)_k z^{-k}.$

3. $\displaystyle\sum_{k=0}^{n}(-1)^k\frac{(2n-k)!}{[(n-k)!]^2(\lambda+1)_k}L_k^{\lambda}(z)=\,_2F_2\!\left(\begin{matrix}-n,\,-n;\,z\\\lambda+1,\,1\end{matrix}\right).$

4. $\displaystyle\sum_{k=0}^{n}\frac{(a)_k}{(n-k)!(\lambda+1)_k}z^{-k}L_k^{\lambda}(z)$

$$=\frac{(a)_n}{n!}z^{-n}\,_3F_3\!\left(\begin{matrix}-n,\,\dfrac{\lambda-a-n+1}{2},\,\dfrac{\lambda-a-n}{2}+1;\,4z\\\lambda+1,\,-a-n+1,\,\lambda-a-n+1\end{matrix}\right).$$

5. $\displaystyle\sum_{k=0}^{n}(-1)^k\frac{(a)_k}{(n-k)!(b)_k(\lambda+1)_k}L_k^{\lambda}(z)=\frac{(b-a)_n}{n!(b)_n}\,_2F_2\!\left(\begin{matrix}-n,\,a;\,z\\a-b-n+1,\,\lambda+1\end{matrix}\right).$

6. $\displaystyle\sum_{k=0}^{n}\frac{(-\lambda-n)_k}{(n-k)!(\lambda+1)_k}(-z)^{-k}L_k^{\lambda}(z)$

$$=\frac{(\lambda+1)_n}{n!}z^{-n}\,_2F_3\!\left(\begin{matrix}-\dfrac{n}{2},\,\dfrac{1-n}{2};\,-z^2\\\dfrac{\lambda+1}{2},\,\dfrac{\lambda}{2}+1,\,\lambda+1\end{matrix}\right).$$

7. $\sum_{k=0}^{n} \dfrac{\sigma_m^{n-k+1}}{(\lambda+1)_k} L_k^\lambda(z) = \dfrac{(-z)^n}{n!(\lambda+1)_n} {}_{m+1}F_{m-1}\!\left(\begin{array}{c}-n,\,-n-\lambda,\,2,\,\ldots,\,2\\ 1,\,\ldots,\,1;\;-z^{-1}\end{array}\right).$

8. $\sum_{k=0}^{n} \dfrac{(-\lambda-n)_k}{(n-k)!(\lambda+1)_k}(-z)^{-k} L_k^\lambda(z)$

$$= \dfrac{(\lambda+1)_n}{n!} z^{-n} {}_2F_3\!\left(\begin{array}{c}-\dfrac{n}{2},\,\dfrac{1-n}{2};\;-z^2\\ \dfrac{\lambda+1}{2},\,\dfrac{\lambda}{2}+1,\,\lambda+1\end{array}\right).$$

9. $\sum_{k=0}^{n}(-1)^k \binom{n}{k}\dfrac{(2k)!}{(\lambda+1)_{2k}} L_{2k}^\lambda(z) = \dfrac{n!(2z)^n}{(\lambda+1)_{2n}} L_n^{\lambda+n}\!\left(\dfrac{z}{2}\right).$

10. $\sum_{k=0}^{n}(-1)^k \binom{n}{k} L_{k+m}^\lambda(z) = (-1)^n L_{m+n}^{\lambda-n}(z).$

11. $\sum_{k=0}^{n}(-1)^k \binom{n}{k}\dfrac{(k+m)!}{(\lambda+m+1)_k} L_{k+m}^\lambda(z) = \dfrac{m!\,z^n}{(\lambda+m+1)_n} L_m^{\lambda+n}(z).$

12. $\sum_{k=0}^{n}(-1)^k \binom{n}{k}\dfrac{(a+m)_k}{(\lambda+m+1)_k} L_{k+m}^\lambda(z)$

$$= \dfrac{(1-a)_n(\lambda+1)_m}{(m+n)!} {}_2F_2\!\left(\begin{array}{c}-m-n,\,a;\;z\\ \lambda+1,\,a-n\end{array}\right).$$

13. $\sum_{k=0}^{n} \sigma_{k+m}^m \dfrac{(-\lambda-n)_k}{(k+m)!} t^k L_{n-k}^\lambda(z)$

$$= \dfrac{t^{-m}}{(\lambda+n+1)_m} \sum_{k=0}^{m}(-1)^k \dfrac{(1-kt)^{m+n}}{k!(m-k)!} L_{m+n}^\lambda\!\left(\dfrac{z}{1-kt}\right).$$

14. $\sum_{k=0}^{n} \dfrac{(-4)^k}{(n-k)!(2k)!}(-\lambda-m)_k L_{m-k}^\lambda(z)$

$$= \dfrac{(-z)^m}{m!\,n!} {}_3F_1\!\left(\begin{array}{c}-m,\,-\lambda-m,\,n+\dfrac{1}{2}\\ \dfrac{1}{2};\;-z^{-1}\end{array}\right) \qquad [m\geq n].$$

15. $\sum_{k=0}^{n} \dfrac{(-4)^k}{(n-k)!(2k+1)!}(-\lambda-m)_k L_{m-k}^\lambda(z)$

$$= \dfrac{(-z)^m}{m!\,n!} {}_3F_1\!\left(\begin{array}{c}-m,\,-\lambda-m,\,n+\dfrac{3}{2}\\ \dfrac{3}{2};\;-z^{-1}\end{array}\right) \qquad [m\geq n].$$

16. $\sum_{k=0}^{n}(-1)^k \binom{n}{k} L_{m-k}^{-m-1/2}(z) = L_m^{-m-n-1/2}(z) \qquad [m\geq n].$

17. $\sum_{k=0}^{[n/2]} \frac{\left(\frac{1}{2}\right)_k}{k!} L_{n-2k}(z) = (-1)^n \frac{2^{-n-1/2}}{n!\sqrt{z}} H_n\left(\sqrt{\frac{z}{2}}\right) H_{n+1}\left(\sqrt{\frac{z}{2}}\right).$

5.10.3. Sums containing $L_{m \pm pk}^{\lambda \pm nk}(z)$

1. $\sum_{k=0}^{n} (-1)^k \frac{(a+k\mu)_{n-1}}{(n-k)!(a+k\mu)_k} L_k^{\lambda+k\mu}(z) = \frac{(-1)^n}{a+n\mu+n-1} L_n^{\lambda-a-n+1}(z).$

2. $\sum_{k=0}^{n} \frac{(-z)^k}{(n-k)!(\lambda+1)_k(\mu+1)_k} L_k^{\lambda+\mu+k}(z) = \frac{n!}{(\lambda+1)_n(\mu+1)_n} L_n^{\lambda}(z) L_n^{\mu}(z).$

3. $\sum_{k=0}^{n} (-1)^k \frac{(a)_k}{(n-k)!(\lambda+1)_{2k}} L_k^{\lambda+k}(z)$

$$= \frac{(a)_n}{n!(\lambda+1)_{2n}} z^n \, {}_3F_1\left(\begin{array}{c}-n, -\lambda-2n, a-\lambda-n\\ 1-a-n;\; z^{-1}\end{array}\right).$$

4. $\sum_{k=0}^{n} \binom{n}{k} \frac{\left(n+\frac{1}{2}\right)_k}{(\lambda+1)_{2k}} (-4z)^k L_k^{\lambda+k}(z)$

$$= \frac{\left(n+\frac{1}{2}\right)_n}{n!(\lambda+1)_{2n}} (2z)^{2n} \, {}_3F_1\left(\begin{array}{c}-2n, -2n, -\lambda-2n\\ -4n;\; -z^{-1}\end{array}\right).$$

5. $\sum_{k=0}^{n} \frac{(a)_k}{(n-k)!\left(a-n+\frac{1}{2}\right)_k (\lambda+1)_{2k}} (-4z)^k L_k^{\lambda+k}(z)$

$$= \frac{(a)_n}{n!\left(\frac{1}{2}-a\right)_n (\lambda+1)_{2n}} (-4z^2)^n \, {}_3F_1\left(\begin{array}{c}-2n, \frac{1}{2}-a-n, -\lambda-2n\\ 1-2a-2n;\; -z^{-1}\end{array}\right).$$

6. $\sum_{k=0}^{n} \frac{\left(\frac{1}{2}-a\right)_k}{(n-k)!(1-a-n)_k (\lambda+1)_{2k}} (-4z)^k L_k^{\lambda+k}(z)$

$$= \frac{\left(\frac{1}{2}-a\right)_n}{n!(a)_n(\lambda+1)_{2n}} (-4z^2)^n \, {}_3F_1\left(\begin{array}{c}-2n, a-n, -\lambda-2n\\ 2a-2n;\; -z^{-1}\end{array}\right).$$

7. $\sum_{k=0}^{2n} \frac{(a)_k}{(n-k)!\left(a-n+\frac{1}{2}\right)_k (\lambda+1)_{2k}} (-4z)^k L_k^{\lambda+k}(z)$

$$= \frac{1}{n!} \, {}_2F_2\left(\begin{array}{c}-2n, 2a;\; z\\ a-n+\frac{1}{2}, \lambda+1\end{array}\right).$$

5.10.3]   5.10. THE LAGUERRE POLYNOMIALS $L_n^\lambda(z)$

8. $\displaystyle\sum_{k=0}^{n}\binom{n}{k}\frac{\left(m+n+\frac{1}{2}\right)_k}{k!\,(\lambda+m+1)_{2k}}(4z)^k L_{k+m}^{\lambda+k}(z)$

$$= \frac{(\lambda+1)_m}{m!}\,{}_2F_2\!\left(\begin{array}{c}-m-2n,\,m+2n+1\\ m+1,\,\lambda+1;\,z\end{array}\right).$$

9. $\displaystyle\sum_{k=0}^{n}\binom{n}{k} L_{m-k}^{\lambda+k}(z) = L_m^{\lambda+n}(z)$ $\quad [m\geq n]$.

10. $\displaystyle\sum_{k=0}^{n}\frac{(-4z)^k}{(n-k)!\,(2k+1)!} L_{m-k}^{\lambda+k}(z) = \frac{(\lambda+1)_m}{m!\,n!}\,{}_2F_2\!\left(\begin{array}{c}-m,\,n+\frac{3}{2}\\ \frac{3}{2},\,\lambda+1;\,z\end{array}\right).$

11. $\displaystyle\sum_{k=0}^{n}\binom{n}{k}\frac{1}{k+2} L_{m-k}^{\lambda+k}(z)$

$$= \frac{1}{(n+1)(n+2)}\left[L_{m+2}^{\lambda-2}(z) + (n+1) L_{m+1}^{\lambda+n}(z) - L_{m+2}^{\lambda+n-1}(z)\right] \quad [m\geq n].$$

12. $\displaystyle\sum_{k=0}^{n}\sigma_{k+m}^{m}\frac{t^k}{(k+m)!} L_{n-k}^{\lambda+k}(z) = (-t)^{-m}\sum_{k=0}^{m}\frac{(-1)^k}{k!\,(m-k)!} L_{m+n}^{\lambda-m}(z-kt).$

13. $\displaystyle\sum_{k=0}^{n}\binom{n}{k} k^r L_{m-k}^{\lambda+k}(z) = n!\sum_{k=1}^{r}\frac{\sigma_r^k}{(n-k)!} L_{m-k}^{\lambda+n}(z)$ $\quad [m\geq n;\,m\geq r]$.

14. $\displaystyle\sum_{k=0}^{n}\binom{n}{k}(m-k)!\,z^k L_{m-k}^{\lambda+k}(z) = (-1)^n (m-n)!\,(-\lambda-m)_n L_{m-n}^{\lambda}(z)$ $\quad [m\geq n]$.

15. $\displaystyle\sum_{k=0}^{n}\binom{n}{k}\frac{(-z)^k}{(\lambda-n+1)_k} L_{m-k}^{\lambda+k}(z) = \frac{(\lambda+1)_m}{(\lambda-n+1)_m} L_m^{\lambda-n}(z)$ $\quad [m\geq n]$.

16. $\displaystyle\sum_{k=0}^{n}\frac{(-4z)^k}{(n-k)!\,(2k)!} L_{m-k}^{n+k-1/2}(z) = (-1)^m\frac{\left(n+\frac{1}{2}\right)_m}{(2m)!\,n!} H_{2m}(\sqrt{z}).$

17. $\displaystyle\sum_{k=1}^{n}\frac{2k+\lambda}{(\lambda+m+1)_k}(-z)^k L_{m-k}^{\lambda+2k}(z)$

$$= -z L_{m-1}^{\lambda+1}(z) - \frac{(-z)^{n+1}}{(\lambda+m+1)_n} L_{m-n-1}^{\lambda+2n+1}(z) \quad [m>n\geq 1].$$

18. $\displaystyle\sum_{k=0}^{n}\frac{(-4z)^k}{(n-k)!\,(2k+1)!} L_{m-k}^{\lambda+k}(z) = \frac{(\lambda+1)_m}{m!\,n!}\,{}_2F_2\!\left(\begin{array}{c}-m,\,n+\frac{3}{2}\\ \lambda+1,\,\frac{3}{2};\,z\end{array}\right)$ $\quad [m\geq n]$.

19. $\displaystyle\sum_{k=0}^{n} \binom{n}{k} \frac{4^{-k}(-\lambda-2m)_k}{\left(\frac{1}{2}-\lambda-2m-n\right)_k} L_{2m-2k}^{\lambda+k}(z)$

$$= \frac{z^{2m}}{(2m)!} {}_3F_1\left(\begin{matrix}-2m, -2\lambda-4m, -\lambda-2m-n\\-2\lambda-4m-2n; -z^{-1}\end{matrix}\right) \quad [m \geq n].$$

20. $\displaystyle\sum_{k=0}^{n} \frac{t^k}{(n-k)!} L_k^{\lambda-k}(z) = t^n L_n^{\lambda-n}\left(z-\frac{1}{t}\right).$

21. $\displaystyle\sum_{k=0}^{n} (a)_k (-z)^{-k} L_k^{\lambda-k}(z) = \frac{(a)_n(-\lambda)_n}{n!} z^{-n} {}_2F_2\left(\begin{matrix}-n, -a-n; z\\1-a-n, \lambda-n+1\end{matrix}\right).$

22. $\displaystyle\sum_{k=0}^{n} (-\lambda-1)_k (-z)^{-k} L_k^{\lambda-k}(z) = (-\lambda)_n (-z)^{-n} L_n^{\lambda-n+1}(z).$

23. $\displaystyle\sum_{k=0}^{n} \sigma_m^{n-k+1} (-z)^{-k} L_k^{\lambda-k}(z) = \frac{(-\lambda)_n}{n!} z^{-n} {}_mF_m\left(\begin{matrix}-n, 2, \ldots, 2; z\\\lambda-n+1, 1, \ldots, 1\end{matrix}\right).$

24. $\displaystyle\sum_{k=0}^{n} \frac{1}{(n-k)!(1-\lambda-n)_k} L_k^{\lambda-k}(z) = \frac{z^n}{n!(\lambda)_n}.$

25. $\displaystyle\sum_{k=0}^{n} \frac{(-1)^k}{(n-k)!(\lambda-n+1)_k} L_k^{\lambda-k}(z) = \frac{(-z)^n}{n!(-\lambda)_n} {}_3F_0\left(\begin{matrix}-\frac{n}{2}, \frac{1-n}{2}, -\lambda\\4z^{-2}\end{matrix}\right).$

26. $\displaystyle\sum_{k=0}^{n-1} \frac{(n-k)^n}{(n-k)!} z^{-k} L_k^{\lambda-k}(z) = \frac{(-\lambda)_n}{n!}(-z)^{-n} \sum_{k=0}^{n} \sigma_n^k \frac{(-n)_k}{(\lambda-n+1)_k} (-z)^k.$

27. $\displaystyle\sum_{k=0}^{n} \frac{(a)_k}{(n-k)!(b)_k} z^{-k} L_k^{\lambda-k}(z) = \frac{(b-a)_n}{n!(b)_n} {}_3F_1\left(\begin{matrix}-n, a, -\lambda; -z^{-1}\\a-b-n+1\end{matrix}\right).$

28. $\displaystyle\sum_{k=0}^{n} \frac{z^{-k}}{(n-k)!(a-k)} L_k^{\lambda-k}(z)$

$$= \frac{(-\lambda)_n}{n!(a-n)}(-z)^{-n} {}_2F_2\left(\begin{matrix}-n, 1; z\\a-n+1, \lambda-n+1\end{matrix}\right).$$

29. $\displaystyle\sum_{k=0}^{n} \frac{z^{-k}}{(n-k)!(k-a)} L_k^{n-k}(z) = \frac{n!}{(-a)_{n+1}}(-z)^{-n} L_n^{a-n}(z).$

30. $\displaystyle\sum_{k=0}^{n} \frac{\left(\frac{1}{2}-a-n\right)_k}{(n-k)!\left(\frac{3}{2}-n\right)_k} z^{-k} L_k^{\lambda-k}(z) = \frac{\left(a+\frac{1}{2}\right)_n (-\lambda)_n}{n!\left(-\frac{1}{2}\right)_n} \frac{(-z)^{-n}}{2a}$

$$\times \left[(2a-1){}_2F_2\left(\begin{matrix}-n, a; z\\a-\frac{1}{2}, \lambda-n+1\end{matrix}\right) + {}_2F_2\left(\begin{matrix}-n, a; z\\a+\frac{1}{2}, \lambda-n+1\end{matrix}\right)\right].$$

5.10. The Laguerre Polynomials $L_n^\lambda(z)$

31. $\displaystyle\sum_{k=0}^{n}\frac{\left(-\lambda-\frac{1}{2}\right)_k}{(n-k)!\left(\frac{3}{2}-n\right)_k}z^{-k}L_k^{\lambda-k}(z)$

$$=\frac{(-\lambda)_n(-z)^{-n}}{2(\lambda-n+1)\left(-\frac{1}{2}\right)_n}\left[(2\lambda+1)L_n^{\lambda-n-1/2}(z)+L_n^{\lambda-n+1/2}(z)\right].$$

32. $\displaystyle\sum_{k=0}^{n}\binom{n}{k}L_{k+m}^{\lambda-k}(z)=L_{m+n}^{\lambda}(z).$

33. $\displaystyle\sum_{k=0}^{n}\binom{n}{k}(a)_k z^{-k}L_{k+m}^{\lambda-k}(z)$

$$=\frac{(m-a+1)_n(-z)^m}{(m+n)!}\,_3F_1\!\left(\begin{matrix}-m-n,\,a-m,\,-\lambda-m\\a-m-n;\,-z^{-1}\end{matrix}\right).$$

34. $\displaystyle\sum_{k=0}^{n}\binom{n}{k}(n-\lambda)_k z^{-k}L_{k+m}^{\lambda-k}(z)=\frac{(\lambda+1)_m(-\lambda)_n z^{-n}}{(\lambda-n+1)_m}L_{m+n}^{\lambda-2n}(z).$

35. $\displaystyle\sum_{k=0}^{n}\binom{n}{k}\frac{(-\lambda-m)_{2k}}{\left(\frac{1}{2}-m-n\right)_k}(4z)^{-k}L_{m-k}^{\lambda-k}(z)$

$$=\frac{(-z)^m}{m!}\,_3F_1\!\left(\begin{matrix}-2m,\,-m-n,\,-\lambda-m\\-2m-2n;\,-z^{-1}\end{matrix}\right)\qquad[m\geq n\geq 1].$$

36. $\displaystyle\sum_{k=0}^{n}\binom{n}{k}\frac{(2k+\lambda)(\lambda)_k}{(\lambda+m+1)_k(\lambda+n+1)_k}z^k L_{m-k}^{\lambda+2k}(z)=\frac{(\lambda)_{m+1}}{(\lambda+n+1)_m}L_m^{\lambda+n}(z)$

$$[m\geq n].$$

37. $\displaystyle\sum_{k=0}^{n}\frac{z^k}{(\lambda+m+1)_k}L_{m-k}^{\lambda+2k}(z)=\frac{(\lambda+1)_m}{m!}$

$$\times\left[{}_2F_2\!\left(\begin{matrix}-m,\,\frac{\lambda}{2};\,z\\\lambda,\,\frac{\lambda}{2}+1\end{matrix}\right)-\frac{(-m)_{n+1}(-z)^{n+1}}{(\lambda+1)_{2n+2}}\,_2F_2\!\left(\begin{matrix}n-m+1,\,n+\frac{\lambda}{2}+1;\,z\\2n+\lambda+2,\,n+\frac{\lambda}{2}+2\end{matrix}\right)\right]$$

$$[m\geq n].$$

38. $\displaystyle\sum_{k=0}^{n}\binom{n}{k}\frac{\left(-\frac{z^2}{4}\right)^k}{\left(\frac{1}{2}-m-n\right)_k(\lambda+m+1)_k}L_{m-k}^{\lambda+2k}(z)$

$$=\frac{(\lambda+1)_m}{m!}\,_2F_2\!\left(\begin{matrix}-2m,\,-m-n;\,z\\-2m-2n,\,\lambda+1\end{matrix}\right)\qquad[m\geq n].$$

39. $\displaystyle\sum_{k=0}^{[n/2]}\frac{\left(\frac{z}{2}\right)^{2k}}{k!\,(a)_k}L_{n-2k}^{\lambda+2k}(z)=\frac{(\lambda+1)_n}{n!}\,_2F_2\!\left(\begin{matrix}-n,\,a-\frac{1}{2};\,2z\\2a-1,\,\lambda+1\end{matrix}\right).$

40. $\displaystyle\sum_{k=0}^{n}(-\lambda)_k(-z)^{-k}L_k^{\lambda-2k}(z) = \frac{(-\lambda)_{2n}}{n!}z^{-n}{}_2F_2\left(\begin{array}{c}-n, \frac{\lambda}{2}-n; z\\ \frac{\lambda}{2}-n+1, \lambda-2n\end{array}\right).$

41. $\displaystyle\sum_{k=0}^{n}\frac{\left(n-\lambda-\frac{1}{2}\right)_k}{(n-k)!}\left(-\frac{4}{z^2}\right)^k L_k^{\lambda-2k}(z) = \left(\frac{1}{2}\right)_n\left(\frac{2}{z}\right)^{2n}L_{2n}^{2\lambda-4n+1}(z).$

42. $\displaystyle\sum_{k=0}^{n}\frac{\left(\frac{1}{2}-a-n\right)_k(-\lambda)_k}{(n-k)!(1-a-2n)_k}\left(-\frac{4}{z^2}\right)^k L_k^{\lambda-2k}(z)$
$$= \frac{(2a)_{2n}(-\lambda)_{2n}}{n!(a)_{2n}}z^{-2n}{}_2F_2\left(\begin{array}{c}-2n, a; z\\ 2a, \lambda-2n+1\end{array}\right).$$

43. $\displaystyle\sum_{k=0}^{n}\frac{(-\lambda)_k}{(n-k)!(a)_k}z^{-k}L_k^{\lambda-2k}(z)$
$$= \frac{(-\lambda)_{2n}}{n!(a)_n}(-z)^{-n}{}_3F_3\left(\begin{array}{c}-n, \frac{a+\lambda-n}{2}, \frac{a+\lambda-n+1}{2}; z\\ \frac{\lambda+1}{2}-n, \frac{\lambda}{2}-n+1, a+\lambda-n\end{array}\right).$$

44. $\displaystyle\sum_{k=0}^{n}\frac{(n-\lambda)_k}{(n-k)!}t^k L_k^{\lambda-2k}(z)$
$$= n!(-t)^n L_n^{\lambda-2n}\left(\frac{\sqrt{t}z-\sqrt{4+tz^2}}{2\sqrt{t}}\right)L_n^{\lambda-2n}\left(\frac{\sqrt{t}z+\sqrt{4+tz^2}}{2\sqrt{t}}\right).$$

45. $\displaystyle\sum_{k=0}^{n}(2k-\lambda)\frac{(-\lambda)_k z^{-k}}{(n-k)!(n-\lambda+1)_k}L_k^{\lambda-2k}(z) = \frac{(-\lambda)_{n+1}}{n!}(-z)^{-n}.$

46. $\displaystyle\sum_{k=0}^{n}\frac{\left(a+n-\frac{1}{2}\right)_k(-\lambda)_k}{(n-k)!(a)_k}\left(-\frac{4}{z^2}\right)^k L_k^{\lambda-2k}(z)$
$$= \frac{1}{n!}{}_3F_1\left(\begin{array}{c}-2n, 2n+2a-1, -\lambda\\ a; -z^{-1}\end{array}\right).$$

47. $\displaystyle\sum_{k=0}^{n}\frac{(2n-2k+a)(-\lambda)_k}{(n-k)!(1-a-n)_k}z^{-k}L_k^{\lambda-2k}(z)$
$$= \frac{(-\lambda)_{2n}az^{-n}}{n!(a)_n}{}_3F_3\left(\begin{array}{c}-n, \frac{\lambda-a}{2}-n, \frac{\lambda-a+1}{2}-n; z\\ \frac{\lambda+1}{2}-n, \frac{\lambda}{2}-n+1, \lambda-a-2n+1\end{array}\right).$$

48. $\displaystyle\sum_{k=0}^{n}(2k-\lambda)\frac{(1-a-n)_k(-\lambda)_k}{(a-\lambda+n)_k}(-z)^{-k}L_k^{\lambda-2k}(z)$
$$= \frac{(a)_n(-\lambda)_{2n+1}(a-\lambda)_n}{n!(a-\lambda)_{2n}}(-z)^{-n}{}_2F_2\left(\begin{array}{c}-n, a-1; z\\ a, \lambda-2n\end{array}\right).$$

5.10.3]   5.10. The Laguerre Polynomials $L_n^\lambda(z)$

49. $\displaystyle\sum_{k=0}^{n} \frac{2k-\lambda}{(\lambda-2k)^2-1} \frac{(-\lambda)_k}{(n-k)!(n-\lambda+1)_k} z^{-k} L_k^{\lambda-2k}(z)$

$$= \frac{(-\lambda)_{n+1}}{n!\left[(\lambda-2n)^2-1\right]} (-z)^{-n} {}_2F_2\left(\begin{array}{c}-n, 1;\ z\\ \frac{\lambda+1}{2}-n, \frac{\lambda+3}{2}-n\end{array}\right).$$

50. $\displaystyle\sum_{k=0}^{n} \frac{\left(\frac{1}{2}-a\right)_k (-\lambda)_k}{(n-k)!(1-a-n)_k} \left(-\frac{4}{z^2}\right)^k L_k^{\lambda-2k}(z)$

$$= \frac{\left(\frac{1}{2}-a\right)_n (-\lambda)_{2n}}{n!(a)_n} \left(-\frac{4}{z^2}\right)^n {}_2F_2\left(\begin{array}{c}-2n, a-n;\ z\\ 2a-2n, \lambda-2n+1\end{array}\right).$$

51. $\displaystyle\sum_{k=0}^{n} \frac{(-\lambda)_k}{(n-k)!(a)_k\left(\frac{3}{2}-a-\lambda-n\right)_k} 2^{-2k} L_k^{\lambda-2k}(z)$

$$= \frac{(-\lambda)_n}{n!(a)_n\left(a+\lambda-\frac{1}{2}\right)_n} \left(\frac{z}{4}\right)^n {}_3F_1\left(\begin{array}{c}-n, 2a+\lambda+n-1, 2-2a-\lambda-n\\ \lambda-n+1;\ -\frac{1}{z}\end{array}\right).$$

52. $\displaystyle\sum_{k=0}^{n} \frac{(-\lambda)_k}{(n-k)!(1-a-n)_k\left(a-\lambda+\frac{1}{2}\right)_k} 2^{-2k} L_k^{\lambda-2k}(z)$

$$= \frac{(-\lambda)_n}{n!(a)_n\left(a-\lambda+\frac{1}{2}\right)_n} \left(\frac{z}{4}\right)^n {}_3F_1\left(\begin{array}{c}-n, 2a-\lambda+n, \lambda-2a-n+1\\ \lambda-n+1;\ -z^{-1}\end{array}\right).$$

53. $\displaystyle\sum_{k=0}^{n}(2k-\lambda) \frac{(a)_k(b)_k(-\lambda)_k}{(n-k)!(1-a-\lambda)_k(1-b-\lambda)_k(n-\lambda+1)_k} z^{-k} L_k^{\lambda-2k}(z)$

$$= \frac{(a)_n(b)_n(-\lambda)_{n+1}}{n!(1-a-\lambda)_n(1-b-\lambda)_n} (-z)^{-n} {}_2F_2\left(\begin{array}{c}-n, 1-a-b-\lambda;\ z\\ 1-a-n, 1-b-n\end{array}\right).$$

54. $\displaystyle\sum_{k=0}^{n}\binom{n}{k} \frac{\left(-\frac{z^2}{4}\right)^k}{\left(\frac{1}{2}-m-n\right)_k(\lambda+m+1)_k} L_{m-k}^{\lambda+2k}(z)$

$$= \frac{(\lambda+1)_m}{m!} {}_2F_2\left(\begin{array}{c}-2m, -m-n;\ z\\ -2m-2n, \lambda+1\end{array}\right) \quad [m \geq n \geq 1].$$

55. $\displaystyle\sum_{k=0}^{n}\binom{n}{k}\left(m+n+\frac{1}{2}\right)_k (-\lambda-m)_k \left(-\frac{4}{z^2}\right)^k L_{k+m}^{\lambda-2k}(z)$

$$= \frac{(-z)^m}{m!} {}_3F_1\left(\begin{array}{c}-m-2n, m+2n+1, -\lambda-m\\ m+1;\ -z^{-1}\end{array}\right).$$

56. $\displaystyle\sum_{k=0}^{n}\binom{2k}{k} \frac{1}{(n-k)!(\lambda+1)_k} L_{2k}^{\lambda-k}(z) = \frac{1}{(\lambda+1)_n} \left[L_n^\lambda(z)\right]^2$ [[34], (42′)].

353

57. $\sum_{k=0}^{n} \frac{(1-a-n)_k}{(n-k)!\,(\lambda+1)_k} z^{-k} L_{2k}^{\lambda-k}(z)$

$$= \frac{(a)_n(-r)_n}{(2n)!} z^{-n} {}_2F_2\!\left(\begin{matrix} -2n,\,1-a-2n \\ a,\,\lambda-n+1;\,-z \end{matrix}\right).$$

58. $\sum_{k=0}^{n} \binom{n}{k} \frac{\left(\lambda+n+\frac{1}{2}\right)_k}{(\lambda+1)_k} 2^{2k} L_{2k}^{\lambda-k}(z)$

$$= \frac{\left(\lambda+\frac{1}{2}\right)_{2n}}{(2\lambda+1)_{2n}} \frac{(4z)^{2n}}{(2n)!} {}_3F_1\!\left(\begin{matrix} -2n,\,-\lambda-2n,\,-2\lambda-2n \\ -2\lambda-4n;\,-z^{-1} \end{matrix}\right).$$

59. $\sum_{k=0}^{n} \frac{(1-2a-n)_k \left(\frac{1}{2}\right)_k}{(n-k)!\,(1-a-n)_k (\lambda+1)_k} z^{-k} L_{2k}^{\lambda-k}(z)$

$$= \frac{(2a)_n(-\lambda)_n}{n!\,(a)_n} (-4z)^{-n} {}_2F_3\!\left(\begin{matrix} -n,\,\frac{1}{2}-a-n;\,\frac{z^2}{4} \\ a+\frac{1}{2},\,\frac{\lambda-n+1}{2},\,\frac{\lambda-n}{2}+1 \end{matrix}\right).$$

60. $\sum_{k=0}^{n} \frac{\lambda-4k}{(n-k)!} \frac{\left(\frac{1}{2}\right)_k \left(-\frac{\lambda}{2}\right)_k \left(n+\frac{1-\lambda}{2}\right)_k}{\left(\frac{1}{2}-n\right)_k \left(n-\frac{\lambda}{2}+1\right)_k} \left(\frac{2}{z}\right)^{2k} L_{2k}^{\lambda-4k}(z)$

$$= 2(-1)^{n+1} \frac{n!\,\left(-\frac{\lambda}{2}\right)_{n+1}}{\left(\frac{1}{2}\right)_n} \left(\frac{2}{z}\right)^{2n} \left[L_n^{(\lambda-1)/2-2n}\!\left(\frac{z}{2}\right)\right]^2.$$

61. $\sum_{k=0}^{n} \binom{2n}{n-k}(2k)!\,z^{-2k} L_{2k}^{-4k}(z)$

$$= 2^{4n-1}(2n)!\,z^{-2n} L_{2n}^{-2n-1/2}\!\left(\frac{z}{2}\right) - \frac{1}{2}\binom{2n}{n}.$$

62. $\sum_{k=0}^{n} \binom{2n+1}{n-k}(2k+1)!\,z^{-2k} L_{2k+1}^{-4k-2}(z)$

$$= 2^{4n+1}(2n+1)!\,z^{-2n} L_{2n+1}^{-2n-3/2}\!\left(\frac{z}{2}\right).$$

63. $\sum_{k=0}^{n} \binom{n}{k} \frac{(-\lambda-2m)_{2k}}{\left(\frac{1}{2}-\lambda-2m-n\right)_k} 4^{-k} L_{2m-2k}^{\lambda+k}(z)$

$$= \frac{z^{2m}}{(2m)!} {}_3F_1\!\left(\begin{matrix} -2m,\,-2\lambda-4m,\,-\lambda-2m-n \\ -2\lambda-4m-2n;\,-z^{-1} \end{matrix}\right) \quad [m \ge n].$$

64. $\sum_{k=0}^{n} \frac{(m+n-2k)!}{(m-k)!\,(n-k)!\,(\lambda+1)_k} z^{2k} L_{m+n-2k}^{\lambda+2k}(z)$

$$= \frac{(\lambda+1)_{m+n}}{(\lambda+1)_m (\lambda+1)_n} L_m^{\lambda}(z)\,L_n^{\lambda}(z) \quad [m \ge n].$$

65. $\displaystyle\sum_{k=0}^{[n/2]} (4k+\lambda) \frac{(a)_k \left(\frac{\lambda}{2}\right)_k}{k! \left(\frac{\lambda}{2} - a + 1\right)_k (\lambda + n + 1)_{2k} (\lambda)_{4k+1}} z^{2k} L_{n-2k}^{\lambda+4k}(z)$

$$= \frac{(\lambda+1)_n}{n!} {}_2F_2\left(\begin{matrix} -n, \frac{\lambda+1}{2} - a;\ z \\ \frac{\lambda+1}{2},\ \lambda - 2a + 1 \end{matrix}\right).$$

66. $\displaystyle\sum_{k=0}^{[n/2]} (-1)^k (2n - 4k + 1) \frac{(n-2k)! \left(-n - \frac{1}{2}\right)_k}{k!} z^{2k} L_{n-2k}^{4k-2n-1}(z)$

$$= 2^{2n+1} \left(\frac{1}{2}\right)_{n+1} L_n^{-n-1}\left(\frac{z}{2}\right).$$

67. $\displaystyle\sum_{k=0}^{[n/3]} (\lambda + 6k) \frac{\left(\frac{\lambda}{3}\right)_k}{k! (\lambda + n + 1)_{3k}} z^{3k} L_{n-3k}^{\lambda+6k}(z) = \frac{(\lambda)_{n+1}}{n!} {}_2F_2\left(\begin{matrix} -n, \frac{\lambda}{3};\ \frac{3z}{4} \\ \frac{\lambda}{2},\ \frac{\lambda+1}{2} \end{matrix}\right).$

68. $\displaystyle\sum_{k=0}^{n} \frac{(1-a-n)_k \left(a + 2n - \frac{1}{2}\right)_k}{(n-k)! (\lambda+1)_k} \left(\frac{4}{z}\right)^k L_{3k}^{\lambda-k}(z)$

$$= \frac{(a)_n \left(a + 2n - \frac{1}{2}\right)_n}{(3n)! (\lambda+1)_{2n}} (2z)^{2n} {}_3F_2\left(\begin{matrix} -3n,\ a - \frac{1}{2},\ 1 - a - 3n,\ -\lambda - 2n \\ 2a - 1,\ 2 - 2a - 6n;\ -\frac{4}{z} \end{matrix}\right).$$

69. $\displaystyle\sum_{k=0}^{n} \frac{(a)_k \left(n - a + \frac{1}{2}\right)_k}{(n-k)! (\lambda+1)_k} \left(-\frac{4}{z^2}\right)^k L_{3k}^{\lambda-2k}(z)$

$$= \frac{(a)_n \left(n - a + \frac{1}{2}\right)_n (-\lambda)_{2n}}{(3n)!} \left(-\frac{4}{z^2}\right)^n$$

$$\times {}_3F_3\left(\begin{matrix} -3n,\ a - 2n,\ \frac{1}{2} - a - n;\ 4z \\ 2a - 4n,\ 1 - 2a - 2n,\ \lambda - 2n + 1 \end{matrix}\right).$$

5.10.4. Sums containing $L_{m \pm pk}^{\lambda \pm nk}(z)$ and special functions

1. $\displaystyle\sum_{k=0}^{n} (-1)^k \binom{n}{k} \psi(k+1)\, L_k^{\lambda}(z)$

$$= (-1)^n \left[\psi(n+1) L_n^{\lambda-n}(z) + \sum_{k=0}^{n-1} \frac{1}{n-k} L_k^{\lambda-n}(z) \right].$$

2. $\displaystyle\sum_{k=0}^{n} \frac{(1-a)_k}{(\lambda+1)_k} \psi(a-k) L_k^{\lambda}(z) = \frac{(1-a)_n}{n! (\lambda+1)_n (n-a+1)} (-z)^n$

$$\times \left[(n - a + 1) \psi(a - n - 1) {}_3F_1\left(\begin{matrix} -n,\ a - n - 1,\ -\lambda - n \\ a - n;\ -z^{-1} \end{matrix}\right) \right.$$

$$\left. - {}_4F_2\left(\begin{matrix} -n,\ a - n - 1,\ a - n - 1,\ -\lambda - n \\ a - n,\ a - n;\ -z^{-1} \end{matrix}\right) \right].$$

3. $\displaystyle\sum_{k=0}^{n} \frac{(-1)^k}{(n-k+1)!(\lambda+1)_k} \psi(a-k) L_k^\lambda(z)$

$\displaystyle = \frac{1}{(\lambda+1)_{n+1}} \psi(a-n-1) \left[\frac{z^{n+1}}{(n+1)!} + (-1)^n L_{n+1}^\lambda(z) \right]$

$\displaystyle + \frac{z^n}{n!(\lambda+1)_n (a-n-1)} {}_4F_2\!\left(\begin{array}{c} -n, -\lambda-n, 1, 1; -\frac{1}{z} \\ a-n, 2 \end{array}\right).$

4. $\displaystyle\sum_{k=0}^{n} \frac{z^{-k}}{(n-k)!} \psi(a-k) L_k^{\lambda-k}(z) = \frac{(-\lambda)_n}{n!} (-z)^{-n} \psi(a-n)$

$\displaystyle - (-1)^n \frac{(-\lambda)_{n-1}}{(n-1)!(a-n)} z^{1-n} {}_3F_3\!\left(\begin{array}{c} 1-n, 1, 1; z \\ a-n+1, \lambda-n+2, 2 \end{array}\right) \quad [m \geq n].$

5. $\displaystyle\sum_{k=0}^{n} \frac{z^{-k}}{(n-k)!} \psi(a-2k) L_k^{\lambda-k}(z) = \frac{(-\lambda)_n}{n!} (-z)^{-n}$

$\displaystyle \times \left[\psi(a-2n) + \frac{nz}{(a-2n)(\lambda-n+1)} {}_3F_3\!\left(\begin{array}{c} 1-n, 1, 1; z \\ \frac{a}{2}-n+1, \lambda-n+2, 2 \end{array}\right) \right.$

$\displaystyle \left. + \frac{nz}{(a-2n+1)(\lambda-n+1)} {}_3F_3\!\left(\begin{array}{c} 1-n, 1, 1; z \\ \frac{a+3}{2}-n, \lambda-n+2, 2 \end{array}\right) \right].$

6. $\displaystyle\sum_{k=0}^{n} \frac{(-1)^k}{(n-k)!} \gamma(a-k, z) L_k^{n-k}(-z) = (-1)^n \frac{z^{a-n}}{a-n} {}_1F_1\!\left(\begin{array}{c} a-2n; -z \\ a-n+1 \end{array}\right).$

7. $\displaystyle\sum_{k=0}^{n} \frac{(-1)^k}{k!} \gamma(\lambda+k, z) L_{n-k}^{\lambda-n+k}(-z) = \frac{(-z)^n}{n!} (1-\lambda)_n \gamma(\lambda-n, z).$

8. $\displaystyle\sum_{k=0}^{n} \frac{(-1)^k}{(n-k)!} \Gamma(\lambda+n-k, z) L_k^{\lambda-k}(-z) = \frac{(1-\lambda)_n}{2n!} z^n \Gamma(\lambda-n, z).$

9. $\displaystyle\sum_{k=0}^{n} \frac{2^k}{(n-k)!\left(\lambda-2n+\frac{3}{2}\right)_{2k}} D_{\lambda-2n+2k+1/2}\!\left(\sqrt{2z}\right) L_k^\lambda(z)$

$\displaystyle = \frac{(-1)^n}{n!\left(-\lambda-\frac{1}{2}\right)_{2n}} D_{\lambda+2n+1/2}\!\left(\sqrt{2z}\right).$

10. $\displaystyle\sum_{k=0}^{n} \frac{\left(w+\sqrt{w^2-1}\right)^k z^k}{k!} P_k(w) L_{n-k}^{\lambda+k}(z)$

$\displaystyle = \frac{2^{-n}(2n)!}{(n!)^3} \left(\frac{z\sqrt{w^2-1}}{\sqrt{w^2-1}-w}\right)^n {}_3F_1\!\left(\begin{array}{c} -n, -n, -\lambda-n \\ \frac{1}{2}-n; \frac{w-\sqrt{w^2-1}}{2z\sqrt{w^2-1}} \end{array}\right).$

5.10.5]   5.10. THE LAGUERRE POLYNOMIALS $L_n^\lambda(z)$

11. $\displaystyle\sum_{k=0}^{n} \frac{2k+1}{(k+n+1)!} z^k P_k(w) L_{n-k}^{2k+1}(z) = \frac{1}{n!} L_n\left(\frac{z-wz}{2}\right).$

12. $\displaystyle\sum_{k=0}^{n} \frac{2k+1}{(k+n+1)!} z^k [P_k(w)]^2 L_{n-k}^{2k+1}(z) = \frac{1}{n!} {}_2F_2\left(\begin{array}{c}-n, \frac{1}{2} \\ 1, 1;\, z-w^2z\end{array}\right).$

13. $\displaystyle\sum_{k=0}^{n} \frac{2k+1}{(k+n+1)!} z^k T_{2k+1}(w) L_{n-k}^{2k+1}(z) = \frac{(-1)^n w}{(2n)!} H_{2n}\left(\sqrt{z-w^2z}\right).$

14. $\displaystyle\sum_{k=0}^{n} \frac{2k+1}{(k+n+1)!} (-z)^k U_{2k}(w) L_{n-k}^{2k+1}(z) = \frac{(-1)^n}{(2n)!} H_{2n}(w\sqrt{z}).$

15. $\displaystyle\sum_{k=0}^{n} (-1)^k \frac{(-\lambda-n)_k}{(2k)!} H_{2k}(w) L_{n-k}^\lambda(z) = \frac{(w^2+z)^n}{\left(\lambda+\frac{1}{2}\right)_n} C_{2n}^{\lambda+1/2}\left(\frac{w}{\sqrt{w^2+z}}\right).$

16. $\displaystyle\sum_{k=0}^{n} (-1)^k \frac{(-\lambda-n)_k}{(2k+1)!} H_{2k+1}(w) L_{n-k}^\lambda(z)$
$$= \frac{(w^2+z)^{n+1/2}}{\left(\lambda+\frac{1}{2}\right)_{n+1}} C_{2n+1}^{\lambda+1/2}\left(\frac{w}{\sqrt{w^2+z}}\right).$$

17. $\displaystyle\sum_{k=0}^{n} \frac{(-z)^k}{(k+1)!} P_k^{(\rho-k,1)}(3) L_{n-k}^{\lambda+k}(z) = \frac{(\lambda+1)_n}{n!} {}_2F_2\left(\begin{array}{c}-n, \rho+2;\, 2z \\ 2, \lambda+1\end{array}\right).$

5.10.5. Sums containing products of $L_{m\pm pk}^{\lambda\pm nk}(z)$

1. $\displaystyle\sum_{k=0}^{n} \frac{(-1)^k}{(n-k)!(\lambda+1)_k} [L_k^\lambda(z)]^2 = \frac{(-1)^n}{(\lambda+1)_n} \binom{2n}{n} L_{2n}^{\lambda-n}(z)$ [[34], (43′)].

2. $\displaystyle\sum_{k=0}^{2n} (-1)^k \frac{k!\, z^{-2k}}{(2n-k)!} [L_k^{n-k-1/2}(z)]^2 = \frac{(-1)^n \pi z^{-4n}}{2^{2n} \Gamma^2\left(\frac{1}{2}-n\right)} L_n^{-n-1/2}(4z).$

3. $\displaystyle\sum_{k=0}^{n} (-1)^k \frac{k!}{(n-k)!} z^{-2k} L_k^{\lambda-k}(z) L_k^{\mu-k}(z)$
$$= (-1)^n z^{-2n} \frac{(-\lambda)_n(-\mu)_n}{n!} {}_3F_3\left(\begin{array}{c}-n, \frac{\lambda+\mu}{2}-n, \frac{\lambda+\mu}{2}-n+1;\, 4z \\ \lambda-n+1, \mu-n+1, \lambda+\mu-2n+1\end{array}\right).$$

4. $\displaystyle\sum_{k=0}^{n} \frac{k!}{(n-k)!} z^{-2k} L_k^{\lambda-k}(z) L_k^{\lambda-k}(-z)$
$$= \frac{(-\lambda)_n^2}{n!} z^{-2n} {}_2F_3\left(\begin{array}{c}-\frac{n}{2}, \frac{1-n}{2};\, -z^2 \\ \frac{\lambda-n+1}{2}, \frac{\lambda-n}{2}+1, \lambda-n+1\end{array}\right).$$

5. $\displaystyle\sum_{k=0}^{n}(\lambda-2k)\frac{k!\,(-\lambda)_k}{(n-k)!\,(n-\lambda+1)_k}(-wz)^{-k}L_k^{\lambda-2k}(w)L_k^{\lambda-2k}(z)$
$$=-(-\lambda)_{n+1}(wz)^{-n}L_n^{\lambda-2n}(w+z).$$

6. $\displaystyle\sum_{k=0}^{n}(\lambda-2k)\frac{(-\lambda)_k^2}{(n-k)!\,(n-\lambda+1)_k}z^{-2k}\left[L_k^{\lambda-2k}(z)\right]^2$
$$=-\frac{(-\lambda)_{n+1}\left(\frac{1}{2}\right)_n}{n!}\left(\frac{2}{z}\right)^{2n}L_{2n}^{\lambda-2n}(z).$$

7. $\displaystyle\sum_{k=0}^{n}\frac{(-\mu-n)_k}{(\lambda+1)_k}L_k^{\lambda}(w)L_{n-k}^{\mu}(z)=\frac{(-z)^n}{(\lambda+1)_n}P_n^{(\lambda,-\lambda-\mu-2n-1)}\left(1+\frac{2w}{z}\right).$

8. $\displaystyle\sum_{k=0}^{n}(-\mu-n)_k(-w)^{-k}L_k^{\lambda-k}(w)L_{n-k}^{\mu}(z)=\frac{(-z)^n}{n!}\,_3F_0\!\left(\begin{array}{c}-n,\,-\lambda,\,-\mu-n\\w^{-1}z^{-1}\end{array}\right).$

9. $\displaystyle\sum_{k=0}^{n}\frac{z^k}{(\lambda+1)_k}L_k^{\lambda}(w)L_{n-k}^{\mu+k}(z)=\frac{(\mu+1)_n}{n!}\,_1F_2\!\left(\begin{array}{c}-n;\,wz\\\lambda+1,\,\mu+1\end{array}\right).$

10. $\displaystyle\sum_{k=0}^{n}\frac{z^{-k}}{(n-k)!}L_m^{\lambda-k}(-z)L_k^{\mu-k}(z)$
$$=\frac{(\lambda-n+1)_m(-\mu)_n}{m!\,n!}(-z)^{-n}\,_2F_2\!\left(\begin{array}{c}-m,\,\mu+1;\,-z\\\lambda-n+1,\,\mu-n+1\end{array}\right).$$

11. $\displaystyle\sum_{k=0}^{n}\frac{(-z)^k}{k!}L_m^{\lambda+k}(z)L_{n-k}^{k-m-1}(-z)=\frac{(-\lambda-m)_n}{n!}L_{m-n}^{\lambda}(z)\qquad[m\geq n].$

12. $\displaystyle\sum_{k=0}^{n}\frac{(-z)^k}{k!}L_m^{\lambda+k}(z)L_{n-k}^{\lambda-n+k}(-z)=\frac{(-1)^n}{n!}(-\lambda-m)_n L_m^{\lambda-n}(z)\qquad[m\geq n].$

13. $\displaystyle\sum_{k=0}^{n}\frac{(-1)^k}{k!}(-\lambda-m)_k L_m^{\lambda-k}(z)L_{n-k}^{m+k}(z)=\binom{m+n}{m}L_{m+n}^{\lambda}(z).$

14. $\displaystyle\sum_{k=0}^{n}\frac{(-1)^k}{k!}(-\lambda-m)_k L_m^{\lambda-k}(z)L_{n-k}^{k-n-\lambda}(z)=\frac{(-z)^n}{n!}L_m^{\lambda+n}(z).$

15. $\displaystyle\sum_{k=0}^{n}\frac{(-z)^k}{k!}L_{m-k}^{\lambda+k}(z)L_{n-k}^{\mu+k}(z)$
$$=\frac{(\lambda+1)_m(\mu+1)_n}{m!\,n!}e^z\,_2F_2\!\left(\begin{array}{c}\lambda+m+1,\,\mu+n+1;\,-z\\\lambda+1,\,\mu+1\end{array}\right)\qquad[m\geq n].$$

16. $\displaystyle\sum_{k=0}^{n}\frac{(-z)^k}{k!}L_{m-k}^{\lambda+k}(z)L_{n-k}^{\lambda+m+k}(z)=\binom{m+n}{m}L_{m+n}^{\lambda}(z)\qquad[m\geq n].$

17. $\displaystyle\sum_{k=0}^{n}\binom{m+k}{k}L_{m+k}^{\lambda-k}(z)L_{n-k}^{k-n-\lambda}(-z)=\frac{(-z)^{n}}{n!}L_{m-n}^{\lambda+n}(z)$ $\quad[m\geq n]$.

18. $\displaystyle\sum_{k=0}^{n}\frac{(-z)^{k}}{k!}L_{m-k}^{\lambda+k}(z)L_{n-k}^{\lambda-n+k}(z)=\binom{m+n}{m}L_{m+n}^{\lambda-n}(z)$ $\quad[m\geq n]$.

19. $\displaystyle\sum_{k=0}^{n}\binom{m+k}{k}L_{m+k}^{\lambda-k}(z)L_{n-k}^{k-m-\lambda-1}(-z)=\frac{(-\lambda-m)_{n}}{n!}L_{m-n}^{\lambda}(z)$ $\quad[m\geq n]$.

20. $\displaystyle\sum_{k=0}^{n}\frac{k!(\lambda-2k)}{(n-k)!}\frac{(-\lambda)_{k}}{(n-\lambda+1)_{k}}(-wz)^{-k}L_{k}^{\lambda-2k}(w)L_{k}^{\lambda-2k}(z)$
$$=-(-\lambda)_{n+1}(wz)^{-n}L_{n}^{\lambda-2n}(w+z).$$

21. $\displaystyle\sum_{k=0}^{n}(2k+\lambda)\frac{(\lambda)_{k}}{(\lambda+n+1)_{k}}\left(-\frac{w}{z}\right)^{k}L_{n-k}^{\lambda+2k}(w)L_{k}^{-\lambda-2k}(z)$
$$=\frac{(\lambda)_{n+1}}{n!}\left(1+\frac{w}{z}\right)^{n}.$$

22. $\displaystyle\sum_{k=0}^{n}(2k-\lambda)\frac{(k!)^{2}(a+n)_{k}(-2\lambda)_{2k}}{(n-k)!(2k)!(1-\lambda+n)_{k}(1-a-\lambda-n)_{k}}$
$$\times z^{-2k}L_{k}^{\lambda-2k}(z)L_{k}^{\lambda-2k}(-z)$$
$$=\frac{(2a)_{2n}(-\lambda)_{n+1}}{(2n)!(a)_{n}(a+\lambda)_{n}}\,{}_{5}F_{2}\!\left(\begin{array}{c}-n,\,n+a,\,-\frac{\lambda}{2},\,\frac{1-\lambda}{2},\,\frac{1}{2}-\lambda\\ a+\frac{1}{2},\,-\lambda;\,4z^{-2}\end{array}\right).$$

23. $\displaystyle\sum_{k=0}^{n}(2k+\lambda)\frac{k!(\lambda)_{k}}{(\lambda+n+1)_{k}}\left(-\frac{z}{w^{2}}\right)^{k}L_{k}^{-\lambda-2k}(-w)L_{k}^{-\lambda-2k}(w)L_{n-k}^{\lambda+2k}(z)$
$$=\frac{(\lambda)_{n+1}}{n!}\,{}_{3}F_{1}\!\left(\begin{array}{c}-n,\,\frac{\lambda}{2},\,\frac{\lambda+1}{2}\\ \lambda;\,4w^{-2}z\end{array}\right).$$

5.10.6. Sums containing $L_{m\pm pk}^{\lambda\pm nk}(\varphi(k,z))$

1. $\displaystyle\sum_{k=0}^{n}(-1)^{k}\binom{n}{k}L_{m}^{\lambda}(w+kz)=0$ $\quad[m<n]$.

2. $\displaystyle\sum_{k=0}^{n}(-1)^{k}\binom{n}{k}L_{n}^{\lambda}(kz)=z^{n}$.

3. $\displaystyle\sum_{k=0}^{n}(-1)^{k}\binom{n}{k}L_{m}^{\lambda}(kz)=\frac{n!(\lambda+1)_{m}}{m!(\lambda+1)_{n}}z^{n}\sum_{k=0}^{m-n}\sigma_{k+n}^{n}\binom{m}{k+n}\frac{(-z)^{k}}{(\lambda+n+1)_{k}}$.

4. $\sum_{k=1}^{n}(-1)^{k}\binom{n}{k}k^{m}L_{m}^{\lambda}\left(\frac{z}{k}\right) = (-1)^{m+1}\frac{z^{m}}{m!}$

$+ (-1)^{m}\frac{n!(\lambda+1)_{m}}{m!(\lambda+1)_{m-n}}z^{m-n}\sum_{k=0}^{m-n}\sigma_{k+n}^{n}\binom{m}{k+n}(n-m-\lambda)_{k}z^{-k}$ $[m \geq n]$.

5. $\sum_{k=0}^{n}(-1)^{k}\binom{n}{k}L_{m+n}^{\lambda}(w+kz) = n!\,z^{n}\sum_{k=0}^{m}\sigma_{k+n}^{n}\frac{(-z)^{k}}{(k+n)!}L_{m-k}^{\lambda+k+n}(w).$

6. $\sum_{k=0}^{n}\frac{(ka+b)^{n-k-1}}{(n-k)!}\left(\frac{a}{z}\right)^{k}L_{k}^{\lambda-k}(kz) = \frac{1}{na+b}\left(\frac{a}{z}\right)^{n}L_{n}^{\lambda-n}\left(-\frac{bz}{a}\right).$

7. $\sum_{k=1}^{n}\frac{k^{k-1}}{k!}z^{k}L_{n-k}^{\lambda+k}(kz) = \frac{(\lambda+2)_{n-1}z}{(n-1)!}$ $[n \geq 1]$.

8. $\sum_{k=0}^{n}\frac{(ka+1)^{n-k-1}}{(n-k)!}z^{-k}L_{k}^{\lambda-k}((ka+1)z) = \frac{(-\lambda)_{n}(-z)^{-n}}{n!(na+1)}.$

9. $\sum_{k=1}^{n}\frac{k^{2k}}{(k+n)!}z^{k}L_{n-k}^{2k}(k^{2}z) = -\frac{1}{2(n!)} + \frac{(-z)^{n}}{2}L_{n}^{-n-1}\left(\frac{1}{z}\right).$

10. $\sum_{k=1}^{n}\frac{k^{2k+2}}{(k+n)!}z^{k}L_{n-k}^{2k}(k^{2}z) = \frac{(-z)^{n}}{2}\left[L_{n-2}^{-n-2}\left(\frac{1}{z}\right) - L_{n-1}^{-n-3}\left(\frac{1}{z}\right)\right]$ $[n \geq 2]$.

11. $\sum_{k=1}^{n}\frac{k^{2k-2}}{(k+n)!}z^{k}L_{n-k}^{2k}(k^{2}z) = \frac{z}{2(n-1)!}.$

12. $\sum_{k=1}^{n}\frac{k^{2k-4}}{(k+n)!}z^{k}L_{n-k}^{2k}(k^{2}z) = -\frac{z^{2}}{8(n-2)!} + \frac{z}{2(n-1)!}.$

13. $\sum_{k=1}^{n}\frac{k^{2k-6}}{(k+n)!}z^{k}L_{n-k}^{2k}(k^{2}z) = \frac{z^{3}}{72(n-3)!} - \frac{5z^{2}}{32(n-2)!} + \frac{z}{2(n-1)!}.$

14. $\sum_{k=1}^{n}\frac{k^{2k}}{k^{2}+a^{2}}\frac{z^{k}}{(k+n)!}L_{n-k}^{2k}(k^{2}z) = \frac{a^{-2}}{2(n!)}\left[{}_{2}F_{2}\left(\begin{array}{c}-n,1;-a^{2}z\\1-ia,1+ia\end{array}\right)-1\right].$

15. $\sum_{k=0}^{n}\frac{(2k+1)^{2k-1}}{(k+n+1)!}z^{k}L_{n-k}^{2k+1}((2k+1)^{2}z) = \frac{1}{n!}.$

16. $\sum_{k=0}^{n}\frac{(2k+1)^{2k-3}}{(k+n+1)!}z^{k}L_{n-k}^{2k+1}((2k+1)^{2}z) = \frac{1}{n!} - \frac{4z}{9(n-1)!}.$

[5.10.7]   5.10. The Laguerre Polynomials $L_n^\lambda(z)$

17. $\displaystyle\sum_{k=0}^{n} \frac{(2k+1)^{2k-5}}{(k+n+1)!} z^k L_{n-k}^{2k+1}((2k+1)^2 z) = \frac{1}{n!} - \frac{40z}{81(n-1)!} + \frac{16z^2}{225(n-2)!}.$

18. $\displaystyle\sum_{k=0}^{n} \frac{(2k+1)^{2k+1}}{(2k+1)^2 + a^2} \frac{z^k}{(k+n+1)!} L_{n-k}^{2k+1}((2k+1)^2 z)$

$$= \frac{1}{n!(1+a^2)} {}_2F_2\left(\begin{matrix}-n,\,1;\,-a^2 z\\ \frac{3-ia}{2},\,\frac{3+ia}{2}\end{matrix}\right).$$

19. $\displaystyle\sum_{k=0}^{n} \frac{(2k+\lambda)^{2k-1}(\lambda)_k}{k!(\lambda+n+1)_k} z^k L_{n-k}^{\lambda+2k}((2k+\lambda)^2 z) = \frac{(\lambda)_{n+1}}{n!\,a^2}.$

20. $\displaystyle\sum_{k=1}^{n} \frac{(-ka)^k (ka+b)^{n-k-1}}{(n-k)!(\lambda+1)_k} L_k^\lambda\!\left(\frac{z}{k}\right) = -\frac{b^{n-1}}{n!}$

$$+ \frac{b^n}{(\lambda+1)_n (na+b)} L_n^\lambda\!\left(-\frac{az}{b}\right).$$

21. $\displaystyle\sum_{k=1}^{n} \frac{k^{n-1}}{k!} (-\lambda-n)_k L_{n-k}^\lambda\!\left(\frac{z}{k}\right) = (-1)^n \frac{\lambda+n}{(n-1)!} z^{n-1}.$

22. $\displaystyle\sum_{k=0}^{n} \frac{(-1)^k}{(n-k)!(\lambda+1)_k} (ka+1)^{n-1} L_k^\lambda\!\left(\frac{z}{ka+1}\right) = \frac{z^n}{n!(na+1)(\lambda+1)_n}.$

23. $\displaystyle\sum_{k=0}^{n} (-1)^k \binom{n}{k} (ka+b)^{m+n} L_{m+n}^\lambda\!\left(\frac{z}{ka+b}\right)$

$$= n!\,(\lambda+m+1)_n (-a)^n b^m \sum_{k=0}^{m} \sigma_{k+n}^n \frac{(-\lambda-m)_k}{(k+n)!} \left(-\frac{a}{b}\right)^k L_{m-k}^\lambda\!\left(\frac{z}{b}\right).$$

5.10.7. Sums containing $L_{m\pm pk}^{\lambda\pm nk}(\varphi(k,z))$ and special functions

1. $\displaystyle\sum_{k=0}^{n} \frac{(k+1)^{n-1}}{(2n-2k)!(\lambda+1)_k} H_{2n-2k}\!\left(\frac{w}{\sqrt{k+1}}\right) L_k^\lambda\!\left(\frac{z}{k+1}\right)$

$$= \frac{\lambda w^{2n+2} z^{-1}}{3(n+1)!\left(\frac{1}{2}\right)_{n+1}} - \frac{2^{2n+2}\lambda z^{-1}(w^2+z)^{n+1}}{(2\lambda-1)_{2n+2}} C_{2n+2}^{\lambda-1/2}\!\left(\frac{w}{\sqrt{w^2+z}}\right).$$

2. $\displaystyle\sum_{k=0}^{n} \frac{(k+1)^{n-1/2}}{(2n-2k+1)!(\lambda+1)_k} H_{2n-2k+1}\!\left(\frac{w}{\sqrt{k+1}}\right) L_k^\lambda\!\left(\frac{z}{k+1}\right)$

$$= \frac{4\lambda w^{2n+3} z^{-1}}{3(n+1)!\left(\frac{5}{2}\right)_n} - \frac{2^{2n+3}\lambda z^{-1}(w^2+z)^{n+3/2}}{(2\lambda-1)_{2n+3}} C_{2n+3}^{\lambda-1/2}\!\left(\frac{w}{\sqrt{w^2+z}}\right).$$

3. $\displaystyle\sum_{k=0}^{n} \frac{(k+1)^{k-1}}{(2k)!} z^k H_{2k}\left(\frac{w}{\sqrt{k+1}}\right) L_{n-k}^{\lambda+k}((n-k)z)$

$$= \sum_{k=0}^{n} \frac{(4w^2 z)^k}{(2k)!(k+1)} L_{n-k}^{\lambda+k}((n+1)z).$$

4. $\displaystyle\sum_{k=0}^{n} \frac{(k+1)^{k-1/2}}{(2k+1)!} z^k H_{2k+1}\left(\frac{w}{\sqrt{k+1}}\right) L_{n-k}^{\lambda+k}((n-k)z)$

$$= 4w \sum_{k=0}^{n} \frac{(4w^2 z)^k}{(2k+2)!} L_{n-k}^{\lambda+k}((n+1)z).$$

5. $\displaystyle\sum_{k=0}^{n-1} \frac{(k+1)^{k-1}}{(2k)!} (n-k)^{n-k} (-\lambda-n)_k H_{2k}\left(\frac{w}{\sqrt{k+1}}\right) L_{n-k}^{\lambda}\left(\frac{z}{n-k}\right)$

$$= (-1)^{n-1} (n+1)^{n-1} \frac{(\lambda+1)_n}{(2n)!} H_{2n}\left(\frac{w}{\sqrt{n+1}}\right)$$

$$+ (n+1)^n \sum_{k=0}^{n} \left(\frac{4w^2}{n+1}\right)^k \frac{(-\lambda-n)_k}{(2k)!(k+1)} L_{n-k}^{\lambda}\left(\frac{z}{n+1}\right).$$

6. $\displaystyle\sum_{k=0}^{n-1} \frac{(k+1)^{k-1/2}}{(2k+1)!} (n-k)^{n-k} (-\lambda-n)_k H_{2k+1}\left(\frac{w}{\sqrt{k+1}}\right) L_{n-k}^{\lambda}\left(\frac{z}{n-k}\right)$

$$= (-1)^{n-1} (n+1)^{n-1/2} \frac{(\lambda+1)_n}{(2n+1)!} H_{2n+1}\left(\frac{w}{\sqrt{n+1}}\right)$$

$$+ 4w(n+1)^n \sum_{k=0}^{n} \left(\frac{4w^2}{n+1}\right)^k \frac{(-\lambda-n)_k}{(2k+2)!} L_{n-k}^{\lambda}\left(\frac{z}{n+1}\right).$$

5.10.8. Sums containing products of $L_{m\pm pk}^{\lambda\pm nk}(\varphi(k,z))$

1. $\displaystyle\sum_{k=0}^{n} (-1)^k \binom{n}{k} [L_m^{\lambda}(w+kz)]^2 = 0$ $[2m < n]$.

2. $\displaystyle\sum_{k=0}^{n} (-1)^k \binom{n}{k} L_m^{\lambda}(\sqrt{k}\, z) L_m^{\lambda}(-\sqrt{k}\, z) = \frac{z^{2m}}{m!} \delta_{m,n}$ $[n \geq m]$.

3. $\displaystyle\sum_{k=1}^{n} (-1)^k \binom{n}{k} k^m L_m^{\lambda}\left(\frac{z}{\sqrt{k}}\right) L_m^{\lambda}\left(-\frac{z}{\sqrt{k}}\right) = (-1)^m \frac{(\lambda+1)_m^2}{m!} \delta_{m,n}$

$$- (-1)^m \frac{z^{2m}}{(m!)^2} \quad [n \geq m].$$

4. $\displaystyle\sum_{k=0}^{n}(k+1)^{n-k-1}(-\mu-n)_k(-w)^{-k}L_k^{\lambda-k}((k+1)w)\,L_{n-k}^{\mu}\left(\dfrac{z}{k+1}\right)$

$\displaystyle =\dfrac{w(-z)^{n+1}}{(n+1)!(\lambda+1)(\mu+n+1)}\left[{}_3F_0\!\left(\begin{matrix}-n-1,-\lambda-1,-\mu-n-1\\ w^{-1}z^{-1}\end{matrix}\right)-1\right].$

5. $\displaystyle\sum_{k=0}^{n}(k+1)^{n-1}\dfrac{(-\lambda-n)_k}{(\mu+1)_k}L_{n-k}^{\lambda}\!\left(\dfrac{w}{k+1}\right)L_k^{\mu}\!\left(\dfrac{z}{k+1}\right)$

$\displaystyle =\dfrac{(-1)^n\mu w^{n+1}z^{-1}}{(n+1)!(\lambda+n+1)}+\dfrac{z^n}{(\mu+1)_n(\lambda+n+1)}P_{n+1}^{(\lambda,-\lambda-\mu-2n-2)}\!\left(1+\dfrac{2w}{z}\right).$

6. $\displaystyle\sum_{k=0}^{n}\dfrac{(k+1)^{k-1}}{(\lambda+1)_k}(-z)^k L_k^{\lambda}\!\left(\dfrac{w}{k+1}\right)L_{n-k}^{\mu+k}((n-k)z)$

$\displaystyle =\sum_{k=0}^{n}\dfrac{(wz)^k}{(k+1)!(\lambda+1)_k}L_{n-k}^{\mu+k}((n+1)z).$

7. $\displaystyle\sum_{k=0}^{n}\dfrac{\left(\dfrac{z}{w}\right)^k}{k+1}L_k^{\lambda-k}((k+1)w)\,L_{n-k}^{\mu+k}((n-k)z)$

$\displaystyle =\sum_{k=0}^{n}\dfrac{(-\lambda)_k}{(k+1)!}\left(-\dfrac{z}{w}\right)^k L_{n-k}^{\mu+k}((n+1)z).$

5.11. The Gegenbauer Polynomials $C_n^{\lambda}(z)$

5.11.1. Sums containing $C_m^{\lambda\pm nk}(z)$

1. $\displaystyle\sum_{k=0}^{n}\dfrac{(-n)_k(n)_k}{k!(k+m)!}C_m^{-k-m}(z)$

$\displaystyle =(-1)^m\dfrac{(-m)_{2n}}{m!(2n)!}(2z)^{m-2n}\,{}_2F_1\!\left(\begin{matrix}n-\dfrac{m}{2},\,n+\dfrac{1-m}{2}\\ 2n+1;\ z^{-2}\end{matrix}\right).$

2. $\displaystyle\sum_{k=0}^{n}(-1)^k\binom{n}{k}\dfrac{(a)_k}{(1-\lambda)_k}C_m^{\lambda-k}(z)$

$\displaystyle =\dfrac{(2z)^m(\lambda)_m(1-\lambda-a-m)_n}{m!(1-\lambda-m)_n}\,{}_3F_2\!\left(\begin{matrix}-\dfrac{m}{2},\,\dfrac{1-m}{2},\,1-\lambda-a-m+n;\ z^{-2}\\ 1-\lambda-m+n,\,1-\lambda-a-m\end{matrix}\right).$

3. $\displaystyle\sum_{k=0}^{n}\dfrac{(\lambda)_k}{(\lambda+m)_k}C_{2m}^{\lambda+k}(z)=(-1)^{m+1}z^{-2}\dfrac{(\lambda)_m}{\left(\dfrac{1}{2}\right)_m}$

$\displaystyle\times\left[P_{m+1}^{(-3/2,\,\lambda+n-1/2)}(1-2z^2)-P_{m+1}^{(-3/2,\,\lambda-3/2)}(1-2z^2)\right].$

4. $\displaystyle\sum_{k=0}^{n}(-1)^k\binom{n}{k}\frac{(\lambda)_k}{(a)_k}C_{2m}^{\lambda+k}(z)$

$$= (-1)^m\frac{(\lambda)_m(a-\lambda-m)_n}{m!\,(a)_n}\,{}_3F_2\left(\begin{array}{c}-m,\,\lambda+m,\,\lambda-a+m+1\\ \lambda-a+m-n+1,\,\frac{1}{2};\,z^2\end{array}\right).$$

5. $\displaystyle\sum_{k=0}^{n}(-1)^k\binom{n}{k}\frac{(\lambda)_k}{(\lambda+m+1)_k}C_{2m+1}^{\lambda+k}(z)$

$$= (-1)^n\frac{2(\lambda)_{m+1}}{\left(\frac{3}{2}\right)_m}z^{2n+1}P_{m-n}^{(\lambda+n-1/2,\,n+1/2)}(2z^2-1)\quad [m\geq n].$$

6. $\displaystyle\sum_{k=0}^{n}\frac{(\lambda)_k}{(\lambda+m+1)_k}C_{2m+1}^{\lambda+k}(z)$

$$= z^{-1}\left[\frac{(\lambda)_{m+1}}{(\lambda+n)_{m+1}}C_{2m+2}^{\lambda+n}(z)-\frac{\lambda+m}{\lambda-1}C_{2m+2}^{\lambda-1}(z)\right].$$

7. $\displaystyle\sum_{k=0}^{n}(-1)^k\binom{n}{k}\frac{(\lambda)_k}{(a)_k}C_{2m+1}^{\lambda+k}(z)$

$$= (-1)^m 2z\frac{(\lambda)_{m+1}(a-\lambda-m-1)_n}{m!\,(a)_n}\,{}_3F_2\left(\begin{array}{c}-m,\,\lambda+m+1,\,\lambda-a+m+2\\ \lambda-a+m-n+2,\,\frac{3}{2};\,z^2\end{array}\right).$$

8. $\displaystyle\sum_{k=0}^{n}(-1)^k\binom{n}{k}\frac{(\lambda)_k}{\left(\lambda+m+\frac{1}{2}\right)_k}C_{2m+1}^{\lambda+k}(z)$

$$= 2^{1-2m}\lambda z\frac{\left(-m-\frac{1}{2}\right)_n(2\lambda+1)_{2m}}{\left(\frac{3}{2}\right)_m\left(\lambda+\frac{1}{2}\right)_{m+n}}P_m^{(\lambda+n-1/2,\,1/2-n)}(2z^2-1).$$

9. $\displaystyle\sum_{k=0}^{n}\binom{n}{k}k^r\frac{\left(\frac{1}{2}-\lambda-m\right)_k}{}(1-\lambda)_k(z^2-1)^{-k}C_{2m+1}^{\lambda-k}(z)$

$$= \frac{2^{2m+1}(\lambda)_{m+1}}{(2m+1)!\,(-\lambda-m)_n}$$

$$\times z(1-z^2)^m\sum_{k=1}^{r}\sigma_r^k(m+n-k)!\,(-n)_k\left(\frac{1}{2}-\lambda-m\right)_k(1-z^2)^{-k}$$

$$\times P_{m+n-k}^{(-\lambda+k-2m-1,\,k-n+1/2)}\left(\frac{z^2+1}{z^2-1}\right)\quad [m+n\geq r].$$

10. $\displaystyle\sum_{k=0}^{n}\binom{n}{k}k^{r}\frac{(\lambda)_{k}}{\left(\lambda+m+\frac{1}{2}\right)_{k}}(z^{2}-1)^{k}C_{2m+1}^{\lambda+k}(z)=\frac{(2\lambda)_{2m+1}z}{(2m+1)!\left(\lambda+\frac{1}{2}\right)_{m+n}}$

$\displaystyle\qquad\times\sum_{k=1}^{r}\sigma_{r}^{k}(m+n-k)!(-n)_{k}(\lambda+m+1)_{k}(1-z^{2})^{k}$

$\displaystyle\qquad\qquad\times P_{m+n-k}^{(\lambda+k-1/2,\,k-n+1/2)}(2z^{2}-1)\qquad [m+n\geq r].$

5.11.2. Sums containing $C_{m\pm pk}^{\lambda}(z)$

1. $\displaystyle\sum_{k=0}^{n}C_{k}^{\lambda}(z)=\frac{(2\lambda)_{2n}}{n!\left(\lambda+\frac{1}{2}\right)_{n}}\left(\frac{z-1}{2}\right)^{n}{}_{3}F_{2}\!\left(\begin{array}{c}-n,\,-n-\lambda,\,\frac{1}{2}-n-\lambda;\\ 1-n-\lambda,\,-2n-2\lambda\end{array}\!\frac{2}{1-z}\right).$

2. $\displaystyle\sum_{k=0}^{n}(-1)^{k}\frac{(2k+\lambda)\left(\frac{1}{2}\right)_{k}}{(n-k)!\left(\lambda+\frac{1}{2}\right)_{k}(\lambda+n+1)_{k}}C_{2k}^{\lambda}(z)=\frac{(\lambda)_{n+1}}{n!\left(\lambda+\frac{1}{2}\right)_{n}}(1-z^{2})^{n}$
$\qquad\qquad\qquad\qquad\qquad\qquad\qquad\qquad\qquad\qquad\qquad\qquad\qquad [n\geq 1].$

3. $\displaystyle\sum_{k=0}^{n}(-1)^{k}\frac{(2k+\lambda)(n)_{k}}{(n-k)!(\lambda-n+1)_{k}(\lambda+n+1)_{k}}C_{2k}^{\lambda}(z)=\frac{(\lambda)_{n+1}}{n!(-\lambda)_{n}}T_{2n}(z).$

4. $\displaystyle\sum_{k=0}^{n}(-1)^{k}\frac{(2k+\lambda+1)\left(\frac{3}{2}\right)_{k}}{(n-k)!\left(\lambda+\frac{1}{2}\right)_{k}(\lambda+n+2)_{k}}C_{2k+1}^{\lambda}(z)$
$\displaystyle\qquad\qquad\qquad\qquad\qquad\qquad\qquad=2\lambda z\frac{(\lambda+1)_{n+1}}{n!\left(\lambda+\frac{1}{2}\right)_{n}}(1-z^{2})^{n}.$

5. $\displaystyle\sum_{k=0}^{n}(-1)^{k}\binom{n}{k}(m-2k)!(m+\lambda-2k)\frac{(-m-\lambda)_{k}(1-m-2\lambda)_{2k}}{(n-m-\lambda+1)_{k}}C_{m-2k}^{\lambda}(z)$
$\displaystyle\qquad=-2^{2n}(m-2n)!(\lambda)_{n}(-m-\lambda)_{n+1}(1-z^{2})^{n}C_{m-2n}^{\lambda+n}(z)\qquad [m\geq 2n].$

6. $\displaystyle\sum_{k=0}^{[n/2]}(-1)^{k}\binom{m}{k}(\lambda+n-2k)\frac{(n-2k)!(-\lambda-n)_{k}(1-2\lambda-n)_{2k}}{(m-n-\lambda+1)_{k}}C_{n-2k}^{\lambda}(z)$
$\displaystyle\qquad=-2^{2m}(n-2m)!(\lambda)_{m}(-\lambda-n)_{m+1}(1-z^{2})^{m}C_{n-2m}^{\lambda+m}(z)\qquad [2m\leq n].$

5.11.3. Sums containing $C_{m\pm pk}^{\lambda\pm nk}(z)$

1. $\displaystyle\sum_{k=0}^{n-1}\frac{2^{-k}(n-k)^{n}}{(n-k)!(1-\lambda)_{k}}(z-1)^{-k}C_{k}^{\lambda-k}(z)$
$\displaystyle\qquad=\frac{\left(\frac{1}{2}-\lambda\right)_{n}}{n!(1-2\lambda)_{n}}\left(\frac{2}{1-z}\right)^{n}\sum_{k=0}^{n}\sigma_{n}^{k}\frac{(-n)_{k}(2\lambda-n)_{k}}{\left(\lambda-n+\frac{1}{2}\right)_{k}}\left(\frac{z-1}{2}\right)^{k}.$

2. $\displaystyle\sum_{k=1}^{n} \frac{k\,\Gamma(2n-k)}{(n-k)!} \frac{(1-2\lambda)_k}{(1-\lambda)_k} \left(\frac{z}{1-z^2}\right)^k C_k^{\lambda-k}(z) = \left(\frac{1}{2}-\lambda\right)_n \left(\frac{4z^2}{z^2-1}\right)^n$

$[n \geq 1]$.

3. $\displaystyle\sum_{k=0}^{n} \frac{(a)_k}{(1-\lambda)_k} 2^{-k}(1-z)^{-k} C_k^{\lambda-k}(z) = \frac{(a+1)_n}{n!}\, {}_3F_2\!\left(\begin{array}{c}-n,\,a,\,\frac{1}{2}-\lambda;\\ a+1,\,1-2\lambda\end{array}\frac{2}{1-z}\right).$

4. $\displaystyle\sum_{k=0}^{n} \frac{k!\,2^{-k}}{(1-\lambda)_k}(1-z)^{-k} C_k^{\lambda-k}(z)$

$= \dfrac{2\lambda(1-z)}{2\lambda+1}\left[1 - \dfrac{(n+1)!}{(-2\lambda)_{n+1}} P_{n+1}^{(-2\lambda-1,\,\lambda-n-3/2)}\!\left(\dfrac{z+3}{z-1}\right)\right].$

5. $\displaystyle\sum_{k=0}^{n} \sigma_{k+m}^{m} \frac{(\lambda)_k}{(k+m)!} t^k C_{n-k}^{\lambda+k}(z)$

$= \dfrac{t^{-m}}{m!(1-\lambda)_m} \displaystyle\sum_{k=0}^{m}(-1)^k \binom{m}{k} C_{m+n}^{\lambda-m}\!\left(\dfrac{kt}{2}+z\right).$

6. $\displaystyle\sum_{k=0}^{n} \binom{n}{k} \frac{(\lambda)_k}{(2\lambda+m)_k} 2^k(z+1)^k C_{m-k}^{\lambda+k}(z)$

$= \left(\dfrac{z+1}{2}\right)^m \dfrac{(2\lambda)_m}{\left(\lambda+\frac{1}{2}\right)_m} P_m^{(\lambda+n-1/2,\,-2\lambda-2m-n)}\!\left(\dfrac{3-z}{1+z}\right).$

7. $\displaystyle\sum_{k=0}^{n} \frac{\sigma_m^{n-k+1}}{(1-\lambda)_k} C_{2k}^{\lambda-k}(z)$

$= (-1)^n \dfrac{(\lambda)_n}{(2n)!}(2z)^{2n}\, {}_{m+1}F_m\!\left(\begin{array}{c}-n,\,\frac{1}{2}-n,\,2,\,\ldots,\,2;\,z^{-2}\\ 1-\lambda-n,\,1,\,\ldots,\,1\end{array}\right).$

8. $\displaystyle\sum_{k=0}^{n} \frac{\left(\frac{1}{2}\right)_k}{(1-\lambda)_k} C_{2k}^{\lambda-k}(z) = -\frac{\left(\frac{3}{2}\right)_n}{2z(1-\lambda)_{n+1}} C_{2n+1}^{\lambda-n-1}(z).$

9. $\displaystyle\sum_{k=0}^{n-1}(-1)^k \frac{(n-k)^n}{(n-k)!(1-\lambda)_k} C_{2k}^{\lambda-k}(z)$

$= \dfrac{(\lambda)_n}{n!\left(\frac{1}{2}\right)_n} z^{2n} \displaystyle\sum_{k=0}^{n}(-1)^k \sigma_n^k \dfrac{(-n)_k\left(\frac{1}{2}-n\right)_k}{(1-\lambda-n)_k} z^{-2k}$ $[n \geq 1]$.

10. $\displaystyle\sum_{k=0}^{n} \frac{t^k}{(n-k)!(1-\lambda)_k} C_{2k}^{\lambda-k}(z) = \frac{(t+1)^n}{(1-\lambda)_n} C_{2n}^{\lambda-n}\!\left(z\sqrt{\frac{t}{t+1}}\right).$

5.11. The Gegenbauer Polynomials $C_n^\lambda(z)$

11. $\displaystyle\sum_{k=0}^{n}(-1)^k\binom{n}{k}\binom{2n}{k}^{-1}\frac{1}{(n-k)!(1-\lambda)_k}C_{2k}^{\lambda-k}(z)$

$$= \frac{n!}{(2n)!}\,{}_3F_2\left(\begin{array}{c}-n,-n,\lambda\\ \frac{1}{2},1;\,z^2\end{array}\right).$$

12. $\displaystyle\sum_{k=0}^{n}(-1)^k\frac{(a)_k}{(n-k)!(b)_k(1-\lambda)_k}C_{2k}^{\lambda-k}(z)$

$$= \frac{(b-a)_n}{n!\,(b)_n}\,{}_3F_2\left(\begin{array}{c}-n,a,\lambda;\,z^2\\ a-b-n+1,\,\frac{1}{2}\end{array}\right).$$

13. $\displaystyle\sum_{k=0}^{n}\frac{\left(\frac{1}{2}\right)_k}{(1-\lambda)_k}(1-z^2)^{-k}C_{2k}^{\lambda-k}(z) = -\frac{\left(\frac{3}{2}\right)_n}{2(-\lambda)_{n+1}z}(1-z^2)^{-n}C_{2n+1}^{\lambda-n}(z).$

14. $\displaystyle\sum_{k=0}^{n}\frac{1}{(n-k)!(1-\lambda)_k}(z^2-1)^{-k}C_{2k}^{\lambda-k}(z) = \frac{\left(\frac{1}{2}-\lambda\right)_n}{n!\left(\frac{1}{2}\right)_n}\left(\frac{z^2}{z^2-1}\right)^n.$

15. $\displaystyle\sum_{k=0}^{n}\frac{\left(\frac{1}{2}-n\right)_k}{(n-k)!(1-\lambda)_k(1-\lambda-n)_k}z^{-2k}C_{2k}^{\lambda-k}(z)$

$$= \frac{\left(\frac{1}{2}\right)_n}{n!(\lambda)_n}z^{-2n}\,{}_4F_3\left(\begin{array}{c}-\frac{n}{2},\frac{1-n}{2},\lambda,\frac{1}{2}-\lambda\\ \frac{1}{4},\frac{1}{2},\frac{3}{4};\,z^4\end{array}\right).$$

16. $\displaystyle\sum_{k=0}^{n}(-1)^k\frac{\left(\frac{1}{2}\right)_k}{(n-k)!(a)_k(1-\lambda)_k}C_{2k}^{\lambda-k}(z)$

$$= \frac{1}{(a)_n}P_n^{(\lambda+a-3/2,\,-a-n+1/2)}(2z^2-1).$$

17. $\displaystyle\sum_{k=0}^{n}\binom{n}{k}\frac{2^{2k}\left(\frac{1}{2}-\lambda\right)_k\left(\frac{1}{2}\right)_k}{(1-2\lambda)_{2k}(n-k+1)}(z^2-1)^{-k}C_{2k}^{\lambda-k}(z) = (-1)^n\frac{(z^2-1)^{-n-1}}{n+1}$

$$\times\left[\frac{\left(\frac{3}{2}\right)_n}{2(1-\lambda)_{n+1}}C_{2n+2}^{\lambda-n-1}(z)+(-1)^n P_{n+1}^{(\lambda-n-3/2,\,-n-3/2)}(2z^2-1)\right].$$

18. $\displaystyle\sum_{k=0}^{n}(-1)^k\frac{(1-a-n)_k}{(n-k)!(1-\lambda)_{2k}}z^{-2k}C_{2k}^{\lambda-2k}(z)$

$$= \frac{(a)_n}{n!(1-\lambda)_n}z^{-2n}\,{}_4F_3\left(\begin{array}{c}-n,\lambda-n,\frac{2a-1}{4},\frac{2a+1}{4}\\ a,a-\frac{1}{2},\frac{1}{2};\,4z^2\end{array}\right).$$

19. $\sum_{k=0}^{n} \dfrac{4k-\lambda+1}{(4k-2\lambda-1)(4k-2\lambda+3)} \dfrac{\left(\frac{1}{2}-\lambda\right)_k \left(\frac{1}{2}\right)_k}{(n-k)!\left(n-\lambda+\frac{3}{2}\right)_k (1-\lambda)_{2k}}$

$\times (1-z^2)^{-k} C_{2k}^{\lambda-2k}(z) = -\dfrac{2(1/2-\lambda)_{n+1}}{n!(4n-2\lambda-1)(4n-2\lambda+3)(1-\lambda)_n} (1-z^2)^{-n}$

$$\times {}_3F_2\left(\begin{matrix}-n, \lambda-n, 1; 1-z^2 \\ \dfrac{2\lambda+1}{4}-n, \dfrac{2\lambda+5}{4}-n\end{matrix}\right).$$

20. $\sum_{k=0}^{n}(-1)^k \binom{n}{k} \dfrac{(\lambda)_k}{\left(\frac{1}{2}-m\right)_k} C_{2m-2k}^{\lambda+k}(z)$

$$= \dfrac{(\lambda)_m}{\left(\frac{1}{2}\right)_m} P_m^{(n+\lambda-1/2,\,-n-1/2)}(2z^2-1) \quad [m \geq n].$$

21. $\sum_{k=0}^{n} \binom{n}{k} k^r \dfrac{(\lambda)_k}{\left(\frac{1}{2}-m\right)_k} (z^2-1)^k C_{2m-2k}^{\lambda+k}(z)$

$$= \dfrac{(\lambda)_m}{\left(\frac{1}{2}\right)_m} \sum_{k=1}^{r} \sigma_r^k (-n)_k (z^2-1)^k P_{m-k}^{(\lambda+k-1/2,\,k-n-1/2)}(2z^2-1)$$

$$[m \geq n;\ m \geq r].$$

22. $\sum_{k=1}^{n} \binom{2n}{n-k} \dfrac{\left(\frac{1}{2}\right)_{2k}}{(k+m)!\left(\frac{1}{2}-m\right)_k} (z^2-1)^k C_{2m-2k}^{2k+1/2}(z)$

$$= \dfrac{1}{2}\binom{2n}{n}\left[\dfrac{n!}{(m+n)!} P_m^{(n,\,-n-1/2)}(2z^2-1) - \dfrac{1}{m!} P_{2m}(z)\right] \quad [m \geq n].$$

23. $\sum_{k=0}^{n} \sigma_{k+m}^{m} \dfrac{(\lambda)_k}{(k+m)!} t^k C_{2n-2k}^{\lambda+k}(z)$

$$= \dfrac{t^{-m}}{m!(1-\lambda)_m} \sum_{k=0}^{m} (-1)^k \binom{m}{k}(1-kt)^{m+n} C_{2m+2n}^{\lambda-m}\left(\dfrac{z}{\sqrt{1-kt}}\right).$$

24. $\sum_{k=0}^{m} \binom{m+n-2k}{m-k} \dfrac{2^{2k}(\lambda)_k^2}{k!(2\lambda)_k}(1-z^2)^k C_{m+n-2k}^{\lambda+k}(z)$

$$= \dfrac{(m+2\lambda)_n}{(2\lambda)_n} C_m^\lambda(z) C_n^\lambda(z) \quad [n \geq m].$$

25. $\sum_{k=0}^{n}(-1)^k \binom{n}{k} \dfrac{(a)_k}{(1-\lambda)_k} C_{2k+2m}^{\lambda-k}(z)$

$$= (-1)^m \dfrac{(m-a+1)_n (\lambda)_m}{(m+n)!} {}_3F_2\left(\begin{matrix}-m-n,\, a-m,\, \lambda+m \\ a-m-n,\, \frac{1}{2};\, z^2\end{matrix}\right).$$

5.11. The Gegenbauer Polynomials $C_n^\lambda(z)$

26. $\displaystyle\sum_{k=0}^n (-1)^k \binom{n}{k} \frac{\left(m+\frac{1}{2}\right)_k}{(1-\lambda)_k} C_{2k+2m}^{\lambda-k}(z)$

$$= (-1)^{m+n} \frac{(\lambda)_m}{\left(\frac{1}{2}\right)_m} P_{m+n}^{(-n-1/2,\,\lambda-1/2)}(1-2z^2).$$

27. $\displaystyle\sum_{k=0}^n (-1)^k \binom{n}{k} \frac{(\lambda+2m+n)_k}{(1-\lambda)_k} C_{2k+2m}^{\lambda-k}(z) = C_{2m+2n}^\lambda(z).$

28. $\displaystyle\sum_{k=0}^n \binom{n}{k} \frac{(a)_k}{(1-\lambda)_k} (z^2-1)^{-k} C_{2k+2m}^{\lambda-k}(z)$

$$= \frac{(1-a+m)_n (\lambda)_m}{(m+n)!} (z^2-1)^m \,{}_3F_2\!\left(\begin{array}{c} -m-n,\, a-m,\, \frac{1}{2}-\lambda-m \\ a-m-n,\, \frac{1}{2};\, \frac{z^2}{z^2-1} \end{array}\right).$$

29. $\displaystyle\sum_{k=0}^n \frac{\left(\frac{3}{2}\right)_k}{(1-\lambda)_k} C_{2k+1}^{\lambda-k}(z) = 2\lambda z P_n^{(3/2,\,\lambda-n-3/2)}(1-2z^2).$

30. $\displaystyle\sum_{k=0}^n \frac{\sigma_m^{n-k+1}}{(1-\lambda)_k} C_{2k+1}^{\lambda-k}(z)$

$$= (-1)^n \frac{(\lambda)_{n+1}}{(2n+1)!} (2z)^{2n+1} \,{}_{m+1}F_m\!\left(\begin{array}{c} -n,\, -\frac{1}{2}-n,\, 2,\, \ldots,\, 2 \\ -\lambda-n,\, 1,\, \ldots,\, 1;\, z^{-2} \end{array}\right).$$

31. $\displaystyle\sum_{k=0}^n \frac{t^k}{(n-k)!(1-\lambda)_k} C_{2k+1}^{\lambda-k}(z) = \frac{t^{-1/2}(t+1)^{n+1/2}}{(1-\lambda)_n} C_{2n+1}^{\lambda-n}\!\left(z\sqrt{\frac{t}{t+1}}\right).$

32. $\displaystyle\sum_{k=0}^n \frac{1}{(n-k)!(1-\lambda)_k} (z^2-1)^{-k} C_{2k+1}^{\lambda-k}(z) = \frac{2\lambda z \left(\frac{1}{2}-\lambda\right)_n}{n!\left(\frac{3}{2}\right)_n} \left(\frac{z^2}{z^2-1}\right)^n.$

33. $\displaystyle\sum_{k=0}^n \frac{\left(\frac{3}{2}\right)_k}{(1-\lambda)_k} (1-z^2)^{-k} C_{2k+1}^{\lambda-k}(z)$

$$= (-1)^n 2\lambda z (1-z^2)^{-n} P_n^{(\lambda-n-1/2,\,3/2)}(2z^2-1).$$

34. $\displaystyle\sum_{k=0}^{n-1} (-1)^k \frac{(n-k)^n}{(n-k)!(1-\lambda)_k} C_{2k+1}^{\lambda-k}(z)$

$$= \frac{2(\lambda)_{n+1}}{n!\left(\frac{3}{2}\right)_n} z^{2n+1} \sum_{k=0}^n (-1)^k \sigma_n^k \frac{(-n)_k \left(-\frac{1}{2}-n\right)_k}{(-\lambda-n)_k} z^{-2k} \quad [n \geq 1].$$

35. $\displaystyle\sum_{k=0}^{n}(-1)^k\binom{n}{k}\binom{2n}{k}^{-1}\frac{1}{(n-k)!(1-\lambda)_k}C_{2k+1}^{\lambda-k}(z)$

$$= 2\lambda z\frac{n!}{(2n)!}\,_3F_2\left(\begin{matrix}-n,-n,\lambda+1\\1,\frac{3}{2};\ z^2\end{matrix}\right).$$

36. $\displaystyle\sum_{k=0}^{n}\binom{n}{k}\frac{\left(\frac{3}{2}\right)_k}{(1-\lambda)_k(n-k+1)}(z^2-1)^{-k}C_{2k+1}^{\lambda-k}(z)=\frac{2\lambda}{n+1}(z^2-1)^{-n-1}$

$$\times\left[\frac{3\left(\frac{5}{2}\right)_n}{4(-\lambda)_{n+2}}C_{2n+3}^{\lambda-n-1}(z)-(-1)^n z P_{n+1}^{(\lambda-n-3/2,-n-1/2)}(2z^2-1)\right].$$

37. $\displaystyle\sum_{k=0}^{n}(-1)^k\frac{(a)_k}{(n-k)!(b)_k(1-\lambda)_k}C_{2k+1}^{\lambda-k}(z)$

$$= 2\lambda z\frac{(b-a)_n}{n!(b)_n}\,_3F_2\left(\begin{matrix}-n,a,\lambda+1;\ z^2\\a-b-n+1,\frac{3}{2}\end{matrix}\right).$$

38. $\displaystyle\sum_{k=0}^{n}(-1)^k\frac{\left(\frac{3}{2}\right)_k}{(n-k)!(a)_k(1-\lambda)_k}C_{2k+1}^{\lambda-k}(z)$

$$= \frac{2\lambda z}{(a)_n}P_n^{(\lambda+a-3/2,-a-n+3/2)}(2z^2-1).$$

39. $\displaystyle\sum_{k=0}^{n}\frac{k!}{(1-\lambda)_k}C_{2k+1}^{\lambda-k}(z) = z^{-1}\left[1-\frac{(n+1)!}{(1-\lambda)_{n+1}}C_{2n+2}^{\lambda-n-1}(z)\right].$

40. $\displaystyle\sum_{k=0}^{n}\frac{\left(-\frac{1}{2}-n\right)_k}{(n-k)!(1-\lambda)_k(-\lambda-n)_k}z^{-2k}C_{2k+1}^{\lambda-k}(z)$

$$= 2\lambda\frac{\left(\frac{3}{2}\right)_n}{n!(\lambda+1)_n}z^{1-2n}\,_4F_3\left(\begin{matrix}-\frac{n}{2},\frac{1-n}{2},\lambda+1,\frac{1}{2}-\lambda\\\frac{3}{4},\frac{5}{4},\frac{3}{2};\ z^4\end{matrix}\right).$$

41. $\displaystyle\sum_{k=0}^{n}\frac{4k-\lambda+1}{(4k-2\lambda-1)(4k-2\lambda+3)}\frac{\left(\frac{1}{2}-\lambda\right)_k\left(\frac{3}{2}\right)_k}{(n-k)!\left(n-\lambda+\frac{3}{2}\right)_k(-\lambda)_{2k+1}}$

$$\times(1-z^2)^{-k}C_{2k+1}^{\lambda-2k}(z) = -\frac{4z\left(\frac{1}{2}-\lambda\right)_{n+1}}{n!(4n-2\lambda-1)(4n-2\lambda+3)(-\lambda)_n}(1-z^2)^{-n}$$

$$\times\,_3F_2\left(\begin{matrix}-n,1,\lambda-n+1;\ 1-z^2\\\frac{2\lambda+1}{4}-n,\frac{2\lambda+5}{4}-n\end{matrix}\right).$$

42. $\sum_{k=0}^{n}(-1)^k\binom{n}{k}\frac{(a)_k}{(1-\lambda)_k}C_{2k+2m+1}^{\lambda-k}(z)$

$$= 2(-1)^m \frac{(m-a+1)_n (\lambda)_{m+1}}{(m+n)!} z\,_3F_2\left(\begin{matrix}-m-n,\, a-m,\, \lambda+m+1\\ a-m-n,\, \tfrac{3}{2};\, z^2\end{matrix}\right).$$

43. $\sum_{k=0}^{n}(-1)^k\binom{n}{k}\frac{(\lambda+2m+n+1)_k}{(1-\lambda)_k}C_{2k+2m+1}^{\lambda-k}(z) = C_{2m+2n+1}^{\lambda}(z).$

44. $\sum_{k=0}^{n}(-1)^k\binom{n}{k}\frac{\left(m+\tfrac{3}{2}\right)_k}{(1-\lambda)_k}C_{2k+2m+1}^{\lambda-k}(z)$

$$= 2(-1)^{m+n}\frac{(\lambda)_{m+1}}{\left(\tfrac{3}{2}\right)_m} z P_{m+n}^{(1/2-n,\,\lambda-1/2)}(1-2z^2).$$

45. $\sum_{k=0}^{n}\binom{n}{k}\frac{(2m+2k)!}{(1-\lambda)_k(\lambda+2m-n+1)_k} 2^{-2k}(z^2-1)^{-k} C_{2k+2m+1}^{\lambda-k}(z)$

$$= \frac{(2m)!(\lambda)_m(1/2-\lambda-m)_n}{(\lambda-n)_m(-\lambda-2m)_n}(1-z^2)^{-n} C_{2m}^{\lambda-n}(z).$$

46. $\sum_{k=0}^{n}\binom{n}{k}\frac{(2m+2k+1)!}{(1-\lambda)_k(\lambda+2m-n+2)_k} 2^{-2k}(z^2-1)^{-k} C_{2k+2m+1}^{\lambda-k}(z)$

$$= \frac{(2m+1)!(\lambda)_{m+1}\left(\tfrac{1}{2}-\lambda-m\right)_n}{(\lambda-n)_{m+1}(-\lambda-2m-1)_n}(1-z^2)^{-n} C_{2m+1}^{\lambda-n}(z).$$

47. $\sum_{k=0}^{n}\binom{n}{k}\frac{(a)_k}{(1-\lambda)_k}(z^2-1)^{-k} C_{2k+2m+1}^{\lambda-k}(z)$

$$= \frac{(1-a+m)_n(\lambda)_{m+1}}{(m+n)!} 2z(z^2-1)^m\,_3F_2\left(\begin{matrix}-m-n,\, a-m,\, \tfrac{1}{2}-\lambda-m\\ a-m-n,\, \tfrac{3}{2};\, \tfrac{z^2}{z^2-1}\end{matrix}\right).$$

48. $\sum_{k=0}^{n}\sigma_{k+m}^{m}\frac{(\lambda)_k}{(k+m)!}t^k C_{2n-2k+1}^{\lambda+k}(z)$

$$= \frac{t^{-m}}{m!(1-\lambda)_m}\sum_{k=0}^{m}(-1)^k\binom{m}{k}(1-kt)^{m+n+1/2} C_{2m+2n+1}^{\lambda-m}\left(\frac{z}{\sqrt{1-kt}}\right).$$

49. $\sum_{k=0}^{n}(-1)^k\binom{n}{k}\frac{(\lambda)_k}{\left(-m-\tfrac{1}{2}\right)_k} C_{2m-2k+1}^{\lambda+k}(z)$

$$= 2z\frac{(\lambda)_{m+1}}{\left(\tfrac{3}{2}\right)_m} P_m^{(n+\lambda-1/2,\,1/2-n)}(2z^2-1) \quad [m\geq n].$$

50. $\displaystyle\sum_{k=0}^{n} \binom{n}{k} k^r \frac{(\lambda)_k}{\left(-m-\frac{1}{2}\right)_k} (z^2-1)^k C_{2m-2k+1}^{\lambda+k}(z)$

$$= 2z \frac{(\lambda)_{m+1}}{\left(\frac{3}{2}\right)_m} \sum_{k=1}^{r} \sigma_r^k (-n)_k (z^2-1)^k P_{m-k}^{(\lambda+k-1/2,\,k-n+1/2)}(2z^2-1)$$

$[m \geq n;\ m \geq r]$.

51. $\displaystyle\sum_{k=1}^{n} \binom{2n}{n-k} \frac{\left(\frac{1}{2}\right)_{2k}}{(k+m)!\left(-m-\frac{1}{2}\right)_k} (z^2-1)^k C_{2m-2k+1}^{2k+1/2}(z)$

$$= \frac{1}{2}\binom{2n}{n}\left[\frac{n!\,z}{(m+n)!} P_m^{(n,\,1/2-n)}(2z^2-1) - \frac{1}{m!} P_{2m+1}(z)\right] \quad [m \geq n].$$

52. $\displaystyle\sum_{k=0}^{[n/2]} (n-2k+\lambda) \frac{(a-\lambda)_k (-n-\lambda)_k}{k!(1-a-n)_k} C_{n-2k}^{\lambda}(z) = \frac{(\lambda)_{n+1}}{(a)_n} C_n^a(z).$

53. $\displaystyle\sum_{k=0}^{[n/3]} \frac{\lambda+n-3k}{k!} \left(-\frac{2\lambda+2n}{3}\right)_k C_{n-3k}^{\lambda}(z) = 3^n \lambda P_n^{\left(\frac{2\lambda-n}{3},\,\frac{\lambda-2n}{3}\right)}\left(\frac{4z-1}{3}\right).$

54. $\displaystyle\sum_{k=0}^{[n/2]} (-1)^k \frac{(n-k)!}{k!} (\lambda)_{2k} \left(\frac{1-z^2}{2}\right)^{2k} C_{2n-4k}^{\lambda+2k}(z)$

$$= \frac{(2\lambda)_{2n}}{\left(\frac{1}{2}\right)_n} \left(\frac{z}{2}\right)^{2n} {}_3F_2\!\left(\begin{array}{c} -n,\,-n-\frac{1}{2},\,-n+\frac{1}{2} \\ -2n-1,\,\lambda+\frac{1}{2};\,2-2z^{-2} \end{array}\right).$$

55. $\displaystyle\sum_{k=0}^{[n/2]} (-1)^k \frac{(n-k)!}{k!} (\lambda)_{2k} \left(\frac{1-z^2}{2}\right)^{2k} C_{2n-4k+1}^{\lambda+2k}(z)$

$$= \frac{2^{-2n}(2\lambda)_{2n+1}}{\left(\frac{3}{2}\right)_n} z^{2n+1}\, {}_3F_2\!\left(\begin{array}{c} -n,\,-n-\frac{1}{2},\,-n-\frac{1}{2} \\ -2n-1,\,\lambda+\frac{1}{2};\,2-2z^{-2} \end{array}\right).$$

56. $\displaystyle\sum_{k=0}^{[n/2]} \frac{(8k+2\lambda-1)(a)_k \left(\frac{2\lambda-1}{4}\right)_k (\lambda)_{4k}}{k!\left(\frac{2\lambda+3}{4}-a\right)_k \left(\lambda+n+\frac{1}{2}\right)_{2k} \left(\frac{1}{2}-n\right)_{2k}} (1-z^2)^{2k} C_{2n-4k}^{\lambda+4k}(z)$

$$= \frac{(2\lambda-1)_{2n+1}}{(2n)!}\, {}_3F_2\!\left(\begin{array}{c} -n,\,n+\lambda,\,\frac{2\lambda+1}{4}-a \\ \frac{2\lambda+1}{4},\,\lambda-2a+\frac{1}{2};\,1-z^2 \end{array}\right).$$

57. $\displaystyle\sum_{k=0}^{[n/2]} \frac{(8k+2\lambda-1)(a)_k \left(\frac{2\lambda-1}{4}\right)_k (\lambda)_{4k}}{k!\left(\frac{2\lambda+3}{4}-a\right)_k \left(\lambda+n+\frac{1}{2}\right)_{2k} \left(-\frac{1}{2}-n\right)_{2k}} (1-z^2)^{2k} C_{2n-4k+1}^{\lambda+4k}(z)$

$$= \frac{(2\lambda-1)_{2n+2}\,z}{(2n+1)!}\, {}_3F_2\!\left(\begin{array}{c} -n,\,n+\lambda+1,\,\frac{2\lambda+1}{4}-a \\ \frac{2\lambda+1}{4},\,\lambda-2a+\frac{1}{2};\,1-z^2 \end{array}\right).$$

5.11.4]   5.11. The Gegenbauer Polynomials $C_n^\lambda(z)$

58. $\displaystyle\sum_{k=0}^{[n/3]} (-1)^k \frac{(2\lambda+12k-1)\left(\frac{2\lambda-1}{6}\right)_k (\lambda)_{6k}}{k!\left(\lambda+n+\frac{1}{2}\right)_{3k}\left(\frac{1}{2}-n\right)_{3k}} (1-z^2)^{3k} C_{2n-6k}^{\lambda+6k}(z)$

$$= \frac{(2\lambda-1)_{2n+1}}{(2n)!} \, {}_3F_2\!\left(\begin{array}{c}-n,\,\frac{2\lambda-1}{6},\,\lambda+n\\ \frac{2\lambda-1}{4},\,\frac{2\lambda+1}{4};\,\frac{3-3z^2}{4}\end{array}\right).$$

59. $\displaystyle\sum_{k=0}^{[n/3]} (-1)^k \frac{(2\lambda+12k-1)\left(\frac{2\lambda-1}{6}\right)_k (\lambda)_{6k}}{k!\left(\lambda+n+\frac{1}{2}\right)_{3k}\left(-\frac{1}{2}-n\right)_{3k}} (1-z^2)^{3k} C_{2n-6k+1}^{\lambda+6k}(z)$

$$= \frac{(2\lambda-1)_{2n+2}}{(2n+1)!} z \, {}_3F_2\!\left(\begin{array}{c}-n,\,\frac{2\lambda-1}{6},\,\lambda+n+1\\ \frac{2\lambda-1}{4},\,\frac{2\lambda+1}{4};\,\frac{3-3z^2}{4}\end{array}\right).$$

5.11.4. Sums containing $C_{m\pm pk}^{\lambda\pm nk}(z)$ and special functions

1. $\displaystyle\sum_{k=0}^{n} \frac{(2^{2n-2k}-1)}{(2n-2k)!(1-\lambda)_{2k}} (4z)^{-2k} B_{2n-2k} C_{2k}^{\lambda-2k}(z)$

$$= -\frac{(4z)^{1-2n}}{2(1-\lambda)_{2n-1}} C_{2n-1}^{\lambda-2n+1}(z) \quad [n\geq 1].$$

2. $\displaystyle\sum_{k=0}^{n} \frac{(4z)^{-2k}}{(2n-2k)!(-\lambda)_{2k+1}} B_{2n-2k} C_{2k+1}^{\lambda-2k}(z) = -\frac{(4z)^{1-2n}}{2(-\lambda)_{2n}} C_{2n}^{\lambda-2n+1}(z).$

3. $\displaystyle\sum_{k=0}^{n} \frac{(a)_k}{(1-\lambda)_k} \psi(k+a)\, C_{2k}^{\lambda-k}(z) = \frac{(a+1)_n}{n!}$

$$\times \left\{\left[\psi(a+n+1)-\frac{1}{a}\right] {}_3F_2\!\left(\begin{array}{c}-n,\,a,\,\lambda;\,z^2\\ a+1,\,\frac{1}{2}\end{array}\right)\right.$$

$$\left. - \frac{2n\lambda z^2}{(a+1)^2} {}_4F_3\!\left(\begin{array}{c}1-n,\,a+1,\,a+1,\,\lambda+1\\ a+2,\,a+2,\,\frac{3}{2};\,z^2\end{array}\right)\right\}.$$

4. $\displaystyle\sum_{k=0}^{n} \frac{(a)_k}{(1-\lambda)_k} \psi(k+a)\, C_{2k+1}^{\lambda-k}(z) = \frac{2\lambda(a+1)_n z}{n!}$

$$\times \left\{\left[\psi(a+n+1)-\frac{1}{a}\right] {}_3F_2\!\left(\begin{array}{c}-n,\,a,\,\lambda+1;\,z^2\\ a+1,\,3/2\end{array}\right)\right.$$

$$\left. - \frac{2n(\lambda+1)z^2}{3(a+1)^2} {}_4F_3\!\left(\begin{array}{c}1-n,\,a+1,\,a+1,\,\lambda+2\\ a+2,\,a+2,\,\frac{5}{2};\,z^2\end{array}\right)\right\}.$$

5. $\sum_{k=0}^{n} \dfrac{2^{-k}(z-1)^{-k}}{(n-k+1)!(1-\lambda)_k} \psi(a-k) C_k^{\lambda-k}(z) = \left(\dfrac{2}{1-z}\right)^{n+1}$

$\times \psi(a-n-1) \left[\dfrac{\left(\frac{1}{2}-\lambda\right)_{n+1}}{(n+1)!(1-2\lambda)_{n+1}} - \dfrac{\left(-\frac{1}{4}\right)^{n+1}}{(1-\lambda)_{n+1}} C_{n+1}^{\lambda-n-1}(z) \right]$

$+ \dfrac{\left(\frac{1}{2}-\lambda\right)_n}{n!(1-2\lambda)_n (a-n-1)} \left(\dfrac{2}{1-z}\right)^n {}_4F_3\!\left(\begin{matrix}-n, 2\lambda-n, 1, 1; \\ \lambda-n+\frac{1}{2}, a-n, 2\end{matrix}\,\dfrac{1-z}{2}\right).$

6. $\sum_{k=0}^{n} \dfrac{(1-a)_k}{(1-\lambda)_k} \dfrac{2^{-k}}{(z-1)^k} \psi(a-k) C_k^{\lambda-k}(z)$

$= \dfrac{\left(\frac{1}{2}-\lambda\right)_n (1-a)_n}{n!(1-2\lambda)_n (a-n-1)} \left(\dfrac{2}{z-1}\right)^n$

$\times \left[(a-n-1)\psi(a-n-1) {}_3F_2\!\left(\begin{matrix}-n, 2\lambda-n, a-n-1, 1 \\ \lambda-n+\frac{1}{2}, a-n;\end{matrix}\,\dfrac{1-z}{2}\right) \right.$

$\left. + {}_4F_3\!\left(\begin{matrix}-n, 2\lambda-n, a-n-1, a-n-1 \\ \lambda-n+\frac{1}{2}, a-n, a-n;\end{matrix}\,\dfrac{1-z}{2}\right) \right].$

7. $\sum_{k=0}^{n} \dfrac{(-n)_k}{(1-\lambda)_k} \left(w-\sqrt{w^2-1}\right)^k P_{n-k}(w) C_{2k}^{\lambda-k}(z)$

$= 2^n \dfrac{\left(\frac{1}{2}\right)_n}{n!} (w^2-1)^{n/2} {}_3F_2\!\left(\begin{matrix}-n, -n, \lambda \\ \frac{1}{2}-n, \frac{1}{2};\end{matrix}\,\dfrac{z^2}{2}-\dfrac{z^2}{2}w(w^2-1)^{-1/2}\right).$

8. $\sum_{k=0}^{n} \dfrac{(-n)_k}{(1-\lambda)_k} \left(w-\sqrt{w^2-1}\right)^k P_{n-k}(w) C_{2k+1}^{\lambda-k}(z)$

$= 2^{n+1} \dfrac{\left(\frac{1}{2}\right)_n}{n!} \lambda z (w^2-1)^{n/2} {}_3F_2\!\left(\begin{matrix}-n, -n, \lambda+1 \\ \frac{1}{2}-n, \frac{3}{2};\end{matrix}\,\dfrac{z^2}{2}-\dfrac{z^2}{2}w(w^2-1)^{-1/2}\right).$

9. $\sum_{k=0}^{n} \dfrac{(2k+1)\left(\frac{3}{2}\right)_{2k}}{(k+n+1)!\left(\frac{1}{2}-n\right)_k} (z^2-1)^k [P_k(w)]^2 C_{2n-2k}^{2k+3/2}(z)$

$= \dfrac{2n+1}{n!} {}_3F_2\!\left(\begin{matrix}-n, n+\frac{3}{2}, \frac{1}{2} \\ 1, 1;\end{matrix}\,(1-w^2)(1-z^2)\right).$

10. $\sum_{k=0}^{n} \dfrac{(2k+1)\left(\frac{3}{2}\right)_{2k}}{(k+n+1)!\left(-\frac{1}{2}-n\right)_k} (z^2-1)^k [P_k(w)]^2 C_{2n-2k+1}^{2k+3/2}(z)$

$= \dfrac{(2n+3)z}{n!} {}_3F_2\!\left(\begin{matrix}-n, n+\frac{5}{2}, \frac{1}{2} \\ 1, 1;\end{matrix}\,(1-w^2)(1-z^2)\right).$

5.11.4] 5.11. The Gegenbauer Polynomials $C_n^\lambda(z)$

11. $\displaystyle\sum_{k=0}^{n} \frac{(4k+1)(2k)!}{\left(n+\frac{3}{2}\right)_k \left(\frac{1}{2}-n\right)_k} (1-z^2)^k P_{2k}\left(\sqrt{\frac{1-w}{2}}\right) P_{2k}\left(\sqrt{\frac{1+w}{2}}\right)$

$\times C_{2n-2k}^{2k+1}(z) = (2n+1)\,{}_4F_3\left(\begin{array}{c} -n,\,n+1,\,\frac{1}{4},\,\frac{3}{4} \\ \frac{1}{2},\,\frac{1}{2},\,1;\,(1-w^2)(1-z^2) \end{array}\right).$

12. $\displaystyle\sum_{k=0}^{n} \frac{(4k+3)(2k+1)!}{\left(n+\frac{5}{2}\right)_k \left(\frac{1}{2}-n\right)_k} (1-z^2)^k P_{2k+1}\left(\sqrt{\frac{1-w}{2}}\right) P_{2k+1}\left(\sqrt{\frac{1+w}{2}}\right)$

$\times C_{2n-2k}^{2k+2}(z) = \dfrac{(2n+3)!}{4(2n)!}\sqrt{1-w^2}\,{}_4F_3\left(\begin{array}{c} -n,\,n+2,\,\frac{3}{4},\,\frac{5}{4} \\ 1,\,\frac{3}{2},\,\frac{3}{2};\,(1-w^2)(1-z^2) \end{array}\right).$

13. $\displaystyle\sum_{k=0}^{n} \frac{(4k+3)(2k+1)!}{\left(n+\frac{5}{2}\right)_k \left(-\frac{1}{2}-n\right)_k} (1-z^2)^k P_{2k+1}\left(\sqrt{\frac{1-w}{2}}\right) P_{2k+1}\left(\sqrt{\frac{1+w}{2}}\right)$

$\times C_{2n-2k+1}^{2k+2}(z) = (n+1)(n+2)(2n+3)\sqrt{1-w^2}\,z$

$\times {}_4F_3\left(\begin{array}{c} -n,\,n+3,\,\frac{3}{4},\,\frac{5}{4} \\ 1,\,\frac{3}{2},\,\frac{3}{2};\,(1-w^2)(1-z^2) \end{array}\right).$

14. $\displaystyle\sum_{k=0}^{n} \frac{\left(\frac{3}{2}\right)_{2k}}{(k+n+1)!\left(\frac{1}{2}-n\right)_k} (1-z^2)^k T_{2k+1}\left(\sqrt{\frac{1-w}{2}}\right) T_{2k+1}\left(\sqrt{\frac{1+w}{2}}\right)$

$\times C_{2n-2k}^{2k+3/2}(z) = \dfrac{(-1)^n}{2\left(\frac{1}{2}\right)_n \sqrt{1-z^2}} P_{2n+1}\left(\sqrt{(1-w^2)(1-z^2)}\right).$

15. $\displaystyle\sum_{k=0}^{n} \frac{\left(\frac{3}{2}\right)_{2k}}{(k+n+1)!\left(-\frac{1}{2}-n\right)_k} (1-z^2)^k$

$\times T_{2k+1}\left(\sqrt{\frac{1-w}{2}}\right) T_{2k+1}\left(\sqrt{\frac{1+w}{2}}\right) C_{2n-2k+1}^{2k+3/2}(z) = \dfrac{(2n+3)\sqrt{1-w^2}\,z}{2n!}$

$\times {}_2F_1\left(\begin{array}{c} -n,\,n+\frac{5}{2} \\ \frac{3}{2};\,(1-w^2)(1-z^2) \end{array}\right).$

16. $\displaystyle\sum_{k=0}^{n} \frac{\left(-n+\frac{1}{2}\right)_k}{(n-k)!(1-\lambda)_k} \left(\frac{2}{z-1}\right)^k H_{2n-2k}(w)$

$\times C_k^{\lambda-k}(z) = 2^{3n} \dfrac{\left(\frac{1}{2}\right)_n \left(\frac{1}{2}-\lambda\right)_n}{n!\,(1-2\lambda)_n} (z-1)^{-n}\,{}_2F_2\left(\begin{array}{c} -n,\,2\lambda-n;\,\frac{w^2-w^2 z}{2} \\ \lambda-n+\frac{1}{2},\,\frac{1}{2} \end{array}\right).$

17. $\displaystyle\sum_{k=0}^{n} \frac{\left(-n-\frac{1}{2}\right)_k}{(n-k)!\,(1-\lambda)_k} \left(\frac{2}{z-1}\right)^k H_{2n-2k+1}(w)\, C_k^{\lambda-k}(z)$

$$= 2^{3n+1} \frac{\left(\frac{3}{2}\right)_n \left(\frac{1}{2}-\lambda\right)_n}{n!\,(1-2\lambda)_n}\, w\,(z-1)^{-n}\,{}_2F_2\!\left(\begin{array}{c}-n,\,2\lambda-n;\ \frac{w^2-w^2z}{2}\\ \lambda-n+\frac{1}{2},\,\frac{3}{2}\end{array}\right).$$

18. $\displaystyle\sum_{k=0}^{n} \frac{\left(\frac{1}{2}-n\right)_k}{(n-k)!\,(1-\lambda)_k} \left(\frac{2}{z}\right)^k H_{2n-2k}(w) C_k^{\lambda-k}(z)$

$$= \frac{(2w)^{2n}}{n!}\,{}_4F_1\!\left(\begin{array}{c}-\frac{n}{2},\,\frac{1-2n}{4},\,\frac{1-n}{2},\,\frac{3-2n}{4}\\ 1-\lambda;\ 4w^{-4}z^{-2}\end{array}\right).$$

19. $\displaystyle\sum_{k=0}^{n} \frac{\left(-\frac{1}{2}-n\right)_k}{(n-k)!\,(1-\lambda)_k} \left(\frac{2}{z}\right)^k H_{2n-2k+1}(w) C_k^{\lambda-k}(z)$

$$= \frac{(2w)^{2n+1}}{n!}\,{}_4F_1\!\left(\begin{array}{c}-\frac{n}{2},\,-\frac{1+2n}{4},\,\frac{1-n}{2},\,\frac{1-2n}{4}\\ 1-\lambda;\ 4w^{-4}z^{-2}\end{array}\right).$$

20. $\displaystyle\sum_{k=0}^{n} \frac{(1-z^2)^{-k}}{(2n-2k)!\,(1-\lambda)_k} H_{2n-2k}(w) C_{2k}^{\lambda-k}(z)$

$$= \frac{\left(\frac{1}{2}-\lambda\right)_n}{n!\,\left(\frac{1}{2}\right)_n}\, z^{2n}(1-z^2)^{-n}\,{}_2F_2\!\left(\begin{array}{c}-n,\,\frac{1}{2}-n;\ w^2(1-z^{-2})\\ \lambda-n+\frac{1}{2},\,\frac{1}{2}\end{array}\right).$$

21. $\displaystyle\sum_{k=0}^{n} \frac{(1-z^2)^{-k}}{(2n-2k+1)!\,(1-\lambda)_k} H_{2n-2k+1}(w) C_{2k}^{\lambda-k}(z)$

$$= 2w\,\frac{\left(\frac{1}{2}-\lambda\right)_n}{n!\,\left(\frac{1}{2}\right)_n}\, z^{2n}(1-z^2)^{-n}\,{}_2F_2\!\left(\begin{array}{c}-n,\,\frac{1}{2}-n;\ w^2(1-z^{-2})\\ \lambda-n+\frac{1}{2},\,\frac{3}{2}\end{array}\right).$$

22. $\displaystyle\sum_{k=0}^{n} \frac{(1-z^2)^{-k}}{(2n-2k)!\,(1-\lambda)_k} H_{2n-2k}(w) C_{2k+1}^{\lambda-k}(z)$

$$= 2\lambda\,\frac{\left(\frac{1}{2}-\lambda\right)_n}{n!\,\left(\frac{3}{2}\right)_n}\, z^{2n+1}(1-z^2)^{-n}\,{}_2F_2\!\left(\begin{array}{c}-n,\,-n-\frac{1}{2};\ w^2(1-z^{-2})\\ \lambda-n+\frac{1}{2},\,\frac{1}{2}\end{array}\right).$$

23. $\displaystyle\sum_{k=0}^{n} \frac{(1-z^2)^{-k}}{(2n-2k+1)!\,(1-\lambda)_k} H_{2n-2k+1}(w) C_{2k+1}^{\lambda-k}(z)$

$$= 4\lambda w\,\frac{\left(\frac{1}{2}-\lambda\right)_n}{n!\,\left(\frac{3}{2}\right)_n}\, z^{2n+1}(1-z^2)^{-n}\,{}_2F_2\!\left(\begin{array}{c}-n,\,-n-\frac{1}{2};\ w^2(1-z^{-2})\\ \lambda-n+\frac{1}{2},\,\frac{3}{2}\end{array}\right).$$

5.11.4]   5.11. THE GEGENBAUER POLYNOMIALS $C_n^\lambda(z)$

24. $\displaystyle\sum_{k=0}^{n} \frac{(-\lambda - n)_k}{(1 - \mu)_k} (-2z)^{-k} L_{n-k}^\lambda(w) C_k^{\mu-k}(z)$

$$= \frac{(-w)^n}{n!} {}_4F_1\left(\begin{array}{c} -\frac{n}{2}, \frac{1-n}{2}, -\frac{\lambda+n}{2}, \frac{1-\lambda-n}{2} \\ 1 - \mu;\ 4w^{-2}z^{-2} \end{array}\right).$$

25. $\displaystyle\sum_{k=0}^{n} \frac{(-\lambda - n)_k}{(1 - \mu)_k} 2^{-k} (1 - z)^{-k} L_{n-k}^\lambda(w)\, C_k^{\mu-k}(z)$

$$= \frac{(\lambda + 1)_n \left(\frac{1}{2} - \mu\right)_n}{n!\,(1 - 2\mu)_n} \left(\frac{2}{1-z}\right)^n {}_2F_2\left(\begin{array}{c} -n,\, 2\mu - n;\ \frac{w - wz}{2} \\ \lambda + 1,\, \mu - n + \frac{1}{2} \end{array}\right).$$

26. $\displaystyle\sum_{k=0}^{n} \frac{(-2)^{-k}}{(1 - \mu)_k} \left(\frac{w}{z-1}\right)^k L_{n-k}^{\lambda+k}(w)\, C_k^{\mu-k}(z)$

$$= \frac{\left(\frac{1}{2} - \mu\right)_n}{n!\,(1 - 2\mu)_n} \left(\frac{2w}{z-1}\right)^n {}_3F_1\left(\begin{array}{c} -n,\, -\lambda - n,\, 2\mu - n \\ \mu - n + \frac{1}{2};\ \frac{z-1}{2w} \end{array}\right).$$

27. $\displaystyle\sum_{k=0}^{n} \frac{(-\lambda - n)_k}{(1 - \mu)_k} (1 - z^2)^{-k} L_{n-k}^\lambda(w) C_{2k}^{\mu-k}(z)$

$$= \frac{(\lambda + 1)_n \left(\frac{1}{2} - \mu\right)_n}{n!\,\left(\frac{1}{2}\right)_n} z^{2n} (z^2 - 1)^{-n} {}_2F_2\left(\begin{array}{c} -n,\, \frac{1}{2} - n;\ w(1 - z^{-2}) \\ \lambda + 1,\, \mu - n + \frac{1}{2} \end{array}\right).$$

28. $\displaystyle\sum_{k=0}^{n} \frac{(-\lambda - n)_k}{(1 - \mu)_k} (1 - z^2)^{-k} L_{n-k}^\lambda(w) C_{2k+1}^{\mu-k}(z)$

$$= 2\mu \frac{(\lambda + 1)_n \left(\frac{1}{2} - \mu\right)_n}{n!\,\left(\frac{3}{2}\right)_n} z^{2n+1} (z^2 - 1)^{-n} {}_2F_2\left(\begin{array}{c} -n,\, -n - \frac{1}{2};\ w(1 - z^{-2}) \\ \lambda + 1,\, \mu - n + \frac{1}{2} \end{array}\right).$$

29. $\displaystyle\sum_{k=0}^{n} \frac{(4k - 2\lambda + 1)\left(\frac{1}{2} - \lambda\right)_k \left(\frac{1}{2}\right)_k}{\left(n - \lambda + \frac{3}{2}\right)_k (1 - \lambda)_{2k}} \left(\frac{w}{z^2 - 1}\right)^k L_{n-k}^{2k-\lambda+1/2}(w) C_{2k}^{\lambda - 2k}(z)$

$$= \frac{2\left(\frac{1}{2} - \lambda\right)_{n+1}}{(1 - \lambda)_n} L_n^{-\lambda}\left(\frac{w}{1 - z^2}\right).$$

30. $\displaystyle\sum_{k=0}^{n} \frac{(4k - 2\lambda + 1)\left(\frac{1}{2} - \lambda\right)_k \left(\frac{3}{2}\right)_k}{\left(n - \lambda + \frac{3}{2}\right)_k (1 - \lambda)_{2k}} \left(\frac{w}{z^2 - 1}\right)^k L_{n-k}^{2k-\lambda+1/2}(w) C_{2k+1}^{\lambda - 2k}(z)$

$$= \frac{4\lambda z \left(\frac{1}{2} - \lambda\right)_{n+1}}{(-\lambda)_n} L_n^{-\lambda - 1}\left(\frac{w}{1 - z^2}\right).$$

377

31. $\sum_{k=0}^{n} \frac{1}{(1-\mu)_k} \left(\frac{w}{1-z^2}\right)^k L_{n-k}^{\lambda+k}(w) C_{2k}^{\mu-k}(z)$

$$= \frac{\left(\frac{1}{2}-\mu\right)_n}{(2n)!} \left(\frac{4wz^2}{1-z^2}\right)^n {}_3F_1\left(\begin{array}{c} -n, -n-\lambda, \frac{1}{2}-n \\ \mu-n+\frac{1}{2}; \frac{z^{-2}-1}{w} \end{array}\right).$$

32. $\sum_{k=0}^{n} \frac{1}{(1-\mu)_k} \left(\frac{w}{1-z^2}\right)^k L_{n-k}^{\lambda+k}(w) C_{2k+1}^{\mu-k}(z)$

$$= \frac{\mu\left(\frac{1}{2}-\mu\right)_n}{(2n)!} (2z)^{2n+1} \left(\frac{w}{1-z^2}\right)^n {}_3F_1\left(\begin{array}{c} -n, -n-\lambda, -n-\frac{1}{2} \\ \mu-n+\frac{1}{2}; \frac{z^{-2}-1}{w} \end{array}\right).$$

33. $\sum_{k=0}^{[n/2]} (\mu)_k (-w)^{-k} L_k^{\lambda-k}(w) C_{n-2k}^{\mu+k}(z)$

$$= \frac{(\mu)_n}{n!} (2z)^n {}_3F_1\left(\begin{array}{c} -\frac{n}{2}, \frac{1-n}{2}, -\lambda \\ 1-\mu-n; -\frac{1}{wz^2} \end{array}\right).$$

34. $\sum_{k=0}^{[n/2]} \frac{(\mu)_k}{(\lambda+1)_k} L_k^{\lambda}(w) C_{n-2k}^{\mu+k}(z) = \frac{(\mu)_n}{n!} (2z)^n {}_2F_2\left(\begin{array}{c} -\frac{n}{2}, \frac{1-n}{2}; \frac{w}{z^2} \\ \lambda+1, 1-\mu-n \end{array}\right).$

35. $\sum_{k=0}^{[n/2]} (2k-n-\lambda)(-n-\lambda)_k L_k^{\lambda-2k+n}(1) C_{n-2k}^{\lambda}(z) = -\frac{(\lambda+1)_n}{n!} H_n(z).$

5.11.5. Sums containing products of $C_{m\pm pk}^{\lambda\pm nk}(z)$

1. $\sum_{k=0}^{n} (-1)^k \frac{(\lambda+k)(\lambda+n)_k}{(n-k)!(\lambda-n+1)_k(2\lambda+n+1)_k} \left[C_k^{\lambda}(z)\right]^2$

$$= \frac{(2\lambda)_{n+1}\left(\frac{1}{2}\right)_n}{2n!(-\lambda)_n\left(\lambda+\frac{1}{2}\right)_n} C_{2n}^{\lambda}(z) \quad [[69], (23)].$$

2. $\sum_{k=0}^{n} (-1)^k \frac{(k+\lambda)k!\left(2\lambda+n-\frac{1}{2}\right)_k}{(n-k)!\left(\frac{3}{2}-n\right)_k (2\lambda)_k (2\lambda+n+1)_k} \left[C_k^{\lambda}(z)\right]^2$

$$= 2^{2n} \lambda(1-2n) \frac{(\lambda)_n (2\lambda+1)_n}{\left(\lambda+\frac{1}{2}\right)_n (4\lambda-1)_{2n}} C_{2n}^{2\lambda-1/2}(z) \quad [[69], (24)].$$

3. $\sum_{k=0}^{n} \frac{(2k+2\lambda-1)(2\lambda-1)_k}{k!\left(\lambda+\frac{1}{2}\right)_k^2 (2\lambda+2k)_{m-k}(2\lambda+2k)_{n-k}} \left(\frac{z^2-1}{4}\right)^k C_{m-k}^{\lambda+k}(z) C_{n-k}^{\lambda+k}(z)$

$$= \frac{2\lambda-1}{(2\lambda)_{m+n}} \binom{m+n}{n} C_{m+n}^{\lambda}(z) \quad [m \geq n; [25], (3)].$$

4. $\displaystyle\sum_{k=0}^{n}\frac{(\mu)_k}{(1-\lambda)_k}\,(1-z^2)^k\,C_{2k}^{\lambda-k}(w)C_{2n-2k+1}^{\mu+k}(z)$

$$= z^{2n+1}\frac{(2\mu)_{2n+1}}{(2n+1)!}\,{}_3F_2\!\left(\begin{array}{c}-n,\,-n-\frac{1}{2},\,\lambda\\ \mu+\frac{1}{2},\,\frac{1}{2};\,w^2(1-z^{-2})\end{array}\right).$$

5. $\displaystyle\sum_{k=0}^{n}\frac{(\mu)_k}{(1-\lambda)_k}\,(1-z^2)^k\,C_{2k}^{\lambda-k}(w)C_{2n-2k}^{\mu+k}(z)$

$$= z^{2n}\frac{(2\mu)_{2n}}{(2n)!}\,{}_3F_2\!\left(\begin{array}{c}-n,\,\frac{1}{2}-n,\,\lambda\\ \mu+\frac{1}{2},\,\frac{1}{2};\,w^2(1-z^{-2})\end{array}\right).$$

6. $\displaystyle\sum_{k=0}^{n}\frac{(\mu)_k}{(1-\lambda)_k}\,(1-z^2)^k\,C_{2k+1}^{\lambda-k}(w)C_{2n-2k}^{\mu+k}(z)$

$$= 2\lambda w z^{2n}\frac{(2\mu)_{2n}}{(2n)!}\,{}_3F_2\!\left(\begin{array}{c}-n,\,\frac{1}{2}-n,\,\lambda+1\\ \mu+\frac{1}{2},\,\frac{3}{2};\,w^2(1-z^{-2})\end{array}\right).$$

7. $\displaystyle\sum_{k=0}^{n}\frac{(\mu)_k}{(1-\lambda)_k}\,(1-z^2)^k\,C_{2k+1}^{\lambda-k}(w)C_{2n-2k+1}^{\mu+k}(z)$

$$= 2\lambda w z^{2n+1}\frac{(2\mu)_{2n+1}}{(2n+1)!}\,{}_3F_2\!\left(\begin{array}{c}-n,\,-n-\frac{1}{2},\,\lambda+1\\ \mu+\frac{1}{2},\,\frac{3}{2};\,w^2(1-z^{-2})\end{array}\right).$$

8. $\displaystyle\sum_{k=0}^{n}\frac{k!(\lambda+k)}{(2\lambda)_k(2\lambda+n+1)_k}\,w^k L_{n-k}^{2\lambda+2k}(w)\left[C_k^{\lambda}(z)\right]^2$

$$= \frac{(2\lambda)_{n+1}}{2n!}\,{}_2F_2\!\left(\begin{array}{c}-n,\,\lambda;\,w-wz^2\\ \lambda+\frac{1}{2},\,2\lambda\end{array}\right).$$

9. $\displaystyle\sum_{k=0}^{n}\frac{(k+\lambda)k!\left(2\lambda+\frac{1}{2}\right)_{2k}}{(2\lambda)_k(2\lambda+n+1)_k\left(\frac{1}{2}-n\right)_k}\,(w^2-1)^k\,C_{2n-2k}^{2\lambda+2k+1/2}(w)\left[C_k^{\lambda}(z)\right]^2$

$$= \frac{\lambda(4\lambda+1)_{2n}}{(2n)!}\,{}_3F_2\!\left(\begin{array}{c}-n,\,\lambda,\,2\lambda+n+\frac{1}{2}\\ \lambda+\frac{1}{2},\,2\lambda;\,(1-w^2)(1-z^2)\end{array}\right).$$

10. $\displaystyle\sum_{k=0}^{n}\frac{(k+\lambda)k!\left(2\lambda+\frac{1}{2}\right)_{2k}}{(2\lambda)_k(2\lambda+n+1)_k\left(-\frac{1}{2}-n\right)_k}\,(w^2-1)^k\,C_{2n-2k+1}^{2\lambda+2k+1/2}(w)\left[C_k^{\lambda}(z)\right]^2$

$$= \frac{(4\lambda)_{2n+2}w}{4(2n+1)!}\,{}_3F_2\!\left(\begin{array}{c}-n,\,\lambda,\,2\lambda+n+\frac{3}{2}\\ \lambda+\frac{1}{2},\,2\lambda;\,(1-w^2)(1-z^2)\end{array}\right).$$

11. $\sum_{k=0}^{n} \dfrac{(2k-2\lambda+1)k!\,(1-2\lambda)_k \left(\frac{3}{2}-2\lambda\right)_{2k}}{(1-\lambda)_k^2 (n-2\lambda+2)_k \left(\frac{1}{2}-n\right)_k} \, 2^{-2k} \left(\dfrac{1-w^2}{1-z^2}\right)^k C_{2n-2k}^{2k-2\lambda+3/2}(w)$

$\times \left[C_k^{\lambda-k}(z)\right]^2 = \dfrac{(1-2\lambda)(3-4\lambda)_{2n}}{(2n)!} \, {}_3F_2\!\left(\begin{matrix} -n,\, n-2\lambda+\frac{3}{2},\, \frac{1}{2}-\lambda \\ 1-\lambda,\, 1-2\lambda;\, \frac{1-w^2}{1-z^2} \end{matrix}\right).$

12. $\sum_{k=0}^{n} \dfrac{(2k-2\lambda+1)k!\,(1-2\lambda)_k \left(\frac{3}{2}-2\lambda\right)_{2k}}{(1-\lambda)_k^2 (n-2\lambda+2)_k \left(-\frac{1}{2}-n\right)_k} \, 2^{-2k} \left(\dfrac{1-w^2}{1-z^2}\right)^k C_{2n-2k+1}^{2k-2\lambda+3/2}(w)$

$\times \left[C_k^{\lambda-k}(z)\right]^2 = \dfrac{(2-4\lambda)_{2n+2}\, w}{2(2n+1)!} \, {}_3F_2\!\left(\begin{matrix} -n,\, n-2\lambda+\frac{5}{2},\, \frac{1}{2}-\lambda \\ 1-\lambda,\, 1-2\lambda;\, \frac{1-w^2}{1-z^2} \end{matrix}\right).$

13. $\sum_{k=0}^{n} \dfrac{(\lambda+2k)(2k)!\,\left(\lambda+\frac{1}{2}\right)_{2k}}{(\lambda+n+1)_k \left(\frac{1}{2}-n\right)_k (2\lambda)_{2k}} (1-w^2)^k$

$\times C_{2n-2k}^{\lambda+2k+1/2}(w)\, C_{2k}^{\lambda}\!\left(\sqrt{\dfrac{1-z}{2}}\right) C_{2k}^{\lambda}\!\left(\sqrt{\dfrac{1+z}{2}}\right)$

$= \dfrac{(2\lambda)_{2n+1}}{2(2n)!} \, {}_4F_3\!\left(\begin{matrix} -n,\, \frac{\lambda}{2},\, \frac{\lambda+1}{2},\, \lambda+n+\frac{1}{2} \\ \lambda,\, \lambda+\frac{1}{2},\, \frac{1}{2};\, (1-w^2)(1-z^2) \end{matrix}\right).$

14. $\sum_{k=0}^{n} \dfrac{(\lambda+2k)(2k)!\,\left(\lambda+\frac{1}{2}\right)_{2k}}{(\lambda+n+1)_k \left(-\frac{1}{2}-n\right)_k (2\lambda)_{2k}} (1-w^2)^k$

$\times C_{2n-2k+1}^{\lambda+2k+1/2}(w)\, C_{2k}^{\lambda}\!\left(\sqrt{\dfrac{1-z}{2}}\right) C_{2k}^{\lambda}\!\left(\sqrt{\dfrac{1+z}{2}}\right)$

$= \dfrac{(2\lambda)_{2n+2}\, w}{2(2n+1)!} \, {}_4F_3\!\left(\begin{matrix} -n,\, \frac{\lambda}{2},\, \frac{\lambda+1}{2},\, \lambda+n+\frac{3}{2} \\ \lambda,\, \lambda+\frac{1}{2},\, \frac{1}{2};\, (1-w^2)(1-z^2) \end{matrix}\right).$

15. $\sum_{k=0}^{n} \dfrac{(\lambda+2k+1)(2k+1)!\,\left(\lambda+\frac{3}{2}\right)_{2k}}{(\lambda+n+2)_k \left(\frac{1}{2}-n\right)_k (2\lambda+1)_{2k}} (1-w^2)^k$

$\times C_{2n-2k}^{\lambda+2k+3/2}(w)\, C_{2k+1}^{\lambda}\!\left(\sqrt{\dfrac{1-z}{2}}\right) C_{2k+1}^{\lambda}\!\left(\sqrt{\dfrac{1+z}{2}}\right)$

$= \dfrac{2\lambda^2(\lambda+1)(2\lambda+3)_{2n}}{(2n)!} \sqrt{1-z^2}\, {}_4F_3\!\left(\begin{matrix} -n,\, \frac{\lambda+1}{2},\, \frac{\lambda}{2}+1,\, \lambda+n+\frac{3}{2} \\ \lambda+\frac{1}{2},\, \lambda+1,\, \frac{3}{2};\, (1-w^2)(1-z^2) \end{matrix}\right).$

16. $\sum_{k=0}^{n} \frac{(\lambda+2k+1)(2k+1)!\left(\lambda+\frac{3}{2}\right)_{2k}}{(\lambda+n+2)_k \left(-\frac{1}{2}-n\right)_k (2\lambda+1)_{2k}} (1-w^2)^k$

$$\times C_{2n-2k+1}^{\lambda+2k+3/2}(w) C_{2k+1}^{\lambda}\left(\sqrt{\frac{1-z}{2}}\right) C_{2k+1}^{\lambda}\left(\sqrt{\frac{1+z}{2}}\right)$$

$$= \frac{\lambda^2 (2\lambda+2)_{2n+2}}{(2n+1)!} w\sqrt{1-z^2}\ {}_4F_3\left(\begin{matrix}-n, \frac{\lambda+1}{2}, \frac{\lambda}{2}+1, \lambda+n+\frac{5}{2} \\ \lambda+\frac{1}{2}, \lambda+1, \frac{3}{2};\ (1-w^2)(1-z^2)\end{matrix}\right).$$

5.11.6. Sums containing $C_{mk+n}^{\lambda k+\mu}(\varphi(k,z))$

1. $\sum_{k=0}^{n}(-1)^k\binom{n}{k} C_m^\lambda(w+kz) = 0$ $\qquad [m<n]$.

2. $\sum_{k=0}^{n}(-1)^k\binom{n}{k} C_m^\lambda(1+kz)$

$$= \frac{n!(2\lambda)_{m+n}}{(m-n)!\left(\lambda+\frac{1}{2}\right)_n}\left(-\frac{z}{2}\right)^n \sum_{k=0}^{m-n} \sigma_{k+n}^n \frac{(n-m)_k(2\lambda+m+n)_k}{(k+n)!\left(\lambda+n+\frac{1}{2}\right)_k}\left(-\frac{z}{2}\right)^k.$$

3. $\sum_{k=0}^{n}(-1)^k\binom{n}{k} C_{m+n}^\lambda(w+kz)$

$$= n!(\lambda)_n(-2z)^n \sum_{k=0}^{m} \sigma_{k+n}^n \frac{(\lambda+n)_k}{(k+n)!}(2z)^k C_{m-k}^{\lambda+k+n}(w).$$

4. $\sum_{k=1}^{n}(-1)^k\binom{n}{k} k^m C_m^\lambda\left(1+\frac{z}{k}\right) = (-1)^m(2\lambda)_m \delta_{m,n} - \frac{(\lambda)_m}{m!}(2z)^m$ $\quad [n\geq m]$.

5. $\sum_{k=1}^{n}(-1)^k\binom{n}{k}(k-z)^m C_m^\lambda\left(\frac{k+z}{k-z}\right) = (-1)^m(2\lambda)_m \delta_{m,n} - \frac{(2\lambda)_m}{m!} z^m$

$\qquad\qquad [n\geq m]$.

6. $\sum_{k=1}^{n}(-1)^k\binom{n}{k} C_{2m}^\lambda(\sqrt{k}\,z) = (-1)^m \frac{m!(\lambda)_{2m}}{(2m)!}(2z)^{2m}\delta_{m,n} - (-1)^m \frac{(\lambda)_m}{m!}$

$\qquad\qquad [n\geq m]$.

7. $\sum_{k=1}^{n}(-1)^k\binom{n}{k} k^{-1/2} C_{2m+1}^\lambda(\sqrt{k}\,z) = (-1)^m \frac{m!(\lambda)_{2m+1}}{(2m+1)!}(2z)^{2m+1}\delta_{m,n}$

$$- (-1)^m \frac{2(\lambda)_{m+1} z}{m!} \quad [n\geq m].$$

8. $\sum_{k=1}^{n}(-1)^k\binom{n}{k} k^m C_{2m}^\lambda\left(\frac{z}{\sqrt{k}}\right) = (\lambda)_m \delta_{m,n} - \frac{(\lambda)_{2m}}{(2m)!}(2z)^{2m}$ $\quad [n\geq m]$.

9. $\displaystyle\sum_{k=1}^{n}(-1)^{k}\binom{n}{k}k^{m+1/2}C_{2m+1}^{\lambda}\left(\frac{z}{\sqrt{k}}\right) = 2(\lambda)_{m+1}z\delta_{m,n}$
$$-\frac{(\lambda)_{2m+1}}{(2m+1)!}(2z)^{2m+1} \quad [n \geq m].$$

10. $\displaystyle\sum_{k=1}^{n}(-1)^{k}\binom{n}{k}C_{2m}^{\lambda}\left(\sqrt{1+kz}\right) = \frac{m!}{(2m)!}(\lambda)_{2m}(-4z)^{m}\delta_{m,n} - \frac{(2\lambda)_{2m}}{(2m)!}$
$$[n \geq m].$$

11. $\displaystyle\sum_{k=1}^{n}\frac{(-1)^{k}}{\sqrt{1+kz}}\binom{n}{k}C_{2m+1}^{\lambda}\left(\sqrt{1+kz}\right) = (-1)^{m}\frac{2^{2m+1}m!}{(2m+1)!}(\lambda)_{2m+1}z^{m}\delta_{m,n}$
$$-\frac{(2\lambda)_{2m+1}}{(2m+1)!} \quad [n \geq m].$$

12. $\displaystyle\sum_{k=1}^{n}(-1)^{k}\binom{n}{k}k^{m}C_{2m}^{\lambda}\left(\frac{z}{\sqrt{k}}\right) = -\frac{(\lambda)_{2m}}{(2m)!}(2z)^{2m}$
$$+\frac{n!\,(\lambda)_{m}(\lambda+m)_{m-n}}{(2m-2n)!}(2z)^{2m-2n}\sum_{k=0}^{m-n}\sigma_{k+n}^{n}\frac{(2n-2m)_{2k}}{(k+n)!(n-2m-\lambda+1)_{k}}(2z)^{-2k}$$
$$[n \geq m].$$

13. $\displaystyle\sum_{k=1}^{n}(-1)^{k}\binom{n}{k}k^{m+1/2}C_{2m+1}^{\lambda}\left(\frac{z}{\sqrt{k}}\right)$
$$= -\frac{(\lambda)_{2m+1}}{(2m+1)!}(2z)^{2m+1} + \frac{n!\,(\lambda)_{m+1}(\lambda+m+1)_{m-n}}{(2m-2n)!}(2z)^{2m-2n+1}$$
$$\times \sum_{k=0}^{m-n}\sigma_{k+n}^{n}\frac{(2n-2m)_{2k}}{(k+n)!(n-2m-\lambda)_{k}(2m-2n-2k+1)}(2z)^{-2k} \quad [n \geq m].$$

14. $\displaystyle\sum_{k=1}^{n}(-1)^{k}\binom{n}{k}k^{m}C_{2m}^{\lambda}\left(\sqrt{1+\frac{z}{k}}\right) = \frac{(-1)^{m}m!}{(2m)!}(2\lambda)_{2m}\delta_{m,n}$
$$-\frac{(\lambda)_{2m}}{(2m)!}(4z)^{m} \quad [n \geq m].$$

15. $\displaystyle\sum_{k=1}^{n}(-1)^{k}\binom{n}{k}\frac{k^{m+1/2}}{\sqrt{k+z}}C_{2m+1}^{\lambda}\left(\sqrt{1+\frac{z}{k}}\right)$
$$= \frac{(-1)^{m}m!}{(2m+1)!}(2\lambda)_{2m+1}\delta_{m,n} - \frac{(\lambda)_{2m+1}}{(2m+1)!}2^{2m+1}z^{m} \quad [n \geq m].$$

16. $\displaystyle\sum_{k=1}^{n}(-1)^{k}\binom{n}{k}(k+z)^{m}C_{2m}^{\lambda}\left(\sqrt{\frac{k}{k+z}}\right) = (-1)^{m}\frac{m!}{(2m)!}(2\lambda)_{2m}\delta_{m,n}$
$$-\frac{(\lambda)_{m}}{m!}(-z)^{m} \quad [n \geq m].$$

5.11.6]   5.11. The Gegenbauer Polynomials $C_n^\lambda(z)$

17. $\displaystyle\sum_{k=1}^{n}(-1)^k\binom{n}{k}k^{-1/2}(k+z)^{m+1/2}C_{2m+1}^{\lambda}\left(\sqrt{\dfrac{k}{k+z}}\right)$

$\qquad = (-1)^m \dfrac{m!}{(2m+1)!}(2\lambda)_{2m+1}\delta_{m,n} - \dfrac{2(\lambda)_{m+1}}{m!}(-z)^m \quad [n\geq m].$

18. $\displaystyle\sum_{k=1}^{n}(-1)^k\binom{n}{k}(k+z)^m C_{2m}^{\lambda}\left(\sqrt{\dfrac{z}{k+z}}\right) = (\lambda)_m\delta_{m,n} - \dfrac{(2\lambda)_{2m}}{(2m)!}z^m$

$\qquad\qquad\qquad\qquad\qquad\qquad\qquad\qquad\qquad\qquad\qquad\qquad [n\geq m].$

19. $\displaystyle\sum_{k=1}^{n}(-1)^k\binom{n}{k}(k+z)^{m+1/2}C_{2m+1}^{\lambda}\left(\sqrt{\dfrac{z}{k+z}}\right)$

$\qquad = (-1)^{m+1}2\sqrt{z}(\lambda)_{m+1}\delta_{m,n} + (-1)^m \dfrac{(2\lambda)_{2m+1}}{(2m+1)!}z^{m+1/2} \quad [n\geq m].$

20. $\displaystyle\sum_{k=1}^{n}(-1)^k\binom{n}{k}k^{m/2}C_m^{\lambda}\left(\dfrac{1}{2}\sqrt{\dfrac{k}{z}}+\dfrac{1}{2}\sqrt{\dfrac{z}{k}}\right) = (-1)^m(\lambda)_m z^{-m/2}\delta_{m,n}$

$\qquad\qquad\qquad\qquad\qquad\qquad\qquad\qquad\qquad - \dfrac{(\lambda)_m}{m!}z^{m/2} \quad [n\geq m].$

21. $\displaystyle\sum_{k=1}^{n}\dfrac{k^{k-1}}{k!}(\lambda)_k(-2z)^k C_{n-k}^{\lambda+k}(1+kz) = -\dfrac{(2\lambda)_{n+1}z}{(n-1)!(2\lambda+1)}.$

22. $\displaystyle\sum_{k=0}^{n}\dfrac{(ka+b)^{n-k-1}}{(n-k)!(1-\lambda)_k}\left(\dfrac{a}{2z}\right)^k C_k^{\lambda-k}(1+kz)$

$\qquad\qquad\qquad\qquad = \dfrac{\left(\dfrac{a}{2z}\right)^n}{(na+b)(1-\lambda)_n} C_n^{\lambda-n}\left(1-\dfrac{bz}{a}\right).$

23. $\displaystyle\sum_{k=0}^{n}\dfrac{2^{-k}}{(n-k)!(1-\lambda)_k}(ka+z+1)^{n-k-1}C_k^{\lambda-k}(ka+z)$

$\qquad\qquad\qquad\qquad = \dfrac{2^n\left(\dfrac{1}{2}-\lambda\right)_n}{n!(na+z+1)(1-2\lambda)_n}.$

24. $\displaystyle\sum_{k=0}^{2n}\dfrac{(2z)^{-k}}{(2n-k)!(1-\lambda)_k}(ka+1)^{2n-k-1}C_k^{\lambda-k}((ka+1)z)$

$\qquad\qquad\qquad\qquad = \dfrac{\left(\dfrac{1}{2}\right)_n z^{-2n}}{(2n)!(2na+1)(1-\lambda)_n}.$

25. $\displaystyle\sum_{k=1}^{n}\dfrac{k^{n-1}}{k!}(\lambda)_k C_{2n-2k}^{\lambda+k}\left(\dfrac{z}{\sqrt{k}}\right) = \dfrac{(\lambda)_{2n-1}}{(2n-2)!}(2z)^{2n-2} \quad [n\geq 1].$

26. $\displaystyle\sum_{k=1}^{n}\dfrac{k^{n-1/2}}{k!}(\lambda)_k C_{2n-2k+1}^{\lambda+k}\left(\dfrac{z}{\sqrt{k}}\right) = \dfrac{(\lambda)_{2n}}{(2n-1)!}(2z)^{2n-1} \quad [n\geq 1].$

27. $\sum_{k=1}^{n} k^{2k-2} \dfrac{\left(\frac{1}{2}\right)_{2k}}{(k+n)!\left(\frac{1}{2}-n\right)_k} z^k C_{2n-2k}^{2k+1/2}\left(\sqrt{1+k^2 z}\right) = -\dfrac{(2n+1)z}{4(n-1)!}.$

28. $\sum_{k=1}^{n} k^{2k-2} \dfrac{\left(\frac{1}{2}\right)_{2k}}{(k+n)!\left(-n-\frac{1}{2}\right)_k} \dfrac{z^k}{\sqrt{1+k^2 z}} C_{2n-2k+1}^{2k+1/2}\left(\sqrt{1+k^2 z}\right)$
$$= -\dfrac{(2n+3)z}{4(n-1)!}.$$

29. $\sum_{k=1}^{n} k^{2k-4} \dfrac{\left(\frac{1}{2}\right)_{2k} z^k}{(k+n)!\left(\frac{1}{2}-n\right)_k} C_{2n-2k}^{2k+1/2}\left(\sqrt{1+k^2 z}\right)$
$$= -\dfrac{(2n+1)z}{4}\left[\dfrac{1}{(n-1)!} + \dfrac{(2n+3)z}{8(n-2)!}\right].$$

30. $\sum_{k=1}^{n} k^{2k-4} \dfrac{\left(\frac{1}{2}\right)_{2k}}{(k+n)!\left(-n-\frac{1}{2}\right)_k} \dfrac{z^k}{\sqrt{1+k^2 z}} C_{2n-2k+1}^{2k+1/2}\left(\sqrt{1+k^2 z}\right)$
$$= -\dfrac{(2n+3)z}{4}\left[\dfrac{1}{(n-1)!} + \dfrac{(2n+5)z}{8(n-2)!}\right].$$

31. $\sum_{k=1}^{n} k^{2k-6} \dfrac{\left(\frac{1}{2}\right)_{2k}}{(k+n)!\left(\frac{1}{2}-n\right)_k} z^k C_{2n-2k}^{2k+1/2}\left(\sqrt{1+k^2 z}\right)$
$$= -\dfrac{(2n+1)(2n+3)(2n+5)z^3}{576(n-3)!} - \dfrac{5(2n+1)(2n+3)z^2}{128(n-2)!} - \dfrac{(2n+1)z}{4(n-1)!}.$$

32. $\sum_{k=1}^{n} k^{2k-6} \dfrac{\left(\frac{1}{2}\right)_{2k}}{(k+n)!\left(-\frac{1}{2}-n\right)_k} \dfrac{z^k}{\sqrt{1+k^2 z}} C_{2n-2k+1}^{2k+1/2}\left(\sqrt{1+k^2 z}\right)$
$$= -\dfrac{(2n+3)(2n+5)(2n+7)z^3}{576(n-3)!} - \dfrac{5(2n+3)(2n+5)z^2}{128(n-2)!} - \dfrac{(2n+3)z}{4(n-1)!}.$$

33. $\sum_{k=1}^{n} \dfrac{k^{2k}}{k^2+a^2} \dfrac{\left(-\frac{1}{2}\right)_{2k}}{(k+n)!\left(\frac{1}{2}-n\right)_k} z^k C_{2n-2k}^{2k+1/2}\left(\sqrt{1+k^2 z}\right)$
$$= -\dfrac{a^{-2}}{2(n!)} - \dfrac{a^{-2}}{(n+1)!(2n-1)z}\left[{}_3F_2\!\left(\begin{matrix}-n-1, n-\frac{1}{2}, 1\\ ia, -ia;\ a^2 z\end{matrix}\right) - 1\right].$$

34. $\sum_{k=1}^{n} \dfrac{k^{2k}}{k^2+a^2} \dfrac{\left(\frac{1}{2}\right)_{2k}}{(k+n)!\left(-\frac{1}{2}-n\right)_k} \dfrac{z^k}{\sqrt{1+k^2 z}} C_{2n-2k+1}^{2k+1/2}\left(\sqrt{1+k^2 z}\right)$
$$= -\dfrac{a^{-2}}{2(n!)} - \dfrac{a^{-2}}{(n+1)!(2n+1)z}\left[{}_3F_2\!\left(\begin{matrix}-n-1, n+\frac{1}{2}, 1\\ ia, -ia;\ a^2 z\end{matrix}\right) - 1\right].$$

5.11. The Gegenbauer Polynomials $C_n^\lambda(z)$

35. $\sum_{k=0}^{n} \frac{(2k+1)^{2k-1} \left(\frac{1}{2}\right)_{2k+1}}{(k+n+1)! \left(\frac{1}{2}-n\right)_k} z^k C_{2n-2k}^{2k+3/2}\left(\sqrt{1+(2k+1)^2 z}\right) = \frac{2n+!}{2(n!)}.$

36. $\sum_{k=0}^{n} \frac{(2k+1)^{2k-1} \left(\frac{3}{2}\right)_{2k}}{(k+n+1)! \left(-\frac{1}{2}-n\right)_k} \frac{z^k}{\sqrt{1+(2k+1)^2 z}}$
$\times C_{2n-2k+1}^{2k+3/2}\left(\sqrt{1+(2k+1)^2 z}\right) = \frac{2n+3}{n!}.$

37. $\sum_{k=0}^{n} \frac{(2k+1)^{2k-3} \left(\frac{1}{2}\right)_{2k+1}}{(k+n+1)! \left(\frac{1}{2}-n\right)_k} z^k C_{2n-2k}^{2k+3/2}\left(\sqrt{1+(2k+1)^2 z}\right)$
$= \frac{2n+1}{2(n!)} \left[1 + \frac{2}{9} n(2n+3)z\right].$

38. $\sum_{k=0}^{n} \frac{(2k+1)^{2k-3} \left(\frac{3}{2}\right)_{2k}}{(k+n+1)! \left(-\frac{1}{2}-n\right)_k} \frac{z^k}{\sqrt{1+(2k+1)^2 z}}$
$\times C_{2n-2k+1}^{2k+3/2}\left(\sqrt{1+(2k+1)^2 z}\right) = (2n+3)\left[\frac{1}{n!} + \frac{2(2n+5)z}{9(n-1)!}\right].$

39. $\sum_{k=0}^{n} \frac{(2k+1)^{2k-5} \left(\frac{1}{2}\right)_{2k+1}}{(k+n+1)! \left(\frac{1}{2}-n\right)_k} z^k C_{2n-2k}^{2k+3/2}\left(\sqrt{1+(2k+1)^2 z}\right)$
$= \frac{2n+1}{2(n!)}\left[1 + \frac{20}{81} n(2n+3)z + \frac{4}{225} n(n+1)(2n+3)(2n+5)z^2\right].$

40. $\sum_{k=0}^{n} \frac{(2k+1)^{2k-5} \left(\frac{3}{2}\right)_{2k}}{(k+n+1)! \left(-n-\frac{1}{2}\right)_k} \frac{z^k}{\sqrt{1+(2k+1)^2 z}}$
$\times C_{2n-2k+1}^{2k+3/2}\left(\sqrt{1+(2k+1)^2 z}\right)$
$= (2n+3)\left[\frac{1}{n!} + \frac{20(2n+5)z}{81(n-1)!} + \frac{4(2n+5)(2n+7)z^2}{225(n-2)!}\right].$

41. $\sum_{k=0}^{n} \frac{(2k+1)^{2k+1}}{(2k+1)^2 + a^2} \frac{\left(\frac{3}{2}\right)_{2k}}{(k+n+1)!(1/2-n)_k} z^k$
$\times C_{2n-2k}^{2k+3/2}\left(\sqrt{1+(2k+1)^2 z}\right) = \frac{2n+1}{n!(1+a^2)} {}_3F_2\left(\begin{array}{c}-n, 1, n+\frac{3}{2}; a^2 z\\ \frac{3-ia}{2}, \frac{3+ia}{2}\end{array}\right).$

42. $\sum_{k=0}^{n} \frac{(2k+1)^{2k+1}}{(2k+1)^2 + a^2} \frac{\left(\frac{3}{2}\right)_{2k}}{(k+n+1)!\left(-\frac{1}{2}-n\right)_k} \frac{z^k}{\sqrt{1+(2k+1)^2 z}}$
$\times C_{2n-2k+1}^{2k+3/2}\left(\sqrt{1+(2k+1)^2 z}\right) = \frac{2n+3}{n!(1+a^2)} {}_3F_2\left(\begin{array}{c}-n, n+\frac{5}{2}, 1\\ \frac{3-ia}{2}, \frac{3+ia}{2}; a^2 z\end{array}\right).$

43. $\sum_{k=0}^{n} \dfrac{(2k+\lambda)^{2k-1}(\lambda)_k \left(\lambda+\dfrac{1}{2}\right)_{2k}}{k!(\lambda+n+1)_k \left(-\dfrac{1}{2}-n\right)_k} \dfrac{z^k}{\sqrt{1+(2k+\lambda)^2 z}}$

$\times C_{2n-2k+1}^{\lambda+2k+1/2}\left(\sqrt{1+(2k+\lambda)^2 z}\right) = \dfrac{2^{-2n}(2\lambda+1)_{2n+1}}{n!\lambda \left(\dfrac{3}{2}\right)_n}.$

44. $\sum_{k=1}^{n} \dfrac{k^{2k+2}\left(\dfrac{1}{2}\right)_{2k}}{(k+n)!\left(\dfrac{1}{2}-n\right)_k} z^k C_{2n-2k}^{2k+1/2}\left(\sqrt{1+k^2 z}\right)$

$= \dfrac{(2n+1)z}{8(n-2)!} \left[(2n+3)\, z\, {}_3F_0\!\left(\genfrac{}{}{0pt}{}{2-n,\, n+\dfrac{5}{2},\, 4}{z}\right) \right.$

$\left. - \dfrac{2}{n-1}\, {}_3F_0\!\left(\genfrac{}{}{0pt}{}{1-n,\, n+\dfrac{3}{2},\, 4}{z}\right) \right] \quad [n \geq 2].$

45. $\sum_{k=1}^{n} \dfrac{k^{2k+2}}{\sqrt{1+k^2 z}} \dfrac{\left(\dfrac{1}{2}\right)_{2k}}{(k+n)!\left(-\dfrac{1}{2}-n\right)_k} z^k C_{2n-2k+1}^{2k+1/2}\left(\sqrt{1+k^2 z}\right)$

$= \dfrac{(2n+3)z}{8(n-2)!} \left[(2n+5)\, z\, {}_3F_0\!\left(\genfrac{}{}{0pt}{}{2-n,\, n+\dfrac{7}{2},\, 4}{z}\right) \right.$

$\left. - \dfrac{2}{n-1}\, {}_3F_0\!\left(\genfrac{}{}{0pt}{}{1-n,\, n+\dfrac{5}{2},\, 4}{z}\right) \right] \quad [n \geq 2].$

46. $\sum_{k=0}^{n} (2k+\lambda)^{2k-1} \dfrac{(\lambda)_k \left(\lambda+\dfrac{1}{2}\right)_{2k} z^k}{k!(\lambda+n+1)_k \left(\dfrac{1}{2}-n\right)_k} C_{2n-2k}^{\lambda+2k+1/2}\left(\sqrt{1+(2k+\lambda)^2 z}\right)$

$= \dfrac{2^{-2n}(2\lambda+1)_{2n}}{n!\lambda \left(\dfrac{1}{2}\right)_n}.$

47. $\sum_{k=1}^{n} \dfrac{(-ka)^k (ka-b)^{n-k-1}}{(n-k)!(1-\lambda)_k} C_{2k}^{\lambda-k}\!\left(\dfrac{z}{\sqrt{k}}\right)$

$= \dfrac{(-1)^n b^{n-1}}{n!} + \dfrac{(-b)^n}{(na-b)(1-\lambda)_n} C_{2n}^{\lambda-n}\!\left(z\sqrt{\dfrac{a}{b}}\right).$

48. $\sum_{k=1}^{n} \dfrac{k^{k+1/2}(-a)^k (ka-b)^{n-k-1}}{(n-k)!(1-\lambda)_k} C_{2k+1}^{\lambda-k}\!\left(\dfrac{z}{\sqrt{k}}\right)$

$= \dfrac{2(-1)^n b^{n-1}\lambda z}{n!} + \dfrac{(-1)^n a^{-1/2} b^{n+1/2}}{(na-b)(1-\lambda)_n} C_{2n+1}^{\lambda-n}\!\left(z\sqrt{\dfrac{a}{b}}\right).$

49. $\sum_{k=0}^{n} \dfrac{(-a)^k (ka+b)^{n-k-1}}{(n-k)!(1-\lambda)_k} (k+z)^k C_{2k}^{\lambda-k}\!\left(\sqrt{\dfrac{z}{k+z}}\right)$

$= \dfrac{(b-az)^n}{(na+b)(1-\lambda)_n} C_{2n}^{\lambda-n}\!\left(\sqrt{\dfrac{az}{az-b}}\right).$

50. $\displaystyle\sum_{k=0}^{n}\frac{(-a)^{k}(ka+b)^{n-k-1}}{(n-k)!(1-\lambda)_{k}}(k+z)^{k+1/2}C_{2k+1}^{\lambda-k}\left(\sqrt{\frac{z}{k+z}}\right)$

$$=\frac{(-1)^{n+1}a^{-1/2}(az-b)^{n+1/2}}{(na+b)(1-\lambda)_{n}}C_{2n+1}^{\lambda-n}\left(\sqrt{\frac{az}{az-b}}\right).$$

51. $\displaystyle\sum_{k=0}^{n}\frac{(-1)^{k}}{(n-k)!(1-\lambda)_{k}}(ka+1)^{n-1}C_{2k}^{\lambda-k}\left(\frac{z}{\sqrt{ka+1}}\right)=\frac{(\lambda)_{n}(2z)^{2n}}{(2n)!(na+1)}.$

52. $\displaystyle\sum_{k=0}^{n}\frac{(-1)^{k}}{(n-k)!(1-\lambda)_{k}}(ka+1)^{n-1/2}C_{2k+1}^{\lambda-k}\left(\frac{z}{\sqrt{ka+1}}\right)$

$$=\frac{(\lambda)_{n+1}(2z)^{2n+1}}{(2n+1)!(na+1)}.$$

53. $\displaystyle\sum_{k=0}^{n}\frac{(ka+1)^{n-k-1}}{(n-k)!(1-\lambda)_{k}}(-z)^{-k}(1+(ka+1)z)^{k}C_{2k}^{\lambda-k}\left(\frac{1}{\sqrt{1+(ka+1)z}}\right)$

$$=\frac{\left(\frac{1}{2}-\lambda\right)_{n}}{(2n)!(na+1)}\left(-\frac{4}{z}\right)^{n}.$$

54. $\displaystyle\sum_{k=0}^{n}\frac{(ka+1)^{n-k-1}}{(n-k)!(1-\lambda)_{k}}(-z)^{-k}(1+(ka+1)z)^{k+1/2}$

$$\times C_{2k+1}^{\lambda-k}\left(\frac{1}{\sqrt{1+(ka+1)z}}\right)=\frac{2\lambda\left(\frac{1}{2}-\lambda\right)_{n}}{(2n+1)!(na+1)}\left(-\frac{4}{z}\right)^{n}.$$

55. $\displaystyle\sum_{k=1}^{n}(-1)^{k}\binom{n}{k}k^{n/2}C_{n}^{\lambda}\left(\frac{k+z}{2\sqrt{kz}}\right)=(\lambda)_{n}\left[(-1)^{n}z^{-n/2}-\frac{z^{n/2}}{n!}\right].$

56. $\displaystyle\sum_{k=0}^{2n}(-1)^{k}\frac{(ka+1)^{2n-k-1}}{(2n-k)!(2\lambda)_{k}}[(ka+1)^{2}+z]^{k/2}C_{k}^{\lambda}\left(\frac{ka+1}{\sqrt{(ka+1)^{2}+z}}\right)$

$$=\frac{\left(\frac{1}{2}\right)_{n}(-z)^{n}}{(2n)!(2na+1)\left(\lambda+\frac{1}{2}\right)_{n}}.$$

57. $\displaystyle\sum_{k=0}^{n}(-1)^{k}\frac{(ka+1)^{n-1}}{(n-k)!(1-\lambda)_{k}}C_{2k}^{\lambda-k}\left(\frac{z}{\sqrt{ka+1}}\right)=\frac{(\lambda)_{n}z^{2n}}{n!(na+1)\left(\frac{1}{2}\right)_{n}}.$

58. $\displaystyle\sum_{k=0}^{n}(-1)^{k}\frac{(ka+1)^{n-1/2}}{(n-k)!(1-\lambda)_{k}}C_{2k+1}^{\lambda-k}\left(\frac{z}{\sqrt{ka+1}}\right)=\frac{2(\lambda)_{n+1}z^{2n+1}}{n!(na+1)\left(\frac{3}{2}\right)_{n}}.$

5.11.7. Sums containing $C_{mk+n}^{\lambda k+\mu}(\varphi(k,z))$ and special functions

1. $\displaystyle\sum_{k=1}^{n}(4k)^k(n-k+1)^{n-k-1/2}\frac{\left(-n-\frac{1}{2}\right)_k}{(n-k)!(1-\lambda)_k}H_{2n-2k+1}\left(\frac{w}{\sqrt{n-k+1}}\right)$

   $\times\, C_{2k}^{\lambda-k}\left(\dfrac{z}{\sqrt{k}}\right)=-\dfrac{(n+1)^{n-1/2}}{n!}H_{2n+1}\left(w\sqrt{\dfrac{1}{n+1}}\right)$

   $+\,(2w)^{2n+1}\displaystyle\sum_{k=0}^{n}\left(\frac{n+1}{w^2}\right)^k\frac{\left(-\frac{1}{2}-n\right)_k}{(n-k+1)!(1-\lambda)_k}C_{2k}^{\lambda-k}\left(\frac{z}{\sqrt{n+1}}\right).$

2. $\displaystyle\sum_{k=1}^{n}2^{2k}k^{k+1/2}(n-k+1)^{n-k-1/2}\frac{\left(-n-\frac{1}{2}\right)_k}{(n-k)!(1-\lambda)_k}H_{2n-2k+1}\left(\frac{w}{\sqrt{n-k+1}}\right)$

   $\times\, C_{2k+1}^{\lambda-k}\left(\dfrac{z}{\sqrt{k}}\right)=-\dfrac{2\lambda z}{n!}(n+1)^{n-1/2}H_{2n+1}\left(w\sqrt{\dfrac{1}{n+1}}\right)$

   $+\,\sqrt{n+1}\,(2w)^{2n+1}\displaystyle\sum_{k=0}^{n}\left(\frac{n+1}{w^2}\right)^k\frac{\left(-\frac{1}{2}-n\right)_k}{(n-k+1)!(1-\lambda)_k}C_{2k+1}^{\lambda-k}\left(\frac{z}{\sqrt{n+1}}\right).$

3. $\displaystyle\sum_{k=1}^{n}2^{2k}k^{k+1/2}(n-k+1)^{n-k-1}\frac{\left(\frac{1}{2}-n\right)_k}{(n-k)!(1-\lambda)_k}H_{2n-2k}\left(\frac{w}{\sqrt{n-k+1}}\right)$

   $\times\, C_{2k+1}^{\lambda-k}\left(\dfrac{z}{\sqrt{k}}\right)=-\dfrac{2\lambda z}{n!}(n+1)^{n-1}H_{2n}\left(\dfrac{w}{\sqrt{n+1}}\right)$

   $+\,\sqrt{n+1}\,(4w^2)^n\displaystyle\sum_{k=0}^{n}\left(\frac{n+1}{w^2}\right)^k\frac{\left(\frac{1}{2}-n\right)_k}{(n-k+1)!(1-\lambda)_k}C_{2k+1}^{\lambda-k}\left(\frac{z}{\sqrt{n+1}}\right).$

4. $\displaystyle\sum_{k=1}^{n}(4k)^k(n-k+1)^{n-k-1}\frac{\left(\frac{1}{2}-n\right)_k}{(n-k)!(1-\lambda)_k}H_{2n-2k}\left(\frac{w}{\sqrt{n-k+1}}\right)$

   $\times\, C_{2k}^{\lambda-k}\left(\dfrac{z}{\sqrt{k}}\right)=-\dfrac{(n+1)^{n-1}}{n!}H_{2n}\left(w\sqrt{\dfrac{1}{n+1}}\right)$

   $-\,(2w)^{2n}\displaystyle\sum_{k=0}^{n}\left(\frac{n+1}{w^2}\right)^k\frac{\left(\frac{1}{2}-n\right)_k}{(n-k+1)!(1-\lambda)_k}C_{2k}^{\lambda-k}\left(\frac{z}{\sqrt{n+1}}\right).$

5. $\displaystyle\sum_{k=0}^{n}(-4)^k(k+1)^{n-1/2}\frac{\left(\frac{1}{2}-n\right)_k}{(n-k)!(1-\lambda)_k}H_{2n-2k}\left(\frac{w}{\sqrt{k+1}}\right)C_{2k+1}^{\lambda-k}\left(\frac{z}{\sqrt{k+1}}\right)$

   $=\dfrac{2^{2n+1}w^{2n+2}z^{-1}}{(n+1)!(2n+1)}\left[1-{}_3F_1\!\left(\begin{array}{c}-n-1,\,-n-\frac{1}{2},\,\lambda\\ \frac{1}{2};\,-w^{-2}z^2\end{array}\right)\right].$

5.11.7]   5.11. The Gegenbauer Polynomials $C_n^\lambda(z)$

6. $\displaystyle\sum_{k=0}^{n}(-4)^k(k+1)^{n-1}\frac{\left(\frac{1}{2}-n\right)_k}{(n-k)!(1-\lambda)_k}H_{2n-2k}\left(\frac{w}{\sqrt{k+1}}\right)C_{2k}^{\lambda-k}\left(\frac{z}{\sqrt{k+1}}\right)$

$$=\frac{2^{2n}w^{2n+2}z^{-2}}{(n+1)!(2n+1)(\lambda-1)}\left[{}_3F_1\left(\begin{array}{c}-n-1,-n-\frac{1}{2},\lambda\\-\frac{1}{2};\,-w^{-2}z^2\end{array}\right)-1\right].$$

7. $\displaystyle\sum_{k=0}^{n}(-4)^k(k+1)^{n-1/2}\frac{\left(-\frac{1}{2}-n\right)_k}{(n-k)!(1-\lambda)_k}$

$$\times H_{2n-2k+1}\left(\frac{w}{\sqrt{k+1}}\right)C_{2k}^{\lambda-k}\left(\frac{z}{\sqrt{k+1}}\right)$$

$$=\frac{2^{2n+1}w^{2n+3}z^{-2}}{(n+1)!(2n+3)(\lambda-1)}\left[{}_3F_1\left(\begin{array}{c}-n-1,-n-\frac{3}{2},\lambda-1\\-\frac{1}{2};\,-w^{-2}z^2\end{array}\right)-1\right].$$

8. $\displaystyle\sum_{k=0}^{n}(-4)^k(k+1)^{n}\frac{\left(-n-\frac{1}{2}\right)_k}{(n-k)!(1-\lambda)_k}H_{2n-2k+1}\left(\frac{w}{\sqrt{k+1}}\right)C_{2k+1}^{\lambda-k}\left(\frac{z}{\sqrt{k+1}}\right)$

$$=\frac{2^{2n+2}w^{2n+3}z^{-1}}{(n+1)!(2n+3)}\left[1-{}_3F_1\left(\begin{array}{c}-n-1,-n-\frac{3}{2},\lambda\\\frac{1}{2};\,-w^{-2}z^2\end{array}\right)\right].$$

9. $\displaystyle\sum_{k=0}^{n}\frac{(k+1)^{-1}}{(1-\mu)_k}\left(\frac{w}{2z}\right)^k L_{n-k}^{\lambda+k}((n-k)w)\,C_k^{\mu-k}(1+(k+1)z)=\frac{\left(\frac{1}{2}-\mu\right)_n}{(1-2\mu)_n}$

$$\times\left(-\frac{2w}{z}\right)^n\sum_{k=0}^{n}\left(-\frac{z}{2w}\right)^k\frac{(2\mu-n)_k}{(n-k+1)!\left(\mu-n+\frac{1}{2}\right)_k}L_k^{\lambda-k+n}((n+1)w).$$

10. $\displaystyle\sum_{k=1}^{n}(-1)^k k^k(n-k+1)^{n-k-1}\frac{(-n-\lambda)_k}{(1-\mu)_k}L_{n-k}^{\lambda}\left(\frac{w}{n-k+1}\right)$

$$\times C_{2k}^{\mu-k}\left(\frac{z}{\sqrt{k}}\right)=-(n+1)^{n-1}L_n^{\lambda}\left(\frac{w}{n+1}\right)$$

$$+(-w)^n\sum_{k=0}^{n}\left(\frac{n+1}{w}\right)^k\frac{(-n-\lambda)_k}{(1-\mu)_k(n-k+1)!}C_{2k}^{\mu-k}\left(\frac{z}{\sqrt{n+1}}\right).$$

11. $\displaystyle\sum_{k=1}^{n}(-1)^k k^{k+1/2}(n-k+1)^{n-k-1}\frac{(-n-\lambda)_k}{(1-\mu)_k}L_{n-k}^{\lambda}\left(\frac{w}{n-k+1}\right)$

$$\times C_{2k+1}^{\mu-k}\left(\frac{z}{\sqrt{k}}\right)=-2\mu z(n+1)^{n-1}L_n^{\lambda}\left(\frac{w}{n+1}\right)$$

$$+\sqrt{n+1}\,w^n\sum_{k=0}^{n}\left(\frac{n+1}{w}\right)^k\frac{(-n-\lambda)_k}{(1-\mu)_k(n-k+1)!}C_{2k+1}^{\mu-k}\left(\frac{z}{\sqrt{n+1}}\right).$$

12. $\displaystyle\sum_{k=0}^{n} \frac{(k+1)^{k-1}}{(1-\mu)_k} w^k L_{n-k}^{\lambda+k}((k+1)w)$

$$\times C_{2k}^{\mu-k}\left(\frac{z}{\sqrt{k+1}}\right) = \frac{(\lambda)_{n+1}(wz^2)^{-1}}{2(n+1)!(\mu-1)}\left[{}_2F_2\!\left(\begin{array}{c}-n-1,\,\mu-1\\ \lambda,\,-\frac{1}{2};\,wz^2\end{array}\right)-1\right].$$

13. $\displaystyle\sum_{k=0}^{n} \frac{(k+1)^{k-1/2}}{(1-\mu)_k} w^k L_{n-k}^{\lambda+k}((k+1)w)\, C_{2k+1}^{\mu-k}\!\left(\frac{z}{\sqrt{k+1}}\right)$

$$= \frac{(\lambda)_{n+1}(wz)^{-1}}{(n+1)!}\left[1 - {}_2F_2\!\left(\begin{array}{c}-n-1,\,\mu\\ \lambda,\,\frac{1}{2};\,wz^2\end{array}\right)\right].$$

14. $\displaystyle\sum_{k=0}^{n} \frac{(k+1)^{k-1}}{(1-\lambda)_k}(-w)^k L_{n-k}^{\lambda+k}((n-k)w)\, C_{2k}^{\lambda-k}\!\left(\frac{z}{\sqrt{k+1}}\right)$

$$= \sum_{k=0}^{n}(4wz^2)^k \frac{(\lambda)_k}{(2k)!(k+1)} L_{n-k}^{\lambda+k}((n+1)w).$$

15. $\displaystyle\sum_{k=0}^{n} \frac{(k+1)^{k-1/2}}{(1-\lambda)_k}(-w)^k L_{n-k}^{\lambda+k}((n-k)w)\, C_{2k+1}^{\lambda-k}\!\left(\frac{z}{\sqrt{k+1}}\right)$

$$= 4\lambda z \sum_{k=0}^{n}(4wz^2)^k \frac{(\lambda+1)_k}{(2k+2)!} L_{n-k}^{\lambda+k}((n+1)w).$$

16. $\displaystyle\sum_{k=0}^{n} \frac{(k+1)^{k-1}}{(1-\lambda)_k} w^k L_{n-k}^{\mu+k}((k+1)w)\, C_{2k}^{\lambda-k}\!\left(\frac{z}{\sqrt{k+1}}\right)$

$$= \frac{(\mu)_{n+1}}{2(n+1)!(\lambda-1)wz^2}\left[{}_2F_2\!\left(\begin{array}{c}-n-1,\,\lambda-1;\,wz^2\\ \mu,\,-\frac{1}{2}\end{array}\right)-1\right].$$

17. $\displaystyle\sum_{k=0}^{n} \frac{(k+1)^{k-1/2}}{(1-\lambda)_k} w^k L_{n-k}^{\lambda+k}((k+1)w)\, C_{2k+1}^{\lambda-k}\!\left(\frac{z}{\sqrt{k+1}}\right)$

$$= \frac{(\lambda)_{n+1}}{wz}\left[\frac{1}{(n+1)!} + \frac{(-1)^n}{(2n+2)!} H_{2n+2}(\sqrt{w}\,z)\right].$$

5.11.8. Sums containing products of $C_{mk+n}^{\lambda k+\mu}(\varphi(k,z))$

1. $\displaystyle\sum_{k=0}^{n}(-1)^k \binom{n}{k}\left[C_m^\lambda(w+kz)\right]^2 = 0$ $\hfill [2m < n].$

2. $\displaystyle\sum_{k=1}^{n}(-1)^k \binom{n}{k} k^m \left[C_m^\lambda\!\left(\sqrt{1+\frac{z}{k}}\right)\right]^2 = (-1)^n \frac{(2\lambda)_n^2}{n!}\delta_{m,n} - \frac{(\lambda)_m^2}{(m!)^2}(4z)^m$

$\hfill [m \le n].$

3. $\sum_{k=0}^{n} (k+1)^{k-1} \frac{(\lambda)_k}{(1-\mu)_k} (-2w)^k C_{n-k}^{\lambda+k}((k+1)w+1) C_{2k}^{\mu-k}\left(\frac{z}{\sqrt{k+1}}\right)$

$$= \frac{(2\lambda-1)_n (2wz^2)^{-1}}{(n+1)!(\mu-1)} \left[1 - {}_3F_2\left(\begin{array}{c} -n-1, 2\lambda+n-1, \mu-1 \\ \lambda-\frac{1}{2}, -\frac{1}{2}; \end{array} -\frac{wz^2}{2}\right)\right].$$

4. $\sum_{k=0}^{n} (k+1)^{k-1/2} \frac{(\lambda)_k}{(1-\mu)_k} (-2w)^k C_{n-k}^{\lambda+k}((k+1)w+1) C_{2k+1}^{\mu-k}\left(\frac{z}{\sqrt{k+1}}\right)$

$$= \frac{(2\lambda-1)_n}{(n+1)!} (wz)^{-1} \left[{}_3F_2\left(\begin{array}{c} -n-1, 2\lambda+n-1, \mu \\ \lambda-\frac{1}{2}, \frac{1}{2}; \end{array} -\frac{wz^2}{2}\right) - 1\right].$$

5. $\sum_{k=0}^{n} (k+1)^{n-1/2} \frac{(\lambda)_k}{(1-\mu)_k} C_{2n-2k}^{\lambda+k}\left(\frac{w}{\sqrt{k+1}}\right) C_{2k+1}^{\mu-k}\left(\frac{z}{\sqrt{k+1}}\right)$

$$= \frac{(\lambda)_{2n+1}}{(2n+2)!} (2w)^{2n+2} z^{-1} \left[1 - {}_3F_2\left(\begin{array}{c} -n-1, -n-\frac{1}{2}, \mu \\ -\lambda-2n, \frac{1}{2}; \end{array} \frac{z^2}{w^2}\right)\right].$$

6. $\sum_{k=0}^{n} (k+1)^{n} \frac{(\lambda)_k}{(1-\mu)_k} C_{2n-2k+1}^{\lambda+k}\left(\frac{w}{\sqrt{k+1}}\right) C_{2k+1}^{\mu-k}\left(\frac{z}{\sqrt{k+1}}\right)$

$$= \frac{(\lambda)_{2n+2}}{(2n+3)!} (2w)^{2n+3} z^{-1} \left[1 - {}_3F_2\left(\begin{array}{c} -n-1, -n-\frac{3}{2}, \mu \\ -\lambda-2n-1, \frac{1}{2}; \end{array} \frac{z^2}{w^2}\right)\right].$$

5.12. The Jacobi Polynomials $P_n^{(\rho,\sigma)}(z)$

5.12.1. Sums containing $P_m^{(\rho\pm pk, \sigma\pm qk)}(z)$

1. $\sum_{k=0}^{n} P_m^{(\rho+k,\sigma)}(z) = \frac{2}{z+1}\left[P_{m+1}^{(\rho+n,\sigma-1)}(z) - P_{m+1}^{(\rho-1,\sigma-1)}(z)\right].$

2. $\sum_{k=0}^{n} (-1)^k \binom{n}{k} P_m^{(\rho+k,\sigma)}(z) = \left(-\frac{1+z}{2}\right)^n P_{m-n}^{(\rho+n,\sigma+n)}(z)$ \qquad $[m \geq n].$

3. $\sum_{k=0}^{n} \binom{n}{k} \frac{(\rho+\sigma+m+1)_k}{(\rho+m+1)_k (k+1)} \left(\frac{z-1}{2}\right)^k P_m^{(\rho+k,\sigma)}(z)$

$$= \frac{2}{(n+1)(\rho+\sigma+m)(1-z)}$$
$$\times \left[(\rho+m) P_m^{(\rho-1,\sigma)}(z) - \frac{(m+n+1)!}{m!(\rho+m+1)_n} P_{m+n+1}^{(\rho-1,\sigma-n-1)}(z)\right].$$

4. $\displaystyle\sum_{k=0}^{n}(-1)^k\binom{n}{k}\frac{(\rho+\sigma+m+1)_k}{(\rho+m+1)_k}P_m^{(\rho+k,\sigma)}(z)$
$$=\frac{(\rho+1)_m(-\sigma-m)_n}{(\rho+1)_n(\rho+n+1)_m}P_m^{(\rho+n,\sigma-n)}(z).$$

5. $\displaystyle\sum_{k=0}^{n}(-1)^k\binom{n}{k}\frac{(k+n-1)!}{(k+m)!}P_m^{(k,\sigma-k)}(z)$
$$=\frac{(n-1)!(\sigma+m+1)_n}{(m+n)!}\left(\frac{z-1}{2}\right)^n P_{m-n}^{(2n,\sigma)}(z) \quad [m\geq n].$$

6. $\displaystyle\sum_{k=0}^{n}(-1)^k\binom{n}{k}P_m^{(\rho-k,\sigma+k)}(z)=P_{m-n}^{(\rho,\sigma+n)}(z) \quad [m\geq n].$

7. $\displaystyle\sum_{k=0}^{n}(-1)^k\binom{n}{k}\frac{(-\sigma-m)_k}{(\rho+m+1)_k}P_m^{(\rho+k,\sigma-k)}(z)=\frac{(\rho+\sigma+m+1)_n}{(\rho+m+1)_n}P_m^{(\rho+n,\sigma)}(z).$

8. $\displaystyle\sum_{k=0}^{n}(-1)^k\binom{n}{k}\frac{(-\sigma-m)_k}{(1-\sigma-n)_k}P_m^{(\rho+k,\sigma-k)}(z)$
$$=\frac{(\rho+\sigma+m+1)_n}{(\sigma)_n}\left(-\frac{z+1}{2}\right)^n P_{m-n}^{(\rho+n,\sigma+n)}(z)$$
$$[m\geq n].$$

9. $\displaystyle\sum_{k=0}^{n}\binom{n}{k}\frac{(-\sigma-m)_k}{(\rho-n+1)_k}\left(\frac{1-z}{1+z}\right)^k P_m^{(\rho+k,\sigma-k)}(z)$
$$=\frac{(m+n)!}{m!(-\rho)_n}\left(-\frac{2}{1+z}\right)^n P_{m+n}^{(\rho-n,\sigma-n)}(z).$$

5.12.2. Sums containing $P_{m\pm nk}^{(\rho\pm pk,\sigma\pm qk)}(z)$

1. $\displaystyle\sum_{k=0}^{n}(-1)^k(2k+\rho+\sigma+1)\frac{(\rho+\sigma+1)_k}{(n-k)!(\rho+1)_k(\rho+\sigma+n+2)_k}P_k^{(\rho,\sigma)}(z)$
$$=\frac{(\rho+\sigma+1)_{n+1}}{n!(\rho+1)_n}\left(\frac{1-z}{2}\right)^n.$$

2. $\displaystyle\sum_{k=0}^{n}(2k+\rho+\sigma+1)\frac{(\rho+\sigma+1)_k^2}{k!(\rho+1)_k}P_k^{(\rho,\sigma)}(z)$
$$=\frac{(n+1)(\rho+\sigma+1)_{n+1}^2}{n!(\rho+1)_{n+1}(\rho+\sigma+1)}P_{n+1}^{(\rho,\sigma)}(z)+\frac{(\rho+\sigma+1)(\rho+\sigma+2)}{2(\rho+1)}$$
$$\times\left[\frac{(\rho+\sigma+3)_n}{n!}\right]^2(1-z)\,_3F_2\!\left(\begin{array}{c}-n,\rho+\sigma+2,\rho+\sigma+n+3\\ \rho+2,\rho+\sigma+3;\end{array}\frac{1-z}{2}\right).$$

3. $\displaystyle\sum_{k=0}^{[n/2]}(-1)^k(2n-4k+1)\frac{(n-2k)!\left(-n-\frac{1}{2}\right)_k(-\rho-n)_{2k}}{k!}P_{n-2k}^{(\rho,-\rho)}(z)$

$$= 2^n\left(\frac{3}{2}\right)_n P_n^{(\rho,-\rho-n)}(2z-1).$$

4. $\displaystyle\sum_{k=0}^{[n/2]}\binom{n}{k}\frac{(n-2k+1)^2(n-2k)!}{(n-k+1)(\rho+1)_{n-2k}}P_{n-2k}^{(\rho,1-\rho)}(z)$

$$= 2^n\frac{n!}{(\rho+1)_n}P_n^{(\rho,1/2-\rho-n)}(2z-1).$$

5. $\displaystyle\sum_{k=0}^{n}\binom{2n+1}{n-k}\frac{(2k+1)!}{(\rho+1)_{2k+1}}P_{2k+1}^{(\rho,-\rho-1)}(z)$

$$= \frac{2^{2n}(2n+1)!}{(\rho+1)_{2n+1}}P_{2n+1}^{(\rho,-\rho-2n-3/2)}(2z-1).$$

6. $\displaystyle\sum_{k=0}^{n}\sigma_{k+m}^m t^k\frac{(-\rho-n)_k}{(k+m)!}P_{n-k}^{(\rho,\sigma+k)}(z)$

$$= \frac{(-1)^{m+n}t^{-m}}{m!(\rho+n+1)_m}\sum_{k=0}^{m}(-1)^k\binom{m}{k}(kt-1)^{m+n}P_{m+n}^{(\rho,\sigma-m)}\left(\frac{kt-z}{kt-1}\right).$$

7. $\displaystyle\sum_{k=0}^{n}\left(\frac{2}{z-1}\right)^k P_k^{(\rho-k,\sigma)}(z) = \left(\frac{2}{z-1}\right)^n P_n^{(\rho-n,\sigma+1)}(z).$

8. $\displaystyle\sum_{k=0}^{n}(-1)^k\frac{(a)_k}{(\sigma+1)_k}P_k^{(\rho-k,\sigma)}(z) = \frac{(a+1)_n}{n!}{}_3F_2\left(\begin{array}{c}-n,a,\rho+\sigma+1\\a+1,\sigma+1;\end{array}\frac{1+z}{2}\right).$

9. $\displaystyle\sum_{k=0}^{n}\frac{1}{(n-k)!(a)_k}P_k^{(\rho-k,\sigma)}(z) = \frac{1}{(a)_n}P_n^{(\rho+a-1,\sigma-a-n+1)}(z).$

10. $\displaystyle\sum_{k=0}^{n}\frac{1}{(n-k)!(\sigma+1)_k}P_k^{(\rho-k,\sigma)}(z) = \frac{(\rho+\sigma+1)_n}{n!(\sigma+1)_n}\left(\frac{z+1}{2}\right)^n.$

11. $\displaystyle\sum_{k=0}^{n}\frac{1}{(n-k)!(\sigma+1)_k}\left(\frac{2}{1-z}\right)^k P_k^{(\rho-k,\sigma)}(z) = \frac{(-\rho)_n}{n!(\sigma+1)_n}\left(\frac{z+1}{z-1}\right)^n.$

12. $\displaystyle\sum_{k=0}^{n}\frac{(-1)^k k!}{(\sigma+1)_k}P_k^{(\rho-k,\sigma)}(z)$

$$= \frac{2\sigma}{\rho+\sigma}(1+z)^{-1}\left[1+(-1)^n\frac{(n+1)!}{(\sigma)_{n+1}}P_{n+1}^{(\rho-n-1,\sigma-1)}(z)\right].$$

13. $$\sum_{k=0}^{n} \frac{(a)_k}{(n-k)!(b)_k(\sigma+1)_k} P_k^{(\rho-k,\sigma)}(z)$$
$$= \frac{(b-a)_n}{n!(b)_n} \, _3F_2\left(\begin{array}{c} -n, a, \rho+\sigma+1; \\ a-b-n+1, \sigma+1 \end{array} \frac{1+z}{2}\right).$$

14. $$\sum_{k=0}^{n} \binom{n}{k}\binom{2n}{k}^{-1} \frac{1}{(n-k)!(\sigma+1)_k} P_k^{(\rho-k,\sigma)}(z)$$
$$= \frac{n!}{(2n)!} \, _3F_2\left(\begin{array}{c} -n, -n, \rho+\sigma+1 \\ \sigma+1, 1; \end{array} \frac{1+z}{2}\right).$$

15. $$\sum_{k=0}^{n} \frac{(-\sigma-n)_k}{(n-k)!(\sigma+1)_k(-\rho-\sigma-n)_k} \left(-\frac{2}{1+z}\right)^k P_k^{(\rho-k,\sigma)}(z)$$
$$= \frac{(\sigma+1)_n}{n!(\rho+\sigma+1)_n} \left(\frac{2}{1+z}\right)^n {}_4F_3\left(\begin{array}{c} -\frac{n}{2}, \frac{1-n}{2}, -\rho, \rho+\sigma+1 \\ \frac{\sigma+1}{2}, \frac{\sigma}{2}+1, \sigma+1; \end{array} \frac{(1+z)^2}{4}\right).$$

16. $$\sum_{k=0}^{n} (2k-\rho) \frac{(-\rho)_k}{(\sigma+1)_k} \left(\frac{2}{1-z}\right)^k P_k^{(\rho-2k,\sigma)}(z)$$
$$= \frac{(-\rho)_{n+1}}{(\sigma+1)_n} \left(\frac{2}{1-z}\right)^n P_n^{(\rho-2n-1,\sigma)}(z).$$

17. $$\sum_{k=0}^{n} (2k-\rho) \frac{(a)_k(-\rho)_k}{(1-a-\rho)_k(-\rho-\sigma)_k} \left(\frac{2}{1-z}\right)^k P_k^{(\rho-2k,\sigma)}(z)$$
$$= \frac{(a)_n(-\rho)_{2n+1}}{n!(1-a-\rho)_n(-\rho-\sigma)_n} \left(\frac{2}{z-1}\right)^n {}_3F_2\left(\begin{array}{c} -n, -n-a, \rho+\sigma-n+1 \\ 1-n-a, \rho-2n; \end{array} \frac{1-z}{2}\right).$$

18. $$\sum_{k=0}^{n} (2k-\rho) \frac{(-n-1)_k(-\rho)_k}{(n-\rho+2)_k(-\rho-\sigma)_k} \left(\frac{2}{1-z}\right)^k P_k^{(\rho-2k,\sigma)}(z)$$
$$= \frac{(-\rho)_{n+2}}{(-\rho-\sigma)_{n+1}} \left(\frac{2}{1-z}\right)^{n+1} \left[1 + (-1)^n (n+1)! \frac{(-\rho)_{n+1}}{(-\rho)_{2n+2}} P_{n+1}^{(\rho-2n-2,\sigma)}(z)\right].$$

19. $$\sum_{k=0}^{n} \frac{(-\rho)_k}{(n-k)!(a)_k(-\rho-\sigma)_k} \left(\frac{2}{z-1}\right)^k P_k^{(\rho-2k,\sigma)}(z)$$
$$= \frac{(-\rho)_{2n}}{n!(a)_n(-\rho-\sigma)_n} \left(\frac{2}{1-z}\right)^n$$
$$\times {}_4F_3\left(\begin{array}{c} -n, \frac{a+\rho-n}{2}, \frac{a+\rho-n+1}{2}, \rho+\sigma-n+1 \\ \frac{\rho+1}{2}-n, \frac{\rho}{2}-n+1, a+\rho-n; \end{array} \frac{1-z}{2}\right).$$

5.12. The Jacobi Polynomials $P_n^{(\rho,\sigma)}(z)$

20. $\sum_{k=0}^{n} \dfrac{(-\rho)_k}{(-\rho-\sigma)_k} \left(\dfrac{2}{1-z}\right)^k P_k^{(\rho-2k,\sigma)}(z)$

$$= \dfrac{(-\rho)_{2n}}{n!(-\rho-\sigma)_n}\left(\dfrac{2}{z-1}\right)^n {}_3F_2\left(\begin{array}{c}-n,\ \frac{\rho}{2}-n,\ \rho+\sigma-n+1\\ \frac{\rho}{2}-n+1,\ \rho-2n;\ \frac{1-z}{2}\end{array}\right).$$

21. $\sum_{k=0}^{n} \sigma_{k+m}^m t^k \dfrac{(\rho+\sigma+n+1)_k}{(k+m)!} P_{n-k}^{(\rho+k,\sigma+k)}(z)$

$$= \dfrac{t^{-m}}{m!(-\rho-\sigma-n)_m}\sum_{k=0}^{m}(-1)^k\binom{m}{k} P_{m+n}^{(\rho-m,\sigma-m)}(2kt+z).$$

22. $\sum_{k=0}^{n} \dfrac{(\rho+\sigma+1)_k}{(n-k)!(1-a-n)_k\left(a+\rho+\sigma+\frac{3}{2}\right)_k}(-2)^{-k}(1+z)^{-k}P_k^{(\rho+k,\sigma-k)}(z)$

$$= \dfrac{(-\sigma)_n(\rho+\sigma+1)_n}{n!(a)_n\left(a+\rho+\sigma+\frac{3}{2}\right)_n} 2^{-n}(1+z)^{-n}$$

$$\times {}_3F_2\left(\begin{array}{c}-n,\ 2a+\rho+\sigma+n+1,\ -2a-\rho-\sigma-n\\ \sigma-n+1,\ -\rho-\sigma-n;\ \frac{1+z}{2}\end{array}\right).$$

23. $\sum_{k=0}^{n} \dfrac{(\rho+\sigma+1)_k}{(n-k)!(a)_k\left(\rho+\sigma-a-n+\frac{5}{2}\right)_k}(-2)^{-k}(1+z)^{-k} P_k^{(\rho+k,\sigma-k)}(z)$

$$= \dfrac{(-\sigma)_n(\rho+\sigma+1)_n}{n!(a)_n\left(a-\rho-\sigma-\frac{3}{2}\right)_n} 2^{-n}(1+z)^{-n}$$

$$\times {}_3F_2\left(\begin{array}{c}-n,\ 2a-\rho-\sigma+n-2,\ \rho+\sigma-2a-n+3\\ \sigma-n+1,\ -\rho-\sigma-n;\ \frac{1+z}{2}\end{array}\right).$$

24. $\sum_{k=0}^{n} \dfrac{\left(\rho+\sigma+n+\frac{1}{2}\right)_k}{(n-k)!(\rho+1)_{2k}} 2^k(z-1)^k P_k^{(\rho+k,\sigma-k)}(z)$

$$= \dfrac{(2n)!}{n!(\rho+1)_{2n}} P_{2n}^{(\rho,\rho+2\sigma)}(z).$$

25. $\sum_{k=0}^{n} \dfrac{\left(a+n-\frac{1}{2}\right)_k(\rho+\sigma+1)_k}{(n-k)!(a)_k(\rho+1)_{2k}} 2^k(z-1)^k P_k^{(\rho+k,\sigma-k)}(z)$

$$= \dfrac{1}{n!} {}_3F_2\left(\begin{array}{c}-2n,\ 2n+2a-1,\ \rho+\sigma+1\\ a,\ \rho+1;\ \frac{1-z}{2}\end{array}\right).$$

26. $\displaystyle\sum_{k=0}^{n} \frac{(a)_k (\rho+\sigma+1)_k}{(n-k)!\,(a-n+\frac{1}{2})_k (\rho+1)_{2k}} 2^k (z-1)^k P_k^{(\rho+k,\sigma-k)}(z)$

$$= (-1)^n \frac{(a)_n (\rho+\sigma+1)_{2n}}{\left(\frac{1}{2}-a\right)_n (\rho+1)_{2n}} (z-1)^{2n}\, {}_3F_2\!\left(\begin{array}{c} -2n,\, \frac{1}{2}-a-n,\, -\rho-2n \\ 1-2a-2n,\, -\rho-\sigma-2n;\ \frac{2}{1-z} \end{array}\right).$$

27. $\displaystyle\sum_{k=0}^{n} \frac{(1-a-n)_k}{(-\rho-\sigma)_k} \left(\frac{2}{1-z}\right)^k P_k^{(\rho-k,\sigma-k)}(z)$

$$= \frac{(a)_n (-\rho)_n}{n!\,(-\rho-\sigma)_n} \left(\frac{2}{1-z}\right)^n {}_3F_2\!\left(\begin{array}{c} -n,\, a-1,\, \rho+\sigma-n+1 \\ a,\, \rho-n+1;\ \frac{1-z}{2} \end{array}\right).$$

28. $\displaystyle\sum_{k=0}^{n} \frac{t^k}{(n-k)!\,(-\rho-\sigma)_k} P_k^{(\rho-k,\sigma-k)}(z) = \frac{t^n}{(-\rho-\sigma)_n} P_n^{(\rho-n,\sigma-n)}\!\left(z - \frac{2}{t}\right).$

29. $\displaystyle\sum_{k=0}^{n} \frac{1}{(n-k)!\,(-\rho-\sigma)_k (n-k+a)} \left(\frac{2}{z-1}\right)^k P_k^{(\rho-k,\sigma-k)}(z)$

$$= \frac{(-\rho)_n}{n!\,a(-\rho-\sigma)_n} \left(\frac{2}{z-1}\right)^n {}_3F_2\!\left(\begin{array}{c} -n,\, 1,\, \rho+\sigma-n+1 \\ a+1,\, \rho-n+1;\ \frac{1-z}{2} \end{array}\right).$$

30. $\displaystyle\sum_{k=0}^{n} \frac{(\rho+\sigma-n+1)_k}{(n-k)!\,(\rho-n+1)_k (-\rho-\sigma)_k} P_k^{(\rho-k,\sigma-k)}(z)$

$$= \frac{(-\rho-\sigma)_n}{n!\,(-\rho)_n} \left(\frac{1-z}{2}\right)^n {}_4F_3\!\left(\begin{array}{c} -\frac{n}{2},\, \frac{1-n}{2},\, -\rho,\, -\sigma;\ \frac{4}{(1-z)^2} \\ -\frac{\rho+\sigma}{2},\, \frac{1-\rho-\sigma}{2},\, -\rho-\sigma \end{array}\right).$$

31. $\displaystyle\sum_{k=0}^{n} \frac{\left(\frac{1}{2}-a-n\right)_k}{(n-k)!\,\left(\frac{3}{2}-n\right)_k (-\rho-\sigma)_k} \left(\frac{2}{z-1}\right)^k P_k^{(\rho-k,\sigma-k)}(z)$

$$= \frac{\left(a+\frac{1}{2}\right)_n (-\rho)_n}{2an!\,\left(-\frac{1}{2}\right)_n (-\rho-\sigma)_n} \left(\frac{2}{1-z}\right)^n$$

$$\times \left[(2a-1)\, {}_3F_2\!\left(\begin{array}{c} -n,\, a,\, \rho+\sigma-n+1 \\ a-\frac{1}{2},\, \rho-n+1;\ \frac{1-z}{2} \end{array}\right) + {}_3F_2\!\left(\begin{array}{c} -n,\, a,\, \rho+\sigma-n+1 \\ a+\frac{1}{2},\, \rho-n+1;\ \frac{1-z}{2} \end{array}\right)\right].$$

32. $\displaystyle\sum_{k=0}^{n} \binom{m}{n-k} \frac{(m-\sigma)_k}{(-\rho-\sigma)_k} \left(\frac{2}{z+1}\right)^k P_k^{(\rho-k,\sigma-k)}(z)$

$$= \frac{(-\sigma)_n}{(-\rho-\sigma)_n} \left(\frac{2}{z+1}\right)^n P_n^{(\rho+m-n,\sigma-m-n)}(z).$$

33. $\displaystyle\sum_{k=0}^{n} \frac{\sigma_m^{n-k+1}}{(-\rho-\sigma)_k}\left(\frac{2}{1-z}\right)^k P_k^{(\rho-k,\sigma-k)}(z)$

$\displaystyle = \frac{(-\rho)_n}{n!(-\rho-\sigma)_n}\left(\frac{2}{z-1}\right)^n {}_{m+1}F_m\left(\begin{array}{c}-n,\rho+\sigma-n+1,2,\ldots,2\\ \rho-n+1,1,\ldots,1;\ \frac{1-z}{2}\end{array}\right)$ $[m \geq 1]$.

34. $\displaystyle\sum_{k=0}^{n} \frac{(a)_k}{(n-k)!(-\rho-\sigma)_{2k}}\left(\frac{2}{z-1}\right)^k P_k^{(\rho-k,\sigma-2k)}(z)$

$\displaystyle = \frac{(a)_n(-\rho)_n}{n!(-\rho-\sigma)_{2n}}\left(\frac{2}{1-z}\right)^n {}_3F_2\left(\begin{array}{c}-n,\rho+\sigma-2n+1,\rho+\sigma+a-n+1\\ \rho-n+1,1-a-n;\ \frac{z-1}{2}\end{array}\right)$.

35. $\displaystyle\sum_{k=0}^{n} \frac{\left(\frac{1}{2}-\rho-n\right)_k (-\sigma)_k}{(n-k)!(1-\rho-2n)_k(-\rho-\sigma)_{2k}}\left(\frac{4}{1+z}\right)^{2k} P_k^{(\rho-k,\sigma-2k)}(z)$

$\displaystyle = \frac{(2\rho)_{2n}(-\sigma)_{2n}}{n!(\rho)_{2n}(-\rho-\sigma)_{2n}}\left(\frac{2}{1+z}\right)^{2n} {}_3F_2\left(\begin{array}{c}-2n,\rho,\rho+\sigma-2n+1;\ \frac{1+z}{2}\\ 2\rho,\sigma-2n+1\end{array}\right)$.

36. $\displaystyle\sum_{k=0}^{n} \frac{\left(\frac{1}{2}-\rho-n\right)_k (-\sigma)_k}{(n-k)!(1-\rho-2n)_k(-\rho-\sigma)_{2k}}\left(\frac{4}{1+z}\right)^{2k} P_k^{(\rho-k,\sigma-2k)}(z)$

$\displaystyle = \frac{(2\rho)_{2n}(-\sigma)_{2n}}{n!(\rho)_{2n}(-\rho-\sigma)_{2n}}\left(\frac{2}{1+z}\right)^{2n} {}_3F_2\left(\begin{array}{c}-2n,\rho,\rho+\sigma-2n+1;\ \frac{1+z}{2}\\ 2\rho,\sigma-2n+1\end{array}\right)$.

37. $\displaystyle\sum_{k=0}^{n}(-1)^k \frac{(a+k\sigma)_{n-1}}{(n-k)!(a+k\sigma)_k} P_k^{(\rho+k\sigma,\tau-k(\sigma+1))}(z)$

$\displaystyle = \frac{(-1)^n}{a+n\sigma+n-1} P_n^{(\rho-a-n+1,a+\tau-1)}(z)$.

38. $\displaystyle\sum_{k=0}^{n} \frac{(a+k\sigma)_{n-1}}{(n-k)!(a+k\sigma)_k}\left(\frac{2}{z-1}\right)^k P_k^{(\rho-k,\tau-k(\sigma+1))}(z)$

$\displaystyle = \frac{1}{a+n\sigma+n-1}\left(\frac{2}{z-1}\right)^n P_n^{(\rho-n,a+\tau-1)}(z)$.

39. $\displaystyle\sum_{k=0}^{n}(-1)^k \binom{n}{k}\frac{(2k)!}{(\rho+1)_{2k}} P_{2k}^{(\rho,\sigma-2k)}(z)$

$\displaystyle = \frac{n!(\rho+\sigma+1)_n}{(\rho+1)_{2n}}(1-z)^n P_n^{(\rho+n,\sigma-n)}\left(\frac{1+z}{2}\right)$.

40. $\displaystyle\sum_{k=0}^{n}\binom{n}{k} P_{k+m}^{(\rho-k,\sigma)}(z) = P_{m+n}^{(\rho,\sigma-n)}(z)$.

41. $\sum_{k=0}^{n} \binom{n}{k} \frac{(a)_k}{(\sigma+m+1)_k} P_{k+m}^{(\rho-k,\sigma)}(z)$

$$= (-1)^m \frac{(m-a+1)_n (\sigma+1)_m}{(m+n)!} \, _3F_2 \left(\begin{array}{c} -m-n,\, a-m,\, \rho+\sigma+m+1 \\ a-m-n,\, \sigma+1;\, \frac{1+z}{2} \end{array} \right).$$

42. $\sum_{k=0}^{n} \binom{n}{k} \frac{(m+1)_k}{(\rho+\sigma+2m-n+2)_k} \left(\frac{2}{1-z} \right)^k P_{k+m}^{(\rho-k,\sigma)}(z)$

$$= \frac{(-\rho-m)_n}{(-\rho-\sigma-2m-1)_n} \left(\frac{2}{1-z} \right)^n P_m^{(\rho-n,\sigma)}(z).$$

43. $\sum_{k=0}^{n} \binom{n}{k} \left(\frac{2}{z+1} \right)^k P_{k+m}^{(\rho-k,\sigma-k)}(z) = \left(\frac{2}{z+1} \right)^n P_{m+n}^{(\rho,\sigma-n)}(z).$

44. $\sum_{k=0}^{n} \binom{n}{k} \frac{(-\sigma-m)_k}{(\rho-n+1)_k} \left(\frac{1-z}{2} \right)^k P_{m-k}^{(\rho+k,\sigma)}(z)$

$$= \frac{(\rho+1)_m}{(\rho-n+1)_m} P_m^{(\rho-n,\sigma)}(z) \quad [m \geq n].$$

45. $\sum_{k=0}^{n} \binom{n}{k} \frac{(a-\rho-\sigma-m+n-1)_k}{(a)_k} \left(\frac{1-z}{2} \right)^k P_{m-k}^{(\rho+k,\sigma)}(z)$

$$= \frac{(\rho+1)_m}{m!} \, _3F_2 \left(\begin{array}{c} -m,\, \rho+\sigma+m-n+1,\, a+n \\ \rho+1,\, a;\, \frac{1-z}{2} \end{array} \right) \quad [m \geq n].$$

46. $\sum_{k=0}^{n} \binom{n}{k} P_{m-k}^{(\rho+k,\sigma)}(z) = P_m^{(\rho+n,\sigma-n)}(z) \quad [m \geq n].$

47. $\sum_{k=0}^{n} \binom{n}{k} k^r P_{m-k}^{(\rho+k,\sigma)}(z)$

$$= \left(\frac{1-z}{2} \right)^m \sum_{k=1}^{r} \sigma_r^k (-n)_k \left(\frac{2}{z-1} \right)^k P_{m-k}^{(k-\rho-\sigma-2m-1,\sigma+k-n)} \left(\frac{z+3}{z-1} \right)$$

$$[m \geq n;\, m \geq r].$$

48. $\sum_{k=0}^{n} (-1)^k \binom{n}{k} (m-k)! (\rho+\sigma+m+1)_k \left(\frac{z+1}{2} \right)^k P_{m-k}^{(\rho+k,\sigma+k)}(z)$

$$= (m-n)! (-\sigma-m)_n P_{m-n}^{(\rho+n,\sigma)}(z) \quad [m \geq n].$$

49. $\sum_{k=0}^{n} \binom{n}{k} \left(\frac{z+1}{2} \right)^k P_{m-k}^{(\rho+k,\sigma+k)}(z) = P_m^{(\rho+n,\sigma)}(z) \quad [m \geq n].$

5.12. The Jacobi Polynomials $P_n^{(\rho,\sigma)}(z)$

50. $\displaystyle\sum_{k=0}^{n} \binom{n}{k} \frac{(\rho+\sigma+m+1)_k}{(\sigma-n+1)_k} \left(\frac{z+1}{2}\right)^k P_{m-k}^{(\rho+k,\sigma+k)}(z)$

$$= \frac{(-\sigma-m)_n}{(-\sigma)_n} P_m^{(\rho+n,\sigma-n)}(z) \quad [m \geq n].$$

51. $\displaystyle\sum_{k=0}^{n} \frac{\left(\frac{1}{2}-a\right)_k (\rho+\sigma+1)_k}{(n-k)!(1-a-n)_k(\rho+1)_{2k}} 2^k (z-1)^k P_k^{(\rho+k,\sigma-k)}(z)$

$$= (-1)^n \frac{\left(\frac{1}{2}-a\right)_n (\rho+\sigma+1)_{2n}}{n!(a)_n(\rho+1)_{2n}} (z-1)^{2n} \, _3F_2\!\left(\begin{array}{c} -2n,\, a-n,\, -\rho-2n \\ 2a-2n,\, -\rho-\sigma-2n;\, \frac{2}{1-z} \end{array}\right).$$

52. $\displaystyle\sum_{k=0}^{n} \frac{(\rho+\sigma+1)_k}{(n-k)!(a)_k \left(\rho+\sigma-a-n-\frac{5}{2}\right)_k} (-2)^{-k} (1+z)^{-k} P_k^{(\rho+k,\sigma-k)}(z)$

$$= \frac{(-\rho)_n(\rho+\sigma+1)_n}{n!(a)_n \left(a-\rho-\sigma-\frac{3}{2}\right)_n} 2^{-n}(1+z)^{-n}$$

$$\times {}_3F_2\!\left(\begin{array}{c} -n,\, 2a-\rho-\sigma+n,\, \rho+\sigma-2a-n+3 \\ \rho-n-1,\, -\rho-\sigma-n;\, \frac{1+z}{2} \end{array}\right).$$

53. $\displaystyle\sum_{k=0}^{n} \binom{n}{k} \frac{(-\sigma-m)_{2k}}{\left(\frac{1}{2}-m-n\right)_k (-\rho-\sigma-m)_k} 2^{-k}(z+1)^{-k} P_{m-k}^{(\rho+k,\sigma-k)}(z)$

$$= \frac{(\rho+\sigma+m+1)_m}{m!} \left(\frac{z+1}{2}\right)^m {}_2F_2\!\left(\begin{array}{c} -2m,\, -m-n,\, -\sigma-m;\, \frac{2}{1+z} \\ -2m-2n,\, -\rho-\sigma-2m \end{array}\right)$$

$$[m \geq n \geq 1].$$

54. $\displaystyle\sum_{k=0}^{n} \binom{n}{k} \frac{\left(m+n+\frac{1}{2}\right)_k (\rho+\sigma+m+1)_k}{(\rho+m+1)_{2k}} 2^k(z-1)^k P_{k+m}^{(\rho+k,\sigma-k)}(z)$

$$= \frac{(\rho+1)_m}{m!} \, _3F_2\!\left(\begin{array}{c} -m-2n,\, m+2n+1,\, \rho+\sigma+m+1 \\ m+1,\, \rho+1;\, \frac{1-z}{2} \end{array}\right).$$

55. $\displaystyle\sum_{k=0}^{n} \frac{(-\sigma)_k}{(n-k)!(1-a-n)_k \left(a-\sigma+\frac{1}{2}\right)_k} (-4)^{-k} P_k^{(\rho+k,\sigma-2k)}(z)$

$$= \frac{(-\sigma)_n(\rho+\sigma+1)_n}{n!(a)_n \left(a-\sigma+\frac{1}{2}\right)_n} \left(\frac{z+1}{8}\right)^n {}_3F_2\!\left(\begin{array}{c} -n,\, 2a-\sigma+n,\, \sigma-2a-n+1 \\ \sigma-n+1,\, -\rho-\sigma-n;\, \frac{2}{z+1} \end{array}\right).$$

56. $\displaystyle\sum_{k=0}^{n} \frac{\left(\frac{1}{2}-a\right)_k (-\sigma)_k}{(n-k)!(1-a-n)_k(-\rho-\sigma)_{2k}} \left(\frac{4}{z+1}\right)^{2k} P_k^{(\rho-k,\sigma-2k)}(z)$

$$= (-1)^n \frac{\left(\frac{1}{2}-a\right)_n (-\sigma)_{2n}}{n!(a)_n(-\rho-\sigma)_{2n}} \left(\frac{4}{z+1}\right)^{2n} {}_3F_2\!\left(\begin{array}{c} -2n,\, a-n,\, \rho+\sigma-2n+1 \\ 2a-2n,\, \sigma-2n+1;\, \frac{z+1}{2} \end{array}\right).$$

57. $\sum_{k=0}^{n} \binom{n}{k} \frac{(\rho+\sigma+m+1)_{2k}}{\left(\frac{1}{2}-m-n\right)_k (\sigma+m+1)_k} \left(\frac{z+1}{4}\right)^{2k} P_{m-k}^{(\rho+k,\sigma+2k)}(z)$

$= (-1)^m \frac{(\sigma+1)_m}{m!} \, _3F_2\left(\begin{array}{c} -2m, -m-n, \rho+\sigma+m+1 \\ -2m-2n, \sigma+1; \frac{z+1}{2} \end{array}\right)$ $[m \geq n \geq 1]$.

58. $\sum_{k=0}^{n} \frac{(-\sigma)_k}{(n-k)!(a)_k \left(\frac{3}{2}-a-\sigma-n\right)_k} (-4)^{-k} P_k^{(\rho+k,\sigma-2k)}(z)$

$= \frac{(-\sigma)_n(\rho+\sigma+1)_n}{n!(a)_n \left(a+\sigma-\frac{1}{2}\right)_n} \left(\frac{1+z}{8}\right)^n \, _3F_2\left(\begin{array}{c} -n, 2a+\sigma+n-1, 2-2a-\sigma-n \\ \sigma-n+1, -\rho-\sigma-n; \frac{2}{1+z} \end{array}\right)$.

59. $\sum_{k=0}^{n} \binom{n}{k} \frac{(-\rho-2m)_k}{\left(\frac{1}{2}-\rho-2m-n\right)_k} 2^{-2k} P_{2m-2k}^{(\rho+k,\sigma+k)}(z)$

$= \frac{(\rho+\sigma+2m+1)_{2m}}{(2m)!} \left(\frac{1-z}{2}\right)^{2m} \, _3F_2\left(\begin{array}{c} -2m, -\rho-2m-n, -2\rho-4m \\ -2\rho-4m-2n, -\rho-\sigma-4m; \frac{2}{1-z} \end{array}\right)$ $[m \geq n]$.

60. $\sum_{k=0}^{n} \frac{(1-2a-n)_k \left(\frac{1}{2}\right)_k}{(n-k)!(1-a-n)_k (\sigma+1)_k (-\rho-\sigma)_k} \left(-\frac{2}{1+z}\right)^k P_{2k}^{(\rho-2k,\sigma-k)}(z)$

$= \frac{(2a)_n (\rho+\sigma+1)_n}{n!(a)_n (\sigma+1)_n} \left(\frac{1+z}{8}\right)^n$

$\times \, _4F_3\left(\begin{array}{c} -n, \frac{1}{2}-a-n, -\frac{\sigma+n}{2}, \frac{1}{2}\frac{\sigma-n}{2} \\ a+\frac{1}{2}, -\frac{\rho+\sigma+n}{2}, \frac{1-\rho-\sigma-n}{2}; \frac{4}{(1+z)^2} \end{array}\right)$.

61. $\sum_{k=0}^{n} \frac{(a)_k \left(\frac{1}{2}\right)_k}{(n-k)! \left(\frac{a-n+1}{2}\right)_k (\sigma+1)_k (-\rho-\sigma)_k} \left(-\frac{2}{1+z}\right)^k P_{2k}^{(\rho-2k,\sigma-k)}(z)$

$= \frac{(a)_n (-\sigma)_n}{n! \left(\frac{1-a-n}{2}\right)_n (-\rho-\sigma)_n} (-2)^{-n}(1+z)^{-n}$

$\times \, _4F_3\left(\begin{array}{c} -n, \frac{a-n}{2}, \frac{\rho+\sigma-n+1}{2}, \frac{\rho+\sigma-n}{2}+1 \\ 1-\frac{a+n}{2}, \frac{\sigma-n+1}{2}, \frac{\sigma-n}{2}+1; \frac{(1+z)^2}{4} \end{array}\right)$.

62. $\sum_{k=0}^{n} \frac{(1-a-n)_k}{(n-k)!(\sigma+1)_k (-\rho-\sigma)_k} \left(-\frac{2}{1+z}\right)^k P_{2k}^{(\rho-2k,\sigma-k)}(z)$

$= \frac{(a)_n (-\sigma)_n}{(2n)!(-\rho-\sigma)_n} \left(-\frac{2}{1+z}\right)^n \, _3F_2\left(\begin{array}{c} -2n, 1-a-2n, \rho+\sigma-n+1 \\ a, \sigma-n+1; -\frac{1+z}{2} \end{array}\right)$.

63. $\displaystyle\sum_{k=0}^{n} \binom{n}{k} \frac{(\rho+\sigma+2m+1)_k}{\left(\rho+\sigma+2m-n+\frac{3}{2}\right)_k} \left(\frac{1+z}{4}\right)^{2k} P_{2m-2k}^{(\rho+k,\sigma+2k)}(z)$

$$= \frac{(\sigma+1)_{2m}}{(2m)!} \,_3F_2\left(\begin{array}{c} -2m,\, \rho+\sigma+2m-n+1,\, 2\rho+2\sigma+4m+2 \\ \sigma+1,\, 2\rho+2\sigma+4m-2n+2;\, \frac{1+z}{2} \end{array}\right) \quad [m \geq n].$$

64. $\displaystyle\sum_{k=0}^{[n/2]} (-1)^k \frac{(n-k)!}{k!} (-\rho-n)_{2k} \left(\frac{1+z}{4}\right)^{2k} P_{n-2k}^{(\rho,\sigma+2k)}(z)$

$$= (\sigma+1)_n \left(\frac{z-1}{2}\right)^n \,_3F_2\left(\begin{array}{c} -n,\, -n-\frac{1}{2},\, -n-\rho \\ -2n-1,\, \sigma+1;\, \frac{2z+2}{z-1} \end{array}\right).$$

65. $\displaystyle\sum_{k=0}^{[n/2]} (-1)^k \frac{(n-k)!}{k!} (\rho+\sigma+n+1)_{2k} \left(\frac{1+z}{4}\right)^{2k} P_{n-2k}^{(\rho+2k,\sigma+2k)}(z)$

$$= (-1)^n (\sigma+1)_n \,_3F_2\left(\begin{array}{c} -n,\, -n-\frac{1}{2},\, \rho+\sigma+n+1 \\ -2n-1,\, \sigma+1;\, z+1 \end{array}\right).$$

66. $\displaystyle\sum_{k=0}^{[n/3]} (\rho+\sigma+2n-6k+1) \frac{(-\rho-n)_{3k}}{k!(-\rho-\sigma-n)_{3k}} \left(-\frac{\rho+\sigma+2n+1}{3}\right)_k$

$\times P_{n-3k}^{(\rho,\sigma)}(z) = -\frac{(-\rho-\sigma-2n-1)_{n+1}}{n!} \left(\frac{1-z}{2}\right)^n$

$$\times \,_3F_2\left(\begin{array}{c} -n,\, -\rho-n,\, -\frac{\rho+\sigma+2n+1}{3} \\ -\frac{\rho+\sigma+2n+1}{2},\, -\frac{\rho+\sigma+2n}{2};\, \frac{3}{2-2z} \end{array}\right).$$

67. $\displaystyle\sum_{k=0}^{[n/3]} (\rho+6k) \frac{\left(\frac{\rho}{3}\right)_k (\rho+\sigma+n+1)_{3k}}{k!(\rho+n+1)_{3k}} \left(\frac{1-z}{2}\right)^{3k} P_{n-3k}^{(\rho+6k,\sigma)}(z)$

$$= \frac{(\rho)_{n+1}}{n!} \,_3F_2\left(\begin{array}{c} -n,\, \frac{\rho}{3},\, \rho+\sigma+n+1 \\ \frac{\rho}{2},\, \frac{\rho+1}{2};\, \frac{3-3z}{8} \end{array}\right).$$

68. $\displaystyle\sum_{k=0}^{n} \frac{(1-a-n)_k \left(a+2n-\frac{1}{2}\right)_k}{(n-k)!(\sigma+1)_{2k}(-\rho-\sigma)_k} \left(\frac{8}{1+z}\right)^k P_{3k}^{(\rho-3k,\sigma-k)}(z)$

$$= \frac{(a)_n \left(a+2n-\frac{1}{2}\right)_n (\rho+\sigma+1)_{2n}}{(3n)!(\sigma+1)_{2n}} (1+z)^{2n}$$

$$\times \,_4F_3\left(\begin{array}{c} -3n,\, a-\frac{1}{2},\, 1-a-3n,\, -\sigma-2n \\ 2a-1,\, 2-2a-6n,\, -\rho-\sigma-2n;\, \frac{8}{1+z} \end{array}\right).$$

69. $\sum_{k=0}^{n} \frac{(1-a-n)_k \left(a+2n-\frac{1}{2}\right)_k}{(n-k)!(\sigma+1)_k(-\rho-\sigma)_{2k}} \left(\frac{4}{1+z}\right)^{2k} P_{3k}^{(\rho-3k,\sigma-2k)}(z)$

$$= \frac{(a)_n \left(a+2n-\frac{1}{2}\right)_n (-\sigma)_{2n}}{(3n)!(-\rho-\sigma)_{2n}} \left(\frac{4}{1+z}\right)^{2n}$$

$$\times {}_4F_3\left(\begin{matrix} -3n,\, a-\frac{1}{2},\, 1-a-3n,\, \rho+\sigma-2n+1 \\ 2a-1,\, 2-2a-6n,\, \sigma-2n+1;\, 2z+2 \end{matrix}\right).$$

5.12.3. Sums containing $P_{n \pm mk}^{(\rho \pm pk, \sigma \pm qk)}(z)$ and special functions

1. $\sum_{k=0}^{n} \frac{1}{(n-k+1)!(-\rho-\sigma)_k} \left(\frac{2}{z-1}\right)^k \psi(a-k) P_k^{(\rho-k,\sigma-k)}(z)$

$$= \frac{1}{(-\rho-\sigma)_{n+1}} \left(\frac{2}{1-z}\right)^{n+1} \psi(a-n-1) \left[\frac{(-\rho)_{n+1}}{(n+1)!} - P_{n+1}^{(\rho-n-1,\sigma-n-1)}(z)\right]$$

$$+ \frac{(-\rho)_n}{n!(-\rho-\sigma)_n(a-n-1)} \left(\frac{2}{1-z}\right)^n {}_4F_3\left(\begin{matrix} -n,\, \rho+\sigma-n+1,\, 1,\, 1; \\ \rho-n+1,\, a-n,\, 2 \end{matrix} \frac{1-z}{2}\right).$$

2. $\sum_{k=0}^{n} \frac{(1-a)_k}{(-\rho-\sigma)_k} \left(\frac{2}{1-z}\right)^k \psi(a-k) P_k^{(\rho-k,\sigma-k)}(z)$

$$= \frac{(-\rho)_n (1-a)_n}{n!(-\rho-\sigma)_n (a-n-1)} \left(\frac{2}{z-1}\right)^n$$

$$\times \left[(a-n-1)\psi(a-n-1) {}_3F_2\left(\begin{matrix} -n,\, \rho+\sigma-n+1,\, a-n-1 \\ \rho-n+1,\, a-n;\, \frac{1-z}{2} \end{matrix}\right)\right.$$

$$\left. + {}_4F_3\left(\begin{matrix} -n,\, \rho+\sigma-n+1,\, a-n-1,\, a-n-1 \\ \rho-n+1,\, a-n,\, a-n;\, \frac{1-z}{2} \end{matrix}\right)\right].$$

3. $\sum_{k=0}^{n} \left(\sqrt{w^2-1}-w\right)^k \frac{(-n)_k}{(\sigma+1)_k} P_{n-k}(w) P_k^{(\rho-k,\sigma)}(z)$

$$= 2^n \frac{\left(\frac{1}{2}\right)_n}{n!} (w^2-1)^{n/2} {}_3F_2\left(\begin{matrix} -n,\, -n,\, \rho+\sigma+1 \\ \sigma+1,\, \frac{1}{2}-n;\, \frac{z+1}{4} - \frac{z+1}{4} w(w^2-1)^{-1/2} \end{matrix}\right).$$

4. $\sum_{k=0}^{n} \frac{(-2i)^k}{(1-z^2)^{k/2}} P_{n-k}\left(\frac{iz}{\sqrt{1-z^2}}\right) P_k^{(\rho-k,\sigma-k)}(z)$

$$= \frac{(-2i)^n}{(1-z^2)^{n/2}} P_n^{(\rho-n-1/2,\sigma-n-1/2)}(z).$$

5. $\displaystyle\sum_{k=0}^{n} \frac{1}{(2n-2k)!(\sigma+1)_k} \left(\frac{2}{z-1}\right)^k H_{2n-2k}(w) P_k^{(\rho-k,\sigma)}(z)$

$$= \frac{(-\rho)_n}{n!(\sigma+1)_n} \left(\frac{1+z}{1-z}\right)^n {}_2F_2\left(\begin{array}{c}-n,\ -n-\sigma;\ \frac{w^2(z-1)}{z+1}\\ \frac{1}{2},\ \rho-n+1\end{array}\right).$$

6. $\displaystyle\sum_{k=0}^{n} \frac{1}{(2n-2k+1)!(\sigma+1)_k} \left(\frac{2}{z-1}\right)^k H_{2n-2k+1}(w) P_k^{(\rho-k,\sigma)}(z)$

$$= \frac{2(-\rho)_n w}{n!(\sigma+1)_n} \left(\frac{1+z}{1-z}\right)^n {}_2F_2\left(\begin{array}{c}-n,\ -n-\sigma;\ \frac{w^2(z-1)}{z+1}\\ \frac{3}{2},\ \rho-n+1\end{array}\right).$$

7. $\displaystyle\sum_{k=0}^{n} \frac{(-w)^k}{(\sigma+1)_k} L_{n-k}^{\lambda+k}(w) P_k^{(\rho-k,\sigma)}(z)$

$$= \frac{(\rho+\sigma+1)_n}{n!(\sigma+1)_n} (-w)^n \left(\frac{z+1}{2}\right)^n {}_3F_1\left(\begin{array}{c}-n,\ -\lambda-n,\ -\sigma-n\\ -\rho-\sigma-n;\ -\frac{2}{w(z+1)}\end{array}\right).$$

8. $\displaystyle\sum_{k=0}^{n} \frac{(-\lambda-n)_k}{(-\rho-\sigma)_k} \left(\frac{2}{1-z}\right)^k L_{n-k}^{\lambda}(w) P_k^{(\rho-k,\sigma-k)}(z)$

$$= \frac{(\lambda+1)_n(-\rho)_n}{n!(-\rho-\sigma)_n} \left(\frac{2}{1-z}\right)^n {}_2F_2\left(\begin{array}{c}-n,\ \rho+\sigma-n+1;\ \frac{w-wz}{2}\\ \lambda+1,\ \rho-n+1\end{array}\right).$$

9. $\displaystyle\sum_{k=0}^{n} \frac{\left(\frac{2w}{1-z}\right)^k}{(-\rho-\sigma)_k} L_{n-k}^{\lambda+k}(w) P_k^{(\rho-k,\sigma-k)}(z)$

$$= \frac{(-\rho)_n}{n!(-\rho-\sigma)_n} \left(\frac{2w}{z-1}\right)^n {}_3F_1\left(\begin{array}{c}-n,\ -\lambda-n,\ \rho+\sigma-n+1\\ \rho-n+1;\ \frac{z-1}{2w}\end{array}\right).$$

10. $\displaystyle\sum_{k=0}^{n} (2k+\rho+\sigma+1) \frac{(\rho+\sigma+1)_k}{(\rho+1)_k(\rho+\sigma+n+2)_k} w^k L_{n-k}^{2k+\rho+\sigma+1}(w) P_k^{(\rho,\sigma)}(z)$

$$= \frac{(\rho+\sigma+1)_{n+1}}{(\rho+1)_n} L_n^{\rho}\left(\frac{w-wz}{2}\right).$$

11. $\displaystyle\sum_{k=0}^{n} \frac{(-\lambda-n)_k}{(\sigma+1)_k} \left(\frac{2}{z-1}\right)^k L_{n-k}^{\lambda}(w) P_k^{(\rho-k,\sigma)}(z)$

$$= \frac{(-\rho)_n(\lambda+1)_n}{n!(\sigma+1)_n} \left(\frac{z+1}{z-1}\right)^n {}_2F_2\left(\begin{array}{c}-n,\ -n-\sigma;\ \frac{w(z-1)}{z+1}\\ \lambda+1,\ \rho-n+1\end{array}\right).$$

12. $\displaystyle\sum_{k=0}^{n} \left(\frac{2}{z-1}\right)^k L_{n-k}^{-\sigma-n-1}(w) P_k^{(\rho-k,\sigma)}(z) = \left(\frac{z+1}{z-1}\right)^n L_n^{\rho-n}\left(w\frac{z-1}{z+1}\right).$

13. $\sum_{k=0}^{n} \frac{1}{(\sigma+1)_k} \left(\frac{2w}{z-1}\right)^k L_{n-k}^{\lambda+k}(w) P_k^{(\rho-k,\sigma)}(z)$

$$= \frac{(-\rho)_n}{n!(\sigma+1)_n} \left(w \frac{1+z}{1-z}\right)^n {}_3F_1 \left(\begin{array}{c} -n, -n-\lambda, -n-\sigma \\ \rho-n+1; \frac{1-z}{w(1+z)} \end{array}\right).$$

14. $\sum_{k=0}^{n} \left(\frac{2}{z-1}\right)^k L_{n-k}^{k-n-\sigma}(w) P_k^{(\rho-k,\sigma-k)}(z) = \left(\frac{z+1}{z-1}\right)^n L_n^{\rho-n}\left(w\frac{z-1}{z+1}\right).$

15. $\sum_{k=0}^{n} 2^{2k} \frac{(\lambda)_k}{(-\rho-\sigma)_k} \left(\frac{1-w}{1-z}\right)^k C_{n-k}^{\lambda+k}(w) P_k^{(\rho-k,\sigma-k)}(z)$

$$= \frac{2^{2n}(\lambda)_n(-\rho)_n}{n!(-\rho-\sigma)_n} \left(\frac{w-1}{1-z}\right)^n {}_3F_2 \left(\begin{array}{c} -n, \frac{1}{2}-\lambda-n, \rho+\sigma-n+1 \\ 1-2\lambda-2n, \rho-n+1; \frac{1-z}{1-w} \end{array}\right).$$

16. $\sum_{k=0}^{n} (w^2-1)^k \frac{(\lambda)_k}{(\sigma+1)_k} C_{2n-2k}^{\lambda+k}(w) P_k^{(\rho-k,\sigma)}(z)$

$$= w^{2n} \frac{(2\lambda)_{2n}}{(2n)!} {}_3F_2 \left(\begin{array}{c} -n, \frac{1}{2}-n, \rho+\sigma+1 \\ \lambda+\frac{1}{2}, \sigma+1; \frac{1}{2}(1-w^{-2})(z+1) \end{array}\right).$$

17. $\sum_{k=0}^{n} (w^2-1)^k \frac{(\lambda)_k}{(\sigma+1)_k} C_{2n-2k+1}^{\lambda+k}(w) P_k^{(\rho-k,\sigma)}(z)$

$$= w^{2n+1} \frac{(2\lambda)_{2n+1}}{(2n+1)!} {}_3F_2 \left(\begin{array}{c} -n, -n-\frac{1}{2}, \rho+\sigma+1 \\ \lambda+\frac{1}{2}, \sigma+1; \frac{1}{2}(1-w^{-2})(z+1) \end{array}\right).$$

18. $\sum_{k=0}^{n} \frac{(2\lambda)_{2k}}{\left(\lambda+\frac{1}{2}\right)_k (2\lambda+n)_k} C_{n-k}^{\lambda+k}(z) P_k^{(\rho-k,\sigma-k)}(z)$

$$= \frac{(2\lambda)_n}{\left(\lambda+\frac{1}{2}\right)_n} P_n^{(\lambda+\rho-1/2,\lambda+\sigma-1/2)}(z).$$

19. $\sum_{k=0}^{n} \frac{(\lambda)_k}{(-\rho-\sigma)_k} \left(\frac{4w}{z-1}\right)^k C_{n-k}^{\lambda+k}(w) P_k^{(\rho-k,\sigma-k)}(z)$

$$= \frac{(\lambda)_n(-\rho)_n}{n!(-\rho-\sigma)_n} \left(\frac{4w}{z-1}\right)^n {}_4F_3 \left(\begin{array}{c} -\frac{n}{2}, \frac{1-n}{2}, \frac{\rho+\sigma-n+1}{2}, \frac{\rho+\sigma-n+2}{2}; \frac{(1-z)^2}{4w^2} \\ \frac{\rho-n+1}{2}, \frac{\rho-n+2}{2}, 1-\lambda-n \end{array}\right).$$

5.12.4. Sums containing products of $P_{m\pm nk}^{(\rho\pm pk,\sigma\pm qk)}(z)$

1. $\displaystyle\sum_{k=0}^{n} \frac{(\rho+\sigma+n+1)_k}{(\mu+1)_k} \left(\frac{1-z}{2}\right)^k P_k^{(\mu,\nu-k)}(w) P_{n-k}^{(\rho+k,\sigma+k)}(z)$

$$= \frac{(\rho+1)_n}{n!} {}_3F_2\left(\begin{matrix} -n, \mu+\nu+1, \rho+\sigma+n+1 \\ \mu+1, \rho+1; \ \frac{(1-w)(1-z)}{4} \end{matrix}\right).$$

2. $\displaystyle\sum_{k=0}^{n} \frac{(\rho+\sigma+n+1)_k}{(-\mu-\nu)_k} \left(\frac{1-z}{1-w}\right)^k P_k^{(\mu-k,\nu-k)}(w) P_{n-k}^{(\rho+k,\sigma+k)}(z)$

$$= \frac{(\rho+1)_n}{n!} {}_3F_2\left(\begin{matrix} -n, -\mu, \rho+\sigma+n+1 \\ -\mu-\nu, \rho+1; \ \frac{1-z}{1-w} \end{matrix}\right).$$

3. $\displaystyle\sum_{k=0}^{n} \frac{(-n-\sigma)_k}{(\nu+1)_k} \left(\frac{1-z}{2}\right)^k P_k^{(\mu-k,\nu)}(w) P_{n-k}^{(\rho-k,\sigma)}(z)$

$$= \frac{(\rho+1)_n}{n!} \left(\frac{1+z}{2}\right)^n {}_3F_2\left(\begin{matrix} -n, -n-\sigma, \mu+\nu+1 \\ \nu+1, \rho+1; \ \frac{(w+1)(z-1)}{2(z+1)} \end{matrix}\right).$$

4. $\displaystyle\sum_{k=0}^{n} (-1)^k \binom{n}{k} \frac{\left(\rho+m+\frac{3}{2}\right)_k}{\left(\frac{1}{2}-\rho-m-n\right)_k} P_{m-k}^{(\rho+k,k+1/2)}(z) P_{m-n+k}^{(\rho+n-k,n-k-1/2)}(z)$

$$= (-1)^m \frac{(-2m)_n \left(\rho+\frac{1}{2}\right)_n (\rho+1)_m (2\rho+2m+2)_n (-\rho-m)_{m-n}}{(m!)^2 \left(\rho+m+\frac{1}{2}\right)_n (2\rho+1)_n}$$

$$\times {}_3F_2\left(\begin{matrix} n-2m, \rho+n+\frac{1}{2}, 2\rho+2m+n+2 \\ \rho+n+1, 2\rho+n+1; \ \frac{1-z}{2} \end{matrix}\right) \quad [m \geq n].$$

5. $\displaystyle\sum_{k=0}^{n} \frac{(\rho+\sigma+m+1)_k}{k!} \left(\frac{z^2-1}{4}\right)^k P_{m-k}^{(\rho+k,\sigma+k)}(z) P_{n-k}^{(\rho+k-n,\sigma+k-n)}(z)$

$$= \binom{m+n}{m} P_{m+n}^{(\rho-n,\sigma-n)}(z).$$

6. $\displaystyle\sum_{k=0}^{n} \frac{(\rho+\sigma+m+1)_k}{k!} \left(\frac{z^2-1}{4}\right)^k P_{m-k}^{(\rho+k,\sigma+k)}(z) P_{n-k}^{(\rho+k+m,\sigma+k-n)}(z)$

$$= \binom{m+n}{m} P_{m+n}^{(\rho,\sigma-n)}(z).$$

5.12.5. Sums containing $P_{m\pm nk}^{(\rho\pm pk,\sigma\pm qk)}(\varphi(k,z))$

1. $\displaystyle\sum_{k=0}^{n} (-1)^k \binom{n}{k} P_m^{(\rho,\sigma)}(w+kz) = 0 \qquad [m < n].$

2. $\displaystyle\sum_{k=0}^{n}(-1)^k\binom{n}{k}P_{m+n}^{(\rho,\sigma)}(w+kz) = n!\,(\rho+\sigma+m+n+1)_n\left(-\frac{z}{2}\right)^n$

$\displaystyle\times\sum_{k=0}^{m}\sigma_{k+n}^n\,\frac{(\rho+\sigma+m+2n+1)_k}{(k+n)!}\left(\frac{z}{2}\right)^k P_{m-k}^{(\rho+k+n,\sigma+k+n)}(w).$

3. $\displaystyle\sum_{k=1}^{n}(-1)^k\binom{n}{k}(k-z)^m\,P_m^{(\rho,\sigma)}\left(\frac{k+z}{k-z}\right) = (-1)^m(\rho+1)_m\delta_{m,n}$

$\displaystyle\hspace{6cm}-\frac{(\sigma+1)_m}{m!}z^m \quad [n\geq m].$

4. $\displaystyle\sum_{k=0}^{n}(-1)^k\binom{n}{k}(ka+z)^{m+n}P_{m+n}^{(\rho,\sigma)}\left(\frac{ka+w}{ka+z}\right)$

$\displaystyle = n!\,(\rho+m+1)_n(-a)^n z^m \sum_{k=0}^{m}\sigma_{k+n}^n\left(-\frac{a}{z}\right)^k\frac{(-\rho-m)_k}{(k+n)!}P_{m-k}^{(\rho,\sigma+k+n)}\left(\frac{w}{z}\right).$

5. $\displaystyle\sum_{k=1}^{n}\frac{k^{k-1}}{k!}(\rho+\sigma+n+1)_k\left(-\frac{z}{2}\right)^k P_{n-k}^{(\rho+k,\sigma+k)}(1+kz)$

$\displaystyle\hspace{5cm}= -\frac{(\rho+2)_{n-1}(\rho+\sigma+n+1)z}{2(n-1)!}.$

6. $\displaystyle\sum_{k=0}^{n}\frac{(ka+b)^{n-k-1}}{(n-k)!(-\rho-\sigma)_k}\left(\frac{2a}{z}\right)^k P_k^{(\rho-k,\sigma-k)}(1+kz)$

$\displaystyle\hspace{4cm}=\frac{\left(\frac{2a}{z}\right)^n}{(na+b)(-\rho-\sigma)_n}P_n^{(\rho-n,\sigma-n)}\left(1-\frac{bz}{a}\right).$

7. $\displaystyle\sum_{k=0}^{n}\frac{(ka+1)^{n-k-1}}{(n-k)!(-\rho-\sigma)_k}\left(\frac{2}{z}\right)^k P_k^{(\rho-k,\sigma-k)}(1+(ka+1)z)$

$\displaystyle\hspace{6cm}=\frac{(-\rho)_n\left(-\frac{2}{z}\right)^n}{n!\,(na+1)(-\rho-\sigma)_n}.$

8. $\displaystyle\sum_{k=1}^{n}\frac{k^{2k}}{k^2+a^2}(\sigma+n+1)_k\frac{\left(-\frac{z}{2}\right)^k}{(k+n)!}P_{n-k}^{(2k,\sigma)}(1+k^2z)$

$\displaystyle = -\frac{a^{-2}}{2(n!)} - \frac{a^{-2}}{(n+1)!(\sigma+n)z}\left[{}_3F_2\!\left(\begin{matrix}-n-1,\sigma+n,1\\ia,-ia;\end{matrix}\frac{a^2z}{2}\right)-1\right].$

9. $\displaystyle\sum_{k=1}^{n}k^{2k}\frac{(\sigma+n+1)_k}{(k+n)!}\left(-\frac{z}{2}\right)^k P_{n-k}^{(2k,\sigma)}(1+k^2z)$

$\displaystyle\hspace{5cm}=\frac{1}{2(n!)}\left[{}_3F_0\!\left(\begin{matrix}-n,\sigma+n+1,1\\ \\ \frac{z}{2}\end{matrix}\right)-1\right].$

10. $\sum_{k=1}^{n} k^{2k-2} \dfrac{(\sigma+n+1)_k}{(k+n)!} \left(-\dfrac{z}{2}\right)^k P_{n-k}^{(2k,\sigma)}(1+k^2 z) = -\dfrac{(\sigma+n+1)z}{4(n-1)!}.$

11. $\sum_{k=1}^{n} k^{2k-4} \dfrac{(\sigma+n+1)_k}{(k+n)!} \left(-\dfrac{z}{2}\right)^k P_{n-k}^{(2k,\sigma)}(1+k^2 z)$
$$= -\dfrac{(\sigma+n+1)(\sigma+n+2)z^2}{32(n-2)!} - \dfrac{(\sigma+n+1)z}{4(n-1)!}.$$

12. $\sum_{k=1}^{n} k^{2k-6} \dfrac{(\sigma+n+1)_k}{(k+n)!} \left(-\dfrac{z}{2}\right)^k P_{n-k}^{(2k,\sigma)}(1+k^2 z)$
$$= -\dfrac{(\sigma+n+1)(\sigma+n+2)(\sigma+n+3)z^3}{576(n-3)!} - \dfrac{5(\sigma+n+1)(\sigma+n+2)z^2}{128(n-2)!}$$
$$- \dfrac{(\sigma+n+1)z}{4(n-1)!}.$$

13. $\sum_{k=0}^{n} (2k+\rho)^{2k-1} \dfrac{(\rho)_k (\sigma+a+n+1)_k}{k!(a+n+1)_k} \left(-\dfrac{z}{2}\right)^k P_{n-k}^{(2k+\rho,\sigma)}(1+(2k+\rho)^2 z)$
$$= \dfrac{(\rho+1)_n}{n!\rho}.$$

14. $\sum_{k=0}^{n} (2k+1)^{2k-3} \dfrac{(\sigma+n+2)_k}{(k+n+1)!} \left(-\dfrac{z}{2}\right)^k P_{n-k}^{(2k+1,\sigma)}(1+(2k+1)^2 z)$
$$= \dfrac{1}{n!} + \dfrac{2(\sigma+n+2)z}{9(n-1)!}.$$

15. $\sum_{k=0}^{n} (2k+1)^{2k-5} \dfrac{(\sigma+n+2)_k}{(k+n+1)!} \left(-\dfrac{z}{2}\right)^k P_{n-k}^{(2k+1,\sigma)}(1+(2k+1)^2 z)$
$$= \dfrac{1}{n!} + \dfrac{20(\sigma+n+2)z}{81(n-1)!} + \dfrac{4(\sigma+n+2)(\sigma+n+3)z^2}{225(n-2)!}.$$

16. $\sum_{k=0}^{n} \dfrac{(2k+1)^{2k+1}}{(2k+1)^2+a^2} \dfrac{(\sigma+n+2)_k}{(k+n+1)!} \left(-\dfrac{z}{2}\right)^k P_{n-k}^{(2k+1,\sigma)}(1+(2k+1)^2 z)$
$$= \dfrac{z^{-1}}{2(n+1)!(\sigma+n+1)a^2} \left[1 - {}_3F_2\left(\begin{array}{c}-n-1,\,1,\,\sigma+n+1\\ \dfrac{1-ia}{2},\,\dfrac{1+ia}{2}\end{array}; \dfrac{a^2 z}{2}\right)\right].$$

17. $\sum_{k=1}^{n} \dfrac{k^{n-1}}{k!} (-\rho-n)_k P_{n-k}^{(\rho,\sigma+k)}\left(1+\dfrac{z}{k}\right)$
$$= -\dfrac{(\rho+\sigma+n+1)_{n-1}(\rho+n)}{(n-1)!} \left(\dfrac{z}{2}\right)^{n-1}.$$

18. $\sum_{k=1}^{n} \dfrac{(-ka)^k (ka+b)^{n-k-1}}{(n-k)!(\rho+1)_k} P_k^{(\rho,\sigma-k)}\left(1+\dfrac{z}{k}\right)$

$$= -\dfrac{b^{n-1}}{n!} + \dfrac{b^n}{(na+b)(\rho+1)_n} P_n^{(\rho,\sigma-n)}\left(1-\dfrac{az}{b}\right).$$

19. $\sum_{k=0}^{n} \dfrac{a^k (ka+b)^{n-k-1}}{(n-k)!(\sigma+1)_k} (z+k)^k P_k^{(\rho-k,\sigma)}\left(\dfrac{z-k}{z+k}\right)$

$$= \dfrac{(az-b)^n}{(na+b)(\sigma+1)_n} P_n^{(\rho-n,\sigma)}\left(\dfrac{az+b}{az-b}\right).$$

20. $\sum_{k=0}^{n} \dfrac{(-1)^k}{(n-k)!(\rho+1)_k} (ka+1)^{n-1} P_k^{(\rho,\sigma-k)}\left(\dfrac{ka+z}{ka+1}\right)$

$$= \dfrac{(\rho+\sigma+1)_n}{n!(na+1)(\rho+1)_n} \left(\dfrac{1-z}{2}\right)^n.$$

5.12.6. Sums containing $P_{n\pm mk}^{(\rho\pm pk,\sigma\pm qk)}(\varphi(k,z))$ and special functions

1. $\sum_{k=1}^{n} (4k)^k (n-k+1)^{n-k-1} \dfrac{\left(\frac{1}{2}-n\right)_k}{(n-k)!(\rho+1)_k} H_{2n-2k}\left(\dfrac{w}{\sqrt{n-k+1}}\right)$

$$\times P_k^{(\rho,\sigma-k)}\left(1+\dfrac{z}{k}\right) = \dfrac{(n+1)^{n-1}}{n!} H_{2n}\left(\dfrac{w}{\sqrt{n+1}}\right)$$

$$+ (-1)^n (2w)^{2n} \sum_{k=0}^{n} \left(\dfrac{n+1}{w^2}\right)^k \dfrac{\left(\frac{1}{2}-n\right)_k}{(n-k+1)!(\rho+1)_k} P_k^{(\rho,\sigma-k)}\left(1+\dfrac{z}{n+1}\right)$$

$$[n \geq 1].$$

2. $\sum_{k=0}^{n} (-4)^k (k+1)^{n-1} \dfrac{\left(\frac{1}{2}-n\right)_k}{(n-k)!(\rho+1)_k}$

$$\times H_{2n-2k}\left(\dfrac{w}{\sqrt{k+1}}\right) P_k^{(\rho,\sigma-k)}\left(1+\dfrac{z}{k+1}\right)$$

$$= \dfrac{\rho(2w)^{2n+2} z^{-1}}{(n+1)!(2n+1)(\rho+\sigma)} \left[{}_3F_1\left(\begin{matrix} -n-1, -n-\frac{1}{2}, \rho+\sigma \\ \rho;\ \dfrac{z}{2w^2} \end{matrix}\right) - 1\right].$$

3. $\sum_{k=0}^{n} (-4)^k (k+1)^{n-1/2} \dfrac{\left(-\frac{1}{2}-n\right)_k}{(n-k)!(\rho+1)_k}$

$$\times H_{2n-2k+1}\left(\dfrac{w}{\sqrt{k+1}}\right) P_k^{(\rho,\sigma-k)}\left(1+\dfrac{z}{k+1}\right)$$

$$= \dfrac{\rho(2w)^{2n+3} z^{-1}}{(n+1)!(2n+3)(\rho+\sigma)} \left[{}_3F_1\left(\begin{matrix} -n-1, -n-\frac{3}{2}, \rho+\sigma \\ \rho;\ \dfrac{z}{2w^2} \end{matrix}\right) - 1\right].$$

5.12.6]   5.12. The Jacobi Polynomials $P_n^{(\rho,\sigma)}(z)$

4. $\displaystyle\sum_{k=0}^{n} \frac{(k+1)^{k-1}}{(\rho+1)_k} w^k L_{n-k}^{k+\rho+\sigma}((k+1)w)\, P_k^{(\rho,\sigma-k)}\!\left(1+\frac{z}{k+1}\right)$

$$= \frac{2\rho}{wz}(\rho+\sigma+1)_n \left[\frac{1}{(\rho)_{n+1}} L_{n+1}^{\rho-1}\!\left(-\frac{wz}{2}\right) - \frac{1}{(n+1)!} \right].$$

5. $\displaystyle\sum_{k=0}^{n} \frac{(k+1)^{k-1}}{(\rho+1)_k} w^k L_{n-k}^{k+\lambda}((k+1)w)\, P_k^{(\rho,\sigma-k)}\!\left(1+\frac{z}{k+1}\right)$

$$= \frac{2\lambda\rho(wz)^{-1}}{(n+1)!(\rho+\sigma)} \left[1 - {}_2F_2\!\left(\begin{matrix}-n-1, \rho+\sigma \\ \lambda, \rho;\end{matrix}\; -\frac{wz}{2}\right) \right].$$

6. $\displaystyle\sum_{k=0}^{n} \frac{(k+1)^{k-1}}{(\rho+1)_k}(-w)^k L_{n-k}^{\lambda+k}((n-k)w)\, P_k^{(\rho,\sigma-k)}\!\left(1+\frac{z}{k+1}\right)$

$$= \sum_{k=0}^{n} \left(-\frac{wz}{2}\right)^k \frac{(\rho+\sigma+1)_k}{(k+1)!(\rho+1)_k} L_{n-k}^{\lambda+k}((n+1)w).$$

7. $\displaystyle\sum_{k=0}^{n} (k+1)^{n-1} \frac{(-\lambda-n)_k}{(\rho+1)_k} L_{n-k}^{\lambda}\!\left(\frac{w}{k+1}\right) P_k^{(\rho,\sigma-k)}\!\left(1+\frac{z}{k+1}\right)$

$$= (-1)^n \frac{2\rho w^{n+1} z^{-1}}{(n+1)!(\lambda+n+1)(\rho+\sigma)} \left[{}_3F_1\!\left(\begin{matrix}-n-1, -\lambda-n-1, \rho+\sigma \\ \rho;\; \frac{z}{2w}\end{matrix}\right) - 1 \right].$$

8. $\displaystyle\sum_{k=0}^{n} (k+1)^{n-1} L_{n-k}^{-\rho-n-1}\!\left(\frac{w}{k+1}\right) P_k^{(\rho,\sigma-k)}\!\left(1+\frac{z}{k+1}\right)$

$$= \frac{2(-1)^n z^{-1}}{\rho+\sigma} \left[\frac{w^{n+1}}{(n+1)!} - \left(\frac{z}{2}\right)^{n+1} L_{n+1}^{-\rho-\sigma-n-1}\!\left(-\frac{2w}{z}\right) \right].$$

9. $\displaystyle\sum_{k=0}^{n} (k+1)^{n-1} \frac{(-\rho-n)_k}{(1-\lambda)_k} C_{2k}^{\lambda-k}\!\left(\frac{w}{\sqrt{k+1}}\right) P_{n-k}^{(\rho,\sigma+k)}\!\left(1+\frac{z}{k+1}\right)$

$$= \frac{2^{-n-2}(\rho+\sigma+1)_{2n+1} w^{-2} z^{n+1}}{(n+1)!(\rho+\sigma+1)_n (\rho+n+1)(\lambda-1)}$$

$$\times \left[1 - {}_3F_2\!\left(\begin{matrix}-n-1, \lambda-1, -\rho-n-1 \\ -\frac{1}{2}, -\rho-\sigma-2n-1;\; -\frac{2w^2}{z}\end{matrix}\right) \right].$$

10. $\displaystyle\sum_{k=0}^{n} (k+1)^{k-1} \frac{(\rho+\sigma+n+1)_k}{(1-\lambda)_k} \left(-\frac{z}{2}\right)^k$

$$\times C_{2k}^{\lambda-k}\!\left(\frac{w}{\sqrt{k+1}}\right) P_{n-k}^{(\rho+k,\sigma+k)}((k+1)z+1)$$

$$= \frac{\rho(\rho+1)_n (w^2 z)^{-1}}{(n+1)!(\lambda-1)(\rho+\sigma+n)} \left[1 - {}_3F_2\!\left(\begin{matrix}-n-1, \lambda-1, \rho+\sigma+n \\ \rho, -\frac{1}{2};\; -\frac{w^2 z}{2}\end{matrix}\right) \right].$$

409

11. $\displaystyle\sum_{k=0}^{n}(k+1)^{n-1/2}\frac{(-\rho-n)_k}{(1-\lambda)_k}C_{2k+1}^{\lambda-k}\left(\frac{w}{\sqrt{k+1}}\right)P_{n-k}^{(\rho,\sigma+k)}\left(1+\frac{z}{k+1}\right)$

$$=\frac{(\rho+\sigma+1)_{2n+1}w^{-1}\left(\frac{z}{2}\right)^{n+1}}{(n+1)!(\rho+\sigma+1)_n(\rho+n+1)}\left[{}_3F_2\left(\begin{array}{c}-n-1,\lambda,-\rho-n-1;\\-\rho-\sigma-2n-1,\frac{1}{2}\end{array}-\frac{2w^2}{z}\right)-1\right].$$

12. $\displaystyle\sum_{k=0}^{n}(k+1)^{k-1/2}\frac{(\rho+\sigma+n+1)_k}{(1-\lambda)_k}\left(-\frac{z}{2}\right)^k$

$$\times C_{2k+1}^{\lambda-k}\left(\frac{w}{\sqrt{k+1}}\right)P_{n-k}^{(\rho+k,\sigma+k)}((k+1)z+1)$$

$$=\frac{2\rho(\rho+1)_n(wz)^{-1}}{(n+1)!(\rho+\sigma+n)}\left[{}_3F_2\left(\begin{array}{c}-n-1,\lambda,\rho+\sigma+n;\\\rho,\frac{1}{2};\end{array}-\frac{w^2z}{2}\right)-1\right].$$

13. $\displaystyle\sum_{k=0}^{n-1}(-1)^k(k+1)^{k-1}(n-k)^{n-k}\frac{(\lambda)_k}{(\rho+1)_k}$

$$\times C_{2n-2k}^{\lambda+k}\left(\frac{w}{\sqrt{n-k}}\right)P_k^{(\rho,\sigma-k)}\left(1+\frac{z}{k+1}\right)$$

$$=(-1)^{n+1}\frac{(n+1)^{n-1}(\lambda)_n}{(\rho+1)_n}P_n^{(\rho,\sigma-n)}\left(1+\frac{z}{n+1}\right)+\frac{(\lambda)_n(\rho+\sigma+1)_n}{(\rho+1)_n}\left(-\frac{z}{2}\right)^n$$

$$\times\sum_{k=0}^{n}\frac{2^k(-\rho-n)_k}{(n-k+1)!(1-\lambda-n)_k(-\rho-\sigma-n)_k}\left(\frac{n+1}{z}\right)^k C_{2k}^{\lambda+n-k}\left(\frac{w}{\sqrt{n+1}}\right).$$

14. $\displaystyle\sum_{k=0}^{n-1}(-1)^k(k+1)^{k-1}(n-k)^{n-k+1/2}\frac{(\lambda)_k}{(\rho+1)_k}$

$$\times C_{2n-2k+1}^{\lambda+k}\left(\frac{w}{\sqrt{n-k}}\right)P_k^{(\rho,\sigma-k)}\left(1+\frac{z}{k+1}\right)$$

$$=2(-1)^{n+1}w\frac{(n+1)^{n-1}(\lambda)_{n+1}}{(\rho+1)_n}P_n^{(\rho,\sigma-n)}\left(1+\frac{z}{n+1}\right)$$

$$+\sqrt{n+1}\frac{(\lambda)_n(\rho+\sigma+1)_n}{(\rho+1)_n}\left(-\frac{z}{2}\right)^n$$

$$\times\sum_{k=0}^{n}\frac{2^k(-\rho-n)_k}{(n-k+1)!(1-\lambda-n)_k(-\rho-\sigma-n)_k}\left(\frac{n+1}{z}\right)^k C_{2k+1}^{\lambda+n-k}\left(\frac{w}{\sqrt{n+1}}\right).$$

5.12.7. Sums containing products of $P_{n\pm mk}^{(\rho\pm pk,\sigma\pm qk)}(\varphi(k,z))$

1. $\displaystyle\sum_{k=0}^{n}(-1)^k\binom{n}{k}\left[P_m^{(\rho,\sigma)}(w+kz)\right]^2=0$ $\quad[2m<n].$

2. $\displaystyle\sum_{k=0}^{n}\frac{(\mu+\nu+n+1)_k}{(\sigma+1)_k}(k+1)^{k-1}\left(\frac{w}{2}\right)^k$

$\times P_{n-k}^{(k+\mu,k+\nu)}((k+1)w+1)P_k^{(\rho-k,\sigma)}\left(\frac{z}{k+1}-1\right)$

$=\dfrac{4\mu\sigma(\mu+1)_n(wz)^{-1}}{(n+1)!(\mu+\nu+n)(\rho+\sigma)}\left[{}_3F_2\left(\begin{array}{c}-n-1,\,\mu+\nu+n,\,\rho+\sigma\\ \mu,\,\sigma;\,-\dfrac{wz}{4}\end{array}\right)-1\right].$

3. $\displaystyle\sum_{k=0}^{n-1}(-1)^k\frac{(-\rho-n)_k}{(\mu+1)_k}(k+1)^{k-1}(n-k)^{n-k}$

$\times P_k^{(\mu,\nu-k)}\left(1+\dfrac{w}{k+1}\right)P_{n-k}^{(\rho,\sigma+k)}\left(1+\dfrac{z}{n-k}\right)$

$=-\dfrac{(n+1)^{n-1}(\rho+1)_n}{(\mu+1)_n}P_n^{(\mu,\nu-n)}\left(1+\dfrac{w}{n+1}\right)$

$+(\rho+1)_n\dfrac{(\mu+\nu+1)_n}{(\mu+1)_n}\left(\dfrac{w}{2}\right)^n$

$\times\displaystyle\sum_{k=0}^{n}\frac{2^k(-\mu-n)_k}{(n-k+1)!(-\mu-\nu-n)_k(\rho+1)_k}\left(\frac{n+1}{w}\right)^k P_k^{(\rho,\sigma+n-k)}\left(1+\dfrac{z}{n+1}\right).$

5.13. The Legendre Function $P_\nu^\mu(z)$

5.13.1. Sums containing $P_{\nu\pm k}^{\mu\pm k}(z)$

1. $\displaystyle\sum_{k=0}^{n}(-2)^{-k}\binom{n}{k}\frac{(\nu-\mu+1)_{2k}}{\left(\nu-n+\frac{3}{2}\right)_k}(1-z^2)^{-k/2}P_{\nu+k}^{\mu-k}(z)$

$=2^{-n}\dfrac{(-\mu-\nu)_{2n}}{\left(-\nu-\frac{1}{2}\right)_n}(1-z^2)^{-n/2}P_{\nu-n}^{\mu-n}(z)\qquad[\arg(1\pm z)<\pi].$

2. $\displaystyle\sum_{k=0}^{n}(-2)^{-k}\binom{n}{k}\frac{(-\mu-\nu)_{2k}}{\left(\frac{1}{2}-\nu-n\right)_k}(1-z^2)^{-k/2}P_{\nu-k}^{\mu-k}(z)$

$=2^{-n}\dfrac{(\nu-\mu+1)_{2n}}{\left(\nu+\frac{1}{2}\right)_n}(1-z^2)^{-n/2}P_{\nu+n}^{\mu-n}(z)\qquad[\arg(1\pm z)<\pi].$

5.14. The Kummer Confluent Hypergeometric Function ${}_1F_1(a;b;z)$

5.14.1. Sums containing ${}_1F_1(a;b;z)$

1. $\displaystyle\sum_{k=0}^{n}(-1)^k\binom{n}{k}\frac{(b-a)_k}{(b)_k}{}_1F_1\left(\begin{array}{c}a;\,z\\ b+k\end{array}\right)=\frac{(a)_n}{(b)_n}{}_1F_1\left(\begin{array}{c}a+n;\,z\\ b+n\end{array}\right).$

2. $\displaystyle\sum_{k=0}^{n}\binom{n}{k}\frac{(b-a)_k}{(b)_k(1-c-n)_k}(-z)^k\,{}_1F_1\!\left(\begin{matrix}a;\ z\\b+k\end{matrix}\right)=e^z\,{}_2F_2\!\left(\begin{matrix}b-a,1-c;\ z\\b,1-c-n\end{matrix}\right).$

3. $\displaystyle\sum_{k=0}^{n}\binom{n}{k}\frac{(-z)^k}{(b)_k}\,{}_1F_1\!\left(\begin{matrix}a;\ z\\b+k\end{matrix}\right)={}_1F_1\!\left(\begin{matrix}a-n\\b;\ z\end{matrix}\right).$

4. $\displaystyle\sum_{k=0}^{n}(-1)^k\binom{n}{k}\frac{(1-b)_k}{(1-c-n)_k}\,{}_1F_1\!\left(\begin{matrix}a;\ z\\b-k\end{matrix}\right)=\frac{(c-b+1)_n}{(c)_n}\,{}_2F_2\!\left(\begin{matrix}a,b-c;\ z\\b,b-c-n\end{matrix}\right).$

5. $\displaystyle\sum_{k=0}^{n}(-1)^k\binom{n}{k}\frac{(1-b)_k}{(2-b-n)_k}\,{}_1F_1\!\left(\begin{matrix}a;\ z\\b-k\end{matrix}\right)$
$$=\frac{(a)_n}{(b-1)_n(b)_n}(-z)^n\,{}_1F_1\!\left(\begin{matrix}a+n;\ z\\b+n\end{matrix}\right).$$

6. $\displaystyle\sum_{k=0}^{n}(-1)^k\binom{n}{k}\frac{(1-b)_k}{(a-b+1)_k}\,{}_1F_1\!\left(\begin{matrix}a;\ z\\b-k\end{matrix}\right)=\frac{(a)_n}{(a-b+1)_n}\,{}_1F_1\!\left(\begin{matrix}a+n\\b;\ z\end{matrix}\right).$

7. $\displaystyle\sum_{k=0}^{n}\binom{n}{k}(1-b)_k z^{-k}\,{}_1F_1\!\left(\begin{matrix}a;\ z\\b-k\end{matrix}\right)=(1-b)_n z^{-n}\,{}_1F_1\!\left(\begin{matrix}a-n;\ z\\b-n\end{matrix}\right).$

5.14.2. Sums containing $_1F_1(a;\ b;\ z)$ and special functions

1. $\displaystyle\sum_{k=0}^{n}\frac{(b-a)_k}{k!(b)_k}(-z)^k L_{n-k}^{b+k-n-1}(-z)\,{}_1F_1\!\left(\begin{matrix}a;\ z\\b+k\end{matrix}\right)$
$$=(-1)^n\frac{(1-b)_n}{n!}\,{}_1F_1\!\left(\begin{matrix}a;\ z\\b-n\end{matrix}\right).$$

2. $\displaystyle\sum_{k=0}^{n}\frac{(b-a)_k}{k!(b)_k}(-z)^k L_{n-k}^{a+k-1}(-z)\,{}_1F_1\!\left(\begin{matrix}a;\ z\\b+k\end{matrix}\right)=\frac{(a)_n}{n!}\,{}_1F_1\!\left(\begin{matrix}a+n\\b;\ z\end{matrix}\right).$

3. $\displaystyle\sum_{k=0}^{n}\frac{(-1)^k}{k!}(1-b)_k L_{n-k}^{k-a}(z)\,{}_1F_1\!\left(\begin{matrix}a;\ z\\b-k\end{matrix}\right)=\frac{(b-a)_n}{n!}\,{}_1F_1\!\left(\begin{matrix}a-n\\b;\ z\end{matrix}\right).$

4. $\displaystyle\sum_{k=0}^{n}\frac{(-1)^k}{k!}(1-b)_k L_{n-k}^{k-b-n+1}(z)\,{}_1F_1\!\left(\begin{matrix}a;\ z\\b-k\end{matrix}\right)=\frac{(b-a)_n}{n!(b)_n}(-z)^n\,{}_1F_1\!\left(\begin{matrix}a;\ z\\b+n\end{matrix}\right).$

5. $\displaystyle\sum_{k=0}^{n}\frac{(-1)^k}{k!}(1-b)_k L_{n-k}^{k-b-n+1}(-z)\,{}_1F_1\!\left(\begin{matrix}a-k;\ z\\b-k\end{matrix}\right)$
$$=\frac{(a)_n}{n!(b)_n}z^n\,{}_1F_1\!\left(\begin{matrix}a+n;\ z\\b+n\end{matrix}\right).$$

6. $\sum_{k=0}^{n} \frac{(1-b)_k}{(n-k)!(a-b+1)_k}(-z)^{-k} L_k^{\lambda-k}(-z) {}_1F_1\left({a;\ z \atop b-k}\right)$

$$= \frac{(1-b)_n(-\lambda)_n}{n!(a-b+1)_n} z^{-n} {}_2F_2\left({a,\ \lambda+1;\ z \atop b-n,\ \lambda-n+1}\right).$$

5.14.3. Sums containing products of ${}_1F_1(a;\ b;\ z)$

1. $\sum_{k=0}^{2n}(-1)^k\binom{2n}{k}\frac{(b-a)_k(1-b-2n)_k}{(b)_k(a-b-2n+1)_k} {}_1F_1\left({a;\ -z \atop b+2n-k}\right) {}_1F_1\left({a;\ z \atop b+k}\right)$

$$= 2^{2n} \frac{(a)_n(b-a)_n\left(\tfrac{1}{2}\right)_n}{(b)_n(b-a)_{2n}} {}_3F_4\left({a+n,\ b-a+n,\ n+\tfrac{1}{2};\ \tfrac{z^2}{4} \atop \tfrac{b}{2}+n,\ \tfrac{b+1}{2}+n,\ b+n,\ \tfrac{1}{2}}\right).$$

5.15. The Tricomi Confluent Hypergeometric Function $\Psi(a;\ b;\ z)$

5.15.1. Sums containing $\Psi(a;\ b;\ z)$

1. $\sum_{k=0}^{n}(-1)^k \binom{n}{k} \Psi\left({a;\ z \atop b+k}\right) = (-1)^n (a)_n \Psi\left({a+n;\ z \atop b+n}\right).$

2. $\sum_{k=0}^{n} \binom{n}{k} \frac{(-z)^k}{(b-a)_k} \Psi\left({a;\ z \atop b+k}\right) = \frac{(-1)^n}{(b-a)_n} \Psi\left({a-n \atop b;\ z}\right).$

3. $\sum_{k=0}^{n} \binom{n}{k}(a)_k \Psi\left({a+k;\ z \atop b+k}\right) = \Psi\left({a;\ z \atop b+n}\right).$

4. $\sum_{k=0}^{n}(-1)^k \binom{n}{k} \Psi\left({a;\ z \atop b-k}\right) = (a)_n \Psi\left({a+n \atop b;\ z}\right).$

5. $\sum_{k=0}^{n} \binom{n}{k}(a-b+1)_k z^{-k} \Psi\left({a;\ z \atop b-k}\right) = z^{-n} \Psi\left({a-n;\ z \atop b-n}\right).$

6. $\sum_{k=0}^{n}(-1)^k \binom{n}{k} \frac{(a-b+1)_k}{(2-b-n)_k} \Psi\left({a;\ z \atop b-k}\right) = \frac{(a)_n}{(b-1)_n} z^n \Psi\left({a+n;\ z \atop b+n}\right).$

7. $\sum_{k=0}^{n} \binom{n}{k} \frac{(a)_k}{(b-n)_k}(-z)^k \Psi\left({a+k;\ z \atop b+k}\right) = \frac{(a-b+1)_n}{(1-b)_n} \Psi\left({a;\ z \atop b-n}\right).$

5.15.2. Sums containing $\Psi(a; b; z)$ and special functions

1. $\displaystyle\sum_{k=0}^{n} \frac{(-z)^k}{k!} L_{n-k}^{k+a-1}(-z)\Psi\left(\begin{matrix}a;\,z\\b+k\end{matrix}\right) = \frac{(a)_n}{n!}(a-b+1)_n \Psi\left(\begin{matrix}a+n\\b;\,z\end{matrix}\right).$

2. $\displaystyle\sum_{k=0}^{n} \frac{(-z)^k}{k!} L_{n-k}^{b-n+k-1}(-z)\Psi\left(\begin{matrix}a;\,z\\b+k\end{matrix}\right) = (-1)^n \frac{(a-b+1)_n}{n!} \Psi\left(\begin{matrix}a;\,z\\b-n\end{matrix}\right).$

3. $\displaystyle\sum_{k=0}^{n} \frac{(-1)^k}{k!} (a-b+1)_k L_{n-k}^{k-n-b+1}(z)\Psi\left(\begin{matrix}a;\,z\\b-k\end{matrix}\right) = \frac{(-z)^n}{n!} \Psi\left(\begin{matrix}a;\,z\\b+n\end{matrix}\right).$

4. $\displaystyle\sum_{k=0}^{n} \frac{(-1)^k}{k!} (a-b+1)_k L_{n-k}^{k-a}(z)\Psi\left(\begin{matrix}a;\,z\\b-k\end{matrix}\right) = \frac{(-1)^n}{n!} \Psi\left(\begin{matrix}a-n\\b;\,z\end{matrix}\right).$

5.16. The Gauss Hypergeometric Function $_2F_1(a,b;\,c;\,z)$

5.16.1. Sums containing $_2F_1(a,b;\,c;\,z)$

1. $\displaystyle\sum_{k=0}^{n} (-1)^k \binom{n}{k} \frac{(d)_k}{(c)_k} \,_2F_1\left(\begin{matrix}a,\,b;\,z\\c+k\end{matrix}\right) = \frac{(c-d)_n}{(c)_n} \,_3F_2\left(\begin{matrix}a,\,b,\,c-d+n\\c+n,\,c-d;\,z\end{matrix}\right).$

2. $\displaystyle\sum_{k=0}^{n} (-1)^k \binom{n}{k} \frac{(b)_k}{(b-a-n+1)_k} \,_2F_1\left(\begin{matrix}a,\,b+k;\,z\\c\end{matrix}\right) = \frac{(a)_n}{(a-b)_n} \,_2F_1\left(\begin{matrix}a+n,\,b\\c;\,z\end{matrix}\right).$

3. $\displaystyle\sum_{k=0}^{n} (-1)^k \binom{n}{k} \frac{(c-a)_k(c-b)_k}{(c)_k(1-n-a-b+c)_k} \,_2F_1\left(\begin{matrix}a,\,b;\,z\\c+k\end{matrix}\right)$
$$= \frac{(a)_n(b)_n}{(c)_n(a+b-c)_n}(1-z)^n \,_2F_1\left(\begin{matrix}a+n,\,b+n\\c+n;\,z\end{matrix}\right).$$

4. $\displaystyle\sum_{k=0}^{n} (-1)^k \binom{n}{k} \frac{(c-a)_k}{(c)_k} \,_2F_1\left(\begin{matrix}a,\,b;\,z\\c+k\end{matrix}\right) = \frac{(a)_n}{(c)_n} \,_2F_1\left(\begin{matrix}a+n,\,b\\c+n;\,z\end{matrix}\right).$

5. $\displaystyle\sum_{k=0}^{n} \binom{n}{k} \frac{(c-b)_k}{(c)_k} \left(\frac{z}{1-z}\right)^k \,_2F_1\left(\begin{matrix}a,\,b;\,z\\c+k\end{matrix}\right) = (1-z)^{-n} \,_2F_1\left(\begin{matrix}a-n,\,b\\c;\,z\end{matrix}\right).$

6. $\displaystyle\sum_{k=0}^{n} \binom{n}{k} \frac{(b)_k}{(c)_k} z^k \,_2F_1\left(\begin{matrix}a+k,\,b+k;\,z\\c+k\end{matrix}\right) = \,_2F_1\left(\begin{matrix}a+n,\,b\\c;\,z\end{matrix}\right).$

7. $\displaystyle\sum_{k=0}^{n} \binom{n}{k} \frac{(b)_k}{(c)_k} (z-1)^k \,_2F_1\left(\begin{matrix}a+k,\,b+k;\,z\\c+k\end{matrix}\right) = \frac{(c-b)_n}{(c)_n} \,_2F_1\left(\begin{matrix}a+n,\,b\\c+n;\,z\end{matrix}\right).$

8. $\sum_{k=0}^{n} \binom{n}{k} \frac{(a)_k (b)_k}{(c)_k (c-n)_k} z^k {}_2F_1\left(\begin{array}{c} a+k, b+k; z \\ c+k \end{array}\right) = {}_2F_1\left(\begin{array}{c} a, b \\ c-n; z \end{array}\right).$

9. $\sum_{k=0}^{n} (-1)^k \binom{n}{k} \frac{(1-c)_k}{(2-c-n)_k} {}_2F_1\left(\begin{array}{c} a, b; z \\ c-k \end{array}\right)$
$$= \frac{(a)_n (b)_n}{(c)_n (c-1)_n} (-z)^n {}_2F_1\left(\begin{array}{c} a+n, b+n \\ c+n; z \end{array}\right).$$

10. $\sum_{k=0}^{n} \binom{n}{k} \frac{(1-c)_k}{(a+b-c-n+1)_k} \left(\frac{1-z}{z}\right)^k {}_2F_1\left(\begin{array}{c} a, b; z \\ c-k \end{array}\right)$
$$= \frac{(1-c)_n}{(c-a-b)_n} (-z)^{-n} {}_2F_1\left(\begin{array}{c} a-n, b-n \\ c-n; z \end{array}\right).$$

11. $\sum_{k=0}^{n} \binom{n}{k} \frac{(1-c)_k}{(b-c+1)_k} \left(\frac{1-z}{z}\right)^k {}_2F_1\left(\begin{array}{c} a, b; z \\ c-k \end{array}\right)$
$$= \frac{(1-c)_n}{(b-c+1)_n} z^{-n} {}_2F_1\left(\begin{array}{c} a-n, b \\ c-n; z \end{array}\right).$$

5.16.2. Sums containing $_2F_1(a, b; c; z)$ and special functions

1. $\sum_{k=0}^{n} \frac{(c-a)_k (c-b)_k}{k! (c)_k} z^k P_{n-k}^{(k-n+c-1, k-n-a-b+c)}(1-2z) {}_2F_1\left(\begin{array}{c} a, b; z \\ c+k \end{array}\right)$
$$= \frac{(1-c)_n}{n!} (z-1)^n {}_2F_1\left(\begin{array}{c} a, b \\ c-n; z \end{array}\right).$$

2. $\sum_{k=0}^{n} \frac{(c-a)_k (c-b)_k}{k! (c)_k} z^k P_{n-k}^{(k-n+c-1, k-a)}(1-2z) {}_2F_1\left(\begin{array}{c} a, b; z \\ c+k \end{array}\right)$
$$= (-1)^n \frac{(1-c)_n}{n!} {}_2F_1\left(\begin{array}{c} a-n, b \\ c-n; z \end{array}\right).$$

3. $\sum_{k=0}^{n} \frac{(a)_k (b)_k}{k! (c)_k} (z-z^2)^k P_{n-k}^{(k-n+c-1, k+b-c)}(1-2z) {}_2F_1\left(\begin{array}{c} a+k, b+k; z \\ c+k \end{array}\right)$
$$= (-1)^n \frac{(1-c)_n}{n!} {}_2F_1\left(\begin{array}{c} a-n, b \\ c-n; z \end{array}\right).$$

4. $\sum_{k=0}^{n} \frac{(a)_k (b)_k}{k! (c)_k} (z-z^2)^k P_{n-k}^{(k-n+c-1, k-n+a+b-c)}(1-2z)$
$$\times {}_2F_1\left(\begin{array}{c} a+k, b+k; z \\ c+k \end{array}\right) = (-1)^n \frac{(1-c)_n}{n!} {}_2F_1\left(\begin{array}{c} a-n, b-n \\ c-n; z \end{array}\right).$$

5. $\displaystyle\sum_{k=0}^{n} \frac{(a)_k (b)_k}{k!\,(c)_k} (z-z^2)^k P_{n-k}^{(k-a+c-1,\,k-n+a+b-c)}(1-2z)$

$$\times {}_2F_1\!\left(\begin{matrix}a+k,\,b+k;\,z\\ c+k\end{matrix}\right) = \frac{(c-a)_n}{n!}\,{}_2F_1\!\left(\begin{matrix}a-n,\,b\\ c;\,z\end{matrix}\right).$$

6. $\displaystyle\sum_{k=0}^{n} \frac{(1-c)_k}{k!}(z-1)^k P_{n-k}^{(k-n-c+1,\,k+b-1)}(1-2z)\,{}_2F_1\!\left(\begin{matrix}a,\,b;\,z\\ c-k\end{matrix}\right)$

$$= \frac{(b)_n (c-a)_n}{n!\,(c)_n}(-z)^n\,{}_2F_1\!\left(\begin{matrix}a,\,b+n\\ c+n;\,z\end{matrix}\right).$$

7. $\displaystyle\sum_{k=0}^{n} \frac{(1-c)_k}{k!}(z-1)^k P_{n-k}^{(k-n-c+1,\,k-n+a+b-c)}(1-2z)\,{}_2F_1\!\left(\begin{matrix}a,\,b;\,z\\ c-k\end{matrix}\right)$

$$= \frac{(c-a)_n (c-b)_n}{n!\,(c)_n} z^n\,{}_2F_1\!\left(\begin{matrix}a,\,b\\ c+n;\,z\end{matrix}\right).$$

8. $\displaystyle\sum_{k=0}^{n} \frac{(1-c)_k}{k!}(z-1)^k P_{n-k}^{(k-n-c+1,\,k+a-1)}(1-2z)\,{}_2F_1\!\left(\begin{matrix}a,\,b;\,z\\ c-k\end{matrix}\right)$

$$= \frac{(a)_n (c-b)_n}{n!\,(c)_n}(-z)^n\,{}_2F_1\!\left(\begin{matrix}a+n,\,b\\ c+n;\,z\end{matrix}\right).$$

9. $\displaystyle\sum_{k=0}^{n} \frac{(1-c)_k}{k!}(z-1)^k P_{n-k}^{(k-a,\,k-n+a+b-c)}(1-2z)\,{}_2F_1\!\left(\begin{matrix}a,\,b;\,z\\ c-k\end{matrix}\right)$

$$= \frac{(c-a)_n}{n!}\,{}_2F_1\!\left(\begin{matrix}a-n,\,b\\ c;\,z\end{matrix}\right).$$

10. $\displaystyle\sum_{k=0}^{n}(-1)^k \frac{(1-c)_k}{k!} P_{n-k}^{(k+a-c,\,k-n-a-b+c)}(1-2z)\,{}_2F_1\!\left(\begin{matrix}a-k,\,b-k;\,z\\ c-k\end{matrix}\right)$

$$= \frac{(a)_n}{n!}(1-z)^n\,{}_2F_1\!\left(\begin{matrix}a+n,\,b\\ c;\,z\end{matrix}\right).$$

11. $\displaystyle\sum_{k=0}^{n}(-1)^k \frac{(1-c)_k}{k!} P_{n-k}^{(k-n-c+1,\,k-a-b+c-n)}(1-2z)\,{}_2F_1\!\left(\begin{matrix}a-k,\,b-k;\,z\\ c-k\end{matrix}\right)$

$$= \frac{(a)_n (b)_n}{n!\,(c)_n}(z-z^2)^n\,{}_2F_1\!\left(\begin{matrix}a+n,\,b+n\\ c+n;\,z\end{matrix}\right).$$

12. $\displaystyle\sum_{k=0}^{n}(-1)^k \frac{(1-c)_k}{k!} P_{n-k}^{(k-n-c+1,\,k-b+c-1)}(1-2z)\,{}_2F_1\!\left(\begin{matrix}a-k,\,b-k;\,z\\ c-k\end{matrix}\right)$

$$= \frac{(a)_n (c-b)_n}{n!\,(c)_n}(-z)^n\,{}_2F_1\!\left(\begin{matrix}a+n,\,b\\ c+n;\,z\end{matrix}\right).$$

13. $\sum_{k=0}^{n} \frac{(1-c)_k}{k!} (z-1)^k P_{n-k}^{(k-n-c+1,\,k-n+a+b-c)} (1-2z) \, _2F_1\!\left(\begin{matrix} a,\,b;\,z \\ c-k \end{matrix}\right)$

$$= \frac{(c-a)_n (c-b)_n}{n!\,(c)_n} (-z)^n \, _2F_1\!\left(\begin{matrix} a,\,b \\ c+n;\,z \end{matrix}\right).$$

5.16.3. Sums containing products of $_2F_1(a,b;c;z)$

1. $\displaystyle\sum_{k=0}^{n} \binom{n}{k} \frac{(a)_k (b)_k}{\left(a+b+\tfrac{1}{2}\right)_k \left(a+b-n+\tfrac{1}{2}\right)_k} z^k$

$\times \, _2F_1\!\left(\begin{matrix} a+k,\,b+k;\,z \\ a+b+k+\tfrac{1}{2} \end{matrix}\right) {}_2F_1\!\left(\begin{matrix} a,\,b;\,z \\ a+b-n+k+\tfrac{1}{2} \end{matrix}\right)$

$$= {}_3F_2\!\left(\begin{matrix} 2a,\,2b,\,a+b;\,z \\ 2a+2b,\,a+b-n+\tfrac{1}{2} \end{matrix}\right).$$

2. $\displaystyle\sum_{k=0}^{n} \binom{n}{k} \frac{(a)_k (b)_k}{\left(a+b-\tfrac{1}{2}\right)_k \left(a+b-n-\tfrac{1}{2}\right)_k} z^k$

$\times \, _2F_1\!\left(\begin{matrix} a+k,\,b+k;\,z \\ a+b+k-\tfrac{1}{2} \end{matrix}\right) {}_2F_1\!\left(\begin{matrix} a,\,b-1;\,z \\ a+b-n+k-\tfrac{1}{2} \end{matrix}\right)$

$$= {}_3F_2\!\left(\begin{matrix} 2a,\,2b-1,\,a+b-1;\,z \\ 2a+2b-2,\,a+b-n-\tfrac{1}{2} \end{matrix}\right).$$

3. $\displaystyle\sum_{k=0}^{n} (-1)^k \binom{n}{k} \frac{(a)_k (b)_k \left(\tfrac{1}{2}-a-b-n\right)_k}{\left(a+b+\tfrac{1}{2}\right)_k (1-a-n)_k (1-b-n)_k}$

$\times \, _2F_1\!\left(\begin{matrix} a+k,\,b+k;\,z \\ a+b+k+\tfrac{1}{2} \end{matrix}\right) {}_2F_1\!\left(\begin{matrix} a+n-k,\,b+n-k;\,z \\ a+b+n-k+\tfrac{1}{2} \end{matrix}\right)$

$$= \frac{(2a)_n (2b)_n (a+b)_n}{(a)_n (b)_n (2a+2b)_n} \, _3F_2\!\left(\begin{matrix} 2a+n,\,2b+n,\,a+b+n;\,z \\ 2a+2b+n,\,a+b+n+\tfrac{1}{2} \end{matrix}\right).$$

4. $\displaystyle\sum_{k=0}^{n} \binom{n}{k} \frac{(a)_k \left(b+\tfrac{1}{2}\right)_k}{(a+b)_k (a+b-n)_k} \left(\frac{z}{1-z}\right)^k \, _2F_1\!\left(\begin{matrix} a-\tfrac{1}{2},\,b;\,z \\ a+b+k \end{matrix}\right)$

$\times \, _2F_1\!\left(\begin{matrix} a,\,b-\tfrac{1}{2};\,z \\ a+b-n+k \end{matrix}\right) = (1-z)^{1/2} \, _3F_2\!\left(\begin{matrix} 2a,\,2b,\,a+b-\tfrac{1}{2};\,z \\ 2a+2b-1,\,a+b-n \end{matrix}\right).$

417

5. $\displaystyle\sum_{k=0}^{n}\binom{n}{k}\frac{(a)_k(b)_k\left(\frac{1}{2}-a-b-n\right)_k}{\left(a+b+\frac{1}{2}\right)_k(1-a-n)_k(1-b-n)_k}(z-1)^k$

$$\times\,{}_2F_1\!\left(\begin{matrix}a+k,\,b+k;\,z\\a+b+k+\frac{1}{2}\end{matrix}\right){}_2F_1\!\left(\begin{matrix}a+\frac{1}{2},\,b+\frac{1}{2};\,z\\a+b+n-k+\frac{1}{2}\end{matrix}\right)$$

$$=\frac{(2a)_n(2b)_n(a+b)_n}{(a)_n(b)_n(2a+2b)_n}(1-z)^{n-1/2}\,{}_3F_2\!\left(\begin{matrix}2a+n,\,2b+n,\,a+b+n;\,z\\2a+2b+n,\,a+b+n+\frac{1}{2}\end{matrix}\right).$$

6. $\displaystyle\sum_{k=0}^{n}(-1)^k\binom{n}{k}\frac{\left(a-\frac{1}{2}\right)_k(b)_k(1-a-b-n)_k}{\left(\frac{3}{2}-a-n\right)_k(1-b-n)_k(a+b)_k}$

$$\times\,{}_2F_1\!\left(\begin{matrix}a,\,b+\frac{1}{2};\,z\\a+b+k\end{matrix}\right){}_2F_1\!\left(\begin{matrix}a,\,b+\frac{1}{2};\,z\\a+b+n-k\end{matrix}\right)$$

$$=\frac{(2a-1)_n(2b)_n\left(a+b-\frac{1}{2}\right)_n}{\left(a-\frac{1}{2}\right)_n(b)_n(2a+2b-1)_n}(1-z)^{n-1}$$

$$\times\,{}_3F_2\!\left(\begin{matrix}2a+n-1,\,2b+n,\,a+b+n-\frac{1}{2}\\2a+2b+n-1,\,a+b+n;\,z\end{matrix}\right).$$

5.17. The Generalized Hypergeometric Function ${}_pF_q((a_p);\,(b_q);\,z)$

5.17.1. Sums containing ${}_pF_q((a_p)\pm mk;\,(b_q)\pm nk;\,z)$

1. $\displaystyle\sum_{k=0}^{n}(-1)^k\binom{n}{k}\frac{(a)_k}{(c)_k}\,{}_pF_{q+1}\!\left(\begin{matrix}(a_p);\,z\\(b_q),\,c+k\end{matrix}\right)$

$$=\frac{(c-a)_n}{(c)_n}\,{}_{p+1}F_{p+2}\!\left(\begin{matrix}(a_p),\,c-a+n;\,z\\(b_q),\,c+n,\,c-a\end{matrix}\right).$$

2. $\displaystyle\sum_{k=0}^{n}(-1)^k\binom{n}{k}\frac{(b+m+n)_k}{(b)_k}\,{}_{p+1}F_{q+1}\!\left(\begin{matrix}-m-n,\,(a_p);\,z\\(b_q),\,b+k\end{matrix}\right)$

$$=(-1)^{m+n}\frac{(m+n)!}{m!(b)_n}\,{}_{p+1}F_{q+1}\!\left(\begin{matrix}-m,\,(a_p);\,z\\(b_q),\,b+n\end{matrix}\right).$$

3. $\displaystyle\sum_{k=0}^{n}\frac{(-n)_k(n)_k}{(k!)^2}\,{}_pF_{q+1}\!\left(\begin{matrix}(a_p);\,z\\(b_q),\,k+1\end{matrix}\right)$

$$=\frac{z^n}{(2n)!}\frac{\prod(a_p)_n}{\prod(b_q)_n}\,{}_pF_{q+1}\!\left(\begin{matrix}(a_p)+n;\,z\\(b_q)+n,\,2n+1\end{matrix}\right).$$

5.17.1]   5.17. The Generalized Hypergeometric Function $_pF_q((a_p); (b_q); z)$

4. $\displaystyle\sum_{k=0}^{n}(-1)^k\binom{n}{k}\frac{(1-c)_k}{(a)_k}\,_pF_{q+1}\!\left(\begin{array}{c}(a_p);\ z\\(b_q),\ c-k\end{array}\right)$

$$=\frac{(a+c-1)_n}{(a)_n}\,_{p+1}F_{q+2}\!\left(\begin{array}{c}(a_p),\ a+c+n-1;\ z\\(b_q),\ c,\ a+c-1\end{array}\right).$$

5. $\displaystyle\sum_{k=0}^{n}(-1)^k\binom{n}{k}\frac{(a)_k}{(b)_k}\,_{p+1}F_q\!\left(\begin{array}{c}(a_p),\ a+k;\ z\\(b_q)\end{array}\right)$

$$=\frac{(b-a)_n}{(b)_n}\,_{p+2}F_{q+1}\!\left(\begin{array}{c}(a_p),\ a,\ a-b+1;\ z\\(b_q),\ a-b-n+1\end{array}\right).$$

6. $\displaystyle\sum_{k=0}^{n}(-1)^k\binom{n}{k}\frac{(a)_k}{(b)_k}\,_{p+1}F_{q+1}\!\left(\begin{array}{c}(a_p),\ k+a;\ z\\(b_q),\ k+b\end{array}\right)$

$$=\frac{(b-a)_n}{(b)_n}\,_{p+1}F_{q+1}\!\left(\begin{array}{c}(a_p),\ a;\ z\\(b_q),\ b+n\end{array}\right).$$

7. $\displaystyle\sum_{k=0}^{n}\binom{n}{k}(-z)^k\frac{\prod(a_p)_k}{\prod(b_q)_k}\,_{p+1}F_q\!\left(\begin{array}{c}(a_p)+k,\ a\\(b_q)+k;\ z\end{array}\right)=\,_{p+1}F_{q+1}\!\left(\begin{array}{c}(a_p),\ a-n;\ z\\(b_q)\end{array}\right).$

8. $\displaystyle\sum_{k=0}^{n}\binom{n}{k}\frac{(-z)^k}{k+1}\frac{\prod(a_p)_k}{\prod(b_q)_k}\,_{p+1}F_q\!\left(\begin{array}{c}(a_p)+k,\ a\\(b_q)+k;\ z\end{array}\right)$

$$=\frac{1}{(n+1)z}\frac{\prod_{j=1}^{q}(b_j-1)}{\prod_{i=1}^{p}(a_i-1)}\left[\,_{p+1}F_q\!\left(\begin{array}{c}a,\ (a_p)-1\\(b_q)-1;\ z\end{array}\right)-\,_{p+1}F_q\!\left(\begin{array}{c}a-n-1,\ (a_p)-1\\(b_q)-1;\ z\end{array}\right)\right].$$

9. $\displaystyle\sum_{k=0}^{n}\binom{n}{k}k^r(-z)^k\frac{\prod(a_p)_k}{\prod(b_q)_k}\,_{p+1}F_q\!\left(\begin{array}{c}-m,\ (a_p)+k\\(b_q)+k;\ z\end{array}\right)$

$$=\sum_{k=1}^{r}\sigma_r^k(-n)_k z^k\frac{\prod(a_p)_k}{\prod(b_q)_k}\,_{p+1}F_q\!\left(\begin{array}{c}-m-n+k,\ (a_p)+k\\(b_q)+k;\ z\end{array}\right).$$

10. $\displaystyle\sum_{k=0}^{n}\sigma_{k+m}^m\frac{(-n)_k}{(k+m)!}z^k\frac{\prod(a_p)_k}{\prod(b_q)_k}\,_{p+1}F_q\!\left(\begin{array}{c}-n+k,\ (a_p)+k\\(b_q)+k;\ az\end{array}\right)$

$$=(-1)^{m(p+q+1)}\frac{z^{-m}}{m!(1+n)_m}\frac{\prod(1-(b_q))_m}{\prod(1-(a_p))_m}\sum_{k=0}^{m}(-1)^k\binom{m}{k}$$

$$\times\,_{p+1}F_q\!\left(\begin{array}{c}-m-n,\ (a_p)-m\\(b_q)-m;\ (k+a)z\end{array}\right).$$

11. $\sum_{k=0}^{n} \frac{\left(-\frac{1}{2}\right)_k}{k!} \left(-\frac{z^2}{4}\right)^k \frac{\prod(a_p)_{2k}}{\prod(b_q)_{2k}} {}_{p+1}F_q\left(\begin{array}{c} -n+k, (a_p)+2k \\ (b_q)+2k; z \end{array}\right)$

$$= {}_{p+2}F_{q+1}\left(\begin{array}{c} -2n, -n-\frac{1}{2}, (a_p) \\ -2n-1, (b_q); z \end{array}\right).$$

12. $\sum_{k=1}^{n} \frac{a+2k-1}{(a)_{2k}(b)_{2k}} z^k \frac{\prod(a_p)_k}{\prod(b_q)_k} {}_pF_{q+3}\left(\begin{array}{c} (a_p)+k; z \\ (b_q)+k, a+2k, b+2k, b-a+1 \end{array}\right)$

$$= \frac{z}{ab(b+1)} \frac{\prod_{i=1}^{p} a_i}{\prod_{j=1}^{q} b_j} {}_{p+1}F_{q+4}\left(\begin{array}{c} (a_p)+1, \frac{b+1}{2}; z \\ (b_q)+1, a+1, b+1, \frac{b+3}{2}, b-a+1 \end{array}\right)$$

$$- \frac{z^{n+1}}{(a)_{2n+1}(b)_{2n+2}} \frac{\prod(a_p)_{n+1}}{\prod(b_q)_{n+1}}$$

$$\times {}_{p+1}F_{q+4}\left(\begin{array}{c} (a_p)+n+1, \frac{b+1}{2}+n; z \\ (b_q)+n+1, a+2n+1, b+2n+1, \frac{b+3}{2}+n, b-a+1 \end{array}\right).$$

13. $\sum_{k=0}^{[(n-1)/2]} \binom{n}{k} {}_{p+2}F_q\left(\begin{array}{c} -n+2k, n-2k, (a_p) \\ (b_q); z \end{array}\right)$

$$= 2^{n-1} {}_{p+2}F_q\left(\begin{array}{c} -n, \frac{1}{2}, (a_p) \\ (b_q); 2z \end{array}\right) - \frac{1+(-1)^n}{4} \binom{n}{[n/2]}.$$

14. $\sum_{k=0}^{[n/2]} \binom{n}{k} \frac{(n-2k+1)^2}{n-k+1} {}_{p+2}F_q\left(\begin{array}{c} -n+2k, n-2k+2, (a_p) \\ (b_q); z \end{array}\right)$

$$= 2^n {}_{p+2}F_q\left(\begin{array}{c} -n, \frac{3}{2}, (a_p) \\ (b_q); 2z \end{array}\right).$$

15. $\sum_{k=0}^{[n/2]} \binom{n}{2k} \frac{\left(\frac{1}{2}\right)_k}{(a)_k} (-4z)^k \frac{\prod(a_p)_k}{\prod(b_q)_k} {}_{p+1}F_q\left(\begin{array}{c} -n+2k, (a_p)+k \\ (b_q)+k; z \end{array}\right)$

$$= {}_{p+2}F_{q+1}\left(\begin{array}{c} -n, (a_p), 1-a-n \\ (b_q), a; -z \end{array}\right).$$

16. $\sum_{k=0}^{n} \frac{1}{(n-k)!(k+n+1)!} {}_pF_{p+2}\left(\begin{array}{c} (a_p); z \\ (b_p), \frac{1}{2}-k, \frac{3}{2}+k \end{array}\right)$

$$= \frac{2^{2n}}{(2n+1)!} {}_pF_{p+2}\left(\begin{array}{c} (a_p); z \\ (b_p), n+\frac{3}{2}, \frac{1}{2} \end{array}\right).$$

5.17. The Generalized Hypergeometric Function $_pF_q((a_p);(b_q);z)$

17. $\sum_{k=0}^{n} \frac{(-z)^k}{(2k)!} \frac{\prod(a_p)_k}{\prod(b_q)_k} {}_pF_{q+1}\left(\begin{array}{c}(a_p)+k;\ z\\(b_q)+k,\ 2k+2\end{array}\right)$

$= 1 - \frac{(-z)^{n+1}}{(2n+2)!} \frac{\prod(a_p)_{n+1}}{\prod(b_q)_{n+1}} {}_{p+1}F_{q+2}\left(\begin{array}{c}(a_p)+n+1,\ n+1;\ z\\(b_q)+n+1,\ n+2,\ 2n+3\end{array}\right).$

18. $\sum_{k=0}^{n} \frac{z^k}{(2k+1)!} \frac{\prod(a_p)_k}{\prod(b_q)_k} {}_pF_{q+1}\left(\begin{array}{c}(a_p)+k;\ z\\(b_q)+k,\ 2k+3\end{array}\right) = {}_pF_{q+1}\left(\begin{array}{c}(a_p);\ z\\(b_q),\ 2\end{array}\right)$

$- \frac{z^{n+1}}{(2n+3)!} \frac{\prod(a_p)_{n+1}}{\prod(b_q)_{n+1}} {}_pF_{q+1}\left(\begin{array}{c}(a_p)+n+1;\ z\\(b_q)+n+1,\ 2n+4\end{array}\right).$

19. $\sum_{k=0}^{n} \binom{2n}{2k}\left(\frac{1}{2}\right)_k\left(-\frac{1}{2}\right)_k (-z^2)^k \frac{\prod(a_p)_{2k}}{\prod(b_q)_{2k}}$

$\times {}_{p+2}F_q\left(\begin{array}{c}-2n+2k,\ k-\frac{1}{2},\ (a_p)+2k\\(b_q)+2k;\ z\end{array}\right) = 1 + nz\frac{\prod_{i=1}^{p} a_i}{\prod_{j=1}^{q} b_j}.$

20. $\sum_{k=0}^{n}(-1)^k\binom{n}{k}\frac{(a+k\mu)_{n-1}(\lambda+k\mu)_k}{(a+k\mu)_k} {}_{p+1}F_{q+1}\left(\begin{array}{c}-k,\ (a_p);\ z\\(b_q),\ \lambda+k\mu\end{array}\right)$

$= \frac{(a-\lambda)_n}{a+n\mu+n-1} {}_{p+1}F_{q+1}\left(\begin{array}{c}-n,\ (a_p);\ z\\(b_q),\ \lambda-a-n+1\end{array}\right).$

5.17.2. Sums containing $_pF_q((a_p)\pm mk;\ (b_q)\pm nk;\ z)$ and special functions

1. $\sum_{k=0}^{n} \frac{(-1)^k}{k!}\binom{n}{k}(a)_k\psi(k+1) {}_{p+1}F_q\left(\begin{array}{c}-k,\ (a_p)\\(b_q);\ z\end{array}\right)$

$= \frac{(1-a)_n}{n!}\psi(n+1) {}_{p+2}F_{q+1}\left(\begin{array}{c}-n,\ (a_p),\ a;\ z\\(b_q),\ a-n\end{array}\right)$

$+ (-1)^n \sum_{k=0}^{n-1} \frac{(a-n)_k}{k!(n-k)} {}_{p+2}F_{q+1}\left(\begin{array}{c}-k,\ (a_p),\ a;\ z\\(b_q),\ a-n\end{array}\right).$

2. $\sum_{k=0}^{n}\binom{n}{k}\frac{(-4z)^k}{\left(\frac{1}{2}\right)_k} B_{2k} \frac{\prod(a_p)_k}{\prod(b_q)_k} {}_{p+1}F_{q+1}\left(\begin{array}{c}-n+k,\ (a_p)+k;\ z\\(b_q)+k,\ \frac{3}{2}\end{array}\right)$

$= {}_{p+1}F_{q+1}\left(\begin{array}{c}-n,\ (a_p);\ z\\(b_q),\ \frac{1}{2}\end{array}\right).$

3. $\sum_{k=0}^{n} \binom{n}{k}(-z)^k B_k(w) \dfrac{\prod(a_p)_k}{\prod(b_q)_k} \,_{p+1}F_{q+1}\!\left(\begin{array}{c}-n+k,\,(a_p)+k,\,1;\,z\\(b_q)+k,\,2\end{array}\right)$

$$= \,_pF_q\!\left(\begin{array}{c}-n,\,(a_p)\\(b_q);\,wz\end{array}\right).$$

5.17.3. Sums containing $_pF_q((a_p)\pm mk;\,(b_q)\pm nk;\,\varphi(k,z))$

1. $\sum_{k=0}^{n}(-1)^k\binom{n}{k}\,_{p+1}F_q\!\left(\begin{array}{c}-m,\,(a_p)\\(b_q);\,w+kz\end{array}\right) = 0$ $\qquad [m<n].$

2. $\sum_{k=1}^{n}\binom{n}{k}k^{k-1}z^k \dfrac{\prod(a_p)_k}{\prod(b_q)_k}\,_{p+1}F_q\!\left(\begin{array}{c}-n+k,\,(a_p)+k\\(b_q)+k;\,kz\end{array}\right) = nz\,\dfrac{\prod_{i=1}^{p}a_i}{\prod_{j=1}^{q}b_j}.$

3. $\sum_{k=1}^{n}\dfrac{k^{2k-4}}{(n-k)!(2k)!}z^k \dfrac{\prod(a_p)_k}{\prod(b_q)_k}\,_{p+1}F_{q+1}\!\left(\begin{array}{c}-n+k,\,(a_p)+k;\,k^2z\\(b_q)+k,\,2k+1\end{array}\right)$

$$= -\dfrac{z^2}{(n-2)!\,8}\,\dfrac{\prod_{i=1}^{p}a_i(a_i+1)}{\prod_{j=1}^{q}b_j(b_j+1)} + \dfrac{z}{(n-1)!\,2}\,\dfrac{\prod_{i=1}^{p}a_i}{\prod_{j=1}^{q}b_j} \qquad [n\ge 2].$$

4. $\sum_{k=0}^{n}\dfrac{(2k+1)^{2k-3}}{(n-k)!(2k+1)!}z^k\dfrac{\prod(a_p)_k}{\prod(b_q)_k}\,_{p+1}F_q\!\left(\begin{array}{c}-n+k,\,(a_p)+k;\,(2k+1)^2z\\(b_q)+k,\,2k+2\end{array}\right)$

$$= \dfrac{1}{n!} - \dfrac{4z}{9(n-1)!}\,\dfrac{\prod_{i=1}^{p}a_i}{\prod_{j=1}^{q}b_j}.$$

5. $\sum_{k=0}^{n}\binom{n}{k}\dfrac{(ka+b)^k}{ka+1}z^k\dfrac{\prod(a_p)_k}{\prod(b_q)_k}\,_{p+1}F_q\!\left(\begin{array}{c}-n+k,\,(a_p)+k\\(b_q)+k;\,(ka+1)z\end{array}\right)$

$$= \,_{p+2}F_{q+1}\!\left(\begin{array}{c}-n,\,(a_p),\,\frac{1}{a};\,(1-b)z\\(b_q),\,\frac{1}{a}+1\end{array}\right).$$

6. $\sum_{k=1}^{n}(-1)^k\binom{n}{k}k^m\,_{p+1}F_q\!\left(\begin{array}{c}-m,\,(a_p);\,\frac{z}{k}\\(b_q)\end{array}\right) = (-1)^m m!\,\delta_{m,n}$

$$- \dfrac{\prod(a_p)_m}{\prod(b_q)_m}(-z)^m \qquad [m\le n].$$

7. $\displaystyle\sum_{k=1}^{n}(-1)^k\binom{n}{k}k^m\,{}_{p+1}F_q\!\left(\begin{matrix}-m,\,(a_p);\\(b_q)\end{matrix}\;\frac{z}{k}\right)=(-1)^{m+1}z^m\frac{\prod(a_p)_m}{\prod(b_q)_m}$

$\displaystyle\qquad+(-1)^m n!\,z^{m-n}\frac{\prod(a_p)_{m-n}}{\prod(b_q)_{m-n}}\sum_{k=0}^{m-n}\sigma_{k+n}^n\binom{m}{k+n}(-1)^{k(p+q+1)}z^k$

$\displaystyle\qquad\qquad\times\frac{\prod(n-m-b_q+1)_k}{\prod(n-m-a_p+1)_k}\qquad[m\geq n].$

8. $\displaystyle\sum_{k=1}^{n}\binom{n}{k}(-ka)^k(ka+b)^{n-k-1}\,{}_{p+1}F_p\!\left(\begin{matrix}-k,\,(a_p)\\(b_p);\,\frac{z}{k}\end{matrix}\right)$

$\displaystyle\qquad=-b^{n-1}+\frac{b^n}{na+b}\,{}_{p+1}F_q\!\left(\begin{matrix}-n,\,(a_p)\\(b_q);\,-\frac{az}{b}\end{matrix}\right).$

9. $\displaystyle\sum_{k=0}^{n}\binom{n}{k}(k+1)^{k-1}(a-k)^{n-k-1}\,{}_{p+1}F_p\!\left(\begin{matrix}-k,\,(a_p)\\(b_q);\,\frac{z}{k+1}\end{matrix}\right)$

$\displaystyle\qquad=\frac{(a+1)^n}{(n+1)(a-n)}\,{}_{p+1}F_p\!\left(\begin{matrix}-n,\,(a_p)\\(b_q);\,\frac{z}{a+1}\end{matrix}\right)$

$\displaystyle\qquad+\frac{(a+1)^n}{(n+1)z}\frac{\prod_{j=1}^{q}(b_j-1)}{\prod_{i=1}^{p}(a_i-1)}\left[1-{}_{p+1}F_p\!\left(\begin{matrix}-n,\,(a_p)-1\\(b_q)-1;\,\frac{z}{a+1}\end{matrix}\right)\right].$

10. $\displaystyle\sum_{k=0}^{n}(-1)^k\binom{n}{k}(ka+1)^{n-1}\,{}_{p+1}F_q\!\left(\begin{matrix}-k,\,(a_p)\\(b_q);\,\frac{z}{ka+1}\end{matrix}\right)=\frac{z^n}{na+1}\frac{\prod(a_p)_n}{\prod(b_q)_n}.$

11. $\displaystyle\sum_{k=0}^{2n}(-1)^k\binom{2n}{k}(ka+1)^{2n-1}\,{}_{p+2}F_q\!\left(\begin{matrix}-\frac{k}{2},\,\frac{1-k}{2},\,(a_p)\\(b_q);\,\frac{z}{(ka+1)^2}\end{matrix}\right)$

$\displaystyle\qquad=\frac{\left(\frac{1}{2}\right)_n}{2na+1}z^n\frac{\prod(a_p)_n}{\prod(b_q)_n}.$

12. $\displaystyle\sum_{k=0}^{2n+1}(-1)^k\binom{2n+1}{k}(ka+1)^{2n}\,{}_{p+2}F_q\!\left(\begin{matrix}-\frac{k}{2},\,\frac{1-k}{2},\,(a_p)\\(b_q);\,\frac{z}{(ka+1)^2}\end{matrix}\right)=0.$

5.17.4. Sums containing $_pF_q((a_p) \pm mk; (b_q) \pm nk; \varphi(k,z))$ and special functions

1. $\displaystyle\sum_{k=0}^{n} \frac{\left(\frac{1}{2}-n\right)_k}{k!(n-k)!} (-4)^k (k+1)^{n-1} H_{2n-2k}\left(\frac{w}{\sqrt{k+1}}\right) {}_{p+1}F_q\left(\begin{matrix}-k, (a_p)\\ (b_q); \end{matrix}\; \frac{z}{k+1}\right)$

$\displaystyle = -\frac{2^{2n+1} w^{2n+2} z^{-1}}{(n+1)!(2n+1)} \frac{\prod_{j=1}^{q}(b_j-1)}{\prod_{i=1}^{p}(a_i-1)} \left[{}_{p+2}F_q\left(\begin{matrix}-n-1, -n-\frac{1}{2}, (a_p)-1\\ (b_q)-1;\; -w^{-2}z\end{matrix}\right) - 1\right].$

2. $\displaystyle\sum_{k=0}^{n} \frac{\left(-n-\frac{1}{2}\right)_k}{k!(n-k)!} (-4)^k (k+1)^{n-1/2}$

$\displaystyle \times H_{2n-2k+1}\left(\frac{w}{\sqrt{k+1}}\right) {}_{p+1}F_q\left(\begin{matrix}-k, (a_p)\\ (b_q); \end{matrix}\; \frac{z}{k+1}\right)$

$\displaystyle = -\frac{2^{2n+2} w^{2n+3} z^{-1}}{(n+1)!(2n+3)} \frac{\prod_{j=1}^{q}(b_j-1)}{\prod_{i=1}^{p}(a_i-1)} \left[{}_{p+2}F_q\left(\begin{matrix}-n-1, -n-\frac{3}{2}, (a_p)-1\\ (b_q)-1;\; -w^{-2}z\end{matrix}\right) - 1\right].$

3. $\displaystyle\sum_{k=0}^{n} \frac{(k+1)^{n-1}}{k!} (-\lambda-n)_k L_{n-k}^{\lambda}\left(\frac{w}{k+1}\right) {}_{p+1}F_q\left(\begin{matrix}-k, (a_p)\\ (b_q); \end{matrix}\; \frac{z}{k+1}\right)$

$\displaystyle = \frac{(-w)^{n+1} z^{-1}}{(n+1)!(\lambda+n+1)} \frac{\prod_{j=1}^{q}(b_j-1)}{\prod_{i=1}^{p}(a_i-1)} \left[{}_{p+2}F_q\left(\begin{matrix}-n-1, -\lambda-n-1, (a_p)-1\\ (b_q)-1;\; -w^{-1}z\end{matrix}\right) - 1\right].$

4. $\displaystyle\sum_{k=0}^{n} \frac{(k+1)^{k-1}}{k!} (-w)^k L_{n-k}^{\lambda+k}((n-k)w) {}_{p+1}F_q\left(\begin{matrix}-k, (a_p)\\ (b_q); \end{matrix}\; \frac{z}{k+1}\right)$

$\displaystyle = \sum_{k=0}^{n} \frac{(wz)^k}{(k+1)!} \frac{\prod(a_p)_k}{\prod(b_q)_k} L_{n-k}^{\lambda+k}((n+1)w).$

5. $\displaystyle\sum_{k=1}^{n} (-k)^k (n-k+1)^{n-k-1} \frac{(-\lambda-n)_k}{k!} L_{n-k}^{\lambda}\left(\frac{w}{n-k+1}\right) {}_{p+1}F_q\left(\begin{matrix}-k, (a_p)\\ (b_q); \end{matrix}\; \frac{z}{k}\right)$

$\displaystyle = -(n+1)^{n-1} L_n^{\lambda}\left(\frac{w}{n+1}\right)$

$\displaystyle + \frac{(-w)^n}{(n+1)!} \sum_{k=0}^{n} \binom{n+1}{k} (-n-\lambda)_k \left(\frac{n+1}{w}\right)^k {}_{p+1}F_q\left(\begin{matrix}-k, (a_p)\\ (b_q); \end{matrix}\; \frac{z}{n+1}\right).$

5.17.5. Sums containing products of $_pF_q((a_p)\pm mk;(b_q)\pm nk;\varphi(k,z))$

1. $\displaystyle\sum_{k=1}^{n}(-1)^k\binom{n+1}{k}k^n\,{}_2F_1\left(\begin{array}{c}-n+k,\,a\\b;\,\frac{n+1}{k}\end{array}\right)\,{}_{p+1}F_q\left(\begin{array}{c}-k,\,(a_p)\\(b_q);\,\frac{z}{k}\end{array}\right)$

$$=\frac{(-1)^{n+1}(n+1)^n(a)_{n+1}}{(a-b+1)(b)_n}$$

$$\times\left[\frac{a-b+1}{n+a}-{}_{p+2}F_{q+1}\left(\begin{array}{c}-n-1,\,(a_p),\,1-b-n\\(b_q),\,-a-n;\,\frac{z}{n+1}\end{array}\right)\right.$$

$$\left.+\frac{(b-1)(b)_n}{(a)_{n+1}}\,{}_{p+1}F_q\left(\begin{array}{c}-n-1,\,(a_p)\\(b_q);\,\frac{z}{n+1}\end{array}\right)\right].$$

2. $\displaystyle\sum_{k=0}^{n-1}\binom{n}{k}(k+1)^{k-1}(n-k)^{n-k}\,{}_{p+1}F_q\left(\begin{array}{c}-k,\,(a_p);\,\frac{w}{k+1}\\(b_q)\end{array}\right)$

$$\times{}_{r+1}F_s\left(\begin{array}{c}-n+k,\,(c_r)\\(d_s);\,\frac{z}{n-k}\end{array}\right)=-(n+1)^{n-1}\,{}_{p+1}F_q\left(\begin{array}{c}-n,\,(a_p)\\(b_q);\,\frac{w}{n+1}\end{array}\right)$$

$$+\frac{(-w)^n}{n+1}\frac{\prod(a_p)_n}{\prod(b_q)_n}\sum_{k=0}^{n+1}\binom{n+1}{k}(-1)^{(p+q+1)k}\left(\frac{n+1}{w}\right)^k$$

$$\times\frac{\prod(1-b_q-n)_k}{\prod(1-a_p-n)_k}\,{}_{r+1}F_s\left(\begin{array}{c}-k,\,(c_r)\\(d_s);\,\frac{z}{n+1}\end{array}\right).$$

5.17.6. Various sums containing $_pF_q((a_p)+mk;(b_q)+nk;z)$

1. $\displaystyle\sum_{k=0}^{n}(-1)^k\binom{n}{k}\frac{(c)_k(e-b)_k}{(e)_k(c-d+1)_k}\,{}_3F_2\left(\begin{array}{c}a,\,b,\,c+k\\d,\,e+k;\,1\end{array}\right)$

$$=\frac{(1-d)_n}{(c-d+1)_n}\,{}_3F_2\left(\begin{array}{c}a-n,\,b,\,c\\d-n,\,e;\,1\end{array}\right)\quad[\mathrm{Re}\,(d+e-a-b)>0].$$

2. $\displaystyle\sum_{k=0}^{n}\binom{n}{k}2^{-2k}(2k)!\,\frac{(a)_k(2a+n)_k}{\left(a+\frac{1}{2}\right)_k(2a)_k\left(\frac{1}{2}\right)_{2k}}$

$$\times{}_4F_3\left(\begin{array}{c}-n+k,\,k+1,\,k+a,\,k+n+2a\\2k+\frac{3}{2},\,k+a+\frac{1}{2},\,k+2a;\,1\end{array}\right)P_{2k}(z)=\left[\frac{n!}{(2a)_n}C_n^a(z)\right]^2$$

[[69], (18)].

3. $\displaystyle\sum_{k=0}^{n}\frac{(k!)^2(a+n)_k}{(n-k)!(2k)!\left(\frac{1}{2}\right)_k\left(a+\frac{1}{2}\right)_k}\,{}_4F_3\left(\begin{array}{c}-n+k,\,k+1,\,k+1,\,k+n+a\\k+\frac{1}{2},\,a+k+\frac{1}{2},\,2k+2;\,1\end{array}\right)$

$$\times[P_k(z)]^2=\frac{(2n)!}{n!(2a)_{2n}}C_{2n}^a(z).$$

4. $\sum_{k=0}^{n} \frac{(k!)^2 (a)_k (2a+n)_k}{(n-k)!(2k)! \left(\frac{1}{2}\right)_k \left(a+\frac{1}{2}\right)_k (2a)_k}$

$\times {}_5F_4\left(\begin{array}{c} -n+k, k+1, k+1, k+a, k+n+2a \\ k+\frac{1}{2}, a+k+\frac{1}{2}, k+2a, 2k+2;\ 1 \end{array}\right) [P_k(z)]^2 = \frac{n!}{(2a)_n^2} [C_n^a(z)]^2.$

5. $\sum_{k=0}^{n} \binom{n}{k} 2^{-2k} \frac{(a)_k (2a+n)_k}{\left(a+\frac{1}{2}\right)_k (2a)_k} {}_4F_3\left(\begin{array}{c} -n+k, k+\frac{3}{2}, k+a, k+n+2a \\ 2k+2, k+a+\frac{1}{2}, k+2a;\ 1 \end{array}\right)$

$\times U_{2k}(z) = \left[\frac{n!}{(2a)_n} C_n^a(z)\right]^2$ [[69], (18)].

6. $\sum_{k=0}^{n} \frac{\left(\frac{3}{2}\right)_k (a+n)_k}{(n-k)!(2k+1)! \left(a+\frac{1}{2}\right)_k} {}_4F_3\left(\begin{array}{c} -n+k, k+\frac{3}{2}, k+2, k+n+a \\ k+1, a+k+\frac{1}{2}, 2k+3;\ 1 \end{array}\right)$

$\times [U_k(z)]^2 = \frac{(2n)!}{n!(2a)_{2n}} C_{2n}^a(z).$

7. $\sum_{k=0}^{n} \binom{n}{k} \frac{2^{-2k}(a)_k(2a+n)_k}{\left(a+\frac{1}{2}\right)_k (2a)_k} {}_4F_3\left(\begin{array}{c} -n+k, k+\frac{3}{2}, k+2, k+a, k+n+2a \\ k+1, a+k+\frac{1}{2}, k+2a, 2k+3;\ 1 \end{array}\right)$

$\times [U_k(z)]^2 = \left[\frac{n!}{(2a)_n} C_n^a(z)\right]^2.$

5.18. Multiple Sums

5.18.1. Sums containing Bessel functions

Condition: $k_i = 0, 1, 2, \ldots$

1. $\sum_{k_1+\ldots+k_{2m}=n} \prod_{i=1}^{2m} \frac{1}{k_i!} J_{\pm k_i \mp 1/2}(z)$

$= \frac{(\pm 1)^n 2^{n-m+1}}{n!\,(\pi z)^{m-1/2}} \sum_{k=1}^{m} (\pm 1)^k \binom{2m}{m-k} k^{n+1/2} J_{n-1/2}(2kz) \quad [n \geq 1].$

2. $\sum_{k_1+\ldots+k_{2m+1}=n} \prod_{i=1}^{2m+1} \frac{1}{k_i!} J_{\pm k_i \mp 1/2}(z)$

$= \frac{(\pm 1)^n}{n!\,(2\pi z)^m} \sum_{k=1}^{m} \binom{2m+1}{m-k} k^{n+1/2} J_{\pm n \mp 1/2}((2k+1)z) \quad [n \geq 1].$

3. $\sum_{k_1+\ldots+k_m=n} \prod_{i=1}^{m} \frac{z_i^{k_i}}{k_i!} K_{k_i-1/2}(z_i) = \frac{\left(\frac{\pi}{2}\right)^{(m-1)/2}}{n!} \frac{(z_1+\ldots+z_m)^{n+1/2}}{(z_1\ldots z_m)^{1/2}}$

$\times K_{n-1/2}(z_1+\ldots+z_m).$

5.18.2. Sums containing orthogonal polynomials

Condition: $k_i = 0, 1, 2, \ldots$

1. $$\sum_{k_1+\ldots+k_m=n} \prod_{i=1}^{m} P_{k_i}(z) = C_n^{m/2}(z).$$

2. $$\sum_{k_1+\ldots+k_m=n} \prod_{i=1}^{m} U_{k_i}(z) = C_n^{m}(z).$$

3. $$\sum_{k_1+\ldots+k_m=n} \prod_{i=1}^{m} \frac{1}{k_i!} H_{2k_i}(z_i) = (-4)^n L_n^{m/2-1}(z_1^2 + \ldots + z_m^2).$$

4. $$\sum_{k_1+\ldots+k_m=n} \prod_{i=1}^{m} \frac{1}{k_i!} H_{2k_i+1}(z_i)$$
$$= (-1)^n 2^{2n+m}(z_1 \ldots z_m)^{1/2} L_n^{3m/2-1}(z_1^2 + \ldots + z_m^2).$$

5. $$\sum_{k_1+\ldots+k_m=n} \prod_{i=1}^{m} L_{k_i}^{\lambda_i}(z_i) = L_n^{\lambda_1+\ldots+\lambda_m+m-1}(z_1 + \ldots + z_m).$$

6. $$\sum_{k_1+\ldots+k_m=n} \prod_{i=1}^{m} \frac{z_i^{k_i}}{k_i!} L_{r_i}^{k_i-r_i}(z_i)$$
$$= \frac{(r_1+\ldots+r_m)!}{n!} \left(\prod_{i=1}^{m} \frac{z_i^{k_i}}{k_i!} \right) u^{n-r} L_r^{n-r}(u)$$
$$[r_i = 0, 1, \ldots, n;\ r = r_1 + \ldots + r_m;\ u = z_1 + \ldots + z_m].$$

7. $$\sum_{k_1+\ldots+k_m=n} \prod_{i=1}^{m} C_{k_i}^{\lambda_i}(z) = C_n^{\lambda_1+\ldots+\lambda_m}(z).$$

8. $$\sum_{k_1+\ldots+k_m=n} \prod_{i=1}^{m} \frac{(-\lambda_i)_{k_i}}{(k_i-2\lambda_i)_{k_i}} C_{k_i}^{\lambda_i}(z)$$
$$= \frac{(-\lambda_1-\ldots-\lambda_m)_n}{(n-2\lambda_1-\ldots-2\lambda_m)_n} C_n^{\lambda_1+\ldots+\lambda_m}(z).$$

9. $$\sum_{k_1+\ldots+k_m=n} \prod_{i=1}^{m} P_{k_i}^{(\rho_i-k_i,\sigma_i-k_i)}(z) = P_n^{(\rho-n,\sigma-n)}(z)$$
$$[\rho = \rho_1 + \ldots + \rho_m;\ \sigma = \sigma_1 + \ldots + \sigma_m].$$

Chapter 6

Infinite Series

6.1. Elementary Functions

6.1.1. Series containing algebraic functions

1. $\displaystyle\sum_{k=0}^{\infty} \frac{\sigma_{k+m}^m}{(k+m)!} z^k \frac{\prod(a_p)_k}{\prod(b_q)_k}$

$= (-1)^{m(p+q+1)} \dfrac{z^{-m}}{m!} \dfrac{\prod(1-(b_q))_m}{\prod(1-(a_p))_m} \displaystyle\sum_{k=0}^{m} (-1)^k \binom{m}{k} {}_pF_q\!\left(\!\begin{array}{c}(a_p)-m;\ kz\\(b_q)-m\end{array}\!\right)$

$[p \leq q]$.

6.1.2. Series containing the exponential function

Notations: $c = \mathbf{K}(k')/\mathbf{K}(k)$, $k' = \sqrt{1-k^2}$.

1. $\displaystyle\sum_{n=1}^{\infty} \frac{1}{n\left(e^{2n\pi c}-1\right)} = -\frac{\pi c}{12} - \frac{1}{6}\ln\frac{2kk'\,\mathbf{K}^3(k)}{\pi^3}$ \hfill [[84], (T1.1)].

2. $\displaystyle\sum_{n=1}^{\infty} \frac{(-1)^n}{n\left(e^{2n\pi c}-1\right)} = -\frac{\pi c}{12} - \frac{1}{12}\ln\frac{k^2}{16k'}$ \hfill [[84], (T1.2)].

3. $\displaystyle\sum_{n=1}^{\infty} \frac{1}{n\left(e^{2n\pi c}+1\right)} = \frac{\pi c}{4} + \frac{1}{2}\ln\frac{k\,\mathbf{K}(k)}{2\pi}$ \hfill [[84], (T1.5)].

4. $\displaystyle\sum_{n=0}^{\infty} \frac{1}{(2n+1)\left[e^{(2n+1)\pi c}+1\right]} = \frac{1}{4}\ln\frac{2\,\mathbf{K}(k)}{\pi}$ \hfill [[84], (T1.6)].

5. $\displaystyle\sum_{n=0}^{\infty} \frac{1}{(2n+1)\left[e^{(2n+1)\pi c}-1\right]} = -\frac{1}{4}\ln\frac{2k'\,\mathbf{K}(k)}{\pi}$ \hfill [[84], (T1.7)].

6. $\displaystyle\sum_{n=1}^{\infty} \frac{(-1)^n}{n\left(e^{2n\pi c}+1\right)} = \frac{\pi c}{4} + \frac{1}{2}\ln\frac{1-k'}{2k}$ \hfill [[84], (T1.8)].

6.1.3. Series containing hyperbolic functions

Notations: $c = \mathbf{K}(k')/\mathbf{K}(k)$, $k' = \sqrt{1-k^2}$.

1. $\displaystyle\sum_{n=1}^{\infty} \frac{\coth(nc\pi) - 1}{n} = -\frac{c\pi}{6} - \frac{1}{3}\ln\frac{2kk'\mathbf{K}^3(k)}{\pi^3}$ [[84], (T1.1)].

2. $\displaystyle\sum_{n=1}^{\infty} (-1)^n \frac{\coth(nc\pi) - 1}{n} = -\frac{\pi c}{6} - \frac{1}{6}\ln\frac{2kk'\mathbf{K}^3(k)}{\pi^3}$ [[84], (T1.2)].

3. $\displaystyle\sum_{n=1}^{\infty} \frac{\operatorname{csch}(nc\pi)}{n} = \frac{c\pi}{12} - \frac{1}{6}\ln\frac{4k'^2}{k}$ [[84], (T1.4)].

4. $\displaystyle\sum_{n=1}^{\infty} (-1)^n \frac{\operatorname{csch}(nc\pi)}{n} = \frac{c\pi}{12} + \frac{1}{6}\ln\frac{kk'}{4}$ [[84], (T1.3)].

5. $\displaystyle\sum_{n=1}^{\infty} \frac{1 - \tanh(nc\pi)}{n} = \frac{c\pi}{2} - \ln\frac{k\mathbf{K}(k)}{2\pi}$ [[84], (T1.5)].

6. $\displaystyle\sum_{n=1}^{\infty} (-1)^n \frac{1 - \tanh(nc\pi)}{n} = \frac{c\pi}{2} + \ln\frac{1-k'}{2k}$ [[84], (T1.8)].

7. $\displaystyle\sum_{n=0}^{\infty} \frac{(-1)^n}{2n+1} \operatorname{csch}[(2n+1)c\pi] = -\frac{1}{4}\ln k'$ [[84], (T1.9)].

8. $\displaystyle\sum_{n=0}^{\infty} \frac{(-1)^n}{2n+1} \operatorname{sech}\frac{(2n+1)c\pi}{2} = \frac{1}{2}\arcsin k$ [[84], (T1.10)].

9. $\displaystyle\sum_{n=1}^{\infty} (-1)^n \operatorname{csch}(nc\pi)\coth(nc\pi) = \frac{2k^2 - 1}{3\pi^2}\mathbf{K}^2(k) - \frac{1}{12}$ [[84], (T1.18)].

10. $\displaystyle\sum_{n=0}^{\infty} \frac{(-1)^n}{(2n+1)^{6m+1}} \operatorname{sech}\frac{(2n+1)\sqrt{3}\pi}{2}$

$\displaystyle = \frac{(-1)^{m+1}}{2}\pi^{6m+1} \sum_{p=0}^{3m} \frac{E_{2p+1}}{(2p+1)!} \frac{B_{6m-2p}}{(6m-2p)!}\cos\frac{(2p+1)\pi}{3}$ [19].

6.1.4. Series containing trigonometric functions

1. $\sum_{k=1}^{\infty} \dfrac{\sin(kx)\cos(ky)}{k^{2m-1}} = (-1)^m \dfrac{x^{2m-1}}{2(2m-1)!} + \sum_{k=0}^{m-2} \dfrac{x^{2k+1}}{(2k+1)!}$

$\times \left[(-1)^{m-1} \dfrac{\pi y^{2m-2k-3}}{2(2m-2k-3)!} + (-1)^k \sum_{j=0}^{m-k-1} (-1)^j \dfrac{y^{2j}}{(2j)!} \zeta(2m-2j-2k-2) \right]$

$$[-\pi < x < \pi; \ |x| < y < 2\pi - |x|; \ [72]].$$

2. $\sum_{k=1}^{\infty} (-1)^k \dfrac{\sin(kx)\cos(ky)}{k^{2m-1}} = (-1)^m \dfrac{x^{2m-1}}{2(2m-1)!} - \sum_{k=0}^{m-2} (-1)^k \dfrac{x^{2k+1}}{(2k+1)!}$

$\times \sum_{j=0}^{m-k-1} (-1)^j \dfrac{y^{2j}}{(2j)!} \left(1 - 2^{2j+2k-2m+3}\right) \zeta(2m-2j-2k-2)$

$$[-\pi < x < \pi; \ |x| - \pi < y < \pi - |x|; \ [72]].$$

3. $\sum_{k=0}^{\infty} \dfrac{(-1)^k \left(k+\frac{1}{2}\right) \sin\sqrt{\left(k+\frac{1}{2}\right)^2 \pi^2 + a^2}}{\left(k+\frac{1}{2}\right)^2 - b^2} \cdot \dfrac{1}{\sqrt{\left(k+\frac{1}{2}\right)^2 \pi^2 + a^2}} = \dfrac{\pi}{2} \sec(b\pi) \dfrac{\sin\sqrt{a^2+b^2\pi^2}}{\sqrt{a^2+b^2\pi^2}}$

$$[[39], (1.10)].$$

4. $\sum_{k=1}^{\infty} \dfrac{(-1)^k}{k^2\sqrt{k^2+a^2}} \sin\left(\sqrt{k^2+a^2}\, x\right) = \dfrac{3-a^2\pi^2}{12a^3} \sin(ax) - \dfrac{x}{4a^2} \cos(ax)$

$$[-\pi < x < \pi].$$

5. $\sum_{k=1}^{\infty} \dfrac{(-1)^k}{k^2} \cos\left(\sqrt{k^2+a^2}\, x\right) = \dfrac{x}{4a} \sin(ax) - \dfrac{\pi^2}{12} \cos(ax) \quad [-\pi < x < \pi].$

6.2. The Psi Function $\psi(z)$

6.2.1. Series containing $\psi(ka+b)$

1. $\sum_{k=1}^{\infty} t^k \psi(k) = \dfrac{t}{t-1} \left[\mathbf{C} + \ln(1-t) \right]$ $\qquad [|t|<1].$

2. $\sum_{k=1}^{\infty} \dfrac{t^k}{k} \psi(k) = \mathbf{C} \ln(1-t) + \dfrac{1}{2} \ln^2(1-t)$ $\qquad [|t|<1].$

3. $\sum_{k=1}^{\infty} \dfrac{t^k}{k^2} \psi(k) = \dfrac{1}{2} \ln t \ln^2(1-t) - \mathbf{C}\operatorname{Li}_2(t) + \ln(1-t)\operatorname{Li}_2(1-t)$

$$- \operatorname{Li}_3(1-t) + \zeta(3) \quad [|t|<1].$$

4. $\sum_{k=1}^{\infty} \frac{1}{k^2} \psi(k) = \zeta(3) - \frac{\pi^2 C}{6}.$

5. $\sum_{k=1}^{\infty} \frac{1}{k(k+1)} \psi(k) = 1 - C$ [[29], (7)].

6. $\sum_{k=1}^{\infty} \frac{1}{(k+a)(k+b)} \psi(k)$
$= \frac{1}{2(a-b)} [\psi'(b+1) - \psi'(a+1) - \psi^2(b+1) + \psi^2(a+1)].$

7. $\sum_{k=2}^{\infty} \frac{t^k}{k(k-1)} \psi(k) = -C[t + \ln(1-t) - t \ln(1-t)]$
$+ \frac{1}{6} [\pi^2 t + 3(t-1) \ln^2(1-t) - 6t \ln t \ln(1-t) - 6t \operatorname{Li}_2(1-t)] \quad [|t| < 1].$

8. $\sum_{k=1}^{\infty} \frac{t^{k+1}}{k(k+1)} \psi(k) = -\frac{1}{2}(1-t) \ln^2(1-t)$
$+ (1-C)[t + (1-t) \ln(1-t)] \quad [|t| < 1].$

9. $\sum_{k=1}^{\infty} \frac{1}{k^2} \psi(k+m) = \zeta(3) + \frac{\pi^2}{6} \psi(m) - \frac{1}{2} \psi''(m)$
$+ C\left[\psi'(m) - \frac{\pi^2}{6}\right] - \sum_{k=1}^{m-1} \frac{1}{k^2} \psi(k).$

10. $\sum_{k=1}^{\infty} \frac{1}{k^3} \psi(k) = \frac{\pi^4}{360} - C\zeta(3).$

11. $\sum_{k=1}^{\infty} \frac{(-1)^k}{k^3} \psi(k) = -\frac{\pi^4}{48} - \frac{\pi^2}{12} \ln^2 2 + \frac{1}{12} \ln^4 2$
$+ \frac{1}{4}(7 \ln 2 + 3C) \zeta(3) + 2 \operatorname{Li}_4\left(\frac{1}{2}\right).$

12. $\sum_{k=1}^{\infty} \frac{1}{k^2(k+1)} \psi(k) = \left(1 - \frac{\pi^2}{6}\right) C + \zeta(3) - 1$ [[58], (10)].

13. $\sum_{k=1}^{\infty} \frac{1}{k(k+1)^2} \psi(k) = \left(\frac{\pi^2}{6} - 2\right) C - \frac{\pi^2}{6} - \zeta(3) + 3$ [[58], (13)].

14. $\sum_{k=1}^{\infty} \frac{1}{k^4} \psi(k) = -\frac{\pi^4 C}{90} - \frac{\pi^2}{6} \zeta(3) + 2\zeta(5).$

15. $\sum_{k=1}^{\infty} \frac{1}{k^5} \psi(k) = \frac{\pi^6}{1260} - \frac{1}{2}\zeta^2(3) - C\zeta(5).$

16. $\sum_{k=1}^{\infty} \frac{1}{k^{n+2}} \psi(k) = \frac{n+2}{2}\zeta(n+3) - C\zeta(n+2)$
$$- \frac{1}{2}\sum_{k=0}^{n} \zeta(k+1)\zeta(n-k+2).$$

17. $\sum_{k=1}^{\infty} \frac{1}{(k+a)^{n+2}} \psi(k)$
$$= \frac{(-1)^{n+1}}{(n+1)!}\left[\frac{1}{2}\psi^{(n+2)}(a+1) - \sum_{k=0}^{n}\binom{n}{k}\psi^{(n-k)}(a+1)\psi^{(k+1)}(a+1)\right].$$

18. $\sum_{k=1}^{\infty} \frac{t^k}{k!}\psi(k) = 2C + e^t\Gamma(0,t) + \Gamma(0,-t) + e^t\ln t + \ln(-t).$

19. $\sum_{k=0}^{\infty} \frac{t^k}{(k!)^2}\psi(k+1) = \frac{1}{2}\ln t\, I_0(2\sqrt{t}) + K_0(2\sqrt{t}).$

20. $\sum_{k=0}^{\infty} \frac{t^k}{k!(k+1)!}\psi(k+1)$
$$= \frac{1}{2t}\left[2 - I_0(2\sqrt{t}) + \sqrt{t}\ln t\, I_1(2\sqrt{t}) - 2\sqrt{t}K_1(2\sqrt{t})\right].$$

21. $\sum_{k=1}^{\infty} \frac{t^k}{k}\psi\left(k+\frac{1}{2}\right) = (C+2\ln 2)\ln(1-t) + \frac{1}{2}\ln^2\frac{1+\sqrt{t}}{1-\sqrt{t}}$ $\qquad [|t|<1].$

22. $\sum_{k=1}^{\infty} \frac{(-1)^k}{k}\psi\left(k+\frac{1}{2}\right) = -\frac{\pi^2}{8} + \ln 2(C+2\ln 2).$

23. $\sum_{k=1}^{\infty} \frac{1}{k^2}\psi\left(k+\frac{1}{2}\right) = \frac{7}{2}\zeta(3) - \frac{\pi^2}{3}\ln 2 - \frac{\pi^2 C}{6}$ $\qquad$ [[29], (15)].

24. $\sum_{k=1}^{\infty} \frac{(-1)^k}{k^2}\psi\left(k+\frac{1}{2}\right) = \frac{7}{2}\zeta(3) + \frac{\pi^2}{6}\ln 2 + \frac{\pi^2 C}{12} - 2\pi G$ $\qquad$ [[29], (19)].

25. $\sum_{k=1}^{\infty} \frac{1}{k(2k+1)}\psi\left(k+\frac{1}{2}\right) = \frac{\pi^2}{6} + 2(\ln 2 - 1)(C+2\ln 2).$

26. $\sum_{k=0}^{\infty} \frac{1}{(k+1)(2k+1)}\psi\left(k+\frac{1}{2}\right) = -\frac{\pi^2}{6} + 2(2-C)\ln 2 - 4\ln^2 2.$

27. $\displaystyle\sum_{k=0}^{\infty} \frac{1}{(k+2)(2k+1)} \psi\left(k+\frac{1}{2}\right)$
$$= \frac{1}{18}\left[4 - \pi^2 + 20\ln 2 - 24\ln^2 2 - 6(2\ln 2 + 1)\,C\right].$$

28. $\displaystyle\sum_{k=1}^{\infty} \frac{1}{(2k-1)^2} \psi\left(k+\frac{1}{2}\right) = \frac{7}{8}\zeta(3) + \frac{\pi^2}{4}\ln 2 - \frac{\pi^2}{8}(C + 2\ln 2).$

29. $\displaystyle\sum_{k=0}^{\infty} \frac{1}{(2k+1)^2} \psi\left(k+\frac{1}{2}\right) = -\frac{1}{8}[\pi^2 C + 7\zeta(3)].$

30. $\displaystyle\sum_{k=0}^{\infty} \frac{1}{(k+2)(2k+1)(2k+3)} \psi\left(k+\frac{1}{2}\right)$
$$= \frac{1}{18}\left[\pi^2 + 14 - 56\ln 2 + 24\ln^2 2 + 12C(\ln 2 - 1)\right].$$

31. $\displaystyle\sum_{k=0}^{\infty} \frac{t^k}{k!} \psi\left(k+\frac{1}{2}\right) = -(C + 2\ln 2)\,e^t + 2te^t\, {}_2F_2\!\left(\begin{array}{c}1,1;\;-t\\ \frac{3}{2},2\end{array}\right).$

32. $\displaystyle\sum_{k=1}^{\infty} \frac{1}{k!\,k} \psi\left(k+\frac{1}{2}\right) = \frac{1}{2}(C + \pi^2) - 2C\ln 2 + \ln 2 - 4\ln^2 2 - 1.$

33. $\displaystyle\sum_{k=1}^{\infty} \frac{\left(\frac{1}{2}\right)_k}{k!\,k} \psi\left(k+\frac{1}{2}\right) = \frac{1}{2}[\pi^2 - 4\ln 2(C + 2\ln 2)].$

34. $\displaystyle\sum_{k=1}^{\infty} \frac{\left(\frac{1}{2}\right)_k}{(a)_k\,k} \psi\left(k+\frac{1}{2}\right) = (C + 2\ln 2)\left[\psi\!\left(a-\frac{1}{2}\right) - \psi(a)\right] + \psi'\!\left(a-\frac{1}{2}\right)$
$$[\mathrm{Re}\,a > 1/2].$$

35. $\displaystyle\sum_{k=0}^{\infty} \frac{\left(\frac{1}{2}\right)_k}{(k+1)!\,(k+1)} \psi\left(k+\frac{1}{2}\right) = 4(\ln 2 - 1)C + 8(2 - 2\ln 2 + \ln^2 2) - \pi^2.$

36. $\displaystyle\sum_{k=0}^{\infty} \frac{\left(\frac{1}{2}\right)_k}{(k+2)!\,(k+2)} \psi\left(k+\frac{1}{2}\right)$
$$= \frac{2}{27}\left[3(7 - 12\ln 2)\,C + 9\pi^2 + 138\ln 2 - 72\ln^2 2 - 154\right].$$

37. $\displaystyle\sum_{k=0}^{\infty} \frac{\left(\frac{1}{2}\right)_k}{(k+2)!\,(2k+3)} \psi\left(k+\frac{1}{2}\right) = \left(\pi - \frac{10}{3}\right)C - \pi + \frac{4}{9}(17 - 15\ln 2).$

38. $\displaystyle\sum_{k=0}^{\infty} \frac{\left(\frac{1}{2}\right)_k}{(k+3)!\,(2k+5)} \psi\left(k+\frac{1}{2}\right) = \left(\frac{23}{15} - \frac{\pi}{2}\right)C + \frac{3\pi}{4} + \frac{46}{15}\ln 2 - \frac{1018}{225}.$

39. $\displaystyle\sum_{k=0}^{\infty} \frac{\left(\frac{1}{2}\right)_k}{k!\,(2k+1)^2} \psi\!\left(k+\frac{1}{2}\right) = -\frac{\pi}{24}\left(\pi^2 + 12\mathbf{C}\ln 2\right).$

40. $\displaystyle\sum_{k=0}^{\infty} \frac{\left(\frac{1}{2}\right)_k}{(k+1)!\,(2k+1)^2} \psi\!\left(k+\frac{1}{2}\right) = [\pi(1-\ln 2) - 2]\,\mathbf{C} - \frac{\pi^3}{12} + 4(1-\ln 2).$

41. $\displaystyle\sum_{k=0}^{\infty} \frac{\left(\frac{1}{2}\right)_k}{(k+1)!\,(2k+1)(2k+3)} \psi\!\left(k+\frac{1}{2}\right) = \frac{\pi}{4} + \left(2 - \frac{3\pi}{4}\right)\mathbf{C} + 4(\ln 2 - 1).$

42. $\displaystyle\sum_{k=0}^{\infty} \frac{\left(\frac{1}{2}\right)_k}{k!\,(k+2)(2k+1)(2k+3)} \psi\!\left(k+\frac{1}{2}\right)$
$\displaystyle\qquad = \frac{1}{288}\left[153\pi + 6(64-21\pi)\mathbf{C} + 768\ln 2 - 1024\right].$

43. $\displaystyle\sum_{k=0}^{\infty} \frac{\left(\frac{1}{2}\right)_k}{(k+1)!\,(k+3)(2k+5)(2k+7)} \psi\!\left(k+\frac{1}{2}\right) = \frac{398}{225} - \frac{29\pi}{96}$
$\displaystyle\qquad + \frac{3}{80}(5\pi - 16)\mathbf{C} - \frac{6}{5}\ln 2.$

44. $\displaystyle\sum_{k=0}^{\infty} \frac{t^k}{k!\left(\frac{1}{2}\right)_k} \psi\!\left(k+\frac{1}{2}\right)$
$\displaystyle\qquad = \frac{1}{2}\left[\cosh(2\sqrt{t})\ln t + 2\sinh(2\sqrt{t})\,\mathrm{shi}(4\sqrt{t}) - 2\cosh(2\sqrt{t})\,\mathrm{chi}(4\sqrt{t})\right].$

45. $\displaystyle\sum_{k=0}^{\infty} \frac{\left(\frac{1}{2}\right)_k^2}{k!\,(k+1)!} \psi\!\left(k+\frac{1}{2}\right) = \frac{4}{\pi}(2 - \mathbf{C} - 4\ln 2).$

46. $\displaystyle\sum_{k=1}^{\infty} \frac{1}{(k+a)(k+a+1)} \psi\!\left(k+\frac{1}{2}\right)$
$\displaystyle\qquad = \frac{1}{a(a+1)(2a+1)}\left[2 + a(2 + \mathbf{C} + 2\ln 2)\right] + \frac{2}{2a+1}\psi(a).$

47. $\displaystyle\sum_{k=1}^{\infty} \frac{t^k}{\left(\frac{1}{2}\right)_k^2} \psi\!\left(k+\frac{1}{2}\right)$
$\displaystyle\qquad = \frac{\pi\sqrt{t}}{2}\left[\ln t\,\mathbf{L}_0(2\sqrt{t}) - 2K_0(2\sqrt{t}) + \frac{1}{\pi^2} G^{42}_{24}\!\left(t\,\bigg|\,\begin{matrix}\frac{1}{2},\frac{1}{2}\\0,0,\frac{1}{2},\frac{1}{2}\end{matrix}\right)\right].$

48. $\displaystyle\sum_{k=0}^{\infty} \frac{t^k}{(2k+1)!} \psi\left(k+\frac{1}{2}\right)$
$= \dfrac{1}{\sqrt{t}} \left[\dfrac{1}{2} \ln \dfrac{t}{4} \sinh\sqrt{t} - 2\operatorname{shi}\left(\sqrt{t}\right) + \cosh\sqrt{t}\,\operatorname{shi}\left(2\sqrt{t}\right)\right.$
$\left. - \sinh\sqrt{t}\,\operatorname{chi}\left(2\sqrt{t}\right)\right].$

49. $\displaystyle\sum_{k=0}^{\infty} \frac{\left(-\frac{1}{2}\right)_k \left(\frac{1}{2}\right)_k}{k!\,(k+1)!} \psi\left(k+\frac{1}{2}\right) = \frac{8}{3\pi}(1 - C - 4\ln 2).$

50. $\displaystyle\sum_{k=0}^{\infty} \frac{\left(\frac{1}{2}\right)_k^2}{k!\,(k+1)!} \psi\left(k+\frac{1}{2}\right) = \frac{4}{\pi}(2 - C - 4\ln 2).$

51. $\displaystyle\sum_{k=0}^{\infty} \frac{\left(\frac{1}{2}\right)_k^2}{[(k+1)!]^2} \psi\left(k+\frac{1}{2}\right) = \frac{4}{\pi}[(\pi - 4)C + 2\pi(\ln 2 - 1) + 4(3 - 4\ln 2)].$

52. $\displaystyle\sum_{k=0}^{\infty} \frac{\left(\frac{1}{2}\right)_k \left(\frac{3}{2}\right)_k}{[(k+1)!]^2} \psi\left(k+\frac{1}{2}\right) = \frac{4}{\pi}[(2 - \pi)C + 2(4 - \pi)(\ln 2 - 1)].$

53. $\displaystyle\sum_{k=0}^{\infty} \frac{\left(\frac{1}{2}\right)_k^2}{(k+1)!\,(k+2)!} \psi\left(k+\frac{1}{2}\right)$
$= \dfrac{4}{27\pi}[3(9\pi - 32)C + 54\pi(\ln 2 - 1) - 384\ln 2 + 304].$

54. $\displaystyle\sum_{k=0}^{\infty} \frac{\left(\frac{1}{2}\right)_k \left(\frac{3}{2}\right)_k}{(k+1)!\,(k+2)!} \psi\left(k+\frac{1}{2}\right)$
$= \dfrac{4}{9\pi}[3(8 - 3\pi)C - 18\pi(\ln 2 - 1) + 8(12\ln 2 - 11)].$

55. $\displaystyle\sum_{k=0}^{\infty} \frac{\left(\frac{1}{2}\right)_k^3}{(k!)^3} \psi\left(k+\frac{1}{2}\right) = \frac{1}{12\pi^3}(\pi - 3C - 6\ln 2)\Gamma^4\left(\frac{1}{4}\right).$

56. $\displaystyle\sum_{k=0}^{\infty} \frac{\left(\frac{1}{2}\right)_k^3}{k!\,[(k+1)!]^2} \psi\left(k+\frac{1}{2}\right)$
$= \dfrac{1}{6\pi^3}\left[(\pi - 6\ln 2 - 3C)\Gamma^4\left(\dfrac{1}{4}\right) + 48(\pi + 6\ln 2 + 3C - 8)\Gamma^4\left(\dfrac{3}{4}\right)\right].$

57. $\displaystyle\sum_{k=0}^{\infty} \frac{\left(\frac{1}{2}\right)_k^3}{[(k+1)!]^3} \psi\left(k+\frac{1}{2}\right) = 8(2 - C - 2\ln 2)$
$+ \dfrac{32}{\pi^3}(3C + 6\ln 2 + \pi - 10)\Gamma^4\left(\dfrac{3}{4}\right).$

58. $\sum_{k=0}^{\infty} \frac{\left(\frac{1}{2}\right)_k^3}{(k+1)![(k+2)!]^2} \psi\left(k+\frac{1}{2}\right) = 8(2 - C - 2\ln 2)$
$$- \frac{4}{243\pi^3}(22 + 15\pi - 45C - 90\ln 2)\Gamma^4\left(\frac{1}{4}\right)$$
$$- \frac{64}{27\pi^3}[78 - 7\pi - 21C - 42\ln 2]\Gamma^4\left(\frac{3}{4}\right).$$

59. $\sum_{k=0}^{\infty} \frac{k!}{(a)_k(k+1)} \psi(k+a) = [(a-1)\psi(a-1) - 1]\psi'(a-1)$
$$- (a-1)\psi''(a-1) \quad [\text{Re } a > 1].$$

60. $\sum_{k=1}^{\infty} \frac{t^k}{k!} \psi(k+a) = \frac{t}{a} e^t \left[a\psi(a)\frac{1-e^{-t}}{t} + {}_2F_2\left({1,1;\ -t \atop a+1,\ 2}\right) \right]$ [[58], (1.1a)].

61. $\sum_{k=0}^{\infty} \frac{(a)_k}{k!} t^k \psi(k+a) = (1-t)^{-a}[\psi(a) - \ln(1-t)]$ [$|t| < 1$].

62. $\sum_{k=0}^{\infty} \frac{(a)_k}{k!(k+b)} t^k \psi(k+a) = \frac{1}{b} \psi(a) {}_2F_1\left({a,\ b \atop b+1;\ t}\right)$
$$+ t^{-b}(1-t)^{1-a} \left\{ (1-t)^{a-1} B(1-a, b) \left[\psi(b-a+1) - \psi(1-a)\right] \right.$$
$$\left. + \frac{1}{1-a} \ln(1-t) {}_2F_1\left({1-a,\ 1-b \atop 2-a;\ 1-t}\right) - \frac{1}{(1-a)^2} {}_3F_2\left({1-a,\ 1-a,\ 1-b \atop 2-a,\ 2-a;\ 1-t}\right) \right\}.$$

63. $\sum_{k=1}^{\infty} \frac{t^k}{(a)_k} \psi(k+a) = \frac{t}{a^2} e^t \left[a^2 t^{-a} \psi(a)\gamma(a,t) + {}_2F_2\left({a,\ a;\ -t \atop a+1,\ a+1}\right) \right]$
$$[[58],\ (1.1b)].$$

64. $\sum_{k=0}^{\infty} \frac{2^{-k} k!}{(a)_k} \psi(k+a) = [(a-1)\psi(a) - 1] \left[\psi\left(\frac{a}{2}\right) - \psi\left(\frac{a-1}{2}\right) \right]$
$$+ \frac{a-1}{2} \left[\psi'\left(\frac{a-1}{2}\right) - \psi'\left(\frac{a}{2}\right) \right].$$

65. $\sum_{k=1}^{\infty} \frac{2^{-k} k!}{k^2 (a)_k} \psi(k+a) = \frac{1}{8} \left[\psi\left(\frac{a+1}{2}\right) - \psi\left(\frac{a}{2}\right) \right] \left[\psi'\left(\frac{a+1}{2}\right) - \psi'\left(\frac{a}{2}\right) \right]$
$$+ \frac{1}{8} \psi(a) \left\{ 4\psi'(a) - \left[\psi\left(\frac{a+1}{2}\right) - \psi\left(\frac{a}{2}\right) \right]^2 \right\}$$
$$+ \frac{1}{8} \left[\zeta\left(3, \frac{a}{2}\right) + \zeta\left(3, \frac{a+1}{2}\right) \right].$$

66. $\displaystyle\sum_{k=0}^{\infty} \frac{(k+n)!}{k!(a)_k} t^k \psi(k+a)$

$\displaystyle = n! \left[\psi(a) {}_1F_1\!\left({n+1 \atop a;\ t}\right) - t^{1-a} e^t L_n^{1-a}(-t)\gamma(a-1,t) \right.$

$\displaystyle \qquad - \sum_{k=1}^{n} \frac{1}{k} L_{n-k}^{k-a+1}(-t) L_{k-1}^{a-k-1}(t) + (a-1)e^t$

$\displaystyle \qquad \left. \times \sum_{k=0}^{n} \frac{(-t)^k}{k!(a+k-1)^2} L_{n-k}^k(-t) {}_2F_2\!\left({a+k-1,\ a+k-1 \atop a+k,\ a+k;\ -t}\right) \right].$

67. $\displaystyle\sum_{k=0}^{\infty} \frac{(a)_k \left(\frac{1}{2}\right)_k}{k!(k+1)!} \psi(k+a)$

$\displaystyle = \frac{2\Gamma\!\left(\frac{3}{2}-a\right)}{\sqrt{\pi}\,\Gamma(2-a)} \left[\frac{1}{1-a} + \pi\cot(a\pi) + 2\psi(a) - \psi\!\left(\frac{3}{2}-a\right) \right]$ [Re $a < 1$].

68. $\displaystyle\sum_{k=0}^{\infty} \frac{(a)_k \left(\frac{1}{2}\right)_k}{(k+1)!(k+2)!} \psi(k+a)$

$\displaystyle = \frac{2}{a-1} \left\{ \psi(a-1) - \frac{4\Gamma\!\left(\frac{7}{2}-a\right)}{3\sqrt{\pi}\,\Gamma(3-a)} \left[\psi(3-a) - \psi\!\left(\frac{7}{2}-a\right) + \psi(a-1)\right] \right\}$

[Re $a < 1$].

69. $\displaystyle\sum_{k=0}^{\infty} \frac{(a)_k^3}{(k!)^3} \psi(k+a) = -\frac{2\Gamma\!\left(-\frac{3a}{2}\right)}{a^2 \Gamma^3\!\left(-\frac{a}{2}\right)} \cos\frac{a\pi}{2}$

$\displaystyle \times \left\{ \pi\tan\frac{a\pi}{2} - 6\psi(a) - 3\left[2\pi\cos\frac{a\pi}{2}\csc\frac{3a\pi}{2} + \psi\!\left(\frac{a}{2}\right) - \psi\!\left(\frac{3a}{2}\right)\right] \right\}$

[Re $a < 1$].

70. $\displaystyle\sum_{k=0}^{\infty} (-1)^k (2k+a) \frac{(a)_k^3}{(k!)^3} \psi(k+a) = \frac{1}{3}\cos(a\pi)$

$\displaystyle + \left[a\psi(a) - \frac{1}{3}\right] {}_3F_2\!\left({a,\ a,\ a \atop 1,\ 1;\ -1}\right) - 2a^3 \psi(a)\, {}_3F_2\!\left({a+1,\ a+1,\ a+1 \atop 2,\ 2;\ -1}\right)$

[Re $a < 1/3$].

71. $\displaystyle\sum_{k=1}^{\infty} \frac{1}{k^2} \psi(2k) = \frac{9}{4}\zeta(3) - \frac{C\pi^2}{6}.$

72. $\displaystyle\sum_{k=1}^{\infty} \frac{1}{k(k+1)} \psi(2k) = \frac{1}{2} - C + 2\ln 2$ \qquad [[62], (B.4)].

73. $\displaystyle\sum_{k=1}^{\infty} \frac{1}{k^2(k+1)} \psi(2k) = \frac{9}{4}\zeta(3) + \left(1 - \frac{\pi^2}{6}\right) C - 2\ln 2 - \frac{1}{2}.$

74. $\displaystyle\sum_{k=0}^{\infty} \frac{(-1)^k}{2k+1} t^{2k+1} \psi(2k+1) = -\left[C + \frac{1}{2}\ln(1+t^2)\right] \arctan t$ $\qquad [|t| < 1]$.

75. $\displaystyle\sum_{k=0}^{\infty} t^{2k} \psi(2k+1) = \frac{1}{t^2-1}\left[C + \frac{1}{2}\ln(1-t^2) - \frac{1}{2t}\ln\frac{1+t}{1-t}\right]$ $\qquad [|t| < 1]$.

76. $\displaystyle\sum_{k=0}^{\infty} (-1)^k t^{2k} \psi(2k+1) = -\frac{1}{1+t^2}\left[C + t \arctan t + \frac{1}{2}\ln(1+t^2)\right]$

$\qquad [|t| < 1]$.

77. $\displaystyle\sum_{k=0}^{\infty} \frac{(-1)^k t^{4k}}{[(2k)!]^2} \psi(2k+1) = \ln t \operatorname{ber}(2t) - \frac{\pi}{4} \operatorname{bei}(2t) + \operatorname{ker}(2t)$ $\qquad [t > 0]$.

78. $\displaystyle\sum_{k=0}^{\infty} \frac{(-1)^k t^{4k}}{[(2k+1)!]^2} \psi(2k+2) = \frac{1}{4t^2}\left[\pi \operatorname{ber}(2t) + 4\ln t \operatorname{bei}(2t) + 4\operatorname{kei}(2t)\right]$

$\qquad [t > 0]$.

79. $\displaystyle\sum_{k=0}^{\infty} \frac{\left(\frac{1}{2}\right)_{2k}}{(2k)!} t^{2k} \psi\left(2k + \frac{1}{2}\right)$

$= \dfrac{2^{-3/2}}{\sqrt{1-t^2}} \left\{ \left(1 - \sqrt{1-t^2}\right)^{1/2} \ln\frac{1+t}{1-t} - \left(1 + \sqrt{1-t^2}\right)^{1/2} \right.$

$\left. \times [2C + 4\ln 2 + \ln(1-t^2)] \right\}$ $\qquad [|t| < 1]$.

80. $\displaystyle\sum_{k=0}^{\infty} \frac{t^{2k+1}}{(2k+1)!} \left(\frac{3}{2}\right)_{2k+1} \psi\left(2k + \frac{3}{2}\right)$

$= \dfrac{2^{-1/2}}{\sqrt{1-t^2}} \left\{ \left(1 + \sqrt{1-t^2}\right)^{1/2} \ln\frac{1+t}{1-t} - \left(1 - \sqrt{1-t^2}\right)^{1/2} \right.$

$\left. \times [2C + 4\ln 2 + \ln(1-t^2)] \right\}$ $\qquad [|t| < 1]$.

81. $\displaystyle\sum_{k=0}^{\infty} \frac{t^k}{k!} \psi(2k+a) = e^t \psi(a) + \frac{t}{a} e^t \,_2F_2\!\left(\begin{matrix}1,1;\,-t\\2,\frac{a}{2}+1\end{matrix}\right)$

$+ \dfrac{t}{a+1} e^t \,_2F_2\!\left(\begin{matrix}1,1;\,-t\\2,\frac{a+3}{2}\end{matrix}\right).$

82. $\displaystyle\sum_{k=0}^{\infty} \frac{t^{2k}}{(2k)!} (a)_{2k} \psi(2k+a) = -\frac{1}{2} \left(t_-^{-a} \ln t_- + t_+^{-a} \ln t_+ \right)$
$$+ \frac{1}{2} \left(t_-^{-a} + t_+^{-a} \right) \psi(a) \quad [t_\pm = 1 \pm t;\ |t| < 1].$$

83. $\displaystyle\sum_{k=0}^{\infty} \frac{t^{2k+1}}{(2k+1)!} (a)_{2k} \psi(2k+a) = \frac{1}{2(1-a)} \left(t_-^{1-a} \ln t_- - t_+^{1-a} \ln t_+ \right)$
$$+ \frac{1}{2(1-a)} \left[\frac{1}{1-a} + \psi(a) \right] \left(t_+^{1-a} - t_-^{1-a} \right) \quad [t_\pm = 1 \pm t;\ |t| < 1].$$

84. $\displaystyle\sum_{k=0}^{\infty} \frac{(-1)^k t^{2k}}{(2k)!} (a)_{2k} \psi(2k+a) = (1+t^2)^{-a/2}$
$$\times \left\{ \cos(au) \left[\psi(a) - \frac{1}{2} \ln(1+t^2) \right] - u \sin(au) \right\} \quad [u = \arctan t;\ |t| < 1].$$

85. $\displaystyle\sum_{k=0}^{\infty} \frac{t^{2k+1}}{(2k+1)!} (a)_{2k+1} \psi(2k+a+1)$
$$= \frac{1}{2} \left[t_+^{-a} \ln t_+ - t_-^{-a} \ln t_- + (t_-^{-a} - t_+^{-a}) \psi(a) \right] \quad [t_\pm = 1 \pm t;\ |t| < 1].$$

86. $\displaystyle\sum_{k=0}^{\infty} \frac{(-1)^k t^{2k+1}}{(2k+1)!} (a)_{2k+1} \psi(2k+a+1)$
$$= \frac{1}{2} (1+t^2)^{-a/2} \left\{ 2u \cos(au) + \sin(au) [2\psi(a) - \ln(1+t^2)] \right\}$$
$$[u = \arctan t;\ |t| < 1].$$

87. $\displaystyle\sum_{k=0}^{\infty} \frac{(-1)^k}{3k+1} \psi(3k+1) = -\frac{\pi^2}{54} - \frac{1}{6} \ln^2 2 - \frac{1}{3\sqrt{3}} \left(\pi + \sqrt{3} \ln 2 \right).$

88. $\displaystyle\sum_{k=0}^{\infty} \frac{1}{(3k+1)(3k+2)} \psi(3k+1) = \frac{\pi}{6\sqrt{3}} (2 - 2C - \ln 3).$

89. $\displaystyle\sum_{k=0}^{\infty} \frac{(a)_{3k}}{(3k)!} \psi(3k+a) = 3^{-a/2-2} \left\{ 3 \cos \frac{a\pi}{6} \left[2\psi(a) - \ln 3 \right] - \pi \sin \frac{a\pi}{6} \right\}$
$$[\operatorname{Re} a < 1].$$

90. $\displaystyle\sum_{k=0}^{\infty} (-1)^k \frac{(a)_{3k}}{(3k)!} \psi(3k+a)$
$$= \frac{1}{3} \left[\left(2^{-a} + 2 \cos \frac{a\pi}{3} \right) \psi(a) - 2^{-a} \ln 2 - \frac{2\pi}{3} \sin \frac{a\pi}{3} \right] \quad [\operatorname{Re} a < 1].$$

91. $\sum_{k=0}^{\infty} \frac{(a)_{3k}}{(3k+1)!} \psi(3k+a)$

$= \frac{3^{-(a+1)/2}}{1-a} \left\{ \cos\frac{(a+1)\pi}{6} \left[\frac{2}{1-a} - \ln 3 + 2\psi(a) \right] - \frac{\pi}{3} \sin\frac{(a+1)\pi}{6} \right\}$
$\hfill [\text{Re}\, a < 1].$

92. $\sum_{k=0}^{\infty} (-1)^k \frac{(a)_{3k}}{(3k+1)!} \psi(3k+a)$

$= \frac{1}{3(a-1)} \left\{ 2^{1-a} \ln 2 + \left[2^{-a} + 2\cos\frac{(a+1)\pi}{3} \right] \left[\frac{1}{a-1} - \psi(a) \right] \right.$
$\left. + \frac{2\pi}{3} \sin\frac{(a+1)\pi}{3} \right\} \hfill [\text{Re}\, a < 2].$

93. $\sum_{k=0}^{\infty} \frac{(a)_{3k}}{(3k+2)!} \psi(3k+a) = -\frac{3^{-a/2}}{(a-1)(a-2)}$

$\times \left\{ \cos\frac{(a+2)\pi}{6} \left[\ln 3 + \frac{4a-6}{(a-1)(a-2)} - 2\psi(a) \right] + \frac{\pi}{3} \sin\frac{(a+2)\pi}{6} \right\}$
$\hfill [\text{Re}\, a < 2].$

94. $\sum_{k=1}^{\infty} \frac{1}{k^2} \psi(nk) = \left(\frac{n^2}{2} + \frac{1}{2n} \right) \zeta(3) - \frac{\pi^2 C}{6} + \pi \sum_{k=1}^{n-1} k\, \text{Cl}_2\left(\frac{2k\pi}{n} \right)$
$\hfill [[61],\,(5)].$

95. $\sum_{k=1}^{\infty} \frac{(-1)^k}{k^2} \psi(nk) = \left(\frac{n^2}{2} - \frac{3}{8n} \right) \zeta(3) + \frac{\pi^2 C}{12} + \pi \sum_{k=1}^{n-1} k\, \text{Cl}_2\left(\frac{2k+1}{n} \pi \right)$
$\hfill [[61],\,(6)].$

96. $\sum_{k=1}^{\infty} \frac{1}{k} \left[\psi\left(\frac{k+1}{2} \right) - \psi\left(\frac{k}{2} \right) \right] = \ln^2 2 + \frac{\pi^2}{6}.$

97. $\sum_{k=1}^{\infty} \frac{1}{k^2} \left[\psi\left(\frac{k+1}{2} \right) - \psi\left(\frac{k}{2} \right) \right] = \frac{13}{4} \zeta(3) - \frac{\pi^2}{3} \ln 2 \hfill [[29],\,(31)].$

98. $\sum_{k=1}^{\infty} \frac{(-1)^k}{k^2} \left[\psi\left(\frac{k+1}{2} \right) - \psi\left(\frac{k}{2} \right) \right] = \frac{\pi^2}{6} \ln 2 - 2\zeta(3) \hfill [[29],\,(32)].$

99. $\sum_{k=1}^{\infty} \frac{1}{k} \left[\psi\left(k + \frac{3}{4} \right) - \psi\left(k + \frac{1}{4} \right) \right] = 8G - 3\pi \ln 2.$

100. $\sum_{k=1}^{\infty} \frac{1}{k} \left[\psi\left(\frac{k}{2} + \frac{3}{4} \right) - \psi\left(\frac{k}{2} + \frac{1}{4} \right) \right] = 4G - \pi \ln 2.$

101. $\sum_{k=1}^{\infty} \frac{1}{k}\left[\psi\left(\frac{3k}{2}+\frac{3}{4}\right)-\psi\left(\frac{3k}{2}+\frac{1}{4}\right)\right] = 12\mathbf{G} - \pi\ln 2 - 2\pi\ln\left(2+\sqrt{3}\right).$

102. $\sum_{k=1}^{\infty} \frac{1}{k}\left[\psi\left(\frac{k}{3}+\frac{3}{4}\right)-\psi\left(\frac{k}{3}+\frac{1}{4}\right)\right] = \frac{8}{3}\mathbf{G} - \pi\ln\frac{9}{8}.$

103. $\sum_{k=1}^{\infty} \frac{1}{k}\left[\psi\left(\frac{k+3}{4}\right)-\psi\left(\frac{k+1}{4}\right)\right] = 4\mathbf{G} - \frac{\pi}{2}\ln 2.$

104. $\sum_{k=1}^{\infty} \frac{1}{k}\left[\psi\left(\frac{k}{5}+\frac{3}{4}\right)-\psi\left(\frac{k}{5}+\frac{1}{4}\right)\right] = \frac{8}{5}\mathbf{G} - \pi\ln\frac{32}{5} + 2\pi\ln\left(1+\sqrt{5}\right).$

105. $\sum_{k=1}^{\infty} \frac{1}{k}\left[\psi\left(\frac{k}{6}+\frac{3}{4}\right)-\psi\left(\frac{k}{6}+\frac{1}{4}\right)\right] = \frac{4}{3}\mathbf{G} + \pi\ln 2.$

106. $\sum_{k=1}^{\infty} \frac{1}{k}\left[\psi\left(\frac{k}{8}+\frac{3}{4}\right)-\psi\left(\frac{k}{8}+\frac{1}{4}\right)\right] = 2\mathbf{G} - \frac{\pi}{4}\ln 2 + \pi\ln\left(1+\sqrt{2}\right).$

107. $\sum_{k=1}^{\infty} \frac{1}{k}\left[\psi\left(\frac{3k}{8}+\frac{3}{4}\right)-\psi\left(\frac{3k}{8}+\frac{1}{4}\right)\right] = \frac{8}{3}\mathbf{G} - \frac{\pi}{4}\ln 2 - \pi\ln\left(1+\sqrt{2}\right)$
$+ \frac{2\pi}{3}\ln\left(2+\sqrt{3}\right).$

108. $\sum_{k=1}^{\infty} \frac{1}{k}\left[\psi\left(\frac{3k}{10}+\frac{3}{4}\right)-\psi\left(\frac{3k}{10}+\frac{1}{4}\right)\right] = \frac{12}{5}\mathbf{G} + 3\pi\ln 2 + 2\pi\ln\left(2+\sqrt{3}\right)$
$- 4\pi\ln\left(1+\sqrt{5}\right).$

109. $\sum_{k=1}^{\infty} \frac{1}{k}\left[\psi\left(\frac{k}{12}+\frac{3}{4}\right)-\psi\left(\frac{k}{12}+\frac{1}{4}\right)\right] = \frac{4}{3}\mathbf{G} + \frac{\pi}{2}\ln 2 + \frac{2\pi}{3}\ln\left(2+\sqrt{3}\right).$

110. $\sum_{k=1}^{\infty} \frac{1}{k}\left[\psi\left(\frac{3k}{16}+\frac{3}{4}\right)-\psi\left(\frac{3k}{16}+\frac{1}{4}\right)\right] = \frac{4}{3}\mathbf{G} - \frac{3\pi}{8}\ln 2 - \frac{2\pi}{3}\ln\left(2+\sqrt{3}\right)$
$+ \pi\ln\left[2\sqrt{2} + (3-\sqrt{2})\sqrt{2+\sqrt{2}}\right].$

111. $\sum_{k=1}^{\infty} \frac{(-1)^k}{k}\left[\psi\left(\frac{mk}{4n+2}+\frac{3}{4}\right)-\psi\left(\frac{mk}{4n+2}+\frac{1}{4}\right)\right] = \frac{4m\mathbf{G}}{2n+1} - 2m\pi\ln 2$
$- 2\pi\sum_{k=0}^{2n}\sum_{p=0}^{m-1}(-1)^k\ln\left|\sin\left(\frac{2k+1}{8n+4}\pi + \frac{2p+1}{4m}\pi\right)\right|.$

112. $\sum_{k=1}^{\infty} \frac{(-1)^k}{k}\left[\psi\left(k+\frac{3}{4}\right)-\psi\left(k+\frac{1}{4}\right)\right] = 8\mathbf{G} - \pi\ln\left(6+4\sqrt{2}\right).$

113. $\sum_{k=1}^{\infty} \frac{(-1)^k}{k} \left[\psi\left(\frac{3k}{2} + \frac{3}{4}\right) - \psi\left(\frac{3k}{2} + \frac{1}{4}\right) \right] = 12G - 2\pi \ln 6.$

114. $\sum_{k=1}^{\infty} \frac{(-1)^k}{k} \left[\psi\left(\frac{k+3}{4}\right) - \psi\left(\frac{k+1}{4}\right) \right] = -\frac{\pi}{2} \ln 2.$

115. $\sum_{k=1}^{\infty} \frac{(-1)^k}{k} \left[\psi\left(\frac{3k}{4} + \frac{3}{4}\right) - \psi\left(\frac{3k}{4} + \frac{1}{4}\right) \right] = \frac{20}{3} G - \frac{\pi}{2} \ln 2$
$$- \frac{4\pi}{3} \ln\left(2 + \sqrt{3}\right).$$

116. $\sum_{k=1}^{\infty} \frac{(-1)^k}{k} \left[\psi\left(\frac{k}{5} + \frac{3}{4}\right) - \psi\left(\frac{k}{5} + \frac{1}{4}\right) \right] = \frac{8}{5} G + 3\pi \ln 2$
$$+ 2\pi \ln\left(1 + \sqrt{2}\right) - 4\pi \ln\left(1 + \sqrt{5}\right).$$

117. $\sum_{k=1}^{\infty} \frac{(-1)^k}{k} \left[\psi\left(\frac{k}{6} + \frac{3}{4}\right) - \psi\left(\frac{k}{6} + \frac{1}{4}\right) \right] = \frac{4}{3} G - 2\pi \ln \frac{3}{2}.$

118. $\sum_{k=1}^{\infty} \frac{(-1)^k}{k} \left[\psi\left(\frac{k}{8} + \frac{3}{4}\right) - \psi\left(\frac{k}{8} + \frac{1}{4}\right) \right] = 2G - \frac{\pi}{4} \ln 2 - \pi \ln\left(1 + \sqrt{2}\right).$

119. $\sum_{k=1}^{\infty} \frac{(-1)^k}{k} \left[\psi\left(\frac{3k}{8} + \frac{3}{4}\right) - \psi\left(\frac{3k}{8} + \frac{1}{4}\right) \right] = \frac{8}{3} G - \frac{\pi}{4} \ln 2$
$$+ \pi \ln\left(1 + \sqrt{2}\right) - \frac{4\pi}{3} \ln\left(2 + \sqrt{3}\right).$$

120. $\sum_{k=1}^{\infty} \frac{(-1)^k}{k} \left[\psi\left(\frac{k}{10} + \frac{3}{4}\right) - \psi\left(\frac{k}{10} + \frac{1}{4}\right) \right] = \frac{4}{5} G + \pi \ln 5$
$$- 2\pi \ln\left(1 + \sqrt{5}\right).$$

121. $\sum_{k=1}^{\infty} \frac{(-1)^k}{k} \left[\psi\left(\frac{3k}{10} + \frac{3}{4}\right) - \psi\left(\frac{3k}{10} + \frac{1}{4}\right) \right] = \frac{12}{5} G - \pi \ln \frac{144}{5}$
$$+ 2\pi \ln\left(1 + \sqrt{5}\right).$$

122. $\sum_{k=1}^{\infty} \frac{(-1)^k}{k} \left[\psi\left(\frac{k}{12} + \frac{3}{4}\right) - \psi\left(\frac{k}{12} + \frac{1}{4}\right) \right] = \frac{\pi}{2} \ln 2 - \frac{2\pi}{3} \ln\left(2 + \sqrt{3}\right).$

123. $\sum_{k=1}^{\infty} \frac{(-1)^k}{k} \left[\psi\left(\frac{3k}{16} + \frac{3}{4}\right) - \psi\left(\frac{3k}{16} + \frac{1}{4}\right) \right] = \frac{4}{3} G + \frac{\pi}{8} \ln 2$
$$+ \frac{4\pi}{3} \ln\left(2 + \sqrt{3}\right) - \pi \ln\left(4 + 2\sqrt{2} + \sqrt{26 + 17\sqrt{2}}\right).$$

124. $\displaystyle\sum_{k=1}^{\infty}\frac{(-1)^k}{k}\left[\psi\left(\frac{\sqrt{3}+2}{2}k+1\right)-\psi\left(\frac{\sqrt{3}+2}{2}k+\frac{1}{2}\right)\right]$
$$=\left(\sqrt{3}-1\right)\frac{\pi^2}{6}-2\ln 2\ln\left(\sqrt{3}+1\right).$$

125. $\displaystyle\sum_{k=1}^{\infty}\frac{(-1)^k}{k}\left[\psi\left(\frac{k}{2\sqrt{3}+4}+1\right)-\psi\left(\frac{k}{2\sqrt{3}+4}+\frac{1}{2}\right)\right]$
$$=\left(1-\sqrt{3}\right)\frac{\pi^2}{6}+2\ln 2\ln\left(\sqrt{3}+1\right)-2\ln^2 2.$$

126. $\displaystyle\sum_{k=1}^{\infty}\frac{(-1)^k}{k}\left[\psi\left(\frac{3-\sqrt{2}}{2}k+1\right)-\psi\left(\frac{3-\sqrt{2}}{2}k+\frac{1}{2}\right)\right]$
$$=\left(3-4\sqrt{2}\right)\frac{\pi^2}{12}+3\ln 2\ln\left(\sqrt{2}+1\right)-\ln^2 2.$$

127. $\displaystyle\sum_{k=1}^{\infty}\frac{(-1)^k}{k}\left[\psi\left(\frac{4+\sqrt{15}}{2}k+1\right)-\psi\left(\frac{4+\sqrt{15}}{2}k+\frac{1}{2}\right)\right]$
$$=\left(\sqrt{15}-2\right)\frac{\pi^2}{6}-2\ln 2\ln\left(\sqrt{3}+\sqrt{5}\right)-2\ln\left(\frac{1+\sqrt{5}}{2}\right)\ln\left(2+\sqrt{3}\right).$$

128. $\displaystyle\sum_{k=1}^{\infty}\frac{(-1)^k}{k}\left[\psi\left(\frac{4-\sqrt{15}}{2}k+1\right)-\psi\left(\frac{4-\sqrt{15}}{2}k+\frac{1}{2}\right)\right]$
$$=\left(2-\sqrt{15}\right)\frac{\pi^2}{6}+2\ln 2\ln\left(\sqrt{3}+\sqrt{5}\right)+2\ln\left(\frac{1+\sqrt{5}}{2}\right)\ln\left(2+\sqrt{3}\right)$$
$$-2\ln^2 2.$$

129. $\displaystyle\sum_{k=1}^{\infty}\frac{(-1)^k}{k}\left[\psi\left(\frac{5+\sqrt{24}}{2}k+1\right)-\psi\left(\frac{5+\sqrt{24}}{2}k+\frac{1}{2}\right)\right]$
$$=\left(\sqrt{\frac{2}{3}}-\frac{5}{12}\right)\pi^2-\frac{3}{2}\ln 2\ln\left(5+2\sqrt{6}\right)-\ln\left(1+\sqrt{2}\right)\ln\left(2+\sqrt{3}\right)-\ln^2 2.$$

130. $\displaystyle\sum_{k=1}^{\infty}\frac{(-1)^k}{k}\left[\psi\left(\frac{5-\sqrt{24}}{2}k+1\right)-\psi\left(\frac{5-\sqrt{24}}{2}k+\frac{1}{2}\right)\right]$
$$=\left(\sqrt{\frac{2}{3}}-\frac{5}{12}\right)\pi^2+\frac{3}{2}\ln 2\ln\left(5+2\sqrt{6}\right)+\ln\left(1+\sqrt{2}\right)\ln\left(2+\sqrt{3}\right)-\ln^2 2.$$

131. $\displaystyle\sum_{k=1}^{\infty}\frac{(-1)^k}{k}\left[\psi\left(\frac{6+\sqrt{35}}{2}k+1\right)-\psi\left(\frac{6+\sqrt{35}}{2}k+\frac{1}{2}\right)\right]$
$$=\left(\sqrt{35}-3\right)\frac{\pi^2}{6}-2\ln 2\ln\left(\sqrt{5}+\sqrt{7}\right)-2\ln\frac{1+\sqrt{5}}{2}\ln\left(8+3\sqrt{7}\right).$$

132. $\sum_{k=1}^{\infty} \frac{(-1)^k}{k} \left[\psi\left(\frac{6-\sqrt{35}}{2} k + 1 \right) - \psi\left(\frac{6-\sqrt{35}}{2} k + \frac{1}{2} \right) \right]$

$= \left(3 - \sqrt{35}\right) \frac{\pi^2}{6} + 2\ln 2 \ln\left(\sqrt{5} + \sqrt{7}\right) + 2\ln \frac{1+\sqrt{5}}{2} \ln\left(8 + 3\sqrt{7}\right) - 2\ln^2 2.$

133. $\sum_{k=1}^{\infty} \frac{(-1)^k}{k} \left[\psi\left(\frac{8+\sqrt{63}}{2} k + 1 \right) - \psi\left(\frac{8+\sqrt{63}}{2} k + \frac{1}{2} \right) \right]$

$= \left(\sqrt{63} - 4\right) \frac{\pi^2}{6} - 2\ln 2 \ln\left(3 + \sqrt{7}\right) - 2\ln \frac{5+\sqrt{21}}{2} \ln\left(2 + \sqrt{3}\right).$

134. $\sum_{k=1}^{\infty} \frac{(-1)^k}{k} \left[\psi\left(\frac{8-\sqrt{63}}{2} k + 1 \right) - \psi\left(\frac{8-\sqrt{63}}{2} k + \frac{1}{2} \right) \right]$

$= \left(4 - \sqrt{63}\right) \frac{\pi^2}{6} + 2\ln 2 \ln\left(3 + \sqrt{7}\right) + 2\ln \frac{5+\sqrt{21}}{2} \ln\left(2 + \sqrt{3}\right) - 2\ln^2 2.$

135. $\sum_{k=1}^{\infty} \frac{(-1)^k}{k} \left[\psi\left(\frac{11+\sqrt{120}}{2} k + 1 \right) - \psi\left(\frac{11+\sqrt{120}}{2} k + \frac{1}{2} \right) \right]$

$= \left(\sqrt{480} - 11\right) \frac{\pi^2}{12} - \ln\left(2 + \sqrt{3}\right) \ln\left(3 + \sqrt{10}\right) - \ln\left(1 + \sqrt{2}\right) \ln\left(4 + \sqrt{15}\right)$

$\quad - \frac{3}{2} \ln 2 \ln\left(11 + \sqrt{120}\right) - \ln \frac{1+\sqrt{5}}{2} \ln\left(5 + \sqrt{24}\right) - \ln^2 2.$

136. $\sum_{k=1}^{\infty} \frac{(-1)^k}{k} \left[\psi\left(\frac{11-\sqrt{120}}{2} k + 1 \right) - \psi\left(\frac{11-\sqrt{120}}{2} k + \frac{1}{2} \right) \right]$

$= \left(11 - \sqrt{480}\right) \frac{\pi^2}{12} + \ln\left(2 + \sqrt{3}\right) \ln\left(3 + \sqrt{10}\right) + \ln\left(1 + \sqrt{2}\right) \ln\left(4 + \sqrt{15}\right)$

$\quad + \frac{3}{2} \ln 2 \ln\left(11 + \sqrt{120}\right) + \ln \frac{1+\sqrt{5}}{2} \ln\left(5 + \sqrt{24}\right) - \ln^2 2.$

137. $\sum_{k=1}^{\infty} \frac{(-1)^k}{k} \left[\psi\left(\frac{12+\sqrt{143}}{2} k + 1 \right) - \psi\left(\frac{12+\sqrt{143}}{2} k + \frac{1}{2} \right) \right]$

$= \left(\sqrt{143} - 6\right) \frac{\pi^2}{6} - 2\ln 2 \ln\left(\sqrt{11} + \sqrt{13}\right) - 2\ln \frac{3+\sqrt{13}}{2} \ln\left(10 + 3\sqrt{11}\right).$

138. $\sum_{k=1}^{\infty} \frac{(-1)^k}{k} \left[\psi\left(\frac{12-\sqrt{143}}{2} k + 1 \right) - \psi\left(\frac{12-\sqrt{143}}{2} k + \frac{1}{2} \right) \right]$

$= \left(6 - \sqrt{143}\right) \frac{\pi^2}{6} + 2\ln 2 \ln\left(\sqrt{11} + \sqrt{13}\right) + 2\ln \frac{3+\sqrt{13}}{2} \ln\left(10 + 3\sqrt{11}\right)$

$\quad - 2\ln^2 2.$

139. $\sum_{k=1}^{\infty} \frac{(-1)^k}{k} \left[\psi\left(\frac{13+\sqrt{168}}{2} k + 1 \right) - \psi\left(\frac{13+\sqrt{168}}{2} k + \frac{1}{2} \right) \right]$

$= \left(\sqrt{672} - 13 \right) \frac{\pi^2}{12} - \frac{1}{2} \ln\left(2 + \sqrt{3} \right) \ln\left(15 + \sqrt{224} \right)$

$\quad - \frac{1}{2} \ln\left(5 + \sqrt{24} \right) \ln\left(8 + \sqrt{63} \right)$

$\quad - \frac{3}{2} \ln 2 \ln\left(13 + \sqrt{168} \right) - \ln \frac{5+\sqrt{21}}{2} \ln\left(1 + \sqrt{2} \right) - \ln^2 2.$

140. $\sum_{k=1}^{\infty} \frac{(-1)^k}{k} \left[\psi\left(\frac{13-\sqrt{168}}{2} k + 1 \right) - \psi\left(\frac{13-\sqrt{168}}{2} k + \frac{1}{2} \right) \right]$

$= \left(13 - \sqrt{672} \right) \frac{\pi^2}{12} + \frac{1}{2} \ln\left(2 + \sqrt{3} \right) \ln\left(15 + \sqrt{224} \right)$

$\quad + \frac{1}{2} \ln\left(5 + \sqrt{24} \right) \ln\left(8 + \sqrt{63} \right)$

$\quad + \frac{3}{2} \ln 2 \ln\left(13 + \sqrt{168} \right) + \ln \frac{5+\sqrt{21}}{2} \ln\left(1 + \sqrt{2} \right) - \ln^2 2.$

141. $\sum_{k=1}^{\infty} \frac{(-1)^k}{k} \left[\psi\left(\frac{14+\sqrt{195}}{2} k + 1 \right) - \psi\left(\frac{14+\sqrt{195}}{2} k + \frac{1}{2} \right) \right]$

$= \left(\sqrt{195} - 7 \right) \frac{\pi^2}{6} - 2 \ln 2 \ln\left(\sqrt{13} + \sqrt{15} \right) - 2 \ln \frac{3+\sqrt{13}}{2} \ln\left(4 + \sqrt{15} \right)$

$\quad - 2 \ln \frac{1+\sqrt{5}}{2} \ln\left(25 + \sqrt{39} \right).$

142. $\sum_{k=1}^{\infty} \frac{(-1)^k}{k} \left[\psi\left(\frac{14-\sqrt{195}}{2} k + 1 \right) - \psi\left(\frac{14-\sqrt{195}}{2} k + \frac{1}{2} \right) \right]$

$= \left(7 - \sqrt{195} \right) \frac{\pi^2}{6} + 2 \ln 2 \ln\left(\sqrt{13} + \sqrt{15} \right) + 2 \ln \frac{3+\sqrt{13}}{2} \ln\left(4 + \sqrt{15} \right) +$

$\quad + 2 \ln \frac{1+\sqrt{5}}{2} \ln\left(25 + \sqrt{39} \right) - 2 \ln^2 2.$

143. $\sum_{k=0}^{\infty} \frac{(-1)^k}{2k+1} \left[\psi\left(\frac{3+\sqrt{8}}{4} (2k+1) + \frac{3}{4} \right) - \psi\left(\frac{3+\sqrt{8}}{4} (2k+1) + \frac{1}{4} \right) \right]$

$= \frac{1}{4} \ln 2 \ln\left(3 + \sqrt{8} \right).$

144. $\sum_{k=0}^{\infty} \frac{(-1)^k}{2k+1} \left[\psi\left(\frac{5+\sqrt{24}}{4} (2k+1) + \frac{3}{4} \right) \right.$

$\quad \left. - \psi\left(\frac{5+\sqrt{24}}{4} (2k+1) + \frac{1}{4} \right) \right]$

$= \frac{1}{2} \ln\left(1 + \sqrt{2} \right) \ln\left(2 + \sqrt{3} \right) - \frac{1}{4} \ln 2 \ln\left(5 + \sqrt{24} \right).$

145. $\displaystyle\sum_{k=0}^{\infty}\frac{(-1)^k}{2k+1}\left[\psi\left(\frac{11+\sqrt{120}}{4}(2k+1)+\frac{3}{4}\right)\right.$
$\left.-\psi\left(\frac{11+\sqrt{120}}{4}(2k+1)+\frac{1}{4}\right)\right]=\frac{1}{4}\ln 2\ln\left(11+\sqrt{120}\right)$
$-\frac{1}{2}\ln\left(1+\sqrt{2}\right)\ln\left(4+\sqrt{15}\right)-\frac{1}{2}\ln\left(2+\sqrt{3}\right)\ln\left(3+\sqrt{10}\right)$
$+\frac{3}{2}\ln\frac{1+\sqrt{5}}{2}\ln\left(5+\sqrt{24}\right).$

146. $\displaystyle\sum_{k=0}^{\infty}\frac{(-1)^k}{2k+1}\left[\psi\left(\frac{13+\sqrt{168}}{4}(2k+1)+\frac{3}{4}\right)\right.$
$\left.-\psi\left(\frac{13+\sqrt{168}}{4}(2k+1)+\frac{1}{4}\right)\right]$
$=\frac{1}{4}\ln\left(2+\sqrt{3}\right)\ln\left(15+\sqrt{224}\right)+\frac{1}{4}\ln\left(5+\sqrt{24}\right)\ln\left(8+\sqrt{63}\right)$
$-\frac{3}{2}\ln\frac{5+\sqrt{21}}{2}\ln\left(1+\sqrt{2}\right)-\frac{1}{4}\ln 2\ln\left(13+\sqrt{168}\right).$

6.2.2. Series containing $\psi(ka+b)$ and trigonometric functions

1. $\displaystyle\sum_{k=0}^{\infty}\frac{t^k}{k!}\left\{\begin{matrix}\sin(kz)\\ \cos(kz)\end{matrix}\right\}\psi\left(k+\frac{1}{2}\right)=-\frac{i^{(1\pm 1)/2}}{2}(C+2\ln 2)\left(e^{u_-}\mp e^{u_+}\right)$
$+i^{(1\pm 1)/2}\left[u_-e^{u_-}{}_2F_2\left(\begin{matrix}1,1;\\ \frac{3}{2},2\end{matrix};-u_-\right)\mp u_+e^{u_+}{}_2F_2\left(\begin{matrix}1,1;\\ \frac{3}{2},2\end{matrix};-u_+\right)\right]\quad\left[u_{\pm}=te^{\pm iz}\right].$

2. $\displaystyle\sum_{k=0}^{\infty}\frac{t^k}{(k!)^2}\left\{\begin{matrix}\sin(kz)\\ \cos(kz)\end{matrix}\right\}\psi(k+1)$
$=\frac{i^{(1\pm 1)/2}}{2}\left[K_0(2u_-)\mp K_0(2u_+)+\ln u_-I_0(2u_-)\mp\ln u_+I_0(2u_+)\right]$
$\left[u_{\pm}=\sqrt{t}\,e^{\pm iz/2}\right].$

6.2.3. Series containing products of $\psi(ka+b)$

1. $\displaystyle\sum_{k=1}^{\infty}t^k\psi^2(k)=\frac{t}{1-t}\left[C^2+2C\ln(1-t)+\ln^2(1-t)+\operatorname{Li}_2(t)\right]\qquad[|t|<1].$

2. $\displaystyle\sum_{k=1}^{\infty}\frac{t^k}{k}\psi^2(k)=\ln(1-t)\left[\operatorname{Li}_2(t)-\frac{\pi^2}{3}-C^2\right]+\ln^2(1-t)(\ln t-C)$
$-\frac{1}{3}\ln^3(1-t)+2\operatorname{Li}_3(1-t)-2\zeta(3)\qquad[|t|<1].$

3. $\displaystyle\sum_{k=1}^{\infty}\frac{(-1)^k}{k}\psi^2(k)=\left(\frac{\pi^2}{12}-C^2\right)\ln 2-C\ln^2 2-\frac{1}{3}\ln^3 2-\frac{1}{4}\zeta(3).$

4. $\sum_{k=1}^{\infty} \frac{1}{k^2} \psi^2(k) = \frac{11\pi^4}{360} + \frac{\pi^2 \mathbf{C}^2}{6} - 2\mathbf{C}\zeta(3)$ [[29], (9)].

5. $\sum_{k=1}^{\infty} \frac{(-1)^k}{k^2} \psi^2(k) = \frac{\pi^2}{480} \left(11\pi^2 - 40\mathbf{C}^2 + 40\ln^2 2\right) - \frac{1}{12}\left[\ln^4 2 + 24\operatorname{Li}_4(\tfrac{1}{2})\right]$
$\qquad - \frac{1}{4}\left(\mathbf{C} + 7\ln 2\right)\zeta(3).$

6. $\sum_{k=1}^{\infty} \frac{1}{k(k+1)} \psi^2(k) = \mathbf{C}^2 - 2\mathbf{C} + \frac{\pi^2}{6} + 1$ [[63], (4.16)].

7. $\sum_{k=1}^{\infty} \frac{1}{k^3} \psi^2(k) = -\frac{\pi^4 \mathbf{C}}{180} + \left(\frac{\pi^2}{6} + \mathbf{C}^2\right)\zeta(3) - \frac{3}{2}\zeta(5).$

8. $\sum_{k=1}^{\infty} \frac{1}{k^4} \psi^2(k) = \frac{37\pi^6}{22680} + \frac{\pi^4 \mathbf{C}^2}{90} + \frac{\pi^2 \mathbf{C}}{3}\zeta(3) - \zeta^2(3) - 4\mathbf{C}\zeta(5).$

9. $\sum_{k=1}^{\infty} \frac{1}{k^5} \psi^2(k) = -\frac{\pi^4}{180}\zeta(3) + \mathbf{C}\left[\zeta^2(3) - \frac{\pi^6}{630}\right] + \left(\frac{\pi^2}{6} + \mathbf{C}^2\right)\zeta(5) - \zeta(7).$

10. $\sum_{k=1}^{\infty} \frac{(-1)^k}{k} \psi^2\left(k+\tfrac{1}{2}\right) = \frac{1}{4}(\mathbf{C} + 2\ln 2)\left[\pi^2 - 4(\mathbf{C} + 2\ln 2)\ln 2\right] - \frac{7}{4}\zeta(3)$
[[29], (19)].

11. $\sum_{k=1}^{\infty} \frac{1}{k^2} \psi^2\left(k+\tfrac{1}{2}\right) = \frac{\pi^4}{8} + \frac{\pi^2}{6}(\mathbf{C} + 2\ln 2)^2 - 7(\mathbf{C} + 2\ln 2)\zeta(3)$
[[29], (22)].

12. $\sum_{k=0}^{\infty} \frac{\left(\tfrac{1}{2}\right)_k}{k!(2k+1)} \psi^2\left(k+\tfrac{1}{2}\right) = \frac{\pi}{4}\left(\pi^2 + 2\mathbf{C}^2\right).$

13. $\sum_{k=0}^{\infty} \frac{\left(\tfrac{1}{2}\right)_k}{k!(2k+1)(2k+3)} \psi^2\left(k+\tfrac{1}{2}\right) = \frac{\pi}{16}\left(\pi^2 + 4\mathbf{C} + 2\mathbf{C}^2 - 6\right).$

14. $\sum_{k=0}^{\infty} \frac{\left(\tfrac{1}{2}\right)_k^2}{k!(k+1)!} \psi^2\left(k+\tfrac{1}{2}\right)$
$\qquad = \frac{2}{3\pi}\left[\pi^2 + 24(2\ln 2 - 1)\mathbf{C} + 6\mathbf{C}^2 + 24(2\ln 2 - 1)^2\right].$

15. $\sum_{k=0}^{\infty} \frac{(a)_k}{k!} \psi^2(k+a) = -\frac{\pi}{a}\csc(a\pi)$ [$\operatorname{Re} a < 0$].

16. $\sum_{k=0}^{\infty} \frac{(a)_k}{k!} \psi(k+a)\psi(k+b) = B(-a, b)[\psi(-a) - \psi(a) - \psi(b-a)]$

$\quad\quad\quad\quad\quad\quad\quad\quad\quad\quad\quad\quad$ [Re $a < 0$; $a, b \neq 0, -1, -2, \ldots$].

17. $\sum_{k=0}^{\infty} \frac{\left(\frac{1}{2}\right)_k}{(k+1)!} \psi\left(k+\frac{1}{2}\right)\psi(k+1) = \pi^2 - 4C + 2C^2 + 8\ln 2(1 - \ln 2)$.

18. $\sum_{k=0}^{\infty} \frac{\left(-\frac{1}{2}\right)_k}{(k+1)!} \psi\left(k-\frac{1}{2}\right)\psi(k+1)$

$\quad\quad\quad\quad = \frac{1}{9}[3\pi^2 - 4C + 6C^2 + 8\ln 2(7 - 3\ln 2) - 56]$.

19. $\sum_{k=0}^{\infty} \frac{\left(\frac{1}{2}\right)_k}{(k+2)!} \psi\left(k+\frac{1}{2}\right)\psi(k+2)$

$\quad\quad\quad\quad = \frac{2}{9}\left[3C^2 + 2C(9\ln 2 - 7) - 3\pi^2 + 56(1 - \ln 2) + 24\ln^2 2\right]$.

20. $\sum_{k=0}^{\infty} \frac{\left(\frac{1}{2}\right)_k}{(k+1)!(k+1)} \psi\left(k+\frac{1}{2}\right)\psi(k+2) = 4\left(\frac{\pi^2}{3} - C^2 - 4\right)(\ln 2 - 1)$

$\quad\quad\quad\quad\quad\quad\quad + C\left[\frac{5\pi^2}{3} - 8(3 - 2\ln 2 + \ln^2 2)\right] + 4[8 - 7\zeta(3)]$.

21. $\sum_{k=0}^{\infty} \frac{\left(\frac{1}{2}\right)_k^2}{k!(k+1)!} \psi^2\left(k+\frac{1}{2}\right)$

$\quad\quad\quad\quad = \frac{2}{3\pi}[6C^2 + 24C(2\ln 2 - 1) + \pi^2 + 24(2\ln 2 - 1)^2]$.

22. $\sum_{k=0}^{\infty} \frac{\left(\frac{1}{2}\right)_k \left(\frac{3}{2}\right)_k}{(k+1)!(k+2)!} \psi\left(k+\frac{1}{2}\right)\psi\left(k+\frac{3}{2}\right)$

$\quad\quad\quad = \frac{4}{9\pi}\{3(3\pi - 8)C^2 + 2[(56 - 96\ln 2) + 9\pi(2\ln 2 - 1)]C$

$\quad\quad\quad\quad\quad - 64(6\ln^2 2 - 7\ln 2 + 1) + 36\pi \ln 2(\ln 2 - 1) - 4\pi^2\}$.

6.2.4. Series containing $\psi'(ka + b)$

1. $\sum_{k=1}^{\infty} t^k \psi'(k) = -\frac{\pi^2}{6} + \text{Li}_2(t) + \frac{1}{1-t}[\ln t \ln(1-t) + \text{Li}_2(1-t)]$ $[|t| < 1]$.

2. $\sum_{k=1}^{\infty} (-1)^k \psi'(k) = -\frac{\pi^2}{8}$.

3. $\displaystyle\sum_{k=1}^{\infty} \frac{t^k}{k}\psi'(k) = 2\zeta(3) - \frac{\pi^2}{6}\ln t + \ln^2 t \ln(1-t) + \ln t \operatorname{Li}_2(t)$
$+ \ln(t-t^2)\operatorname{Li}_2(1-t) - 2\operatorname{Li}_3(1-t)$ $\quad [|t|<1]$.

4. $\displaystyle\sum_{k=1}^{\infty} \frac{1}{k}\psi'(k) = 2\zeta(3)$ $\hfill [[63],\,(4.12)]$.

5. $\displaystyle\sum_{k=1}^{\infty} \frac{(-1)^k}{k}\psi'(k) = \frac{1}{4}\zeta(3) - \frac{\pi^2}{4}\ln 2$.

6. $\displaystyle\sum_{k=1}^{\infty} \frac{1}{k(k+1)}\psi'(k) = 1$ $\hfill [[63],\,(4.14)]$.

7. $\displaystyle\sum_{k=1}^{\infty} \frac{(k!)^2}{(2k+2)!} t^k \psi'(k) = -\frac{7}{2} - \frac{\pi^2}{12} + 3\frac{\sqrt{4-t}}{\sqrt{t}}\arcsin\frac{\sqrt{t}}{2}$
$+ \frac{3t+\pi^2+6}{3t}\arcsin^2\frac{\sqrt{t}}{2} - \frac{2}{3t}\arcsin^4\frac{\sqrt{t}}{2}$ $\quad [|t|<4]$.

8. $\displaystyle\sum_{k=0}^{\infty} \frac{1}{(2k+1)^2}\psi'\!\left(k+\frac{1}{2}\right) = \frac{5\pi^4}{96}$.

9. $\displaystyle\sum_{k=0}^{\infty} \frac{\left(\frac{1}{2}\right)_k}{k!} t^k \psi'\!\left(k+\frac{1}{2}\right) = \frac{\pi^2}{2\sqrt{1-t}} - \frac{2}{\sqrt{1-t}}\arcsin^2\sqrt{t}$ $\quad [|t|<1]$.

10. $\displaystyle\sum_{k=0}^{\infty} \frac{\left(\frac{1}{2}\right)_k}{k!(2k+1)} t^k \psi'\!\left(k+\frac{1}{2}\right) = \frac{\pi^2}{2\sqrt{t}}\arcsin\sqrt{t} - \frac{2}{3\sqrt{t}}\arcsin^3\sqrt{t}$
$\hfill [|t|<1]$.

11. $\displaystyle\sum_{k=0}^{\infty} \frac{\left(\frac{1}{2}\right)_k}{(k+1)!(2k+1)} t^k \psi'\!\left(k+\frac{1}{2}\right) = \frac{8+\pi^2}{t}\left(\sqrt{1-t}-1\right)$
$+ \frac{8+\pi^2}{\sqrt{t}}\arcsin\sqrt{t} - \frac{4}{t}\sqrt{1-t}\arcsin^2\sqrt{t} - \frac{4}{3\sqrt{t}}\arcsin^3\sqrt{t}$ $\quad [|t|<1]$.

12. $\displaystyle\sum_{k=0}^{\infty} \frac{(a)_k}{k!}\psi'(k+a) = \frac{\pi}{a}\csc(a\pi)$ $\hfill [\operatorname{Re} a < 1;\ a\neq 0,-1,-2,\ldots]$.

13. $\displaystyle\sum_{k=0}^{\infty} \frac{(a)_k}{k!}\psi(k+a)\psi'(k+a) = -\frac{\pi}{a}\csc(a\pi)\left[2\mathbf{C}+\psi(-a)\right]$
$\hfill [\operatorname{Re} a < 1;\ a\neq 0,-1,-2,\ldots]$.

6.3. The Hurwitz Zeta Function $\zeta(s, z)$

6.3.1. Series containing $\zeta(k, z)$

1. $\displaystyle\sum_{k=2}^{\infty} \frac{t^k}{k+n} \zeta(k, z) = \sum_{k=0}^{n} \binom{n}{k} t^{-k} \left.\frac{\partial \zeta(s, z-t)}{\partial s}\right|_{s=-k} - t^{-n} \left.\frac{\partial \zeta(s, z)}{\partial s}\right|_{s=-n}$

$\displaystyle - \sum_{k=0}^{n-1} \frac{t^{-k}}{n-k} \zeta(-k, z) - \frac{t}{n+1} [\psi(n+1) - \psi(z) + \mathbf{C}] \quad [|t| < |z|; \; [50]].$

2. $\displaystyle\sum_{k=2}^{\infty} \frac{(k-1)!}{(k+n)!} t^k \zeta(k, z) = \frac{(-t)^{-n}}{n!} \left.\frac{\partial \zeta(s, z-t)}{\partial s}\right|_{s=-n}$

$\displaystyle - \frac{1}{n!} \sum_{k=0}^{n} \binom{n}{k} (-t)^{-k} \left.\frac{\partial \zeta(s, z)}{\partial s}\right|_{s=-k}$

$\displaystyle - \frac{1}{n!} \sum_{k=0}^{n-1} \binom{n}{k} \frac{(-t)^{-k}}{k+1} B_{k+1}(z) [\psi(n+1) - \psi(k+1)]$

$\displaystyle + \frac{t}{(n+1)!} [\psi(n+1) - \psi(z) + \mathbf{C}] \quad [|t| < |z|; \; [50]].$

6.4. The Sine Si(z) and Cosine ci(z) Integrals

6.4.1. Series containing Si$(\varphi(k)x)$

1. $\displaystyle\sum_{k=1}^{\infty} \frac{(-1)^k}{k(k+1)} \mathrm{Si}((2k+1)x) = \mathrm{Si}(x) - 2\sin x \qquad [-\pi/2 < x < \pi/2].$

2. $\displaystyle\sum_{k=1}^{\infty} \frac{(-1)^k}{k(k^2 a^2 + b^2)} \mathrm{Si}(kx) = -\frac{x}{2b^2} + \frac{\pi}{2b^2} \operatorname{csch}\frac{b\pi}{a} \operatorname{shi}\left(\frac{bx}{a}\right) \qquad [-\pi \leq x \leq \pi].$

3. $\displaystyle\sum_{k=1}^{\infty} \frac{(-1)^k}{k(k^2 a^2 - b^2)} \mathrm{Si}(kx) = \frac{x}{2b^2} - \frac{\pi}{2b^2} \csc\frac{b\pi}{a} \mathrm{Si}\left(\frac{bx}{a}\right) \qquad [-\pi \leq x \leq \pi].$

4. $\displaystyle\sum_{k=2}^{\infty} \frac{(-1)^k}{k(k^2-1)} \mathrm{Si}(kx) = \frac{x}{2} + \frac{1}{2} \sin x - \frac{3}{4} \mathrm{Si}(x) \qquad [-\pi \leq x \leq \pi].$

5. $\displaystyle\sum_{k=0}^{\infty} \frac{(2k+1)^{-1}}{((2k+1)^2 a^2 + b^2)} \mathrm{Si}((2k+1)x)$

$\displaystyle = \frac{\pi}{4b^2} \left[\tanh\frac{b\pi}{2a} \operatorname{shi}\left(\frac{bx}{a}\right) - \operatorname{chi}\left(\frac{bx}{a}\right) + \ln\frac{bx}{a} + \mathbf{C} \right] \qquad [0 \leq x \leq \pi].$

6. $\displaystyle\sum_{k=1}^{\infty} \frac{1}{k(k+1)(2k+1)} \operatorname{Si}((2k+1)x) = 3\operatorname{Si}(x) + \pi \operatorname{ci}(x) - 2\sin x$
$$- \pi \ln x - \pi C \qquad [0 < x < \pi/2].$$

7. $\displaystyle\sum_{k=1}^{\infty} \frac{(-1)^k}{\sqrt{k^2 + a^2}} \operatorname{Si}\left(\sqrt{k^2 + a^2}\,x\right) = -\frac{1}{2a}\operatorname{Si}(ax) \qquad [-\pi < x < \pi].$

8. $\displaystyle\sum_{k=1}^{\infty} \frac{(-1)^k}{k^2\sqrt{k^2 + a^2}} \operatorname{Si}\left(\sqrt{k^2 + a^2}\,x\right)$
$$= \frac{1}{12a^3}[(3 - \pi^2 a^2)\operatorname{Si}(ax) - 3\sin(ax)] \qquad [-\pi < x < \pi].$$

9. $\displaystyle\sum_{k=0}^{\infty} \frac{(-1)^k}{(2k+1)\sqrt{(2k+1)^2 + a^2}} \operatorname{Si}\left(\sqrt{(2k+1)^2 + a^2}\,x\right) = \frac{\pi}{4a}\operatorname{Si}(ax)$
$$[-\pi/2 < x < \pi/2].$$

6.4.2. Series containing $\operatorname{ci}(\varphi(k)x)$

1. $\displaystyle\sum_{k=1}^{\infty} \frac{(-1)^k}{k^2 + a^2}\left[\operatorname{ci}\left(\sqrt{k^2 + a^2}\,x\right) - \frac{1}{2}\ln(k^2 + a^2)\right]$
$$= \frac{1}{2a^2}\operatorname{csch}(a\pi)[\sinh(a\pi)\ln a + a\pi(C + \ln x) - \sinh(a\pi)\operatorname{ci}(ax)]$$
$$[0 < a < \pi].$$

2. $\displaystyle\sum_{k=0}^{\infty} \frac{(-1)^k}{(2k+1)^3}[\operatorname{ci}((2k+1)x) - \ln(2k+1)] = \frac{\pi^3}{32}(C + \ln x) - \frac{\pi x^2}{16}$
$$[0 < x < \pi/2].$$

3. $\displaystyle\sum_{k=0}^{\infty} \frac{(-1)^k}{(4k^2 - 1)(2k+3)}[\operatorname{ci}((2k+1)x) - \ln(2k+1)]$
$$= -\frac{\pi}{16}\left[C + \ln\frac{x}{2} + \operatorname{ci}(2x)\right].$$

4. $\displaystyle\sum_{k=0}^{\infty}(-1)^k \frac{(2k+1)^{-1}}{(2k+1)^2 + a^2}\left[\operatorname{ci}\left(\sqrt{(2k+1)^2 + a^2}\,x\right) - \frac{1}{2}\ln((2k+1)^2 + a^2)\right]$
$$= -\frac{\pi}{4a^2}[C + \ln(ax) - \operatorname{ci}(ax)] + \frac{\pi(e^{a\pi/2} - 1)^2}{4a^2(e^{a\pi} + 1)}(C + \ln x) \qquad [0 < a < \pi].$$

5. $\displaystyle\sum_{k=0}^{\infty} \frac{(2k+1)^{-2}}{(2k+1)^2 a^2 + b^2}[\operatorname{ci}((2k+1)x) - \ln(2k+1)]$
$$= \frac{\pi}{8b^3}\left\{b(\pi C - 2x + \pi \ln x) + 2a\operatorname{shi}\left(\frac{bx}{a}\right) + 2a\tanh\frac{b\pi}{2a}\left[\ln\frac{b}{a} - \operatorname{chi}\left(\frac{bx}{a}\right)\right]\right\}$$
$$[0 < x < \pi].$$

6. $\displaystyle\sum_{k=0}^{\infty} \frac{(2k+1)^{-2}}{(2k+1)^2 a^2 - b^2} [\operatorname{ci}((2k+1)x) - \ln(2k+1)]$

$= -\dfrac{\pi}{8b^3} \left\{ 2a\operatorname{Si}\left(\dfrac{bx}{a}\right) + b(\pi C - 2x + \pi \ln x) + 2a \tan \dfrac{b\pi}{2a} \left[\ln \dfrac{b}{a} - \operatorname{ci}\left(\dfrac{bx}{a}\right)\right] \right\}$

$[0 < x < \pi].$

6.4.3. Series containing $\operatorname{Si}(kx)$ and trigonometric functions

1. $\displaystyle\sum_{k=1}^{\infty} \frac{\cos(ky)}{k^{2m-1}} \operatorname{Si}(kx) = (-1)^m \frac{x^{2m-1}}{2(2m-1)!(2m-1)} + \sum_{k=0}^{m-2} \frac{x^{2k+1}}{(2k+1)!(2k+1)}$

$\times \left[(-1)^{m-1} \dfrac{\pi y^{2m-2k-3}}{2(2m-2k-3)!} + (-1)^k \displaystyle\sum_{j=0}^{m-k-1} (-1)^j \dfrac{y^{2j}}{(2j)!} \zeta(2m-2j-2k-2) \right]$

$[m \geq 1;\ 0 < x < \pi;\ x < y < 2\pi - x].$

2. $\displaystyle\sum_{k=1}^{\infty} (-1)^k \frac{\cos(ky)}{k^{2m-1}} \operatorname{Si}(kx) = (-1)^m \frac{x^{2m-1}}{2(2m-1)!(2m-1)}$

$- \displaystyle\sum_{k=0}^{m-2} \frac{(-1)^k x^{2k+1}}{(2k+1)!(2k+1)} \sum_{j=0}^{m-k-1} (-1)^j \frac{y^{2j}}{(2j)!} \left(1 - 2^{2j+2k-2m+3}\right)$

$\times \zeta(2m-2j-2k-2) \qquad [m \geq 1;\ -\pi < x < \pi;\ |x| - \pi < y < \pi - |x|].$

6.4.4. Series containing products of $\operatorname{Si}(kx)$

1. $\displaystyle\sum_{k=1}^{\infty} \frac{(-1)^k}{k^{2m}} \operatorname{Si}(kx) \operatorname{Si}(ky) = (-1)^m \frac{x^{2m-1} y}{2(2m-1)!(2m-1)}$

$- \displaystyle\sum_{k=0}^{m-2} \frac{(-1)^k x^{2k+1}}{(2k+1)!(2k+1)} \sum_{j=0}^{m-k-1} (-1)^j \frac{y^{2j+1}}{(2j+1)!(2j+1)}$

$\times \left(1 - 2^{2j+2k-2m+3}\right) \zeta(2m-2j-2k-2)$

$[m \geq 1;\ -\pi < x < \pi;\ |x| - \pi < y < \pi - |x|].$

2. $\displaystyle\sum_{k=1}^{\infty} \frac{(-1)^k}{k^n} \prod_{i=1}^{n} \operatorname{Si}(kx_i) = -\frac{1}{2} \prod_{i=1}^{n} x_i \qquad \left[x_i > 0;\ \displaystyle\sum_{i=1}^{n} x_i < \pi\right].$

6.5. The Fresnel Integrals $S(x)$ and $C(x)$

6.5.1. Series containing $S(\varphi(k)x)$, $C(\varphi(k)x)$ and algebraic functions

1. $\displaystyle\sum_{k=2}^{\infty} \frac{(-1)^k}{k^{3/2}(k^2-1)} S(kx) = \frac{1}{6}\sqrt{\frac{x}{2\pi}} (2x + 3\sin x) - S(x) \qquad [-\pi \leq x \leq \pi].$

2. $\sum_{k=1}^{\infty} \dfrac{(-1)^k}{k^{3/2}\left(k^2 a^2 + b^2\right)} S(kx) = -\dfrac{2}{3b^2 \sqrt{\pi}} \left(\dfrac{x}{2}\right)^{3/2}$
$\quad + \dfrac{a^{1/2} \pi}{(2b)^{5/2}} \operatorname{csch} \dfrac{b\pi}{a} \left[\operatorname{erfi}\left(\sqrt{\dfrac{bx}{a}}\right) - \operatorname{erf}\left(\sqrt{\dfrac{bx}{a}}\right)\right] \quad [-\pi < x < \pi].$

3. $\sum_{k=1}^{\infty} \dfrac{(-1)^k}{k^{3/2}\left(k^2 a^2 - b^2\right)} S(kx) = \dfrac{2}{3b^2 \sqrt{\pi}} \left(\dfrac{x}{2}\right)^{3/2} \dfrac{a^{1/2} \pi}{2b^{5/2}} \csc \dfrac{b\pi}{a} S\left(\dfrac{bx}{a}\right)$
$\quad [-\pi < x < \pi].$

4. $\sum_{k=1}^{\infty} \dfrac{(2k+1)^{-3/2}}{k(k+1)} S((2k+1)x) = 4S(x) + \pi C(x) - \sqrt{\dfrac{2x}{\pi}}(\pi + \sin x)$
$\quad [0 \le x \le \pi].$

5. $\sum_{k=0}^{\infty} (-1)^k \dfrac{(2k+1)^{-1/2}}{(2k-1)(2k+3)} S((2k+1)x) = -\dfrac{\pi}{2^{7/2}} S(2x) \quad [-\pi/2 < x < \pi/2].$

6. $\sum_{k=0}^{\infty} \dfrac{(-1)^k}{(2k+1)^{5/2}} S((2k+1)x) = \dfrac{1}{6}\sqrt{\dfrac{\pi x^3}{2}} \quad [-\pi/2 < x < \pi/2].$

7. $\sum_{k=1}^{\infty} \dfrac{(-1)^k}{k(k+1)(2k+1)^{1/2}} S((2k+1)x) = 2S(x) - \sqrt{\dfrac{2x}{\pi}} \sin x$
$\quad [-\pi/2 < x < \pi/2].$

8. $\sum_{k=1}^{\infty} \dfrac{(-1)^k}{(k^2 + a^2)^{3/4}} S\left(\sqrt{k^2 + a^2}\, x\right) = \dfrac{1}{2a^{3/2}} S(ax) \quad [-\pi/2 < x < \pi/2].$

9. $\sum_{k=1}^{\infty} \dfrac{(-1)^k}{k^2(k^2 + a^2)^{3/4}} S\left(\sqrt{k^2 + a^2}\, x\right)$
$\quad = \dfrac{1}{24 a^{7/2}} \left[(9 - 2\pi^2 a^2) S(ax) - 3\sqrt{\dfrac{2ax}{\pi}} \sin(ax)\right] \quad [-\pi < x < \pi].$

10. $\sum_{k=0}^{\infty} \dfrac{(-1)^k}{(2k+1)((2k+1)^2 + a^2)^{3/4}} S\left(\sqrt{(2k+1)^2 + a^2}\, x\right) = \dfrac{\pi}{4a^{3/2}} S(ax)$
$\quad [-\pi/2 < x < \pi/2].$

11. $\sum_{k=1}^{\infty} \dfrac{(-1)^k}{k^{1/2}\left(k^2 a^2 + b^2\right)} C(kx)$
$\quad = -\dfrac{1}{b^2}\sqrt{\dfrac{x}{2\pi}} + \dfrac{\pi}{2^{5/2} a^{1/2} b^{3/2}} \operatorname{csch} \dfrac{b\pi}{a} \left[\operatorname{erf}\left(\sqrt{\dfrac{bx}{a}}\right) + \operatorname{erfi}\left(\sqrt{\dfrac{bx}{a}}\right)\right]$
$\quad [-\pi < x < \pi].$

12. $\displaystyle\sum_{k=1}^{\infty} \frac{(-1)^k}{k^{1/2}\left(k^2 a^2 - b^2\right)} C(kx) = \frac{1}{b^2}\sqrt{\frac{x}{2\pi}} - \frac{\pi}{2a^{1/2}b^{3/2}} \csc\frac{b\pi}{a} C\left(\frac{bx}{a}\right)$

$[-\pi < x < \pi].$

13. $\displaystyle\sum_{k=2}^{\infty} \frac{(-1)^k}{k^{1/2}\left(k^2 - 1\right)} C(kx) = \sqrt{\frac{x}{8\pi}}\,(\cos x + 2) - \frac{1}{2}C(x)$ $\quad [-\pi < x < \pi].$

14. $\displaystyle\sum_{k=1}^{\infty} \frac{(2k+1)^{-1/2}}{k(k+1)} C((2k+1)x) = 2C(x) - \pi S(x) - \sqrt{\frac{2x}{\pi}}\cos x$

$[0 \le x \le \pi].$

15. $\displaystyle\sum_{k=0}^{\infty} (-1)^k \frac{(2k+1)^{1/2}}{(2k-1)(2k+3)} C((2k+1)x) = -\frac{\pi}{2^{7/2}} C(2x)$ $\quad [0 \le x \le \pi].$

16. $\displaystyle\sum_{k=0}^{\infty} \frac{(-1)^k}{(2k+1)^{3/2}} C((2k+1)x) = \frac{1}{2}\sqrt{\frac{\pi x}{2}}$ $\quad [-\pi/2 < x < \pi/2].$

17. $\displaystyle\sum_{k=1}^{\infty} (-1)^k \frac{(2k+1)^{1/2}}{k(k+1)} C((2k+1)x) = -\sqrt{\frac{2x}{\pi}}\cos x$ $\quad [-\pi/2 < x < \pi/2].$

18. $\displaystyle\sum_{k=1}^{\infty} \frac{(-1)^k}{(k^2+a^2)^{1/4}} C\!\left(\sqrt{k^2+a^2}\,x\right) = -\frac{1}{2a^{1/2}} C(ax)$ $\quad [-\pi/2 < x < \pi/2].$

19. $\displaystyle\sum_{k=1}^{\infty} \frac{(-1)^k}{k^2(k^2+a^2)^{1/4}} C\!\left(\sqrt{k^2+a^2}\,x\right)$

$= \dfrac{1}{24a^{5/2}}\left[(2\pi^2 a^2 - 3)C(ax) - 3\sqrt{\dfrac{2ax}{\pi}}\cos(ax)\right]$ $\quad [-\pi < x < \pi].$

20. $\displaystyle\sum_{k=0}^{\infty} \frac{(-1)^k}{(2k+1)((2k+1)^2+a^2)^{1/4}} C\!\left(\sqrt{(2k+1)^2+a^2}\,x\right) = \frac{\pi}{4a^{1/2}} C(ax)$

$[-\pi/2 < x < \pi/2].$

6.5.2. Series containing $S(\varphi(k)x)$, $C(\varphi(k)x)$ and trigonometric functions

1. $\displaystyle\sum_{k=1}^{\infty} \frac{(-1)^k}{k^{5/2}} \sin(kx) S(ky) = -\frac{x}{3}\sqrt{\frac{y^3}{2\pi}}$ $\quad [x, y > 0;\ x+y < \pi].$

2. $\displaystyle\sum_{k=1}^{\infty} \frac{(-1)^k}{k^{3/2}} \cos(kx) S(ky) = -\frac{1}{3}\sqrt{\frac{y^3}{2\pi}}$ $\quad [x, y > 0;\ x+y < \pi].$

3. $\sum_{k=1}^{\infty} \dfrac{\cos(kx)}{k^{2m-1/2}} S(ky) = (-1)^m \dfrac{y^{2m-1/2}}{(2m-1)!(4m-1)\sqrt{2\pi}}$

$+ \sqrt{\dfrac{2y^3}{\pi}} \sum_{k=0}^{m-2} \dfrac{y^{2k}}{(2k+1)!(4k+3)}$

$\times \left[(-1)^{m-1} \dfrac{\pi x^{2m-2k-3}}{2(2m-2k-3)!} + (-1)^k \sum_{j=0}^{m-k-1} (-1)^j \dfrac{x^{2j}}{(2j)!} \zeta(2m-2j-2k-2) \right]$

$[m \geq 1;\ 0 < y < \pi;\ y < x < 2\pi - y]$.

4. $\sum_{k=1}^{\infty} (-1)^k \dfrac{\cos(kx)}{k^{2m-1/2}} S(ky) = (-1)^m \dfrac{y^{2m-1/2}}{(2m-1)!(4m-1)\sqrt{2\pi}}$

$- \sqrt{\dfrac{2y^3}{\pi}} \sum_{k=0}^{m-2} \dfrac{(-1)^k y^{2k}}{(2k+1)!(4k+3)} \sum_{j=0}^{m-k-1} (-1)^j \dfrac{x^{2j}}{(2j)!} \left(1 - 2^{2j+2k-2m+3}\right)$

$\times \zeta(2m-2j-2k-2) \quad [m \geq 1;\ -\pi < y < \pi;\ |y| - \pi < x < \pi - |y|]$.

5. $\sum_{k=1}^{\infty} \dfrac{(-1)^k}{k^{3/2}} \sin(kx) C(ky) = -x \sqrt{\dfrac{y}{2\pi}}$ $\quad [x, y > 0;\ x + y < \pi]$.

6. $\sum_{k=1}^{\infty} \dfrac{(-1)^k}{k^{1/2}} \cos(kx) C(ky) = -\sqrt{\dfrac{y}{2\pi}}$ $\quad [x, y > 0;\ x + y < \pi]$.

7. $\sum_{k=1}^{\infty} \dfrac{\cos(kx)}{k^{2m-3/2}} C(ky) = (-1)^m \dfrac{y^{2m-3/2}}{(2m-2)!(4m-3)\sqrt{2\pi}}$

$+ \sqrt{\dfrac{2y}{\pi}} \sum_{k=0}^{m-2} \dfrac{y^{2k}}{(2k)!(4k+1)}$

$\times \left[(-1)^m \dfrac{\pi x^{2m-2k-3}}{2(2m-2k-3)!} + (-1)^k \sum_{j=0}^{m-k-1} (-1)^j \dfrac{x^{2j}}{(2j)!} \zeta(2m-2j-2k-2) \right]$

$[m \geq 2;\ 0 < y < \pi;\ y < x < 2\pi - y]$.

8. $\sum_{k=1}^{\infty} (-1)^k \dfrac{\cos(kx)}{k^{2m-3/2}} C(ky) = (-1)^m \dfrac{y^{2m-3/2}}{(2m-2)!(4m-3)\sqrt{2\pi}}$

$- \sqrt{\dfrac{2y}{\pi}} \sum_{k=0}^{m-2} \dfrac{(-1)^k y^{2k}}{(2k)!(4k+1)} \sum_{j=0}^{m-k-1} (-1)^j \dfrac{x^{2j}}{(2j)!} \left(1 - 2^{2j+2k-2m+3}\right)$

$\times \zeta(2m-2j-2k-2) \quad [m \geq 2;\ -\pi < y < \pi;\ |y| - \pi < x < \pi - |y|]$.

9. $\displaystyle\sum_{k=2}^{\infty} \frac{(-1)^k}{k^{1/2}(k^2-1)} [\sin(kx)S(kx) + \cos(kx)C(kx)]$
$= \dfrac{1}{2}(x\cos x - \sin x)S(x) - \dfrac{1}{2}(\cos x + x\sin x)C(x) + 3\sqrt{\dfrac{x}{8\pi}}$
$\qquad\qquad [-\pi/2 < x < \pi/2].$

10. $\displaystyle\sum_{k=1}^{\infty} \frac{(-1)^k}{k^{1/2}(k^2+a^2)} [\sin(kx)S(kx) + \cos(kx)C(kx)] = -\dfrac{1}{b^2}\sqrt{\dfrac{x}{2\pi}}$
$+ \dfrac{\pi}{2^{5/2}a^{1/2}b^{3/2}} \operatorname{csch}\dfrac{b\pi}{a} \left[e^{bx/a}\operatorname{erf}\left(\sqrt{\dfrac{bx}{a}}\right) + e^{-bx/a}\operatorname{erfi}\left(\sqrt{\dfrac{bx}{a}}\right) \right]$
$\qquad\qquad [-\pi < x < \pi].$

11. $\displaystyle\sum_{k=1}^{\infty} \frac{(-1)^k}{k^{3/2}(k^2+a^2)} [\cos(kx)S(kx) - \sin(kx)C(kx)] = \dfrac{4}{3\sqrt{\pi}\,b^2}\left(\dfrac{x}{2}\right)^{3/2}$
$- \dfrac{a^{1/2}\pi}{(2b)^{5/2}} \operatorname{csch}\dfrac{b\pi}{a} \left[e^{bz/a}\operatorname{erf}\left(\sqrt{\dfrac{bx}{a}}\right) - e^{-bz/a}\operatorname{erfi}\left(\sqrt{\dfrac{bx}{a}}\right) \right]$
$\qquad\qquad [-\pi < x < \pi].$

12. $\displaystyle\sum_{k=2}^{\infty} \frac{(-1)^k}{k^{1/2}(k^2-1)} [\sin(kx)S(kx) + \cos(kx)C(kx)]$
$= \dfrac{1}{2}(x\cos x - \sin x)S(x) - \dfrac{1}{2}(\cos x + x\sin x)C(x) + 3\sqrt{\dfrac{x}{8\pi}}$
$\qquad\qquad [-\pi/2 < x < \pi/2].$

13. $\displaystyle\sum_{k=2}^{\infty} \frac{(-1)^k}{k^{3/2}(k^2-1)} [\cos(kx)S(kx) - \sin(kx)C(kx)]$
$= -\dfrac{1}{2}(x\sin x + 2\cos x)S(x) + \dfrac{1}{2}(2\sin x - x\cos x)C(x) - \dfrac{4}{3\sqrt{\pi}}\left(\dfrac{x}{2}\right)^{3/2}$
$\qquad\qquad [-\pi < x < \pi].$

14. $\displaystyle\sum_{k=1}^{\infty} (-1)^k \frac{(2k+1)^{1/2}}{k(k+1)}$
$\times [\sin((2k+1)x)S((2k+1)x) + \cos((2k+1)x)C((2k+1)x)]$
$= 2x\left[\sin x\, C(x) - \cos x\, S(x) - \dfrac{1}{\sqrt{2\pi x}}\right] \quad [-\pi/2 < x < \pi/2].$

15. $\displaystyle\sum_{k=1}^{\infty} \frac{(2k+1)^{-3/2}}{k(k+1)}$
$\times [\cos((2k+1)x)S((2k+1)x) - \sin((2k+1)x)C((2k+1)x)]$
$= [4\cos x + (2x-\pi)\cos x]S(x) - [4\sin x + (\pi-2x)\cos x]C(x) + \sqrt{2\pi x}$
$\qquad\qquad [0 < x < \pi/2].$

16. $\displaystyle\sum_{k=1}^{\infty} \frac{(2k+1)^{-1/2}}{k(k+1)}$
$\times [\sin((2k+1)x)S((2k+1)x) + \cos((2k+1)x)C((2k+1)x)]$
$= [2\sin x + (\pi - 2x)\cos x] S(x) + [2\cos x + (2x-\pi)\sin x] C(x) - \sqrt{\dfrac{2x}{\pi}}$
$\hfill [0 < x < \pi/2].$

17. $\displaystyle\sum_{k=0}^{\infty} \frac{(-1)^k}{(2k+1)^{3/2}}$
$\times [\sin((2k+1)x)S((2k+1)x) + \cos((2k+1)x)C((2k+1)x)] = \sqrt{\dfrac{\pi x}{8}}$
$\hfill [-\pi/2 < x < \pi/2].$

18. $\displaystyle\sum_{k=0}^{\infty} (-1)^k \frac{(2k+1)^{1/2}}{(2k-1)(2k+3)}$
$\times [\sin((2k+1)x)S((2k+1)x) + \cos((2k+1)x)C((2k+1)x)]$
$= -\dfrac{\pi}{2^{5/2}} [\sin(2x)S(2x) + \cos(2x)C(2x)] \quad [-\pi/2 < x < \pi/2].$

19. $\displaystyle\sum_{k=0}^{\infty} (-1)^k \frac{(2k+1)^{-1/2}}{(2k-1)(2k+3)}$
$\times [\cos((2k+1)x)S((2k+1)x) - \sin((2k+1)x)C((2k+1)x)]$
$= -\dfrac{\pi}{2^{7/2}} [\cos(2x)S(2x) - \sin(2x)C(2x)] \quad [-\pi/2 < x < \pi/2].$

20. $\displaystyle\sum_{k=1}^{\infty} (-1)^k \frac{(2k+1)^{-1/2}}{k(k+1)}$
$\times [\cos((2k+1)x)S((2k+1)x) - \sin((2k+1)x)C((2k+1)x)]$
$= 2(\cos x + x\sin x) S(x) - 2(\sin x - x\cos x) C(x) \quad [-\pi/2 < x < \pi/2].$

21. $\displaystyle\sum_{k=0}^{\infty} \frac{(-1)^k}{(2k+1)^{5/2}}$
$\times [\cos((2k+1)x)S((2k+1)x) - \sin((2k+1)x)C((2k+1)x)]$
$= -\dfrac{1}{3}\sqrt{\dfrac{\pi x^3}{2}} \quad [-\pi/2 < x < \pi/2].$

6.5.3. Series containing $S(kx)$, $C(kx)$ and $\operatorname{Si}(kx)$

1. $\displaystyle\sum_{k=1}^{\infty} \frac{(-1)^k}{k^{5/2}} \operatorname{Si}(kx) S(ky) = -\dfrac{x}{3}\sqrt{\dfrac{y^3}{2\pi}} \qquad [x, y > 0;\ x+y < \pi].$

2. $\displaystyle\sum_{k=1}^{\infty} \frac{(-1)^k}{k^{3/2}} \operatorname{Si}(kx) C(ky) = -x\sqrt{\dfrac{y}{2\pi}} \qquad [x, y > 0;\ x+y < \pi].$

6.5.4. Series containing products of $S(kx)$ and $C(kx)$

1. $\sum_{k=1}^{\infty} \frac{(-1)^k}{k^{2m+1}} S(kx) S(ky) = (-1)^m \frac{x^{2m-1/2} y^{3/2}}{3(2m-1)!(4m-1)\pi}$

$- \frac{2}{\pi} \sum_{k=0}^{m-2} \frac{(-1)^k x^{2k+3/2}}{(2k+1)!(4k+3)} \sum_{j=0}^{m-k-1} (-1)^j \frac{y^{2j+3/2}}{(2j+1)!(4j+3)} \left(1 - 2^{2j+2k-2m+3}\right)$

$\times \zeta(2m-2j-2k-2) \quad [m \geq 1;\ -\pi < x < \pi;\ |x| - \pi < y < \pi - |x|].$

2. $\sum_{k=1}^{\infty} \frac{(-1)^k}{k^{2m}} S(kx) C(ky) = (-1)^m \frac{x^{2m-1/2} y^{1/2}}{(2m-1)!(4m-1)\pi}$

$- \frac{2}{\pi} \sum_{k=0}^{m-2} \frac{(-1)^k x^{2k+3/2}}{(2k+1)!(4k+3)} \sum_{j=0}^{m-k-1} (-1)^j \frac{y^{2j+1/2}}{(2j)!(4j+1)} \left(1 - 2^{2j+2k-2m+3}\right)$

$\times \zeta(2m-2j-2k-2) \quad [m \geq 1;\ -\pi < x < \pi;\ |x| - \pi < y < \pi - |x|].$

3. $\sum_{k=1}^{\infty} \frac{(-1)^k}{k^{2m-1}} C(kx) C(ky) = (-1)^m \frac{x^{2m-3/2} y^{1/2}}{(2m-2)!(4m-3)\pi}$

$- \frac{2}{\pi} \sum_{k=0}^{m-2} \frac{(-1)^k x^{2k+1/2}}{(2k)!(4k+1)} \sum_{j=0}^{m-k-1} (-1)^j \frac{y^{2j+1/2}}{(2j)!(4j+1)} \left(1 - 2^{2j+2k-2m+3}\right)$

$\times \zeta(2m-2j-2k-2) \quad [m \geq 2;\ -\pi < x < \pi;\ |x| - \pi < y < \pi - |x|].$

4. $\sum_{k=1}^{\infty} \frac{(-1)^k}{k^{3n/2}} \prod_{i=1}^{n} S(kx_i) = -\frac{3^{-n}}{2} \left(\frac{2}{\pi}\right)^{n/2} \prod_{i=1}^{n} x_i^{3/2} \quad \left[x_i > 0;\ \sum_{i=1}^{n} x_i < \pi\right].$

5. $\sum_{k=1}^{\infty} \frac{(-1)^k}{k^{n/2}} \prod_{i=1}^{n} C(kx_i) = -\frac{1}{2} \left(\frac{2}{\pi}\right)^{n/2} \prod_{i=1}^{n} x_i^{1/2} \quad \left[x_i > 0;\ \sum_{i=1}^{n} x_i < \pi\right].$

6.6. The Incomplete Gamma Function $\gamma(\nu, z)$

6.6.1. Series containing $\gamma(\nu \pm k, z)$

1. $\sum_{k=0}^{\infty} \frac{t^k}{k!(\nu)_k} \gamma(\nu + k, z) = \frac{z^\nu}{\nu} e^{-z} \Phi_3(1;\ \nu + 1;\ z, tz).$

2. $\sum_{k=0}^{\infty} \frac{1}{(\nu)_k} \gamma(\nu + k, z) = (\nu - 1) z^{\nu-1} e^{-z} - (\nu - 1)(\nu - z - 1)\gamma(\nu - 1, z).$

3. $$\sum_{k=0}^{\infty} \frac{(-z)^k}{(k!)^2} \gamma(k+1, z)$$
$$= \frac{e^{-z}}{2} [1 + (2z-1) J_0(2z) - \pi z J_0(2z) \mathbf{H}_1(2z) + \pi z J_1(2z) \mathbf{H}_0(2z)].$$

4. $$\sum_{k=0}^{\infty} \frac{(1-\nu)_k}{k!} t^k \gamma(\nu - k, z) = e^{-t} \gamma(\nu, z-t) \qquad [|t| < |z|].$$

5. $$\sum_{k=0}^{\infty} \frac{t^k}{k!(\nu)_k} \gamma(\nu + k, z) = \Gamma(\nu) (2t)^{-\nu} \int_0^{2\sqrt{tz}} x^\nu e^{-x^2/(4t)} I_{\nu-1}(x) dx.$$

6. $$\sum_{k=0}^{\infty} \frac{2^{-2k}}{k!(a)_k} \gamma(\nu + 2k, z) = \frac{z^\nu}{\nu} {}_2F_2\left(\begin{array}{c} a - \frac{1}{2}, \nu; -2z \\ 2a - 1, \nu + 1 \end{array}\right).$$

7. $$\sum_{k=0}^{\infty} \frac{2^k}{(\nu+1)_k (k+2\nu)} P_k^{(-k-\nu,\nu)}(0) \gamma(3\nu + k, z) = \frac{\nu}{6} \gamma^3(\nu, z).$$

8. $$\sum_{k=0}^{\infty} \frac{(-1)^k}{k!} E_k(w) \gamma(k+1, z) = \frac{2}{w}(1 - e^{-wz}) + \psi\left(\frac{w}{2}\right) - \psi\left(\frac{w+1}{2}\right)$$
$$+ 2e^{-wz} \Phi(-e^{-z}, 1, w).$$

6.6.2. Series containing products of $\gamma(\nu + k, z)$

1. $$\sum_{k=0}^{\infty} \frac{1}{k!(\nu)_k} \gamma^2(\nu + k, z) = \frac{z^{2\nu}}{\nu^2} {}_2F_2\left(\begin{array}{c} \nu, \nu + \frac{1}{2}; -4z \\ \nu + 1, 2\nu + 1 \end{array}\right).$$

2. $$\sum_{k=0}^{\infty} \frac{1}{k!(\nu)_k} \gamma(\nu + k, z) \gamma(\nu + k, -z) = \frac{e^{i\pi\nu}}{\nu^2} z^{2\nu} {}_1F_2\left(\begin{array}{c} \frac{\nu}{2}; -z^2 \\ \frac{\nu}{2} + 1, \nu + 1 \end{array}\right).$$

6.7. The Parabolic Cylinder Function $D_\nu(z)$

6.7.1. Series containing $D_{\nu \pm nk}(z)$ and elementary functions

1. $$\sum_{k=0}^{\infty} \frac{t^k}{k!} D_{\nu+2k}(z) = (1 + 2t)^{-(\nu+1)/2} \exp\frac{tz^2}{2(2t+1)} D_\nu\left(\frac{z}{\sqrt{2t+1}}\right)$$
$$[|t| < 1/2; \ [80], (3.1)].$$

2. $$\sum_{k=0}^{\infty} \frac{(-\nu)_{2k}}{k!} t^k D_{\nu-2k}(z) = (1 - 2t)^{\nu/2} \exp\frac{tz^2}{2(1-2t)} D_\nu\left(\frac{z}{\sqrt{1-2t}}\right)$$
$$[|t| < 1/2; \ [80], (4.2)].$$

6.8. The Bessel Functions $J_\nu(z)$ and $Y_\nu(z)$

6.8.1. Series containing $J_{nk+\nu}(z)$

1. $\displaystyle\sum_{k=0}^{\infty} J_{2k+\nu}(z) = \frac{\left(\frac{z}{2}\right)^\nu}{\Gamma(\nu+1)}\,{}_1F_2\left(\begin{array}{c}\frac{\nu}{2};\, -\frac{z^2}{4}\\ \frac{\nu}{2}+1,\,\nu\end{array}\right).$

2. $\displaystyle\sum_{k=0}^{\infty}(-1)^k (2k+\nu)^2 J_{2k+\nu}(z)$
$\displaystyle = \frac{\nu z^{\nu-1}}{2^\nu \Gamma(\nu-2)}\left[\frac{\sin z}{\nu-2}\,{}_3F_4\left(\begin{array}{c}\frac{\nu}{2}-1,\,\frac{2\nu-1}{4},\,\frac{2\nu+1}{4}\\ \frac{\nu}{2},\,\nu-\frac{1}{2},\,\nu,\,\frac{1}{2};\,-z^2\end{array}\right)\right.$
$\displaystyle \left. - \frac{z\cos z}{\nu-1}\,{}_3F_4\left(\begin{array}{c}\frac{\nu-1}{2},\,\frac{2\nu+1}{4},\,\frac{2\nu+3}{4}\\ \frac{\nu+1}{2},\,\nu,\,\nu+\frac{1}{2},\,\frac{3}{2};\,-z^2\end{array}\right)\right].$

3. $\displaystyle\sum_{k=0}^{\infty} \frac{\left(\frac{z}{2}\right)^k}{k!(k+a)} J_{k+\nu}(z) = \frac{\left(\frac{z}{2}\right)^\nu}{a\Gamma(\nu+1)}\,{}_1F_2\left(\begin{array}{c}1;\, -\frac{z^2}{4}\\ a+1,\,\nu+1\end{array}\right).$

4. $\displaystyle\sum_{k=0}^{\infty}(2k+a)\frac{(a)_k}{k!} J_{2k+\nu}(z) = \frac{a}{\Gamma(\nu+1)}\left(\frac{z}{2}\right)^\nu\,{}_2F_3\left(\begin{array}{c}\frac{\nu-a}{2},\,\frac{\nu-a+1}{2};\,-\frac{z^2}{4}\\ \frac{\nu+1}{2},\,\frac{\nu}{2}+1,\,\nu-a+1\end{array}\right).$

5. $\displaystyle\sum_{k=0}^{\infty}(2k+\nu)\frac{(\nu-\mu+1)_k}{(\mu)_k} J_{2k+\nu}(z) = \frac{\left(\frac{z}{2}\right)^\nu}{\Gamma(\nu)}\,{}_1F_2\left(\begin{array}{c}\mu-1;\,-\frac{z^2}{4}\\ \mu,\,\nu\end{array}\right).$

6. $\displaystyle\sum_{k=0}^{\infty}(\pm t)^k \frac{\left(\nu+\frac{1}{2}\right)_k}{k!(2\nu+1)_k} J_{k+\nu}(z)$
$\displaystyle = \Gamma(\nu+1)\left(\frac{4}{t}\right)^\nu J_\nu\!\left(\frac{z-\sqrt{z^2-2tz}}{2}\right) J_{\pm\nu}\!\left(\frac{z+\sqrt{z^2-2tz}}{2}\right)\quad [|t|<|z|/2].$

7. $\displaystyle\sum_{k=0}^{\infty}(-1)^k (2k+\nu)\frac{(\nu-\mu-1)_{2k}}{(\nu+\mu+2)_{2k}} J_{2k+\nu}(z) = \frac{(\mu+\nu)(\mu+\nu+1)z^{\nu-1}}{2^\nu \Gamma(\nu)}$
$\displaystyle \times \left[\frac{\sin z}{\mu+\nu}\,{}_3F_4\left(\begin{array}{c}\frac{2\nu-1}{4},\,\frac{2\nu+1}{4},\,\frac{\mu+\nu}{2};\,-z^2\\ \nu-\frac{1}{2},\,\nu,\,\frac{\mu+\nu}{2}+1,\,\frac{1}{2}\end{array}\right)\right.$
$\displaystyle \left. - \frac{z\cos z}{\mu+\nu+1}\,{}_3F_4\left(\begin{array}{c}\frac{2\nu+1}{4},\,\frac{2\nu+3}{4},\,\frac{\mu+\nu+1}{2};\,-z^2\\ \frac{\nu+1}{2},\,\nu+\frac{1}{2},\,\frac{\mu+\nu+3}{2},\,\frac{3}{2}\end{array}\right)\right].$

8. $\displaystyle\sum_{k=0}^{\infty}(2k+\nu)\frac{(\nu)_k \left(\frac{1}{2}\right)_k}{k!\left(\nu+\frac{1}{2}\right)_k} J_{2k+\nu}(z) = \frac{\left(\frac{z}{2}\right)^\nu}{\Gamma(\nu)}\,{}_1F_2\left(\begin{array}{c}\frac{\nu}{2};\,-\frac{z^2}{4}\\ \frac{\nu+1}{2},\,\nu+\frac{1}{2}\end{array}\right).$

9. $\displaystyle\sum_{k=0}^{\infty}(4k+1)\frac{\left(\frac{1}{2}-\nu\right)_k\left(\frac{1}{2}+\nu\right)_k\left(\frac{1}{2}\right)_k}{k!\,(1-\nu)_k(1+\nu)_k}J_{2k+1/2}(z)$

$$= \nu\sqrt{2\pi z}\,\csc(\nu\pi)\,J_{-\nu}\left(\frac{z}{2}\right)J_{\nu}\left(\frac{z}{2}\right).$$

10. $\displaystyle\sum_{k=0}^{\infty}(-1)^k(2k+\nu)\frac{(2\nu)_{2k}(a)_{2k}(b)_{2k}}{(2k)!\,(2\nu-a+1)_{2k}(2\nu-b+1)_{2k}}J_{2k+\nu}(z)$

$$= \frac{\left(\frac{z}{2}\right)^{\nu}}{2\Gamma(\nu)}\left[e^{-iz}{}_2F_2\!\left(\begin{array}{c}\nu+\frac{1}{2},\,2\nu-a-b+1;\,2iz\\ 2\nu-a+1,\,2\nu-b+1\end{array}\right)\right.$$
$$\left. + e^{iz}{}_2F_2\!\left(\begin{array}{c}\nu+\frac{1}{2},\,2\nu-a-b+1;\,-2iz\\ 2\nu-a+1,\,2\nu-b+1\end{array}\right)\right].$$

11. $\displaystyle\sum_{k=0}^{\infty}(2k+\nu)\frac{(a)_k(\nu)_k}{k!\,(\nu-a+1)_k}J_{4k+2\nu}(z)$

$$= \frac{2^{-2\nu-1}z^{2\nu}}{\Gamma(2\nu)}\,{}_1F_2\!\left(\begin{array}{c}\nu-a+\frac{1}{2};\,-\frac{z^2}{4}\\ \nu+\frac{1}{2},\,2\nu-2a+1\end{array}\right).$$

12. $\displaystyle\sum_{k=0}^{\infty}(6k+\nu)\frac{\left(\frac{\nu}{3}\right)_k}{k!}J_{6k+\nu}(z) = \frac{\left(\frac{z}{2}\right)^{\nu}}{\Gamma(\nu)}\,{}_1F_2\!\left(\begin{array}{c}\frac{\nu}{3};\,-\frac{3z^2}{16}\\ \frac{\nu}{2},\,\frac{\nu+1}{2}\end{array}\right).$

6.8.2. Series containing two Bessel functions $J_{nk+\nu}(z)$

1. $\displaystyle\sum_{k=1}^{\infty}J_{k+\nu}^2(z) = \frac{\left(\frac{z}{2}\right)^{2\nu+2}}{\Gamma^2(\nu+2)}\,{}_2F_3\!\left(\begin{array}{c}\nu+1,\,\nu+\frac{3}{2};\,-z^2\\ \nu+2,\,\nu+2,\,2\nu+2\end{array}\right).$

2. $\displaystyle\sum_{k=0}^{\infty}(-1)^k J_{2k+1}(w)\,J_{2n(2k+1)}(z) = \frac{1}{2\pi}\int_0^{\pi}\sin\left(w\cos(2nx)\right)\cos\left(z\sin x\right)dx.$

3. $\displaystyle\sum_{k=0}^{\infty}(2k+1)\,J_{k+1/2}^2(z) = \frac{2z}{\pi}.$

4. $\displaystyle\sum_{k=0}^{\infty}(2k+3)\,J_{k+3/2}^2(z) = \frac{1}{\pi z}\left[\cos(2z)+2z^2-1\right].$

5. $\displaystyle\sum_{k=0}^{\infty}(k+\nu)\,J_{k+\nu}(w)\,J_{k+\nu}(z) = \frac{wz}{2(z-w)}\left[J_{\nu-1}(w)J_{\nu}(z)-J_{\nu}(w)J_{\nu-1}(z)\right].$

6. $\displaystyle\sum_{k=0}^{\infty}(k+\nu)J_{k+\nu}^2(z)$
$$= \frac{z^2}{4}[J_{\nu-1}^2(z) + J_\nu^2(z) - J_{\nu-1}(z)J_{\nu+1}(z) - J_{\nu-2}(z)J_\nu(z)].$$

7. $\displaystyle\sum_{k=2}^{\infty} k^2(k^2-1)^2 J_k^2(z) = \frac{9z^4}{16} + \frac{5z^6}{32}.$

8. $\displaystyle\sum_{k=0}^{\infty} \frac{(-1)^k}{(2k-1)(2k+3)} J_k(z) J_{k+1}(z) = \frac{\pi}{8z}[J_0(2z)\,\mathbf{H}_1(2z) - J_1(2z)\,\mathbf{H}_0(2z)].$

9. $\displaystyle\sum_{k=0}^{\infty} \frac{(-1)^k}{(2k+1)^3} J_{\nu-k-1/2}(z) J_{\nu+k+1/2}(z) = \frac{1}{8}\int_0^\pi x(\pi-x) J_{2\nu}(2z\sin x)\,dx.$

10. $\displaystyle\sum_{k=0}^{\infty} \frac{(\pm 1)^k}{k!} J_{\nu-k}(z) J_{\nu+k}(z) = \frac{2}{\pi}\int_0^{\pi/2} e^{\pm\cos(2x)} \cos(\sin 2x)\, J_{2\nu}(2z\cos x)\,dx.$

11. $\displaystyle\sum_{k=0}^{\infty}(-1)^k \frac{(2\nu)_k}{k!} J_{k+\nu}^2(z) = \frac{\left(\frac{z}{2}\right)^{2\nu}}{\Gamma^2(\nu+1)}\,_1F_2\!\left(\begin{matrix}\nu+\frac{1}{2};\,-z^2\\ \nu+1,\,\nu+1\end{matrix}\right).$

12. $\displaystyle\sum_{k=0}^{\infty}(-1)^k (k+\nu)\frac{(2\nu)_k}{k!} J_{k+\nu}(w) J_{k+\nu}(z) = \frac{2^{-\nu}}{\Gamma(\nu)}\left(\frac{wz}{w+z}\right)^\nu J_\nu(w+z).$

13. $\displaystyle\sum_{k=0}^{\infty} \frac{(a)_k}{(1-a)_k} J_k^2(z) = \frac{1}{2} J_0^2(z) + \frac{1}{2}\,_1F_2\!\left(\begin{matrix}\frac{1}{2}-a;\,-z^2\\ 1,\,1-a\end{matrix}\right).$

14. $\displaystyle\sum_{k=0}^{\infty}(-1)^k \frac{(a)_k}{(1-a)_k} J_k^2(z) = \frac{1}{2} J_0^2(z) + \frac{1}{2}\,_1F_2\!\left(\begin{matrix}\frac{1}{2};\,-z^2\\ 1,\,1-a\end{matrix}\right).$

15. $\displaystyle\sum_{k=0}^{\infty} \frac{(a)_k}{(2-a)_k} J_{k+1/2}^2(z) = \frac{2z}{\pi}\,_2F_3\!\left(\begin{matrix}1,\,\frac{3}{2}-a;\,-z^2\\ \frac{3}{2},\,\frac{3}{2},\,2-a\end{matrix}\right).$

16. $\displaystyle\sum_{k=0}^{\infty}(-1)^k (2k+1)\frac{(a)_k}{(2-a)_k} J_{k+1/2}^2(z) = \frac{\Gamma(2-a)}{\sqrt{\pi}} z^{a-1/2}\,\mathbf{H}_{1/2-a}(2z).$

17. $\displaystyle\sum_{k=0}^{\infty}(2k+a)\frac{(a)_k}{k!} J_{k+\nu}^2(z)$
$$= \frac{a}{\Gamma^2(\nu+1)}\left(\frac{z}{2}\right)^{2\nu}\,_2F_3\!\left(\begin{matrix}\nu-\frac{a}{2},\,\nu+\frac{1-a}{2};\,-z^2\\ \nu+1,\,\nu+1,\,2\nu-a+1\end{matrix}\right).$$

18. $\displaystyle\sum_{k=0}^{\infty} \frac{k+\nu}{4(k+\nu)^2-1} \frac{(2\nu)_k}{k!} J_{k+\nu}^2(z)$

$$= \frac{\left(\frac{z}{2}\right)^{2\nu}}{(4\nu^2-1)\Gamma(\nu)\Gamma(\nu+1)} {}_1F_2\left(\begin{array}{c} 1;\ -z^2 \\ \nu+1,\ \nu+\frac{3}{2} \end{array}\right).$$

19. $\displaystyle\sum_{k=0}^{\infty} (-1)^k \frac{k+\nu}{4(k+\nu)^2-1} \frac{(2\nu)_k}{k!} J_{k+\nu}^2(z)$

$$= \frac{\left(\frac{z}{2}\right)^{2\nu}}{(4\nu^2-1)\Gamma(\nu)\Gamma(\nu+1)} {}_1F_2\left(\begin{array}{c} \nu+\frac{1}{2};\ -z^2 \\ \nu+1,\ \nu+\frac{3}{2} \end{array}\right).$$

20. $\displaystyle\sum_{k=0}^{\infty} (-1)^k \frac{(1-a)_k}{(a)_k} J_{\nu-k}(z) J_{\nu+k}(z)$

$$= \frac{1}{2} J_\nu^2(z) + 2^{2a-3} z^{2\nu} \frac{\Gamma^2(a)\Gamma\left(a+\nu-\frac{1}{2}\right)}{\sqrt{\pi}\,\Gamma(2a-1)\Gamma(a+\nu)\Gamma(2\nu+1)} {}_1F_2\left(\begin{array}{c} a+\nu-\frac{1}{2};\ -z^2 \\ a+\nu,\ 2\nu+1 \end{array}\right).$$

21. $\displaystyle\sum_{k=0}^{\infty} (-1)^k \frac{(a)_k}{k!} J_{k+\mu}(z) J_{\nu-k}(z) = \frac{\Gamma(\mu+\nu-a+1)}{\Gamma(\nu+1)\Gamma(\mu+\nu+1)\Gamma(\mu-a+1)}$

$$\times \left(\frac{z}{2}\right)^{\mu+\nu} {}_2F_3\left(\begin{array}{c} \frac{\mu+\nu-a+1}{2},\ \frac{\mu+\nu-a}{2}+1;\ -z^2 \\ \nu+1,\ \mu+\nu+1,\ \mu-a+1 \end{array}\right).$$

22. $\displaystyle\sum_{k=0}^{\infty} (-1)^k \frac{\left(\frac{1}{2}-\nu\right)_k}{\left(\frac{3}{2}+\nu\right)_k} J_{\nu+k}(z) J_{\nu-k-1}(z) = \frac{\Gamma\left(\nu+\frac{3}{2}\right)}{2\Gamma(\nu+1)} z^{-1/2} J_{2\nu-1/2}(2z).$

23. $\displaystyle\sum_{k=0}^{\infty} \frac{(\mu+\nu)_k (\mu+\nu-a+1)_k}{k!\,(a)_k} J_{k+\mu}(z) J_{k+\nu}(z)$

$$= \frac{\left(\frac{z}{2}\right)^{\mu+\nu}}{\Gamma(\mu+1)\Gamma(\nu+1)} {}_2F_3\left(\begin{array}{c} \frac{\mu+\nu+1}{2},\ a-\frac{\mu+\nu}{2} \\ a,\ \mu+1,\ \nu+1;\ -z^2 \end{array}\right).$$

24. $\displaystyle\sum_{k=0}^{\infty} \frac{(2\nu)_k (2\nu-a+1)_k}{k!\,(a)_k} J_{k+\nu}^2(z) = \frac{\left(\frac{z}{2}\right)^{2\nu}}{\Gamma^2(\nu+1)} {}_2F_3\left(\begin{array}{c} \nu+\frac{1}{2},\ a-\nu;\ -z^2 \\ a,\ \nu+1,\ \nu+1 \end{array}\right).$

25. $\displaystyle\sum_{k=0}^{\infty} (-1)^k (k+\nu) \frac{(2\nu)_k (2\nu-a+1)_k}{k!\,(a)_k} J_{k+\nu}^2(z) = \frac{\left(\frac{z}{2}\right)^{2\nu}}{\nu\Gamma^2(\nu)} {}_1F_2\left(\begin{array}{c} \nu+\frac{1}{2};\ -z^2 \\ a,\ \nu+1 \end{array}\right).$

26. $\sum_{k=0}^{\infty} (-1)^k \dfrac{(\mu-\nu)_k (\mu-\nu-a+1)_k}{k!\,(a)_k} J_{\mu+k}(z)\, J_{\nu-k}(z)$

$$= \dfrac{\Gamma(a)\,\Gamma\!\left(\dfrac{\mu-\nu}{2}+1\right)\Gamma\!\left(a+\dfrac{3\nu-\mu}{2}\right)}{\Gamma(\nu+1)\,\Gamma(\mu-\nu+1)\,\Gamma\!\left(\dfrac{\mu+\nu}{2}+1\right)\Gamma(a+\nu)\,\Gamma\!\left(a+\dfrac{\nu-\mu}{2}\right)} \left(\dfrac{z}{2}\right)^{\mu+\nu}$$

$$\times\ {}_2F_3\!\left(\begin{array}{c}\dfrac{\mu+\nu+1}{2},\ a+\dfrac{3\nu-\mu}{2};\ -z^2\\[2pt] \nu+1,\ \mu+\nu+1,\ \nu+a\end{array}\right).$$

27. $\sum_{k=0}^{\infty} (-1)^k J_{2k+1}(w)\, J_{2n(2k+1)}(z) = \dfrac{1}{2\pi}\int_0^{\pi} \sin(w\cos(2nx))\cos(z\sin x)\,dx.$

28. $\sum_{k=1}^{\infty} \dfrac{\Gamma^2\!\left(\dfrac{2k+1}{4}\right)}{\Gamma^2\!\left(\dfrac{2k+3}{4}\right)} J_k^2(z) = -\dfrac{1}{4\pi^2}\Gamma^4\!\left(\dfrac{1}{4}\right) J_0^2(z)$

$$+ \dfrac{4}{\pi}\int_0^{\pi/2} J_0(2z\sin x)\,\mathbf{K}(\cos x)\,dx.$$

29. $\sum_{k=1}^{\infty} (-1)^k \dfrac{\Gamma^2\!\left(\dfrac{2k+1}{4}\right)}{\Gamma^2\!\left(\dfrac{2k+3}{4}\right)} J_k^2(z) = -\dfrac{1}{4\pi^2}\Gamma^4\!\left(\dfrac{1}{4}\right) J_0^2(z)$

$$+ \dfrac{4}{\pi}\int_0^1 \dfrac{1}{\sqrt{1-x^2}} J_0(2zx)\,\mathbf{K}(x)\,dx.$$

6.8.3. Series containing three Bessel functions $J_{nk+\nu}(z)$

1. $\sum_{k=0}^{\infty} (-1)^k J_{k+\nu}(z)\, J_{\nu-k}(z)\, J_{2k}(2z)$

$$= \dfrac{1}{2} J_\nu^2(z) J_0(2z) + \dfrac{\left(\dfrac{z}{2}\right)^{2\nu}}{2\Gamma^2(\nu+1)}\,{}_3F_4\!\left(\begin{array}{c}\nu+\dfrac{1}{2},\ \nu+\dfrac{1}{4},\ \nu+\dfrac{3}{4};\ -4z^2\\[2pt] \nu+1,\ 2\nu+\dfrac{1}{2},\ 2\nu+1,\ \dfrac{1}{2}\end{array}\right).$$

2. $\sum_{k=0}^{\infty} (-1)^k J_{k+\nu}(z)\, J_{\nu-k-1}(z)\, J_{2k+1}(2z)$

$$= \dfrac{2\left(\dfrac{z}{2}\right)^{2\nu}}{\Gamma(\nu)\Gamma(\nu+1)}\,{}_3F_4\!\left(\begin{array}{c}\nu+\dfrac{1}{2},\ \nu+\dfrac{1}{4},\ \nu+\dfrac{3}{4};\ -4z^2\\[2pt] \nu+1,\ 2\nu,\ 2\nu+\dfrac{1}{2},\ \dfrac{3}{2}\end{array}\right).$$

3. $\sum_{k=0}^{\infty} J_{k+1/2}^2(w)\, J_{2k+1}(z) = \dfrac{1}{\pi}\int_0^1 \dfrac{\sin(zx)}{\sqrt{1-x^2}}\,\mathbf{H}_0(2wx)\,dx.$

4. $\displaystyle\sum_{k=0}^{\infty} J_k(w) J_{k+1}(w) J_{2k+1}(z) = \frac{1}{\pi} \int_0^1 \frac{\sin(zx)}{\sqrt{1-x^2}} J_1(2wx)\, dx.$

6.8.4. Series containing four Bessel functions $J_{nk+\nu}(z)$

1. $\displaystyle\sum_{k=0}^{\infty} J_k^4(z) = \frac{1}{2} J_0^4(z) + \frac{1}{2}\, {}_2F_3\!\left(\begin{array}{c}\frac{1}{2},\frac{1}{2};\ -4z^2\\ 1,1,1\end{array}\right).$

2. $\displaystyle\sum_{k=0}^{\infty} J_k^2(z) J_{k+1}^2(z) = \frac{z^2}{4}\, {}_2F_3\!\left(\begin{array}{c}\frac{3}{2},\frac{3}{2};\ -4z^2\\ 2,2,3\end{array}\right).$

3. $\displaystyle\sum_{k=0}^{\infty} J_{k+1/2}^2(w) J_{k+1/2}^2(z) = \frac{1}{\pi}\int_0^1 \frac{1}{\sqrt{1-x^2}}\, \mathbf{H}_0(2wx)\, \mathbf{H}_0(2zx)\, dx.$

4. $\displaystyle\sum_{k=0}^{\infty} J_{k+1/2}^2(w) J_k(z) J_{k+1}(z) = \frac{1}{\pi}\int_0^1 \frac{1}{\sqrt{1-x^2}}\, J_1(2zx)\, \mathbf{H}_0(2wx)\, dx.$

5. $\displaystyle\sum_{k=0}^{\infty} (-1)^k J_{k+1/2}^2(w) J_{(2k+1)n}^2(z) = \frac{1}{2\pi}\int_0^\pi J_0\!\left(2z\sin\frac{x}{2}\right) \mathbf{H}_0(2w\cos(nx))\, dx.$

6. $\displaystyle\sum_{k=0}^{\infty} J_{3k+3/2}^2(w) J_{2k+2}^2(z) = -\frac{1}{2\pi}\int_0^\pi J_0\!\left(2z\sin\frac{3x}{2}\right) \mathbf{H}_0(2w\cos(2x))\, dx.$

7. $\displaystyle\sum_{k=0}^{\infty} J_k^2(w) J_{nk}^2(z) = \frac{1}{2} J_0^2(w) J_0^2(z) + \frac{1}{\pi}\int_0^{\pi/2} J_0(2z\sin x) J_0(2w\sin(nx))\, dx.$

8. $\displaystyle\sum_{k=1}^{\infty} J_{\mu+k}(w) J_{\mu-k}(w) J_{\nu+k}(z) J_{\nu-k}(z)$

$\displaystyle\qquad = -\frac{1}{2} J_\mu^2(w) J_\nu^2(z) + \frac{1}{\pi}\int_0^1 \frac{1}{\sqrt{1-x^2}}\, J_{2\mu}(2wx) J_{2\nu}(2zx)\, dx.$

9. $\displaystyle\sum_{k=0}^{\infty} J_{\mu+k+1/2}(w) J_{\mu-k-1/2}(w) J_{\nu+k+1/2}(z) J_{\nu-k-1/2}(z)$

$\displaystyle\qquad = \frac{1}{\pi}\int_0^1 \frac{1}{\sqrt{1-x^2}}\, J_{2\mu}(2wx) J_{2\nu}(2zx)\, dx.$

10. $\sum_{k=1}^{\infty}(-1)^{k}J_{\mu+k}(w)J_{\mu-k}(w)J_{\nu+k}(z)J_{\nu-k}(z)$

$$= -\frac{1}{2}J_{\mu}^{2}(w)J_{\nu}^{2}(z) + \frac{1}{\pi}\int_{0}^{\pi/2}J_{2\mu}(2w\sin x)J_{2\nu}(2z\cos x)\,dx.$$

11. $\sum_{k=1}^{\infty}J_{k}(z_{1})J_{k}(z_{2})J_{k}(z_{3})J_{k}(z_{4}) = -\frac{1}{2}J_{0}(z_{1})J_{0}(z_{2})J_{0}(z_{3})J_{0}(z_{4})$

$$+ \frac{1}{2\pi}\int_{0}^{\pi}J_{0}\left(\sqrt{z_{1}^{2}+z_{2}^{2}-2z_{1}z_{2}\cos x}\right)J_{0}\left(\sqrt{z_{3}^{2}+z_{4}^{2}-2z_{3}z_{4}\cos x}\right)dx.$$

6.8.5. Series containing $J_{k+\nu}(z)$ and $\psi(z)$

1. $\sum_{k=0}^{\infty}\frac{\left(\frac{z}{2}\right)^{k}}{(k+1)!}\psi(k+a)J_{k}(z) = \frac{4}{z^{2}} + \frac{2}{z}\psi(a-1)J_{1}(z)$

$$- \Gamma(a-1)\left(\frac{2}{z}\right)^{a}J_{a-2}(z).$$

2. $\sum_{k=0}^{\infty}(-1)^{k}(4k+1)\psi(2k+1)J_{2k+1/2}(z) = -\sqrt{\frac{z}{2\pi}}\sin z\,\text{Si}(2z)$

$$- \sqrt{\frac{z}{2\pi}}\cos z\,[\text{ci}(2z) - \ln(2z) + \text{C}].$$

3. $\sum_{k=0}^{\infty}(-1)^{k}(4k+3)\psi(2k+2)J_{2k+3/2}(z) = \sqrt{\frac{z}{2\pi}}\cos z\,\text{Si}(2z)$

$$- \sqrt{\frac{z}{2\pi}}\sin z\,[\text{ci}(2z) - \ln(2z) + \text{C}].$$

6.8.6. Series containing $J_{\nu}(\varphi(k,x))$

1. $\sum_{k=1}^{\infty}k^{m-2n-1/2}J_{1/2-m}(kx)$

$$= \frac{(-1)^{m+n+1}}{(2n+1)!\sqrt{\pi}}2^{m-1/2}x^{2n-m+1/2}\sum_{k=0}^{2n+1}B_{k}\binom{2n+1}{k}\left(\frac{k-1}{2}-n\right)_{m}\left(\frac{2\pi}{x}\right)^{k}$$

$$[m \leq 2n;\ 0 < x \leq 2\pi].$$

2. $\sum_{k=1}^{\infty}k^{m-2n-1/2}J_{m+1/2}(kx)$

$$= \frac{(-1)^{n+1}}{(2n+1)!\sqrt{\pi}}2^{m-1/2}x^{2n-m+1/2}\sum_{k=0}^{2n+1}B_{k}\binom{2n+1}{k}\left(\frac{k}{2}-n\right)_{m}\left(\frac{2\pi}{x}\right)^{k}$$

$$[m \leq 2n;\ 0 < x \leq 2\pi].$$

3. $\displaystyle\sum_{k=1}^{\infty} k^{m-2n+1/2} J_{-m-1/2}(kx)$

$$= \frac{(-1)^{n+1}}{(2n)!\sqrt{\pi}} 2^{m-1/2} x^{2n-m-1/2} \sum_{k=0}^{2n} B_k \binom{2n}{k} \left(\frac{k+1}{2} - n\right)_m \left(\frac{2\pi}{x}\right)^k$$

$$[m \leq 2n - 1;\ 0 < x \leq 2\pi].$$

4. $\displaystyle\sum_{k=1}^{\infty} (-1)^k k^{m-2n+1/2} J_{m-1/2}(kx) = \frac{(-1)^{n+1}}{(2n)!} 2^{m+2n-1/2} \pi^{2n-1/2} x^{-m-1/2}$

$$\times \sum_{k=0}^{2n} \frac{B_{2n-k}}{2^k} \binom{2n}{k} \sum_{p=0}^{k} \binom{k}{p} \left(-\frac{p}{2}\right)_m \left(\frac{x}{\pi}\right)^p \quad [m \leq 2n - 1;\ 0 < x \leq \pi].$$

5. $\displaystyle\sum_{k=1}^{\infty} (-1)^k k^{m-2n+1/2} J_{-m-1/2}(kx)$

$$= \frac{(-1)^{m+n+1}}{(2n)!} 2^{m+2n-1/2} \pi^{2n-1/2} x^{-m-1/2}$$

$$\times \sum_{k=0}^{2n} \frac{B_{2n-k}}{2^k} \binom{2n}{k} \sum_{p=0}^{k} \binom{k}{p} \left(\frac{1-p}{2}\right)_m \left(\frac{x}{\pi}\right)^p \quad [m \leq 2n - 1;\ 0 < x \leq \pi].$$

6. $\displaystyle\sum_{k=1}^{\infty} (-1)^k k^{m-2n-1/2} J_{1/2-m}(kx) = \frac{(-1)^{m+n+1}}{(2n+1)!} 2^{m+2n+1/2} \pi^{2n+1/2} x^{-m-1/2}$

$$\times \sum_{k=0}^{2n+1} \frac{B_{2n-k+1}}{2^k} \binom{2n+1}{k} \sum_{p=0}^{k} \binom{k}{p} \left(-\frac{p}{2}\right)_m \left(\frac{x}{\pi}\right)^p \quad [m < 2n;\ 0 < x \leq \pi].$$

7. $\displaystyle\sum_{k=1}^{\infty} (-1)^k k^{m-2n-1/2} J_{m+1/2}(kx) = \frac{(-1)^{n+1}}{(2n+1)!} 2^{m+2n+1/2} \pi^{2n+1/2} x^{-m-1/2}$

$$\times \sum_{k=0}^{2n+1} \frac{B_{2n-k+1}}{2^k} \binom{2n+1}{k} \sum_{p=0}^{k} \binom{k}{p} \left(\frac{1-p}{2}\right)_m \left(\frac{x}{\pi}\right)^p \quad [m \leq 2n;\ 0 < x \leq \pi].$$

8. $\displaystyle\sum_{k=1}^{\infty} k^{m-2n+1/2} J_{m-1/2}(kx)$

$$= \frac{(-1)^{n+1}}{(2n)!\sqrt{\pi}} 2^{m-1/2} x^{2n-m-1/2} \sum_{k=0}^{2n} B_k \binom{2n}{k} \left(\frac{k}{2} - n\right)_m \left(\frac{2\pi}{x}\right)^k$$

$$[m \leq 2n - 1;\ 0 < x \leq 2\pi].$$

9. $\displaystyle\sum_{k=2}^{\infty} (-1)^k \frac{k^{-\nu}}{k^2 - 1} J_\nu(kx) = \frac{\left(\frac{x}{2}\right)^\nu}{2\Gamma(\nu+1)} - \frac{1}{4} J_\nu(x) - \frac{x}{2} J_{\nu+1}(x)$

$$[-\pi \leq x \leq \pi;\ \operatorname{Re}\nu > -2].$$

10. $$\sum_{k=1}^{\infty} \frac{(-1)^k}{(k^2+a^2)^{\nu/2}} J_\nu\left(\sqrt{k^2+a^2}\,x\right) = -\frac{a^{-\nu}}{2} J_\nu(ax)$$
$$[-\pi < x < \pi;\ \operatorname{Re}\nu > -1/2].$$

11. $$\sum_{k=0}^{\infty} (2k+1)^{m-2n-1/2} J_{1/2-m}((2k+1)x)$$
$$= \frac{(-1)^{m+n}}{(2n+1)!} 2^{m-3/2} \pi^{2n+1/2} x^{-m-1/2}$$
$$\times \sum_{k=0}^{2n} B_{2n-k+1}\binom{2n+1}{k}(2-2^{2n-k+2})\left(-\frac{k}{2}\right)_m \left(\frac{x}{\pi}\right)^k \quad [m \le 2n;\ 0 \le x \le \pi].$$

12. $$\sum_{k=0}^{\infty} (2k+1)^{m-2n-1/2} J_{m+1/2}((2k+1)x)$$
$$= \frac{(-1)^n}{(2n+1)!} 2^{m-3/2} \pi^{2n+1/2} x^{-m-1/2}$$
$$\times \sum_{k=0}^{2n} B_{2n-k+1}\binom{2n+1}{k}(2-2^{2n-k+2})\left(\frac{1-k}{2}\right)_m \left(\frac{x}{\pi}\right)^k$$
$$[m \le 2n;\ 0 \le x \le \pi].$$

13. $$\sum_{k=1}^{\infty} \frac{(2k+1)^{-\nu}}{k(k+1)} J_\nu((2k+1)x) = J_\nu(x) + 2x\,J_{\nu+1}(x) - \pi\,\mathbf{H}_\nu(x)$$
$$[-\pi/2 \le x \le \pi/2;\ \operatorname{Re}\nu > -3/2].$$

14. $$\sum_{k=1}^{\infty} (-1)^k \frac{(2k+1)^{1-\nu}}{k(k+1)} J_\nu((2k+1)x) = 2x\,J_{\nu+1}(x) - J_\nu(x)$$
$$[-\pi/2 \le x \le \pi/2;\ \operatorname{Re}\nu > -3/2].$$

15. $$\sum_{k=0}^{\infty} \frac{(2k+1)^{-\nu}}{(2k+1)^2 a^2 \pm b^2} J_\nu((2k+1)x)$$
$$= \frac{\pi a^{\nu-1}}{4b^{\nu+1}}\left[\begin{Bmatrix}\tanh(b\pi/(2a))\\ \tan(b\pi/(2a))\end{Bmatrix} I_\nu\left(\frac{bx}{a}\right) - \begin{Bmatrix}\mathbf{L}_\nu(bx/a)\\ \mathbf{H}_\nu(bx/a)\end{Bmatrix}\right]$$
$$[0 < x < \pi;\ \operatorname{Re}\nu > -3/2].$$

16. $$\sum_{k=0}^{\infty} \frac{(-1)^k}{2k+1}[(2k+1)^2+a^2]^{-\nu/2} J_\nu\left(\sqrt{(2k+1)^2+a^2}\,x\right) = \frac{\pi a^{-\nu}}{4} J_\nu(ax)$$
$$[0 < x < \pi/2;\ \operatorname{Re}\nu > -3/2].$$

17. $\sum_{k=0}^{\infty} \frac{(-1)^k}{\left(k+\frac{1}{2}\right)^{2n-1}} \left[\left(k+\frac{1}{2}\right)^2 \pi^2 + a^2\right]^{-\nu/2} J_\nu\left(\sqrt{\left(k+\frac{1}{2}\right)^2 \pi^2 + a^2}\right)$

$= (-1)^{n-1} \frac{\pi^{2n-1}}{2a^\nu} \sum_{k=0}^{n-1} \frac{(2a)^{-k}}{k!(2n-2k-2)!} E_{2n-2k-2} J_{\nu+k}(a)$ $[\operatorname{Re}\nu > 1/2 - 2n]$.

6.8.7. Series containing $J_\nu(kx)$ and trigonometric functions

1. $\sum_{k=1}^{\infty} \frac{(-1)^k}{k^{\nu+1}} \sin(kx) J_\nu(ky) = -\frac{xy^\nu}{2^{\nu+1}\Gamma(\nu+1)}$

$$[x, y > 0;\ x+y < \pi;\ \operatorname{Re}\nu > -3/2].$$

2. $\sum_{k=1}^{\infty} \frac{(-1)^k}{k^\nu} \cos(kx) J_\nu(ky) = -\frac{\left(\frac{y}{2}\right)^\nu}{2\Gamma(\nu+1)}$

$$[x, y > 0;\ x+y < \pi;\ \operatorname{Re}\nu > -1/2].$$

3. $\sum_{k=1}^{\infty} \frac{\cos(kx)}{k^{2m+\nu-2}} J_\nu(ky)$

$= \frac{\left(\frac{y}{2}\right)^\nu}{\Gamma(\nu+1)} \left\{ \frac{(-1)^m}{2(m-1)!(\nu+1)_{m-1}} \left(\frac{y}{2}\right)^{2m-2} + \sum_{k=0}^{m-2} \frac{\left(\frac{y}{2}\right)^{2k}}{k!(\nu+1)_k} \right.$

$\left. \times \left[\frac{(-1)^{m-1}\pi}{2(2m-2k-3)!} x^{2m-2k-3} + (-1)^k \sum_{j=0}^{m-k-1} \frac{(-x^2)^j}{(2j)!} \zeta(2m-2j-2k-2) \right] \right\}$

$$[m \geq 1;\ 0 < y < \pi;\ y < x < 2\pi - y].$$

4. $\sum_{k=1}^{\infty} \frac{(-1)^k}{k^{2m+\nu-2}} \cos(kx) J_\nu(ky) = \frac{(-1)^m}{2(m-1)!\Gamma(m+\nu)} \left(\frac{y}{2}\right)^{2m+\nu-2}$

$- \sum_{k=0}^{m-2} \frac{(-1)^k}{k!\Gamma(k+\nu+1)} \left(\frac{y}{2}\right)^{2k+\nu}$

$\times \sum_{j=0}^{m-k-1} \frac{(-1)^j}{(2j)!} x^{2j} \left(1 - 2^{2j+2k-2m+3}\right) \zeta(2m-2j-2k-2)$

$$[m \geq 1;\ \operatorname{Re}\nu > 3/2 - 2m;\ -\pi < y < \pi;\ |y| - \pi < x < \pi - |y|].$$

6.8.8. Series containing products of $J_\nu(\varphi(k,x))$

1. $\sum_{k=1}^{\infty} \frac{(-1)^k}{k^{2m+\mu+\nu}} J_\mu(kx) J_\nu(ky) = \frac{(-1)^{m+1}}{2m!\Gamma(m+\mu+1)\Gamma(\nu+1)} \left(\frac{x}{2}\right)^{2m+\mu} \left(\frac{y}{2}\right)^\nu$

$- \sum_{k=0}^{m-1} \frac{(-1)^k}{k!\Gamma(k+\mu+1)} \left(\frac{x}{2}\right)^{2k+\mu}$

$$\times \sum_{j=0}^{m-k} \frac{(-1)^j}{j!\Gamma(j+\nu+1)} \left(\frac{y}{2}\right)^{2j+\nu} \left(1 - 2^{2j+2k-2m+1}\right) \zeta(2m - 2j - 2k)$$

$$[-\pi < x < \pi;\ |x| - \pi < y < \pi - |x|;\ \text{Re}(\mu + \nu) > -2m - 1].$$

2. $\displaystyle\sum_{k=2}^{\infty} (-1)^k \frac{k^{-2\nu}}{k^2 - 1} J_\nu^2(kx) = \frac{\left(\frac{x}{2}\right)^{2\nu}}{2\Gamma^2(\nu+1)} - \frac{1}{4} J_\nu^2(x)$

$$- J_{\nu+1}(x)\left[2(\nu+1) J_{\nu+1}(x) - x J_{\nu+2}(x)\right]$$

$$[-\pi \le x \le \pi;\ \text{Re}\,\nu > -1].$$

3. $\displaystyle\sum_{k=1}^{\infty} \frac{(-1)^k}{(k^2 + a^2)^\nu} J_\nu^2\left(\sqrt{k^2 + a^2}\,x\right) = -\frac{a^{-2\nu}}{2} J_\nu^2(ax)$

$$[-\pi < x < \pi;\ \text{Re}\,\nu > -1/2].$$

4. $\displaystyle\sum_{k=1}^{\infty} \frac{(-1)^k}{k^2(k^2+a^2)^\nu} J_\nu^2\left(\sqrt{k^2+a^2}\,x\right)$

$$= \frac{a^{-2\nu-2}}{12} \left\{ 6 J_{\nu+1}(ax)\left[2(\nu+1) J_{\nu+1}(ax) - ax J_{\nu+2}(ax)\right] - \pi^2 a^2 J_\nu^2(ax) \right\}$$

$$[-\pi < x < \pi;\ \text{Re}\,\nu > -3/2].$$

5. $\displaystyle\sum_{k=0}^{\infty} \frac{(-1)^k}{2k+1}\left[(2k+1)^2 + a^2\right]^{-\nu} J_\nu^2\left(\sqrt{(2k+1)^2 + a^2}\,x\right) = \frac{\pi a^{-2\nu}}{4} J_\nu^2(ax)$

$$[0 < x < \pi/2;\ \text{Re}\,\nu > -1].$$

6.8.9. Series containing products of $J_\nu(kx)$ and trigonometric functions

1. $\displaystyle\sum_{k=1}^{\infty} \frac{(-1)^k}{k^{\mu+\nu+1}} \sin(kx) J_\mu(ky) J_\nu(ky) = -\frac{x\left(\frac{y}{2}\right)^{\mu+\nu}}{2\Gamma(\mu+1)\Gamma(\nu+1)}$

$$[x, y > 0;\ x + y < \pi;\ \text{Re}(\mu+\nu) > -1].$$

2. $\displaystyle\sum_{k=1}^{\infty} \frac{(-1)^k}{k^{\mu+\nu}} \cos(kx) J_\mu(ky) J_\nu(ky) = -\frac{\left(\frac{y}{2}\right)^{\mu+\nu}}{2\Gamma(\mu+1)\Gamma(\nu+1)}$

$$[x, y > 0;\ x + y < \pi;\ \text{Re}(\mu+\nu) > -1].$$

3. $\displaystyle\sum_{k=1}^{\infty} \frac{\cos(kx)}{k^{2m+\mu+\nu-2}} J_\mu(ky) J_\nu(ky) = \frac{\left(\frac{y}{2}\right)^{\mu+\nu}}{\Gamma(\mu+1)\Gamma(\nu+1)}$

$$\times \left\{ \frac{(-1)^m 2^{2m-3}(\mu+\nu+1)_{2m-2}}{(\mu+1)_{m-1}(\nu+1)_{m-1}(\mu+\nu+1)_{m-1}} y^{2m-2} \right.$$

$$+ \sum_{k=0}^{m-2} \frac{(\mu+\nu+1)_{2k}}{k!(\mu+1)_k(\nu+1)_k(\mu+\nu+1)_k} \left(\frac{y}{2}\right)^{2k} \left[(-1)^{m-1}\frac{\pi x^{2m-2k-3}}{2(2m-2k-3)!}\right.$$

$$+ (-1)^k \sum_{j=0}^{m-k-1} (-1)^j \frac{x^{2j}}{(2j)!} \zeta(2m - 2j - 2k - 2) \Bigg] \Bigg\}$$

$$[m \geq 1;\ 0 < y < \pi/2;\ y < x/2 < \pi - y].$$

4. $\displaystyle\sum_{k=1}^{\infty} \frac{(-1)^k}{k^{2m+\mu+\nu-2}} \cos(kx) J_\mu(ky) J_\nu(ky) = \frac{\left(\frac{y}{2}\right)^{\mu+\nu}}{\Gamma(\mu+1)\Gamma(\nu+1)}$

$$\times \Bigg\{ \frac{(-1)^m (\mu + \nu + 1)_{2m-2}}{2(m-1)!(\mu+1)_{m-1}(\nu+1)_{m-1}(\mu+\nu+1)_{m-1}} \left(\frac{y}{2}\right)^{2m-2}$$

$$- \sum_{k=0}^{m-2} \frac{(-1)^k (\mu+\nu+1)_{2k}}{k!(\mu+1)_k(\nu+1)_k(\mu+\nu+1)_k} \left(\frac{y}{2}\right)^{2k}$$

$$\times \sum_{j=0}^{m-k-1} \frac{(-x^2)^j}{(2j)!} \left(1 - 2^{2j+2k-2m+3}\right) \zeta(2m - 2j - 2k - 2) \Bigg\}$$

$$[m \geq 1;\ -\pi/2 < y < \pi/2;\ 2|y| - \pi < x < \pi - 2|y|;\ \operatorname{Re}(\mu+\nu) > 1 - 2m].$$

6.8.10. Series containing $J_\nu(kx)$ and $\operatorname{Si}(kx)$

1. $\displaystyle\sum_{k=1}^{\infty} \frac{(-1)^k}{k^{2m+\mu+\nu-1}} \operatorname{Si}(kx) J_\mu(ky) J_\nu(ky)$

$$= \frac{(-1)^m \Gamma(2m+\mu+\nu-1) x}{2(m-1)! \Gamma(m+\mu) \Gamma(m+\nu) \Gamma(m+\mu+\nu)} \left(\frac{y}{2}\right)^{2m+\mu+\nu-2}$$

$$- \sum_{k=0}^{m-2} \frac{(-1)^k \Gamma(2k+\mu+\nu+1)}{k! \Gamma(2k+\mu+\nu+1) \Gamma(k+\mu+1) \Gamma(k+\nu+1)} \left(\frac{y}{2}\right)^{2k+\mu+\nu}$$

$$\times \sum_{j=0}^{m-k-1} \frac{(-1)^j}{(2j+1)!(2j+1)} x^{2j+1} \left(1 - 2^{2j+2k-2m+3}\right) \zeta(2m-2j-2k-2)$$

$$[m \geq 1;\ -\pi < x < \pi;\ |x| - \pi < 2y < \pi - |x|].$$

6.8.11. Series containing $J_\nu(kx)$, $S(kx)$ and $C(kx)$

1. $\displaystyle\sum_{k=1}^{\infty} \frac{(-1)^k}{k^{\nu+1/2}} C(kx) J_\nu(ky) = -\frac{2^{-\nu-1/2} x^{1/2} y^\nu}{\sqrt{\pi}\, \Gamma(\nu+1)}$

$$[x, y > 0;\ x + y < \pi;\ \operatorname{Re}\nu > -1].$$

2. $\displaystyle\sum_{k=1}^{\infty} \frac{(-1)^k}{k^{\nu+3/2}} S(kx) J_\nu(ky) = -\frac{2^{-\nu-1/2} x^{3/2} y^\nu}{3\sqrt{\pi}\, \Gamma(\nu+1)}$

$$[x, y > 0;\ x + y < \pi;\ \operatorname{Re}\nu > -2].$$

6.8.12. Series containing $J_{k\mu+\nu}(\varphi(k,z))$

Notation: $\Delta(z) = \left|\dfrac{z}{1+\sqrt{1-z^2}} e^{\sqrt{1-z^2}}\right|$.

1. $\displaystyle\sum_{k=1}^{\infty} \frac{1}{k^2+a^2} J_{2k}(kz) = \frac{1}{2a^2}\left[{}_1F_2\!\left(\begin{matrix}1;\\ 1-ia,\, 1+ia\end{matrix}\,\frac{a^2z^2}{4}\right) - 1\right]$ $\quad [\Delta(z/2) < 1]$.

2. $\displaystyle\sum_{k=1}^{\infty} \frac{k^{-(k+\nu)/2}(k+a)^{k-1}}{k!}\left(\frac{z}{2}\right)^k J_{k+\nu}(\sqrt{k}\,z) = -\frac{\left(\frac{z}{2}\right)^\nu}{a\,\Gamma(\nu+1)} + a^{-\nu/2-1} I_\nu(\sqrt{a}\,z)$.

3. $\displaystyle\sum_{k=1}^{\infty} \frac{k^{(k-\nu)/2-1}}{k!}\left(\frac{z}{2}\right)^k J_{k+\nu}(\sqrt{k}\,z) = \frac{\left(\frac{z}{2}\right)^{\nu+2}}{\Gamma(\nu+2)}$.

4. $\displaystyle\sum_{k=1}^{\infty} (-1)^k J_{k/2}^2(kz) = -\frac{1}{2} + \frac{1}{2\sqrt{1-4z^2}} - \frac{\arcsin(2z)}{\pi\sqrt{1-4z^2}}$ $\quad [\Delta(z) < 1]$.

5. $\displaystyle\sum_{k=1}^{\infty} k^2 J_k^2(kz) = \frac{z^2(z^2+4)}{16(1-z^2)^{7/2}}$ $\quad [\Delta(z) < 1]$.

6. $\displaystyle\sum_{k=1}^{\infty} \frac{1}{k^2} J_k^2(kz) = \frac{z^2}{4}$ $\quad [\Delta(z) < 1]$.

7. $\displaystyle\sum_{k=1}^{\infty} \frac{1}{k^4} J_k^2(kz) = \frac{z^2}{4} - \frac{3z^4}{64}$ $\quad [\Delta(z) < 1]$.

8. $\displaystyle\sum_{k=1}^{\infty} \frac{1}{k^6} J_k^2(kz) = \frac{5z^6}{1152} - \frac{15z^4}{256} + \frac{z^2}{4}$ $\quad [\Delta(z) < 1]$.

9. $\displaystyle\sum_{k=0}^{\infty} \frac{1}{(2k+1)^2} J_{k+1/2}^2((2k+1)z) = \frac{2z}{\pi}$ $\quad [\Delta(2z) < 1]$.

10. $\displaystyle\sum_{k=0}^{\infty} \frac{1}{(2k+1)^4} J_{k+1/2}^2((2k+1)z) = \frac{2z}{\pi}\left(1 - \frac{8z^2}{27}\right)$ $\quad [\Delta(2z) < 1]$.

11. $\displaystyle\sum_{k=0}^{\infty} \frac{1}{(2k+1)^6} J_{k+1/2}^2((2k+1)z) = \frac{2z}{\pi}\left(\frac{128z^4}{3375} - \frac{80z^3}{243} + 1\right)$ $\quad [\Delta(2z) < 1]$.

12. $\displaystyle\sum_{k=1}^{\infty} J_{k-1/2}(kz)\, J_{k+1/2}(kz) = \frac{\arcsin z}{\pi z\sqrt{1-z^2}} - \frac{1}{\pi}$ $\quad [\Delta(z) < 1]$.

13. $\displaystyle\sum_{k=1}^{\infty} \frac{1}{k^2 - a^2} J_{k-1/2}(kz) J_{k+1/2}(kz)$

$$= \frac{1}{\pi a^2} - \frac{1}{\pi a^2} {}_2F_3\left(\begin{array}{c} 1, 1; \ -a^2 z^2 \\ 1+a, \ 1-a, \ \frac{3}{2} \end{array}\right) \qquad [\Delta(z) < 1].$$

14. $\displaystyle\sum_{k=1}^{\infty} \frac{(-1)^k}{k^2} J_{k/2}^2(kz) = \frac{z^2}{4} - \frac{2z}{\pi}$ \hfill $[\Delta(z) < 1]$.

15. $\displaystyle\sum_{k=1}^{\infty} \frac{(-1)^k}{k^2 - a^2} J_{k/2}^2(kz)$

$$= \frac{1}{2a^2} - \frac{\pi}{4a} \csc\frac{\pi a}{2} J_{-a/2}(az) J_{a/2}(az) + \frac{2z}{\pi(a^2-1)} {}_2F_3\left(\begin{array}{c} 1, 1; \ -a^2 z^2 \\ \frac{3}{2}, \ \frac{3-a}{2}, \ \frac{3+a}{2} \end{array}\right)$$
$$[\Delta(z) < 1].$$

16. $\displaystyle\sum_{k=0}^{\infty} \frac{(2\nu)_k}{k!} (k+\nu)^{-2\nu-1} J_{k+\nu}^2((k+\nu)z)) = \frac{\left(\frac{z}{2}\right)^{2\nu}}{\nu\Gamma^2(\nu+1)}$ \hfill $[\Delta(2z) < 1]$.

6.8.13. Various series containing $J_\nu(z)$

1. $\displaystyle\sum_{k=1}^{\infty} J_k(w_1) J_k(w_2) J_0(kz)$

$$= -\frac{1}{2} J_0(w_1) J_0(w_2) + \frac{1}{\pi} \int_0^z (z^2 - x^2)^{-1/2} J_0\left(\sqrt{w_1^2 + w_2^2 - 2w_1 w_2 \cos x}\right) dx.$$

2. $\displaystyle\sum_{k=0}^{\infty} \frac{1}{(2k+1)^{\nu-1}} J_{k+1/2}^2(w) J_\nu((2k+1)z)$

$$= \frac{2(2z)^{-\nu}}{\sqrt{\pi}\,\Gamma\left(\nu - \frac{1}{2}\right)} \int_0^z x(z^2 - x^2)^{\nu-3/2} \mathbf{H}_0(2w \sin x) \, dx.$$

3. $\displaystyle\sum_{k=0}^{\infty} \frac{(-1)^k}{(2k+1)^\nu} J_{k+1/2}^2(w) J_\nu((2k+1)z)$

$$= \frac{(2z)^{-\nu}}{\sqrt{\pi}\,\Gamma\left(\nu + \frac{1}{2}\right)} \int_0^z (z^2 - x^2)^{\nu-1/2} \mathbf{H}_0(2w \cos x) \, dx.$$

4. $\displaystyle\sum_{k=1}^{\infty} \frac{(-1)^k}{k^{n\nu}} \prod_{i=1}^{n} J_\nu(kx_i) = -\frac{2^{-n\nu-1}}{\Gamma^n(\nu+1)} \prod_{i=1}^{n} x_i^\nu$

$$\left[x_i > 0; \ \sum_{i=1}^{n} x_i < \pi/2; \ \mathrm{Re}\,\nu > -1/2\right].$$

6.8.14. Series containing $Y_{k+\nu}(z)$

1. $\sum_{k=0}^{\infty} J_{k+1/2}(w) J_{2k+1}(z) Y_{k+1/2}(w) = -\dfrac{1}{\pi} \displaystyle\int_0^1 \dfrac{1}{\sqrt{1-x^2}} \sin(zx) J_0(2wx) \, dx.$

2. $\sum_{k=0}^{\infty} J^2_{k+1/2}(w) J_{k+1/2}(z) Y_{k+1/2}(z) = -\dfrac{1}{\pi} \displaystyle\int_0^1 \dfrac{1}{\sqrt{1-x^2}} J_0(2zx) \mathbf{H}_0(2wx) \, dx.$

3. $\sum_{k=0}^{\infty} J_k(z) J_{k+1/2}(z) J_{k+1}(z) Y_{k+1/2}(z) = \dfrac{1}{4\pi z}[J_0(4z) - 1].$

4. $\sum_{k=1}^{\infty} J_k(x_1) J_k(x_2) Y_k(x_3) Y_k(x_4) = -\dfrac{1}{2} J_0(x_1) J_0(x_2) Y_0(x_3) Y_0(x_4)$
$+ \dfrac{1}{2\pi} \displaystyle\int_0^{\pi} Y_0\!\left(\sqrt{x_1^2 + x_3^2 - 2x_1 x_3 \cos x}\right) Y_0\!\left(\sqrt{x_2^2 + x_4^2 - 2x_2 x_4 \cos x}\right) dx$
$[x_1 < x_3, x_2 < x_4].$

6.9. The Modified Bessel Function $I_\nu(z)$

6.9.1. Series containing $I_{nk+\nu}(z)$

1. $\sum_{k=1}^{\infty} k \, I_k(z) = \dfrac{z}{2}[I_0(z) + I_1(z)].$

2. $\sum_{k=0}^{\infty} \dfrac{(a)_k}{k!} I_{k+\nu}(z) = \dfrac{\left(\frac{z}{2}\right)^\nu}{\Gamma(\nu+1)} e^z \, {}_2F_2\!\left(\begin{array}{c} \nu + \frac{1-a}{2}, \nu - \frac{a}{2} + 1 \\ \nu+1, \, 2\nu - a + 1; \, -2z \end{array}\right).$

3. $\sum_{k=0}^{\infty} \dfrac{k+\nu}{(2k+\nu)^2 - 1} \dfrac{(2\nu)_k}{k!} I_{k+\nu}(z)$
$= \dfrac{2^{-2\nu-3/2} z^{-1/2}}{(2\nu-1)\Gamma(\nu)} e^{(2\nu-3)\,\mathrm{sgn}\,(\mathrm{Im}\,z)\pi i/2 - z} \gamma\!\left(\nu + \dfrac{1}{2}, -2z\right).$

4. $\sum_{k=0}^{\infty} (-1)^k (k+\nu) \dfrac{(a)_k (b)_k (2\nu)_k}{k!(2\nu - a + 1)_k (2\nu - b + 1)_k} I_{k+\nu}(z)$
$= \dfrac{\left(\frac{z}{2}\right)^\nu}{\Gamma(\nu)} e^{-z} \, {}_2F_2\!\left(\begin{array}{c} \nu + \frac{1}{2}, \, 2\nu - a - b + 1; \, 2z \\ 2\nu - a + 1, \, 2\nu - b + 1 \end{array}\right).$

5. $\sum_{k=0}^{\infty} \dfrac{(\mu)_k}{k!} I_{k+\nu}(z) = \dfrac{\left(\frac{z}{2}\right)^\nu e^z}{\Gamma(\nu+1)} \, {}_2F_2\!\left(\begin{array}{c} \nu + \frac{1-\mu}{2}, \nu - \frac{\mu}{2} + 1 \\ \nu + 1, \, 2\nu - \mu + 1; \, -2z \end{array}\right).$

6. $\displaystyle\sum_{k=1}^{\infty}(-1)^k\frac{\Gamma^2\left(\frac{2k+1}{4}\right)}{\Gamma^2\left(\frac{2k+3}{4}\right)}I_{k/2}(z)$

$$= -\frac{1}{4\pi^2}\Gamma^4\left(\frac{1}{4}\right)I_0(z) + \frac{4}{\pi}\int_0^{\pi/2} e^{z\cos(4x)}\,\text{erfc}\left(\sqrt{2z}\cos(2x)\right)\mathbf{K}(\cos x)\,dx.$$

7. $\displaystyle\sum_{k=0}^{\infty}(-1)^k I_{2k+\nu}(z) = \frac{\left(\frac{z}{2}\right)^\nu}{\Gamma(\nu+1)}\,{}_1F_2\!\left(\begin{matrix}\frac{\nu}{2};\ \frac{z^2}{4}\\ \frac{\nu}{2}+1,\ \nu\end{matrix}\right).$

8. $\displaystyle\sum_{k=0}^{\infty}(-1)^k(2k+\nu)\frac{(\nu-\mu+1)_k}{(\mu)_k}I_{2k+\nu}(z) = \frac{\left(\frac{z}{2}\right)^\nu}{\Gamma(\nu)}\,{}_1F_2\!\left(\begin{matrix}\mu-1;\ \frac{z^2}{4}\\ \mu,\ \nu\end{matrix}\right).$

9. $\displaystyle\sum_{k=0}^{\infty}(-1)^k(2k+\nu)\frac{(\nu)_k\left(\frac{2\nu+1}{4}\right)_k\left(\frac{1}{2}\right)_k}{k!\left(\nu+\frac{1}{2}\right)_k\left(\frac{2\nu+3}{4}\right)_k}I_{2k+\nu}(z)$

$$= 2^{\nu-1}z^{1/2}\frac{\Gamma^2\left(\frac{2\nu+3}{4}\right)}{\Gamma(\nu)}I_{(2\nu-1)/4}^2\!\left(\frac{z}{2}\right).$$

10. $\displaystyle\sum_{k=0}^{\infty}(2k+\nu)\frac{(a)_k(\nu)_k}{k!\,(\nu-a+1)_k}I_{2k+\nu}(z) = \frac{\left(\frac{z}{2}\right)^\nu}{\Gamma(\nu)}e^{-z}\,{}_1F_1\!\left(\begin{matrix}\nu-a+\frac{1}{2};\ 2z\\ 2\nu-2a+1\end{matrix}\right).$

11. $\displaystyle\sum_{k=0}^{\infty}\sigma_{k+m}^m\frac{t^k}{(k+m)!}I_{k+\nu}(z)$

$$= (-t)^{-m}z^{(\nu-m)/2}\sum_{k=0}^{m}(-1)^k\frac{(2kt+z)^{(m-\nu)/2}}{k!\,(m-k)!}I_{\nu-m}\!\left(\sqrt{z(2kt+z)}\right).$$

6.9.2. Series containing $I_{k+\nu}(z)$ and $\psi(z)$

1. $\displaystyle\sum_{k=0}^{\infty}\frac{\left(-\frac{z}{2}\right)^k}{k!}\psi(k+a)I_{k+\nu}(z) = \frac{\left(\frac{z}{2}\right)^\nu}{\Gamma(\nu+1)}\psi(a)$

$$-\frac{\left(\frac{z}{2}\right)^{\nu+2}}{\Gamma(\nu+2)a}\,{}_2F_3\!\left(\begin{matrix}1,1;\ \frac{z^2}{4}\\ a+1,\ \nu+2,\ 2\end{matrix}\right).$$

2. $\displaystyle\sum_{k=0}^{\infty}\frac{\left(-\frac{z}{2}\right)^k}{(k+1)!}\psi(k+a)I_k(z) = -\frac{4}{z^2} + \frac{2}{z}\psi(a-1)I_1(z)$

$$+\Gamma(a-1)\left(\frac{2}{z}\right)^a I_{a-2}(z).$$

3. $$\sum_{k=0}^{\infty} \frac{\left(-\frac{z}{2}\right)^k}{k!} \psi(2k+a) I_{k+\nu}(z) = \frac{\left(\frac{z}{2}\right)^\nu}{\Gamma(\nu+1)} \psi(a)$$
$$- \frac{\left(\frac{z}{2}\right)^{\nu+2}}{\Gamma(\nu+2)} \left[\frac{1}{a} {}_2F_3\left(\begin{matrix}1,1;\frac{z^2}{4}\\ \frac{a}{2}+1, \nu+2, 2\end{matrix}\right) + \frac{1}{a+1} {}_2F_3\left(\begin{matrix}1,1;\frac{z^2}{4}\\ \frac{a+3}{2}, \nu+2, 2\end{matrix}\right) \right].$$

6.9.3. Series containing products of $I_{nk+\nu}(z)$

1. $$\sum_{k=2}^{\infty} (-1)^k k^2 (k^2-1)^2 I_k^2(z) = \frac{9z^4}{16} - \frac{5z^6}{32}.$$

2. $$\sum_{k=1}^{\infty} (-1)^k \frac{\Gamma^2\left(\frac{2k+1}{4}\right)}{\Gamma^2\left(\frac{2k+3}{4}\right)} I_k^2(z) = -\frac{1}{4\pi^2} \Gamma^4\left(\frac{1}{4}\right) I_0^2(z)$$
$$+ \frac{4}{\pi} \int_0^{\pi/2} I_0(2z \sin x) \, \mathbf{K}(\cos x) \, dx.$$

3. $$\sum_{k=0}^{\infty} \frac{(a)_k}{(1-a)_k} I_k^2(z) = \frac{1}{2} I_0^2(z) + \frac{1}{2} {}_1F_2\left(\begin{matrix}\frac{1}{2};z^2\\ 1, 1-a\end{matrix}\right).$$

4. $$\sum_{k=0}^{\infty} (-1)^k \frac{(a)_k}{(1-a)_k} I_k^2(z) = \frac{1}{2} I_0^2(z) + \frac{1}{2} {}_1F_2\left(\begin{matrix}\frac{1}{2}-a;z^2\\ 1, 1-a\end{matrix}\right).$$

5. $$\sum_{k=0}^{\infty} (-1)^k (4k+1) \frac{\left(\frac{1}{2}-a\right)_k \left(\frac{1}{2}+a\right)_k \left(\frac{1}{2}\right)_k}{k!(1-a)_k(1+a)_k} I_{k+1/4}^2(z)$$
$$= \frac{2^{7/2} z^{1/2}}{\Gamma^2\left(\frac{1}{4}\right)} {}_2F_3\left(\begin{matrix}\frac{1}{2},\frac{3}{4};z^2\\ 1-a, 1+a, \frac{5}{4}\end{matrix}\right).$$

6. $$\sum_{k=0}^{\infty} (-1)^k (2k+1) I_{k+1/2}^2(z) = \frac{2z}{\pi}.$$

7. $$\sum_{k=0}^{\infty} (-1)^k \frac{(a)_k}{(2-a)_k} I_{k+1/2}^2(z) = \frac{2z}{\pi} {}_2F_3\left(\begin{matrix}\frac{3}{2}-a, 1;z^2\\ 2-a, \frac{3}{2}, \frac{3}{2}\end{matrix}\right).$$

8. $$\sum_{k=0}^{\infty} (2k+1) \frac{(a)_k}{(2-a)_k} I_{k+1/2}^2(z) = \frac{\Gamma(2-a)}{\sqrt{\pi}} z^{a-1/2} \mathbf{L}_{1/2-a}(2z).$$

9. $$\sum_{k=0}^{\infty} \frac{1}{(2k-1)(2k+3)} I_k(z) I_{k+1}(z) = \frac{\pi}{8z} [I_0(2z) \mathbf{L}_1(2z) - I_1(2z) \mathbf{L}_0(2z)].$$

10. $\displaystyle\sum_{k=0}^{\infty}(-1)^k \frac{(\nu)_k}{(2-\nu)_k} I_k(z) I_{k+1}(z) = \frac{z}{2} {}_1F_2\left(\begin{array}{c}\frac{3}{2}-\nu;\ z^2\\ 2-\nu,\ 2\end{array}\right).$

11. $\displaystyle\sum_{k=0}^{\infty}(-1)^k(2k+3)\, I_{k+3/2}^2(z) = \frac{1}{\pi z}\left[\cosh(2z) - 2z^2 - 1\right].$

12. $\displaystyle\sum_{k=1}^{\infty}(-1)^k I_{k+\nu}^2(z) = -\frac{\left(\frac{z}{2}\right)^{2\nu+2}}{\Gamma^2(\nu+2)}\, {}_2F_3\left(\begin{array}{c}\nu+1,\ \nu+\frac{3}{2};\ z^2\\ \nu+2,\ \nu+2,\ 2\nu+2\end{array}\right).$

13. $\displaystyle\sum_{k=0}^{\infty}\frac{k+\nu}{4(k+\nu)^2-1}\,\frac{(2\nu)_k}{k!}\, I_{k+\nu}^2(z)$

$$= \frac{\left(\frac{z}{2}\right)^{2\nu}}{(4\nu^2-1)\Gamma(\nu)\Gamma(\nu+1)}\, {}_1F_2\left(\begin{array}{c}\nu+\frac{1}{2};\ z^2\\ \nu+1,\ \nu+\frac{3}{2}\end{array}\right).$$

14. $\displaystyle\sum_{k=0}^{\infty}(-1)^k\frac{k+\nu}{4(k+\nu)^2-1}\,\frac{(2\nu)_k}{k!}\, I_{k+\nu}^2(z)$

$$= \frac{\left(\frac{z}{2}\right)^{2\nu}}{(4\nu^2-1)\Gamma(\nu)\Gamma(\nu+1)}\, {}_1F_2\left(\begin{array}{c}1;\ z^2\\ \nu+1,\ \nu+\frac{3}{2}\end{array}\right).$$

15. $\displaystyle\sum_{k=0}^{\infty}(-1)^k \frac{(2\nu)_k(2\nu-a+1)_k}{k!\,(a)_k}\, I_{k+\nu}^2(z) = \frac{\left(\frac{z}{2}\right)^{2\nu}}{\Gamma^2(\nu+1)}\, {}_2F_3\left(\begin{array}{c}\nu+\frac{1}{2},\ a-\nu;\ z^2\\ a,\ \nu+1,\ \nu+1\end{array}\right).$

16. $\displaystyle\sum_{k=0}^{\infty}(k+\nu)\frac{(2\nu)_k(2\nu-a+1)_k}{k!\,(a)_k}\, I_{k+\nu}^2(z) = \frac{\left(\frac{z}{2}\right)^{2\nu}}{\nu\Gamma^2(\nu)}\, {}_1F_2\left(\begin{array}{c}\nu+\frac{1}{2};\ z^2\\ a,\ \nu+1\end{array}\right).$

17. $\displaystyle\sum_{k=0}^{\infty}(-1)^k(k+\nu)\frac{(2\nu)_k(a)_k(b)_k}{k!\,(2\nu-a+1)_k(2\nu-b+1)_k}\, I_{k+\nu}^2(z)$

$$= \frac{\left(\frac{z}{2}\right)^{2\nu}}{\nu\Gamma^2(\nu)}\, {}_2F_3\left(\begin{array}{c}\nu+\frac{1}{2},\ 2\nu-a-b+1;\ z^2\\ \nu+1,\ 2\nu-a+1,\ 2\nu-b+1\end{array}\right).$$

18. $\displaystyle\sum_{k=0}^{\infty}(-1)^k \frac{(a)_k}{k!}\, I_{k+\mu}(z) I_{\nu-k}(z)$

$$= \frac{\Gamma(\mu+\nu-a+1)}{\Gamma(\nu+1)\Gamma(\mu+\nu+1)\Gamma(\mu-a+1)}\left(\frac{z}{2}\right)^{\mu+\nu}$$

$$\times {}_2F_3\left(\begin{array}{c}\frac{\mu+\nu-a+1}{2},\ \frac{\mu+\nu-a}{2}+1;\ z^2\\ \nu+1,\ \mu+\nu+1,\ \mu-a+1\end{array}\right).$$

19. $\sum_{k=0}^{\infty}(k+\nu)\frac{(2\nu)_k}{k!}I_{k+\nu}(w)I_{k+\nu}(z) = \frac{2^{-\nu}}{\Gamma(\nu)}\left(\frac{wz}{w+z}\right)^{\nu}I_{\nu}(w+z).$

20. $\sum_{k=0}^{\infty}(-1)^k\frac{\left(\frac{1}{2}-\nu\right)_k}{\left(\frac{3}{2}+\nu\right)_k}I_{\nu+k}(z)I_{\nu-k-1}(z) = \frac{\Gamma\left(\nu+\frac{3}{2}\right)}{2\Gamma(\nu+1)}z^{-1/2}I_{2\nu-1/2}(2z).$

21. $\sum_{k=0}^{\infty}I_{k+\nu}(z)I_{\nu-k}(z)I_{2k}(2z)$

$= \frac{1}{2}I_{\nu}^2(z)I_0(2z) + \frac{\left(\frac{z}{2}\right)^{2\nu}}{2\Gamma^2(\nu+1)}\,{}_3F_4\left(\begin{array}{c}\nu+\frac{1}{4},\nu+\frac{1}{2},\nu+\frac{3}{4};\,4z^2\\ \nu+1,2\nu+\frac{1}{2},2\nu+1,\frac{1}{2}\end{array}\right).$

22. $\sum_{k=0}^{\infty}I_{k+\nu}(z)I_{\nu-k-1}(z)I_{2k+1}(2z)$

$= \frac{2\left(\frac{z}{2}\right)^{2\nu}}{\Gamma(\nu)\Gamma(\nu+1)}\,{}_3F_4\left(\begin{array}{c}\nu+\frac{1}{4},\nu+\frac{1}{2},\nu+\frac{3}{4};\,4z^2\\ \nu+1,2\nu,2\nu+\frac{1}{2},\frac{3}{2}\end{array}\right).$

23. $\sum_{k=0}^{\infty}I_k^2(z)I_{k+1}^2(z) = \frac{z^2}{4}\,{}_2F_3\left(\begin{array}{c}\frac{3}{2},\frac{3}{2};\,4z^2\\ 2,2,3\end{array}\right).$

24. $\sum_{k=0}^{\infty}I_k^2(w)I_{nk}^2(z) = \frac{1}{2}I_0^2(w)I_0^2(z) + \frac{1}{\pi}\int_0^{\pi/2}I_0(2z\sin x)I_0(2w\sin(nx))\,dx.$

25. $\sum_{k=1}^{\infty}(-1)^k J_0(kz)I_k(w_1)I_k(w_2)$

$= -\frac{1}{2}I_0(w_1)I_0(w_2) + \frac{1}{\pi}\int_0^z (z^2-x^2)^{-1/2}I_0\left(\sqrt{w_1^2+w_2^2-2w_1w_2\cos x}\right)dx.$

26. $\sum_{k=0}^{\infty}(-1)^k J_{nk}^2(w)I_k^2(z) = \frac{1}{2}J_0^2(w)I_0^2(z)$

$+ \frac{1}{\pi}\int_0^{\pi/2} J_0(2w\sin x)I_0(2z\sin(nx))\,dx.$

27. $\sum_{k=1}^{\infty} J_k(x_1) Y_k(x_2) I_k(x_3) K_k(x_4) = -\frac{1}{2} J_0(x_1) Y_0(x_2) I_0(x_3) K_0(x_4)$

$+ \frac{1}{2\pi} \int_0^{\pi} Y_0\left(\sqrt{x_1^2 + x_2^2 - 2x_1 x_2 \cos x}\right) K_0\left(\sqrt{x_3^2 + x_4^2 - 2x_3 x_4 \cos x}\right) dx$

$[x_1 < x_3;\ x_2 < x_4].$

6.9.4. Series containing $I_{nk+\mu}((nk+\nu)z)$

Notation: $\Delta(z) = \left| \dfrac{z}{1+\sqrt{1+z^2}} e^{\sqrt{1+z^2}} \right|.$

1. $\sum_{k=1}^{\infty} \dfrac{(-1)^k}{k^2} I_{2k}(kz) = -\dfrac{z^2}{8}$ \hspace{2em} $[\Delta(z/2) < 1].$

2. $\sum_{k=0}^{\infty} \dfrac{(-1)^k}{(2k+1)^6} I_{2k+1}((2k+1)z) = \dfrac{z^5}{450} + \dfrac{5z^3}{81} + \dfrac{z}{2}$ \hspace{2em} $[\Delta(z) < 1].$

3. $\sum_{k=0}^{\infty} \dfrac{(-1)^k}{k^2+a^2} I_{2k}(kz) = \dfrac{2}{a^2 z^2}\left[1 + \dfrac{z^2}{4} - {}_1F_2\!\left(\begin{matrix}1;\ -\dfrac{a^2 z^2}{4} \\ -ia,\ ia\end{matrix}\right)\right]$ \hspace{1em} $[\Delta(z/2) < 1].$

6.9.5. Series containing products of $I_{nk+\nu}((nk+\nu)z)$

Notation: $\Delta(z) = \left| \dfrac{z}{1+\sqrt{1+z^2}} e^{\sqrt{1+z^2}} \right|.$

1. $\sum_{k=1}^{\infty} (-1)^k k^2 I_k^2(kz) = \dfrac{z^2(z^2-4)}{16(z^2+1)^{7/2}}$ \hspace{2em} $[\Delta(z) < 1].$

2. $\sum_{k=1}^{\infty} \dfrac{(-1)^k}{k^2} I_k^2(kz) = -\dfrac{z^2}{4}$ \hspace{2em} $[\Delta(z) < 1].$

3. $\sum_{k=1}^{\infty} \dfrac{(-1)^k}{k^4} I_k^2(kz) = -\dfrac{3z^4}{64} - \dfrac{z^2}{4}$ \hspace{2em} $[\Delta(z) < 1].$

4. $\sum_{k=1}^{\infty} \dfrac{(-1)^k}{k^6} I_k^2(kz) = -\dfrac{5z^6}{1152} - \dfrac{15z^4}{256} - \dfrac{z^2}{4}$ \hspace{2em} $[\Delta(z) < 1].$

5. $\sum_{k=0}^{\infty} \dfrac{(-1)^k}{(2k+1)^2} I_{k+1/2}^2((2k+1)z) = \dfrac{2z}{\pi}$ \hspace{2em} $[\Delta(z) < 1].$

6. $\sum_{k=0}^{\infty} \dfrac{(-1)^k}{(2k+1)^4} I_{k+1/2}^2((2k+1)z) = \dfrac{2z}{\pi}\left(\dfrac{8z^2}{27} + 1\right)$ \hspace{2em} $[\Delta(z) < 1].$

7. $\sum_{k=0}^{\infty} \frac{(-1)^k}{(2k+1)^6} I_{k+1/2}^2((2k+1)z) = \frac{2z}{\pi}\left(\frac{128z^4}{3375} + \frac{80z^3}{243} + 1\right)$ $[\Delta(z) < 1]$.

8. $\sum_{k=1}^{\infty} (-1)^k I_{k-1/2}(kz) I_{k+1/2}(kz) = \frac{1}{\pi z\sqrt{z^2+1}} \ln\left(z + \sqrt{z^2+1}\right) - \frac{1}{\pi}$

$[\Delta(z) < 1]$.

9. $\sum_{k=1}^{\infty} \frac{(-1)^k}{k^2 - a^2} I_{k-1/2}(kz) I_{k+1/2}(kz) = \frac{1}{\pi a^2} - \frac{1}{\pi a^2} {}_2F_3\left(\begin{array}{c} 1, 1; \, a^2z^2 \\ 1+a, 1-a, \frac{3}{2} \end{array}\right)$

$[\Delta(z) < 1]$.

6.10. The Struve Functions $\mathbf{H}_\nu(z)$ and $\mathbf{L}_\nu(z)$

6.10.1. Series containing $\mathbf{H}_{k+\nu}(z)$ and $\mathbf{L}_{k+\nu}(z)$

1. $\sum_{k=0}^{\infty} \frac{\left(\frac{z}{2}\right)^k}{k!} \mathbf{H}_k(z) = \frac{2}{\pi} \operatorname{shi}(z)$.

2. $\sum_{k=0}^{\infty} \frac{\left(-\frac{z}{2}\right)^k}{k!} \mathbf{L}_k(z) = \frac{2}{\pi} \operatorname{Si}(z)$.

3. $\sum_{k=0}^{\infty} \frac{\left(\pm\frac{z}{2}\right)^k}{\left(\frac{3}{2}\right)_k} \left\{\begin{array}{c} \mathbf{H}_{k+\nu}(z) \\ \mathbf{L}_{k+\nu}(z) \end{array}\right\} = \frac{z^{\nu+1}}{2^\nu\sqrt{\pi}\,\Gamma\left(\nu+\frac{3}{2}\right)} {}_2F_5\left(\begin{array}{c} \frac{1}{2}, 1; \, \frac{z^4}{256} \\ \frac{2\nu+3}{4}, \frac{2\nu+5}{4}, \frac{3}{4}, \frac{5}{4}, \frac{3}{2} \end{array}\right)$.

4. $\sum_{k=0}^{\infty} \frac{\left(\frac{z}{2}\right)^k}{k!} \mathbf{H}_{k+1/2}(z)$

$= \frac{z^{3/2}}{\sqrt{2\pi}} \left\{ 2I_0(z) - \frac{2}{z} I_1(z) + \pi\left[I_0(z)\mathbf{L}_1(z) - I_1(z)\mathbf{L}_0(z)\right]\right\}$.

5. $\sum_{k=0}^{\infty} \frac{\left(-\frac{z}{2}\right)^k}{k!} \mathbf{L}_{k+1/2}(z)$

$= \frac{z^{3/2}}{\sqrt{2\pi}} \left\{ 2J_0(z) - \frac{2}{z} J_1(z) + \pi\left[J_1(z)\mathbf{H}_0(z) - J_0(z)\mathbf{H}_1(z)\right]\right\}$.

6. $\sum_{k=0}^{\infty} \frac{\left(\frac{z}{2}\right)^k}{(k+1)!} \mathbf{H}_{k+1/2}(z) = -\sqrt{\frac{8}{\pi z^3}} \sin z$

$+ \sqrt{\frac{8}{\pi z}} \left\{I_0(z) - \frac{\pi}{2}\left[I_1(z)\mathbf{L}_0(z) - I_0(z)\mathbf{L}_1(z)\right]\right\}$.

7. $\sum_{k=0}^{\infty} \dfrac{\left(-\dfrac{z}{2}\right)^k}{(k+1)!} \mathbf{L}_{k+1/2}(z) = \sqrt{\dfrac{8}{\pi z^3}} \sinh z$

$\qquad - \sqrt{\dfrac{8}{\pi z}} \left\{ J_0(z) + \dfrac{\pi}{2} [J_1(z) \mathbf{H}_0(z) - J_0(z) \mathbf{H}_1(z)] \right\}.$

6.10.2. Series containing $\mathbf{H}_\nu(\varphi(k)x)$

1. $\sum_{k=1}^{\infty} (-1)^k \dfrac{k^{-\nu-1}}{k^2 a^2 + b^2} \mathbf{H}_\nu(kx) = \dfrac{\pi a^\nu}{2 b^{\nu+2}} \operatorname{csch}\left(\dfrac{b\pi}{a}\right) \mathbf{L}_\nu\left(\dfrac{bx}{a}\right)$

$\qquad - \dfrac{\left(\dfrac{x}{2}\right)^{\nu+1}}{\sqrt{\pi} b^2 \Gamma\left(\nu + \dfrac{3}{2}\right)} \qquad [-\pi \le x \le \pi; \ \operatorname{Re} \nu > -7/2].$

2. $\sum_{k=1}^{\infty} (-1)^k \dfrac{k^{-\nu-1}}{k^2 a^2 - b^2} \mathbf{H}_\nu(kx) = \dfrac{\left(\dfrac{x}{2}\right)^{\nu+1}}{\sqrt{\pi} b^2 \Gamma\left(\nu + \dfrac{3}{2}\right)}$

$\qquad - \dfrac{\pi a^\nu}{2 b^{\nu+2}} \csc\left(\dfrac{b\pi}{a}\right) \mathbf{H}_\nu\left(\dfrac{bx}{a}\right) \qquad [-\pi \le x \le \pi; \ \operatorname{Re} \nu > -7/2].$

3. $\sum_{k=2}^{\infty} (-1)^k \dfrac{k^{-\nu-1}}{k^2 - 1} \mathbf{H}_\nu(kx) = \dfrac{(x/2)^{\nu+1}}{\sqrt{\pi} \Gamma\left(\nu + \dfrac{3}{2}\right)} - \left(\nu + \dfrac{3}{4}\right) \mathbf{H}_\nu(x) + \dfrac{x}{2} \mathbf{H}_{\nu-1}(x)$

$\qquad [-\pi \le x \le \pi; \ \operatorname{Re} \nu > -7/2].$

4. $\sum_{k=0}^{\infty} (-1)^k \dfrac{(2k+1)^{-\nu}}{(2k-1)(2k+3)} \mathbf{H}_\nu((2k+1)x) = -\dfrac{\pi}{2^{\nu+3}} \mathbf{H}_\nu(2x)$

$\qquad [-\pi/2 \le x \le \pi/2; \ \operatorname{Re} \nu > -5/2].$

5. $\sum_{k=0}^{\infty} \dfrac{(-1)^k}{(2k+1)^{\nu+2}} \mathbf{H}_\nu((2k+1)x) = \dfrac{2^{-\nu-2} \sqrt{\pi}}{\Gamma\left(\nu + \dfrac{3}{2}\right)} x^{\nu+1}$

$\qquad [-\pi/2 < x < \pi/2; \ \operatorname{Re} \nu > -7/2].$

6. $\sum_{k=1}^{\infty} (-1)^k \dfrac{(2k+1)^{-\nu}}{k(k+1)} \mathbf{H}_\nu((2k+1)x) = (4\nu+1) \mathbf{H}_\nu(x) - 2z \mathbf{H}_{\nu-1}(x)$

$\qquad [-\pi/2 \le x \le \pi/2; \ \operatorname{Re} \nu > -5/2].$

7. $\sum_{k=1}^{\infty} \dfrac{(-1)^k}{(k^2 + a^2)^{(\nu+1)/2}} \mathbf{H}_\nu\left(\sqrt{k^2 + a^2}\, x\right) = -\dfrac{a^{-\nu-1}}{2} \mathbf{H}_\nu(ax)$

$\qquad [-\pi \le x \le \pi; \ \operatorname{Re} \nu > -3/2].$

8. $\displaystyle\sum_{k=1}^{\infty} \frac{(-1)^k}{k^2(k^2+a^2)^{(\nu+1)/2}} \mathbf{H}_\nu\left(\sqrt{k^2+a^2}\,x\right)$

$\displaystyle = \frac{a^{-\nu-3}}{12}\{[3(2\nu+1) - \pi^2 a^2]\mathbf{H}_\nu(ax) - 3ax\,\mathbf{H}_{\nu-1}(ax)\}$

$[-\pi \le x \le \pi;\ \mathrm{Re}\,\nu > -7/2].$

6.10.3. Series containing $\mathbf{H}_\nu(kx)$ and trigonometric functions

1. $\displaystyle\sum_{k=1}^{\infty} \frac{(-1)^k}{k^{\nu+2}} \sin(kx)\,\mathbf{H}_\nu(ky) = -\frac{x}{\sqrt{\pi}\,\Gamma\!\left(\nu+\frac{3}{2}\right)} \left(\frac{y}{2}\right)^{\nu+1}$

$[x,y > 0;\ x+y < \pi;\ \mathrm{Re}\,\nu > -3/2].$

2. $\displaystyle\sum_{k=1}^{\infty} \frac{(-1)^k}{k^{2m+\nu}} \sin(kx)\,\mathbf{H}_\nu(ky) = \frac{(-1)^m x\left(\frac{y}{2}\right)^{2m+\nu-1}}{2\Gamma\!\left(m+\frac{1}{2}\right)\Gamma\!\left(m+\nu+\frac{1}{2}\right)}$

$\displaystyle - \sum_{k=0}^{m-2} \frac{(-1)^k\left(\frac{y}{2}\right)^{2k+\nu+1}}{\Gamma\!\left(k+\frac{3}{2}\right)\Gamma\!\left(k+\nu+\frac{3}{2}\right)} \sum_{j=0}^{m-k-1} \frac{(-1)^j}{(2j+1)!} x^{2j+1}$

$\times \left(1 - 2^{2j+2k-2m+3}\right)\zeta(2m-2j-2k-2)$

$[-\pi < x < \pi;\ |x| - \pi < y < \pi - |x|;\ \mathrm{Re}\,\nu > -2m - 1/2].$

3. $\displaystyle\sum_{k=1}^{\infty} \frac{(-1)^k}{k^{\nu+1}} \cos(kx)\,\mathbf{H}_\nu(ky) = -\frac{\left(\frac{y}{2}\right)^{\nu+1}}{\sqrt{\pi}\,\Gamma\!\left(\nu+\frac{3}{2}\right)}$

$[x,y > 0;\ x+y < \pi;\ \mathrm{Re}\,\nu > -3/2].$

4. $\displaystyle\sum_{k=1}^{\infty} \frac{\cos(kx)}{k^{2m+\nu-1}} \mathbf{H}_\nu(ky)$

$\displaystyle = \frac{\left(\frac{y}{2}\right)^{\nu+1}}{\sqrt{\pi}\,\Gamma\!\left(\nu+\frac{3}{2}\right)} \Bigg\{(-1)^m \frac{\left(\frac{y}{2}\right)^{2m-2}}{\left(\frac{3}{2}\right)_{m-1}\left(\nu+\frac{3}{2}\right)_{m-1}} + \sum_{k=0}^{m-2} \frac{\left(\frac{y}{2}\right)^{2k}}{\left(\frac{3}{2}\right)_k\left(\nu+\frac{3}{2}\right)_k}$

$\displaystyle \times \left[(-1)^{m-1}\frac{\pi x^{2m-2k-3}}{(2m-2k-3)!} + 2(-1)^k \sum_{j=0}^{m-k-1} \frac{(-x^2)^j}{(2j)!}\zeta(2m-2j-2k-2)\right]\Bigg\}$

$[0 < y < \pi;\ y < x < 2\pi - x;\ \mathrm{Re}\,\nu > 3/2 - 2m].$

5. $\displaystyle\sum_{k=1}^{\infty} (-1)^k \frac{\cos(kx)}{k^{2m+\nu-1}} \mathbf{H}_\nu(ky) = \frac{\left(\frac{y}{2}\right)^{\nu+1}}{\sqrt{\pi}\,\Gamma\!\left(\nu+\frac{3}{2}\right)}$

$\displaystyle \times \left\{\frac{(-1)^m\left(\frac{y}{2}\right)^{2m+\nu-1}}{2\Gamma\!\left(m+\frac{1}{2}\right)\Gamma\!\left(m+\nu+\frac{1}{2}\right)} - \sqrt{\pi} \sum_{k=0}^{m-2} \frac{(-1)^k\left(\frac{y}{2}\right)^{2k+\nu+1}}{\Gamma\!\left(k+\frac{3}{2}\right)\Gamma\!\left(k+\nu+\frac{3}{2}\right)}\right.$

$$\times \sum_{j=0}^{m-k-1} \frac{(-x^2)^j}{(2j)!} \left(1 - 2^{2j+2k-2m+3}\right) \zeta(2m - 2j - 2k - 2) \Bigg\}$$

$$[-\pi < y < \pi;\; |y| - \pi < x < \pi - |y|;\; \operatorname{Re}\nu > 3/2 - 2m].$$

6.10.4. Series containing $\mathbf{H}_\nu(kx)$ and $\operatorname{Si}(kx)$

1. $\displaystyle\sum_{k=1}^{\infty} \frac{(-1)^k}{k^{2m+\nu}} \operatorname{Si}(kx)\, \mathbf{H}_\nu(ky)$

$$= \frac{(-1)^m x \left(\frac{y}{2}\right)^{2m+\nu-1}}{2\Gamma\left(m + \frac{1}{2}\right)\Gamma\left(m + \nu + \frac{1}{2}\right)} - \sum_{k=0}^{m-2} \frac{(-1)^k \left(\frac{y}{2}\right)^{2k+\nu+1}}{\Gamma\left(k + \frac{3}{2}\right)\Gamma\left(k + \nu + \frac{3}{2}\right)}$$

$$\times \sum_{j=0}^{m-k-1} (-1)^j \frac{x^{2j+1}}{(2j+1)!\,(2j+1)} \left(1 - 2^{2j+2k-2m+3}\right) \zeta(2m - 2j - 2k - 2)$$

$$[m \geq 1;\; -\pi < y < \pi;\; |y| - \pi < x < \pi - |y|].$$

6.10.5. Series containing $\mathbf{H}_\nu(kx)$, $S(kx)$ and $C(kx)$

1. $\displaystyle\sum_{k=1}^{\infty} \frac{(-1)^k}{k^{\nu+5/2}} S(kx)\, \mathbf{H}_\nu(ky) = -\frac{2^{-\nu-1/2} x^{3/2} y^{\nu+1}}{3\pi \Gamma\left(\nu + \frac{3}{2}\right)}$

$$[x, y > 0;\; x + y < \pi;\; \operatorname{Re}\nu > -3].$$

2. $\displaystyle\sum_{k=1}^{\infty} \frac{(-1)^k}{k^{2m+\nu+1/2}} S(kx)\, \mathbf{H}_\nu(ky)$

$$= \frac{(-1)^m x^{3/2} \left(\frac{y}{2}\right)^{2m+\nu-1}}{3\sqrt{2\pi}\, \Gamma\left(m + \frac{1}{2}\right)\Gamma\left(m + \nu + \frac{1}{2}\right)} - \sqrt{\frac{2}{\pi}} \sum_{k=0}^{m-2} \frac{(-1)^k \left(\frac{y}{2}\right)^{2k+\nu+1}}{\Gamma\left(k + \frac{3}{2}\right)\Gamma\left(k + \nu + \frac{3}{2}\right)}$$

$$\times \sum_{j=0}^{m-k-1} (-1)^j \frac{x^{2j+3/2}}{(2j+1)!\,(4j+3)} \left(1 - 2^{2j+2k-2m+3}\right) \zeta(2m - 2j - 2k - 2)$$

$$[m \geq 1;\; -\pi < y < \pi;\; |y| - \pi < x < \pi - |y|].$$

3. $\displaystyle\sum_{k=1}^{\infty} \frac{(-1)^k}{k^{\nu+3/2}} C(kx)\, \mathbf{H}_\nu(ky) = -\frac{2^{-\nu-1/2} x^{1/2} y^{\nu+1}}{\pi \Gamma\left(\nu + \frac{3}{2}\right)}$

$$[x, y > 0;\; x + y < \pi;\; \operatorname{Re}\nu > -2].$$

4. $\displaystyle\sum_{k=1}^{\infty} \frac{(-1)^k}{k^{2m+\nu-1/2}} C(kx)\, \mathbf{H}_\nu(ky)$

$$= \frac{(-1)^m x^{1/2} \left(\frac{y}{2}\right)^{2m+\nu-1}}{\sqrt{2\pi}\, \Gamma\left(m + \frac{1}{2}\right)\Gamma\left(m + \nu + \frac{1}{2}\right)} - \sqrt{\frac{2}{\pi}} \sum_{k=0}^{m-2} \frac{(-1)^k \left(\frac{y}{2}\right)^{2k+\nu+1}}{\Gamma\left(k + \frac{3}{2}\right)\Gamma\left(k + \nu + \frac{3}{2}\right)}$$

$$\times \sum_{j=0}^{m-k-1} (-1)^j \frac{x^{2j+1/2}}{(2j)!\,(4j+1)} \left(1 - 2^{2j+2k-2m+3}\right) \zeta(2m - 2j - 2k - 2)$$

$$[m \geq 1;\ -\pi < y < \pi;\ |y| - \pi < x < \pi - |y|].$$

6.10.6. Series containing $\mathbf{H}_\nu(\varphi(k)x)$ and $J_\mu(kx)$

1. $\displaystyle\sum_{k=1}^{\infty} \frac{(-1)^k}{k^{\mu+\nu+1}} J_\mu(kx)\, \mathbf{H}_\nu(ky) = -\frac{\left(\frac{x}{2}\right)^\mu \left(\frac{y}{2}\right)^{\nu+1}}{\sqrt{\pi}\,\Gamma(\mu+1)\Gamma\!\left(\nu+\frac{3}{2}\right)}$

$$[x, y > 0;\ x + y < \pi;\ \mathrm{Re}\,(\mu+\nu) > -2].$$

2. $\displaystyle\sum_{k=0}^{\infty} \frac{1}{(2k+1)^\nu} J_k(w)\, J_{k+1}(w)\, \mathbf{H}_\nu((2k+1)z)$

$$= \frac{(2z)^{-\nu}}{\sqrt{\pi}\,\Gamma\!\left(\nu+\frac{1}{2}\right)} \int_0^z (z^2 - x^2)^{\nu-1/2}\, J_1(2w\sin x)\, dx.$$

3. $\displaystyle\sum_{k=0}^{\infty} \frac{1}{(2k+1)^\nu} J_{k+1/2}^2(w)\, \mathbf{H}_\nu((2k+1)z)$

$$= \frac{(2z)^{-\nu}}{\sqrt{\pi}\,\Gamma\!\left(\nu+\frac{1}{2}\right)} \int_0^z (z^2 - x^2)^{\nu-1/2}\, \mathbf{H}_0(2w\sin x)\, dx.$$

4. $\displaystyle\sum_{k=1}^{\infty} \frac{(-1)^k}{k^{2m+\mu+\nu-1}} J_\mu(kx)\, \mathbf{H}_\nu(ky)$

$$= \frac{(-1)^m \left(\frac{x}{2}\right)^\mu \left(\frac{y}{2}\right)^{2m+\nu-1}}{\Gamma\!\left(m+\frac{1}{2}\right)\Gamma(\mu+1)\Gamma\!\left(m+\nu+\frac{1}{2}\right)} - \sum_{k=0}^{m-2} \frac{(-1)^k \left(\frac{y}{2}\right)^{2k+\nu+1}}{\Gamma\!\left(k+\frac{3}{2}\right)\Gamma\!\left(k+\nu+\frac{3}{2}\right)}$$

$$\times \sum_{j=0}^{m-k-1} (-1)^j \frac{\left(\frac{x}{2}\right)^{2j+\mu}}{j!\,\Gamma(j+\mu+1)} \left(1 - 2^{2j+2k-2m+3}\right) \zeta(2m - 2j - 2k - 2)$$

$$[m \geq 1;\ -\pi < y < \pi;\ |y| - \pi < x < \pi - |y|].$$

6.10.7. Series containing product of $\mathbf{H}_\nu(kx)$

1. $\displaystyle\sum_{k=1}^{\infty} \frac{(-1)^k}{k^{2m+\mu+\nu}} \mathbf{H}_\mu(kx)\, \mathbf{H}_\nu(ky)$

$$= \frac{(-1)^m \pi^{-1/2} \left(\frac{x}{2}\right)^{2m+\mu-1} \left(\frac{y}{2}\right)^{\nu+1}}{\Gamma\!\left(m+\frac{1}{2}\right)\Gamma\!\left(\nu+\frac{3}{2}\right)\Gamma\!\left(m+\mu+\frac{1}{2}\right)} - \sum_{k=0}^{m-2} \frac{(-1)^k \left(\frac{x}{2}\right)^{2k+\mu+1}}{\Gamma\!\left(k+\frac{3}{2}\right)\Gamma\!\left(k+\mu+\frac{3}{2}\right)}$$

$$\times \sum_{j=0}^{m-k-1} (-1)^j \frac{\left(\frac{y}{2}\right)^{2j+\nu+1}}{\Gamma\left(j+\frac{3}{2}\right)\Gamma\left(j+\nu+\frac{3}{2}\right)} \left(1 - 2^{2j+2k-2m+3}\right)$$
$$\times \zeta(2m - 2j - 2k - 2) \quad [m \geq 1;\ -\pi < x < \pi;\ |x| - \pi < y < \pi - |x|].$$

2. $\displaystyle\sum_{k=1}^{\infty} \frac{(-1)^k}{k^{n\nu+\nu}} \prod_{i=1}^{n} \mathbf{H}_\nu(kx_i) = -\frac{2^{-n\nu-1}}{\pi^{n/2}\Gamma^n\left(\nu + \frac{3}{2}\right)} \prod_{i=1}^{n} x_i^{\nu+1}$

$$\left[x_i > 0;\ \sum_{i=1}^{n} x_i < \pi;\ \nu > -1\right].$$

6.11. The Legendre Polynomials $P_n(z)$

6.11.1. Series containing $P_{nk+m}(z)$

1. $\displaystyle\sum_{k=1}^{\infty} \frac{2k+1}{k(k+1)} P_k(x) = \ln\frac{2}{1-x} - 1 \qquad [-1 \leq x < 1;\ [47]].$

2. $\displaystyle\sum_{k=1}^{\infty} \frac{2k+1}{k^2(k+1)^2} P_k(x) = 1 - \frac{\pi^2}{6} + \mathrm{Li}_2\left(\frac{1+x}{2}\right) \qquad [-1 \leq x \leq 1;\ [47]].$

3. $\displaystyle\sum_{k=1}^{\infty} \frac{2k+1}{k^3(k+1)^3} P_k(x) = \frac{\pi^2}{6} - 2 + 2\zeta(3) + \ln\frac{1-x}{2}\mathrm{Li}_2\left(\frac{1-x}{2}\right)$
$$- \mathrm{Li}_2\left(\frac{1+x}{2}\right) - 2\mathrm{Li}_3\left(\frac{1-x}{2}\right) \quad [-1 \leq x < 1;\ [47],\ (15)].$$

6.11.2. Series containing $P_{nk+m}(z)$ and Bessel functions

1. $\displaystyle\sum_{k=0}^{\infty} (2k+1) J_{2k+1}(w) P_k(z) = \frac{w}{2} J_0\left(w\sqrt{\frac{1-z}{2}}\right).$

2. $\displaystyle\sum_{k=0}^{\infty} (-1)^k (4k+1) J_{2k+1/2}(w) P_{2k}(z) = \sqrt{\frac{2w}{\pi}} \cos(wz).$

3. $\displaystyle\sum_{k=0}^{\infty} (-1)^k (4k+3) J_{2k+3/2}(w) P_{2k+1}(z) = \sqrt{\frac{2w}{\pi}} \sin(wz).$

4. $\displaystyle\sum_{k=0}^{\infty} (4k+1) \frac{\left(\frac{1}{2}\right)_k}{k!} J_{2k+1/2}(w) P_{2k}(z) = \sqrt{\frac{2w}{\pi}} J_0\left(w\sqrt{1-z^2}\right).$

5. $\displaystyle\sum_{k=0}^{\infty} (4k+3) \frac{\left(\frac{3}{2}\right)_k}{k!} J_{2k+3/2}(w) P_{2k+1}(z) = z\sqrt{\frac{2w^3}{\pi}} J_0\left(w\sqrt{1-z^2}\right).$

6. $\sum_{k=0}^{\infty}(2k+1)J_{k+1/2}^2(w)P_k(z) = \dfrac{\sqrt{2}}{\pi\sqrt{1-z}}\sin\left(w\sqrt{2-2z}\right).$

7. $\sum_{k=0}^{\infty}(-1)^k(4k+1)J_{k+\nu}(w)J_{k-\nu+1/2}(w)P_{2k}(z)$

$$= \dfrac{1}{\Gamma(\nu+1)\Gamma\left(\frac{3}{2}-\nu\right)}\sqrt{\dfrac{w}{2}}\;{}_2F_3\left(\begin{array}{c}\frac{3}{4},\frac{5}{4};\;-w^2z^2\\ \frac{1}{2},\frac{3}{2}-\nu,1+\nu\end{array}\right).$$

8. $\sum_{k=0}^{\infty}(4k+1)\dfrac{\left(\frac{1}{2}\right)_k}{k!}J_{k+\nu}(w)J_{k-\nu+1/2}(w)P_{2k}(z)$

$$= \dfrac{1}{\Gamma(\nu+1)\Gamma\left(\frac{3}{2}-\nu\right)}\sqrt{\dfrac{w}{2}}\;{}_2F_3\left(\begin{array}{c}\frac{3}{4},\frac{5}{4};\;w^2z^2-w^2\\ 1,\frac{3}{2}-\nu,1+\nu\end{array}\right).$$

9. $\sum_{k=0}^{\infty}(-1)^k(4k+1)J_{k+1/4}^2(w)P_{2k}(z)$

$$= \dfrac{1}{\sqrt{z}}J_{1/4}(wz)\left[J_{1/4}(wz)-2wzJ_{5/4}(wz)\right].$$

10. $\sum_{k=0}^{\infty}(-1)^k(4k+3)J_{k+3/4}^2(w)P_{2k+1}(z)$

$$= \dfrac{1}{\sqrt{z}}J_{3/4}(wz)\left[3J_{3/4}(wz)-2wzJ_{7/4}(wz)\right].$$

11. $\sum_{k=0}^{\infty}(2k+1)I_{k+1/2}(w)P_k(z) = \sqrt{\dfrac{2w}{\pi}}\,e^{wz}.$

12. $\sum_{k=0}^{\infty}(4k+1)I_{2k+1/2}(w)P_{2k}(z) = \sqrt{\dfrac{2w}{\pi}}\,\cosh(wz).$

13. $\sum_{k=0}^{\infty}(4k+3)I_{2k+3/2}(w)P_{2k+1}(z) = \sqrt{\dfrac{2w}{\pi}}\,\sinh(wz).$

14. $\sum_{k=0}^{\infty}(-1)^k(4k+1)\dfrac{\left(\frac{1}{2}\right)_k}{k!}I_{k+1/4}(w)P_{2k}(z)$

$$= \dfrac{2^{5/4}}{\pi}\Gamma\left(\dfrac{3}{4}\right)w^{1/4}e^{-w}\;{}_1F_1\left(\begin{array}{c}\frac{3}{4};\;2w-2wz^2\\ 1\end{array}\right).$$

15. $\displaystyle\sum_{k=0}^{\infty}(-1)^k(4k+3)\frac{\left(\frac{3}{2}\right)_k}{k!}I_{k+3/4}(w)P_{2k+1}(z)$

$$=\frac{2^{3/4}}{\pi}\Gamma\left(\frac{1}{4}\right)w^{3/4}ze^{-w}{}_1F_1\left(\begin{array}{c}\frac{5}{4};\ 2w-2wz^2\\ 1\end{array}\right).$$

6.11.3. Series containing products of $P_{nk+m}(z)$

1. $\displaystyle\sum_{k=1}^{\infty}\frac{2k+1}{k(k+1)}P_k(x)P_k(y)=\ln\frac{4}{(1-x)(1+y)}-1 \qquad [-1\le x\le y<1].$

2. $\displaystyle\sum_{k=1}^{\infty}\frac{2k+1}{k^2(k+1)^2}P_k(x)P_k(y)=1-\ln\frac{1+y}{2}\ln\frac{(1-x)(1-y)}{4}+\mathrm{Li}_2\left(\frac{1+x}{2}\right)$

$$-\mathrm{Li}_2\left(\frac{1+y}{2}\right) \qquad [-1\le x\le y<1].$$

3. $\displaystyle\sum_{k=0}^{\infty}(2k+1)J_{2k+1}(w)[P_k(z)]^2=\frac{w}{2}J_0^2\left(\frac{w}{2}\sqrt{1-z^2}\right).$

4. $\displaystyle\sum_{k=0}^{\infty}(2k+1)J_{k+\nu}(w)J_{k-\nu+1}(w)[P_k(z)]^2$

$$=\frac{w\sin(\nu\pi)}{2\nu(1-\nu)\pi}{}_2F_3\left(\begin{array}{c}\frac{1}{2},\frac{3}{2};\ w^2z^2-w^2\\ 1,\ \nu+1,\ 2-\nu\end{array}\right).$$

5. $\displaystyle\sum_{k=0}^{\infty}(2k+1)J_{k+1/2}^2(w)[P_k(z)]^2=\frac{2w}{\pi}I_0\left(2w\sqrt{z^2-1}\right)$

$$-wI_1\left(2w\sqrt{z^2-1}\right)\mathbf{L}_0\left(2w\sqrt{z^2-1}\right)$$
$$+wI_0\left(2w\sqrt{z^2-1}\right)\mathbf{L}_1\left(2w\sqrt{z^2-1}\right).$$

6.11.4. Series containing $P_{nk+m}(\varphi(k,z))$

1. $\displaystyle\sum_{k=1}^{\infty}\frac{k^n}{\sqrt{k+z}}t^k P_{2n}\left(\sqrt{1+\frac{z}{k}}\right)$

$$=\sum_{k=0}^{n}\binom{n}{k}\frac{\left(\frac{1}{2}-n\right)_k}{k!}(-z)^k\Phi\left(t,\ k-n+\frac{1}{2},\ z\right)-\frac{\left(n+\frac{1}{2}\right)_n}{n!}z^{n-1/2}$$

$$[|t|<1].$$

2. $\displaystyle\sum_{k=0}^{\infty}\frac{(k+1)^{k/2-1}}{k!}t^k(k+z+1)^{k/2}P_k\left(\frac{2k+z+2}{2\sqrt{(k+1)(k+z+1)}}\right)$

$$=e^{-w-wz/2}\left[I_0\left(\frac{wz}{2}\right)+I_1\left(\frac{wz}{2}\right)\right] \qquad [t=-we^w;\ |we^{w+1}|<1].$$

6.12. The Chebyshev Polynomials $T_k(z)$ and $U_k(z)$

6.12.1. Series containing $T_{nk+m}(\varphi(k,z))$

1. $\displaystyle\sum_{k=0}^{\infty}\frac{t^k}{k!}T_k(z)=e^{tz}\cos\left(t\sqrt{1-z^2}\right)$

2. $\displaystyle\sum_{k=0}^{\infty}\frac{1}{(2k+1)\left[(2k+1)^2-a^2\right]^{n+1/2}}T_{2n+1}\left(\frac{2k+1}{\sqrt{(2k+1)^2-a^2}}\right)$
$$=\frac{a^{-2n-1}}{4}\sum_{k=0}^{2n}\frac{\left(\frac{a}{2}\right)^k}{k!}\left[(-1)^k\psi^{(k)}\left(\frac{1+a}{2}\right)-\psi^{(k)}\left(\frac{1-a}{2}\right)\right].$$

3. $\displaystyle\sum_{k=1}^{\infty}\frac{(-1)^k}{(k^2a^2-b^2)^{n+1/2}}\sin kx\, T_{2n+1}\left(\frac{ka}{\sqrt{k^2a^2-b^2}}\right)$
$$=-\frac{\pi}{(2n)!2a}\mathrm{D}_b^{2n}\left[\sin\frac{bx}{a}\csc\frac{b\pi}{a}\right]\quad [-\pi<x<\pi].$$

6.12.2. Series containing $T_{nk+m}(z)$ and Bessel functions

1. $\displaystyle\sum_{k=0}^{\infty}(-1)^k J_{2k}(w)T_{2k}(z)=\frac{1}{2}\left[\cos(wz)+J_0(w)\right].$

2. $\displaystyle\sum_{k=0}^{\infty}(-1)^k J_{2k+1}(w)T_{2k+1}(z)=\frac{1}{2}\sin(wz).$

3. $\displaystyle\sum_{k=0}^{\infty}(2k+1)J_{2k+1}(w)T_{2k+1}(z)=\frac{wz}{2}\cos\left(w\sqrt{1-z^2}\right).$

4. $\displaystyle\sum_{k=0}^{\infty}(-1)^k J_k^2(w)T_{2k}(z)=\frac{1}{2}[J_0(2wz)+J_0^2(wz)].$

5. $\displaystyle\sum_{k=0}^{\infty}(-1)^k J_{k+1/2}^2(w)T_{2k+1}(z)=\frac{1}{2}\mathbf{H}_0(2wz).$

6. $\displaystyle\sum_{k=0}^{\infty}(2k+1)J_{k+1/2}^2(w)T_{2k+1}(z)=wz\,\mathbf{H}_{-1}\left(2w\sqrt{1-z^2}\right).$

7. $\displaystyle\sum_{k=0}^{\infty}(-1)^k J_k(w)J_{k+1}(w)T_{2k+1}(z)=\frac{1}{2}J_1(2wz).$

8. $\displaystyle\sum_{k=0}^{\infty}I_k(w)T_k(z)=\frac{1}{2}\left[e^{wz}+I_0(w)\right].$

9. $\sum_{k=0}^{\infty} I_{2k}(w)T_{2k}(z) = \frac{1}{2}\left[\cosh(wz) + I_0(w)\right].$

10. $\sum_{k=0}^{\infty} I_{2k+1}(w)T_{2k+1}(z) = \frac{1}{2}\sinh(wz).$

11. $\sum_{k=0}^{\infty} I_{k+1/2}^2(w)T_{2k+1}(z) = \frac{1}{2}\mathbf{L}_0(2wz).$

12. $\sum_{k=0}^{\infty} I_{k/2}^2(w)T_k(z) = \frac{1}{2}\left[I_0(2wz) + I_0^2(w) + \mathbf{L}_0(2wz)\right].$

13. $\sum_{k=0}^{\infty} I_k^2(w)T_{2k}(z) = \frac{1}{2}\left[I_0(2wz) + I_0^2(w)\right].$

6.12.3. Series containing $U_{nk+m}(\varphi(k,z))$

1. $\sum_{k=0}^{\infty} \frac{t^k}{k!} U_k(z) = e^{tz}\left[\cos\left(t\sqrt{1-z^2}\right) + \frac{z}{\sqrt{1-z^2}}\sin\left(t\sqrt{1-z^2}\right)\right].$

2. $\sum_{k=1}^{\infty} \frac{(-1)^k}{(k^2a^2 - b^2)^{n+3/2}} \sin kx \, U_{2n+1}\left(\frac{ka}{\sqrt{k^2a^2 - b^2}}\right)$
$= -\frac{\pi}{(2n+1)!2ab} D_b^{2n+1}\left[\sin\frac{bx}{a} \csc\frac{b\pi}{a}\right] \quad [-\pi < x < \pi].$

3. $\sum_{k=1}^{\infty} \frac{(-1)^k}{k(k^2a^2 - b^2)^{n+3/2}} \cos kx \, U_{2n+1}\left(\frac{ka}{\sqrt{k^2a^2 - b^2}}\right)$
$= -\frac{\pi}{(2n+1)!2b} D_b^{2n+1}\left[\frac{1}{b}\cos\frac{bx}{a} \csc\frac{b\pi}{a}\right] - (n+1)ab^{-2n-4} \quad [-\pi \leq x \leq \pi].$

4. $\sum_{k=0}^{\infty} \frac{1}{(2k+1)\left[(2k+1)^2 - a^2\right]^{n+3/2}} U_{2n+1}\left(\frac{2k+1}{\sqrt{(2k+1)^2 - a^2}}\right)$
$= -\frac{a^{-2n-3}}{4}\sum_{k=0}^{2n+1} \frac{\left(\frac{a}{2}\right)^k}{k!}\left[(-1)^k\psi^{(k)}\left(\frac{1+a}{2}\right) - \psi^{(k)}\left(\frac{1-a}{2}\right)\right].$

6.12.4. Series containing $U_{nk+m}(z)$ and Bessel functions

1. $\sum_{k=0}^{\infty} (k+1)J_{2k+2}(w)U_k(z) = \frac{w}{8}\sqrt{\frac{2}{1-z}}\sin\left(w\sqrt{\frac{1-z}{2}}\right).$

2. $\sum_{k=0}^{\infty} J_{2k+1}(w)U_{2k}(z) = \frac{1}{2\sqrt{1-z^2}}\sin\left(w\sqrt{1-z^2}\right).$

3. $\displaystyle\sum_{k=0}^{\infty}(-1)^{k}(2k+1)J_{2k+1}(w)U_{2k}(z) = \frac{w}{2}\cos(wz)$.

4. $\displaystyle\sum_{k=0}^{\infty}(-1)^{k}(k+1)J_{2k+2}(w)U_{2k+1}(z) = \frac{w}{4}\sin(wz)$.

5. $\displaystyle\sum_{k=0}^{\infty}(k+1)J_{2k+2}(w)U_{2k+1}(z) = \frac{wz}{4\sqrt{1-z^2}}\sin\left(w\sqrt{1-z^2}\right)$.

6. $\displaystyle\sum_{k=0}^{\infty}(k+1)J_{k+\nu}(w)J_{k-\nu+2}(w)U_k(z)$
$$= \frac{w^2\sin(\nu\pi)}{4\pi\nu(\nu-1)(\nu-2)}\,{}_1F_2\!\left(\begin{array}{c}2;\\3-\nu,1+\nu\end{array}\frac{w^2(z-1)}{2}\right).$$

7. $\displaystyle\sum_{k=0}^{\infty}(k+1)J_k(w)J_{k+2}(w)U_k(z)$
$$= \frac{1}{2^{3/2}(1-z)}\left[\sqrt{2}\,J_2\!\left(w\sqrt{2-2z}\right) - w\sqrt{1-z}\,J_3\!\left(w\sqrt{2-2z}\right)\right].$$

8. $\displaystyle\sum_{k=0}^{\infty}J_{k+1/2}^2(w)U_{2k}(z) = \frac{1}{2\sqrt{1-z^2}}\,\mathbf{H}_0\!\left(2w\sqrt{1-z^2}\right)$.

9. $\displaystyle\sum_{k=0}^{\infty}(-1)^k(2k+1)J_{k+1/2}^2(w)U_{2k}(z) = w\,\mathbf{H}_{-1}(2wz)$.

10. $\displaystyle\sum_{k=0}^{\infty}(-1)^k(2k+1)J_{k+\nu}(w)J_{k-\nu+1}(w)U_{2k}(z) = \frac{w\sin(\nu\pi)}{2\pi\nu(\nu^2-1)(\nu-2)}$
$$\times\left[(2+\nu-\nu^2)\,{}_1F_2\!\left(\begin{array}{c}1;\\2-\nu,1+\nu\end{array}-w^2z^2\right) - 2w^2z^2\,{}_1F_2\!\left(\begin{array}{c}2;\\3-\nu,2+\nu\end{array}-w^2z^2\right)\right].$$

11. $\displaystyle\sum_{k=0}^{\infty}(-1)^k(2k+1)J_k(w)J_{k+1}(w)U_{2k}(z) = \frac{1}{2z}J_1(2wz) - wJ_2(2wz)$.

12. $\displaystyle\sum_{k=0}^{\infty}J_{k+\nu}(w)J_{k-\nu+1}(w)U_{2k}(z) = \frac{w\sin(\nu\pi)}{2\pi\nu(1-\nu)}\,{}_1F_2\!\left(\begin{array}{c}1;\\2-\nu,1+\nu\end{array}w^2z^2-w^2\right)$.

13. $\displaystyle\sum_{k=0}^{\infty}J_{k-1/2}(w)J_{k+3/2}(w)U_{2k}(z)$
$$= \frac{1}{2w(1-z^2)}\left[w\sqrt{1-z^2}\,\mathbf{H}_0\!\left(2w\sqrt{1-z^2}\right) - \mathbf{H}_1\!\left(2w\sqrt{1-z^2}\right)\right].$$

14. $\displaystyle\sum_{k=0}^{\infty}J_k(w)J_{k+1}(w)U_{2k}(z) = \frac{1}{2\sqrt{1-z^2}}J_1\!\left(2w\sqrt{1-z^2}\right)$.

15. $\displaystyle\sum_{k=0}^{\infty} (k+1) J_{k+1}^2(w) U_{2k+1}(z) = \frac{wz}{2\sqrt{1-z^2}} J_1\left(2w\sqrt{1-z^2}\right).$

16. $\displaystyle\sum_{k=0}^{\infty} (-1)^k (k+1) J_{k+1}^2(w) U_{2k+1}(z) = \frac{w}{2} J_1(2wz).$

17. $\displaystyle\sum_{k=0}^{\infty} (-1)^k (2k+1) J_{k+1/2}^2(w) U_{2k}(z) = w\,\mathbf{H}_{-1}(2wz).$

18. $\displaystyle\sum_{k=0}^{\infty} (-1)^k (k+1) J_{k+1}^2(w) U_{2k+1}(z) = \frac{w}{2} J_1(2wz).$

19. $\displaystyle\sum_{k=0}^{\infty} (2k+1)\, I_{k+1/2}(w) U_{2k}(z) = \sqrt{\frac{2w}{\pi}}\, e^{-w} + 2wz e^{2wz^2 - w}\,\mathrm{erf}\left(\sqrt{2w}\,z\right).$

20. $\displaystyle\sum_{k=0}^{\infty} (k+1)\, I_{k+1}(w) U_k(z) = \frac{w}{2} e^{wz}.$

21. $\displaystyle\sum_{k=0}^{\infty} (2k+1)\, I_{2k+1}(w) U_{2k}(z) = \frac{w}{2}\cosh(wz).$

22. $\displaystyle\sum_{k=0}^{\infty} (k+1)\, I_{2k+2}(w) U_{2k+1}(z) = \frac{w}{4}\sinh(wz).$

23. $\displaystyle\sum_{k=0}^{\infty} (-1)^k (k+1)\, I_{k+1}(w) U_{2k+1}(z) = wz e^{w - 2wz^2}.$

24. $\displaystyle\sum_{k=0}^{\infty} (k+1) I_{(k+1)/2}^2(w) U_k(z) = w[I_1(2wz) + \mathbf{L}_{-1}(2wz)].$

25. $\displaystyle\sum_{k=0}^{\infty} (2k+1)\, I_{k+1/2}^2(w) U_{2k}(z) = w\,\mathbf{L}_{-1}(2wz).$

26. $\displaystyle\sum_{k=0}^{\infty} (k+1) I_{k+1}^2(w) U_{2k+1}(z) = \frac{w}{2} I_1(2wz).$

6.13. Hermite Polynomials $H_n(z)$

6.13.1. Series containing $H_{nk+m}(z)$ and Bessel functions

1. $\displaystyle\sum_{k=0}^{\infty} \frac{\left(\mp\frac{w}{2}\right)^k}{(2k)!} \left\{\begin{array}{c} J_{\nu+k}(w) \\ I_{\nu+k}(w) \end{array}\right\} H_{2k}(z) = \frac{\left(\frac{w}{2}\right)^\nu}{\Gamma(\nu+1)}\, _0F_2\left(\begin{array}{c} \mp\frac{w^2 z^2}{4} \\ \nu+1, \frac{1}{2} \end{array}\right).$

2. $\sum_{k=0}^{\infty} \frac{\left(\mp \frac{w}{2}\right)^k}{(2k+1)!} \left\{ \begin{matrix} J_{\nu+k}(w) \\ I_{\nu+k}(w) \end{matrix} \right\} H_{2k+1}(z) = \frac{2\left(\frac{w}{2}\right)^\nu z}{\Gamma(\nu+1)} \, {}_0F_2\left(\begin{matrix} \mp \frac{w^2 z^2}{4} \\ \nu+1, \frac{3}{2} \end{matrix} \right).$

6.13.2. Series containing products of $H_{nk+m}(z)$

1. $\sum_{k=0}^{\infty} \frac{t^k}{(2k)!} H_{2k}(w) H_{2k}(z)$

$$= (1-4t)^{-1/2} \exp\left(\frac{4tw^2 + 4tz^2}{4t-1}\right) \cosh\left(\frac{4\sqrt{t}\,wz}{1-4t}\right) \quad [|t| < 1/4].$$

2. $\sum_{k=0}^{\infty} \frac{t^k}{(2k+1)!} H_{2k+1}(w) H_{2k+1}(z)$

$$= t^{-1/2}(1-4t)^{-1/2} \exp\left(\frac{4tw^2 + 4tz^2}{4t-1}\right) \sinh\left(\frac{4\sqrt{t}\,wz}{1-4t}\right) \quad [|t| < 1/4].$$

3. $\sum_{k=0}^{\infty} \frac{t^k}{(k!)^2} [H_{2k}(z)]^2$

$$= \frac{4}{\pi} e^{2z^2} \int_0^\infty \int_0^\infty e^{-x-y} \cos(2xz) \cos(2yz) I_0(8\sqrt{t}\,xy)\,dxdy$$

$$[|t| < 1/16;\ [34]].$$

4. $\sum_{k=0}^{\infty} \frac{t^k}{(k!)^2} [H_{2k+1}(z)]^2$

$$= \frac{16}{\pi} e^{2z^2} \int_0^\infty \int_0^\infty xy e^{-x^2-y^2} \sin(2xz) \sin(2yz) I_0(8\sqrt{t}\,xy)\,dxdy$$

$$[|t| < 1/16;\ [34]].$$

6.13.3. Series containing $H_{nk+m}(\varphi(k,z))$

1. $\sum_{k=0}^{\infty} \frac{(k+1)^{k-1}}{(2k)!} t^k H_{2k}\left(\frac{z}{\sqrt{k+1}}\right)$

$$= \frac{e^{-w}}{2wz^2} [2z\sqrt{w} \sinh(2z\sqrt{w}) - \cosh(2z\sqrt{w}) + 1]$$

$$[t = we^w;\ |we^{w+1}| < 1].$$

2. $\sum_{k=0}^{\infty} \frac{(k+1)^{k-1/2}}{(2k+1)!} t^k H_{2k+1}\left(\frac{z}{\sqrt{k+1}}\right) = \frac{e^{-w}}{wz} [\cosh(2z\sqrt{w}) - 1]$

$$[t = we^w;\ |we^{w+1}| < 1].$$

6.13.4. Series containing $H_{nk+m}(\varphi(k,z))$ and special functions

1. $\displaystyle\sum_{k=0}^{\infty} \frac{(-1)^k}{(2k)!}(k+1)^{-3/2}\gamma\left(k+\frac{1}{2},(k+1)w\right)H_{2k}\left(\frac{z}{\sqrt{k+1}}\right)$
$$= w^{-1/2}z^{-2}[1-\cos(2\sqrt{w}\,z)].$$

2. $\displaystyle\sum_{k=0}^{\infty} \frac{(-1)^k}{(2k+1)!}(k+1)^{-2}\gamma\left(k+\frac{3}{2},(k+1)w\right)H_{2k+1}\left(\frac{z}{\sqrt{k+1}}\right)$
$$= 2w^{1/2}z^{-1}\left[1-\frac{\sin(2\sqrt{w}\,z)}{2\sqrt{w}\,z}\right].$$

3. $\displaystyle\sum_{k=0}^{\infty} \frac{(k+1)^{(k-\nu)/2-1}}{(2k)!}\left(-\frac{w}{2}\right)^k J_{k+\nu}(\sqrt{k+1}\,w)\,H_{2k}\left(\frac{z}{\sqrt{k+1}}\right)$
$$= \frac{\left(\frac{w}{2}\right)^{\nu-2}z^{-2}}{2\Gamma(\nu)}\left[{}_0F_2\!\left(\begin{array}{c}-\frac{w^2z^2}{4}\\\nu,-\frac{1}{2}\end{array}\right)-1\right].$$

4. $\displaystyle\sum_{k=0}^{\infty} \frac{(k+1)^{(k-\nu-1)/2}}{(2k+1)!}\left(-\frac{w}{2}\right)^k J_{k+\nu}(\sqrt{k+1}\,w)\,H_{2k+1}\left(\frac{z}{\sqrt{k+1}}\right)$
$$= \frac{\left(\frac{w}{2}\right)^{\nu-2}z^{-1}}{\Gamma(\nu)}\left[1-{}_0F_2\!\left(\begin{array}{c}-\frac{w^2z^2}{4}\\\nu,\frac{1}{2}\end{array}\right)\right].$$

5. $\displaystyle\sum_{k=0}^{\infty} \frac{(k+1)^{(k-\nu)/2-1}}{(2k)!}\left(\frac{w}{2}\right)^k I_{k+\nu}(\sqrt{k+1}\,w)\,H_{2k}\left(\frac{z}{\sqrt{k+1}}\right)$
$$= \frac{\left(\frac{w}{2}\right)^{\nu-2}z^{-2}}{2\Gamma(\nu)}\left[1-{}_0F_2\!\left(\begin{array}{c}\frac{w^2z^2}{4}\\\nu,-\frac{1}{2}\end{array}\right)\right].$$

6. $\displaystyle\sum_{k=0}^{\infty} \frac{(k+1)^{(k-\nu-1)/2}}{(2k+1)!}\left(\frac{w}{2}\right)^k I_{k+\nu}(\sqrt{k+1}\,w)\,H_{2k+1}\left(\frac{z}{\sqrt{k+1}}\right)$
$$= \frac{\left(\frac{w}{2}\right)^{\nu-2}z^{-1}}{\Gamma(\nu)}\left[{}_0F_2\!\left(\begin{array}{c}\frac{w^2z^2}{4}\\\nu,\frac{1}{2}\end{array}\right)-1\right].$$

6.13.5. Series containing products of $H_{nk+m}(\varphi(k,z))$

1. $\displaystyle\sum_{k=1}^{\infty} \frac{\cos(kx)}{k^{2m-2}} H_{2n}(\sqrt{k}\,y)\,H_{2n}(i\sqrt{k}\,y)$
$$= (-1)^{n+1}\frac{[(2n)!]^2}{n!}\frac{2^{6m-7}\left(n+\frac{1}{2}\right)_{m-1}y^{4m-4}}{(n-m+1)!(4m-4)!}$$

$$+ \frac{[(2n)!]^2}{n!} \sum_{k=0}^{m-2} \frac{2^{6k} y^{4k}}{(n-k)!(4k)!} \left(n+\frac{1}{2}\right)_k$$

$$\times \left[(-1)^{m-1} \frac{\pi x^{2m-2k-3}}{2(2m-2k-3)!} + (-1)^k \sum_{j=0}^{m-k-1} (-1)^j \frac{x^{2j}}{(2j)!} \zeta(2m-2j-2k-2)\right]$$

$$\left[m \geq 1;\; 0 < y < \sqrt{\pi};\; y^2 < x < 2\pi - y^2\right].$$

2. $\displaystyle\sum_{k=1}^{\infty} \frac{\cos(kx)}{k^{2m-1}} H_{2n+1}(\sqrt{k}\,y) H_{2n+1}(i\sqrt{k}\,y)$

$$= (-1)^m 4i \frac{[(2n+1)!]^2}{n!} \frac{2^{8m-8}\left(n+\frac{3}{2}\right)_{m-1} y^{4m-2}}{(n-m+1)!(4m-2)!}$$

$$+ 4i \frac{[(2n+1)!]^2}{n!} y^2 \sum_{k=0}^{m-2} \frac{2^{6k} y^{4k}}{(n-k)!(4k+2)!} \left(n+\frac{3}{2}\right)_k$$

$$\times \left[(-1)^{m-1} \frac{\pi x^{2m-2k-3}}{(2m-2k-3)!} + 2(-1)^k \sum_{j=0}^{m-k-1} (-1)^j \frac{x^{2j}}{(2j)!} \zeta(2m-2j-2k-2)\right]$$

$$\left[m \geq 1;\; 0 < y < \sqrt{\pi};\; y^2 < x < 2\pi - y^2\right].$$

6.14. The Laguerre Polynomials $L_n^\lambda(z)$

6.14.1. Series containing $L_{nk+m}^{\lambda \pm lk}(z)$

1. $\displaystyle\sum_{k=0}^{\infty} \frac{z^k}{(\lambda+n+1)_k} L_n^{\lambda+k}(z) = \frac{\lambda+n}{n} L_{n-1}^\lambda(z)$ $\hspace{2em} [n \geq 1].$

2. $\displaystyle\sum_{k=0}^{\infty} \frac{(ka+b)^m}{(\lambda+m+1)_k} z^k L_m^{\lambda+k}(z)$

$$= \frac{(\lambda+1)_m}{m!} b^m \sum_{k=0}^{m} \binom{m}{k}\left(\frac{a}{b}\right)^k \sum_{j=0}^{k} \sigma_k^j \frac{j!}{(\lambda+1)_j} z^j \,{}_1F_1\left(\begin{matrix}-m+j+1\\ \lambda+j+1;\,1-z^2\end{matrix}\right)$$

$$[|z| < 1].$$

3. $\displaystyle\sum_{k=0}^{\infty} \frac{(-\lambda-n)_k}{(k+n)!} t^k L_n^{\lambda-k}(z) = (n+\lambda+1)_n(-t)^n$

$$\times \left[(1-t)^{\lambda+n} L_n^{\lambda+n}(z-tz) - (-t)^n \sum_{k=0}^{n-1} \frac{(-\lambda-2n)_k}{k!} t^k L_n^{\lambda+n-k}(z)\right]$$

$$[|t| < 1].$$

4. $\displaystyle\sum_{k=0}^{\infty} \frac{t^k}{(\lambda+1)_k} L_k^\lambda(z) = \Gamma(\lambda+1)(tz)^{-\lambda/2} e^t J_\lambda(2\sqrt{tz}).$

495

5. $\sum_{k=0}^{\infty} \dfrac{z^k}{(\lambda+2)_k} L_k^{\lambda+k}(z) = e^z.$

6. $\sum_{k=0}^{\infty} \dfrac{(a)_{2k}}{(a+1)_k (\lambda+1)_{2k}} z^k L_k^{\lambda+k}(z) = {}_1F_1\!\left(\begin{array}{c} a;\, z \\ \lambda+1 \end{array}\right).$

7. $\sum_{k=0}^{\infty} \dfrac{(a)_k (b)_k}{\left(a+b+\tfrac{1}{2}\right)_k (\lambda+1)_{2k}} (4z)^k L_k^{\lambda+k}(z) = {}_2F_2\!\left(\begin{array}{c} 2a,\, 2b \\ a+b+\tfrac{1}{2};\, z \end{array}\right).$

8. $\sum_{k=0}^{\infty} \binom{k+n}{k} t^k L_{k+n}^{\lambda}(z) = (1-t)^{-\lambda-n-1} e^{tz/(t-1)} L_n^{\lambda}\!\left(\dfrac{z}{1-t}\right)$ $[|t|<1].$

9. $\sum_{k=0}^{\infty} \dfrac{t^k}{k!} L_k^k(z) = e^{2t} I_0(2\sqrt{t(t-z)})$ [34].

10. $\sum_{k=0}^{\infty} \dfrac{(k+n)!}{k!\,(a)_k} z^k L_{k+n}^{\lambda+k}(z) = (\lambda+1)_n e^z {}_2F_2\!\left(\begin{array}{c} \lambda+n+1,\, \lambda-a+n+2 \\ \lambda+1,\, a;\, z \end{array}\right).$

11. $\sum_{k=0}^{\infty} \dfrac{(a)_k \left(\tfrac{1}{2}-a+n\right)_k}{k!\,(\lambda+n+1)_{2k}} (4z)^k L_{k+n}^{\lambda+k}(z) = \dfrac{(\lambda+1)_n}{n!} {}_2F_2\!\left(\begin{array}{c} 2a-n,\, n-2a+1 \\ n+1,\, \lambda+1;\, z \end{array}\right).$

6.14.2. Series containing $L_{nk+m}^{\lambda\pm lk}(z)$ and special functions

1. $\sum_{k=0}^{\infty} \dfrac{1}{(\lambda+1)_k} \gamma(k+\lambda+1,\, w) L_k^{\lambda}(z) = \Gamma(\lambda+1) \left(\dfrac{w}{z}\right)^{(\lambda+1)/2} J_{\lambda+1}(2\sqrt{wz}).$

2. $\sum_{k=0}^{\infty} \dfrac{1}{(\lambda+1)_k} \gamma(\nu+k,\, w) L_k^{\lambda}(z) = \dfrac{w^\nu}{\nu} {}_1F_2\!\left(\begin{array}{c} \nu;\, -wz \\ \lambda+1,\, \nu+1 \end{array}\right).$

3. $\sum_{k=0}^{\infty} \dfrac{1}{k!} \gamma(k+\nu,\, z) L_n^{\lambda+k}(z) = \dfrac{(\lambda+1)_n}{n!\,\nu} z^\nu {}_2F_2\!\left(\begin{array}{c} -n,\, 1;\, z \\ \lambda+1,\, \nu+1 \end{array}\right).$

4. $\sum_{k=0}^{\infty} \dfrac{\left(\pm\tfrac{w}{2}\right)^k}{(\lambda+1)_k} \left\{\begin{array}{c} J_{\nu+k}(w) \\ I_{\nu+k}(w) \end{array}\right\} L_k^{\lambda}(z) = \dfrac{\left(\tfrac{w}{2}\right)^\nu}{\Gamma(\nu+1)} {}_0F_2\!\left(\begin{array}{c} \mp\tfrac{w^2 z}{4} \\ \lambda+1,\, \nu+1 \end{array}\right).$

5. $\sum_{k=0}^{\infty} (2k-\lambda)(-\lambda)_k (-z)^{-k} J_{k+\mu}(w) J_{k-\lambda-\mu}(w) L_k^{\lambda-2k}(z)$

$$= -\dfrac{\lambda \left(\tfrac{2}{w}\right)^\lambda}{\Gamma(\mu+1)\Gamma(1-\lambda-\mu)} {}_2F_2\!\left(\begin{array}{c} \tfrac{1-\lambda}{2},\, 1-\tfrac{\lambda}{2};\, \tfrac{w^2}{z} \\ \mu+1,\, 1-\lambda-\mu \end{array}\right).$$

6. $\sum_{k=0}^{\infty} t^k P_k(w) L_k(z) = v^{-1/2} e^{t(t-w)z/v} J_0\left(\frac{tz\sqrt{1-w^2}}{v}\right)$

$$[v = 1 - 2tw + t^2;\ |t| < 1].$$

6.14.3. Series containing products of $L_{nk+m}^{\lambda \pm lk}(z)$

1. $\sum_{k=0}^{\infty} \frac{t^k}{k!} L_m^{\lambda+k}(w) L_n^{\mu+k}(z) = e^t \sum_{k=0}^{m} \frac{t^k}{k!} L_{m-k}^{\lambda+k}(w-t) L_{n-k}^{\mu+k}(z-t)$ $\quad [m \le n]$.

2. $\sum_{k=0}^{\infty} \frac{k!}{(\lambda+1)_k} t^k L_k^\lambda(w) L_k^\lambda(z)$

$= \Gamma(\lambda+1) \frac{(txy)^{-\lambda/2}}{1-t} \exp\left[\frac{t(w+z)}{t-1}\right] I_\lambda\left(\frac{2\sqrt{txy}}{1-t}\right)$ $\quad [|t| < 1;\ [14],\ 2.5(17)]$.

3. $\sum_{k=0}^{\infty} t^k \left[L_k^\lambda(z)\right]^2$

$= z^{-\lambda} e^{2z} \int_0^\infty \int_0^\infty (xy)^{\lambda/2} e^{-x-y} J_\lambda(2\sqrt{xz}) J_\lambda(2\sqrt{yz}) I_0(2z\sqrt{txy})\, dx$

$$[|t| < 1;\ [34]].$$

6.14.4. Series containing products of $L_n^\lambda(kx)$

1. $\sum_{k=1}^{\infty} \frac{\cos(ky)}{k^{2m-2}} L_n^\lambda(-kx) L_n^\lambda(kx)$

$= -\frac{(\lambda+1)_n^2}{(n!)^2} \frac{(2x)^{2m-2}}{2(2m-1)!} \frac{(-n)_{m-1}(\lambda+n+1)_{m-1}\left(\frac{3}{2}\right)_{m-1}}{2(2m-1)!(\lambda+1)_{m-1}(\lambda+1)_{2m-2}}$

$+ \frac{(\lambda+1)_n^2}{(n!)^2} \sum_{k=0}^{m-2} \frac{x^{2k}}{k!} \frac{(-n)_k(\lambda+n+1)_k}{(\lambda+1)_k(\lambda+1)_{2k}} \left[(-1)^{m-k-1} \frac{\pi y^{2m-2k-3}}{2(2m-2k-3)!}\right.$

$\left. + \sum_{j=0}^{m-k-1} (-1)^j \frac{y^{2j}}{(2j)!} \zeta(2m-2j-2k-2)\right]$

$$[m \ge 1;\ 0 < x < \pi;\ x < y < 2\pi - x].$$

6.14.5. Series containing $L_{nk+m}^{\lambda \pm lk}(\varphi(k, z))$

1. $\sum_{k=0}^{\infty} \frac{t^k}{k+1} L_k^{\lambda-k}((k+1)z) = \frac{ze^{-w}}{(\lambda+1)w} \left[\left(1+\frac{w}{z}\right)^{\lambda+1} - 1\right]$

$$\left[t = we^w/z;\ |we^{w+1}|,\ |w/z| < 1\right].$$

2. $\sum_{k=0}^{\infty} \dfrac{(k+1)^{k-1}}{k!} t^k L_k\left(\dfrac{z}{k+1}\right) = \dfrac{e^{-w}}{\sqrt{wz}} I_1(2\sqrt{wz})$ $\qquad [t = -we^w;\ |we^{w+1}| < 1]$.

6.14.6. Series containing $L_{nk+m}^{\lambda \pm lk}(\varphi(k, z))$ and special functions

1. $\sum_{k=0}^{\infty} \dfrac{(k+1)^{-\nu-1}}{(\lambda+1)_k} \gamma(k+\nu, (k+1)w) L_k^{\lambda}\left(\dfrac{z}{k+1}\right)$

$$= \dfrac{\lambda w^{\nu-1} z^{-1}}{1-\nu}\left[{}_1F_2\!\left(\!\begin{array}{c}\nu-1;\ -wz\\ \lambda,\ \nu\end{array}\!\right) - 1\right].$$

2. $\sum_{k=0}^{\infty} \dfrac{(k+1)^{-\lambda-2}}{(\lambda+1)_k} \gamma(k+\lambda+1, (k+1)w) L_k^{\lambda}\left(\dfrac{z}{k+1}\right)$

$$= w^\lambda z^{-1}\left[1 - \Gamma(\lambda+1)(wz)^{-\lambda/2} J_\lambda(2\sqrt{wz})\right].$$

3. $\sum_{k=0}^{\infty} \dfrac{(k+1)^{(k-\nu)/2-1}}{(\lambda+1)_k}\left(\dfrac{w}{2}\right)^k J_{k+\nu}(\sqrt{k+1}\,w) L_k^{\lambda}\left(\dfrac{z}{k+1}\right)$

$$= \dfrac{\lambda\left(\dfrac{w}{2}\right)^{\nu-2} z^{-1}}{\Gamma(\nu)}\left[1 - {}_0F_2\!\left(\!\begin{array}{c}-\dfrac{w^2 z}{4}\\ \lambda,\ \nu\end{array}\!\right)\right].$$

4. $\sum_{k=0}^{\infty} \dfrac{(k+1)^{(k-\nu)/2-1}}{(\lambda+1)_k}\left(-\dfrac{w}{2}\right)^k I_{k+\nu}(\sqrt{k+1}\,w) L_k^{\lambda}\left(\dfrac{z}{k+1}\right)$

$$= \dfrac{\lambda\left(\dfrac{w}{2}\right)^{\nu-2} z^{-1}}{\Gamma(\nu)}\left[{}_0F_2\!\left(\!\begin{array}{c}\dfrac{w^2 z}{4}\\ \lambda,\ \nu\end{array}\!\right) - 1\right].$$

6.14.7. Series containing products of $L_{mk+n}^{\lambda \pm lk}(\varphi(k, z))$

1. $\sum_{k=0}^{\infty} \dfrac{(k+1)^{k-1}}{(\lambda+n+1)_k} t^k L_n^{\lambda+k}((k+1)w) L_k^{\lambda+n}\left(\dfrac{z}{k+1}\right)$

$$= (tz)^{-1} \dfrac{(\lambda)_{n+1}}{n!}\left[1 - \Gamma(\lambda)(wz)^{(1-\lambda)/2} J_{\lambda-1}(2\sqrt{wz})\right] \qquad [t = we^{-w}].$$

2. $\sum_{k=0}^{\infty} \dfrac{(k+1)^{k-1}}{(\mu+1)_k} t^k L_n^{\lambda+k}((k+1)w) L_k^{\mu}\left(\dfrac{z}{k+1}\right)$

$$= (tz)^{-1} \dfrac{(\lambda)_n \mu}{n!}\left[1 - {}_1F_2\!\left(\!\begin{array}{c}\lambda+n;\ -wz\\ \mu,\ \lambda\end{array}\!\right)\right] \qquad [t = we^{-w}].$$

6.15. The Gegenbauer Polynomials $C_n^\lambda(z)$

6.15.1. Series containing $C_{nk+m}^{\lambda\pm lk}(z)$

1. $\displaystyle\sum_{k=0}^{\infty} \frac{(\lambda)_k}{k!} t^k C_n^{\lambda+k}(z) = (1-t)^{-\lambda-n/2} C_n^\lambda\left(\frac{z}{\sqrt{1-t}}\right)$ $\quad [|t|<1]$.

2. $\displaystyle\sum_{k=0}^{\infty} \frac{(1-2\lambda-n)_{2k}}{k!\,(1-\lambda)_k} t^k C_n^{\lambda-k}(z)$
$= (1-4t)^{\lambda-1/2}(1-4t+4tz^2)^{n/2} C_n^\lambda\left(\frac{z}{\sqrt{1-4t+4tz^2}}\right)$ $\quad [|t|<1/4]$.

3. $\displaystyle\sum_{k=0}^{\infty}(ka+b)^m \frac{(\lambda)_k}{\left(\lambda+m+\frac{1}{2}\right)_k}(1-z^2)^k C_{2m}^{\lambda+k}(z)$
$= \frac{(2\lambda)_{2m}}{(2m)!} b^m \sum_{k=0}^{m}\binom{m}{k}\left(\frac{a}{b}\right)^k \sum_{j=0}^{k} \sigma_k^j \frac{j!(\lambda+m)_j}{\left(\lambda+\frac{1}{2}\right)_j}(1-z^2)^j$
$\times {}_2F_1\left(\begin{array}{c}-m+j+1,\lambda+j+m\\ \lambda+j+\frac{1}{2};\,1-z^2\end{array}\right)$ $\quad [|z|<1]$.

4. $\displaystyle\sum_{k=0}^{\infty}(ka+b)^m \frac{(\lambda)_k}{\left(\lambda+m+\frac{1}{2}\right)_k}(1-z^2)^k C_{2m+1}^{\lambda+k}(z)$
$= \frac{(2\lambda)_{2m+1}}{(2m+1)!} b^m z \sum_{k=0}^{m}\binom{m}{k}\left(\frac{a}{b}\right)^k \sum_{j=0}^{k} \sigma_k^j \frac{j!(\lambda+m+1)_j}{\left(\lambda+\frac{1}{2}\right)_j}(1-z^2)^j$
$\times {}_2F_1\left(\begin{array}{c}-m+j+1,\lambda+j+m+1\\ \lambda+j+\frac{1}{2};\,1-z^2\end{array}\right)$ $\quad [|z|<1]$.

5. $\displaystyle\sum_{k=0}^{\infty} \frac{t^k}{(1-\lambda)_k} C_k^{\lambda-k}(z) = \Gamma(1-\lambda) t^\lambda e^{-2tz} I_{-\lambda}(2t)$.

6. $\displaystyle\sum_{k=0}^{\infty} \frac{t^k}{(1-\lambda)_k} C_{2k}^{\lambda-k}(z) = e^t {}_1F_1\left(\begin{array}{c}\lambda;\,-tz^2\\ \frac{1}{2}\end{array}\right)$.

7. $\displaystyle\sum_{k=0}^{\infty} \frac{t^k}{(k+n)!} C_{2k}^{-k-n}(z) = \frac{(-1)^n}{(2n)!} e^t H_{2n}(iz\sqrt{t})$.

8. $\displaystyle\sum_{k=0}^{\infty} \frac{(a)_k}{(1-\lambda)_k} t^k C_{2k}^{\lambda-k}(z) = (1-t)^{-a} {}_2F_1\left(\begin{array}{c}a,\lambda\\ \frac{1}{2};\,\frac{tz^2}{t-1}\end{array}\right)$ $\quad [|t|<1]$.

9. $\sum_{k=0}^{\infty} \dfrac{t^k}{(1-\lambda)_k} C_{2k+1}^{\lambda-k}(z) = 2\lambda z e^t \, {}_1F_1\left(\begin{array}{c}\lambda+1\\ \frac{3}{2};\end{array} -tz^2\right).$

10. $\sum_{k=0}^{\infty} \dfrac{t^k}{(k+n+1)!} C_{2k+1}^{-k-n-1}(z) = \dfrac{(-1)^n i}{(2n+1)!} t^{-1/2} e^t H_{2n+1}(iz\sqrt{t}).$

11. $\sum_{k=0}^{\infty} \dfrac{t^k}{\left(\frac{3}{2}\right)_k} C_{2k+1}^{-k-1/2}(z) = -\dfrac{1}{2}\sqrt{\dfrac{\pi}{t}} e^t \operatorname{erf}(\sqrt{t}\, z).$

12. $\sum_{k=0}^{\infty} \dfrac{t^k}{\left(\frac{3}{2}\right)_k (k+1)} C_{2k+1}^{-k-1/2}(i) = -\dfrac{\pi i}{4t} \operatorname{erfi}^2(\sqrt{t}).$

13. $\sum_{k=0}^{\infty} \binom{k+n}{k} \dfrac{(1-2\lambda-n)_k}{(1-\lambda)_k} t^k C_{k+n}^{\lambda-k}(z)$
$$= [(1+2tz)^2 - 4t^2]^{\lambda-1/2} C_n^{\lambda}(z - 2t + 2tz^2).$$

6.15.2. Series containing $C_{nk+m}^{\lambda\pm lk}(z)$ and special functions

1. $\sum_{k=0}^{\infty} (-1)^k (2k+\lambda) J_{2k+\lambda}(w) C_{2k}^{\lambda}(z) = \dfrac{\left(\frac{w}{2}\right)^{\lambda}}{\Gamma(\lambda)} \cos(wz).$

2. $\sum_{k=0}^{\infty} (-1)^k (2k+\lambda+1) J_{2k+\lambda+1}(w) C_{2k+1}^{\lambda}(z) = \dfrac{\left(\frac{w}{2}\right)^{\lambda}}{\Gamma(\lambda)} \sin(wz).$

3. $\sum_{k=0}^{\infty} (2k+\lambda) \dfrac{\left(\frac{1}{2}\right)_k}{\left(\lambda+\frac{1}{2}\right)_k} J_{2k+\lambda}(w) C_{2k}^{\lambda}(z)$
$$= \sqrt{\dfrac{w}{2}} (z^2-1)^{(1-2\lambda)/4} \dfrac{\Gamma\left(\lambda+\frac{1}{2}\right)}{\Gamma(\lambda)} I_{\lambda-1/2}\left(w\sqrt{z^2-1}\right).$$

4. $\sum_{k=0}^{\infty} (2k+\lambda+1) \dfrac{\left(\frac{3}{2}\right)_k}{\left(\lambda+\frac{1}{2}\right)_k} J_{2k+\lambda+1}(w) C_{2k+1}^{\lambda}(z)$
$$= \sqrt{\dfrac{w^3}{2}} z(z^2-1)^{(1-2\lambda)/4} \dfrac{\Gamma\left(\lambda+\frac{1}{2}\right)}{\Gamma(\lambda)} I_{\lambda-1/2}\left(w\sqrt{z^2-1}\right).$$

5. $\sum_{k=0}^{\infty} (4k-2\lambda+1) \dfrac{\left(\frac{1}{2}-\lambda\right)_k \left(\frac{1}{2}\right)_k}{(1-\lambda)_{2k}} (z^2-1)^{-k} J_{2k-\lambda+1/2}(w) C_{2k}^{\lambda-2k}(z)$
$$= \sqrt{2w}\,(z^2-1)^{-\lambda/2} \dfrac{\Gamma(1-\lambda)}{\Gamma\left(\frac{1}{2}-\lambda\right)} I_{-\lambda}\left(\dfrac{w}{\sqrt{z^2-1}}\right).$$

6.15.2]	6.15. The Gegenbauer Polynomials $C_n^\lambda(z)$

6. $\sum_{k=0}^{\infty}(4k - 2\lambda + 1) \dfrac{\left(\frac{1}{2}-\lambda\right)_k \left(\frac{3}{2}\right)_k}{(1-\lambda)_{2k}} (z^2 - 1)^{-k} J_{2k-\lambda+1/2}(w) C_{2k+1}^{\lambda-2k}(z)$

$$= -\sqrt{2w^3} z \, (z^2 - 1)^{-(\lambda+1)/2} \dfrac{\Gamma(1-\lambda)}{\Gamma\left(\frac{1}{2}-\lambda\right)} I_{-\lambda-1}\left(\dfrac{w}{\sqrt{z^2-1}}\right).$$

7. $\sum_{k=0}^{\infty}(2k + \lambda) \dfrac{\left(\frac{1}{2}\right)_k}{\left(\lambda+\frac{1}{2}\right)_k} J_{k+\nu}(w) \, J_{k+\lambda-\nu}(w) C_{2k}^{\lambda}(z)$

$$= \dfrac{\lambda \left(\frac{w}{2}\right)^{\lambda}}{\Gamma(\nu+1)\Gamma(\lambda-\nu+1)} {}_2F_3\left(\begin{array}{c}\frac{\lambda+1}{2}, \frac{\lambda}{2}+1; \, w^2 z^2 - w^2 \\ \lambda+\frac{1}{2}, \nu+1, \lambda-\nu+1 \end{array}\right).$$

8. $\sum_{k=0}^{\infty}(2k + \lambda + 1) \dfrac{\left(\frac{3}{2}\right)_k}{\left(\lambda+\frac{1}{2}\right)_k} J_{k+\nu}(w) \, J_{k+\lambda-\nu+1}(w) C_{2k+1}^{\lambda}(z)$

$$= \dfrac{2^{-\lambda}\lambda(\lambda+1) w^{\lambda+1} z}{\Gamma(\nu+1)\Gamma(\lambda-\nu+2)} {}_2F_3\left(\begin{array}{c}\frac{\lambda}{2}+1, \frac{\lambda+3}{2}; \, w^2 z^2 - w^2 \\ \lambda+\frac{1}{2}, \nu+1, \lambda-\nu+2 \end{array}\right).$$

9. $\sum_{k=0}^{\infty}(4k - 2\lambda + 1) \dfrac{\left(\frac{1}{2}-\lambda\right)_k \left(\frac{1}{2}\right)_k}{(1-\lambda)_{2k}} (z^2 - 1)^{-k} J_{k+\nu}(w) \, J_{k-\lambda-\nu+1/2}(w)$

$$\times C_{2k}^{\lambda-2k}(z) = \dfrac{(1-2\lambda)\left(\frac{w}{2}\right)^{1/2-\lambda}}{\Gamma(\nu+1)\Gamma\left(\frac{3}{2}-\lambda-\nu\right)} {}_2F_3\left(\begin{array}{c}\frac{3-2\lambda}{4}, \frac{5-2\lambda}{4}; \, \frac{w^2}{z^2-1} \\ 1-\lambda, \nu+1, \frac{3}{2}-\lambda-\nu \end{array}\right).$$

10. $\sum_{k=0}^{\infty}(4k - 2\lambda + 1) \dfrac{\left(\frac{1}{2}-\lambda\right)_k \left(\frac{3}{2}\right)_k}{(-\lambda)_{2k+1}} (z^2 - 1)^{-k} J_{k+\nu}(w) \, J_{k-\lambda-\nu+1/2}(w)$

$$\times C_{2k+1}^{\lambda-2k}(z) = \dfrac{2^{\lambda+1/2}(2\lambda-1) w^{1/2-\lambda} z}{\Gamma(\nu+1)\Gamma\left(\frac{3}{2}-\lambda-\nu\right)} {}_2F_3\left(\begin{array}{c}\frac{3-2\lambda}{4}, \frac{5-2\lambda}{4}; \, \frac{w^2}{z^2-1} \\ -\lambda, \nu+1, \frac{3}{2}-\lambda-\nu \end{array}\right).$$

11. $\sum_{k=0}^{\infty}(k + \lambda) \, I_{k+\lambda}(w) C_k^{\lambda}(z) = \dfrac{\left(\frac{w}{2}\right)^{\lambda}}{\Gamma(\lambda)} e^{wz}.$

12. $\sum_{k=0}^{\infty}(2k + \lambda) \, I_{2k+\lambda}(w) C_{2k}^{\lambda}(z) = \dfrac{\left(\frac{w}{2}\right)^{\lambda}}{\Gamma(\lambda)} \cosh(wz).$

13. $\sum_{k=0}^{\infty}(2k + \lambda + 1) \, I_{2k+\lambda+1}(w) C_{2k+1}^{\lambda}(z) = \dfrac{\left(\frac{w}{2}\right)^{\lambda}}{\Gamma(\lambda)} \sinh(wz).$

14. $$\sum_{k=0}^{\infty}(-1)^k(2k+\lambda)\frac{\left(\frac{1}{2}\right)_k}{\left(\lambda+\frac{1}{2}\right)_k}I_{2k+\lambda}(w)C_{2k}^{\lambda}(z)$$

$$=\sqrt{\frac{w}{2}}\,(z^2-1)^{(1-2\lambda)/4}\frac{\Gamma\left(\lambda+\frac{1}{2}\right)}{\Gamma(\lambda)}J_{\lambda-1/2}\left(w\sqrt{z^2-1}\right).$$

15. $$\sum_{k=0}^{\infty}(-1)^k(2k+\lambda+1)\frac{\left(\frac{3}{2}\right)_k}{\left(\lambda+\frac{1}{2}\right)_k}I_{2k+\lambda+1}(w)C_{2k+1}^{\lambda}(z)$$

$$=\sqrt{\frac{w^3}{2}}\,z(z^2-1)^{(1-2\lambda)/4}\frac{\Gamma\left(\lambda+\frac{1}{2}\right)}{\Gamma(\lambda)}J_{\lambda-1/2}\left(w\sqrt{z^2-1}\right).$$

16. $$\sum_{k=0}^{\infty}(4k-2\lambda+1)\frac{\left(\frac{1}{2}-\lambda\right)_k\left(\frac{1}{2}\right)_k}{(1-\lambda)_{2k}}(1-z^2)^{-k}I_{k+(1-2\lambda)/4}(w)C_{2k}^{\lambda-2k}(z)$$

$$=\frac{2^{(2\lambda+7)/4}}{\Gamma\left(\frac{1-2\lambda}{4}\right)}w^{(1-2\lambda)/4}e^{-w}\,{}_1F_1\!\left(\begin{array}{c}\frac{3-2\lambda}{4};\ \frac{2w}{1-z^2}\\ 1-\lambda\end{array}\right).$$

17. $$\sum_{k=0}^{\infty}(4k-2\lambda+1)\frac{\left(\frac{1}{2}-\lambda\right)_k\left(\frac{3}{2}\right)_k}{(1-\lambda)_{2k}}(1-z^2)^{-k}I_{k+(1-2\lambda)/4}(w)C_{2k+1}^{\lambda-2k}(z)$$

$$=\frac{2^{(2\lambda+11)/4}}{\Gamma\left(\frac{1-2\lambda}{4}\right)}\lambda z w^{(1-2\lambda)/4}e^{-w}\,{}_1F_1\!\left(\begin{array}{c}\frac{3-2\lambda}{4}\\ -\lambda;\ \frac{2w}{1-z^2}\end{array}\right).$$

18. $$\sum_{k=0}^{\infty}(k+\lambda)\,I_{(k+\lambda)/2}^2(w)C_k^{\lambda}(z)=2^{2-\lambda}w^{\lambda}$$

$$\times\left[\frac{1}{\lambda\Gamma^2\left(\frac{\lambda}{2}\right)}\,{}_1F_2\!\left(\begin{array}{c}\frac{\lambda+1}{2};\ w^2z^2\\ \frac{\lambda}{2}+1,\frac{1}{2}\end{array}\right)+\frac{\lambda wz}{(\lambda+1)\Gamma^2\left(\frac{\lambda+1}{2}\right)}\,{}_1F_2\!\left(\begin{array}{c}\frac{\lambda}{2}+1;\ w^2z^2\\ \frac{\lambda+3}{2},\frac{3}{2}\end{array}\right)\right].$$

6.15.3. Series containing products of $C_{nk+m}^{\lambda\pm lk}(z)$

1. $$\sum_{k=0}^{\infty}\frac{k!\,(1-2\lambda)_k}{(1-\lambda)_k^2}t^k C_k^{\lambda-k}(w)\,C_k^{\lambda-k}(z)=(u_-u_+)^{\lambda-1/2}\,{}_2F_1\!\left(\begin{array}{c}\frac{1}{2}-\lambda,\frac{1}{2}-\lambda\\ 1-2\lambda;\ \frac{16t}{u_-u_+}\end{array}\right)$$

$$\left[u_-=1-4t(w+1)(z-1),\ u_+=1-4t(w-1)(z+1);\ |16tu_-^{-1}u_+^{-1}|<1\right].$$

2. $\displaystyle\sum_{k=0}^{\infty}(-4)^{k}k!\,(2k+\lambda)\,\frac{\left(\frac{1}{2}\right)_{k}}{(2\lambda)_{2k}}\,J_{2k+\lambda}(w)$

$\times C_{2k}^{\lambda}\left(\sqrt{\dfrac{1+z}{2}}\right)C_{2k}^{\lambda}\left(\sqrt{\dfrac{1-z}{2}}\right)=2^{\lambda-1}w^{1/2}(1-z^{2})^{(1-2\lambda)/4}\,\dfrac{\Gamma\!\left(\lambda+\frac{1}{2}\right)}{\Gamma(\lambda)}$

$\times \cos\left(\dfrac{w}{2}\sqrt{1-z^{2}}\right)J_{\lambda-1/2}\!\left(\dfrac{w}{2}\sqrt{1-z^{2}}\right).$

3. $\displaystyle\sum_{k=0}^{\infty}(-4)^{k}k!\,(2k+\lambda+1)\,\frac{\left(\frac{3}{2}\right)_{k}}{(2\lambda)_{2k+1}}\,J_{2k+\lambda+1}(w)$

$\times C_{2k+1}^{\lambda}\left(\sqrt{\dfrac{1+z}{2}}\right)C_{2k+1}^{\lambda}\left(\sqrt{\dfrac{1-z}{2}}\right)=2^{\lambda-1}w^{1/2}(1-z^{2})^{(1-2\lambda)/4}\,\dfrac{\Gamma\!\left(\lambda+\frac{1}{2}\right)}{\Gamma(\lambda)}$

$\times \sin\left(\dfrac{w}{2}\sqrt{1-z^{2}}\right)J_{\lambda-1/2}\!\left(\dfrac{w}{2}\sqrt{1-z^{2}}\right).$

4. $\displaystyle\sum_{k=0}^{\infty}(2k+\lambda)\,\frac{(2k)!}{(2\lambda)_{2k}}\,I_{k+\lambda/2}(w)C_{2k}^{\lambda}\left(\sqrt{\dfrac{1+z}{2}}\right)C_{2k}^{\lambda}\left(\sqrt{\dfrac{1-z}{2}}\right)$

$=\dfrac{2^{1-\lambda/2}}{\Gamma\!\left(\frac{\lambda}{2}\right)}w^{\lambda/2}e^{-w}\,{}_{3}F_{3}\!\left(\begin{array}{c}\frac{\lambda}{2},\frac{\lambda+1}{2},\frac{\lambda+1}{2};\,2w-2wz^{2}\\ \frac{1}{2},\lambda,\lambda+\frac{1}{2}\end{array}\right).$

5. $\displaystyle\sum_{k=0}^{\infty}(2k+\lambda+1)\,\frac{(2k+1)!}{(2\lambda+1)_{2k}}\,I_{k+(\lambda+1)/2}(w)$

$\times C_{2k+1}^{\lambda}\left(\sqrt{\dfrac{1+z}{2}}\right)C_{2k+1}^{\lambda}\left(\sqrt{\dfrac{1-z}{2}}\right)=\dfrac{2^{(3-\lambda)/2}\lambda^{2}}{\Gamma\!\left(\frac{\lambda+1}{2}\right)}w^{(\lambda+1)/2}\sqrt{1-z^{2}}\,e^{-w}$

$\times {}_{3}F_{3}\!\left(\begin{array}{c}\frac{\lambda+1}{2},\frac{\lambda}{2}+1,\frac{\lambda}{2}+1;\,2w-2wz^{2}\\ \frac{3}{2},\lambda+\frac{1}{2},\lambda+1\end{array}\right).$

6.15.4. Series containing $C_{nk+m}^{\lambda\pm lk}(\varphi(k,z))$

1. $\displaystyle\sum_{k=0}^{\infty}\frac{(k+1)^{-1}}{(1-\lambda)_{k}}t^{k}C_{k}^{\lambda-k}(1+(k+1)z)=\dfrac{2\lambda ze^{-w}}{(2\lambda+1)w}\left[1-{}_{1}F_{1}\!\left(\begin{array}{c}-\lambda-\frac{1}{2}\\ -2\lambda;\,-\frac{2w}{z}\end{array}\right)\right]$

$\left[t=\dfrac{w}{2z}e^{w};\,|we^{w+1}|<1\right].$

2. $\displaystyle\sum_{k=0}^{\infty}\frac{(k+1)^{k-1}}{(1-\lambda)_{k}}t^{k}C_{2k}^{\lambda-k}\!\left(\dfrac{z}{\sqrt{k+1}}\right)=\dfrac{e^{-w}}{2(\lambda-1)wz^{2}}\left[1-{}_{1}F_{1}\!\left(\begin{array}{c}\lambda-1\\ -\frac{1}{2};\,wz^{2}\end{array}\right)\right]$

$\left[t=-we^{w};\,|we^{w+1}|<1\right].$

3. $\displaystyle\sum_{k=0}^{\infty} \frac{(k+1)^{k-1/2}}{(1-\lambda)_k} t^k C_{2k+1}^{\lambda-k}\left(\frac{z}{\sqrt{k+1}}\right) = \frac{e^{-w}}{wz}\left[{}_1F_1\left(\begin{array}{c}\lambda;\ wz^2\\ \frac{1}{2}\end{array}\right) - 1\right]$

$$[t = -we^w;\ |we^{w+1}| < 1].$$

4. $\displaystyle\sum_{k=0}^{\infty} \frac{(k+1)^{-1}}{(1-\lambda)_k} t^k (k+z)^k C_{2k}^{\lambda-k}\left(\sqrt{\frac{z-1}{z+k}}\right)$

$$= \frac{e^{-w}}{(2\lambda+1)w(z-1)}\left[1 - {}_1F_1\left(\begin{array}{c}-\lambda-\frac{1}{2}\\ -\frac{1}{2};\ w(1-z)\end{array}\right)\right]$$

$$[t = -we^w;\ |we^{w+1}| < 1].$$

5. $\displaystyle\sum_{k=0}^{\infty} \frac{(k+z)^{k+1/2}}{(1-\lambda)_k(k+1)} t^k C_{2k+1}^{\lambda-k}\left(\sqrt{\frac{z-1}{z+k}}\right)$

$$= \frac{2\lambda e^{-w}}{(2\lambda+1)w\sqrt{z-1}}\left[{}_1F_1\left(\begin{array}{c}-\lambda-\frac{1}{2}\\ \frac{1}{2};\ w(1-z)\end{array}\right) - 1\right]$$

$$[t = -we^w;\ |we^{w+1}| < 1].$$

6.15.5. Series containing $C_{nk+m}^{\lambda \pm lk}(\varphi(k,z))$ and special functions

1. $\displaystyle\sum_{k=0}^{\infty} \frac{(k+1)^{-(k+\nu)/2-1}}{(1-\lambda)_k}\left(-\frac{w}{4z}\right)^k J_{k+\nu}(\sqrt{k+1}\,w) C_k^{\lambda-k}((k+1)z+1)$

$$= \frac{2\lambda\left(\frac{w}{2}\right)^{\nu-2} z}{(2\lambda+1)\Gamma(\nu)}\left[{}_1F_2\left(\begin{array}{c}-\lambda-\frac{1}{2};\ \frac{w^2 z^{-1}}{2}\\ -2\lambda,\ \nu\end{array}\right) - 1\right].$$

2. $\displaystyle\sum_{k=0}^{\infty} \frac{(k+1)^{(k-\nu)/2-1}}{(1-\lambda)_k}\left(\frac{w}{2}\right)^k J_{k+\nu}(\sqrt{k+1}\,w) C_{2k}^{\lambda-k}\left(\frac{z}{\sqrt{k+1}}\right)$

$$= \frac{\left(\frac{w}{2}\right)^{\nu-2} z^{-2}}{2(\lambda-1)\Gamma(\nu)}\left[{}_1F_2\left(\begin{array}{c}\lambda-1;\ -\frac{w^2 z^2}{4}\\ -\frac{1}{2},\ \nu\end{array}\right) - 1\right].$$

3. $\displaystyle\sum_{k=0}^{\infty} \frac{(k+1)^{(k-\nu-1)/2}}{(1-\lambda)_k}\left(\frac{w}{2}\right)^k J_{k+\nu}(\sqrt{k+1}\,w) C_{2k+1}^{\lambda-k}\left(\frac{z}{\sqrt{k+1}}\right)$

$$= \frac{\left(\frac{w}{2}\right)^{\nu-2} z^{-1}}{\Gamma(\nu)}\left[1 - {}_1F_2\left(\begin{array}{c}\lambda;\ -\frac{w^2 z^2}{4}\\ \frac{1}{2},\ \nu\end{array}\right)\right].$$

6.16. The Jacobi Polynomials $P_n^{(\rho,\sigma)}(z)$

6.16.1. Series containing $P_{m\pm nk}^{(\rho\pm pk,\sigma\pm qk)}(z)$

1. $\displaystyle\sum_{k=0}^{\infty} \frac{t^k}{(\sigma+1)_k} P_k^{(\rho-k,\sigma)}(z) = e^{-t} {}_1F_1\left(\begin{matrix}\rho+\sigma+1\\\sigma+1;\end{matrix}\frac{t(z+1)}{2}\right).$

2. $\displaystyle\sum_{k=0}^{\infty} \frac{(a)_k}{(\sigma+1)_k} t^k P_k^{(\rho-k,\sigma)}(z) = (1+t)^{-a} {}_2F_1\left(\begin{matrix}a,\rho+\sigma+1\\\sigma+1;\end{matrix}\frac{t(1+z)}{2(t+1)}\right)$ $[|t|<1].$

3. $\displaystyle\sum_{k=0}^{\infty}(ka+b)^m \frac{(\rho+\sigma+m+1)_k}{(\rho+m+1)_k}\left(\frac{1-z}{2}\right)^k P_m^{(\rho+k,\sigma)}(z)$

$$= \frac{(\rho+1)_m}{m!} b^m \sum_{k=0}^{m}\binom{m}{k}\left(\frac{a}{b}\right)^k$$

$\times \displaystyle\sum_{j=0}^{k} \sigma_k^j \frac{j!(\rho+\sigma+m+1)_j}{(\rho+1)_j}\left(\frac{1-z}{2}\right)^j {}_2F_1\left(\begin{matrix}-m+j+1,\rho+\sigma+j+m+1\\\rho+j+1;\end{matrix}\frac{1-z}{2}\right)$

$$[|1-z|<2].$$

4. $\displaystyle\sum_{k=0}^{\infty} \frac{t^k}{(\rho+1)_k} P_k^{(\rho,-\rho-k)}(z) = \rho\left(\frac{2}{tz-t}\right)^\rho e^{t(z+1)/2} \gamma\left(\rho,\frac{tz-t}{2}\right).$

5. $\displaystyle\sum_{k=0}^{\infty} \frac{t^k}{(-\rho-\sigma)_k} P_k^{(\rho-k,\sigma-k)}(z) = e^{-t(z+1)/2} {}_1F_1\left(\begin{matrix}-\sigma;\\-\rho-\sigma\end{matrix}t\right).$

6. $\displaystyle\sum_{k=0}^{\infty} \frac{(a)_k}{(-\rho-\sigma)_k} t^k P_k^{(\rho-k,\sigma-k)}(z)$

$$= \left(1+\frac{tz-t}{2}\right)^{-a} {}_2F_1\left(\begin{matrix}a,-\rho\\-\rho-\sigma;\end{matrix}\frac{2t}{t-tz-2}\right) \quad [|t|<1].$$

7. $\displaystyle\sum_{k=0}^{\infty}\binom{k+n}{k} t^k P_{k+n}^{(\rho-k,\sigma-k)}(z)$

$$= \left(1+\frac{t(z+1)}{2}\right)^\rho \left(1+\frac{t(z-1)}{2}\right)^\sigma P_n^{(\rho,\sigma)}\left(z+\frac{tz^2-t}{2}\right).$$

8. $\displaystyle\sum_{k=0}^{\infty} \frac{(-\rho-n)_k}{k!} t^k P_n^{(\rho-k,\sigma+k)}(z) = (1-t)^\rho P_n^{(\rho,\sigma)}(t+z-tz)$ $[|t|<1].$

9. $\displaystyle\sum_{k=0}^{\infty} \frac{(a)_k(\frac{1}{2}-a+n)_k(\rho+\sigma+n+1)_k}{k!\,(\rho+n+1)_{2k}} 2^k (1-z)^k P_{k+n}^{(\rho+k,\sigma-k)}(z)$

$$= \frac{(\rho+1)_n}{n!} {}_3F_2\left(\begin{matrix} 2a-n,\, n-2a+1,\, \rho+\sigma+n+1 \\ n+1,\, \rho+1;\; \frac{1-z}{2} \end{matrix}\right).$$

10. $\displaystyle\sum_{k=0}^{\infty} \frac{(a)_k\left(\frac{1}{2}-a+n\right)_k(-\rho-n)_k}{k!\,(-\rho-\sigma-n)_{2k}} \left(\frac{4}{z-1}\right)^{2k} P_{k+n}^{(\rho-2k,\sigma-k)}(z)$

$$= \frac{(\rho+\sigma+1)_{2n}}{n!\,(\rho+\sigma+1)_n} \left(\frac{z-1}{2}\right)^n {}_3F_2\left(\begin{matrix} 2a-n,\, n-2a+1,\, -\rho-n \\ n+1,\, -\rho-\sigma-2n;\; \frac{2}{1-z} \end{matrix}\right).$$

6.16.2. Series containing $P_{m\pm nk}^{(\rho\pm pk,\,\sigma\pm qk)}(z)$ and special functions

1. $\displaystyle\sum_{k=0}^{\infty} (2k+\rho+\sigma+1)\, \frac{(\rho+\sigma+1)_k}{(\rho+1)_k} J_{k+a}(w)\, J_{k+\rho+\sigma-a+1}(w)\, P_k^{(\rho,\sigma)}(z)$

$$= \frac{(\rho+\sigma+1)\left(\frac{w}{2}\right)^{\rho+\sigma+1}}{\Gamma(a+1)\Gamma(\rho+\sigma-a+2)} {}_2F_3\left(\begin{matrix} \frac{\rho+\sigma}{2}+1,\, \frac{\rho+\sigma+3}{2};\; \frac{w^2 z-w^2}{2} \\ a+1,\, \rho+1,\, \rho+\sigma-a+2 \end{matrix}\right).$$

2. $\displaystyle\sum_{k=0}^{\infty} (2k-\rho)\, \frac{(-\rho)_k}{(-\rho-\sigma)_k} \left(\frac{2}{1-z}\right)^k J_{k+a}(w)\, J_{k-\rho-a}(w)\, P_k^{(\rho-2k,\,\sigma)}(z)$

$$= -\frac{\rho\left(\frac{2}{w}\right)^{\rho}}{\Gamma(a+1)\Gamma(1-\rho-a)} {}_2F_3\left(\begin{matrix} \frac{1-\rho}{2},\, 1-\frac{\rho}{2};\; \frac{2w^2}{z-1} \\ a+1,\, -\rho-\sigma,\, 1-\rho-a \end{matrix}\right).$$

3. $\displaystyle\sum_{k=0}^{\infty} (2k-\rho)\, \frac{(-\rho)_k}{(-\rho-\sigma)_k} \left(\frac{2}{z-1}\right)^k I_{k-\rho/2}(w)\, P_k^{(\rho-2k,\,\sigma)}(z)$

$$= \frac{2}{\Gamma\left(-\frac{\rho}{2}\right)} \left(\frac{2}{w}\right)^{\rho/2} e^{-w}\, {}_1F_1\left(\begin{matrix} \frac{1-\rho}{2};\; \frac{4w}{1-z} \\ -\rho-\sigma \end{matrix}\right).$$

4. $\displaystyle\sum_{k=0}^{\infty} \frac{(\rho+\sigma+n+1)_k}{(1-\lambda)_k} \left(\frac{1-z}{2}\right)^k C_{2k}^{\lambda-k}(w)\, P_n^{(\rho+k,\,\sigma)}(z)$

$$= \frac{(\rho+1)_n}{n!} \left(\frac{1+z}{2}\right)^{-\rho-\sigma-n-1} {}_3F_2\left(\begin{matrix} \lambda,\, \rho+n+1,\, \rho+\sigma+n+1 \\ \frac{1}{2},\, \rho+1;\; \frac{w^2(z-1)}{z+1} \end{matrix}\right)$$

$$[|1-z|<2].$$

5. $\displaystyle\sum_{k=0}^{\infty} \frac{(\rho+\sigma+n+1)_k}{(1-\lambda)_k} \left(\frac{1-z}{2}\right)^k C_{2k+1}^{\lambda-k}(w) P_n^{(\rho+k,\sigma)}(z)$

$= 2\lambda w \dfrac{(\rho+1)_n}{n!} \left(\dfrac{1+z}{2}\right)^{-\rho-\sigma-n-1} {}_3F_2\left(\begin{array}{c} \lambda+1,\ \rho+n+1,\ \rho+\sigma+n+1 \\ \frac{3}{2},\ \rho+1;\ \dfrac{w^2(z-1)}{z+1} \end{array}\right)$

$$[|1-z|<2].$$

6.16.3. Series containing products of $P_{m\pm nk}^{(\rho\pm pk,\sigma\pm qk)}(z)$

1. $\displaystyle\sum_{k=0}^{\infty} (-1)^k (2k+\rho+\sigma+1) \frac{k!\,(\rho+\sigma+1)_k}{(\rho+1)_k (\sigma+1)_k}$

$\times J_{k+a}(w) J_{k+\rho+\sigma-a+1}(w) P_k^{(\rho,\sigma)}(z) P_k^{(\rho,\sigma)}(-z) = \dfrac{(\rho+\sigma+1)\left(\frac{w}{2}\right)^{\rho+\sigma+1}}{\Gamma(a+1)\Gamma(\rho+\sigma-a+2)}$

$\times {}_4F_5\left(\begin{array}{c} \dfrac{\rho+\sigma+1}{2},\ \dfrac{\rho+\sigma}{2}+1,\ \dfrac{\rho+\sigma}{2}+1,\ \dfrac{\rho+\sigma+3}{2};\ w^2z^2 - w^2 \\ a+1,\ \rho+1,\ \sigma+1,\ \rho+\sigma+1,\ \rho+\sigma-a+2 \end{array}\right).$

6.16.4. Series containing $P_{m\pm nk}^{(\rho\pm pk,\sigma\pm qk)}(\varphi(k,z))$

Notation: $t = -we^w$, $|we^{w+1}| < 1$.

1. $\displaystyle\sum_{k=0}^{\infty} \frac{(k+1)^{-1}}{(-\rho-\sigma)_k} \left(-\frac{2t}{z}\right)^k P_k^{(\rho-k,\sigma-k)}(1+(k+1)z)$

$= \dfrac{(\rho+\sigma+1)ze^{-w}}{2(\rho+1)w} \left[1 - {}_1F_1\left(\begin{array}{c} -\rho-1;\ -\dfrac{2w}{z} \\ -\rho-\sigma-1 \end{array}\right)\right].$

2. $\displaystyle\sum_{k=0}^{\infty} \frac{(k+1)^{k-1}}{k!\,(\rho+1)_k} t^k P_k^{(\rho,\sigma-k)}\left(1+\frac{z}{k+1}\right)$

$= \dfrac{2\rho e^{-w}}{(\rho+\sigma)wz} \left[1 - {}_1F_1\left(\begin{array}{c} \rho+\sigma;\ -\dfrac{wz}{2} \\ \rho \end{array}\right)\right].$

6.16.5. Series containing $P_{m\pm nk}^{(\rho\pm pk,\sigma\pm qk)}(\varphi(k,z))$ and special functions

1. $\displaystyle\sum_{k=0}^{\infty} \frac{(k+1)^{-(k+\nu)/2-1}}{(-\rho-\sigma)_k} \left(-\frac{w}{z}\right)^k J_{k+\nu}(\sqrt{k+1}\,w) P_k^{(\rho-k,\sigma-k)}((k+1)z-1)$

$= \dfrac{(\rho+\sigma+1)\left(\frac{w}{2}\right)^{\nu-2} z}{2(\sigma+1)\Gamma(\nu)} \left[1 - {}_1F_2\left(\begin{array}{c} -\sigma-1;\ -\dfrac{w^2 z^{-1}}{2} \\ -\rho-\sigma-1,\ \nu \end{array}\right)\right].$

2. $\sum_{k=0}^{\infty} \frac{(k+1)^{(k-\nu)/2-1}}{(\sigma+1)_k} \left(-\frac{w}{2}\right)^k J_{k+\nu}(\sqrt{k+1}\, w) P_k^{(\rho-k,\sigma)}\left(\frac{z}{k+1}-1\right)$

$$= \frac{2\sigma \left(\frac{w}{2}\right)^{\nu-2} z^{-1}}{(\rho+\sigma)\Gamma(\nu)} \left[1 - {}_1F_2\left(\begin{matrix}\rho+\sigma;\\ \sigma, \nu\end{matrix}\ -\frac{w^2 z}{8}\right)\right].$$

3. $\sum_{k=0}^{\infty} \frac{(k+1)^{-(k+\nu)/2-1}}{(\sigma+1)_k} \left(\frac{w}{2}\right)^k$

$\times (z-k-1)^k J_{k+\nu}(\sqrt{k+1}\, w) P_k^{(\rho-k,\sigma)}\left(\frac{z+k+1}{z-k-1}\right)$

$$= \frac{\sigma \left(\frac{w}{2}\right)^{\nu-2} z^{-1}}{(\rho+1)\Gamma(\nu)} \left[{}_1F_2\left(\begin{matrix}-\rho-1;\\ \sigma, \nu\end{matrix}\ -\frac{w^2 z}{4}\right) - 1\right].$$

4. $\sum_{k=0}^{\infty} \frac{(k+1)^{-(k+\nu)/2-1}}{(-\rho-\sigma)_k} \left(\frac{w}{z}\right)^k I_{k+\nu}(\sqrt{k+1}\, w) P_k^{(\rho-k,\sigma-k)}((k+1)z-1)$

$$= \frac{(\rho+\sigma+1)\left(\frac{w}{2}\right)^{\nu-2} z}{2(\sigma+1)\Gamma(\nu)} \left[{}_1F_2\left(\begin{matrix}-\sigma-1;\ \frac{w^2 z^{-1}}{2}\\ -\rho-\sigma-1, \nu\end{matrix}\right) - 1\right].$$

5. $\sum_{k=0}^{\infty} \frac{(k+1)^{(k-\nu)/2-1}}{(\sigma+1)_k} \left(\frac{w}{z}\right)^k I_{k+\nu}(\sqrt{k+1}\, w) P_k^{(\rho-k,\sigma)}\left(\frac{z}{k+1}-1\right)$

$$= \frac{2\sigma \left(\frac{w}{2}\right)^{\nu-2} z^{-1}}{(\rho+\sigma)\Gamma(\nu)} \left[{}_1F_2\left(\begin{matrix}\rho+\sigma;\ \frac{w^2 z}{8}\\ \sigma, \nu\end{matrix}\right) - 1\right].$$

6. $\sum_{k=0}^{\infty} \frac{(k+1)^{-(k+\nu)/2-1}}{(\sigma+1)_k} \left(-\frac{w}{2}\right)^k (z-k-1)^k I_{k+\nu}(\sqrt{k+1}\, w)$

$\times P_k^{(\rho-k,\sigma)}\left(\frac{z+k+1}{z-k-1}\right) = \frac{\sigma\left(\frac{w}{2}\right)^{\nu-2} z^{-1}}{(\rho+1)\Gamma(\nu)} \left[1 - {}_1F_2\left(\begin{matrix}-\rho-1;\ \frac{w^2 z}{4}\\ \sigma, \nu\end{matrix}\right)\right].$

6.17. The Generalized Hypergeometric Function ${}_pF_q((a_p); (b_q); z)$

6.17.1. Series containing ${}_pF_q((a_p(k)); (b_q(k)); z)$

1. $\sum_{k=0}^{\infty} \frac{t^k}{k!\,(b)_k} {}_1F_1\left(\begin{matrix}a;\ z\\ b+k\end{matrix}\right) = \Phi_3(a;\ b;\ z, t).$

2. $\sum_{k=0}^{\infty} \frac{(-z^2)^k}{k!\,(k+1)!} {}_1F_1\left(\begin{matrix}2;\ z\\ k+2\end{matrix}\right) = 1 + 2z\, J_0(2z) - J_1(2z)$

$$+ \pi z \left[J_1(2z)\, \mathbf{H}_0(2z) - J_0(2z)\, \mathbf{H}_1(2z)\right].$$

3. $\sum_{k=0}^{\infty} (-1)^k \dfrac{(a)_k (b-a)_k \left(b-\dfrac{1}{2}\right)_k \left(\dfrac{1}{2}\right)_k}{k!\,(b)_k (2b-1)_{4k}} (2z)^{2k} {}_1F_1\!\left(\begin{array}{c}2a+2k;\ z\\ 2b+4k\end{array}\right)$

$$= \left[{}_1F_1\!\left(\begin{array}{c}a;\ \dfrac{z}{2}\\ b\end{array}\right)\right]^2.$$

4. $\sum_{k=0}^{\infty} (-1)^k \dfrac{\left(a+\dfrac{1}{2}\right)_k \left(b-a+\dfrac{1}{2}\right)_k \left(b-\dfrac{1}{2}\right)_k \left(\dfrac{1}{2}\right)_k}{k!\,(b)_k (2b-1)_{4k}} (2z)^{2k} {}_1F_1\!\left(\begin{array}{c}2a+2k;\ z\\ 2b+4k\end{array}\right)$

$$= {}_1F_1\!\left(\begin{array}{c}a-\dfrac{1}{2}\\ b;\ \dfrac{z}{2}\end{array}\right) {}_1F_1\!\left(\begin{array}{c}a+\dfrac{1}{2}\\ b;\ \dfrac{z}{2}\end{array}\right).$$

5. $\sum_{k=0}^{\infty} \dfrac{(b)_k}{k!\,(c)_k} t^k {}_2F_1\!\left(\begin{array}{c}a,\ b+k;\ z\\ c+k\end{array}\right) = \Phi_1(b,a;\ c;\ z,t)$ $\qquad [|z|<1].$

6. $\sum_{k=0}^{\infty} \dfrac{t^k}{k!} {}_3F_2\!\left(\begin{array}{c}-k,\ k+1,\ z\\ 1,\ 1;\ 1\end{array}\right) = {}_1F_1\!\left(\begin{array}{c}z;\ -t\\ 1\end{array}\right) {}_1F_1\!\left(\begin{array}{c}1-z\\ 1;\ t\end{array}\right).$

7. $\sum_{k=0}^{\infty} \dfrac{z^k}{(k!)^2} {}_3F_2\!\left(\begin{array}{c}-k,\ \dfrac{1}{2},\ 1;\ 1\\ k+1,\ \dfrac{3}{2}\end{array}\right) = \dfrac{1}{4}\{-\pi I_1(4\sqrt{z})\,\mathbf{L}_0(4\sqrt{z})$

$$+ [1+I_0(4\sqrt{z})]\,[2+\pi\,\mathbf{L}_1(4\sqrt{z})]\}.$$

8. $\sum_{k=0}^{\infty} \dfrac{t^{2k+1}}{k!\,(2k+1)} {}_3F_2\!\left(\begin{array}{c}-k,\ -k-\dfrac{1}{2},\ a;\ z\\ \dfrac{1}{2}-k,\ a+1\end{array}\right) = \dfrac{\sqrt{\pi}\,a}{2(t^2z)^a}\,\mathrm{erfi}\,(t)\,\gamma(a,\,t^2z)$ $\qquad [9].$

9. $\sum_{k=0}^{\infty} \dfrac{(-z)^k}{k!\,(a)_k} {}_1F_2\!\left(\begin{array}{c}\dfrac{1}{2};\ z\\ a+k,\ \dfrac{3}{2}\end{array}\right) = \dfrac{\sqrt{\pi}}{2}\Gamma(a)\,z^{(1-2a)/4}\,\mathbf{H}_{a-3/2}(2\sqrt{z}).$

10. $\sum_{k=1}^{\infty} \dfrac{k}{(2k+1)!} t^k {}_1F_2\!\left(\begin{array}{c}k+\dfrac{1}{2};\ z\\ k+\dfrac{3}{2},\ \dfrac{3}{2}\end{array}\right)$

$$= \dfrac{1}{8}\sqrt{\dfrac{t}{z}}\left[\dfrac{\sinh(\sqrt{t}+2\sqrt{z})}{\sqrt{t}+2\sqrt{z}} - \dfrac{\sinh(\sqrt{t}-2\sqrt{z})}{\sqrt{t}-2\sqrt{z}}\right].$$

11. $\sum_{k=0}^{\infty} \dfrac{(-z)^k}{(k!)^2} {}_1F_2\!\left(\begin{array}{c}a;\ z\\ k+1,\ a+1\end{array}\right) = \Gamma(a+1)\,z^{-a/2}\,J_a(2\sqrt{z}).$

12. $\sum_{k=0}^{\infty} \dfrac{(-z)^k}{k!\,(a)_k} {}_1F_2\!\left(\begin{array}{c}\dfrac{1}{2};\ z\\ k+a,\ \dfrac{3}{2}\end{array}\right) = \dfrac{1}{2}\Gamma(a)\sqrt{\pi}\,z^{(1-2a)/4}\,\mathbf{H}_{a-3/2}(2\sqrt{z}).$

13. $\displaystyle\sum_{k=1}^{\infty} \frac{2k-1}{(2k)!} t^k {}_1F_2\!\left(\begin{array}{c} k;\ z \\ k+1,\ \frac{3}{2} \end{array}\right)$

$\displaystyle = \frac{t}{2(4z-t)} \left[2\cosh\sqrt{t}\cosh(2\sqrt{z}) - \sqrt{\frac{t}{z}}\sinh\sqrt{t}\sinh(2\sqrt{z}) - 2 \right].$

14. $\displaystyle\sum_{k=1}^{\infty} \frac{t^k}{(2k)!} {}_1F_2\!\left(\begin{array}{c} k;\ z \\ k+1,\ \frac{1}{2} \end{array}\right)$

$\displaystyle = \frac{t}{t-4z} \left[\cosh\sqrt{t}\cosh(2\sqrt{z}) - 2\sqrt{\frac{z}{t}}\sinh\sqrt{t}\sinh(2\sqrt{z}) - 1 \right].$

15. $\displaystyle\sum_{k=1}^{\infty} \frac{(4z)^k}{(2k)!} {}_1F_2\!\left(\begin{array}{c} k;\ z \\ k+1,\ \frac{1}{2} \end{array}\right) = \frac{1}{2}\sinh^2(2\sqrt{z}).$

16. $\displaystyle\sum_{k=1}^{\infty} \frac{2k-1}{(2k)!} (4z)^k\, {}_1F_2\!\left(\begin{array}{c} k;\ z \\ k+1,\ \frac{3}{2} \end{array}\right) = \frac{1}{2}\sinh^2(2\sqrt{z}).$

17. $\displaystyle\sum_{k=0}^{\infty} \frac{t^k}{(2k+1)!} {}_1F_2\!\left(\begin{array}{c} k+\frac{1}{2};\ z \\ k+\frac{3}{2},\ \frac{1}{2} \end{array}\right) = \frac{1}{2}\left[\frac{\sinh(\sqrt{t}-2\sqrt{z})}{\sqrt{t}-2\sqrt{z}} + \frac{\sinh(\sqrt{t}+2\sqrt{z})}{\sqrt{t}+2\sqrt{z}} \right].$

18. $\displaystyle\sum_{k=0}^{\infty} \frac{(4z)^k}{(2k+1)!} {}_1F_2\!\left(\begin{array}{c} k+\frac{1}{2};\ z \\ k+\frac{3}{2},\ \frac{1}{2} \end{array}\right) = \frac{1}{2} + \frac{1}{8\sqrt{z}}\sinh(4\sqrt{z}).$

19. $\displaystyle\sum_{k=0}^{\infty} \frac{(4z)^k}{k!(k+1)!} {}_1F_2\!\left(\begin{array}{c} 1;\ z \\ \frac{k}{2}+1,\ \frac{k+3}{2} \end{array}\right) = \frac{\sinh(4\sqrt{z})}{4\sqrt{z}}.$

20. $\displaystyle\sum_{k=0}^{\infty} \frac{(-4z)^k}{k!(k+2)!} {}_1F_2\!\left(\begin{array}{c} 1;\ z \\ \frac{k+3}{2},\ \frac{k}{2}+2 \end{array}\right) = \frac{1}{8z}\left[1 - J_0(4\sqrt{z})\right].$

21. $\displaystyle\sum_{k=0}^{\infty} \frac{(-z)^k}{k!\left(\frac{1}{2}\right)_k} {}_1F_2\!\left(\begin{array}{c} -\frac{1}{2};\ z \\ \frac{1}{2},\ k+\frac{1}{2} \end{array}\right) = 1 - \pi\sqrt{z}\,\mathbf{H}_0(2\sqrt{z}).$

22. $\displaystyle\sum_{k=0}^{\infty} \frac{(-16\sqrt{z})^k}{(k!)^2} {}_1F_4\!\left(\begin{array}{c} 1;\ -z \\ \frac{k+1}{4},\ \frac{k+2}{4},\ \frac{k+3}{4},\ \frac{k}{4}+1 \end{array}\right) = \frac{1}{2} J_0(8z^{1/4})$

$\displaystyle \hspace{8cm} + \frac{1}{2}\cos(4\sqrt{2}\,z^{1/4}).$

23. $\displaystyle\sum_{k=0}^{\infty} \frac{(-16\sqrt{z})^k}{k!(k+2)!} {}_1F_4\!\left(\begin{array}{c} 1;\ z \\ \frac{k+3}{4},\ \frac{k+4}{4},\ \frac{k+5}{4},\ \frac{k+6}{4} \end{array}\right) = \frac{\sin^2(4z^{1/4})}{32z^{1/2}}.$

6.17.1]   6.17. The Generalized Hypergeometric Function $_pF_q((a_p);(b_q);z)$

24. $\displaystyle\sum_{k=0}^{\infty} \frac{(-z^2)^k}{k!\,(k+1)!}\, _2F_2\!\left(\begin{matrix}2,\,2;\,z\\1,\,k+2\end{matrix}\right) = 2z + (4z^2+1)\,J_0(2z) - 2z\,J_1(2z)$
$$+ 2\pi z^2[J_1(2z)\,\mathbf{H}_0(2z) - J_0(2z)\,\mathbf{H}_1(2z)].$$

25. $\displaystyle\sum_{k=0}^{\infty} \frac{\left(\frac{z}{4}\right)^k}{\left(\frac{3}{2}\right)_k}\, _2F_2\!\left(\begin{matrix}k+\frac{1}{2},\,k+1;\,z\\k+\frac{3}{2},\,2k+2\end{matrix}\right) = \frac{1}{2}\sqrt{\frac{\pi}{z}}\,\mathrm{erfi}\,(\sqrt{z}).$

26. $\displaystyle\sum_{k=0}^{\infty} \frac{(-z)^k}{k!\,(k+1)!}\, _2F_3\!\left(\begin{matrix}1,\,1;\,z\\2,\,2,\,k+2\end{matrix}\right) = \frac{\pi}{2z}Y_0(2\sqrt{z})$
$$- \frac{1}{z}\left(\mathbf{C} + \frac{1}{2}\ln z\right) J_0(2\sqrt{z}).$$

27. $\displaystyle\sum_{k=0}^{\infty} \frac{(-z)^k}{k!\,(k+1)!}\, _2F_3\!\left(\begin{matrix}\frac{1}{2},\,1;\,z\\k+2,\,\frac{3}{2},\,2\end{matrix}\right) = \frac{1}{z}\left[J_0(2\sqrt{z}) - \cos(2\sqrt{z})\right].$

28. $\displaystyle\sum_{k=0}^{\infty} \frac{(-4z)^k}{(2k+1)!\,(2k+3)}\, _2F_3\!\left(\begin{matrix}\frac{1}{2},\,1;\,z\\k+\frac{5}{2},\,\frac{3}{2},\,2\end{matrix}\right)$
$$= \frac{1}{4z^{3/2}}\left[\sin(2\sqrt{z}) - \pi\sqrt{z}\,\mathbf{H}_{-1}(2\sqrt{z})\right].$$

29. $\displaystyle\sum_{k=0}^{\infty} \frac{(4z)^k}{k!\,(2a-1)_k}\, _2F_3\!\left(\begin{matrix}a,\,1;\,z\\a+\frac{k-1}{2},\,a+\frac{k}{2},\,a-1\end{matrix}\right)$
$$= 2^{3-2a}\Gamma(2a-2)z^{3/2-a}I_{2a-3}(4\sqrt{z}).$$

30. $\displaystyle\sum_{k=0}^{\infty} \frac{(4z)^k}{k!\,(k+2)!}\, _2F_3\!\left(\begin{matrix}1,\,1;\,z\\ \frac{k+3}{2},\,\frac{k}{2}+2,\,2\end{matrix}\right)$
$$= \frac{1}{8z}\{[\ln(2\sqrt{z}) + \mathbf{C}]\,I_0(4\sqrt{z}) + K_0(4\sqrt{z})\}.$$

31. $\displaystyle\sum_{k=0}^{\infty} \frac{(-4z)^k}{k!\,(k+2)!}\, _2F_3\!\left(\begin{matrix}1,\,1;\,-z\\ \frac{k+3}{2},\,\frac{k}{2}+2,\,2\end{matrix}\right)$
$$= \frac{1}{16z}\{\pi Y_0(4\sqrt{z}) - 2[\ln(2\sqrt{z}) + \mathbf{C}]\,J_0(4\sqrt{z})\}.$$

32. $\displaystyle\sum_{k=0}^{\infty} \frac{(-4z)^k}{k!\,(k+2)!}\, _2F_3\!\left(\begin{matrix}1,\,1;\,z\\ \frac{k+3}{2},\,\frac{k}{2}+2,\,2\end{matrix}\right) = -\frac{1}{2\pi z}\int_0^1 \frac{\cos(4x\sqrt{z})\ln(2x)}{\sqrt{1-x^2}}\,dx.$

33. $\displaystyle\sum_{k=0}^{\infty} \frac{(-4z)^k}{(k!)^2}\, _2F_3\!\left(\begin{matrix}\frac{1}{2},\,1;\,z\\ \frac{k+1}{2},\,\frac{k}{2}+1,\,\frac{3}{2}\end{matrix}\right) = \frac{1}{4}\{\pi J_1(4\sqrt{z})\,\mathbf{H}_0(4\sqrt{z})$
$$+ [1 + J_0(4\sqrt{z})][2 - \pi\,\mathbf{H}_1(4\sqrt{z})]\}.$$

34. $\displaystyle\sum_{k=0}^{\infty} \frac{(-4z)^k}{k!\,(k+1)!}\, {}_2F_3\!\left(\begin{array}{c}\frac{1}{2},\, 1;\, z \\ \frac{k}{2}+1,\, \frac{k+3}{2},\, \frac{3}{2}\end{array}\right) = \frac{\pi}{8\sqrt{z}}\, \mathbf{H}_0(4\sqrt{z}).$

35. $\displaystyle\sum_{k=0}^{\infty} \frac{(4z)^k}{k!\,(k+2)!}\, {}_2F_3\!\left(\begin{array}{c}2,\, 2;\, z \\ \frac{k+3}{2},\, \frac{k}{2}+2,\, 1\end{array}\right) = \frac{\sinh(4\sqrt{z})}{8\sqrt{z}}.$

36. $\displaystyle\sum_{k=0}^{\infty} \frac{(4z)^k}{k!\,(k+2)!}\, {}_2F_3\!\left(\begin{array}{c}a,\, 1;\, z \\ \frac{k+3}{2},\, \frac{k}{2}+2,\, 3-a\end{array}\right) = \frac{a-2}{8(a-1)z}\, I_0(4\sqrt{z})$
$\qquad\qquad + \dfrac{2^{a-4}}{a-1}\,\Gamma(3-a)\, z^{(a-3)/2}\, I_{1-a}(4\sqrt{z}).$

37. $\displaystyle\sum_{k=0}^{\infty} \frac{(-4z)^k}{k!\,(2a+1)_k}\, {}_2F_3\!\left(\begin{array}{c}\frac{1}{2},\, 1;\, z \\ \frac{k+1}{2}+a,\, \frac{k}{2}+a+1,\, \frac{3}{2}\end{array}\right)$
$= 2^{2-2a}(2a-1)\,\pi\, z^{-a}\,\Gamma(2a+1) \displaystyle\int_0^{\pi/2} \frac{1}{x}\, J_{2a-1}(x)\, \mathbf{H}_0(4\sqrt{z}-x)\, dx$

$\qquad\qquad\qquad\qquad\qquad\qquad\qquad [\operatorname{Re} a > 1/2].$

38. $\displaystyle\sum_{k=0}^{\infty} \frac{(4z)^k}{k!\,(k+2)!}\, {}_2F_3\!\left(\begin{array}{c}\frac{1}{2},\, 1;\, z \\ \frac{k+3}{2},\, \frac{k}{2}+2,\, \frac{5}{2}\end{array}\right)$
$= \dfrac{3}{32z^{3/2}}\, \bigl[4\sqrt{z}\, I_0(4\sqrt{z}) - \sinh(4\sqrt{z})\bigr].$

39. $\displaystyle\sum_{k=0}^{\infty} \frac{(-4z)^k}{k!\,(k+2)!}\, {}_2F_3\!\left(\begin{array}{c}\frac{1}{2},\, 1;\, z \\ \frac{k+3}{2},\, \frac{k}{2}+2,\, \frac{5}{2}\end{array}\right) = \frac{3}{8z}\left[J_0(4\sqrt{z}) + \frac{\pi}{2}\,\mathbf{H}_1(4\sqrt{z}) - 1\right].$

40. $\displaystyle\sum_{k=0}^{\infty} \frac{(4z)^k}{k!\,(k+2)!}\, {}_2F_3\!\left(\begin{array}{c}\frac{1}{2},\, 1;\, z \\ \frac{k+3}{2},\, \frac{k}{2}+2,\, \frac{5}{2}\end{array}\right)$
$= \dfrac{3}{32z^{3/2}}\, \bigl[4z^{1/2} I_0(4\sqrt{z}) - \sinh(4\sqrt{z})\bigr].$

41. $\displaystyle\sum_{k=0}^{\infty} \frac{(16z)^k}{k!\,(k+1)!}\, {}_2F_5\!\left(\begin{array}{c}1,\, \frac{5}{4};\, z \\ \frac{k+2}{4},\, \frac{k+3}{4},\, \frac{k+4}{4},\, \frac{k+5}{4},\, \frac{1}{4}\end{array}\right) = \frac{1}{2} + \frac{1}{2}\, I_0(8\sqrt{z}).$

42. $\displaystyle\sum_{k=0}^{\infty} \frac{(4z)^k}{k!\,(a+1)_k}\, {}_3F_4\!\left(\begin{array}{c}a,\, \frac{a}{2}+1,\, b;\, z \\ \frac{k+a+1}{2},\, \frac{k+a+2}{2},\, \frac{a}{2},\, a-b+1\end{array}\right)$
$= 2^{b-a}\,\Gamma(a-b+1)\, z^{(b-a)/2}\, I_{a-b}(4\sqrt{z}).$

43. $\sum_{k=0}^{\infty} \dfrac{(-4z)^k}{k!\,(2a+2)_k}\, {}_3F_4\!\left(\begin{array}{c} a,\ a+\frac{3}{2},\ 1;\ z \\ \frac{k+2}{2}+a,\ \frac{k+3}{2}+a,\ a+\frac{1}{2},\ a+2 \end{array}\right)$

$$= \frac{2^{-2a-2}z^{-a-1/2}}{2a+1}\,\Gamma(2a+3)\,J_{2a+1}(4\sqrt{z}) + \frac{a}{2a+1}\,{}_1F_2\!\left(\begin{array}{c} a+1;\ -4z \\ a+2,\ 2a+2 \end{array}\right).$$

44. $\sum_{k=0}^{\infty} \dfrac{(-4z)^k}{k!\,(k+2)!}\, {}_3F_4\!\left(\begin{array}{c} \frac{2}{3},\ 1,\ \frac{4}{3};\ z \\ \frac{k+3}{2},\ \frac{k}{2}+2,\ \frac{5}{3},\ \frac{7}{3} \end{array}\right) = \dfrac{1}{z} J_0(4\sqrt{z}) - \dfrac{1}{z}\,{}_1F_2\!\left(\begin{array}{c} 1;\ -4z \\ \frac{2}{3},\ \frac{4}{3} \end{array}\right).$

45. $\sum_{k=0}^{\infty} \dfrac{(4z)^k}{k!\,(k+1)!}\, {}_3F_4\!\left(\begin{array}{c} 1,\ \frac{3}{2},\ \frac{3}{2};\ z \\ \frac{k}{2}+1,\ \frac{k+3}{2},\ \frac{1}{2},\ \frac{1}{2} \end{array}\right) = \cosh(4\sqrt{z}).$

46. $\sum_{k=0}^{\infty} \dfrac{(4z)^k}{k!\,(k+2)!}\, {}_3F_4\!\left(\begin{array}{c} 1-a,\ 1+a,\ 1;\ z \\ \frac{k+3}{2},\ \frac{k}{2}+2,\ 2-a,\ 2+a \end{array}\right)$

$$= \frac{1-a^2}{8a^2 z}\left[I_0(4\sqrt{z}) - \frac{2a}{\sin(a\pi)} \int_0^{\pi/2} \frac{\cos(2ax)\cosh(4\sqrt{z}\sin x)}{\sqrt{1-x^2}}\,dx\right].$$

47. $\sum_{k=0}^{\infty} \dfrac{(4z)^k}{(k+1)!\,(2k+1)}\, {}_4F_4\!\left(\begin{array}{c} k+\frac{1}{2},\ k+1,\ 1,\ \frac{3}{2};\ z \\ k+\frac{3}{2},\ \frac{k}{2}+1,\ \frac{k+3}{2},\ \frac{1}{2} \end{array}\right) = \dfrac{1}{4}\sqrt{\dfrac{\pi}{z}}\,\operatorname{erfi}(2\sqrt{z}).$

48. $\sum_{k=0}^{\infty} \dfrac{(a)_k(b)_k}{\left(a+b+\frac{1}{2}\right)_k} (4z)^k \dfrac{\prod(a_p)_k}{\prod(b_q)_k}\, {}_{p+1}F_q\!\left(\begin{array}{c} -k,\ (a_p)+k \\ (b_q)+k;\ z \end{array}\right)$

$$= {}_{p+2}F_{q+1}\!\left(\begin{array}{c} (a_p),\ 2a,\ 2b \\ (b_q),\ a+b+\frac{1}{2};\ z \end{array}\right).$$

49. $\sum_{k=0}^{\infty} \dfrac{(n+a)_k}{k!} t^k\, {}_{p+1}F_{q+1}\!\left(\begin{array}{c} -n-k,\ (a_p) \\ a,\ (b_q);\ z \end{array}\right)$

$$= (1-t)^{-n-a} \sum_{k=0}^{n} \binom{n}{k} \dfrac{(t-1)^{-k}z^k}{(a)_k} \dfrac{\prod(a_p)_k}{\prod(b_q)_k}\, {}_pF_q\!\left(\begin{array}{c} (a_p)+k;\ \frac{tz}{t-1} \\ (b_q)+k \end{array}\right) \quad [|t|<1].$$

50. $\sum_{k=0}^{\infty} \dfrac{(1-b)_k}{k!} t^k\, {}_pF_{q+1}\!\left(\begin{array}{c} (a_p);\ z \\ (b_q),\ b-k \end{array}\right) = (1-t)^{b-1}\, {}_pF_{q+1}\!\left(\begin{array}{c} (a_p);\ z-tz \\ (b_q),\ b \end{array}\right)$

$$[|t|<1].$$

51. $\sum_{k=0}^{\infty} \dfrac{(1-a)_k}{k!} t^k\, {}_{p+1}F_{q+1}\!\left(\begin{array}{c} -n-k,\ (a_p) \\ a-k,\ (b_q);\ z \end{array}\right)$

$$= (1-t)^{a-1} \sum_{k=0}^{n} \binom{n}{k} \dfrac{(t-1)^k z^k}{(a)_k} \dfrac{\prod(a_p)_k}{\prod(b_q)_k}\, {}_pF_q\!\left(\begin{array}{c} (a_p)+k \\ (b_q)+k;\ tz \end{array}\right) \quad [|t|<1].$$

52. $\displaystyle\sum_{k=0}^{\infty} \frac{(-z)^k}{(b)_k} \frac{\prod (a_p)_k}{\prod (b_q)_k} {}_pF_q\!\left(\begin{matrix}(a_p)+k;\ z\\(b_q)+k\end{matrix}\right) = {}_{p+1}F_{q+1}\!\left(\begin{matrix}(a_p),\ b-1;\ z\\(b_q),\ b\end{matrix}\right).$

53. $\displaystyle\sum_{k=0}^{\infty} \frac{(-z)^k}{k!(k+b)} \frac{\prod (a_p)_k}{\prod (b_q)_k} {}_pF_q\!\left(\begin{matrix}(a_p)+k;\ z\\(b_q)+k\end{matrix}\right) = \frac{1}{b}\, {}_{p+1}F_{q+1}\!\left(\begin{matrix}(a_p),\ 1;\ z\\(b_q),\ b+1\end{matrix}\right).$

54. $\displaystyle\sum_{k=0}^{\infty} \frac{\left(-\tfrac{1}{2}\right)_k}{k!\left(a+\tfrac{1}{2}\right)_k} (-z)^k \frac{\prod (a_p)_k}{\prod (b_q)_k} {}_pF_q\!\left(\begin{matrix}(a_p)+k;\ z\\(b_q)+k\end{matrix}\right)$

$\displaystyle = \frac{1}{2a}\left[(2a-1)\, {}_{p+1}F_{q+1}\!\left(\begin{matrix}(a_p),\ a;\ z\\(b_q),\ a-\tfrac{1}{2}\end{matrix}\right) + {}_{p+1}F_{p+1}\!\left(\begin{matrix}(a_p),\ a;\ z\\(b_q),\ a+\tfrac{1}{2}\end{matrix}\right)\right].$

55. $\displaystyle\sum_{k=0}^{\infty} \frac{\sigma_{k+m}^m}{(k+m)!} z^k \frac{\prod (a_p)_k}{\prod (b_q)_k} {}_pF_q\!\left(\begin{matrix}(a_p)+k;\ az\\(b_q)+k\end{matrix}\right)$

$\displaystyle = (-1)^{m(p+q+1)} \frac{z^{-m}}{m!} \frac{\prod (1-(b_q))_m}{\prod (1-(a_p))_m} \sum_{k=0}^{m} (-1)^k \binom{m}{k}\, {}_pF_q\!\left(\begin{matrix}(a_p)-m;\ (k+a)z\\(b_q)-m\end{matrix}\right).$

56. $\displaystyle\sum_{k=0}^{\infty} \frac{z^k}{(k+1)!} P_k^{(\rho-k,\,1)}(3) \frac{\prod (a_p)_k}{\prod (b_q)_k} {}_pF_q\!\left(\begin{matrix}(a_p)+k;\ z\\(b_q)+k\end{matrix}\right)$

$\displaystyle \qquad\qquad\qquad\qquad = {}_{p+1}F_{q+1}\!\left(\begin{matrix}(a_p),\ \rho+2;\ 2z\\(b_q),\ 2\end{matrix}\right).$

57. $\displaystyle\sum_{k=0}^{\infty} \frac{(a)_k}{k!} z^k \frac{\prod (a_p)_k}{\prod (b_q)_k} {}_{p+1}F_q\!\left(\begin{matrix}(a_p)+k,\ 1;\ z\\(b_q)+k\end{matrix}\right) = {}_{p+1}F_q\!\left(\begin{matrix}(a_p),\ a+1\\(b_q);\ z\end{matrix}\right).$

58. $\displaystyle\sum_{k=0}^{\infty} (ka+b)^m z^k \frac{\prod (a_p)_k}{\prod (b_q)_k} {}_{p+1}F_q\!\left(\begin{matrix}(a_p)+k,\ a;\ z\\(b_q)+k\end{matrix}\right)$

$\displaystyle = b^m \sum_{k=0}^{m} \binom{m}{k}\left(\frac{a}{b}\right)^k \sum_{j=0}^{k} \sigma_k^j j!\, z^j \frac{\prod (a_p)_j}{\prod (b_q)_j}\, {}_{p+1}F_q\!\left(\begin{matrix}(a_p)+j,\ a+j+1\\(b_q)+j;\ z\end{matrix}\right)$

$\qquad\qquad\qquad\qquad\qquad\qquad\qquad\qquad\qquad\qquad [|z|<1].$

59. $\displaystyle\sum_{k=0}^{\infty} \frac{z^k}{k!(a)_k} \frac{\prod (a_p)_k}{\prod (b_q)_k}\, {}_pF_{q+1}\!\left(\begin{matrix}(a_p)+k;\ z\\(b_q)+k,\ b\end{matrix}\right)$

$\displaystyle \qquad\qquad = {}_{p+2}F_{q+3}\!\left(\begin{matrix}(a_p),\ \dfrac{a+b-1}{2},\ \dfrac{a+b}{2};\ 4z\\(b_q),\ a,\ b,\ a+b-1\end{matrix}\right).$

6.17. The Generalized Hypergeometric Function $_pF_q((a_p);(b_q);z)$

60. $\displaystyle\sum_{k=0}^{\infty} \frac{(-z)^k}{k!(b)_k} \frac{\prod(a_p)_k}{\prod(b_q)_k} \, _pF_{q+1}\!\left(\begin{array}{c}(a_p)+k;\ z\\(b_q)+k,\ b\end{array}\right)$

$$=\, _{2p}F_{2q+3}\!\left(\begin{array}{c}\left(\frac{a_p}{2}\right),\left(\frac{a_p+1}{2}\right);\ -\dfrac{z^2}{4^{q-p+1}}\\ \left(\frac{b_q}{2}\right),\left(\frac{b_q+1}{2}\right),\frac{b}{2},\frac{b+1}{2},b\end{array}\right).$$

61. $\displaystyle\sum_{k=0}^{\infty} \frac{(-z)^k}{k!} \frac{\prod(a_p)_k}{\prod(b_q)_k} \, _{p+1}F_{q+1}\!\left(\begin{array}{c}(a_p)+k,\ a;\ z\\(b_q)+k,\ b\end{array}\right)$

$$=\, _{p+1}F_{q+1}\!\left(\begin{array}{c}(a_p),\ b-a;\ -z\\(b_q),\ b\end{array}\right).$$

62. $\displaystyle\sum_{k=0}^{\infty} \frac{(a)_k}{k!(b)_k}(-z)^k \frac{\prod(a_p)_k}{\prod(b_q)_k} \, _{p+1}F_{q+1}\!\left(\begin{array}{c}(a_p)+k,\ a;\ z\\(b_q)+k,\ b\end{array}\right)$

$$=\, _{2p+2}F_{2q+3}\!\left(\begin{array}{c}\frac{(a_p)}{2},\frac{(a_p)+1}{2},\ a,\ b-a;\ \dfrac{z^2}{4^{q-p+1}}\\ \frac{(b_q)}{2},\frac{(b_q)+1}{2},\frac{b}{2},\frac{b+1}{2},b\end{array}\right).$$

63. $\displaystyle\sum_{k=0}^{\infty} \frac{(a)_k(b)_k}{k!\left(a+b+\frac{1}{2}\right)_k}z^k \frac{\prod(a_p)_k}{\prod(b_q)_k} \, _{p+2}F_{q+1}\!\left(\begin{array}{c}(a_p)+k,\ a,\ b;\ z\\(b_q)+k,\ a+b+\frac{1}{2}\end{array}\right)$

$$=\, _{p+3}F_{q+2}\!\left(\begin{array}{c}(a_p),\ 2a,\ 2b,\ a+b;\ z\\(b_q),\ a+b+\frac{1}{2},\ 2a+2b\end{array}\right).$$

64. $\displaystyle\sum_{k=0}^{\infty} \frac{(b)_k\left(\frac{1}{2}-a-b\right)_k}{k!(1-a)_k}(4z)^k \frac{\prod(a_p)_k}{\prod(b_q)_k} \, _{p+1}F_q\!\left(\begin{array}{c}(a_p)+k,\ a-k\\(b_q)+k;\ z\end{array}\right)$

$$=\, _{p+2}F_{q+1}\!\left(\begin{array}{c}(a_p),\ a+2b,\ 1-a-2b\\(b_q),\ 1-a;\ z\end{array}\right).$$

65. $\displaystyle\sum_{k=0}^{\infty} \frac{(b)_k}{k!(k+a)} \, _{p+1}F_{q+1}\!\left(\begin{array}{c}k+a,\ (a_p);\ z\\k+a+1,\ (b_q)\end{array}\right)$

$$=\mathrm{B}(a,1-b)\, _{p+1}F_{q+1}\!\left(\begin{array}{c}a,\ (a_p);\ z\\a-b+1,\ (b_q)\end{array}\right) \quad [\mathrm{Re}\,b<1].$$

66. $\displaystyle\sum_{k=0}^{\infty} \frac{(-4z)^k}{(k!)^2} \frac{\prod(a_p)_k}{\prod(b_q)_k} \, _{p+2}F_{q+3}\!\left(\begin{array}{c}(a_p)+k,\ \frac{1}{2},\ 1;\ z\\(b_q)+k,\ \frac{k+1}{2},\frac{k+2}{2},\frac{3}{2}\end{array}\right)$

$$=\frac{1}{2}+\frac{1}{2}\, _{p+1}F_{q+2}\!\left(\begin{array}{c}(a_p),\frac{1}{2};\ -4z\\(b_q),1,\frac{3}{2}\end{array}\right)-\frac{8z}{3}\frac{\prod_{i=1}^{p}a_i}{\prod_{j=1}^{q}b_j}\, _{p+1}F_{q+2}\!\left(\begin{array}{c}(a_p)+1,1;\ -4z\\(b_q)+1,\frac{3}{2},\frac{5}{2}\end{array}\right).$$

67. $$\sum_{k=0}^{\infty} \frac{(-4z)^k}{k!(2a)_k} \frac{\prod(a_p)_k}{\prod(b_q)_k} {}_{p+1}F_{q+2}\left(\begin{array}{c}(a_p)+k,\,1;\,z\\(b_q)+k,\,\frac{k}{2}+a,\,\frac{k+1}{2}+a\end{array}\right)$$
$$= {}_{p+1}F_{q+2}\left(\begin{array}{c}(a_p),\,a-\frac{1}{2};\,-4z\\(b_q),\,a+\frac{1}{2},\,2a-1\end{array}\right).$$

68. $$\sum_{k=0}^{\infty} \frac{(-4z)^k}{k!(k+2)!} \frac{\prod(a_p)_k}{\prod(b_q)_k} {}_{p+1}F_{q+2}\left(\begin{array}{c}(a_p)+k,\,1;\,z\\(b_q)+k,\,\frac{k+3}{2},\,\frac{k}{2}+2\end{array}\right)$$
$$= \frac{1}{8z} \frac{\prod\limits_{j=1}^{q}(b_j-1)}{\prod\limits_{i=1}^{p}(a_i-1)}\left[1 - {}_pF_{q+1}\left(\begin{array}{c}(a_p)-1;\,-4z\\(b_q)-1,\,1\end{array}\right)\right].$$

69. $$\sum_{k=0}^{\infty} \frac{(-4z)^k}{k!(2a+1)_k} \frac{\prod(a_p)_k}{\prod(b_q)_k} {}_{p+2}F_{q+3}\left(\begin{array}{c}(a_p)+k,\,a+1,\,2a;\,z\\(b_q)+k,\,\frac{k+1}{2}+a,\,\frac{k}{2}+a+1,\,a\end{array}\right) = 1.$$

70. $$\sum_{k=0}^{\infty} \frac{(4z)^k}{k!(k+2)!} \frac{\prod(a_p)_k}{\prod(b_q)_k} {}_{p+2}F_{q+3}\left(\begin{array}{c}(a_p)+k,\,a,\,1;\,z\\(b_q)+k,\,\frac{k+3}{2},\,\frac{k}{2}+2,\,3-a\end{array}\right)$$
$$= \frac{a-2}{8(a-1)z} \frac{\prod\limits_{j=1}^{q}(b_j-1)}{\prod\limits_{i=1}^{p}(a_i-1)}\left[{}_pF_{q+1}\left(\begin{array}{c}(a_p)-1;\,4z\\(b_q)-1,\,1\end{array}\right) - {}_pF_{q+1}\left(\begin{array}{c}(a_p)-1;\,4z\\(b_q)-1,\,2-a\end{array}\right)\right].$$

71. $$\sum_{k=0}^{\infty} \frac{(-4z)^k}{k!(k+2)!} \frac{(a_p)_k}{(b_q)_k} {}_{p+2}F_{q+3}\left(\begin{array}{c}(a_p)+k,\,2,\,2;\,z\\(b_q)+k,\,\frac{k+3}{2},\,\frac{k}{2}+2,\,1\end{array}\right) = \frac{1}{2}.$$

72. $$\sum_{k=0}^{\infty} \frac{(-4z)^k}{k!(k+1)!} {}_{p+1}F_{q+2}\left(\begin{array}{c}(a_p)+k,\,1;\,z\\(b_q)+k,\,\frac{k}{2}+1,\,\frac{k+3}{2}\end{array}\right)$$
$$= {}_{p+1}F_{q+2}\left(\begin{array}{c}(a_p),\,\frac{1}{2};\,-4z\\(b_q),\,1,\,\frac{3}{2}\end{array}\right).$$

73. $$\sum_{k=0}^{\infty} \frac{(-16z)^k}{k!(k+1)!(a-1)_k(a)_k} \frac{\prod(a_p)_k}{\prod(b_q)_k}$$
$$\times {}_{p+2}F_{q+5}\left(\begin{array}{c}(a_p)+k,\,1,\,\frac{3}{2};\,z\\(b_q)+k,\,\frac{a+k}{2},\,\frac{a+k+1}{2},\,\frac{k+2}{2},\,\frac{k+3}{2},\,\frac{1}{2}\end{array}\right)$$
$$= {}_pF_{q+3}\left(\begin{array}{c}(a_p);\,-4z\\(b_q),\,a-1,\,\frac{a}{2},\,\frac{a+1}{2}\end{array}\right).$$

74. $\displaystyle\sum_{k=0}^{\infty} \frac{(16z)^k}{(k!)^2 (b)_k^2} \frac{\prod (a_p)_k}{\prod (b_q)_k}$

$\times {}_{p+2}F_{q+5}\left(\begin{array}{c}(a_p)+k, a, 1; z \\ (b_q)+k, \frac{k+b}{2}, \frac{k+b+1}{2}, \frac{k+1}{2}, \frac{k+2}{2}, 1-a\end{array}\right)$

$= \dfrac{1}{2}{}_{p+2}F_{q+5}\left(\begin{array}{c}(a_p), \frac{b-a}{2}, \frac{b-a+1}{2}; 16z \\ (b_q), 1-a, b, \frac{b}{2}, \frac{b+1}{2}, b-a\end{array}\right) + \dfrac{1}{2}{}_pF_{q+3}\left(\begin{array}{c}(a_p); 16z \\ (b_q), b, b, 1\end{array}\right).$

75. $\displaystyle\sum_{k=0}^{\infty} \frac{(-4z)^k}{k!(2a-1)_k} \frac{\prod (a_p)_k}{\prod (b_q)_k}$

$\times {}_{p+3}F_{q+4}\left(\begin{array}{c}(a_p)+k, 1, a, b; z \\ (b_q)+k, \frac{k-1}{2}+a, \frac{k}{2}+a, a-1, 2a-b-1\end{array}\right)$

$= {}_pF_{q+1}\left(\begin{array}{c}(a_p), 2a-b-2; -4z \\ (b_q), 2a-2, 2a-b-1\end{array}\right).$

76. $\displaystyle\sum_{k=0}^{\infty} \frac{(-4z)^k}{k!(2a+3)_k} \frac{\prod (a_p)_k}{\prod (b_q)_k}$

$\times {}_{p+3}F_{q+4}\left(\begin{array}{c}(a_p)+k, a, a+2, 1; z \\ (b_q)+k, \frac{k+3}{2}+a, \frac{k}{2}+a+2, a+1, a+3\end{array}\right)$

$= {}_{p+1}F_{q+2}\left(\begin{array}{c}(a_p), a+2; -4z \\ (b_q), a+3, 2a+2\end{array}\right).$

77. $\displaystyle\sum_{k=0}^{\infty} \frac{(4z)^k}{k!(a+1)_k} \frac{\prod (a_p)_k}{\prod (b_q)_k}$

$\times {}_{p+3}F_{q+4}\left(\begin{array}{c}(a_p)+k, a, \frac{a}{2}+1, b; z \\ (b_q)+k, \frac{k+a+1}{2}, \frac{k+a+2}{2}, \frac{a}{2}, a-b+1\end{array}\right)$

$= {}_pF_{q+1}\left(\begin{array}{c}(a_p); 4z \\ (b_q), a-b+1\end{array}\right).$

78. $\displaystyle\sum_{k=0}^{\infty} \frac{(-4z)^k}{k!(2a+1)_k} \frac{\prod (a_p)_k}{\prod (b_q)_k}$

$\times {}_{p+4}F_{q+5}\left(\begin{array}{c}(a_p)+k, \frac{1}{2}, 2a, a+1, 2a-\frac{1}{2}; z \\ (b_q)+k, \frac{k+1}{2}+a, \frac{k+2}{2}+a, \frac{3}{2}, a, 2a+\frac{1}{2}\end{array}\right)$

$= {}_{p+1}F_{q+2}\left(\begin{array}{c}(a_p), 1; -4z \\ (b_q), \frac{3}{2}, 2a+\frac{1}{2}\end{array}\right).$

79. $\displaystyle\sum_{k=0}^{\infty} \frac{(-4z)^k}{k!(k+1)!} \frac{\prod(a_p)_k}{\prod(b_q)_k} {}_{p+4}F_{q+5}\left(\begin{array}{c}(a_p)+k, \frac{1}{3}, \frac{2}{3}, 1, \frac{3}{2}; z\\ (b_q)+k, \frac{k+2}{2}, \frac{k+3}{2}, \frac{1}{2}, \frac{4}{3}, \frac{5}{3}\end{array}\right)$

$\qquad = {}_{p+1}F_{q+2}\left(\begin{array}{c}(a_p), 1; -4z\\ (b_q), \frac{4}{3}, \frac{5}{3}\end{array}\right).$

80. $\displaystyle\sum_{k=0}^{\infty} \frac{(-z)^k}{k!(k+a)} \frac{\prod(2a_p)_{2k}}{\prod(2b_q)_{2k}} {}_{2p}F_{2q}\left(\begin{array}{c}(a_p)+k, (a_p)+k+\frac{1}{2}; z\\ (b_q)+k, (b_q)+k+\frac{1}{2}\end{array}\right)$

$\qquad = \frac{1}{a} {}_{2p+1}F_{2q+1}\left(\begin{array}{c}(a_p), (a_p)+\frac{1}{2}, 1; z\\ (b_q), (b_q)+\frac{1}{2}, a+1\end{array}\right).$

81. $\displaystyle\sum_{k=0}^{\infty} \frac{(a)_{2k}}{k!(b)_k}(-z)^k \frac{\prod(a_p)_k}{\prod(b_q)_k} {}_{p+1}F_q\left(\begin{array}{c}(a_p)+k, a+2k\\ (b_q)+k; z\end{array}\right)$

$\qquad = {}_{p+2}F_{q+1}\left(\begin{array}{c}(a_p), a, a-b+1\\ (b_q), b; -z\end{array}\right).$

82. $\displaystyle\sum_{k=0}^{\infty} \frac{\left(\frac{a}{2}\right)_k (b)_k}{k!(2b)_k}(-4z)^k \frac{\prod(a_p)_k}{\prod(b_q)_k} {}_{p+1}F_q\left(\begin{array}{c}(a_p)+k, a+2k\\ (b_q)+k; z\end{array}\right)$

$\qquad = {}_{2p+2}F_{2q+1}\left(\begin{array}{c}\frac{(a_p)}{2}, \frac{(a_p)+1}{2}, \frac{a}{2}, \frac{a+1}{2}-b\\ \frac{(b_q)}{2}, \frac{(b_q)+1}{2}, b+\frac{1}{2}; \frac{z^2}{4^{q-p}}\end{array}\right).$

83. $\displaystyle\sum_{k=0}^{\infty} \frac{(a)_{2k}(b)_k}{k!(c)_k(a+b-c+1)_k}(-z)^k \frac{\prod(a_p)_k}{\prod(b_q)_k} {}_{p+1}F_q\left(\begin{array}{c}(a_p)+k, a+2k\\ (b_q)+k; z\end{array}\right)$

$\qquad = {}_{p+2}F_{q+1}\left(\begin{array}{c}(a_p), a, a-c+1, c-b\\ (b_q), c, a+b-c+1; z\end{array}\right).$

84. $\displaystyle\sum_{k=0}^{\infty} \frac{(a)_k}{k!}(-z)^k \frac{\prod(a_p)_k}{\prod(b_q)_k} {}_{p+2}F_{q+1}\left(\begin{array}{c}(a_p)+k, 2a+2k, a\\ (b_q)+k, 2a; z\end{array}\right)$

$\qquad = {}_{3p+2}F_{3q+1}\left(\begin{array}{c}\frac{(a_p)}{3}, \frac{(a_p)+1}{3}, \frac{(a_p)+2}{3}, \frac{a}{2}, \frac{a+1}{2}\\ \frac{(b_q)}{3}, \frac{(b_q)+1}{3}, \frac{(b_q)+2}{3}, a+\frac{1}{2}; -\frac{z^3}{27^{q-p}}\end{array}\right).$

85. $\displaystyle\sum_{k=0}^{\infty} \frac{(a)_{2k}}{k!(b)_k}(-z)^k \frac{\prod(a_p)_k}{\prod(b_q)_k} {}_{p+1}F_q\left(\begin{array}{c}(a_p)+k, a+2k\\ (b_q)+k; z\end{array}\right)$

$\qquad = {}_{p+2}F_{q+1}\left(\begin{array}{c}(a_p), a, a-b+1\\ (b_q), b; -z\end{array}\right).$

6.17.1]   6.17. The Generalized Hypergeometric Function $_pF_q((a_p);(b_q);z)$

86. $\sum_{k=1}^{\infty} \frac{(-z)^k}{(a)_{2k}} \frac{\prod(a_p)_k}{\prod(b_q)_k} {}_pF_{q+1}\left(\begin{array}{c}(a_p)+k;\ z\\(b_q)+k,\ a+2k\end{array}\right)$

$$= -\frac{z}{a(a+1)} \frac{\prod_{i=1}^{p} a_i}{\prod_{j=1}^{q} b_j} {}_{p+1}F_{q+2}\left(\begin{array}{c}(a_p)+1,\ \frac{a+1}{2};\ z\\(b_p)+1,\ \frac{a+3}{2},\ a+1\end{array}\right).$$

87. $\sum_{k=0}^{\infty} \frac{(2k+a-1)(a-1)_k(b)_k(c)_k}{k!\,(a-b)_k(a-c)_k(a-1)_{2k+1}}$

$\times (-z)^k \frac{\prod(a_p)_k}{\prod(b_q)_k} {}_pF_{q+1}\left(\begin{array}{c}(a_p)+k;\ z\\(b_q)+k,\ a+2k\end{array}\right) = {}_{p+1}F_{q+2}\left(\begin{array}{c}(a_p),\ a-b-c;\ z\\(b_q),\ a-b,\ a-c\end{array}\right).$

88. $\sum_{k=0}^{\infty} \frac{2k+a-1}{(2k+a-2)(2k+a)} \frac{(a-1)_k}{k!\,(a)_{2k}} (-z)^k \frac{\prod(a_p)_k}{\prod(b_p)_k} {}_pF_{q+1}\left(\begin{array}{c}(a_p)+k;\ z\\(b_q)+k,\ a+2k\end{array}\right)$

$$= \frac{a-1}{a(a-2)} {}_{p+1}F_{q+2}\left(\begin{array}{c}(a_p),\ 1;\ z\\(b_q),\ \frac{a}{2},\ \frac{a}{2}+1\end{array}\right).$$

89. $\sum_{k=0}^{\infty} \frac{(2k+a-1)(a-b)_k}{(a-1)_{2k+1}(b)_k} (-z)^k \frac{\prod(a_p)_k}{\prod(b_p)_k} {}_pF_{p+1}\left(\begin{array}{c}(a_p)+k;\ z\\(b_p)+k,\ a+2k\end{array}\right)$

$$= {}_{p+1}F_{p+2}\left(\begin{array}{c}(a_p),\ b-1;\ z\\(b_p),\ a-1,\ b\end{array}\right).$$

90. $\sum_{k=0}^{\infty} \frac{(a)_k}{k!\,(b)_{2k}} (-z)^k \frac{\prod(a_p)_k}{\prod(b_q)_k} {}_pF_{p+1}\left(\begin{array}{c}(a_p)+k;\ z\\(b_q)+k,\ b+2k\end{array}\right)$

$$= {}_{p+2}F_{p+3}\left(\begin{array}{c}(a_p),\ \frac{b-a}{2},\ \frac{b-a+1}{2};\ z\\(b_q),\ \frac{b}{2},\ \frac{b+1}{2},\ b-a\end{array}\right).$$

91. $\sum_{k=0}^{\infty} \frac{(a-1)_k(a-b)_k}{k!\,(a)_{2k}(b)_k} (-z)^k \frac{\prod(a_p)_k}{\prod(b_q)_k} {}_pF_{q+1}\left(\begin{array}{c}(a_p)+k;\ z\\(b_q)+k,\ a+2k\end{array}\right)$

$$= {}_{p+1}F_{p+2}\left(\begin{array}{c}(a_p),\ \frac{1-a}{2}+b;\ z\\(b_q),\ \frac{a+1}{2},\ b\end{array}\right).$$

92. $\sum_{k=0}^{\infty} \frac{2k+a-1}{(2k+a)(2k+a-2)} \frac{(a-1)_k}{k!\,(a)_{2k}} (-z)^k \frac{\prod(a_p)_k}{\prod(b_q)_k} {}_pF_{q+1}\left(\begin{array}{c}(a_p)+k;\ z\\(b_q)+k,\ a+2k\end{array}\right)$

$$= \frac{a-1}{a(a-2)} {}_{p+1}F_{q+2}\left(\begin{array}{c}(a_p),\ 1;\ z\\(b_q),\ \frac{a}{2},\ \frac{a}{2}+1\end{array}\right).$$

93. $\displaystyle\sum_{k=0}^{\infty} \frac{(2k+b)(b)_k}{k!(a)_{2k}} \frac{\prod(a_p)_k}{\prod(b_q)_k} (-z)^k {}_pF_{q+1}\left(\begin{matrix}(a_p)+k;\ z\\(b_p)+k,\ a+2k\end{matrix}\right)$

$$= b\,{}_{p+2}F_{q+3}\left(\begin{matrix}(a_p),\ \frac{a-b-1}{2},\ \frac{a-b}{2};\ z\\(b_p),\ \frac{a}{2},\ \frac{a+1}{2},\ a-b\end{matrix}\right).$$

94. $\displaystyle\sum_{k=0}^{\infty} \frac{(2k+a-1)(a-1)_k(a-b)_k}{k!(a)_{2k}(b)_k} z^k \frac{\prod(a_p)_k}{\prod(b_p)_k} {}_pF_{p+1}\left(\begin{matrix}(a_p)+k;\ z\\(b_p)+k,\ a+2k\end{matrix}\right)$

$$= (a-1)\,{}_pF_{p+1}\left(\begin{matrix}(a_p);\ z\\(b_p),\ b\end{matrix}\right).$$

95. $\displaystyle\sum_{k=0}^{\infty} \frac{(1-a)_k}{(2k)!(a)_k(b)_{2k}} z^k \frac{\prod(a_p)_k}{\prod(b_q)_k} {}_pF_{q+3}\left(\begin{matrix}(a_p)+k;\ z\\(b_q)+k,\ b+2k,\ 2k+1,\ b\end{matrix}\right)$

$$= \frac{1}{2}\,{}_{p+2}F_{q+5}\left(\begin{matrix}(a_p),\ \frac{a+b-1}{2},\ \frac{a+b}{2};\ z\\(b_q),\ a,\ b,\ \frac{b}{2},\ \frac{b+1}{2},\ a+b-1\end{matrix}\right) + \frac{1}{2}\,{}_pF_{q+3}\left(\begin{matrix}(a_p);\ z\\(b_q),\ b,\ b,\ 1\end{matrix}\right).$$

96. $\displaystyle\sum_{k=0}^{\infty} \frac{(-z)^k}{(2k)!(b)_{2k}} \frac{\prod(a_p)_k}{\prod(b_q)_k} {}_pF_{q+3}\left(\begin{matrix}(a_p)+k;\ z\\(b_q)+k,\ b+2k,\ 2k+2,\ b-1\end{matrix}\right)$

$$= {}_pF_{q+3}\left(\begin{matrix}(a_p);\ \frac{z}{4}\\(b_q),\ \frac{b}{2},\ \frac{b+1}{2},\ b-1\end{matrix}\right).$$

97. $\displaystyle\sum_{k=0}^{\infty} \frac{\left(\frac{z}{2}\right)^{2k}}{k!(a)_k} \frac{\prod(a_p)_{2k}}{\prod(b_q)_{2k}} {}_pF_q\left(\begin{matrix}(a_p)+2k;\ z\\(b_q)+2k\end{matrix}\right) = {}_{p+1}F_{q+1}\left(\begin{matrix}(a_p),\ a-\frac{1}{2};\ 2z\\(b_q),\ 2a-1\end{matrix}\right).$

98. $\displaystyle\sum_{k=0}^{\infty} \frac{(-z^2)^k}{k!(k+1)!} \frac{\prod(a_p)_{2k}}{\prod(b_q)_{2k}} {}_pF_{q+1}\left(\begin{matrix}(a_p)+2k,\ 2;\ z\\(b_q)+2k,\ k+2\end{matrix}\right)$

$$= 1 + z\frac{\prod\limits_{i=1}^{p} a_i}{\prod\limits_{j=1}^{q} b_j}\left[2\,{}_{2p}F_{2q+1}\left(\begin{matrix}\frac{(a_p)+1}{2},\ \frac{(a_p)}{2}+1,\ \frac{1}{2};\ -z^2\\\frac{(b_q)+1}{2},\ \frac{(b_q)}{2}+1,\ 1,\ \frac{3}{2}\end{matrix}\right)\right.$$

$$\left. - {}_{2p}F_{2q+1}\left(\begin{matrix}\frac{(a_p)+1}{2},\ \frac{(a_p)}{2}+1;\ -z^2\\\frac{(b_q)+1}{2},\ \frac{(b_q)}{2}+1,\ 2\end{matrix}\right)\right].$$

99. $\displaystyle\sum_{k=0}^{\infty} \frac{(a)_k(b)_k}{k!\left(a+b+\frac{1}{2}\right)_k}\left(-\frac{z^2}{4}\right)^k \frac{\prod(a_p)_{2k}}{\prod(b_q)_{2k}} {}_{p+1}F_q\left(\begin{matrix}(a_p)+2k,\ a+k\\(b_q)+2k;\ z\end{matrix}\right)$

$$= {}_{p+2}F_{q+1}\left(\begin{matrix}(a_p),\ 2a,\ a+b\\(b_q),\ 2a+2b;\ z\end{matrix}\right).$$

6.17. The Generalized Hypergeometric Function $_pF_q((a_p);(b_q);z)$

100. $\displaystyle\sum_{k=0}^{\infty}\frac{(a)_{3k}}{k!\,(b)_k\left(a-b+\frac{3}{2}\right)_k}\frac{\prod(a_p)_{2k}}{\prod(b_q)_{2k}}\left(\frac{z}{2}\right)^{2k}\,_{p+1}F_q\!\left(\begin{array}{c}(a_p)+2k,\,a+3k\\(b_q)+2k;\,z\end{array}\right)$

$$= {}_{p+3}F_{q+2}\!\left(\begin{array}{c}(a_p),\,a,\,b-\frac{1}{2},\,a-b+1\\(b_q),\,2b-1,\,2a-2b+2;\,4z\end{array}\right)\quad [9].$$

101. $\displaystyle\sum_{k=0}^{\infty}\frac{z^{2k}}{(4k+2)!\,(b)_{4k}}\frac{\prod(a_p)_{2k}}{\prod(b_q)_{2k}}\,_pF_{q+3}\!\left(\begin{array}{c}(a_p)+2k;\,z\\(b_q)+2k,\,b+4k,\,4k+3,\,b-2\end{array}\right)$

$$= \frac{(b-2)(b-1)}{4z}\frac{\prod_{j=1}^{q}(b_j-1)}{\prod_{i=1}^{p}(a_i-1)}\left[\,_{p+2}F_{q+5}\!\left(\begin{array}{c}(a_p)-1,\,\frac{2b-5}{4},\,\frac{2b-3}{4};\,z\\(b_q)-1,\,\frac{b}{2}-1,\,\frac{b-1}{2},\,b-\frac{5}{2},\,b-2,\,\frac{1}{2}\end{array}\right)\right.$$

$$\left.- {}_pF_{q+3}\!\left(\begin{array}{c}(a_p)-1;\,\frac{z}{4}\\(b_q)-1,\,\frac{b}{2}-1,\,\frac{b-1}{2},\,b-2\end{array}\right)\right].$$

102. $\displaystyle\sum_{k=0}^{\infty}\frac{(6k+a)\left(\frac{a}{3}\right)_k}{k!\,(a)_{6k+1}}(-z^3)^k\frac{\prod(a_p)_{3k}}{\prod(b_q)_{3k}}\,_pF_{q+1}\!\left(\begin{array}{c}(a_p)+3k;\,z\\(b_q)+3k,\,a+6k+1\end{array}\right)$

$$= {}_{p+1}F_{q+2}\!\left(\begin{array}{c}(a_p),\,\frac{a}{3};\,\frac{3z}{4}\\(b_q),\,\frac{a}{2},\,\frac{a+1}{2}\end{array}\right).$$

6.17.2. Series containing $_pF_q((a_p(k));(b_q(k));z)$ and trigonometric functions

1. $\displaystyle\sum_{k=1}^{\infty}\frac{1}{k}\frac{\sin(k\nu\pi)}{\Gamma(\mu+k\nu)\Gamma(\mu-k\nu)}\,_pF_{q+2}\!\left(\begin{array}{c}(a_p);\,z\\(b_q),\,\mu+k\nu,\,\mu-k\nu\end{array}\right)$

$$= \frac{\pi(1-\nu)}{2\Gamma^2(\mu)}\,_pF_{q+2}\!\left(\begin{array}{c}(a_p);\,z\\(b_q),\,\mu,\,\mu\end{array}\right)\quad [\mathrm{Re}\,\mu>1/2;\,0<\nu<1].$$

2. $\displaystyle\sum_{k=1}^{\infty}\frac{(-1)^k}{k}\frac{\sin(k\nu\pi)}{\Gamma(\mu+k\nu)\Gamma(\mu-k\nu)}\,_pF_{q+2}\!\left(\begin{array}{c}(a_p);\,z\\(b_q),\,\mu+k\nu,\,\mu-k\nu\end{array}\right)$

$$= -\frac{\pi\nu}{2\Gamma^2(\mu)}\,_pF_{q+2}\!\left(\begin{array}{c}(a_p);\,z\\(b_q),\,\mu,\,\mu\end{array}\right)\quad [\mathrm{Re}\,\mu>0;\,0<\nu<1/2].$$

3. $\displaystyle\sum_{k=0}^{\infty}\frac{(-1)^k}{2k+1}\frac{\cos[(2k+1)\nu\pi]}{\Gamma(\mu+(2k+1)\nu)\Gamma(\mu-(2k+1)\nu)}$

$$\times\,_pF_{p+2}\!\left(\begin{array}{c}(a_p);\,z\\(b_q),\,\mu+(2k+1)\nu,\,\mu-(2k+1)\nu\end{array}\right) = \frac{\pi}{4\Gamma^2(\mu)}\,_pF_{p+2}\!\left(\begin{array}{c}(a_p);\,z\\(b_q),\,\mu,\,\mu\end{array}\right)$$

$$[\mathrm{Re}\,\mu>1/2;\,-1/4<\nu<1/4].$$

4. $\sum_{k=0}^{\infty} \frac{(-1)^k}{(2k+1)^2} \frac{\sin\left[(2k+1)\nu\pi\right]}{\Gamma(\mu+(2k+1)\nu)\Gamma(\mu-(2k+1)\nu)}$

$\times {}_pF_{q+2}\left(\begin{matrix}(a_p);\ z\\(b_q),\ \mu+(2k+1)\nu,\ \mu-(2k+1)\nu\end{matrix}\right) = \frac{\pi^2\nu}{4\Gamma^2(\mu)}\ {}_pF_{q+2}\left(\begin{matrix}(a_p);\ z\\(b_q),\ \mu,\ \mu\end{matrix}\right)$

$[\operatorname{Re}\mu > 1/2;\ -1/4 < \nu < 1/4].$

5. $\sum_{k=0}^{\infty} \frac{(-1)^k}{(2k-1)(2k+3)} \frac{\sin\left[(2k+1)\nu\pi\right]}{\Gamma(\mu+(2k+1)\nu)\Gamma(\mu-(2k+1)\nu)}$

$\times {}_pF_{q+2}\left(\begin{matrix}(a_p);\ z\\(b_q),\ \mu+(2k+1)\nu,\ \mu-(2k+1)\nu\end{matrix}\right)$

$= -\frac{\pi\sin(2\nu\pi)}{8\Gamma(\mu+2\nu)\Gamma(\mu-2\nu)}\ {}_pF_{q+2}\left(\begin{matrix}(a_p);\ z\\(b_q),\ \mu+2\nu,\ \mu-2\nu\end{matrix}\right)$

$[\operatorname{Re}\mu > 0;\ -1/4 < \nu < 1/4].$

6. $\sum_{k=0}^{\infty} \frac{1}{(2k-1)(2k+3)} \frac{\cos\left[(2k+1)\nu\pi\right]}{\Gamma(\mu+(2k+1)\nu)\Gamma(\mu-(2k+1)\nu)}$

$\times {}_pF_{q+2}\left(\begin{matrix}(a_p);\ z\\(b_q),\ \mu+(2k+1)\nu,\ \mu-(2k+1)\nu\end{matrix}\right)$

$= -\frac{\pi\sin(2\nu\pi)}{8\Gamma(\mu+2\nu)\Gamma(\mu-2\nu)}\ {}_pF_{q+2}\left(\begin{matrix}(a_p);\ z\\(b_q),\ \mu+2\nu,\ \mu-2\nu\end{matrix}\right)$

$[\operatorname{Re}\mu > 1/2;\ 0 < \nu < 1/2].$

7. $\sum_{k=0}^{\infty} \frac{(2k+1)}{(2k-1)(2k+3)} \frac{\sin\left[(2k+1)\nu\pi\right]}{\Gamma(\mu+(2k+1)\nu)\Gamma(\mu-(2k+1)\nu)}$

$\times {}_pF_{q+2}\left(\begin{matrix}(a_p);\ z\\(b_q),\ \mu+(2k+1)\nu,\ \mu-(2k+1)\nu\end{matrix}\right)$

$= \frac{\pi\cos(2\nu\pi)}{4\Gamma(\mu+2\nu)\Gamma(\mu-2\nu)}\ {}_pF_{q+2}\left(\begin{matrix}(a_p);\ z\\(b_q),\ \mu+2\nu,\ \mu-2\nu\end{matrix}\right)$

$[\operatorname{Re}\mu > 1/2;\ 0 < \nu < 1/2].$

8. $\sum_{k=0}^{\infty} \frac{(-1)^k(2k+1)}{(2k-1)(2k+3)} \frac{\cos\left[(2k+1)\nu\pi\right]}{\Gamma(\mu+(2k+1)\nu)\Gamma(\mu-(2k+1)\nu)}$

$\times {}_pF_{q+2}\left(\begin{matrix}(a_p);\ z\\(b_q),\ \mu+(2k+1)\nu,\ \mu-(2k+1)\nu\end{matrix}\right)$

$= -\frac{\pi\cos(2\nu\pi)}{4\Gamma(\mu+2\nu)\Gamma(\mu-2\nu)}\ {}_pF_{q+2}\left(\begin{matrix}(a_p);\ z\\(b_q),\ \mu+2\nu,\ \mu-2\nu\end{matrix}\right)$

$[\operatorname{Re}\mu > 1/2;\ -1/4 < \nu < 1/4].$

6.17. The Generalized Hypergeometric Function $_pF_q((a_p);(b_q);z)$

9. $\displaystyle\sum_{k=1}^{\infty} \frac{1}{(k^2a^2-b^2)} \frac{\cos(k\nu\pi)}{\Gamma(\mu+k\nu)\Gamma(\mu-k\nu)} \,_pF_{q+2}\!\left(\!\begin{array}{c}(a_p);\,z\\(b_q),\,\mu+k\nu,\,\mu-k\nu\end{array}\!\right)$

$= -\dfrac{\pi\csc\dfrac{b\pi}{a}\cos\dfrac{(1-\nu)b\pi}{a}}{2ab\,\Gamma\!\left(\mu+\dfrac{b\nu}{a}\right)\Gamma\!\left(\mu-\dfrac{b\nu}{a}\right)}\,_pF_{q+2}\!\left(\!\begin{array}{c}(a_p);\,z\\(b_q),\,\mu+\dfrac{b\nu}{a},\,\mu-\dfrac{b\nu}{a}\end{array}\!\right)$

$\quad + \dfrac{1}{2\Gamma^2(\mu)b^2}\,_pF_{q+2}\!\left(\!\begin{array}{c}(a_p);\,z\\(b_q),\,\mu,\,\mu\end{array}\!\right)$ $\quad[\mathrm{Re}\,\mu>0;\ 0<\nu<1]$.

10. $\displaystyle\sum_{k=1}^{\infty} \frac{(-1)^k}{k^2a^2-b^2} \frac{\cos(k\nu\pi)}{\Gamma(\mu+k\nu)\Gamma(\mu-k\nu)} \,_pF_{q+2}\!\left(\!\begin{array}{c}(a_p);\,z\\(b_q),\,\mu+k\nu,\,\mu-k\nu\end{array}\!\right)$

$= -\dfrac{\pi\csc\dfrac{b\pi}{a}\cos\dfrac{b\nu\pi}{a}}{2ab\,\Gamma\!\left(\mu+\dfrac{b\nu}{a}\right)\Gamma\!\left(\mu-\dfrac{b\nu}{a}\right)}\,_pF_{q+2}\!\left(\!\begin{array}{c}(a_p);\,z\\(b_q),\,\mu+\dfrac{b\nu}{a},\,\mu-\dfrac{b\nu}{a}\end{array}\!\right)$

$\quad + \dfrac{1}{2\Gamma^2(\mu)b^2}\,_pF_{q+2}\!\left(\!\begin{array}{c}(a_p);\,z\\(b_q),\,\mu,\,\mu\end{array}\!\right)$ $\quad[\mathrm{Re}\,\mu>-1/2;\ 0<\nu<1]$.

11. $\displaystyle\sum_{k=1}^{\infty} \frac{k}{(k^2a^2-b^2)} \frac{\sin(k\nu\pi)}{\Gamma(\mu-k\nu)\Gamma(\mu+k\nu)} \,_pF_{q+2}\!\left(\!\begin{array}{c}(a_p);\,z\\(b_q),\,\mu+k\nu,\,\mu-k\nu\end{array}\!\right)$

$= \dfrac{\pi\csc\dfrac{b\pi}{a}\sin\dfrac{(1-\nu)b\pi}{a}}{2a^2\,\Gamma\!\left(\mu+\dfrac{b\nu}{a}\right)\Gamma\!\left(\mu-\dfrac{b\nu}{a}\right)}\,_pF_{q+2}\!\left(\!\begin{array}{c}(a_p);\,z\\(b_q),\,\mu+\dfrac{b\nu}{a},\,\mu-\dfrac{b\nu}{a}\end{array}\!\right)$

$\quad[\mathrm{Re}\,\mu>1/2;\ 0<\nu<1]$.

12. $\displaystyle\sum_{k=1}^{\infty} \frac{(-1)^k k}{k^2a^2-b^2} \frac{\sin(k\nu\pi)}{\Gamma(\mu-k\nu)\Gamma(\mu+k\nu)} \,_pF_{q+2}\!\left(\!\begin{array}{c}(a_p);\,z\\(b_q),\,\mu+k\nu,\,\mu-k\nu\end{array}\!\right)$

$= -\dfrac{\pi\csc\dfrac{b\pi}{a}\sin\dfrac{\nu b\pi}{a}}{2a^2\,\Gamma\!\left(\mu+\dfrac{b\nu}{a}\right)\Gamma\!\left(\mu-\dfrac{b\nu}{a}\right)}\,_pF_{q+2}\!\left(\!\begin{array}{c}(a_p);\,z\\(b_q),\,\mu+\dfrac{b\nu}{a},\,\mu-\dfrac{b\nu}{a}\end{array}\!\right)$

$\quad[\mathrm{Re}\,\mu>0;\ 0<\nu<1/2]$.

13. $\displaystyle\sum_{k=0}^{\infty} \frac{(-1)^k}{[(2k+1)^2a^2-b^2]} \frac{\sin[(2k+1)\nu\pi]}{\Gamma(\mu+(2k+1)\nu)\Gamma(\mu-(2k+1)\nu)}$

$\qquad \times\,_pF_{q+2}\!\left(\!\begin{array}{c}(a_p);\,z\\(b_q),\,\mu+(2k+1)\nu,\,\mu-(2k+1)\nu\end{array}\!\right)$

$= \dfrac{\pi\sec\dfrac{b\pi}{2a}\sin\dfrac{b\nu\pi}{a}}{4ab\,\Gamma\!\left(\mu+\dfrac{b\nu}{a}\right)\Gamma\!\left(\mu-\dfrac{b\nu}{a}\right)}\,_pF_{q+2}\!\left(\!\begin{array}{c}(a_p);\,z\\(b_q),\,\mu+\dfrac{b\nu}{a},\,\mu-\dfrac{b\nu}{a}\end{array}\!\right)$

$\quad[\mathrm{Re}\,\mu>0;\ -1/4<\nu<1/4]$.

14. $\sum_{k=0}^{\infty} \frac{(-1)^k (2k+1)}{[(2k+1)^2 a^2 - b^2]} \frac{\cos[(2k+1)\nu\pi]}{\Gamma(\mu + (2k+1)\nu)\Gamma(\mu - (2k+1)\nu)}$

$$\times \, _pF_{q+2}\left(\begin{matrix}(a_p); \, z \\ (b_q), \, \mu + (2k+1)\nu, \, \mu - (2k+1)\nu\end{matrix}\right)$$

$$= \frac{\pi \sec\frac{b\pi}{2a} \cos\frac{b\nu\pi}{a}}{4a^2 \Gamma\left(\mu + \frac{b\nu}{a}\right)\Gamma\left(\mu - \frac{b\nu}{a}\right)} \, _pF_{q+2}\left(\begin{matrix}(a_p); \, z \\ (b_q), \, \mu + \frac{b\nu}{a}, \, \mu - \frac{b\nu}{a}\end{matrix}\right)$$

$$[\mathrm{Re}\,\mu > 1/2; \; -1/4 < \nu < 1/4].$$

6.17.3. Series containing $_pF_q((a_p(k)); (b_q(k)); z)$ and special functions

1. $\sum_{k=0}^{\infty} \frac{(-z)^k}{k!} \psi(k+a) \frac{\prod(a_p)_k}{\prod(b_q)_k} {}_pF_q\left(\begin{matrix}(a_p)+k; \, z \\ (b_q)+k\end{matrix}\right)$

$$= \psi(a) - \frac{z}{a} \frac{\prod_{i=1}^{p} a_i}{\prod_{j=1}^{q} b_j} {}_{p+2}F_{q+2}\left(\begin{matrix}(a_p)+1,\, 1,\, 1;\, z \\ (b_q)+1,\, a+1,\, 2\end{matrix}\right).$$

2. $\sum_{k=1}^{\infty} \frac{(-z)^k}{(a)_k} \psi(k+a) \frac{\prod(a_p)_k}{\prod(b_q)_k} {}_pF_q\left(\begin{matrix}(a_p)+k;\, z \\ (b_q)+k\end{matrix}\right) =$

$$-\frac{z}{a^2} \frac{\prod_{i=1}^{p} a_i}{\prod_{j=1}^{q} b_j} \left[a\psi(a)\, {}_{p+1}F_{q+1}\left(\begin{matrix}(a_p)+1,\, a;\, z \\ (b_q)+1,\, a+1\end{matrix}\right)\right.$$

$$\left. + \, {}_{p+2}F_{q+2}\left(\begin{matrix}(a_p)+1,\, a,\, a;\, z \\ (b_q)+1,\, a+1,\, a+1\end{matrix}\right)\right].$$

3. $\sum_{k=0}^{\infty} \frac{(-z)^k}{k!} \psi(2k+a) \frac{\prod(a_p)_k}{\prod(b_q)_k} {}_pF_q\left(\begin{matrix}(a_p)+k;\, z \\ (b_q)+k\end{matrix}\right)$

$$= \psi(a) - z \frac{\prod_{i=1}^{p} a_i}{\prod_{j=1}^{q} b_j} \left[\frac{1}{a}\, {}_{p+2}F_{q+2}\left(\begin{matrix}(a_p)+1,\, 1,\, 1;\, z \\ (b_q)+1,\, 2,\, \frac{a}{2}+1\end{matrix}\right)\right.$$

$$\left. + \frac{1}{a+1}\, {}_{p+2}F_{q+2}\left(\begin{matrix}(a_p)+1,\, 1,\, 1;\, z \\ (b_q)+1,\, 2,\, \frac{a+3}{2}\end{matrix}\right)\right].$$

4. $\sum_{k=0}^{\infty} \frac{(16z)^k}{(2k)!} B_{2k} \frac{\prod(a_p)_k}{\prod(b_q)_k} {}_pF_{q+1}\left(\begin{matrix}(a_p)+k;\, z \\ (b_q)+k,\, \frac{3}{2}\end{matrix}\right) = {}_pF_{q+1}\left(\begin{matrix}(a_p);\, z \\ (b_q),\, \frac{1}{2}\end{matrix}\right).$

5. $\displaystyle\sum_{k=0}^{\infty} (2^{2k}-1) \frac{(16z)^k}{(2k)!} B_{2k} \frac{\prod (a_p)_k}{\prod (b_q)_k} {}_pF_{q+1}\left(\begin{array}{c}(a_p)+k;\ z\\ (b_q)+k,\ \frac{1}{2}\end{array}\right)$

$$= 4z \frac{\prod_{i=1}^{p} a_i}{\prod_{j=1}^{q} b_j} {}_pF_{q+1}\left(\begin{array}{c}(a_p)+1;\ z\\ (b_q)+1,\ \frac{3}{2}\end{array}\right).$$

6. $\displaystyle\sum_{k=0}^{\infty} \frac{z^k}{k!} B_k(w) \frac{\prod (a_p)_k}{\prod (b_q)_k} {}_{p+1}F_{q+1}\left(\begin{array}{c}(a_p)+k,\ 1;\ z\\ (b_q)+k,\ 2\end{array}\right) = {}_pF_q\left(\begin{array}{c}(a_p);\ wz\\ (b_q)\end{array}\right).$

7. $\displaystyle\sum_{k=0}^{\infty} \frac{z^k}{\left(\frac{1}{2}\right)_{2k}} P_{2k}(w) \frac{\prod (a_p)_k}{\prod (b_q)_k} {}_pF_{q+1}\left(\begin{array}{c}(a_p)+k;\ z\\ (b_q)+k,\ 2k+\frac{3}{2}\end{array}\right)$

$$= {}_pF_{q+1}\left(\begin{array}{c}(a_p);\ w^2 z\\ (b_q),\ \frac{1}{2}\end{array}\right).$$

8. $\displaystyle\sum_{k=0}^{\infty} \frac{z^k}{\left(\frac{1}{2}\right)_{2k+1}} P_{2k+1}(w) \frac{\prod (a_p)_k}{\prod (b_q)_k} {}_pF_{q+1}\left(\begin{array}{c}(a_p)+k;\ z\\ (b_q)+k,\ 2k+\frac{5}{2}\end{array}\right)$

$$= 2w\, {}_pF_{q+1}\left(\begin{array}{c}(a_p);\ w^2 z\\ (b_q),\ \frac{3}{2}\end{array}\right).$$

9. $\displaystyle\sum_{k=0}^{\infty} \frac{z^k}{(2k)!} T_{2k}(w) \frac{\prod (a_p)_k}{\prod (b_q)_k} {}_pF_{q+1}\left(\begin{array}{c}(a_p)+k;\ z\\ (b_q)+k,\ 2k+1\end{array}\right)$

$$= \frac{1}{2}\, {}_pF_{q+1}\left(\begin{array}{c}(a_p);\ w^2 z\\ (b_q),\ \frac{1}{2}\end{array}\right) + \frac{1}{2}\, {}_pF_{q+1}\left(\begin{array}{c}(a_p);\ z\\ (b_q),\ 1\end{array}\right).$$

10. $\displaystyle\sum_{k=0}^{\infty} \frac{z^k}{(2k+1)!} T_{2k+1}(w) \frac{\prod (a_p)_k}{\prod (b_q)_k} {}_pF_{q+1}\left(\begin{array}{c}(a_p)+k;\ z\\ (b_q)+k,\ 2k+2\end{array}\right)$

$$= w\, {}_pF_{q+1}\left(\begin{array}{c}(a_p);\ w^2 z\\ (b_q),\ \frac{3}{2}\end{array}\right).$$

11. $\displaystyle\sum_{k=0}^{\infty} \frac{z^k}{(2k)!} U_{2k}(w) \frac{\prod (a_p)_k}{\prod (b_q)_k} {}_pF_{q+1}\left(\begin{array}{c}(a_p)+k;\ z\\ (b_q)+k,\ 2k+2\end{array}\right)$

$$= {}_pF_{q+1}\left(\begin{array}{c}(a_p);\ w^2 z\\ (b_q),\ \frac{1}{2}\end{array}\right).$$

12. $\displaystyle\sum_{k=0}^{\infty} \frac{z^k}{(2k+1)!} U_{2k+1}(w) \frac{\prod (a_p)_k}{\prod (b_q)_k} \,{}_pF_{q+1}\!\left(\begin{array}{c} (a_p)+k;\ z \\ (b_q)+k,\, 2k+3 \end{array}\right)$

$$= 2w\, {}_pF_{q+1}\!\left(\begin{array}{c} (a_p);\ w^2 z \\ (b_q),\, \frac{3}{2} \end{array}\right).$$

13. $\displaystyle\sum_{k=0}^{\infty} \frac{(-z)^k}{(\lambda+1)_k} L_k^\lambda(w) \frac{\prod (a_p)_k}{\prod (b_q)_k} \,{}_{p+1}F_q\!\left(\begin{array}{c} (a_p)+k;\ z \\ (b_q)+k \end{array}\right) = {}_pF_{q+1}\!\left(\begin{array}{c} (a_p);\ wz \\ (b_q),\, \lambda+1 \end{array}\right).$

14. $\displaystyle\sum_{k=0}^{\infty} \frac{(b)_k (c-a)_k}{(c)_k (\lambda+1)_k} z^k L_k^\lambda(w) \,{}_2F_1\!\left(\begin{array}{c} a,\, b+k;\ z \\ c+k \end{array}\right)$

$$= (1-z)^{-b} \,{}_2F_2\!\left(\begin{array}{c} b,\, c-a;\ \frac{wz}{z-1} \\ c,\, \lambda+1 \end{array}\right) \quad [|z|<1].$$

15. $\displaystyle\sum_{k=0}^{\infty} \frac{z^{k/2}}{(\lambda)_k} C_k^\lambda(w) \frac{\prod (a_p)_{k/2}}{\prod (b_q)_{k/2}} \,{}_pF_{q+1}\!\left(\begin{array}{c} (a_p)+\frac{k}{2};\ z \\ (b_q)+\frac{k}{2},\, \lambda+k+1 \end{array}\right)$

$$= {}_pF_{q+1}\!\left(\begin{array}{c} (a_p);\ w^2 z \\ (b_q),\, \frac{1}{2} \end{array}\right) + 2w z^{1/2} \frac{\prod \Gamma[(b_q)] \prod \Gamma\!\left[(a_p)+\frac{1}{2}\right]}{\prod \Gamma[(a_p)] \prod \Gamma\!\left[(b_q)+\frac{1}{2}\right]}$$

$$\times {}_pF_{q+1}\!\left(\begin{array}{c} (a_p)+\frac{1}{2};\ w^2 z \\ (b_q)+\frac{1}{2},\, \frac{3}{2} \end{array}\right).$$

16. $\displaystyle\sum_{k=0}^{\infty} \frac{z^k}{(\lambda)_{2k}} C_{2k}^\lambda(w) \frac{\prod (a_p)_k}{\prod (b_q)_k} \,{}_pF_{q+1}\!\left(\begin{array}{c} (a_p)+k;\ z \\ (b_q)+k,\, \lambda+2k+1 \end{array}\right)$

$$= {}_pF_{q+1}\!\left(\begin{array}{c} (a_p);\ w^2 z \\ (b_q),\, \frac{1}{2} \end{array}\right).$$

17. $\displaystyle\sum_{k=0}^{\infty} \frac{z^k}{(\lambda)_{2k+1}} C_{2k+1}^\lambda(w) \frac{\prod (a_p)_k}{\prod (b_q)_k} \,{}_pF_{q+1}\!\left(\begin{array}{c} (a_p)+k;\ z \\ (b_q)+k,\, \lambda+2k+2 \end{array}\right)$

$$= 2w\, {}_pF_{q+1}\!\left(\begin{array}{c} (a_p);\ w^2 z \\ (b_q),\, \frac{3}{2} \end{array}\right).$$

18. $\displaystyle\sum_{k=0}^{\infty} \frac{\left(\frac{z}{2w}\right)^k}{(1-\lambda)_k} C_k^{\lambda-k}(w) \frac{\prod (a_p)_k}{\prod (b_q)_k} \,{}_pF_q\!\left(\begin{array}{c} (a_p)+k;\ z \\ (b_q)+k \end{array}\right)$

$$= {}_pF_{q+1}\!\left(\begin{array}{c} \frac{(a_p)}{2},\, \frac{(a_p)+1}{2};\ \frac{z^2}{4^{q-p+1} w^2} \\ \frac{(b_q)}{2},\, \frac{(b_q)+1}{2},\, 1-\lambda \end{array}\right).$$

19. $\sum_{k=0}^{\infty} \dfrac{(b)_k (c-a)_k}{(c)_k (1-\lambda)_k} z^k C_{2k}^{\lambda-k}(w) \, _2F_1\!\left(\begin{array}{c} a,\, b+k;\ z \\ c+k \end{array}\right)$

$$= (1-z)^{-b} \, _3F_2\!\left(\begin{array}{c} b,\, c-a,\, \lambda \\ \dfrac{1}{2},\, c; \ \dfrac{w^2 z}{z-1} \end{array}\right) \qquad [|z|<1].$$

20. $\sum_{k=0}^{\infty} \dfrac{(b)_k (c-a)_k}{(c)_k (1-\lambda)_k} z^k C_{2k+1}^{\lambda-k}(w) \, _2F_1\!\left(\begin{array}{c} a,\, b+k;\ z \\ c+k \end{array}\right)$

$$= 2\lambda w (1-z)^{-b} \, _3F_2\!\left(\begin{array}{c} b,\, c-a,\, \lambda+1 \\ \dfrac{3}{2},\, c; \ \dfrac{w^2 z}{z-1} \end{array}\right) \qquad [|z|<1].$$

21. $\sum_{k=0}^{\infty} \dfrac{(b)_k (c-a)_k}{(c)_k (\rho+1)_k} z^k P_k^{(\rho,\sigma-k)}(w) \, _2F_1\!\left(\begin{array}{c} a,\, b+k;\ z \\ c+k \end{array}\right)$

$$= (1-z)^{-b} \, _3F_2\!\left(\begin{array}{c} b,\, c-a,\, \rho+\sigma+1 \\ c,\, \rho+1; \ \dfrac{(w-1)z}{2(1-z)} \end{array}\right) \qquad [|z|<1].$$

6.17.4. Series containing products of $_pF_q((a_p(k));(b_q(k));z)$

1. $\sum_{k=0}^{\infty} \dfrac{(-z)^k}{k!} \dfrac{\prod (c_r)_k}{\prod (d_s)_k} \, _{p+1}F_q\!\left(\begin{array}{c} -k,\, (a_p) \\ (b_q);\ w \end{array}\right) {}_r F_s\!\left(\begin{array}{c} (c_r)+k;\ z \\ (d_s)+k \end{array}\right)$

$$= {}_{p+r}F_{q+s}\!\left(\begin{array}{c} (a_p),\, (c_r);\ wz \\ (b_q),\, (d_s) \end{array}\right).$$

2. $\sum_{k=0}^{\infty} (-1)^{k(r+s+1)} \binom{n}{k} \dfrac{\left(\dfrac{w}{z}\right)^k}{(1-a)_k} \dfrac{\prod (a_p)_k \prod (1-d_s)_k}{\prod (b_q)_k \prod (1-c_r)_k}$

$$\times \, _{p+1}F_{q+1}\!\left(\begin{array}{c} -n+k,\, (a_p)+k;\ w \\ 2k+a+1,\, (b_q)+k \end{array}\right)$$

$$\times \, _{r+2}F_{s+1}\!\left(\begin{array}{c} -k,\, a-k,\, (c_r)-k;\ z \\ 1-a-2k,\, (d_s)-k \end{array}\right)$$

$$= {}_{p+s+1}F_{q+r+1}\!\left(\begin{array}{c} -n,\, (a_p),\, 1-(d_s);\ (-1)^{(r+s)} \dfrac{w}{z} \\ 1-a,\, (b_q),\, 1-(c_r) \end{array}\right).$$

6.17.5. Series containing $_pF_{p+1}((a_p);(b_{p+1});\varphi(k,x))$

Notation: $d = \operatorname{Re}\left(\sum_{j=1}^{p+1} b_j - \sum_{i=1}^{p} a_i\right).$

1. $\sum_{k=1}^{\infty} (-1)^k \, _pF_{p+1}\!\left(\begin{array}{c} (a_p);\ -k^2 x \\ (b_{p+1}) \end{array}\right) = -\dfrac{1}{2} \qquad \left[\operatorname{Re} b_j > 1;\ d > 1/2;\ 0 < x < \pi^2/4\right].$

2. $\displaystyle\sum_{k=1}^{\infty} \frac{1}{k^{2n}} {}_pF_{p+1}\left(\begin{array}{c}(a_p);\ -k^2x\\(b_{p+1})\end{array}\right) = \frac{(-1)^{n+1}}{(2n+1)!\sqrt{\pi}}$

$\displaystyle\times \sum_{k=0}^{2n+1} \binom{2n+1}{k} B_k (2\pi)^k (4x)^{n-k/2} \Gamma\left[\left(n + \frac{3-k}{2}\right)\right]$

$\displaystyle\times \frac{\prod \Gamma[(a_p) + n - k/2] \prod \Gamma[(b_{p+1})]}{\prod \Gamma[(a_p)] \prod \Gamma[(b_{p+1}) + n - k/2]}$

$\left[\operatorname{Re} b_j > -n - 1/2;\ \operatorname{Re} a_i > 2n - 2 + (1 \pm 1)/2;\ d > 2n + 1/2;\ 0 < x < \pi^2\right].$

3. $\displaystyle\sum_{k=1}^{\infty} \frac{(-1)^k}{k^{2n}} {}_pF_{p+1}\left(\begin{array}{c}(a_p);\ -k^2x\\(b_{p+1})\end{array}\right) = \frac{(-1)^{n+1}}{2(2n)!} \pi^{2n-1/2}$

$\displaystyle\times \sum_{k=0}^{2n} \binom{2n}{k} 2^k B_k \sum_{j=0}^{2n-k} \binom{2n-k}{j} \pi^{-j} (4x)^{j/2} \Gamma\left(\frac{j+1}{2}\right)$

$\displaystyle\times \frac{\prod \Gamma[(a_p) + j/2] \prod \Gamma[(b_{p+1})]}{\prod \Gamma[(a_p)] \prod \Gamma[(b_{p+1}) + j/2]}$

$\left[\operatorname{Re} b_j > -n - 1/2;\ d > 1/2 - 2n;\ 0 < x < \pi^2/4\right].$

4. $\displaystyle\sum_{k=0}^{\infty} \frac{1}{k^2 a^2 \pm b^2} {}_pF_{p+1}\left(\begin{array}{c}(a_p);\ -k^2x\\(b_{p+1})\end{array}\right)$

$\displaystyle = \mp \frac{1}{2b^2} \pm \frac{\pi}{2ab} \left\{\begin{array}{c}\coth(b\pi/a)\\ \cot(b\pi/a)\end{array}\right\} {}_pF_{p+1}\left(\begin{array}{c}(a_p);\ \pm\frac{b^2 x}{a^2}\\(b_{p+1})\end{array}\right)$

$\displaystyle \pm \frac{\sqrt{\pi x}}{a^2} \frac{\prod \Gamma[(a_p)] \prod \Gamma\left[(b_{p+1}) + \frac{1}{2}\right]}{\prod \Gamma\left[(a_p) + \frac{1}{2}\right] \prod \Gamma[(b_{p+1})]} {}_{p+1}F_{p+2}\left(\begin{array}{c}(a_p) + \frac{1}{2},\ 1;\ \pm\frac{b^2 x}{a^2}\\(b_{p+1}) + \frac{1}{2},\ \frac{3}{2}\end{array}\right)$

$\left[\operatorname{Re} b_j > 1/2;\ d > -3/2;\ 0 < x < \pi^2\right].$

5. $\displaystyle\sum_{k=1}^{\infty} \frac{(-1)^k}{k^2 a^2 \pm b^2} {}_pF_{p+1}\left(\begin{array}{c}(a_p);\ -k^2x\\(b_{p+1})\end{array}\right)$

$\displaystyle = \mp \frac{1}{2b^2} \pm \frac{\pi}{2ab} \left\{\begin{array}{c}\operatorname{csch}(b\pi/a)\\ \csc(b\pi/a)\end{array}\right\} {}_pF_{p+1}\left(\begin{array}{c}(a_p);\ \pm\frac{b^2 x}{a^2}\\(b_{p+1})\end{array}\right)$

$\left[\operatorname{Re} b_j > 1/2;\ d > -3/2;\ 0 < x < \pi^2/4\right].$

6. $\displaystyle\sum_{k=2}^{\infty}\frac{(-1)^k}{k^2-1}\,_pF_{p+1}\!\left(\begin{matrix}(a_p);\ -k^2x\\(b_{p+1})\end{matrix}\right)=\frac{1}{2}-\frac{1}{4}\,_pF_{p+1}\!\left(\begin{matrix}(a_p);\ -x\\(b_{p+1})\end{matrix}\right)$

$\displaystyle -\,x\,\frac{\prod_{i=1}^{p}a_i}{\prod_{j=1}^{p+1}b_j}\,_pF_{p+1}\!\left(\begin{matrix}(a_p)+1;\ -x\\(b_{p+1})+1\end{matrix}\right)\quad\left[\mathrm{Re}\,b_j>1/2;\ d>-3/2;\ 0<x<\pi^2/4\right].$

7. $\displaystyle\sum_{k=1}^{\infty}\frac{\cos(ky)}{k^{2m-2}}\,_pF_{p+1}\!\left(\begin{matrix}(a_p);\ -k^2x\\(b_{p+1})\end{matrix}\right)=(-1)^m\frac{(4x)^{m-1}}{2(m-1)!}\frac{\prod(a_p)_{m-1}}{\prod(b_{p+1})_{m-1}}$

$\displaystyle +\sum_{k=0}^{m-2}\frac{x^k}{k!}\frac{\prod(a_p)_k}{\prod(b_{p+1})_k}\Bigg[(-1)^{m-1}\frac{\pi y^{2m-2k-3}}{2(2m-2k-3)!}$

$\displaystyle +(-1)^k\sum_{j=0}^{m-k-1}\frac{(-y^2)^j}{(2j)!}\zeta(2m-2j-2k-2)\Bigg]$

$\left[\mathrm{Re}\,b_j>3/2-m;\ d>2m-5/2;\ m\geq 1;\ 0<x<\pi^2/4;\ 2\sqrt{x}<y<2\pi-2\sqrt{x}\right].$

8. $\displaystyle\sum_{k=1}^{\infty}\,_pF_{p+1}\!\left(\begin{matrix}(a_p);\ -(k^2+a^2)x\\(b_{p+1})\end{matrix}\right)=-\frac{1}{2}\,_pF_{p+1}\!\left(\begin{matrix}(a_p);\ -a^2x\\(b_{p+1})\end{matrix}\right)$

$\displaystyle +\frac{1}{2}\sqrt{\frac{\pi}{x}}\frac{\prod\Gamma\!\left[(a_p)-\frac{1}{2}\right]\prod\Gamma[(b_{p+1})]}{\prod\Gamma[(a_p)]\prod\Gamma\!\left[(b_{p+1})-\frac{1}{2}\right]}\,_pF_{p+1}\!\left(\begin{matrix}(a_p)-\frac{1}{2};\ -a^2x\\(b_{p+1})-\frac{1}{2}\end{matrix}\right)$

$\left[\mathrm{Re}\,b_j>1/2;\ d>1/2;\ 0<x<\pi^2\right].$

9. $\displaystyle\sum_{k=1}^{\infty}(-1)^k\,_pF_{p+1}\!\left(\begin{matrix}(a_p);\ -(k^2+a^2)x\\(b_{p+1})\end{matrix}\right)=-\frac{1}{2}\,_pF_{p+1}\!\left(\begin{matrix}(a_p);\ -a^2x\\(b_{p+1})\end{matrix}\right)$

$\left[\mathrm{Re}\,b_j>1/2;\ d>1/2;\ 0<x<\pi^2/4\right].$

10. $\displaystyle\sum_{k=1}^{\infty}\frac{(-1)^k}{k^2}\,_pF_{p+1}\!\left(\begin{matrix}(a_p);\ -(k^2+a^2)x\\(b_{p+1})\end{matrix}\right)$

$\displaystyle =-\frac{\pi^2}{12}\,_pF_{p+1}\!\left(\begin{matrix}(a_p);\ -a^2x\\(b_{p+1})\end{matrix}\right)+\frac{x}{2}\frac{\prod_{i=1}^{p}a_i}{\prod_{j=1}^{p}b_j}\,_pF_{p+1}\!\left(\begin{matrix}(a_p)+1;\ -a^2x\\(b_{p+1})+1\end{matrix}\right)$

$\left[\mathrm{Re}\,b_j>-1/2;\ d>-3/2;\ 0<x<\pi^2/4\right].$

11. $\displaystyle\sum_{k=1}^{\infty} \frac{(-1)^k}{k^2} {}_pF_{p+1}\left(\begin{array}{c}(a_p); -(k^2+a^2)x \\ (b_{p+1})\end{array}\right) = -\frac{\pi^2}{12} {}_pF_{p+1}\left(\begin{array}{c}(a_p); -a^2 x \\ (b_{p+1})\end{array}\right)$

$\displaystyle + \frac{x}{2} \frac{\prod_{i=1}^{p} a_i}{\prod_{j=1}^{p+1} b_j} {}_pF_{p+1}\left(\begin{array}{c}(a_p)+1; -a^2 x \\ (b_{p+1})+1\end{array}\right)$ $\quad [\operatorname{Re} b_j > -1/2;\ d > -3/2;\ 0 < x < \pi^2/4]$.

12. $\displaystyle\sum_{k=0}^{\infty} \frac{1}{(2k+1)^2} {}_pF_{p+1}\left(\begin{array}{c}(a_p); -(2k+1)^2 x \\ (b_{p+1})\end{array}\right)$

$\displaystyle = \frac{\pi^2}{8} - \frac{\sqrt{\pi x}}{2} \frac{\prod \Gamma\left[(a_p)+\frac{1}{2}\right] \prod \Gamma[(b_{p+1})]}{\prod \Gamma[(a_p)] \prod \Gamma\left[(b_{p+1})+\frac{1}{2}\right]}$

$\quad [\operatorname{Re} b_j > -1/2;\ d > -3/2;\ 0 < x < \pi^2/4]$.

13. $\displaystyle\sum_{k=0}^{\infty} \frac{(-1)^k}{(2k+1)^{2n+1}} {}_pF_{p+1}\left(\begin{array}{c}(a_p); -(2k+1)^2 x \\ (b_{p+1})\end{array}\right)$

$\displaystyle = \frac{(-1)^n}{(2n)!} 2^{2n-2} \sqrt{\pi}\, x^n \sum_{k=0}^{2n} \binom{2n}{k} E_k\left(\frac{\pi}{4\sqrt{x}}\right)^k \Gamma\left(n + \frac{1-k}{2}\right)$

$\displaystyle \times \frac{\prod \Gamma\left[(a_p)+n-\frac{k}{2}\right] \prod \Gamma[(b_{p+1})]}{\prod \Gamma[(a_p)] \prod \Gamma\left[(b_{p+1})+n-\frac{k}{2}\right]}$

$\quad [\operatorname{Re} b_j > -n;\ d > -2n - 1/2;\ 0 < x < \pi^2/16]$.

14. $\displaystyle\sum_{k=0}^{\infty} \frac{1}{(2k-1)(2k+3)} {}_pF_{p+1}\left(\begin{array}{c}(a_p); -(2k+1)^2 x \\ (b_{p+1})\end{array}\right)$

$\displaystyle = -\frac{\sqrt{\pi x}}{2} \frac{\prod \Gamma\left[(a_p)+\frac{1}{2}\right] \prod \Gamma[(b_{p+1})]}{\prod \Gamma[(a_p)] \prod \Gamma\left[(b_{p+1})+\frac{1}{2}\right]} {}_pF_{p+1}\left(\begin{array}{c}(a_p)+\frac{1}{2}, 1;\ -4x \\ (b_{p+1})+\frac{1}{2}, \frac{3}{2}\end{array}\right)$

$\quad [\operatorname{Re} b_j > -1/2;\ d > -3/2;\ 0 < x < \pi^2/4]$.

15. $\displaystyle\sum_{k=0}^{\infty} (-1)^k \frac{2k+1}{(2k-1)(2k+3)} {}_pF_{p+1}\left(\begin{array}{c}(a_p); -(2k+1)^2 x \\ (b_{p+1})\end{array}\right)$

$\displaystyle = -\frac{\pi}{4} {}_pF_{p+1}\left(\begin{array}{c}(a_p);\ -4x \\ (b_{p+1})\end{array}\right)$ $\quad [\operatorname{Re} b_j > 0;\ d > -1/2;\ 0 < x < \pi^2/16]$.

16. $\displaystyle\sum_{k=0}^{\infty} \frac{(-1)^k}{(2k+3)(4k^2-1)} {}_pF_{p+1}\left(\begin{array}{c}(a_p); -(2k+1)^2 x \\ (b_{p+1})\end{array}\right)$

$\displaystyle = -\frac{\pi}{16}\left[1 + {}_pF_{p+1}\left(\begin{array}{c}(a_p);\ -4x \\ (b_{p+1})\end{array}\right)\right]$ $\quad [\operatorname{Re} b_j > -1;\ d > -1/2;\ |x| < \pi^2/16]$.

6.17. The Generalized Hypergeometric Function $_pF_q((a_p); (b_q); z)$

17. $\displaystyle\sum_{k=1}^{\infty} \frac{1}{k(k+1)} {}_pF_{p+1}\!\left(\begin{array}{c}(a_p);\ -(2k+1)^2 x \\ (b_{p+1})\end{array}\right) = {}_pF_{p+1}\!\left(\begin{array}{c}(a_p);\ -x \\ (b_{p+1})\end{array}\right) -$

$\displaystyle -2\sqrt{\pi x}\, \frac{\prod \Gamma\!\left[(a_p)+\frac{1}{2}\right] \prod \Gamma[(b_{p+1})]}{\prod \Gamma[(a_p)] \prod \Gamma\!\left[(b_{p+1})+\frac{1}{2}\right]}\, {}_{p+1}F_{p+2}\!\left(\begin{array}{c}(a_p)+\frac{1}{2},\ 1;\ -x \\ (b_{p+1})+\frac{1}{2},\ \frac{3}{2}\end{array}\right)$

$\displaystyle + 4x\, \frac{\prod_{i=1}^{p} a_i}{\prod_{j=1}^{p+1} b_j}\, {}_pF_{p+1}\!\left(\begin{array}{c}(a_p)+1;\ -x \\ (b_{p+1})+1\end{array}\right) \quad [\mathrm{Re}\,b_j > -1/2;\ d > -3/2;\ 0 < x < \pi^2/4].$

18. $\displaystyle\sum_{k=1}^{\infty} (-1)^k \frac{2k+1}{k(k+1)} {}_pF_{p+1}\!\left(\begin{array}{c}(a_p);\ -(2k+1)^2 x \\ (b_{p+1})\end{array}\right) = {}_pF_{p+1}\!\left(\begin{array}{c}(a_p);\ -x \\ (b_{p+1})\end{array}\right)$

$\displaystyle - 2\, {}_{p+1}F_{p+2}\!\left(\begin{array}{c}(a_p),\ \frac{3}{2};\ -x \\ (b_{p+1}),\ \frac{1}{2}\end{array}\right) \quad [\mathrm{Re}\,b_j > -n;\ d > -1/2;\ 0 < x < \pi^2/16].$

19. $\displaystyle\sum_{k=0}^{\infty} \frac{1}{(2k+1)^2 a^2 \pm b^2}\, {}_pF_{p+1}\!\left(\begin{array}{c}(a_p);\ -(2k+1)^2 x \\ (b_{p+1})\end{array}\right)$

$\displaystyle = \frac{\pi}{4ab}\left\{\begin{array}{c}\tanh[b\pi/(2a)] \\ \tan[b\pi/(2a)]\end{array}\right\} {}_pF_{p+1}\!\left(\begin{array}{c}(a_p);\ \pm\frac{b^2 x}{a^2} \\ (b_{p+1})\end{array}\right)$

$\displaystyle - \frac{\sqrt{\pi x}}{2a^2}\, \frac{\prod \Gamma\!\left[(a_p)+\frac{1}{2}\right] \prod \Gamma[(b_{p+1})]}{\prod \Gamma[(a_p)] \prod \Gamma\!\left[(b_{p+1})+\frac{1}{2}\right]}\, {}_{p+1}F_{p+2}\!\left(\begin{array}{c}(a_p)+\frac{1}{2},\ 1;\ \pm\frac{b^2 x}{a^2} \\ (b_{p+1})+\frac{1}{2},\ \frac{3}{2}\end{array}\right)$

$[\mathrm{Re}\,b_j > -1/2;\ d > -3/2;\ 0 \le x \le \pi^2/4].$

20. $\displaystyle\sum_{k=0}^{\infty} \frac{(-1)^k}{2k+1}\, {}_pF_{p+1}\!\left(\begin{array}{c}(a_p);\ -((2k+1)^2+a^2)x \\ (b_{p+1})\end{array}\right) = \frac{\pi}{4}\, {}_pF_{p+1}\!\left(\begin{array}{c}(a_p);\ -a^2 x \\ (b_{p+1})\end{array}\right)$

$[\mathrm{Re}\,b_j > 0;\ d > -1/2;\ 0 < x < \pi^2/16].$

6.17.6. Series containing $_pF_{p+1}((a_p(k));\ (b_{p+1}(k));\ \varphi(k)z)$

Notation: $\Delta(z) = \left|\dfrac{z^{1/2}}{1+\sqrt{1-z}}\, e^{\sqrt{1-z}}\right|.$

1. $\displaystyle\sum_{k=1}^{\infty} \frac{k^{2k}}{(2k)!}\, z^k\, \frac{\prod (a_p)_k}{\prod (b_p)_k}\, {}_pF_{p+1}\!\left(\begin{array}{c}(a_p)+k;\ -k^2 z \\ (b_p)+k,\ 2k+1\end{array}\right)$

$\displaystyle = \frac{1}{2}\, {}_{p+1}F_p\!\left(\begin{array}{c}(a_p),\ 1;\ z \\ (b_p)\end{array}\right) - \frac{1}{2} \quad [\Delta(z) < 1].$

2. $\displaystyle\sum_{k=1}^{\infty} \frac{k^{2k-2}}{(2k)!} z^k \frac{\prod (a_p)_k}{\prod (b_p)_k} {}_pF_{p+1}\!\left(\begin{matrix}(a_p)+k;\ -k^2 z\\(b_p)+k,\ 2k+1\end{matrix}\right) = \frac{z}{2}\, \frac{\prod_{i=1}^{p} a_i}{\prod_{j=1}^{p} b_j}$ $\quad[\Delta(z)<1]$.

3. $\displaystyle\sum_{k=1}^{\infty} \frac{k^{2k-4}}{(2k)!} z^k \frac{\prod (a_p)_k}{\prod (b_p)_k} {}_pF_{p+1}\!\left(\begin{matrix}(a_p)+k;\ -k^2 z\\(b_p)+k,\ 2k+1\end{matrix}\right)$

$$= \frac{z}{2}\, \frac{\prod_{i=1}^{p} a_i}{\prod_{j=1}^{p} b_j} - \frac{z^2}{8}\, \frac{\prod_{i=1}^{p} a_i(a_i+1)}{\prod_{j=1}^{p} b_j(b_j+1)} \quad [\Delta(z)<1].$$

4. $\displaystyle\sum_{k=1}^{\infty} \frac{k^{2k-6}}{(2k)!} z^k \frac{\prod (a_p)_k}{\prod (b_p)_k} {}_{p+1}F_{p+2}\!\left(\begin{matrix}(a_p)+k;\ -k^2 z\\(b_p)+k,\ 2k+1\end{matrix}\right)$

$$= \frac{z}{2}\, \frac{\prod_{i=1}^{p} a_i}{\prod_{j=1}^{p} b_j} - \frac{5z^2}{32}\, \frac{\prod_{i=1}^{p} a_i(a_i+1)}{\prod_{j=1}^{p} b_j(b_j+1)} + \frac{z^3}{72}\, \frac{\prod_{i=1}^{p} a_i(a_i+1)(a_i+2)}{\prod_{j=1}^{p} b_j(b_j+1)(b_j+2)} \quad [\Delta(z)<1].$$

5. $\displaystyle\sum_{k=1}^{\infty} \frac{k^{2k}}{(2k)!(k^2-a^2)} z^k \frac{\prod (a_p)_k}{\prod (b_p)_k} {}_pF_{p+1}\!\left(\begin{matrix}(a_p)+k;\ -k^2 z\\(b_p)+k,\ 2k+1\end{matrix}\right)$

$$= \frac{1}{2a^2}\left[1 - {}_{p+1}F_{p+2}\!\left(\begin{matrix}(a_p),\ 1;\ -a^2 z\\(b_p),\ 1-a,\ 1+a\end{matrix}\right)\right] \quad [\Delta(z)<1].$$

6. $\displaystyle\sum_{k=0}^{\infty} \frac{(2k+1)^{2k-3}}{(2k+1)!} z^k \frac{\prod (a_p)_k}{\prod (b_p)_k} {}_pF_{p+1}\!\left(\begin{matrix}(a_p)+k;\ -(2k+1)^2 z\\(b_p)+k,\ 2k+2\end{matrix}\right)$

$$= 1 - \frac{4z}{9}\, \frac{\prod_{i=1}^{p} a_i}{\prod_{j=1}^{p} b_j} \quad [|\Delta(2z)|<1].$$

7. $\displaystyle\sum_{k=0}^{\infty} \frac{(2k+1)^{2k-5}}{(2k+1)!} z^k \frac{\prod (a_p)_k}{\prod (b_p)_k} {}_pF_{p+1}\!\left(\begin{matrix}(a_p)+k;\ -(2k+1)^2 z\\(b_p)+k,\ 2k+2\end{matrix}\right)$

$$= 1 - \frac{40z}{81}\, \frac{\prod_{i=1}^{p} a_i}{\prod_{j=1}^{p} b_j} + \frac{16z^2}{225}\, \frac{\prod_{i=1}^{p} a_i(a_i+1)}{\prod_{j=1}^{p} b_j(b_j+1)} \quad [|\Delta(2z)|<1].$$

6.17.7]   6.17. THE GENERALIZED HYPERGEOMETRIC FUNCTION $_pF_q((a_p);(b_q);z)$

8. $\displaystyle\sum_{k=0}^{\infty} \frac{(2k+1)^{2k+1}}{(2k+1)![(2k+1)^2+a^2]} z^k \frac{\prod(a_p)_k}{\prod(b_p)_k} {}_pF_{p+1}\left(\begin{array}{c}(a_p)+k;\ -(2k+1)^2z\\(b_p)+k,\ 2k+2\end{array}\right)$

$\displaystyle = \frac{1}{a^2+1} {}_{p+1}F_{p+2}\left(\begin{array}{c}(a_p),\ 1;\ a^2z\\(b_p),\ \frac{3-ia}{2},\ \frac{3+ia}{2}\end{array}\right)$ $[|\Delta(2z)|<1]$.

9. $\displaystyle\sum_{k=0}^{\infty} \frac{(k+1)^{k-1}}{k!} t^k {}_{p+1}F_q\left(\begin{array}{c}-k,\ (a_p)\\(b_q);\ \frac{z}{k+1}\end{array}\right)$

$\displaystyle = (tz)^{-1} \frac{\prod_{j=1}^{q}(b_j-1)}{\prod_{i=1}^{p}(a_i-1)} \left[{}_pF_q\left(\begin{array}{c}(a_p)-1;\ wz\\(b_q)-1\end{array}\right) - 1\right]$ $[t=-we^w;\ |we^{w+1}|<1]$.

6.17.7. Series containing $_pF_q((a_p(k));(b_q(k));\varphi(k)z)$ and special functions

1. $\displaystyle\sum_{k=0}^{\infty} \frac{(k+1)^{(k-\nu)/2-1}}{k!} \left(\frac{w}{2}\right)^k J_{k+\nu}(\sqrt{k+1}\,w) {}_{p+1}F_q\left(\begin{array}{c}-k,\ (a_p);\ \frac{z}{k+1}\\(b_q)\end{array}\right)$

$\displaystyle = \frac{\left(\frac{w}{2}\right)^{\nu-2} z^{-1}}{\Gamma(\nu)} \frac{\prod_{j=1}^{q}(b_j-1)}{\prod_{i=1}^{p}(a_i-1)} \left[1 - {}_pF_{q+1}\left(\begin{array}{c}(a_p)-1;\ -\frac{wz^2}{4}\\(b_q)-1,\ \nu\end{array}\right)\right]$.

2. $\displaystyle\sum_{k=0}^{\infty} \frac{(k+1)^{(k-\nu)/2-1}}{k!} \left(-\frac{w}{2}\right)^k I_{k+\nu}(\sqrt{k+1}\,w) {}_{p+1}F_q\left(\begin{array}{c}-k,\ (a_p);\ \frac{z}{k+1}\\(b_q)\end{array}\right)$

$\displaystyle = \frac{\left(\frac{w}{2}\right)^{\nu-2} z^{-1}}{\Gamma(\nu)} \frac{\prod_{j=1}^{q}(b_j-1)}{\prod_{i=1}^{p}(a_i-1)} \left[{}_pF_{q+1}\left(\begin{array}{c}(a_p)-1;\ \frac{wz^2}{4}\\(b_q)-1,\ \nu\end{array}\right) - 1\right]$.

3. $\displaystyle\sum_{k=0}^{\infty} \frac{(k+1)^{k-1}}{(2k)!} z^k H_{2k}\left(\frac{w}{\sqrt{k+1}}\right) \frac{\prod(a_p)_k}{\prod(b_q)_k} {}_pF_q\left(\begin{array}{c}(a_p)+k;\ (k+1)z\\(b_q)+k\end{array}\right)$

$\displaystyle = \frac{w^{-2}z^{-1}}{2} \frac{\prod_{j=1}^{q}(b_j-1)}{\prod_{i=1}^{p}(a_i-1)} \left[1 - {}_pF_{q+1}\left(\begin{array}{c}(a_p)-1;\ w^2z\\(b_q)-1,\ -\frac{1}{2}\end{array}\right)\right]$.

4. $\displaystyle\sum_{k=0}^{\infty} \frac{(k+1)^{k-1/2}}{(2k+1)!} z^k H_{2k+1}\left(\frac{w}{\sqrt{k+1}}\right) \frac{\prod(a_p)_k}{\prod(b_q)_k} {}_pF_q\left(\begin{array}{c}(a_p)+k;\ (k+1)z\\(b_q)+k\end{array}\right)$

$$= (wz)^{-1} \frac{\prod\limits_{j=1}^{q}(b_j-1)}{\prod\limits_{i=1}^{p}(a_i-1)} \left[{}_pF_{q+1}\left(\begin{array}{c}(a_p)-1;\ w^2z\\(b_q)-1,\ \frac{1}{2}\end{array}\right) - 1\right].$$

5. $\displaystyle\sum_{k=0}^{\infty} \frac{(k+1)^{k-1}}{(\lambda+1)_k} (-z)^k L_k^{\lambda-k}\left(\frac{w}{k+1}\right) \frac{\prod(a_p)_k}{\prod(b_q)_k} {}_pF_q\left(\begin{array}{c}(a_p)+k;\ (k+1)z\\(b_q)+k\end{array}\right)$

$$= \lambda(wz)^{-1} \frac{\prod\limits_{j=1}^{q}(b_j-1)}{\prod\limits_{i=1}^{p}(a_i-1)} \left[{}_pF_{q+1}\left(\begin{array}{c}(a_p)-1;\ wz\\(b_q)-1,\ \lambda\end{array}\right) - 1\right].$$

6. $\displaystyle\sum_{k=0}^{\infty} (k+1)^{-1} \frac{\left(\frac{1}{2}-\lambda\right)_k}{(1-2\lambda)_{2k}} \left(\frac{z}{2w}\right)^k$

$$\times \frac{\prod(a_p)_k}{\prod(b_q)_k} C_k^{\lambda-k}(1+(k+1)w)\, {}_pF_q\left(\begin{array}{c}(a_p)+k;\ (k+1)z\\(b_q)+k\end{array}\right)$$

$$= \frac{2\lambda w}{(2\lambda+1)z} \frac{\prod\limits_{i=1}^{q}(b_i-1)}{\prod\limits_{i=1}^{p}(a_i-1)} \left[1 - {}_{p+1}F_{q+1}\left(\begin{array}{c}(a_p)-1,\ -\lambda-\frac{1}{2};\ -2w^{-1}z\\(b_q)-1,\ -2\lambda\end{array}\right)\right]$$

$$[|ze^{z+1}|<1].$$

7. $\displaystyle\sum_{k=0}^{\infty} \frac{(k+1)^{k-1}}{(1-\lambda)_k} (-z)^k C_{2k}^{\lambda-k}\left(\frac{w}{\sqrt{k+1}}\right) \frac{\prod(a_p)_k}{\prod(b_q)_k} {}_pF_q\left(\begin{array}{c}(a_p)+k;\ (k+1)z\\(b_q)+k\end{array}\right)$

$$= \frac{w^{-2}z^{-1}}{2(\lambda-1)} \frac{\prod\limits_{j=1}^{q}(b_j-1)}{\prod\limits_{i=1}^{p}(a_i-1)} \left[1 - {}_{p+1}F_{q+1}\left(\begin{array}{c}(a_p)-1,\ \lambda-1;\ w^2z\\(b_q)-1,\ -\frac{1}{2}\end{array}\right)\right].$$

8. $\displaystyle\sum_{k=0}^{\infty} \frac{(k+1)^{k-1/2}}{(1-\lambda)_k} (-z)^k C_{2k+1}^{\lambda-k}\left(\frac{w}{\sqrt{k+1}}\right) \frac{\prod(a_p)_k}{\prod(b_q)_k} {}_pF_q\left(\begin{array}{c}(a_p)+k;\ (k+1)z\\(b_q)+k\end{array}\right)$

$$= (wz)^{-1} \frac{\prod\limits_{j=1}^{q}(b_j-1)}{\prod\limits_{i=1}^{p}(a_i-1)} \left[{}_{p+1}F_{q+1}\left(\begin{array}{c}(a_p)-1,\ \lambda;\ w^2z\\(b_q)-1,\ \frac{1}{2}\end{array}\right) - 1\right].$$

6.17. The Generalized Hypergeometric Function $_pF_q((a_p);(b_q);z)$

9. $\displaystyle\sum_{k=0}^{\infty} \frac{(k+1)^{k-1}}{(\sigma+1)_k} z^k P_k^{(\rho-k,\sigma)}\left(\frac{w}{k+1}-1\right) \frac{\prod(a_p)_k}{\prod(b_q)_k}\, _pF_q\!\left(\begin{array}{c}(a_p)+k;\ (k+1)z\\ (b_q)+k\end{array}\right)$

$\displaystyle = \frac{2\sigma(wz)^{-1}}{\rho+\sigma} \frac{\prod_{j=1}^{q}(b_j-1)}{\prod_{i=1}^{p}(a_i-1)} \left[_{p+1}F_{q+1}\!\left(\begin{array}{c}(a_p)-1,\ \rho+\sigma\\ (b_q)-1,\ \sigma;\ \frac{wz}{2}\end{array}\right)-1\right].$

10. $\displaystyle\sum_{k=0}^{\infty} \frac{(k+1)^{-1}}{(-\lambda-\sigma)_k} \left(\frac{2z}{w}\right)^k$

$\displaystyle \times P_k^{(\rho-k,\sigma-k)}((k+1)w-1) \frac{\prod(a_p)_k}{\prod(b_q)_k}\, _pF_q\!\left(\begin{array}{c}(a_p)+k;\ (k+1)z\\ (b_q)+k\end{array}\right)$

$\displaystyle = \frac{(\rho+\sigma+1)w}{2(\sigma+1)z} \frac{\prod_{j=1}^{q}(b_j-1)}{\prod_{i=1}^{p}(a_i-1)} \left[_{p+1}F_{q+1}\!\left(\begin{array}{c}(a_p)-1,\ -\sigma-1;\ 2w^{-1}z\\ (b_q)-1,\ -\rho-\sigma-1\end{array}\right)-1\right].$

11. $\displaystyle\sum_{k=0}^{\infty} \frac{(k+1)^{-1}(w-k-1)^k}{(\sigma+1)_k}(-z)^k$

$\displaystyle \times P_k^{(\rho-k,\sigma)}\left(\frac{w+k+1}{w-k-1}\right) \frac{\prod(a_p)_k}{\prod(b_q)_k}\, _pF_q\!\left(\begin{array}{c}(a_p)+k;\ (k+1)z\\ (b_q)+k\end{array}\right)$

$\displaystyle = \frac{\sigma(wz)^{-1}}{\rho+1} \frac{\prod_{j=1}^{q}(b_j-1)}{\prod_{i=1}^{p}(a_i-1)} \left[1-\,_{p+1}F_{q+1}\!\left(\begin{array}{c}(a_p)-1,\ -\rho-1;\ wz\\ (b_q)-1,\ \sigma\end{array}\right)\right].$

12. $\displaystyle\sum_{k=0}^{\infty} \frac{(k+1)^{-1}}{(-\rho-\sigma)_k} \left(\frac{2z}{w}\right)^k P_k^{(\rho-k,\sigma-k)}(1+(k+1)w)$

$\displaystyle \times \frac{\prod(a_p)_k}{\prod(b_q)_k}\, _pF_q\!\left(\begin{array}{c}(a_p)+k;\ (k+1)z\\ (b_q)+k\end{array}\right)$

$\displaystyle = \frac{\rho+\sigma+1}{\rho+1} \frac{w}{2z} \frac{\prod_{i=1}^{q}(b_i-1)}{\prod_{i=1}^{p}(a_i-1)} \left[1-\,_{p+1}F_{q+1}\!\left(\begin{array}{c}(a_p)-1,\ -\rho-1;\ -2w^{-1}z\\ (b_q)-1,\ -\rho-\sigma-1\end{array}\right)\right]$

$[|ze^{z+1}|<1].$

13. $\sum_{k=0}^{\infty} \frac{(k+1)^{k-1}}{(\rho+1)_k} (-z)^k \frac{\prod (a_p)_k}{\prod (b_q)_k}$

$\times P_k^{(\rho, \sigma-k)} \left(1 + \frac{w}{k+1}\right) {}_pF_q\left(\begin{array}{c}(a_p)+k;\ (k+1)z\\(b_q)+k\end{array}\right)$

$= \frac{2\rho}{(\rho+\sigma)wz} \frac{\prod\limits_{i=1}^{q}(b_i-1)}{\prod\limits_{i=1}^{p}(a_i-1)} \left[1 - {}_{p+1}F_{q+1}\left(\begin{array}{c}(a_p)-1,\ \rho+\sigma\\(b_q)-1,\ \rho;\ -\frac{wz}{2}\end{array}\right)\right]$ $\quad [|ze^{z+1}|<1].$

6.17.8. Series containing products of ${}_pF_q((a_p(k));\ (b_q(k));\ \varphi(k)z)$

1. $\sum_{k=0}^{\infty} \frac{(-z)^k}{k!} (k+1)^{k-1} \frac{\prod (c_r)_k}{\prod (d_s)_k}$

$\times {}_{p+1}F_q\left(\begin{array}{c}-k,\ (a_p)\\(b_q);\ \frac{w}{k+1}\end{array}\right) {}_rF_s\left(\begin{array}{c}(c_r)+k;\ (k+1)z\\(d_s)+k\end{array}\right)$

$= (wz)^{-1} \frac{\prod\limits_{i=1}^{q}(b_i-1) \prod\limits_{j=1}^{s}(d_j-1)}{\prod\limits_{i=1}^{p}(a_i-1) \prod\limits_{j=1}^{r}(d_j-1)} \left[{}_{p+r}F_{q+s}\left(\begin{array}{c}(a_p)-1,\ (c_r)-1;\ wz\\(b_q)-1,\ (d_s)-1\end{array}\right) - 1\right]$

$[|ze^{z+1}|<1].$

Chapter 7

The Connection Formulas

7.1. Elementary Functions

7.1.1. Trigonometric functions

1. $\sin nz = \sin z \sum_{k=0}^{[(n-1)/2]} \frac{(1-n)_{2k}}{k!(1-n)_k} 2^{n-2k-1} \cos^{n-2k-1} z.$

2. $\sin 2nz = n \cos z \sum_{k=0}^{n-1} \frac{(1-n)_k(1+n)_k}{(2k+1)!} 2^{2k+1} \sin^{2k+1} z.$

3. $\sin(2n+1)z = (2n+1) \sum_{k=0}^{n} \frac{(-n)_k(n+1)_k}{(2k+1)!} 2^{2k} \sin^{2k+1} z.$

4. $\cos nz = \frac{1}{2} \sum_{k=0}^{[n/2]} \frac{(-n)_{2k}}{k!(1-n)_k} 2^{n-2k} \cos^{n-2k} z.$

5. $\cos 2nz = \sum_{k=0}^{n} \frac{(-n)_k(n)_k}{(2k)!} 2^{2k} \sin^{2k} z.$

6. $\cos(2n+1)z = \cos z \sum_{k=0}^{n} \frac{(-n)_k(n+1)_k}{(2k)!} 2^{2k} \sin^{2k} z.$

7.2. Special Functions

7.2.1. The psi function $\psi(z)$

1. $\psi'\left(\frac{1}{4}\right) = \pi^2 + 8G.$

2. $\psi'\left(\frac{3}{4}\right) = \pi^2 - 8G.$

3. $\psi^{(n)}(z) = (-1)^{n+1} n! \, z^{-n-1} {}_{n+2}F_{n+1}\left(\begin{matrix}1, z, z, \ldots, z; 1 \\ z+1, z+1, \ldots, z+1\end{matrix}\right)$ $[n \geq 1].$

7.2.2. The incomplete gamma functions $\Gamma(\nu, z)$ and $\gamma(\nu, z)$

1. $\Gamma(\nu - n, z) = \dfrac{(-1)^n}{(1-\nu)_n} \left[\Gamma(\nu, z) - z^{\nu-1} e^{-z} \displaystyle\sum_{k=0}^{n-1} (1-\nu)_k (-z)^{-k} \right]$.

2. $\Gamma(\nu + n, z) = (\nu)_n \Gamma(\nu, z) + z^{\nu+n-1} e^{-z} \displaystyle\sum_{k=0}^{n-1} (1-\nu-n)_k (-z)^{-k}$.

3. $\gamma\left(n + \dfrac{1}{2}, z\right) = \sqrt{\pi} \left(\dfrac{1}{2}\right)_n \operatorname{erf}(\sqrt{z})$
$\qquad\qquad - \dfrac{2}{2n+1} z^{n+1/2} e^{-z} \displaystyle\sum_{k=1}^{n} \left(-n - \dfrac{1}{2}\right)_k (-z)^{-k}$.

4. $\gamma\left(\dfrac{1}{2} - n, z\right) = \dfrac{(-1)^n}{\left(\dfrac{1}{2}\right)_n} \sqrt{\pi} \operatorname{erf}(\sqrt{z}) + z^{-n-1/2} e^{-z} \displaystyle\sum_{k=1}^{n} \dfrac{z^k}{\left(\dfrac{1}{2} - n\right)_k}$.

7.2.3. The parabolic cylinder function $D_\nu(z)$

1. $D_{\nu+n}(z) = 2^{-n/2} \displaystyle\sum_{k=0}^{n} \binom{n}{k} 2^{k/2} (-\nu)_k H_{n-k}\left(\dfrac{z}{\sqrt{2}}\right) D_{\nu-k}(z)$.

2. $D_{\nu-n}(z) = \dfrac{1}{(-\nu)_n} \left(\dfrac{i}{\sqrt{2}}\right)^n \displaystyle\sum_{k=0}^{n} \binom{n}{k} (-\sqrt{2}i)^k H_{n-k}\left(\dfrac{iz}{\sqrt{2}}\right) D_{\nu+k}(z)$.

3. $D_n(z) = 2^{-n/2} e^{-z^2/4} H_n\left(\dfrac{z}{\sqrt{2}}\right)$.

4. $D_{-n-1}(z) = \dfrac{2^{(1-n)/2} i^n}{n!} e^{-z^2/4} \displaystyle\sum_{k=1}^{n} \binom{n}{k} (-i)^k H_{k-1}\left(\dfrac{z}{\sqrt{2}}\right) H_{n-k}\left(\dfrac{iz}{\sqrt{2}}\right)$
$\qquad\qquad + \dfrac{2^{-(n+1)/2} \sqrt{\pi}}{n!} i^n e^{z^2/4} \operatorname{erfc}\left(\dfrac{z}{\sqrt{2}}\right) H_n\left(\dfrac{iz}{\sqrt{2}}\right)$.

5. $D_{2n-1/2}(z)$
$\qquad = (-2)^n n! \sqrt{\dfrac{z}{2\pi}} \displaystyle\sum_{k=0}^{n} \dfrac{\left(-\dfrac{z^2}{8}\right)^k}{k!} L_{n-k}^{k-1/2}\left(\dfrac{z^2}{4}\right) \displaystyle\sum_{m=0}^{k} \binom{k}{m} K_{2m-k+1/4}\left(\dfrac{z^2}{4}\right)$.

6. $D_{2n-3/2}(z) = (-2)^n n! \sqrt{\dfrac{z^3}{2\pi}} \displaystyle\sum_{k=0}^{n} \dfrac{\left(-\dfrac{z^2}{8}\right)^k}{k!} L_{n-k}^{k-1/2}\left(\dfrac{z^2}{4}\right)$
$\qquad\qquad \times \displaystyle\sum_{m=0}^{k} \binom{k}{m} \left[K_{2m-k+3/4}\left(\dfrac{z^2}{4}\right) - K_{2m-k+1/4}\left(\dfrac{z^2}{4}\right) \right]$.

7. $D_{-2n-1/2}(z)$
$$= \frac{2^{n-1/2}n!}{\left(\frac{1}{2}\right)_{2n}}\sqrt{\frac{z}{\pi}}\sum_{k=0}^{n}\frac{\left(-\frac{z^2}{8}\right)^k}{k!}L_{n-k}^{k-1/2}\left(-\frac{z^2}{4}\right)\sum_{m=0}^{k}\binom{k}{m}K_{2m-k+1/4}\left(\frac{z^2}{4}\right).$$

8. $D_{-2n-3/2}(z) = \dfrac{2^{n-1/2}n!}{\left(\frac{3}{2}\right)_{2n}}\sqrt{\dfrac{z^3}{\pi}}\sum_{k=0}^{n}\dfrac{\left(-\frac{z^2}{8}\right)^k}{k!}L_{n-k}^{k+1/2}\left(-\dfrac{z^2}{4}\right)$
$$\times \sum_{m=0}^{k}\binom{k}{m}\left[K_{2m-k+3/4}\left(\frac{z^2}{4}\right) - K_{2m-k+1/4}\left(\frac{z^2}{4}\right)\right].$$

7.2.4. The Bessel functions $J_\nu(z)$, $H_\nu^{(1)}(z)$, $H_\nu^{(2)}(z)$, $I_\nu(z)$ and $K_\nu(z)$

1. $J_{\nu+n}(z) = (\nu)_n\left(\dfrac{2}{z}\right)^n {}_2F_3\left(\begin{matrix}-\frac{n}{2}, \frac{1-n}{2}; -z^2 \\ -n, \nu, 1-\nu-n\end{matrix}\right)J_\nu(z)$

$\qquad - (\nu+1)_{n-1}\left(\dfrac{2}{z}\right)^{n-1} {}_2F_3\left(\begin{matrix}\frac{1-n}{2}, 1-\frac{n}{2}; -z^2 \\ 1-n, \nu+1, 1-\nu-n\end{matrix}\right)J_{\nu-1}(z)$
$$[n \geq 1;\ [7],\ 2.7.5.2.(23)].$$

2. $J_{\nu-n}(z) = (-\nu)_n\left(-\dfrac{2}{z}\right)^n {}_2F_3\left(\begin{matrix}-\frac{n}{2}, \frac{1-n}{2}; -z^2 \\ -n, -\nu, \nu-n+1\end{matrix}\right)J_\nu(z)$

$\qquad - (1-\nu)_{n-1}\left(-\dfrac{2}{z}\right)^{n-1} {}_2F_3\left(\begin{matrix}\frac{1-n}{2}, 1-\frac{n}{2}; -z^2 \\ 1-n, 1-\nu, \nu-n+1\end{matrix}\right)J_{\nu+1}(z) \quad [n \geq 1].$

3. $J_{n+1/2}(z) = (-1)^n\sqrt{\dfrac{2}{\pi z}}\sum_{k=0}^{n}(-1)^k\dfrac{(n+k)!}{k!(n-k)!}(2z)^{-k}\sin\left(z + \dfrac{n-k}{2}\pi\right).$

4. $J_{-n-1/2}(z) = \sqrt{\dfrac{2}{\pi z}}\sum_{k=0}^{n}(-1)^k\dfrac{(n+k)!}{k!(n-k)!}(2z)^{-k}\cos\left(z - \dfrac{n+k}{2}\pi\right).$

5. $J_{n-1/2}(mz) = m^{-n-1/2}\left(\dfrac{\pi z}{2}\right)^{(m-1)/2}\sum_{k=0}^{[(m-1)/2]}(-1)^{k+n}\binom{m}{2k}$

$\qquad \times \displaystyle\sum_{p_1+\ldots+p_m=n}(-1)^{p_{2k+1}+\ldots+p_m}\dfrac{n!}{p_1!\ldots p_m!}\prod_{i=1}^{2k}J_{1/2-p_i}(z)\prod_{j=2k+1}^{m}J_{p_j-1/2}(z).$

6. $J_{n-1/2}(2z) = 2^{-n-1}(\pi z)^{1/2}\sum_{k=0}^{n}\binom{n}{k}[J_{k-1/2}(z)J_{n-k-1/2}(z)$
$$- (-1)^n J_{1/2-k}(z)J_{k-n+1/2}(z)].$$

7. $J_{1/2-n}(mz) = m^{-n-1/2}\left(\dfrac{\pi z}{2}\right)^{(m-1)/2} \displaystyle\sum_{k=0}^{[(m-1)/2]} (-1)^k \binom{m}{2k+1}$

$\times \displaystyle\sum_{p_1+\ldots+p_m=n} (-1)^{p_{2k+2}+\ldots+p_m} \dfrac{n!}{p_1!\ldots p_m!} \prod_{i=1}^{2k+1} J_{1/2-p_i}(z) \prod_{j=2k+2}^{m} J_{p_j-1/2}(z).$

8. $J_{n+1/2}^2(z) + J_{-n-1/2}^2(z) = \dfrac{4}{\pi^2} K_{n+1/2}(iz) K_{n+1/2}(-iz) \qquad [|\arg z| < \pi/2].$

9. $\qquad = \dfrac{(n!)^2}{2^{2n-1}\pi} z^{-2n-1} L_n^{-2n-1}(2iz) L_n^{-2n-1}(-2iz).$

10. $J_{n-1/2}(x+iy) = \sqrt{\dfrac{\pi}{2}}\, \dfrac{x^{n+1/2} y^{1/2}}{(x^2+y^2)^{(2n+1)/4}}$

$\times \Bigg\{ \displaystyle\sum_{k=0}^{n} \binom{n}{k}\left(\dfrac{y}{x}\right)^k \bigg[(-1)^k \cos\left(\dfrac{2n+1}{2} \arctan \dfrac{y}{x}\right) J_{n-k-1/2}(x) I_{k-1/2}(y)$

$\qquad - (-1)^n \sin\left(\dfrac{2n+1}{2} \arctan \dfrac{y}{x}\right) J_{k-n+1/2}(x) I_{1/2-k}(y) \bigg]$

$\qquad - i\bigg[(-1)^k \sin\left(\dfrac{2n+1}{2} \arctan \dfrac{y}{x}\right) J_{n-k-1/2}(x) I_{k-1/2}(y)$

$\qquad + (-1)^n \cos\left(\dfrac{2n+1}{2} \arctan \dfrac{y}{x}\right) J_{k-n-1/2}(x) I_{1/2-k}(y)\bigg]\Bigg\}.$

11. $J_{1/2-n}(x+iy) = \sqrt{\dfrac{\pi}{2}}\, \dfrac{x^{n+1/2} y^{1/2}}{(x^2+y^2)^{(2n+1)/4}}$

$\times \Bigg\{ \displaystyle\sum_{k=0}^{n} \binom{n}{k}\left(\dfrac{y}{x}\right)^k \bigg[\cos\left(\dfrac{2n+1}{2} \arctan \dfrac{y}{x}\right) J_{k-n+1/2}(x) I_{k-1/2}(y) +$

$\qquad + (-1)^{n-k} \sin\left(\dfrac{2n+1}{2} \arctan \dfrac{y}{x}\right) J_{n-k-1/2}(x) I_{1/2-k}(y)\bigg]$

$\qquad + i\bigg[-\sin\left(\dfrac{2n+1}{2} \arctan \dfrac{y}{x}\right) J_{k-n+1/2}(x) I_{k-1/2}(y)$

$\qquad + (-1)^{n-k} \cos\left(\dfrac{2n+1}{2} \arctan \dfrac{y}{x}\right) J_{n-k-1/2}(x) I_{1/2-k}(y)\bigg]\Bigg\}.$

12. $H^{(1)}_{n-1/2}(\sqrt{z})$

$= \sqrt{\dfrac{2}{\pi}}\, z^{-1/4} e^{i(\sqrt{z}-n\pi/2)} \displaystyle\sum_{k=0}^{n-1} (-1)^k \dfrac{(n+k-1)!}{k!(n-k-1)!}\, \dfrac{1}{(2i\sqrt{z})^k} \qquad [n \geq 1].$

13. $H^{(2)}_{n-1/2}(\sqrt{z}) = \sqrt{\dfrac{2}{\pi}}\, z^{-1/4} e^{-i(\sqrt{z}-n\pi/2)} \displaystyle\sum_{k=0}^{n-1} \dfrac{(n+k-1)!}{k!(n-k-1)!}\, \dfrac{1}{(2i\sqrt{z})^k}$

$\qquad\qquad\qquad [n \geq 1].$

14. $I_{\nu+n}(z) = (\nu)_n \left(-\dfrac{2}{z}\right)^n {}_2F_3\left(\begin{matrix} -\dfrac{n}{2}, \dfrac{1-n}{2}; z^2 \\ -n, \nu, 1-\nu-n \end{matrix}\right) I_\nu(z)$

$+ (\nu+1)_{n-1} \left(-\dfrac{2}{z}\right)^{n-1} {}_2F_3\left(\begin{matrix} \dfrac{1-n}{2}, 1-\dfrac{n}{2}; z^2 \\ 1-n, \nu+1, 1-\nu-n \end{matrix}\right) I_{\nu-1}(z) \quad [n \geq 1].$

15. $I_{\nu-n}(z) = (-\nu)_n \left(-\dfrac{2}{z}\right)^n {}_2F_3\left(\begin{matrix} -\dfrac{n}{2}, \dfrac{1-n}{2}; z^2 \\ -n, -\nu, \nu-n+1 \end{matrix}\right) I_\nu(z)$

$+ (1-\nu)_{n-1} \left(-\dfrac{2}{z}\right)^{n-1} {}_2F_3\left(\begin{matrix} \dfrac{1-n}{2}, 1-\dfrac{n}{2}; z^2 \\ 1-n, 1-\nu, \nu-n+1 \end{matrix}\right) I_{\nu+1}(z) \quad [n \geq 1].$

16. $I^2_{n+1/2}(z) - I^2_{-n-1/2}(z) = (-1)^n \dfrac{4i}{\pi^2} K_{n+1/2}(z) K_{n+1/2}(-z)$

$\qquad\qquad [0 < \arg z < \pi].$

17. $\quad = -\dfrac{(n!)^2}{2^{2n-1}\pi} z^{-2n-1} L_n^{-2n-1}(2z) L_n^{-2n-1}(-2z).$

18. $K_{\nu+n}(z) = (\nu)_n \left(\dfrac{2}{z}\right)^n {}_2F_3\left(\begin{matrix} -\dfrac{n}{2}, \dfrac{1-n}{2}; z^2 \\ -n, \nu, 1-\nu-n \end{matrix}\right) K_\nu(z)$

$+ (\nu+1)_{n-1} \left(\dfrac{2}{z}\right)^{n-1} {}_2F_3\left(\begin{matrix} \dfrac{1-n}{2}, 1-\dfrac{n}{2}; z^2 \\ 1-n, \nu+1, 1-\nu-n \end{matrix}\right) K_{\nu-1}(z) \quad [n \geq 1].$

19. $K_{\nu-n}(z) = (-\nu)_n \left(\dfrac{2}{z}\right)^n {}_2F_3\left(\begin{matrix} -\dfrac{n}{2}, \dfrac{1-n}{2}; z^2 \\ -n, -\nu, \nu-n+1 \end{matrix}\right) K_\nu(z)$

$+ (1-\nu)_{n-1} \left(\dfrac{2}{z}\right)^{n-1} {}_2F_3\left(\begin{matrix} \dfrac{1-n}{2}, 1-\dfrac{n}{2}; z^2 \\ 1-n, 1-\nu, \nu-n+1 \end{matrix}\right) K_{\nu+1}(z) \quad [n \geq 1].$

20. $K_{n+1/2}(z) = n! \sqrt{\dfrac{\pi}{2z}} (-2z)^{-n} e^{-z} L_n^{-2n-1}(2z).$

21. $K_{n-1/2}(mz)$

$\quad = m^{-n-1/2} \left(\dfrac{2\pi}{z}\right)^{(m-1)/2} \displaystyle\sum_{p_1+\ldots+p_m=n} \dfrac{n!}{p_1!\ldots p_m!} \prod_{i=1}^m K_{p_i-1/2}(z).$

22. $K_{n-1/2}(2z) = \pm(-2)^{-n} (\pi z)^{1/2} \displaystyle\sum_{k=0}^n (-1)^k \binom{n}{k} I_{\pm n \mp k \mp 1/2}(z) K_{k-1/2}(z).$

23. $\quad = 2^{-n} \left(\dfrac{z}{\pi}\right)^{1/2} \displaystyle\sum_{k=0}^n \binom{n}{k} K_{k-1/2}(z) K_{n-k-1/2}(z).$

24. $J_{\nu+n}(\sqrt[4]{z}) K_\nu(\sqrt[4]{z}) = \dfrac{1}{4\sqrt{\pi}}(\nu)_n \Bigg\{ 2^{-n/2} {}_2F_3\left(\begin{matrix}-\frac{n}{2}, \frac{1-n}{2}; -\sqrt{z}\\ -n, \nu, -\nu-n+1\end{matrix}\right)$

$\times G_{04}^{30}\left(\dfrac{z}{64}\bigg| -\dfrac{n}{4}, \dfrac{2-n}{4}, \dfrac{2\nu-n}{4}, -\dfrac{2\nu+n}{4}\right)$

$- \dfrac{2^n z^{-n/4}}{\nu}(1-\delta_{n,0}) {}_2F_3\left(\begin{matrix}\frac{1-n}{2}, 1-\frac{n}{2}; -\sqrt{z}\\ 1-n, \nu+1, -\nu-n+1\end{matrix}\right)$

$\times \left[G_{04}^{30}\left(\dfrac{z}{64}\bigg| 0, \dfrac{1}{2}, \dfrac{\nu}{2}, 1-\dfrac{\nu}{2}\right) + G_{04}^{30}\left(\dfrac{z}{64}\bigg| 0, \dfrac{1}{2}, \dfrac{1+\nu}{2}, \dfrac{1-\nu}{2}\right)\right]\Bigg\}.$

25. $J_{\nu-n}(\sqrt[4]{z}) K_\nu(\sqrt[4]{z})$

$= (-1)^n \dfrac{2^{n-2}}{\sqrt{\pi}}(-\nu)_n z^{-n/4} \Bigg\{ \left[{}_2F_3\left(\begin{matrix}-\frac{n}{2}, \frac{1-n}{2}; -\sqrt{z}\\ -n, -\nu, \nu-n+1\end{matrix}\right)\right.$

$\left. - (1-\delta_{n,0}) {}_2F_3\left(\begin{matrix}\frac{1-n}{2}, 1-\frac{n}{2}; -\sqrt{z}\\ 1-n, 1-\nu, \nu-n+1\end{matrix}\right)\right] G_{04}^{30}\left(\dfrac{z}{64}\bigg| 0, \dfrac{1}{2}, \dfrac{\nu}{2}, -\dfrac{\nu}{2}\right)$

$+ \dfrac{1-\delta_{n,0}}{\nu} {}_2F_3\left(\begin{matrix}\frac{1-n}{2}, 1-\frac{n}{2}; -\sqrt{z}\\ 1-n, 1-\nu, \nu-n+1\end{matrix}\right)$

$\times \left[G_{04}^{30}\left(\dfrac{z}{64}\bigg| 0, \dfrac{1}{2}, \dfrac{\nu}{2}, 1-\dfrac{\nu}{2}\right) + G_{04}^{30}\left(\dfrac{z}{64}\bigg| 0, \dfrac{1}{2}, \dfrac{1+\nu}{2}, \dfrac{1-\nu}{2}\right)\right]\Bigg\}.$

7.2.5. The Struve functions $\mathbf{H}_\nu(z)$ and $\mathbf{L}_\nu(z)$

1. $\mathbf{H}_{-n}(z) = (-1)^n \mathbf{H}_n(z) - \dfrac{(-1)^n}{\pi} \sum_{k=0}^{n-1} \dfrac{\left(\frac{1}{2}\right)_k}{\left(\frac{1}{2}\right)_{n-k}} \left(\dfrac{z}{2}\right)^{n-2k-1}.$

2. $\mathbf{H}_{n+1/2}(z) = Y_{n+1/2}(z) + \dfrac{\left(\frac{z}{2}\right)^{n-1/2}}{n!\sqrt{\pi}} \sum_{k=0}^{n} (-n)_k \left(\dfrac{1}{2}\right)_k \left(-\dfrac{4}{z^2}\right)^k.$

3. $\mathbf{H}_{-n-1/2}(z) = (-1)^n J_{n+1/2}(z).$

4. $\mathbf{H}_{\nu+n}(z) = (\nu)_n \left(\dfrac{2}{z}\right)^n {}_2F_3\left(\begin{matrix}-\frac{n}{2}, \frac{1-n}{2}; -z^2\\ -n, \nu, 1-\nu-n\end{matrix}\right) \mathbf{H}_\nu(z)$

$- (\nu+1)_{n-1}\left(\dfrac{2}{z}\right)^{n-1} {}_2F_3\left(\begin{matrix}\frac{1-n}{2}, 1-\frac{n}{2}; -z^2\\ 1-n, \nu+1, 1-\nu-n\end{matrix}\right) \mathbf{H}_{\nu-1}(z)$

$+ \dfrac{\left(\frac{z}{2}\right)^{\nu+n-1}}{\sqrt{\pi}} \sum_{k=0}^{n-1} \dfrac{(1-\nu-n)_k}{\Gamma\left(\nu-k+n+\frac{1}{2}\right)} \left(-\dfrac{4}{z^2}\right)^k {}_2F_3\left(\begin{matrix}-\frac{k}{2}, \frac{1-k}{2}; -z^2\\ -k, \nu-k+n, 1-\nu-n\end{matrix}\right)$

$[n \geq 1].$

5. $\mathbf{H}_{\nu-n}(z) = (-\nu)_n \left(-\dfrac{2}{z}\right)^n {}_2F_3\left(\begin{array}{c}-\dfrac{n}{2}, \dfrac{1-n}{2}; -z^2 \\ -n, -\nu, \nu-n+1\end{array}\right) \mathbf{H}_\nu(z)$

$\quad - (1-\nu)_{n-1} \left(-\dfrac{2}{z}\right)^{n-1} {}_2F_3\left(\begin{array}{c}\dfrac{1-n}{2}, 1-\dfrac{n}{2}; -z^2 \\ 1-n, 1-\nu, \nu-n+1\end{array}\right) \mathbf{H}_{\nu+1}(z)$

$\quad + \dfrac{\left(\dfrac{z}{2}\right)^{\nu-n+1}}{\sqrt{\pi}} \sum_{k=0}^{n-1} \dfrac{(\nu-n+1)_k}{\Gamma\left(\nu+k-n+\dfrac{5}{2}\right)} {}_2F_3\left(\begin{array}{c}-\dfrac{k}{2}, \dfrac{1-k}{2}; -z^2 \\ -k, n-k-\nu, \nu-n+1\end{array}\right) \quad [n \geq 1].$

6. $\mathbf{L}_{-n}(z) = \mathbf{L}_n(z) + \dfrac{1}{\pi} \sum_{k=0}^{n-1} (-1)^k \dfrac{\left(\dfrac{1}{2}\right)_k}{\left(\dfrac{1}{2}\right)_{n-k}} \left(\dfrac{z}{2}\right)^{n-2k-1}.$

7. $\mathbf{L}_{n+1/2}(z) = I_{-n-1/2}(z) - \dfrac{\left(\dfrac{z}{2}\right)^{n-1/2}}{n!\sqrt{\pi}} \sum_{k=0}^{n} (-n)_k \left(\dfrac{1}{2}\right)_k \left(\dfrac{4}{z^2}\right)^k.$

8. $\mathbf{L}_{-n-1/2}(z) = I_{n+1/2}(z).$

9. $\mathbf{L}_{\nu+n}(z) = (\nu)_n \left(-\dfrac{2}{z}\right)^n {}_2F_3\left(\begin{array}{c}-\dfrac{n}{2}, \dfrac{1-n}{2}; z^2 \\ -n, \nu, 1-\nu-n\end{array}\right) \mathbf{L}_\nu(z)$

$\quad + (\nu+1)_{n-1} \left(-\dfrac{2}{z}\right)^{n-1} {}_2F_3\left(\begin{array}{c}\dfrac{1-n}{2}, 1-\dfrac{n}{2}; z^2 \\ 1-n, \nu+1, 1-\nu-n\end{array}\right) \mathbf{L}_{\nu-1}(z)$

$\quad - \dfrac{\left(\dfrac{z}{2}\right)^{\nu+n-1}}{\sqrt{\pi}} \sum_{k=0}^{n-1} \dfrac{(1-\nu-n)_k}{\Gamma\left(\nu-k+n+\dfrac{1}{2}\right)} \left(\dfrac{4}{z^2}\right)^k {}_2F_3\left(\begin{array}{c}-\dfrac{k}{2}, \dfrac{1-k}{2}; z^2 \\ -k, \nu-k+n, 1-\nu-n\end{array}\right)$

$\quad [n \geq 1].$

10. $\mathbf{L}_{\nu-n}(z) = (-\nu)_n \left(-\dfrac{2}{z}\right)^n {}_2F_3\left(\begin{array}{c}-\dfrac{n}{2}, \dfrac{1-n}{2}; z^2 \\ -n, -\nu, \nu-n+1\end{array}\right) \mathbf{L}_\nu(z)$

$\quad + (1-\nu)_{n-1} \left(-\dfrac{2}{z}\right)^{n-1} {}_2F_3\left(\begin{array}{c}\dfrac{1-n}{2}, 1-\dfrac{n}{2}; z^2 \\ 1-n, 1-\nu, \nu-n+1\end{array}\right) \mathbf{L}_{\nu+1}(z)$

$\quad + \dfrac{\left(\dfrac{z}{2}\right)^{\nu-n+1}}{\sqrt{\pi}} \sum_{k=0}^{n-1} \dfrac{(\nu-n+1)_k}{\Gamma\left(\nu+k-n+\dfrac{5}{2}\right)} {}_2F_3\left(\begin{array}{c}-\dfrac{k}{2}, \dfrac{1-k}{2}; z^2 \\ -k, n-k-\nu, \nu-n+1\end{array}\right) \quad [n \geq 1].$

7.2.6. The Anger $\mathbf{J}_\nu(z)$ and Weber $\mathbf{E}_\nu(z)$ functions

1. $\mathbf{J}_{-\nu}(z) = \mathbf{J}_\nu(-z).$

2. $\mathbf{J}_n(z) = J_n(z).$

3. $\mathbf{J}_{\nu+n}(z) = (\nu+1)_{n-1}\left(\dfrac{2}{z}\right)^{n-1} {}_2F_3\left(\begin{array}{c}\frac{1-n}{2}, 1-\frac{n}{2}; -z^2\\ 1-n, \nu+1, 1-n-\nu\end{array}\right)\mathbf{J}_\nu(z)$

$-(\nu+2)_{n-2}\left(\dfrac{2}{z}\right)^{n-2} {}_2F_3\left(\begin{array}{c}1-\frac{n}{2}, \frac{3-n}{2}; -z^2\\ 2-n, \nu+2, 1-n-\nu\end{array}\right)\mathbf{J}_{\nu+1}(z)$

$+(-1)^n\dfrac{\sin(\nu\pi)}{\pi}\sum_{k=0}^{n-2}(1-\nu-n)_k\left(\dfrac{2}{z}\right)^{k+1} {}_2F_3\left(\begin{array}{c}-\frac{k}{2}, \frac{1-k}{2}; -z^2\\ -k, \nu-k+n, 1-n-\nu\end{array}\right)$

$[n \geq 2]$.

4. $\mathbf{J}_{\nu-n}(z) = (-\nu)_n\left(-\dfrac{2}{z}\right)^{n} {}_2F_3\left(\begin{array}{c}-\frac{n}{2}, \frac{1-n}{2}; -z^2\\ -n, -\nu, \nu-n+1\end{array}\right)\mathbf{J}_\nu(z)$

$-(1-\nu)_{n-1}\left(-\dfrac{2}{z}\right)^{n-1} {}_2F_3\left(\begin{array}{c}\frac{1-n}{2}, 1-\frac{n}{2}; -z^2\\ 1-n, 1-\nu, \nu-n+1\end{array}\right)\mathbf{J}_{\nu+1}(z)$

$-(-1)^n\dfrac{\sin(\nu\pi)}{\pi}\sum_{k=0}^{n-1}(\nu-n+1)_k\left(-\dfrac{2}{z}\right)^{k+1} {}_2F_3\left(\begin{array}{c}-\frac{k}{2}, \frac{1-k}{2}; -z^2\\ k, n-k-\nu, \nu-n+1\end{array}\right)$

$[n \geq 1]$.

5. $\mathbf{E}_{-\nu}(z) = -\mathbf{E}_\nu(-z)$.

6. $\mathbf{E}_n(z) = \dfrac{2z^{n-1}}{(2n-1)!!\,\pi}\sum_{k=0}^{[(n-1)/2]}\left(\dfrac{1}{2}\right)_k\left(\dfrac{1}{2}-n\right)_k\left(-\dfrac{4}{z^2}\right)^k - \mathbf{H}_n(z)$.

7. $\mathbf{E}_{-n}(z) = (-1)^n\dfrac{2z^{n-1}}{(2n-1)!!\,\pi}\sum_{k=0}^{[(n-1)/2]}\left(\dfrac{1}{2}\right)_k\left(\dfrac{1}{2}-n\right)_k\left(-\dfrac{4}{z^2}\right)^k$

$-(-1)^n\mathbf{H}_{-n}(z)$.

8. $\mathbf{E}_{n+1/2}(z) = (-1)^n\mathbf{J}_{-n-1/2}(z) = (-1)^n\mathbf{J}_{n+1/2}(-z)$.

9. $\mathbf{E}_{-n+1/2}(z) = (-1)^n\mathbf{J}_{n-1/2}(z) = (-1)^n\mathbf{J}_{-n+1/2}(-z)$.

10. $\mathbf{E}_{\nu+n}(z) = (\nu+1)_{n-1}\left(\dfrac{2}{z}\right)^{n-1} {}_2F_3\left(\begin{array}{c}\frac{1-n}{2}, 1-\frac{n}{2}; -z^2\\ 1-n, \nu+1, 1-n-\nu\end{array}\right)\mathbf{E}_\nu(z)$

$-(\nu+2)_{n-2}\left(\dfrac{2}{z}\right)^{n-2} {}_2F_3\left(\begin{array}{c}1-\frac{n}{2}, \frac{3-n}{2}; -z^2\\ 2-n, \nu+2, 1-n-\nu\end{array}\right)\mathbf{E}_{\nu+1}(z)$

$+\dfrac{1}{\pi}\sum_{k=0}^{n-2}\left[1+(-1)^{k+n}\cos(\nu\pi)\right](1-\nu-n)_k$

$\times\left(-\dfrac{2}{z}\right)^{k+1} {}_2F_3\left(\begin{array}{c}-\frac{k}{2}, \frac{1-k}{2}; -z^2\\ -k, \nu-k+n, 1-n-\nu\end{array}\right)$ $[n \geq 2]$.

11. $E_{\nu-n}(z) = (-\nu)_n \left(-\dfrac{2}{z}\right)^n {}_2F_3\left(\begin{array}{c}-\dfrac{n}{2}, \dfrac{1-n}{2}; -z^2 \\ -n, -\nu, \nu-n+1\end{array}\right) E_\nu(z)$

$- (1-\nu)_{n-1}\left(-\dfrac{2}{z}\right)^{n-1} {}_2F_3\left(\begin{array}{c}\dfrac{1-n}{2}, 1-\dfrac{n}{2}; -z^2 \\ 1-n, 1-\nu, \nu-n+1\end{array}\right) E_{\nu+1}(z)$

$- \dfrac{1}{\pi} \sum_{k=0}^{n-1} \left[1 + (-1)^{k+n}\cos(\nu\pi)\right] (\nu-n+1)_k \left(\dfrac{2}{z}\right)^{k+1}$

$\times {}_2F_3\left(\begin{array}{c}-\dfrac{k}{2}, \dfrac{1-k}{2}; -z^2 \\ -k, n-k-\nu, \nu-n+1\end{array}\right) \quad [n \geq 1].$

12. $J_{n+1/2}(z) = \dfrac{(-1)^n}{2\pi} \sum_{k=0}^{n-1} (-1)^k \left(\dfrac{2n-2k+3}{4}\right)_k \left(\dfrac{2}{z}\right)^{k+1}$

$- \dfrac{1}{2} \sum_{k=1}^{n} \binom{n}{k} \left(\dfrac{2}{z}\right)^{k-1} \sum_{m=0}^{k-1} \binom{k-1}{m} \left(\dfrac{3}{4}\right)_{k-m-1} \left(\dfrac{z}{2}\right)^m$

$\times \{J_{m-1/2}(z) \left[(-1)^{k+n} J_{k-n-1/2}(z) + J_{n-k+1/2}(z)\right]$
$- (-1)^m J_{1/2-m}(z) \left[(-1)^{k+n} J_{k-n-1/2}(z) - J_{n-k+1/2}(z)\right]\}$
$+ [J_{n+1/2}(z) - (-1)^n J_{-n-1/2}(z)] S(z) + [J_{n+1/2}(z)$
$\qquad\qquad + (-1)^n J_{-n-1/2}(z)] C(z).$

13. $J_{1/2-n}(z) = \dfrac{(-1)^n}{2\pi} \sum_{k=0}^{n-1} \left(\dfrac{2n-2k+1}{4}\right)_k \left(\dfrac{2}{z}\right)^{k+1}$

$- \dfrac{1}{2} \sum_{k=1}^{n} \binom{n}{k} \left(\dfrac{2}{z}\right)^{k-1} \sum_{m=0}^{k-1} \binom{k-1}{m} \left(\dfrac{3}{4}\right)_{k-m-1} \left(\dfrac{z}{2}\right)^m$

$\times \{J_{m-1/2}(z) \left[(-1)^k J_{1/2-n+k}(z) + (-1)^n J_{n-k-1/2}(z)\right]$
$+ (-1)^m J_{1/2-m}(z) \left[(-1)^k J_{1/2-n+k}(z) - (-1)^n J_{n-k-1/2}(z)\right]\}$
$+ [J_{1/2-n}(z) - (-1)^n J_{n-1/2}(z)] S(z) + [J_{1/2-n}(z) + (-1)^n J_{n-1/2}(z)] C(z).$

7.2.7. The Airy functions $\text{Ai}(z)$ and $\text{Bi}(z)$

1. $\text{Ai}\left(e^{\pi i/6} z\right) = \dfrac{1}{\pi}\sqrt{\dfrac{z}{6}} \left[\ker_{1/3}\left(\dfrac{2}{3} z^{3/2}\right) - \ker_{1/3}\left(\dfrac{2}{3} z^{3/2}\right)\right]$

$+ \dfrac{i}{\pi}\sqrt{\dfrac{z}{6}} \left[\ker_{1/3}\left(\dfrac{2}{3} z^{3/2}\right) + \ker_{1/3}\left(\dfrac{2}{3} z^{3/2}\right)\right] \quad [\text{Re } z > 0].$

2. $\text{Bi}\left(e^{\pi i/6} z\right) = \dfrac{1}{2}\sqrt{\dfrac{z}{6}} \left[2\,\text{ber}_{-1/3}\left(\dfrac{2}{3} z^{3/2}\right) - 2\,\text{bei}_{-1/3}\left(\dfrac{2}{3} z^{3/2}\right)\right.$
$\left. + (1+\sqrt{3})\,\text{ber}_{1/3}\left(\dfrac{2}{3} z^{3/2}\right) - (1-\sqrt{3})\,\text{bei}_{1/3}\left(\dfrac{2}{3} z^{3/2}\right)\right]$
$+ \dfrac{i}{2}\sqrt{\dfrac{z}{6}} \left[2\,\text{ber}_{-1/3}\left(\dfrac{2}{3} z^{3/2}\right) + 2\,\text{bei}_{-1/3}\left(\dfrac{2}{3} z^{3/2}\right)\right.$
$\left. + (1-\sqrt{3})\,\text{ber}_{1/3}\left(\dfrac{2}{3} z^{3/2}\right) + (1+\sqrt{3})\,\text{bei}_{1/3}\left(\dfrac{2}{3} z^{3/2}\right)\right] \quad [\text{Re } z > 0].$

7.2.8. The Kelvin functions $\text{ber}_\nu(z)$, $\text{bei}_\nu(z)$, $\text{ker}_\nu(z)$ and $\text{kei}_\nu(z)$

1. $\text{ber}_{1/2}(z) = \sqrt{\dfrac{2}{\pi z}} \left(\sin \dfrac{3\pi}{8} \sinh \dfrac{z}{\sqrt{2}} \cos \dfrac{z}{\sqrt{2}} - \cos \dfrac{3\pi}{8} \cosh \dfrac{z}{\sqrt{2}} \sin \dfrac{z}{\sqrt{2}} \right).$

2. $\text{bei}_{1/2}(z) = \sqrt{\dfrac{2}{\pi z}} \left(\sin \dfrac{3\pi}{8} \cosh \dfrac{z}{\sqrt{2}} \sin \dfrac{z}{\sqrt{2}} + \cos \dfrac{3\pi}{8} \sinh \dfrac{z}{\sqrt{2}} \cos \dfrac{z}{\sqrt{2}} \right).$

3. $\text{ber}_{-1/2}(z) = \sqrt{\dfrac{2}{\pi z}} \left(\cos \dfrac{3\pi}{8} \cosh \dfrac{z}{\sqrt{2}} \cos \dfrac{z}{\sqrt{2}} + \sin \dfrac{3\pi}{8} \sinh \dfrac{z}{\sqrt{2}} \sin \dfrac{z}{\sqrt{2}} \right).$

4. $\text{bei}_{-1/2}(z) = \sqrt{\dfrac{2}{\pi z}} \left(\cos \dfrac{3\pi}{8} \sinh \dfrac{z}{\sqrt{2}} \sin \dfrac{z}{\sqrt{2}} - \sin \dfrac{3\pi}{8} \cosh \dfrac{z}{\sqrt{2}} \cos \dfrac{z}{\sqrt{2}} \right).$

5. $\text{ker}_{1/2}(z) = \sqrt{\dfrac{\pi}{2z}}\, e^{-z/\sqrt{2}} \cos\left(\dfrac{z}{\sqrt{2}} + \dfrac{3\pi}{8} \right).$

6. $\text{kei}_{1/2}(z) = -\sqrt{\dfrac{\pi}{2z}}\, e^{-z/\sqrt{2}} \sin\left(\dfrac{z}{\sqrt{2}} + \dfrac{3\pi}{8} \right).$

7. $\text{ker}_{-1/2}(z) = \sqrt{\dfrac{\pi}{2z}}\, e^{-z/\sqrt{2}} \cos\left(\dfrac{z}{\sqrt{2}} - \dfrac{\pi}{8} \right).$

8. $\text{kei}_{-1/2}(z) = -\sqrt{\dfrac{\pi}{2z}}\, e^{-z/\sqrt{2}} \sin\left(\dfrac{z}{\sqrt{2}} - \dfrac{\pi}{8} \right).$

9. $\text{ber}_{\nu+n}(z) = (\nu)_n \left(-\dfrac{2}{z} \right)^n$

$\times {}_4F_7\left(\begin{array}{c} -\dfrac{n}{4},\, \dfrac{1-n}{4},\, \dfrac{2-n}{4},\, \dfrac{3-n}{4};\, -\dfrac{z^4}{16} \\ -\dfrac{n}{2},\, \dfrac{1-n}{2},\, \dfrac{\nu}{2},\, \dfrac{1+\nu}{2},\, \dfrac{1-n-\nu}{2},\, 1-\dfrac{n+\nu}{2},\, \dfrac{1}{2} \end{array} \right)$

$\times \left[\cos \dfrac{n\pi}{4} \text{ber}_\nu(z) - \sin \dfrac{n\pi}{4} \text{bei}_\nu(z) \right]$

$- \dfrac{1}{\nu} (\nu)_n \left(-\dfrac{2}{z} \right)^{n-1} {}_4F_7\left(\begin{array}{c} \dfrac{1-n}{4},\, \dfrac{2-n}{4},\, \dfrac{3-n}{4},\, 1-\dfrac{n}{4};\, -\dfrac{z^4}{16} \\ \dfrac{1-n}{2},\, 1-\dfrac{n}{2},\, \dfrac{1+\nu}{4},\, 1+\dfrac{\nu}{2},\, \dfrac{1-n-\nu}{2},\, 1-\dfrac{n+\nu}{2},\, \dfrac{1}{2} \end{array} \right)$

$\times \left[\sin \dfrac{(n+1)\pi}{4} \text{ber}_{\nu-1}(z) + \cos \dfrac{(n+1)\pi}{4} \text{bei}_{\nu-1}(z) \right]$

$+ \dfrac{n-1}{\nu(\nu+n-1)} (\nu)_n \left(-\dfrac{2}{z} \right)^{n-2}$

$\times {}_4F_7\left(\begin{array}{c} \dfrac{2-n}{4},\, \dfrac{3-n}{4},\, 1-\dfrac{n}{4},\, \dfrac{5-n}{4};\, -\dfrac{z^4}{16} \\ \dfrac{1-n}{2},\, 1-\dfrac{n}{2},\, \dfrac{1+\nu}{2},\, 1+\dfrac{\nu}{2},\, 1-\dfrac{n+\nu}{2},\, \dfrac{3-n-\nu}{2},\, \dfrac{3}{2} \end{array} \right)$

$\times \left[\sin \dfrac{n\pi}{4} \text{ber}_\nu(z) + \cos \dfrac{n\pi}{4} \text{bei}_\nu(z) \right] - \dfrac{n-2}{\nu(\nu+1)(\nu+n-1)} (\nu)_n \left(-\dfrac{2}{z} \right)^{n-3}$

$$\times {}_4F_7\left(\begin{matrix}\frac{3-n}{4},1-\frac{n}{4},\frac{5-n}{4},\frac{6-n}{4};-\frac{z^4}{16}\\1-\frac{n}{2},\frac{3-n}{2},1+\frac{\nu}{2},\frac{3+\nu}{2},1-\frac{n+\nu}{2},\frac{3-n-\nu}{2},\frac{3}{2}\end{matrix}\right)$$

$$\times\left[\cos\frac{(n+1)\pi}{4}\,\mathrm{ber}_{\nu-1}(z)-\sin\frac{(n+1)\pi}{4}\,\mathrm{bei}_{\nu-1}(z)\right].$$

10. $\mathrm{bei}_{\nu+n}(z)=(\nu)_n\left(-\dfrac{2}{z}\right)^n$

$$\times {}_4F_7\left(\begin{matrix}-\frac{n}{4},\frac{1-n}{4},\frac{2-n}{4},\frac{3-n}{4};-\frac{z^4}{16}\\-\frac{n}{2},\frac{1-n}{2},\frac{\nu}{2},\frac{1+\nu}{2},\frac{1-n-\nu}{2},1-\frac{n+\nu}{2},\frac{1}{2}\end{matrix}\right)$$

$$\times\left[\sin\frac{n\pi}{4}\,\mathrm{ber}_\nu(z)+\cos\frac{n\pi}{4}\,\mathrm{bei}_\nu(z)\right]$$

$$+\frac{(\nu+1)_n}{n+\nu}\left(-\frac{2}{z}\right)^{n-1}{}_4F_7\left(\begin{matrix}\frac{1-n}{4},\frac{2-n}{4},\frac{3-n}{4},1-\frac{n}{4};-\frac{z^4}{16}\\\frac{1-n}{2},1-\frac{n}{2},\frac{1+\nu}{2},1+\frac{\nu}{2},\frac{1-n-\nu}{2},1-\frac{n+\nu}{2},\frac{1}{2}\end{matrix}\right)$$

$$\times\left[\cos\frac{(n+1)\pi}{4}\,\mathrm{ber}_{\nu-1}(z)-\sin\frac{(n+1)\pi}{4}\,\mathrm{bei}_{\nu-1}(z)\right]$$

$$+\frac{n-1}{(n+\nu)(n+\nu-1)}(\nu+1)_n\left(-\frac{2}{z}\right)^{n-2}$$

$$\times {}_4F_7\left(\begin{matrix}\frac{2-n}{4},\frac{3-n}{4},1-\frac{n}{4},\frac{5-n}{4};-\frac{z^4}{16}\\\frac{1-n}{2},\frac{3-2n}{4},1-\frac{n}{2},\frac{1+\nu}{2},1+\frac{\nu}{2},\frac{1-n-\nu}{2},1-\frac{n+\nu}{2},\frac{3}{2}\end{matrix}\right)$$

$$\times\left[\cos\frac{n\pi}{4}\,\mathrm{ber}_\nu(z)-\sin\frac{n\pi}{4}\,\mathrm{bei}_\nu(z)\right]$$

$$-\frac{n-2}{(n+\nu)[(n+\nu)^2-1]}(\nu+2)_n\left(-\frac{2}{z}\right)^{n-3}$$

$$\times {}_4F_7\left(\begin{matrix}\frac{3-n}{4},1-\frac{n}{4},\frac{5-n}{4},\frac{6-n}{4};-\frac{z^4}{16}\\1-\frac{n}{2},\frac{3-n}{2},1+\frac{\nu}{2},\frac{3+\nu}{2},1-\frac{n+\nu}{2},\frac{3-n-\nu}{2},\frac{3}{2}\end{matrix}\right)$$

$$\times\left[\sin\frac{(n+1)\pi}{4}\,\mathrm{ber}_{\nu-1}(z)+\cos\frac{(n+1)\pi}{4}\,\mathrm{bei}_{\nu-1}(z)\right].$$

11. $\mathrm{ber}_{\nu-n}(z)=(-\nu)_n\left(\dfrac{2}{z}\right)^n$

$$\times {}_4F_7\left(\begin{matrix}-\frac{n}{4},\frac{1-n}{4},\frac{2-n}{4},\frac{3-n}{4};-\frac{z^4}{16}\\-\frac{n}{2},\frac{1-n}{2},-\frac{\nu}{2},\frac{1-\nu}{2},\frac{1-n+\nu}{2},1+\frac{\nu-n}{2},\frac{1}{2}\end{matrix}\right)$$

$$\times\left[\cos\frac{n\pi}{4}\,\mathrm{ber}_\nu(z)-\sin\frac{n\pi}{4}\,\mathrm{bei}_\nu(z)\right]$$

$$-\frac{2}{\nu}(-\nu)_n\left(\frac{2}{z}\right)^{n-1}{}_4F_7\left(\begin{matrix}\frac{1-n}{4},\frac{2-n}{4},\frac{3-n}{4},1-\frac{n}{4};-\frac{z^4}{16}\\\frac{1-n}{2},1-\frac{n}{2},\frac{1-\nu}{2},1-\frac{\nu}{2},\frac{1-n+\nu}{2},1+\frac{\nu-n}{2},\frac{1}{2}\end{matrix}\right)$$

$$\times \left[\sin\frac{(n+1)\pi}{4}\operatorname{ber}_{\nu+1}(z)+\cos\frac{(n+1)\pi}{4}\operatorname{bei}_{\nu+1}(z)\right]$$

$$+\frac{n-1}{\nu(n-\nu-1)}(-\nu)_n\left(\frac{2}{z}\right)^{n-2}$$

$$\times {}_4F_7\left(\begin{array}{c}\frac{2-n}{4},\frac{3-n}{4},1-\frac{n}{4},\frac{5-n}{4};-\frac{z^4}{16}\\ \frac{1-n}{2},1-\frac{n}{2},\frac{1-\nu}{2},1-\frac{\nu}{2},1+\frac{\nu-n}{2},\frac{3-n+\nu}{2},\frac{3}{2}\end{array}\right)$$

$$\times \left[\sin\frac{n\pi}{4}\operatorname{ber}_\nu(z)+\cos\frac{n\pi}{4}\operatorname{bei}_\nu(z)\right]+\frac{n-2}{\nu(\nu-1)(\nu-n+1)}(-\nu)_n\left(\frac{2}{z}\right)^{n-3}$$

$$\times {}_4F_7\left(\begin{array}{c}\frac{3-n}{4},1-\frac{n}{4},\frac{5-n}{4},\frac{6-n}{4};-\frac{z^4}{16}\\ 1-\frac{n}{2},\frac{3-n}{2},1-\frac{\nu}{2},\frac{3-\nu}{2},1+\frac{\nu-n}{2},\frac{3-n+\nu}{2},\frac{3}{2}\end{array}\right)$$

$$\times \left[\cos\frac{(n+1)\pi}{4}\operatorname{ber}_{\nu+1}(z)-\sin\frac{(n+1)\pi}{4}\operatorname{bei}_{\nu+1}(z)\right].$$

12. $\operatorname{bei}_{\nu-n}(z)=(-\nu)_n\left(\frac{2}{z}\right)^n$

$$\times {}_4F_7\left(\begin{array}{c}-\frac{n}{4},\frac{1-n}{4},\frac{2-n}{4},\frac{3-n}{4};-\frac{z^4}{16}\\ -\frac{n}{2},\frac{1-n}{2},-\frac{\nu}{2},\frac{1-\nu}{2},\frac{1-n+\nu}{2},1+\frac{\nu-n}{2},\frac{1}{2}\end{array}\right)$$

$$\times \left[\sin\frac{n\pi}{4}\operatorname{ber}_\nu(z)+\cos\frac{n\pi}{4}\operatorname{bei}_\nu(z)\right]$$

$$-\frac{(1-\nu)_n}{\nu-n}\left(\frac{2}{z}\right)^{n-1}{}_4F_7\left(\begin{array}{c}\frac{1-n}{4},\frac{2-n}{4},\frac{3-n}{4},1-\frac{n}{4};-\frac{z^4}{16}\\ \frac{1-n}{2},1-\frac{n}{2},\frac{1-\nu}{2},1-\frac{\nu}{2},\frac{1-n+\nu}{2},1+\frac{\nu-n}{2},\frac{1}{2}\end{array}\right)$$

$$\times \left[\cos\frac{(n+1)\pi}{4}\operatorname{ber}_{\nu+1}(z)-\sin\frac{(n+1)\pi}{4}\operatorname{bei}_{\nu+1}(z)\right]$$

$$+\frac{n-1}{(n-\nu)(n-\nu-1)}(1-\nu)_n\left(\frac{2}{z}\right)^{n-2}$$

$$\times {}_4F_7\left(\begin{array}{c}\frac{2-n}{4},\frac{3-n}{4},1-\frac{n}{4},\frac{5-n}{4};-\frac{z^4}{16}\\ \frac{1-n}{2},1-\frac{n}{2},\frac{1-\nu}{2},1-\frac{\nu}{2},1+\frac{\nu-n}{2},\frac{3-n+\nu}{2},\frac{3}{2}\end{array}\right)$$

$$\times \left[\cos\frac{n\pi}{4}\operatorname{ber}_\nu(z)-\sin\frac{n\pi}{4}\operatorname{bei}_\nu(z)\right]$$

$$-\frac{n-2}{(n-\nu)[(n-\nu)^2-1]}(2-\nu)_n\left(\frac{2}{z}\right)^{n-3}$$

$$\times {}_4F_7\left(\begin{array}{c}\frac{3-n}{4},1-\frac{n}{4},\frac{5-n}{4},\frac{6-n}{4};-\frac{z^4}{16}\\ 1-\frac{n}{2},\frac{3-n}{2},1-\frac{\nu}{2},\frac{3-\nu}{2},1+\frac{\nu-n}{2},\frac{3-n+\nu}{2},\frac{3}{2}\end{array}\right)$$

$$\times \left[\sin\frac{(n+1)\pi}{4}\operatorname{ber}_{\nu+1}(z)+\cos\frac{(n+1)\pi}{4}\operatorname{bei}_{\nu+1}(z)\right].$$

13. $\operatorname{ker}_{n+1/2}(z)=(-1)^n n!\sqrt{\pi}\,(2z)^{-n-1/2}e^{-z/\sqrt{2}}$

$$\times\left[\cos\left(\frac{z}{\sqrt{2}}+\frac{6n+3}{8}\pi\right)\sum_{k=0}^{[n/2]}\frac{(-1)^k(\sqrt{2}z)^{2k}}{(2k)!}L_{n-2k}^{2k-2n-1}(\sqrt{2}z)\right.$$

$$- \sin\left(\frac{z}{\sqrt{2}} + \frac{6n+3}{8}\pi\right) \sum_{k=0}^{[(n-1)/2]} \frac{(-1)^k (\sqrt{2}z)^{2k+1}}{(2k+1)!} L_{n-2k-1}^{2k-2n}(\sqrt{2}z) \Bigg]$$

$$[|\arg z| < \pi].$$

14. $\text{kei}_{n+1/2}(z) = (-1)^{n+1} n! \sqrt{\pi} (2z)^{-n-1/2} e^{-z/\sqrt{2}}$

$$\times \Bigg[\sin\left(\frac{z}{\sqrt{2}} + \frac{6n+3}{8}\pi\right) \sum_{k=0}^{[n/2]} \frac{(-1)^k (\sqrt{2}z)^{2k}}{(2k)!} L_{n-2k}^{2k-2n-1}(\sqrt{2}z)$$

$$+ \cos\left(\frac{z}{\sqrt{2}} + \frac{6n+3}{8}\pi\right) \sum_{k=0}^{[(n-1)/2]} \frac{(-1)^k (\sqrt{2}z)^{2k+1}}{(2k+1)!} L_{n-2k-1}^{2k-2n}(\sqrt{2}z) \Bigg]$$

$$[|\arg z| < \pi].$$

7.2.9. The Legendre polynomials $P_n(z)$

1. $P_n(z) = (-1)^n P_n(-z)$.

2. $\quad = C_n^{1/2}(z)$.

3. $\quad = P_n^{(0,0)}(z)$.

4. $\quad = \left(\dfrac{z-1}{2}\right)^n P_n^{(0,-2n-1)}\left(\dfrac{3+z}{1-z}\right)$.

5. $\quad = (-2)^n (z^2-1)^{n/2} P_n^{(-n-1/2,-n-1/2)}\left(\dfrac{z}{\sqrt{z^2-1}}\right)$.

6. $P_{2n}(z) = (z^2-1)^n P_n^{(-1/2,-2n-1/2)}\left(\dfrac{1+z^2}{1-z^2}\right)$.

7. $\quad = z^{2n} P_n^{(0,-2n-1/2)}\left(\dfrac{2}{z^2}-1\right)$.

8. $P_{2n+1}(z) = z(z^2-1)^n P_n^{(1/2,-2n-3/2)}\left(\dfrac{1+z^2}{1-z^2}\right)$.

9. $\quad = z^{2n+1} P_n^{(0,-2n-3/2)}\left(\dfrac{2}{z^2}-1\right)$.

7.2.10. The Chebyshev polynomials $T_n(z)$ and $U_n(z)$

1. $T_n(z) = (-1)^n T_n(-z)$.

2. $\quad = (-1)^n T_{2n}\left(\sqrt{\dfrac{1-z}{2}}\right)$.

3. $\quad = T_{2n}\left(\sqrt{\dfrac{1+z}{2}}\right)$.

4. $\quad = U_n(z) - z U_{n-1}(z)$ \hfill $[n \geq 1]$.

5. $\quad = \dfrac{n}{2} \lim\limits_{\lambda \to 0} \dfrac{1}{\lambda} C_n^\lambda(z).$

6. $\quad = \dfrac{n!}{\left(\frac{1}{2}\right)_n} P_n^{(-1/2,\,-1/2)}(z).$

7. $T_{2n}(z) = T_n(2z^2 - 1).$

8. $\quad = (-1)^n T_{2n}\!\left(\sqrt{1-z^2}\right).$

9. $\quad = 2T_n^2(z) - 1.$

10. $\quad = (-1)^{n+1}$
$\quad\quad + 2T_{2n}\!\left(\sqrt{\dfrac{1}{2} + \dfrac{1}{2}(1-z^2)^{1/2}}\right) T_{2n}\!\left(\sqrt{\dfrac{1}{2} - \dfrac{1}{2}(1-z^2)^{1/2}}\right).$

11. $\quad = \dfrac{n!}{\left(n+\frac{1}{2}\right)_n} z^{2n} C_{2n}^{1/2-2n}\!\left(\dfrac{\sqrt{z^2-1}}{z}\right).$

12. $\quad = \dfrac{n!}{\left(\frac{1}{2}\right)_n} z^{2n} P_n^{(-1/2,\,-2n)}\!\left(\dfrac{2}{z^2}-1\right).$

13. $T_{2n+1}(z) = 2T_{2n+1}\!\left(\sqrt{\dfrac{1+\sqrt{1-z^2}}{2}}\right) T_{2n+1}\!\left(\sqrt{\dfrac{1-\sqrt{1-z^2}}{2}}\right).$

14. $\quad = (-1)^n z U_{2n}\!\left(\sqrt{1-z^2}\right).$

15. $\quad = \dfrac{n!}{\left(n+\frac{3}{2}\right)_n} z^{2n+1} C_{2n}^{-1/2-2n}\!\left(\dfrac{\sqrt{z^2-1}}{z}\right).$

16. $\quad = \dfrac{n!}{\left(\frac{1}{2}\right)_n} z P_n^{(-1/2,\,1/2)}(2z^2 - 1).$

17. $\quad = \dfrac{n!}{\left(\frac{1}{2}\right)_n} z^{2n+1} P_n^{(-1/2,\,-2n-1)}\!\left(\dfrac{2}{z^2}-1\right).$

18. $T_n^2(z) = \dfrac{1}{2}[T_{2n}(z) + 1].$

19. $\quad = 1 - (1-z^2)U_{n-1}^2(z)$ $\quad\quad\quad\quad\quad\quad\quad\quad [n \geq 1].$

20. $U_n(z) = (-1)^n U_n(-z).$

21. $\quad = C_n^1(z).$

22. $\quad = \dfrac{(n+1)!}{\left(\frac{3}{2}\right)_n} P_n^{(1/2,\,1/2)}(z).$

23. $U_{2n}(z) = \dfrac{(-1)^n}{\sqrt{1-z^2}} T_{2n+1}\left(\sqrt{1-z^2}\right).$

24. $\quad = -\dfrac{n!(2n+1)}{2\left(n+\dfrac{1}{2}\right)_{n+1}} z^{2n+1}(z^2-1)^{-1/2} C_{2n+1}^{-1/2-2n}\left(\sqrt{1-\dfrac{1}{z^2}}\right).$

25. $\quad = \dfrac{n!}{\left(\dfrac{1}{2}\right)_n} P_n^{(1/2,-1/2)}(2z^2-1).$

26. $\quad = \dfrac{(n!)^2}{(2n)!}(2z)^{2n} P_n^{(1/2,-2n-1)}\left(\dfrac{2}{z^2}-1\right).$

27. $U_{2n+1}(z) = 2z U_n(2z^2-1).$

28. $\quad = (-1)^n \dfrac{z}{\sqrt{1-z^2}} U_{2n+1}\left(\sqrt{1-z^2}\right).$

29. $\quad = -\dfrac{2(n+1)!}{(4n+3)\left(n+\dfrac{3}{2}\right)_n} z^{2n+2}(z^2-1)^{-1/2} C_{2n+1}^{-3/2-2n}\left(\sqrt{1-\dfrac{1}{z^2}}\right).$

30. $\quad = \dfrac{n!(n+1)!}{(2n+1)!}(2z)^{2n+1} P_n^{(1/2,-2n-2)}\left(\dfrac{2}{z^2}-1\right).$

31. $U_n^2(z) = \dfrac{1}{1-z^2}[1 - T_{n+1}^2(z)].$

32. $T_{2n}(z)T_{2n+1}\left(\sqrt{1-z^2}\right) + z T_{2n+1}(z)U_{2n-1}\left(\sqrt{1-z^2}\right)$
$$= (-1)^n \sqrt{1-z^2} \quad [n \geq 1].$$

7.2.11. The Hermite polynomials $H_n(z)$

1. $H_n(z) = (-1)^n H_n(-z).$

2. $H_{2n}(z) = (-1)^n 2^{2n} n! L_n^{-1/2}(z^2).$

3. $H_{2n+1}(z) = (-1)^n 2^{2n+1} n! z L_n^{1/2}(z^2).$

4. $H_{2n}(z_1 z_2 \ldots z_m)$
$$= (2n)! \sum_{k_1=0}^{n} \sum_{k_2=0}^{k_1} \cdots \sum_{k_{m-1}=0}^{k_{m-2}} \dfrac{z_1^{2k_1}(z_1^2-1)^{n-k_1}}{(n-k_1)!} \dfrac{z_2^{2k_2}(z_2^2-1)^{k_1-k_2}}{(k_1-k_2)!} \cdots$$
$$\times \dfrac{z_{m-1}^{2k_{m-1}}(z_{m-1}^2-1)^{k_{m-2}-k_{m-1}}}{(k_{m-2}-k_{m-1})!} \dfrac{1}{(2k_{m-1})!} H_{2k_{m-1}}(z_m).$$

5. $H_{2n+1}(z_1 z_2 \ldots z_m) = (2n+1)! \, z_1 z_2 \ldots z_{m-1}$

$$\times \sum_{k_1=0}^{n} \sum_{k_2=0}^{k_1} \ldots \sum_{k_{m-1}=0}^{k_{m-2}} \frac{z_1^{2k_1}(z_1^2-1)^{n-k_1}}{(n-k_1)!} \frac{z_2^{2k_2}(z_2^2-1)^{k_1-k_2}}{(k_1-k_2)!} \ldots$$

$$\times \frac{z_{m-1}^{2k_{m-1}}(z_{m-1}^2-1)^{k_{m-2}-k_{m-1}}}{(k_{m-2}-k_{m-1})!} \frac{1}{(2k_{m-1}+1)!} H_{2k_{m-1}+1}(z_m).$$

7.2.12. The Laguerre polynomials $L_n^\lambda(z)$

1. $L_n^{-1/2}(z) = \dfrac{(-1)^n}{2^{2n} n!} H_{2n}(\sqrt{z})$.

2. $L_n^{1/2}(z) = \dfrac{(-1)^n}{2^{2n+1} n! \sqrt{z}} H_{2n+1}(\sqrt{z})$.

3. $L_n^{-m}(z) = \dfrac{(n-m)!}{n!}(-z)^m L_{n-m}^m(z)$ \qquad\qquad $[1 \leq m \leq n]$.

4. $L_n^{-n}(z) = \dfrac{(-z)^n}{n!}$.

5. $L_n^{1-n}(z) = \dfrac{(-1)^n}{n!} z^{n-1}(z-n)$.

6. $L_n^{-n-1}(z) = \dfrac{(-1)^n}{n!} e^z \Gamma(n+1, z)$.

7. $\quad = (-1)^n \sum_{k=0}^{n} \dfrac{z^k}{k!}$.

8. $L_n^{-2n-1}(z) = \dfrac{(-z)^n}{n!} \sqrt{\dfrac{z}{\pi}} e^{z/2} K_{n+1/2}\!\left(\dfrac{z}{2}\right)$.

9. $L_n^{-2n-1}(z) L_n^{-2n-1}(-z) = \dfrac{\pi}{4(n!)^2} z^{2n+1} \left[I_{-n-1/2}^2\!\left(\dfrac{z}{2}\right) - I_{n+1/2}^2\!\left(\dfrac{z}{2}\right) \right]$.

10. $L_n^\lambda(z_1 + z_2 + \ldots + z_m)$

$$= \sum_{k_1=0}^{n} \sum_{k_2=0}^{k_1} \ldots \sum_{k_{m-1}=0}^{k_{m-2}} L_{n-k_1}^{\lambda-\lambda_1-1}(z_1) L_{k_1-k_2}^{\lambda_1-\lambda_2-1}(z_2) \ldots$$

$$\times L_{k_{m-2}-k_{m-1}}^{\lambda_{m-2}-\lambda_{m-1}-1}(z_{m-1}) L_{k_{m-1}}^{\lambda_{m-1}}(z_m).$$

11. $L_n^\lambda(z_1 z_2 \ldots z_m)$

$$= (\lambda+1)_n \sum_{k_1=0}^{n} \sum_{k_2=0}^{k_1} \ldots \sum_{k_{m-1}=0}^{k_{m-2}} \frac{z_1^{k_1}(1-z_1)^{n-k_1}}{(n-k_1)!} \frac{z_2^{k_2}(1-z_2)^{k_1-k_2}}{(k_1-k_2)!} \ldots$$

$$\times \frac{z_{m-1}^{k_{m-1}}(1-z_{m-1})^{k_{m-1}}}{(k_{m-2}-k_{m-1})!} \frac{1}{(\lambda+1)_{k_{m-1}}} L_{k_{m-1}}^\lambda(z_m).$$

7.2.13. The Gegenbauer polynomials $C_n^\lambda(z)$

1. $C_n^\lambda(z) = (-1)^n C_n^\lambda(-z)$.

2. $ = (-2)^{-n} \dfrac{(2\lambda)_n}{\left(\lambda + \frac{1}{2}\right)_n} (z^2 - 1)^{n/2} C_n^{1/2-\lambda-n}\left(\dfrac{z}{\sqrt{z^2-1}}\right)$.

3. $ = \dfrac{(2\lambda)_n}{\left(\lambda + \frac{1}{2}\right)_n} P_n^{(\lambda-1/2,\,\lambda-1/2)}(z)$.

4. $ = 2^n \dfrac{(\lambda)_n}{(2\lambda+n)_n}(z+1)^n P_n^{(\lambda-1/2,\,-2\lambda-2n)}\left(\dfrac{3-z}{1+z}\right)$.

5. $ = \dfrac{(2\lambda)_n}{\left(\lambda + \frac{1}{2}\right)_n}\left(\dfrac{z+1}{2}\right)^n P_n^{(\lambda-1/2,\,-2\lambda-2n)}\left(\dfrac{3-z}{1+z}\right)$.

6. $ = (2z^2 + 2z\sqrt{z^2-1} - 1)^{-n/2} P_n^{(2\lambda-1,\,-\lambda-n)}\left(4z^2 + 4z\sqrt{z^2-1} - 3\right)$.

7. $ = (-2)^n (z^2-1)^{n/2} P_n^{(-\lambda-n,\,-\lambda-n)}\left(\dfrac{z}{\sqrt{z^2-1}}\right)$.

8. $C_{2n}^\lambda(z) = (-1)^n \dfrac{(\lambda)_n}{\left(\frac{1}{2}\right)_n} P_n^{(-1/2,\,\lambda-1/2)}(1 - 2z^2)$.

9. $\phantom{C_{2n}^\lambda(z)} = \dfrac{(\lambda)_n}{\left(\frac{1}{2}\right)_n}(z^2-1)^n P_n^{(-1/2,\,-\lambda-2n)}\left(\dfrac{1+z^2}{1-z^2}\right)$.

10. $\phantom{C_{2n}^\lambda(z)} = \dfrac{n!}{(2n)!}(\lambda)_n (2z)^{2n} P_n^{(\lambda-1/2,\,-\lambda-2n)}\left(\dfrac{2}{z^2} - 1\right)$.

11. $C_{2n+1}^\lambda(z) = 2(-1)^n \dfrac{(\lambda)_{n+1}}{\left(\frac{3}{2}\right)_n} z P_n^{(1/2,\,\lambda-1/2)}(1-2z^2)$.

12. $\phantom{C_{2n+1}^\lambda(z)} = \dfrac{2(\lambda)_{n+1}}{\left(\frac{3}{2}\right)_n} z(z^2-1)^n P_n^{(1/2,\,-\lambda-2n-1)}\left(\dfrac{1+z^2}{1-z^2}\right)$.

13. $\phantom{C_{2n+1}^\lambda(z)} = \dfrac{n!}{(2n+1)!}(\lambda)_{n+1}(2z)^{2n+1} P_n^{(\lambda-1/2,\,-\lambda-2n-1)}\left(\dfrac{2}{z^2}-1\right)$.

14. $C_{2n}^{-m-n}(z) = C_{2m}^{-m-n}(z)$.

15. $C_{2n+1}^{-m-n}(z) = C_{2m-1}^{-m-n}(z)$ \hfill $[m \geq 1]$.

16. $\displaystyle\lim_{\lambda \to 0} \dfrac{1}{\lambda} C_n^\lambda(z) = \dfrac{2}{n} T_n(z)$ \hfill $[n \geq 1]$.

17. $C_n^{1/2}(z) = P_n(z)$.

18. $C_n^{-1/2}(z) = \dfrac{1}{n-1}[zP_{n-1}(z) - P_n(z)]$ $\hspace{2em}[n \geq 2]$.

19. $C_n^1(z) = U_n(z)$.

20. $C_n^{1/2-n}(z) = (-2)^n \dfrac{\left(\frac{1}{2}\right)_n}{n!}(z^2-1)^{n/2}T_n\left(\dfrac{z}{\sqrt{z^2-1}}\right)$.

21. $C_n^{-n-1/2}(z) = 2^{n-1}\dfrac{\left(\frac{3}{2}\right)_n}{(n+1)!}[(1-z)^{n+1} + (-1)^n(1+z)^{n+1}]$.

22. $C_n^{-n}(z) = (-2)^n(z^2-1)^{n/2}P_n\left(\dfrac{z}{\sqrt{z^2-1}}\right)$ $\hspace{2em}[n \geq 1]$.

23. $C_{2n}^{1/2-n}(z) = \dfrac{\left(\frac{1}{2}\right)_n}{n!}(1-z^2)^n$.

24. $C_{2n+1}^{1/2-n}(z) = \dfrac{\left(\frac{1}{2}\right)_n}{n!}z(1-z^2)^n$.

25. $C_{2n}^{1/2-2n}(z) = \dfrac{\left(n+\frac{1}{2}\right)_n}{n!}(1-z^2)^n T_{2n}\left(\dfrac{1}{\sqrt{1-z^2}}\right)$.

26. $\phantom{C_{2n}^{1/2-2n}(z)} = \dfrac{\left(n+\frac{1}{2}\right)_n}{n!}(1-z^2)^n T_n\left(\dfrac{1+z^2}{1-z^2}\right)$.

27. $C_{2n}^{-1/2-2n}(z) = 2^{2n-1}\dfrac{\left(\frac{3}{2}\right)_{2n}}{(2n+1)!}[(1-z)^{2n+1} + (1+z)^{2n+1}]$.

28. $C_{2n}^{m-n+1/2}(z) = \dfrac{m!\left(\frac{1}{2}-m\right)_n}{n!\left(\frac{1}{2}-n\right)_m}(1-z^2)^{n-m}C_{2m}^{n-m+1/2}(z)$.

29. $C_{2n+1}^{m-n+1/2}(z) = \dfrac{m!\left(-\frac{1}{2}-m\right)_{n+1}}{n!\left(-\frac{1}{2}-n\right)_{m+1}}(1-z^2)^{n-m}C_{2m+1}^{n-m+1/2}(z)$.

30. $C_{2n}^{1/2-2n}(z) = \dfrac{\left(\frac{1}{2}\right)_{2n}}{\left(\frac{1}{2}\right)_n^2}(-z)^n C_n^{1/2-n}\left(\dfrac{1+z^2}{2z}\right)$.

31. $C_{2n+1}^{-1/2-2n}(z) = -\dfrac{2\left(n+\frac{1}{2}\right)_{n+1}}{n!(2n+1)}z(1-z^2)^n U_{2n}\left(\dfrac{1}{\sqrt{1-z^2}}\right)$.

32. $C_{2n+1}^{-3/2-2n}(z) = 2^{2n}\dfrac{\left(\frac{3}{2}\right)_{2n+1}}{(2n+2)!}[(1-z)^{2n+2} - (1+z)^{2n+2}]$.

33. $C_{2n}^{1/4-n}(z) = \left(\dfrac{z}{2}\right)^{2n} C_n^{1/4-n}\left(1 - \dfrac{2}{z^2}\right).$

34. $C_{2n+1}^{-n-1/4}(z) = -\left(\dfrac{z}{2}\right)^{2n+1} C_n^{-n-1/4}\left(1 - \dfrac{2}{z^2}\right).$

35. $\left[C_n^{-n-1/2}(z)\right]^2 = 2^{2n-1} \dfrac{\left(\frac{3}{2}\right)_n^2}{[(n+1)!]^2} (z^2-1)^{n+1} \left[T_{n+1}\left(\dfrac{z^2+1}{z^2-1}\right) - 1\right].$

36. $C_n^\lambda(z_1 + z_2 + \ldots + z_m)$

$$= 2^n (\lambda)_n \sum_{k_1=0}^{n} \sum_{k_2=0}^{k_1} \cdots \sum_{k_{m-1}=0}^{k_{m-2}} \dfrac{z_1^{n-k_1}}{(n-k_1)!} \dfrac{z_2^{k_1-k_2}}{(k_1-k_2)!} \cdots$$

$$\times \dfrac{z_{m-1}^{k_{m-2}-k_{m-1}}}{(k_{m-2}-k_{m-1})!} \dfrac{(-2)^{-k_{m-1}}}{(1-\lambda-n)_{k_{m-1}}} C_{k_{m-1}}^{\lambda+n-k_{m-1}}(z_m).$$

37. $C_{2n}^\lambda(z_1 z_2 \ldots z_m) = (-1)^n (\lambda)_n$

$$\times \sum_{k_1=0}^{n} \sum_{k_2=0}^{k_1} \cdots \sum_{k_{m-1}=0}^{k_{m-2}} \dfrac{z_1^{2k_1}(1-z_1^2)^{n-k_1}}{(n-k_1)!} \dfrac{z_2^{2k_2}(1-z_2^2)^{k_1-k_2}}{(k_1-k_2)!} \cdots$$

$$\times \dfrac{z_{m-1}^{2k_{m-1}}(1-z_{m-1}^2)^{k_{m-2}-k_{m-1}}}{(k_{m-2}-k_{m-1})!} \dfrac{1}{(1-\lambda-n)_{k_{m-1}}} C_{2k_{m-1}}^{\lambda+n-k_{m-1}}(z_m).$$

38. $C_{2n+1}^\lambda(z_1 z_2 \ldots z_m) = (-1)^n (\lambda)_n z_1 z_2 \ldots z_{m-1}$

$$\times \sum_{k_1=0}^{n} \sum_{k_2=0}^{k_1} \cdots \sum_{k_{m-1}=0}^{k_{m-2}} \dfrac{z_1^{2k_1}(1-z_1^2)^{n-k_1}}{(n-k_1)!} \dfrac{z_2^{2k_2}(1-z_2^2)^{k_1-k_2}}{(k_1-k_2)!} \cdots$$

$$\times \dfrac{z_{m-1}^{2k_{m-1}}(1-z_{m-1}^2)^{k_{m-2}-k_{m-1}}}{(k_{m-2}-k_{m-1})!} \dfrac{1}{(1-\lambda-n)_{k_{m-1}}} C_{2k_{m-1}+1}^{\lambda+n-k_{m-1}}(z_m).$$

7.2.14. The Jacobi polynomials $P_n^{(\rho,\sigma)}(z)$

1. $P_n^{(\rho,\sigma)}(z) = (-1)^n P_n^{(\sigma,\rho)}(-z).$

2. $\quad = \left(\dfrac{1-z}{2}\right)^n P_n^{(-\rho-\sigma-2n-1,\sigma)}\left(\dfrac{z+3}{z-1}\right).$

3. $\quad = (-1)^n \left(\dfrac{1+z}{2}\right)^n P_n^{(-\rho-\sigma-2n-1,\rho)}\left(\dfrac{z-3}{z+1}\right).$

4. $P_n^{(\rho,\rho)}(z) = \dfrac{(\rho+1)_n}{(2\rho+1)_n} C_n^{\rho+1/2}(z).$

5. $\quad = (-1)^n \left(\dfrac{z^2-1}{4}\right)^{n/2} C_n^{-\rho-n}\left(\dfrac{z}{\sqrt{z^2-1}}\right).$

6. $P_n^{(1/2,\sigma)}(z) = \dfrac{(-1)^n}{\sqrt{2(1-z)}} \dfrac{\left(\frac{3}{2}\right)_n}{\left(\sigma+\frac{1}{2}\right)_{n+1}} C_{2n+1}^{\sigma+1/2}\left(\sqrt{\dfrac{1-z}{2}}\right).$

7. $\quad = -2^{-n-1} \dfrac{\left(\frac{3}{2}\right)_n}{(\sigma+n+1)_{n+1}} \dfrac{(z+1)^{n+1/2}}{(z-1)^{1/2}} C_{2n+1}^{-\sigma-2n-1}\left(\sqrt{\dfrac{z-1}{z+1}}\right).$

8. $P_n^{(-1/2,\sigma)}(z) = (-1)^n \dfrac{\left(\frac{1}{2}\right)_n}{(\sigma+1/2)_n} C_{2n}^{\sigma+1/2}\left(\sqrt{\dfrac{1-z}{2}}\right).$

9. $\quad = \dfrac{\left(\frac{1}{2}\right)_n}{(\sigma+n+1)_n} \left(\dfrac{z+1}{2}\right)^n C_{2n}^{-\sigma-2n}\left(\sqrt{\dfrac{z-1}{z+1}}\right).$

10. $P_n^{(\rho,-\rho-2n-1/2)}(z) = \dfrac{(2n)!(\rho+1)_n}{n!(2\rho+1)_{2n}} \left(\dfrac{z+1}{2}\right)^n C_{2n}^{\rho+1/2}\left(\sqrt{\dfrac{2}{z+1}}\right).$

11. $\quad = \dfrac{(2n)!(\rho+1)_n}{n!(\rho+1)_{2n}} \left(\dfrac{1-z}{8}\right)^n C_{2n}^{-\rho-2n}\left(\sqrt{\dfrac{2}{1-z}}\right).$

12. $P_n^{(\rho,-\rho-2n-3/2)}(z) = \dfrac{(2n+1)!(\rho+1)_n}{n!(2\rho+1)_{2n+1}} \left(\dfrac{z+1}{2}\right)^{n+1/2} C_{2n+1}^{\rho+1/2}\left(\sqrt{\dfrac{2}{z+1}}\right).$

13. $\quad = -\dfrac{(2n+1)!(\rho+1)_n}{n!(\rho+1)_{2n+1}} \left(\dfrac{1-z}{8}\right)^{n+1/2} C_{2n+1}^{-\rho-2n-1}\left(\sqrt{\dfrac{2}{1-z}}\right).$

14. $P_n^{(\rho,\ 2\rho\ 2n-1)}(z) = \dfrac{(\rho+1)_n}{(2\rho+1)_n} \left(\dfrac{z+1}{2}\right)^n C_n^{\rho+1/2}\left(\dfrac{3-z}{1+z}\right).$

15. $\quad = (-1)^n \left(\dfrac{1-z}{2}\right)^{n/2} C_n^{-\rho-n}\left(\dfrac{3-z}{2^{3/2}\sqrt{1-z}}\right).$

16. $P_n^{(\rho,-n-(\rho+1)/2)}(z) = 2^{-3n} \dfrac{(\rho+1)_n}{\left(\frac{\rho}{2}+1\right)_n} (1-z)^n C_n^{-n-\rho/2}\left(\dfrac{z+3}{z-1}\right).$

17. $P_n^{(\rho,m-n)}(z) = \dfrac{m!}{n!} \dfrac{\Gamma(n+\rho+1)}{\Gamma(m+\rho+1)} \left(\dfrac{z+1}{2}\right)^{n-m} P_m^{(\rho,n-m)}(z).$

18. $P_n^{(\rho,-\rho-m-n)}(z) = \dfrac{(m-1)!}{n!} \dfrac{\Gamma(n+\rho+1)}{\Gamma(m+\rho)} P_{m-1}^{(\rho,-\rho-m-n)}(z) \qquad [m \geq 1].$

19. $P_n^{(\rho,m)}(z) = \dfrac{(m+n)!}{n!(\rho+n+1)_m} \left(\dfrac{2}{z+1}\right)^m P_{m+n}^{(\rho,-m)}(z).$

20. $P_n^{(-n,\rho)}(z) = \dfrac{(\rho+1)_n}{n!} \left(\dfrac{z-1}{2}\right)^n.$

21. $P_n^{(0,0)}(z) = P_n(z).$

22. $P_n^{(0,-1)}(z) = \frac{1}{2}[P_{n-1}(z) + P_n(z)]$ $\qquad$ [$n \geq 1$].

23. $P_n^{(0,1)}(z) = \frac{1}{1-z}[P_n(z) - P_{n+1}(z)]$.

24. $P_n^{(0,1/2)}(z) = \left(\frac{2}{z+1}\right)^{1/2} P_{2n+1}\left(\sqrt{\frac{z+1}{2}}\right)$.

25. $P_n^{(0,-1/2)}(z) = P_{2n}\left(\sqrt{\frac{z+1}{2}}\right)$.

26. $P_n^{(1/2,1/2)}(z) = \frac{\left(\frac{3}{2}\right)_n}{(n+1)!} U_n(z)$.

27. $P_n^{(-1/2,-1/2)}(z) = \frac{\left(\frac{1}{2}\right)_n}{n!} T_n(z)$.

28. $P_n^{(-1/2,1/2)}(z) = (-1)^n P_n^{(1/2,-1/2)}(-z)$.

29. $\qquad = (-1)^n \frac{\left(\frac{1}{2}\right)_n}{n!} U_{2n}\left(\sqrt{\frac{1-z}{2}}\right)$.

30. $\qquad = \frac{\left(\frac{1}{2}\right)_n}{n!} \left(\frac{2}{z+1}\right)^{1/2} T_{2n+1}\left(\sqrt{\frac{z+1}{2}}\right)$.

31. $P_n^{(0,-n-1/2)}(z) = \left(\frac{z+1}{2}\right)^{n/2} P_n\left(\frac{z+3}{2^{3/2}\sqrt{z+1}}\right)$.

32. $P_n^{(0,-2n-1)}(z) = (-1)^n \left(\frac{z+1}{2}\right)^n P_n\left(\frac{z-3}{z+1}\right)$.

33. $P_n^{(0,-2n-1/2)}(z) = \left(\frac{z+1}{2}\right)^n P_{2n}\left(\sqrt{\frac{2}{z+1}}\right)$.

34. $P_n^{(0,-2n-3/2)}(z) = \left(\frac{z+1}{2}\right)^{n+1/2} P_{2n+1}\left(\sqrt{\frac{2}{z+1}}\right)$.

35. $P_n^{(1/2,-2n-1)}(z) = \frac{(2n)!}{(n!)^2} \left(\frac{z+1}{8}\right)^n U_{2n}\left(\sqrt{\frac{2}{z+1}}\right)$.

36. $P_n^{(1/2,-2n-3/2)}(z) = (-2)^{-n} \frac{(-z-1)^{n+1/2}}{(1-z)^{1/2}} P_{2n+1}\left(\sqrt{\frac{z-1}{z+1}}\right)$.

37. $P_n^{(1/2,-2n-2)}(z) = \frac{(2n+1)!}{n!(n+1)!} \left(\frac{z+1}{8}\right)^{n+1/2} U_{2n+1}\left(\sqrt{\frac{2}{z+1}}\right)$.

38. $\quad = (-1)^n \dfrac{\left(\frac{3}{2}\right)_n}{(n+1)!} \left(\dfrac{z+1}{2}\right)^n U_n\left(\dfrac{z-3}{z+1}\right).$

39. $P_n^{(-1/2,-2n)}(z) = (-1)^n \dfrac{\left(\frac{1}{2}\right)_n}{n!} \left(\dfrac{z+1}{2}\right)^n T_n\left(\dfrac{z-3}{z+1}\right).$

40. $\quad = \dfrac{\left(\frac{1}{2}\right)_n}{n!} \left(\dfrac{z+1}{2}\right)^n T_{2n}\left(\sqrt{\dfrac{2}{z+1}}\right).$

41. $P_n^{(-1/2,-2n-1/2)}(z) = (-1)^n \left(\dfrac{z+1}{2}\right)^n P_{2n}\left(\sqrt{\dfrac{z-1}{z+1}}\right).$

42. $P_n^{(-1/2,-2n-1)}(z) = (-1)^n \dfrac{\left(\frac{3}{2}\right)_n}{n!(2n+1)} \left(\dfrac{z+1}{2}\right)^n U_{2n}\left(\sqrt{\dfrac{z-1}{z+1}}\right).$

43. $\quad = \dfrac{\left(\frac{1}{2}\right)_n}{n!} \left(\dfrac{z+1}{2}\right)^{n+1/2} T_{2n+1}\left(\sqrt{\dfrac{2}{z+1}}\right).$

44. $P_n^{(-n-1,-n-1)}(z) = \left(-\dfrac{1}{2}\right)^n (z^2-1)^{n/2} U_n\left(\dfrac{z}{\sqrt{z^2-1}}\right).$

45. $P_n^{(-n-1/2,-n-1/2)}(z) = \left(\dfrac{1}{2}\right)^n (z^2-1)^{n/2} P_n\left(\dfrac{z}{\sqrt{z^2-1}}\right).$

46. $P_{2n}^{(-2n-1,-2n-1)}(z) = 2^{-2n}(1-z^2)^{n+1/2} T_{2n+1}\left(\dfrac{1}{\sqrt{1-z^2}}\right).$

47. $P_{2n+1}^{(-2n-2,-2n-2)}(z) = -2^{-2n-1} z(1-z^2)^{n+1/2} U_{2n+1}\left(\dfrac{1}{\sqrt{1-z^2}}\right).$

48. $(1+z) P_n^{(\rho,-\rho)}(z) P_n^{(-\rho-1,\rho+1)}(z) + (1-z) P_n^{(-\rho,\rho)}(z) P_n^{(\rho+1,-\rho-1)}(z)$
$$= \dfrac{2(-\rho)_n (\rho+1)_n}{(n!)^2}.$$

49. $P_n^{(\rho,\sigma)}(z_1 + z_2 + \ldots + z_m)$
$$= \sum_{k_1=0}^{n} \sum_{k_2=0}^{k_1} \cdots \sum_{k_{m-1}=0}^{k_{m-2}} \dfrac{\left(\frac{z_1}{2}\right)^{n-k_1} \left(\frac{z_2}{2}\right)^{k_1-k_2} \cdots \left(\frac{z_{m-1}}{2}\right)^{k_{m-2}-k_{m-1}}}{(n-k_1)!(k_1-k_2)! \cdots (k_{m-2}-k_{m-1})!}$$
$$\times (\rho+\sigma+n+1)_{n-k_{m-1}} P_{k_{m-1}}^{(\rho+n-k_{m-1},\sigma+n-k_{m-1})}(z_m).$$

7.2.15. The polynomials of the imaginary argument

1. $P_n(iz) = \left(-\dfrac{i}{2}\right)^n (1+z^2)^{n/2} C_n^{-n}\left(\dfrac{z}{\sqrt{1+z^2}}\right)$ $\quad [n \geq 1].$

2. $T_{2n}(iz) = (-1)^n T_{2n}\left(\sqrt{1+z^2}\right).$

3. $T_{2n+1}(iz) = (-1)^n iz U_{2n}\left(\sqrt{1+z^2}\right)$.

4. $U_{2n}(iz) = \dfrac{(-1)^n}{\sqrt{1+z^2}} T_{2n+1}\left(\sqrt{1+z^2}\right)$.

5. $U_{2n+1}(iz) = (-1)^n \dfrac{iz}{\sqrt{1+z^2}} U_{2n+1}\left(\sqrt{1+z^2}\right)$.

6. $H_{2n}(iz) = (-4)^n n!\, L_n^{-1/2}(-z^2)$.

7. $H_{2n+1}(iz) = (-1)^n 2^{2n+1} n!\, iz\, L_n^{1/2}(-z^2)$.

8. $C_n^\lambda(iz) = (-2i)^n \dfrac{(\lambda)_n}{(2\lambda+n)_n} (1+z^2)^{n/2} C_n^{1/2-\lambda-n}\left(\dfrac{z}{\sqrt{1+z^2}}\right)$.

7.2.16. The complete elliptic integral $K(z)$

1. $K\left(\dfrac{1}{\sqrt{2}}\right) = \dfrac{1}{4\sqrt{\pi}} \Gamma^2\left(\dfrac{1}{4}\right)$.

2. $K\left(\sqrt{\dfrac{1-\sqrt{2}}{2}}\right) = \dfrac{2^{-11/4}}{\pi^{1/2}} \Gamma\left(\dfrac{1}{8}\right) \Gamma\left(\dfrac{3}{8}\right)$.

3. $K\left(\sqrt{\dfrac{2-\sqrt{3}}{4}}\right) = \dfrac{3^{-1/4} \pi^{-1/2}}{4} \Gamma\left(\dfrac{1}{6}\right) \Gamma\left(\dfrac{1}{3}\right)$.

4. $K\left(\sqrt{\dfrac{4-3\sqrt{2}}{8}}\right) = \dfrac{2^{-9/4}}{\pi^{1/2}} \Gamma^2\left(\dfrac{1}{4}\right)$.

5. $K\left(\sqrt{2}-1\right) = \dfrac{(1+\sqrt{2})^{1/2}}{2^{13/4} \pi^{1/2}} \Gamma\left(\dfrac{1}{8}\right) \Gamma\left(\dfrac{3}{8}\right)$ [48].

6. $K\left(\dfrac{\sqrt{3}-1}{2\sqrt{2}}\right) = \dfrac{3^{1/4}}{2^{7/3} \pi} \Gamma^3\left(\dfrac{1}{3}\right)$ [48].

7. $K\left(3-2\sqrt{2}\right) = \dfrac{1+\sqrt{2}}{2^{7/2} \pi^{1/2}} \Gamma^2\left(\dfrac{1}{4}\right)$ [48].

8. $K\left(\sqrt{2\sqrt{2}-2}\right) = \dfrac{(2+\sqrt{2})^{1/2}}{8\pi^{1/2}} \Gamma\left(\dfrac{1}{8}\right) \Gamma\left(\dfrac{3}{8}\right)$.

9. $K\left(\sqrt{3-2\sqrt{2}}\right) = \dfrac{(2+\sqrt{2})^{1/2}}{2^{7/2} \pi^{1/2}} \Gamma\left(\dfrac{1}{8}\right) \Gamma\left(\dfrac{3}{8}\right)$.

10. $K\left(2\sqrt{3\sqrt{2}-4}\right) = \dfrac{(3+2\sqrt{2})^{1/2}}{2^{5/2} \pi^{1/2}} \Gamma^2\left(\dfrac{1}{4}\right)$.

11. $K\left(\sqrt{12\sqrt{2}-16}\right) = \dfrac{2+\sqrt{2}}{8\pi^{1/2}}\Gamma^2\left(\dfrac{1}{4}\right)$.

12. $K\left(\sqrt{17-12\sqrt{2}}\right) = \dfrac{(3+2\sqrt{2})^{1/2}}{2^{7/2}\pi^{1/2}}\Gamma^2\left(\dfrac{1}{4}\right)$.

13. $K\left((\sqrt{3}-\sqrt{2})(2-\sqrt{3})\right)$
$$= \dfrac{[(2-\sqrt{2})(3+\sqrt{6})(1+\sqrt{3})]^{1/2}}{48\pi^{1/2}}\Gamma\left(\dfrac{1}{24}\right)\Gamma\left(\dfrac{11}{24}\right) \quad [48].$$

14. $K\left(\dfrac{3-\sqrt{7}}{4\sqrt{2}}\right) = \dfrac{7^{-1/4}}{4\pi}\Gamma\left(\dfrac{1}{7}\right)\Gamma\left(\dfrac{2}{7}\right)\Gamma\left(\dfrac{4}{7}\right)$ [48].

15. $K\left(\dfrac{(\sqrt{3}-1)(\sqrt{2}-\sqrt[4]{3})}{2}\right) = \dfrac{3^{-3/4}(1+\sqrt{3})}{2^{5/2}\pi^{1/2}}\Gamma^2\left(\dfrac{1}{4}\right)$ [48].

16. $K\left(\dfrac{3-2\sqrt{2}}{(1+\sqrt[4]{2})^4}\right) = \dfrac{\left(1+\sqrt[4]{2}\right)^2}{2^{9/2}\pi^{1/2}}\Gamma^2\left(\dfrac{1}{4}\right)$ [48].

17. $K\left(\dfrac{(3-2\sqrt[4]{5})(\sqrt{5}-2)}{\sqrt{2}}\right) = \dfrac{(2+\sqrt{5})}{20\pi^{1/2}}\Gamma^2\left(\dfrac{1}{4}\right)$ [48].

18. $K\left((5-2\sqrt{6})(3-2\sqrt{2})\right) = \dfrac{3^{1/4}(\sqrt{3}+2\sqrt{2}+1)}{2^{29/6}\pi}\Gamma^3\left(\dfrac{1}{3}\right)$ [49].

19. $K\left((\sqrt{2}-1)^3(2-\sqrt{3})^2\right)$
$$= \dfrac{2^{-13/4}(1+\sqrt{2})^{1/2}(\sqrt{6}+\sqrt{2}-1)}{3\pi^{1/2}}\Gamma\left(\dfrac{1}{8}\right)\Gamma\left(\dfrac{3}{8}\right) \quad [49].$$

20. $K\left(\dfrac{(2-\sqrt{3})(3-\sqrt{5})(\sqrt{3}-\sqrt{5})}{8\sqrt{2}}\right)$
$$= \dfrac{3^{-1/4}5^{-7/12}}{4\pi}\Gamma\left(\dfrac{1}{15}\right)\Gamma\left(\dfrac{4}{15}\right)\Gamma\left(\dfrac{2}{3}\right) \quad [49].$$

21. $K\left(\sqrt{\dfrac{5\sqrt{6}-3\sqrt{2}-8}{5\sqrt{6}-3\sqrt{2}+8}}\right)$
$$= \dfrac{3^{-1/4}}{16\pi^{1/2}}\left(2\sqrt{2}+\sqrt{3}+2\sqrt{6}+6\right)^{1/2}\Gamma\left(\dfrac{1}{6}\right)\Gamma\left(\dfrac{1}{3}\right).$$

7.2.17. The complete elliptic integral $E(z)$

1. $E\left(\dfrac{1}{\sqrt{2}}\right) = \dfrac{1}{8\sqrt{\pi}}\left[\Gamma^2\left(\dfrac{1}{4}\right)+4\Gamma^2\left(\dfrac{3}{4}\right)\right]$.

2. $E(3 - 2\sqrt{2}) = \dfrac{1}{8}\sqrt{\dfrac{2}{\pi}}\left[\Gamma^2\left(\dfrac{1}{4}\right) + 4(\sqrt{2} - 1)\,\Gamma^2\left(\dfrac{3}{4}\right)\right].$

3. $E\left(\sqrt{3 - 2\sqrt{2}}\right) = \dfrac{(\sqrt{2} - 1)^{1/2}}{2^{15/4}\pi^{1/2}}\left[(1 + \sqrt{2})\Gamma\left(\dfrac{1}{8}\right)\Gamma\left(\dfrac{3}{8}\right) + 8\Gamma\left(\dfrac{5}{8}\right)\Gamma\left(\dfrac{7}{8}\right)\right].$

4. $E\left(\sqrt{2\sqrt{2} - 2}\right) = \dfrac{(2 - \sqrt{2})^{1/2}}{2^{7/2}\pi^{1/2}}\left[\Gamma\left(\dfrac{1}{8}\right)\Gamma\left(\dfrac{3}{8}\right) + 8\Gamma\left(\dfrac{5}{8}\right)\Gamma\left(\dfrac{7}{8}\right)\right].$

5. $E\left(2\sqrt{3\sqrt{2} - 4}\right) = \dfrac{2 - \sqrt{2}}{8\pi^{1/2}}\left[\Gamma^2\left(\dfrac{1}{4}\right) + 8\Gamma^2\left(\dfrac{3}{4}\right)\right].$

6. $E\left(\sqrt{\dfrac{1 - \sqrt{2}}{2}}\right) = \dfrac{2^{-5/4}}{\pi^{1/2}}\left[\dfrac{1 + \sqrt{2}}{8}\Gamma\left(\dfrac{1}{8}\right)\Gamma\left(\dfrac{3}{8}\right) + \Gamma\left(\dfrac{5}{8}\right)\Gamma\left(\dfrac{7}{8}\right)\right].$

7. $E\left(\sqrt{\dfrac{2 - \sqrt{3}}{4}}\right) = \dfrac{3^{-3/4}(1 + \sqrt{3})}{8\pi^{1/2}}\Gamma\left(\dfrac{1}{6}\right)\Gamma\left(\dfrac{1}{3}\right) + \dfrac{3^{1/4}}{4\pi^{1/2}}\Gamma\left(\dfrac{2}{3}\right)\Gamma\left(\dfrac{5}{6}\right).$

8. $E\left(\sqrt{\dfrac{4 - 3\sqrt{2}}{8}}\right) = \dfrac{2^{-15/4}}{\pi^{1/2}}\left[(1 + \sqrt{2})\,\Gamma^2\left(\dfrac{1}{4}\right) + 4\Gamma^2\left(\dfrac{3}{4}\right)\right].$

9. $E\left(\dfrac{3 - \sqrt{7}}{4\sqrt{2}}\right) = \dfrac{7^{-3/4}(2 + \sqrt{7})}{8\pi}\Gamma\left(\dfrac{1}{7}\right)\Gamma\left(\dfrac{2}{7}\right)\Gamma\left(\dfrac{4}{7}\right)$
$$+ \dfrac{7^{1/4}}{8\pi}\Gamma\left(\dfrac{3}{7}\right)\Gamma\left(\dfrac{5}{7}\right)\Gamma\left(\dfrac{6}{7}\right).$$

7.2.18. The Legendre function $P_\nu^\mu(z)$

1. $P_{n-1/2}^\mu(z) = 2^n\left(\dfrac{3 - 2\mu - 2n}{4}\right)_n (1 - z^2)^{n/2}$
$$\times \sum_{k=0}^n \binom{n}{k}\dfrac{(-z^2)^k}{\left(\dfrac{2\mu - 2n + 1}{4}\right)_k}\left[\delta_{k,0}P_{-1/2}^{\mu-n}(z) + (2z)^{-k}(1 - z^2)^{-k/2}\right.$$
$$\left.\times \sum_{p=0}^{k-1}\dfrac{(k + p - 1)!}{p!\,(k - p - 1)!}(2z)^{-p}(1 - z^2)^{p/2}\,P_{-1/2}^{\mu+k-n-p}(z)\right].$$

2. $= 2^n\dfrac{\left(\dfrac{3 + 2\mu - 2n}{4}\right)_n}{\left(\dfrac{1}{2} - \mu\right)_n\left(\dfrac{1}{2} + \mu\right)_n}(1 - z^2)^{n/2}\sum_{k=0}^n\binom{n}{k}\dfrac{z^{2k}}{\left(\dfrac{1 - 2\mu - 2n}{4}\right)_k}$
$$\times \left[\delta_{k,0}P_{-1/2}^{\mu+n}(z) + (-2z)^{-k}(1 - z^2)^{-k/2}\right.$$
$$\left.\times \sum_{p=0}^{k-1}\dfrac{(k + p - 1)!}{p!\,(k - p - 1)!}\left(\dfrac{1}{2} - \mu - n\right)_{k-p}^2 (2z)^{-p}(1 - z^2)^{p/2}\,P_{-1/2}^{\mu-k+n+p}(z)\right].$$

3. $P_{-n-1/2}^\mu(z) = P_{n-1/2}^\mu(z).$

Chapter 8

Representations of Hypergeometric Functions and of the Meijer G Function

8.1. The Hypergeometric Functions

8.1.1. The Gauss hypergeometric function $_2F_1(a, b; c; z)$

1. $_2F_1\!\left(\begin{matrix}a-n,\,b\\c;\,z\end{matrix}\right) = \sum_{k=0}^{n}(-z)^k\binom{n}{k}\frac{(b)_k}{(c)_k}\,_2F_1\!\left(\begin{matrix}a,\,b+k\\c+k;\,z\end{matrix}\right).$

2. $_2F_1\!\left(\begin{matrix}a+j,\,b+m\\c+n;\,z\end{matrix}\right) = \dfrac{(-1)^m (c)_n (1-z)^{c-a-b-j+n}}{(a)_j (b)_m (c-a)_n (c-b)_n}$

   $\times \displaystyle\sum_{k=0}^{m}(-1)^k\binom{m}{k}(j+k)!\,(a-b-m)_{m-k}\sum_{p=0}^{j+k}\dfrac{z^p}{p!}(1-z)^{p-k}$

   $\times P_{j+k-p}^{(a-k+p-1,\,c-a-b-j-k+n+p)}(1-2z)\,\mathrm{D}^{n+p}\!\left[(1-z)^{a+b-c}\,_2F_1\!\left(\begin{matrix}a,\,b\\c;\,z\end{matrix}\right)\right].$

3. $_2F_1\!\left(\begin{matrix}a+j,\,b-m\\c+n;\,z\end{matrix}\right) = \dfrac{m!\,(c)_n (1-z)^{c-a-b-j+n}}{(a)_j (c-b+n)_m (c-a)_n (c-b)_n}$

   $\times \displaystyle\sum_{k=0}^{m}\dfrac{(j+k)!}{k!}P_{m-k}^{(c-a-b+k+n,\,a+b-c+j+k-m-n)}(1-2z)\sum_{p=0}^{j+k}\dfrac{z^p(1-z)^p}{p!}$

   $\times P_{j+k-p}^{(a-k+p-1,\,c-a-b-j-k+n+p)}(1-2z)\,\mathrm{D}^{n+p}\!\left[(1-z)^{a+b-c}\,_2F_1\!\left(\begin{matrix}a,\,b\\c;\,z\end{matrix}\right)\right].$

4. $_2F_1\!\left(\begin{matrix}a-j,\,b-m\\c+n;\,z\end{matrix}\right) = \dfrac{(-1)^{j+m}(c)_n(1-z)^{c-a-b+j+m+n}}{(c-a+n)_j(c-b+n)_m(c-a)_n(c-b)_n}$

   $\times \displaystyle\sum_{k=0}^{m}\binom{m}{k}(b-a-m)_{m-k}\sum_{p=0}^{j+k}\binom{j+k}{p}(a-c-j-n+1)_{j+k-p}(-z)^p$

   $\times \mathrm{D}^{n+p}\!\left[(1-z)^{a+b-c}\,_2F_1\!\left(\begin{matrix}a,\,b\\c;\,z\end{matrix}\right)\right].$

5. $\,_2F_1\!\left(\begin{matrix} a+j, b+m \\ c-n;\ z \end{matrix}\right) = \dfrac{(-1)^{j+m+n} z^{n-c+1}}{(a)_j (b)_m (1-c)_n} \sum_{k=0}^{m} \binom{m}{k}(a-b-m)_{m-k}$

$$\times \sum_{p=0}^{j+k} \binom{j+k}{p}(c-a-j-n)_{j+k-p}(-z)^p\, D^{n+p}\!\left[z^{c-1}\,_2F_1\!\left(\begin{matrix} a, b \\ c;\ z \end{matrix}\right)\right].$$

6. $\,_2F_1\!\left(\begin{matrix} a+j, b-m \\ c-n;\ z \end{matrix}\right) = \dfrac{(-1)^{j+n} m!\, z^{n-c+1}}{(a)_j (c-b-n)_m (1-c)_n}$

$$\times \sum_{k=0}^{m} \dfrac{(z-1)^k}{k!}\, P_{m-k}^{(c-a-b+k-n,\, a+b-c+j+k-m+n)}(1-2z)$$

$$\times \sum_{p=0}^{j+k}(-z)^p\binom{j+k}{p}(c-a-j-n)_{j+k-p}\, D^{n+p}\!\left[z^{c-1}\,_2F_1\!\left(\begin{matrix} a, b \\ c;\ z \end{matrix}\right)\right].$$

7. $\,_2F_1\!\left(\begin{matrix} a-j, b-m \\ c-n;\ z \end{matrix}\right) = \dfrac{(-1)^{m+n} z^{n-c+1}(1-z)^m}{(c-a-n)_j (c-b-n)_m (1-c)_n}$

$$\times \sum_{k=0}^{m}(z-1)^{-k}\binom{m}{k}(j+k)!\,(b-a-m)_{m-k}\sum_{p=0}^{j+k}\dfrac{z^p(1-z)^p}{p!}$$

$$\times P_{j+k-p}^{(p-a-k,\, a+b-c-j-k+n+p)}(1-2z)\, D^{n+p}\!\left[z^{c-1}\,_2F_1\!\left(\begin{matrix} a, b \\ c;\ z \end{matrix}\right)\right].$$

8. $\,_2F_1\!\left(\begin{matrix} a, a+m+\tfrac{1}{2} \\ n+\tfrac{1}{2};\ z \end{matrix}\right) = \dfrac{2^{n-1} m!\,\left(\tfrac{1}{2}\right)_n z^{-n/2}(1-z)^{n-m-2a}}{\left(a+\tfrac{1}{2}\right)_m (-2a)_{2n}}$

$$\times \sum_{k=0}^{m} \dfrac{z^k (1-z)^k}{k!}\, P_{m-k}^{(k+a-1/2,\, n+k-m-2a)}(1-2z)$$

$$\times \Bigg[(-1)^{k+n}(-2a)_{k+n}(2\sqrt{z})^{-k}(1+\sqrt{z})^{2a-k-n}$$

$$+(-1)^{k+n}\sum_{p=1}^{k+n-1}\dfrac{(k+n+p-1)!}{p!(k+n-p-1)!}(-2a)_{k+n-p}$$

$$\times (2\sqrt{z})^{-k-p}(1+\sqrt{z})^{2a-k-n+p}+(-2a)_{k+n}(2\sqrt{z})^{-k}(1-\sqrt{z})^{2a-k-n}$$

$$+\sum_{p=1}^{k+n-1}(-1)^p\dfrac{(k+n+p-1)!}{p!(k+n-p-1)!}(-2a)_{k+n-p}(2\sqrt{z})^{-k-p}(1-\sqrt{z})^{2a-k-n+p}\Bigg].$$

9. $\,_2F_1\!\left(\begin{matrix} a, a+m+\tfrac{1}{2} \\ \tfrac{1}{2}-n;\ z \end{matrix}\right) = \dfrac{(-1)^m}{2\left(a+\tfrac{1}{2}\right)_m \left(\tfrac{1}{2}\right)_n}$

$$\times \sum_{k=0}^{m}\binom{m}{k}(-m-n-a)_{m-k}\sum_{p=0}^{k+n} z^p \binom{k+n}{p}\left(\tfrac{1}{2}\right)_{k+n-p}$$

8.1.1]   8.1. THE HYPERGEOMETRIC FUNCTIONS

$$\times \left[(2a)_p (2\sqrt{z})^{-p} (1+\sqrt{z})^{-p-2a} \right.$$

$$+ \sum_{r=1}^{p-1} \frac{(p+r-1)!}{r!(p-r-1)!} (2a)_{p-r} (2\sqrt{z})^{(-p-r)/2} (1+\sqrt{z})^{r-p-2a}$$

$$+ (-1)^p (2a)_p (2\sqrt{z})^{-p} (1-\sqrt{z})^{-p-2a}$$

$$\left. + (-1)^p \sum_{r=1}^{p-1} (-1)^r \frac{(p+r-1)!}{r!(p-r-1)!} (2a)_{p-r} (2\sqrt{z})^{(-p-r)/2} (1-\sqrt{z})^{r-p-2a} \right].$$

10. $\displaystyle {}_2F_1\!\left(\begin{array}{c} a,\, a+m+\frac{1}{2} \\ 2a+n+1;\, z \end{array}\right) = \frac{(-1)^m 2^{2a} z^{-n} n!\, (2a+1)_n}{\left(a+\frac{1}{2}\right)_m (a+1)_n \left(a-m+\frac{1}{2}\right)_n}$

$$\times \sum_{k=0}^{n} \frac{(z-1)^k}{k!} P_{n-k}^{(k-n-a+1/2,\, k+m-n-1/2)}(1-2z) \sum_{p=0}^{k+m} \binom{k+m}{p}$$

$$\times \left(\frac{1}{2}-a-m\right)_{k+m-p} (-z)^p \left[(2a)_p (2\sqrt{1-z})^{-p} (1+\sqrt{1-z})^{-2a-p} \right.$$

$$\left. + \sum_{r=1}^{p-1} \frac{(p+r-1)!}{r!(p-r-1)!} (2a)_{p-r} (2\sqrt{1-z})^{-p-r} (1+\sqrt{1-z})^{-2a-p+r} \right].$$

11. $\displaystyle {}_2F_1\!\left(\begin{array}{c} a,\, a+m+\frac{1}{2} \\ 2a-n+1;\, z \end{array}\right) = \frac{(-1)^m 2^{2a}}{\left(a+\frac{1}{2}\right)_m (-2a)_n}$

$$\times \sum_{k=0}^{n} \binom{n}{k} \left(-a-\frac{1}{2}\right)_{n-k} \sum_{p=0}^{k+m} \binom{k+m}{p}$$

$$\times \left(\frac{1}{2}-a-m\right)_{k+m-p} (-z)^p \left[(2a)_p (2\sqrt{1-z})^{-p} (1+\sqrt{1-z})^{-2a-p} \right.$$

$$\left. + \sum_{r=1}^{p-1} \frac{(p+r-1)!}{r!(p-r-1)!} (2a)_{p-r} (2\sqrt{1-z})^{-p-r} (1+\sqrt{1-z})^{-2a-p+r} \right].$$

12. $\displaystyle {}_2F_1\!\left(\begin{array}{c} j+1,\, m+1 \\ n+2;\, z \end{array}\right) = \frac{(-1)^{n+1}(n+1) z^{-n-1}(1-z)^n}{j!\, n!} \sum_{k=0}^{m} \binom{m}{k} \frac{(j+k)!}{k!}$

$$\times \sum_{p=0}^{j+k} (-1)^p \frac{(n+p)!}{p!} (1-z)^{p-k-j} P_{j+k-p}^{(p-k,\, n-j-k+p)}(1-2z)$$

$$\times \left[\ln(1-z) - \sum_{r=1}^{n+p} \frac{1}{r} \left(\frac{z}{z-1}\right)^r \right].$$

13. $\displaystyle {}_2F_1\left(\begin{matrix} j+1,\, m+1 \\ n+\frac{3}{2};\, z \end{matrix}\right)$

$$= \frac{(-1)^n(2n+1)z^{-n-1/2}(1-z)^{n-j-1/2}}{j!\,\left(\frac{1}{2}\right)_n} \sum_{k=0}^{m} \binom{m}{k}\frac{(j+k)!}{k!}(1-z)^{-k}$$

$$\times \sum_{p=0}^{j+k} \frac{(z-1)^p}{p!} P_{j+k-p}^{(p-k,\,n-j-k+p-1/2)}(1-2z) \left[\left(\frac{1}{2}\right)_{n+p} \arcsin\sqrt{z}\right.$$

$$\left. - (n+p)!\,\frac{i}{2}\sum_{r=1}^{n+p} \frac{\left(\frac{1}{2}\right)_{n+p-r}}{(n+p-r)!\,r}\left(\frac{z}{z-1}\right)^{r/2} P_{r-1}\left(\frac{2z-1}{2\sqrt{z(z-1)}}\right)\right].$$

14. $\displaystyle {}_2F_1\left(\begin{matrix} j+1,\, m+1 \\ n+\frac{3}{2};\, -z \end{matrix}\right)$

$$= \frac{(2n+1)z^{-n-1/2}(1+z)^{n-j-1/2}}{j!\,\left(\frac{1}{2}\right)_n} \sum_{k=0}^{m} \binom{m}{k}\frac{(j+k)!}{k!}(1+z)^{-k}$$

$$\times \sum_{p=0}^{j+k} \frac{(-1)^p}{p!}(1+z)^p P_{j+k-p}^{(p-k,\,n-j-k+p-1/2)}(1+2z)$$

$$\times \left[\left(\frac{1}{2}\right)_{n+p} \ln\left(\sqrt{z}+\sqrt{1+z}\right) - \frac{(n+p)!}{2}\sum_{r=1}^{n+p} \frac{\left(\frac{1}{2}\right)_{n+p-r}}{(n+p-r)!\,r}\left(\frac{z}{1+z}\right)^{r/2}\right.$$

$$\left. \times P_{r-1}\left(\frac{2z+1}{2\sqrt{z(1+z)}}\right)\right].$$

15. $\displaystyle {}_2F_1\left(\begin{matrix} j+1,\, m+1 \\ \frac{3}{2}-n;\, z \end{matrix}\right) = \frac{(-1)^{j+m+n}z^{n-1/2}(1-z)^{-n-1/2}}{j!\,m!\,\left(-\frac{1}{2}\right)_n}$

$$\times \sum_{k=0}^{m}\binom{m}{k}(-m)_{m-k}\sum_{p=0}^{j+k}\binom{j+k}{p}\left(\frac{1}{2}-j-n\right)_{j+k-p}\left(\frac{z}{z-1}\right)^p$$

$$\times \left[\left(\frac{1}{2}\right)_{n+p}\arcsin\sqrt{z} + \frac{1}{2}\sum_{r=1}^{n+p}\binom{n+p}{r}(r-1)!\left(\frac{1}{2}\right)_{n+p-r} i^{r-1}\left(\frac{z}{1-z}\right)^{-r/2}\right.$$

$$\left. \times P_{r-1}\left(\frac{2z-1}{2i\sqrt{z(1-z)}}\right)\right].$$

8.1. THE HYPERGEOMETRIC FUNCTIONS

16. $_2F_1\left(\begin{array}{c}j+1,\, m+1\\ \frac{3}{2}-n;\, -z\end{array}\right) = \dfrac{(-1)^j z^{n-1/2}(1+z)^{-n-1/2}}{j!\left(-\frac{1}{2}\right)_n}$

$$\times \sum_{k=0}^{m} \frac{(-1)^k}{k!} \sum_{p=0}^{j+k} \binom{j+k}{p}\left(\frac{1}{2}-j-n\right)_{j+k-p}\left(\frac{z}{1+z}\right)^p$$

$$\times \left[\left(\frac{1}{2}\right)_{n+p} \ln\left(\sqrt{z}+\sqrt{1+z}\right) - \frac{(n+p)!}{2}\sum_{r=1}^{n+p}\frac{\left(\frac{1}{2}\right)_{n+p-r}}{(n+p-r)!\, r}\left(\frac{z}{1+z}\right)^{-r/2}\right.$$

$$\left.\times P_{r-1}\left(\frac{2z+1}{2\sqrt{z(1+z)}}\right)\right].$$

17. $_2F_1\left(\begin{array}{c}j+\frac{1}{2},\, m+1\\ n+2;\, z\end{array}\right) = \dfrac{2(-1)^m(n+1)z^{-n-1}}{m!\left(\frac{1}{2}\right)_j\left(\frac{3}{2}\right)_n}\sum_{k=0}^{m}(-1)^k\binom{m}{k}(j+k)!$

$$\times \left(-m-\frac{1}{2}\right)_{m-k}\sum_{p=0}^{k+j}\frac{(n+p)!}{p!}(1-z)^{-j-k}P_{j+k-p}^{(p-k-1/2,\, n-j-k+p+1/2)}(1-2z)$$

$$\times \left[P_{n+p}^{(-n-p-1,\, -n-p-1/2)}(1-2z) - (-1)^{n+p}(1-z)^{n+p+1/2}\right].$$

18. $_2F_1\left(\begin{array}{c}\frac{1}{2}-j,\, m+1\\ n+2;\, z\end{array}\right) = \dfrac{2(n+1)j!\, z^{-n-1}(1-z)^{-m}}{m!\left(\frac{3}{2}\right)_{j+n}}$

$$\times \sum_{k=0}^{j}\frac{(k+m)!}{k!}P_{j-k}^{(k+n+1/2,\, k-j+m-n-1/2)}(1-2z)$$

$$\times \sum_{p=0}^{k+m}\frac{(n+p)!}{p!}(1-z)^{-j-k}P_{k+m-p}^{(p-k,\, n-k-m+p+1/2)}(1-2z)$$

$$\times \left[P_{n+p}^{(-n-p-1,\, -n-p-1/2)}(1-2z) - (-1)^{n+p}(1-z)^{n+p+1/2}\right].$$

19. $_2F_1\left(\begin{array}{c}j+1,\, m+1\\ \frac{3}{2}-n;\, -z\end{array}\right) = \dfrac{(-1)^j z^{n-1/2}(1+z)^{-n-1/2}}{j!\left(-\frac{1}{2}\right)_n}$

$$\times \sum_{k=0}^{m} \frac{(-1)^k}{k!} \sum_{p=0}^{j+k} \binom{j+k}{p}\left(\frac{1}{2}-j-n\right)_{j+k-p}\left(\frac{z}{1+z}\right)^p$$

$$\times \left[\left(\frac{1}{2}\right)_{n+p} \ln\left(\sqrt{z}+\sqrt{1+z}\right) - \frac{(n+p)!}{2}\sum_{r=1}^{n+p}\frac{\left(\frac{1}{2}\right)_{n+p-r}}{(n+p-r)!\, r}\left(\frac{z}{1+z}\right)^{-r/2}\right.$$

$$\left.\times P_{r-1}\left(\frac{2z+1}{2\sqrt{z(1+z)}}\right)\right].$$

20. $\displaystyle {}_2F_1\left(\begin{array}{c}j+\frac{1}{2},\, m+1\\ n+\frac{3}{2};\, -z\end{array}\right)$

$$= \frac{(2n+1)\left(\frac{3}{2}\right)_m z^{-n}(1+z)^{n-j}}{n!\left(\frac{1}{2}\right)_j} \sum_{k=0}^{m} \frac{(j+k)!}{(m-k)!(2k+1)!}\left(\frac{4}{1+z}\right)^k$$

$$\times \sum_{p=0}^{j+k}(-1)^p \frac{(1+z)^p}{p!} P_{j+k-p}^{(p-k-1/2,\,n-j-k+p)}(1+2z)\left[\left(\frac{1}{2}\right)_{n+p} z^{-1/2}\arctan\sqrt{z}\right.$$

$$\left.+\frac{(n+p)!}{2}\sum_{r=1}^{n+p}(-1)^r \frac{\left(\frac{1}{2}\right)_{n+p-r}}{(n+p-r)!\,r}(z+1)^{-r}P_{r-1}^{(1/2-r,\,-r)}(1+2z)\right].$$

21. $\displaystyle {}_2F_1\left(\begin{array}{c}j+\frac{1}{2},\, m+1\\ \frac{3}{2}-n;\, z\end{array}\right)= \frac{(-1)^{j+n}\left(\frac{3}{2}\right)_m}{2\left(\frac{1}{2}\right)_j\left(-\frac{1}{2}\right)_n}(1-z)^{-n}$

$$\times \sum_{k=0}^{m}\frac{(-4)^k}{(m-k)!(2k+1)!}\sum_{p=0}^{j+k}\binom{j+k}{p}(n+p-1)!(1-j-n)_{j+k-p}(z-1)^{-p}$$

$$\times P_{n+p-1}^{(1/2-n-p,\,-n-p)}(1-2z).$$

22. $\displaystyle {}_2F_1\left(\begin{array}{c}\frac{1}{2}-j,\, m+1\\ \frac{3}{2}+n;\, z\end{array}\right)= \frac{(-1)^{j+n}(2n+1)}{2(j+n)!}z^{-n}(1-z)^{j-m+n}$

$$\times \sum_{k=0}^{m}\frac{(z-1)^k}{k!}P_{m-k}^{(k-n,\,j+k-m+n)}(1-2z)\sum_{p=0}^{j+k}\binom{j+k}{p}(-j-n)_{j+k-p}$$

$$\times \left[\left(\frac{1}{2}\right)_{n+p}z^{-1/2}\ln\frac{1+\sqrt{z}}{1-\sqrt{z}}+\sum_{r=1}^{n+p}\binom{n+p}{r}(r-1)!\left(\frac{1}{2}\right)_{n+p-r}(z-1)^{-r}\right.$$

$$\left.\times P_{r-1}^{(1/2-r,\,-r)}(1-2z)\right].$$

23. $\displaystyle {}_2F_1\left(\begin{array}{c}\frac{1}{2}-j,\, m+1\\ \frac{3}{2}+n;\, -z\end{array}\right)= \frac{(-1)^j(2n+1)}{(j+n)!}z^{-n}(1+z)^{j-m+n}$

$$\times \sum_{k=0}^{m}\frac{(-1)^k}{k!}(z+1)^k P_{m-k}^{(k-n,\,j+k-m+n)}(1+2z)\sum_{p=0}^{j+k}\binom{j+k}{p}(-j-n)_{j+k-p}$$

$$\times \left[\left(\frac{1}{2}\right)_{n+p}z^{-1/2}\arctan\sqrt{z}+\frac{(n+p)!}{2}\sum_{r=1}^{n+p}(-1)^r\frac{(1+z)^{-r}}{(n+p-r)!\,r}\left(\frac{1}{2}\right)_{n+p-r}\right.$$

$$\left.\times P_{r-1}^{(1/2-r,\,-r)}(1+2z)\right].$$

24. $\displaystyle {}_2F_1\left(\begin{array}{c}\frac{1}{2}-j,\, m+1\\ \frac{3}{2}-n;\, z\end{array}\right) = \frac{(-1)^j(1-z)^{j-m}}{2\,(j!\,m!)} \sum_{k=0}^{n} \binom{n}{k} \frac{(k+m)!}{\left(\frac{3}{2}-n\right)_k}(1-z)^{-k}$

$\displaystyle \times \sum_{p=0}^{k+m} \frac{(z-1)^p}{p!} P_{k+m-p}^{(p-k,\,j-k-m+p)}(1-2z)\left[\left(\frac{1}{2}-j\right)_{j+p} z^{-1/2}\ln\frac{1+\sqrt{z}}{1-\sqrt{z}}\right.$

$\displaystyle \left. + (j+p)!\sum_{r=1}^{j+p}\frac{\left(\frac{1}{2}-j\right)_{j+p-r}}{(j+p-r)!\,r}(z-1)^{-r} P_{r-1}^{(1/2-r,\,-r)}(1-2z)\right].$

25. $\displaystyle {}_2F_1\left(\begin{array}{c}\frac{1}{2}-j,\, m+1\\ \frac{3}{2}-n;\, -z\end{array}\right) = \frac{(-1)^j(1+z)^{j-m}}{j!\,m!}$

$\displaystyle \times \sum_{k=0}^{n}\binom{n}{k}(k+m)!\,(1+z)^{-k}\sum_{p=0}^{k+m}\frac{(-1)^p}{p!}(1+z)^p P_{k+m-p}^{(p-k,\,j-k-m+p)}(1+2z)$

$\times$

$\displaystyle \times \left[\left(\frac{1}{2}-j\right)_{j+p} z^{-1/2}\arctan\sqrt{z} + \frac{(j+p)!}{2}\sum_{r=1}^{m+p}(-1)^r\frac{\left(\frac{1}{2}-j\right)_{j+p-r}}{(j+p-r)!\,r}(z+1)^{-r}\right.$

$\displaystyle \left. \times P_{r-1}^{(1/2-r,\,-r)}(1+2z)\right].$

26. $\displaystyle {}_2F_1\left(\begin{array}{c}j+\frac{1}{2},\, m+\frac{1}{2}\\ n+\frac{3}{2};\, z\end{array}\right) = \frac{(-1)^j\left(\frac{3}{2}\right)_n}{\left(\frac{1}{2}\right)_j\left(\frac{1}{2}\right)_m(n!)^2} z^{-(n+1)/2}(z-1)^{-j+n/2}$

$\displaystyle \times \sum_{k=0}^{m}\binom{m}{k}\frac{(j+k)!}{k!}(1-z)^{-k}\sum_{p=0}^{j+k}\frac{(p+n)!}{p!}(z^2-z)^{p/2}$

$\displaystyle \times P_{j+k-p}^{(p-k-1/2,\,n-j-k+p+1/2)}(1-2z)\left[\arcsin\sqrt{z}\, P_{n+p}\!\left(\frac{2z-1}{2\sqrt{z^2-z}}\right)\right.$

$\displaystyle \left. -\frac{i}{2}\sum_{r=1}^{n+p}\frac{1}{r}P_{n+p-r}\!\left(\frac{2z-1}{2\sqrt{z^2-z}}\right)P_{r-1}\!\left(\frac{2z-1}{2\sqrt{z^2-z}}\right)\right].$

27. $\displaystyle {}_2F_1\left(\begin{array}{c}j+\frac{1}{2},\, m+\frac{1}{2}\\ n+\frac{3}{2};\, -z\end{array}\right) = \frac{(-1)^{m+n}\left(\frac{3}{2}\right)_n}{\left(\frac{1}{2}\right)_j\left(\frac{1}{2}\right)_m(n!)^2} z^{-(n+1)/2}(1+z)^{-j+n/2}$

$\displaystyle \times \sum_{k=0}^{m}\binom{m}{k}\frac{(j+k)!}{k!}(1+z)^{-k}\sum_{p=0}^{j+k}(-1)^p\frac{(p+n)!}{p!}$

$\displaystyle \times (z^2+z)^{p/2} P_{j+k-p}^{(p-k-1/2,\,n-j-k+p+1/2)}(1+2z)$

$$\times \left[-\ln\left(\sqrt{z}+\sqrt{1+z}\right) P_{n+p}\left(\frac{2z+1}{2\sqrt{z^2+z}}\right) \right.$$

$$\left. + \frac{1}{2} \sum_{r=1}^{n+p} \frac{1}{r} P_{n+p-r}\left(\frac{2z+1}{2\sqrt{z^2+z}}\right) P_{r-1}\left(\frac{2z+1}{2\sqrt{z^2+z}}\right) \right].$$

28. $\;{}_2F_1\left(\begin{array}{c} j+\frac{1}{2}, \frac{1}{2}-m \\ n+\frac{3}{2};\, z \end{array}\right) = \dfrac{m!\left(\frac{3}{2}\right)_n}{\left(\frac{1}{2}\right)_j (n+1)_m (n!)^2} z^{-(n+1)/2}(1-z)^{-j+n/2}$

$$\times \sum_{k=0}^{m} \frac{(j+k)!}{k!} P_{m-k}^{(k+n+1/2,\, j+k-m-n-1/2)}(1-2z) \sum_{p=0}^{j+k} \frac{(p+n)!}{p!}$$

$$\times i^{p+n}(z-z^2)^{p/2} P_{j+k-p}^{(p-k-1/2,\, n-j-k+p+1/2)}(1-2z)$$

$$\times \left[\arcsin\sqrt{z}\, P_{n+p}\left(\frac{2z-1}{2i\sqrt{z(1-z)}}\right) - \frac{i}{2} \sum_{r=1}^{n+p} \frac{1}{r} P_{n+p-r}\left(\frac{2z-1}{2i\sqrt{z(1-z)}}\right) \right.$$

$$\left. \times P_{r-1}\left(\frac{2z-1}{2i\sqrt{z(1-z)}}\right) \right].$$

29. $\;{}_2F_1\left(\begin{array}{c} j+\frac{1}{2}, \frac{1}{2}-m \\ n+\frac{3}{2};\, -z \end{array}\right) = \dfrac{m!\left(\frac{3}{2}\right)_n z^{-(n+1)/2}(1+z)^{-j+n/2}}{\left(\frac{1}{2}\right)_j (n+1)_m (n!)^2} \sum_{k=0}^{m} \frac{(j+k)!}{k!}$

$$\times P_{m-k}^{(k+n+1/2,\, j+k-m-n-1/2)}(1+2z) \sum_{p=0}^{j+k} (-1)^p \frac{(n+p)!}{p!}$$

$$\times (z+z^2)^{p/2} P_{j+k-p}^{(p-k-1/2,\, n-j-k+p+1/2)}(1+2z) \left[\ln\left(\sqrt{z}+\sqrt{1+z}\right) \right.$$

$$\left. \times P_{n+p}\left(\frac{2z+1}{2\sqrt{z+z^2}}\right) - \frac{1}{2} \sum_{r=1}^{n+p} \frac{1}{r} P_{n+p-r}\left(\frac{2z+1}{2\sqrt{z+z^2}}\right) P_{r-1}\left(\frac{2z+1}{2\sqrt{z+z^2}}\right) \right].$$

30. $\;{}_2F_1\left(\begin{array}{c} j-\frac{1}{2}, m-\frac{1}{2} \\ n+\frac{3}{2};\, z \end{array}\right) = \dfrac{(-1)^{j+m}\left(\frac{3}{2}\right)_n}{(j+n)!(m+n)!}$

$$\times z^{-(n+1)/2}(1-z)^{j+m+n/2} \sum_{k=0}^{m-k} \binom{m}{k}(-m)_{m-k}$$

$$\times \sum_{p=0}^{j+k} (-1)^p \binom{j+k}{p}(n+p)!(-j-n)_{j+k-p}\, i^{p+n} \left(\frac{z}{1-z}\right)^{-p/2}$$

$$\times \left[\arcsin\sqrt{z}\,P_{n+p}\!\left(\frac{2z-1}{2i\sqrt{z(1-z)}}\right) - \frac{i}{2}\sum_{r=1}^{n+p}\frac{1}{r}P_{n+p-r}\!\left(\frac{2z-1}{2i\sqrt{z(1-z)}}\right)\right.$$

$$\left. \times P_{r-1}\!\left(\frac{2z-1}{2i\sqrt{z(1-z)}}\right)\right].$$

31. $\;{}_2F_1\!\left(\begin{array}{c}j-\tfrac{1}{2},\,m-\tfrac{1}{2}\\ n+\tfrac{3}{2};\,-z\end{array}\right) = \dfrac{(-1)^j m!\,\left(\tfrac{3}{2}\right)_n}{(j+n)!\,(m+n)!}\,z^{-(n+1)/2}(1+z)^{j+m+n/2}$

$$\times \sum_{k=0}^{m}(-1)^k\binom{m}{k}\sum_{p=0}^{j+k}\binom{j+k}{p}(n+p)!\,(-j-n)_{j+k-p}\!\left(\frac{z}{1+z}\right)^{p/2}$$

$$\times \left[\ln(\sqrt{z}+\sqrt{1+z})\,P_{n+p}\!\left(\frac{2z+1}{2\sqrt{z^2+z}}\right) - \frac{1}{2}\sum_{r=1}^{n+p}\frac{1}{r}P_{n+p-r}\!\left(\frac{2z+1}{2\sqrt{z^2+z}}\right)\right.$$

$$\left. \times P_{r-1}\!\left(\frac{2z+1}{2\sqrt{z^2+z}}\right)\right].$$

32. $\;{}_2F_1\!\left(\begin{array}{c}j+\tfrac{1}{2},\,m+\tfrac{1}{2}\\ \tfrac{3}{2}-n;\,z\end{array}\right) = \dfrac{(-1)^{j+1}m!}{2\left(\tfrac{1}{2}\right)_j\left(\tfrac{1}{2}\right)_m\left(-\tfrac{1}{2}\right)_n}(-z)^{-1/2}\!\left(\frac{z}{z-1}\right)^{n/2}$

$$\times \sum_{k=0}^{m}\frac{(-1)^k}{k!}\binom{m}{k}\sum_{p=0}^{j+k}i^{n+p-1}\binom{j+k}{p}(n+p-1)!\,(1-j-n)_{j+k-p}$$

$$\times \left(\frac{z}{z-1}\right)^{p/2}P_{n+p-1}\!\left(\frac{1-2z}{2\sqrt{z^2-z}}\right).$$

33. $\;{}_2F_1\!\left(\begin{array}{c}j+\tfrac{1}{2},\,m+\tfrac{1}{2}\\ \tfrac{3}{2}-n;\,-z\end{array}\right) = \dfrac{(-1)^{j+1}m!\,z^{(n-1)/2}(1+z)^{-n/2}}{2\left(\tfrac{1}{2}\right)_j\left(\tfrac{1}{2}\right)_m\left(-\tfrac{1}{2}\right)_n}\sum_{k=0}^{m}\frac{(-1)^k}{k!}\binom{m}{k}$

$$\times \sum_{p=0}^{j+k}\binom{j+k}{p}(n+p-1)!\,(1-j-n)_{j+k-p}\!\left(\frac{z}{1+z}\right)^{p/2}P_{n+p-1}\!\left(\frac{1+2z}{2\sqrt{z^2+z}}\right).$$

34. $\;{}_2F_1\!\left(\begin{array}{c}j+\tfrac{1}{2},\,\tfrac{1}{2}-m\\ \tfrac{3}{2}-n;\,z\end{array}\right) = \dfrac{(-1)^{n+1}z^{-1/2}(1-z)^{-j}}{m!\left(\tfrac{1}{2}\right)_j\left(-\tfrac{1}{2}\right)_n}$

$$\times \sum_{k=n-1}^{n}\binom{n}{k}(j+k)!\,(1-z)^{-k}\sum_{p=0}^{j+k}\frac{(m+p)!}{p!}P_{j+k-p}^{(p-k-1/2,\,m-j-k+p+1/2)}(1-2z)$$

$$\times \left[\arcsin\sqrt{z}\,P_{m+p}^{(-p-1/2,\,-m-p-1/2)}(1-2z) + \frac{1}{2}\sum_{r=1}^{m+p}\frac{i^{r+1}}{r}(z-z^2)^{r/2}\right.$$

$$\left. \times P_{m+p-r}^{(r-p-1/2,\,r-m-p-1/2)}(1-2z)\,P_{r-1}\!\left(\frac{2z-1}{2i\sqrt{z(1-z)}}\right)\right].$$

35. $_2F_1\left(\begin{array}{c}j+\frac{1}{2},\frac{1}{2}-m\\ \frac{3}{2}-n;\;-z\end{array}\right)=\dfrac{(-1)^{n+1}z^{-1/2}(1-z)^{-j}}{m!\left(\frac{1}{2}\right)_j\left(-\frac{1}{2}\right)_n}$

$\times\displaystyle\sum_{k=n-1}^{n}\binom{n}{k}(j+k)!\sum_{p=0}^{j+k}\dfrac{(m+p)!}{p!}(1+z)^{-k}P_{j+k-p}^{(p-k-1/2,\,m-j-k+p+1/2)}(1+2z)$

$\times\left[\ln\left(\sqrt{z}+\sqrt{1+z}\right)P_{m+p}^{(-p-1/2,\,-m-p-1/2)}(1+2z)\right.$

$+\dfrac{1}{2}\displaystyle\sum_{r=1}^{m+p}\dfrac{(-1)^r}{r}(z+z^2)^{r/2}P_{m+p-r}^{(r-p-1/2,\,r-m-p-1/2)}(1+2z)$

$\left.\times P_{r-1}\left(\dfrac{1+2z}{2\sqrt{z(1+z)}}\right)\right].$

36. $_2F_1\left(\begin{array}{c}\frac{1}{2}-j,\frac{1}{2}-m\\ \frac{3}{2}-n;\;z\end{array}\right)=\dfrac{(-1)^{m+n}n!\,z^{-1/2}(1-z)^{m-n}}{j!\,m!\left(-\frac{1}{2}\right)_n}$

$\times\displaystyle\sum_{k=0}^{n}\dfrac{(1-z)^k}{k!}P_{n-k}^{(k-n+1/2,\,j+k+m-n+1/2)}(1-2z)$

$\times\displaystyle\sum_{p=0}^{k+m}(-1)^{k-p}\binom{k+m}{p}(j+p)!(-m)_{k+m-p}(1-z)^{-p}$

$\times\left[\arcsin\sqrt{z}\,P_{j+p}^{(-p-1/2,\,-j-p-1/2)}(1-2z)+\dfrac{1}{2}\displaystyle\sum_{r=1}^{j+p}\dfrac{i^{r-1}}{r}(z-z^2)^{r/2}\right.$

$\left.\times P_{j+p-r}^{(r-p-1/2,\,r-j-p-1/2)}(1-2z)P_{r-1}\left(\dfrac{2z-1}{2i\sqrt{z(1-z)}}\right)\right].$

37. $_2F_1\left(\begin{array}{c}\frac{1}{2}-j,\frac{1}{2}-m\\ \frac{3}{2}-n;\;-z\end{array}\right)=\dfrac{(-1)^{m+n}n!\,z^{-1/2}(1+z)^{m-n}}{j!\,m!\left(-\frac{1}{2}\right)_n}$

$\times\displaystyle\sum_{k=0}^{n}\dfrac{(1+z)^k}{k!}P_{n-k}^{(k-n+1/2,\,j+k+m-n+1/2)}(1+2z)$

$\times\displaystyle\sum_{p=0}^{k+m}(-1)^{k-p}\binom{k+m}{p}(j+p)!(-m)_{k+m-p}(1+z)^{-p}$

$\times\left[\ln\left(\sqrt{z}+\sqrt{1+z}\right)P_{j+p}^{(-p-1/2,\,-j-p-1/2)}(1+2z)\right.$

$$+\frac{1}{2}\sum_{r=1}^{j+p}\frac{(-1)^{r-1}}{r}(z+z^2)^{r/2}P_{j+p-r}^{(r-p-1/2,\,r-j-p-1/2)}(1+2z)$$

$$\times P_{r-1}\left(\frac{2z+1}{2\sqrt{z(1+z)}}\right)\Bigg].$$

38. $\displaystyle {}_2F_1\left(\genfrac{}{}{0pt}{}{a,\,b}{\frac{1}{2};\,z}\right) = \frac{1}{2\pi^{1/2}}\frac{\Gamma\left(a+\frac{1}{2}\right)\Gamma\left(b+\frac{1}{2}\right)}{\Gamma\left(a+b+\frac{1}{2}\right)}$

$$\times\left[{}_2F_1\left(\genfrac{}{}{0pt}{}{2a,\,2b}{a+b+\frac{1}{2};\,\frac{1+\sqrt{z}}{2}}\right) + {}_2F_1\left(\genfrac{}{}{0pt}{}{2a,\,2b}{a+b+\frac{1}{2};\,\frac{1-\sqrt{z}}{2}}\right)\right].$$

39. $\displaystyle {}_2F_1\left(\genfrac{}{}{0pt}{}{a,\,b}{\frac{3}{2};\,z}\right) = \frac{1}{4(\pi z)^{1/2}}\frac{\Gamma\left(a-\frac{1}{2}\right)\Gamma\left(b-\frac{1}{2}\right)}{\Gamma\left(a+b-\frac{1}{2}\right)}$

$$\times\left[{}_2F_1\left(\genfrac{}{}{0pt}{}{2a-1,\,2b-1}{a+b-\frac{1}{2};\,\frac{1+\sqrt{z}}{2}}\right) - {}_2F_1\left(\genfrac{}{}{0pt}{}{2a-1,\,2b-1}{a+b-\frac{1}{2};\,\frac{1-\sqrt{z}}{2}}\right)\right].$$

40. $\displaystyle {}_2F_1\left(\genfrac{}{}{0pt}{}{a,\,1-a}{b;\,z}\right) = (1-z)^{b-1}\,{}_2F_1\left(\genfrac{}{}{0pt}{}{\frac{b-a}{2},\,\frac{b+a-1}{2}}{c;\,4z(1-z)}\right)$ [[64], (3.31)].

41. $\displaystyle {}_2F_1\left(\genfrac{}{}{0pt}{}{a,\,b}{2b;\,-\frac{4z}{(1-z)^2}}\right) = (1-z)^{2a}\,{}_2F_1\left(\genfrac{}{}{0pt}{}{a,\,a-b+\frac{1}{2}}{b+\frac{1}{2};\,z^2}\right)$ [[18], (4.10)].

42. $\displaystyle {}_2F_1\left(\genfrac{}{}{0pt}{}{a,\,b;\,\frac{z^2}{4(z-1)}}{a+b+\frac{1}{2}}\right) = (1-z)^a\,{}_2F_1\left(\genfrac{}{}{0pt}{}{2a,\,a+b;\,z}{2a+2b}\right).$

43. $\displaystyle {}_2F_1\left(\genfrac{}{}{0pt}{}{a,\,b;\,4z(1-z)}{a+b+\frac{1}{2}}\right) = (1-z)^{1/2-a-b}\,{}_2F_1\left(\genfrac{}{}{0pt}{}{a-b+\frac{1}{2},\,b-a+\frac{1}{2}}{a+b+\frac{1}{2};\,z}\right)$

[[64], (3.31)].

44. $\displaystyle {}_2F_1\left(\genfrac{}{}{0pt}{}{-n,\,a}{2a;\,z}\right) = (-1)^n\frac{n!}{(2a)_n}(1-z)^{n/2}C_n^a\left(\frac{z-2}{2\sqrt{1-z}}\right).$

45. $\displaystyle \lim_{a\to 0}{}_2F_1\left(\genfrac{}{}{0pt}{}{-n,\,a}{2a;\,z}\right) = (-1)^n(1-z)^{n/2}T_n\left(\frac{z-2}{2\sqrt{1-z}}\right).$

46. $\displaystyle {}_2F_1\left(\genfrac{}{}{0pt}{}{-n,\,1}{2;\,z}\right) = \frac{2}{(n+1)z}\left[1+(-1)^n(1-z)^{(n+1)/2}T_{n+1}\left(\frac{z-2}{2\sqrt{1-z}}\right)\right].$

47. $\displaystyle = \frac{(-1)^n}{n+1}(1-z)^{n/2}U_n\left(\frac{z-2}{2\sqrt{1-z}}\right).$

48. $\displaystyle {}_2F_1\left(\genfrac{}{}{0pt}{}{-n,\,b}{\frac{1}{2};\,z}\right) = \frac{n!}{\left(b+\frac{1}{2}\right)_n}(1-z)^n C_{2n}^{1/2-b-n}\left(\sqrt{\frac{z}{z-1}}\right).$

49. $\displaystyle {}_2F_1\left(\genfrac{}{}{0pt}{}{-n,\,b}{\frac{3}{2};\,z}\right)$
$\displaystyle = (-1)^n \frac{n!}{2\left(b-\frac{1}{2}\right)_{n+1}} z^{-1/2}(z-1)^{n+1/2} C_{2n+1}^{1/2-b-n}\left(\sqrt{\frac{z}{z-1}}\right).$

50. $\displaystyle {}_2F_1\left(\genfrac{}{}{0pt}{}{-n,\,n}{1;\,z}\right) = \frac{1}{2}\left[P_n(1-2z)+P_{n-1}(1-2z)\right]$ $[n\geq 1].$

51. $\displaystyle {}_2F_1\left(\genfrac{}{}{0pt}{}{-n,\,n}{\frac{3}{2};\,z}\right) = \frac{(-1)^n}{1-4n^2}\left[T_{2n}(\sqrt{z})+\frac{2n(1-z)}{\sqrt{z}}U_{2n-1}(\sqrt{z})\right]$ $[n\geq 1].$

52. $\displaystyle {}_2F_1\left(\genfrac{}{}{0pt}{}{-n,\,\frac{1}{2}-n}{\frac{1}{2};\,z}\right) = (z-1)^n T_{2n}\left(\sqrt{\frac{z}{z-1}}\right).$

53. $\displaystyle {}_2F_1\left(\genfrac{}{}{0pt}{}{-n,\,\frac{1}{2}-n}{1;\,z}\right) = (1-z)^n P_{2n}\left(\frac{1}{\sqrt{1-z}}\right).$

54. $\displaystyle {}_2F_1\left(\genfrac{}{}{0pt}{}{-n,\,\frac{1}{2}-n}{\frac{3}{2};\,z}\right) = \frac{1}{2n+1} z^{-1/2}(z-1)^{n+1/2} T_{2n+1}\left(\sqrt{\frac{z}{z-1}}\right).$

55. $\displaystyle {}_2F_1\left(\genfrac{}{}{0pt}{}{-n,\,-n-\frac{1}{2}}{\frac{1}{2};\,z}\right) = (z-1)^n U_{2n}\left(\sqrt{\frac{z}{z-1}}\right).$

56. $\displaystyle {}_2F_1\left(\genfrac{}{}{0pt}{}{-n,\,-n-\frac{1}{2}}{\frac{3}{2};\,z}\right) = \frac{1}{2(n+1)} z^{-1/2}(z-1)^{n+1/2} U_{2n+1}\left(\sqrt{\frac{z}{z-1}}\right).$

57. $\displaystyle {}_2F_1\left(\genfrac{}{}{0pt}{}{-n,\,\frac{1}{2}-n}{\frac{1}{2}-2n;\,z}\right) = \frac{(2n)!}{\left(\frac{1}{2}\right)_{2n}} \left(\frac{z}{4}\right)^n P_{2n}\left(\frac{1}{\sqrt{z}}\right).$

58. $\displaystyle {}_2F_1\left(\genfrac{}{}{0pt}{}{-n,\,-n-\frac{1}{2}}{-\frac{1}{2}-2n;\,z}\right) = 2^{-2n}\frac{(2n+1)!}{\left(\frac{3}{2}\right)_{2n}} z^{n+1/2} P_{2n+1}\left(\frac{1}{\sqrt{z}}\right).$

59. $\displaystyle {}_2F_1\left(\genfrac{}{}{0pt}{}{-n,\,a;\,z}{\frac{a-n+1}{2}}\right) = \frac{(-1)^n n!}{\left(\frac{1-a-n}{2}\right)_n}(z^2-z)^{n/2} C_n^{(1-a-n)/2}\left(\frac{2z-1}{2\sqrt{z^2-z}}\right).$

8.1.1]   8.1. The Hypergeometric Functions

60. $\;{}_2F_1\left(\begin{array}{c}-n,\,-n\\ \frac{1}{2}-n;\,z\end{array}\right)=\dfrac{(-1)^n n!}{\left(\frac{1}{2}\right)_n}(z^2-z)^{n/2}P_n\left(\dfrac{2z-1}{2\sqrt{z^2-z}}\right).$

61. $\;{}_2F_1\left(\begin{array}{c}-n,\,-n-\frac{1}{2}\\ -2n;\,z\end{array}\right)=\left(-\dfrac{z}{4}\right)^n U_{2n}\left(\sqrt{1-\dfrac{1}{z}}\right).$

62. $\;{}_2F_1\left(\begin{array}{c}-\frac{1}{4},\,\frac{1}{4}\\ 1;\,z\end{array}\right)=\dfrac{2}{\pi}(\sqrt{z}+1)^{1/2}\,\mathrm{E}\left(\sqrt{\dfrac{2z^{1/2}}{z^{1/2}+1}}\right).$

63. $\;{}_2F_1\left(\begin{array}{c}\frac{1}{8},\,\frac{1}{8}\\ \frac{3}{4};\,-z\end{array}\right)=\dfrac{\Gamma^2\left(\frac{3}{4}\right)}{\pi^{3/2}}(\sqrt{z}+\sqrt{z+1})^{-1/4}$
$\times\left[\mathrm{K}\left(\sqrt{\dfrac{1}{2}+\dfrac{2^{-1/2}z^{1/4}}{(\sqrt{z}+\sqrt{z+1})^{1/2}}}\right)+\mathrm{K}\left(\sqrt{\dfrac{1}{2}-\dfrac{2^{-1/2}z^{1/4}}{(\sqrt{z}+\sqrt{z+1})^{1/2}}}\right)\right].$

64. $\;{}_2F_1\left(\begin{array}{c}\frac{1}{8},\,\frac{3}{8}\\ \frac{1}{2};\,z\end{array}\right)=\dfrac{2^{1/4}}{\pi^{3/2}}\Gamma\left(\dfrac{5}{8}\right)\Gamma\left(\dfrac{7}{8}\right)$
$\times\Bigg\{\left[2^{1/2}+(1+\sqrt{z})^{1/2}\right]^{-1/2}\mathrm{K}\left(\dfrac{2^{1/2}(1+\sqrt{z})^{1/4}}{\sqrt{2^{1/2}+(1+\sqrt{z})^{1/2}}}\right)$
$+\left[2^{1/2}+(1-\sqrt{z})^{1/2}\right]^{-1/2}\mathrm{K}\left(\dfrac{2^{1/2}(1-\sqrt{z})^{1/4}}{\sqrt{2^{1/2}+(1-\sqrt{z})^{1/2}}}\right)\Bigg\}.$

65. $\;{}_2F_1\left(\begin{array}{c}\frac{1}{8},\,\frac{3}{8}\\ 1;\,z\end{array}\right)$
$=\dfrac{2^{5/4}}{\pi}\left(\sqrt{2}+\sqrt{1-\sqrt{1-z}}\right)^{-1/2}\mathrm{K}\left(\dfrac{2^{1/2}(1-\sqrt{1-z})^{1/4}}{\left(\sqrt{2}+\sqrt{1-\sqrt{1-z}}\right)^{1/2}}\right).$

66. $\;{}_2F_1\left(\begin{array}{c}\frac{1}{8},\,\frac{3}{8}\\ 1;\,-z\end{array}\right)$
$=\dfrac{2^{5/4}}{\pi}(z+2+2\sqrt{z+1})^{-1/8}\mathrm{K}\left(\sqrt{\dfrac{1}{2}-2^{-1/2}\left(\dfrac{1+\sqrt{z+1}}{z+2+2\sqrt{z+1}}\right)^{1/2}}\right).$

67. $\;{}_2F_1\left(\begin{array}{c}\frac{1}{8},\,\frac{5}{8}\\ \frac{3}{4};\,z\end{array}\right)=\dfrac{\Gamma^2\left(\frac{3}{4}\right)}{\pi^{3/2}}(\sqrt{z}+1)^{-1/4}$
$\times\left[\mathrm{K}\left(\sqrt{\dfrac{1}{2}+\dfrac{2^{-1/2}z^{1/4}}{(\sqrt{z}+1)^{1/2}}}\right)+\mathrm{K}\left(\sqrt{\dfrac{1}{2}-\dfrac{2^{-1/2}z^{1/4}}{(\sqrt{z}+1)^{1/2}}}\right)\right].$

68. $= \dfrac{2^{1/4}}{\pi^{3/2} z^{1/8}} \Gamma^2\left(\dfrac{3}{4}\right) u^{1/2} \left[\mathbf{K}\left(\sqrt{\dfrac{1}{2}+u}\right) + \mathbf{K}\left(\sqrt{\dfrac{1}{2}-u}\right) \right]$

$\left[u = z^{1/4} \left[(1+\sqrt{1-z})^{1/2} + (1-\sqrt{1-z})^{1/2} \right]^{-1} \right].$

69. $\;{}_2F_1\!\left(\begin{array}{c}\frac{1}{8},\frac{5}{8}\\ 1;\,z\end{array}\right) = \dfrac{2^{5/4}}{\pi}(1+\sqrt{1-z})^{-1/4}\,\mathbf{K}\!\left(\sqrt{\dfrac{1}{2}-\dfrac{2^{-1/2}(1-z)^{1/4}}{(1+\sqrt{1-z})^{1/2}}}\right).$

70. $\;{}_2F_1\!\left(\begin{array}{c}\frac{1}{4},\frac{1}{4}\\ \frac{1}{2};\,z\end{array}\right) = \dfrac{\Gamma^2\!\left(\frac{3}{4}\right)}{\pi^{3/2}}\left[\mathbf{K}\!\left(\sqrt{\dfrac{1+\sqrt{z}}{2}}\right) + \mathbf{K}\!\left(\sqrt{\dfrac{1-\sqrt{z}}{2}}\right)\right].$

71. $\;{}_2F_1\!\left(\begin{array}{c}\frac{1}{4},\frac{1}{4}\\ \frac{1}{2};\,-z\end{array}\right) = \dfrac{\Gamma^2\!\left(\frac{3}{4}\right)}{\pi^{3/2}}(z+1)^{-1/4}$

$\times \left[\mathbf{K}\!\left(\sqrt{\dfrac{1}{2}+\dfrac{1}{2}\sqrt{\dfrac{z}{z+1}}}\right) + \mathbf{K}\!\left(\sqrt{\dfrac{1}{2}-\dfrac{1}{2}\sqrt{\dfrac{z}{z+1}}}\right)\right].$

72. $\;{}_2F_1\!\left(\begin{array}{c}\frac{1}{4},\frac{1}{4}\\ \frac{3}{4};\,z\end{array}\right) = \dfrac{\Gamma^2\!\left(\frac{3}{4}\right)}{\pi^{3/2}}\left(2\sqrt{z^2-z}-2z+1\right)^{-1/4}$

$\times \left[\mathbf{K}\!\left(\sqrt{\dfrac{1}{2}+\dfrac{(z^2-z)^{1/4}}{(2\sqrt{z^2-z}-2z+1)^{1/2}}}\right) + \mathbf{K}\!\left(\sqrt{\dfrac{1}{2}-\dfrac{(z^2-z)^{1/4}}{(2\sqrt{z^2-z}-2z+1)^{1/2}}}\right)\right]$

$[\operatorname{Re} z < 1/2].$

73. $\;{}_2F_1\!\left(\begin{array}{c}\frac{1}{4},\frac{1}{4}\\ \frac{3}{4};\,-z\end{array}\right) = \dfrac{\Gamma^2\!\left(\frac{3}{4}\right)}{\pi^{3/2}}\left(\sqrt{z}+\sqrt{z+1}\right)^{-1/2}$

$\times \left[\mathbf{K}\!\left(\sqrt{\dfrac{1}{2}+\dfrac{(z^2+z)^{1/4}}{\sqrt{z}+\sqrt{z+1}}}\right) + \mathbf{K}\!\left(\sqrt{\dfrac{1}{2}-\dfrac{(z^2+z)^{1/4}}{\sqrt{z}+\sqrt{z+1}}}\right)\right] \quad [\operatorname{Re} z > -1/2].$

74. $\;{}_2F_1\!\left(\begin{array}{c}\frac{1}{4},\frac{1}{4}\\ 1;\,z\end{array}\right) = \dfrac{2}{\pi}\,\mathbf{K}\!\left(\sqrt{\dfrac{1}{2}-\dfrac{1}{2}(1-z)^{1/2}}\right).$

75. $\;{}_2F_1\!\left(\begin{array}{c}\frac{1}{4},\frac{1}{4}\\ 1;\,-z\end{array}\right) = \dfrac{2^{3/2}}{\pi}(1+\sqrt{z+1})^{-1/2}\,\mathbf{K}\!\left(\dfrac{\sqrt{z}}{1+\sqrt{z+1}}\right).$

76. $\;{}_2F_1\!\left(\begin{array}{c}\frac{1}{4},\frac{1}{2}\\ \frac{3}{4};\,z\end{array}\right) = \dfrac{\pi^{-3/2}\Gamma^2\!\left(\frac{3}{4}\right)}{(\sqrt{z}+1)^{1/2}}$

$\times \left[\mathbf{K}\!\left(\sqrt{\dfrac{1}{2}+\dfrac{z^{1/4}}{z^{1/2}+1}}\right) + \mathbf{K}\!\left(\sqrt{\dfrac{1}{2}-\dfrac{z^{1/4}}{z^{1/2}+1}}\right)\right] \quad [0 < z < 1].$

8.1. The Hypergeometric Functions

77. $\displaystyle {}_2F_1\left(\genfrac{}{}{0pt}{}{\frac{1}{4},\frac{1}{2}}{1;\,z}\right) = \frac{2^{3/2}}{\pi\left(\sqrt{1-z}+1\right)^{1/2}}\,\mathbf{K}\left(\sqrt{\frac{1}{2} - \frac{(1-z)^{1/4}}{z} + \frac{(1-z)^{3/4}}{z}}\right).$

78. $\displaystyle {}_2F_1\left(\genfrac{}{}{0pt}{}{\frac{1}{4},\frac{1}{2}}{1;\,-z}\right) = \frac{8}{\pi u_+}\,\mathbf{K}\left(\frac{u_-}{u_+}\right) \quad \left[u_\pm = 1 + \sqrt[4]{z+1} \pm \sqrt{2}\left(1+\sqrt{z+1}\right)^{1/2}\right].$

79. $\displaystyle {}_2F_1\left(\genfrac{}{}{0pt}{}{\frac{1}{4},\frac{3}{4}}{1;\,z}\right) = \frac{2}{\pi}(1+\sqrt{z})^{-1/2}\,\mathbf{K}\left(\frac{2^{1/2} z^{1/4}}{(\sqrt{z}+1)^{1/2}}\right).$

80. $\displaystyle \phantom{{}_2F_1\left(\genfrac{}{}{0pt}{}{\frac{1}{4},\frac{3}{4}}{1;\,z}\right)} = \frac{2}{\pi}(1-z)^{-1/4}\,\mathbf{K}\left(\sqrt{\frac{1-(1-z)^{-1/2}}{2}}\right).$

81. $\displaystyle {}_2F_1\left(\genfrac{}{}{0pt}{}{\frac{1}{4},\frac{3}{4}}{1;\,-z}\right) = \frac{2}{\pi}(z+1)^{-1/4}\,\mathbf{K}\left(\sqrt{\frac{\sqrt{z+1}-1}{2\sqrt{z+1}}}\right).$

82. $\displaystyle {}_2F_1\left(\genfrac{}{}{0pt}{}{\frac{3}{8},\frac{3}{8}}{\frac{5}{4};\,-z}\right) = \frac{\Gamma^2\left(\frac{1}{4}\right)}{(2\pi)^{3/2} z^{1/4}}\left(\sqrt{z}+\sqrt{z+1}\right)^{-1/4}$
$\displaystyle \times \left[\mathbf{K}\left(\sqrt{\frac{1}{2}+\frac{2^{-1/2} z^{1/4}}{(\sqrt{z}+\sqrt{z+1})^{1/2}}}\right) - \mathbf{K}\left(\sqrt{\frac{1}{2}-\frac{2^{-1/2} z^{1/4}}{(\sqrt{z}+\sqrt{z+1})^{1/2}}}\right)\right].$

83. $\displaystyle {}_2F_1\left(\genfrac{}{}{0pt}{}{\frac{3}{8},\frac{7}{8}}{1;\,z}\right) = \frac{2^{5/4}}{\pi}(1-z)^{-1/4}\left(1+\sqrt{1-z}\right)^{-1/4}$
$\displaystyle \times \mathbf{K}\left(\sqrt{\frac{1}{2} - \frac{2^{-1/2}(1-z)^{1/2}}{\left(1-z+\sqrt{1-z}\right)^{1/2}}}\right).$

84. $\displaystyle {}_2F_1\left(\genfrac{}{}{0pt}{}{\frac{3}{8},\frac{7}{8}}{\frac{5}{4};\,z}\right) = \frac{\Gamma^2\left(\frac{1}{4}\right)}{(2\pi)^{3/2} z^{1/4}}\left(\sqrt{z}+1\right)^{-1/4}$
$\displaystyle \times \left[\mathbf{K}\left(\sqrt{\frac{1}{2}+\frac{2^{-1/2} z^{1/4}}{(\sqrt{z}+1)^{1/2}}}\right) - \mathbf{K}\left(\sqrt{\frac{1}{2}-\frac{2^{-1/2} z^{1/4}}{(\sqrt{z}+1)^{1/2}}}\right)\right].$

85. $\displaystyle {}_2F_1\left(\genfrac{}{}{0pt}{}{\frac{1}{2},\frac{1}{2}}{\frac{3}{4};\,-z}\right) = \frac{\Gamma^2\left(\frac{3}{4}\right)}{\pi^{3/2}}(z+1)^{-1/4}\left(\sqrt{z}+\sqrt{z+1}\right)^{-1/2}$
$\displaystyle \times \left[\mathbf{K}\left(\sqrt{\frac{1}{2}+\frac{(z^2+z)^{1/4}}{\sqrt{z}+\sqrt{z+1}}}\right) + \mathbf{K}\left(\sqrt{\frac{1}{2}-\frac{(z^2+z)^{1/4}}{\sqrt{z}+\sqrt{z+1}}}\right)\right]$
$[\operatorname{Re} z > -1/2].$

86. $\displaystyle {}_2F_1\left(\genfrac{}{}{0pt}{}{\frac{1}{2},\frac{1}{2}}{1;\,z}\right) = \frac{2}{\pi}\,\mathbf{K}\left(\sqrt{z}\right).$

87. $\,_2F_1\!\left(\!\begin{array}{c}\frac{1}{2},\frac{1}{2}\\ 1;\,-z\end{array}\!\right) = \dfrac{4}{\pi\left(\sqrt{z+1}+1\right)}\,\mathbf{K}\!\left(\dfrac{\sqrt{z+1}-1}{\sqrt{z+1}+1}\right).$

88. $\,_2F_1\!\left(\!\begin{array}{c}\frac{1}{2},\frac{3}{4}\\ 1;\,z\end{array}\!\right) = \dfrac{2^{3/2}}{\pi}(1-z)^{-1/4}\left(1+\sqrt{1-z}\right)^{-1/2}$
$$\times \mathbf{K}\!\left(\sqrt{\dfrac{1}{2}-\dfrac{(1-z)^{1/4}}{1+\sqrt{1-z}}}\right).$$

89. $\,_2F_1\!\left(\!\begin{array}{c}\frac{1}{2},\frac{3}{4}\\ 1;\,-z\end{array}\!\right) = \dfrac{8}{\pi u_+\sqrt{z+1}}\,\mathbf{K}\!\left(\dfrac{u_-}{u_+}\right)$
$$\left[u_\pm = 1+\sqrt[4]{z+1}\pm\sqrt{2}\left(1+\sqrt{z+1}\right)^{1/2}\right].$$

90. $\,_2F_1\!\left(\!\begin{array}{c}\frac{5}{8},\frac{7}{8}\\ 1;\,z\end{array}\!\right) = \dfrac{2^{5/4}}{\pi}(1-z)^{-1/2}\left[2^{1/2}+\left(1-\sqrt{1-z}\right)^{1/2}\right]^{-1/2}$
$$\times \mathbf{K}\!\left(\dfrac{2^{1/2}(1-\sqrt{1-z})^{1/4}}{\sqrt{2^{1/2}+\left(1-\sqrt{1-z}\right)^{1/2}}}\right).$$

91. $\,_2F_1\!\left(\!\begin{array}{c}\frac{3}{4},\frac{3}{4}\\ \frac{1}{2};\,-z\end{array}\!\right) = \dfrac{\Gamma^2\!\left(\frac{1}{4}\right)}{4\pi^{3/2}}(z+1)^{-3/4}$
$$\times\left[2\,\mathbf{E}\!\left(\sqrt{\dfrac{1}{2}+\dfrac{1}{2}\sqrt{\dfrac{z}{z+1}}}\right)+2\,\mathbf{E}\!\left(\sqrt{\dfrac{1}{2}-\dfrac{1}{2}\sqrt{\dfrac{z}{z+1}}}\right)\right.$$
$$\left.-\mathbf{K}\!\left(\sqrt{\dfrac{1}{2}+\dfrac{1}{2}\sqrt{\dfrac{z}{z+1}}}\right)-\mathbf{K}\!\left(\sqrt{\dfrac{1}{2}-\dfrac{1}{2}\sqrt{\dfrac{z}{z+1}}}\right)\right]\quad[|\arg z|<\pi].$$

92. $\,_2F_1\!\left(\!\begin{array}{c}\frac{3}{4},\frac{3}{4}\\ 1;\,z\end{array}\!\right) = \dfrac{2}{\pi\sqrt{1-z}}\,\mathbf{K}\!\left(\sqrt{\dfrac{1}{2}-\dfrac{1}{2}\sqrt{1-z}}\right).$

93. $\,_2F_1\!\left(\!\begin{array}{c}\frac{3}{4},\frac{3}{4}\\ 1;\,-z\end{array}\!\right) = \dfrac{2^{3/2}}{\pi\sqrt{z+1}\left(\sqrt{z+1}+1\right)^{1/2}}\,\mathbf{K}\!\left(\dfrac{\sqrt{z}}{\sqrt{z+1}+1}\right).$

94. $\,_2F_1\!\left(\!\begin{array}{c}\frac{3}{4},\frac{3}{4}\\ \frac{5}{4};\,-z\end{array}\!\right) = \dfrac{\Gamma^2\!\left(\frac{1}{4}\right)}{4\pi^{3/2}(z^2+z)^{1/4}}\left(\sqrt{z}+\sqrt{z+1}\right)^{-1/2}$
$$\times\left[\mathbf{K}\!\left(\sqrt{\dfrac{1}{2}+\dfrac{(z^2+z)^{1/4}}{\sqrt{z}+\sqrt{z+1}}}\right)-\mathbf{K}\!\left(\sqrt{\dfrac{1}{2}-\dfrac{(z^2+z)^{1/4}}{\sqrt{z}+\sqrt{z+1}}}\right)\right]\quad[\operatorname{Re} z>0].$$

95. $\,_2F_1\!\left(\!\begin{array}{c}\frac{3}{4},\frac{3}{4}\\ \frac{3}{2};\,z\end{array}\!\right) = \dfrac{\Gamma^2\!\left(\frac{1}{4}\right)}{2\pi^{3/2}z^{1/2}}\left[\mathbf{K}\!\left(\sqrt{\dfrac{1+\sqrt{z}}{2}}\right)-\mathbf{K}\!\left(\sqrt{\dfrac{1-\sqrt{z}}{2}}\right)\right].$

96. $\displaystyle {}_2F_1\left(\begin{array}{c}\frac{3}{4},\frac{3}{4}\\\frac{3}{2};\,-z\end{array}\right) = \frac{\Gamma^2\left(\frac{1}{4}\right)}{2\pi^{3/2}z^{1/2}}(z+1)^{-1/4}$

$\displaystyle \times \left[\mathbf{K}\left(\sqrt{\frac{1}{2}+\frac{1}{2}\sqrt{\frac{z}{z+1}}}\right) - \mathbf{K}\left(\sqrt{\frac{1}{2}-\frac{1}{2}\sqrt{\frac{z}{z+1}}}\right)\right]$ $[|\arg z|<\pi]$.

97. $\displaystyle {}_2F_1\left(\begin{array}{c}a,\,b\\c;\,-1\end{array}\right) = 2^{-a}\,{}_2F_1\left(\begin{array}{c}a,\,c-b\\c;\,\frac{1}{2}\end{array}\right).$

98. $\displaystyle {}_2F_1\left(\begin{array}{c}a,\,a+b+n+1\\a+\frac{b}{2}+n+1;\,-1\end{array}\right) = 2^{-a}\,{}_2F_1\left(\begin{array}{c}a,\,-\frac{b}{2};\,\frac{1}{2}\\a+\frac{b}{2}+n+1\end{array}\right).$

99. $\displaystyle {}_2F_1\left(\begin{array}{c}m,\,a\\a+n;\,-1\end{array}\right) = \frac{2^{-m}}{(m-1)!}\sum_{k=0}^{m-1}\binom{m-1}{k}\frac{(n)_k(1-n)_{m-k-1}}{\mathrm{B}(k+n,\,a-k)}$

$\displaystyle \times \sum_{p=0}^{k+n-1}(-1)^p\binom{k+n-1}{p}\left[\psi\left(\frac{a-k+p+1}{2}\right)-\psi\left(\frac{a-k+p}{2}\right)\right]$ $[m\geq 1]$.

100. $\displaystyle {}_2F_1\left(\begin{array}{c}a,\,\frac{a-2-\sqrt{2-a}}{2}\\\frac{a+4-\sqrt{2-a}}{2};\,-1\end{array}\right) = 2^{-a-1}\left[2+a\left(3+\sqrt{2-a}\right)\right].$

101. $\displaystyle {}_2F_1\left(\begin{array}{c}a,\,\frac{a-3-\sqrt{7-3a}}{2}\\\frac{a+5-\sqrt{7-3a}}{2};\,-1\end{array}\right) = \frac{2^{-a-1}}{3}\left[6+a\left(15-a+4\sqrt{7-3a}\right)\right].$

102. $\displaystyle {}_2F_1\left(\begin{array}{c}\frac{1}{4},\frac{1}{4}\\1;\,-1\end{array}\right) = 2^{-7/4}\pi^{-3/2}\Gamma\left(\frac{1}{8}\right)\Gamma\left(\frac{3}{8}\right).$

103. $\displaystyle {}_2F_1\left(\begin{array}{c}-n,\,-2n-\frac{2}{3}\\\frac{4}{3};\,-8\end{array}\right) = (-27)^n\frac{\left(\frac{5}{6}\right)_n}{\left(\frac{3}{2}\right)_n}$ [[38], (3.7)].

104. $\displaystyle {}_2F_1\left(\begin{array}{c}-n,\,-2n+\frac{2}{3}\\\frac{2}{3};\,-8\end{array}\right) = \frac{(-4)^n(2n)_n}{\left(\frac{1}{3}\right)_n\left(\frac{2}{3}\right)_n}\left[\left(\frac{1}{6}\right)_n + \frac{1}{2}\left(\frac{1}{2}\right)_n\right]$

$[n\geq 1;\,[38],\,(3.8)]$.

105. $\displaystyle {}_2F_1\left(\begin{array}{c}-n,\,-2n+\frac{4}{3}\\\frac{1}{3};\,-8\end{array}\right) = \frac{(-4)^n(2n-1)_n}{\left(-\frac{1}{3}\right)_n\left(\frac{1}{3}\right)_n}\left[\left(-\frac{1}{6}\right)_n + \frac{1}{2}\left(-\frac{1}{2}\right)_n\right]$

$[n\geq 1;\,[38],\,(3.9)]$.

106. $\displaystyle {}_2F_1\left(\begin{matrix}-n,\ \frac{1-3n}{6}\\ \frac{2}{3};\ -8\end{matrix}\right) = (-1)^{(n+1)/2}3^{(3n-1)/2}\delta_{1,n-2[n/2]}$

$\hspace{6cm} + (-1)^{n/2}3^{3n/2}\delta_{0,n-2[n/2]}\quad [[38],\ (3.12)].$

107. $\displaystyle {}_2F_1\left(\begin{matrix}-n,\ \frac{2-3n}{6}\\ \frac{1}{3};\ -8\end{matrix}\right) = (-1)^{(n+1)/2}3^{(3n-1)/2}\delta_{1,n-2[n/2]}$

$\hspace{3cm} + (-1)^{n/2}3^{3n/2-1}\left[1 + 2\dfrac{\left(\frac{1}{2}\right)_{n/2}}{\left(\frac{1}{6}\right)_{n/2}}\right]\delta_{0,n-2[n/2]}\quad [n \geq 2;\ [38],\ (3.13)].$

108. $\displaystyle {}_2F_1\left(\begin{matrix}-2n,\ -n-\frac{1}{6}\\ \frac{4}{3};\ -8\end{matrix}\right) = (-1)^n\dfrac{3^{3n}}{2n+1}\hspace{3cm}[[38],\ (3.14)].$

109. $\displaystyle {}_2F_1\left(\begin{matrix}-2n-1,\ -n-\frac{5}{6}\\ \frac{5}{3};\ -8\end{matrix}\right) = (-1)^{n+1}\dfrac{3^{3n+2}}{2n+3}\hspace{2cm}[[38],\ (3.15)].$

110. $\displaystyle {}_2F_1\left(\begin{matrix}-2n-1,\ -n-\frac{7}{6}\\ \frac{7}{3};\ -8\end{matrix}\right) = (-1)^{n+1}\dfrac{5\cdot 3^{3n+2}}{(2n+3)(2n+5)}\hspace{1cm}[[38],\ (3.16)].$

111. $\displaystyle {}_2F_1\left(\begin{matrix}\frac{1}{4},\ \frac{1}{4}\\ \frac{3}{4};\ -\frac{1}{8}\end{matrix}\right) = \dfrac{3^{3/4}(1+\sqrt{3})}{2^{5/4}\pi^2}\Gamma\!\left(\frac{1}{3}\right)\Gamma^2\!\left(\frac{3}{4}\right)\Gamma\!\left(\frac{7}{6}\right).$

112. $\displaystyle {}_2F_1\left(\begin{matrix}\frac{1}{4},\ \frac{1}{4}\\ 1;\ -\frac{1}{8}\end{matrix}\right) = 2^{-5/4}\pi^{-3/2}\Gamma^2\!\left(\frac{1}{4}\right).$

113. $\displaystyle {}_2F_1\left(\begin{matrix}\frac{3}{4},\ \frac{3}{4}\\ 1;\ -\frac{1}{8}\end{matrix}\right) = \dfrac{2^{1/4}}{3\pi^{3/2}}\Gamma^2\!\left(\frac{1}{4}\right).$

114. $\displaystyle {}_2F_1\left(\begin{matrix}\frac{3}{4},\ \frac{3}{4}\\ \frac{5}{4};\ -\frac{1}{8}\end{matrix}\right) = \dfrac{\sqrt{3}-1}{2^{37/12}3^{1/4}\pi^{5/2}}\Gamma^2\!\left(\frac{1}{4}\right)\Gamma^3\!\left(\frac{1}{3}\right).$

115. $\displaystyle {}_2F_1\left(\begin{matrix}a,\ a+\frac{1}{2}\\ \frac{4a+5}{6};\ \frac{1}{9}\end{matrix}\right) = \sqrt{\pi}\left(\frac{3}{4}\right)^a\dfrac{\Gamma\!\left(\frac{4a+5}{6}\right)}{\Gamma\!\left(\frac{2a+3}{6}\right)\Gamma\!\left(\frac{2a+5}{6}\right)}\quad [[51],\ (1.1)].$

116. $\displaystyle {}_2F_1\left(\begin{matrix}a,\ \frac{1-a}{2}\\ \frac{3a+5}{6};\ \frac{1}{9}\end{matrix}\right) = \left(\frac{3}{4}\right)^{a/2}\dfrac{\Gamma\!\left(\frac{3a+5}{6}\right)\Gamma\!\left(\frac{2}{3}\right)}{\Gamma\!\left(\frac{3a+4}{6}\right)\Gamma\!\left(\frac{5}{6}\right)}\quad [[51],\ (1.2)].$

8.1.1] 8.1. The Hypergeometric Functions

117. $\;_2F_1\left(\begin{array}{c}a,\,1-\dfrac{a}{2}\\ \dfrac{3a+4}{6};\,\dfrac{1}{9}\end{array}\right)=\sqrt{\pi}\left(\dfrac{3}{4}\right)^{a/2}\dfrac{\Gamma\left(\dfrac{3a+4}{6}\right)}{\Gamma\left(\dfrac{a+1}{2}\right)\Gamma\left(\dfrac{2}{3}\right)}$ [[51], (1.3)].

118. $\;_2F_1\left(\begin{array}{c}a,\,a+\dfrac{1}{4}\\ \dfrac{5}{4}-2a;\,\dfrac{1}{9}\end{array}\right)=\dfrac{3^{5a}\,\Gamma\left(\dfrac{5}{4}-2a\right)\Gamma\left(\dfrac{2}{3}\right)\Gamma\left(\dfrac{13}{12}\right)}{2^{6a}\,\Gamma\left(\dfrac{2}{3}-a\right)\Gamma\left(\dfrac{13}{12}-a\right)\Gamma\left(\dfrac{5}{4}\right)}$ [[51], (1.4)].

119. $\;_2F_1\left(\begin{array}{c}a,\,a+\dfrac{1}{4}\\ \dfrac{9}{4}-2a;\,\dfrac{1}{9}\end{array}\right)=\dfrac{3^{5a}\,\Gamma\left(\dfrac{9}{4}-2a\right)\Gamma\left(\dfrac{4}{3}\right)\Gamma\left(\dfrac{17}{12}\right)}{2^{6a}\,\Gamma\left(\dfrac{4}{3}-a\right)\Gamma\left(\dfrac{17}{12}-a\right)\Gamma\left(\dfrac{9}{4}\right)}$ [[51], (1.5)].

120. $\;_2F_1\left(\begin{array}{c}-n,\,-n+\dfrac{1}{4}\\ 2n+\dfrac{5}{4};\,\dfrac{1}{9}\end{array}\right)=\dfrac{\left(\dfrac{5}{4}\right)_{2n}}{\left(\dfrac{2}{3}\right)_n\left(\dfrac{13}{12}\right)_n}\left(\dfrac{2^6}{3^5}\right)^n$ [[38], (6.5)].

121. $\;_2F_1\left(\begin{array}{c}-n,\,-n+\dfrac{1}{4}\\ 2n+\dfrac{9}{4};\,\dfrac{1}{9}\end{array}\right)=\dfrac{\left(\dfrac{9}{4}\right)_{2n}}{\left(\dfrac{4}{3}\right)_n\left(\dfrac{17}{12}\right)_n}\left(\dfrac{2^6}{3^5}\right)^n$ [[38], (6.6)].

122. $\;_2F_1\left(\begin{array}{c}-\dfrac{1}{4},\,\dfrac{1}{4}\\ 1;\,\dfrac{1}{9}\end{array}\right)=\dfrac{3^{-1/2}}{2\pi^{3/2}}\left[\Gamma^2\left(\dfrac{1}{4}\right)+4\Gamma^2\left(\dfrac{3}{4}\right)\right].$

123. $\;_2F_1\left(\begin{array}{c}\dfrac{1}{4},\,\dfrac{1}{2}\\ \dfrac{3}{4};\,\dfrac{1}{9}\end{array}\right)=\dfrac{3^{1/4}(1+\sqrt{3})}{8\pi^2}\,\Gamma\left(\dfrac{1}{6}\right)\Gamma\left(\dfrac{1}{3}\right)\Gamma^2\left(\dfrac{3}{4}\right)$ [[49], (7.4)].

124. $\;_2F_1\left(\begin{array}{c}\dfrac{1}{4},\,\dfrac{3}{4}\\ 1;\,\dfrac{1}{9}\end{array}\right)=\dfrac{3^{1/2}}{4\pi^{3/2}}\,\Gamma^2\left(\dfrac{1}{4}\right).$

125. $\;_2F_1\left(\begin{array}{c}a,\,\dfrac{1}{2};\,\dfrac{1}{4}\\ -2a\pm n+\dfrac{3}{2}\end{array}\right)$

$=\dfrac{2^{\pm n+3/2}}{3^{\pm n+1}}\,\Gamma\left[\begin{array}{ccc}\dfrac{1}{2}-a,&\dfrac{\pm 2n+3}{4}-a,&\dfrac{\pm 2n+5}{4}-a\\ \dfrac{\pm 2n+3}{6}-a,&\dfrac{\pm 2n+5}{6}-a,&\dfrac{\pm 2n+7}{6}-a\end{array}\right]R_1(\pm n)$

$-\,2^{3/2}(-3)^{\pm n-2}\,\Gamma\left[\begin{array}{ccc}1-a,&\dfrac{\pm 2n+3}{4}-a,&\dfrac{\pm 2n+5}{4}-a\\ \dfrac{\pm n+1}{2}-a,&\dfrac{\pm n}{2}-a+1,&\dfrac{3}{2}-a\end{array}\right]R_2(\pm n),$

$R_1(1)=R_2(0)=0,\;R_1(0)=R_2(1)=1,$

$R_1(n)=(-1)^n n\displaystyle\sum_{k=[n/3]}^{[n/2]}\left(\dfrac{3}{2}\right)^{4k}\dfrac{(k-1)!}{(n-2k)!(3k-n)!}\dfrac{\left(\dfrac{1}{2}-a\right)_k}{(1-a)_k}$ $[n>1],$

$R_2(n)={}_4F_3\left(\begin{array}{c}-\dfrac{n-1}{3},\,-\dfrac{n-2}{3},\,-\dfrac{n}{3}+1,\,1-a\\ -\dfrac{n}{2}+1,\,-\dfrac{n-3}{2},\,\dfrac{3}{2}-a;\,1\end{array}\right)$ $[n>1],$

581

$$R_1(-n) = {}_4F_3\left(\begin{array}{c}-\dfrac{n}{3},\ -\dfrac{n-1}{3},\ -\dfrac{n-2}{3},\ a\\ -\dfrac{n-1}{2},\ 1-\dfrac{n}{2},\ \dfrac{1}{2}-a;\ 1\end{array}\right) \qquad [n \geq 1],$$

$$R_2(-n) = (-1)^n(n+1)\sum_{k=[(n+1)/3]}^{[(n+1)/2]}\left(\dfrac{3}{2}\right)^{4k}\dfrac{(k-1)!}{(n-2k+1)!\,(3k-n-1)!}\dfrac{\left(a-\dfrac{1}{2}\right)_k}{(a)_k}$$
$$[n \geq 1;\ [83]].$$

126. ${}_2F_1\left(\begin{array}{c}m+\dfrac{1}{2},\ a;\ \dfrac{1}{4}\\ -2a+n+\dfrac{3}{2}\end{array}\right)$

$$= \left(\dfrac{4}{3}\right)^m \dfrac{m!}{(1/2)_m}\sum_{k=0}^{m}\dfrac{2^{-2k}}{k!}P_{m-k}^{(k-1/2,\,-3a-m+k+n+1)}\left(\dfrac{1}{2}\right)$$

$$\times\ \dfrac{(-2a+n+1)_k\left(-3a+n+\dfrac{3}{2}\right)_k}{\left(-2a+n+\dfrac{3}{2}\right)_k}\ {}_2F_1\left(\begin{array}{c}\dfrac{1}{2},\ a;\ \dfrac{1}{4}\\ -2a+k+n+\dfrac{3}{2}\end{array}\right).$$

127. ${}_2F_1\left(\begin{array}{c}\dfrac{1}{2}-m,\ a;\ \dfrac{1}{4}\\ -2a+n+\dfrac{3}{2}\end{array}\right) = \dfrac{2^{-2m}(-3)^m}{(-2a+n+1)_m}\sum_{k=0}^{m}\binom{m}{k}(-3)^{-k}$

$$\times\ \dfrac{(-2a+n+1)_k(2a-m-n)_{m-k}\left(-3a+n+\dfrac{3}{2}\right)_k}{\left(-2a+n+\dfrac{3}{2}\right)_k}\ {}_2F_1\left(\begin{array}{c}\dfrac{1}{2},\ a;\ \dfrac{1}{4}\\ -2a+k+n+\dfrac{3}{2}\end{array}\right).$$

128. ${}_2F_1\left(\begin{array}{c}\dfrac{1}{3},\ \dfrac{1}{2}\\ \dfrac{5}{6};\ \dfrac{1}{4}\end{array}\right) = \dfrac{2^{5/3}}{3}$ \hfill $[[66],\ (A7)].$

129. ${}_2F_1\left(\begin{array}{c}\dfrac{1}{2},\ \dfrac{2}{3}\\ \dfrac{7}{6};\ \dfrac{1}{4}\end{array}\right) = \dfrac{2^{-8/3}}{\sqrt{3}\,\pi^3}\Gamma^6\left(\dfrac{1}{3}\right)$ \hfill $[[66],\ (A8)].$

130. ${}_2F_1\left(\begin{array}{c}a,\ b;\ \dfrac{1}{2}\\ \dfrac{a+b+n}{2}\end{array}\right) = \dfrac{\sqrt{\pi}\,\Gamma\left(\dfrac{a+b+n}{2}\right)}{\left(\dfrac{b-a-n}{2}\right)_n}\sum_{k=0}^{n}\binom{n}{k}(-2)^{-k}(a)_k$

$$\times\left[\dfrac{1}{\Gamma\left(\dfrac{a+k+1}{2}\right)\Gamma\left(\dfrac{b+k-n}{2}\right)} + \dfrac{1}{\Gamma\left(\dfrac{a+k}{2}\right)\Gamma\left(\dfrac{b+k-n}{2}+1\right)}\right].$$

131. $_2F_1\left(\begin{matrix}a,\, n-a\\b;\,\frac{1}{2}\end{matrix}\right) = 2^{n-b}\sqrt{\pi}\,\frac{\Gamma(b)}{(-a)_n}\sum_{k=0}^n \binom{n}{k}(-2)^{-k}(a+b-n)_k$

$$\times\left[\frac{1}{\Gamma\left(\frac{a+b+k-n}{2}\right)\Gamma\left(\frac{b-a+k-n+1}{2}\right)}\right.$$

$$\left.+\frac{1}{\Gamma\left(\frac{a+b+k-n+1}{2}\right)\Gamma\left(\frac{b-a+k-n}{2}\right)}\right].$$

132. $_2F_1\left(\begin{matrix}a,\,-n-a\\b;\,\frac{1}{2}\end{matrix}\right) = 2^{-b}\sqrt{\pi}\,n!\,\frac{\Gamma(b)}{(a+b)_n}\sum_{k=0}^n \frac{1}{k!}P_{n-k}^{(a+k,\,k-b-n)}(0)$

$$\times\left[\frac{1}{\Gamma\left(\frac{a+b-k}{2}\right)\Gamma\left(\frac{b-a-k+1}{2}\right)}+\frac{1}{\Gamma\left(\frac{a+b-k+1}{2}\right)\Gamma\left(\frac{b-a-k}{2}\right)}\right].$$

133. $_2F_1\left(\begin{matrix}a,\,a-2b+m\\a-b+n;\,\frac{1}{2}\end{matrix}\right) = \frac{\sqrt{\pi}\,\Gamma(a-b+n)}{(-b)_n(b-m)_n(a-2b)_m}$

$$\times\sum_{k=0}^n 2^{-k}\binom{n}{k}(a)_k(a-2b)_k(b-a-m)_{n-k}$$

$$\times\sum_{p=0}^m 2^{-p}\binom{m}{p}(a+k)_p(a-2b+k)_m$$

$$\times\left[\frac{1}{\Gamma\left(\frac{k+p+a}{2}-b\right)\Gamma\left(\frac{k+p+a+1}{2}\right)}+\frac{1}{\Gamma\left(\frac{k+p+a+1}{2}-b\right)\Gamma\left(\frac{k+p+a}{2}\right)}\right].$$

134. $_2F_1\left(\begin{matrix}a,\,a-2b-n\\a-b;\,\frac{1}{2}\end{matrix}\right)$

$$= n!\sqrt{\pi}\,\frac{\Gamma(a-b)}{(b)_n}\sum_{k=0}^n \frac{2^{-2k}}{k!}(a)_k(a-2b)_k P_{n-k}^{(b+k-1,\,a-b+k-n)}(0)$$

$$\times\left[\frac{1}{\Gamma\left(\frac{k+a+1}{2}\right)\Gamma\left(\frac{k+a}{2}-b\right)}+\frac{1}{\Gamma\left(\frac{k+a}{2}\right)\Gamma\left(\frac{k+a+1}{2}-b\right)}\right].$$

135. $_2F_1\left(\begin{matrix}m+1,\,n+1\\\frac{s}{t}+r;\,\frac{1}{2}\end{matrix}\right) = (-1)^{r-1}2^{m+n+1}\left(\frac{s}{t}+r-1\right)$

$$\times\sum_{j=0}^m \frac{(-2)^{-j}}{j!}\left(\frac{s}{t}+r-n-1\right)_j P_{m-j}^{(j,\,s/t+j+r-m-n-2)}(0)$$

$$\times\sum_{k=0}^n \frac{(-2)^{-k}}{k!}\left(\frac{s}{t}+j+r-1\right)_k P_{n-k}^{(k,\,s/t+j+k+r-n-2)}(0)$$

$$\times \left\{ \frac{\pi}{2} \csc \frac{s\pi}{t} - 2 \sum_{q=0}^{[t/2]-1} \cos\left[(2q+1)\frac{s\pi}{t}\right] \ln \sin\left[(2q+1)\frac{s\pi}{2t}\right] \right.$$

$$\left. - \sum_{p=0}^{j+k+r-2} \frac{(-1)^p}{p+s/t} \right\} \qquad [s, t = 1, 2, \ldots;\ m = 1, 2, \ldots, n-1;\ n = 2, 3, \ldots].$$

136. $\ {}_2F_1\left(\begin{matrix} -\frac{1}{4}, \frac{1}{4} \\ 1;\ \frac{1}{2} \end{matrix}\right) = \frac{2^{-5/2}}{\pi^{3/2}}\left[\Gamma\left(\frac{1}{8}\right)\Gamma\left(\frac{3}{8}\right) + 8\Gamma\left(\frac{5}{8}\right)\Gamma\left(\frac{7}{8}\right)\right].$

137. $\ {}_2F_1\left(\begin{matrix} \frac{1}{6}, \frac{1}{3} \\ \frac{1}{2};\ \frac{1}{2} \end{matrix}\right) = \frac{[(\sqrt{3}-\sqrt{2})(\sqrt{2}+2)]^{1/2}}{4} \frac{\Gamma^2\left(\frac{1}{24}\right)\Gamma\left(\frac{1}{3}\right)}{\Gamma\left(\frac{1}{12}\right)\Gamma^2\left(\frac{1}{6}\right)}$ $[[49],\ (5.19)].$

138. $\ {}_2F_1\left(\begin{matrix} \frac{2}{3}, \frac{5}{6} \\ \frac{3}{2};\ \frac{1}{2} \end{matrix}\right) = \frac{[(\sqrt{6}-2)(5\sqrt{2}-7)]^{1/2}}{2^{13/4}\pi} \frac{\Gamma\left(\frac{1}{24}\right)\Gamma\left(\frac{1}{6}\right)\Gamma\left(\frac{1}{3}\right)}{\Gamma\left(\frac{13}{24}\right)}$ $[[49],\ (6.11)].$

139. $\ {}_2F_1\left(\begin{matrix} \frac{1}{4}, \frac{1}{4} \\ \frac{1}{2};\ \frac{3}{4} \end{matrix}\right) = \frac{3^{3/4}(1+\sqrt{3})}{2\pi^2}\Gamma\left(\frac{1}{3}\right)\Gamma^2\left(\frac{3}{4}\right)\Gamma\left(\frac{7}{6}\right)$ $[[49],\ (3.12)].$

140. $\ {}_2F_1\left(\begin{matrix} \frac{1}{4}, \frac{1}{2} \\ \frac{3}{4};\ \frac{3}{4} \end{matrix}\right) = \left(\frac{4}{3}\right)^{3/4}$ $[[49],\ (7.6)].$

141. $\ {}_2F_1\left(\begin{matrix} \frac{1}{2}, \frac{3}{4} \\ \frac{5}{4};\ \frac{3}{4} \end{matrix}\right) = \frac{1}{12\pi^2}\Gamma^4\left(\frac{1}{4}\right)$ $[[49],\ (7.9)].$

142. $\ {}_2F_1\left(\begin{matrix} \frac{3}{4}, \frac{3}{4} \\ \frac{3}{2};\ \frac{3}{4} \end{matrix}\right) = \frac{\sqrt{3}-1}{2^{7/3} 3^{1/4} \pi^{5/2}}\Gamma^2\left(\frac{1}{4}\right)\Gamma^3\left(\frac{1}{3}\right)$ $[[49],\ (6.5)].$

143. $\ {}_2F_1\left(\begin{matrix} a,\ a+\frac{1}{2} \\ 4a+2;\ \frac{8}{9} \end{matrix}\right) = \sqrt{\pi}\ \frac{3^a \Gamma\left(\frac{4a+5}{6}\right)}{\Gamma\left(\frac{2a+3}{6}\right)\Gamma\left(\frac{2a+5}{6}\right)}$ $[[51],\ (3.1)].$

144. $\ {}_2F_1\left(\begin{matrix} a,\ 1-2a \\ \frac{2}{3};\ \frac{8}{9} \end{matrix}\right) = \frac{2}{3^a}\sin\left(\frac{5\pi}{6} - \pi a\right)$ $[[51],\ (3.2)].$

145. $\ {}_2F_1\left(\begin{matrix} a,\ 2-2a \\ \frac{4}{3};\ \frac{8}{9} \end{matrix}\right) = \sqrt{\pi}\ \frac{3^{-a}\Gamma\left(\frac{1}{6}\right)}{2\Gamma\left(a+\frac{1}{6}\right)\Gamma\left(\frac{3}{2}-a\right)}$ $[[51],\ (3.3)].$

146. $\ {}_2F_1\left(\begin{matrix} a,\ a+\frac{1}{4} \\ 4a+1;\ \frac{8}{9} \end{matrix}\right) = \frac{108^{a/3}\Gamma\left(\frac{4a+7}{12}\right)\Gamma\left(\frac{2a+5}{6}\right)\Gamma\left(\frac{2}{3}\right)\Gamma\left(\frac{3}{4}\right)}{\Gamma\left(\frac{4a+9}{12}\right)\Gamma\left(\frac{a+2}{3}\right)\Gamma\left(\frac{7}{12}\right)\Gamma\left(\frac{5}{6}\right)}$ $[[51],\ (3.4)].$

8.1.1]   8.1. The Hypergeometric Functions

147. $\displaystyle {}_2F_1\left(\begin{array}{c} a,\, a-\frac{1}{4} \\ \frac{4a+1}{3};\, \frac{8}{9} \end{array}\right) = \frac{108^{a/3}\Gamma\left(\frac{4a+1}{12}\right)\Gamma\left(\frac{2a+5}{6}\right)\Gamma\left(\frac{1}{4}\right)\Gamma\left(\frac{2}{3}\right)}{\Gamma\left(\frac{4a+3}{12}\right)\Gamma\left(\frac{a+2}{3}\right)\Gamma\left(\frac{1}{12}\right)\Gamma\left(\frac{5}{6}\right)}$ [[51], (3.5)].

148. $\displaystyle {}_2F_1\left(\begin{array}{c} -\frac{1}{4},\, \frac{1}{4} \\ 1;\, \frac{8}{9} \end{array}\right) = \frac{3^{-1/2}}{(2\pi)^{3/2}}\left[\Gamma^2\left(\frac{1}{4}\right) + 8\Gamma^2\left(\frac{3}{4}\right)\right].$

149. $\displaystyle {}_2F_1\left(\begin{array}{c} \frac{1}{8},\, \frac{3}{8} \\ \frac{1}{2};\, \frac{8}{9} \end{array}\right) = \frac{(\sqrt{2}-1)^{3/2}(\sqrt{3}-1)(\sqrt{6}+\sqrt{2}-1)^{1/2}}{2^{11/3}3^{1/2}\pi^3}$

$$\times \Gamma^2\left(\frac{1}{24}\right)\Gamma\left(\frac{1}{4}\right)\Gamma^2\left(\frac{7}{8}\right)\Gamma\left(\frac{11}{12}\right)$$ [[49], (4.12)].

150. $\displaystyle {}_2F_1\left(\begin{array}{c} \frac{1}{4},\, \frac{3}{4} \\ 1;\, \frac{8}{9} \end{array}\right) = \frac{3^{1/2}}{(2\pi)^{3/2}}\Gamma^2\left(\frac{1}{4}\right).$

151. $\displaystyle {}_2F_1\left(\begin{array}{c} \frac{5}{8},\, \frac{7}{8} \\ \frac{3}{2};\, \frac{8}{9} \end{array}\right) = \frac{3^{1/2}(2\sqrt{3}-\sqrt{6}-1)^{1/2}}{8\pi^2}$

$$\times \cos\frac{\pi}{24}\cos\frac{\pi}{8}\Gamma\left(\frac{1}{24}\right)\Gamma\left(\frac{1}{8}\right)\Gamma\left(\frac{3}{8}\right)\Gamma\left(\frac{11}{24}\right)$$ [[49], (6.8)].

152. $\displaystyle {}_2F_1\left(\begin{array}{c} \frac{1}{6},\, \frac{1}{3} \\ \frac{1}{2};\, \frac{25}{27} \end{array}\right) = \frac{3^{3/2}}{4}$ [[49], (5.15)].

153. $\displaystyle {}_2F_1\left(\begin{array}{c} \frac{2}{3},\, \frac{5}{6} \\ \frac{3}{2};\, \frac{25}{27} \end{array}\right) = \frac{3^{5/2}}{160\pi^2}\Gamma^2\left(\frac{1}{6}\right)\Gamma^2\left(\frac{1}{3}\right)$ [[49], (6.12)].

154. $\displaystyle {}_2F_1\left(\begin{array}{c} \frac{3}{8},\, \frac{7}{8} \\ \frac{5}{4};\, \frac{48}{49} \end{array}\right) = \frac{2^{-7/2}7^{3/4}}{3\pi^2}\Gamma^4\left(\frac{1}{4}\right).$

155. $\displaystyle {}_2F_1\left(\begin{array}{c} \frac{1}{4},\, \frac{1}{4} \\ \frac{1}{2};\, \frac{63}{64} \end{array}\right) = \frac{(4+\sqrt{7})^{1/2}}{2^{3/2}7^{1/4}\pi^{5/2}}\Gamma\left(\frac{1}{7}\right)\Gamma\left(\frac{2}{7}\right)\Gamma\left(\frac{4}{7}\right)\Gamma^2\left(\frac{3}{4}\right)$ [[49], (3.13)].

156. $\displaystyle {}_2F_1\left(\begin{array}{c} \frac{3}{4},\, \frac{3}{4} \\ \frac{3}{2};\, \frac{63}{64} \end{array}\right) = \frac{\sqrt{7}-1}{3\cdot 7^{3/4}\pi^{5/2}}\Gamma\left(\frac{1}{7}\right)\Gamma\left(\frac{2}{7}\right)\Gamma\left(\frac{4}{7}\right)\Gamma^2\left(\frac{1}{4}\right)$ [[49], (6.6)].

157. $\displaystyle {}_2F_1\left(\begin{array}{c} \frac{1}{8},\, \frac{3}{8} \\ \frac{1}{2};\, \frac{49}{81} \end{array}\right) = \frac{8\sqrt{3}(1+\sqrt{2})^{1/2}}{\pi^{5/2}}\Gamma^2\left(\frac{7}{8}\right)\Gamma^3\left(\frac{5}{4}\right)$ [[49], (4.11)].

158. $\displaystyle {}_2F_1\left(\begin{array}{c} \frac{1}{4},\, \frac{1}{2} \\ \frac{3}{4};\, \frac{49}{81} \end{array}\right) = \frac{3(1+\sqrt{7})}{2^4 7^{1/4}\pi^{5/2}}\Gamma\left(\frac{1}{7}\right)\Gamma\left(\frac{2}{7}\right)\Gamma\left(\frac{4}{7}\right)\Gamma^2\left(\frac{3}{4}\right)$ [[49], (7.5)].

Ch. 8. Representations of Hypergeometric and Meijer G Functions   [8.1.1]

159. $\,_2F_1\left(\begin{matrix}\frac{5}{8},\frac{7}{8}\\ \frac{3}{2};\frac{49}{81}\end{matrix}\right)=\dfrac{3^{5/2}\left(\sqrt{2}-1\right)^{3/2}}{112\pi^{3/2}}\Gamma^2\left(\dfrac{1}{8}\right)\Gamma\left(\dfrac{1}{4}\right)$ [[49], (6.7)].

160. $\,_2F_1\left(\begin{matrix}\frac{1}{8},\frac{3}{8}\\ \frac{1}{2};\frac{80}{81}\end{matrix}\right)=\dfrac{3^{1/2}\left(3\sqrt{5}+4\sqrt{2}+1\right)^{1/2}}{2^{3/10}5^{1/4}\pi^3}$

$\times \sin^2\dfrac{3\pi}{40}\cos\dfrac{3\pi}{20}\Gamma^2\left(\dfrac{7}{40}\right)\Gamma\left(\dfrac{3}{20}\right)\Gamma\left(\dfrac{13}{20}\right)\Gamma^2\left(\dfrac{37}{40}\right)$ [[49], (4.13)].

161. $\,_2F_1\left(\begin{matrix}\frac{1}{4},\frac{1}{2}\\ \frac{3}{4};\frac{80}{81}\end{matrix}\right)=\dfrac{9}{5}$ [[49], (7.7)].

162. $\,_2F_1\left(\begin{matrix}\frac{1}{2},\frac{3}{4}\\ \frac{5}{4};\frac{80}{81}\end{matrix}\right)=\dfrac{9}{8\cdot 5^{5/4}\pi^2}\Gamma^4\left(\dfrac{1}{4}\right)$ [[49], (7.10)].

163. $\,_2F_1\left(\begin{matrix}\frac{5}{8},\frac{7}{8}\\ \frac{3}{2};\frac{80}{81}\end{matrix}\right)=\dfrac{3^{5/2}\left(3\sqrt{5}-4\sqrt{2}-1\right)^{1/2}}{2^{7/2}5^{5/4}\pi^2}$

$\times \cos\dfrac{\pi}{40}\cos\dfrac{9\pi}{40}\Gamma\left(\dfrac{1}{40}\right)\Gamma\left(\dfrac{9}{40}\right)\Gamma\left(\dfrac{11}{40}\right)\Gamma\left(\dfrac{19}{40}\right)$ [[49], (6.9)].

164. $\,_2F_1\left(\begin{matrix}\frac{1}{12},\frac{5}{12}\\ \frac{1}{2};\frac{98}{125}\end{matrix}\right)=\dfrac{1+\sqrt{2}}{2^{11/4}\pi^2}\left(\dfrac{5}{3}\right)^{1/4}\Gamma\left(\dfrac{1}{8}\right)\Gamma\left(\dfrac{3}{8}\right)\Gamma^2\left(\dfrac{3}{4}\right)$ [48].

165. $\,_2F_1\left(\begin{matrix}\frac{1}{12},\frac{5}{12}\\ \frac{1}{2};\frac{121}{125}\end{matrix}\right)=\dfrac{\sqrt[4]{15}\left(1+\sqrt{3}\right)}{2^{17/6}\pi^{5/2}}\Gamma^3\left(\dfrac{1}{3}\right)\Gamma^2\left(\dfrac{3}{4}\right)$ [48].

166. $\,_2F_1\left(\begin{matrix}\frac{1}{6},\frac{1}{3}\\ \frac{1}{2};\frac{121}{125}\end{matrix}\right)=\dfrac{3^{1/2}\left(\sqrt{5}-1\right)^{1/2}}{2^{11/2}5^{1/4}\pi^4}$

$\times\Gamma\left(\dfrac{1}{30}\right)\Gamma\left(\dfrac{1}{15}\right)\Gamma\left(\dfrac{4}{15}\right)\Gamma\left(\dfrac{19}{30}\right)\Gamma^2\left(\dfrac{2}{3}\right)\Gamma^2\left(\dfrac{5}{6}\right)$ [[49], (5.20)].

167. $\,_2F_1\left(\begin{matrix}\frac{2}{3},\frac{5}{6}\\ \frac{3}{2};\frac{121}{125}\end{matrix}\right)=\dfrac{5^{5/4}\left(5\sqrt{5}-11\right)^{1/2}}{11\cdot 2^{9/2}\pi^2}\Gamma\left(\dfrac{1}{30}\right)\Gamma\left(\dfrac{19}{30}\right)\Gamma\left(\dfrac{1}{15}\right)\Gamma\left(\dfrac{4}{15}\right)$

[[49], (6.13)].

168. $\,_2F_1\left(\begin{matrix}\frac{1}{12},\frac{5}{12}\\ \frac{1}{2};\frac{1323}{1331}\end{matrix}\right)=\dfrac{3\sqrt[4]{11}}{4}$ [48].

169. $\,_2F_1\left(\begin{matrix}\frac{1}{8},\frac{3}{8}\\ \frac{1}{2};\frac{2400}{2401}\end{matrix}\right)=\dfrac{2\sqrt{7}}{3}$ [[49], (4.10)].

170. $_2F_1\left(\begin{matrix}\frac{5}{8},\frac{7}{8}\\\frac{3}{2};\frac{2400}{2401}\end{matrix}\right)=\frac{49}{480\pi^2}\sqrt{\frac{7}{3}}\Gamma^2\left(\frac{1}{8}\right)\Gamma^2\left(\frac{3}{8}\right)$ [[49], (6.10)].

171. $_2F_1\left(\begin{matrix}\frac{3}{8},\frac{7}{8}\\\frac{5}{4};\frac{25920}{25921}\end{matrix}\right)=\frac{5^{-5/4}161^{3/4}}{24\pi^2}\Gamma^4\left(\frac{1}{4}\right){}_2F_1\left(\begin{matrix}a,\frac{4a+1}{6}\\\frac{2a+5}{6};17-12\sqrt{2}\end{matrix}\right)$

$$=48^{-a/2}(3+2\sqrt{2})^a\frac{\sqrt{\pi}\,\Gamma\left(\frac{2a+5}{6}\right)}{\Gamma\left(\frac{a+3}{6}\right)\Gamma\left(\frac{a+5}{6}\right)}.$$

172. $_2F_1\left(\begin{matrix}a,\frac{4a+1}{6}\\\frac{4a+1}{3};12\sqrt{2}-16\end{matrix}\right)=3^{-a/2}(3+2\sqrt{2})^a\frac{\sqrt{\pi}\,\Gamma\left(\frac{2a+5}{6}\right)}{\Gamma\left(\frac{a+3}{6}\right)\Gamma\left(\frac{a+5}{6}\right)}.$

173. $_2F_1\left(\begin{matrix}-\frac{1}{2},-\frac{1}{2}\\1;3-2\sqrt{2}\end{matrix}\right)=\frac{2^{-7/4}}{\pi^{3/2}}(\sqrt{2}-1)^{1/2}\left[\Gamma\left(\frac{1}{8}\right)\Gamma\left(\frac{3}{8}\right)+8\Gamma\left(\frac{5}{8}\right)\Gamma\left(\frac{7}{8}\right)\right].$

174. $_2F_1\left(\begin{matrix}-\frac{1}{4},\frac{1}{4}\\1;\frac{57-40\sqrt{2}}{49}\end{matrix}\right)$

$$=\frac{(3+\sqrt{2})^{-1/2}}{2^{9/4}\pi^{3/2}}\left[(1+\sqrt{2})\Gamma\left(\frac{1}{8}\right)\Gamma\left(\frac{3}{8}\right)+8\Gamma\left(\frac{5}{8}\right)\Gamma\left(\frac{7}{8}\right)\right].$$

175. $_2F_1\left(\begin{matrix}-\frac{1}{4},\frac{1}{4}\\1;\frac{249-176\sqrt{2}}{441}\end{matrix}\right)$

$$=\frac{3^{-1/2}}{2\pi^{3/2}}(1+2\sqrt{2})^{-1/2}\left[(1+\sqrt{2})\Gamma^2\left(\frac{1}{4}\right)+4\Gamma^2\left(\frac{3}{4}\right)\right].$$

176. $_2F_1\left(\begin{matrix}\frac{1}{12},\frac{5}{12}\\\frac{1}{2};\frac{3514+988\sqrt{2}}{17^3}\end{matrix}\right)$

$$=\frac{3^{-3/4}(1+\sqrt{6})}{8\pi^2}\left(\frac{5+2\sqrt{2}}{2+\sqrt{3}}\right)^{1/4}\Gamma\left(\frac{1}{24}\right)\Gamma\left(\frac{11}{24}\right)\Gamma^2\left(\frac{3}{4}\right)\quad[48].$$

177. $_2F_1\left(\begin{matrix}\frac{1}{8},\frac{3}{8}\\\frac{1}{2};\frac{25}{2401}(16\sqrt{2}-13)^2\end{matrix}\right)=\frac{3}{8}(6\sqrt{2}+4)^{1/2}$ [[49], (4.9)].

178. $_2F_1\left(\begin{matrix}\frac{1}{8},\frac{3}{8}\\1;\frac{32}{2401}(325\sqrt{2}-457)\end{matrix}\right)=\frac{(3+\sqrt{2})^{1/2}}{2^{11/4}\pi^{3/2}}\Gamma\left(\frac{1}{8}\right)\Gamma\left(\frac{3}{8}\right).$

179. $_2F_1\left(\begin{matrix}\frac{1}{6}, \frac{1}{3}\\ \frac{1}{2};\end{matrix} \frac{3^4\left(17\sqrt{6}-22\right)^2}{2\cdot 5^6}\right) = \frac{5}{8}\left(\sqrt{6}+1\right)$ [[49], (5.17)].

180. $_2F_1\left(\begin{matrix}\frac{1}{4}, \frac{1}{4}\\ \frac{1}{2};\end{matrix} -96+56\sqrt{3}\right) = \frac{\sqrt{6}+\sqrt{2}}{3^{3/4}}$ [[49], (3.9)].

181. $_2F_1\left(\begin{matrix}\frac{1}{4}, \frac{1}{4}\\ 1;\end{matrix} -56-40\sqrt{2}\right) = \frac{\left(2-\sqrt{2}\right)^{1/2}}{4\pi^{3/2}}\Gamma\left(\frac{1}{8}\right)\Gamma\left(\frac{3}{8}\right).$

182. $_2F_1\left(\begin{matrix}\frac{1}{4}, \frac{1}{4}\\ 1;\end{matrix} 40\sqrt{2}-56\right) = \frac{1}{2^{9/4}\pi^{3/2}}\left(1+\sqrt{2}\right)^{1/2}\Gamma\left(\frac{1}{8}\right)\Gamma\left(\frac{3}{8}\right).$

183. $_2F_1\left(\begin{matrix}\frac{1}{4}, \frac{1}{4}\\ 1;\end{matrix} -16(26-15\sqrt{3})\right) = \frac{3^{-1/4}}{4\pi^{3/2}}\left(2+\sqrt{3}\right)^{1/2}\Gamma\left(\frac{1}{6}\right)\Gamma\left(\frac{1}{3}\right).$

184. $_2F_1\left(\begin{matrix}\frac{1}{4}, \frac{1}{4}\\ 1;\end{matrix} \frac{7-5\sqrt{2}}{8}\right) = \frac{\left(1+\sqrt{2}\right)^{1/4}}{4\pi^{3/2}}\Gamma\left(\frac{1}{8}\right)\Gamma\left(\frac{3}{8}\right).$

185. $_2F_1\left(\begin{matrix}\frac{1}{4}, \frac{1}{4}\\ 1;\end{matrix} \frac{15\sqrt{3}-26}{16}\right) = \frac{3^{-1/4}}{(2\pi)^{3/2}}\left(2+\sqrt{3}\right)^{1/4}\Gamma\left(\frac{1}{6}\right)\Gamma\left(\frac{1}{3}\right).$

186. $_2F_1\left(\begin{matrix}\frac{1}{4}, \frac{1}{4}\\ 1;\end{matrix} \frac{140-99\sqrt{2}}{32}\right) = \frac{1}{2^{15/8}\pi^{3/2}}\left(1+\sqrt{2}\right)^{1/2}\Gamma^2\left(\frac{1}{4}\right).$

187. $_2F_1\left(\begin{matrix}\frac{1}{4}, \frac{3}{4}\\ 1;\end{matrix} \frac{8}{49}(5\sqrt{2}-1)\right) = \frac{\left(2+3\sqrt{2}\right)^{1/2}}{4\pi^{3/2}}\Gamma\left(\frac{1}{8}\right)\Gamma\left(\frac{3}{8}\right).$

188. $_2F_1\left(\begin{matrix}\frac{1}{4}, \frac{3}{4}\\ 1;\end{matrix} \frac{16}{441}(12-11\sqrt{2})\right) = \frac{\left(12-3\sqrt{2}\right)^{1/2}}{2^{11/4}\pi^{3/2}}\Gamma^2\left(\frac{1}{4}\right).$

189. $_2F_1\left(\begin{matrix}\frac{1}{4}, \frac{3}{4}\\ 1;\end{matrix} \frac{249-176\sqrt{2}}{441}\right) = \frac{3^{1/2}}{8\pi^{3/2}}\left(1+2\sqrt{2}\right)^{1/2}\Gamma^2\left(\frac{1}{4}\right).$

190. $_2F_1\left(\begin{matrix}\frac{3}{8}, \frac{7}{8}\\ 1;\end{matrix} \frac{64}{13225}(153\sqrt{3}-266)\right)$
$= \frac{2^{-5/2}}{11\pi^{3/2}}\left(\frac{5}{3}\right)^{3/4}\left(29861+18884\sqrt{3}\right)^{1/4}\Gamma\left(\frac{1}{6}\right)\Gamma\left(\frac{1}{3}\right).$

191. $\displaystyle {}_2F_1\left(\begin{matrix}\frac{1}{2},\frac{1}{2}\\ 1;\ 17-12\sqrt{2}\end{matrix}\right) = \frac{(3+2\sqrt{2})^{1/2}}{2^{5/2}\pi^{3/2}}\Gamma^2\left(\frac{1}{4}\right).$

192. $\displaystyle {}_2F_1\left(\begin{matrix}\frac{3}{4},\frac{3}{4}\\ 1;\ 40\sqrt{2}-56\end{matrix}\right) = \frac{2^{-9/4}}{7\pi^{3/2}}\left(137+97\sqrt{2}\right)^{1/2}\Gamma\left(\frac{1}{8}\right)\Gamma\left(\frac{3}{8}\right).$

193. $\displaystyle {}_2F_1\left(\begin{matrix}\frac{3}{4},\frac{3}{4}\\ 1;\ \frac{7-5\sqrt{2}}{8}\end{matrix}\right) = \frac{2^{-1/2}}{7\pi^{3/2}}\left(31+41\sqrt{2}\right)^{1/4}\Gamma\left(\frac{1}{8}\right)\Gamma\left(\frac{3}{8}\right).$

194. $\displaystyle {}_2F_1\left(\begin{matrix}\frac{3}{4},\frac{3}{4}\\ 1;\ \frac{15\sqrt{3}-26}{16}\end{matrix}\right) = \frac{2^{1/2}3^{-3/4}}{11\pi^{3/2}}\left(962+551\sqrt{3}\right)^{1/4}\Gamma\left(\frac{1}{6}\right)\Gamma\left(\frac{1}{3}\right).$

195. $\displaystyle {}_2F_1\left(\begin{matrix}\frac{3}{4},\frac{3}{4}\\ 1;\ \frac{140-99\sqrt{2}}{32}\end{matrix}\right) = \frac{2^{1/2}}{21\pi^{3/2}}\left(782+\frac{1107}{\sqrt{2}}\right)^{1/4}\Gamma^2\left(\frac{1}{4}\right).$

196. $\displaystyle {}_2F_1\left(\begin{matrix}\frac{3}{4},\frac{3}{4}\\ \frac{3}{2};\ 56\sqrt{3}-96\end{matrix}\right) = \frac{2^{-7/2}}{3\pi^2}\left(7+4\sqrt{3}\right)^{3/4}\Gamma^4\left(\frac{1}{4}\right).$

8.1.2. The hypergeometric function ${}_3F_2(a_1,a_2,a_3;b_1,b_2;z)$

1. $\displaystyle {}_3F_2\left(\begin{matrix}a,b,1-b;\ z\\ \frac{a+b}{2},\frac{a-b+1}{2}\end{matrix}\right) = \frac{1+8z}{3(1-4z)^a}\ {}_3F_2\left(\begin{matrix}\frac{a}{3},\frac{a+1}{3},\frac{a+2}{3};\ -\frac{27z}{(1-4z)^3}\\ \frac{a+b}{2},\frac{a-b+1}{2}\end{matrix}\right)$

$\displaystyle +\frac{2}{3(1-4z)^{a-1}}\ {}_3F_2\left(\begin{matrix}\frac{a-1}{3},\frac{a}{3},\frac{a+1}{3};\ -\frac{27z}{(1-4z)^3}\\ \frac{a+b}{2},\frac{a-b+1}{2}\end{matrix}\right)\quad [-1/8<z<1/4;\ [52],(6)].$

2. $\displaystyle {}_3F_2\left(\begin{matrix}a,b,3-b;\ z\\ \frac{a+b+1}{2},\frac{a-b}{2}+2\end{matrix}\right) = \frac{(a+b-1)(b-a-2)}{12(b-1)(b-2)z(1-4z)^{a-1}}$

$\displaystyle \times\left[\frac{1+8z}{1-4z}\ {}_3F_2\left(\begin{matrix}\frac{a}{3},\frac{a+1}{3},\frac{a+2}{3};\ -\frac{27z}{(1-4z)^3}\\ \frac{a+b-1}{2},\frac{a-b}{2}+1\end{matrix}\right)\right.$

$\displaystyle \left. -{}_3F_2\left(\begin{matrix}\frac{a-1}{3},\frac{a}{3},\frac{a+1}{3};\ -\frac{27z}{(1-4z)^3}\\ \frac{a+b-1}{2},\frac{a-b}{2}+1\end{matrix}\right)\right]\quad [-1/8<z<1/4;\ [52],(6)].$

3. $\displaystyle {}_3F_2\left(\begin{matrix}a+1,-a,b\\ a,c;\ z\end{matrix}\right) = (1-z)^{-b}\ {}_3F_2\left(\begin{matrix}b,a+c-1,\frac{a+c+1}{2}\\ c,\frac{a+c-1}{2};\ \frac{z}{z-1}\end{matrix}\right)$

$[|\arg(1-z)|<\pi].$

4. $\displaystyle {}_3F_2\!\left(\begin{array}{c} a,\, a+\frac{1}{2},\, b;\ -\frac{4z}{(1-z)^2} \\ c,\, 2a+b-c+1 \end{array}\right) = (1-z)^{2a}\, {}_3F_2\!\left(\begin{array}{c} 2a,\, 2a-c+1,\, c-b;\ z \\ c,\, 2a+b-c+1 \end{array}\right).$

5. $\displaystyle {}_3F_2\!\left(\begin{array}{c} a,\, a+\frac{1}{3},\, a+\frac{2}{3};\ \frac{27z^2}{4(1-z)^3} \\ b,\, 3a-b+\frac{3}{2} \end{array}\right)$

$$= (1-z)^{3a}\, {}_3F_2\!\left(\begin{array}{c} 3a,\, b-\frac{1}{2},\, 3a-b+1;\ 4z \\ 2b-1,\, 6a-2b+2 \end{array}\right).$$

6. $\displaystyle {}_3F_2\!\left(\begin{array}{c} -n,\, -n,\, -n;\ z \\ \frac{1}{2}-n,\, -2n \end{array}\right) = \left(-\frac{z}{4}\right)^n \left[\frac{n!}{\left(\frac{1}{2}\right)_n} P_n\!\left(\sqrt{1-\frac{1}{z}}\right)\right]^2.$

7. $\displaystyle {}_3F_2\!\left(\begin{array}{c} \frac{1}{4},\, \frac{1}{4},\, \frac{1}{4} \\ \frac{1}{2},\, \frac{3}{4};\ -z \end{array}\right) = \frac{\Gamma^4\!\left(\frac{3}{4}\right)}{\pi^3}\left(\sqrt{z}+\sqrt{z+1}\right)^{-1/2}$

$$\times \left[\mathbf{K}\!\left(\sqrt{\frac{1}{2}+\frac{2^{-1/2}z^{1/4}}{(\sqrt{z}+\sqrt{z+1})^{1/2}}}\right) + \mathbf{K}\!\left(\sqrt{\frac{1}{2}-\frac{2^{-1/2}z^{1/4}}{(\sqrt{z}+\sqrt{z+1})^{1/2}}}\right)\right]^2.$$

8. $\displaystyle {}_3F_2\!\left(\begin{array}{c} \frac{1}{2},\, \frac{1}{2},\, \frac{1}{2} \\ 1,\, 1;\ z \end{array}\right) = \frac{4}{\pi^2}\, \mathbf{K}^2\!\left(\sqrt{\frac{1-\sqrt{1-z}}{2}}\right).$

9. $\displaystyle {}_3F_2\!\left(\begin{array}{c} \frac{1}{2},\, \frac{1}{2},\, \frac{1}{2} \\ 1,\, 1;\ -z \end{array}\right) = \frac{8}{\pi^2(\sqrt{z+1}+1)}\, \mathbf{K}^2\!\left(\frac{\sqrt{z}}{\sqrt{z+1}+1}\right).$

10. $\displaystyle {}_3F_2\!\left(\begin{array}{c} \frac{1}{2},\, \frac{1}{2},\, 1 \\ \frac{3}{2},\, \frac{3}{2};\ z \end{array}\right) = \frac{1}{2\sqrt{z}}\left[\operatorname{Li}_2(\sqrt{z}) - \operatorname{Li}_2(-\sqrt{z})\right].$

11. $\displaystyle {}_3F_2\!\left(\begin{array}{c} \frac{1}{2},\, \frac{1}{2},\, 1 \\ \frac{3}{2},\, \frac{3}{2};\ -z \end{array}\right) = \frac{1}{2\sqrt{z}}\,[2\arctan\sqrt{z}\,\ln\sqrt{z}$

$$+ \operatorname{Cl}_2(2\arctan\sqrt{z}) + \operatorname{Cl}_2(\pi - 2\arctan\sqrt{z})].$$

12. $\displaystyle {}_3F_2\!\left(\begin{array}{c} \frac{1}{2},\, \frac{2}{3},\, \frac{4}{3};\ -\frac{27}{4}z(1-z)^{-3} \\ \frac{3}{2},\, \frac{3}{2} \end{array}\right) = \frac{(1-z)^{3/2}}{\sqrt{z}}\arcsin\sqrt{z}$

$$[|z|,\, |27z(1-z)^{-3}/4| < 1].$$

13. $_3F_2\left(\begin{array}{c}\frac{1}{2},1,1\\ \frac{3}{4},\frac{5}{4};-z\end{array}\right) = \frac{1}{z^{1/4}(z+1)^{1/4}}\left\{\cos\frac{\pi+2\arctan\sqrt{z}}{4}\ln(z_++z_-)\right.$

$\left.+\sin\frac{\pi+2\arctan\sqrt{z}}{4}\arcsin\frac{(\sqrt{z}-\sqrt{z+1}+1)^{1/2}}{\sqrt{2}}\right\}$

$\left[z_\pm = 2^{-1/2}(\sqrt{z}+\sqrt{z+1}\pm 1)^{1/2}\right].$

14. $_3F_2\left(\begin{array}{c}\frac{1}{2},1,1;z\\ \frac{3}{2},\frac{3}{2}\end{array}\right) = \frac{1}{4\sqrt{z}}\left[i\pi^2 + 4\arcsin\sqrt{z}\ln\frac{1-e^{i\arcsin\sqrt{z}}}{1+e^{i\arcsin\sqrt{z}}}\right.$

$\left.+4i\,\mathrm{Li}_2\left(-e^{i\arcsin\sqrt{z}}\right) - 4i\,\mathrm{Li}_2\left(e^{i\arcsin\sqrt{z}}\right)\right].$

15. $_3F_2\left(\begin{array}{c}\frac{1}{2},1,1\\ \frac{3}{2},\frac{3}{2};-z\end{array}\right) = \frac{1}{4\sqrt{z}}\left[\pi^2 + 4\ln(\sqrt{z}+\sqrt{z+1})\ln\frac{\sqrt{z}+\sqrt{z+1}-1}{\sqrt{z}+\sqrt{z+1}+1}\right.$

$\left.+4\,\mathrm{Li}_2\left(-\frac{1}{\sqrt{z}+\sqrt{z+1}}\right) - 4\,\mathrm{Li}_2\left(\frac{1}{\sqrt{z}+\sqrt{z+1}}\right)\right].$

16. $_3F_2\left(\begin{array}{c}1,1,\frac{3}{2}\\ \frac{5}{4},\frac{7}{4};-z\end{array}\right) = \frac{3}{2z^{3/4}(z+1)^{1/4}}\left\{\sin\frac{\pi+2\arctan\sqrt{z}}{4}\ln(z_++z_-)\right.$

$\left.-\cos\frac{\pi+2\arctan\sqrt{z}}{4}\arcsin\frac{(\sqrt{z}-\sqrt{z+1}+1)^{1/2}}{\sqrt{2}}\right\}$

$\left[z_\pm = 2^{-1/2}(\sqrt{z}+\sqrt{z+1}\pm 1)^{1/2}\right].$

17. $= \frac{\Gamma^4\left(\frac{3}{4}\right)}{\pi^3}\left(1+2z-2\sqrt{z^2+z}\right)^{1/4}$

$\times\left[\mathbf{K}\left(\sqrt{\frac{1}{2}+\frac{1}{2^{1/2}}\left(\sqrt{z^2+z}-z\right)^{1/2}}\right)\right.$

$\left.+\mathbf{K}\left(\sqrt{\frac{1}{2}-\frac{1}{2^{1/2}}\left(\sqrt{z^2+z}-z\right)^{1/2}}\right)\right]^2.$

18. $_3F_2\left(\begin{array}{c}\frac{3}{4},\frac{3}{4},\frac{3}{4}\\ \frac{5}{4},\frac{3}{2};-z\end{array}\right) = \frac{\Gamma^4\left(\frac{1}{4}\right)}{8\pi^3 z^{1/2}}\left(\sqrt{z}+\sqrt{z+1}\right)^{-1/2}$

$\times\left[\mathbf{K}\left(\sqrt{\frac{1}{2}+\frac{2^{-1/2}z^{1/4}}{(\sqrt{z}+\sqrt{z+1})^{1/2}}}\right) - \mathbf{K}\left(\sqrt{\frac{1}{2}-\frac{2^{-1/2}z^{1/4}}{(\sqrt{z}+\sqrt{z+1})^{1/2}}}\right)\right]^2.$

19. $\quad = \dfrac{\Gamma^4\left(\frac{1}{4}\right)}{8\pi^3} \dfrac{(1+2z-2\sqrt{z^2+z})^{3/4}}{\sqrt{z^2+z}-z}$

$$\times \left[\mathbf{K}\left(\sqrt{\dfrac{1}{2}+\dfrac{1}{2^{1/2}}\left(\sqrt{z^2+z}-z\right)^{1/2}}\right)\right.$$

$$\left.-\mathbf{K}\left(\sqrt{\dfrac{1}{2}-\dfrac{1}{2^{1/2}}\left(\sqrt{z^2+z}-z\right)^{1/2}}\right)\right]^2.$$

20. $\quad {}_3F_2\left(\begin{array}{c}\frac{1}{4},\frac{1}{2},\frac{3}{4}\\1,1;\,z\end{array}\right) = \dfrac{2^{5/2}}{\pi^2}\left(2-z+2\sqrt{1-z}\right)^{-1/4}$

$$\times \mathbf{K}^2\left(\sqrt{\dfrac{1}{2}-\left(\dfrac{1-\sqrt{1-z}}{2z}\right)^{1/2}}\right).$$

21. $\quad {}_3F_2\left(\begin{array}{c}\frac{1}{2},\frac{1}{2},\frac{1}{2}\\\frac{3}{4},\frac{5}{4};\,-z\end{array}\right) = \dfrac{1}{2^{1/2}\pi z^{1/4}}\left(\sqrt{z+1}+\sqrt{z}\right)^{-1/2}$

$$\times \left[\mathbf{K}^2\left(\sqrt{\dfrac{1}{2}+\dfrac{2^{-1/2}z^{1/4}}{(\sqrt{z}+\sqrt{z+1})^{1/2}}}\right) - \mathbf{K}^2\left(\sqrt{\dfrac{1}{2}-\dfrac{2^{-1/2}z^{1/4}}{(\sqrt{z}+\sqrt{z+1})^{1/2}}}\right)\right].$$

22. $\quad {}_3F_2\left(\begin{array}{c}\frac{1}{2},\frac{1}{2},\frac{1}{2}\\1,\frac{3}{2};\,z\end{array}\right) = \dfrac{2}{\pi\sqrt{z}}\int_0^{\sqrt{z}}\mathbf{K}(x)\,dx.$

23. $\quad {}_3F_2\left(\begin{array}{c}\frac{1}{2},\frac{1}{2},\frac{1}{2}\\1,\frac{3}{2};\,-z\end{array}\right) = \dfrac{2}{\pi\sqrt{z}}\int_0^{\sqrt{z/(z+1)}}\dfrac{\mathbf{K}(x)}{1-x^2}\,dx.$

24. $\quad {}_3F_2\left(\begin{array}{c}\frac{1}{2},\frac{1}{2},\frac{1}{2}\\\frac{3}{2},\frac{3}{2};\,z\end{array}\right) = \dfrac{1}{2\sqrt{z}}\operatorname{Cl}_2(2\arcsin\sqrt{z}) + \dfrac{\arcsin\sqrt{z}}{\sqrt{z}}\ln(2\sqrt{z}).$

25. $\quad {}_3F_2\left(\begin{array}{c}\frac{1}{2},\frac{1}{2},\frac{1}{2}\\\frac{3}{2},\frac{3}{2};\,-z\end{array}\right) = \dfrac{1}{\sqrt{z}}\left[\dfrac{\pi^2}{12} - \dfrac{1}{2}\ln^2(\sqrt{z}+\sqrt{1+z}) +\right.$

$$\left.+\ln\left(\sqrt{z}+\sqrt{1+z}\right)\ln\left(1+\sqrt{z}+\sqrt{1+z}\right) + \operatorname{Li}_2(-\sqrt{z}-\sqrt{1+z})\right.$$

$$\left.- \operatorname{Li}_2(1-\sqrt{z}-\sqrt{1+z})\right].$$

26. $\quad {}_3F_2\left(\begin{array}{c}1,1,1\\\frac{3}{2},\frac{3}{2};\,z\end{array}\right) = \dfrac{1}{\sqrt{z}}\int_0^{\arcsin\sqrt{z}}\dfrac{x\,dx}{\sqrt{z-\sin^2 x}}.$

8.1.2]   8.1. THE HYPERGEOMETRIC FUNCTIONS

27. $\ _3F_2\left(\begin{matrix}1,1,1\\\frac{3}{2},\frac{3}{2};\end{matrix}-z\right)=\frac{1}{\sqrt{z}}\int_0^{\ln(\sqrt{z}+\sqrt{z+1})}\frac{x\,dx}{\sqrt{z-\sinh^2 x}}.$

28. $\ _3F_2\left(\begin{matrix}a,b,c\\d,e;1\end{matrix}\right)=\frac{\Gamma(d)\,\Gamma(e)\,\Gamma(d+e-a-b-c)}{\Gamma(c)\,\Gamma(d+e-a-c)\,\Gamma(d+e-b-c)}$
$$\times\ _3F_2\left(\begin{matrix}d-c,e-c,d+e-a-b-c\\d+e-a-c,d+e-b-c;1\end{matrix}\right)\quad[\operatorname{Re} c,\ \operatorname{Re}(d+e-a-b-c)>0].$$

29. $\ _3F_2\left(\begin{matrix}a-n,b-n,c-n\\d-n,e-n;1\end{matrix}\right)$
$$=\frac{1}{(1-d)_n(1-e)_n}\sum_{k=0}^{n}(-1)^k\binom{n}{k}(1-d)_k(e-c)_k$$
$$\times(c-n)_{n-k}(d-a-b)_{n-k}\ _3F_2\left(\begin{matrix}a,b,c-k;1\\d-k,e\end{matrix}\right).$$

30. $\ _3F_2\left(\begin{matrix}a,b,c;1\\2a+b+1,a+2b+1\end{matrix}\right)=\frac{2(a+b)-c}{2(a+b)}\ _3F_2\left(\begin{matrix}a+1,b+1,c;1\\2a+b+1,a+2b+1\end{matrix}\right)$
$$[\operatorname{Re}(2a+2b-c)>0;\ [56]].$$

31. $\ _3F_2\left(\begin{matrix}a,b,c;1\\a+1,2a-b-c+\frac{bc}{a+1}+3\end{matrix}\right)$
$$=\frac{(a+1)(a-b+2)(a-c+2)}{(a+2)(a^2+3a-b-ab-c-ac+bc+2)}$$
$$\times\ _3F_2\left(\begin{matrix}a+2,b,c;1\\a+3,2a-b-c+\frac{bc}{a+1}+3\end{matrix}\right)\quad\left[\operatorname{Re}\left(2a-2b-2c+\frac{b}{a+1}\right)>-4;\ [56]\right].$$

32. $\ _3F_2\left(\begin{matrix}a,b,c;1\\a+1,c+\frac{a(a-c+1)}{b-1}+1\end{matrix}\right)=\frac{(a-b+2)(a^2+a-c-ac+bc)}{(a+1)(a^2+2a-ab-c-ac+bc)}$
$$\times\ _3F_2\left(\begin{matrix}a+1,b-1,c;1\\a+2,c+\frac{a(a-c+1)}{b-1}\end{matrix}\right)\quad[\operatorname{Re}(a(a-c+1)/(b-1)+b)<2;\ [56]].$$

33. $\ _3F_2\left(\begin{matrix}a,b,c;1\\a+1,\frac{bc}{-a+b+c}\end{matrix}\right)$
$$=\frac{a^2+b+c+bc-a(b+c+1)}{-a+b+c+bc}\ _3F_2\left(\begin{matrix}a,b+1,c+1;1\\a+1,\frac{bc}{-a+b+c}+2\end{matrix}\right)$$
$$[\operatorname{Re}[bc/(-a+b+c)+b+c]<2;\ [56]].$$

593

34. $\displaystyle {}_3F_2\left(\begin{array}{c} a,\, b,\, c;\ 1 \\ a+1,\, d \end{array}\right)$

$$= \frac{(a-b+1)(a-c+1)(a-c+2)(d-1)}{(a+1)(c-1)(a-d+1)(a-d+2)}\, {}_3F_2\left(\begin{array}{c} a+1,\, b,\, c-1 \\ a+2,\, d-1;\ 1 \end{array}\right)$$

$$- \frac{a(a-b+1) + (c-1)(b-d+1)}{(c-1)(a-d+1)(a-d+2)}\, \frac{\Gamma(d)\Gamma(d-b-c+1)}{\Gamma(d-b)\Gamma(d-c)}$$

$$[\operatorname{Re}[d-b-c] > -1;\ [56]].$$

35. $\displaystyle {}_3F_2\left(\begin{array}{c} 1,\, a,\, 1-a \\ \tfrac{3}{2},\, b;\ 1 \end{array}\right) = \frac{2^{2-2b}\cos[(a-b)\pi]\,\Gamma(2b-1)\,\Gamma(a-b+1)}{(2a-1)\Gamma(a+b-1)}$

$$+ \frac{2^{3-2b}(b-1)}{(2a-1)(a-b+1)}\, {}_2F_1\left(\begin{array}{c} 3-2b,\, a-b+1 \\ a-b+2;\ -1 \end{array}\right)\quad [\operatorname{Re}b > 1/2].$$

36. $\displaystyle {}_3F_2\left(\begin{array}{c} a,\, a,\, 1 \\ 2a,\, 2a;\ 1 \end{array}\right) = \frac{2^{6a-6}\sqrt{\pi}\,(2a-1)^2\csc^2(a\pi)}{21a}\,\frac{\Gamma^3\!\left(a-\tfrac{1}{2}\right)}{\Gamma^3(a)}$

$$+ \frac{2a-1}{a-1}\, {}_3F_2\left(\begin{array}{c} a,\, a,\, 1;\ 1 \\ 2a,\, 2-a \end{array}\right)\quad [1/2 < \operatorname{Re}a < 1;\ [46],\,(3,\,4)].$$

37. $\displaystyle {}_3F_2\left(\begin{array}{c} a,\, a+\tfrac{1}{2},\, 1;\ 1 \\ 2a+\tfrac{1}{2},\, 4a \end{array}\right)$

$$= \frac{2^{18a-7}3^{3/2-6a}\sqrt{\pi}\,(4a-1)^2\csc(2a\pi)}{2a-1}\,\frac{\Gamma^3\!\left(2a-\tfrac{1}{2}\right)}{\Gamma\!\left(2a-\tfrac{1}{3}\right)\Gamma(2a)\,\Gamma\!\left(2a+\tfrac{1}{3}\right)}$$

$$+ \frac{4a-1}{2a-1}\, {}_3F_2\left(\begin{array}{c} 2a,\, 6a-1,\, 1;\ \tfrac{1}{2} \\ 4a,\, 2-2a \end{array}\right)\quad [\operatorname{Re}a > 1/4;\ [46],\,(7,\,8)].$$

38. $\displaystyle {}_3F_2\left(\begin{array}{c} a,\, \tfrac{1}{2},\, 1;\ 1 \\ a+\tfrac{1}{2},\, 2a \end{array}\right) = \frac{2^{4a-2}\csc(a\pi)}{2a-1}\,\frac{\Gamma^3\!\left(a+\tfrac{1}{2}\right)}{\Gamma(a)\,\Gamma\!\left(2a-\tfrac{1}{2}\right)}$

$$+ \frac{3(2a-1)}{4(a-1)}\, {}_3F_2\left(\begin{array}{c} a,\, 2a-\tfrac{1}{2},\, 1;\ \tfrac{1}{4} \\ a+\tfrac{1}{2},\, 2-a \end{array}\right)\quad [\operatorname{Re}a > 1/2;\ [46],\,(7,\,8)].$$

39. $\displaystyle {}_3F_2\left(\begin{array}{c} \tfrac{1}{2},\, \tfrac{1}{2},\, 1 \\ a,\, a;\ 1 \end{array}\right) = 2^{2a-3}\sqrt{\pi}\,\sec^2(a\pi)\,\frac{\Gamma^2(a)\,\Gamma(a-1)}{\Gamma^3\!\left(a-\tfrac{1}{2}\right)}$

$$+ \frac{2(a-1)}{2a-3}\, {}_3F_2\left(\begin{array}{c} \tfrac{1}{2},\, 1,\, a-\tfrac{1}{2} \\ a,\, \tfrac{5}{2}-a;\ 1 \end{array}\right)\quad [1 < \operatorname{Re}a < 3/2;\ [46],\,(1,\,2)].$$

40. $\displaystyle {}_3F_2\left(\begin{array}{c} -n,\, a,\, a+\tfrac{1}{2} \\ b,\, b+\tfrac{1}{2};\ 1 \end{array}\right) = \frac{n!\,(2b-2a)_n}{(2b)_{2n}}\, P_n^{(2b+n-1,\,2a-2b-2n)}(3).$

8.1. The Hypergeometric Functions

41. $\displaystyle {}_3F_2\!\left(\begin{array}{c}-n,\,a,\,b;\,1\\ \frac{a+b+1}{2},\,c\end{array}\right)=\frac{\left(\frac{1-a-b}{2}+c\right)_n}{(c)_n}\,{}_3F_2\!\left(\begin{array}{c}-n,\,\frac{a-b+1}{2},\,\frac{b-a+1}{2};\,1\\ \frac{a+b+1}{2},\,\frac{1-a-b}{2}+c\end{array}\right).$

42. $\displaystyle {}_3F_2\!\left(\begin{array}{c}-n,\,a,\,b;\,1\\ 2a,\,b-a-n+1\end{array}\right)=\frac{(a)_n(2a-b)_n}{(2a)_n(a-b)_n}.$

43. $\displaystyle {}_3F_2\!\left(\begin{array}{c}-n,\,a,\,b;\,1\\ a-n+\frac{1}{2},\,b+\frac{1}{2}\end{array}\right)=\frac{\left(b-a+\frac{1}{2}\right)_n\left(\frac{1}{2}\right)_n}{\left(\frac{1}{2}-a\right)_n\left(b+\frac{1}{2}\right)_n}.$

44. $\displaystyle {}_3F_2\!\left(\begin{array}{c}-n,\,a,\,\frac{1}{2}-a-n;\,1\\ b,\,b+\frac{1}{2}\end{array}\right)$
$\displaystyle =\frac{(2b-2a)_{2n}}{(2b)_{2n}}\,{}_3F_2\!\left(\begin{array}{c}-n,\,a,\,1-2b-2n;\,1\\ a-b-n+\frac{1}{2},\,a-b-n+1\end{array}\right).$

45. $\displaystyle {}_3F_2\!\left(\begin{array}{c}-n,\,\frac{1}{2}-n,\,a;\,1\\ b,\,b+\frac{1}{2}\end{array}\right)=\frac{(2n)!}{(2a-2b+1)(2b)_{2n}}$
$\displaystyle \times\left[2a\,P_{2n}^{(2b-a-1,\,2a-2b-2n+1)}(3)-(2n+1)P_{2n+1}^{(2b-a-2,\,2a-2b-2n)}(3)\right]$
[[55], (3.18)].

46. $\displaystyle {}_3F_2\!\left(\begin{array}{c}-m-n,\,a,\,b;\,1\\ c-m,\,a+b-c\end{array}\right)$
$\displaystyle =\frac{(a-c)_{m+1}(b-c)_{m+1}}{(-c)_{m+1}(a+b-c)_{m+1}}\,{}_3F_2\!\left(\begin{array}{c}-n+1,\,a,\,b;\,1\\ c+1,\,a+b-c+m+1\end{array}\right)\quad[n\geq 1].$

47. $\displaystyle {}_3F_2\!\left(\begin{array}{c}-n,\,a,\,b;\,1\\ \frac{a-n}{2},\,\frac{a-n+1}{2}\end{array}\right)=\frac{(-2)^n n!}{(1-a)_n}\,P_n^{(-n-2b,\,2b-a)}(0).$

48. $\displaystyle {}_3F_2\!\left(\begin{array}{c}-n,\,a,\,b;\,1\\ \frac{b-n}{2},\,a+\frac{b-n}{2}+1\end{array}\right)=\frac{(b+n)\left(a-\frac{b+n}{2}+1\right)_n}{(b-n)\left(-a-\frac{b+n}{2}\right)_n}.$

49. $\displaystyle {}_3F_2\!\left(\begin{array}{c}-n,\,a,\,b;\,1\\ \frac{a-n}{2},\,\frac{a-n+1}{2}\end{array}\right)$
$\displaystyle =\frac{n!}{(1-a)_n}\left[P_n^{(2b-a-1,\,a-n-1)}(3)+2P_{n-1}^{(2b-a,\,a-n)}(3)\right]\quad[n\geq 1].$

50. $\displaystyle {}_3F_2\!\left(\begin{array}{c}-n,\,a,\,b;\,1\\ \frac{b-n+1}{2},\,\frac{b-n}{2}+1\end{array}\right)=\frac{n!\,b}{(b+n)(-b)_n}\,P_n^{(2a-b-1,\,b-n-1)}(3).$

51. $_3F_2\left(\begin{matrix}-n, a, b; 1\\ b+2, \dfrac{a-b+an-bn+ab-n-1}{b}\end{matrix}\right)$
$$=\dfrac{(n+1)(b+1)\left(\dfrac{b-a-an+n+1}{b}\right)_n}{(b+n+1)\left(\dfrac{2b-a-ab-an+n+1}{b}\right)_n}\qquad [[38],\,(1.9)].$$

52. $_3F_2\left(\begin{matrix}-n, \frac{1}{4}, \frac{1}{2}; 1\\ \dfrac{5-4n}{8}, \dfrac{9-4n}{8}\end{matrix}\right)=\dfrac{1-4n}{1+4n}.$

53. $_3F_2\left(\begin{matrix}-2n, 2n+1, \frac{1}{2}\\ 1-2a, 1+2a; 1\end{matrix}\right) = {_3F_2}\left(\begin{matrix}-n, n+\frac{1}{2}, \frac{1}{2}\\ 1-a, 1+a; 1\end{matrix}\right).$

54. $_3F_2\left(\begin{matrix}-2n, a, b; 1\\ 2a, \frac{b}{2}-n\end{matrix}\right) = \dfrac{\left(1+\frac{b}{2}\right)_n}{\left(1-\frac{b}{2}\right)_n}\,{_3F_2}\left(\begin{matrix}-n, a+n, \frac{b+1}{2}; 1\\ a+\frac{1}{2}, \frac{b}{2}+1\end{matrix}\right).$

55. $_3F_2\left(\begin{matrix}-2n, a, b; 1\\ 2a, \frac{b}{2}-n\end{matrix}\right) = \dfrac{\left(a-\frac{b}{2}\right)_n\left(\frac{1}{2}\right)_n}{\left(a+\frac{1}{2}\right)_n\left(1-\frac{b}{2}\right)_n}.$

56. $_3F_2\left(\begin{matrix}-2n, a, b; 1\\ 2a, \frac{b+1}{2}-n\end{matrix}\right) = \dfrac{\left(a+\frac{1-b}{2}\right)_n\left(\frac{1}{2}\right)_n}{\left(a+\frac{1}{2}\right)_n\left(\frac{1-b}{2}\right)_n}.$

57. $_3F_2\left(\begin{matrix}-2n-1, a, b; 1\\ 2a, \frac{b}{2}-n\end{matrix}\right) = 0.$

58. $_3F_2\left(\begin{matrix}-\frac{n}{2}, \frac{1-n}{2}, a\\ a+\frac{1}{2}, b; 1\end{matrix}\right) = \dfrac{(a)_n}{(2a)_n}\,{_3F_2}\left(\begin{matrix}-n, a, 1-b-n\\ 1-a-n, b; -1\end{matrix}\right)\qquad [[55],\,(3.14)].$

59. $_3F_2\left(\begin{matrix}-\frac{n}{2}, \frac{1-n}{2}, a\\ b, b+\frac{1}{2}; 1\end{matrix}\right)$
$$=\dfrac{n!}{(2b)_n}\left[P_n^{(2b-a-1,\,2a-2b-n-1)}(3) + P_{n-1}^{(2b-a,\,2a-2b-n)}(3)\right]\qquad [n\geq 1].$$

60. $_3F_2\left(\begin{matrix}-\frac{n}{2}, \frac{1-n}{2}, a; 1\\ b, a-b-n+\frac{3}{2}\end{matrix}\right) = \dfrac{\left(b-\frac{1}{2}\right)_n(2b-2a-1)_n}{(2b-1)_n\left(b-a-\frac{1}{2}\right)_n}.$

61. $_3F_2\left(\begin{matrix}-\frac{2n}{3}, \frac{1-2n}{3}, \frac{2-2n}{3}\\ a, \frac{1}{2}-n; 1\end{matrix}\right) = \dfrac{n!}{(a)_n}\left(\dfrac{2}{3}\right)^{2n} P_n^{(n+2a-2,\,2-3a-3n)}\left(\dfrac{1}{2}\right).$

8.1. The Hypergeometric Functions

62. $\displaystyle {}_3F_2\left(\begin{array}{c}-\frac{n}{2},\frac{1-n}{2},a;\,1\\ \frac{a-n+1}{3},\frac{a-n+2}{3}\end{array}\right)=\frac{2^{2n}n!}{(1-a)_n}P_n^{(-a/3-2n/3,\,a-n-1)}\left(\frac{5}{4}\right).$

63. $\displaystyle {}_3F_2\left(\begin{array}{c}-n,\frac{n}{2}+a,\frac{n+1}{2}+a\\ \frac{2a+1}{3},\frac{2a+2}{3};\,1\end{array}\right)=\frac{2^n n!}{(2a)_n}P_n^{(-4a/3-n,\,2a-1)}\left(\frac{5}{4}\right)$ [[38], (3.6)].

64. $\displaystyle {}_3F_2\left(\begin{array}{c}-\frac{2n}{3},\frac{1-2n}{3},\frac{2-2n}{3}\\ \frac{1}{2}-n,\,a;\,1\end{array}\right)=\left(\frac{2}{3}\right)^{2n}\frac{n!}{(a)_n}P_n^{(2a+n-2,\,2-3a-3n)}\left(\frac{1}{2}\right)$

[[38], (3.17)].

65. $\displaystyle {}_3F_2\left(\begin{array}{c}-\frac{n}{3},\frac{1-n}{3},\frac{2-n}{3}\\ a,\,2-2a-n;\,1\end{array}\right)=\left(-\frac{4}{3}\right)^n\frac{n!}{(2a-1)_n}P_n^{(1/2-a-n,\,3a-5/2)}\left(\frac{1}{2}\right)$

[[38], (3.18)].

66. $\displaystyle {}_3F_2\left(\begin{array}{c}-\frac{n}{3},\frac{1-n}{3},\frac{2-n}{3};\,1\\ a,\,a+\frac{1}{3}\end{array}\right)=\frac{3^{n/2}n!}{(3a-1)_n}C_n^{a-1/3}\left(\frac{\sqrt{3}}{2}\right)$ [[38], (3.19)].

67. $\displaystyle {}_3F_2\left(\begin{array}{c}-n,\frac{2n+1}{4},\frac{2n+3}{4}\\ \frac{1}{2},\frac{5}{6};\,1\end{array}\right)=\frac{(-2)^n n!}{\left(\frac{5}{6}\right)_n}C_{2n}^{1/6-n}\left(\frac{3}{2\sqrt{2}}\right).$

68. $\displaystyle {}_3F_2\left(\begin{array}{c}-\frac{2n}{3},\frac{1-2n}{3},\frac{2-2n}{3}\\ \frac{1}{2}-n,\frac{1}{2};\,1\end{array}\right)=\frac{3^{-n}n!\left(-\frac{1}{2}\right)_n}{\left(-\frac{1}{2}\right)_{2n}}C_{2n}^{n-1/2}\left(\frac{2}{\sqrt{3}}\right).$

69. $\displaystyle {}_3F_2\left(\begin{array}{c}-\frac{2n}{3},\frac{1-2n}{3},\frac{2-2n}{3}\\ \frac{1}{2}-n,\frac{3}{2};\,1\end{array}\right)=\frac{3^{-n-1/2}n!\left(\frac{3}{2}\right)_n}{2\left(\frac{5}{2}\right)_{2n}}C_{2n+1}^{n+3/2}\left(\frac{2}{\sqrt{3}}\right).$

70. $\displaystyle {}_3F_2\left(\begin{array}{c}-\frac{n}{3},\frac{1-n}{3},\frac{2-n}{3}\\ \frac{5}{6}-n,\frac{7}{6}-n;\,1\end{array}\right)=(-i)^n\frac{3^{n/2}\left(\frac{1}{2}\right)_n\left(-\frac{1}{2}\right)_{2n}}{\left(-\frac{1}{2}\right)_{3n}}T_n(i\sqrt{3}).$

71. $\displaystyle {}_3F_2\left(\begin{array}{c}-mn,\,mn,\,1;\,1\\ 1-ma,\,1+ma\end{array}\right)={}_3F_2\left(\begin{array}{c}-n,\,n,\,1;\,1\\ 1-a,\,1+a\end{array}\right)$ $[m=1,2,\ldots]$.

72. $\displaystyle {}_3F_2\left(\begin{array}{c}-n,\,a,\,b;\,1\\ \frac{b-n+1}{2},\,a+\frac{b-n+1}{2}\end{array}\right)=(-1)^n\frac{\left(a+\frac{1-b-n}{2}\right)_n}{\left(a+\frac{1+b-n}{2}\right)_n}.$

73. $\displaystyle {}_3F_2\left(\begin{array}{c}-n,\,a,\,b;\,1\\ \frac{a+b+1}{2},\,\frac{a+b+1}{2}-n\end{array}\right)=\frac{\left(\frac{a-b+1}{2}\right)_n\left(\frac{b-a+1}{2}\right)_n}{\left(\frac{a+b+1}{2}\right)_n\left(\frac{1-a-b}{2}\right)_n}.$

74. $_3F_2\left(\begin{matrix} a, 2a - \frac{1}{2}, \frac{1}{2} \\ a + \frac{1}{2}, 2a; \ -1 \end{matrix}\right) = \frac{2^{1/2-2a}\pi \Gamma^2\left(a + \frac{1}{2}\right)}{\Gamma^2\left(\frac{4a+3}{8}\right)\Gamma^2\left(\frac{4a+5}{8}\right)}.$

75. $_3F_2\left(\begin{matrix} \frac{1}{2} + a, \frac{1}{2} - a, \frac{1}{2} \\ 1 + a, 1 - a; \ -1 \end{matrix}\right) = \frac{2^{-1/2}\pi^2 a \csc(a\pi)}{\Gamma\left(\frac{5+4a}{8}\right)\Gamma\left(\frac{5-4a}{8}\right)\Gamma\left(\frac{7+4a}{8}\right)\Gamma\left(\frac{7-4a}{8}\right)}.$

76. $_3F_2\left(\begin{matrix} -n, a, b; \ -1 \\ 1 - a - n, 1 - b - n \end{matrix}\right) = \frac{(2a)_n}{(a)_n} {}_3F_2\left(\begin{matrix} -\frac{n}{2}, \frac{1-n}{2}, a; \ 1 \\ a + \frac{1}{2}, 1 - b - n \end{matrix}\right).$

77. $_3F_2\left(\begin{matrix} -n, 2a, 1 + a \\ a, b; \ -1 \end{matrix}\right) = \frac{(b - 2a)_n}{(b)_n} {}_3F_2\left(\begin{matrix} -n, a + \frac{1}{2}, 2a - b + 1; \ 1 \\ a + \frac{1-b-n}{2}, a - \frac{b+n}{2} + 1 \end{matrix}\right).$

78. $_3F_2\left(\begin{matrix} \frac{3}{2}, \frac{3}{2}, \frac{3}{2} \\ 2, 2; \ -1 \end{matrix}\right) = \frac{1}{4\sqrt{2}\pi^3}\Gamma^2\left(\frac{1}{8}\right)\Gamma^2\left(\frac{3}{8}\right) - \frac{4}{\pi}.$

79. $_3F_2\left(\begin{matrix} \frac{1}{2}, \frac{1}{2}, \frac{1}{2} \\ \frac{3}{2}, \frac{3}{2}; \ -\frac{1}{4} \end{matrix}\right) = \frac{\pi^2}{10}.$

80. $_3F_2\left(\begin{matrix} \frac{1}{2}, \frac{1}{2}, \frac{1}{2} \\ \frac{3}{2}, \frac{3}{2}; \ -\frac{1}{8} \end{matrix}\right) = \frac{1}{6\sqrt{2}}\left(\pi^2 - 3\ln^2 2\right).$

81. $_3F_2\left(\begin{matrix} \frac{3}{2}, \frac{3}{2}, \frac{3}{2} \\ 2, 2; \ -\frac{1}{8} \end{matrix}\right) = -\frac{64\sqrt{2}}{3\pi} + \frac{4\sqrt{2}}{3\pi^3}\Gamma^4\left(\frac{1}{4}\right).$

82. $_3F_2\left(\begin{matrix} \frac{3}{2}, \frac{3}{2}, \frac{3}{2} \\ 2, 2; \ \frac{1}{64} \end{matrix}\right) = \frac{4096}{21\pi} - \frac{320}{21\sqrt{7}\pi^4}\Gamma^2\left(\frac{1}{7}\right)\Gamma^2\left(\frac{2}{7}\right)\Gamma^2\left(\frac{4}{7}\right).$

83. $_3F_2\left(\begin{matrix} \frac{1}{2}, 1, 1 \\ \frac{1}{4}, \frac{3}{4}; \ \frac{1}{16} \end{matrix}\right) = \frac{16}{15} + \frac{\pi\sqrt{3}}{27} - \frac{2\sqrt{5}}{25}\ln\frac{1+\sqrt{5}}{2}.$

84. $_3F_2\left(\begin{matrix} a, b, 1 - b; \ \frac{1}{4} \\ \frac{a+b}{2}, \frac{a-b+1}{2} \end{matrix}\right) = \frac{2^{2a/3}\Gamma\left(\frac{a+b}{2}\right)\Gamma\left(\frac{a-b+1}{2}\right)}{3\Gamma(a)}$

$\times \left[\frac{\Gamma\left(\frac{a}{3}\right)}{\Gamma\left(\frac{a+3b}{6}\right)\Gamma\left(\frac{a-3b+3}{6}\right)} + \frac{2^{1/3}\Gamma\left(\frac{a+2}{3}\right)}{\Gamma\left(\frac{a+3b+2}{6}\right)\Gamma\left(\frac{a-3b+5}{6}\right)}\right]$ [[52], (ii)].

85. $${}_3F_2\left(\begin{array}{c} a, b, 1-b;\ \frac{1}{4} \\ \frac{a+b+1}{2},\ \frac{a-b}{2}+1 \end{array}\right) = \frac{2^{2a/3}\Gamma\left(\frac{a}{3}+1\right)\Gamma\left(\frac{a+b+1}{2}\right)\Gamma\left(\frac{a-b}{2}+1\right)}{\Gamma\left(\frac{a+3b+3}{6}\right)\Gamma\left(\frac{a-3b}{6}+1\right)\Gamma(a+1)}$$
[[52], (i)].

86. $${}_3F_2\left(\begin{array}{c} a, b, 3-b;\ \frac{1}{4} \\ \frac{a+b+1}{2},\ \frac{a-b}{2}+2 \end{array}\right) = \frac{2^{2(a+2)/3}\Gamma\left(\frac{a+b+1}{2}\right)\Gamma\left(\frac{a-b}{2}+2\right)}{3(b-1)(b-2)\Gamma(a)}$$

$$\times\left[-\frac{2^{2/3}\Gamma\left(\frac{a}{3}\right)}{\Gamma\left(\frac{a+3b-3}{6}\right)\Gamma\left(\frac{a-3b+6}{6}\right)} + \frac{\Gamma\left(\frac{a+2}{3}\right)}{\Gamma\left(\frac{a+3b-1}{6}\right)\Gamma\left(\frac{a-3b+8}{6}\right)}\right]$$ [[52], (iii)].

87. $${}_3F_2\left(\begin{array}{c} -3n-m, a, 1-a;\ \frac{1}{4} \\ \frac{a-3n-m+1}{2},\ 1-\frac{a+3n+m}{2} \end{array}\right) = \frac{(3n)!\,\delta_{0,m}}{2^{2n}n!\left(\frac{a+n}{2}\right)_n\left(\frac{n-a+1}{2}\right)_n}$$
$[m=0,1,2;\ [52], (iv)]$.

88. $${}_3F_2\left(\begin{array}{c} -3n-m, a, 1-a;\ \frac{1}{4} \\ \frac{a-3n-m}{2},\ \frac{1-a-3n-m}{2} \end{array}\right) = \frac{(3n)!\,\delta_{0,m}}{2^{2n}n!\left(\frac{a+n+1}{2}\right)_n\left(\frac{n-a+2}{2}\right)_n}$$
$$+\frac{(3n+2)!\,\delta_{2,m}}{2^{2n+1}n!\left(\frac{a+n+1}{2}\right)_{n+1}\left(\frac{n-a}{2}+1\right)_{n+1}}$$ $[m=0,1,2;\ [52], (v)]$.

89. $${}_3F_2\left(\begin{array}{c} a,\ \frac{4}{3}-a,\ 1-3a;\ \frac{1}{4} \\ \frac{1}{3}-a,\ \frac{3}{2}-2a \end{array}\right) = \frac{3^{3a}\Gamma\left(\frac{3}{2}-2a\right)\Gamma\left(\frac{5}{3}\right)}{2^{4a-1}\sqrt{\pi}\,\Gamma\left(\frac{5}{3}-2a\right)}$$ [[52], (vi)].

90. $${}_3F_2\left(\begin{array}{c} a, 1, 2;\ \frac{1}{4} \\ \frac{a}{2}+1,\ \frac{a+3}{2} \end{array}\right)$$
$$= \frac{a(a+1)}{6}\left[\psi\left(\frac{a+3}{6}\right)-\psi\left(\frac{a}{6}\right)+\psi\left(\frac{a+2}{6}\right)-\psi\left(\frac{a+5}{6}\right)\right]$$ [[52], (vii)].

91. $${}_3F_2\left(\begin{array}{c} a, 2a-\frac{1}{2}, 1;\ \frac{1}{4} \\ a+\frac{1}{2},\ 2-a \end{array}\right) = -\frac{2^{4a}(a-1)\csc(a\pi)}{3(2a-1)^2}\frac{\Gamma^3\left(a+\frac{1}{2}\right)}{\Gamma(a)\Gamma\left(2a-\frac{1}{2}\right)}$$
$$+\frac{4(a-1)}{3(2a-1)}\,{}_3F_2\left(\begin{array}{c} a,\ \frac{1}{2},\ 1;\ 1 \\ a+\frac{1}{2},\ 2a \end{array}\right)$$ $[\mathrm{Re}\,a>1/2;\ [46], (7,8)]$.

92. $${}_3F_2\left(\begin{array}{c} \frac{1}{2}, 1, 1;\ \frac{1}{4} \\ \frac{3}{2},\ \frac{3}{2} \end{array}\right) = \frac{8}{3}\mathbf{G} - \frac{\pi}{3}\ln\left(2+\sqrt{3}\right).$$

93. $${}_3F_2\left(\begin{array}{c} \frac{1}{2}, \frac{1}{2}, \frac{1}{2};\ \frac{1}{4} \\ \frac{3}{2}, \frac{3}{2};\ \frac{1}{4} \end{array}\right) = \frac{1}{16\sqrt{3}}\left[\zeta\left(2,\frac{1}{6}\right)-\zeta\left(2,\frac{1}{3}\right)+\zeta\left(2,\frac{2}{3}\right)-\zeta\left(2,\frac{5}{6}\right)\right].$$

94. $_3F_2\left(\begin{matrix}\frac{1}{2},1,1\\ \frac{3}{2},\frac{3}{2};\frac{1}{4}\end{matrix}\right) = \frac{1}{3}\left[8G - \pi\ln\left(2+\sqrt{3}\right)\right]$ [[52], (viii)].

95. $_3F_2\left(\begin{matrix}1,1,1\\ \frac{3}{2},2;\frac{1}{4}\end{matrix}\right) = \frac{\pi^2}{9}$.

96. $_3F_2\left(\begin{matrix}\frac{3}{2},\frac{3}{2},\frac{3}{2}\\ 2,2;\frac{1}{4}\end{matrix}\right) = \frac{64}{3\pi} - \frac{4}{3\sqrt{3}\pi^3}\Gamma^2\left(\frac{1}{6}\right)\Gamma^2\left(\frac{1}{3}\right)$.

97. $_3F_2\left(\begin{matrix}a,1,1;\frac{1}{2}\\ \frac{a}{2}+1,2\end{matrix}\right) = \frac{a}{2(a-1)}\left[\psi\left(\frac{a}{2}\right) + 2\ln 2 + C\right]$.

98. $_3F_2\left(\begin{matrix}1,1,1\\ 2,a;\frac{1}{2}\end{matrix}\right) = \frac{1-a}{4}$

$\times\left\{\left[\psi\left(\frac{a-1}{2}\right) - \psi\left(\frac{a}{2}\right)\right]^2 - \zeta\left(2,\frac{a-1}{2}\right) - \zeta\left(2,\frac{a}{2}\right)\right\}$ [[28], (4.18)].

99. $_3F_2\left(\begin{matrix}a,3a-1,1;\frac{1}{2}\\ 2a,2-a\end{matrix}\right) = -2^{9a-7}3^{3/2-3a}\sqrt{\pi}\,\frac{\csc(a\pi)(2a-1)\Gamma^3\left(a-\frac{1}{2}\right)}{\Gamma\left(a-\frac{1}{3}\right)\Gamma(a-1)\Gamma\left(a+\frac{1}{3}\right)}$

$+ \frac{a-1}{2a-1}\,_3F_2\left(\begin{matrix}\frac{a}{2},\frac{a+1}{2},1;1\\ 2a,a+\frac{1}{2}\end{matrix}\right)$ [Re $a > 1/4$; [46], (7, 8)].

100. $_3F_2\left(\begin{matrix}\frac{1}{2},\frac{1}{2},\frac{1}{2}\\ \frac{3}{2},\frac{3}{2};\frac{1}{2}\end{matrix}\right) = \frac{\pi}{2^{5/2}}\ln 2 + \frac{G}{\sqrt{2}}$.

101. $_3F_2\left(\begin{matrix}\frac{1}{2},\frac{1}{2},\frac{1}{2}\\ \frac{3}{2},\frac{3}{2};\frac{3}{4}\end{matrix}\right) = \frac{\pi}{3\sqrt{3}}\ln 3$

$+ \frac{1}{72}\left[\zeta\left(2,\frac{1}{6}\right) - \zeta\left(2,\frac{1}{3}\right) + \zeta\left(2,\frac{2}{3}\right) - \zeta\left(2,\frac{5}{6}\right)\right]$.

102. $_3F_2\left(\begin{matrix}-3n-1, a, -a-3n-\frac{1}{2}\\ 2a, -2a-6n-1;4\end{matrix}\right) = 0$ [[27], (2.8)].

103. $_3F_2\left(\begin{matrix}-3n-2, a, -a-3n-\frac{3}{2}\\ 2a, -2a-6n-3;4\end{matrix}\right) = 0$ [[27], (2.7)].

104. $_3F_2\left(\begin{matrix}\frac{1}{4},\frac{1}{3},\frac{2}{3}\\ \frac{1}{2},\frac{5}{4};27(17-12\sqrt{2})\end{matrix}\right) = \frac{1+\sqrt{2}}{16\sqrt{\pi}}\Gamma^2\left(\frac{1}{4}\right)$.

8.1.3. The hypergeometric function $_4F_3(a_1, a_2, a_3, a_4; b_1, b_2, b_3; z)$

1. $_4F_3\left(\begin{array}{c} a, a+\frac{1}{3}, a+\frac{2}{3}, b \\ \frac{3a}{2}, \frac{3a+1}{2}, b+1; -\frac{27z}{4(1-z)^3} \end{array}\right) = (1-z)^{3a} {}_2F_1\left(\begin{array}{c} 1, 3a-2b \\ b+1; z \end{array}\right)$

$\left[|z|, |27z(1-z)^{-3}/4| < 1; [[38], (5.13)]\right]$.

2. $_4F_3\left(\begin{array}{c} a, a+\frac{1}{3}, a+\frac{2}{3}, a-n \\ \frac{3a}{2}, \frac{3a+1}{2}, a-n+1; -\frac{27z}{4(1-z)^3} \end{array}\right)$
$= -\frac{(3n-1)!\,a}{(-a)_n(a)_{2n}} z^{3n-1}(1-z)^{3a-3n} P_{3n-1}^{(-3n, 1-a-2n)}\left(\frac{2}{z}-1\right)$

$\left[n \geq 1; |z|, |27z(1-z)^{-3}/4| < 1\right]$.

3. $_4F_3\left(\begin{array}{c} n+\frac{1}{2}, n+\frac{5}{6}, n+\frac{7}{6}, \frac{1}{2} \\ \frac{6n+3}{4}, \frac{6n+5}{4}, \frac{3}{2}; -\frac{27z}{4(1-z)^3} \end{array}\right)$
$= -\frac{(3n-1)!}{\left(\frac{3}{2}\right)_{3n-1}} \frac{(1-z)^{3/2}}{\sqrt{z}} C_{6n-1}^{1/2-3n}(\sqrt{z})$ $\left[n \geq 1; |z|, |27z(1-z)^{-3}/4| < 1\right]$.

4. $_4F_3\left(\begin{array}{c} n-\frac{1}{2}, n-\frac{1}{6}, n+\frac{1}{6}, -\frac{1}{2} \\ \frac{6n-3}{4}, \frac{6n-1}{4}, \frac{1}{2}; -\frac{27z}{4(1-z)^3} \end{array}\right) = \frac{(3n-1)!}{\left(\frac{3}{2}\right)_{3n-1}} (1-z)^{-3/2} C_{6n-2}^{1/2-3n}(\sqrt{z})$

$\left[n \geq 1; |z|, |27z(1-z)^{-3}/4| < 1\right]$.

5. $_4F_3\left(\begin{array}{c} -n, -n, -n, \frac{1}{2}-n \\ -2n, -2n, 1; z \end{array}\right)$
$= \frac{(n!)^2}{(2n)!} \left(\frac{z}{4}\right)^n P_n\left(\frac{\sqrt{1-z}-3}{\sqrt{1-z}+1}\right) P_n\left(\frac{\sqrt{1-z}+3}{\sqrt{1-z}-1}\right)$.

6. $_4F_3\left(\begin{array}{c} -n, -n, -n, \frac{1}{2}-n \\ -2n, \frac{1}{2}-2n, \frac{1}{2}; z \end{array}\right)$
$= \frac{(n!)^2}{\left(\frac{1}{2}\right)_{2n}} \left(\frac{z}{4}\right)^n P_{2n}\left(\sqrt{\frac{\sqrt{1-z}+1}{\sqrt{1-z}-1}}\right) P_{2n}\left(\sqrt{\frac{\sqrt{1-z}-1}{\sqrt{1-z}+1}}\right)$.

7. $_4F_3\left(\begin{array}{c} -n, n+\frac{1}{2}, \frac{1}{4}, \frac{3}{4} \\ \frac{1}{2}, \frac{1}{2}, 1; z \end{array}\right)$
$= (-1)^n \frac{n!}{\left(\frac{1}{2}\right)_n} P_{2n}\left(\sqrt{\frac{1-\sqrt{1-z}}{2}}\right) P_{2n}\left(\sqrt{\frac{1+\sqrt{1-z}}{2}}\right)$.

8. $_4F_3\left(\begin{array}{c}-n, n+\frac{3}{2}, \frac{3}{4}, \frac{5}{4}\\ 1, \frac{3}{2}, \frac{3}{2}; z\end{array}\right)$

$$= (-1)^n \frac{n!2}{\left(\frac{3}{2}\right)_n \sqrt{z}} P_{2n+1}\left(\sqrt{\frac{1-\sqrt{1-z}}{2}}\right) P_{2n+1}\left(\sqrt{\frac{1+\sqrt{1-z}}{2}}\right).$$

9. $_4F_3\left(\begin{array}{c}-n, \frac{1}{4}-n, \frac{1}{2}-n, \frac{3}{4}-n\\ \frac{1}{2}-2n, \frac{1}{2}-2n, 1; z\end{array}\right)$

$$= \frac{(2n)!}{\left(\frac{1}{2}\right)_{2n}} \left(\frac{z}{16}\right)^n P_{2n}\left(\sqrt{\frac{2}{1+\sqrt{1-z}}}\right) P_{2n}\left(\sqrt{\frac{2}{1-\sqrt{1-z}}}\right).$$

10. $_4F_3\left(\begin{array}{c}-n, -\frac{1}{2}-n, -\frac{1}{4}-n, \frac{1}{4}-n\\ -\frac{1}{2}-2n, -\frac{1}{2}-2n, 1; z\end{array}\right)$

$$= \frac{n!}{\left(n+\frac{3}{2}\right)_n} \left(\frac{z}{4}\right)^{n+1/2} P_{2n+1}\left(\sqrt{\frac{2}{1+\sqrt{1-z}}}\right) P_{2n+1}\left(\sqrt{\frac{2}{1-\sqrt{1-z}}}\right).$$

11. $_4F_3\left(\begin{array}{c}-n, \frac{1-2n}{4}, \frac{3-2n}{4}, \frac{1}{2}\\ \frac{1}{2}-n, \frac{1}{2}-n, 1; z\end{array}\right)$

$$= \frac{n!}{\left(\frac{1}{2}\right)_n} \left(\frac{z}{4}\right)^{n/2} P_n\left(\frac{3+\sqrt{1-z}}{2^{3/2}\sqrt{1+\sqrt{1-z}}}\right) P_n\left(\frac{3-\sqrt{1-z}}{2^{3/2}\sqrt{1-\sqrt{1-z}}}\right).$$

12. $_4F_3\left(\begin{array}{c}-n, \frac{1}{4}-n, \frac{1}{2}-n, \frac{3}{4}-n\\ -2n, \frac{1}{2}-2n, \frac{3}{2}; z\end{array}\right)$

$$= \frac{1}{2n+1} \left(\frac{z}{16}\right)^n U_{2n}\left(\sqrt{\frac{2}{1+\sqrt{1-z}}}\right) U_{2n}\left(\sqrt{\frac{2}{1-\sqrt{1-z}}}\right).$$

13. $_4F_3\left(\begin{array}{c}-n, -\frac{1}{2}-n, -\frac{1}{4}-n, \frac{1}{4}-n\\ -\frac{1}{2}-2n, -1-2n, \frac{3}{2}; z\end{array}\right)$

$$= \frac{1}{2n+2} \left(\frac{z}{16}\right)^{n+1/2} U_{2n+1}\left(\sqrt{\frac{2}{1+\sqrt{1-z}}}\right) U_{2n+1}\left(\sqrt{\frac{2}{1-\sqrt{1-z}}}\right).$$

14. $_4F_3\left(\begin{array}{c}-n, \frac{1}{4}-n, \frac{1}{2}-n, \frac{3}{4}-n\\ \frac{1}{2}-2n, 1-2n, \frac{1}{2}; z\end{array}\right)$

$$= 2\left(\frac{z}{16}\right)^n T_n\left(\frac{3-\sqrt{1-z}}{1+\sqrt{1-z}}\right) T_n\left(\frac{3+\sqrt{1-z}}{1-\sqrt{1-z}}\right) \quad [n \geq 1].$$

15. $${}_4F_3\left(\begin{matrix}-n,\, -\frac{1}{4}-n,\, -\frac{1}{2}-n,\, \frac{1}{4}-n \\ -\frac{1}{2}-2n,\, -2n,\, \frac{1}{2};\, z\end{matrix}\right)$$
$$= 2^{-4n-1}z^{n+1/2}T_{2n+1}\left(\sqrt{\frac{2}{1-\sqrt{1-z}}}\right)T_{2n+1}\left(\sqrt{\frac{2}{1+\sqrt{1-z}}}\right).$$

16. $${}_4F_3\left(\begin{matrix}-\frac{n}{2},\, \frac{1-n}{2},\, a,\, a+\frac{1}{2} \\ \frac{1}{4},\, \frac{1}{2},\, \frac{3}{4};\, z\end{matrix}\right) = \frac{n!}{2(1-2a)_n}$$
$$\times\left[C_{2n}^{2a-n}\left(z^{1/4}\right) + (-4)^n\frac{(2a-n)_{2n}}{(4a)_{2n}}(1+z^{1/2})^n\right.$$
$$\left.\times C_{2n}^{1/2-2a-n}\left(\frac{z^{1/4}}{\sqrt{1+z^{1/2}}}\right)\right].$$

17. $${}_4F_3\left(\begin{matrix}-\frac{n}{2},\, \frac{1-n}{2},\, \frac{n}{2},\, \frac{n+1}{2};\, z \\ \frac{1}{4},\, \frac{1}{2},\, \frac{3}{4}\end{matrix}\right)$$
$$= \frac{1}{2}\left[(-1)^n T_{2n}(z^{1/4}) + T_{2n}\left(\sqrt{1+z^{1/2}}\right)\right].$$

18. $${}_4F_3\left(\begin{matrix}-\frac{n}{2},\, \frac{1-n}{2},\, \frac{n}{2}+1,\, \frac{n+3}{2} \\ \frac{3}{4},\, \frac{5}{4},\, \frac{3}{2};\, z\end{matrix}\right)$$
$$= \frac{z^{-1/2}}{4(n+1)^2}\left[(-1)^n T_{2n+2}(z^{1/4}) + T_{2n+2}\left(\sqrt{1+z^{1/2}}\right)\right].$$

19. $${}_4F_3\left(\begin{matrix}-\frac{n}{2},\, \frac{1-n}{2},\, \frac{n+1}{2},\, \frac{n}{2}+1 \\ \frac{1}{2},\, \frac{3}{4},\, \frac{5}{4};\, z\end{matrix}\right)$$
$$= \frac{1}{2(2n+1)}\left[(-1)^n z^{-1/4}T_{2n+1}(z^{1/4}) + U_{2n}\left(\sqrt{1+z^{1/2}}\right)\right].$$

20. $${}_4F_3\left(\begin{matrix}-\frac{n}{2},\, \frac{1-n}{2},\, \frac{n+3}{2},\, \frac{n}{2}+2 \\ \frac{5}{4},\, \frac{3}{2},\, \frac{7}{4};\, z\end{matrix}\right)$$
$$= \frac{3z^{-3/4}}{4(n+1)(n+2)(2n+3)}\left[(-1)^n T_{2n+3}(z^{1/4}) + z^{1/4}U_{2n+2}\left(\sqrt{1+z^{1/2}}\right)\right].$$

21. $${}_4F_3\left(\begin{matrix}\frac{1}{4},\, \frac{1}{4},\, \frac{3}{4},\, \frac{3}{4} \\ \frac{1}{2},\, 1,\, 1;\, z\end{matrix}\right) = \frac{4}{\pi^2}\mathbf{K}\left(\sqrt{\frac{1}{2}-\frac{k}{2}}\right)\mathbf{K}\left(\sqrt{\frac{1}{2}-\frac{1}{2k}}\right)$$
$$\left[k = 1 - 2z - 2\sqrt{z^2-z}\right].$$

22. $${}_4F_3\left(\begin{matrix}\frac{1}{2},\, \frac{1}{2},\, \frac{5}{6},\, \frac{7}{6} \\ \frac{3}{4},\, \frac{5}{4},\, \frac{3}{2};\, -\frac{27z}{4(1-z)^3}\end{matrix}\right) = \frac{(1-z)^{3/2}}{2\sqrt{z}}\ln\frac{1+\sqrt{z}}{1-\sqrt{z}} \qquad [-1/2 < z < 1].$$

23. $\displaystyle {}_4F_3\left(\begin{array}{c}\frac{1}{2},\frac{1}{2},1,1\\ \frac{3}{4},\frac{5}{4},\frac{3}{2};\end{array}-z\right) = \frac{\left(1+2z-2\sqrt{z^2+z}\right)^{1/2}}{4\left(\sqrt{z^2+z}-z\right)}$

$$\times \arcsin\sqrt{2\sqrt{z^2+z}+2z}\, \ln \frac{1+\sqrt{2\sqrt{z^2+z}-2z}}{1-\sqrt{2\sqrt{z^2+z}-2z}}.$$

24. $\displaystyle {}_4F_3\left(\begin{array}{c}\frac{3}{4},\frac{3}{4},\frac{13}{12},\frac{17}{12}\\ \frac{9}{8},\frac{13}{8},\frac{7}{4};\end{array} -\frac{27z}{4(1-z)^3}\right) = \frac{3(1-z)^{9/4}}{4z^{3/4}}\left(\ln\frac{1+z^{1/4}}{1-z^{1/4}} - 2\arctan z^{1/4}\right)$

$$\left[|z|,\ |27z(1-z)^{-3}/4| < 1\right].$$

25. $\displaystyle {}_4F_3\left(\begin{array}{c}1,1,1,1\\ \frac{3}{2},2,2;\ z\end{array}\right)$

$$= \frac{i}{3z}\left[2\arcsin^3\sqrt{z} - 6i\arcsin^2\sqrt{z}\,\ln\left(1-e^{-2i\arcsin\sqrt{z}}\right)\right.$$
$$\left. + 6\arcsin\sqrt{z}\,\mathrm{Li}_2\left(e^{-2i\arcsin\sqrt{z}}\right) - 3i\,\mathrm{Li}_3\left(e^{-2i\arcsin\sqrt{z}}\right) + 3i\zeta(3)\right].$$

26. $\displaystyle {}_4F_3\left(\begin{array}{c}1,1,1,1\\ \frac{3}{2},2,2;\ -z\end{array}\right) = \frac{1}{3z}\left\{-2\ln^3\left(\sqrt{z}+\sqrt{z+1}\right)\right.$

$$+ 6\ln^2\left(\sqrt{z}+\sqrt{z+1}\right)\left[i\pi + \ln\left(2z+2\sqrt{z+z^2}\right)\right]$$
$$+ 6\ln\left(\sqrt{z}+\sqrt{z+1}\right)\mathrm{Li}_2\left(1+2z+2\sqrt{z+z^2}\right)$$
$$\left. - 3\,\mathrm{Li}_3\left(1+2z+2\sqrt{z+z^2}\right) + 3\zeta(3)\right\}.$$

27. $\displaystyle {}_4F_3\left(\begin{array}{c}1,1,1,\frac{3}{2}\\ \frac{5}{4},\frac{7}{4},2;\ -z\end{array}\right) = -\frac{3}{z}\arcsin^2\frac{\left(\sqrt{z}-\sqrt{z+1}+1\right)^{1/2}}{\sqrt{2}}$

$$+\frac{3}{z}\ln^2\left[\left(\sqrt{z}+\sqrt{z+1}+\sqrt{2}\sqrt{z+\sqrt{z}\sqrt{z+1}}\right)^{1/2}\right].$$

28. $\displaystyle {}_4F_3\left(\begin{array}{c}1,1,\frac{4}{3},\frac{5}{3}\\ \frac{3}{2},2,2;\ -\frac{27z}{4(1-z)^3}\end{array}\right) = -\frac{(1-z)^3}{z}\ln(1-z)$

$$\left[|z|,\ |27z(1-z)^{-3}/4| < 1\right].$$

8.1. The Hypergeometric Functions

29. $${}_4F_3\left(\begin{matrix} a, a+\frac{1}{2}, b, b+\frac{1}{2}; 1 \\ a-b+\frac{1}{2}, a-b+1, \frac{1}{2} \end{matrix}\right)$$
$$= \frac{\Gamma(2a-2b+1)}{2}\left[\frac{2^{-4b}}{\sqrt{\pi}}\frac{\Gamma\left(\frac{1}{2}-2b\right)}{\Gamma(2a-4b+1)} + \frac{2^{-2a}\sqrt{\pi}}{\Gamma\left(a+\frac{1}{2}\right)\Gamma(a-2b+1)}\right]$$
$$[\text{Re } b < 1/4;\ [31],\ (2.2)].$$

30. $${}_4F_3\left(\begin{matrix} a, a+\frac{1}{2}, b, b+\frac{1}{2}; 1 \\ a-b+1, a-b+\frac{3}{2}, \frac{3}{2} \end{matrix}\right)$$
$$= \frac{\Gamma(2a-2b+2)}{(2a-1)(2b-1)}\left[\frac{2^{1-4b}}{\sqrt{\pi}}\frac{\Gamma\left(\frac{3}{2}-2b\right)}{\Gamma(2a-4b+2)} - \frac{2^{-2a}\sqrt{\pi}}{\Gamma(a)\Gamma\left(a-2b+\frac{3}{2}\right)}\right]$$
$$[\text{Re } b < 1/4;\ [31],\ (2.3)].$$

31. $${}_4F_3\left(\begin{matrix} a, b, 1, 1; 1 \\ \frac{a}{2}+1, 2b-1, 2 \end{matrix}\right) = \frac{a}{2(a-1)}$$
$$\times \left[2\ln 2 + \mathbf{C} + \psi\left(\frac{a}{2}\right) + \psi\left(b-\frac{1}{2}\right) - \psi\left(b-\frac{a}{2}\right)\right] \quad [\text{Re}(2b-a) > 0].$$

32. $${}_4F_3\left(\begin{matrix} a, b, c, 2c-2; 1 \\ c-1, 2c-b-1, d \end{matrix}\right) = \frac{\Gamma(d)\Gamma(d-a-2b)}{\Gamma(d-a)\Gamma(d-2b)}\, {}_3F_2\left(\begin{matrix} a, -b, 2c-2b-2; 1 \\ 2c-b-1, d-2b \end{matrix}\right).$$

33. $${}_4F_3\left(\begin{matrix} a, 1, 1, 1; 1 \\ 2a-1, \frac{3}{2}, 2 \end{matrix}\right) = \frac{\pi^2}{8} + \frac{1}{4}\psi'\left(a-\frac{1}{2}\right) \qquad [\text{Re } a > 1/2].$$

34. $${}_4F_3\left(\begin{matrix} a, b, 1, 1; 1 \\ \frac{a}{2}+1, 2b-1, 2 \end{matrix}\right)$$
$$= \frac{a}{2(a-1)}\left[\psi\left(\frac{a}{2}\right) + \psi\left(b-\frac{1}{2}\right) - \psi\left(b-\frac{a}{2}\right) + 2\ln 2 + \mathbf{C}\right].$$

35. $${}_4F_3\left(\begin{matrix} a, 1, 1, 1; 1 \\ 3-a, 2, 2 \end{matrix}\right) = \frac{a-2}{2(a-1)}\left[\frac{\pi^2}{6} - \psi'(2-a)\right] \qquad [a < 2].$$

36. $$= \zeta(3) \qquad [a = 1].$$

37. $${}_4F_3\left(\begin{matrix} -n, a, b, b+\frac{1}{2}; 1 \\ c, \frac{a-n}{2}, \frac{a-n+1}{2} \end{matrix}\right) = \frac{(2b-a+1)_n}{(1-a)_n}\, {}_3F_2\left(\begin{matrix} -n, 2b, 2b-c+1; -1 \\ 2b-a+1, c \end{matrix}\right)$$
$$[[55],\ (3.30)].$$

38. $${}_4F_3\left(\begin{matrix} -n, a, b, b+\frac{1}{2}; 1 \\ a-n+\frac{1}{2}, c, 2b-c+1 \end{matrix}\right)$$
$$= \frac{(c-2b)_n}{(c)_n}\, {}_3F_2\left(\begin{matrix} -2n, a-n, 2b; 1 \\ 2a-2n, 2b-c-n+1 \end{matrix}\right) \quad [[55],\ (3.14)].$$

39. $\displaystyle {}_4F_3\left(\begin{array}{c}-n,\,a,\,b,\,c;\,1\\2a,\,\frac{b+c}{2},\,\frac{b+c+1}{2}\end{array}\right)=\frac{(c)_{2n}}{(b+c)_{2n}}{}_4F_3\left(\begin{array}{c}-n,\,\frac{1}{2}-a-n,\,\frac{b}{2},\,\frac{b+1}{2};\,1\\a+\frac{1}{2},\,\frac{1-c}{2}-n,\,1-\frac{c}{2}-n\end{array}\right)$

[[55], (3.12)].

40. $\displaystyle {}_4F_3\left(\begin{array}{c}-n,\,a,\,b,\,c;\,1\\a-n+\frac{1}{2},\,\frac{b+c}{2},\,\frac{b+c+1}{2}\end{array}\right)={}_3F_2\left(\begin{array}{c}-2n,\,2a,\,c;\,1\\a-n+\frac{1}{2},\,b+c\end{array}\right)$ [[55], (3.12)].

41. $\displaystyle {}_4F_3\left(\begin{array}{c}-n,\,a,\,b,\,b+\frac{1}{2};\,1\\1-a-n,\,\frac{1}{2}-b-n,\,1-b-n\end{array}\right)$
$$=\frac{(4b)_{2n}}{(2b)_{2n}}{}_4F_3\left(\begin{array}{c}-n,\,\frac{1}{2}-a-n,\,2b;\,1\\1-2a-2n,\,2b+\frac{1}{2}\end{array}\right)\quad [[55],\,(3.12)].$$

42. $\displaystyle {}_4F_3\left(\begin{array}{c}-n,\,a,\,b,\,c;\,1\\a+b+\frac{1}{2},\,\frac{c-n}{2},\,\frac{c-n+1}{2}\end{array}\right)$
$$=\frac{\left(a+b-c+\frac{1}{2}\right)_n}{(1-a)_n}{}_3F_2\left(\begin{array}{c}-n,\,a-b+\frac{1}{2},\,b-a+\frac{1}{2};\,1\\a+b+\frac{1}{2},\,c-a-b-n+\frac{1}{2}\end{array}\right)\quad [[55],\,(3.16)].$$

43. $\displaystyle {}_4F_3\left(\begin{array}{c}-n,\,a,\,b,\,b+\frac{1}{2};\,1\\1-a-n,\,1-b-n,\,\frac{3}{2}-b-n\end{array}\right)$
$$=\frac{(4b-1)_{2n}}{(2b-1)_{2n}}{}_4F_3\left(\begin{array}{c}-n,\,\frac{1}{2}-a-n,\,2b-1;\,1\\1-2a-2n,\,2b-\frac{1}{2}\end{array}\right)\quad [[55],\,(3.12)].$$

44. $\displaystyle {}_4F_3\left(\begin{array}{c}-n,\,a,\,b,\,1-b\\-b,\,c,\,d;\,1\end{array}\right)=\frac{(d-a)_n}{(d)_n}{}_4F_3\left(\begin{array}{c}-n,\,a,\,c-b-1,\,\frac{c-b+1}{2}\\c,\,\frac{c-b-1}{2},\,a-d-n+1;\,1\end{array}\right).$

45. $\displaystyle {}_4F_3\left(\begin{array}{c}-n,\,1,\,1,\,1;\,1\\b,\,2,\,2\end{array}\right)$
$$=\frac{1-b}{n+1}\left\{[\psi(n+2)+\mathbf{C}]\,\psi(b-1)-\sum_{k=0}^{n}\frac{1}{k+1}\psi(k+b)\right\}.$$

46. $\displaystyle {}_4F_3\left(\begin{array}{c}-n,\,a,\,b,\,b+\frac{1}{2};\,1\\\frac{a-n}{2},\,\frac{a-n+1}{2},\,c\end{array}\right)=\frac{(2b-a+1)_n}{(1-a)_n}{}_3F_2\left(\begin{array}{c}-n,\,2b,\,2b-c+1;\,-1\\2b-a+1,\,c\end{array}\right)$

[[55], (3.30)].

47. $\displaystyle {}_4F_3\left(\begin{array}{c}-n,\,a,\,b,\,a+b+\frac{1}{2};\,1\\\frac{a-n}{2},\,\frac{a-n+1}{2},\,a+2b+1\end{array}\right)$
$$=\frac{(a+2b+1)_n}{(1-a)_n}{}_4F_3\left(\begin{array}{c}-n,\,a,\,a+b+1,\,2a+2b;\,1\\a+b,\,a+2b+1,\,a+2b+1\end{array}\right)\quad [[55],\,(3.20)].$$

48. $_4F_3\left(\begin{matrix}-n,\,a,\,a+\frac{1}{2},\,b;\,1\\ \frac{b-n+1}{2},\,\frac{b-n}{2}+1,\,c\end{matrix}\right)$

$$=\frac{b(2a-b)_n}{(b+n)(-b)_n}{}_4F_3\left(\begin{matrix}-n,\,a+\frac{1}{2},\,2a-1,\,2a-c;\,-1\\ a-\frac{1}{2},\,2a-b,\,c\end{matrix}\right)\qquad [[55],\,(3.18)].$$

49. $_4F_3\left(\begin{matrix}-n,\,a,\,b,\,c;\,1\\ a+b+\frac{1}{2},\,\frac{c-n}{2},\,\frac{c-n+1}{2}\end{matrix}\right)$

$$=\frac{\left(a+b-c+\frac{1}{2}\right)_n}{(1-c)_n}{}_3F_2\left(\begin{matrix}-n,\,a-b+\frac{1}{2},\,b-a+\frac{1}{2};\,1\\ a+b+\frac{1}{2},\,c-a-b-n+\frac{1}{2}\end{matrix}\right)\qquad [[55],\,(3.6)].$$

50. $_4F_3\left(\begin{matrix}-n,\,a,\,b,\,c;\,1\\ a+b+\frac{1}{2},\,\frac{c-n}{2},\,\frac{c-n+1}{2}\end{matrix}\right)={}_3F_2\left(\begin{matrix}-n,\,2a,\,2b;\,1\\ a+b+\frac{1}{2},\,c-n\end{matrix}\right)\qquad [[55],\,(3.6)].$

51. $_4F_3\left(\begin{matrix}-n,\,\frac{1}{2}-n,\,a,\,b;\,1\\ \frac{a+b+1}{2},\,\frac{a+b}{2}+1,\,c\end{matrix}\right)$

$$=\frac{a(a+1)_{2n}}{(a-b)(a+b+1)_{2n}}{}_4F_3\left(\begin{matrix}-2n-1,\,\frac{1}{2}-n,\,b,\,-c-2n;\,-1\\ -n-\frac{1}{2},\,-a-2n,\,c\end{matrix}\right)\qquad [[55],\,(3.18)].$$

52. $_4F_3\left(\begin{matrix}-\frac{n}{2},\,\frac{1-n}{2},\,a,\,b;\,1\\ \frac{a+b}{2},\,\frac{a+b+1}{2},\,c\end{matrix}\right)=\frac{(b)_n}{(a+b)_n}{}_3F_2\left(\begin{matrix}-n,\,a,\,1-c-n;\,-1\\ c,\,1-b-n\end{matrix}\right)$

$$[[55],\,(3.30)].$$

53. $_4F_3\left(\begin{matrix}-n,\,\frac{1}{2}-n,\,\frac{1}{6},\,\frac{1}{2};\,1\\ \frac{1}{3}-n,\,\frac{2}{3}-n,\,\frac{7}{6}\end{matrix}\right)=\frac{(2n)!\left(\frac{3}{2}\right)_{3n}}{(3n)!\left(\frac{3}{2}\right)_{2n}}{}_3F_2\left(\begin{matrix}-n,\,\frac{1}{6},\,\frac{1}{2};\,\frac{1}{4}\\ 2n+\frac{3}{2},\,\frac{7}{6}\end{matrix}\right).$

54. $_4F_3\left(\begin{matrix}-n,\,\frac{1}{6},\,\frac{1}{2},\,\frac{2}{3};\,1\\ \frac{4-3n}{9},\,\frac{7-3n}{9},\,\frac{10-3n}{9}\end{matrix}\right)=\frac{n!}{\left(-\frac{1}{3}\right)_n(6n+1)}P_n^{(1/6,\,-n-2/3)}\left(\frac{1}{2}\right).$

55. $_4F_3\left(\begin{matrix}-n,\,\frac{1}{6},\,\frac{1}{2},\,\frac{5}{3};\,1\\ \frac{7-3n}{9},\,\frac{10-3n}{9},\,\frac{13-3n}{9}\end{matrix}\right)=-\frac{n!}{4\left(-\frac{4}{3}\right)_n(6n-5)(6n+1)}$

$$\times\left[20P_n^{(1/6,\,-n-2/3)}\left(\frac{1}{2}\right)+3P_{n-1}^{(7/6,\,-n+1/3)}\left(\frac{1}{2}\right)\right]\qquad [n\geq 1].$$

56. $_4F_3\left(\begin{matrix}-\frac{n}{2},\,\frac{1-n}{2},\,a,\,b;\,1\\ a-b+1,\,b-\frac{n}{2},\,b+\frac{1-n}{2}\end{matrix}\right)$

$$=\frac{(2a-2b+1)_n}{(1-2b)_n}{}_3F_2\left(\begin{matrix}-n,\,a,\,2a-2b+n+1;\,1\\ a-b+1,\,2a-2b+1\end{matrix}\right)\qquad [[55],\,(3.12)].$$

57. $_4F_3\left(\begin{matrix}-\frac{n}{2},\frac{1-n}{2},a,-a;\,1\\b,1-b-n,\frac{1}{2}\end{matrix}\right)=\frac{1}{2(b)_n}[(a+b)_n+(b-a)_n].$

58. $_4F_3\left(\begin{matrix}-\frac{n}{2},\frac{1-n}{2},a,b;\,1\\a-b+1,b+\frac{1-n}{2},b-\frac{n}{2}+1\end{matrix}\right)$
$=\frac{(2a-2b)_n}{(-2b)_n}\,_3F_2\left(\begin{matrix}-n,a,2a-2b+n;\,1\\a-b,2a-2b+1\end{matrix}\right)$ [[55], (3.12)].

59. $_4F_3\left(\begin{matrix}-2n,2n+2,\frac{1}{2},1;\,1\\1-2a,1+2a,\frac{3}{2}\end{matrix}\right)=\,_4F_3\left(\begin{matrix}-n,n+1,\frac{1}{2},1;\,1\\1-a,1+a,\frac{3}{2}\end{matrix}\right).$

60. $_4F_3\left(\begin{matrix}-2n,2n+2a,\frac{1}{2},1;\,1\\a+\frac{1}{2},1-2i,1+2i\end{matrix}\right)=\,_4F_3\left(\begin{matrix}-n,n+a,\frac{1}{2},1;\,1\\a+\frac{1}{2},1-i,1+i\end{matrix}\right).$

61. $_4F_3\left(\begin{matrix}-\frac{n}{2},\frac{1-n}{2},a,a+\frac{1}{2};\,1\\b,c,c+\frac{1}{2}\end{matrix}\right)$
$=\frac{(2c-2a)_n}{(2c)_n}\,_3F_2\left(\begin{matrix}-n,2a,b-\frac{1}{2};\,2\\2b-1,2a-2c-n+1\end{matrix}\right)$ [[55], (3.4)].

62. $_4F_3\left(\begin{matrix}-\frac{n}{2},\frac{1-n}{2},a,a+n;\,1\\1-n,a+\frac{n+1}{2},a+\frac{n}{2}+1\end{matrix}\right)=\frac{2(a+n)_n}{(2a+n+1)_n}.$

63. $_4F_3\left(\begin{matrix}-\frac{n}{2},\frac{1-n}{2},a,b;\,1\\a+b+\frac{1}{2},c,1-c-n\end{matrix}\right)=\frac{(a+c)_n}{(c)_n}\,_3F_2\left(\begin{matrix}-n,2a,a+b;\,1\\2a+2b,a+c\end{matrix}\right)$

[[55], (3.8)].

64. $_4F_3\left(\begin{matrix}-\frac{n}{2},\frac{1-n}{2},a,b;\,1\\\frac{a+b+1}{2},\frac{a+3b}{2},1-b-n\end{matrix}\right)$
$=\frac{(b)_n}{(a+b)_n}\,_4F_3\left(\begin{matrix}-n,a,b,1-\frac{a+3b}{2}-n;\,1\\\frac{a+3b}{2},1-b-n,1-b-n\end{matrix}\right)$ [[55], (3.20)].

65. $_4F_3\left(\begin{matrix}-\frac{n}{3},\frac{1-n}{3},\frac{2-n}{3},a;\,1\\\frac{n+1}{3}+a,\frac{n+2}{3}+a,\frac{n}{3}+a+1\end{matrix}\right)$
$=\frac{(2n)!(3a+1)_n}{(3a+1)_{2n}\left(\frac{1}{2}\right)_n}P_n^{(a-1/2,-n-1/2)}\left(\frac{1}{2}\right).$

66. $_4F_3\left(\begin{array}{c}-\frac{n}{3},\ \frac{1-n}{3},\ \frac{2-n}{3},\ a;\ 1\\ \frac{2+a-n}{3},\ \frac{3+a-n}{3},\ \frac{4+a-n}{3}\end{array}\right)$

$$= \frac{(2a-1)_n}{(-a-1)_n}\ _2F_1\left(\begin{array}{c}-n,\ 2a+n-1;\ \frac{3}{4}\\ a-\frac{1}{2}\end{array}\right).$$

67. $\quad = \dfrac{(-1)^n n!}{2(-a-1)_n}\left[3C^a_{n-1}\left(\dfrac{1}{2}\right) + \dfrac{2a+n-2}{a-1}C^{a-1}_n\left(\dfrac{1}{2}\right)\right]$ $\quad [n \geq 1]$.

68. $_4F_3\left(\begin{array}{c}-\frac{n}{3},\ \frac{1-n}{3},\ \frac{2-n}{3},\ \frac{1}{2};\ 1\\ \frac{3-2n}{6},\ \frac{5-2n}{6},\ \frac{7-2n}{6}\end{array}\right) = \dfrac{(-1)^n n!}{\left(-\frac{1}{2}\right)_n} P_n\left(\dfrac{1}{2}\right).$

69. $_4F_3\left(\begin{array}{c}-\frac{n}{3},\ \frac{1-n}{3},\ \frac{2-n}{3},\ \frac{1}{2};\ 1\\ \frac{2n+5}{6},\ \frac{2n+7}{6},\ \frac{2n+9}{6}\end{array}\right) = \dfrac{12^{n/2}n!\left(\frac{5}{2}\right)_n}{\left(\frac{5}{2}\right)_{2n}} P_n\left(\dfrac{7}{4\sqrt{3}}\right).$

70. $_4F_3\left(\begin{array}{c}-\frac{n}{3},\ \frac{1-n}{3},\ \frac{2-n}{3},\ \frac{3}{2};\ 1\\ \frac{7-2n}{6},\ \frac{9-2n}{6},\ \frac{11-2n}{6}\end{array}\right)$

$$= \dfrac{(-1)^n n!}{2\left(-\frac{5}{2}\right)_n}\left[3C^{3/2}_{n-1}\left(\dfrac{1}{2}\right) + 2(n+1)P_n\left(\dfrac{1}{2}\right)\right] \quad [n \geq 1].$$

71. $_4F_3\left(\begin{array}{c}-\frac{n}{3},\ \frac{1-n}{3},\ \frac{2-n}{3},\ -n;\ 1\\ \frac{1-2n}{3},\ \frac{2-2n}{3},\ 1-\frac{2n}{3}\end{array}\right) = \dfrac{(-i)^n 2^{1-2n} 3^{n/2} n!}{\left(\frac{1}{2}\right)_n} P_n\left(\dfrac{i}{\sqrt{3}}\right)$ $\quad [n \geq 1]$.

72. $_4F_3\left(\begin{array}{c}-\frac{1+2n}{3},\ -\frac{2n}{3},\ \frac{1-2n}{3},\ -\frac{1}{2}-2n\\ -\frac{1+8n}{6},\ \frac{1-8n}{6},\ \frac{3-8n}{6};\ 1\end{array}\right) = \dfrac{3^n \left(\frac{3}{2}\right)^2_{2n}}{(4n+1)\left(\frac{3}{2}\right)_{4n}} U_{2n}\left(\dfrac{2}{\sqrt{3}}\right).$

73. $_4F_3\left(\begin{array}{c}-\frac{1+2n}{3},\ -\frac{2n}{3},\ \frac{1-2n}{3},\ -\frac{3}{2}-2n\\ -\frac{3+8n}{6},\ -\frac{1+8n}{6},\ \frac{1-8n}{6};\ 1\end{array}\right) = \dfrac{3^{n-1}(4n+3)\left(\frac{3}{2}\right)^2_{2n}}{(n+1)\left(\frac{5}{2}\right)_{4n}} U_n\left(\dfrac{5}{3}\right).$

74. $_4F_3\left(\begin{array}{c}\frac{1}{2},\ 1,\ 1,\ 1\\ \frac{3}{2},\ \frac{3}{2},\ \frac{3}{2};\ 1\end{array}\right) = \dfrac{7}{2}\zeta(3) - \pi\mathbf{G}.$

75. $_4F_3\left(\begin{array}{c}1,\ 1,\ 1,\ 1\\ \frac{3}{2},\ \frac{3}{2},\ 2;\ 1\end{array}\right) = 2\pi\mathbf{G} - \dfrac{7}{2}\zeta(3).$

76. $_4F_3\left(\begin{array}{c}1,\ 1,\ 1,\ 1\\ \frac{3}{2},\ 2,\ 2;\ 1\end{array}\right) = \dfrac{\pi^2}{2}\ln 2 - \dfrac{7}{4}\zeta(3).$

77. $_4F_3\left(\begin{matrix}1, \frac{3}{2}, \frac{3}{2}, \frac{3}{2}\\ 2, \frac{5}{2}, \frac{5}{2}; 1\end{matrix}\right) = 9\pi \ln 2 - 18.$

78. $_4F_3\left(\begin{matrix}a, b, c, 2c-2; -1\\ c-1, 2c-a-1, 2c-b-1\end{matrix}\right) = \frac{\Gamma(2c-a-1)\Gamma(2c-b-1)}{\Gamma(2c-1)\Gamma(2c-a-b-1)}$
$[\operatorname{Re}(c-a-b) > 1/2].$

79. $_4F_3\left(\begin{matrix}a, a, a, \frac{a}{2}+1\\ \frac{a}{2}, 1, 1; -1\end{matrix}\right) = \frac{\sin(a\pi)}{a\pi}.$

80. $_4F_3\left(\begin{matrix}a, \frac{1}{2}, \frac{1}{2}, \frac{5}{4}; -1\\ \frac{3}{2}-a, \frac{1}{4}, 1\end{matrix}\right) = \frac{2\Gamma\left(\frac{3}{2}-a\right)}{\sqrt{\pi}\,\Gamma(1-a)}$ $[\operatorname{Re} a < 3/4].$

81. $_4F_3\left(\begin{matrix}\frac{1}{2}, \frac{1}{2}, \frac{1}{2}, \frac{5}{4}\\ \frac{1}{4}, 1, 1; -1\end{matrix}\right) = \frac{2}{\pi}.$

82. $_4F_3\left(\begin{matrix}\frac{1}{3}, \frac{1}{2}, \frac{2}{3}, \frac{6}{5}\\ \frac{1}{5}, 1, 1; -\frac{9}{16}\end{matrix}\right) = \frac{4}{\sqrt{3}\,\pi}.$

83. $_4F_3\left(\begin{matrix}a, a+\frac{1}{2}, \frac{6-a}{5}, \frac{1}{2}\\ \frac{1-a}{5}, \frac{3}{2}-2a, 1; -\frac{1}{4}\end{matrix}\right) = \frac{2\Gamma\left(\frac{3}{2}-2a\right)}{\sqrt{\pi}(1-a)\Gamma(1-2a)}$ $[[45], (3.1)].$

84. $_4F_3\left(\begin{matrix}\frac{1}{8}, \frac{1}{2}, \frac{5}{8}, \frac{47}{40}\\ \frac{7}{40}, 1, \frac{5}{4}; -\frac{1}{4}\end{matrix}\right) = \frac{2\sqrt{2}}{7\pi^{3/2}}\Gamma^2\left(\frac{1}{4}\right).$

85. $_4F_3\left(\begin{matrix}\frac{1}{4}, \frac{1}{2}, \frac{3}{4}, \frac{23}{20}\\ \frac{3}{20}, 1, 1; -\frac{1}{4}\end{matrix}\right) = \frac{8}{3\pi}$ $[[46], (5, 6)].$

86. $_4F_3\left(\begin{matrix}1, 1, 1, 1\\ \frac{3}{2}, 2, 2; -\frac{1}{4}\end{matrix}\right) = \frac{4}{5}\zeta(3).$

87. $_4F_3\left(\begin{matrix}a+1, 3a, 3a, 1-3a\\ a, 3a+\frac{1}{2}, 1; -\frac{1}{8}\end{matrix}\right) = \frac{2^{3a}\Gamma\left(3a+\frac{1}{2}\right)}{\sqrt{\pi}\,\Gamma(3a+1)}$ $[[45], (2.3)].$

88. $_4F_3\left(\begin{matrix}\frac{1}{2}, \frac{1}{2}, \frac{1}{2}, \frac{7}{6}\\ \frac{1}{6}, 1, 1; -\frac{1}{8}\end{matrix}\right) = \frac{2\sqrt{2}}{\pi}$ $[[43], (17)].$

89. $_4F_3\left(\begin{matrix}1, 1, 1, 1\\ \frac{3}{2}, 2, 2; -\frac{1}{8}\end{matrix}\right) = \zeta(3) - \frac{2}{3}\ln^3 2.$

8.1. The Hypergeometric Functions

90. $_4F_3\left(\begin{array}{c} \frac{1}{3}, \frac{1}{2}, \frac{2}{3}, \frac{58}{51} \\ \frac{7}{51}, 1, 1; -\frac{1}{16} \end{array}\right) = \dfrac{12\sqrt{3}}{7\pi}$ [[43], (9)].

91. $_4F_3\left(\begin{array}{c} \frac{1}{3}, \frac{1}{2}, \frac{2}{3}, \frac{10}{9} \\ \frac{1}{9}, 1, 1; -\frac{1}{80} \end{array}\right) = \dfrac{4}{\pi}\sqrt{\dfrac{3}{5}}$ [[43], (10)].

92. $_4F_3\left(\begin{array}{c} \frac{1}{4}, \frac{1}{2}, \frac{3}{4}, \frac{31}{28} \\ \frac{3}{28}, 1, 1; -\frac{1}{48} \end{array}\right) = \dfrac{16}{3\sqrt{3}\,\pi}$ [[43], (2)].

93. $_4F_3\left(\begin{array}{c} \frac{1}{4}, \frac{1}{2}, \frac{3}{4}, \frac{283}{260} \\ \frac{23}{260}, 1, 1; -\frac{1}{324} \end{array}\right) = \dfrac{72}{23\pi}$ [[43], (3)].

94. $_4F_3\left(\begin{array}{c} \frac{1}{6}, \frac{1}{2}, \frac{5}{6}, \frac{5681}{5418} \\ \frac{263}{5418}, 1, 1; -\frac{1}{512000} \end{array}\right) = \dfrac{640}{263\pi}\sqrt{\dfrac{5}{3}}$ [[43], (6)].

95. $_4F_3\left(\begin{array}{c} \frac{1}{4}, \frac{1}{2}, \frac{3}{4}, \frac{299}{280} \\ \frac{19}{280}, 1, 1; \frac{1}{9801} \end{array}\right) = \dfrac{18\sqrt{11}}{19\pi}$ [[43], (25)].

96. $_4F_3\left(\begin{array}{c} \frac{1}{4}, \frac{1}{2}, \frac{3}{4}, \frac{43}{40} \\ \frac{3}{40}, 1, 1; \frac{1}{2401} \end{array}\right) = \dfrac{49}{9\sqrt{3}\,\pi}$ [[43], (24)].

97. $_4F_3\left(\begin{array}{c} \frac{1}{4}, \frac{1}{2}, \frac{3}{4}, \frac{11}{10} \\ \frac{1}{10}, 1, 1; \frac{1}{81} \end{array}\right) = \dfrac{9}{2\sqrt{2}\,\pi}$ [[43], (23)].

98. $_4F_3\left(\begin{array}{c} \frac{1}{2}, \frac{1}{2}, \frac{1}{2}, \frac{47}{42} \\ \frac{5}{42}, 1, 1; \frac{1}{64} \end{array}\right) = \dfrac{16}{5\pi}$ [[43], (21)].

99. $_4F_3\left(\begin{array}{c} a, a+\frac{1}{2}, 1-2a, \frac{5-2a}{4} \\ \frac{1-2a}{4}, \frac{3}{2}-2a, 1; \frac{1}{9} \end{array}\right) = \dfrac{3^{2a}\Gamma\left(\frac{3}{2}-2a\right)}{2^{4a-1}\sqrt{\pi}\,\Gamma(2-2a)}$ [[45], (3.2)].

100. $_4F_3\left(\begin{array}{c} \frac{1}{4}, \frac{1}{2}, \frac{3}{4}, \frac{9}{8} \\ \frac{1}{8}, 1, 1; \frac{1}{9} \end{array}\right) = \dfrac{2\sqrt{3}}{\pi}$ [[43], (22)].

101. $_4F_3\left(\begin{array}{c} \frac{1}{2}, \frac{1}{2}, \frac{1}{2}, a+1 \\ a, \frac{6a+3}{4}, \frac{6a+3}{4}; \frac{1}{4} \end{array}\right) = \dfrac{2\Gamma^2\left(\frac{6a+3}{4}\right)}{3a\,\Gamma^2\left(\frac{6a+1}{4}\right)}$ [[45], (2.1)].

102. $_4F_3\left(\begin{array}{c} a+1, 3a, 1-3a, \frac{1}{2} \\ a, 3a+\frac{1}{2}, 1; \frac{1}{4} \end{array}\right) = \dfrac{2\Gamma\left(3a+\frac{1}{2}\right)}{\sqrt{\pi}\,\Gamma(3a+1)}$ [[45], (2.2)].

103. $${}_4F_3\left(\begin{matrix}\frac{1}{2},\frac{1}{2},\frac{1}{2},\frac{1}{2}\\ \frac{3}{2},\frac{3}{2},\frac{3}{2};\frac{1}{4}\end{matrix}\right)=\frac{7\pi^3}{216}.$$

104. $${}_4F_3\left(\begin{matrix}\frac{1}{2},\frac{1}{2},\frac{1}{2},\frac{7}{6}\\ \frac{1}{6},1,1;\frac{1}{4}\end{matrix}\right)=\frac{4}{\pi}$$ [[43], (20)].

105. $${}_4F_3\left(\begin{matrix}1,1,1,1\\ \frac{3}{2},2,2;\frac{1}{4}\end{matrix}\right)$$
$$=-\frac{8}{3}\zeta(3)+\frac{\pi}{12\sqrt{3}}\left[\zeta\left(2,\frac{1}{6}\right)-\zeta\left(2,\frac{1}{3}\right)+\zeta\left(2,\frac{2}{3}\right)-\zeta\left(2,\frac{5}{6}\right)\right].$$

106. $${}_4F_3\left(\begin{matrix}\frac{1}{4},\frac{1}{2},\frac{3}{4},\frac{8}{7}\\ \frac{1}{7},1,1;\frac{32}{81}\end{matrix}\right)=\frac{9}{2\pi}$$ [[43], (29)].

107. $${}_4F_3\left(\begin{matrix}1,1,1,1\\ \frac{3}{2},2,2;\frac{1}{2}\end{matrix}\right)=\frac{\pi^2}{8}\ln 2-\frac{35}{16}\zeta(3)+\pi\mathbf{G}.$$

108. $${}_4F_3\left(\begin{matrix}1,1,1,1\\ \frac{3}{2},2,2;\frac{3}{4}\end{matrix}\right)$$
$$=\frac{4\pi^2}{27}\ln 3-\frac{52}{27}\zeta(3)+\frac{\pi}{27\sqrt{3}}\left[\zeta\left(2,\frac{1}{6}\right)-\zeta\left(2,\frac{1}{3}\right)+\zeta\left(2,\frac{2}{3}\right)-\zeta\left(2,\frac{5}{6}\right)\right].$$

109. $${}_4F_3\left(\begin{matrix}-n,-n,-\frac{2n}{3}+1,\frac{1}{2}\\ -\frac{2n}{3},1,1;4\end{matrix}\right)=0$$ $[n\neq 3n]$.

110. $${}_4F_3\left(\begin{matrix}\frac{1}{6},\frac{2}{9},\frac{5}{9},\frac{8}{9}\\ \frac{1}{3},\frac{5}{6},\frac{7}{6};108(56\sqrt{3}-97)\end{matrix}\right)=\frac{(2+\sqrt{3})^{2/3}}{2}\sqrt{\frac{3}{\pi}}\Gamma\left(\frac{4}{3}\right)\Gamma\left(\frac{7}{6}\right).$$

8.1.4. The hypergeometric function ${}_5F_4((a_1,\ldots,a_5);(b_1,\ldots,b_4);z)$

1. $${}_5F_4\left(\begin{matrix}\frac{1}{2},\frac{1}{2},1,1,1\\ \frac{3}{4},1,\frac{5}{4},\frac{3}{2};-z\end{matrix}\right)=\frac{2}{\sqrt{z}}\arcsin\frac{\left(\sqrt{z}-\sqrt{z+1}+1\right)^{1/2}}{\sqrt{2}}$$
$$\times\ln\left[\left(\sqrt{z}+\sqrt{z+1}+\sqrt{2}\sqrt{z+\sqrt{z}\sqrt{z+1}}\right)^{1/2}\right].$$

8.1. The Hypergeometric Functions

2. $_5F_4\left(\begin{array}{c} a,\,b,\,c,\,2c-2,\,-\frac{1}{2};\,1 \\ c-1,\,2c-\frac{1}{2},\,2c-a-1,\,2c-b-1 \end{array}\right)$

$$= \frac{\Gamma\left(2c-\frac{1}{2}\right)\Gamma(2c-a-1)\Gamma(2c-b-1)\Gamma\left(2c-a-b-\frac{1}{2}\right)}{\Gamma(2c-1)\Gamma\left(2c-a-\frac{1}{2}\right)\Gamma\left(2c-b-\frac{1}{2}\right)\Gamma(2c-a-b-1)}$$

$$[\operatorname{Re}(2c-a-b) > 1/2].$$

3. $_5F_4\left(\begin{array}{c} a,\,b,\,b+1,\,a-b-1,\,2a-2;\,1 \\ a-1,\,a+b,\,2a-b-2,\,2a-b-1 \end{array}\right)$

$$= \frac{4^{b-a+1}\sqrt{\pi}\,\Gamma(a+b)\,\Gamma(2a-b-2)\Gamma(2a-b-1)}{\Gamma^2(a)\,\Gamma\left(a-b-\frac{1}{2}\right)\Gamma(2a-2)} \quad [\operatorname{Re}(a-b) > 1].$$

4. $_5F_4\left(\begin{array}{c} a,\,a,\,b,\,b,\,2a+2b-1;\,1 \\ a+2b,\,a+2b,\,2a+b,\,2a+b \end{array}\right)$

$$= \frac{\Gamma^2(a+2b)\Gamma^2(2a+b)}{\Gamma^2(a+b)\Gamma^2(2a+2b)}\,_3F_2\left(\begin{array}{c} 2a,\,2b,\,a+b;\,1 \\ 2a+2b,\,2a+2b \end{array}\right) \quad [\operatorname{Re}(a+b) > 0].$$

5. $_5F_4\left(\begin{array}{c} a,\,b,\,c,\,2c-2,\,\frac{1}{2};\,1 \\ c-1,\,2c-\frac{3}{2},\,2c-a-1,\,2c-b-1 \end{array}\right)$

$$= \frac{\Gamma\left(2c-\frac{3}{2}\right)\Gamma(2c-a-1)\Gamma(2c-b-1)\Gamma\left(2c-a-b-\frac{3}{2}\right)}{\Gamma(2c-1)\Gamma\left(2c-a-\frac{3}{2}\right)\Gamma\left(2c-b-\frac{3}{2}\right)\Gamma(2c-a-b-1)}$$

$$[\operatorname{Re}(2c-a-b) > 3/2].$$

6. $_5F_4\left(\begin{array}{c} -n,\,a,\,b,\,c,\,c+\frac{1}{2};\,1 \\ \frac{a-n}{2},\,\frac{a-n+1}{2},\,a+b,\,2c-a+1 \end{array}\right)$

$$= \frac{(2c-a-1)_n}{(1-a)_n}\,_4F_3\left(\begin{array}{c} -n,\,a,\,2c,\,2c-a-b+1;\,1 \\ a+b,\,2c-a+1,\,2c-a+1 \end{array}\right) \quad [[55],\,(3.20)].$$

7. $_5F_4\left(\begin{array}{c} -n,\,a,\,b,\,a+b+\frac{1}{2};\,1 \\ \frac{a-n}{2},\,\frac{a-n+1}{2},\,a+2b+1 \end{array}\right)$

$$= \frac{(a+2b+1)_n}{(1-a)_n}\,_4F_3\left(\begin{array}{c} -n,\,a,\,a+b+1,\,2a+2b;\,1 \\ a+b,\,a+2b+1,\,a+2b+1 \end{array}\right) \quad [[55],\,(3.20)].$$

8. $_5F_4\left(\begin{array}{c} -n,\,a,\,a+\frac{1}{2},\,\frac{1}{6},\,\frac{1}{2};\,1 \\ \frac{7}{6},\,\frac{2a-n}{3},\,\frac{2a-n+1}{3},\,\frac{2a-n+2}{3} \end{array}\right) = \frac{\left(\frac{3}{2}-2a\right)_n}{(1-2a)_n}\,_3F_2\left(\begin{array}{c} -n,\,\frac{1}{6},\,\frac{1}{2};\,\frac{1}{4} \\ \frac{7}{6},\,\frac{3}{2}-2a \end{array}\right).$

9. $_5F_4\left(\begin{array}{c}-n, a, a+\frac{1}{2}, b, b+\frac{1}{3}; 1\\ \frac{2a-n}{3}, \frac{2a-n+1}{3}, \frac{2a-n+2}{3}, 2b+\frac{5}{6}\end{array}\right)$

$$= \frac{(3b-2a+1)_n}{(1-2a)_n} {}_3F_2\left(\begin{array}{c}-n, \frac{1}{3}-b, 3b; \frac{1}{4}\\ 2b+\frac{5}{6}, 3b-a+1\end{array}\right) \quad [[55], (3.26)].$$

10. $_5F_4\left(\begin{array}{c}-n, a, b, b+\frac{1}{3}, b+\frac{2}{3}; 1\\ \frac{2a-n-1}{3}, \frac{2a-n}{3}, \frac{2a-n+1}{3}, 3b-a+2\end{array}\right)$

$$= \frac{(3b-2a+2)_n}{(2-2a)_n} {}_4F_3\left(\begin{array}{c}-n, 3b, 3b-2a+3, 2a-3b-2; \frac{1}{4}\\ a-\frac{1}{2}, 3b-a+2, c, 3b-2a+2\end{array}\right) \quad [[55], (3.26)].$$

11. $_5F_4\left(\begin{array}{c}-\frac{n}{2}, \frac{1-n}{2}, a, b, c; 1\\ 1-a-n, a+b, \frac{a+c}{2}, \frac{a+c+1}{2}\end{array}\right)$

$$= \frac{(a)_n}{(a+c)_n} {}_4F_3\left(\begin{array}{c}-n, a, 1-a-b, c; 1\\ 1-a-n, 1-a-n, a+b\end{array}\right) \quad [[55], (3.20)].$$

12. $_5F_4\left(\begin{array}{c}-\frac{n}{2}, \frac{1-n}{2}, a, b+\frac{1}{3}, b+\frac{2}{3}; 1\\ \frac{a-n}{3}, \frac{a-n+1}{3}, \frac{a-n+2}{3}, 2b+\frac{3}{2}\end{array}\right)$

$$= \frac{(3b-a+1)_n}{(1-a)_n} {}_4F_3\left(\begin{array}{c}-n, b-\frac{1}{2}, 2b+1, 3b; 4\\ 2b-1, 3b-a+1, 4b+2\end{array}\right) \quad [[55], (3.28)].$$

13. $_5F_4\left(\begin{array}{c}-\frac{n}{2}, \frac{1-n}{2}, a, b, b+\frac{1}{3}; 1\\ \frac{a-n}{3}, \frac{a-n+1}{3}, \frac{a-n+2}{3}, 2b+\frac{5}{6}\end{array}\right)$

$$= \frac{(3b-a)_n}{(1-a)_n} {}_4F_3\left(\begin{array}{c}-n, b-\frac{5}{6}, 2b+\frac{1}{3}, 3b-1; 4\\ 2b-\frac{5}{3}, 3b-a, 4b+\frac{2}{3}\end{array}\right) \quad [[55], (3.28)].$$

14. $_5F_4\left(\begin{array}{c}-2n, -2n, -2n, -2n, 3n+1\\ -5n, 1, 1, 1; 1\end{array}\right) = (-1)^n \frac{[(3n)!]^3 (4n)!}{[n!]^4 [(2n)!]^2 (5n)!}.$

15. $_5F_4\left(\begin{array}{c}a, b, c, d, \frac{d}{2}+1; -1\\ \frac{d}{2}, d-b+1, d-c+1, d-a+1\end{array}\right) = \frac{\Gamma(d-c+1)\Gamma(d-a+1)}{\Gamma(d+1)\Gamma(d-c-a+1)}$

$$\times {}_3F_2\left(\begin{array}{c}a, c, \frac{d+1}{2}-b; 1\\ \frac{d+1}{2}, d-b+1\end{array}\right) \quad [\mathrm{Re}\,(d-a-c) > -1].$$

16. $_5F_4\left(\begin{array}{c}\frac{5}{6}, 1, 1, \frac{7}{6}, \frac{17}{10}\\ \frac{7}{10}, \frac{3}{2}, \frac{3}{2}, \frac{3}{2}; -\frac{9}{16}\end{array}\right) = \frac{8}{21} \ln \frac{27}{4} \quad [[43], (30)].$

8.1. The Hypergeometric Functions

17. $\displaystyle {}_5F_4\left(\begin{matrix} a+1, a+\frac{1}{10}, a+\frac{7}{20}, a+\frac{3}{5}, 1 \\ a, a+\frac{17}{20}, a+\frac{17}{20}, a+\frac{17}{20}; -\frac{1}{4} \end{matrix}\right) = \frac{20a-3}{25a}$

$\displaystyle \times {}_3F_2\left(\begin{matrix} a+\frac{7}{20}, \frac{1}{2}, 1; 1 \\ a+\frac{17}{20}, 2a+\frac{7}{10} \end{matrix}\right)$ [Re $a > 3/20$; [46], (7, 8)].

18. $\displaystyle {}_5F_4\left(\begin{matrix} \frac{1}{2}, \frac{3}{4}, 1, 1, \frac{7}{5} \\ \frac{2}{5}, \frac{5}{4}, \frac{5}{4}, \frac{5}{4}; -\frac{1}{4} \end{matrix}\right) = \frac{1}{64\pi}\Gamma^4\left(\frac{1}{4}\right)$ [[43]].

19. $\displaystyle {}_5F_4\left(\begin{matrix} \frac{3}{4}, 1, 1, \frac{5}{4}, \frac{33}{20} \\ \frac{13}{20}, \frac{3}{2}, \frac{3}{2}, \frac{3}{2}; -\frac{1}{4} \end{matrix}\right) = \frac{16}{13}\ln 2$ [[43]].

20. $\displaystyle {}_5F_4\left(\begin{matrix} a+1, a+\frac{1}{3}, a+\frac{1}{3}, a+\frac{1}{3}, 1 \\ a, a+\frac{5}{6}, a+\frac{5}{6}, a+\frac{5}{6}; -\frac{1}{8} \end{matrix}\right) = \frac{6a-1}{9a}$

$\displaystyle \times {}_3F_2\left(\begin{matrix} \frac{3a+1}{6}, \frac{3a+4}{6}, 1 \\ a+\frac{5}{6}, a+\frac{5}{6}, a+\frac{5}{6}; 1 \end{matrix}\right)$ [Re $a > 1/6$; [46], (5, 6)].

21. $\displaystyle {}_5F_4\left(\begin{matrix} 1, 1, 1, 1, \frac{5}{3} \\ \frac{2}{3}, \frac{3}{2}, \frac{3}{2}, \frac{3}{2}; -\frac{1}{8} \end{matrix}\right) = G.$

22. $\displaystyle {}_5F_4\left(\begin{matrix} a+1, a+\frac{16}{231}, a+\frac{31}{77}, a+\frac{170}{231}, 1 \\ a, a+\frac{139}{154}, a+\frac{139}{154}, a+\frac{139}{154}; -\frac{27}{512} \end{matrix}\right) = \frac{32(154a-15)}{5929a}$

$\displaystyle \times {}_3F_2\left(\begin{matrix} \frac{a}{2}+\frac{31}{154}, \frac{a}{2}+\frac{54}{77}, 1; 1 \\ a+\frac{139}{154}, 2a+\frac{62}{77} \end{matrix}\right)$ [Re $a > 15/154$; [46], (7, 8)].

23. $\displaystyle {}_5F_4\left(\begin{matrix} \frac{1}{6}, \frac{1}{2}, \frac{5}{6}, \frac{169}{154} \\ \frac{15}{154}, 1, 1; -\frac{27}{512} \end{matrix}\right) = \frac{32\sqrt{2}}{15\pi}$ [[46], (9, 10)].

24. $\displaystyle {}_5F_4\left(\begin{matrix} \frac{2}{3}, 1, 1, \frac{4}{3}, \frac{123}{77} \\ \frac{46}{77}, \frac{3}{2}, \frac{3}{2}, \frac{3}{2}; -\frac{27}{512} \end{matrix}\right) = \frac{128}{92}\ln 2$ [[46], (9, 10)].

25. $\displaystyle {}_5F_4\left(\begin{matrix} \frac{5}{6}, 1, 1, \frac{7}{6}, \frac{167}{102} \\ \frac{65}{102}, \frac{3}{2}, \frac{3}{2}, \frac{3}{2}; -\frac{1}{16} \end{matrix}\right) = \frac{216}{65}\ln\frac{4}{3}$ [[43], (9)].

26. $\displaystyle {}_5F_4\left(\begin{array}{c} a+1,\, a+\frac{1}{7},\, a+\frac{11}{28},\, a+\frac{9}{14},\, 1 \\ a,\, a+\frac{25}{28},\, a+\frac{25}{28},\, a+\frac{25}{28};\, -\frac{1}{48} \end{array}\right)$

$\displaystyle = 2^{6a+19/14}3^{a-17/28}\sec\left(\frac{28a-3}{28}\pi\right)\frac{\Gamma^3\left(a+\frac{25}{28}\right)}{7a\,\Gamma\left(a+\frac{11}{28}\right)\Gamma\left(2a+\frac{2}{7}\right)}$

$\displaystyle +\frac{3(28a-3)^2}{49a(28a-17)}\,{}_3F_2\left(\begin{array}{c} a+\frac{11}{28},\, 2a+\frac{2}{7},\, 1 \\ \frac{45}{28}-a,\, 2a+\frac{11}{14};\, \frac{3}{4} \end{array}\right)$ [[46], (12)].

27. $\displaystyle {}_5F_4\left(\begin{array}{c} \frac{3}{4},\, 1,\, 1,\, \frac{5}{4},\, \frac{45}{28} \\ \frac{17}{28},\, \frac{3}{2},\, \frac{3}{2},\, \frac{3}{2};\, -\frac{1}{48} \end{array}\right) = \frac{32}{17}\ln\frac{27}{16}$ [[43], (2)].

28. $\displaystyle {}_5F_4\left(\begin{array}{c} \frac{1}{2},\, \frac{3}{4},\, 1,\, 1,\, \frac{19}{14} \\ \frac{5}{14},\, \frac{5}{4},\, \frac{5}{4},\, \frac{5}{4};\, -\frac{1}{48} \end{array}\right) = \frac{3^{-5/4}}{10\sqrt{2}\,\pi}\Gamma^4\left(\frac{1}{4}\right).$

29. $\displaystyle {}_5F_4\left(\begin{array}{c} \frac{5}{6},\, 1,\, 1,\, \frac{7}{6},\, \frac{29}{18} \\ \frac{11}{18},\, \frac{3}{2},\, \frac{3}{2},\, \frac{3}{2};\, -\frac{1}{80} \end{array}\right) = \frac{24}{11}\ln\frac{3^9}{2^2 5^5}$ [[43], (10)].

30. $\displaystyle {}_5F_4\left(\begin{array}{c} \frac{3}{4},\, 1,\, 1,\, \frac{5}{4},\, \frac{413}{260} \\ \frac{153}{260},\, \frac{3}{2},\, \frac{3}{2},\, \frac{3}{2};\, -\frac{1}{324} \end{array}\right) = \frac{144}{17}\ln\frac{9}{8}$ [[43], (3)].

31. $\displaystyle {}_5F_4\left(\begin{array}{c} \frac{3}{4},\, 1,\, 1,\, \frac{5}{4},\, \frac{1007}{644} \\ \frac{363}{644},\, \frac{3}{2},\, \frac{3}{2},\, \frac{3}{2};\, -\frac{1}{25920} \end{array}\right) = \frac{3456}{121}\ln\frac{2^{18}}{3^4 5^5}$ [[43], (40)].

32. $\displaystyle {}_5F_4\left(\begin{array}{c} \frac{3}{4},\, 1,\, 1,\, \frac{5}{4},\, \frac{439}{280} \\ \frac{159}{280},\, \frac{3}{2},\, \frac{3}{2},\, \frac{3}{2};\, \frac{1}{9801} \end{array}\right) = \frac{594\sqrt{11}}{53}\left(-\frac{\pi}{2}+4\arcsin\frac{7}{18}\right)$ [[43], (25)].

33. $\displaystyle {}_5F_4\left(\begin{array}{c} \frac{3}{4},\, 1,\, 1,\, \frac{5}{4},\, \frac{63}{40} \\ \frac{23}{40},\, \frac{3}{2},\, \frac{3}{2},\, \frac{3}{2};\, \frac{1}{2401} \end{array}\right) = \frac{2401}{69\sqrt{3}}\left(-\frac{\pi}{6}+4\arcsin\frac{1}{7}\right)$ [[43], (24)].

34. $\displaystyle {}_5F_4\left(\begin{array}{c} \frac{3}{4},\, 1,\, 1,\, \frac{5}{4},\, \frac{8}{5} \\ \frac{3}{5},\, \frac{3}{2},\, \frac{3}{2},\, \frac{3}{2};\, \frac{1}{81} \end{array}\right) = \frac{27}{4\sqrt{2}}\left(\frac{\pi}{2}-4\arcsin\frac{1}{3}\right)$ [[43], (23)].

35. $\displaystyle {}_5F_4\left(\begin{array}{c} a+1,\, a+\frac{8}{21},\, a+\frac{8}{21},\, a+\frac{8}{21},\, 1 \\ a,\, a+\frac{37}{42},\, a+\frac{37}{42},\, a+\frac{37}{42};\, \frac{1}{64} \end{array}\right) = \frac{8(42a-5)}{441a}$

$\displaystyle \times {}_3F_2\left(\begin{array}{c} a+\frac{8}{21},\, a+\frac{8}{21},\, 1;\, 1 \\ 2a+\frac{16}{21},\, 2a+\frac{16}{21} \end{array}\right)$ [Re $a > 5/42$; [46], (3, 4)].

36. $_5F_4\left(\begin{array}{c} 1,1,1,1,\frac{34}{21} \\ \frac{13}{21},\frac{3}{2},\frac{3}{2},\frac{3}{2};\frac{1}{64} \end{array}\right) = \frac{4\pi^2}{39}$ [[43], (21)].

37. $_5F_4\left(\begin{array}{c} \frac{3}{4},1,1,\frac{5}{4},\frac{13}{8} \\ \frac{5}{8},\frac{3}{2},\frac{3}{2},\frac{3}{2};\frac{1}{9} \end{array}\right) = \frac{\sqrt{3}\pi}{5}$ [[46], (11)].

38. $_5F_4\left(\begin{array}{c} a+1, a+\frac{1}{8}, a+\frac{3}{8}, a+\frac{5}{8}, 1 \\ a, a+\frac{7}{8}, a+\frac{7}{8}, a+\frac{7}{8}; \frac{1}{9} \end{array}\right)$

$= 2^{8a-5} 3^{2a+1/4} \csc\left(\frac{8a+1}{4}\pi\right) \frac{\Gamma\left(a-\frac{1}{8}\right)\Gamma^3\left(a+\frac{7}{8}\right)}{a\pi\Gamma\left(4a-\frac{1}{2}\right)}$

$+ \frac{9(8a-1)^2}{64a(8a-3)} {}_3F_2\left(\begin{array}{c} a+\frac{3}{8}, \frac{1}{2}, 1; \frac{3}{4} \\ a+\frac{7}{8}, \frac{7}{4} - 2a \end{array}\right)$ [[46], (11)].

39. $_5F_4\left(\begin{array}{c} a+1, a+\frac{1}{3}, a+\frac{1}{3}, a+\frac{1}{3}, 1 \\ a, a+\frac{5}{6}, a+\frac{5}{6}, a+\frac{5}{6}; \frac{1}{4} \end{array}\right) = \frac{2(6a-1)}{9a} {}_3F_2\left(\begin{array}{c} \frac{1}{2}, \frac{1}{2}, 1; 1 \\ a+\frac{5}{6}, a+\frac{5}{6} \end{array}\right)$

[Re $a > 1/6$; [46], (1, 2)].

40. $_5F_4\left(\begin{array}{c} \frac{1}{2},\frac{1}{2},\frac{1}{2},\frac{1}{2},\frac{1}{2} \\ \frac{3}{2},\frac{3}{2},\frac{3}{2},\frac{3}{2};\frac{1}{4} \end{array}\right)$

$= \frac{\pi}{12}\zeta(3) + \frac{1}{1024\sqrt{3}}\left[\zeta\left(4,\frac{1}{6}\right) - \zeta\left(4,\frac{1}{3}\right) + \zeta\left(4,\frac{2}{3}\right) - \zeta\left(4,\frac{5}{6}\right)\right].$

41. $_5F_4\left(\begin{array}{c} 1,1,1,1,1 \\ \frac{3}{2}, 2, 2, 2; \frac{1}{4} \end{array}\right) = \frac{17\pi^4}{1620}.$

42. $_5F_4\left(\begin{array}{c} 1,1,1,1,\frac{5}{3} \\ \frac{2}{3},\frac{3}{2},\frac{3}{2},\frac{3}{2};\frac{1}{4} \end{array}\right) = \frac{\pi^2}{8}$ [[43]].

43. $_5F_4\left(\begin{array}{c} \frac{3}{4}, 1, 1, \frac{5}{4}, \frac{23}{14} \\ \frac{9}{14},\frac{3}{2},\frac{3}{2},\frac{3}{2};\frac{32}{81} \end{array}\right) = \frac{9}{4\sqrt{2}}\left(\frac{\pi}{2} - 2\arcsin\frac{1}{3}\right)$ [[43], (29)].

44. $_5F_4\left(\begin{array}{c} -n, a, a+\frac{1}{2}, \frac{a}{2}+1, \frac{2a+1}{3}+n \\ -3n, \frac{a}{2}, 2a+3n+1, \frac{1}{2}; 9 \end{array}\right) = \frac{\left(\frac{2a+2}{3}\right)_n \left(\frac{2a}{3}+1\right)_n}{\left(\frac{1}{3}\right)_n \left(\frac{2}{3}\right)_n}$

[[38], (6.3)].

8.1.5. The hypergeometric function $_6F_5(a_1,\ldots,a_6;b_1,\ldots,b_5;z)$

1. $_6F_5\left(\begin{array}{c} a, a+\frac{1}{2}, b, b+\frac{1}{2}, \frac{a+b}{2}, \frac{a+b+1}{2}; z \\ \frac{1}{2}, a+b, a+b+\frac{1}{2}, \frac{2a+2b+1}{4}, \frac{2a+2b+3}{4} \end{array}\right)$

$= {}_4F_3^2\left(\begin{array}{c} \frac{a}{2}, \frac{a+1}{2}, \frac{b}{2}, \frac{b+1}{2}; z \\ \frac{2a+2b+1}{4}, \frac{2a+2b+3}{4}, \frac{1}{2} \end{array}\right) + \frac{4a^2b^2z}{(2a+2b+1)^2}$

$\times {}_4F_3^2\left(\begin{array}{c} \frac{a+1}{2}, \frac{a}{2}+1, \frac{b+1}{2}, \frac{b}{2}+1; z \\ \frac{2a+2b+3}{4}, \frac{2a+2b+5}{4}, \frac{3}{2} \end{array}\right)$ [42].

2. $_6F_5\left(\begin{array}{c} -\frac{n}{3}, \frac{1-n}{3}, \frac{2-n}{3}, a, b, b+\frac{1}{2}; 1 \\ \frac{a+2b}{3}, \frac{a+2b+1}{3}, \frac{a+2b+2}{3}, c, \frac{3}{2}-c-n \end{array}\right)$

$= \frac{(2b)_n}{(a+2b)_n} {}_4F_3\left(\begin{array}{c} -n, a, 2c+n-1, 2-2c-n; \frac{1}{4} \\ 1-2b-n, c, \frac{3}{2}-c-n \end{array}\right)$ [[55], (3.26)].

3. $_6F_5\left(\begin{array}{c} a, b, c, \frac{c+3}{2}, 1, 1; 1 \\ \frac{c+1}{2}, c+1, c-a+2, c-b+2, 2 \end{array}\right) = \frac{c(c-a+1)(c-b+1)}{(a-1)(b-1)(c+1)}$

$\times [-\psi(c) + \psi(c-a+1) + \psi(c-b+1) - \psi(c-a-b+2)]$.

4. $_6F_5\left(\begin{array}{c} -n, a, a+\frac{1}{2}, b, b+\frac{1}{3}, b+\frac{2}{3}; 1 \\ \frac{2a-n}{3}, \frac{2a-n+1}{3}, \frac{2a-n+2}{3}, c, 3b-c+\frac{3}{2} \end{array}\right)$

$= \frac{(3b-2a+1)_n}{(1-2a)_n} {}_4F_3\left(\begin{array}{c} -n, 3b, 3b-2c+2, 2c-3b-1; \frac{1}{4} \\ 3b-2a+1, c, 3b-c+\frac{3}{2} \end{array}\right)$ [[55], (3.26)].

5. $_6F_5\left(\begin{array}{c} -n, \frac{1}{3}-n, \frac{2}{3}-n, a, b, b+\frac{1}{2}; 1 \\ \frac{a+2b}{3}, \frac{a+2b+1}{3}, \frac{a+2b+2}{3}, c, \frac{3}{2}-c-3n \end{array}\right)$

$= \frac{(a)_{3n}}{(a+2b)_{3n}} {}_4F_3\left(\begin{array}{c} -3n, 2b, c-\frac{1}{2}, 1-c-3n; 4 \\ 1-a-3n, 2c-1, 2-2c-6n \end{array}\right)$ [[55], (3.28)].

6. $_6F_5\left(\begin{array}{c} a, b, c, d, e, \frac{e}{2}+1; -1 \\ \frac{e}{2}, e-a+1, e-b+1, e-c+1, e-d+1 \end{array}\right)$

$= \frac{\Gamma(e-a+1)\Gamma(e-d+1)}{\Gamma(e+1)\Gamma(e-d-a+1)} {}_3F_2\left(\begin{array}{c} a, d, e-b-c+1; 1 \\ e-b+1, e-c+1 \end{array}\right)$

$[\operatorname{Re}(3e-2b-2c-2d-2a+3), \operatorname{Re}(e-d-a+1) > 0]$.

8.1.6] 8.1. THE HYPERGEOMETRIC FUNCTIONS

7. $\,_6F_5\left(\begin{array}{c} a, b, c, d, e, 2e-2; \; -1 \\ e-1, 2e-a-1, 2e-b-1, 2e-c-1, 2e-d-1 \end{array}\right)$

$$= \frac{\Gamma(2e-c-1)\Gamma(2e-d-1)}{\Gamma(2e-1)\Gamma(2e-c-d-1)} \,_3F_2\left(\begin{array}{c} c, d, 2e-a-b-1; \; 1 \\ 2e-a-1, 2e-b-1 \end{array}\right)$$

$$[\mathrm{Re}\,(2e-c-d) > 1; \; \mathrm{Re}\,(3e-a-b-c-d) > 3/2].$$

8. $\,_6F_5\left(\begin{array}{c} a, a+\frac{1}{2}, 2a, 1-2a, \frac{9a+16-\sqrt{25a^2+8a+4}}{14}, \frac{9a+16+\sqrt{25a^2+8a+4}}{14} \\ a+\frac{3}{4}, a+\frac{5}{4}, \frac{9a+2-\sqrt{25a^2+8a+4}}{14}, \frac{9a+2+\sqrt{25a^2+8a+4}}{14} \end{array}; \; -\frac{1}{48}\right)$

$$= \frac{2^{4a+1}\Gamma\left(2a+\frac{3}{2}\right)}{3^{2a}\sqrt{\pi}\,\Gamma(2a+2)} \qquad [[45], (3.3)].$$

9. $\,_6F_5\left(\begin{array}{c} \frac{1}{4}, \frac{1}{2}, \frac{3}{4}, \frac{3}{4}, \frac{39-2\sqrt{11}}{28}, \frac{39+2\sqrt{11}}{28} \\ 1, \frac{9}{8}, \frac{13}{8}, \frac{11-2\sqrt{11}}{28}, \frac{11+2\sqrt{11}}{28} \end{array}; \; \frac{1}{64}\right) = \frac{10\sqrt{2}}{33\pi^{3/2}}\Gamma^2\left(\frac{1}{4}\right)$ [42].

10. $\,_6F_5\left(\begin{array}{c} a, a, 1-a, \frac{1}{2}, \frac{7a+24-\sqrt{28a^2+9}}{21}, \frac{7a+24+\sqrt{28a^2+9}}{21} \\ 1, \frac{2a+3}{4}, \frac{2a+5}{4}, \frac{7a+3-\sqrt{28a^2+9}}{21}, \frac{7a+3+\sqrt{28a^2+9}}{21} \end{array}; \; \frac{1}{64}\right)$

$$= \frac{4\Gamma\left(a+\frac{3}{2}\right)}{\sqrt{\pi}\,(a+2)\Gamma(a+1)} \qquad [[45], (2.4)].$$

11. $\,_6F_5\left(\begin{array}{c} \frac{1}{4}, \frac{1}{4}, \frac{3}{4}, \frac{3}{4}, \frac{16-\sqrt{7}}{12}, \frac{16+\sqrt{7}}{12} \\ 1, 1, \frac{3}{2}, \frac{4-\sqrt{7}}{12}, \frac{4+\sqrt{7}}{12} \end{array}; \; \frac{1}{4}\right) = \frac{16}{3\sqrt{2\pi}}$ [45].

12. $\,_6F_5\left(\begin{array}{c} 1, 1, 1, 1, 1, 1 \\ \frac{3}{2}, 2, 2, 2, 2; \; \frac{1}{4} \end{array}\right)$

$$= \frac{2\pi^2}{9}\zeta(3) - \frac{38}{3}\zeta(5) + \frac{\pi}{192\sqrt{3}}\left[\zeta\left(4, \frac{1}{6}\right) - \zeta\left(4, \frac{1}{3}\right) + \zeta\left(4, \frac{2}{3}\right) - \zeta\left(4, \frac{5}{6}\right)\right].$$

8.1.6. The hypergeometric function $\,_7F_6(a_1, \ldots, a_7; b_1, \ldots, b_6; z)$

1. $\,_7F_6\left(\begin{array}{c} -n, a, b, c, \frac{a}{3}+1, 1-b, a-c+n+\frac{1}{2}; \; 1 \\ \frac{a}{3}, \frac{a-b}{2}+1, \frac{a+b+1}{2}, a-2c+1, a+2n+1, 2c-a-2n \end{array}\right)$

$$= \frac{(a+1)_{2n}\left(\frac{a+b+1}{2}-c\right)_n\left(\frac{a-b}{2}-c+1\right)_n}{\left(\frac{a-b}{2}+1\right)_n\left(\frac{a+b+1}{2}\right)_n(a-2c+1)_{2n}} \qquad [[38], (1.7)].$$

2. $_7F_6\left(\begin{matrix} -2n-m,\, a,\, b,\, c,\, \frac{2a}{3}+1,\, a-b+\frac{1}{2},\, 2a-c+2n+m+1;\, 1 \\ \frac{2a}{3},\, 2b,\, 2a-2b+1,\, a-\frac{c}{2}+1,\, \frac{c+1}{2}-n-\frac{m}{2},\, 1+a+n+\frac{m}{2} \end{matrix}\right)$

$$= 2^{-2n}\frac{(2n)!\,(a+1)_n \left(b+\frac{1-c}{2}\right)_n (a-b-\frac{c}{2}+1)_n}{n!\left(b+\frac{1}{2}\right)_n (a-b+1)_n \left(a-\frac{c}{2}+1\right)_n \left(\frac{1-c}{2}\right)_n}\,\delta_{0,m}$$

$$[m=0,1;\,[38],\,(1.8)].$$

3. $_7F_6\left(\begin{matrix} a,\, 1-a,\, \frac{1}{2},\, \frac{1}{2},\, \frac{1}{2},\, \frac{6a+21-\sqrt{36a^2-28a+1}}{20},\, \frac{6a+21+\sqrt{36a^2-28a+1}}{20} \\ a+\frac{1}{2},\, a+\frac{1}{2},\, 1,\, 1,\, \frac{6a+1-\sqrt{36a^2-28a+1}}{20},\, \frac{6a+1+\sqrt{36a^2-28a+1}}{20}; -\frac{1}{4} \end{matrix}\right)$

$$= \frac{4\Gamma^2\left(a+\frac{1}{2}\right)}{\pi a\,\Gamma^2(a)} \quad [[45],\,(5.1)].$$

4. $_7F_6\left(\begin{matrix} \frac{1}{2},\, \frac{1}{2},\, \frac{1}{2},\, \frac{1}{2},\, \frac{1}{2},\, \frac{12-i}{10},\, \frac{12+i}{10} \\ 1,\,1,\,1,\,1,\, \frac{2-i}{10},\, \frac{2+i}{10};\, -\frac{1}{4} \end{matrix}\right) = \frac{8}{\pi^2}$ $\quad [[45],\,(5.4)].$

5. $_7F_6\left(\begin{matrix} \frac{1}{4},\, \frac{1}{3},\, \frac{1}{2},\, \frac{2}{3},\, \frac{3}{4},\, \frac{9}{8}-\frac{i}{24}\sqrt{\frac{17}{7}},\, \frac{9}{8}+\frac{i}{24}\sqrt{\frac{17}{7}} \\ 1,\,1,\,1,\,1,\, \frac{1}{8}-\frac{i}{24}\sqrt{\frac{17}{7}},\, \frac{1}{8}+\frac{i}{24}\sqrt{\frac{17}{7}};\, -\frac{1}{48} \end{matrix}\right) = \frac{48}{5\pi^2}$ $\quad [[43],\,(15)].$

6. $_7F_6\left(\begin{matrix} \frac{1}{2},\, \frac{1}{2},\, \frac{1}{2},\, \frac{1}{2},\, \frac{1}{2},\, \frac{91}{82}-\frac{4i}{41}\sqrt{\frac{2}{5}},\, \frac{91}{82}+\frac{4i}{41}\sqrt{\frac{2}{5}} \\ 1,\,1,\,1,\,1,\, \frac{9}{82}-\frac{4i}{41}\sqrt{\frac{2}{5}},\, \frac{9}{82}+\frac{4i}{41}\sqrt{\frac{2}{5}};\, -\frac{1}{1024} \end{matrix}\right) = \frac{128}{13\pi^2}$ $\quad [[44],\,(1\text{--}1)].$

7. $_7F_6\left(\begin{matrix} \frac{1}{6},\, \frac{1}{4},\, \frac{1}{2},\, \frac{3}{4},\, \frac{5}{6},\, \frac{1779-i\sqrt{5279}}{1640},\, \frac{1779+i\sqrt{5279}}{1640} \\ 1,\,1,\,1,\,1,\, \frac{139-i\sqrt{5279}}{1640},\, \frac{139+i\sqrt{5279}}{1640};\, -\frac{1}{1024} \end{matrix}\right) = \frac{256}{15\sqrt{3}\,\pi^2}$ $\quad [[43],\,(14)].$

8. $_7F_6\left(\begin{matrix} \frac{1}{8},\, \frac{3}{8},\, \frac{1}{2},\, \frac{5}{8},\, \frac{7}{8},\, \frac{259-i\sqrt{89}}{240},\, \frac{259+i\sqrt{89}}{240} \\ 1,\,1,\,1,\,1,\, \frac{19-i\sqrt{89}}{240},\, \frac{19+i\sqrt{89}}{240};\, \frac{1}{2401} \end{matrix}\right) = \frac{56\sqrt{7}}{15\pi^2}$ $\quad [[44],\,(2\text{--}5)].$

9. $_7F_6\left(\begin{matrix} \frac{1}{8},\, \frac{1}{2},\, \frac{1}{2},\, \frac{1}{2},\, \frac{5}{8},\, \frac{59}{48}-\frac{1}{48}\sqrt{\frac{77}{5}},\, \frac{59}{48}-\frac{1}{48}\sqrt{\frac{77}{5}} \\ 1,\,1,\, \frac{5}{4},\, \frac{5}{4},\, \frac{11}{48}-\frac{1}{48}\sqrt{\frac{77}{5}},\, \frac{11}{48}+\frac{1}{48}\sqrt{\frac{77}{5}};\, \frac{1}{16} \end{matrix}\right) = \frac{2}{11\pi^3}\Gamma^2\left(\frac{1}{4}\right)$

$$[[43],\,(20)].$$

8.1.8] 8.1. The Hypergeometric Functions

10. $_7F_6\left(\begin{array}{c}\frac{1}{8}, \frac{1}{2}, \frac{1}{2}, \frac{1}{2}, \frac{5}{8}, \frac{295-\sqrt{385}}{240}, \frac{295+\sqrt{385}}{240}\\ 1, 1, \frac{5}{4}, \frac{5}{4}, \frac{55-\sqrt{385}}{240}, \frac{55+\sqrt{385}}{240}; \frac{1}{16}\end{array}\right) = \frac{2}{11\pi^3}\Gamma^4\left(\frac{1}{4}\right)$ [42].

11. $_7F_6\left(\begin{array}{c}\frac{1}{4}, \frac{1}{2}, \frac{1}{2}, \frac{1}{2}, \frac{3}{4}, \frac{137-i\sqrt{71}}{120}, \frac{137+i\sqrt{71}}{120}\\ 1, 1, 1, 1, \frac{17-i\sqrt{71}}{120}, \frac{17+i\sqrt{71}}{120}; \frac{1}{16}\end{array}\right) = \frac{32}{3\pi^2}$ [[43], (27)].

8.1.7. The hypergeometric function $_8F_7(a_1, \ldots, a_8; b_1, \ldots, b_7; z)$

1. $_8F_7\left(\begin{array}{c}-n, -n-\frac{1}{2}, a, a+\frac{1}{2}, b, b+\frac{1}{2}, -a-b-n-\frac{1}{4}, -a-b-n+\frac{1}{4}; 1\\ \frac{1}{2}, -a-n, \frac{1}{2}-a-n, -b-n, \frac{1}{2}-b-n, a+b+\frac{1}{4}, a+b+\frac{3}{4}\end{array}\right)$
$= \frac{(4a+1)_{2n}(4b+1)_{2n}(2a+2b+1)_{2n}}{(2a+1)_{2n}(2b+1)_{2n}(4a+4b+1)_{2n}}$.

2. $_8F_7\left(\begin{array}{c}1, 1, 1, 1, \frac{3}{2}, \frac{9}{4}, \frac{9-i\sqrt{3}}{4}, \frac{9+i\sqrt{3}}{4}\\ \frac{5}{4}, 2, \frac{5}{2}, \frac{5}{2}, \frac{5}{2}, \frac{5-i\sqrt{3}}{4}, \frac{5+i\sqrt{3}}{4}; 1\end{array}\right) = \frac{27}{10}\zeta(3) - \frac{54}{35}$.

3. $_8F_7\left(\begin{array}{c}\frac{a}{6}+1, \frac{a}{3}, \frac{b}{3}, \frac{b+1}{3}, \frac{b+2}{3}, \frac{c}{3}, \frac{c+1}{3}, \frac{c+2}{3}; -1\\ \frac{a}{6}, \frac{a-b+1}{3}, \frac{a-b+2}{3}, \frac{a-b}{3}+1, \frac{a-c+1}{3}, \frac{a-c+2}{3}, \frac{a-c}{3}+1\end{array}\right)$
$= \frac{\Gamma(a-b+1)\Gamma(a-c+1)}{\Gamma(a+1)\Gamma(a-b-c+1)} {}_3F_2\left(\begin{array}{c}\frac{a}{3}, b, c; \frac{3}{4}\\ \frac{a}{2}, \frac{a+1}{2}\end{array}\right)$ [Re $(5a-6b-6c) > -3$].

4. $_8F_7\left(\begin{array}{c}\frac{3}{4}, 1, 1, 1, 1, \frac{5}{4}, \frac{197-i\sqrt{71}}{120}, \frac{197+i\sqrt{71}}{120}\\ \frac{3}{2}, \frac{3}{2}, \frac{3}{2}, \frac{3}{2}, \frac{3}{2}, \frac{77-i\sqrt{71}}{120}, \frac{77+I\sqrt{71}}{120}; -\frac{1}{4}\end{array}\right) = \frac{8\pi^2}{75}$ [43].

5. $_8F_7\left(\begin{array}{c}1, 1, 1, 1, 1, 1, \frac{17-i}{10}, \frac{17+i}{10}\\ \frac{3}{2}, \frac{3}{2}, \frac{3}{2}, \frac{3}{2}, \frac{3}{2}, \frac{7-i}{10}, \frac{7+i}{10}; -\frac{1}{4}\end{array}\right) = \frac{7}{10}\zeta(3)$.

8.1.8. The hypergeometric function $_{10}F_9(a_1, \ldots, a_{10}; b_1, \ldots, b_9; z)$

1. $_{10}F_9\left(\begin{array}{c}\frac{1}{2}, \frac{1}{2}, \frac{1}{2}, \frac{1}{2}, \frac{1}{2}, \frac{1}{2}, \frac{1}{2}, \frac{7}{6}, \frac{16-i\sqrt{3}}{14}, \frac{16+i\sqrt{3}}{14}\\ \frac{1}{6}, 1, 1, 1, 1, 1, 1, \frac{2-i\sqrt{3}}{14}, \frac{2+i\sqrt{3}}{14}; \frac{1}{64}\end{array}\right) = \frac{32}{\pi^3}$ [[44], (4–1)].

8.1.9. The Kummer confluent hypergeometric function $_1F_1(a; b; z)$

1. $_1F_1\!\left(\begin{matrix} a+m;\ z \\ b+n \end{matrix}\right)$
$$= \frac{(-1)^n m!\,(b)_n}{(a)_m (b-a)_n} e^z \sum_{k=0}^{m} \frac{z^k}{k!} L^{a+k-1}_{m-k}(-z)\, \mathrm{D}^{k+n}_z\!\left[e^{-z}\, _1F_1\!\left(\begin{matrix} a;\ z \\ b \end{matrix}\right)\right].$$

2. $_1F_1\!\left(\begin{matrix} a-m;\ z \\ b+n \end{matrix}\right) = \frac{(-1)^{m+n}(b)_n}{(b-a+n)_m (b-a)_n} e^z$
$$\times \sum_{k=0}^{m} (-z)^k \binom{m}{k}(a-b-m-n+1)_{m-k}\, \mathrm{D}^{k+n}_z\!\left[e^{-z}\, _1F_1\!\left(\begin{matrix} a;\ z \\ b \end{matrix}\right)\right].$$

3. $_1F_1\!\left(\begin{matrix} a+m;\ z \\ b-n \end{matrix}\right) = \frac{z^{1-a}}{(a)_m (1-b)_n}$
$$\times \sum_{k=0}^{n} \binom{n}{k}(a-b)_{n-k}(-z)^k\, \mathrm{D}^{k+m}\!\left[z^{a+m-1}\, _1F_1\!\left(\begin{matrix} a;\ z \\ b \end{matrix}\right)\right].$$

4. $_1F_1\!\left(\begin{matrix} a-m;\ z \\ b-n \end{matrix}\right)$
$$= \frac{(-1)^n n!\, z^{a-b+1}}{(b-a)_m (1-b)_n} e^z \sum_{k=0}^{n} \frac{z^k}{k!} L^{a+k-n}_{n-k}(-z)\, \mathrm{D}^{k+m}\!\left[z^{b-a+m-1} e^{-z}\, _1F_1\!\left(\begin{matrix} a;\ z \\ b \end{matrix}\right)\right].$$

5. $_1F_1\!\left(\begin{matrix} a;\ z \\ 2a+n \end{matrix}\right) = \Gamma\!\left(a-\frac{1}{2}\right)\!\left(\frac{z}{4}\right)^{1/2-a} e^{z/2} \sum_{k=0}^{n} (-1)^k \binom{n}{k}$
$$\times \frac{(2a-1)_k}{(2a+n)_k}\!\left(a+k-\frac{1}{2}\right) I_{a+k-1/2}\!\left(\frac{z}{2}\right).$$

6. $_1F_1\!\left(\begin{matrix} a;\ z \\ 2a-n \end{matrix}\right) = \Gamma\!\left(a-n-\frac{1}{2}\right)\!\left(\frac{z}{4}\right)^{n-a+1/2} e^{z/2} \sum_{k=0}^{n} \binom{n}{k}$
$$\times \frac{(2a-2n-1)_k}{(2a-n)_k}\!\left(a+k-n-\frac{1}{2}\right) I_{a+k-n-1/2}\!\left(\frac{z}{2}\right).$$

7. $_1F_1\!\left(\begin{matrix} m+1 \\ b;\ z \end{matrix}\right) = (b-1)z^{1-b} e^z \sum_{k=0}^{m} \frac{(-1)^k}{k!} L^k_{m-k}(-z)\gamma(b+k-1,z).$

8. $_1F_1\!\left(\begin{matrix} m+1;\ z \\ n+2 \end{matrix}\right)$
$$= (-1)^n (n+1) z^{-n-1} \sum_{k=0}^{m} \frac{(k+n)!}{k!} L^k_{m-k}(-z)[(-1)^{k+n} e^z - L^{-k-n-1}_{k+n}(z)].$$

9. $_1F_1\!\left(\begin{matrix} \frac{1}{6};\ z \\ \frac{1}{3} \end{matrix}\right) = \frac{3^{1/6}}{2} \Gamma\!\left(\frac{2}{3}\right) e^{z/2} \left[\sqrt{3}\,\mathrm{Ai}\!\left(\left(\frac{3z}{4}\right)^{2/3}\right) + \mathrm{Bi}\!\left(\left(\frac{3z}{4}\right)^{2/3}\right)\right].$

8.1.10]					8.1. The Hypergeometric Functions

10. $\,_1F_1\!\left(\begin{array}{c}\frac{5}{6};\,z\\ \frac{2}{3}\end{array}\right) = 2^{-5/3}3^{5/6}\Gamma\!\left(\frac{4}{3}\right)e^{z/2}$

$$\times \left[-3^{5/6}z^{1/3}\,\mathrm{Ai}\!\left(\left(\frac{3z}{4}\right)^{2/3}\right) - 2^{2/3}\sqrt{3}\,\mathrm{Ai}'\!\left(\left(\frac{3z}{4}\right)^{2/3}\right)\right.$$
$$\left. + (3z)^{1/3}\,\mathrm{Bi}\!\left(\left(\frac{3z}{4}\right)^{2/3}\right) + 2^{2/3}\,\mathrm{Bi}'\!\left(\left(\frac{3z}{4}\right)^{2/3}\right)\right].$$

8.1.10. The Tricomi confluent hypergeometric function $\Psi(a;\,b;\,z)$

1. $\Psi\!\left(\begin{array}{c}a+m;\,z\\ b+n\end{array}\right) = \dfrac{(-1)^n n!\,z^{-a-n+1}}{(a)_m(a-b+1)_m}$

$$\times \sum_{k=0}^{n}\frac{z^k}{k!}L_{n-k}^{k-a-n+1}(z)\,\mathrm{D}^{k+m}\!\left[z^{a+m-1}\Psi\!\left(\begin{array}{c}a;\,z\\ b\end{array}\right)\right].$$

2. $\Psi\!\left(\begin{array}{c}a-m;\,z\\ b+n\end{array}\right) = (-1)^{m+n}(b-a+n)_m\,e^z$

$$\times \sum_{k=0}^{m}\binom{m}{k}\frac{z^k}{(b-a+n)_k}\,\mathrm{D}^{k+n}\!\left[e^{-z}\Psi\!\left(\begin{array}{c}a;\,z\\ b\end{array}\right)\right].$$

3. $\Psi\!\left(\begin{array}{c}a+m;\,z\\ b-n\end{array}\right) = \dfrac{(a-b)_n\,z^{-a+1}}{(a)_m(a-b+1)_m(a-b+m+1)_n}$

$$\times \sum_{k=0}^{n}\binom{n}{k}\frac{z^k}{(b-a-n+1)_k}\,\mathrm{D}^{k+m}\!\left[z^{a+m-1}\Psi\!\left(\begin{array}{c}a;\,z\\ b\end{array}\right)\right].$$

4. $\Psi\!\left(\begin{array}{c}a-m;\,z\\ b-n\end{array}\right) = \dfrac{(-1)^{m+n}m!\,z^{n-b+1}}{(a-b+1)_n}\sum_{k=0}^{m}\frac{z^k}{k!}L_{m-k}^{k-a}(z)\,\mathrm{D}^{k+n}\!\left[z^{b-1}\Psi\!\left(\begin{array}{c}a;\,z\\ b\end{array}\right)\right].$

5. $\Psi\!\left(\begin{array}{c}a;\,z\\ a+n\end{array}\right) = (-1)^{n-1}(n-1)!\,z^{1-a-n}L_{n-1}^{1-a-n}(z)$ $\qquad [n\geq 1].$

6. $\Psi\!\left(\begin{array}{c}a;\,z\\ a-n\end{array}\right) = (-1)^n$

$$\times \left[e^z\Gamma(1-a,z)L_n^{a-n-1}(-z) - z^{1-a}\sum_{k=1}^{n}\frac{1}{k}L_{n-k}^{a+k-n-1}(-z)L_{k-1}^{1-a-k}(z)\right].$$

7. $\Psi\!\left(\begin{array}{c}a;\,z\\ 2a+n\end{array}\right) = \dfrac{(-1)^n}{\sqrt{\pi}}\,n!\,z^{1/2-a-n}e^{z/2}$

$$\times \sum_{k=0}^{n}L_{n-k}^{k-n-a+1}(z)\sum_{p=0}^{k}\frac{(-z/4)^p}{p!}L_{k-p}^{p-k-1/2}\!\left(-\frac{z}{2}\right)\sum_{r=0}^{p}\binom{p}{r}K_{a-p+2r-1/2}\!\left(\frac{z}{2}\right).$$

623

8. $\Psi\begin{pmatrix} a;\ z \\ 2a-n \end{pmatrix} = \dfrac{(-1)^n n!}{\sqrt{\pi}\,(1-a)_n}\,z^{1/2-a}e^{z/2}$

$$\times \sum_{k=0}^{n} \dfrac{(-z/4)^k}{k!}\,L_{n-k}^{a+k-n-1/2}\!\left(-\dfrac{z}{2}\right) \sum_{p=0}^{k} \binom{k}{p} K_{a-k+2p-1/2}\!\left(\dfrac{z}{2}\right).$$

9. $\Psi\begin{pmatrix} a;\ z \\ n+\tfrac12 \end{pmatrix} = 2^a n!(-z)^{-n}e^{z/2}$

$$\times \sum_{k=0}^{n} \dfrac{(-1)^k}{k!}\,L_{n-k}^{k-n-a+1}(z) \sum_{p=0}^{k} \binom{k}{p}(1-a)_{k-p}\,z^p$$

$$\times \left[(-1)^p \delta_{p,0} D_{-2a}(\sqrt{2z}) + (2z)^{-p/2} \sum_{r=0}^{p-1} \dfrac{\Gamma(p+r)}{r!\,\Gamma(p-r)}\,(8z)^{-r/2}(2a)_{p-r}\right.$$

$$\left.\times D_{r-p-2a}(\sqrt{2z})\right].$$

10. $\Psi\begin{pmatrix} a;\ z \\ \tfrac12 - n \end{pmatrix} = \dfrac{2^a e^{z/2}}{\left(a+\tfrac12\right)_n}\sum_{p=0}^{n}\binom{n}{p}\left(\dfrac{1}{2}\right)_{n-p} z^p$

$$\times \left[(-1)^p \delta_{p,0} D_{-2a}(\sqrt{2z}) + (2z)^{-p/2}\sum_{r=0}^{p-1}\dfrac{\Gamma(p+r)}{r!\,\Gamma(p-r)}\,(8z)^{-r/2}(2a)_{p-r}\right.$$

$$\left.\times D_{r-p-2a}(\sqrt{2z})\right].$$

11. $\Psi\begin{pmatrix} m+1 \\ b;\ z \end{pmatrix} = \dfrac{1}{(2-b)_m}\left[z^{1-b}e^z\Gamma(b-1,z)L_m^{1-b}(-z)\right.$

$$\left. -\sum_{k=1}^{m}\dfrac{1}{k}\,L_{m-k}^{k-b+1}(-z)L_{k-1}^{b-k-1}(z)\right].$$

12. $\Psi\begin{pmatrix} m+1;\ z \\ n+1 \end{pmatrix} = -\dfrac{1}{(m!)^2}\sum_{k=0}^{n}\binom{n}{k}(k+m)!(-z)^{-k}$

$$\times \left[e^z \operatorname{Ei}(-z)L_{k+m}^{-k}(-z) + \sum_{p=1}^{k+m}\dfrac{(-z)^p}{p}\,L_{k+m-p}^{p-k}(-z)L_{p-1}^{-p}(z)\right] \quad [m>n].$$

13. $\Psi\begin{pmatrix} m+\frac{1}{2}; z \\ n+\frac{1}{2} \end{pmatrix} = \frac{2^{2m} n! (-z)^{-n}}{(2m)!} \sum_{k=0}^{n} \frac{(k+m)!}{k!} L_{n-k}^{k-n+1/2}(z)$

$\times \left[\sqrt{\pi} e^z \operatorname{erfc}(\sqrt{z}) L_{k+m}^{-k-1/2}(-z) + 2 \sum_{p=1}^{k+m} \frac{(-z)^p}{p!} L_{k+m-p}^{p-k-1/2}(-z) \right.$

$\left. \times \sum_{r=0}^{p-1} \frac{\Gamma(p+r)}{r! \Gamma(p-r)} (2\sqrt{z})^{-p-r} H_{p-r-1}(\sqrt{z}) \right] \quad [m > n].$

14. $\Psi\begin{pmatrix} m+\frac{1}{2}; z \\ \frac{1}{2} - n \end{pmatrix} = \frac{1}{\left(\frac{1}{2} - n\right)_{m+n}} \left[e^z \Gamma\left(n+\frac{1}{2}, z\right) L_{m+n}^{-n-1/2}(-z) \right.$

$\left. - z^{n+1/2} \sum_{k=1}^{m+n} \frac{1}{k} L_{m+n-k}^{k-n-1/2}(-z) L_{k-1}^{n-k+1/2}(z) \right].$

15. $\Psi\begin{pmatrix} \frac{1}{2} - m; z \\ n+\frac{1}{2} \end{pmatrix} = (-1)^m 2^{1-n} (n)_m z^{-n/2} \sum_{k=0}^{m} \binom{m}{k} \frac{\left(-\frac{\sqrt{z}}{2}\right)^k}{(n)_k}$

$\times \sum_{r=0}^{k+n-1} \frac{\Gamma(k+n+r)}{r! \Gamma(k+n-r)} (2\sqrt{z})^{-r} H_{k+n-r-1}(\sqrt{z}) \quad [n \geq 1].$

16. $\Psi\begin{pmatrix} \frac{1}{2} - m; z \\ \frac{1}{2} - n \end{pmatrix}$

$= \frac{(-1)^{m+n} m!}{n!} \sum_{k=0}^{m} \frac{(k+n)!}{k!} L_{m-k}^{k-1/2}(z) [\sqrt{\pi} e^z \operatorname{erfc}(\sqrt{z}) L_{k+n}^{-k-n-1/2}(-z)$

$+ 2 \sum_{p=1}^{k+n} \frac{(-z)^p}{p!} L_{k+n-p}^{p-k-n-1/2}(-z) \sum_{r=0}^{p-1} \frac{\Gamma(p+r)}{r! \Gamma(p-r)} (2\sqrt{z})^{-p-r} H_{p-r-1}(\sqrt{z})]$

$\quad [m > n].$

17. $\Psi\begin{pmatrix} \frac{1}{6}; z \\ \frac{1}{3} \end{pmatrix} = 2^{2/3} 3^{1/6} \sqrt{\pi} e^{z/2} \operatorname{Ai}\left(\left(\frac{3z}{4}\right)^{2/3}\right).$

8.1.11. The hypergeometric function $_1F_2(a_1; b_1, b_2; z)$

1. $_1F_2\begin{pmatrix} 1; z \\ b, b+\frac{1}{2} \end{pmatrix}$

$= \frac{2b-1}{2\sqrt{z}} \sinh(2\sqrt{z}) \,_1F_2\begin{pmatrix} b-1; z \\ b, \frac{1}{2} \end{pmatrix} - 2(b-1) \cosh(2\sqrt{z}) \,_1F_2\begin{pmatrix} b-\frac{1}{2}; z \\ b+\frac{1}{2}, \frac{3}{2} \end{pmatrix}.$

2. $= \dfrac{1-2b}{b}\sqrt{z}\sinh(2\sqrt{z})\,{}_1F_2\!\left(\begin{matrix}b;\,z\\b+1,\,\frac{3}{2}\end{matrix}\right) + \cosh(2\sqrt{z})\,{}_1F_2\!\left(\begin{matrix}b-\frac{1}{2};\,z\\b+\frac{1}{2},\,\frac{1}{2}\end{matrix}\right).$

3. $= \dfrac{2^{-2b}e^{-b\pi i}z^{1/2-b}}{\Gamma(1-2b)}$
$\times \left\{\Gamma(3-2b)\left[e^{-2\sqrt{z}-b\pi i}\Gamma(2b-2,\,-2\sqrt{z}) - e^{2\sqrt{z}+b\pi i}\Gamma(2b-2,\,2\sqrt{z})\right]\right.$
$\left. + \pi[\csc(b\pi)\sinh(2\sqrt{z}) + i\sec(b\pi)\cosh(2\sqrt{z})]\right\}\quad [z>0].$

4. ${}_1F_2\!\left(\begin{matrix}a;\,z\\a+1,\,\frac{1}{2}\end{matrix}\right)$
$= \cosh(2\sqrt{z})\,{}_1F_2\!\left(\begin{matrix}1;\,z\\a+\frac{1}{2},\,a+1\end{matrix}\right) - \dfrac{2\sqrt{z}}{2a+1}\sinh(2\sqrt{z})\,{}_1F_2\!\left(\begin{matrix}1;\,z\\a+1,\,a+\frac{3}{2}\end{matrix}\right).$

5. ${}_1F_2\!\left(\begin{matrix}a;\,z\\a+1,\,\frac{3}{2}\end{matrix}\right)$
$= \dfrac{1}{2a-1}\left[\dfrac{a}{\sqrt{z}}\sinh(2\sqrt{z})\,{}_1F_2\!\left(\begin{matrix}1;\,z\\a,\,a+\frac{1}{2}\end{matrix}\right) - \cosh(2\sqrt{z})\,{}_1F_2\!\left(\begin{matrix}1;\,z\\a+\frac{1}{2},\,a+1\end{matrix}\right)\right].$

6. ${}_1F_2\!\left(\begin{matrix}j+1;\,\pm z\\\frac{3}{2}-m,\,c\end{matrix}\right) = (-1)^j\dfrac{\sqrt{\pi}}{2}\dfrac{\Gamma(c)}{j!\left(-\frac{1}{2}\right)_m}\sum_{i=0}^{j}\binom{j}{i}\left(\dfrac{1}{2}-j-m\right)_{j-i}$
$\times \sum_{k=0}^{i+m}(-1)^k\binom{i+m}{k}\left(c-\dfrac{3}{2}\right)_{i-k+m} z^{(2k-2c+1)/4}\left\{\begin{matrix}\mathbf{L}_{c-k-3/2}(2\sqrt{z})\\\mathbf{H}_{c-k-3/2}(2\sqrt{z})\end{matrix}\right\}.$

7. ${}_1F_2\!\left(\begin{matrix}j+1;\,\pm z\\m+\frac{5}{2},\,c\end{matrix}\right) = \dfrac{\left(\frac{3}{2}\right)_{m+1}}{j!\,m!}\Gamma(c)\sum_{k=0}^{m}(-1)^k\binom{m}{k}\sum_{i=0}^{k}(-k)_i$
$\times \left[\dfrac{\sqrt{\pi}}{2}\sum_{p=0}^{j}(-1)^{j-p}\binom{j}{p}(c-j-1)_{j-p}z^{(2p-2c+2i-1)/4}\left\{\begin{matrix}\mathbf{L}_{c-i-p-5/2}(2\sqrt{z})\\\mathbf{H}_{c-i-p-5/2}(2\sqrt{z})\end{matrix}\right\}\right.$
$\left. - \dfrac{(-1)^j(i-j+1)_j}{(2k-2i+1)\Gamma(c-i-1)}z^{-i-1}\right].$

8. ${}_1F_2\!\left(\begin{matrix}n-\frac{1}{2};\,\pm z\\n+1,\,2n+1\end{matrix}\right)$
$= \dfrac{2^{2n+1}(n!)^2}{2n+1}z^{1/2-n}\left[\sqrt{z}\left\{\begin{matrix}I_{n-1}(\sqrt{z})\\J_{n-1}(\sqrt{z})\end{matrix}\right\}^2 - \left\{\begin{matrix}I_{n-1}(\sqrt{z})I_n(\sqrt{z})\\J_{n-1}(\sqrt{z})J_n(\sqrt{z})\end{matrix}\right\}\right.$
$\left. - \sqrt{z}\left\{\begin{matrix}I_{n-2}(\sqrt{z})I_n(\sqrt{z})\\J_{n-2}(\sqrt{z})J_n(\sqrt{z})\end{matrix}\right\} + \dfrac{1-2n}{2\sqrt{z}}\left\{\begin{matrix}I_n(\sqrt{z})\\J_n(\sqrt{z})\end{matrix}\right\}^2\right].$

8.1.11] 8.1. The Hypergeometric Functions

9. $\displaystyle {}_1F_2\left(\begin{array}{c} j+\frac{1}{2};\ \pm z \\ j+m+\frac{3}{2},\ n+1 \end{array}\right) = \frac{n!}{m!}\left(j+\frac{1}{2}\right)_{m+1}\sum_{k=0}^{m}(-1)^k\binom{m}{k}\frac{z^{-j-k}}{\left(\frac{1}{2}-j-k\right)_n}$

$\displaystyle \times\Bigg[\sum_{p=0}^{2j+2k-1}(\pm 1)^{j+k}\left(\frac{1}{2}-j-k\right)_p z^{(2j+2k-p-1)/2}\left\{\begin{array}{c} I_{p+1}(2\sqrt{z}) \\ J_{p+1}(2\sqrt{z}) \end{array}\right\}$

$\displaystyle + (-1)^{j+k}\left(\frac{1}{2}-j-k\right)_{2j+2k}\left(2\left\{\begin{array}{c} I_0(2\sqrt{z}) \\ J_0(2\sqrt{z}) \end{array}\right\}\right.$

$\displaystyle +\pi\left\{\begin{array}{c} I_0(2\sqrt{z})\,\mathbf{L}_1(2\sqrt{z}) - I_1(2\sqrt{z})\,\mathbf{L}_0(2\sqrt{z}) \\ J_0(2\sqrt{z})\,\mathbf{H}_1(2\sqrt{z}) + J_1(2\sqrt{z})\,\mathbf{H}_0(2\sqrt{z}) \end{array}\right\}$

$\displaystyle -\frac{2}{\sqrt{z}}\sum_{r=0}^{j+k-1}(\mp 1)^r\left\{\begin{array}{c} I_{2r+1}(2\sqrt{z}) \\ J_{2r+1}(2\sqrt{z}) \end{array}\right\}\Bigg)$

$\displaystyle -\sum_{i=0}^{n-1}(\pm 1)^{j+k}\left(\frac{1}{2}-j-k\right)_i z^{(2j+2k-i-1)/2}\left\{\begin{array}{c} I_{i+1}(2\sqrt{z}) \\ J_{i+1}(2\sqrt{z}) \end{array}\right\}\Bigg].$

10. $\displaystyle {}_1F_2\left(\begin{array}{c} j+\frac{1}{4};\ \pm z \\ j+m+\frac{5}{4},\ n+\frac{1}{2} \end{array}\right) = \frac{\left(j+\frac{1}{4}\right)_{m+1}}{m!}\Gamma\!\left(n+\frac{1}{2}\right)\sum_{k=0}^{m}(-1)^k\binom{m}{k}$

$\displaystyle \times \frac{1}{\left(\frac{1}{4}-j-k\right)_n}\Bigg[(\pm 1)^{j+k}(4j+4k-1)!!\,2^{-4j-4k-1/2}z^{-j-k-1/4}$

$\displaystyle \times\left\{\begin{array}{c} \operatorname{erf}(\sqrt{2}\,z^{1/4})+\operatorname{erfi}(\sqrt{2}\,z^{1/4}) \\ 2^{3/2}C(2\sqrt{z}) \end{array}\right\} + \frac{1}{\sqrt{\pi z}}\left\{\begin{array}{c} \sinh(2\sqrt{z}) \\ \sin(2\sqrt{z}) \end{array}\right\}$

$\displaystyle \times \sum_{p=0}^{j+k-1}\frac{(4j+4k-1)!!}{(4j+4k-4p-1)!!}(\pm 16 z)^{-p}$

$\displaystyle -\frac{4}{\sqrt{\pi}}\left\{\begin{array}{c} \cosh(2\sqrt{z}) \\ \cos(2\sqrt{z}) \end{array}\right\}\sum_{p=0}^{j+k-1}\frac{(4j+4k-1)!!}{(4j+4k-4p-3)!!}(\pm 16 z)^{-p-1}$

$\displaystyle -\sum_{k=0}^{n-1}\left(\frac{1}{4}-j-k\right)_i z^{-(2i+1)/4}\left\{\begin{array}{c} I_{i+1/2}(2\sqrt{z}) \\ J_{i+1/2}(2\sqrt{z}) \end{array}\right\}\Bigg].$

11. $\displaystyle {}_1F_2\left(\begin{array}{c} j+\frac{3}{4};\ \pm z \\ j+m+\frac{7}{4},\ n+\frac{1}{2} \end{array}\right) = \frac{\left(j+\frac{3}{4}\right)_{m+1}}{m!}\Gamma\!\left(n+\frac{1}{2}\right)\sum_{k=0}^{m}(-1)^k\binom{m}{k}$

$\displaystyle \times\frac{1}{\left(-j-k-\frac{1}{4}\right)_n}\Bigg[(\pm 1)^{j+k}(4j+4k+1)!!\,2^{-4j-4k-5/2}z^{-j-k-3/4}$

$\displaystyle \times\left\{\begin{array}{c} \operatorname{erf}(\sqrt{2}\,z^{1/4})-\operatorname{erfi}(\sqrt{2}\,z^{1/4}) \\ -2^{3/2}S(2\sqrt{z}) \end{array}\right\} + \frac{1}{\sqrt{\pi z}}\left\{\begin{array}{c} \sinh(2\sqrt{z}) \\ \sin(2\sqrt{z}) \end{array}\right\}$

$$\times \sum_{p=0}^{j+k} \frac{(4j+4k+1)!!}{(4j+4k-4p+1)!!} (\pm 16z)^{-p}$$

$$-\frac{4}{\sqrt{\pi}} \left\{ \begin{matrix} \cosh(2\sqrt{z}) \\ \cos(2\sqrt{z}) \end{matrix} \right\} \sum_{p=0}^{j+k-1} \frac{(4j+4k+1)!!}{(4j+4k-4p-1)!!} (\pm 16z)^{-p-1}$$

$$- \sum_{i=0}^{n-1} \left(-j-k-\frac{1}{4} \right)_i z^{-(2i+1)/4} \left\{ \begin{matrix} I_{i+1/2}(2\sqrt{z}) \\ J_{i+1/2}(2\sqrt{z}) \end{matrix} \right\} \Bigg].$$

12. $${}_1F_2 \left(\begin{matrix} j+\frac{1}{4};\ \pm z \\ j+m+\frac{5}{4},\ \frac{1}{2}-n \end{matrix} \right) = \frac{\left(j+\frac{1}{4}\right)_{m+1}}{m!} \Gamma\left(\frac{1}{2}-n\right) \sum_{k=0}^{m} (-1)^k \binom{m}{k}$$

$$\times \Bigg[\sum_{i=0}^{n-1} \left(\frac{1}{4} -j-k-n \right)_i z^{(2n-2i-1)/4} \left\{ \begin{matrix} I_{i-n+1/2}(2\sqrt{z}) \\ J_{i-n+1/2}(2\sqrt{z}) \end{matrix} \right\}$$

$$+ (-1)^n \left(j+k+\frac{3}{4} \right)_n$$

$$\times \left(\frac{1}{\sqrt{\pi z}} \left\{ \begin{matrix} \sinh(2\sqrt{z}) \\ \sin(2\sqrt{z}) \end{matrix} \right\} \sum_{p=0}^{j+k-1} \frac{(4j+4k-1)!!}{(4j+4k-4p-1)!!} (\pm 16z)^{-p} \right.$$

$$- \frac{4}{\sqrt{\pi}} \left\{ \begin{matrix} \cosh(2\sqrt{z}) \\ \cos(2\sqrt{z}) \end{matrix} \right\} \sum_{p=0}^{j+k-1} \frac{(4j+4k-1)!!}{(4j+4k-4p-3)!!} (\pm 16z)^{-p-1}$$

$$+ (\pm 1)^{j+k} (4j+4k-1)!!\, 2^{-4j-4k-1/2} z^{-j-k-1/4}$$

$$\times \left\{ \begin{matrix} \operatorname{erf}(\sqrt{2}\, z^{1/4}) + \operatorname{erfi}(\sqrt{2}\, z^{1/4}) \\ 2^{3/2} C(2\sqrt{z}) \end{matrix} \right\} \Bigg) \Bigg].$$

13. $${}_1F_2 \left(\begin{matrix} j+\frac{3}{4};\ \pm z \\ j+m+\frac{7}{4},\ \frac{1}{2}-n \end{matrix} \right) = \frac{\left(j+\frac{3}{4}\right)_{m+1}}{m!} \Gamma\left(\frac{1}{2}-n\right) \sum_{k=0}^{m} (-1)^k \binom{m}{k}$$

$$\times \Bigg[\sum_{i=0}^{n-1} \left(-\frac{1}{4} -j-k-n \right)_i z^{(2n-2i-1)/4} \left\{ \begin{matrix} I_{i-n+1/2}(2\sqrt{z}) \\ J_{i-n+1/2}(2\sqrt{z}) \end{matrix} \right\}$$

$$+ (-1)^n \left(j+k+\frac{5}{4} \right)_n$$

$$\times \left(\frac{1}{\sqrt{\pi z}} \left\{ \begin{matrix} \sinh(2\sqrt{z}) \\ \sin(2\sqrt{z}) \end{matrix} \right\} \sum_{p=0}^{j+k} \frac{(4j+4k+1)!!}{(4j+4k-4p+1)!!} (\pm 16z)^{-p} \right.$$

$$- \frac{4}{\sqrt{\pi}} \left\{ \begin{matrix} \cosh(2\sqrt{z}) \\ \cos(2\sqrt{z}) \end{matrix} \right\} \sum_{p=0}^{j+k-1} \frac{(4j+4k+1)!!}{(4j+4k-4p-1)!!} (\pm 16z)^{-p-1}$$

$$+ (\pm 1)^{j+k} (4j+4k+1)!!\, 2^{-4j-4k-5/2} z^{-j-k-3/4}$$

$$\times \left\{ \begin{matrix} \operatorname{erf}(\sqrt{2}\, z^{1/4}) - \operatorname{erfi}(\sqrt{2}\, z^{1/4}) \\ -2^{3/2} S(2\sqrt{z}) \end{matrix} \right\} \Bigg) \Bigg].$$

8.1. The Hypergeometric Functions

14. $\displaystyle {}_1F_2\left(\begin{array}{c} a;\ \pm z \\ a-m+\frac{1}{2},\ 2a-n \end{array}\right) = \frac{\Gamma^2\left(a+\frac{1}{2}\right)}{\left(\frac{1}{2}-a\right)_m (1-2a)_n} \left(\frac{2}{\sqrt{z}}\right)^{2a-1}$

$\displaystyle \times \sum_{k=0}^n \binom{n}{k} \left(\frac{1}{2}-a-m\right)_{n-k} \sum_{i=0}^{k+m} (\mp 1)^i \binom{k+m}{i} \left(\frac{1}{2}-a\right)_{k+m-i} \left(\frac{\sqrt{z}}{2}\right)^i$

$\displaystyle \times \sum_{p=0}^i \binom{i}{p} \left\{ \begin{array}{c} I_{a+p-1/2}(\sqrt{z})\, I_{a+i-p-1/2}(\sqrt{z}) \\ J_{a+p-1/2}(\sqrt{z})\, J_{a+i-p-1/2}(\sqrt{z}) \end{array} \right\}.$

15. $\displaystyle {}_1F_2\left(\begin{array}{c} m+\frac{1}{2};\ \pm z \\ a-n,\ 2-a \end{array}\right) = (-1)^{m+n} \frac{\pi(1-a)\csc(a\pi)}{\left(\frac{1}{2}\right)_m (1-a)_n} \left(\frac{\sqrt{z}}{2}\right)^n$

$\displaystyle \times \sum_{k=0}^m (\mp 1)^k \binom{m}{k} \left(a-m-n-\frac{1}{2}\right)_{m-k} \left(\frac{\sqrt{z}}{2}\right)^k \sum_{i=0}^{k+n} (\pm 1)^{i+n} \binom{k+n}{i}$

$\displaystyle \times \left\{ \begin{array}{c} I_{n+k-i-a+1}(\sqrt{z})\, I_{a-i-1}(\sqrt{z}) \\ J_{n+k-i-a+1}(\sqrt{z})\, J_{a-i-1}(\sqrt{z}) \end{array} \right\}.$

16. $\displaystyle {}_1F_2\left(\begin{array}{c} a;\ \pm z \\ a-n,\ b \end{array}\right) = \Gamma(b) z^{(1-b)/2} \sum_{k=0}^n (\pm 1) \binom{n}{k}^k \frac{z^{k/2}}{(a-n)_k} \left\{ \begin{array}{c} I_{b+k-1}(2\sqrt{z}) \\ J_{b+k-1}(2\sqrt{z}) \end{array} \right\}.$

8.1.12. The hypergeometric function ${}_2F_2(a_1, a_2;\ b_1, b_2;\ z)$

1. $\displaystyle {}_2F_2\left(\begin{array}{c} a+1,\ 2a \\ a,\ b;\ z \end{array}\right) = e^z\, {}_2F_2\left(\begin{array}{c} 2a-b+2,\ b-2a-1 \\ 2a-b+1,\ b;\ -z \end{array}\right)$ [[30], (12)].

2. $\displaystyle {}_2F_2\left(\begin{array}{c} \frac{n+1}{2},\ \frac{n}{2}+1 \\ n+1,\ 1;\ z \end{array}\right) = e^{z/2} \sum_{k=0}^n \binom{n}{k} I_k\left(\frac{z}{2}\right).$

8.1.13. The hypergeometric function ${}_2F_3(a_1, a_2;\ b_1, b_2, b_3;\ z)$

1. $\displaystyle {}_2F_3\left(\begin{array}{c} a,\ a+\frac{1}{2};\ z \\ b,\ b+\frac{1}{2},\ \frac{1}{2} \end{array}\right) = \cosh(2\sqrt{z})\ {}_2F_3\left(\begin{array}{c} b-a,\ b-a+\frac{1}{2};\ z \\ b,\ b+\frac{1}{2},\ \frac{1}{2} \end{array}\right)$

$\displaystyle + \frac{a-b}{b} 2\sqrt{z} \sinh(2\sqrt{z})\ {}_2F_3\left(\begin{array}{c} b-a+\frac{1}{2},\ b-a+1;\ z \\ b+\frac{1}{2},\ b+1,\ \frac{3}{2} \end{array}\right)$ [[25], (12)].

2. $\displaystyle = \frac{1}{2}\left[{}_1F_1\left(\begin{array}{c} 2a;\ -2\sqrt{z} \\ 2b \end{array}\right) + {}_1F_1\left(\begin{array}{c} 2a;\ 2\sqrt{z} \\ 2b \end{array}\right) \right].$

3. $\displaystyle {}_2F_3\left(\begin{matrix}a,\,a+\frac{1}{2};\,z\\b,\,b+\frac{1}{2},\,\frac{3}{2}\end{matrix}\right)=\frac{2b-1}{2(2a-1)\sqrt{z}}\sinh(2\sqrt{z})\,{}_2F_3\left(\begin{matrix}b-a,\,b-a+\frac{1}{2};\,z\\b-\frac{1}{2},\,b,\,\frac{1}{2}\end{matrix}\right)$

$\displaystyle\qquad+\frac{2(a-b)}{2a-1}\cosh(2\sqrt{z})\,{}_2F_3\left(\begin{matrix}b-a+\frac{1}{2},\,b-a+1;\,z\\b,\,b+\frac{1}{2},\,\frac{3}{2}\end{matrix}\right)$ [[25], (12)].

4. $\displaystyle\qquad=\frac{2b-1}{4(2a-1)\sqrt{z}}\left[{}_1F_1\left(\begin{matrix}2a-1;\,2\sqrt{z}\\2b-1\end{matrix}\right)-{}_1F_1\left(\begin{matrix}2a-1;\,-2\sqrt{z}\\2b-1\end{matrix}\right)\right].$

5. $\displaystyle {}_2F_3\left(\begin{matrix}-n,\,n+\frac{1}{2};\,z\\\frac{1}{4},\,\frac{1}{2},\,\frac{3}{4}\end{matrix}\right)=\left[\frac{n!}{(2n)!}\right]^2 H_{2n}\!\left(\sqrt[4]{4z}\right) H_{2n}\!\left(i\sqrt[4]{4z}\right).$

6. $\displaystyle {}_2F_3\left(\begin{matrix}-n,\,n+\frac{3}{2};\,z\\\frac{3}{4},\,\frac{5}{4},\,\frac{3}{2}\end{matrix}\right)=\left[\frac{n!}{(2n+1)!}\right]^2 \frac{1}{8i\sqrt{z}} H_{2n+1}\!\left(\sqrt[4]{4z}\right) H_{2n+1}\!\left(i\sqrt[4]{4z}\right).$

7. $\displaystyle {}_2F_3\left(\begin{matrix}n,\,n+\frac{1}{2};\,z\\n+1,\,n+1,\,2n+1\end{matrix}\right)$

$\displaystyle\qquad=(n!)^2\left(-\frac{4}{z}\right)^n\left[1+I_0^2(\sqrt{z})+(-1)^n I_n^2(\sqrt{z})-2\sum_{k=0}^{n}(-1)^k I_k^2(\sqrt{z})\right].$

8. $\displaystyle {}_2F_3\left(\begin{matrix}\frac{1}{2},\,\frac{1}{2};\,z\\1,\,1,\,1\end{matrix}\right)=\frac{1}{\pi}\int_0^z\frac{1}{\sqrt{x(z-x)}}\,I_0^2(\sqrt{x})\,dx.$

9. $\displaystyle {}_2F_3\left(\begin{matrix}\frac{1}{2},\,\frac{1}{2};\,z\\1,\,1,\,\frac{3}{2}\end{matrix}\right)=\frac{1}{2\sqrt{z}}\int_0^z\frac{1}{\sqrt{x}}\,I_0^2(\sqrt{x})\,dx.$

10. $\displaystyle {}_2F_3\left(\begin{matrix}\frac{1}{2},\,\frac{1}{2};\,z\\\frac{3}{2},\,\frac{3}{2},\,\frac{3}{2}\end{matrix}\right)=\frac{1}{4\sqrt{z}}\int_0^z\frac{\operatorname{shi}(2\sqrt{x})}{x}\,dx.$

11. $\displaystyle {}_2F_3\left(\begin{matrix}\frac{1}{2},\,1;\,z\\\frac{3}{2},\,\frac{3}{2},\,\frac{3}{2}\end{matrix}\right)=\frac{\pi}{8\sqrt{z}}\int_0^z\frac{1}{x}\,\mathbf{L}_0(2\sqrt{x})\,dx.$

12. $\displaystyle {}_2F_3\left(\begin{matrix}\frac{1}{2},\,1;\,z\\\frac{3}{2},\,\frac{3}{2},\,2\end{matrix}\right)=\frac{2\sqrt{z}\operatorname{shi}(2\sqrt{z})-\cosh(2\sqrt{z})+1}{2z}.$

13. $\displaystyle {}_2F_3\left(\begin{matrix}\frac{1}{2},\,1;\,z\\\frac{3}{2},\,2,\,2\end{matrix}\right)=\frac{1}{z}[1+(4z-1)I_0(2\sqrt{z})-2\sqrt{z}\,I_1(2\sqrt{z})]$

$\displaystyle\qquad+2\pi[I_0(2\sqrt{z})\,\mathbf{L}_1(2\sqrt{z})-I_1(2\sqrt{z})\,\mathbf{L}_0(2\sqrt{z})].$

14. $_2F_3\left(\begin{matrix}1,1;z\\2,2,2\end{matrix}\right) = \dfrac{1}{z}\int_0^z \dfrac{I_0(2\sqrt{x})-1}{x}\,dx.$

15. $_2F_3\left(\begin{matrix}1,1;z\\ \frac{3}{2},\frac{3}{2},\frac{3}{2}\end{matrix}\right) = \dfrac{\pi}{8\sqrt{z}}\int_0^z \dfrac{1}{\sqrt{x(z-x)}}\,\mathbf{L}_0(2\sqrt{x})\,dx.$

16. $_2F_3\left(\begin{matrix}1,1;z\\ \frac{3}{2},\frac{3}{2},2\end{matrix}\right) = \dfrac{1}{4\sqrt{z}}\int_0^z \dfrac{\cosh(2\sqrt{x})-1}{x\sqrt{z-x}}\,dx.$

17. $_2F_3\left(\begin{matrix}1,1;z\\ \frac{3}{2},2,2\end{matrix}\right) = \dfrac{\operatorname{chi}(2\sqrt{z})-\ln(2\sqrt{z})-C}{z}.$

8.1.14. The hypergeometric function $_3F_0(a_1,a_2,a_3;z)$

1. $_3F_0\left(\begin{matrix}-n,\frac{1}{2},1\\ z\end{matrix}\right) = n!\sqrt{\pi}\,z^{(2n-1)/4}\left[I_{-n-1/2}\left(\dfrac{2}{\sqrt{z}}\right) - \mathbf{L}_{n+1/2}\left(\dfrac{2}{\sqrt{z}}\right)\right].$

2. $_3F_0\left(\begin{matrix}-n,\frac{1}{2},1\\ -z\end{matrix}\right) = n!\sqrt{\pi}\,z^{(2n-1)/4}\left[\mathbf{H}_{n+1/2}\left(\dfrac{2}{\sqrt{z}}\right) - Y_{n+1/2}\left(\dfrac{2}{\sqrt{z}}\right)\right].$

8.1.15. The hypergeometric function $_5F_0(a_1,a_2,\ldots,a_5;z)$

1. $_5F_0\left(\begin{matrix}-\frac{n}{2},\frac{1-n}{2},\frac{1}{4},\frac{3}{4},1\\ z\end{matrix}\right) = 2^{-n-1/2}n!\sqrt{\pi}\,z^{(2n-1)/8}$
$\times\left[\mathbf{H}_{n+1/2}(4z^{-1/4}) - \mathbf{L}_{n+1/2}(4z^{-1/4}) + I_{-n-1/2}(4z^{-1/4}) - Y_{n+1/2}(4z^{-1/4})\right].$

2. $_5F_0\left(\begin{matrix}-\frac{n}{2},\frac{1-n}{2},\frac{3}{4},1,\frac{5}{4}\\ z\end{matrix}\right) = n!\sqrt{\pi}\left(\dfrac{z}{16}\right)^{(2n-3)/8}$
$\times\left[\mathbf{H}_{n+3/2}(4z^{-1/4}) + \mathbf{L}_{n+3/2}(4z^{-1/4}) - Y_{n+3/2}(4z^{-1/4}) - I_{-n-3/2}(4z^{-1/4})\right].$

8.1.16. The hypergeometric function $_4F_1(a_1,\ldots,a_4;b_1;z)$

1. $_4F_1\left(\begin{matrix}-n,n+1,a,1-a\\ \frac{1}{2};z\end{matrix}\right)$
$= \dfrac{1}{\sqrt{z}}\left[I_{a+n}\left(\dfrac{1}{\sqrt{z}}\right)K_{a-n-1}\left(\dfrac{1}{\sqrt{z}}\right) + I_{a-n-1}\left(\dfrac{1}{\sqrt{z}}\right)K_{a+n}\left(\dfrac{1}{\sqrt{z}}\right)\right].$

2. $_4F_1\left(\begin{matrix}-n,n+1,a,1-a\\ \frac{1}{2};-z\end{matrix}\right)$
$= (-1)^n\dfrac{\pi}{2\sqrt{z}}\left[J_{a+n}\left(\dfrac{1}{\sqrt{z}}\right)Y_{a-n-1}\left(\dfrac{1}{\sqrt{z}}\right) - J_{a-n-1}\left(\dfrac{1}{\sqrt{z}}\right)Y_{a+n}\left(\dfrac{1}{\sqrt{z}}\right)\right].$

3. $\,_4F_1\left(\begin{array}{c}-n,\,\frac{1}{4}-n,\,\frac{1}{2}-n,\,\frac{3}{4}-n\\ \frac{1}{2}-2n;\,z\end{array}\right) = \left(-\frac{z}{64}\right)^n H_{2n}\left(\sqrt[4]{\frac{4}{z}}\right) H_{2n}\left(i\sqrt[4]{\frac{4}{z}}\right).$

4. $\,_4F_1\left(\begin{array}{c}-n,\,-\frac{1}{2}-n,\,-\frac{1}{4}-n,\,\frac{1}{4}-n\\ -\frac{1}{2}-2n;\,z\end{array}\right)$

$$= (-1)^{n+1} i \left(\frac{z}{64}\right)^{n+1/2} H_{2n+1}\left(\sqrt[4]{\frac{4}{z}}\right) H_{2n+1}\left(i\sqrt[4]{\frac{4}{z}}\right).$$

8.1.17. The hypergeometric function $\,_6F_1(a_1,\ldots,a_6;\,b_1;\,z)$

1. $\,_6F_1\left(\begin{array}{c}-\frac{n}{2},\,\frac{1-n}{2},\,\frac{n+1}{2},\,\frac{n}{2}+1,\,\frac{1}{4},\,\frac{3}{4}\\ \frac{1}{2};\,-z\end{array}\right)$

$$= \frac{\pi}{\sqrt{2}\,z^{1/4}} \Big[\operatorname{bei}^2_{n+1/2}(2z^{-1/4}) - 2\operatorname{ber}_{n+1/2}(2z^{-1/4})\operatorname{bei}_{n+1/2}(2z^{-1/4})$$
$$- \operatorname{ber}^2_{n+1/2}(2z^{-1/4}) + \operatorname{bei}^2_{-n-1/2}(2z^{-1/4})$$
$$- 2\operatorname{ber}_{-n-1/2}(2z^{-1/4})\operatorname{bei}_{-n-1/2}(2z^{-1/4}) - \operatorname{ber}^2_{-n-1/2}(2z^{-1/4})\Big].$$

2. $\,_6F_1\left(\begin{array}{c}-\frac{n}{2},\,\frac{1-n}{2},\,\frac{n+3}{2},\,\frac{n}{2}+2,\,\frac{3}{4},\,\frac{5}{4}\\ \frac{3}{2};\,-z\end{array}\right) = -\frac{4\sqrt{2}\,\pi}{(n+1)(n+2)z^{3/4}}$

$$\times \Big[\operatorname{bei}^2_{n+3/2}(2z^{-1/4}) + 2\operatorname{ber}_{n+3/2}(2z^{-1/4})\operatorname{bei}_{n+3/2}(2z^{-1/4})$$
$$- \operatorname{ber}^2_{n+3/2}(2z^{-1/4}) + \operatorname{bei}^2_{-n-3/2}(2z^{-1/4})$$
$$+ 2\operatorname{ber}_{-n-3/2}(2z^{-1/4})\operatorname{bei}_{-n-3/2}(2z^{-1/4}) - \operatorname{ber}^2_{-n-3/2}(2z^{-1/4})\Big].$$

8.1.18. The hypergeometric function $\,_8F_3(a_1,\ldots,a_8;\,b_1,b_2,b_3;\,z)$

1. $\,_8F_3\left(\begin{array}{c}-\frac{n}{4},\,\frac{1-n}{4},\,\frac{2-n}{4},\,\frac{3-n}{4},\,\frac{n+1}{4},\,\frac{n+2}{4},\,\frac{n+3}{4},\,\frac{n}{4}+1\\ \frac{1}{4},\,\frac{1}{2},\,\frac{3}{4};\,-z\end{array}\right) = \frac{\sqrt{\pi}}{z^{1/8}}$

$$\times \left\{\left[\sin\frac{3\pi}{8}\sinh(\sqrt{2}\,z^{-1/4})\sin\left(\frac{n\pi}{2}-\sqrt{2}\,z^{-1/4}\right)\right.\right.$$
$$\left.+\cos\frac{3\pi}{8}\cosh(\sqrt{2}\,z^{-1/4})\cos\left(\frac{n\pi}{2}-\sqrt{2}\,z^{-1/4}\right)\right]\operatorname{ber}_{-n-1/2}(2z^{-1/4})$$
$$+\left[\cos\frac{3\pi}{8}\sinh(\sqrt{2}\,z^{-1/4})\sin\left(\frac{n\pi}{2}-\sqrt{2}\,z^{-1/4}\right)\right.$$
$$\left.-\sin\frac{3\pi}{8}\cosh(\sqrt{2}\,z^{-1/4})\cos\left(\frac{n\pi}{2}-\sqrt{2}\,z^{-1/4}\right)\right]\operatorname{bei}_{-n-1/2}(2z^{-1/4})$$
$$-\left[\cos\frac{3\pi}{8}\cosh(\sqrt{2}\,z^{-1/4})\sin\left(\frac{n\pi}{2}+\sqrt{2}\,z^{-1/4}\right)\right.$$
$$\left.\left.+\sin\frac{3\pi}{8}\sinh(\sqrt{2}\,z^{-1/4})\cos\left(\frac{n\pi}{2}+\sqrt{2}\,z^{-1/4}\right)\right]\operatorname{ber}_{n+1/2}(2z^{-1/4})\right.$$

$$+ \left[\sin\frac{3\pi}{8}\cosh\left(\sqrt{2}\,z^{-1/4}\right)\sin\left(\frac{n\pi}{2}+\sqrt{2}\,z^{-1/4}\right)\right.$$
$$\left.-\cos\frac{3\pi}{8}\sinh\left(\sqrt{2}\,z^{-1/4}\right)\cos\left(\frac{n\pi}{2}+\sqrt{2}\,z^{-1/4}\right)\right]\mathrm{bei}_{n+1/2}\left(2z^{-1/4}\right)\Big]\Big\}.$$

2. $\displaystyle {}_8F_3\left(\begin{matrix}-\frac{n}{4},\frac{1-n}{4},\frac{2-n}{4},\frac{3-n}{4},\frac{n+5}{4},\frac{n+6}{4},\frac{n+7}{4},\frac{n}{4}+2\\ \frac{3}{4},\frac{5}{4},\frac{3}{2};\,-z\end{matrix}\right)$

$$= \frac{32\sqrt{\pi}}{(n+1)(n+2)(n+3)(n+4)z^{5/8}}$$
$$\times\Big\{\left[\sin\frac{3\pi}{8}\cosh\left(\sqrt{2}\,z^{-1/4}\right)\cos\left(\frac{n\pi}{2}-\sqrt{2}\,z^{-1/4}\right)\right.$$
$$\left.-\cos\frac{3\pi}{8}\sinh\left(\sqrt{2}\,z^{-1/4}\right)\sin\left(\frac{n\pi}{2}-\sqrt{2}\,z^{-1/4}\right)\right]\mathrm{ber}_{-n-5/2}\left(2z^{-1/4}\right)$$
$$+\left[\cos\frac{3\pi}{8}\cosh\left(\sqrt{2}\,z^{-1/4}\right)\cos\left(\frac{n\pi}{2}-\sqrt{2}\,z^{-1/4}\right)\right.$$
$$\left.+\sin\frac{3\pi}{8}\sinh\left(\sqrt{2}\,z^{-1/4}\right)\sin\left(\frac{n\pi}{2}-\sqrt{2}\,z^{-1/4}\right)\right]\mathrm{bei}_{-n-5/2}\left(2z^{-1/4}\right)$$
$$+\left[\cos\frac{3\pi}{8}\sinh\left(\sqrt{2}\,z^{-1/4}\right)\cos\left(\frac{n\pi}{2}+\sqrt{2}\,z^{-1/4}\right)\right.$$
$$\left.-\sin\frac{3\pi}{8}\cosh\left(\sqrt{2}\,z^{-1/4}\right)\sin\left(\frac{n\pi}{2}+\sqrt{2}\,z^{-1/4}\right)\right]\mathrm{ber}_{n+5/2}\left(2z^{-1/4}\right)$$
$$-\left[\sin\frac{3\pi}{8}\sinh\left(\sqrt{2}\,z^{-1/4}\right)\cos\left(\frac{n\pi}{2}+\sqrt{2}\,z^{-1/4}\right)\right.$$
$$\left.+\cos\frac{3\pi}{8}\cosh\left(\sqrt{2}\,z^{-1/4}\right)\sin\left(\frac{n\pi}{2}+\sqrt{2}\,z^{-1/4}\right)\right]\mathrm{bei}_{n+5/2}\left(2z^{-1/4}\right)\Big]\Big\}.$$

8.1.19. The hypergeometric function $\,{}_0F_3(b_1,b_2,b_3;z)$

1. $\displaystyle {}_0F_3\left(\begin{matrix}\,\\1,1,1\\\frac{1}{2},\frac{1}{2},\frac{1}{2}\end{matrix}\,;\,\frac{z}{}\right)$

$$= z^{1/4}\,\mathrm{ber}_1(2^{3/2}z^{1/4})\left[2\,\mathrm{ker}_0(2^{3/2}z^{1/4})-2\,\mathrm{kei}_0(2^{3/2}z^{1/4})-\pi\,\mathrm{bei}_0(2^{3/2}z^{1/4})\right]$$
$$+ z^{1/4}\,\mathrm{bei}_1(2^{3/2}z^{1/4})\left[2\,\mathrm{ker}_0(2^{3/2}z^{1/4})+2\,\mathrm{kei}_0(2^{3/2}z^{1/4})+\pi\,\mathrm{ber}_0(2^{3/2}z^{1/4})\right]$$
$$+ z^{1/4}\,\mathrm{bei}_0(2^{3/2}z^{1/4})\left[2\,\mathrm{ker}_1(2^{3/2}z^{1/4})-2\,\mathrm{kei}_1(2^{3/2}z^{1/4})-\pi\,\mathrm{bei}_1(2^{3/2}z^{1/4})\right]$$
$$+ z^{1/4}\,\mathrm{ber}_0(2^{3/2}z^{1/4})\left[2\,\mathrm{ker}_1(2^{3/2}z^{1/4})+2\,\mathrm{kei}_1(2^{3/2}z^{1/4})+\pi\,\mathrm{ber}_1(2^{3/2}z^{1/4})\right].$$

2. $\displaystyle {}_0F_3\left(\begin{matrix}\,\\1,1,1\\\frac{1}{2},\frac{1}{2},\frac{1}{2}\end{matrix}\,;\,\frac{-z}{}\right)$

$$= -\left(\frac{z}{4}\right)^{1/4}\left[2J_1\left(2^{3/2}z^{1/4}\right)K_0\left(2^{3/2}z^{1/4}\right)-2J_0\left(2^{3/2}z^{1/4}\right)K_1\left(2^{3/2}z^{1/4}\right)\right.$$
$$\left.+\pi Y_0\left(2^{3/2}z^{1/4}\right)I_1\left(2^{3/2}z^{1/4}\right)+\pi Y_1\left(2^{3/2}z^{1/4}\right)I_0\left(2^{3/2}z^{1/4}\right)\right].$$

3. $\displaystyle {}_0F_3\left(\begin{matrix}\,\\1,1,3\\\frac{1}{2},\frac{1}{2},\frac{1}{2}\end{matrix}\,;\,\frac{-z}{}\right)$

$$= \frac{1}{2}\left[2J_1\left(2^{3/2}z^{1/4}\right)K_1\left(2^{3/2}z^{1/4}\right)-\pi Y_1\left(2^{3/2}z^{1/4}\right)I_1\left(2^{3/2}z^{1/4}\right)\right].$$

4. $_0F_3\left(\begin{matrix}-z\\ \frac{1}{2},\frac{5}{6},\frac{7}{6}\end{matrix}\right) = \frac{2^{-4/3}\pi}{3^{2/3}z^{1/6}}\left[\operatorname{Ai}\left(-2^{1/3}3^{2/3}z^{1/6}\right)\operatorname{Bi}\left(2^{1/3}3^{2/3}z^{1/6}\right)\right.$
$$\left. - \operatorname{Ai}\left(2^{1/3}3^{2/3}z^{1/6}\right)\operatorname{Bi}\left(-2^{1/3}3^{2/3}z^{1/6}\right)\right].$$

5. $_0F_3\left(\begin{matrix}-z\\ \frac{2}{3},\frac{7}{6},\frac{4}{3}\end{matrix}\right) = \frac{3^{1/3}\Gamma^2\left(\frac{4}{3}\right)}{2^{8/3}z^{1/3}}\left[\operatorname{Bi}\left(2^{1/3}3^{2/3}z^{1/6}\right) - \sqrt{3}\operatorname{Ai}\left(2^{1/3}3^{2/3}z^{1/6}\right)\right]$
$$\times\left[\sqrt{3}\operatorname{Ai}\left(-2^{1/3}3^{2/3}z^{1/6}\right) - \operatorname{Bi}\left(-2^{1/3}3^{2/3}z^{1/6}\right)\right].$$

6. $_0F_3\left(\begin{matrix}z\\ \frac{4}{3},\frac{3}{2},\frac{5}{3}\end{matrix}\right) = \frac{4\pi}{3^{5/2}z^{1/2}}\left[\operatorname{ber}_{-1/3}(2^{3/2}z^{1/4})\operatorname{ber}_{1/3}(2^{3/2}z^{1/4})\right.$
$$\left. + \operatorname{bei}_{-1/3}(2^{3/2}z^{1/4})\operatorname{bei}_{1/3}(2^{3/2}z^{1/4})\right].$$

7. $_0F_3\left(\begin{matrix}-z\\ \frac{4}{3},\frac{3}{2},\frac{5}{3}\end{matrix}\right) = \frac{2^{-1/3}\pi}{3^{13/6}z^{2/3}}\left[3\operatorname{Ai}\left(2^{1/3}3^{2/3}z^{1/6}\right)\operatorname{Ai}\left(-2^{1/3}3^{2/3}z^{1/6}\right)\right.$
$$\left. - \operatorname{Bi}\left(2^{1/3}3^{2/3}z^{1/6}\right)\operatorname{Bi}\left(-2^{1/3}3^{2/3}z^{1/6}\right)\right].$$

8. $_0F_3\left(\begin{matrix}-z\\ \frac{3}{2},\frac{3}{2},\frac{3}{2}\end{matrix}\right)$
$$= \frac{1}{8\sqrt{z}}\left[2J_0\left(2^{3/2}z^{1/4}\right)K_0\left(2^{3/2}z^{1/4}\right) + \pi Y_0\left(2^{3/2}z^{1/4}\right)I_0\left(2^{3/2}z^{1/4}\right)\right].$$

8.1.20. The hypergeometric function $_0F_7(b_1,\ldots,b_7;z)$

1. $_0F_7\left(\begin{matrix}-z\\ \frac{1}{8},\frac{1}{4},\frac{3}{8},\frac{1}{2},\frac{5}{8},\frac{3}{4},\frac{7}{8}\end{matrix}\right) = \sinh\left(4\sqrt{2}\,z^{1/8}\sin\frac{\pi}{8}\right)\sinh\left(4\sqrt{2}\,z^{1/8}\cos\frac{\pi}{8}\right)$
$$\times \sin\left(4\sqrt{2}\,z^{1/8}\sin\frac{\pi}{8}\right)\sin\left(4\sqrt{2}\,z^{1/8}\cos\frac{\pi}{8}\right)$$
$$+ \cosh\left(4\sqrt{2}\,z^{1/8}\sin\frac{\pi}{8}\right)\cosh\left(4\sqrt{2}\,z^{1/8}\cos\frac{\pi}{8}\right)$$
$$\times \cos\left(4\sqrt{2}\,z^{1/8}\sin\frac{\pi}{8}\right)\cos\left(4\sqrt{2}\,z^{1/8}\cos\frac{\pi}{8}\right).$$

2. $_0F_7\left(\begin{matrix}-z\\ \frac{5}{8},\frac{3}{4},\frac{7}{8},\frac{9}{8},\frac{5}{4},\frac{11}{8},\frac{3}{2}\end{matrix}\right) = \frac{3}{512\sqrt{z}}$
$$\times\left\{\sinh\left(4\sqrt{2}\,z^{1/8}\sin\frac{\pi}{8}\right)\cosh\left(4\sqrt{2}\,z^{1/8}\cos\frac{\pi}{8}\right)\right.$$
$$\times \sin\left(4\sqrt{2}\,z^{1/8}\cos\frac{\pi}{8}\right)\cos\left(4\sqrt{2}\,z^{1/8}\sin\frac{\pi}{8}\right)$$
$$- \sinh\left(4\sqrt{2}\,z^{1/8}\cos\frac{\pi}{8}\right)\cosh\left(4\sqrt{2}\,z^{1/8}\sin\frac{\pi}{8}\right)$$
$$\left.\times \sin\left(4\sqrt{2}\,z^{1/8}\sin\frac{\pi}{8}\right)\cos\left(4\sqrt{2}\,z^{1/8}\cos\frac{\pi}{8}\right)\right\}.$$

8.1.21. The hypergeometric function $_2F_5(a_1, a_2; b_1, \ldots, b_5; z)$

1. $_2F_5\left(\begin{array}{c} a, a+\frac{1}{2}; z \\ a+\frac{1}{4}, a+\frac{3}{4}, 2a-\frac{1}{2}, 2a, \frac{1}{2} \end{array}\right) = \frac{1}{2}\Gamma\left(2a-\frac{1}{2}\right)\Gamma\left(2a+\frac{1}{2}\right)z^{1/2-a}$
$\times \left[J_{2a-3/2}(2z^{1/4})J_{2a-1/2}(2z^{1/4}) + I_{2a-3/2}(2z^{1/4})I_{2a-1/2}(2z^{1/4})\right].$

2. $_2F_5\left(\begin{array}{c} a, a+\frac{1}{2}; -z \\ a+\frac{1}{4}, a+\frac{3}{4}, 2a-\frac{1}{2}, 2a, \frac{1}{2} \end{array}\right) = \Gamma\left(2a-\frac{1}{2}\right)\Gamma\left(2a+\frac{1}{2}\right)z^{1/2-a}$
$\times \left\{\sin(3a\pi)\left[\mathrm{bei}_{2a-3/2}(2z^{1/4})\,\mathrm{bei}_{2a-1/2}(2z^{1/4})\right.\right.$
$\left.- \mathrm{ber}_{2a-3/2}(2z^{1/4})\,\mathrm{ber}_{2a-1/2}(2z^{1/4})\right]$
$+ \cos(3a\pi)\left[\mathrm{ber}_{2a-3/2}(2z^{1/4})\,\mathrm{bei}_{2a-1/2}(2z^{1/4})\right.$
$\left.\left.+ \mathrm{bei}_{2a-3/2}(2z^{1/4})\,\mathrm{ber}_{2a-1/2}(2z^{1/4})\right]\right\}$ [[81], (13)].

3. $_2F_5\left(\begin{array}{c} a, a+\frac{1}{2}; z \\ a+\frac{1}{4}, a+\frac{3}{4}, 2a-1, 2a-\frac{1}{2}, \frac{3}{2} \end{array}\right)$
$= \frac{1}{4(2a-1)}\Gamma\left(2a-\frac{1}{2}\right)\Gamma\left(2a+\frac{1}{2}\right)z^{1/2-a}$
$\times \left[I_{2a-5/2}(2z^{1/4})I_{2a-3/2}(2z^{1/4}) - J_{2a-5/2}(2z^{1/4})J_{2a-3/2}(2z^{1/4})\right].$

4. $_2F_5\left(\begin{array}{c} a, a+\frac{1}{2}; -z \\ a+\frac{1}{4}, a+\frac{3}{4}, 2a-1, 2a-\frac{1}{2}, \frac{3}{2} \end{array}\right)$
$= \frac{1}{2(2a-1)}\Gamma\left(2a-\frac{1}{2}\right)\Gamma\left(2a+\frac{1}{2}\right)z^{1/2-a}$
$\times \left\{\sin(3a\pi)\left[\mathrm{ber}_{2a-5/2}(2z^{1/4})\,\mathrm{ber}_{2a-3/2}(2z^{1/4})\right.\right.$
$\left.- \mathrm{bei}_{2a-5/2}(2z^{1/4})\,\mathrm{bei}_{2a-3/2}(2z^{1/4})\right]$
$- \cos(3a\pi)\left[\mathrm{ber}_{2a-5/2}(2z^{1/4})\,\mathrm{bei}_{2a-3/2}(2z^{1/4})\right.$
$\left.\left.+ \mathrm{bei}_{2a-5/2}(2z^{1/4})\,\mathrm{ber}_{2a-3/2}(2z^{1/4})\right]\right\}$ [[81], (14)].

5. $_2F_5\left(\begin{array}{c} \frac{1}{4}, \frac{3}{4}; -z \\ \frac{1}{2}-a, 1-a, a+\frac{1}{2}, a+1, \frac{1}{2} \end{array}\right) = 2a\pi\csc(2a\pi)$
$\times \left[\mathrm{ber}_{2a}(2z^{1/4})\,\mathrm{ber}_{-2a}(2z^{1/4}) - \mathrm{bei}_{2a}(2z^{1/4})\,\mathrm{bei}_{-2a}(2z^{1/4})\right].$

6. $_2F_5\left(\begin{array}{c} \frac{3}{4}, \frac{5}{4}; -z \\ \frac{3}{2}-a, 1-a, a+1, a+\frac{3}{2}, \frac{3}{2} \end{array}\right) = a(1-4a^2)\pi\csc(2a\pi)z^{-1/2}$
$\times \left[\mathrm{ber}_{2a}(2z^{1/4})\,\mathrm{bei}_{-2a}(2z^{1/4}) + \mathrm{ber}_{-2a}(2z^{1/4})\,\mathrm{bei}_{2a}(2z^{1/4})\right].$

8.1.22. The hypergeometric function $_4F_7(a_1, \ldots, a_4; b_1, \ldots, b_7; z)$

1. $_4F_7\left(\begin{matrix} a, a+\frac{1}{4}, a+\frac{1}{2}, a+\frac{3}{4}; z \\ 2a, 2a+\frac{1}{2}, b, b+\frac{1}{2}, 2a-b+\frac{1}{2}, 2a-b+1, \frac{1}{2} \end{matrix}\right)$

$$= \frac{1}{2}\Gamma(2b)\,\Gamma(4a-2b+1)\,z^{1/4-a}$$

$$\times \left[J_{2b-1}\!\left(2z^{1/4}\right)J_{4a-2b}\!\left(2z^{1/4}\right) + I_{2b-1}\!\left(2z^{1/4}\right)I_{4a-2b}\!\left(2z^{1/4}\right)\right] \quad [[81],\,(11)].$$

2. $_4F_7\left(\begin{matrix} a, a+\frac{1}{4}, a+\frac{1}{2}, a+\frac{3}{4}; -z \\ 2a, 2a+\frac{1}{2}, b, b+\frac{1}{2}, 2a-b+\frac{1}{2}, 2a-b+1, \frac{1}{2} \end{matrix}\right)$

$$= \Gamma(2b)\,\Gamma(4a-2b+1)\,z^{1/4-a}$$

$$\times \left\{\cos\frac{3(4a-1)\pi}{4}\left[\mathrm{ber}_{2b-1}\!\left(2z^{1/4}\right)\mathrm{ber}_{4a-2b}\!\left(2z^{1/4}\right)\right.\right.$$
$$\left. - \mathrm{bei}_{2b-1}\!\left(2z^{1/4}\right)\mathrm{bei}_{4a-2b}\!\left(2z^{1/4}\right)\right]$$
$$+ \sin\frac{3(4a-1)\pi}{4}\left[\mathrm{ber}_{2b-1}\!\left(2z^{1/4}\right)\mathrm{bei}_{4a-2b}\!\left(2z^{1/4}\right)\right.$$
$$\left.\left. + \mathrm{bei}_{2b-1}\!\left(2z^{1/4}\right)\mathrm{ber}_{4a-2b}\!\left(2z^{1/4}\right)\right]\right\} \quad [[81],\,(11)].$$

3. $_4F_7\left(\begin{matrix} a, a+\frac{1}{4}, a+\frac{1}{2}, a+\frac{3}{4}; z \\ 2a, 2a\;\frac{1}{2}, b, b+\frac{1}{2}, 2a-b+\frac{1}{2}, 2a-b+1, \frac{3}{2} \end{matrix}\right)$

$$= \frac{\Gamma(2b)\,\Gamma(4a-2b+1)}{2(4a-1)}\,z^{1/4-a}$$

$$\times \left[I_{2b-2}\!\left(2z^{1/4}\right)I_{4a-2b-1}\!\left(2z^{1/4}\right) - J_{2b-2}\!\left(2z^{1/4}\right)J_{4a-2b-1}\!\left(2z^{1/4}\right)\right]$$
$$[[81],\,(12)].$$

4. $_4F_7\left(\begin{matrix} a, a+\frac{1}{4}, a+\frac{1}{2}, a+\frac{3}{4}; -z \\ 2a, 2a-\frac{1}{2}, b, b+\frac{1}{2}, 2a-b+\frac{1}{2}, 2a-b+1, \frac{3}{2} \end{matrix}\right)$

$$= \frac{\Gamma(2b)\,\Gamma(4a-2b+1)}{4a-1}\,z^{1/4-a}$$

$$\times \left\{\sin\frac{3(4a-3)\pi}{4}\left[\mathrm{bei}_{2b-2}\!\left(2z^{1/4}\right)\mathrm{bei}_{4a-2b-1}\!\left(2z^{1/4}\right)\right.\right.$$
$$\left. - \mathrm{ber}_{2b-2}\!\left(2z^{1/4}\right)\mathrm{ber}_{4a-2b-1}\!\left(2z^{1/4}\right)\right]$$
$$+ \cos\frac{3(4a-3)\pi}{4}\left[\mathrm{ber}_{2b-2}\!\left(2z^{1/4}\right)\mathrm{bei}_{4a-2b-1}\!\left(2z^{1/4}\right)\right.$$
$$\left.\left. + \mathrm{bei}_{2b-2}\!\left(2z^{1/4}\right)\mathrm{ber}_{4a-2b-1}\!\left(2z^{1/4}\right)\right]\right\} \quad [[81],\,(12)].$$

8.1.23. The generalized hypergeometric function $_pF_q((a_p); (b_q); z)$

Notation: $K_{n-1} = k_1 + k_2 + \ldots + k_{n-1}$.

1. $_pF_q\left(\begin{array}{c}(a_{p-n}), (a_n); z \\ (b_{q-n}), (a_n + m_n)\end{array}\right)$

$$= \frac{1}{\prod_{i=0}^{n} B(a_i, m_i)} \sum_{k_1=0}^{m_1-1} \cdots \sum_{k_n=0}^{m_n-1} \prod_{i=0}^{n} \frac{(-1)^{k_i}}{a_i + k_i} \binom{m_i - 1}{k_i}$$

$$\times \;_pF_q\left(\begin{array}{c}(a_{p-n}), (a_n) + k_i; z \\ (b_{q-n}), (a_n) + k_i + 1\end{array}\right) \quad [m_i = 1, 2, \ldots].$$

2. $_{p+1}F_q\left(\begin{array}{c}(a_p), 1 \\ (b_q); z\end{array}\right) = z^{-1} \frac{\prod_{j=0}^{q}(b_j - 1)}{\prod_{i=0}^{p}(a_i - 1)} \left[_{p+1}F_q\left(\begin{array}{c}(a_p) - 1, 1 \\ (b_q) - 1; z\end{array}\right) - 1\right].$

3. $_{p+1}F_p\left(\begin{array}{c}(a)_{p+1}; z \\ (b)_p\end{array}\right) = \frac{\prod_{j=1}^{p} \Gamma(b_j)}{\prod_{h=1}^{p+1} \Gamma(a_h)} \sum_{k=1}^{p+1} (e^{\pi i} z^{-1})^{a_k}$

$$\times \frac{\Gamma(a_k)}{\prod_{j=1}^{p} \Gamma(b_h - a_k)} \prod_{\substack{h=1 \\ h \neq k}}^{p+1} \Gamma(a_h - a_k) \;_{p+1}F_p\left(\begin{array}{c}1 + a_k - (b)_p, a_k; z \\ 1 + a_k - (a)'_{p+1}\end{array}\right)$$

$[(a)'_{p+1} = (a_1, a_2, \ldots, a_{k-1}, a_{k+1}, \ldots, a_{p+1}); \; 0 < \arg z < 2\pi].$

4. $_pF_q\left(\begin{array}{c}(a_p); z_1 + z_2 + \ldots + z_n \\ (b_q)\end{array}\right)$

$$= \sum_{k_1=0}^{\infty} \sum_{k_2=0}^{\infty} \cdots \sum_{k_{n-1}=0}^{\infty} \left(\prod_{i=1}^{n-1} \frac{z_i^{k_i}}{k_i!}\right) \frac{\prod(a_p)_{K_{n-1}}}{\prod(b_q)_{K_{n-1}}} \;_pF_q\left(\begin{array}{c}(a_p) + K_{n-1}; z_n \\ (b_q) + K_{n-1}\end{array}\right)$$

$$\left[\begin{array}{c}K_{n-1} = k_1 + k_2 + \ldots + k_{n-1}; \; |z_1| + |z_2| + \ldots + |z_n| < 1, \\ \text{if } p = q + 1 \text{ and } a_i \neq 0, -1, -2, \ldots \text{ for } i = 1, \ldots, n\end{array}\right].$$

5. $_{p+1}F_q\left(\begin{array}{c}-m, (a_p); z_1 + z_2 + \ldots + z_n \\ (b_q)\end{array}\right)$

$$= m! \sum_{k_1=0}^{n} \sum_{k_2=0}^{n-k_1} \cdots \sum_{k_{n-1}=0}^{n-K_{n-2}} \left(\prod_{i=1}^{n-1} \frac{(-z_i)^{k_i}}{k_i!}\right) \frac{\prod(a_p)_{K_{n-1}}}{\prod(b_q)_{K_{n-1}}} \frac{1}{(m - K_{n-1})!}$$

$$\times \;_pF_q\left(\begin{array}{c}-m + K_{n-1}, (a_p) + K_{n-1}; z_n \\ (b_q) + K_{n-1}\end{array}\right).$$

6. $${}_pF_q\left(\begin{matrix}(a_p);\\ (b_q)\end{matrix}\bigg| z_1z_2\ldots z_n\right) = e^{-z_1-z_2-\ldots-z_{n-1}}\sum_{k_1=0}^{\infty}\sum_{k_2=0}^{\infty}\ldots\sum_{k_{n-1}=0}^{\infty}\left(\prod_{i=1}^{n-1}\frac{z_i^{k_i}}{k_i!}\right)$$
$$\times\;{}_{p+n-1}F_q\left(\begin{matrix}-k_1,-k_2,\ldots,-k_{n-1},(a_p)\\ (b_q);\;(-1)^{n-1}z_n\end{matrix}\right).$$

7. $${}_{p+1}F_q\left(\begin{matrix}-m,(a_p);\;z_1z_2\ldots z_n\\ (b_q)\end{matrix}\right) = m!\,(z_1z_2\ldots z_{n-1})^m$$
$$\times\sum_{k_1=0}^{n}\sum_{k_2=0}^{n-k_1}\ldots\sum_{k_{n-1}=0}^{n-k_1-\ldots-k_{n-1}}\sum_{k_{n-1}=0}^{n-K_{n-2}}\left(\prod_{i=1}^{n-1}\frac{z_i^{-K_i}(1-z_i)^{k_i}}{k_i!}\right)\frac{1}{(m-K_{n-1})!}$$
$$\times\;{}_pF_q\left(\begin{matrix}-m+K_{n-1},(a_p)\\ (b_q);\;z_n\end{matrix}\right).$$

8. $${}_{p+2}F_{p+1}\left(\begin{matrix}-n,a,2,\ldots,2;\;1\\ b,1,\ldots,1\end{matrix}\right) = \frac{n!}{(b)_n}\sum_{k=0}^{p}(-1)^k\frac{(a)_k(b-a)_{n-k}}{(n-k)!}S_{p+1}^{k+1}.$$

9. $${}_pF_p\left(\begin{matrix}a+1,a+1,\ldots,a+1\\ a,a,\ldots,a;\end{matrix}\bigg| z\right) = \sum_{k=0}^{p}\binom{p}{k}\frac{1}{a^k}\,\mathrm{D}_t^k\left[e^{ze^t}\right]\bigg|_{t=0}.$$

8.2. The Meijer Function $G_{p,q}^{m,n}\left(z\bigg|\begin{matrix}(a_p)\\(b_q)\end{matrix}\right)$

8.2.1. General formulas

1. $$G_{p+1,q+2}^{m+1,n+1}\left(z\bigg|\begin{matrix}a,(a_p)\\ a,(b_q),b\end{matrix}\right) = (-1)^{a-b}G_{p,q+1}^{m+1,n}\left(z\bigg|\begin{matrix}(a_p)\\ b,(b_q)\end{matrix}\right)$$
$$-(-1)^{a-b}\sum_{k=1}^{a-b}\mathop{\mathrm{Res}}_{s=-a+k}\left[\frac{\Gamma(b+s)\prod_{i=1}^{m}\Gamma(b_i+s)\prod_{j=1}^{n}\Gamma(1-a_i-s)}{\prod_{i=n+1}^{p}\Gamma(a_i+s)\prod_{j=m+1}^{q}\Gamma(1-b_j-s)}z^{-s}\right]$$
$$\left[\begin{matrix}q\geq 1;\;0\leq n\leq p\leq q;\;0\leq m\leq q;\;a-b=1,2,\ldots;\\ a_k-b\neq 1,2,\ldots\text{ for }k=1,\ldots,n;\;[15],(2.1)\end{matrix}\right].$$

2. $$G_{p+1,q+2}^{m+1,n+1}\left(z\bigg|\begin{matrix}a,(a_p)\\ a,(b_q),b\end{matrix}\right) = (-1)^{a-b}G_{p,q+1}^{m+1,n}\left(z\bigg|\begin{matrix}(a_p)\\ b,(b_q)\end{matrix}\right)$$
$$\left[\begin{matrix}q\geq 1;\;0\leq n\leq p\leq q;\;0\leq m\leq q;\;a-b=0,-1,-2,\ldots;\\ a_k-b\neq 1,2,\ldots\text{ for }k=1,\ldots,n;\;[15],(2.2)\end{matrix}\right].$$

3. $$G_{p+2,q+1}^{m,n+1}\left(z\bigg|\begin{matrix}a,(a_p),b\\ (b_q),b\end{matrix}\right) = (-1)^{a-b}G_{p,q+1}^{m+1,n}\left(z\bigg|\begin{matrix}(a_p),a\\ (b_q)\end{matrix}\right)$$
$$\left[\begin{matrix}q\geq 1,0\leq n\leq p\leq q,0\leq m\leq q;\;a-b=0,-1,-2,\ldots;\\ a_k-b\neq 1,2,\ldots\text{ for }k=1,\ldots,n;\;[15],(2.4)\end{matrix}\right].$$

8.2. The Meijer Function $G_{p,q}^{m,n}\left(z \left| \begin{matrix} (a_p) \\ (b_q) \end{matrix} \right.\right)$

4. $G_{2p,2q+2}^{2m+1,2n}\left(z \left| \begin{matrix} (a_p), (a_p)+\frac{1}{2} \\ 0, (b_q), (b_q)+\frac{1}{2}, \frac{1}{2} \end{matrix} \right.\right)$

$$= 2^{-\tau}\pi^{-\rho}G_{p,q+1}^{m+1,n}\left(2^{\delta}e^{\pi i/2}z^{1/2} \left| \begin{matrix} (2a_p) \\ 0, (2b_q) \end{matrix} \right.\right)$$

$$+ 2^{-\tau}\pi^{-\rho}G_{p,q+1}^{m+1,n}\left(2^{\delta}e^{-\pi i/2}z^{1/2} \left| \begin{matrix} (2a_p) \\ 0, (2b_q) \end{matrix} \right.\right)$$

$$\left[\begin{matrix} \delta = q-p+1, \rho = (p+q+1)/2 - m - n, \\ \tau = 2\sum_{j=1}^{q} b_j - 2\sum_{i=1}^{a} a_i + p - m - n + 1; \ [25], (21) \end{matrix}\right].$$

5. $G_{2p,2q+2}^{2m+1,2n}\left(z \left| \begin{matrix} (a_p), (a_p)+\frac{1}{2} \\ 1/2, (b_q), (b_q)+\frac{1}{2}, 0 \end{matrix} \right.\right)$

$$= 2^{-\tau}\pi^{-\rho}iG_{p,q+1}^{m+1,n}\left(2^{\delta}e^{\pi i/2}z^{1/2} \left| \begin{matrix} (2a_p) \\ 0, (2b_q) \end{matrix} \right.\right)$$

$$- 2^{-\tau}\pi^{-\rho}iG_{p,q+1}^{m+1,n}\left(2^{\delta}e^{-\pi i/2}z^{1/2} \left| \begin{matrix} (2a_p) \\ 0, (2b_q) \end{matrix} \right.\right)$$

$$\left[\begin{matrix} \delta = q-p+1; \rho = (p+q+1)/2 - m - n; \\ \tau = 2\sum_{j=1}^{q} b_j - 2\sum_{i=1}^{a} a_i + p - m - n + 1; \ [25], (21) \end{matrix}\right].$$

6. $G_{p,q+1}^{m+1,n}\left(z \left| \begin{matrix} (a_p) \\ 0, (b_q) \end{matrix} \right.\right)$

$$= 2^{\sigma}\pi^{\rho}G_{2p,2q+2}^{2m+1,2n}\left(\frac{e^{\pm\pi i}z^2}{4^{\delta}} \left| \begin{matrix} (a_p)/2, (a_p)/2+1/2 \\ 0, (b_q)/2, (b_q)/2+1/2, 1/2 \end{matrix} \right.\right)$$

$$- 2^{\sigma-\delta}\pi^{\rho}zG_{2p,2q+2}^{2m+1,2n}\left(\frac{e^{\pm\pi i}z^2}{4^{\delta}} \left| \begin{matrix} (a_p)/2, (a_p)/2-1/2 \\ 0, (b_q)/2, (b_q)/2+1/2, -1/2 \end{matrix} \right.\right)$$

$$\left[\begin{matrix} \delta = q-p+1; \rho = (p+q+1)/2 - m - n; \\ \sigma = \sum_{j=1}^{q} b_j - \sum_{i=1}^{a} a_i + p - m - n; \ [20], (19) \end{matrix}\right].$$

8.2.2. Various Meijer G functions

1. $G_{22}^{12}\left(z \left| \begin{matrix} \frac{3}{4}, \frac{3}{4} \\ \frac{1}{2}, 0 \end{matrix} \right.\right) = \frac{2}{(z+1)^{1/4}}$

$$\times \left[\mathbf{K}\left(\sqrt{\frac{1}{2} + \frac{1}{2}\sqrt{\frac{z}{z+1}}}\right) - \mathbf{K}\left(\sqrt{\frac{1}{2} - \frac{1}{2}\sqrt{\frac{z}{z+1}}}\right)\right] \quad [|\arg(z+1)| < \pi].$$

2. $G_{48}^{12}\left(z \left|\begin{array}{c} \frac{1}{4}, \frac{3}{4}, \frac{1}{6}, \frac{5}{6} \\ 0, \frac{1}{6}, \frac{5}{6}, -a, \frac{1}{2}, a, \frac{1}{2} - a, a + \frac{1}{2} \end{array}\right.\right)$
$= \dfrac{1}{2\sqrt{2}\,\pi^{5/2}} \left[\operatorname{ber}_{2a}\left(2\sqrt[4]{z}\right) \operatorname{ber}_{-2a}\left(2\sqrt[4]{z}\right) - \operatorname{bei}_{2a}\left(2\sqrt[4]{z}\right) \operatorname{bei}_{-2a}\left(2\sqrt[4]{z}\right)\right].$

3. $G_{33}^{12}\left(z \left|\begin{array}{c} \frac{3}{4}, \frac{3}{4}, \frac{5}{4} \\ 0, 0, \frac{5}{4} \end{array}\right.\right) = -\dfrac{\sqrt{2}}{\pi^2} \Gamma^2\left(\dfrac{1}{4}\right) \mathbf{K}\left(\sqrt{\dfrac{1}{2} - \dfrac{1}{2}\sqrt{1-z}}\right)$ $\quad [|z| < 1].$

4. $G_{13}^{20}\left(z \left|\begin{array}{c} \frac{3}{2} \\ 1, 1, 0 \end{array}\right.\right) = -\sqrt{\pi}\, z [J_0(\sqrt{z}) Y_0(\sqrt{z}) + J_1(\sqrt{z}) Y_1(\sqrt{z})].$

5. $G_{22}^{20}\left(z \left|\begin{array}{c} \frac{3}{4}, \frac{3}{4} \\ 0, 0 \end{array}\right.\right) = \dfrac{i\,\Gamma^2\left(\frac{1}{4}\right)}{\pi^2 z^{1/4}} \theta(1 - |z|) \left[\delta_{\operatorname{Im} z, 0} + \operatorname{sgn}(\operatorname{Im} z)\right]$

$\times \left[\mathbf{K}\left(\sqrt{\dfrac{1}{2} - \dfrac{1}{2}\sqrt{\dfrac{z-1}{z}}}\right) - \mathbf{K}\left(\sqrt{\dfrac{1}{2} + \dfrac{1}{2}\sqrt{\dfrac{z-1}{z}}}\right)\right]$ $\quad [z \notin (-1, 0)].$

6. $G_{15}^{20}\left(z \left|\begin{array}{c} \frac{1}{2} \\ 0, 0, \frac{1}{2}, \frac{1}{2}, \frac{1}{2} \end{array}\right.\right) = \dfrac{4}{\pi^2} \operatorname{ker}\left(4z^{1/4}\right) - \dfrac{1}{\pi} \operatorname{bei}\left(2z^{1/4}\right).$

7. $G_{24}^{20}\left(z \left|\begin{array}{c} a, \frac{1}{2} \\ -a - \frac{1}{2}, a + \frac{1}{2}, a, \frac{1}{2} \end{array}\right.\right) = \dfrac{1}{\pi} \cos(a\pi)\, I_{-2a-1}(2\sqrt{z})$
$+ \dfrac{2}{\pi^2} \sin(a\pi)\, K_{2a+1}(2\sqrt{z}).$

8. $G_{24}^{20}\left(z \left|\begin{array}{c} a, 0 \\ -a - \frac{1}{2}, a + \frac{1}{2}, a, 0 \end{array}\right.\right) = \dfrac{1}{\pi} \sin(a\pi)\, I_{-2a-1}(2\sqrt{z})$
$+ \dfrac{2}{\pi^2} \cos(a\pi)\, K_{-2a-1}(2\sqrt{z}).$

9. $G_{24}^{20}\left(z \left|\begin{array}{c} a, \frac{1}{2} \\ \frac{1}{2} - a, a - \frac{1}{2}, a, 0 \end{array}\right.\right) = \dfrac{1}{\pi^{5/2}} \left[\pi^2 \cos(a\pi)\, I_{1/2-a}^2(\sqrt{z})\right.$
$+ 2\pi I_{1/2-a}(\sqrt{z}) K_{1/2-a}(\sqrt{z}) + 2\cos(a\pi)\, K_{1/2-a}^2(\sqrt{z})\big].$

10. $G_{26}^{20}\left(z \left|\begin{array}{c} a, a + \frac{1}{2} \\ \frac{1}{2}, a + \frac{1}{4}, 0, -a - \frac{1}{4}, a, a + \frac{1}{2} \end{array}\right.\right)$
$= \dfrac{1}{\pi^{5/2}} \operatorname{ber}_{2a+1/2}\left(2^{3/2} z^{1/4}\right) \left[\pi \sin \dfrac{(12a+1)\pi}{4} \operatorname{ber}_{-2a-1/2}\left(2^{3/2} z^{1/4}\right)\right.$
$+ \pi \cos \dfrac{(12a+3)\pi}{4} \operatorname{bei}_{-2a-1/2}\left(2^{3/2} z^{1/4}\right)$

$$+ 2\sin\frac{(4a+1)\pi}{4}\ker_{-2a-1/2}\left(2^{3/2}z^{1/4}\right)$$
$$- 2\cos\frac{(4a+1)\pi}{4}\mathrm{kei}_{-2a-1/2}\left(2^{3/2}z^{1/4}\right)\bigg]$$
$$+ \frac{1}{\pi^{5/2}}\mathrm{bei}_{2a+1/2}\left(2^{3/2}z^{1/4}\right)\bigg[\pi\sin\frac{(12a+1)\pi}{4}\mathrm{bei}_{-2a-1/2}\left(2^{3/2}z^{1/4}\right)$$
$$- \pi\cos\frac{(12a+1)\pi}{4}\mathrm{ber}_{-2a-1/2}\left(2^{3/2}z^{1/4}\right)$$
$$+ 2\cos\frac{(4a+1)\pi}{4}\ker_{-2a-1/2}\left(2^{3/2}z^{1/4}\right)$$
$$+ 2\sin\frac{(4a+1)\pi}{4}\mathrm{kei}_{-2a-1/2}\left(2^{3/2}z^{1/4}\right)\bigg].$$

11. $G_{26}^{20}\left(z \left| \begin{array}{c} a,\, a+\frac{1}{2} \\ 0,\, a+\frac{1}{4},\, \frac{1}{2},\, -a-\frac{1}{4},\, a,\, a+\frac{1}{2} \end{array}\right.\right)$
$$= \frac{1}{\pi^{5/2}}\mathrm{ber}_{2a+1/2}\left(2^{3/2}z^{1/4}\right)\bigg[\pi\cos\frac{(12a+1)\pi}{4}\mathrm{ber}_{-2a-1/2}\left(2^{3/2}z^{1/4}\right)$$
$$- \pi\sin\frac{(12a+1)\pi}{4}\mathrm{bei}_{-2a-1/2}\left(2^{3/2}z^{1/4}\right)$$
$$- 2\cos\frac{(4a+1)\pi}{4}\ker_{-2a-1/2}\left(2^{3/2}z^{1/4}\right)$$
$$- 2\sin\frac{(4a+1)\pi}{4}\mathrm{kei}_{-2a-1/2}\left(2^{3/2}z^{1/4}\right)\bigg]$$
$$+ \frac{1}{\pi^{5/2}}\mathrm{bei}_{2a+1/2}\left(2^{3/2}z^{1/4}\right)\bigg[\pi\sin\frac{(12a+1)\pi}{4}\mathrm{ber}_{-2a-1/2}\left(2^{3/2}z^{1/4}\right)$$
$$+ \pi\cos\frac{(12a+1)\pi}{4}\mathrm{bei}_{-2a-1/2}\left(2^{3/2}z^{1/4}\right)$$
$$+ 2\sin\frac{(4a+1)\pi}{4}\ker_{-2a-1/2}\left(2^{3/2}z^{1/4}\right)$$
$$- 2\cos\frac{(4a+1)\pi}{4}\mathrm{kei}_{-2a-1/2}\left(2^{3/2}z^{1/4}\right)\bigg].$$

12. $G_{35}^{20}\left(z \left| \begin{array}{c} \frac{1}{6},\, \frac{1}{2},\, \frac{5}{6} \\ 0,\, \frac{1}{3},\, \frac{1}{6},\, \frac{1}{2},\, \frac{5}{6} \end{array}\right.\right) = \frac{3^{1/6}}{2\pi^2}\,\mathrm{Ai}\left(-\sqrt[3]{9z}\right).$

13. $G_{35}^{20}\left(z \left| \begin{array}{c} \frac{1}{6},\, \frac{1}{2},\, \frac{5}{6} \\ 0,\, \frac{2}{3},\, \frac{1}{6},\, \frac{1}{3},\, \frac{1}{2} \end{array}\right.\right) = \frac{1}{\pi^{3/2}}\left(\frac{3}{2}\right)^{1/3}\mathrm{Ai}\left(-\sqrt[3]{\frac{9z}{4}}\right)\mathrm{Bi}\left(-\sqrt[3]{\frac{9z}{4}}\right).$

14. $G_{37}^{20}\left(z \left| \begin{array}{c} 0,\, 0,\, \frac{1}{2} \\ a,\, a+\frac{1}{2},\, 0,\, 0,\, \frac{1}{2},\, \frac{1}{2}-a,\, -a \end{array}\right.\right)$
$$= \frac{\sin(4a\pi)}{4\pi^3}\left[\mathrm{bei}_{4a}\left(4z^{1/4}\right) - \tan(2a\pi)\,\mathrm{ber}_{4a}\left(4z^{1/4}\right)\right].$$

15. $G_{13}^{21}\left(z \left| \begin{array}{c} \frac{1}{2} \\ 0, n, 0 \end{array}\right.\right) = (-1)^n \dfrac{\sqrt{\pi}}{2^{n-1}} z^{n/2} \sum_{k=0}^{n} (-1)^k \binom{n}{k} I_{n-k}(\sqrt{z}) K_k(\sqrt{z})$.

16. $G_{13}^{21}\left(z \left| \begin{array}{c} 1 \\ \frac{1}{2} - n, n + \frac{1}{2}, 0 \end{array}\right.\right)$
$= 2(-1)^n \pi \sqrt{z}\, [K_0(2\sqrt{z})\, \mathbf{L}_{-1}(2\sqrt{z}) + K_1(2\sqrt{z})\, \mathbf{L}_0(2\sqrt{z})]$
$\qquad\qquad\qquad\qquad\qquad + 4(-1)^n \sum_{k=0}^{n-1} (-1)^k K_{2k+1}(2\sqrt{z})$.

17. $G_{13}^{21}\left(z \left| \begin{array}{c} \frac{1}{2} \\ \frac{1}{2}, \frac{1}{2}, 0 \end{array}\right.\right) = \dfrac{1}{\sqrt{\pi}} \left[e^{-2\sqrt{z}}\, \mathrm{Ei}\,(2\sqrt{z}) - e^{2\sqrt{z}}\, \mathrm{Ei}\,(-2\sqrt{z}) \right] \quad [z > 0]$.

18. $G_{13}^{21}\left(z \left| \begin{array}{c} \frac{5}{6} \\ 0, \frac{2}{3}, \frac{1}{3} \end{array}\right.\right) = 2^{5/3} 3^{1/3} \pi^{3/2}\, \mathrm{Ai}\left(\sqrt[3]{\dfrac{9z}{4}}\right) \mathrm{Bi}\left(\sqrt[3]{\dfrac{9z}{4}}\right)$.

19. $G_{13}^{21}\left(z \left| \begin{array}{c} \frac{3}{2} \\ \frac{1}{3}, \frac{2}{3}, 0 \end{array}\right.\right)$
$= 8\pi^{3/2}\left\{ -3\left(\dfrac{3z^2}{2}\right)^{1/3} \mathrm{Ai}\left(\left(\dfrac{9z}{4}\right)^{1/3}\right) \mathrm{Bi}\left(\left(\dfrac{9z}{4}\right)^{1/3}\right) + \mathrm{Ai}'\left(\left(\dfrac{9z}{4}\right)^{1/3}\right) \right.$
$\left. \times \mathrm{Bi}\left(\left(\dfrac{9z}{4}\right)^{1/3}\right) + (18z)^{1/3}\, \mathrm{Ai}'\left(\left(\dfrac{9z}{4}\right)^{1/3}\right) \mathrm{Bi}'\left(\left(\dfrac{9z}{4}\right)^{1/3}\right) \right\} + 4\sqrt{\pi}$.

20. $G_{13}^{21}\left(z \left| \begin{array}{c} \frac{3}{2} \\ \frac{1}{3}, \frac{2}{3}, 0 \end{array}\right.\right)$
$= 8\pi^{3/2}\left\{ -3\left(\dfrac{3z^2}{2}\right)^{1/3} \mathrm{Ai}\left(\left(\dfrac{9z}{4}\right)^{1/3}\right) \mathrm{Bi}\left(\left(\dfrac{9z}{4}\right)^{1/3}\right) + \mathrm{Ai}\left(\left(\dfrac{9z}{4}\right)^{1/3}\right) \right.$
$\left. \times \mathrm{Bi}'\left(\left(\dfrac{9z}{4}\right)^{1/3}\right) + (18z)^{1/3}\, \mathrm{Ai}'\left(\left(\dfrac{9z}{4}\right)^{1/3}\right) \mathrm{Bi}'\left(\left(\dfrac{9z}{4}\right)^{1/3}\right) \right\} - 4\sqrt{\pi}$.

21. $G_{23}^{21}\left(z \left| \begin{array}{c} 1, n + \frac{3}{2} \\ 1, 2n + 1, 0 \end{array}\right.\right) = \dfrac{z^{n+1} e^{-z/2}}{(2n+1)\sqrt{\pi}} \left[K_n\left(\dfrac{z}{2}\right) - K_{n+1}\left(\dfrac{z}{2}\right) \right]$
$\qquad\qquad\qquad\qquad\qquad\qquad\qquad + \dfrac{2^{2n+1} n!}{(2n+1)\sqrt{\pi}}$.

22. $G_{23}^{21}\left(z \left| \begin{array}{c} a+1, a \\ 0, b, a \end{array}\right.\right) = \dfrac{e^{-z}}{a} \left[z \Psi\left(\begin{array}{c} a-b+1 \\ 2-b;\ z \end{array}\right) - \Psi\left(\begin{array}{c} a-b \\ 1-b;\ z \end{array}\right) \right]$.

23. $G_{23}^{21}\left(z \left| \begin{array}{c} 1, a \\ 1, b, 0 \end{array}\right.\right)$
$= \dfrac{\Gamma(b)}{\Gamma(a)} + \dfrac{e^{-z}}{a-1}\left[(b-a)z\Psi\binom{a-b+1}{3-b;\,z} - (b-1)\Psi\binom{a-b;\,z}{2-b}\right].$

24. $G_{24}^{21}\left(z \left| \begin{array}{c} 1, -n \\ \frac{1}{2} - n,\, n+\frac{1}{2},\, 0,\, -n \end{array}\right.\right) = \pi\sqrt{z}\,Y_1(2\sqrt{z})\,\mathbf{H}_0(2\sqrt{z})$

$- 2\sqrt{z}\,Y_2(2\sqrt{z}) + \pi\left[\sqrt{z}\,Y_2(2\sqrt{z}) - Y_1(2\sqrt{z})\right]\mathbf{H}_1(2\sqrt{z}) - 2\sum_{k=1}^{n-1} Y_{2k+1}(2\sqrt{z})$

$[n \geq 1].$

25. $G_{33}^{21}\left(z \left| \begin{array}{c} 1, 0, 1 \\ \frac{1}{2}, \frac{1}{2}, 0 \end{array}\right.\right) = -\theta(1-z)\,\dfrac{2\sqrt{z}}{\pi}\,\mathbf{K}\left(\sqrt{1-z}\right).$

26. $G_{35}^{21}\left(z \left| \begin{array}{c} \frac{1}{2}, 0, 1 \\ \frac{1}{2}, \frac{1}{2}, 0, 0, 1 \end{array}\right.\right)$
$= \dfrac{2}{\pi^{5/2}}\left[\sinh(2\sqrt{z})\,\mathrm{chi}\,(2\sqrt{z}) - \cosh(2\sqrt{z})\,\mathrm{shi}\,(2\sqrt{z})\right].$

27. $G_{22}^{22}\left(z \left| \begin{array}{c} \frac{3}{4}, \frac{3}{4} \\ 0, 0 \end{array}\right.\right)$
$= 2\Gamma^2\left(\tfrac{1}{4}\right) z^{-1/4}\left[\mathbf{K}\left(\sqrt{\tfrac{1}{2} - \tfrac{1}{2}\sqrt{1-\tfrac{1}{z}}}\right) + \mathbf{K}\left(\sqrt{\tfrac{1}{2} - \tfrac{1}{2}\sqrt{1-\tfrac{1}{z}}}\right)\right].$

28. $G_{23}^{22}\left(z \left| \begin{array}{c} 1, n+\frac{3}{2} \\ 1, 2n+1, 0 \end{array}\right.\right) = (-1)^n\,\dfrac{\sqrt{\pi}\,z^{n+1}e^{z/2}}{2n+1}\left[K_n\left(\tfrac{z}{2}\right) + K_{n+1}\left(\tfrac{z}{2}\right)\right]$
$- (-1)^n\,\dfrac{2^{2n+1}n!\sqrt{\pi}}{2n+1}.$

29. $G_{23}^{22}\left(z \left| \begin{array}{c} 1, n+1 \\ n+1, n+1, 0 \end{array}\right.\right)$
$= \left[\dfrac{z^n}{n} + (-1)^n n!\,(\mathbf{C} + \ln z) + n!\sum_{k=1}^{n-1}(-1)^k\,\dfrac{z^{n-k}}{(n-k)!\,(n-k)}\right]$
$- e^z\left[z^n + n!\sum_{k=1}^{n}(-1)^k\,\dfrac{z^{n-k}}{(n-k)!}\right]\mathrm{Ei}\,(-z)\quad [z > 0].$

30. $G_{23}^{22}\left(z \left| \begin{array}{c} 1, a \\ 1, b, 0 \end{array}\right.\right) = \Gamma(1-a)\Gamma(b) - \Gamma(1-a)\Gamma(b-a+1)$
$\times\left[(b+z-1)\Psi\binom{2-a;\,z}{2-b} - (a-2)z\Psi\binom{3-a;\,z}{3-b}\right].$

31. $G^{22}_{34}\left(z \;\middle|\; \begin{matrix} 1, 1, \frac{1}{2} \\ 1, 1, 0, \frac{1}{2} \end{matrix}\right) = \frac{1}{\pi}\left[\mathbf{C} + \ln z - e^{-z}\operatorname{Ei}(z)\right]$ $\qquad [|\arg z| < \pi].$

32. $G^{22}_{46}\left(z \;\middle|\; \begin{matrix} \frac{1}{4}, \frac{3}{4}, 0, a \\ -a - \frac{1}{2}, a + \frac{1}{2}, 0, -a, a, a+1 \end{matrix}\right)$
$\qquad = \frac{\sqrt{2}}{\pi^2}\cosh(\sqrt{z})\left[\pi\sin(a\pi)I_{-2a-1}(\sqrt{z}) + 2\cos(a\pi)K_{2a+1}(\sqrt{z})\right].$

33. $G^{22}_{46}\left(z \;\middle|\; \begin{matrix} \frac{1}{4}, \frac{3}{4}, 0, a \\ \frac{1}{2} - a, a + \frac{1}{2}, 0, -a, a, a \end{matrix}\right)$
$\qquad = \frac{\sqrt{2}}{\pi^2}\sinh(\sqrt{z})\left[2\cos(a\pi)K_{2a}(\sqrt{z}) - \pi\sin(a\pi)I_{-2a}(\sqrt{z})\right].$

34. $G^{22}_{46}\left(z \;\middle|\; \begin{matrix} \frac{1}{4}, \frac{3}{4}, \frac{1}{2}, a \\ \frac{1}{2} - a, a + \frac{1}{2}, \frac{1}{2}, -a, a, a \end{matrix}\right)$
$\qquad = \frac{\sqrt{2}}{\pi^2}\sinh(\sqrt{z})\left[2\sin(a\pi)K_{2a}(\sqrt{z}) - \pi\cos(a\pi)I_{-2a}(\sqrt{z})\right].$

35. $G^{22}_{46}\left(z \;\middle|\; \begin{matrix} \frac{1}{4}, \frac{3}{4}, \frac{1}{2}, a \\ -a - \frac{1}{2}, a + \frac{1}{2}, \frac{1}{2}, -a, a, a+1 \end{matrix}\right)$
$\qquad = \frac{\sqrt{2}}{\pi^2}\cosh(\sqrt{z})\left[\pi\cos(a\pi)I_{-2a-1}(\sqrt{z}) + 2\sin(a\pi)K_{2a+1}(\sqrt{z})\right].$

36. $G^{23}_{33}\left(z \;\middle|\; \begin{matrix} \frac{1}{2}, \frac{1}{2}, \frac{1}{2} \\ 0, 0, 0 \end{matrix}\right) = 4\sqrt{\pi}\,\mathbf{K}\!\left(\sqrt{\frac{1}{2} - \frac{1}{2}\sqrt{1-z}}\right)\mathbf{K}\!\left(\sqrt{\frac{1}{2} + \frac{1}{2}\sqrt{1-z}}\right).$

37. $G^{30}_{04}\left(z \;\middle|\; 0, \frac{1}{6}, \frac{1}{3}, \frac{2}{3}\right) = 2^{5/3}3^{1/3}\pi^{3/2}\operatorname{Ai}\!\left(\sqrt[6]{324z}\right)\operatorname{Bi}\!\left(-\sqrt[6]{324z}\right).$

38. $G^{30}_{13}\left(z \;\middle|\; \begin{matrix} \frac{1}{2} \\ 0, 0, n \end{matrix}\right) = \frac{2^{1-n}}{\sqrt{\pi}} z^{n/2} \sum_{k=0}^{n}\binom{n}{k} K_k(\sqrt{z}) K_{n-k}(\sqrt{z}).$

39. $G^{30}_{13}\left(z \;\middle|\; \begin{matrix} 1 \\ -n, 0, n+1 \end{matrix}\right) = -2(-1)^n K_0(2\sqrt{z})$
$\qquad\qquad\qquad\qquad\qquad + 4(-1)^n\sum_{k=1}^{n}(-1)^k K_{2k}(2\sqrt{z}).$

8.2.2]   8.2. THE MEIJER FUNCTION $G_{p,q}^{m,n}\left(z \left| \begin{matrix}(a_p)\\(b_q)\end{matrix}\right.\right)$

40. $G_{15}^{30}\left(z \left| \begin{matrix}\frac{1}{4}\\0, \frac{1}{2}, 1, 0, \frac{1}{4}\end{matrix}\right.\right) = \frac{2^{3/2}}{\sqrt{\pi}} z^{1/4}$
$\times \left[\operatorname{ber}_1\left(2^{3/2}z^{1/4}\right)\operatorname{kei}_0\left(2^{3/2}z^{1/4}\right) - \operatorname{ber}_0\left(2^{3/2}z^{1/4}\right)\operatorname{ker}_1\left(2^{3/2}z^{1/4}\right)\right.$
$\left. - \operatorname{bei}_0\left(2^{3/2}z^{1/4}\right)\operatorname{kei}_1\left(2^{3/2}z^{1/4}\right) - \operatorname{bei}_1\left(2^{3/2}z^{1/4}\right)\operatorname{ker}_0\left(2^{3/2}z^{1/4}\right)\right].$

41. $G_{15}^{30}\left(z \left| \begin{matrix}\frac{3}{4}\\0, \frac{1}{2}, 1, 0, \frac{3}{4}\end{matrix}\right.\right) = \frac{2^{3/2}z^{1/4}}{\sqrt{\pi}}$
$\times \left[\operatorname{bei}_0\left(2^{3/2}z^{1/4}\right)\operatorname{ker}_1\left(2^{3/2}z^{1/4}\right) - \operatorname{ber}_1\left(2^{3/2}z^{1/4}\right)\operatorname{ker}_0\left(2^{3/2}z^{1/4}\right)\right.$
$\left. - \operatorname{ber}_0\left(2^{3/2}z^{1/4}\right)\operatorname{kei}_1\left(2^{3/2}z^{1/4}\right) - \operatorname{bei}_1\left(2^{3/2}z^{1/4}\right)\operatorname{kei}_0\left(2^{3/2}z^{1/4}\right)\right].$

42. $G_{23}^{30}\left(z \left| \begin{matrix}1, 1\\0, 0, a\end{matrix}\right.\right) = \Gamma(a)\left[\psi(a) - \ln z\right] + \frac{z^a}{a^2}{}_2F_2\left(\begin{matrix}a, a; -z\\a+1, a+1\end{matrix}\right).$

43. $G_{15}^{30}\left(z \left| \begin{matrix}\frac{1}{4}\\0, \frac{1}{2}, 1, 0, \frac{3}{4}\end{matrix}\right.\right) = \frac{2^{3/2}}{\sqrt{\pi}} z^{1/4}$
$\times \left[\operatorname{ber}_1\left(2^{3/2}z^{1/4}\right)\operatorname{ker}\left(2^{3/2}z^{1/4}\right) - \operatorname{bei}\left(2^{3/2}z^{1/4}\right)\operatorname{ker}_1\left(2^{3/2}z^{1/4}\right)\right.$
$\left. - \operatorname{ber}\left(2^{3/2}z^{1/4}\right)\operatorname{kei}_1\left(2^{3/2}z^{1/4}\right) - \operatorname{bei}_1\left(2^{3/2}z^{1/4}\right)\operatorname{kei}\left(2^{3/2}z^{1/4}\right)\right].$

44. $G_{15}^{30}\left(z \left| \begin{matrix}\frac{3}{4}\\0, \frac{1}{2}, 1, 0, \frac{3}{4}\end{matrix}\right.\right) = \frac{2^{3/2}}{\sqrt{\pi}} z^{1/4}$
$\times \left[\operatorname{bei}\left(2^{3/2}z^{1/4}\right)\operatorname{ker}_1\left(2^{3/2}z^{1/4}\right) - \operatorname{ber}\left(2^{3/2}z^{1/4}\right)\operatorname{kei}_1\left(2^{3/2}z^{1/4}\right)\right.$
$\left. - \operatorname{ber}_1\left(2^{3/2}z^{1/4}\right)\operatorname{ker}\left(2^{3/2}z^{1/4}\right) - \operatorname{bei}_1\left(2^{3/2}z^{1/4}\right)\operatorname{kei}\left(2^{3/2}z^{1/4}\right)\right].$

45. $G_{26}^{30}\left(z \left| \begin{matrix}\frac{3}{4}, \frac{5}{4}\\\frac{1}{2}, \frac{1}{2}, 1, 0, \frac{1}{2}, 1\end{matrix}\right.\right)$
$= \frac{2^{5/2}z^{1/2}}{\pi^{3/2}}\left\{\operatorname{ber}_0\left(2z^{1/4}\right)\operatorname{ker}_0\left(2z^{1/4}\right) + \operatorname{ber}_1\left(2z^{1/4}\right)\operatorname{ker}_1\left(2z^{1/4}\right)\right.$
$- \operatorname{bei}_0\left(2z^{1/4}\right)\left[\pi\operatorname{ber}_0\left(2z^{1/4}\right) + \operatorname{kei}_0\left(2z^{1/4}\right)\right]$
$\left. - \operatorname{bei}_1\left(2z^{1/4}\right)\left[\pi\operatorname{ber}_1\left(2z^{1/4}\right) + \operatorname{kei}_1\left(2z^{1/4}\right)\right]\right\}.$

46. $G_{35}^{30}\left(z \left| \begin{matrix}\frac{1}{2}, \frac{1}{2}, \frac{1}{2}\\0, -a, a, \frac{1}{2}, \frac{1}{2}\end{matrix}\right.\right) = \frac{1}{\pi^{5/2}}\left\{2K_a^2(\sqrt{z}) - \pi^2 I_{-a}(\sqrt{z}) I_a(\sqrt{z})\right.$
$\left. + \pi \sin(a\pi)\left[I_a(\sqrt{z}) - I_{-a}(\sqrt{z})\right]K_a(\sqrt{z})\right\}.$

47. $G^{30}_{35}\left(z \left| \begin{array}{c} \frac{1}{2}, \frac{1}{2}, a \\ 0, \frac{1}{2} - a, a - \frac{1}{2}, \frac{1}{2}, a \end{array} \right.\right)$
$$= \frac{\sin(a\pi)}{\pi^{5/2}} \left[2K^2_{1/2-a}(\sqrt{z}) - \pi^2 I^2_{1/2-a}(\sqrt{z}) \right] \quad [a \neq 0, 1, 2, \ldots].$$

48. $G^{31}_{15}\left(z \left| \begin{array}{c} \frac{3}{4} \\ \frac{1}{4}, \frac{3}{4}, \frac{3}{4}, 0, \frac{1}{2} \end{array} \right.\right) = -\sqrt{\frac{2}{\pi}} e^{-7\pi i/4} \left[\frac{i\pi}{2} \sin\left(4e^{\pi i/4} z^{1/4}\right) \right.$
$- \sin\left(4e^{\pi i/4} z^{1/4}\right) \operatorname{ci}\left(4e^{\pi i/4} z^{1/4}\right) + \sinh\left(4e^{\pi i/4} z^{1/4}\right) \operatorname{chi}\left(4e^{\pi i/4} z^{1/4}\right)$
$\left. - \cosh\left(4e^{\pi i/4} z^{1/4}\right) \operatorname{shi}\left(4e^{\pi i/4} z^{1/4}\right) + \cos\left(4e^{\pi i/4} z^{1/4}\right) \operatorname{Si}\left(4e^{\pi i/4} z^{1/4}\right) \right].$

49. $G^{31}_{24}\left(z \left| \begin{array}{c} 1, \frac{3}{2} \\ 1, 1, 1, 0 \end{array} \right.\right) = \frac{2z}{\sqrt{\pi}} \left[K^2_0(\sqrt{z}) - K^2_1(\sqrt{z}) \right] + \frac{2}{\sqrt{\pi}}.$

50. $G^{31}_{24}\left(z \left| \begin{array}{c} 0, 1 \\ 0, 0, 0, \frac{1}{2} \end{array} \right.\right) = \frac{2}{\sqrt{\pi}} \left[\operatorname{chi}^2(2\sqrt{z}) - \operatorname{shi}^2(2\sqrt{z}) \right].$

51. $G^{31}_{24}\left(z \left| \begin{array}{c} \frac{3}{2}, \frac{1}{2} \\ 0, \frac{1}{3}, \frac{2}{3}, \frac{1}{2} \end{array} \right.\right) = 8\pi^{3/2} \left\{ -\frac{3^{4/3}}{2^{1/3}} z^{2/3} \operatorname{Ai}^2\left(\left(\frac{9z}{4}\right)^{1/3}\right) \right.$
$\left. + \operatorname{Ai}\left(\left(\frac{9z}{4}\right)^{1/3}\right) \operatorname{Ai}'\left(\left(\frac{9z}{4}\right)^{1/3}\right) + (18z)^{1/3} \left[\operatorname{Ai}'\left(\left(\frac{9z}{4}\right)^{1/3}\right)\right]^2 \right\}.$

52. $G^{32}_{26}\left(z \left| \begin{array}{c} \frac{1}{4}, \frac{3}{4} \\ 0, 0, \frac{1}{2}, 0, \frac{1}{2}, \frac{1}{2} \end{array} \right.\right)$
$$= 4\sqrt{2\pi} \left[\operatorname{ber}\left(2z^{1/4}\right) \operatorname{ker}\left(2z^{1/4}\right) - \operatorname{bei}\left(2z^{1/4}\right) \operatorname{kei}\left(2z^{1/4}\right) \right].$$

53. $G^{32}_{26}\left(z \left| \begin{array}{c} \frac{1}{4}, \frac{3}{4} \\ 0, \frac{1}{2}, \frac{1}{2}, 0, 0, \frac{1}{2} \end{array} \right.\right) =$
$$-4\sqrt{2\pi} \left[\operatorname{ber}\left(2z^{1/4}\right) \operatorname{kei}\left(2z^{1/4}\right) + \operatorname{bei}\left(2z^{1/4}\right) \operatorname{ker}\left(2z^{1/4}\right) \right].$$

54. $G^{32}_{33}\left(z \left| \begin{array}{c} \frac{1}{2}, \frac{1}{2}, \frac{1}{2} \\ 0, 0, 0 \end{array} \right.\right)$
$$= 2\sqrt{\pi} \left[\mathbf{K}^2\left(\sqrt{\frac{1}{2} - \frac{1}{2}\sqrt{1-z}}\right) + \mathbf{K}^2\left(\sqrt{\frac{1}{2} + \frac{1}{2}\sqrt{1-z}}\right) \right].$$

8.2.2]   8.2. THE MEIJER FUNCTION $G_{p,q}^{m,n}\left(z \mid \begin{matrix} (a_p) \\ (b_q) \end{matrix}\right)$

55. $G_{24}^{40}\left(z \mid \begin{matrix} \frac{1}{2}, 1 \\ a+b, a-b, b-a, -a-b \end{matrix}\right) = \frac{2}{(a+b)\sqrt{\pi}} K_{2a}(\sqrt{z}) K_{2b}(\sqrt{z})$

$\qquad + \frac{\sqrt{z}}{(a^2 - b^2)\sqrt{\pi}} [K_{2a-1}(\sqrt{z}) K_{2b}(\sqrt{z}) - K_{2a}(\sqrt{z}) K_{2b-1}(\sqrt{z})].$

56. $G_{35}^{40}\left(z \mid \begin{matrix} 0, \frac{1}{6}, \frac{2}{3} \\ \frac{1}{3} - a, -a, a, a + \frac{1}{3}, 0 \end{matrix}\right)$

$\qquad = \frac{3^{1/6}\sqrt{\pi}}{2^{2/3}} \left[e^{ia\pi} H_{2a}^{(1)}(\sqrt{z}) \operatorname{Ai}\left(\frac{3^{2/3} e^{-i\pi/3}}{2^{2/3}} z^{1/3} \right) \right.$

$\qquad \left. + e^{-ia\pi} H_{2a}^{(2)}(\sqrt{z}) \operatorname{Ai}\left(\frac{3^{2/3} e^{i\pi/3}}{2^{2/3}} z^{1/3} \right) \right].$

57. $G_{35}^{40}\left(z \mid \begin{matrix} \frac{1}{6}, \frac{1}{2}, \frac{2}{3} \\ \frac{1}{3} - a, -a, a, a + \frac{1}{3}, \frac{1}{2} \end{matrix}\right)$

$\qquad = \frac{3^{1/6}\sqrt{\pi}\, i}{2^{2/3}} \left[e^{ia\pi} H_{2a}^{(1)}(\sqrt{z}) \operatorname{Ai}\left(\frac{3^{2/3} e^{-i\pi/3}}{2^{2/3}} z^{1/3} \right) \right.$

$\qquad \left. - e^{-ia\pi} H_{2a}^{(2)}(\sqrt{z}) \operatorname{Ai}\left(\frac{3^{2/3} e^{i\pi/3}}{2^{2/3}} z^{1/3} \right) \right].$

58. $G_{37}^{40}\left(z \mid \begin{matrix} \frac{1}{4}, \frac{1}{2}, \frac{3}{4} \\ \frac{1}{2} - a, -a, a, a + \frac{1}{2}, 0, \frac{1}{2}, \frac{1}{2} \end{matrix}\right)$

$= \frac{\sqrt{2}}{\pi^{3/2}} \{ \operatorname{ber}_{2a}(2z^{1/4}) [\cos(4a\pi) \operatorname{ker}_{-2a}(2z^{1/4}) + \sin(4a\pi) \operatorname{kei}_{-2a}(2z^{1/4})]$

$\qquad - \operatorname{bei}_{2a}(2z^{1/4}) [\pi \operatorname{ber}_{-2a}(2z^{1/4}) - \sin(4a\pi) \operatorname{ker}_{-2a}(2z^{1/4})$

$\qquad + \cos(4a\pi) \operatorname{kei}_{-2a}(2z^{1/4})] + \operatorname{ber}_{-2a}(2z^{1/4})$

$\qquad \times [\cos(4a\pi) \operatorname{ker}_{2a}(2z^{1/4}) - \sin(4a\pi) \operatorname{kei}_{2a}(2z^{1/4})] - \operatorname{bei}_{-2a}(2z^{1/4})$

$\qquad \times [\pi \operatorname{ber}_{2a}(2z^{1/4}) + \sin(4a\pi) \operatorname{ker}_{2a}(2z^{1/4}) + \cos(4a\pi) \operatorname{kei}_{2a}(2z^{1/4})] \}.$

59. $G_{37}^{40}\left(z \mid \begin{matrix} 0, \frac{1}{4}, \frac{3}{4} \\ \frac{1}{2} - a, -a, a, a + \frac{1}{2}, 0, 0, \frac{1}{2} \end{matrix}\right)$

$= \frac{\sqrt{2}}{\pi^{3/2}} \{ \operatorname{ber}_{2a}(2z^{1/4}) [\cos(4a\pi) \operatorname{kei}_{-2a}(2z^{1/4}) - \sin(4a\pi) \operatorname{ker}_{-2a}(2z^{1/4})]$

$\qquad + \operatorname{bei}_{2a}(2z^{1/4}) [\cos(4a\pi) \operatorname{ker}_{-2a}(2z^{1/4}) + \sin(4a\pi) \operatorname{kei}_{-2a}(2z^{1/4})]$

$\qquad + \operatorname{ber}_{-2a}(2z^{1/4}) [\pi \operatorname{ber}_{2a}(2z^{1/4}) + \sin(4a\pi) \operatorname{ker}_{2a}(2z^{1/4})$

$\qquad + \cos(4a\pi) \operatorname{kei}_{2a}(2z^{1/4})] - \operatorname{bei}_{-2a}(2z^{1/4})$

$\qquad \times [\pi \operatorname{bei}_{2a}(2z^{1/4}) - \cos(4a\pi) \operatorname{ker}_{2a}(2z^{1/4}) + \sin(4a\pi) \operatorname{kei}_{2a}(2z^{1/4})] \}.$

60. $G_{59}^{40}\left(z \left|\begin{array}{c} 0, \frac{1}{2}, \frac{1}{2}, a, a+\frac{1}{2} \\ \frac{1}{2}-b, -b, b, b+\frac{1}{2}, 0, \frac{1}{2}, \frac{1}{2}, a, a+\frac{1}{2} \end{array}\right.\right)$
$$= \frac{1}{4\pi^3}\{\sin[2(a-b)\pi]\operatorname{ber}_{4b}(4z^{1/4}) + \sin[2(a+b)\pi]\operatorname{ber}_{-4b}(4z^{1/4})$$
$$+ (\sin[2(a-b)\pi]\operatorname{bei}_{4b}(4z^{1/4}) - \sin[2(a+b)\pi]\operatorname{bei}_{-4b}(4z^{1/4}))\tan(2b\pi)\}.$$

61. $G_{59}^{40}\left(z \left|\begin{array}{c} 0, 0, \frac{1}{2}, a, a+\frac{1}{2} \\ \frac{1}{2}-b, -b, b, b+\frac{1}{2}, 0, 0, \frac{1}{2}, a, a+\frac{1}{2} \end{array}\right.\right)$
$$= \frac{1}{4\pi^3}\{(\sin[2(a-b)\pi]\operatorname{ber}_{4b}(4z^{1/4})$$
$$- \sin[2(a+b)\pi]\operatorname{ber}_{-4b}(4z^{1/4}))\tan(2b\pi)$$
$$- \sin[2(a-b)\pi]\operatorname{bei}_{4b}(4z^{1/4}) - \sin[2(a+b)\pi]\operatorname{bei}_{-4b}(4z^{1/4})\}.$$

62. $G_{24}^{41}\left(z \left|\begin{array}{c} 0, 1 \\ 0, 0, 0, \frac{1}{2} \end{array}\right.\right) = 2\sqrt{\pi}\,[\operatorname{si}^2(2\sqrt{z}) - \operatorname{ci}^2(2\sqrt{z})].$

63. $G_{46}^{42}\left(z \left|\begin{array}{c} 1, 1, \frac{1}{4}, \frac{3}{4} \\ \frac{1}{2}, \frac{1}{2}, 1, 1, \frac{1}{4}, \frac{3}{4} \end{array}\right.\right) = \frac{1}{2\pi^2} G_{24}^{42}\left(z \left|\begin{array}{c} 1, 1 \\ \frac{1}{2}, \frac{1}{2}, 1, 1 \end{array}\right.\right) - 2\sqrt{z}\,K_0(2\sqrt{z}).$

64. $G_{6,10}^{43}\left(z \left|\begin{array}{c} 0, \frac{1}{4}, \frac{3}{4}, \frac{1}{2}, a, a+\frac{1}{2} \\ a-\frac{1}{4}, a+\frac{1}{4}, b, b+\frac{1}{2}, \frac{1}{4}-a, \frac{3}{4}-a, a, a+\frac{1}{2}, \frac{1}{2}-b, -b \end{array}\right.\right)$
$$= -\frac{\sqrt{2}}{\pi^{3/2}}\{\operatorname{ber}_{2a+2b-1/2}(2z^{1/4})[\pi\cos(4b\pi)\operatorname{bei}_{2b-2a+1/2}(2z^{1/4})$$
$$- \pi\sin(4a\pi)\operatorname{ber}_{2b-2a+1/2}(2z^{1/4})$$
$$+ 2\cos(4a\pi)\operatorname{ker}_{2b-2a+1/2}(2z^{1/4}) + 2\sin(4a\pi)\operatorname{kei}_{2b-2a+1/2}(2z^{1/4})]$$
$$\operatorname{bei}_{2a+2b-1/2}(2z^{1/4})[\pi\cos(4b\pi)\operatorname{ber}_{2b-2a+1/2}(2z^{1/4})$$
$$+ \pi\sin(4b\pi)\operatorname{bei}_{2b-2a+1/2}(2z^{1/4})$$
$$+ 2\sin(4a\pi)\operatorname{ker}_{2b-2a+1/2}(2z^{1/4}) - 2\cos(4a\pi)\operatorname{kei}_{2b-2a+1/2}(2z^{1/4})]\}.$$

65. $G_{6,10}^{43}\left(z \left|\begin{array}{c} \frac{1}{4}, \frac{1}{2}, \frac{3}{4}, 0, a, a+\frac{1}{2} \\ a-\frac{1}{4}, a+\frac{1}{4}, b, b+\frac{1}{2}, \frac{1}{4}-a, \frac{3}{4}-a, a, a+\frac{1}{2}, \frac{1}{2}-b, -b \end{array}\right.\right)$
$$= \frac{\sqrt{2}}{\pi^{3/2}}\{\operatorname{ber}_{2a+2b-1/2}(2z^{1/4})[\cos(4a\pi)\operatorname{kei}_{2b-2a+1/2}(2z^{1/4})$$
$$- \sin(4a\pi)\operatorname{ker}_{2b-2a+1/2}(2z^{1/4}) - 2\operatorname{bei}_{2a+2b-1/2}(2z^{1/4})$$
$$\times [\cos(4a\pi)\operatorname{ker}_{2b-2a+1/2}(2z^{1/4}) + \sin(4a\pi)\operatorname{kei}_{2b-2a+1/2}(2z^{1/4})$$
$$+ \pi\operatorname{ber}_{2b-2a+1/2}(2z^{1/4})[\cos(4b\pi)\operatorname{ber}_{2a+2b-1/2}(2z^{1/4})$$

$+ \sin(4b\pi) \operatorname{bei}_{2a+2b-1/2}(2z^{1/4}) + \pi \operatorname{bei}_{2b-2a+1/2}(2z^{1/4})$

$\times \left[\sin(4b\pi) \operatorname{ber}_{2a+2b-1/2}(2z^{1/4}) - \cos(4b\pi) \operatorname{bei}_{2a+2b-1/2}(2z^{1/4})\right]\}.$

66. $G_{48}^{50}\left(z \left|\begin{array}{c} \frac{1}{4}, \frac{3}{4}, a, a+\frac{1}{2} \\ \frac{1}{2}, \frac{1}{4}-a, \frac{3}{4}-a, a-\frac{1}{4}, a+\frac{1}{4}, 0, a, a+\frac{1}{2} \end{array}\right.\right)$

$= \dfrac{\sqrt{2}}{\pi^{5/2}} \{2\pi \operatorname{bei}_{1/2-2a}(2z^{1/4}) \left[\pi \cos(4a\pi) \operatorname{ber}_{1/2-2a}(2z^{1/4})\right.$

$\left. - \operatorname{kei}_{1/2-2a}(2z^{1/4})\right] + 2 \operatorname{ker}_{1/2-2a}(2z^{1/4}) \left[\pi \operatorname{ber}_{1/2-2a}(2z^{1/4})\right.$

$\left. - 2\cos(4a\pi)\operatorname{kei}_{1/2-2a}(2z^{1/4})\right] + \sin(4a\pi)\left[\pi^2 \operatorname{ber}^2_{1/2-2a}(2z^{1/4})\right.$

$\left.\left. - \pi^2 \operatorname{bei}^2_{1/2-2a}(2z^{1/4}) + 2\operatorname{ker}^2_{1/2-2a}(2z^{1/4}) - 2\operatorname{kei}^2_{1/2-2a}(2z^{1/4})\right]\right\}.$

67. $G_{26}^{52}\left(z \left|\begin{array}{c} \frac{1}{4}, \frac{3}{4} \\ 0, \frac{1}{2}-a, -a, a, a+\frac{1}{2}, \frac{1}{2} \end{array}\right.\right)$

$= 8\sqrt{2}\pi^{3/2} \sec(2a\pi) \{\cos(4a\pi)\left[\operatorname{ker}^2_{2a}(2z^{1/4}) - \operatorname{kei}^2_{2a}(2z^{1/4})\right]$

$- 2\sin(4a\pi) \operatorname{ker}_{2a}(2z^{1/4}) \operatorname{kei}_{2a}(2z^{1/4})$

$- \pi\left[\operatorname{ber}_{2a}(2z^{1/4})\operatorname{kei}_{2a}(2z^{1/4}) + \operatorname{bei}_{2a}(2z^{1/4})\operatorname{ker}_{2a}(2z^{1/4})\right]\}.$

68. $G_{26}^{52}\left(z \left|\begin{array}{c} \frac{1}{4}, \frac{3}{4} \\ \frac{1}{2}, \frac{1}{2}-a, -a, a, a+\frac{1}{2}, 0 \end{array}\right.\right)$

$= 8\sqrt{2}\pi^{3/2}\sec(2a\pi)\{\sin(4a\pi)\left[\operatorname{ker}^2_{2a}(2z^{1/4}) - \operatorname{kei}^2_{2a}(2z^{1/4})\right]$

$+ 2\cos(4a\pi)\operatorname{ker}_{2a}(2z^{1/4})\operatorname{kei}_{2a}(2z^{1/4})$

$+ \pi\left[\operatorname{ber}_{2a}(2z^{1/4})\operatorname{ker}_{2a}(2z^{1/4}) - \operatorname{bei}_{2a}(2z^{1/4})\operatorname{kei}_{2a}(2z^{1/4})\right]\}.$

69. $G_{37}^{60}\left(z \left|\begin{array}{c} \frac{1}{4}, \frac{3}{4}, a \\ 0, \frac{1}{2}, \frac{1-a}{2}, -\frac{a}{2}, \frac{a}{2}, \frac{a+1}{2}, a \end{array}\right.\right)$

$= -8\sqrt{\dfrac{2}{\pi}} \operatorname{ker}_a(2z^{1/4}) \operatorname{kei}_a(2z^{1/4}).$

70. $G_{37}^{60}\left(z \left|\begin{array}{c} \frac{1}{4}, \frac{3}{4}, a \\ 0, \frac{1}{2}, \frac{1-2a}{4}, \frac{3-a}{4}, \frac{2a+1}{4}, \frac{2a+1}{4}, a \end{array}\right.\right)$

$= 4\sqrt{\dfrac{2}{\pi}} \left[\operatorname{ker}^2_{a-1/2}(2z^{1/4}) - \operatorname{kei}^2_a(2z^{1/4})\right].$

71. $G_{59}^{60}\left(z \,\middle|\, \begin{matrix} 0, 0, \frac{1}{4}, \frac{1}{2}, \frac{3}{4} \\ a, a+\frac{1}{2}, \frac{1}{2}-b, -b, b, b+\frac{1}{2}, 0, \frac{1}{2}-a, -a \end{matrix}\right)$

$= \dfrac{\sqrt{2}}{\pi^{3/2}} \{\operatorname{ber}_{2a+2b}(2z^{1/4}) \,[\pi \cos(4a\pi) \operatorname{ber}_{2a-2b}(2z^{1/4})$

$\quad + \pi \sin(4a\pi) \operatorname{bei}_{2a-2b}(2z^{1/4}) - \sin(4b\pi) \operatorname{ker}_{2a-2b}(2z^{1/4})$

$\quad + \cos(4b\pi) \operatorname{kei}_{2a-2b}(2z^{1/4})] + \operatorname{bei}_{2a+2b}(2z^{1/4})$

$\quad \times [\pi \sin(4a\pi) \operatorname{ber}_{2a-2b}(2z^{1/4}) - \pi \cos(4a\pi) \operatorname{bei}_{2a-2b}(2z^{1/4})$

$\quad + \cos(4b\pi) \operatorname{ker}_{2a-2b}(2z^{1/4}) + \sin(4b\pi) \operatorname{kei}_{2a-2b}(2z^{1/4})]$

$+ \operatorname{ber}_{2a-2b}(2z^{1/4}) [\sin(4b\pi) \operatorname{ker}_{2a+2b}(2z^{1/4}) + \cos(4b\pi) \operatorname{kei}_{2a+2b}(2z^{1/4})]$

$+ \operatorname{bei}_{2a-2b}(2z^{1/4}) [\cos(4b\pi) \operatorname{ker}_{2a+2b}(2z^{1/4}) - \sin(4b\pi) \operatorname{kei}_{2a+2b}(2z^{1/4})]\}.$

72. $G_{59}^{60}\left(z \,\middle|\, \begin{matrix} 0, \frac{1}{4}, \frac{1}{2}, \frac{1}{2}, \frac{3}{4} \\ a, a+\frac{1}{2}, \frac{1}{2}-b, -b, b, b+\frac{1}{2}, \frac{1}{2}, \frac{1}{2}-a, -a \end{matrix}\right)$

$= \dfrac{\sqrt{2}}{\pi^{3/2}} \{\operatorname{ber}_{2a+2b}(2z^{1/4}) \,[\pi \sin(4a\pi) \operatorname{ber}_{2a-2b}(2z^{1/4})$

$\quad - \pi \cos(4a\pi) \operatorname{bei}_{2a-2b}(2z^{1/4}) + \cos(4b\pi) \operatorname{ker}_{2a-2b}(2z^{1/4})$

$\quad + \sin(4b\pi) \operatorname{kei}_{2a-2b}(2z^{1/4})] - \operatorname{bei}_{2a+2b}(2z^{1/4})$

$\quad \times [\pi \cos(4a\pi) \operatorname{ber}_{2a-2b}(2z^{1/4}) + \pi \sin(4a\pi) \operatorname{bei}_{2a-2b}(2z^{1/4})$

$\quad - \sin(4b\pi) \operatorname{ker}_{2a-2b}(2z^{1/4}) + \cos(4b\pi) \operatorname{kei}_{2a-2b}(2z^{1/4})]$

$+ \operatorname{ber}_{2a-2b}(2z^{1/4}) [\cos(4b\pi) \operatorname{ker}_{2a+2b}(2z^{1/4}) - \sin(4b\pi) \operatorname{kei}_{2a+2b}(2z^{1/4})]$

$- \operatorname{bei}_{2a-2b}(2z^{1/4}) [\sin(4b\pi) \operatorname{ker}_{2a+2b}(2z^{1/4}) + \cos(4b\pi) \operatorname{kei}_{2a+2b}(2z^{1/4})]\}.$

73. $G_{48}^{63}\left(z \,\middle|\, \begin{matrix} 0, \frac{1}{4}, \frac{3}{4}, \frac{1}{2} \\ a, a+\frac{1}{2}, \frac{1}{2}-b, -b, b, b+\frac{1}{2}, \frac{1}{2}-a, -a \end{matrix}\right) =$

$\qquad -8\sqrt{2}\,\pi^{5/2} \csc(4b\pi) \,\{\operatorname{ber}_{2a+2b}(2z^{1/4})$

$\times [\cos(4b\pi) \operatorname{ker}_{2a-2b}(2z^{1/4}) + \sin(4b\pi) \operatorname{kei}_{2a-2b}(2z^{1/4})]$

$+ \operatorname{bei}_{2a+2b}(2z^{1/4}) [\sin(4b\pi) \operatorname{ker}_{2a-2b}(2z^{1/4}) - \cos(4b\pi) \operatorname{kei}_{2a-2b}(2z^{1/4})]$

$+ \operatorname{ber}_{2a-2b}(2z^{1/4}) [\sin(4b\pi) \operatorname{kei}_{2a+2b}(2z^{1/4}) - \cos(4b\pi) \operatorname{ker}_{2a+2b}(2z^{1/4})]$

$+ \operatorname{bei}_{2a-2b}(2z^{1/4}) [\sin(4b\pi) \operatorname{ker}_{2a+2b}(2z^{1/4}) + \cos(4b\pi) \operatorname{kei}_{2a+2b}(2z^{1/4})]\}.$

74. $G_{48}^{63}\left(z \,\middle|\, \begin{matrix} \frac{1}{4}, \frac{1}{2}, \frac{3}{4}, 0 \\ a, a+\frac{1}{2}, \frac{1}{2}-b, -b, b, b+\frac{1}{2}, \frac{1}{2}-a, -a \end{matrix}\right)$

$= 8\sqrt{2}\,\pi^{5/2} \csc(4b\pi) \,\{\operatorname{ber}_{2a+2b}(2z^{1/4})$

$\times [\sin(4b\pi) \operatorname{ker}_{2a-2b}(2z^{1/4}) - \cos(4b\pi) \operatorname{kei}_{2a-2b}(2z^{1/4})]$

$- \operatorname{bei}_{2a+2b}(2z^{1/4}) [\cos(4b\pi) \operatorname{ker}_{2a-2b}(2z^{1/4}) + \sin(4b\pi) \operatorname{kei}_{2a-2b}(2z^{1/4})]$

[8.2.2]   8.2. The Meijer Function $G_{p,q}^{m,n}\left(z \left| \begin{matrix} (a_p) \\ (b_q) \end{matrix}\right.\right)$

$+ \operatorname{ber}_{2a-2b}\left(2z^{1/4}\right)\left[\sin(4b\pi)\operatorname{ker}_{2a+2b}\left(2z^{1/4}\right) + \cos(4b\pi)\operatorname{kei}_{2a+2b}\left(2z^{1/4}\right)\right]$
$+ \operatorname{bei}_{2a-2b}\left(2z^{1/4}\right)\left[\cos(4b\pi)\operatorname{ker}_{2a+2b}\left(2z^{1/4}\right) - \sin(4b\pi)\operatorname{kei}_{2a+2b}\left(2z^{1/4}\right)\right]\}.$

75. $G_{5,9}^{8,0}\left(z \left| \begin{matrix} 0, \frac{1}{4}, \frac{1}{2}, \frac{3}{4}, a \\ \frac{1-a}{2}, -\frac{a}{2}, \frac{a}{2}, \frac{a+1}{2}, \frac{1}{2}-b, -b, b, b+\frac{1}{2}, a \end{matrix}\right.\right) =$
$-4\sqrt{\dfrac{2}{\pi}}\left[\operatorname{ker}_{a+2b}\left(2z^{1/4}\right)\operatorname{kei}_{a-2b}\left(2z^{1/4}\right) + \operatorname{kei}_{a+2b}\left(2z^{1/4}\right)\operatorname{ker}_{a-2b}\left(2z^{1/4}\right)\right].$

76. $G_{5,9}^{8,0}\left(z \left| \begin{matrix} 0, \frac{1}{4}, \frac{1}{2}, \frac{3}{4}, a \\ \frac{1-2a}{4}, \frac{3-2a}{4}, \frac{2a-1}{4}, \frac{2a+1}{4}, \frac{1}{2}-b, -b, b, b+\frac{1}{2}, a \end{matrix}\right.\right)$
$= 4\sqrt{\dfrac{2}{\pi}}\left[\operatorname{ker}_{a+2b-1/2}\left(2z^{1/4}\right)\operatorname{ker}_{a-2b-1/2}\left(2z^{1/4}\right)\right.$
$\left. - \operatorname{kei}_{a+2b-1/2}\left(2z^{1/4}\right)\operatorname{kei}_{a-2b-1/2}\left(2z^{1/4}\right)\right].$

77. $G_{5,9}^{8,0}\left(z \left| \begin{matrix} \frac{1}{8}, \frac{3}{8}, \frac{5}{8}, \frac{7}{8}, a \\ \frac{1-2a}{4}, \frac{3-2a}{4}, \frac{1-a}{2}, -\frac{a}{2}, \frac{a}{2}, \frac{a+1}{2}, \frac{2a+1}{4}, \frac{2a+3}{4}, a \end{matrix}\right.\right)$
$= 2\sqrt{2}\,e^{-\sqrt{2}z^{1/4}}\left[\sin\left(\sqrt{2}z^{1/4}\right)\operatorname{ker}_{2a}\left(2z^{1/4}\right) - \cos\left(\sqrt{2}z^{1/4}\right)\operatorname{kei}_{2a}\left(2z^{1/4}\right)\right].$

78. $G_{5,9}^{8,0}\left(z \left| \begin{matrix} \frac{1}{8}, \frac{3}{8}, \frac{5}{8}, \frac{7}{8}, a \\ \frac{1-2a}{4}, \frac{3-2a}{4}, \frac{1-a}{2}, 1-\frac{a}{2}, \frac{a}{2}, \frac{a+1}{2}, \frac{2a-1}{4}, \frac{2a+1}{4}, a \end{matrix}\right.\right)$
$= 2\sqrt{2}\,e^{-\sqrt{2}z^{1/4}}\left[\cos\left(\sqrt{2}z^{1/4}\right)\operatorname{ker}_{2a-1}\left(2z^{1/4}\right)\right.$
$\left. + \sin\left(\sqrt{2}z^{1/4}\right)\operatorname{kei}_{2a-1}\left(2z^{1/4}\right)\right].$

79. $G_{5,9}^{8,4}\left(z \left| \begin{matrix} \frac{1}{8}, \frac{3}{8}, \frac{5}{8}, \frac{7}{8}, a \\ \frac{1-2a}{4}, \frac{3-2a}{4}, \frac{1-a}{2}, -\frac{a}{2}, \frac{a}{2}, \frac{a+1}{2}, \frac{2a+1}{4}, \frac{2a+3}{4}, a \end{matrix}\right.\right) =$
$-16\sqrt{2}\,\pi^4 e^{\sqrt{2}z^{1/4}}\sec(2a\pi)\left[\sin\left(\sqrt{2}z^{1/4}\right)\operatorname{ker}_{2a}\left(2z^{1/4}\right)\right.$
$\left. + \cos\left(\sqrt{2}z^{1/4}\right)\operatorname{kei}_{2a}\left(2z^{1/4}\right)\right].$

80. $G_{5,9}^{8,4}\left(z \left| \begin{matrix} \frac{1}{8}, \frac{3}{8}, \frac{5}{8}, \frac{7}{8}, a \\ \frac{1-2a}{4}, \frac{3-2a}{4}, \frac{1-a}{2}, 1-\frac{a}{2}, \frac{a}{2}, \frac{a+1}{2}, \frac{2a-1}{4}, \frac{2a+1}{4}, a \end{matrix}\right.\right)$
$= 16\sqrt{2}\,\pi^4 e^{\sqrt{2}z^{1/4}}\sec(2a\pi)\left[\sin\left(\sqrt{2}z^{1/4}\right)\operatorname{kei}_{2a-1}\left(2z^{1/4}\right)\right.$
$\left. - \cos\left(\sqrt{2}z^{1/4}\right)\operatorname{ker}_{2a-1}\left(2z^{1/4}\right)\right].$

8.3. Representation in Terms of Hypergeometric Functions

8.3.1. Elementary functions

1. $\dfrac{1}{1+\sqrt{1+z}} = \dfrac{1}{2} {}_2F_1\!\left(\begin{matrix}\frac{1}{2},\,1\\2;\,-z\end{matrix}\right).$

2. $\left(1+\sqrt{1+z}\right)^{-1/2} = 2^{-1/2}\,{}_2F_1\!\left(\begin{matrix}\frac{1}{4},\,\frac{3}{4}\\\frac{3}{2};\,-z\end{matrix}\right).$

3. $(1-\sqrt{z})^{-1/2} + (1+\sqrt{z})^{-1/2} = 2\,{}_2F_1\!\left(\begin{matrix}\frac{1}{4},\,\frac{3}{4}\\\frac{1}{2};\,z\end{matrix}\right).$

4. $(1-\sqrt{z})^{-3/2} + (1+\sqrt{z})^{-3/2} = 2\,{}_2F_1\!\left(\begin{matrix}\frac{3}{4},\,\frac{5}{4}\\\frac{1}{2};\,z\end{matrix}\right).$

5. $(1+z)^{\nu} = {}_1F_0(-\nu;\,z).$

6. $\sinh z = z\,{}_3F_2\!\left(\begin{matrix}\frac{iz}{\pi},\,\frac{iz}{\pi},\,\frac{iz}{\pi}\\1,\,1;\,-1\end{matrix}\right) + \dfrac{2z^3}{\pi^2}\,{}_3F_2\!\left(\begin{matrix}\frac{iz}{\pi}+1,\,\frac{iz}{\pi}+1,\,\frac{iz}{\pi}+1\\2,\,2;\,-1\end{matrix}\right)$
   $[\operatorname{Re}(iz) < 2\pi/3].$

7. $\cosh z = -\left(iz - \dfrac{\pi}{2}\right) z\,{}_3F_2\!\left(\begin{matrix}\frac{iz}{\pi}-\frac{1}{2},\,\frac{iz}{\pi}-\frac{1}{2},\,\frac{iz}{\pi}-\frac{1}{2}\\1,\,1;\,-1\end{matrix}\right)$
   $+ \dfrac{2}{\pi^2}\left(iz - \dfrac{\pi}{2}\right)^3\,{}_3F_2\!\left(\begin{matrix}\frac{iz}{\pi}+\frac{1}{2},\,\frac{iz}{\pi}+\frac{1}{2},\,\frac{iz}{\pi}+\frac{1}{2}\\2,\,2;\,-1\end{matrix}\right)$ $[\operatorname{Re}(iz) < 7\pi/6].$

8. $\sinh\sqrt{z} = \sqrt{z}\,{}_0F_1\!\left(\begin{matrix}\frac{3}{2};\,\frac{z}{4}\end{matrix}\right).$

9. $\cosh\sqrt{z} = {}_0F_1\!\left(\begin{matrix}\frac{1}{2};\,\frac{z}{4}\end{matrix}\right).$

10. $\sinh\sqrt[4]{z} = z^{1/4}\,{}_0F_3\!\left(\begin{matrix}\frac{z}{256}\\\frac{1}{2},\,\frac{3}{4},\,\frac{5}{4}\end{matrix}\right) + \dfrac{z^{3/4}}{6}\,{}_0F_3\!\left(\begin{matrix}\frac{z}{256}\\\frac{5}{4},\,\frac{3}{2},\,\frac{7}{4}\end{matrix}\right).$

11. $\cosh\sqrt[4]{z} = {}_0F_3\!\left(\begin{matrix}\frac{z}{256}\\\frac{1}{4},\,\frac{1}{2},\,\frac{3}{4}\end{matrix}\right) + \dfrac{z^{1/2}}{2}\,{}_0F_3\!\left(\begin{matrix}\frac{z}{256}\\\frac{3}{4},\,\frac{5}{4},\,\frac{3}{2}\end{matrix}\right).$

12. $\operatorname{sech} z = \dfrac{4\pi}{\pi^2+4z^2}\,{}_4F_3\!\left(\begin{matrix}1,\,\frac{3}{2},\,\frac{1}{2}-\frac{iz}{\pi},\,\frac{1}{2}+\frac{iz}{\pi}\\\frac{1}{2},\,\frac{3}{2}-\frac{iz}{\pi},\,\frac{3}{2}+\frac{iz}{\pi};\,-1\end{matrix}\right).$

8.3.1]   8.3. Representation in Terms of Hypergeometric Functions

13. $\operatorname{csch} z = \dfrac{2}{z} {}_3F_2\left(\begin{array}{c} 1, -\frac{iz}{\pi}, \frac{iz}{\pi}; -1 \\ 1-\frac{iz}{\pi}, 1+\frac{iz}{\pi} \end{array}\right) - \dfrac{1}{z}.$

14. $\tanh z = \dfrac{8z}{\pi^2+4z^2} {}_3F_2\left(\begin{array}{c} 1, \frac{1}{2}-\frac{iz}{\pi}, \frac{1}{2}+\frac{iz}{\pi} \\ \frac{3}{2}-\frac{iz}{\pi}, \frac{3}{2}+\frac{iz}{\pi}; 1 \end{array}\right).$

15. $\coth z = \dfrac{2}{z} {}_3F_2\left(\begin{array}{c} 1, -\frac{iz}{\pi}, \frac{iz}{\pi} \\ 1-\frac{iz}{\pi}, 1+\frac{iz}{\pi}; 1 \end{array}\right) - \dfrac{1}{z}.$

16. $\sin z = z\,{}_3F_2\left(\begin{array}{c} \frac{z}{\pi}, \frac{z}{\pi}, \frac{z}{\pi} \\ 1, 1; -1 \end{array}\right) - \dfrac{2z^3}{\pi^2}\,{}_3F_2\left(\begin{array}{c} \frac{z}{\pi}+1, \frac{z}{\pi}+1, \frac{z}{\pi}+1 \\ 2, 2; -1 \end{array}\right)$ $[\operatorname{Re} z < 2\pi/3]$.

17. $\cos z = -\left(z-\dfrac{\pi}{2}\right){}_3F_2\left(\begin{array}{c} \frac{z}{\pi}-\frac{1}{2}, \frac{z}{\pi}-\frac{1}{2}, \frac{z}{\pi}-\frac{1}{2} \\ 1, 1; -1 \end{array}\right)$

$\quad - \dfrac{2}{\pi^2}\left(z-\dfrac{\pi}{2}\right)^3 {}_3F_2\left(\begin{array}{c} \frac{z}{\pi}+\frac{1}{2}, \frac{z}{\pi}+\frac{1}{2}, \frac{z}{\pi}+\frac{1}{2} \\ 2, 2; -1 \end{array}\right)$ $[\operatorname{Re} z < 7\pi/6]$.

18. $\sin\sqrt{z} = \sqrt{z}\,{}_0F_1\left(\dfrac{3}{2}; -\dfrac{z}{4}\right).$

19. $\cos\sqrt{z} = {}_0F_1\left(\dfrac{1}{2}; -\dfrac{z}{4}\right).$

20. $\sin\sqrt[4]{z} = z^{1/4}\,{}_0F_3\left(\begin{array}{c} \frac{z}{256} \\ \frac{1}{2}, \frac{3}{4}, \frac{5}{4} \end{array}\right) - \dfrac{z^{3/4}}{6}\,{}_0F_3\left(\begin{array}{c} \frac{z}{256} \\ \frac{5}{4}, \frac{3}{2}, \frac{7}{4} \end{array}\right).$

21. $\cos\sqrt[4]{z} = {}_0F_3\left(\begin{array}{c} \frac{z}{256} \\ \frac{1}{4}, \frac{1}{2}, \frac{3}{4} \end{array}\right) - \dfrac{z^{1/2}}{2}\,{}_0F_3\left(\begin{array}{c} \frac{z}{256} \\ \frac{3}{4}, \frac{5}{4}, \frac{3}{2} \end{array}\right).$

22. $\sinh\sqrt[4]{z}\,\sin\sqrt[4]{z} = z^{1/2}\,{}_0F_3\left(\begin{array}{c} -\frac{z}{64} \\ \frac{3}{4}, \frac{5}{4}, \frac{3}{2} \end{array}\right).$

23. $\cosh\sqrt[4]{z}\,\cos\sqrt[4]{z} = {}_0F_3\left(\begin{array}{c} -\frac{z}{64} \\ \frac{1}{4}, \frac{1}{2}, \frac{3}{4} \end{array}\right).$

24. $\left\{\begin{array}{c} \sinh\sqrt[4]{z}\,\cos\sqrt[4]{z} \\ \cosh\sqrt[4]{z}\,\sin\sqrt[4]{z} \end{array}\right\} = z^{1/4}\,{}_0F_3\left(\begin{array}{c} -\frac{z}{64} \\ \frac{1}{2}, \frac{3}{4}, \frac{5}{4} \end{array}\right) \mp \dfrac{z^{3/4}}{3}\,{}_0F_3\left(\begin{array}{c} -\frac{z}{64} \\ \frac{5}{4}, \frac{3}{2}, \frac{7}{4} \end{array}\right).$

25. $\sec z = \dfrac{4\pi}{\pi^2-4z^2}\,{}_4F_3\left(\begin{array}{c} 1, \frac{3}{2}, \frac{1}{2}-\frac{z}{\pi}, \frac{1}{2}+\frac{z}{\pi} \\ \frac{1}{2}, \frac{3}{2}-\frac{z}{\pi}, \frac{3}{2}+\frac{z}{\pi}; -1 \end{array}\right).$

26. $\csc z = \dfrac{2}{z}\,{}_3F_2\!\left(\begin{array}{c}1,\,-\frac{z}{\pi},\,\frac{z}{\pi};\,-1\\1-\frac{z}{\pi},\,1+\frac{z}{\pi}\end{array}\right) - \dfrac{1}{z}.$

27. $\tan z = \dfrac{8z}{\pi^2 - 4z^2}\,{}_3F_2\!\left(\begin{array}{c}1,\,\frac{1}{2}-\frac{z}{\pi},\,\frac{1}{2}+\frac{z}{\pi}\\\frac{3}{2}-\frac{z}{\pi},\,\frac{3}{2}+\frac{z}{\pi};\,1\end{array}\right).$

28. $\cot z = \dfrac{2}{z}\,{}_3F_2\!\left(\begin{array}{c}1,\,-\frac{z}{\pi},\,\frac{z}{\pi}\\1-\frac{z}{\pi},\,1+\frac{z}{\pi};\,1\end{array}\right) - \dfrac{1}{z}.$

29. $\ln(1+z) = z\,{}_2F_1\!\left(\begin{array}{c}1,\,1\\2;\,-z\end{array}\right).$

30. $\ln(\sqrt{z}+\sqrt{1+z}) = \sqrt{z}\,{}_2F_1\!\left(\begin{array}{c}\frac{1}{2},\,\frac{1}{2}\\\frac{3}{2};\,-z\end{array}\right).$

31. $\dfrac{1}{\sqrt{1+z}}\ln(\sqrt{z}+\sqrt{1+z}) = \sqrt{z}\,{}_2F_1\!\left(\begin{array}{c}1,\,1\\\frac{3}{2};\,-z\end{array}\right).$

32. $\ln^2(\sqrt{z}+\sqrt{1+z}) = z\,{}_3F_2\!\left(\begin{array}{c}1,\,1,\,1;\,-z\\\frac{3}{2},\,2\end{array}\right).$

33. $\ln\dfrac{1+\sqrt{z}}{1-\sqrt{z}} = 2\sqrt{z}\,{}_2F_1\!\left(\begin{array}{c}\frac{1}{2},\,1\\\frac{3}{2};\,z\end{array}\right).$

34. $\arcsin\sqrt{z} = \sqrt{z}\,{}_2F_1\!\left(\begin{array}{c}\frac{1}{2},\,\frac{1}{2}\\\frac{3}{2};\,z\end{array}\right).$

35. $\dfrac{1}{\sqrt{1-z}}\arcsin\sqrt{z} = \sqrt{z}\,{}_2F_1\!\left(\begin{array}{c}1,\,1\\\frac{3}{2};\,z\end{array}\right).$

36. $\arcsin^2\sqrt{z} = z\,{}_3F_2\!\left(\begin{array}{c}1,\,1,\,1;\,z\\\frac{3}{2},\,2\end{array}\right).$

37. $\arctan\sqrt{z} = \sqrt{z}\,{}_2F_1\!\left(\begin{array}{c}\frac{1}{2},\,1\\\frac{3}{2};\,-z\end{array}\right).$

38. $\sin(\nu\arcsin\sqrt{z}) = \nu\sqrt{z}\,{}_2F_1\!\left(\begin{array}{c}\frac{1-\nu}{2},\,\frac{1+\nu}{2}\\\frac{3}{2};\,z\end{array}\right).$

8.3.2]   8.3. Representation in Terms of Hypergeometric Functions

39. $\cos(\nu \arcsin \sqrt{z}) = {}_2F_1\left(\begin{array}{c} -\dfrac{\nu}{2}, \dfrac{\nu}{2} \\ \dfrac{1}{2}; z \end{array}\right).$

40. $\dfrac{1}{\sqrt{1-z}} \sin(\nu \arcsin \sqrt{z}) = \nu\sqrt{z}\, {}_2F_1\left(\begin{array}{c} 1-\dfrac{\nu}{2}, 1+\dfrac{\nu}{2} \\ \dfrac{3}{2}; z \end{array}\right).$

41. $\dfrac{1}{\sqrt{1-z}} \cos(\nu \arcsin \sqrt{z}) = {}_2F_1\left(\begin{array}{c} \dfrac{1-\nu}{2}, \dfrac{1+\nu}{2} \\ \dfrac{1}{2}; z \end{array}\right).$

42. $(1+z)^{-\nu/2} \sin(\nu \arctan \sqrt{z}) = \nu\sqrt{z}\, {}_2F_1\left(\begin{array}{c} \dfrac{1+\nu}{2}, 1+\dfrac{\nu}{2} \\ \dfrac{3}{2}; -z \end{array}\right).$

43. $(1+z)^{-\nu/2} \cos(\nu \arctan \sqrt{z}) = {}_2F_1\left(\begin{array}{c} \dfrac{\nu}{2}, \dfrac{1+\nu}{2} \\ \dfrac{1}{2}; -z \end{array}\right).$

8.3.2. Special functions

1. $\psi^{(n)}(z) = n!(-z)^{-n-1}\, {}_{n+2}F_{n+1}\left(\begin{array}{c} 1, z, z, \ldots, z; 1 \\ z+1, z+1, \ldots, z+1 \end{array}\right).$

2. $\Phi(z, n, a) = a^{-n}\, {}_{n+1}F_n\left(\begin{array}{c} 1, a, a, \ldots, a; z \\ a+1, a+1, \ldots, a+1 \end{array}\right).$

3. $\mathrm{Li}_n(z) = z\, {}_{n+1}F_n\left(\begin{array}{c} 1, 1, \ldots, 1; z \\ 2, 2, \ldots, 2 \end{array}\right).$

4. $\mathrm{Si}(\sqrt{z}) = z^{1/2}\, {}_1F_2\left(\begin{array}{c} \dfrac{1}{2}; -\dfrac{z}{4} \\ \dfrac{3}{2}, \dfrac{3}{2} \end{array}\right).$

5. $\mathrm{ci}(\sqrt{z}) - \ln\sqrt{z} - \mathbf{C} = -\dfrac{z}{4}\, {}_2F_3\left(\begin{array}{c} 1, 1; -\dfrac{z}{4} \\ \dfrac{3}{2}, 2, 2 \end{array}\right).$

6. $\mathrm{Ei}(-z) - \ln z - \mathbf{C} = -z\, {}_2F_2\left(\begin{array}{c} 1, 1; -z \\ 2, 2 \end{array}\right).$

7. $e^z \mathrm{Ei}(-z) = -\Psi\left(\begin{array}{c} 1; z \\ 1 \end{array}\right).$

8. $\mathrm{erf}(\sqrt{z}) = \dfrac{2z^{1/2}}{\sqrt{\pi}}\, {}_1F_1\left(\begin{array}{c} \dfrac{1}{2}; -z \\ \dfrac{3}{2} \end{array}\right).$

9. $e^z \mathrm{erf}(\sqrt{z}) = \dfrac{2z^{1/2}}{\sqrt{\pi}}\, {}_1F_1\left(\begin{array}{c} 1; z \\ \dfrac{3}{2} \end{array}\right).$

CH. 8. REPRESENTATIONS OF HYPERGEOMETRIC AND MEIJER G FUNCTIONS [8.3.2

10. $\left\{\begin{matrix}\operatorname{erf}(\sqrt[4]{z})\\ \operatorname{erfi}(\sqrt[4]{z})\end{matrix}\right\} = \dfrac{2z^{1/4}}{\sqrt{\pi}}\,{}_1F_2\!\left(\begin{matrix}\tfrac{1}{4};\ \tfrac{z}{4}\\ \tfrac{1}{2},\ \tfrac{5}{4}\end{matrix}\right) \mp \dfrac{2z^{3/4}}{3\sqrt{\pi}}\,{}_1F_2\!\left(\begin{matrix}\tfrac{3}{4};\ \tfrac{z}{4}\\ \tfrac{3}{2},\ \tfrac{7}{4}\end{matrix}\right).$

11. $\left\{\begin{matrix}e^{\sqrt{z}}\operatorname{erf}(\sqrt[4]{z})\\ e^{-\sqrt{z}}\operatorname{erfi}(\sqrt[4]{z})\end{matrix}\right\} = \dfrac{2}{\sqrt{\pi}}z^{1/4}\,{}_1F_2\!\left(\begin{matrix}1;\ \tfrac{z}{4}\\ \tfrac{3}{4},\ \tfrac{5}{4}\end{matrix}\right) \pm \dfrac{4}{3\sqrt{\pi}}z^{3/4}\,{}_1F_2\!\left(\begin{matrix}1;\ \tfrac{z}{4}\\ \tfrac{5}{4},\ \tfrac{7}{4}\end{matrix}\right).$

12. $e^{z}\operatorname{erfc}(\sqrt{z}) = \dfrac{1}{\sqrt{\pi}}\Psi\!\left(\begin{matrix}\tfrac{1}{2};\ z\\ \tfrac{1}{2}\end{matrix}\right).$

13. $\operatorname{erf}(\sqrt[4]{z})\operatorname{erfi}(\sqrt[4]{z}) = \dfrac{4z^{1/2}}{\pi}\,{}_2F_3\!\left(\begin{matrix}\tfrac{1}{2},\ 1;\ \tfrac{z}{4}\\ \tfrac{3}{4},\ \tfrac{5}{4},\ \tfrac{3}{2}\end{matrix}\right).$

14. $S(\sqrt{z}) = \dfrac{1}{3}\sqrt{\dfrac{2}{\pi}}z^{3/4}\,{}_1F_2\!\left(\begin{matrix}\tfrac{3}{4};\ -\tfrac{z}{4}\\ \tfrac{3}{2},\ \tfrac{7}{4}\end{matrix}\right).$

15. $C(\sqrt{z}) = \sqrt{\dfrac{2}{\pi}}z^{1/4}\,{}_1F_2\!\left(\begin{matrix}\tfrac{1}{4};\ -\tfrac{z}{4}\\ \tfrac{1}{2},\ \tfrac{5}{4}\end{matrix}\right).$

16. $\sin\sqrt{z}\,S(\sqrt{z}) + \cos\sqrt{z}\,C(\sqrt{z}) = \sqrt{\dfrac{2}{\pi}}z^{1/4}\,{}_1F_2\!\left(\begin{matrix}1;\ -\tfrac{z}{4}\\ \tfrac{3}{4},\ \tfrac{5}{4}\end{matrix}\right).$

17. $\sin\sqrt{z}\,C(\sqrt{z}) - \cos\sqrt{z}\,S(\sqrt{z}) = \dfrac{2}{3}\sqrt{\dfrac{2}{\pi}}z^{3/4}\,{}_1F_2\!\left(\begin{matrix}1;\ -\tfrac{z}{4}\\ \tfrac{5}{4},\ \tfrac{7}{4}\end{matrix}\right).$

18. $S(\sqrt{z},\nu) = \Gamma(\nu)\sin\dfrac{\nu\pi}{2} - \dfrac{z^{(\nu+1)/2}}{\nu+1}\,{}_1F_2\!\left(\begin{matrix}\tfrac{\nu+1}{2};\ -\tfrac{z}{4}\\ \tfrac{\nu+3}{2},\ \tfrac{3}{2}\end{matrix}\right).$

19. $C(\sqrt{z},\nu) = \Gamma(\nu)\cos\dfrac{\nu\pi}{2} - \dfrac{z^{\nu/2}}{\nu}\,{}_1F_2\!\left(\begin{matrix}\tfrac{\nu}{2};\ -\tfrac{z}{4}\\ \tfrac{\nu}{2}+1,\ \tfrac{1}{2}\end{matrix}\right).$

20. $\gamma(\nu, z) = \dfrac{z^{\nu}}{\nu}\,{}_1F_1\!\left(\begin{matrix}\nu;\ -z\\ \nu+1\end{matrix}\right).$

21. $e^{z}\gamma(\nu, z) = \dfrac{z^{\nu}}{\nu}\,{}_1F_1\!\left(\begin{matrix}1;\ z\\ \nu+1\end{matrix}\right).$

22. $e^{z}\Gamma(\nu, z) = z^{\nu}\Psi\!\left(\begin{matrix}1;\ z\\ \nu+1\end{matrix}\right).$

23. $\phantom{e^{z}\Gamma(\nu, z)} = \Psi\!\left(\begin{matrix}1-\nu;\ z\\ 1-\nu\end{matrix}\right).$

24. $D_\nu(\sqrt{z}) = 2^{\nu/2} e^{-z/4} \Psi\begin{pmatrix} -\frac{\nu}{2}; \frac{z}{2} \\ \frac{1}{2} \end{pmatrix}$.

25. $\begin{Bmatrix} J_\nu(\sqrt{z}) \\ I_\nu(\sqrt{z}) \end{Bmatrix} = \dfrac{\left(\frac{\sqrt{z}}{2}\right)^\nu}{\Gamma(\nu+1)} {}_0F_1\left(\nu+1; \mp\dfrac{z}{4}\right)$.

26. $\begin{Bmatrix} J_\nu(\sqrt[4]{z}) \\ I_\nu(\sqrt[4]{z}) \end{Bmatrix} = \dfrac{z^{\nu/4}}{2^\nu \Gamma(\nu+1)} {}_0F_3\left(\begin{matrix} \frac{z}{256} \\ \frac{1}{2}, \frac{\nu+1}{2}, \frac{\nu}{2}+1 \end{matrix}\right)$

$$\mp \dfrac{z^{(\nu+2)/4}}{2^{\nu+2} \Gamma(\nu+2)} {}_0F_3\left(\begin{matrix} \frac{z}{256} \\ \frac{3}{2}, \frac{\nu}{2}+1, \frac{\nu+3}{2} \end{matrix}\right).$$

27. $e^z I_\nu(z) = \dfrac{\left(\frac{z}{2}\right)^\nu}{\Gamma(\nu+1)} {}_1F_1\left(\begin{matrix} \nu+\frac{1}{2}; 2z \\ 2\nu+1 \end{matrix}\right)$.

28. $e^{iz} J_\nu(z) = \dfrac{\left(\frac{z}{2}\right)^\nu}{\Gamma(\nu+1)} {}_1F_1\left(\begin{matrix} \nu+\frac{1}{2}; 2iz \\ 2\nu+1 \end{matrix}\right)$.

29. $\begin{Bmatrix} \sin\sqrt{z} J_\nu(\sqrt{z}) \\ \sinh\sqrt{z} I_\nu(\sqrt{z}) \end{Bmatrix} = \dfrac{z^{(\nu+1)/2}}{2^\nu \Gamma(\nu+1)} {}_2F_3\left(\begin{matrix} \frac{2\nu+3}{4}, \frac{2\nu+5}{4}; \mp z \\ \frac{3}{2}, \nu+1, \nu+\frac{3}{2} \end{matrix}\right)$.

30. $\begin{Bmatrix} \cos\sqrt{z} J_\nu(\sqrt{z}) \\ \cosh\sqrt{z} I_\nu(\sqrt{z}) \end{Bmatrix} = \dfrac{z^{\nu/2}}{2^\nu \Gamma(\nu+1)} {}_2F_3\left(\begin{matrix} \frac{2\nu+1}{4}, \frac{2\nu+3}{4}; \mp z \\ \frac{1}{2}, \nu+\frac{1}{2}, \nu+1 \end{matrix}\right)$.

31. $\begin{Bmatrix} J_\nu^2(\sqrt{z}) \\ I_\nu^2(\sqrt{z}) \end{Bmatrix} = \dfrac{\left(\frac{z}{4}\right)^\nu}{\Gamma^2(\nu+1)} {}_1F_2\left(\begin{matrix} \nu+\frac{1}{2}; \mp z \\ \nu+1, 2\nu+1 \end{matrix}\right)$.

32. $\begin{Bmatrix} J_\mu(\sqrt{z}) J_\nu(\sqrt{z}) \\ I_\mu(\sqrt{z}) I_\nu(\sqrt{z}) \end{Bmatrix} = \dfrac{\left(\frac{\sqrt{z}}{2}\right)^{\mu+\nu}}{\Gamma(\mu+1)\Gamma(\nu+1)} {}_2F_3\left(\begin{matrix} \frac{\mu+\nu+1}{2}, \frac{\mu+\nu}{2}+1; \mp z \\ \mu+1, \nu+1, \mu+\nu+1 \end{matrix}\right)$.

33. $J_\nu(\sqrt[4]{z}) I_\nu(\sqrt[4]{z}) = \dfrac{\left(\frac{\sqrt{z}}{4}\right)^\nu}{\Gamma^2(\nu+1)} {}_0F_3\left(\begin{matrix} -\frac{z}{64} \\ \frac{\nu+1}{2}, \frac{\nu}{2}+1, \nu+1 \end{matrix}\right)$.

34. $J_\nu(\sqrt[4]{z}) I_{-\nu}(\sqrt[4]{z}) = \dfrac{\sin\nu\pi}{\nu\pi} {}_0F_3\left(\begin{matrix} -\frac{z}{64} \\ 1-\frac{\nu}{2}, 1+\frac{\nu}{2}, \frac{1}{2} \end{matrix}\right)$

$$+ \dfrac{\sin\nu\pi}{2(1-\nu^2)\pi} z^{1/2} {}_0F_3\left(\begin{matrix} -\frac{z}{64} \\ \frac{3-\nu}{2}, \frac{3+\nu}{2}, \frac{3}{2} \end{matrix}\right).$$

35. $J_{n+1/2}^2\left(\dfrac{1}{\sqrt{z}}\right) + Y_{n+1/2}^2\left(\dfrac{1}{\sqrt{z}}\right) = \dfrac{2\sqrt{z}}{\pi}\,{}_3F_0\!\left(\begin{matrix}-n,\,n+1,\,\tfrac{1}{2}\\ -z\end{matrix}\right).$

36. $I_{-n-1/2}^2\left(\dfrac{1}{\sqrt{z}}\right) - I_{n+1/2}^2\left(\dfrac{1}{\sqrt{z}}\right) = (-1)^n\dfrac{2\sqrt{z}}{\pi}\,{}_3F_0\!\left(\begin{matrix}-n,\,n+1,\,\tfrac{1}{2}\\ z\end{matrix}\right).$

37. $e^z K_{n+1/2}(z) = \dfrac{(2n)!\sqrt{\pi}}{n!}(2z)^{-n-1/2}\,{}_1F_1\!\left(\begin{matrix}-n;\,2z\\ -2n\end{matrix}\right).$

38. $\left\{\begin{matrix}\mathbf{H}_\nu(\sqrt{z})\\ \mathbf{L}_\nu(\sqrt{z})\end{matrix}\right\} = \dfrac{2^{-\nu} z^{(\nu+1)/2}}{\sqrt{\pi}\,\Gamma\!\left(\nu+\tfrac{3}{2}\right)}\,{}_1F_2\!\left(\begin{matrix}1;\,\mp\tfrac{z}{4}\\ \nu+\tfrac{3}{2},\,\tfrac{3}{2}\end{matrix}\right).$

39. $\mathbf{H}_{n+1/2}(\sqrt{z}) - Y_{n+1/2}(\sqrt{z}) = \dfrac{\left(\tfrac{\sqrt{z}}{2}\right)^{n-1/2}}{n!\sqrt{\pi}}\,{}_3F_0\!\left(\begin{matrix}-n,\,\tfrac{1}{2},\,1\\ -\tfrac{4}{z}\end{matrix}\right).$

40. $\mathbf{L}_{n+1/2}(\sqrt{z}) - I_{-n-1/2}(\sqrt{z}) = \dfrac{\left(\tfrac{\sqrt{z}}{2}\right)^{n-1/2}}{n!\sqrt{\pi}}\,{}_3F_0\!\left(\begin{matrix}-n,\,\tfrac{1}{2},\,1\\ \tfrac{4}{z}\end{matrix}\right).$

41. $\mathrm{Ai}(z) = \dfrac{3^{-2/3}}{\Gamma\!\left(\tfrac{2}{3}\right)}\,{}_0F_1\!\left(\begin{matrix}\tfrac{2}{3};\,\tfrac{z^3}{9}\end{matrix}\right) - \dfrac{3^{-1/3}z}{\Gamma\!\left(\tfrac{1}{3}\right)}\,{}_0F_1\!\left(\begin{matrix}\tfrac{4}{3};\,\tfrac{z^3}{9}\end{matrix}\right).$

42. $\mathrm{Bi}(z) = \dfrac{3^{-1/6}}{\Gamma\!\left(\tfrac{2}{3}\right)}\,{}_0F_1\!\left(\begin{matrix}\tfrac{2}{3};\,\tfrac{z^3}{9}\end{matrix}\right) + \dfrac{3^{1/6}z}{\Gamma\!\left(\tfrac{1}{3}\right)}\,{}_0F_1\!\left(\begin{matrix}\tfrac{4}{3};\,\tfrac{z^3}{9}\end{matrix}\right).$

43. $\mathrm{Ai}'(z) = \dfrac{3^{-2/3}z^2}{2\Gamma\!\left(\tfrac{2}{3}\right)}\,{}_0F_1\!\left(\begin{matrix}\tfrac{5}{3};\,\tfrac{z^3}{9}\end{matrix}\right) - \dfrac{3^{-1/3}}{\Gamma\!\left(\tfrac{1}{3}\right)}\,{}_0F_1\!\left(\begin{matrix}\tfrac{1}{3};\,\tfrac{z^3}{9}\end{matrix}\right).$

44. $\mathrm{Bi}'(z) = \dfrac{3^{-1/6}z^2}{2\Gamma\!\left(\tfrac{2}{3}\right)}\,{}_0F_1\!\left(\begin{matrix}\tfrac{5}{3};\,\tfrac{z^3}{9}\end{matrix}\right) + \dfrac{3^{1/6}}{\Gamma\!\left(\tfrac{1}{3}\right)}\,{}_0F_1\!\left(\begin{matrix}\tfrac{1}{3};\,\tfrac{z^3}{9}\end{matrix}\right).$

45. $\mathrm{Ai}^2(z) = \dfrac{\Gamma\!\left(\tfrac{1}{6}\right)}{2^{5/3}3^{5/6}\pi^{3/2}}\,{}_1F_2\!\left(\begin{matrix}\tfrac{1}{6};\,\tfrac{4z^3}{9}\\ \tfrac{1}{3},\,\tfrac{2}{3}\end{matrix}\right) - \dfrac{z}{\sqrt{3}\,\pi}\,{}_1F_2\!\left(\begin{matrix}\tfrac{1}{2};\,\tfrac{4z^3}{9}\\ \tfrac{2}{3},\,\tfrac{4}{3}\end{matrix}\right)$
$$+ \dfrac{\Gamma\!\left(\tfrac{5}{6}\right)z^2}{2^{4/3}3^{1/6}\pi^{3/2}}\,{}_1F_2\!\left(\begin{matrix}\tfrac{5}{6};\,\tfrac{4z^3}{9}\\ \tfrac{4}{3},\,\tfrac{5}{3}\end{matrix}\right).$$

46. $\mathrm{Bi}^2(z) = \dfrac{3^{1/6}}{2^{5/3}\pi^{3/2}}\Gamma\!\left(\tfrac{1}{6}\right)\,{}_1F_2\!\left(\begin{matrix}\tfrac{1}{6};\,\tfrac{4z^3}{9}\\ \tfrac{1}{3},\,\tfrac{2}{3}\end{matrix}\right) + \dfrac{3^{1/2}z}{\pi}\,{}_1F_2\!\left(\begin{matrix}\tfrac{1}{2};\,\tfrac{4z^3}{9}\\ \tfrac{2}{3},\,\tfrac{4}{3}\end{matrix}\right)$
$$+ \dfrac{3^{5/6}z^2}{2^{4/3}\pi^{3/2}}\Gamma\!\left(\tfrac{5}{6}\right)\,{}_1F_2\!\left(\begin{matrix}\tfrac{5}{6};\,\tfrac{4z^3}{9}\\ \tfrac{4}{3},\,\tfrac{5}{3}\end{matrix}\right).$$

8.3.2]   8.3. Representation in Terms of Hypergeometric Functions

47. $\operatorname{Ai}(z)\operatorname{Ai}(-z) = \dfrac{3^{-4/3}}{\Gamma^2\left(\frac{2}{3}\right)}\,_0F_3\left(\begin{array}{c}-\frac{z^6}{324}\\ \frac{1}{3},\frac{2}{3},\frac{5}{6}\end{array}\right) - \dfrac{3^{-8/3}z^2}{\Gamma^2\left(\frac{4}{3}\right)}\,_0F_3\left(\begin{array}{c}-\frac{z^6}{324}\\ \frac{2}{3},\frac{7}{6},\frac{4}{3}\end{array}\right)$

$\qquad + \dfrac{3^{-3/2}z^4}{4\pi}\,_0F_3\left(\begin{array}{c}-\frac{z^6}{324}\\ \frac{4}{3},\frac{3}{2},\frac{5}{3}\end{array}\right).$

48. $\operatorname{Bi}(z)\operatorname{Bi}(-z) = \dfrac{3^{-1/3}}{\Gamma^2\left(\frac{2}{3}\right)}\,_0F_3\left(\begin{array}{c}-\frac{z^6}{324}\\ \frac{1}{3},\frac{2}{3},\frac{5}{6}\end{array}\right) - \dfrac{3^{-5/3}z^2}{\Gamma^2\left(\frac{4}{3}\right)}\,_0F_3\left(\begin{array}{c}-\frac{z^6}{324}\\ \frac{2}{3},\frac{7}{6},\frac{4}{3}\end{array}\right)$

$\qquad - \dfrac{3^{-1/2}z^4}{4\pi}\,_0F_3\left(\begin{array}{c}-\frac{z^6}{324}\\ \frac{4}{3},\frac{3}{2},\frac{5}{3}\end{array}\right).$

49. $\operatorname{Ai}(z)\operatorname{Bi}(-z) = \dfrac{3^{-5/6}}{\Gamma^2\left(\frac{2}{3}\right)}\,_0F_3\left(\begin{array}{c}-\frac{z^6}{324}\\ \frac{1}{3},\frac{2}{3},\frac{5}{6}\end{array}\right) + \dfrac{3^{-13/6}z^2}{\Gamma^2\left(\frac{4}{3}\right)}\,_0F_3\left(\begin{array}{c}-\frac{z^6}{324}\\ \frac{2}{3},\frac{7}{6},\frac{4}{3}\end{array}\right)$

$\qquad - \dfrac{z}{\pi}\,_0F_3\left(\begin{array}{c}-\frac{z^6}{324}\\ \frac{1}{2},\frac{5}{6},\frac{7}{6}\end{array}\right).$

50. $\operatorname{ber}_\nu(z) = \cos\dfrac{3\nu\pi}{4}\,\dfrac{\left(\frac{z}{2}\right)^\nu}{\Gamma(\nu+1)}\,_0F_3\left(\begin{array}{c}-\frac{z^4}{256}\\ \frac{\nu+1}{2},\frac{\nu}{2}+1,\frac{1}{2}\end{array}\right)$

$\qquad - \sin\dfrac{3\nu\pi}{4}\,\dfrac{\left(\frac{z}{2}\right)^{\nu+2}}{\Gamma(\nu+2)}\,_0F_3\left(\begin{array}{c}-\frac{z^4}{256}\\ \frac{\nu}{2}+1,\frac{\nu+3}{2},\frac{3}{2}\end{array}\right).$

51. $\operatorname{bei}_\nu(z) = \sin\dfrac{3\nu\pi}{4}\,\dfrac{\left(\frac{z}{2}\right)^\nu}{\Gamma(\nu+1)}\,_0F_3\left(\begin{array}{c}-\frac{z^4}{256}\\ \frac{\nu+1}{2},\frac{\nu}{2}+1,\frac{1}{2}\end{array}\right)$

$\qquad + \cos\dfrac{3\nu\pi}{4}\,\dfrac{\left(\frac{z}{2}\right)^{\nu+2}}{\Gamma(\nu+2)}\,_0F_3\left(\begin{array}{c}-\frac{z^4}{256}\\ \frac{\nu}{2}+1,\frac{\nu+3}{2},\frac{3}{2}\end{array}\right).$

52. $\operatorname{ber}_\nu^2(z) = \dfrac{z^{2\nu}}{2^{2\nu+1}\Gamma^2(\nu+1)}$

$\qquad\times\left[\cos\dfrac{3\nu\pi}{2}\,_2F_5\left(\begin{array}{c}\frac{2\nu+1}{4},\frac{2\nu+3}{4};-\frac{z^4}{16}\\ \frac{\nu+1}{2},\frac{\nu}{2}+1,\nu+\frac{1}{2},\nu+1,\frac{1}{2}\end{array}\right) - \dfrac{z^2}{2(\nu+1)}\sin\dfrac{3\nu\pi}{2}\right.$

$\qquad\left.\times\,_2F_5\left(\begin{array}{c}\frac{2\nu+3}{4},\frac{2\nu+5}{4};-\frac{z^4}{16}\\ \frac{\nu}{2}+1,\frac{\nu+3}{2},\nu+1,\nu+\frac{3}{2},\frac{3}{2}\end{array}\right) + {}_0F_3\left(\begin{array}{c}\frac{z^4}{64}\\ \frac{\nu+1}{2},\frac{\nu}{2}+1,\nu+1\end{array}\right)\right].$

53. $\operatorname{bei}_\nu^2(z) = \dfrac{z^{2\nu}}{2^{2\nu+1}\Gamma^2(\nu+1)}$

$\times \left[-\cos\dfrac{3\nu\pi}{2} {}_2F_5\left(\begin{matrix} \dfrac{2\nu+1}{4}, \dfrac{2\nu+3}{4}; -\dfrac{z^4}{16} \\ \dfrac{\nu+1}{2}, \dfrac{\nu}{2}+1, \nu+\dfrac{1}{2}, \nu+1, \dfrac{1}{2} \end{matrix} \right) + \dfrac{z^2}{2(\nu+1)} \sin\dfrac{3\nu\pi}{2} \right.$

$\left. \times {}_2F_5\left(\begin{matrix} \dfrac{2\nu+3}{4}, \dfrac{2\nu+5}{4}; -\dfrac{z^4}{16} \\ \dfrac{\nu}{2}+1, \dfrac{\nu+3}{2}, \nu+1, \nu+\dfrac{3}{2}, \dfrac{3}{2} \end{matrix} \right) + {}_0F_3\left(\begin{matrix} \dfrac{z^4}{64} \\ \dfrac{\nu+1}{2}, \dfrac{\nu}{2}+1, \nu+1 \end{matrix} \right) \right].$

54. $\operatorname{ber}_\nu(z)\operatorname{bei}_\nu(z) = \dfrac{z^{2\nu}}{2^{2\nu+1}\Gamma^2(\nu+1)}$

$\times \left[\sin\dfrac{3\nu\pi}{2} {}_2F_5\left(\begin{matrix} \dfrac{2\nu+1}{4}, \dfrac{2\nu+3}{4}; -\dfrac{z^4}{16} \\ \dfrac{\nu+1}{2}, \dfrac{\nu}{2}+1, \nu+\dfrac{1}{2}, \nu+1, \dfrac{1}{2} \end{matrix} \right) \right.$

$\left. + \dfrac{z^2}{2(\nu+1)} \cos\dfrac{3\nu\pi}{2} {}_2F_5\left(\begin{matrix} \dfrac{2\nu+3}{4}, \dfrac{2\nu+5}{4}; -\dfrac{z^4}{16} \\ \dfrac{\nu}{2}+1, \dfrac{\nu+3}{2}, \nu+1, \nu+\dfrac{3}{2}, \dfrac{3}{2} \end{matrix} \right) \right].$

55. $\mathbf{E}_\nu(\sqrt{z}) = \dfrac{1}{\nu\pi}(1-\cos\nu\pi)\,{}_1F_2\left(\begin{matrix} 1; -\dfrac{z}{4} \\ 1-\dfrac{\nu}{2}, 1+\dfrac{\nu}{2} \end{matrix} \right)$

$- \dfrac{\sqrt{z}}{\pi(1-\nu^2)}(1+\cos\nu\pi)\,{}_1F_2\left(\begin{matrix} 1; -\dfrac{z}{4} \\ \dfrac{3-\nu}{2}, \dfrac{3+\nu}{2} \end{matrix} \right).$

56. $\mathbf{J}_\nu(\sqrt{z}) = \dfrac{\sin\nu\pi}{\nu\pi}\,{}_1F_2\left(\begin{matrix} 1; -\dfrac{z}{4} \\ 1-\dfrac{\nu}{2}, 1+\dfrac{\nu}{2} \end{matrix} \right) + \dfrac{\sin\nu\pi}{\pi(1-\nu^2)}\sqrt{z}\,{}_1F_2\left(\begin{matrix} 1; -\dfrac{z}{4} \\ \dfrac{3-\nu}{2}, \dfrac{3+\nu}{2} \end{matrix} \right).$

57. $P_{2n}(\sqrt{z}) = (-1)^n \dfrac{\left(\frac{1}{2}\right)_n}{n!}\,{}_2F_1\left(\begin{matrix} -n, n+\dfrac{1}{2} \\ \dfrac{1}{2}; z \end{matrix} \right).$

58. $P_{2n+1}(\sqrt{z}) = (-1)^n \dfrac{\left(\frac{3}{2}\right)_n}{n!}\sqrt{z}\,{}_2F_1\left(\begin{matrix} -n, n+\dfrac{3}{2} \\ \dfrac{3}{2}; z \end{matrix} \right).$

59. $P_n\left(\dfrac{1}{\sqrt{z}}\right) = \dfrac{\left(\frac{1}{2}\right)_n}{n!}\left(\dfrac{2}{\sqrt{z}}\right)^n {}_2F_1\left(\begin{matrix} -\dfrac{n}{2}, \dfrac{1-n}{2} \\ \dfrac{1}{2}-n; z \end{matrix} \right).$

60. $P_n(1+z) = {}_2F_1\left(\begin{matrix} -n, n+1 \\ 1; -\dfrac{z}{2} \end{matrix} \right).$

61. $P_n\left(1+\dfrac{1}{z}\right) = \dfrac{\left(\frac{1}{2}\right)_n}{n!}\left(\dfrac{2}{z}\right)^n {}_2F_1\left(\begin{matrix} -n, -n \\ -2n; -2z \end{matrix} \right).$

8.3. Representation in Terms of Hypergeometric Functions

62. $(1-z)^n P_n\left(\dfrac{1+z}{1-z}\right) = {}_2F_1\left(\begin{matrix}-n,\,-n\\1;\,z\end{matrix}\right).$

63. $P_{2n}(\sqrt{1+z}) = {}_2F_1\left(\begin{matrix}-n,\,n+\tfrac{1}{2}\\1;\,-z\end{matrix}\right).$

64. $\dfrac{1}{\sqrt{1+z}}\,P_{2n+1}(\sqrt{1+z}) = {}_2F_1\left(\begin{matrix}-n,\,n+\tfrac{3}{2}\\1;\,-z\end{matrix}\right).$

65. $P_n\left(\dfrac{1+z}{2\sqrt{z}}\right) = \dfrac{(2n)!}{(n!)^2}(4\sqrt{z})^{-n}\,{}_2F_1\left(\begin{matrix}-n,\,\tfrac{1}{2}\\\tfrac{1}{2}-n;\,z\end{matrix}\right).$

66. $(1+z)^{n/2}P_n\left(\dfrac{2+z}{2\sqrt{1+z}}\right) = {}_2F_1\left(\begin{matrix}-n,\,\tfrac{1}{2}\\1;\,-z\end{matrix}\right).$

67. $P_{2n}\left(\sqrt{1+\dfrac{1}{z}}\right) = \dfrac{\left(\tfrac{1}{2}\right)_{2n}}{(2n)!}\left(\dfrac{4}{z}\right)^n\,{}_2F_1\left(\begin{matrix}-n,\,-n\\\tfrac{1}{2}-2n;\,-z\end{matrix}\right).$

68. $\dfrac{1}{\sqrt{1+z}}\,P_{2n+1}\left(\sqrt{1+\dfrac{1}{z}}\right) = \dfrac{\left(\tfrac{1}{2}\right)_{2n+1}}{(2n+1)!}\left(\dfrac{4}{z}\right)^{n+1/2}{}_2F_1\left(\begin{matrix}-n,\,-n\\-\tfrac{1}{2}-2n;\,-z\end{matrix}\right).$

69. $(z-1)^n P_{2n}\left(\sqrt{\dfrac{z}{z-1}}\right) = \dfrac{\left(\tfrac{1}{2}\right)_n}{n!}\,{}_2F_1\left(\begin{matrix}-n,\,-n\\\tfrac{1}{2};\,z\end{matrix}\right).$

70. $(z-1)^{n+1/2}P_{2n+1}\left(\sqrt{\dfrac{z}{z-1}}\right) = \dfrac{\left(\tfrac{3}{2}\right)_n}{n!}\sqrt{z}\,{}_2F_1\left(\begin{matrix}-n,\,-n\\\tfrac{3}{2};\,z\end{matrix}\right).$

71. $(1+z)^{n/2}P_n\left(\dfrac{1}{\sqrt{1+z}}\right) = {}_2F_1\left(\begin{matrix}-\tfrac{n}{2},\,\tfrac{1-n}{2}\\1;\,-z\end{matrix}\right).$

72. $\left[P_n(\sqrt{1+z})\right]^2 = {}_3F_2\left(\begin{matrix}-n,\,n+1,\,\tfrac{1}{2}\\1,\,1;\,-z\end{matrix}\right).$

73. $\left[P_n\left(\sqrt{1+\dfrac{1}{z}}\right)\right]^2 = \left[\dfrac{\left(\tfrac{1}{2}\right)_n}{n!}\right]^2\left(\dfrac{4}{z}\right)^n\,{}_3F_2\left(\begin{matrix}-n,\,-n,\,-n\\\tfrac{1}{2}-n,\,-2n;\,-z\end{matrix}\right).$

74. $P_{2n}\left(\sqrt{\dfrac{1+\sqrt{1-z}}{2}}\right)P_{2n}\left(\sqrt{\dfrac{1-\sqrt{1-z}}{2}}\right)$

$\qquad\qquad = (-1)^n\dfrac{\left(\tfrac{1}{2}\right)_n}{n!}\,{}_4F_3\left(\begin{matrix}-n,\,n+\tfrac{1}{2},\,\tfrac{1}{4},\,\tfrac{3}{4}\\\tfrac{1}{2},\,\tfrac{1}{2},\,1;\,z\end{matrix}\right).$

75. $P_{2n+1}\left(\sqrt{\dfrac{1+\sqrt{1-z}}{2}}\right) P_{2n+1}\left(\sqrt{\dfrac{1-\sqrt{1-z}}{2}}\right)$
$$= (-1)^n \dfrac{\left(\tfrac{3}{2}\right)_n \sqrt{z}}{2\,(n!)}\; {}_4F_3\left(\begin{array}{c} -n,\, n+\tfrac{3}{2},\, \tfrac{3}{4},\, \tfrac{5}{4} \\ 1,\, \tfrac{3}{2},\, \tfrac{3}{2};\, z \end{array}\right).$$

76. $P_{2n+1}\left(\sqrt{\dfrac{2}{1+\sqrt{1-z}}}\right) P_{2n+1}\left(\sqrt{\dfrac{2}{1-\sqrt{1-z}}}\right)$
$$= (-1)^n \dfrac{\left(n+\tfrac{3}{2}\right)_n}{n!}\left(\dfrac{4}{z}\right)^{n+1/2} {}_4F_3\left(\begin{array}{c} -n,\, -\tfrac{1}{2}-n,\, -\tfrac{1}{4}-n,\, \tfrac{1}{4}-n \\ -\tfrac{1}{2}-2n,\, -\tfrac{1}{2}-2n,\, 1;\, z \end{array}\right).$$

77. $P_{2n}\left(\sqrt{\dfrac{2}{1+\sqrt{1-z}}}\right) P_{2n}\left(\dfrac{2}{1-\sqrt{1-z}}\right)$
$$= \dfrac{\left(n+\tfrac{1}{2}\right)_n}{n!}\left(\dfrac{4}{z}\right)^n {}_4F_3\left(\begin{array}{c} -n,\, \tfrac{1}{4}-n,\, \tfrac{1}{2}-n,\, \tfrac{3}{4}-n \\ \tfrac{1}{2}-2n,\, \tfrac{1}{2}-2n,\, 1;\, z \end{array}\right).$$

78. $P_n\left(\dfrac{3+\sqrt{1-z}}{2^{3/2}\sqrt{1+\sqrt{1-z}}}\right) P_n\left(\dfrac{3-\sqrt{1-z}}{2^{3/2}\sqrt{1-\sqrt{1-z}}}\right)$
$$= \dfrac{\left(\tfrac{1}{2}\right)_n}{n!}\left(\dfrac{4}{z}\right)^{n/2} {}_4F_3\left(\begin{array}{c} -n,\, \tfrac{1-2n}{4},\, \tfrac{3-2n}{4},\, \tfrac{1}{2};\, z \\ \tfrac{1}{2}-n,\, \tfrac{1}{2}-n,\, 1 \end{array}\right).$$

79. $T_{2n}(\sqrt{z}) = (-1)^n\, {}_2F_1\left(\begin{array}{c} -n,\, n \\ \tfrac{1}{2};\, z \end{array}\right).$

80. $T_{2n+1}(\sqrt{z}) = (-1)^n (2n+1)\sqrt{z}\, {}_2F_1\left(\begin{array}{c} -n,\, n+1 \\ \tfrac{3}{2};\, z \end{array}\right).$

81. $T_n\left(\dfrac{1}{\sqrt{z}}\right) = 2^{n-1} z^{-n/2}\, {}_2F_1\left(\begin{array}{c} -\tfrac{n}{2},\, \tfrac{1-n}{2} \\ 1-n;\, z \end{array}\right).$

82. $T_n(1+z) = {}_2F_1\left(\begin{array}{c} -n,\, n \\ \tfrac{1}{2};\, -\tfrac{z}{2} \end{array}\right).$

83. $T_n\left(1+\dfrac{1}{z}\right) = 2^{n-1} z^{-n}\, {}_2F_1\left(\begin{array}{c} -n,\, \tfrac{1}{2}-n \\ 1-2n;\, -2z \end{array}\right).$

84. $(1-z)^n T_n\left(\dfrac{1+z}{1-z}\right) = {}_2F_1\left(\begin{array}{c} -n,\, \tfrac{1}{2}-n \\ \tfrac{1}{2};\, z \end{array}\right).$

8.3. Representation in Terms of Hypergeometric Functions

85. $T_{2n}(\sqrt{1+z}) = {}_2F_1\left(\begin{matrix} -n, n \\ \frac{1}{2}; -z \end{matrix}\right).$

86. $\dfrac{1}{\sqrt{1+z}} T_{2n+1}(\sqrt{1+z}) = {}_2F_1\left(\begin{matrix} -n, n+1 \\ \frac{1}{2}; -z \end{matrix}\right).$

87. $T_{2n}\left(\sqrt{1+\dfrac{1}{z}}\right) = 2^{2n-1} z^{-n} \, {}_2F_1\left(\begin{matrix} -n, \frac{1}{2}-n \\ 1-2n; -z \end{matrix}\right).$

88. $\dfrac{1}{\sqrt{1+z}} T_{2n+1}\left(\sqrt{1+\dfrac{1}{z}}\right) = 2^{2n} z^{-n-1/2} \, {}_2F_1\left(\begin{matrix} -n, \frac{1}{2}-n \\ -2n; -z \end{matrix}\right).$

89. $(z-1)^n T_{2n}\left(\sqrt{\dfrac{z}{z-1}}\right) = {}_2F_1\left(\begin{matrix} -n, \frac{1}{2}-n \\ \frac{1}{2}; z \end{matrix}\right).$

90. $(z-1)^{n+1/2} T_{2n+1}\left(\sqrt{\dfrac{z}{z-1}}\right) = (2n+1)\sqrt{z} \, {}_2F_1\left(\begin{matrix} -n, \frac{1}{2}-n \\ \frac{3}{2}; z \end{matrix}\right).$

91. $(1+z)^{n/2} T_n\left(\dfrac{1}{\sqrt{1+z}}\right) = {}_2F_1\left(\begin{matrix} -\frac{n}{2}, \frac{1-n}{2} \\ \frac{1}{2}; -z \end{matrix}\right).$

92. $\left[T_n(\sqrt{1+z})\right]^2 = \dfrac{1}{2}\left[1 + {}_2F_1\left(\begin{matrix} -n, n \\ \frac{1}{2}; -z \end{matrix}\right)\right].$

93. $T_n\left(\dfrac{3-\sqrt{1-z}}{1+\sqrt{1-z}}\right) T_n\left(\dfrac{3+\sqrt{1-z}}{1-\sqrt{1-z}}\right)$
$= 2^{4n-1} z^{-n} \, {}_4F_3\left(\begin{matrix} -n, \frac{1}{4}-n, \frac{1}{2}-n, \frac{3}{4}-n \\ \frac{1}{2}-2n, 1-2n, \frac{1}{2}; z \end{matrix}\right) \quad [n \geq 1].$

94. $U_{2n}(\sqrt{z}) = (-1)^n \, {}_2F_1\left(\begin{matrix} -n, n+1 \\ \frac{1}{2}; z \end{matrix}\right).$

95. $U_{2n+1}(\sqrt{z}) = (-1)^n 2(n+1)\sqrt{z} \, {}_2F_1\left(\begin{matrix} -n, n+2 \\ \frac{3}{2}; z \end{matrix}\right).$

96. $U_n\left(\dfrac{1}{\sqrt{z}}\right) = 2^n z^{-n/2} \, {}_2F_1\left(\begin{matrix} -\frac{n}{2}, \frac{1-n}{2} \\ -n; z \end{matrix}\right).$

97. $U_n(1+z) = (n+1) \, {}_2F_1\left(\begin{matrix} n, n+2 \\ \frac{3}{2}; -\frac{z}{2} \end{matrix}\right).$

98. $U_n\left(1+\dfrac{1}{z}\right) = \left(\dfrac{2}{z}\right)^n {}_2F_1\left(\begin{array}{c}-n,\ -n-\frac{1}{2}\\-2n-1;\ -2z\end{array}\right).$

99. $(1-z)^n U_n\left(\dfrac{1+z}{1-z}\right) = (n+1)\,{}_2F_1\left(\begin{array}{c}-n,\ -n-\frac{1}{2}\\ \frac{3}{2};\ z\end{array}\right).$

100. $U_{2n}(\sqrt{1+z}) = (2n+1)\,{}_2F_1\left(\begin{array}{c}-n,\ n+1\\ \frac{3}{2};\ -z\end{array}\right).$

101. $\dfrac{1}{\sqrt{1+z}} U_{2n+1}(\sqrt{1+z}) = 2(n+1)\,{}_2F_1\left(\begin{array}{c}-n,\ n+2\\ \frac{3}{2};\ -z\end{array}\right).$

102. $U_{2n}\left(\sqrt{1+\dfrac{1}{z}}\right) = 2^{2n} z^{-n}\,{}_2F_1\left(\begin{array}{c}-n,\ -n-\frac{1}{2}\\ -2n;\ -z\end{array}\right).$

103. $\dfrac{1}{\sqrt{1+z}} U_{2n+1}\left(\sqrt{1+\dfrac{1}{z}}\right) = 2^{2n+1} z^{-n-1/2}\,{}_2F_1\left(\begin{array}{c}-n,\ -n-\frac{1}{2}\\ -2n-1;\ -z\end{array}\right).$

104. $(z-1)^n U_{2n}\left(\sqrt{\dfrac{z}{z-1}}\right) = {}_2F_1\left(\begin{array}{c}-n,\ -n-\frac{1}{2}\\ \frac{1}{2};\ z\end{array}\right).$

105. $(z-1)^{n+1/2} U_{2n+1}\left(\sqrt{\dfrac{z}{z-1}}\right) = 2(n+1)\sqrt{z}\,{}_2F_1\left(\begin{array}{c}-n,\ -n-\frac{1}{2}\\ \frac{3}{2};\ z\end{array}\right).$

106. $(1+z)^{n/2} U_n\left(\dfrac{1}{\sqrt{1+z}}\right) = (n+1)\,{}_2F_1\left(\begin{array}{c}-\frac{n}{2},\ \frac{1-n}{2}\\ \frac{3}{2};\ -z\end{array}\right).$

107. $\left[U_n(\sqrt{1+z})\right]^2 = (n+1)^2\,{}_3F_2\left(\begin{array}{c}-n,\ n+2,\ 1\\ \frac{3}{2},\ 2;\ -z\end{array}\right).$

108. $U_{2n}\left(\sqrt{\dfrac{2}{1+\sqrt{1-z}}}\right) U_{2n}\left(\sqrt{\dfrac{2}{1-\sqrt{1-z}}}\right)$
$= (2n+1)\left(\dfrac{z}{16}\right)^{-n}{}_4F_3\left(\begin{array}{c}-n,\ \frac{1}{4}-n,\ \frac{1}{2}-n,\ \frac{3}{4}-n\\ -2n,\ \frac{1}{2}-2n,\ \frac{3}{2};\ z\end{array}\right).$

109. $U_{2n+1}\left(\sqrt{\dfrac{2}{1+\sqrt{1-z}}}\right) U_{2n+1}\left(\sqrt{\dfrac{2}{1-\sqrt{1-z}}}\right)$
$= 2(n+1)\left(\dfrac{z}{16}\right)^{-n-1/2}{}_4F_3\left(\begin{array}{c}-n,\ -\frac{1}{2}-n,\ -\frac{1}{4}-n,\ \frac{1}{4}-n\\ -\frac{1}{2}-2n,\ -1-2n,\ \frac{3}{2};\ z\end{array}\right).$

8.3. Representation in Terms of Hypergeometric Functions   [8.3.2]

110. $H_n(\sqrt{z}) = 2^n \sqrt{z}\, \Psi\!\left(\begin{matrix}\frac{1-n}{2};\ z\\ \frac{3}{2}\end{matrix}\right).$

111. $H_{2n}(\sqrt{z}) = (-1)^n \dfrac{(2n)!}{n!}\, {}_1F_1\!\left(\begin{matrix}-n;\ z\\ \frac{1}{2}\end{matrix}\right).$

112. $H_{2n+1}(\sqrt{z}) = (-1)^n \dfrac{(2n+1)!}{n!}\, 2\sqrt{z}\, {}_1F_1\!\left(\begin{matrix}-n;\ z\\ \frac{3}{2}\end{matrix}\right).$

113. $H_n\!\left(\dfrac{1}{\sqrt{z}}\right) = 2^n z^{-n/2}\, {}_2F_0\!\left(\begin{matrix}-\frac{n}{2},\ \frac{1-n}{2}\\ -z\end{matrix}\right).$

114. $e^{-z} H_{2n}(\sqrt{z}) = (-1)^n 2^{2n} \left(\dfrac{1}{2}\right)_n {}_1F_1\!\left(\begin{matrix}n+\frac{1}{2}\\ \frac{1}{2};\ -z\end{matrix}\right).$

115. $e^{-z} H_{2n+1}(\sqrt{z}) = (-1)^n 2^{2n+1} \left(\dfrac{3}{2}\right)_n \sqrt{z}\, {}_1F_1\!\left(\begin{matrix}n+\frac{3}{2}\\ \frac{3}{2};\ -z\end{matrix}\right).$

116. $H_{2n}(\sqrt[4]{z})\, H_{2n}(i\sqrt[4]{z}) = \left[\dfrac{(2n)!}{n!}\right]^2 {}_2F_3\!\left(\begin{matrix}-n,\ n+\frac{1}{2};\ \frac{z}{4}\\ \frac{1}{4},\ \frac{1}{2},\ \frac{3}{4}\end{matrix}\right).$

117. $H_{2n+1}(\sqrt[4]{z})\, H_{2n+1}(i\sqrt[4]{z}) = 4i\left[\dfrac{(2n+1)!}{n!}\right]^2 \sqrt{z}\, {}_2F_3\!\left(\begin{matrix}-n,\ n+\frac{3}{2}\\ \frac{3}{4},\ \frac{5}{4},\ \frac{3}{2};\ \frac{z}{4}\end{matrix}\right).$

118. $H_{2n}(z^{-1/4})\, H_{2n}(iz^{-1/4}) = \left(-\dfrac{16}{z}\right)^n {}_4F_1\!\left(\begin{matrix}-n,\ \frac{1}{4}-n,\ \frac{1}{2}-n,\ \frac{3}{4}-n\\ \frac{1}{2}-2n;\ 4z\end{matrix}\right).$

119. $H_{2n+1}(z^{-1/4})\, H_{2n+1}(iz^{-1/4})$
$= (-1)^n 2^{4n+2} i z^{-n-1/2}\, {}_4F_1\!\left(\begin{matrix}-n,\ -\frac{1}{2}-n,\ -\frac{1}{4}-n,\ \frac{1}{4}-n\\ -\frac{1}{2}-2n;\ 4z\end{matrix}\right).$

120. $L_n^\lambda(z) = \dfrac{(\lambda+1)_n}{n!}\, {}_1F_1\!\left(\begin{matrix}-n;\ z\\ \lambda+1\end{matrix}\right).$

121. $= \dfrac{(-1)^n}{n!}\, \Psi\!\left(\begin{matrix}-n;\ z\\ \lambda+1\end{matrix}\right).$

122. $L_n^\lambda\!\left(\dfrac{1}{z}\right) = \dfrac{(-z)^{-n}}{n!}\, {}_2F_0\!\left(\begin{matrix}-n,\ -\lambda-n\\ -z\end{matrix}\right).$

123. $e^{-z} L_n^\lambda(z) = \dfrac{(\lambda+1)_n}{n!}\, {}_1F_1\!\left(\begin{matrix}\lambda+n+1\\ \lambda+1;\ -z\end{matrix}\right).$

124. $L_n^\lambda(\sqrt{z}) L_n^\lambda(-\sqrt{z}) = \left[\dfrac{(\lambda+1)_n}{n!}\right]^2 {}_2F_3\left(\begin{array}{c} -n, \lambda+n+1; \frac{z}{4} \\ \frac{\lambda+1}{2}, \frac{\lambda}{2}+1, \lambda+1 \end{array}\right).$

125. $L_n^\lambda\left(\dfrac{1}{\sqrt{z}}\right) L_n^\lambda\left(-\dfrac{1}{\sqrt{z}}\right) = \dfrac{(-z)^{-n}}{(n!)^2} {}_4F_1\left(\begin{array}{c} -n, -\frac{\lambda}{2}-n, \frac{1-\lambda}{2}-n, -\lambda-n \\ -\lambda-2n;\ 4z \end{array}\right).$

126. $C_{2n}^\lambda(\sqrt{z}) = (-1)^n \dfrac{(\lambda)_n}{n!} {}_2F_1\left(\begin{array}{c} -n, \lambda+n \\ \frac{1}{2};\ z \end{array}\right).$

127. $C_{2n+1}^\lambda(\sqrt{z}) = (-1)^n \dfrac{(\lambda)_{n+1}}{n!} 2\sqrt{z}\, {}_2F_1\left(\begin{array}{c} -n, \lambda+n+1 \\ -\frac{3}{2};\ z \end{array}\right).$

128. $C_n^\lambda\left(\dfrac{1}{\sqrt{z}}\right) = \dfrac{(\lambda)_n}{n!} 2^n z^{-n/2} {}_2F_1\left(\begin{array}{c} -\frac{n}{2}, \frac{1-n}{2} \\ 1-\lambda-n;\ z \end{array}\right).$

129. $C_n^\lambda(1+z) = \dfrac{(2\lambda)_n}{n!} {}_2F_1\left(\begin{array}{c} -n, 2\lambda+n \\ \lambda+\frac{1}{2};\ -\frac{z}{2} \end{array}\right).$

130. $C_n^\lambda\left(1+\dfrac{1}{z}\right) = \dfrac{(\lambda)_n}{n!}\left(\dfrac{2}{z}\right)^n {}_2F_1\left(\begin{array}{c} -n, \frac{1}{2}-\lambda-n \\ 1-2\lambda-2n;\ -2z \end{array}\right).$

131. $(1-z)^n C_n^\lambda\left(\dfrac{1+z}{1-z}\right) = \dfrac{(2\lambda)_n}{n!} {}_2F_1\left(\begin{array}{c} -n, \frac{1}{2}-\lambda-n \\ \lambda+\frac{1}{2};\ z \end{array}\right).$

132. $C_{2n}^\lambda(\sqrt{1+z}) = \dfrac{(2\lambda)_{2n}}{(2n)!} {}_2F_1\left(\begin{array}{c} -n, \lambda+n \\ \lambda+\frac{1}{2};\ -z \end{array}\right).$

133. $\dfrac{1}{\sqrt{1+z}} C_{2n+1}^\lambda(\sqrt{1+z}) = \dfrac{(2\lambda)_{2n+1}}{(2n+1)!} {}_2F_1\left(\begin{array}{c} -n, \lambda+n+1 \\ \lambda+\frac{1}{2};\ -z \end{array}\right).$

134. $C_n^\lambda\left(\dfrac{z+1}{2\sqrt{z}}\right) = \dfrac{(\lambda)_n}{n!} z^{-n/2} {}_2F_1\left(\begin{array}{c} -n, \lambda \\ 1-\lambda-n;\ z \end{array}\right).$

135. $(1+z)^{n/2} C_n^\lambda\left(\dfrac{1}{\sqrt{1+z}}\right) = \dfrac{(2\lambda)_n}{n!} {}_2F_1\left(\begin{array}{c} -\frac{n}{2}, \frac{1-n}{2} \\ \lambda+\frac{1}{2};\ -z \end{array}\right).$

136. $C_{2n}^\lambda\left(\sqrt{1+\dfrac{1}{z}}\right) = \dfrac{(\lambda)_{2n}}{(2n)!} 2^{2n} z^{-n} {}_2F_1\left(\begin{array}{c} -n, \frac{1}{2}-\lambda-n \\ 1-\lambda-2n;\ -z \end{array}\right).$

8.3. Representation in Terms of Hypergeometric Functions

137. $\dfrac{1}{\sqrt{1+z}} C_{2n+1}^\lambda\left(\sqrt{1+\dfrac{1}{z}}\right)$

$$= \dfrac{(\lambda)_{2n+1}}{(2n+1)!} 2^{2n+1} z^{-n-1/2} {}_2F_1\left(\begin{array}{c} -n,\ \frac{1}{2}-\lambda-n \\ -\lambda-2n;\ -z \end{array}\right).$$

138. $\left[C_n^\lambda(\sqrt{1+z})\right]^2 = \left[\dfrac{(2\lambda)_n}{n!}\right]^2 {}_3F_2\left(\begin{array}{c} -n,\ \lambda,\ 2\lambda+n;\ -z \\ \lambda+\frac{1}{2},\ 2\lambda \end{array}\right).$

139. $P_n^{(\rho,\sigma)}(1+z) = \dfrac{(\rho+1)_n}{n!} {}_2F_1\left(\begin{array}{c} -n,\ \rho+\sigma+n+1 \\ \rho+1;\ -\frac{z}{2} \end{array}\right).$

140. $P_n^{(\rho,\sigma)}\left(1+\dfrac{1}{z}\right) = \dfrac{(\rho+\sigma+n+1)_n}{n!}(2z)^{-n}\ {}_2F_1\left(\begin{array}{c} -n,\ -\rho-n;\ -2z \\ -\rho-\sigma-2n \end{array}\right).$

141. $(1-z)^n P_n^{(\rho,\sigma)}\left(\dfrac{1+z}{1-z}\right) = \dfrac{(\rho+1)_n}{n!} {}_2F_1\left(\begin{array}{c} -n,\ -\sigma-n \\ \rho+1;\ z \end{array}\right).$

142. $P_n^{(\rho,\sigma)}(\sqrt{1+z})P_n^{(\rho,\sigma)}(-\sqrt{1+z})$

$$= (-1)^n \dfrac{(\rho+1)_n(\sigma+1)_n}{(n!)^2} {}_4F_3\left(\begin{array}{c} -n,\ \frac{\rho+\sigma+1}{2},\ \frac{\rho+\sigma}{2}+1,\ \rho+\sigma+n+1;\ -z \\ \rho+1,\ \sigma+1,\ \rho+\sigma+1 \end{array}\right).$$

143. $\left\{\begin{array}{l} \mathbf{K}(\sqrt{z}) \\ \mathbf{E}(\sqrt{z}) \end{array}\right\} = \dfrac{\pi}{2} {}_2F_1\left(\begin{array}{c} \pm\frac{1}{2},\ \frac{1}{2} \\ 1;\ z \end{array}\right).$

144. $\dfrac{1}{1-z}\mathbf{E}(\sqrt{z}) = \dfrac{\pi}{2} {}_2F_1\left(\begin{array}{c} \frac{1}{2},\ \frac{3}{2} \\ 1;\ z \end{array}\right).$

145. $\mathbf{D}(\sqrt{z}) = \dfrac{\pi}{4} {}_2F_1\left(\begin{array}{c} \frac{1}{2},\ \frac{3}{2} \\ 2;\ z \end{array}\right).$

References

Monographs

1. Prudnikov A.P., Brychkov Yu.A., Marichev O.I., *Integrals and Series. V. 1. Elementary Functions*, New York–London: Gordon & Breach, 1986.
2. —, —, —, *Integrals and Series. V. 2. Special Functions*, New York–London: Gordon & Breach, 1986.
3. —, —, —, *Integrals and Series. V. 3. More Special Functions*, New York–London: Gordon & Breach, 1990.
4. —, —, —, *Integrals and Series. V. 4. Laplace Transforms*, New York–London: Gordon & Breach, 1992.
5. —, —, —, *Integrals and Series. V. 5. Inverse Laplace Transforms*, New York–London: Gordon & Breach, 1992.
6. Bateman H., Erdélyi A., *Higher Transcendental Functions, V. 1*, New York–Toronto: London, McGraw-Hill, 1953.
7. —, —, *Higher Transcendental Functions, V. 2*, New York–Toronto: London, McGrow-Hill, 1953.
8. Abramowitz M., Stegun I.A. (ed.), *Handbook of Mathematical Functions with Formulas, Graphs, and Mathematical Tables*, New York: Dover, 1992.
9. Khadzhi P.I., *The Error Function*, Institute of Applied Physics, Acad. Sci. Mold. SSR, Kishinev, 1971.
10. Luke Y.L., *The Special Functions and their Approximations, V. 1*, New York: Academic Press, 1969.
11. Magnus W., Oberhettinger F., Soni R.P., *Formulas and Theorems for the Special Functions of Mathematical Physics*, New York–Berlin: Springer-Verlag, 1966.
12. Rainville E.D., *Special Functions*, New York: Macmillan, 1963.
13. Schwatt I.J., *An Introduction to the Operations with Series*, New York: Chelsea, 1924.
14. Srivastava H.M., Manocha H.L., *A Treatise on Generating Functions*, New York: Halsted Press, 1984.

Journal papers

15. Adamchik V., The evaluation of integrals of Bessel functions via G-function identities, *J. Comput. Appl. Math.*, 64 (1995), N 3, 283–290.
16. —, A class of logarithmic integrals, In: *Proceedings of ISSAC*, 1997, New York: Academic Press, 2001, 1–8.
17. Al-Saqabi B.N., Kalla S.L., Srivastava H.M., A certain family of of infinite series associated with digamma functions, *J. Math. Anal. Appl.*, 159 (1991), N 1, 361–372.

18. Bailey W. N., Products of generalized hypergeometric series, *Proc. London Math. Soc.*, (2) 28(1928), 242–254.

19. Berndt B., Analytic Eisenstein series, theta-functions, and series relations in the spirit of Ramanujan, *J. Reine Angew. Math.*, 303/304 (1978), 332–335.

20. Brown B. M., Eastham M. S. P., A note on the Dixon formula for a finite hypergeometric series, *J. Comp. Appl. Math.*, 194 (2006), N 1, 173–175.

21. Brychkov Yu. A., Evaluation of some classes of definite and indefinite integrals, *Integral Transforms Spec. Func.*, 13 (2002), N 2, 163–167.

22. —, On some new series of special functions, *Appl. Math. Comput.*, 187 (2007), N 1, 101–104.

23. —, On some formulas for Weber $\mathbf{E}_\nu(z)$ and Anger $\mathbf{J}_\nu(z)$ functions, *Integral Transforms Spec. Func.*, 18 (2007), N 1–2, 15–26.

24. —, Geddes K. O., Differentiation of hypergeometric functions with respect to parameters. In: *Proc. Int. Conf. on Abstract and Applied Analysis, Hanoi, Vietnam, 13–17 Aug. 2002*, Singapore: World Scientific P. C., 2004, 15–28.

25. Carlitz L., The product of two ultraspherical polynomials, *Proc. Glasgow Math. Assoc.*, 5 (1961), 76–79.

26. Carlson B. C., Some extensions of Lardner's relations between $_0F_3$ and Bessel functions, *SIAM J. Math. Anal.*, 1 (1970), N 2, 232–242.

27. Chen Kung-Yu, Liu Shuoh-Jung, Srivastava H. M., Some double-series identities and associated generating function relationships, *Appl. Math. Lett.*, 19 (2006), N 9, 887–893.

28. Choi Junesang, Srivastava H. M., Explicit evaluation of Euler and related sums, *The Ramanujan J.*, 10 (2005), N 1, 51–70.

29. De Doelder P. J., On some series containing $\psi(x) - \psi(y)$ and $(\psi(x) - \psi(y))^2$ for certain values of x and y, *J. Comput. Appl. Math.*, 37 (1991), N 1–3, 125–141.

30. Exton H., On the reducibility of the Kampé de Fériet function, *J. Comp. Appl. Math.*, 83 (1997), N 1, 119–121.

31. —, Some new summation formula for the generalized hypergeometric function of higher order, *J. Comp. Appl. Math.*, 79 (1997), N 2, 183–187.

32. —, New hypergeometric transformations, *J. Comp. Appl. Math.*, 92 (1998), N 2, 135–137.

33. Feldheim E., Alcuni risultati sulle funzioni di Whittaker e del cilindro parabolico, *Atti Acc. Sci. Torino*, 76 (1941), 541–555.

34. —, Relations entre les polynomes de Jacobi, Laguerre et Hermite, *Acta Math.*, 75 (1943), 117–138.

35. Fox C., The expression of hypergeometric series in terms of similar series, *Proc. London Math. Soc.*, 26 (1927), 201–210.

36. Froehlich J., Parameter derivatives of the Jacobi polynomials and the Gaussian hypergeometric function, *Integral Transforms Spec. Func.*, 2 (1994), N 4, 252–266.

37. Furtlehner C., Ouvry S., Integrals involving four Macdonald functions and their relation to $7\zeta(3)/2$, http://arXiv:math-ph/0306004 v2 1 Apr. 2004.

38. Gessel I., Stanton D., Strange evaluation of hypergeometric series, *SIAM J. Math. Anal.*, 13 (1982), N 2, 295–307.

39. Gosper R. W., Ismail M. E. H., Zhang R., On some strange summation formulas, *Illinois J. Math.*, 37 (1993), N 2, 240–277.

40. Grosjean C. C., Some new integrals arising from mathematical physics. III, *Simon Stevin*, 43 (1969), 3–46.

41. —, Some new integrals arising from mathematical physics. IV, *Simon Stevin*, 45 (1971/72), N 3–4, 321–383.
42. Guillera J., Some binomial series obtained by the WZ-method, *Adv. Appl. Math.*, 29 (2002), 599–603.
43. —, Series closely related to Ramanujan formulas for pi, 2002, 1–8, http://personal.auna.com/jguillera.
44. —, About a new kind of Ramanujan-type series, *Exp. Math.*, 12 (2003), N 4, 507–510.
45. —, Generators of some Ramanujan formulas, *Ramanujan J.*, 11 (2006), N 1, 41–48.
46. —, Hypergeometric identities for 10 extended Ramanujan-type series, *The Ramanujan J.*, 15 (2008), N 2, 219–234.
47. Gustavsson J., Some sums of Legendre and Jacobi polynomials, *Math. Bohem.*, 126 (2001), N 1, 141–149.
48. Joyce G. S., Zucker I. J., Special values of the hypergeometric functions, *Math. Proc. Cambridge Philos. Soc.*, 109 (1991), N 2, 257–261.
49. —, —, Special values of the hypergeometric functions. II, *Math. Proc. Cambridge Philos. Soc.*, 131 (2001), N 2, 309–319.
50. Kanemitsu Sh., Kumagai H., Yoshimoto M., Sums involving Hurwitz Zeta function, *Ramanujan J.*, 5 (2001), N 1, 5–19.
51. Karlsson P. W., On two hypergeometric summation formulas conjectured by Gosper, *Simon Stevin*, 60 (1966), N 4, 329–337.
52. —, Clausen hypergeometric series with variable 1/4, *J. Math. Anal. Appl.*, 196 (1995), N 1, 172–180.
53. Kelber Ju., Schnittler Ch., Übensee H., Integralformeln für Airyfunktionen, *Wiss. Z. TH Ilmenau*, 31 (1985), N 4, 147–156.
54. Koepf W., Identities for families of orthogonal polynomials and special functions, *Integral Transforms Spec. Func.*, 5 (1997), N 1–2, 69–102.
55. Krattenthaler C., Srinivasa Rao K., Automatic generation of hypergeometric identities by the beta integral method, *J. Comp. Appl. Math.*, 160 (2003), N 1–2, 159–173.
56. —, Rivoal T., How can we escape Thomae's relations, *J. Math. Soc. Japan*, 58 (2006), 183–210.
57. Laurenzi B. J., Moment integrals of power of Airy functions, *Z. Angew. Math. Phys.*, 44 (1993), N 5, 891–908.
58. Miller A. R., Summations for certain series containing the digamma function, *J. Phys. A: Math. Gen.*, 39 (2006), 3011–3020.
59. Miller J., Adamchik V. S., Derivatives of the Hurwitz zeta function for rational arguments, *J. Comput. Appl. Math.*, 100 (1998), N 2, 201–206.
60. Muzaffar H., Williams K. S., A restricted Epstein zeta function and the evaluation of some definite integrals, *Acta Arith.*, 104 (2002), N 1, 23–66.
61. Ogreid O. M., Osland P., Summing one- and two-dimensional series related to the Euler series, *J. Comput. Appl. Math.*, 98 (1998), N 2, 245–271.
62. —, —, More series related to the Euler series, *J. Comput. Appl. Math.*, 136 (2001), N 1–2, 389–403.
63. —, —, Some infinite series related to Feynman diagrams, *J. Comput. Appl. Math.*, 140 (2002), N 1–2, 659–671.
64. Rahman M., Verma A., Quadratic transformation formulas for basic hypergeometric series, *Trans. Amer. Math. Soc.*, 335 (1993), 277–302.

65. Reid W. H., Integral representations for products of Airy functions, *ZAMP*, 46 (1995), N 2, 159–170.

66. —, Integral representations for products of Airy functions. Part 2. Cubic products, *ZAMP*, 48 (1997), N 4, 646–655.

67. —, Integral representations for products of Airy functions. Part 3. Quartic products, *ZAMP*, 48 (1997), N 4, 656–664.

68. Rottbrand K., Finite-sum rules for MacDonald's function and Hankel's symbols, *Integral Transform. Spec. Func.*, 10 (2000), N 2, 115–124.

69. Sánchez-Ruiz J., Linearization and connection formulae involving squares of Gegenbauer polynomials, *Appl. Math. Letters*, 14 (2001), N 3, 261–267.

70. Lin Shy-Der, Srivastava H. M., Wang Pin-Yu, Some families of hypergeometric transformations and generating functions, *Mathematical and computer modelling*, 36 (2002), 445–459.

71. Srivastava H. M., Remarks on some series expansions associated with certain products of the incomplete gamma functions, *Comput. Math. Appl.*, 46 (2003), N 12, 1749–1759.

72. Stanković M. S., Vidanović M. V., Tričković S. B., Some series over the product of two trigonometric functions und series involving Bessel functions, *Z. Anal. Anwendungen*, 20 (2001), N 1, 235–246.

73. Szmytkowski R., On the derivative of the Legendre function of the first kind with respect to its degree, *J. Phys. A: Math. Gen.*, 39 (2006), 15147–15172.

74. Toscano L., Sulle derivate dei polinomi di Laguerre e del tipo ultrasferico rispetto al parametro, *Boll. Un. Mat. Ital.*, 8 (1953), 193–195.

75. —, Le funzioni del cilindro parabolico come caso limite delle funzioni ipergeometriche, *Boll. Un. Mat. Ital.* (3), 9 (1954), 29–38.

76. —, Su una formula limite tra funzioni ipergeometriche di Kummer e funzioni del cilindro parabolico, *Boll. Un. Mat. Ital.*, 10 (1955), 239–243.

77. —, Contributo alla formazione del formulario dei polinomi ultrasferici, *Giornale di matematiche di Battaglini* (5), 89 (1961), 85–109.

78. —, Formule di derivazione su particolari funzioni ipergeometriche di Gauss i di Kummer, *Matematiche* (Catania), 20 (1965), 156–167.

79. —, Formule di derivazione per le funzioni ipergeometriche di Gauss, *Riv. Mat. Univ. Parma* (2), 8 (1967), 163–180.

80. —, Sulle funzioni del cilindro parabolico, *Matematiche* (Catania), 26 (1971), 104–126.

81. Tremblay R., Fugère B. J., Products of two restricted hypergeometric functions, *J. Math. Anal. Appl.*, 198 (1996), N 3, 844–852.

82. Tricomi F., Sulla formula d'inversione di Widder, *Rend. R. Acc. Naz. Lincei*, 25 (1936).

83. Vidunas R., A generalization of Kummer's identity. http://arXiv.org/abs/math.CA/0005095.

84. Zucker I. J., Some infinite series of exponential and hyperbolic functions, *SIAM J. Math. Anal.*, 15 (1984), N 2, 406–413.

85. —, Joyce G. S., Delves R. T., On the evaluation of the integral $\int_0^{\pi/4} \ln\left(\cos^{m/n}\theta + \sin^{m/n}\theta\right) d\theta$, *Ramanujan J.*, 2 (1998), N 3, 317–326.

Index of Notations for Functions and Constants

$\operatorname{Ai}(z) = \dfrac{1}{\pi}\sqrt{\dfrac{z}{3}}\, K_{1/3}\!\left(\dfrac{2}{3} z^{3/2}\right)$ is the Airy function

$\arccos z$, $\operatorname{arccot} z$, $\arcsin z$, $\arctan z$ are inverse trigonometric functions

B_n are the Bernoulli numbers

$B_n(z)$ are the Bernoulli polynomials

$\operatorname{bei}_\nu(z)$, $\operatorname{ber}_\nu(z)$, $\operatorname{bei}(z) \equiv \operatorname{bei}_0(z)$, $\operatorname{ber}(z) \equiv \operatorname{ber}_0(z)$ are the Kelvin functions

$\operatorname{Bi}(z) = \sqrt{\dfrac{z}{3}}\left[I_{-1/3}\!\left(\dfrac{2}{3} z^{3/2}\right) + I_{1/3}\!\left(\dfrac{2}{3} z^{3/2}\right)\right]$ is the Airy function

$C = -\psi(1) = 0{,}5772156649\ldots$ is the Euler constant

$C(z) = \dfrac{1}{\sqrt{2\pi}}\displaystyle\int_0^z \dfrac{\cos t}{\sqrt{t}}\, dt$ is the Fresnel cosine integral

$C(z,\nu) = \displaystyle\int_z^\infty t^{\nu-1} \cos t\, dt\quad [\operatorname{Re}\nu < 1]$ is the generalized Fresnel cosine integral

$C_n^\lambda(z) = \dfrac{(2\lambda)_n}{n!}\, {}_2F_1\!\left(\begin{matrix}-n,\, n+2\lambda\\ \lambda+\tfrac{1}{2};\end{matrix}\ \dfrac{1-z}{2}\right)$ are the Gegenbauer polynomials

$\operatorname{chi}(z) = C + \ln z + \displaystyle\int_0^z \dfrac{\cosh t - 1}{t}\, dt$ is the hyperbolic cosine integral

$\operatorname{ci}(z) = -\displaystyle\int_z^\infty \dfrac{\cos t}{t}\, dt$ is the cosine integral

$\operatorname{Cl}_2(z) = -\displaystyle\int_0^z \ln\!\left(2\sin\dfrac{t}{2}\right) dt$ is the Clausen integral

$D = \dfrac{d}{dz},\ D_a = \dfrac{d}{da}$

$D(k) = \displaystyle\int_0^{\pi/2} \dfrac{\sin^2 t\, dt}{\sqrt{1 - k^2 \sin^2 t}}$ is the complete elliptic integral

$D_\nu(z) = 2^{\nu/2} e^{-z^2/4} \Psi\left(-\frac{\nu}{2}, \frac{1}{2}; \frac{z^2}{2}\right)$ is the parabolic cylinder function

$\mathbf{E}(k) = \int_0^{\pi/2} \sqrt{1 - k^2 \sin^2 t}\, dt$ is the complete elliptic integral of the second kind

E_n are the Euler numbers

$E_n(z)$ are the Euler polynomials

$\mathbf{E}_\nu(z) = \dfrac{1}{\pi} \int_0^\pi \sin(\nu t - z \sin t)\, dt$ is the Weber function

$\mathrm{Ei}\,(z) = \int_{-\infty}^{z} \dfrac{e^t}{t}\, dt$ is the exponential integral

$\mathrm{erf}\,(z) = \dfrac{2}{\sqrt{\pi}} \int_0^z e^{-t^2}\, dt$ is the error function

$\mathrm{erfc}\,(z) = 1 - \mathrm{erf}\,(z) = \dfrac{2}{\sqrt{\pi}} \int_z^\infty e^{-t^2}\, dt$ is the complementary error function

$\mathrm{erfi}(z) = \dfrac{2}{\sqrt{\pi}} \int_0^z e^{t^2}\, dt$ is the error function of imaginary argument

$${}_2F_1\!\left(\begin{matrix} a,\ b;\ z \\ c \end{matrix}\right) \equiv {}_2F_1\!\left(\begin{matrix} a,\ b \\ c;\ z \end{matrix}\right) \equiv {}_2F_1(a,\ b;\ c;\ z) = \sum_{k=0}^{\infty} \frac{(a)_k (b)_k}{(c)_k} \frac{z^k}{k!} \qquad [|z| < 1],$$

$$= \frac{\Gamma(c)}{\Gamma(a)\Gamma(c-b)} \int_0^1 t^{b-1}(1-t)^{c-b-1}(1-tz)^{-a}\, dt \quad [\mathrm{Re}\, c > \mathrm{Re}\, b > 0;\ |\arg(1-z)| < \pi]$$

is the Gauss hypergeometric function

$${}_pF_q\!\left(\begin{matrix} (a_p);\ z \\ (b_q) \end{matrix}\right) \equiv {}_pF_q\!\left(\begin{matrix} (a_p) \\ (b_q);\ z \end{matrix}\right) \equiv {}_pF_q((a_p);\ (b_q);\ z)$$

$$\equiv {}_pF_q(a_1, \ldots, a_p;\ b_1, \ldots, b_q;\ z) = \sum_{k=0}^{\infty} \frac{(a_1)_k (a_2)_k \ldots (a_p)_k}{(b_1)_k (b_2)_k \ldots (b_q)_k} \frac{z^k}{k!}$$

is the generalized hypergeometric function

${}_1F_1\!\left(\begin{matrix} a;\ z \\ b \end{matrix}\right) \equiv {}_1F_1\!\left(\begin{matrix} a \\ b;\ z \end{matrix}\right) \equiv {}_1F_1(a;\ b;\ z) = \sum\limits_{k=0}^{\infty} \dfrac{(a)_k z^k}{(b)_k k!}$ is the Kummer confluent hypergeometric function

$G = \sum\limits_{k=0}^{\infty} \dfrac{(-1)^k}{(2k+1)^2} = 0{,}9159655942\ldots$ is the Catalan constant

$$G_{pq}^{mn}\!\left(z \left| \begin{matrix} (a_p) \\ (b_q) \end{matrix}\right.\right) \equiv G_{pq}^{mn}\!\left(z \left| \begin{matrix} a_1, \ldots, a_p \\ b_1, \ldots, b_q \end{matrix}\right.\right)$$

$$= \frac{1}{2\pi i} \int_L \frac{\Gamma(b_1 + s) \ldots \Gamma(b_m + s)\Gamma(1 - a_1 - s) \ldots \Gamma(1 - a_n - s)}{\Gamma(a_{n+1} + s) \ldots \Gamma(a_p + s)\Gamma(1 - b_{m+1} - s) \ldots \Gamma(1 - b_q - s)} z^{-s}\, ds,$$

Index of Notations for Functions and Constants

$L = L_{\pm\infty}, L_{i\infty}$, is the Meijer G function

$$\mathbf{H}_\nu(z) = \frac{2\left(\frac{z}{2}\right)^{\nu+1}}{\sqrt{\pi}\,\Gamma\left(\nu+\frac{3}{2}\right)}\, {}_1F_2\left(\begin{array}{c}1;\ -\frac{z^2}{4}\\ \frac{3}{2},\nu+\frac{3}{2}\end{array}\right)$$ is the Struve function

$H_\nu^{(1)}(z) = J_\nu(z) + iY_\nu(z)$, $H_\nu^{(2)}(z) = J_\nu(z) - iY_\nu(z)$ are the Hankel functions of the first and second kind (the Bessel functions of the third kind)

$H_n(z) = (-1)^n e^{z^2} \dfrac{d^n}{dz^n} e^{-z^2}$ are the Hermite polynomials

$I_\nu(z) = \dfrac{1}{\Gamma(\nu+1)}\left(\dfrac{z}{2}\right)^\nu {}_0F_1\left(\nu+1;\ \dfrac{z^2}{4}\right) = e^{-\nu\pi i/2} J_\nu\left(e^{\pi i/2}z\right)$ is the modified Bessel function of the first kind

$J_\nu(z) = \dfrac{1}{\Gamma(\nu+1)}\left(\dfrac{z}{2}\right)^\nu {}_0F_1\left(\nu+1;\ -\dfrac{z^2}{4}\right)$ is the Bessel function of the first kind

$\mathbf{J}_\nu(z) = \dfrac{1}{\pi}\displaystyle\int_0^\pi \cos(\nu t - z\sin t)\,dt$ is the Anger function

$\mathbf{K}(k) = \displaystyle\int_0^{\pi/2} \dfrac{dt}{\sqrt{1-k^2\sin^2 t}}$ is the complete elliptic integral of the first kind

$K_\nu(z) = \dfrac{\pi[I_{-\nu}(z) - I_\nu(z)]}{2\sin\nu\pi}$ $[\nu \neq n]$, $K_n(z) = \lim\limits_{\nu\to n} K_\nu(z)$ $[n=0,\pm 1,\pm 2,\ldots]$ is the Macdonald function (the modified Bessel function of the third kind)

$\mathrm{kei}_\nu(z), \mathrm{ker}_\nu(z), \mathrm{kei}(z) = \mathrm{kei}_0(z), \mathrm{ker}(z) = \mathrm{ker}_0(z)$ are the Kelvin functions

$\mathbf{L}_\nu(z) = e^{-(\nu+1)\pi i/2} \mathbf{H}_\nu\left(e^{\pi i/2}z\right)$ is the modified Struve function

$L_n(z) = L_n^0(z)$ are the Laguerre polynomials

$L_n^\lambda(z) = \dfrac{z^{-\lambda}e^z}{n!} \dfrac{d^n}{dz^n}\left(z^{n+\lambda}e^{-z}\right)$ are the generalized Laguerre polynomials

$\mathrm{Li}_\nu(z) = \displaystyle\sum_{k=1}^\infty \dfrac{z^k}{k^\nu}$ $[|z|<1]$,

$= \dfrac{z}{\Gamma(\nu)} \displaystyle\int_0^\infty \dfrac{t^{\nu-1}\,dt}{e^t - z}$ $[\mathrm{Re}\,\nu > 0;\ |\arg(1-z)| < \pi]$ is the polylogarithm of the order ν

$\mathrm{Li}_2(z)$ is the Euler dilogarithm

$M_{\varkappa,\mu}(z) = z^{\mu+1/2} e^{-z/2}\, {}_1F_1\left(\begin{array}{c}\mu-\varkappa+\frac{1}{2}\\ 2\mu+1;\ z\end{array}\right)$ is the Whittaker confluent hypergeometric function

$P_n(z) = \dfrac{2^{-n}}{n!} \dfrac{d^n}{dz^n}(z^2-1)^n$ are the Legendre polynomials

$P_\nu(z) \equiv P_\nu^0(z) = {}_2F_1\left(\begin{array}{c}-\nu, 1+\nu\\ 1;\ \frac{1-z}{2}\end{array}\right)$ $[|\arg(1+z)| < \pi]$ is the Legendre function of the first kind

INDEX OF NOTATIONS FOR FUNCTIONS AND CONSTANTS

$$P_\nu^\mu(z) = \frac{1}{\Gamma(1-\mu)} \left(\frac{z+1}{z-1}\right)^{\mu/2} {}_2F_1\left(\begin{array}{c} -\nu, \nu+1 \\ 1-\mu; \dfrac{1-z}{2} \end{array}\right)$$
$$[|\arg(z \pm 1)| < \pi;\ \mu \neq m;\ m = 1, 2, \ldots]$$

$$P_\nu^m(z) = (z^2 - 1)^{m/2} \left(\frac{d}{dz}\right)^m P_\nu(z) \qquad [|\arg(z-1)| < \pi;\ m = 1, 2, \ldots]$$

$$P_\nu^\mu(x) = \frac{1}{\Gamma(1-\mu)} \left(\frac{1+x}{1-x}\right)^{\mu/2} {}_2F_1\left(\begin{array}{c} -\nu, \nu+1 \\ 1-\mu; \dfrac{1-x}{2} \end{array}\right)$$
$$[-1 < x < 1;\ \mu \neq m;\ m = 1, 2, \ldots]$$

$$P_\nu^m(x) = (-1)^m (1-x^2)^{m/2} \left(\frac{d}{dx}\right)^m P_\nu(x) \qquad [-1 < x < 1;\ m = 1, 2, \ldots]$$

is the associated Legendre function of the first kind

$$P_n^{(\rho,\sigma)}(z) = \frac{(-1)^n}{2^n n!} (1-z)^{-\rho}(1+z)^{-\sigma} \frac{d^n}{dz^n} \left[(1-z)^{\rho+n}(1+z)^{\sigma+n}\right]$$

$$= \frac{(\rho+1)_n}{n!} {}_2F_1\left(\begin{array}{c} -n, \rho+\sigma+n+1 \\ \rho+1; \dfrac{1-z}{2} \end{array}\right) \text{ are the Jacobi polynomials}$$

$Q_\nu(z) \equiv Q_\nu^0(z)$ is the Legendre function of the second kind

$$Q_\nu^\mu(z) = \frac{e^{i\mu\pi}\sqrt{\pi}}{2^{\nu+1}} \Gamma\!\left[\begin{array}{c}\mu+\nu+1 \\ \nu+3/2\end{array}\right] z^{-\mu-\nu-1}(z^2-1)^{\mu/2}\, {}_2F_1\!\left(\begin{array}{c} \dfrac{\mu+\nu+1}{2}, \dfrac{\mu+\nu}{2}+1 \\ \nu+\dfrac{3}{2};\ \dfrac{1}{z^2} \end{array}\right)$$

$$[|\arg z|, |\arg(z \pm 1)| < \pi;\ \nu + 1/2,\ \mu + \nu \neq -1, -2, -3, \ldots]$$

$$Q_{-n-3/2}^\mu(z) = \frac{e^{i\mu\pi}\sqrt{\pi}\,\Gamma(\mu+n+3/2)}{2^{n+3/2}(n+1)!}$$
$$\times z^{-\mu-n-3/2}(z^2-1)^{\mu/2}\, {}_2F_1\!\left(\begin{array}{c} \dfrac{2\mu+2n+3}{4}, \dfrac{2\mu+2n+5}{4} \\ n+2;\ \dfrac{1}{z^2} \end{array}\right)$$

$$[|\arg(z \pm 1)|, |\arg z| < \pi;\ \mu + \nu \neq -1, -2, -3, \ldots]$$

$$Q_\nu^\mu(x) = \frac{e^{-i\mu\pi}}{2}\left[e^{-\mu\pi/2}Q_\nu^\mu(x+i0) + e^{i\mu\pi/2}Q_\nu^\mu(x-i0)\right]$$

$$= \frac{\pi}{2\sin\mu\pi}\left[P_\nu^\mu(x)\cos\mu\pi - \Gamma\!\left[\begin{array}{c}\nu+\mu+1\\ \nu-\mu+1\end{array}\right] P_\nu^{-\mu}(x)\right]$$
$$[-1 < x < 1;\ \mu \neq \pm m;\ \mu + \nu \neq -1, -2, -3, \ldots],$$

$$= (-1)^m (1-x^2)^{m/2} \left(\frac{d}{dx}\right)^m Q_\nu(x) \qquad [\mu = m;\ \nu \neq -m-1, -m-2, \ldots],$$

$$= (-1)^m \Gamma\!\left[\begin{array}{c}\nu-m+1\\ \mu+m+1\end{array}\right] Q_\nu^m(x) \qquad [\mu = -m;\ \nu \neq -m-1, -m-2, \ldots]$$

is the associated Legendre function of the second kind

$$S(z) = \frac{1}{\sqrt{2\pi}} \int_0^z \frac{\sin t}{\sqrt{t}}\, dt \text{ is the Fresnel cosine integral}$$

$$S(z, \nu) = \int_z^\infty t^{\nu-1} \sin t\, dt \quad [\operatorname{Re}\nu < 1] \text{ is the generalized Fresnel sine integral}$$

$$S_n^{(m)} = \sum_{k=0}^{n-m} (-1)^k \binom{n+k-1}{n-m+k}\binom{2n-m}{n-m-k} \sigma_{n-m+k}^k \text{ are the Stirling numbers of the first kind}$$

Index of Notations for Functions and Constants

$$\operatorname{sgn} x = \begin{cases} 1, & x > 0, \\ 0, & x = 0, \\ -1, & x < 0 \end{cases}$$

$\operatorname{shi}(z) = \int_0^z \dfrac{\sinh t}{t}\, dt = -i\operatorname{Si}(iz)$ is the hyperbolic sine integral

$\operatorname{Si}(z) = \int_0^z \dfrac{\sin t}{t}\, dt$ is the sine integral

$\operatorname{si}(z) = \operatorname{Si}(z) - \dfrac{\pi}{2} = -\int_z^{\infty} \dfrac{\sin t}{t}\, dt$ is the sine integral

$T_n(z) = \cos(n \arccos z) = {}_2F_1\left(\begin{array}{c} -n,\, n \\ \frac{1}{2}; \end{array} \dfrac{1-z}{2}\right)$ are the Chebyshev polynomials of the first kind

$U_n(z) = \dfrac{\sin[(n+1)\arccos z]}{\sqrt{1-z^2}} = (n+1)\,{}_2F_1\left(\begin{array}{c} -n,\, n+2 \\ \frac{3}{2}; \end{array} \dfrac{1-z}{2}\right)$ are the Chebyshev polynomials of the second kind

$W_{\varkappa,\mu}(z) = z^{\mu+1/2} e^{-z/2}\Psi\left(\begin{array}{c} \mu - \varkappa + \frac{1}{2} \\ 2\mu+1;\ z \end{array}\right)$ is the Whittaker confluent hypergeometric function

$Y_\nu(z) = \dfrac{\cos\nu\pi\, J_\nu(z) - J_{-\nu}(z)}{\sin\nu\pi}$ $[\nu \neq n]$, $Y_n(z) = \lim\limits_{\nu\to n} Y_\nu(z)$ $[n = 0, \pm 1, \pm 2, \ldots]$ is the Neumann function (the Bessel function of the second kind)

$\mathrm{B}(\alpha,\beta) = \dfrac{\Gamma(\alpha)\Gamma(\beta)}{\Gamma(\alpha+\beta)}$ is the beta function

$\mathrm{B}_z(\alpha,\beta) = \int_0^z t^{\alpha-1}(1-t)^{\beta-1}\, dt$ $[\operatorname{Re}\alpha > 1;\ z < 1]$ is the incomplete beta function

$\Gamma(z) = \int_0^{\infty} t^{z-1} e^{-t}\, dt$ $[\operatorname{Re} z > 0]$ is the gamma function

$\Gamma(\nu, z) = \int_z^{\infty} t^{\nu-1} e^{-t}\, dt$ is the complementary incomplete gamma function

$\gamma(\nu, z) = \Gamma(\nu) - \Gamma(\nu, z) = \int_0^z t^{\nu-1} e^{-t}\, dt$ $[\operatorname{Re}\nu > 0]$ is the incomplete gamma function

$\Gamma[(a_p)] = \prod\limits_{k=1}^p \Gamma(a_k)$

$\Delta(k, a) = \dfrac{a}{k},\ \dfrac{a+1}{k},\ \ldots,\ \dfrac{a+k-1}{k}$

$\Delta(k, (a_p)) = \dfrac{(a_p)}{k},\ \dfrac{(a_p)+1}{k},\ \ldots,\ \dfrac{(a_p)+k-1}{k}$

$\delta_{m,n} = \begin{cases} 0, & m \neq n, \\ 1, & m = n \end{cases}$ is the Kronecker symbol

$$\zeta(z) = \sum_{k=1}^{\infty} \frac{1}{k^z} \quad [\operatorname{Re} z > 1] \text{ is the Riemann zeta function}$$

$$\zeta(z, v) = \sum_{k=0}^{\infty} \frac{1}{(v+k)^z} \quad [\operatorname{Re} z > 1;\ v \neq 0, -1, -2, \ldots] \text{ is the Hurwitz zeta function}$$

$$\theta(x) = \begin{cases} 1, & x \geq 0, \\ 0, & x < 0 \end{cases} \text{ is the Heaviside function}$$

$$\Xi_1(a, a', b;\ c;\ w, z) = \sum_{k,l=0}^{\infty} \frac{(a)_k (a')_l (b)_k}{(c)_{k+l}} \frac{w^k z^l}{k!\, l!} \qquad [|w| < 1]$$

$$\Xi_2(a, b;\ c;\ w, z) = \sum_{k,l=0}^{\infty} \frac{(a)_k (b)_k}{(c)_{k+l}} \frac{w^k z^l}{k!\, l!} \qquad [|w| < 1]$$

$$\sigma_n^m = \frac{1}{m!} \sum_{k=0}^{m} (-1)^{m-k} \binom{m}{k} k^n \text{ are the Stirling numbers of the second kind}$$

$$\Phi(z, s, v) = \sum_{k=0}^{\infty} \frac{z^k}{(v+k)^s} \qquad [|z| < 1;\ v \neq 0, -1, -2, \ldots]$$

$$\Phi_1(a, b;\ c;\ w, z) = \sum_{k,l=0}^{\infty} \frac{(a)_{k+l} (b)_k}{(c)_{k+l}} \frac{w^k z^l}{k!\, l!} \qquad [|w| < 1]$$

$$\Phi_2(b, b';\ c;\ w, z) = \sum_{k,l=0}^{\infty} \frac{(b)_k (b')_l}{(c)_{k+l}} \frac{w^k z^l}{k!\, l!},$$

$$\Phi_3(b;\ c;\ w, z) = \sum_{k,l=0}^{\infty} \frac{(b)_k}{(c)_{k+l}} \frac{w^k z^l}{k!\, l!}$$

$$\Psi\begin{pmatrix} a;\ z \\ b \end{pmatrix} \equiv \Psi\begin{pmatrix} a \\ b;\ z \end{pmatrix} \equiv \Psi(a;\ b;\ z)$$
$$= \frac{\Gamma(1-b)}{\Gamma(1+a-b)} {}_1F_1\begin{pmatrix} a;\ z \\ b \end{pmatrix} + \frac{\Gamma(b-1)}{\Gamma(a)} z^{1-b} {}_1F_1\begin{pmatrix} 1+a-b \\ 2-b;\ z \end{pmatrix}$$

is the Tricomi confluent hypergeometric function

$$\psi_1(z) = \frac{4}{\pi^2}\, \mathbf{K}^2(x),$$

$$\psi_2(z) = \frac{4}{\pi^2} \left\{ \mathbf{K}^2(x) - \frac{x^2}{1-x^2} [\mathbf{K}(x) - \mathbf{D}(x)]^2 \right\},$$

$$\psi_3(z) = \frac{4}{\pi^2} \left\{ 3\,\mathbf{K}^2(x) - \frac{4(1-x^2+x^4)}{(1-x^2)^2} [\mathbf{K}(x) - \mathbf{D}(x)]^2 \right.$$
$$\left. + \frac{1}{(1-x^2)^2} [2(1-2x^2)\,\mathbf{D}(x) - (1-3x^2)\,\mathbf{K}(x)]^2 \right\} \quad \left[x = \left(\frac{1-\sqrt{1-z}}{2} \right)^{1/2} \right]$$

$$\Psi_1(a, b;\ c, c';\ w, z) = \sum_{k,l=0}^{\infty} \frac{(a)_{k+l} (b)_k}{(c)_k (c')_l} \frac{w^k z^l}{k!\, l!} \qquad [|w| < 1]$$

$$\Psi_2(a;\ c, c';\ w, z) = \sum_{k,l=0}^{\infty} \frac{(a)_{k+l}}{(c)_k (c')_l} \frac{w^k z^l}{k!\, l!}$$

$$\psi(z) = [\ln \Gamma(z)]' = \frac{\Gamma'(z)}{\Gamma(z)} \text{ is the psi function}$$

Index of Notations for Symbols

$(a) = a_1, a_2, \ldots, a_A;\quad (a_p) = a_1, a_2, \ldots, a_p$

$(a_p - b_p) = a_1 - b_1, a_2 - b_2, \ldots, a_p - b_p$

$(a) + s = a_1 + s, a_2 + s, \ldots, a_A + s;\quad (a_p) + s = a_1 + s, \ldots, a_p + s$

$(a)' - a_j = a_1 - a_j, \ldots, a_{j-1} - a_j, a_{j+1} - a_j, \ldots, a_A - a_j \qquad [1 \le j \le A]$

$(a_p)' - a_j = a_1 - a_j, \ldots, a_{j-1} - a_j, a_{j+1} - a_j, \ldots, a_p - a_j \qquad [1 \le j \le p]$

$(a)_k = a(a+1)\ldots(a+k-1) \quad [k = 1, 2, 3, \ldots],\quad (a)_0 = 1$ is the Pochhammer symbol

$n! = 1 \cdot 2 \cdot 3 \ldots (n-1)n = (1)_n,\quad 0! = 1! = (-1)! = 1$

$(2n)!! = 2 \cdot 4 \cdot 6 \ldots (2n-2)2n = 2^n n!$

$(2n+1)!! = 1 \cdot 3 \cdot 5 \ldots (2n+1) = \dfrac{2^{n+1}}{\sqrt{\pi}} \Gamma\left(n + \dfrac{3}{2}\right) = \left(\dfrac{3}{2}\right)_n 2^n$

$n!! = \begin{cases} (2k)!!, & n = 2k, \\ (2k+1)!!, & n = 2k+1, \end{cases} \quad 0!! = (-1)!! = 1$

$\binom{n}{k} = \dfrac{n(n-1)\ldots(n-k+1)}{k!} = \dfrac{n!}{k!(n-k)!} = \dfrac{(-1)^k(-n)_k}{k!},\quad \binom{n}{0} = 1$

$\operatorname{Re} a, \operatorname{Re} b > c$ means $\operatorname{Re} a > c$ and $\operatorname{Re} b > c$

$[x] = n \, [n \le x < n+1,\, n = 0, \pm 1, \pm 2, \ldots]$ is the integer part of x

$x_+^\lambda = \begin{cases} x^\lambda, & x > 0, \\ 0, & x < 0 \end{cases}$

$\prod (a_p)_k = \prod_{j=1}^{p} (a_j)_k$

$\prod ((a_p) + b)_k = \prod_{j=1}^{p} (a_j + b)_k$

$\prod_{k=m}^{n} a_k = a_m a_{m+1} \ldots a_n \qquad [n \ge m],$
$\qquad\qquad\, = 1 \qquad\qquad\qquad\;\; [n < m]$

$\sum_{k=m}^{n} a_k = a_m + a_{m+1} + \ldots + a_n \qquad [n \ge m],$
$\qquad\qquad\, = 0 \qquad\qquad\qquad\qquad\quad [n < m]$

Index of Notations for Symbols

$$\prod(a_p)_k = \prod_{j=1}^{p}(a_j)_k$$

$$\prod((a_p)+b)_k = \prod_{j=1}^{p}(a_j+b)_k$$

$$\prod_{k=m}^{n} a_k = a_m a_{m+1} \ldots a_n \quad [n \geq m],$$
$$\qquad\qquad = 1 \qquad\qquad\quad [n < m]$$

$$\prod_{k=1}^{\infty} a_k(z) = \lim_{n \to \infty} \prod_{k=1}^{n} a_k(z)$$

$$\sum_{k=m}^{n} a_k = a_m + a_{m+1} + \ldots + a_n \quad [n \geq m],$$
$$\qquad\qquad = 0 \qquad\qquad\qquad\quad [n < m]$$

$$\sum_{k=1}^{\infty} a_k(z) = \lim_{n \to \infty} \sum_{k=1}^{n} a_k(z)$$